Algae

Algae

Linda E. Graham
University of Wisconsin

Lee W. Wilcox
University of Wisconsin

Prentice Hall

Upper Saddle River, NJ 07458

Graham, Linda E.
    Algae / Linda E. Graham, Lee Warren Wilcox.
            p.   cm.
    Includes bibliographical references.
    ISBN: 0-13-660333-5
    1. Algae. 2. Algology   I. Wilcox, Lee Warre.   II. Title.
QK566.G735   2000   99-24517
579.8--dc21   CIP

Aquisitions Editor: Teresa Ryu
Editorial Assistant: Lisa Tarabojkia
Production Editor: Kim Dellas
Manufacturing Manager: Trudy Pisciotti
Art Director: Jayne Conte
Cover Design: Lee Wilcox and Bruce Kenselaar
Cover Photo: Linda E. Graham

Any Photos that do not have credit are copyrighted
by the authors.

Printed in the United States of America

10 9 8 7 6 5 4 3

ISBN 0-13-660333-5

Prentice-Hall International (UK) Limited, *London*
Prentice-Hall of Australia Pty. Limited, *Sydney*
Prentice-Hall Canada Inc., *Toronto*
Prentice-Hall Hispanoamericana, S.A., *Mexico City*
Prentice-Hall of India Private Limited, *New Delhi*
Prentice-Hall of Japan Inc., *Tokyo*
Prentice Hall (*Singapore*) Pte. Ltd.,
Editora Prentice-Hall do Brasil, Ltda., *Rio de Janeiro*

Brief Contents

Contents

5 Algal Diversity and Relationships— Taxonomy, Systematics, and Phylogeny   80

6 Cyanobacteria (Chloroxybacteria)   97

7 Endosymbiosis and the Origin of Eukaryotic Algae—With a Focus on Glaucophytes, Chlorarachniophytes, and Apicomplexans   132

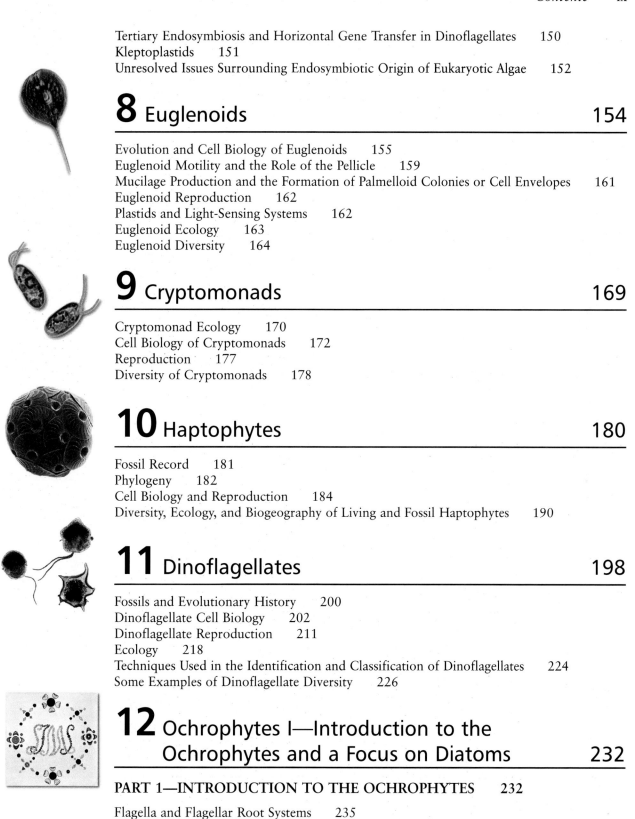

13 Ochrophytes II—Raphidophyceans, Chrysophyceans, Synurophyceans, and Eustigmatophyceans   269

14 Ochrophytes III—Pelagophyceans, Silicoflagellates, Pedinellids, and Related Forms   294

15 Ochrophytes IV—Chrysomeridaleans, Phaeothamniophyceans, Tribophyceans, and Phaeophyceans 301

16 Red Algae 343

17 Green Algae I—Introduction and Prasinophyceans 397

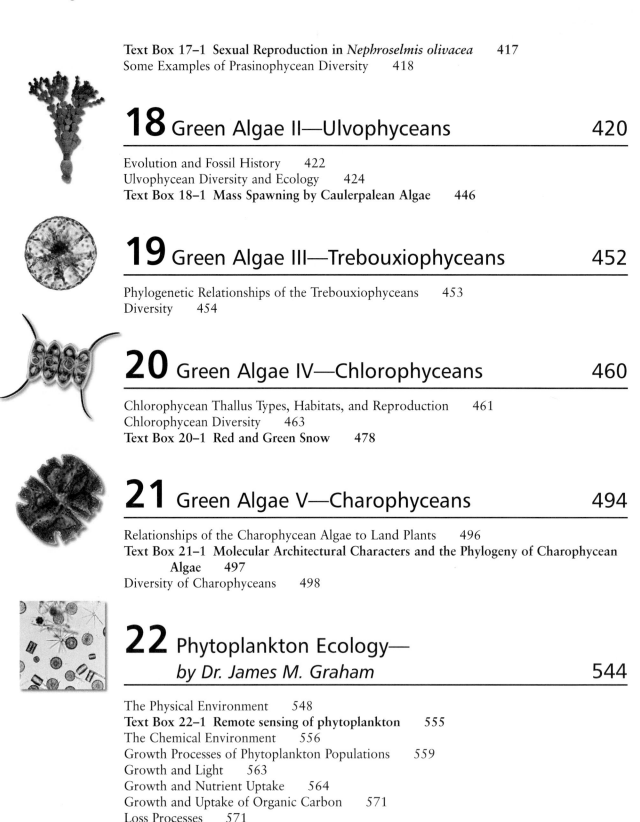

23 Macroalgal and Periphyton Ecology 603

Preface

This book is designed for use by undergraduate and graduate students in courses on the algae and aquatic ecology, as well as by researchers and professionals in the fields of aquatic ecology and technological applications of algae. This text includes extensive photographic illustrations and provides detailed descriptions of representative algal genera. Professor Paul Silva, University of California-Berkeley, graciously researched the etymologies of the generic names of these representative forms that are included with the descriptions. Terms defined in the glossary appear in boldface in the text.

This book covers freshwater, marine, and terrestrial forms. For the student this provides the widest possible background, and thus greater flexibility in research options and when entering the job market. Our wide ecological treatment allows discussion of some important algal transmigrations. Furthermore, considerations of algal biodiversity and evolutionary topics also require coverage of forms that occupy a wide range of algal habitats.

Distinctive features of this book include a series of five introductory chapters designed to stimulate student interest and to provide an overview of the importance of algae. A survey of algal habitats, general characteristics, nutritional variation, and life-history types constitutes Chapter 1. Chapter 2 is focused on the roles of algae in global biogeochemistry and algal influences on present and past climates and atmospheric chemistry. Chapter 3, which considers biotic associations involving algae, includes discussion of herbivory interactions, algal food quality, predatory algae, pathogens of algae, algae as pathogens, and herbivory- and pathogen-defense adaptations. A chapter on technological applications (Chapter 4) includes discussion of algae grown for use as food and in the production of industrially useful materials, mariculture of economically useful algae, and applications of algae in effluent treatment and space technologies. We have, in Chapter 5, provided a primer on modern approaches to algal systematics, including coverage of major molecular systematic techniques and procedures commonly used to evaluate the significance of phylogenies. Already essential in studies of algal evolution, molecular phylogenetics will increasingly be added to the repertoire of algal ecologists. Operating under the assumption that students may be familiar with major algal groups but not with generic names, we have routinely coupled descriptive modifiers with generic names of algae in the five introductory chapters.

Another distinctive feature of this text is a chapter (Chapter 7) on the topic of endosymbiosis—an extremely widespread and ecologically important natural phenomenon that is the mechanism by which the eukaryotic algal groups arose. We believe knowledge of this subject to be essential for understanding differences in physiological and ecological behavior among protist groups. Chapter 7 provides an introduction to the eukaryotic algae. A core group of 15 chapters (Chapters 6 and 8–21) focuses on one or more groups of related algae. Each chapter includes group-specific structural, physiological, evolutionary, and ecological information. This book concludes with two chapters that provide brief synthetic treatments of phytoplankton ecology (Chapter 22), contributed by Dr. James Graham, and periphyton and seaweed ecology (Chapter 23).

Throughout this book we have included examples of new findings and approaches in algal molecular biology. Many instances of recently discovered algal forms are provided to demonstrate that algal biodiversity is incompletely known and that unknown forms await discovery by the curious and prepared investigator. We have provided somewhat more detail regarding the ultrastructure of flagellar apparatuses of algae than is found in many introductory texts, because such information is often essential (as are pigment and molecular sequence data) for the detection and classification of new forms. There is also an emphasis on algae of extreme habitats, in view of recent widespread interest in exotic biodiversity and its possible relevance to extraterrestrial biology (exobiology or astrobiology).

We have not provided taxonomic keys to the algal genera described due to space constraints and because we believe that the correlative use of more comprehensive, locally relevant taxonomic keys is a better alternative. Further, we have provided class, order, and family classifications only when these are supported by both classical and molecular data. References are provided to classification systems currently in use for the major groups, but we have avoided detailing those that have as yet not been tested by application of molecular phylogenetic methods. Colloquial names referring to members of phyla (divisions) are given the suffix "phyte" and members of classes, "phycean."

We have arranged algal groups in order of their phylogenetic divergence (antiquity), as inferred primarily from nuclear SSU rDNA gene sequences. The red algae are a possible exception, their phylogenetic position being currently controversial. We have placed the chapter on red algae between chapters covering brown and marine green algae for the convenience of users in coastal regions where brown, red, and green algae often co-occur. Each chapter is designed to stand alone—the content not depending on that found in other chapters (except Chapter 1)—so that Chapters 2 through 23 may be read in any order.

We have tried to incorporate both very recent and older classic references to research literature from around the world, but have been unable to provide a comprehensive review of the literature due to space limits. Further, we have attempted to reference work accomplished by investigators throughout the world, though few works in languages other than English are cited. Although there is much useful information on algae available on the internet, we have chosen not to include website addresses because of their volatility. We have commonly placed literature citations at the end of a series of relevant sentences to facilitate uninterrupted flow of thought. We would be pleased to receive recommendations of critical literature citations that might be added to a later edition.

Numerous phycologists from around the world graciously contributed original photographs to this project. Contributors are cited in the figure captions. As a token of our appreciation for the use of these images as well as the use of line art, we pledge a substantial contribution arising from book royalties to the International Phycological Congress for use by students for travel to congress meetings.

We are also grateful to Kandis Elliot, UW-Madison Department of Botany artist, who provided technical advice; the staff at Prentice Hall, particularly our editor, Teresa Ryu, and production editor, Kim Dellas; the UW-Madison Biology Library staff; and Professors Jane Gray, University of Oregon, and Ron Hoham, Colgate University, who provided reviews of limited material at our request. The following people were commissioned by Prentice Hall to review one or more book chapters, and we are very appreciative of their helpful efforts:

Robert Bell, University of Wisconsin, Stevens Point
David J. Chapman, University of California, Santa Barbara
David B. Czarnecki, Loras College
W. Marshall Darley, University of Georgia
Walter Dodds, Kansas State University
Greta A. Fryxell, University of Texas, Austin
David Garbary, St. Francis Xavier University
Margaret Ginzberg, Cornell University
Arthur Grossman, Carnegie Institution
Donald Kapraun, University of North Carolina, Wilmington
Louise A. Lewis, University of Connecticut
Carla K. Oldham-Ott, The University of Akron
Donald W. Ott, The University of Akron
Christopher Peterson, Loyola University, Chicago
Laurie L. Richardson, Florida International University
James Rosowski, University of Nebraska
Carol J. Slocum, Stockton College
John W. Stiller, University of Washington
F. R. Trainor, University of Connecticut
D. Reid Wiseman, College of Charleston
Mary K. Wicksten, Texas A&M University
Peggy A. Winter, University of West Florida

Chapter 1

Pediastrum

Introduction to the Algae

Occurrence, Relationships, Nutrition, Definition, General Features

From tiny single-celled species one micrometer in diameter to giant seaweeds over 50 meters long, algae are abundant and ancient organisms that can be found in virtually every ecosystem in the biosphere. For billions of years algae have exerted profound effects on our planet and its biota, and they continue to do so today. Still, in many habitats algae often go unnoticed unless environmental conditions become favorable for the development of conspicuous and sometimes massive proliferations of their numbers—a situation often brought about by human activity. People from many cultures, ancient and modern, have used algae for a variety of purposes. With the advent of biotechnology, algae are poised to play greater, albeit often subtle, roles in the day-to-day lives of human beings. In the following passages we provide a brief overview of algal habitats and activities that demonstrates algae occur in both expected and highly surprising places. This survey will set the stage for a circumscription of the algae, i.e., a definition for this enigmatic group of organisms.

An Overview of the Occurrence and Activities of Algae

Algae in the Marine Habitat

On land the largest and most striking plants are the trees. Together with their herbaceous relatives, their foliage makes green the most conspicuous color of the biosphere. Underwater there are "trees" of similar height that are less widely appreciated because most humans spend little time in their realm. Brown undulating forests of 50-meter-long giant kelps, as tall and crowded as their terrestrial counterparts, dominate significant stretches of submerged temperate coastlines (Fig. 1–1). Like trees, kelps use pho-

Figure 1–1 Kelp forest off the Chilean coast. The predominant alga pictured is *Macrocystis*. (Photograph courtesy R. Searles)

tosynthesis to convert the energy of sunlight into chemical energy, but the green of their chlorophyll is masked by large amounts of brown pigments. These **accessory pigments** aid in the collection of light not absorbed directly by chlorophyll molecules and channel the light

Figure 1–2 Nearshore underwater marine algae (primarily the calcified brown alga *Padina*) in the Bahamas. A gorgonian coral is in the foreground.

energy to chlorophyll *a*—the only pigment that is able to effectively convert the energy of absorbed light into high energy bonds of organic molecules. This is necessary because as light passes through water, the longer wavelengths are filtered out first, such that eventually all that remains is a faint blue-green light that cannot be absorbed by chlorophyll.

Brown seaweeds are not limited to temperate waters, as they also form luxuriant thickets beneath polar ice sheets rarely noticed by anyone but phycologists or algologists—scientists who study these and other algae. The depth record for algae is held by dark purple-colored crusts of yet unnamed red algae discovered in tropical waters by phycologists using submersibles. These organisms live at depths greater than 250 meters, where the light intensity is only 0.0005% that of surface light. The accessory pigments of these algae—whose role is the same as that for those found in the kelps—are essential for the survival of photosynthetic organisms in such low-irradiance environments. In contrast, algae that live in high-irradiance habitats typically have pigments that help protect against photodamage. It is the composition and amounts of accessory and protective photosynthetic pigments that give algae their wide variety of colors and, for several algal groups, their common names such as the brown algae, red algae, and green algae. (We should caution, however, that attempting to identify a particular alga by color alone could be problematic, since, for example, there are red-colored green algae and brown or purple-colored red algae; other characteristics and features must also be considered.)

Figure 1–3 A small decorator crab with various attached algae.

The rocky or sandy shallows of temperate and tropical oceans harbor a vast array of brown, red, and green algal growths that may form thin and sometimes slippery films on rocks; diaphanous, lacy, or fleshy forms attached by holdfasts; or miniature jointed shrubs armored with limestone (Fig. 1–2). Myriad smaller algae, like the epiphytes found on rain forest trees, attach themselves to, or actually grow within, larger seaweeds, rocks, corals, and shells. Algae share the tidal zone with numerous invertebrate animals such as barnacles and snails, which often compete with them for space or consume them. Occasionally small clumps of seaweeds may appear to crawl slowly across the ocean floor or along a coral reef—closer inspection reveals "decorator" crabs that have adorned themselves with a fashionable selection of brown, green, or red algae as a camouflage (Fig. 1–3).

Tropical fringes are typically populated with a breathtakingly diverse array of submersed reef-forming corals, whose very existence and form are dependent upon intracellular tenants—microscopic golden algal cells known as **zooxanthellae**—that generate food and oxygen in exchange for metabolic by-products (carbon dioxide and ammonia) released by the coral cells. Zooxanthellae allow corals to thrive in the typically low-nutrient conditions of tropical waters. Because of their obligate association with these photosynthetic algae, reef-building corals are limited to shallow, well-illuminated waters less than 20 meters or

so in depth. Beneficial algae also occur within the cells and tissues of a wide variety of other marine animals such as nudibranchs, anemones, giant clams, ascidians, and sponges, as well as inside the cells of radiolarians and foraminiferans, which are but two types of the multitudinous simple organisms known as **protists**, an informal group to which the algae also belong.

Sandy tropical shallows may also contain extensive microbial mats composed of an interwoven community of cyanobacteria (also known as chloroxybacteria, blue-green algae, or cyanophytes), diatoms, and other microorganisms. In a few places—notably Shark Bay, Australia and tidal channels close to Exuma Island in the Bahamas—generations of calcium carbonate-depositing, sediment-trapping, cyanobacteria have built layered hummocks up to two meters high (Fig. 1–4). These hummocks represent modern versions of more widespread fossil formations known as **stromatolites**, which are commonly associated with the occurrence of earth's earliest life-forms.

In addition to these conspicuous marine algal communities with their relatively large seaweeds, coral formations, or algal aggregations, the surrounding ocean waters—occupying approximately 70% of the Earth's surface—teem with some 5000 species of tiny floating or swimming emerald, ruby, topaz, and turquoise jewels known as **phytoplankton** (Fig. 1–5). Although individually visible to humans only with the aid of a microscope, large populations can give ocean waters green or rusty hues. Color variations reflect differences in the types and amounts of blue-green, red, orange, and golden accessory pigments accompanying the green of chlorophyll. Like those giving larger seaweeds their brown, purple, or red coloration, these variously colored pigments also

Figure 1–4 Modern-day stromatolites in Shark Bay, Australia. (Photograph courtesy A. Knoll)

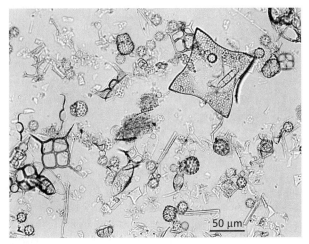

Figure 1–5 Sedimented phytoplankton from the late Cretaceous and early Cenozoic Arctic Ocean, including silicoflagellate and diatom remains. (Specimen courtesy D. Clark)

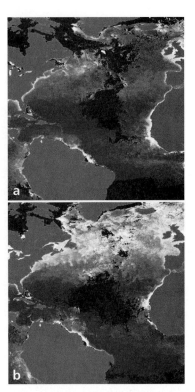

Figure 1–6 Two NASA satellite images of the North Atlantic taken in (a) winter and (b) spring. Brighter areas represent higher concentrations of chlorophyll and, hence, phytoplankton. Totally black areas are regions for which data were not collected.

assist in harvesting light for photosynthesis and in photoprotection. Enormous variation in phytoplankton shape and size has resulted from multiple adaptive solutions to two important problems: sinking to depths where the low levels of light limit photosynthesis and growth, and herbivory—the consumption of algae by animals and protists.

Populations of marine phytoplankton can become so large that they are detectable by satellite remote sensing technology. Such **blooms** are in fact one of the more dramatic vegetational features of the planet when viewed from space (Fig. 1–6). Collectively, marine microalgae have been modifying the earth's atmosphere for more than 2.7 billion years (Buick, 1992), and they continue to exert a powerful influence on modern atmospheric chemistry and biogeochemical cycling of carbon, sulfur, nitrogen, phosphorus, and other elements (Chapter 2). Hundreds of millions of years' worth of past phytoplankton growth and sedimentation have generated important oil and limestone deposits. Algal plankton also form the base of marine food chains, supporting both microbial and animal plankton (zooplankton), upon which economically important marine fisheries and ecologically significant marine mammal and bird populations are dependent.

The Algae of Freshwaters

Freshwater lakes, ponds, and streams contain similar botanical gardens of planktonic microalgae and attached forms (**periphyton**), which are often themselves festooned with epiphytes (Fig. 1–7). Although not exhibiting the phenomenal size range of their marine relatives, freshwater algae nonetheless display a wide diversity of form and function. As in the

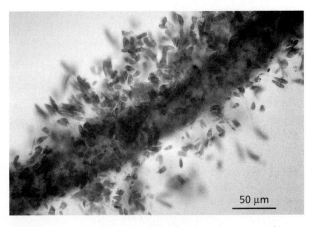

Figure 1–7 A specimen of the freshwater green alga *Oedogonium* with large numbers of epiphytic diatoms.

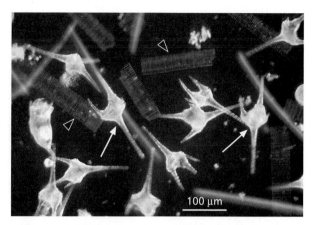

Figure 1–9 Freshwater phytoplankton from a southern Wisconsin lake. The predominant organisms in this sample include the dinoflagellate *Ceratium hirundinella* (arrows) and the filamentous diatom *Fragilaria* (arrowheads).

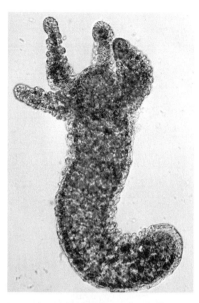

Figure 1–8 *Hydra* containing endosymbiotic green algae known as zoochlorellae.

oceans, it is not uncommon to find that certain photosynthetic freshwater algae colonize the cells and tissues of protozoa or coelenterates like the familiar *Hydra* (Fig. 1–8). Cyanobacteria living within the tissues of water ferns (see Fig. 3–31) can be a major contributor to the nitrogen economy of rice cultivation in paddies and thus influence the nutrition of millions of human beings. Freshwater phytoplankton (Fig. 1–9) and periphyton (also known as benthic algae) form the base of the aquatic food chain, without which freshwater fisheries could not exist. In

addition to oceanic and freshwater environments, some algae have adapted to extreme habitats such as hot springs and brine lakes.

Algal Blooms

Conspicuous blooms of microscopic algae (Fig. 1–10) occur in marine and freshwaters, often in response to pollution with nutrients such as nitrogen and/or phosphate. Nutrient pollution can usually be traced to human activities, such as discharge of effluents containing sewage or industrial wastes, or the use of agricultural fertilizers. Surface algal scums or weedy

Figure 1–10 Algal bloom in a freshwater marshland consisting primarily of the green alga *Spirogyra*.

Figure 1–11 The darkened areas on this Colorado snowfield are reddish patches of "snow algae." Although members of the green algae, they are red because of the high levels of photoprotective pigments within their cells.

Figure 1–13 *Umbilicaria*, a large lichen with a membranous thallus, growing on a shaded cliff with the fern *Polypodium*.

shoreline growths often have pernicious effects on aquatic ecosystems. Water transparency may become so reduced that organisms such as corals, aquatic plants, and periphyton no longer receive sufficient light for photosynthesis. It has been estimated that 50% or more of marine and freshwater algal blooms produce poisons that affect neuromuscular systems, are toxic to the liver, or are carcinogenic to vertebrates. These toxins can cause massive fish kills, death of birds, cattle, dogs and other animals, and serious illness, or death, in humans (Carmichael, 1997). One recently discovered bloom-forming marine microalga is a deadly "ambush predator"—single-celled dinoflagellates known as *Pfiesteria* secrete a highly toxic compound into the water to kill fish, after which they consume the decaying flesh. *Pfiesteria* was apparently detected only recently because normally small populations had relatively cryptic effects. The recent appearance of larger

and more harmful populations of this alga in Chesapeake Bay is correlated with increases in water pollution associated with local agricultural activities (Burkholder, et al., 1992) (see Chapter 3).

Terrestrial Algae

A considerable number of algae have adapted to life on land, such as those occurring in the snows of mountain ranges (Fig. 1–11), in "cryptobiotic crusts" typical of desert and grassland soils, or embedded within surfaces of rocks in deserts (Fig. 1–12), polar regions, and other biomes. The activities of soil and rock algae are thought to enhance soil formation and water retention, increase the availability of nutrients for plants growing nearby, and minimize soil erosion (Johansen, 1993).

Several species of terrestrial algae, together with fungi, form the distinctive life-forms known as lichens (Fig. 1–13). Lichens are ecologically important because of their role as pioneers in early stages of succession, where they help to convert rock into soil, slowly dissolving it with excreted acids. Lichens also help to stabilize fragile desert soils and are used as living barometers of air quality because of their sensitivity to air pollution.

Some terrestrial algae occur in surprising places. For example, algae can impart a greenish cast to the

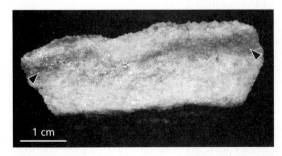

Figure 1–12 A band of cryptoendolithic algae (arrowheads) growing within sandstone. (Specimen courtesy R. Bell)

Figure 1–14 Microalgal production facility. Algal cultures are grown in successively larger volumes and finally cultivated for biomass harvest in long raceways stirred by paddle wheels (foreground). (Photograph courtesy J. Kranzfelder)

fur of giant sloths and sometimes live within the hollow hairs of polar bears. Also, the pink color of flamingos is due to a red-colored algal accessory (carotenoid) pigment consumed as they feed. Algae also occur regularly within the tissues of various plants, including bryophytes, cycads, and the tropical flowering plant *Gunnera*, where they are found in the petioles of its giant leaves.

Human Uses of Algae

For millennia people throughout the world have collected algae for food, fodder, or fertilizer. More recently algae have begun to play important roles in biotechnology. For example, they have been used to absorb excess nutrients from effluents, thereby reducing nutrient pollution in lakes and streams. Algae also generate industrially useful biomolecules (Fig. 1–14), and serve as a human food source, either directly (Fig. 1–15) or indirectly, by supporting aquaculture of shrimp and other aquatic animals. Algae are increasingly being cropped in lab-based bioreactors, outdoor production ponds, and engineered offshore environments.

Algae have provided science with uniquely advantageous model systems for the study of photosynthesis

Figure 1–15 Harvesting the red alga *Porphyra*, grown on nets in an aquaculture operation for use as food. (Photograph courtesy B. Waaland)

and other molecular, biochemical, and cellular-level phenomena of wider importance. Examples include Melvin Calvin's elucidation of the light-independent ("dark") reactions of photosynthesis in the green alga *Chlorella*. Studies of algae have been essential to our understanding of basic photosynthetic processes, and they continue to break new conceptual ground. The relative simplicity, antiquity, and vast diversity of algae, coupled with excellent fossil records in some cases, have also made algae invaluable systems for deciphering organismal and organellar evolution and ecosystem function, and for understanding the effects of human disturbance upon the biosphere (see Chapter 4).

What Are the Algae?

Phylogenetic Relationships of Algae

In order to take a closer look in subsequent chapters at the essential roles of algae in nature and to consider the biotechnological utility of algae, it is first necessary to define these organisms. This is not an easy task, because algae clearly do not form a cohesive group. In the parlance of the field of biology known as **systematics**, which endeavors to understand how organisms are related to one another and to organize them into groups based upon their evolutionary (**phylogenetic**) relationships, we would say that algae are not **monophyletic**. The origin of a monophyletic group (or clade) can be traced to a single hypothetical common ancestor. The organisms commonly regarded as algae, however, appear to have originated on multiple occasions.

Evolutionary relationships among organisms can be visually represented as a **phylogenetic tree**. Various features, including cell structure, biochemistry, molecular (amino acid and nucleic acid) sequences, and molecular architectural data, can be used to infer such phylogenies. The phylogenetic tree depicted in Figure 1–16 shows the major algal groups and their relationships as deduced from nuclear-encoded small subunit (18S) ribosomal RNA gene sequences. As we will see in a more complete discussion of algal systematics in Chapter 5, 18S rDNA sequences offer particular advantages for discerning organismal relationships, particularly in highly diverse, ancient groups such as the algae. Molecular sequence information has provided evidence for the existence of eight to nine major clades or phyla (divisions) of algae. These are the cyanobacteria (chloroxybacteria) classified among the Eubacteria, and the eukaryotic phyla Glaucophyta

(glaucophytes, sometimes classified with the red algae), Euglenophyta (euglenoids), Cryptophyta (cryptomonads), Haptophyta (haptophytes), Dinophyta (dinoflagellates), Ochrophyta (a diverse array of tiny flagellates, diatoms, chrysophyceans, brown algae and a host of other groups), Rhodophyta (red algae), and Chlorophyta (green algae). Summaries of the major features of each of these groups are found at the end of this chapter and in Table 1–1. As more information, particularly nucleic acid sequence data, becomes available, it is possible that the composition of these groups, their classification as phyla, or concepts of relative divergence times may change. However, inasmuch as the emerging molecular-based phylogenies correspond well with concepts of relationships based on the cellular structure and biochemistry of algal cells, it is likely that many of the relationships described here are substantially accurate. Two groups of plastid-containing protists, the apicomplexans and chlorarachniophytes (discussed in Chapter 7), may, after additional study, be defined as algal phyla.

A closer examination of Figure 1–16 reveals that some algal clades are actually more closely related to particular, well-defined groups of non-photosynthetic protozoa than to other groups of algae. One example is the close relationship of euglenoids (euglenids) with kinetoplastid protozoa (organisms with an unusual mitochondrion-associated DNA-containing body, the kinetoplast). This suggests that the familiar laboratory organism *Euglena*, which is characterized by bright green chloroplasts, has a closer kinship to *Trypanosoma*, the agent of human diseases such as sleeping sickness, than to other green-colored algae. Another striking example—golden-colored, plastid-containing dinoflagellates such as *Peridinium*, are more closely related to ciliate protozoa (e.g., *Paramecium*) and to malarial and toxoplasmodial parasites (also known as **apicomplexans**) than to some other algae with golden chloroplasts, such as diatoms. How is this possible? An explanation for these apparent anomalies relates both to the phenomenon known as **endosymbiosis**, in which one or more endosymbiotic organisms (**endosymbionts**) live inside the cells of a **host** organism, as well as to the widespread **horizontal transfer** (movement from one organism to another) of genes involved in photosynthesis. These topics will be covered in more detail in Chapter 7.

Eukaryotic cells are defined by the presence of a double membrane-bound nucleus and, in most cases, other organelles such as mitochondria and chloroplasts (plastids). The plastid is the site of photosyn-

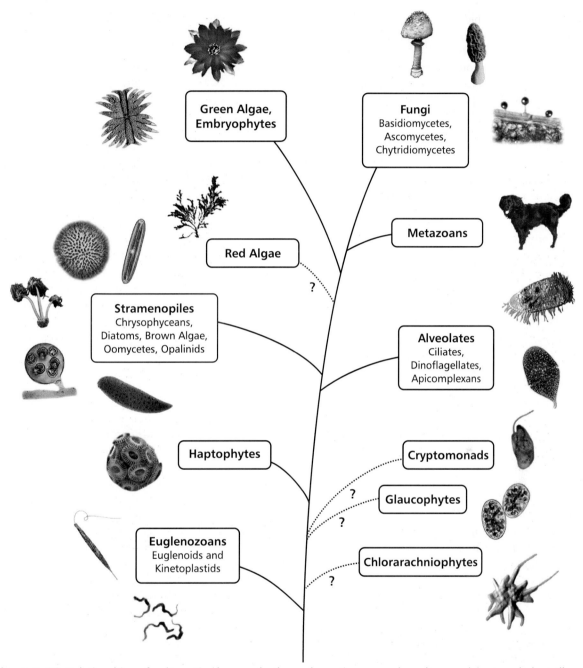

Figure 1–16 Relationships of eukaryotic algae and other eukaryotic groups, based on nuclear-encoded small subunit ribosomal RNA sequencing studies described by Schlegel (1994) and additional molecular sequence information. As new data become available, this phylogeny may change, and the placement of groups labeled with question marks may be better resolved.

thesis in eukaryotic algae. The term chloroplast, while appropriate for all algae because they all contain the green pigment chlorophyll, is applied principally to the usually green-colored plastids of euglenoids, green algae, and land plants. In view of the compelling evidence presented in Chapter 7, mitochondria and plastids are now considered to have originated by the endosymbiotic incorporation of bacteria into host cells. The transformation of free-living bacteria into mitochondria and plastids has been accompanied by

Table 1–1 Predominant photosynthetic pigments, storage products, and cell wall components for the major algal groups

	photosynthetic pigments	storage products	cell covering
Cyanobacteria (Chloroxybacteria)	chlorophyll *a* phycocyanin allophycocyanin phycoerythrin β-carotene xanthophylls	cyanophycin granules, cyanophytan starch (glycogen)	peptidoglycan
prochlorophytes	chlorophyll *a, b* β-carotene xanthophylls	cyanophytan starch (glycogen)	peptidoglycan
Glaucophyta	chlorophyll *a* phycocyanin allophycocyanin β-carotene xanthophylls	starch	some cellulosic
Euglenophyta	chlorophyll *a, b* β-carotene, other carotenes and xanthophylls	paramylon	proteinaceous pellicle beneath plasma membrane
Cryptophyta	chlorophyll *a, c* phycocyanin phycoerythrin α, β-carotene xanthophylls	starch	proteinaceous periplast beneath plasma membrane
Haptophyta	chlorophyll *a, c* β-carotene xanthophylls	chrysolaminaran	$CaCO_3$ scales common
Dinophyta	chlorophyll *a, c* β-carotene xanthophylls	starch	cellulosic plates in vesicles beneath plasma membrane
Ochrophyta (diatoms, chrysophytes, brown algae, etc.)	chlorophyll *a, c* β-carotene xanthophylls	chrysolaminaran, lipids	some naked; some with silica/organic scales; cellulose, alginates in some
eustigmatophytes	chlorophyll *a* β-carotene xanthophylls	chrysolaminaran	present, but unknown composition
Rhodophyta	chlorophyll *a* phycocyanin phycoerythrin allophycocyanin α, β-carotene xanthophylls	floridean starch	cellulose, sulfated polysaccharides; some calcified
Chlorophyta	chlorophyll *a, b* β-carotene, other carotenes and xanthophylls	starch	wall of cellulose/other polymers; scales on some; some naked; some calcified

the transfer of genetic function from one cellular compartment to another, such that the molecular biology of organelles has been dramatically influenced by endosymbiosis. Among the algae occurrence of multiple endosymbiotic events has generated particularly diverse cellular anatomies. Endosymbiosis has been a major force in the origin of the various algal groups and has been such a recurrent theme in algal evolution that it will be explored in much greater detail in Chapter 7 and subsequent chapters.

As previously mentioned and illustrated by Figure 1–16, the organisms known as algae do not fall within a single group linked by a common ancestor, so we cannot give the set of all organisms considered to be algae a formal name within the rules of biological nomenclature. There can be no relationship-based kingdom or phylum (division) that at the same time includes all of the algae and excludes organisms such as trypanosomes, oomycetes, ciliates and opalinids—organisms not generally considered to be algae. So, in contrast to monophyletic groups (such as the Kingdoms Fungi, Plantae, and Animalia), there are no defining features for the algae as a whole. However, sets of characteristics do define each of the algal phyla (pp. 19–21, Table 1–1).

Variations in Algal Nutrition—Relevance to Defining the Algae

As an alternative to a phylogenetic definition for the algae, we might attempt to define this group of organisms in terms of their ecologically important functions, most notably photosynthesis and **carbon fixation**—the transformation of dissolved inorganic carbon, such as carbon dioxide or bicarbonate ion, into an organic form. Therefore presence of chlorophyll—specifically chlorophyll *a*, which occurs in all photosynthetic algae—might be considered as a criterion for algae. Unfortunately, possession of chlorophyll and associated accessory pigments, along with the ability to perform photosynthesis, does not characterize all of the organisms that are considered to be algae. This has become particularly obvious in light of modern systematic investigations coupled with recent findings in studies of algal nutrition. Most algal groups contain colorless members that are devoid of chlorophyll or plastids but which are indisputably related to pigmented forms. Moreover, most major algal groups do not consist exclusively of obligate **photoautotrophs**. In other words, they have members that do not depend entirely upon the photosynthetic

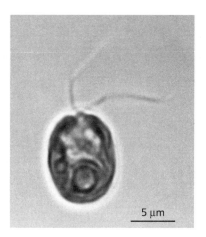

Figure 1–17 The motile unicellular green alga, *Chlamydomonas reinhardtii*. A number of *Chlamydomonas*-like flagellates can use acetate as a carbon source.

apparatus and the availability of light to produce organic carbon from CO_2.

The majority of algal groups contain **heterotrophic** species that obtain organic carbon from the external environment either by ingesting particles by a process known as **phagotrophy**, or through uptake of dissolved organic compounds, an ability termed **osmotrophy**. Many kinds of pigmented algae, including multicellular forms, are able to supplement photosynthesis with import of sugars, acetate, or other small organic compounds using cell membrane-based transport proteins similar to those found in osmotrophic protozoa and bacteria. Examples of osmotrophs include the acetate flagellates—relatives of the green alga *Chlamydomonas* that can use acetate as their sole source of carbon and energy (Fig. 1–17). Some algae, referred to as **auxotrophs**, are incapable of synthesizing certain essential vitamins and hence must import them. Only three vitamins are known to be required by auxotrophic algae—biotin, thiamine, and cobalamin (B_{12}).

A variety of other **flagellates**—motile, single-celled or colonial algae distinguished by possession of one or more flagella per cell—accomplish phagotrophy through intricate cellular feeding apparatuses similar to those of protozoa. Some dinoflagellates, euglenoids, and chrysophyceans are renowned for their ability to ingest whole planktonic cells (as well as smaller particulate organic materials). For example, the chrysophycean *Poterioochromonas malhamenhis* can graze algal prey two to three times its diameter (Zhang, et al.,

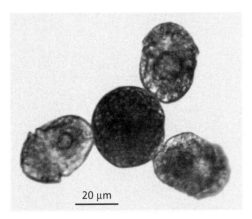

Figure 1–18 Three cells of the mixotrophic dinoflagellate, *Amphidinium cryophilum*, feeding upon another dinoflagellate. (Photograph by L. Wilcox and G. Wedemayer)

1996). Dinoflagellates so commonly exhibit phagotrophy that Jacobson and Anderson (1996) concluded that dinoflagellates are "overwhelmingly and fundamentally a phagotrophic lineage."

Numerous algae exhibit a mixed mode of nutrition; that is, photosynthesis in addition to osmotrophy and/or phagotrophy—an ability termed **mixotrophy**. In the literature the term mixotrophy has sometimes been applied exclusively to pigmented phagotrophic algae, in which case the multicellular green, red, and brown algae would be said to lack mixotrophic representatives. However in this text we will follow the definition of mixotrophy used in a major limnology text (Cole, 1994) because of its functional and ecological relevance—mixotrophy is the ability of chlorophyll-containing algae to utilize "both organic and inorganic carbon sources." Such a definition includes both phagotrophy and osmotrophy, and inasmuch as there is considerable evidence that osmotrophy occurs in multicellular green, red, and brown algae, mixotrophy can be said to characterize at least some members of all major algal groups. An example of a mixotrophic alga is the dinoflagellate *Amphidinium cryophilum*, a cold-water organism that is active during winter beneath ice and snow cover (Fig. 1–18). Observations suggest that when light levels are adequate, it actively photosynthesizes; under low light levels it feeds upon other dinoflagellates (Wedemayer, et al., 1982). An example of mixotrophy in a nonflagellate alga is the uptake of dissolved amino acids by rhizoids of the green seaweed *Caulerpa* (Chisholm, et al., 1996) (Fig. 1–19). Mixotrophic algae may modulate their different nutritional modes as the environment changes, sometimes functioning primarily as photoautotrophs (producers)—by "turning up" expression of the photosynthetic apparatus—and at other times as heterotrophs (consumers). Mixotrophs do not always behave as might be expected, however: some increase grazing rates as light levels decrease, while others decrease grazing under the same conditions (Jones, 1997). The mechanisms by which nutritional mode is regulated and the ecological significance of nutritional modulation by algae are not adequately understood at present and are potentially fertile areas for future research.

The widespread occurrence of mixotrophy among algae reflects the involvement of endosymbiosis in the evolutionary history of particular algal groups. Mixotrophs have retained feeding mechanisms originally present in host cells, despite their having gained photosynthetic capabilities through the acquisition of autotrophic endosymbionts (Tuchman, 1996). The frequent occurrence of mixotrophy among algae suggests that it is highly adaptive in aquatic habitats and therefore needs to be considered when making assessments of algal abundance and function in natural systems. Instantaneous and highly integrative ecosystem-level measures of algal abundance such as chlorophyll *a* levels, or water-sample measures of activity such as conversion of radiolabeled inorganic carbon to cellular organic compounds, can provide useful estimates of algal function under well-defined conditions. However because many algae are so nutritionally versatile, and since colorless members of the algae are not uncommon, such measures can generate an incomplete picture of **primary productivity** (conversion of carbon dioxide to organics) and **remineralization** (conversion of organic compounds to inorganic molecules). Mixotrophic algae acting as autotrophs at one time may undergo rapid transition to heterotrophy; instantaneous measures would not detect such shifts. In view of algal nutritional versatility, the accuracy of a commonly used estimator of algal productivity—particle size coupled with pigment composition—will vary depending upon the taxonomic composition of algal communities (Tang, 1996). Using this approach, productivity would be underestimated if algae were supplementing the carbon fixed through photosynthesis with that obtained through osmotrophy or phagotrophy.

The foregoing overview of algal diversity and nutritional variability leads to the concept of algae as a heterogeneous aggregation of remotely related

Figure 1–19 *Caulerpa*, a green seaweed capable of auto- and osmotrophy. The form of this macroalgal species is quite similar to that of certain gorgonian corals. (Photograph courtesy Ronald J. Stephenson)

groups that have several ecophysiological attributes in common: (1) most are photosynthetic oxygenic producers of organic compounds; (2) most are aquatic, occurring in oceans, lakes, ponds, wetlands, rivers, and streams; and (3) most are smaller and less complex than land plants. Exceptions to the first two of these generalities have been noted—there are heterotrophic and mixotrophic algae, and algae may occur in nonaquatic habitats. We need now to address the features of algae that separate them from plants.

Distinguishing Algae from Plants

Algae are often conceptually and systematically linked to the land plants because they include at least some forms—such as seaweeds—that look much like plants, and as with plants, are commonly sessile (attached or rooted in place), oxygenic (oxygen-producing) photosynthesizers. In contrast, fungi, though sessile, are invariably heterotrophic. Also, although fungi possess

cell walls, theirs (composed of chitin) are biochemically distinct from those of plants and algae. While some animals may contain symbiotic algal cells, animals (metazoa) can be distinguished from algae and other protists in that all are multicellular and obligately heterotrophic.

Photosynthetic algae and plants typically require sunlight as well as similar types and amounts of inorganic nutrients such as carbon dioxide, phosphorus, and fixed nitrogen. Since both groups store food in the form of organic carbon compounds, they are vulnerable to herbivorous predators and parasitic microorganisms such as viruses, bacteria, and fungi. Photosynthetic algae and plants must also contend with taller photosynthetic competitors and epiphytes, both of which reduce light available for photosynthesis. Therefore similarities exist in the means by which plants and algae obtain sunlight and in the uptake of inorganic nutrients at the cell membrane, as well as in the defense strategies they employ, such as production of anti-herbivory, allelopathic, or antimicrobial compounds.

Many seaweeds are very plantlike in appearance, having root, stem, and leaf analogs in the form of anchoring holdfasts, stipes, and blades. Certain kelps are noted for their internal phloemlike conducting tissues. However, as the phylogeny illustrated by Figure 1–16 suggests, brown, red, and even green seaweeds are only remotely related to land plants. Kelps, for example, are much more closely related to smaller brown seaweeds, their unicellular cousins among the golden algae, and heterotrophic pseudofungi such as oomycetes, than to land plants. Thus similarities in macroscopic structure between seaweeds and land plants are due to **parallel evolution**.

In modern classification schemes based upon molecular phylogenetic systematics, the plant kingdom (Kingdom Plantae) includes only the **embryophytes** (land plants), which comprise the bryophytes and vascular plants (Fig. 1–16). Even the most plantlike seaweeds are classified in the Kingdom Protista (Raven, et al., 1999). Embryophytes, though sharing many features with green algae (such as photosynthetic pigments and storage products), nevertheless are characterized by a suite of characters not found among the green algae. These include a multicellular, diploid embryo stage that is developmentally and nutritionally dependent on parental gametophyte tissues for at least some time during early development. This feature is the source of the term embryophyte (Margulis and Schwartz, 1988). Other unique and defining features (**autapomorphies**) of embryophytes include tissue-generating apical meristems and preprophase microtubule bands in cells preparing to divide—features absent from the closest green algal relatives of embryophytes (Graham, 1996). Thus there are phylogenetic reasons for distinguishing algae from land plants.

A Definition for the Algae

In summary, we can say that algae are (with numerous exceptions) aquatic organisms that (with frequent exceptions) are photosynthetic, oxygenic autotrophs that are (except for the kelps) typically smaller and less structurally complex than land plants. This rather inelegant definition allows us to include the cyanobacteria (chloroxybacteria), which, although **prokaryotic**, resemble some members of other (eukaryotic) algal groups in terms of overall structure and ecosystem function. Cyanobacteria are distinguished from other bacteria mainly by their possession of the biochemical apparatus necessary for oxygen production (some other bacteria are photosynthetic but do not generate molecular oxygen). In addition to cyanobacteria, the algae consist of several groups of remotely related protists, many members of which have acquired plastids through endosymbiosis. Although it is possible to clearly distinguish algae from plants, animals, and fungi, they cannot, as a whole, be separated from the rest of the protists.

General Characteristics of Algae

As should now be apparent, it is difficult to make broad statements about the ancient and diverse assemblage of organisms referred to as algae. Attempts to list the salient features of a particular algal group is likewise difficult, since for virtually every characteristic considered there are at least some exceptions. In the next portion of this chapter we shall nevertheless attempt to outline some general characteristics of the major algal groups, including pigment composition, types of storage products, and the nature of the cell covering. We will also introduce some of the common types of algal growth forms in addition to the basic modes of sexual and asexual reproduction in algae. In Chapters 6 and 8–21 we will cover these topics in more detail with respect to individual algal groups and consider some of the exceptions to the generalizations made here, which often provide unique insights into the creative ways algae have adapted to diverse habitats and selective pressures.

Range of Morphological Diversity in Algae

A great deal of variation exists in the morphology of the algal **thallus** (the algal body), the most commonly encountered forms of which are described briefly in the following paragraphs and illustrated in Figure 1–20.

Unicells and colonies

Many algae occur as solitary cells (unicells) while others may be made up of several to many individual cells held together loosely or in a highly organized fashion. Some unicellular algae are nonmotile, while others possess one (or more) of the various means of locomotion found among the algae. As mentioned earlier, some algae have locomotory structures known as flagella. Such flagellates can be either unicellular or colonial. A **colony** is an assemblage of individual cells in which there may be either a variable number of cells or a predictable number and arrangement of cells that remain constant throughout the life of the individual. A colony of the latter type is referred to as a **coenobium**. Depending on the organism, cells in coenobia may be either flagellated or nonmotile.

Filaments

A common growth form among the algae is the **filament**, where daughter cells remain attached to each other following cell division forming a chain of cells. Filaments may be unbranched or branched and may be **uniseriate** (a single series of cells) or **multiseriate** (pluriseriate), where a few to many individual filaments fuse together to form a larger, more complex structure. Linear colonies, formed by some diatoms, for example, can be distinguished from true filaments by the fact that cells of the former each possess their own individual walls, whereas adjacent cells of true filaments share a wall.

Coenocytic or siphonaceous forms

Less common are algae with a **coenocytic** or **siphonous** growth habit. Such organisms basically consist of one large multinucleate cell, without cross walls.

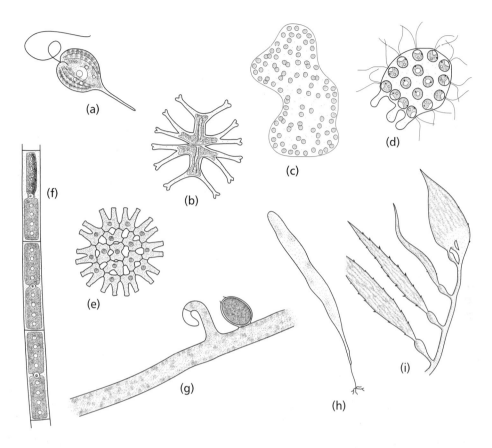

Figure 1–20 Algal growth habits include motile (a—the euglenoid *Phacus*) and nonmotile (b—the green alga *Micrasterias*) unicells; non-coenobic colonies or aggregations (c—the cyanobacterium *Microcystis*); motile (d—the green *Platydorina*); and nonmotile (e—the green *Pediastrum*) coenobia; filaments (f—the green *Mougeotia*); siphonous (g—the tribophycean *Vaucheria*); pseudoparenchymatous multicellular types (h—the green *Monostroma*); and parenchymatous forms (i—the brown kelp *Macrocystis*). (h: After Bold and Wynne, 1985)

Parenchymatous and pseudoparenchymatous algae

Parenchyma is a term used to describe plant (or algal) tissue that is composed of relatively undifferentiated, isodiametric cells generated by a **meristem**. It results from cell divisions occurring in three directions, which gives rise to a three-dimensional form. Pseudoparenchymatous algae have thalli that superficially resemble parenchyma, but which are actually composed of appressed filaments or amorphous cell aggregates. Evolutionarily, parenchymatous growth habits are thought to represent the most highly derived state, with pseudoparenchymatous forms representing an intermediate condition between filamentous and parenchymatous conditions. Parenchymatous and pseudoparenchymatous algae assume a wide range of shapes (sheets, tubes, stem- and leaf-like arrangements, etc.) and sizes (microscopic to lengths of 50 m or greater).

Algal Reproduction

Algae reproduce by a variety of means, both sexual and asexual. In sexual reproduction, **plasmogamy**—fusion of haploid reproductive cells (**gametes**)—is followed by **karyogamy** (nuclear fusion), to form a diploid zygote. The homologous chromosomes contributed by each of the two gametes pair and at some point are partitioned into haploid cells through the process of meiosis. Asexual reproduction is a means by

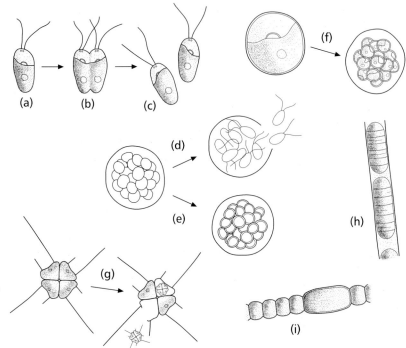

Figure 1–21 Examples of asexual reproduction in the algae include (a)–(c) cellular bisection; (d) zoospore, (e) aplanospore, and (f) autospore production; (g) autocolony formation; (h) fragmentation; and (i) akinetes.

which an individual organism can produce additional copies of itself without such unions of cytoplasmic and nuclear materials or meiosis.

Asexual reproduction

A number of the more common processes and structures involved in asexual reproduction in the algae are introduced here and illustrated in Figure 1–21. Additional discussion is found in Chapters 6 and 8–21.

Cellular bisection. In many unicellular algae reproduction is simply by longitudinal or transverse cell division (Fig. 1–21a, b, c). In multicellular algae (or colonies with indeterminate numbers of cells), this process would lead to growth of the individual, i.e., an increase in the size and the number of its cells. In unicells cell division leads to population growth.

Zoospore and aplanospore formation. **Zoospores** are flagellate reproductive cells that may be produced within vegetative cells or in specialized cells, depending on the organism (Fig. 1–21d). Zoospores contain all of the components necessary to form a new individual. Sometimes, rather than forming flagella, the spores begin their development before being released from the parental cell or sporangium (Fig. 1–21e). These non-motile spores (which possess the ontogenetic possibility of developing into zoospores) are termed **aplanospores**.

Autospore or monospore production. **Autospores** and **monospores** are also nonmotile spores, but unlike aplanospores, lack the capacity to develop into zoospores. They typically look like miniature versions of the parental cell in which they form (Fig. 1–21f). In green algae, such cells are known as autospores; they are termed monospores in red algae.

Autocolony formation. In coenobia, each cell goes through a consistent number of successive divisions giving rise to a miniature version—an **autocolony**—of the original coenobium (Fig. 1–21g). Depending on the organism, autocolonies may be formed from nonmotile or motile cells that arrange themselves in a pattern identical to that of the parental cells.

Fragmentation. Some multicellular and non-coenobic colonial algae give rise to new individuals through **fragmentation** (Fig. 1–21h). In colonies and some filamentous species this can be a simple, more or less random process whereby an individual breaks into two or more pieces, each of which can continue to grow. In other algae the process is more specialized, with, for example, predictably occurring weak links in filaments or elaborate budlike propagules in some multicellular algae.

Akinetes. An **akinete** is a specialized cell that develops a thickened cell wall and is enlarged, compared to typical vegetative cells (Fig. 1–21i). It is usually a resistant structure with large amounts of stored food reserves that allow the alga to survive harsh environmental conditions, germinating when they improve. Rather than a means of producing additional copies of the individual during active growth, akinetes represent a type of survival mechanism.

Sexual reproduction

Gametes and gamete fusion. Gametes may be nearly identical to vegetative cells in appearance or vastly different, depending upon the alga. In isogamous sexual reproduction (**isogamy**), motile gametes that are more or less identical in size and shape fuse with each other. In **anisogamy**, two motile gametes of different size or behavior pair, while in **oogamy**, a flagellate or nonflagellate cell fuses with a larger immobile egg. Among algae, oogamous sexual reproduction is thought to represent the derived condition, and isogamy and anisogamy, the basal and intermediate conditions, respectively.

In anisogamous and oogamous species, the two types of gametes may be produced by the same individual, in which case the species is termed **monecious**. If they are produced by separate individuals, it is said to be **dioecious**. When gametes from the same individual are able to fuse and produce viable offspring, the organism is termed **homothallic** (self-fertile). If such gametes are incompatible, then two individuals of different genetic makeup are required for successful mating, and the organism is termed **heterothallic** (self-sterile).

Life-history types. Algae exhibit an amazing diversity of life histories. The three principal types are illustrated in Figures 1–22, 1–23, and 1–24, and summarized here. The primary differences between them include the point where meiosis occurs and the type of cells it produces, and whether or not there is more than one free-living stage present in the life cycle. Characteristics of the three types are:

1. The major portion (vegetative phase) of the life cycle is spent in the haploid state, with meiosis taking place upon germination of the zygote (**zygotic meiosis**) (Fig. 1–22).

2. The vegetative phase is diploid, with meiosis giving rise to the haploid gametes (**gametic meiosis**) (Fig. 1–23).

3. Two or three multicellular phases occur—the **gametophyte** (typically haploid) and one (or more, in the case of many red algae) **sporophytes** (typically diploid). The gametophyte produces gametes through mitosis, and the sporophyte produces spores through meiosis (**sporic meiosis**). This type of life cycle illustrates the phenomenon of **alternation of generations**. Alternation of generations in the algae can be **isomorphic**, in which the gametophyte and sporophyte are structurally identical, or **heteromorphic**, where gametophyte and sporophyte phases are dissimilar (Fig. 1–24).

The first two life-cycle types are sometimes termed **haplobiontic** (one type of free-living individual) and the final type, **diplobiontic** (two free-living stages). The first type can also be termed **haplontic** (a single, predominant haploid phase) and the second **diplontic** (a single, predominant diploid phase). Using this terminology, the third type is referred to as a **diplohaplontic** life cycle (two phases, haploid and diploid) (Bold and Wynne, 1985). To avoid the use of what can be confusing, similar-sounding terms (haplobiontic, haplontic, haploid, etc.), we will distinguish these life-cycle types by the nature of meiosis (z=zygotic, g=gametic, and s=sporic), with the realization that these terms are inconsistent in that zygotic refers to the place where meiosis occurs, while gametic and sporic refer to the nature of the meiotic products. We have included an icon in each of the life-cycle diagrams in this text to indicate its type.

The third type of life cycle also typifies the land plants (sporic, alternation of generations, diplobiontic, diplohaplontic). Careful comparison of alternation of generations in the algae reveals fundamental differences among different algal groups, as well as critical differences from the life cycle of land plants. It is quite clear that this type of life cycle has evolved multiple times. Therefore the possession of this life-history pattern by plants and a particular algal group should not be taken as evidence of a close relationship between the two. In fact, as we shall see in Chapters 17 and 21, the evidence is compelling that the green algal progenitors of land plants lacked alternation of generations and that origin of this trait in plants coincided with the evolution of the multicellular plant embryo.

A final comment on algal reproduction is that in some algae, one or both gametes may develop into haploid individuals if they do not happen to fuse with another gamete to form a zygote. This phenomenon is referred to as **parthenogenesis**.

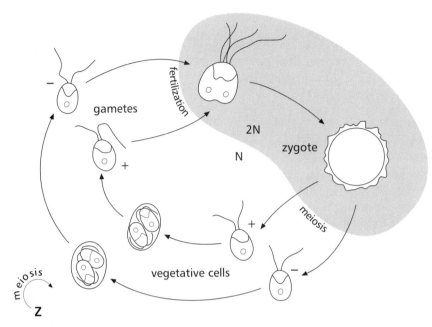

Figure 1–22 Zygotic meiosis in the green unicellular flagellate *Chlamydomonas*.

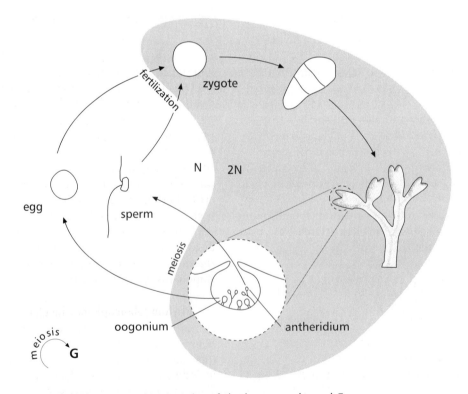

Figure 1–23 Gametic meiosis in a monecious species of the brown rockweed *Fucus*.

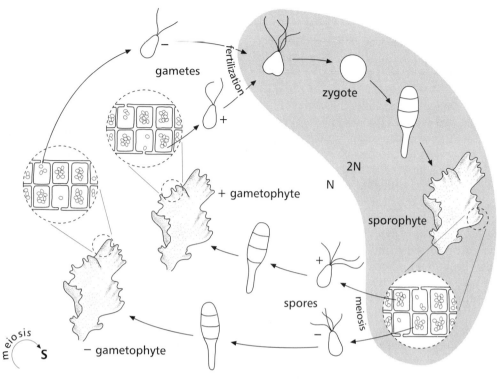

Figure 1–24 Sporic meiosis in the green alga *Ulva*. Note that there are two free-living multicellular stages, one haploid and one diploid (alternation of generations).

Biochemical and Structural Features of Algae

A number of characteristics have been traditionally used in distinguishing among the major algal groups. Prominent among these are the types of photosynthetic pigments, nature of the cell covering, and the type(s) of storage reserves present. More extensive discussion of each of these topics will be given in Chapters 6 and 8–21. We have, however, included a brief summary of these features in Table 1–1, as a reference for subsequent chapters.

Summaries of the Nine Algal Phyla Treated in This Book

Phylum Cyanobacteria (chloroxybacteria, blue-green algae) is a well-defined group of eubacteria. Cyanobacteria include unicellular and filamentous forms, some having specialized cells. Uniquely among bacteria, cyanobacteria produce oxygen as a by-product of photosynthesis. Chlorophyll *a* and accessory and protective pigments—phycobilins and carot-

enoids—are present, associated with membranous thylakoids. Some members of the group—the prochlorophytes—also possess chlorophyll *b*. The photosynthetic storage products include an α-1,4-glucan known as cyanophytan starch. Among autotrophs, cyanobacterial cells are unique in being prokaryotic in organization, hence typical eukaryotic flagella and organelles (chloroplasts, mitochondria, and nuclei) are lacking. Cyanobacteria are common and diverse in both freshwaters and the sea. Given the difficulties in applying species concepts to prokaryotes, the number of species has been difficult to determine. Sexual reproduction of the typical eukaryotic type, involving gamete fusion, is not present.

Phylum Glaucophyta (the glaucophytes) includes several eukaryotes having blue-green plastids (known as cyanelles or cyanellae) that differ from other plastids and resemble cyanobacteria in several ways, including the possession of a thin peptidoglycan wall. The cyanelles/plastids possess chlorophyll *a* and phycobilins, as well as carotenoids. Granules of true starch (an α-1,4-glucan) are produced in the cytoplasm. There are about nine genera, all freshwater. At least one of these is a flagellate, and nonflagellate forms (unicells and colonies) typically produce flagellate asexual reproductive cells. Sexual reproduction is unknown. This group has sometimes been included within the red algae (Rhodophyta).

Phylum Euglenophyta includes the euglenoid flagellates, which occur as unicells or colonies. There are about 40 genera, two thirds of which are heterotrophic, some having colorless plastids and some lacking plastids altogether. One third have green plastids with chlorophyll *a* and the accessory pigment chlorophyll *b* as well as the carotenoids that are typical of green algae, and are capable of photosynthesis. Cell walls are lacking, but there is a protein-rich pellicle beneath the cell membrane. One to several flagella may be present, and nonflagellate cells can undergo a type of motion involving changes in cell shape. The storage material is not starch but rather a β-1,3-linked glucan known as paramylon, which occurs as granules in the cytoplasm of pigmented as well as most colorless forms. Most of the 900 or so species are freshwater, and sexual reproduction is not known.

Phylum Cryptophyta contains the unicellular cryptomonad flagellates, with 12–23 genera. A few are colorless, but most possess variously colored plastids with chlorophyll *a*. Chlorophyll *c*, carotenoids, and phycobilins constitute the accessory pigments. Alloxanthin is a xanthophyll that is unique to cryptomonads. There is not a typical cell wall. Rather, rigid proteinaceous plates of various shapes occur beneath the cell membrane. Cells can be recognized by their typical flattened asymmetrical shape and the two anterior, slightly unequal flagella. The storage carbohydrate is starch, located in a space between plastid membranes. There are about 100 freshwater species and about 100 marine species. There is some evidence for sexual reproduction.

Phylum Haptophyta (haptophytes or prymnesiophytes) comprises unicellular flagellates or nonflagellate unicells or colonies that have flagellate life-history stages. The photosynthetic pigments include chlorophyll *a*, and accessory and photoprotective pigments including chlorophyll *c* and carotenoids such as fucoxanthin. Species vary in the form of chlorophyll *c* and presence or absence and form of fucoxanthin. There is a β-1,3-glucan storage material. Two flagella and a nearby structure known

as a haptonema characterize the apices of flagellates. Many species, known as **coccolithophorids**, produce calcium carbonate-rich scales called **coccoliths**. The 300 species are primarily marine. Sexual reproduction, known in some cases to involve heteromorphic alternation of generations, is widespread.

Phylum Dinophyta encompasses the dinoflagellates, mostly unicellular flagellates having two dissimilar flagella. About one half of the 550 genera are colorless heterotrophs; the rest possess plastids that vary significantly in pigment composition and type of **Rubisco**, the enzyme responsible for photosynthetic carbon fixation. Pigmentation is usually golden brown, reflecting the common occurrence of the unique accessory xanthophyll peridinin, but green and other colors are known. True starch granules occur in the cytoplasm. The cell covering is a peripheral layer of membrane-bound vesicles, which in many cases enclose cellulosic plates. Of the 2000–4000 species, the vast majority are marine; only about 220 freshwater forms are recognized. Symbiotic dinoflagellates known as zooxanthellae occur in reef-forming corals and other marine invertebrates. Sexual reproduction is known.

Phylum Ochrophyta (also known as chromophytes) includes diatoms, raphidophyceans, chrysophyceans, synurophyceans, eustigmatophyceans, pelagophyceans, silicoflagellates, pedinellids, tribophyceans, phaeophyceans (brown algae), and some other groups. Members range in size from microscopic unicells to giant kelps having considerable tissue differentiation. Chlorophyll *a* is present in most ochrophytes, but some colorless heterotrophic forms also occur. In the pigmented forms, dominant accessory and photoprotective pigments may include chlorophyll *c* and carotenoids such as fucoxanthin or vaucheriaxanthin. The food reserve is cytoplasmic lipid droplets and/or a soluble carbohydrate—the β-1,3-glucan chrysolaminaran or laminaran—which occurs in cytoplasmic vacuoles. There are usually two heteromorphic flagella, one bearing many distinctive three-piece hairs known as mastigonemes. Cell coverings vary widely and include silica scales and enclosures as well as cellulose cell walls. There are more than 250 genera and 10,000 species of extant diatoms alone. Some groups of the Ochrophyta are primarily freshwater, some are primarily marine, and some, such as diatoms, are common in both fresh and salt water. The brown algae known as the giant kelps are the largest of all the algae. Sexual reproduction is common, and several types of life cycles occur.

Phylum Rhodophyta, or the red algae, has members that occur as unicells, simple filaments, or complex filamentous aggregations. Pigments are present in all except certain parasitic forms, and include chlorophyll *a* together with accessory phycobilins and carotenoids. Flagella are not present (unless the glaucophyte algae are included within this phylum). The cytoplasmic carbohydrate food reserve is granular floridean starch, an α-1,4-glucan. Cell walls are loosely constructed of cellulose and sulfated polygalactans, and some are impregnated with calcium carbonate. The calcified red algae known as corallines are widespread and ecologically significant in coral reef systems. The 4000–6000 species are primarily marine, favoring warm tropical waters. Sexual reproduction is common, as is alternation of generations. A triphasic life history characterizes most red algae and is unique to this group.

Phylum Chlorophyta (green algae or chlorophytes) have unicellular or multicellular thalli. Some are flagellates, and others produce reproductive cells, the majority of which are biflagellate. In addition to chlorophyll *a*, the pigments chlorophyll *b*, β-carotene, and other carotenoids occur in plastids. Uniquely, starch is produced within plastids of green algae (and land plants). Cell walls of some are cellulosic as in land plants, but the walls of other green algae are composed of different polymers, and some are calcified. Early divergent flagellates (the prasinophyceans) and one multicellular clade (the ulvophycean green seaweeds) are primarily marine, whereas other groups are primarily terrestrial or freshwater. One of the freshwater lineages (the charophyceans) gave rise to the land plants (embryophytes). The common and frequently calcified macrophyte stoneworts (the charaleans) are members of the charophycean lineage. There are about 17,000 species. Sexual reproduction is common, and all three major types of life cycle occur.

Chapter 2

The Roles of Algae in Biogeochemistry

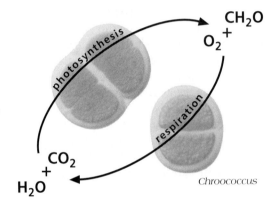

Chroococcus

Biogeochemistry is the study of chemical interactions between the atmosphere, hydrosphere (aquatic systems), lithosphere (crustal minerals), and the biosphere (living organisms). Algae have played a significant role in the earth's biogeochemistry since the origin of cyanobacteria—probably over three billion years ago—and they play an indispensable role in global biogeochemistry today. Algae are particularly important in global cycling of the elements N, O, S, P, and C, and are becoming ever more significant as human population growth results in massive alteration of biogeochemical cycling of these and other elements. Human-induced changes in the global nitrogen cycle, for example, have already had multiple consequences, including dramatic increases in the frequency, extent, and duration of harmful algal blooms during the past two decades. Such algal blooms cause fish kills through anoxia and toxins, cause human illness, and contribute to the loss of natural biodiversity in aquatic systems. In attempting to feed ever larger populations, humans have doubled the natural rate of nitrogen input into terrestrial systems, adding as much fixed nitrogen as do all natural sources combined. Runoffs from agricultural lands and feedlots, together with sewage effluents, have significantly increased nitrate levels in streams, rivers, estuaries, and coastal oceans around the world. For example, there has been a doubling of nitrate in the Mississippi River (U.S.A.) since 1965, with the consequent increase in nuisance algal blooms in the Gulf of Mexico. Nitrogen influx is considered to be the single greatest threat to coastal ecosystems throughout the world (Vitousek, et al., 1997a, b).

Algae are of additional biogeochemical concern because some generate volatile halogen-containing compounds that destroy stratospheric ozone and thus may be responsible (together with industry-generated halocarbons) for increases in deleterious ultraviolet (UV) penetration to the earth's surface. Algae are also a major force in the global sulfur cycle because many species produce volatile sulfur compounds that are active in cloud formation and consequently have a cooling effect on climate. Since the industrial age, algal production of sulfur (some 12–58 trillion g yr^{-1}) has been overtaken by human inputs of sulfur compounds to the atmosphere (some 69–103 trillion g yr^{-1}) (Malin and Kirst, 1997). Scientists are interested in possible deleterious feedback effects of such climate perturbations, as well as nutrient changes such as the nitrogen cycle alterations mentioned above, on natural systems.

Other human impacts on aquatic ecosystems that influence algal populations and global biogeochemistry include changes in phosphate concentrations of natural waters and increases in atmospheric carbon dioxide levels, which are correlated with measurable increases in global temperature. These types of environmental alterations are expected to change the character of algal communities and have deleterious food-chain effects (see Chapters 22 and 23). Some experts (Riebesell, et al., 1993) have suggested that algae may be able to reduce atmospheric carbon dioxide levels through photosynthetic carbon fixation followed by sedimentation of algal-produced organic carbon to the deep ocean, but others (Falkowski and Raven, 1997) think that this is unlikely to occur on a biogeochemical scale. Hence a major current controversy concerns the extent to which algae may or may not be able to "draw down" excess carbon dioxide, which would help to buffer possible climatic effects (global warming). This particular issue hinges on whether or not algal growth is limited by the availability of carbon dioxide. To better understand such issues and the varied biogeochemical roles played by algae, we will begin with a discussion of the chemical factors that limit algal growth.

Limiting Factors

For growth to occur algae require a wide range of mineral (inorganic) nutrients in addition to light, including the potentially growth-limiting nutrients carbon dioxide, combined ("fixed") nitrogen, phosphate, iron, and in the case of diatoms and some other algae, silica (Table 2–1). The list of minerals required for algal growth is similar to that for higher plants. Limiting factors are necessary resources that may, if not present in sufficient quantities, limit the growth of algal and plant populations. The concept of limiting factors, originally developed by Liebig for higher land plants, is known as the law of the minimum. Liebig observed that at any single point in time, higher plant growth is typically limited by the single mineral nutrient whose environmental concentration was closest to the critical minimal level required. When the level of

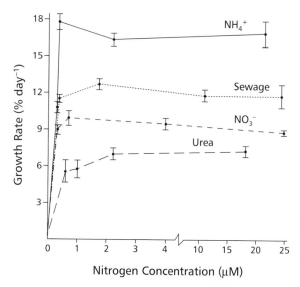

Figure 2–1 An example of results obtained from laboratory analysis of growth of an algal species (the red seaweed *Agardhiella subulata*) in response to differences in the concentrations of a variety of combined nitrogen sources in the external medium. Growth rates first exhibit rapid (logarithmic) increase, then reach a plateau stage during which increases in nitrogen concentration do not change growth rate. At the onset of the plateau stage, the algal growth rate has become limited by environmental levels of some other required resource. (After de Boer, et al., 1978 by permission of the *Journal of Phycology*)

this limiting mineral was increased, growth increased concomitantly until another mineral became limiting. Although algae differ widely in the levels of chemical factors that limit their growth, and nutrient ratios are also important, in general the concept of limiting nutrients that Liebig developed for plants may also be applied to algae.

Determination of the concentrations at which major nutrients limit algal growth requires laboratory analysis of an individual alga's growth at varying resource levels (Fig. 2–1). Such analysis can often help to explain why a particular alga is found in one type of aquatic environment but not another. Comparisons of the nutrient concentration, or ratios such as N/P, at which different algal species are limited can be used to predict which organisms should predominate under stable resource conditions (Chapter 22). However aquatic habitats may differ in the kinds of nutrients that are limiting, patches of nutrient-limited water (as well as more nutrient-rich patches) may occur within water bodies, algal species have differing

Table 2–1 Elements commonly required by algae

element	examples of function/location in algal cells
N	amino acids, nucleotides, chlorophyll, phycobilins
P	ATP, DNA, phospholipids
Cl	oxygen-production in photosynthesis, trichloroethylene, perchloroethylene
S	some amino acids, nitrogenase, thylakoid lipids, CoA, carrageenan, agar, DMSP, biotin
Si	diatom frustules, silicoflagellate skeletons, synurophyte scales and stomatocyst walls, walls of the ulvophyte *Cladophora*
Na	nitrate reductase
Ca	alginates, calcium carbonate, calmodulin
Mg	chlorophyll
Fe	ferredoxin, cytochromes, nitrogenase, nitrate and nitrite reductase, catalase, glutamate synthetase
K	agar and carrageenan, osmotic regulation (ionic form), cofactor for many enzymes
Mo	nitrate reductase, nitrogenase
Mn	oxygen-evolving complex of photosystem II, wall-like lorica of some euglenoids and the chlorophyte *Dysmorphococcus*
Zn	carbonic anhydrase, Cu/Zn superoxide dismutase, alcohol dehydrogenase, glutamic dehydrogenase
Cu	plastocyanin, Cu/Zn superoxide dismutase, cytochrome oxidase
Co	vitamin B_{12}
V	bromoperoxidase, some nitrogenases
Br, I	halogenated compounds with antimicrobial, anti-herbivore, or allelopathic functions

capacities to store excess amounts of nutrients, and nutrient levels may change rapidly—all of which make predicting the outcome of algal competition for limiting resources difficult.

Release of algal growth from nutrient limitation is commonly correlated with the sudden development of algal blooms. Phosphorus is the mineral nutrient that is most commonly limiting in **oligotrophic** freshwater lakes (Vitousek, et al., 1997a, b). Oligotrophic lakes are characterized by low nutrient levels, low pro-

ductivity, and high species diversity. Phosphate (PO_4^{3-}) levels are low in oligotrophic lakes because this ion readily binds Al^{3+}, Fe^{3+}, and Ca^{2+}, forming highly insoluble complexes in soils and lake sediments. Unless drained by volcanic soils, oligotrophic lakes usually receive sufficient nitrogen from surface and ground water that they are not nitrogen-limited (Lampert and Sommer, 1997). Further, freshwaters often contain large numbers and biomasses of nitrogen-fixing cyanobacteria (Chapter 6) (Falkowski and Raven,

1997). As phosphate is added to an oligotrophic lake, the higher concentrations allow for greater aquatic plant and algal growth (i.e., increased primary productivity), a process known as **eutrophication**. When lakes become rich in phosphorus and exhibit increased plant and algal biomass and reduced biodiversity, they are described as **eutrophic**. Hypereutrophic lakes are exceedingly rich in nutrients and typically characterized by repeated and enduring algal blooms. Such changes in nutrient status often have undesirable effects on the appearance, fisheries production, and recreational use of lakes and other waters. When phosphorus is so abundant that it no longer limits algal growth in lakes, another nutrient (usually nitrogen, though sometimes iron) becomes limiting. During algal blooms in alkaline lakes, and in acidic softwater lakes of pH 5 or lower, inorganic carbon may limit algal growth (Hein, 1997; Turner, et al., 1991; Fairchild and Sherman, 1993).

For reasons that will be discussed later, inorganic carbon levels are unlikely to limit the growth of most marine phytoplankton or seaweeds (Raven, 1997b, c). Rather, algal growth in oligotrophic marine waters is most commonly limited by the availability of nitrogen (Vitousek, et al., 1997a, b) and, indirectly, by iron, which influences nitrogen fixation rates and hence the availability of nitrogen. The relative scarcity of combined nitrogen in marine waters has been attributed to the low ratio of fixed nitrogen to phosphate in land-derived sediments, and the fact that combined nitrogen is less rapidly regenerated from decomposing organic matter than is phosphate (Cole, 1994). There is considerable evidence that in the oceans the amount of fixed nitrogen depends on the ratio of nitrogen fixation to denitrification, which in turn depends on the oxidation state of the ocean and the supply of trace elements such as iron (Falkowski, 1997). The strongest effect of nitrogen deficiency in algae is the decline in photosynthetic pigments (chlorophylls and phycobilins) that contain nitrogen. By this and other means nitrogen levels affect the extent of carbon fixation on geological timescales (reviewed by Turpin, 1991). Exceptions to the general rule of ocean nitrogen-limitation include some tropical lagoons where sand binds phosphate (in which case phosphate is limiting), and eutrophic estuaries and seas receiving high nutrient loads. In areas of the open ocean known as high-nutrient, low chlorophyll (HNLC) regions, such as the arctic and equatorial Pacific Ocean and the Southern Ocean surrounding Antarctica, there is increasing evidence that neither nitrate, phosphate, nor silica limit

algal productivity, but rather that iron levels represent the major control on phytoplankton growth.

Every year oceanic phytoplankton produce some 40×10^{15} g of fixed carbon, with an additional 4×10^{15} g generated by coastal macroalgae (seaweeds) and periphyton. Currently this is 50–100% of the primary productivity of terrestrial ecosystems (Raven, 1997b, c). As marine ecosystems become further nitrogen-enriched, productivity and algal biomass is likely to increase until it becomes limited by another nutrient, such as iron. Hence oceanographers and climatologists are interested in interactions between iron, nitrogen, carbon, and other nutrients. They seek to understand natural nitrogen fixation by cyanobacteria (and some other eubacteria) and its relationship to carbon fixation, and how carbon fixation might change in ways that influence climate. Biologists hope to predict how nitrogen pollution will affect algae and other components of aquatic ecosystems. Consequently, we will explore these major algal nutrients in more detail, beginning with nitrogen utilization by algae and the production of fixed nitrogen by cyanobacteria.

Algae and the Nitrogen Cycle

Combined nitrogen is required by algae to manufacture amino acids, nucleic acids, chlorophyll, and other nitrogen-containing organic compounds. Some algae are capable of taking up organic forms of combined nitrogen, such as urea and amino acids (see review by Tuchman, 1996). For example, radiolabeled glycine was demonstrated to enter cells of the diatom *Phaeodactylum tricornutum*, and this organism can remove 14 different amino acids from dilute solution (Lu and Stephens, 1984). The coccolithophorid *Emiliania huxleyi* is able to take up small amides (such as acetamide), from which ammonium is cleaved within the cell, and some algae can use extracellular enzymes to cleave ammonium from dissolved amino acids (Palenik and Henson, 1997).

Inorganic sources of combined nitrogen include ammonium (NH_4^+) and nitrate (NO_3^-), which are taken up at the algal cell membrane. Most algae are able to import ammonium or nitrate, but ammonium is often preferred because it can be used in a more direct fashion than nitrate in the biosynthesis of amino acids. Conversion of nitrate to ammonium requires energy and the enzyme nitrate reductase (NR) (Fig. 2–2). Nitrate reductase includes iron and molybdenum as cofactors, and iron concentrations may be low in neutral to alka-

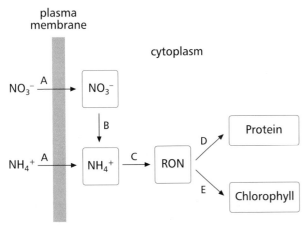

Figure 2–2 A comparison of algal cell utilization of nitrate versus ammonium. Following uptake involving membrane transport systems (A), ammonium can be converted directly to reduced organic nitrogen (RON), then utilized in protein and chlorophyll synthesis, via enzyme systems C, D, and E. In contrast, nitrate utilization requires an additional enzymatic step (B). (After Lobban and Harrison, 1994)

line natural waters where iron is bound in low solubility compounds, notably $Fe(OH)_3$. In contrast, ammonium uptake is independent of iron availability. However, aquatic nitrifying bacteria readily convert ammonium to nitrate, often limiting the availability of ammonium to algae. Consequently there has been selective pressure for the evolution of nitrate reductases, which occur in a wide range of algal groups. Cyanobacteria contain a distinctive form of nitrate reductase, and at least two other forms of this enzyme occur in eukaryotic algae. As would be expected, nitrate reductase activity is usually inducible by low ammonium levels, while NR expression is repressed by high levels of ammonium. Ecologists find that measuring NR activity or gene expression is a useful way of estimating rates of nitrate incorporation into algae, which are otherwise difficult to measure. Algal nitrate reductases and their ecological utility are reviewed by Berges (1997).

Cyanobacteria are unique among oxygenic photosynthesizers in their ability to fix nitrogen, that is, to convert gaseous N_2 into ammonium, a type of metabolism known as **diazotrophy**. Because N_2 is abundant, making up some 70% of the atmosphere, this ability provides cyanobacteria with a competitive advantage over other algae and plants growing in nitrogen-limited waters. The ability to fix nitrogen is one of the factors contributing to development of

undesirable cyanobacterial blooms in fresh and marine waters. The cyanobacterium *Trichodesmium* is a common bloom-former in the tropical North Atlantic Ocean and the Caribbean. Its success is due in part to the ability to fix as much as 30 mg of nitrogen m^{-2} day^{-1} (Carpenter and Romans, 1991; Capone, et al., 1997). In freshwaters rich in phosphate, nuisance cyanobacterial blooms of N_2-fixing *Anabaena* and *Aphanizomenon* are common. Some 40–60% of the nitrogen fixed by cyanobacteria can be excreted from the cells into the open water, thus providing this essential nutrient to other algae and aquatic plants. Cyanobacteria-containing marine and freshwater diatoms, the freshwater fern *Azolla*, terrestrial plants such as the liverwort *Blasia*, various hornworts, cycads, wheat plants, and many types of lichens, also benefit from the fixed nitrogen excreted from their cyanobacterial symbionts. More details about cyanobacterial nitrogen fixation, nitrogen-fixation symbioses, and their ecological significance can be found in Chapters 6, 7, and 22. Here we will consider further the biogeochemical history of cyanobacterial nitrogen fixation, cyanobacterial roles in the origin of an oxygen-rich atmosphere, and how oxygen-rich atmospheres have influenced nitrogen fixation and the origin of eukaryotic cells.

Cyanobacteria and the Origin of an Oxygen-Rich Atmosphere

Although fossils outwardly resembling cyanobacteria are known from deposits that are 3.5 billion years or more in age, the earliest strong evidence for the occurrence of oxygenic photosynthesis (diagnostic for cyanobacteria but not similar-appearing organisms) consists of geochemical signatures approximately 2.7 billion years old (Buick, 1992) (see also Chapter 6). It is thought that at the time when cyanobacteria first appeared, the earth's atmosphere was rich in carbon dioxide (10–100 times present levels) and that oxygen was sparse (about 10^{-8} times that of present levels) (Raven, 1997b). Cyanobacterial photosynthesis would not have been limited by the availability of carbon dioxide, but oxygen-requiring eukaryotes would not have been able to exist. Thus the cyanobacteria dominated earth's biosphere for at least one billion years and probably more. Oxygen derived from their photosynthesis gradually accumulated in the atmosphere, eventually reaching modern levels, about 21%

(Fig. 2–3). The rise to dominance of cyanobacteria—the earliest known oxygenic photosynthesizers—has been described as the single most significant evolutionary event in the history of life on earth, for without it, subsequent origin of eukaryotic life would have been impossible. An oxygen-rich atmosphere made aerobic (oxygen-using) respiration possible. The use of oxygen as an electron acceptor resulted in an 18-fold increase in respiration efficiency, necessary for survival of most eukaryotes. An oxygen-rich atmosphere—poisonous to all anaerobic (non-oxygen–using) prokaryotes—also caused a radical change in community dominance by favoring organisms possessing aerobic metabolism. Atmospheric change allowed the biosynthesis of sterols, a process that requires gaseous oxygen. Sterols render cell membranes flexible, facilitating endosymbiotic incorporation within host archaeal cells of the eubacterial cells ancestral to mitochondria and chloroplasts (the latter arising from cyanobacteria) (see Chapter 7 for more details regarding the origin of eukaryotes). Finally, an oxygen-rich atmosphere was necessary for generation of the stratospheric ozone shield that protects life on earth's surface from the damaging effects of ultraviolet radiation.

This allowed algae and other organisms to colonize surface waters and the land surface, habitats previously rendered sterile by UV.

Cyanobacteria, having altered the earth's atmosphere, had to adapt to the increased oxygen levels. Their ability to fix nitrogen—inherited from ancestral bacteria—was and is based on the enzyme **nitrogenase**, which is poisoned by oxygen. As the atmosphere became enriched in oxygen, cyanobacteria had to evolve methods for protecting this critical enzyme. Cyanobacteria have been described as the champions of the prokaryotic world in terms of their diverse adaptations for reducing oxygen toxicity to nitrogenase (Paerl and Pinkney, 1996). As a result, cyanobacteria are widely studied by genetic engineers seeking to incorporate nitrogen fixation capacities (nitrogenase and oxygen-protection) into crop plants. Additional details regarding cyanobacterial oxygen-protection adaptations will be found in Chapter 6.

Algae and the Carbon Cycle

As mentioned earlier, cyanobacteria and their descendants—plastids of algae and plants—have been generating oxygen for at least 2.7 billion years (Buick, 1992), but this process alone does not explain how oxygen accumulated in the earth's atmosphere or how it remains at such high levels today. The explanation involves numerous aspects of the carbon cycle, including the ability of algae and plants to produce organic compounds that are resistant to microbial breakdown and thus are buried in sediments, and the ability of algae to generate calcium carbonate.

Algae and Organic Carbon Burial

Consider the reactions that describe oxygenic photosynthesis and aerobic (oxygen-using) respiration (Fig. 2–4). If aerobic breakdown of organic carbon occurred at the same rate as oxygen liberation and the formation of organic compounds in photosynthesis, the accumulation of atmospheric oxygen would not occur. It turns out, however, that moribund algae, or organic carbon compounds produced by algae, can sink through the water column to the bottom without being completely decomposed. Thus algal organic carbon (and that of other marine organisms) can be buried in deep anoxic ocean sediments where there are very slow rates of turnover and thus be sequestered from microbial oxidation for up to thousands of years.

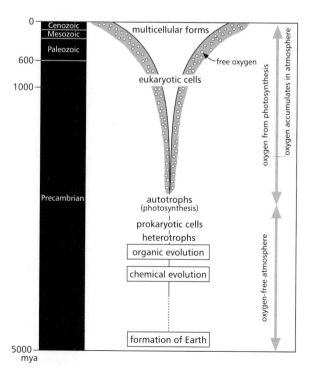

Figure 2–3 Diagram illustrating that atmospheric oxygen increase is correlated with the appearance and radiation of eukaryotes. (mya=millions of years ago)

$$6H_2O + 6CO_2 \underset{\text{Respiration}}{\overset{\text{Photosynthesis}}{\rightleftarrows}} C_6H_{12}O_6 + 6O_2$$

Figure 2–4 The reactants in photosynthesis are the products of cellular respiration and vice versa. At the ecosystem level, if respiration were to occur at the same rate as photosynthesis, atmospheric oxygen and carbon dioxide levels would not change. However burial of organic carbon in sediments can reduce the rate at which organic carbon can be respired back to carbon dioxide. Organic carbon burial is one process that helps to explain large increases in atmospheric oxygen levels that have occurred during the formation of the earth's atmosphere.

When this occurs, global photosynthetic oxygen production can exceed the rate of oxygen consumption, allowing atmospheric oxygen to accumulate.

Various organic compounds produced by algae are known to resist decomposition and are thus termed **refractory carbon**. For example, many dinoflagellates (Chapter 11), some eustigmatophyceans (Chapter 13), and certain green algae (Chapters 20 and 21) are known to produce materials that are resistant to chemical breakdown and decay. Some green algae produce phenol-containing materials similar to the sporopollenin of higher plant spore walls. Sporopollenin is considered to be the single most decay-resistant biopolymer known and is responsible for the persistence of plant spore remains in the fossil record. Several green algae and eustigmatophyceans produce decay-resistant cell wall compounds known as **algaenans**, which are biochemically distinguishable from sporopollenin. Algaenans are polymers of unbranched hydrocarbons (high-molecular–weight lipids) that lack phenolic groups (Gelin, et al., 1997). A number of fossil fuel deposits are known to be derived at least in part from algaenans (Chapter 20). Dinoflagellates possess a very resistant phenolic polymer that is distinct from sporopollenin and algaenans (Kokinos, et al., 1998).

Organic carbon originating from algal exudates or decomposing algae can be converted by aquatic bacteria into fairly decay-resistant colloids 1–1000 μm in diameter, a size lying at the boundary between soluble carbon and particles. Some 30–50% of so-called dissolved organic carbon—perhaps 250 gigatons on a global basis—may occur in such colloidal form (Wells, 1998). Much of this—perhaps 20% of total dissolved organic carbon in surface seawater (which represents 30% of total global annual primary productivity)—occurs as acylpolysaccharides (APS) that appear to be resistant to further bacterial decomposition. APS are branched carbohydrates having two acetate groups for every five sugars; they occur in all fresh and marine waters that have been examined and in deep sea deposits, suggesting long-term survival. APS also aggregate with other materials such as bacteria, zooplankton remains, and fecal pellets, to form larger (>500 μm), readily sinking particles known as **marine snow** (discussed more intensively in Chapter 3). The global reserve of APS is thought to be 10–15 gigatons, with phytoplankton representing the only known source (Aluwihare, et al., 1997).

It has been experimentally demonstrated that APS are generated from algal exudates. Investigators first filtered seawater to remove materials as small as APS, then inoculated the water with the diatom *Thalassiosira weissflogii*, which is known to produce organic exudates. Next they removed the algae and added bacterial communities isolated from seawater by filtration. After an incubation period, during which the bacteria apparently metabolized the algal exudate, the bacteria were filtered out and the seawater examined for colloidal APS organics, which were detected (Aluwihare, et al., 1997). Other experiments have demonstrated production of decay-resistant organic compounds from diatoms. An analysis of the fate of organic carbon compounds produced by large-volume diatom cultures showed that 25–35% of these materials had not been converted back to carbon dioxide after 2.5 years (Fry, et al., 1996).

The biogeochemistry of the ocean is considered to be primarily driven by downward movement of this type of organic carbon from sunlit surface waters. Though much of algal photosynthetic productivity is recycled by grazing zooplankton and microbes (Chapter 3), 10–30% of open ocean organic compounds escape the surface as dissolved organics, APS colloids, or particles. This is known as the biological pump, and represents a mechanism by which atmospheric carbon dioxide can be drawn down on a long-term basis. As the result of operation of the biological pump for billions of years, the earth's sedimentary rocks hold more than 10^7 gigatons of organic carbon of marine origin (Reimers, 1998). Rates of organic carbon sedimentation are influenced by the "rain rate" (the rate at which organic carbon sinks through the water column, which is a function of algal productivity) and the amount of time that organic carbon is exposed to oxygen (and aerobic decomposers). Over long time periods, and as a result of thermal and microbial changes, such organic sediments may be transformed into hydrocarbon-rich

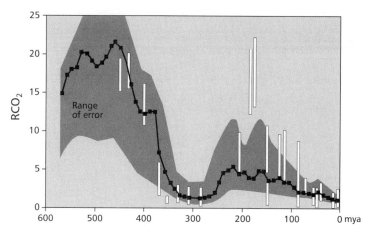

Figure 2–5 A model of current views regarding changes in earth atmospheric carbon dioxide levels through time (black line). The model is supported by measurements of stable carbon isotope ratio measurements (white bars) (see Text Box 2–1) extending back about 500 million years. According to this model, carbon dioxide levels were considerably higher in the remote past when algae were the dominant autotrophs. However considerable uncertainty (represented by the dark gray area) surrounds quantitative estimates of ancient atmospheric carbon dioxide levels. (Redrawn with permission from Berner, R. A., The rise of land plants and their effect on weathering and CO_2. *Science* 276:544–546. ©1997 American Association for the Advancement of Science)

rocks—sources of fossil fuels such as coal, oil, or natural gas (methane). In the past there have been periods of unusually great organic sedimentation and fossil hydrocarbon formation. For example, during the Mesozoic era (65–250 million years ago) warm temperatures, sluggish ocean circulation, and low seawater oxygen content contributed to a particularly high organic rain rate, which resulted in the economically significant North Sea oil deposits. We know that algae were involved in the formation of this oil because fossils, consisting in part of calcareous scales typical of haptophyte algae (coccolithophorids), are abundant in the North Sea deposits (Tucker and Wright, 1990).

The Role of Carbon Burial in Past Atmospheric Chemistry Changes

Paleoclimatologists are interested in understanding the relationships between atmospheric carbon dioxide levels and climate in the past, as a key to being able to predict the effects of changes in atmospheric chemistry on climates of the present and future. They have used stable carbon isotope methods (see Text Box 2–1) to estimate levels of carbon dioxide present in the past. Such studies reveal that 500 million years ago the earth's atmosphere contained about 16 times as much carbon dioxide as it presently does (about 0.04%, or 350–360 parts per million). If the atmosphere con-

tained 0.64% carbon dioxide today, earth would be considerably hotter because of operation of the "greenhouse effect," the absorption of solar heat that would otherwise be radiated into space. Carbon dioxide and other greenhouse gases are the atmospheric components that absorb heat; the higher their concentration, the more heat is absorbed. A dramatic drawdown of carbon dioxide that occurred between 500 and 300 million years ago (Fig. 2–5) has been essential to development of modern climates and organismal communities. The particularly steep carbon dioxide decrease that occurred during the late Paleozoic era (some 300 million years ago) has been attributed to the rise of large populations of woody land plants (Berner, 1997). Algae are believed to have contributed substantially to change in atmospheric composition, particularly 400–500 million years ago, before the evolution of woody plants. Mechanisms by which algae can decrease atmospheric carbon dioxide concentration include burial of organic carbon, as discussed earlier, and sedimentation of inorganic calcium carbonates, the subject of the next section.

The Role of Algae in Carbonate Formation

Beginning with cyanobacteria in Precambrian times, various groups of algae have sequestered very large

Text Box 2–1 Stable Isotopic Measurements

There are two stable (i.e., nonradioactive) isotopes of carbon on earth, ^{12}C and ^{13}C. ^{12}C is far more abundant than ^{13}C, accounting for approximately 99% of stable carbon. Photosynthetic organisms discriminate between these isotopes, preferentially using ^{12}C. Discrimination primarily occurs at the active site of Rubisco during carbon fixation (carboxylation). Sedimentary carbonates, algal biomass, or other samples to be tested are first dried and then combusted to CO_2 which is fed to an isotope ratio mass spectrometer. For each sample, $\delta\,^{13}C$ is calculated according to the following formula:

$$\delta\,^{13}C = \frac{R_{sample} - R_{standard}}{R_{standard}} \times 1000$$

where R_{sample} is the $^{13}C{:}^{12}C$ ratio of the sample and $R_{standard}$ is the $^{13}C{:}^{12}C$ ratio of a standard—a calcium carbonate deposit from the Cretaceous "Pee Dee" formation. The standard has a $^{13}C{:}^{12}C$ ratio of 0.011237. Algae that are able to use bicarbonate as a source of dissolved inorganic carbon (DIC) in photosynthesis are characterized by $\delta\,^{13}C$ values that are significantly less negative than those of algae that appear to be restricted to use of dissolved CO_2. A further calculation of Δ (overall discrimination) is often made to compensate for variations in the $\delta\,^{13}C$ of the source carbon:

$$\Delta = \frac{(\delta\ air\ [or\ other\ C\ source] - \delta_{sample})}{(1 + \delta_{sample})^{-1}}$$

CO_2 resulting from decomposition of organic material may be depleted in ^{13}C in comparison to air, whereas DIC in environments characterized by high photosynthetic rates may be relatively rich in ^{13}C (see Maberly, et al., 1992 and sources cited therein).

amounts of carbon from the atmosphere as carbonate sediments and rocks. In fact, most of the world's carbon has been transformed into limestone, primarily by algae (McConnaughey, 1994), contributing significantly to an ancient drawdown of atmospheric carbon dioxide. Such carbonates are also important because they contain at least 40% of the world's known hydrocarbon reserves (Tucker and Wright, 1990). This is not surprising, since relatively heavy calcified algal remains sink readily, carrying associated organic compounds into the deep ocean.

Modern algae known to generate carbonates include various cyanobacteria as well as green and brown seaweeds that precipitate calcium carbonate on their surfaces, the freshwater green algae known as stoneworts because of surficial carbonate encrustations, some red algae (including tough coral-like forms known as corallines) with carbonates in their cell walls, and the haptophyte algae known as coccolithophorids. Coccolithophorids bear miniscule calcium carbonate scales—coccoliths—that are produced within the algal cells, then transported through the cell

membrane to the cell exterior (Fig. 2–6) (Chapter 10). Coccolithophorid algae alone account for 20–40% of modern biologically produced carbonates, and fossil remains of their coccoliths (mentioned earlier) indicate that coccolithophorids were abundant in the past as well. Substantial fossil $CaCO_3$ deposits were generated by cyanobacteria (Chapter 6), and green seaweeds such as *Halimeda* (Fig. 2–7). Mounds of calcareous *Halimeda* remains up to 52 m thick have been recorded from the shelf behind the northern Great Barrier Reef (Australia) and other locations (Tucker and Wright, 1990).

The process by which algae produce calcium carbonate is known as **calcification,** but the exact mechanism by which they accomplish and control calcium carbonate deposition is unknown. Moreover, the advantages of algal calcification are poorly understood, though calcification may protect coralline red algae from herbivory (Chapter 16). In the cases of corallines, the stonewort *Chara,* and coccolithophorids, calcification appears to be a mechanism assisting in the acquisition of carbon dioxide for photosynthesis (McConnaughey and Whelan, 1996). This would be helpful in today's relatively carbon dioxide–depleted environment in overcoming (inorganic) carbon-limitation of photosynthesis, our next topic.

Impact of Past Atmospheric Composition Change on Algal Photosynthesis

Rubisco (Ribulose bisphosphate carboxylase/oxygenase) is the enzyme that converts inorganic carbon (CO_2) to reduced (organic) carbon in all oxygen-evolving photosynthetic organisms. Rubisco catalyzes the following reaction:

$$RuBP + CO_2 + H_2O \rightarrow 2PGA$$

The enzyme is known as a carboxylase because it uses CO_2 as a reactant in the production of PGA (phosphoglyceric acid) (see reaction above), and an oxygenase because it can also use O_2 in the reaction:

$$RuBP + O_2 \rightarrow PGA + phosphoglycolate$$

The phosphoglycolate can then be lost from the cell. This reaction leads to the metabolic pathway known as **photorespiration** and is energetically undesirable because it results in a net loss of photosynthetically

Figure 2–7 A portion of a calcareous mound formed millions of years ago by growth of the green seaweed *Halimeda*. Individual calcified segments of the algal thallus can be distinguished.

generated organic carbon. Carboxylation is favored by high levels of carbon dioxide in relation to oxygen, whereas phosphoglycolate formation occurs in relatively high oxygen, low carbon dioxide conditions. Photorespiration was not as serious a problem for Precambrian (older than 590 million years) algae, when the atmosphere (and water) contained higher carbon dioxide:oxygen ratios than at present. However as oxygenic photosynthesis raised atmospheric O_2 levels (Fig. 2–3), and atmospheric carbon dioxide drawdown occurred (Fig. 2–5), algae had to acquire physiological adaptations allowing them to cope with the changed conditions. Some cyanobacteria and eukaryotic algae (including relatives and ancestors of land plants) responded by moving from aquatic to terrestrial habitats, the advantage being that gases such as CO_2 diffuse 10^4 times more readily in air than in water, thus providing an increased supply of scarce CO_2 for photosynthesis. Other adaptations exhibited by modern algae include methods for reducing the effects of photorespiration, evolutionary change in Rubisco, acquisition of carbon-concentrating mechanisms, and means for using an alternate source of inorganic carbon, the bicarbonate ion (HCO_3^-), which is more abundant in most aquatic systems than dissolved CO_2.

Algal (and plant) cells can recover some of the energy in phosphoglycolate (which would otherwise be lost to the cell) by oxidizing it to glyoxylate. Algae differ in glycolate-oxidizing mechanisms, reflecting

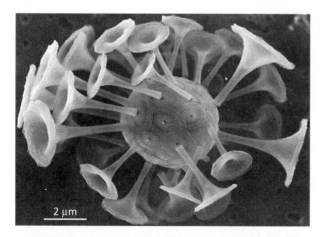

Figure 2–6 The coccolithophorid *Discosphaera tubifera* from the central North Atlantic. Members of this group of primarily unicellular marine phytoplankton have a covering of often elaborate calcified scales, or coccoliths. Certain ocean sediments are rich in the coccolith remains of these algae. (From Kleijne, 1992)

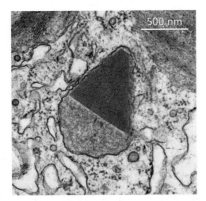

Figure 2–8 Transmission electron micrograph of a peroxisome, the site of glycolate oxidation in some green algae. (Reprinted with permission from Graham, L. G., and Y. Kaneko. 1991. Subcellular structures of relevance to the origin of land plants [Embryophytes] from green algae. *Critical Reviews in Plant Science* 10:323–42. ©CRC Press, Boca Raton, Florida)

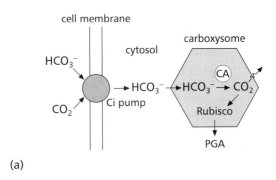

(a)

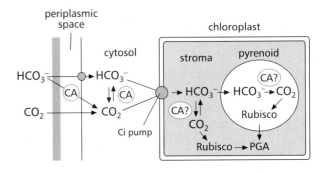

(b)

Figure 2–9 Models of carbonic anhydrase (CA) function in cells of (a) cyanobacteria and (b) eukaryotic cells. Because bicarbonate is a charged ion, uptake of bicarbonate across cell or chloroplast membranes would require transporter molecules and energy, whereas carbon dioxide can diffuse freely across membranes. Transformation of extracellular bicarbonate into CO_2 in the periplasmic space is therefore advantageous for algal cells, as is cytoplasmic entrapment of CO_2 by conversion to bicarbonate. Similarly, chloroplast CA may function to trap bicarbonate within the organelle and to regenerate CO_2 in the vicinity of Rubisco when it is needed for fixation. It should be noted that these models are largely untested. (After Badger and Price, 1994, with permission of *Plant Physiology and Plant Molecular Biology*, Volume 45. ©1994 by Annual Reviews—www.annualreviews.org)

alternate evolutionary responses to the same problem. Glycolate oxidase, located in cell organelles known as **peroxisomes** (Fig. 2–8), is utilized by reds, browns, tribophyceans, raphidophyceans, eustigmatophyceans, and charophycean greens (and their descendants, the land plants). A different enzyme, glycolate dehydrogenase (located in mitochondria), oxidizes glycolate in cyanobacteria, euglenoids, cryptomonads, dinoflagellates, diatoms, and non-charophycean green algae (Raven, 1997b, c). Glycolate-oxidizing systems were surveyed in many groups of algae by Suzuki, et al. (1991).

Another way algae have responded to altered atmospheric conditions is by evolutionary change in Rubisco, such that the enzyme's active site has a higher affinity for carbon dioxide, thus decreasing the impact of photorespiration (Newman and Cattolico, 1990). A measure of the binding strength of Rubisco for CO_2 is the K_m (CO_2) or half-saturation constant—the concentration of substrate at which the catalytic reaction occurs at one half its maximal rate. The lower the value of K_m, the greater the affinity of the enzyme for its substrate. Cyanobacteria, the first oxygenic photosynthesizers, have the highest K_m (CO_2) of all the algae, 80–330 micromoles (μmol). The relatively low CO_2-binding capacity of their Rubisco reflects their evolution in early CO_2-rich atmospheric conditions. Later-appearing eukaryotic algae have K_m (CO_2) values ranging from 45–70 μmol, and those for land plants, which diverged even more recently (Fig. 1–16), fall between

10 and 25 μmol. Further, algae vary in relative sensitivity to O_2-binding by Rubisco, a quantity known as τ (Gr. *tau*). Those having a high value of τ are less susceptible to oxygenase activity, and include reds, cryptomonads, and ochrophytes. Algal groups having a lower τ value are more susceptible to oxygenase activity, and include cyanobacteria and dinoflagellates. Green algae fall within the middle of this range (Raven, 1997b).

A third adaptation to reduced CO_2 and increased O_2 levels was the development in algae of carbon concentrating mechanisms (CCMs). These vary among algae, but all serve to increase the supply of CO_2 to Rubisco. CCMs frequently involve the use of bicarbonate ion, which is often much more abundant than dissolved carbon dioxide. At pH 8–9 (characteristic of many lakes and the ocean) the concentration of bicarbonate (about 2 mM) is 200 times greater than that of dissolved carbon dioxide (only about 12–15 μm). Only at low pH (around 6) does dissolved CO_2 become more abundant than bicarbonate (see Fig. 22–14). CCMs allow algal cells to concentrate inorganic carbon to a level several times that of the external environment. For example, the dinoflagellate *Peridinium gatunense* can concentrate inorganic carbon by 20–80 times that of the surrounding water (Berman-Frank and Erez, 1996). CCMs are thought to be considerably less energetically expensive than glycolate production and disposal (Raven, 1997b). They are so valuable that even algae characterized by Rubisco with high affinity for CO_2 and low susceptibility to photorespiration, have them. CCMs are absolutely essential for cyanobacteria, because the CO_2 affinity of their Rubisco is so low, and they are particularly vulnerable to the effects of photorespiration (see review of cyanobacterial CCMs by Kaplan, et al., 1998). The majority of green algae are also highly dependent upon CCMs (Badger and Price, 1994). Algal CCMs may involve cell membrane inorganic carbon transporters, enzymes that interconvert CO_2 and bicarbonate, calcification-linked processes, and specialized cellular structures—**carboxysomes**—in cyanobacteria and **pyrenoids** in many eukaryotic algae (see below).

Carbon Concentration Mechanisms of Cyanobacteria

Cyanobacteria are thought to take up either carbon dioxide (if available) by passive diffusion or active transport, or bicarbonate by the use of a membrane-bound transporter or pump (Fig. 2–9a). The nature of the hypothesized CO_2 or bicarbonate transporters is not yet known. The use of energy would be required to pump bicarbonate ions into cells because it is charged, but once inside cells, bicarbonate is much less likely to diffuse out than is carbon dioxide. Hence the cytoplasm may contain a store of bicarbonate, but before this inorganic carbon can be used in photosynthesis, it must be converted to carbon dioxide, because the latter is the only form of dissolved inorganic carbon that can be used by Rubisco in carbon

fixation. Bicarbonate ions are converted to CO_2 via the reaction:

$$HCO_3^- + H^+ \rightarrow CO_2 + H_2O$$

However, the uncatalyzed rate of this reaction is too slow to supply carbon dioxide to algal photosynthesis, so cyanobacteria use a zinc-containing enzyme known as **carbonic anhydrase**, which dramatically speeds the reaction. In fact, carbonic anhydrase (CA) is noteworthy among enzymes for its almost unbelievably high turnover rate—a feature that greatly facilitates provision of CO_2 to algal photosynthesis (Stemler, 1997). In cyanobacteria, CA is located

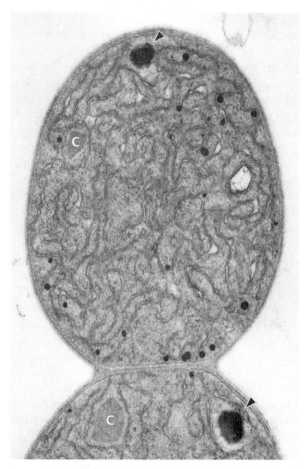

Figure 2–10 Carboxysomes (C) are polyhedral inclusions in the cells of cyanobacteria, such as *Cylindrospermum*, viewed here by transmission electron microscopy. The enzymes carbonic anhydrase and Rubisco have been localized to the carboxysomes of other cyanobacteria. Arrowheads indicate cyanophycin particles, which represent a store of combined nitrogen.

within carboxysomes—polygonal protein structures that can be readily identified in electron micrographs (Fig. 2–10). Carboxysomes are also the location of Rubisco in cyanobacterial cells; co-localization of these two enzymes facilitates the capture of CO_2 (generated by CA) by Rubisco before it can diffuse out of the cell. The CCMs of cyanobacteria have recently been reviewed by Price, et al. (1998).

Carbon Concentration Mechanisms of Eukaryotic Algae

Eukaryotic algae possess a wide range of mechanisms for acquiring inorganic carbon for photosynthesis. These mechanisms are often species specific, defying attempts to generalize, but membrane transporters and carbonic anhydrase are thought to be frequently involved (Fig. 2–9b). There is physiological evidence for the active transport of CO_2 and/or bicarbonate ion in various microalgae (including the common green algae *Chlamydomonas* and *Scenedesmus*), as well as in common macroalgae such as the green *Ulva* and red *Gracilaria* (Beer, 1994; Badger and Price, 1994; Colman and Rotatore, 1995; Sukenik, et al., 1997). Another bicarbonate transporter (inorganic carbon pump) is hypothesized to be located in the plastid envelope (Fig. 2–9b). As in cyanobacterial cells, plastid bicarbonate must be transformed into CO_2 before it can be used in photosynthesis. This is accomplished by carbonic anhydrase located within the plastid (Fig. 2–9b). The location of plastid CA is debated and may vary among algae (Raven, 1997b, c). It has been suggested that in some algae structures known as pyrenoids (Fig. 2–11) play a role in CCMs. Pyrenoids are the structural equivalents of cyanobacterial carboxysomes (Badger and Price, 1994). However, pyrenoids do not occur in the plastids of many algae that have CCMs. The green alga *Coccomyxa*, for example, lacks pyrenoids but contains substantial amounts of chloroplast CA (Badger and Price, 1994). Hence the role of pyrenoids in CCMs is unclear. In some cases, there is good evidence that CA is located within thylakoids (Stemler, 1997). There may be two distinct forms of plastid CA, one soluble and the other membrane-bound (Amoroso, et al., 1996). The diversity and roles of plastids in CCMs of algae were reviewed by Badger, et al. (1998), Moroney and Chen (1998), and Moroney and Somanchi (1998).

Many eukaryotic algae are known to use abundant bicarbonate by production of external carbonic anhydrase. Its role is to catalyze the conversion of extra-

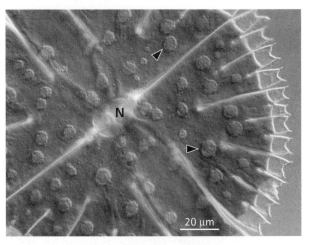

Figure 2–11 Pyrenoids (arrowheads) in the chloroplast of the green alga *Micrasterias*, as viewed by differential interference contrast light microscopy. Many kinds of eukaryotic algae contain pyrenoids in their plastids; these are often visualized at the light microscopic level by staining the starch shell that often surrounds pyrenoids with a solution of iodine and potassium iodide. The nucleus (N) of this unicellular alga is also evident.

cellular bicarbonate into carbon dioxide, which then moves into the cell via passive or active transport (Fig. 2–9b). Immunolocalization at the transmission electron microscopic level (Fig. 2–12) has revealed that external CA is located in the periplasmic space—the region between the cell membrane and the cell wall. Evidence for external CA has been found in a wide variety of algae, including green microalgae such as *Chlamydomonas*, *Chlorella*, *Dunaliella*, and *Scenedesmus* (Badger and Price, 1994), dinoflagellates (Berman-Frank, et al., 1995), various diatoms (Nimer, et al., 1997; John-McKay and Colman, 1997) and macroalgae such as the common red seaweed *Porphyra* (Mercado, et al., 1997; Iglesias-Rodriguez and Merrett, 1997). Gene sequencing and inhibitor studies have revealed that this periplasmic CA is quite different from the CA found in plastids, the former more closely resembling CAs of animal cells (Suzuki, et al., 1994).

In some eukaryotic algae, external CA is constitutive (always present), whereas in other cases it is induced by low levels of environmental carbon dioxide. Still other species completely lack external CA, even when grown in a medium having a low concentration of inorganic carbon. In a study of 18 species of marine phytoplankton, Nimer, et al. (1997) found that extracellular CA was constitutive in four of the five dinofla-

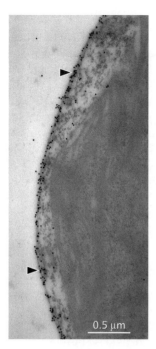

Figure 2–12 Carbonic anhydrase immunolocalized to the cell wall and periplasmic space (the region between the cell wall and the cell membrane—arrowheads) of the green alga *Chlamydomonas reinhardtii*. Here, a thin section of algae prepared for examination by transmission electron microscopy was treated with an antibody to the enzyme carbonic anhydrase. A second antibody labeled with small gold particles was then used as a method for visualizing the location of the first antibody and the antigen to which it is bound. The distribution of gold particles thus reflects the distribution of carbonic anhydrase. (Unpublished micrograph by P. Arancibia, L. Graham, and W. Russin. The antibody was contributed by J. Coleman.)

gellates examined and that external CA was inducible in several ecologically important algae such as the diatom *Skeletonema costatum* and the coccolithophorid *Emiliania huxleyii*. These workers concluded that most diatoms are "CO_2 users" that utilize bicarbonate only indirectly, via external CA. Considerable variation among species has also been observed. While external CA could not be induced in the diatom *Chaetoceros compressum* (Nimer, et al., 1997), there is evidence that the related *C. calcitrans* does utilize bicarbonate, either by active transport or external CA (Korb, et al., 1997). Similarly, two species of the diatom *Thalassiosira* exhibited inducible external CA, but *T. pseudonana* did not (Nimer, et al., 1997). A study of 11 species of the diatom *Phaeodactylum* revealed that only eight of them could produce external CA (John-McKay and Colman,

1997). One of these, *P. tricornutum*, apparently cannot import bicarbonate directly but instead relies upon external CA to generate CO_2 (Laws, et al., 1997). Algal CAs have been reviewed by Sültemeyer, et al. (1993, 1998), and Suzuki, et al. (1994).

Yet another way algae can generate CO_2 from bicarbonate is to excrete acid (H^+) across cell membranes; the acid then reacts with bicarbonate to yield carbon dioxide and water, whereupon the CO_2 can move into cells and then the plastid (McConnaughey, 1998). Such processes are typically linked to calcification. In the stonewort green algae, import of hydrogen ions from the external medium creates alkaline bands where calcification occurs, and compensatory export of hydrogen ions at other regions of the cell surface generates acid bands where carbon dioxide can be produced from bicarbonate (Fig. 2–13a, b). This CO_2 may move into cells via ATP-mediated active transport (Mimura, et al., 1993). In the coccolithophorids, calcification occurs in intracellular vesicles across the surface of which hydrogen ions may be exchanged with calcium cations (Fig. 2–13c). Interaction of protons with bicarbonate then generates carbon dioxide that can be used in photosynthesis. Some open ocean coccolithophorids (*Coccolithus pelagicus* and *Gephryocapsa oceanica*), which do not encounter much variation in bicarbonate concentrations, rely primarily on this means of obtaining CO_2; they lack extracellular carbonic anhydrase (Nimer, et al., 1997). Other coccolithophorids, such as *Emiliania huxleyii*, that occupy coastal regions where influx of fresh water may result in variation in bicarbonate concentration, may use both external CA and calcification to acquire CO_2. At the surfaces of coralline red algae, exchange of protons with calcium ions (Fig. 2–13d) is hypothesized. The hydrogen ions then react with bicarbonate to release CO_2 for photosynthesis. Reactions that link algal calcification to CO_2 acquisition are shown in Figure 2–14.

In summary, the balance of evidence available to date suggests that most common marine phytoplankton are able to utilize bicarbonate either directly or indirectly, and thus are unlikely to significantly drawdown atmospheric CO_2 levels. If atmospheric CO_2 levels continue to increase, the extent to which algae are limited by dissolved carbon dioxide levels would decrease and the utility of their CCMs would decline, but even for dinoflagellates and cyanobacteria (which are extremely reliant on CCMs), atmospheric levels of CO_2 would have to increase several-fold before they would be able to rely on diffusive CO_2 supplies. Thus nitrogen, phosphorus, iron, silica, and light are prob-

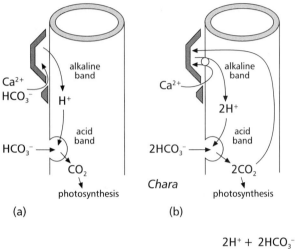

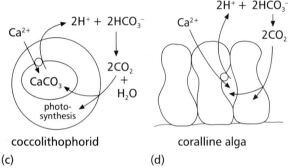

Figure 2–13 Models of the calcification process in three groups of algae: the green algal genus *Chara*, the coccolithophorids, and red algae known as corallines. Calcification of *Chara* (alternate possibilities are shown in (a) and (b) and corallines (d) occurs externally, whereas this process occurs within intracellular vesicles of coccolithophorids (c). In each case Ca^{2+} transport across a membrane is thought to be linked to export of H^+, which participates in the extracellular conversion of bicarbonate (HCO_3^-) to CO_2 that is subsequently taken up by cells for use in photosynthesis. Calcification can be viewed as an adaptation that allows algae living in alkaline waters to obtain sufficient inorganic carbon for photosynthesis. It should be noted that these models represent hypotheses that need further testing. (Reprinted from *Earth Science Reviews* Vol. 42, McConnaughey, T. A., and J. F. Whelan. Calcification generates protons for nutrient and bicarbonate uptake. Pages 95–118. ©1996, with permission from Elsevier Science)

ably more commonly limiting to algal growth than is inorganic carbon (Raven, 1997b).

Algal Use of Organic Carbon

A growing body of evidence (most recently reviewed by Tuchman, 1996, but see also Neilson and Lewin,

1974) indicates that algae more commonly take up and utilize organic carbon than has been generally recognized. Examples of obligate osmotrophs (consumers of dissolved organic carbon) include the colorless chrysophycean *Paraphysomonas*, and the colorless relatives of green algae, *Prototheca* and *Polytoma*. But there is also strong evidence that many pigmented algae are facultative osmotrophs, particularly when light-limited. For example, several species of the freshwater or soil diatom *Navicula* are able to grow with glucose as the sole carbon source in the light or in the dark (Lewin, 1953).

In addition, some algae occupying organic-rich habitats utilize dissolved organic carbon but only in the light. An example is the diatom *Cocconeis*, which was isolated from sediments of Biscayne Bay, Florida, that were rich in decaying sea grasses (Bunt, 1969). The green algae *Chlorella* (which may occur in organic-rich habitats such as soils and sugar mill effluents) and *Coleochaete* (a member of periphyton communities) are able to utilize glucose but only in the light, and in the latter case, only when dissolved inorganic carbon is limiting to growth (Graham, et al., 1994). Lewitus and Kana (1994, 1995) demonstrated that axenic (bacteria-free) cultures of several estuarine phytoplankton species, including the green alga *Closterium* (a relative of *Coleochaete*), two cryptomonads, two ochrophytes, and a coccolithophorid, utilized glucose in mitochondrial respiration at low light levels and via chlororespiration (a form of respiration that occurs in chloroplasts) at high irradiance. The first of these processes allows algae to overcome growth restrictions resulting from reduced light, while the latter provides the algae with greater ability to use saturating levels of light in photosynthesis. Many algae possess active glucose-specific transport systems; in some cases these are constitutive (expressed under most conditions), whereas in other cases the systems are inducible by darkness or high levels of environmental glucose (Carthew and Hellebust, 1983).

Tuchman (1996) pointed out that all groups of photosynthetic algae had ancestors that were heterotrophic; thus the capacity for uptake of organic compounds should be widespread. The green alga *Chlorella* has at least three well-characterized cell membrane hexose transporter proteins (Stadler, et al., 1995) that are homologous to members of the "major facilitator superfamily" of transporters found in plants (*Arabidopsis*), animals (including humans), fungi, other protists, cyanobacteria, and other bacteria (Griffith, et al., 1992). Transporter protein members of this

Calcification: $CO_2 + Ca^{2+} + H_2O \longrightarrow CaCO_3 + 2H^+$

HCO_3^- utilization: $2H^+ + 2HCO_3^- \longrightarrow 2CO_2 + 2H_2O$

Photosynthesis: $CO_2 + H_2O \longrightarrow CH_2O + O_2$

Net: $2HCO_3^- + Ca^{2+} \longrightarrow CaCO_3 + CH_2O + O_2$

Figure 2–14 Reactions linking the calcification process in algae to dissolved carbon utilization and photosynthesis (McConnaughey, 1998).

family distinctively possess 12 α-helical transmembrane regions (which are embedded with the cell membrane) connected by internal and external loops (Fig. 2–15). Will and Tanner (1996) used molecular techniques to demonstrate in *Chlorella* that the likely site for substrate recognition and binding occurs at the first external loop. Other workers have shown that glucose uptake by *Chlorella* suppresses expression of carbonic anhydrase, and hence the CCM, even when

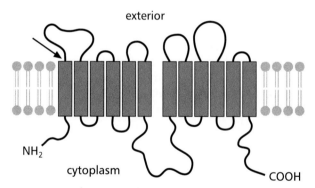

Figure 2–15 Model of the possible conformation of one of several well-studied hexose transporter proteins that occur in the cell membranes of the green alga *Chlorella*. Hexose transporters similar to those of *Chlorella* are found in all groups of organisms from bacteria to higher plants and animals. Members of this large family of homologous proteins are characterized by 12 transmembrane regions (dark gray boxes) consisting of α–helically coiled regions rich in hydrophobic amino acids, separated by loops of more hydrophilic amino acid sequences that extend both into the cytoplasm and outward to the exterior environment. The arrow indicates the site where sugar is thought to bind. The ubiquity of such transporters in prokaryotes and eukaryotes suggest that they are likely to be of widespread occurrence in algae. (After *Botanica Acta* 106 (1993):277–286. Sauer, N., and W. Tanner. Molecular biology of sugar transporters in plants. Georg Thieme Verlag, Stuttgart)

ambient CO_2 levels were low (Shiraiwa and Umino, 1991; Villarejo, et al., 1997).

Iron Limitation of Algal Growth

Iron was probably a much more common ion in primordial anoxic oceans than in today's oxygen-rich surface waters. Thus early cells most likely had an abundant supply of Fe^{2+}, and became dependent on iron as a cofactor for many enzymes, including cytochrome oxidase, iron-superoxide dismutase, catalase, glutamate synthetase, nitrate and nitrite reductase, and nitrogenase, as well as ferredoxin, an electron carrier required for nitrogen fixation. Each nitrogenase complex contains 32–36 iron atoms. As the atmosphere and oceans increased in oxygen content, dissolved iron levels decreased by 6–7 orders of magnitude, as the result of oxidation. Modern seawater iron concentrations are low because iron is poorly soluble in oxygenated seawater. It quickly becomes oxidized to insoluble iron hydroxides, only becoming soluble again under reducing or anoxic conditions in the deep sediments. As dead organisms are decomposed, small amounts of iron are released. It is estimated that each atom of iron in seawater has been recycled through the biota an average of 170 times before being lost to the sediments in sinking particles (Hutchins, 1995). Modern algae have acquired adaptations that allow them to harvest and store iron when it is available.

When iron is relatively abundant, some algae can store it in protein aggregations known as **phytoferritin**. Certain small cyanobacteria, some dinoflagellates, and certain diatoms are known to harvest iron from low concentrations in seawater by producing surface iron-binding organic molecules, known as **siderophores**. The large filamentous surface aggregates formed by the cyanobacterium *Trichodesmium* can apparently capture

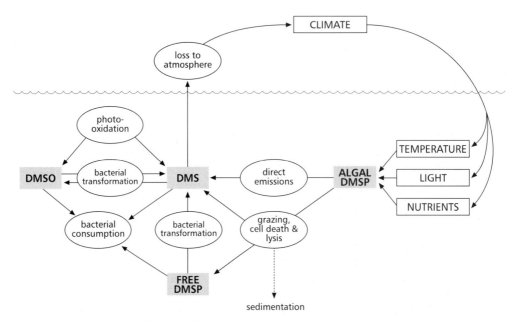

Figure 2–16 The marine dimethyl sulfide (DMS) cycle. DMS is regarded as an important atmospheric climate-modifying molecule. It arises from dimethylsulfonioproprionate (DMSP) generated by marine phytoplankton and released from cells. In seawater, both extracellular DMSP and DMS can be oxidized to dimethylsulfoxide (DMSO), which is not climatically active. (After Malin, et al., 1992 by permission of the *Journal of Phycology*)

iron-rich dust entering the ocean from the atmosphere (Hutchins, 1995). This may explain its abundance in the relatively iron-rich Arabian and Caribbean Seas and the Indian Ocean (Falkowski and Raven, 1997).

Martin and Fitzwater (1988) suggested that the high nutrient-low chlorophyll (HNLC) regions of the ocean could be fertilized with iron in order to increase phytoplankton growth and thus enhance organic carbon sedimentation. This was viewed as a possible engineering solution to the problem of increasing atmospheric levels of carbon dioxide; increased algal growth was hypothesized to draw down inorganic carbon. In order to test this idea the large-scale "Iron Experiment" or "Ironex" was conducted. Ships were used to distribute 450 kilograms of iron within a 64-km² region of the equatorial Pacific Ocean south of the Galapagos Islands. The results demonstrated that growth of phytoplankton, especially diatoms, could indeed be increased by iron fertilization, as chlorophyll levels rose by a factor of two or three and phytoplankton biomass increased by four times (Martin, et al., 1994; de Baar, et al., 1995). However, contrary to expectation, dissolved inorganic carbon levels decreased only slightly. It has been suggested that zooplankton herbivory, loss of iron to colloid formation, and/or development of another micronutrient limitation might explain these

results (Cullen, 1995). Studies of iron effects on ocean phytoplankton are also being conducted in naturally enriched waters of the Galapagos plume and the Antarctic polar jet (Cullen, 1995).

Algae and the Sulfur Cycle

Sulfur is required by algae for biosynthesis of two amino acids—cysteine (where S plays an important role in covalent disulfide bond formation and hence tertiary protein structure) and methionine. In addition, it is required for S-containing thylakoid lipids. Uptake of sulfate requires energy (is ATP-dependent). In marine waters sulfate is abundant and hence does not limit algal growth; whereas in freshwaters anaerobic bacteria can reduce sulfate levels, converting it to H_2S Thus, sulfate may sometimes be limiting to algal growth in this habitat (Falkowski and Raven, 1997). Some freshwater algae possess high-affinity sulfate transport systems (Yildiz-Fitnat, et al., 1994).

The growth of some marine phytoplankton is related to climatically relevant aspects of the global carbon cycle. Haptophytes such as the widespread *Emiliania huxleyii* and *Phaeocystis*, together with dinoflagellates, some diatoms, and green prasinophyceans that grow in a wide range of habitats from the poles to the tropics, generate dimethylsulfonio-proprionate (DMSP) for use in osmoregulation or as a cryoprotectant in cold conditions. The DMSP can then be converted to volatile dimethyl sulfide (DMS), which is released from cells and oxidized to DMSO in the atmosphere to form sulfate aerosols (Fig. 2–16). These naturally produced sulfate aerosols, together with those generated by human activities, are believed to exert significant cooling effects on regional and global climates. By serving as cloud condensation nuclei, the aerosols promote increases in cloud cover and hence increase the earth's albedo—the fraction of incident radiation (including infrared radiation or heat) reflected back into space.

Methionine is the metabolic precursor for biosynthesis of DMSP, and high levels of DMSP and DMS are associated with regions of high algal productivity. DMS emissions also reflect seasonal patterns of phytoplankton growth. Culture studies have revealed that most DMS is released from algal cells when they are senescent. DMSP can be converted to DMS by bacteria, but algae also possess the necessary enzymes. The role of algae in DMSP and DMS production has been reviewed by Malin and Kirst (1997).

Algal Production of Halocarbons

A variety of marine phytoplankton and macroalgae produce volatile halocarbons—organic compounds that contain halogens such as chlorine or bromine. These are generated by a vanadium-containing enzyme known as bromoperoxidase, which occurs widely in algae. Algal halocarbons are of concern because they are oxidized in the atmosphere to hydroxyl, chlorine, and bromine radicals that can destroy stratospheric ozone (as can halocarbons originating from human activities). Methyl bromide (CH_3Br) is probably the single most abundant halocarbon in the atmosphere. Brominated halocarbons are regarded as 4–100 times more active than chlorine compounds in ozone destruction. The global input of methyl bromide to the atmosphere is estimated to be 97–300 Gg yr^{-1} (1 Gg = 10^9 grams); about 50% of this is natural production and 50% is man-made (arising from pesticides and antifreeze). Macroalgae are estimated to generate only 0.6% of coastal production of halocarbons, thus phytoplankton are considered to be a more significant source. A study of methyl bromide production by phytoplankton revealed that 13 of 19 species tested could generate halocarbon and that the haptophyte *Phaeocystis* was the most prolific producer, generating 1.7–30 pg (picograms) CH_3Br per μg chlorophyll *a* per day (Sœmundsdóttir and Matrai, 1998). Marine red, green, and brown macroalgae generate compounds such as trichloroethylene and perchloroethylene.

Recommended Books

Falkowski, P. G., and J. A. Raven. 1997. *Aquatic Photosynthesis*. Blackwell Science, Malden, MA.

Lucas, W. J., and J. A. Berry (editors). 1985. *Inorganic Carbon Uptake by Aquatic Photosynthetic Organisms*. American Society of Plant Physiology, Rockville, MD.

Chapter 3

Algae in Biotic Associations

Bosmina

(Photograph courtesy C. Taylor)

Casual observation of seaweeds cast adrift on the beach or a brief survey of microscopic algae trapped under the cover glass of a microscope slide may leave the impression that algae are relatively independent of other biological organisms. Actually algae are involved in an amazing array of biotic interactions. Closer study of the seaweeds might reveal their roles as hosts for other algae or fungi that live within their tissues or as substrates for attachment of a variety of smaller algae. Longer-term observations might show how the seaweeds serve as food for sea urchin, abalone, or amphipod scavengers and, while rotting, support the growth of a wide range of microorganisms.

With careful observation microscopic algae can be seen to be actors in multitudinous small-scale dramas. Some are violently consumed in whirlwinds created by voracious zooplankton having indiscriminate tastes, while other microalgae are delicately selected and eaten by more finicky planktonic gourmets. Certain microalgae pack intracellular arsenals of spearlike projectiles, ready to be used in self-defense. Others produce chemical repellents that give them a foul taste to zooplankton consumers, who summarily reject them. Yet other algae are themselves consumers, ingesting bacteria and other organisms, some of which may be related algae. Some phytoplankters appear to have shielded themselves against disease by virtue of protective, nearly indestructible surface barriers that resist microbial attack, whereas the internal structure of others may be ravaged by viral, bacterial, protistan, or fungal infections. Certain microalgae appear to encourage the growth of useful bacteria on their surfaces, while others seem to reap similar benefits by living inside protozoa or on the surfaces of zooplankton. In this chapter we will survey important ways in which macroalgae (seaweeds) and microalgae are associated with other organisms—bacteria, protists, fungi, and higher plants and animals, as well as with viruses.

40

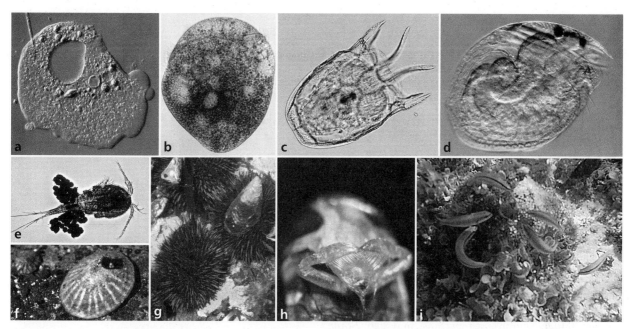

Figure 3–1 Examples of herbivores that consume algae as major portions of their diets include: (a) amoebae, (b) ciliates, (c) rotifers, (d) cladocerans, (e) copepods (dark bodies to sides of animal are egg sacs), (f) limpets, (g) sea urchins, (h) water boatmen (note fore legs, which are modified to scoop up algae), and (i) fish. (g: Courtesy C. Taylor; h: Specimen provided by the University of Wisconsin Insect Research Collection)

Types and Importance of Algal Biotic Associations

Of particular interest to many aquatic ecologists are the multiple roles of algae in food webs dominated by microbes—known as detrital food webs—as well as in those dominated by herbivorous protists and animals. Here **herbivory** is defined as consumption of algae and bacteria, a common use of the term in aquatic ecology. The evolutionary lineages of many algal groups and their herbivorous attackers extend back into the past at least 500 million years further than those of terrestrial plants and their herbivores. In other words, aquatic herbivory interactions have existed more than twice as long as terrestrial ones; this great span of evolutionary time has contributed to the complexity of algal food web associations. Recently much attention has been focused upon algae that function as herbivores or predators. Our increasing knowledge of these forms has revealed that modern aquatic food web complexity is considerably greater than previously realized.

Many ecologists are concerned with algal production of toxic compounds that affect the quality of fisheries and human health. Pathogenic microbes that attack algae (viruses, bacteria, fungi, and funguslike protists) are of concern to ecologists seeking to understand processes that result in loss of algal biomass in natural waters, and by aquatic farmers (aquaculturists) who grow algal crops in aquaculture systems or who use algae as a food source for cultivation of shellfish. Chemical compounds produced by algae as microbial defenses hold great promise for development as pharmaceuticals. A few algae can themselves be pathogenic or allergenic to humans and thus are of medical interest. Algae are also involved in numerous kinds of mutualistic relationships—symbiotic associations with other organisms that confer advantages to the partners or at least cause no harm. Some of these mutualisms are of considerable ecological significance, and all provide insight into endosymbiosis as an evolutionary process (see Chapter 7). An appreciation for the variety of ways in which algae interact with other organisms is also important to people who seek to model ecosystem structure and function, manage natural waters or mariculture operations, or develop strategies for restoration of perturbed or dysfunctional aquatic systems.

In this chapter we will consider the following algal biotic associations: algae as food for herbivores; algal herbivory defenses; pathogens of algae (viruses, bacteria, fungi); algae as parasites, pathogens, or allergens; and algal mutualisms (symbioses). Algae that live inside the cells of other eukaryotes are described in Chapter 7, as are cases where cells of various types of organisms exist endosymbiotically within algal cells.

Algae in Herbivore-Dominated Food Webs

Diverse types of organisms feed preferentially on algae, ranging from microscopic protists (including flagellates, amoebae, and ciliates) and rotifers, such as *Keratella*, to crustacean zooplankton (including cladocerans such as *Daphnia* and *Bosmina*) and copepods (including *Diaptomus* and *Cyclops*), which are barely visible to the naked eye, to somewhat larger "mesograzers" such as oligochaete worms, freshwater dipteran larvae, and marine amphipods, to macroscopic gastropods (such as the common intertidal limpet *Acmaea*), crabs, sea urchins, and fish. Some representative grazers are shown in Figure 3–1.

Grazing can have significant biogeochemical effects, as shown by the case of predation by the dinoflagellate *Oxyrrhis* on the haptophyte *Emiliania*. Laboratory experiments revealed that only ingested cells of *Emiliania* were able to produce DMS, a volatile sulfur compound of climatic significance (see Chapter 2). Even low levels of grazing increased DMS production by 30–400% (Wolfe and Steinke, 1996; Wolfe, et al., 1997). Lakes that are dominated by planktivorous (zooplankton-eating) fish contribute more significantly to CO_2 drawdown (Chapter 2) than do lakes dominated by piscivorous fish (those that eat the planktivores). When zooplankton are removed, algal plankton is less constrained (Schindler, et al., 1997).

As one might expect, protists and metazoan animals feed on algae whose sizes fall within a range that can be accommodated by the grazers' size, oral aperture, ingestion apparatus, or other aspects of its feeding mechanism. Two major zooplankton feeding modes include **filter feeding** and **raptorial feeding**. Filter feeders such as *Daphnia* and *Bosmina* (Fig. 3–1d) are typically much larger than their food, which is concentrated from large volumes of water in a relatively indiscriminate fashion. However they can, by various mechanisms, often reject food that clogs the filtering apparatus or is otherwise unsuitable. Raptorial feeders (for example, copepods) (Fig. 3–1e) can be closer in size to that of food particles, which are often quite selectively removed from among particle mixtures after having been "tasted." Raptorial feeders

(a)

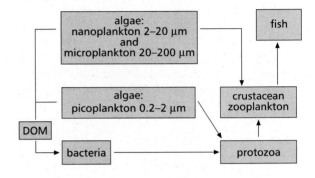

(b)

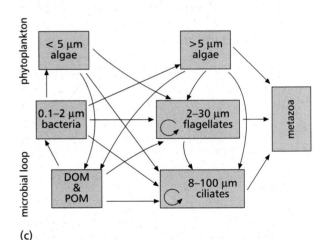

(c)

Figure 3–2 Changing concepts of aquatic microbial food webs through time. (a) In the mid-1960s through the early 1970s prevailing views of food webs were relatively simple, involving linear transfer of food energy from algal producers to consumers (zooplankton) to secondary consumers (fish). (b) In the 1970s researchers realized that various size classes of phytoplankton were consumed by differently sized herbivores, and that bacteria were part of the food web as well. (c) Bacteria and very small algae (picoplankton) were recognized as important members of a previously under-appreciated microbial loop. Food web models arising in the 1980s emphasized diversity of connections among members of the microbial community. (After Graham, 1991)

can discriminate between algal food that is higher or lower in nutritional quality, more or less digestible, and toxic versus non-toxic.

Figure 3–3 *Dinobryon*, an example of a mixotrophic algal flagellate. Individual biflagellate cells are arrayed in a dichotomously branched, treelike colony. Cells contain golden-brown plastids, but are also capable of ingesting particulate prey. (Photograph courtesy C. Taylor)

In view of the strong size relationship between algal food and planktonic grazers, ecologists have grouped phytoplankton into size classes. The **picoplankton** include organisms that are 0.2–2 μm in diameter, such as bacteria (including some cyanobacteria) and certain very small eukaryotic algae; **nanoplankton** range from 2–20 μm, **microplankton** 20–200 μm; and **mesoplankton** from 0.2–2 mm in diameter. (In some older literature the term

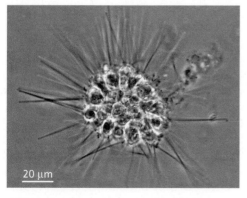

Figure 3–4 *Chrysosphaerella*, another colonial, mixotrophic flagellate whose cells, like those of *Dinobryon*, contain golden plastids.

nanoplankton was used to cover the size ranges now known as picoplankton, nanoplankton, and a portion of the organisms now classified as microplankton.) Our views of the ecological roles of these size classes of algae in food webs has changed over time (Fig. 3–2).

In the next sections we will consider the diversity of protist and zooplankton herbivores in more detail. Herbivory by mesograzers and macrograzers and algal strategies for coping with them are covered in Chapter 23, where the focus is on macroalgal and periphyton ecology.

Protistan herbivory

Flagellates. Various protists, including algal species, amoebae, and ciliates can be extremely important grazers in both fresh and marine waters. Small flagellates and ciliates can ingest as much as 100% of their own cell volume within one hour, and double their populations in about three hours. In one lake study small heterotrophic flagellates, such as bodonids and choanoflagellates, accounted for 84% of total protistan grazing on the picoplankton, primarily cyanobacteria such as *Synechococcus* (Carrias, et al., 1996). Marine heterotrophic flagellates are considered to be the most important link between picoplankton and the zooplankton. The freshwater flagellate *Pseudospora* has an amoeboid stage that preferentially consumes colonial green algae such as *Eudorina* from the inside. This protozoan can reduce populations of such algae by 90% within one or two weeks (Canter and Lund, 1968).

A variety of freshwater and marine euglenoids, cryptomonads, chrysophyceans, dinoflagellates, and prasinophycean green algae ingest cells of other organisms as food (i.e., they are phagotrophic). Many of these algae contain chloroplasts and carry out photosynthesis in addition to phagocytosis (particle feeding) and/or uptake of dissolved organic compounds from the environment (osmotrophy). This mixed mode of feeding is known as mixotrophy (see Chapter 1). As much as 30% of the phytoplankton in oligotrophic lakes is mixotrophic. Examples include the flagellate unicellular or colonial chrysophycean genera, *Dinobryon* (Fig. 3–3), *Uroglena*, *Chrysosphaerella* (Fig. 3–4), *Catenochrysis*, *Chromulina*, *Ochromonas*, and *Chrysococcus*, which feed on bacterioplankton. Certain colorless euglenoid flagellates feed upon related pigmented euglenoids. *Peranema trichophorum*, for example, uses a rodlike cellular ingestion apparatus to push similarly sized green *Euglena* cells into its upper

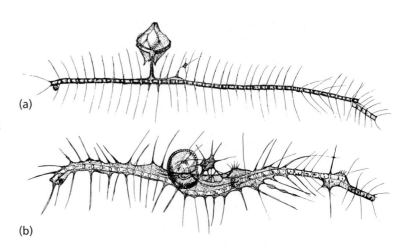

Figure 3–5 Illustration of a pallium, a feeding structure that can be extended from the cells of certain dinoflagellates. (a) *Protoperidinium spinulosum* (above) is shown extending its pallium around the filamentous diatom *Chaetoceros* (below). (b) Within the extended pallium, the diatom cell contents are liquified and then taken into the feeding cell. (From Jacobson and Anderson, 1986 by permission of the *Journal of Phycology*)

end, which can dilate pythonlike to accommodate such large food particles. Similarly, many dinoflagellates, both colorless and pigmented, are capable of feeding upon large prey, some of which include other dinoflagellates and various algal species. For example, the dinoflagellate *Heterocapsa* kills the dinoflagellate

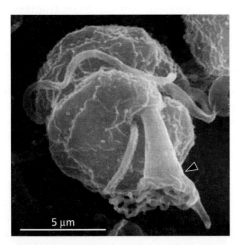

Figure 3–6 One life-history stage of the fish-eating dinoflagellate *Pfiesteria*, viewed with scanning electron microscopy (SEM). SEM preparations are often necessary to visualize details necessary to identify and classify microscopic algae, and to better observe feeding structures, such as the peduncle (arrowhead). Such preparations are made by using chemicals to retard structural changes in cells, then slowly dehydrating cells and drying them in such a way that cell structure does not collapse. As a final step, dried cells are coated with a thin layer of metal—commonly gold—to prevent buildup of electrons on the cell surface while the specimen is being examined in the scanning electron microscope. (From Lewitus, et al., 1999 by permission of the *Journal of Phycology*)

Gyrodinium upon contact. A wide variety of feeding structures are found among the dinoflagellates (see Chapter 11 for more details). As an example, the widespread and common marine *Protoperidinium depressum* casts a cellular "fishing net" termed a **pallium**—basically an extension of the cell membrane that contains an extracellular vacuole—into the water to collect bacteria and larger phytoplankton (Fig. 3–5). Using transmission electron microscopy (TEM), Jacobson and Anderson (1996) found that many photosynthetic walled marine dinoflagellates, including *Alexandrium ostenfeldii*, *Gonyaulax diegensis*, *Ceratium longipipes*, and *Prorocentrum micans*, contained food vacuoles whose contents were full of recognizable remains of partially digested protozoan ciliates and other dinoflagellates.

Some dinoflagellates practice the phytoplanktonic equivalent of mammoth hunting—the single-celled *Pfiesteria piscicida* (Fig. 3–6) consumes pieces of fish far larger than itself. The presence of fish induces the dinoflagellate cells to release potent neurotoxins that kill fish, then the dinoflagellates extrude feeding tubes to consume fish tissues as they slough off. This alga is thought to be responsible for major kill events, decimating populations of up to one million fish, including commercially valuable species. *Pfiesteria*, like some other dinoflagellates (see Chapter 11) can transform into very different looking life stages, including amoebae (Burkholder, et al., 1992; Steidinger, et al., 1996). The amoeboid stages most likely possess distinctive feeding mechanisms of their own. The physiological ecology of *Pfiesteria piscicida* and its predatory behavior were reviewed by Burkholder, et al. (1998).

Amoebae. Marine and freshwater amoebae can harvest larger unicellular and filamentous algae by engulfing them with pseudopodia—cellular processes

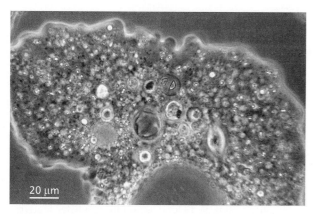

Figure 3–7 An amoeba that is actively feeding by extending pseudopodia. A variety of pigmented algae are present in its cytoplasm.

that actively extend from cell surfaces (Fig. 3–7). *Amoeba proteus* has been observed to feed upon flagellate algae such as the cryptomonad *Chilomonas* and the euglenoid *Euglena*. *Amoeba discoides* engulfs green flagellates such as *Chlamydomonas* and *Pandorina*, but rejects living cells of the nonflagellate greens *Chlorella* and *Staurastrum* (Ho and Alexander, 1974). Shelled (thecate) amoebae also ingest algal prey (Fig. 3–8). *Thecamoeba verrucosa* can ingest, and wind up within its cells, filaments of *Oscillatoria* that are several hundred micrometers in length. Algal food particles adhere to the axopodia (thin processes that extend

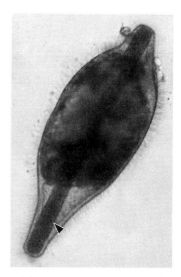

Figure 3–9 This is one of many ciliates known to consume algae via a cytostome (cell mouth). In this individual, the remains of an algal meal—a cyanobacterial filament (arrowhead)—are visible within a cytoplasmic food vacuole.

from the cell surfaces) of heliozoans, in a form of passive filter feeding. The food is then transported inside the cell. Algal food is similarly harvested by pseudopodia extending from mineral-shelled marine radiolarians and foraminifera (Grell, 1973).

Ciliates. Ciliates are protists that are typically covered with cilia at some point in their life cycle and are characterized by having two types of nuclei in their cells. They are important consumers in both marine and freshwaters. Ciliates less than 30 μm in largest dimension primarily eat picoplankton, while those greater than 50 μm consume mostly nanoplankton or larger algae (Rassoulzadegan, et al., 1988). Simek, et al. (1996) have recently defined three ecological groups of ciliate herbivores: (1) raptorial feeders like *Urotricha*, which primarily feed on nanoplankton; (2) highly efficient filter feeders like *Vorticella* and *Cyclidium*; and (3) coarse filter feeders and detritivores such as *Coleps* (Fig. 3–9), which eat living or dead algae.

Most ciliates possess a cell "mouth" (cytostome), a surface region specialized for food intake that leads to a cytoplasm-filled pharynx. The pharynx of raptorial ciliates, such as *Frontonia* and *Nassula*, is reinforced by rodlike structures to form a gullet (cytopharyngeal basket) that helps them ingest large diatoms, dinoflagellates, and filamentous cyanobacteria. Filter feeders, including *Paramecium* and *Tetrahymena*, create water currents that carry food to the cell mouth. Frequently, as in *Vorticella* (Fig. 3–10), cilia in

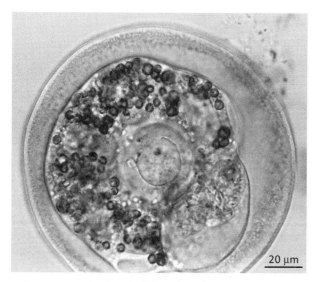

Figure 3–8 A testate (shelled or thecate) amoeba that has fed upon algae. Remains of the algal food can be observed within the cytoplasm through the translucent test wall.

Figure 3–10 *Vorticella* is a sessile (attached) ciliate that feeds upon algae and other particles by using undulating membranes and cilia to generate water currents. This particular species contains endosymbiotic green algae.

the mouth region are fused into undulating membranes that help to generate food-collection currents. Food is enclosed in cell vacuoles that form at the base of the cell mouth, then digested as the vacuoles move through the cell. Suctorians, sessile ciliates including *Tokophyra* (Fig. 3–11), ingest food material through retractable tentacles, incorporating it into vacuoles at the base of the tentacles. The marine and freshwater ciliates known as tintinnids have several undulating food-collecting anterior membranes. They can remove as much as 41% of the standing crop of algal chlorophyll *a* produced in a day and can consume as much as 52% of the nanoplankton (Verity, 1985).

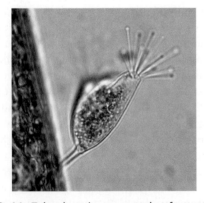

Figure 3–11 *Tokophyra* is an example of a suctorian ciliate; these protozoa attach themselves to solid substrata, then use tentacles to snare food particles (including algae) from the surrounding water. (Photograph courtesy C. Taylor)

Crustacean grazers

Cladocerans such as *Daphnia* and *Bosmina* (Fig. 3–1d) are very important grazers in freshwaters, while copepods (though present in freshwaters) are particularly significant in marine food webs. Cladocerans reproduce asexually most or all year-round (in warmer waters) with generation times on the order of weeks. Their populations can increase rapidly if appropriate algal food is available. Maximum body sizes are generally 0.3–3 mm in length. In contrast, copepods reproduce sexually, and generation times are longer.

Cladocerans and some copepods are filter feeders. They collect algae by sieving water through fine meshes of a filtering apparatus, sweeping them into the mouth region (Fig. 3–12). They are poor at particle selection, but if large algae clog the filter apparatus, the offending material can be removed with an abdominal claw. Peak abundance of cladocerans typically occurs in late spring and summer when their preferred foods—small naked green algae, cryptomonads, and small diatoms—are most abundant. Cladocerans can filter about five times the volume of a eutrophic lake in one day, but filtering rates are lower in oligotrophic waters (Porter, 1977). Because the water volumes filtered by cladocerans may contain some algal food that is low in nutrients or contains toxins, their ingestion of such food may be unavoidable. The situation facing filter-feeding zooplankton has been compared to that of cattle grazing in fields occupied by a mixture of edible and toxic plants—it is difficult for the herbivores to recognize and reject toxic food. However recent evidence suggests that the cladoceran *Daphnia* can detect the odor of edible algae such as the green alga *Scenedesmus* (van Gool and Ringelberg, 1996) and use this ability to find patches of water that are richer in edible algae, thus effectively avoiding patches of less desirable algae.

Copepods (Fig. 3–1e) and some cladocerans are raptors; they grasp large algae such as chain-forming diatoms like candy canes and break off pieces (Porter, 1977). Marine herbivorous copepods, such as *Acartia*, feed selectively on nanoplankton including coccolithophorids, and they actively discriminate against toxic or low quality algal food. Freshwater copepods will select and consume filaments of the cyanobacterium *Oscillatoria tenuis*, whereas they reject the nearly morphologically identical, but toxic *Planktothrix* (*Oscillatoria*) *rubescens* (DeMott and Moxter, 1991). Studies conducted by videotaping copepod feeding have shown that particle size is the most important

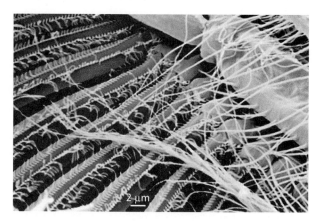

Figure 3–12 Components of the filtering apparatus of the common cladoceran *Daphnia*, viewed with SEM. The spacing of the mesh determines the size of particles that can be retained and consumed.

factor influencing food capture, and that selection occurs after capture. Particles are manipulated and tasted, with chemical factors playing a more important role in the selection phase than does size (DeMott and Moxter, 1991). The upper end of the size range of algae that can be consumed is determined by the opening width of the mandibles or the carapace gape. This is about 50 μm for copepods and some large cladocerans (Lampert and Sommer, 1997).

Algal Food Quality

Experiments have shown that neither the marine copepod *Diaptomus* nor the freshwater cladoceran *Daphnia* absolutely require algae as food. However in both cases more viable young were produced per unit of food biomass ingested, when high levels of algae were consumed (Sanders, et al., 1996), suggesting that algae possess desirable qualities as zooplankton food. Algae that support good growth and reproduction of herbivores are termed "high quality" food. Although rotifers cannot discriminate between high- and low-quality food, they grow better on the former (Rothhaupt, 1995).

Whether or not a particular alga is considered to be high in food quality depends upon two factors, accessibility of the food to a particular herbivore (which is a function of the alga's perceived size and shape) and chemical content. Algal species vary widely in their content of mineral nutrients and essential organic compounds such as vitamins or fatty acids, and some match the nutrient requirements of herbi-

vores much better than do others (Sterner, et al., 1992). Algal species that have high levels of essential biochemical constituents (and that are also consumed at high rates) are considered high-quality foods. Cryptomonads, for example, are superior food for zooplankton because these unicellular, wall-less flagellates are readily ingested and digested, and because they contain relatively high proportions of two essential fatty acids (eicosapentaenoic and docosahexaenoic acids), and lack toxins. Algae that may be high-quality food under some conditions may be poor food under others. For example, the green flagellate *Chlamydomonas* is high-quality food when grown with adequate N and P, but possesses decreased amounts of essential polyunsaturated fatty acids (and is thus of lesser food quality) when P-limited (Weers and Gulati, 1997). Low-quality algae are ingested at low rates or avoided altogether and/or are low in essential herbivore nutrients or produce toxic compounds. The cyanobacterium *Planktothrix rubescens* is an example of a low-quality algal food for zooplankton: its filamentous shape makes ingestion difficult, and it contains high levels of microcystin, a potent neurotoxin. Poor food quality can cause a decreased herbivore growth rate, body size, and reproduction rate, or prevent growth and development altogether.

Assessment of algal food quality is often conducted in laboratory microcosms—relatively small volume containers that are inoculated with the alga to be tested and the herbivore of interest. Good growth and reproduction of the herbivore are taken as evidence for high food quality; reduced growth and reproduction suggest (but do not prove) the opposite. Ianora, et al. (1996) used microcosm experiments of this kind to establish that the dinoflagellate *Prorocentrum minimum* and the chrysophycean *Isochrysis galbana* are high-quality food sources for the copepod *Acartia clausi*. However these workers found that axenic cultures of four diatom species that are abundant in marine waters (and generally considered to be high-quality food for copepods) did not support good reproduction of *Acartia* and suggested that the diatoms produced inhibitory chemicals. On the other hand, Jónasdóttir and Kiorboe (1996) found that diatoms did not reduce reproduction of a related copepod species and suggested that the reduced reproduction was due to low oxygen concentration in the microcosms. In view of the prevailing paradigm stating that diatoms constitute good food for copepods, and to attempt to resolve this controversy, 15 laboratories in 12 countries ran simultaneous exper-

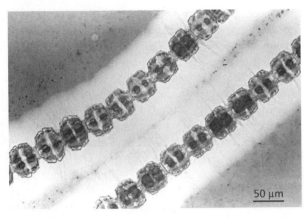

Figure 3–13 The often extensive mucilaginous sheaths of various algae, such as the green alga *Desmidium majus*, are thought to reduce the chance of ingestion by herbivores. Such sheaths are often transparent and difficult to visualize unless a contrast method is used. Here, dilute india ink has been added to the preparation. The ink particles cannot penetrate the sheath, and thus reveal its limits. The presence of such a sheath can have the effect of making the alga appear larger to potential herbivores. (Photograph by M. Fisher and L. Wilcox)

iments on the food quality of diatoms for freshwater, estuarine, and marine copepods (Ban, et al., 1997). The results provided strong evidence that 16 of the 17 diatoms tested were inferior food because they produce a compound that blocked embryo develop-

Figure 3–14 *Scenedesmus*, a freshwater green alga, is an example of algae that can produce spiny cell extensions and long, bristlelike structures, which may make the organism appear large to herbivores and help reduce sedimentation.

ment in 16 different copepods. The copepods that were fed diatoms exhibited significantly reduced egg production and egg viability compared to controls. The authors of this report suggested that the inhibitory compound (whose chemical structure is unknown) may be an example of an anti-herbivory defense, our next topic.

Algal Defenses

Phytoplankton Defenses

Phytoplankton and seaweeds (see Chapter 23) possess a variety of mechanisms that prevent herbivory from completely decimating their populations under natural conditions. Among phytoplankton these include regulation of population size and seasonal occurrence, modifications of organismal structure or reproduction, and production of chemical deterrents. Some members of nearly all major algal groups couple very small cell size with extremely rapid cell division (on the order of hours to days) to generate large populations of cryptic cells. Since it is difficult for herbivores to find and consume all such algal cells, a few persist to generate new populations when growth conditions allow. Growth in the cold season, when populations of herbivores are relatively low, can be another successful herbivore-avoidance strategy.

There are a variety of ways in which phytoplankton structure may deter herbivore selection and ingestion. In fact, the enormous diversity in algal form is due in part to the powerful influence that herbivory has had upon the course of algal speciation over hundreds of millions of years. Many phytoplankton species produce extensive gelatinous coatings (Fig. 3–13) or spiny projections from the cell surface (Fig. 3–14), or assume elongated shapes or colonial morphologies (Fig. 3–15). Such structural features have the effect of making relatively small cells appear larger to herbivores or may clog their ingestion apparatus. The common, highly edible phytoplankton green algal species *Scenedesmus subspicatus* can be induced to form larger colonies composed of cells with tougher walls and more spiny projections (Fig. 3–16) when exposed to either living *Daphnia* or medium in which the *Daphnia* has been grown. This is an example of an inducible defense mechanism; similar responses have long been known from zooplankton prey that have been exposed to predators (Hessen and van Donk, 1993; van Donk, et al., 1999).

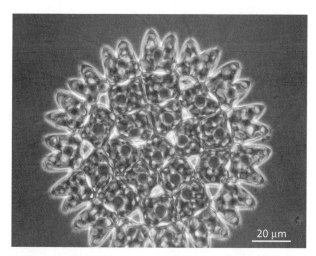

Figure 3–15 *Pediastrum*, a green alga that occurs commonly in freshwaters, is characterized by an elaborate, lacy colonial structure. Cells of some *Pediastrum* coenobia, such as this one, also possess long, slender bristles. As with *Scenedesmus*, such structural features likely reduce settling in the water column and may also function in herbivore avoidance.

Some phytoplankton appear to be protected from digestion by mucilage and tough, digestion-resistant cell walls. These forms not only survive passage through a zooplankter's gut but actually benefit from the oppor-

tunity to obtain mineral nutrients during gut passage. An example is the colonial green genus *Sphaerocystis* (Fig. 3–17), which was documented to benefit in this way from ingestion (Porter, 1977). The unicellular green algae *Chlamydomonas reinhardtii* and *Selenastrum capricornutum*, which are readily digested by zooplankton when the algae are not P-limited, become much more resistant to digestion when they are limited by phosphorus. It is hypothesized that P-limitation triggers a chemical change in their cell walls that allows them to survive passage through zooplankton and that this represents a defense mechanism that reduces the effects of grazing on algal populations when growth rates are low (van Donk, et al., 1997).

Phytoplankton that employ increased cell size as a defensive mechanism may, as a consequence, possess a reduced ability to remain suspended in the water column. Therefore large-celled algae typically possess floatation or suspension adaptations that offset a tendency to sink below euphotic (well-lighted) surface waters. Cyanobacteria have **gas vesicles** that provide a buoyancy adjustment mechanism (see Chapter 6 for more details). Various large dinoflagellates (Fig. 3–18), colonial green algae, and chrysophyceans can swim by virtue of their flagella. In the case of the colonial species, flagellar motion of the individual cells is coordinated. For some algal forms motility may also confer herbivore-avoidance advantages, as well. The mucilage that surrounds cells of some phytoplankton species, and which joins together individual cells of many colonial forms, can reduce accessibility of algae to herbivores as well as slow sinking, since it is composed primarily of water held in a glycoprotein

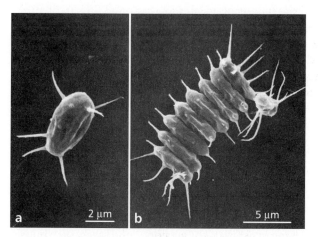

Figure 3–16 Putative inducible morphological defenses are exhibited by *Scenedesmus subspicatus* in cultures that contain exudates from the cladoceran herbivore *Daphnia*, or the animals themselves. The alga was primarily one or two-celled when not exposed to *Daphnia* exudate (a), whereas eight-celled colonies became significantly more abundant when the alga was exposed to *Daphnia* or its exudate (b). (From Hessen and van Donk, 1993)

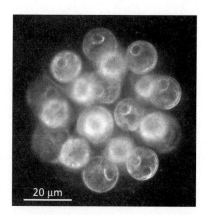

Figure 3–17 *Sphaerocystis*, a colonial green alga found in freshwaters, can survive ingestion by zooplankton by virtue of copious mucilage and a tough cell wall.

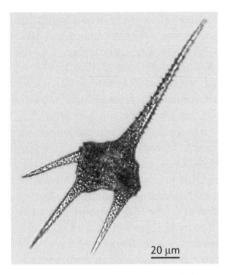

20 μm

Figure 3–18 The dinoflagellate *Ceratium hirundinella* is an example of a relatively large algal cell that is resistant to herbivory but which readily sediments if it is unable to continue swimming by means of its two flagella. (Photograph courtesy C. Taylor)

matrix. Cell surface projections, mentioned above in the context of herbivore avoidance, may also serve to increase buoyancy. In general, it is difficult to establish whether sinking resistance or herbivore avoidance has been the primary selective factor influencing production or retention of flagellar and gas vesicle-based motility, mucilage, and surface projections of phytoplankton species. Algal adaptations that reduce sedimentation rates are discussed further in Chapter 22.

A number of cases are known of toxic phytoplankton taxa that are very closely related to nontoxic forms, and such data support the concept that chemical defenses are adaptive (DeMott and Moxter, 1991). An exception to the generally accepted defensive role of dinoflagellate toxins is the case of *Pfiesteria* toxins; these poison fish, which are then consumed by the dinoflagellate. This is probably an example of evolutionary exaptation—the modification of an ancestral feature (defensive toxins) for use in a new way (active predation behavior).

There is also evidence that the bioluminescence exhibited by many marine dinoflagellates has a defensive function. When the algal cells are agitated, they give off a flash of blue-green light that results from reaction of the substrate luciferin with the enzyme luciferase, as in fireflies and various luminous bacteria (Sweeney, 1987) (see Chapter 11 for more details). In the case of dinoflagellates, biolu-

minescence appears to deter copepod predation. Two possible protective mechanisms have been suggested: a direct "startle" effect on the herbivores themselves and a more indirect effect—increased predation upon copepods that have fed upon glowing dinoflagellates and thus rendered more visible to their predators. An experimental text of the latter hypothesis showed that bioluminescent dinoflagellates increased the foraging efficiency of stickleback fish on copepods. This is viewed as an example of how prey (dinoflagellates) take advantage of predation interactions at a higher trophic level (Abrahams and Townsend, 1993).

Many species of phytoplankton and seaweeds produce toxic or defensive compounds. In the case of **endotoxins**, these compounds are retained within cells or tissues and released only in the event of herbivore damage or death of the algae. In other cases exotoxins are released into the environment from apparently healthy living cells. There is a great deal of variation among algal taxa and geographical location in the occurrence of chemical defenses. Potent toxins are known to be produced by some 20 species of marine phytoplankton. While particular ochrophytes, haptophytes, diatoms, cyanobacteria, and dinoflagellates are toxic, many are not. Toxic dinoflagellates are almost entirely marine. In freshwaters the primary algal toxin-producing species are cyanobacteria. The ecological and economic effects of these phytoplanktonic toxins are so great that they will be considered in some detail in the next section.

Cyanobacterial Toxins

Various cyanobacteria produce one or more toxins belonging to three general classes: lipopolysaccharide endotoxins, hepatotoxin tumor promoters, and neurotoxins. These affect not only zooplankton and other herbivores, but also humans and aquacultured food organisms. Neurotoxins produced by cyanobacteria in drinking water can poison dogs and farm animals, waterfowl, and fish. Signs of poisoning include staggering, gasping, convulsions, and cyanosis (bluish appearance of the skin). Death by respiratory arrest may occur within minutes to hours after toxin ingestion.

Incidents of animal poisonings are associated with large populations or blooms of cyanobacteria. Some 50–75% of cyanobacterial blooms generate toxins, and some experts believe that all cyanobacterial blooms should be regarded as toxic until disproved (Codd, et al., 1995). It is unwise to allow pets or livestock to consume waters for which there is evidence of cyanobacterial bloom-level populations. The factors involved in excessive growth of cyanobacteria in freshwaters and possible methods for prevention of toxic blooms are discussed in Chapters 6 and 22. Although cyanobacterial toxins are generally regarded as defense-related, it has been proposed that they might serve additional cellular functions, such as nitrogen storage. The structure, function, detection, risk assessment, and control of cyanobacterial toxins has been reviewed by Carmichael (1997) and Sivonen (1996).

More than forty cyanobacterial species are either demonstrated or suspected toxin producers. Establishment of a cyanobacterium's ability to produce toxins requires cultivation of clonal, axenic cultures—labgrown populations that have originated from single isolated cells or filaments and which do not contain any other cellular organisms (bacteria or eukaryotes). Toxins can be identified by high performance liquid chromatography and/or mass spectrometry. Toxicity assessments are conducted by injecting algal extracts or purified toxins into mice and determining the LD_{50}, that is, the dosage that results in death (Lethal Dose) of 50% of the animals. With such methods (Codd, et al., 1995), cyanobacteria that have been established to produce toxins include some species or strains of *Anabaena*, *Aphanizomenon*, *Cylindrospermopsis*, *Microcystis*, *Nodularia*, *Nostoc*, and *Planktothrix*. The structure and biosynthesis of cyanobacterial toxins was reviewed by Rinehart, et al. (1994).

Lipopolysaccharides (LPS) are endotoxins that occur within the cell envelopes of *Anabaena*, *Synechococcus*, and *Microcystis*. They are related to toxic LPS produced by pathogenic bacteria (*Salmonella*), but seem to have lower toxicity. The cyanobacterial LPS are implicated in cases of fever and inflammation in humans who have bathed or showered in water that contains cyanobacterial blooms.

Hepatotoxins cause most cases of animal poisonings that are associated with cyanobacterial growths in drinking water. Wild and domestic animals, including livestock and dogs, may experience weakness, heavy breathing, pallor, cold extremities, vomiting , diarrhea, and massive bleeding in the liver that can lead to death within 2–24 hours after ingestion. Some human deaths and cases of illness have been confirmed, and health experts are also concerned that long-term ingestion of drinking water containing hepatotoxins may lead to increases in human liver cancer, because laboratory analyses indicate that hepatotoxins have tumor-promoting activity.

Hepatotoxins are cyclic peptides. Microcystin is a heptapeptide (Fig. 3–19a), and nodularin is a pentapeptide. Microcystin can constitute as much as 1.2% of *Microcystis* protein and can occur in more than 45 chemical variations. This toxin is also produced by strains of *Anabaena*, *Nostoc*, and *Oscillatoria*; the level of the toxin in the cyanobacteria can be quite variable. The molecular basis for hepatotoxin toxicity is irreversible inhibition of liver-cell protein phosphatases. The specificity and potency of the cyanobacterial hepatotoxins are about the same as those of the marine algal toxin okadaic acid, which causes diarrhetic shellfish poisoning and has also been established to possess tumor-promoting activity (Codd, et al., 1995).

Microcystin can persist in freshwaters for two or more weeks before being degraded by bacteria. In some parts of the world untreated natural waters are used for drinking, a cause for concern when cyanobacterial blooms are present. Significant increases in the rate of liver cancer in a Chinese village has been correlated with contamination of surface drinking water supplies by microcystin (see references in Carmichael, 1997). Municipal water treatment plants remove variable amounts (sometimes up to 98%) of microcystins from lake water. Although ozonation, treatment with activated charcoal, and chlorination (Tsuji, et al., 1997) can remove microcystin, in some cases, as much as 20% may escape the treatment process (Carmichael, 1997). Relatively simple yet very sensitive ELISA (Enzyme-Linked Immunosorbent Assay) methods have been developed for detection of microcystins in water, microorganisms, and animal tissues. The ELISA procedure is useful in determining how much microcystin remains in treated water. Recurring, serious toxic algal blooms, particularly in the brackish water of the Baltic Sea and in Australian waters, have led several countries to consider establishing maximum acceptable microcystin levels for drinking water (Carmichael, 1997). Researchers have recently isolated a new species of the bacterium *Sphingomonas* that has the ability to break down microcystin, using this compound as a sole carbon and nitrogen source. A bacterial enzyme, a protease appropriately named microcystinase, can open the ring structure of microcystin, thereby reducing toxicity 160-fold. Additional bacterial proteases further break down microcystin to harmless constituents. Enzymes such as these may someday prove useful in water treatment processes (Bourne, et al., 1996).

Neurotoxins produced by cyanobacteria act to block neuromuscular activity. Anatoxin-a is a secondary amine alkaloid produced by certain strains of *Anabaena flos-aquae*, some benthic species of *Planktothrix*, and perhaps other cyanobacteria. **Saxitoxins** (Fig. 3–19b) are cell membrane sodium-channel blockers so potent that lab mice are killed in minutes at an LD_{50} intraperitoneally injected dose of 10 μg kg^{-1}. Significantly, saxitoxins are also produced by marine phytoplankton, which are discussed in the next section.

Toxins Produced by Marine Dinoflagellates and Diatoms

The incidence of ecologically, economically, and medically serious poisoning episodes attributed to marine phytoplankton-produced toxins seems to be increasingly dramatically on a worldwide basis. Some experts contend that this reflects greater monitoring or increased human awareness and use of marine resources, while others view this increase as evidence of negative human influences upon the natural environment, such as pollution of water with sewage or other materials. The toxins produced by marine dinoflagellates and diatoms have been reviewed by Plumley (1997) and Baden, et al. (1998). Here we will provide a brief summary of the chemical nature of the major dinoflagellate and diatom toxins, their mechanisms of action, and their poisonous effects on humans, marine mammals, shorebirds, and fish. A few other marine phytoplankton, including ochrophytes and haptophytes, also produce toxins, which are discussed in Chapters 10 and 13.

During the warm season of the year there are nearly always reports from throughout the world of dramatic poisoning events such as the 1987 deaths of 14 humpback whales on the Cape Cod (U.S.A.) coast, or the 1991 deaths of numerous brown pelicans near Monterey Bay, California. In the latter case the killer was identified as domoic acid (see below), produced by marine diatoms, which had been eaten by anchovies, which in turn were consumed by the pelicans (Fritz, et al., 1992). The Cape Cod whales died from the effects of saxitoxin (described below) produced by a bloom of the dinoflagellate *Alexandrium tamarense*, which had become concentrated in the organs of mackerel eaten by the whales (Anderson, 1994).

Massive fish kills can result when fish swim through blooms of toxic phytoplankton, exposing them to algal neurotoxins that paralyze their respiratory systems, leading to asphyxiation. Tons of dead fish can wash up on beaches, creating a deleterious effect on recreation and tourism. Fish grown in cages in aquaculture operations are particularly vulnerable to algal toxins because they are unable to escape their presence. Ciguatera fish poisoning affects humans who harvest fish from subtropical and tropical coral reef lagoons. Toxin-producing dinoflagellates such as *Ostreopsis*, which grow attached to red or brown seaweeds, are inadvertently consumed by herbivorous fish, which are in turn eaten by humans (Faust, et al., 1996).

As a result of consuming shellfish whose tissues have concentrated phytoplankton toxins, humans can also suffer the effects of neurotoxic, paralytic, diarrhetic, or amnesiac (involving serious short-term memory loss) poisoning. People who frequent beaches or waters where phytoplankton blooms occur can suffer respiratory illnesses and eye irritation from water- and spray-borne toxins. Laboratory workers and fisherman exposed to the toxin of the dinoflagellate *Pfiesteria piscicida* have experienced memory loss and learning deficits, as have laboratory rats exposed to the same toxin (Levin, et al., 1997). For these reasons governments around the world have established systems for monitoring the occurrence of coastal algal blooms and toxin content of shellfish, but such monitoring varies greatly in effectiveness. Development of techniques for rapid and accurate identification of developing toxic blooms is a high priority for coastal nations, states, and provinces. Some recent molecular systematic approaches that may be helpful are described in Chapter 5.

Saxitoxins (Fig. 3–19b) comprise a family of 18 or more compounds that bind to cell-membrane sodium channels, thus blocking the flux of sodium in and out of animal nerve and muscle cells. This uncouples nerve-muscle communication, resulting in paralysis. Saxitoxins are responsible for the occurrence in humans of paralytic shellfish poisoning. Brevitoxins (Fig. 3–19e) are larger and differently structured molecules that also affect sodium-channel function, likewise causing paralytic poisoning.

Okadaic acid (OA) (Fig. 3–19c) and dinophysis-toxin-1 (DTX–1) (named for the dinoflagellate *Dinophysis*, which produces it) are responsible for diarrhetic shellfish poisoning (DSP) in humans. These toxins (like the hepatotoxins of cyanobacteria) are potent inhibitors of serine- and threonine-specific phosphatases, which are involved in many essential cellular processes in eukaryotes. These compounds thus have the potential to affect not only animals, but also other eukaryotes. One study showed that OA

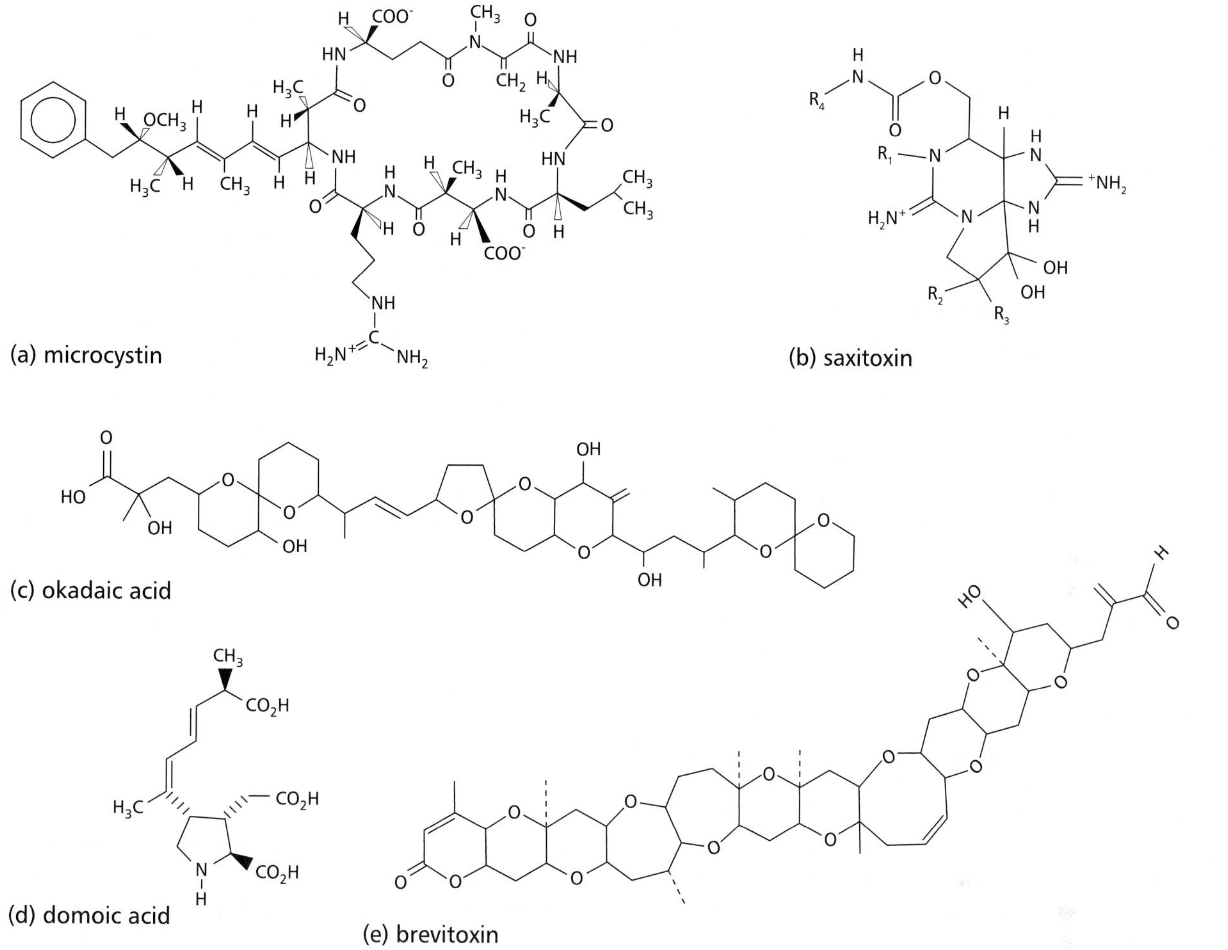

Figure 3–19 The chemical structures of some algal toxins. (a) Microcystin, a highly toxic cyclic peptide produced by various cyanobacteria. It is named for the common bloom-forming genus *Microcystis*, which can produce this compound. Enzymes that break the ring structure reduce toxicity substantially. (b) Saxitoxin, one of a group of cell membrane sodium channel blockers that have highly toxic effects upon animal cells, are produced by certain cyanobacteria and marine dinoflagellates. (c) Okadaic acid, a compound produced by certain dinoflagellates, is toxic to eukaryotes in general. (d) Domoic acid, a neurotoxin produced by the diatom *Pseudo-nitzschia multiseries*. (e) Brevitoxin, produced by certain dinoflagellates, like saxitoxin, affects animal nerve and muscle cell membrane sodium channels, but has a very different chemical structure. (a: After Bourne, et al., 1996; b, c, e: After Taylor, 1990; d: After Baden et al., 1998 with kind permission of Kluwer Academic Publishers)

could inhibit the growth of all non-DSP related microalgae at micromolar concentrations. However a strain of the dinoflagellate *Prorocentrum lima*, which is known to produce both OA and DTX–1, was not affected at much higher toxin levels. This finding suggests that OA and DTX–1 may have broader spectrum allelopathic effects, and they may retard the growth of organisms that might compete with the toxin formers (Windust, et al., 1996).

Domoic acid (DA) (Fig. 3–19d) is a neurotoxin (produced by the diatom *Pseudo-nitzschia multiseries*) that causes amnesiac shellfish poisoning—also known as toxic encephalopathy—in humans. Toxin-producing diatoms are apparently consumed by shellfish such as mussels, which are not themselves affected, but whose tissues may concentrate the toxin to levels that are poisonous to people (Perl, et al., 1990). Domoic acid binds to kainate receptors in the central nervous system, causing depolarization of neurons, which leads to neuron degeneration and death. Memory loss in human victims results from lesions in the hippocampus, which is rich in kainate receptors.

Pan, et al. (1996) discovered that phosphorus-limitation enhances DA production by *Pseudo-nitzschia multiseries*, and noted that other authors have reported similar P-limitation enhancement of saxitoxin production by the dinoflagellate *Alexandrium tamarense* and the haptophyte *Chrysochromulina polylepis*. These findings are of concern in relation to unperturbed marine waters (which are typically N-limited), because combined nitrogen levels have been rising in coastal waters as a result of pollution. Thus more marine habitats are becoming P-limited (Pan, et al., 1996). Human-generated N-pollution may cause undesirable positive feedback effects in the development of toxic phytoplankton blooms, such that the greater the pollution, the greater the chances that phytoplankton blooms will produce high levels of toxins. The physiology and bloom dynamics of domoic acid-producing *Pseudo-nitzschia* species were reviewed by Bates, et al. (1998).

There is significant current interest in understanding the evolutionary origin of phytoplankton toxins, as this may shed light on the potential for development of new toxic species. Only one or two unique enzymes are required to generate domoic acid from harmless chemical precursors. Thus evolutionary origin of this toxin is relatively well understood. In contrast, other toxins have more complex pathways whose origins are much less clear. The occurrence of neurotoxic compounds of similar structure and/or physiological effect in a relatively few taxa of unrelated groups of algae as well as various bacteria has led some experts to suspect that the toxins are produced by bacterial cells or encoded by plasmids or genes of bacterial origin that have become located within the cells or genomes of some algae. Although an extensive search has not yet turned up toxin-bearing plasmids, bacteria isolated from toxigenic dinoflagellates are able to produce toxins (Plumley, 1997). The linked genes for saxitoxin synthesis have been localized to the chromosomes of dinoflagellates (Plumley, 1997).

Detrital Food Webs and Pathogens of Algae

Detritus originates from the breakdown of terrestrial plants, soils, and animal wastes, as well as from aquatic plants, animals, and algae. Phytoplankton, periphyton, and macroalgae (along with aquatic plants) are major detritus-producers in both freshwater and marine ecosystems. Amorphous, mucilaginous suspended aggregates of detritus in the oceans is called marine snow. Marine snow includes polysaccharide-rich polymers originating as algal and bacterial exudates, remains of algae such as the haptophyte *Phaeocystis*, and numerous other components. Marine waters usually contain one to ten "snow" particles per liter. The particles are typically greater than 500 μm in size. Algae included in such aggregates sink much faster than were they not. Sinking rates of marine snow can vary from 1–400 m day^{-1}. Transport of particulate organic matter from the surface to the ocean floor is part of the process that draws down atmospheric CO_2 (see Chapter 2) (Kies, 1995).

As described in Chapter 2, algal remains that reach deep anoxic regions of the ocean may escape complete decomposition and, after very long time periods, may contribute to the formation of fossil carbon deposits. However, during transport to deeper waters, algal remains may be consumed by detrital feeders or used by microorganisms for their growth. Marine snow is consumed by many types of animals: ostracods, amphipods, copepods, salps, fish, and others. There is also circumstantial evidence that krill consume marine snow. Aggregations of algal and other small particles in marine snow allows such animals to consume foods that would individually be too small. It is estimated that more than 99% of oceanic phytoplankton productivity is respired in the upper ocean (Duarte and Cebrián, 1996). Planktonic microalgae are associated with a variety of decomposer organisms, such as bacteria and fungi. Entire microbial communities may exist within and upon aggregations of the algal remains contained in fecal pellets produced by zooplankton, mesograzers, or herbivorous fish. These microbes decompose organic materials and accomplish other biogeochemically significant conversions such as nitrogen fixation, nitrification (conversion of ammonia to nitrate), denitrification (conversion of nitrate to N_2 gas), methanogenesis, and methane oxidation (Paerl and Pinckney, 1996). Microbes also help to recycle mineral nutrients that are necessary for algal growth in subsequent seasons. In Lake Tahoe (U.S.A.), for example, sinking detritus originating from spring phytoplankton blooms was the chief source of nitrate (via microbial nitrification) available for the next year's bloom (Paerl, et al., 1975).

Microbial populations can be consumed by protists who are then eaten by zooplankton, which are themselves consumed by fish or planktivorous marine mammals. This indirect method by which algal fixed carbon contributes to population growth of animals higher in food webs is known as the microbial loop (Fig. 3–2). In general, little is known about the relative quantity of algal carbon that is circulated via the microbial loop versus direct herbivory, or in comparison to the amount of algal carbon that survives in sediments; these proportions most likely vary widely among aquatic environments. In the next section we will focus on microbial saprophytes and pathogens that have been observed to attack dead or living algae: the zoosporic fungi known as chytrids; oomycetes (which are no longer classified with fungi, but rather with ochrophytes); parasitic dinoflagellates; bacteria; and viruses. Saprophytes and pathogens of macroalgae are covered in Chapter 23.

Fungal Parasites of Phytoplankton

Fungal parasites of phytoplankton are important because they have the potential to greatly reduce algal populations in a host-specific manner, i.e., most chytrid species attack only one or a few closely related algal species. For this reason some experts believe that fungal epidemics might play a role in ending or preventing the formation of algal blooms or in the timing and pattern of phytoplankton succession (see Chapter 22). Chytrids (early divergent aquatic fungi) produce zoospores that seem to locate host algae through chemical cues. The zoospores then settle, invade the algal cell, and germinate into a nonmotile stage that uses algal cell components for growth and production of more zoospores, which are released into the water. In the last stages of an epidemic, sexually formed chytrid resting stages are produced. Chytrids have been observed to attack freshwater cyanobacteria, dinoflagellates, green algae such as *Oedogonium* (Fig. 3–20) and *Staurastrum paradoxum*, chrysophyceans, and diatoms (including the common and abundant species *Fragilaria crotonensis*, *Tabellaria fenestrata*, and *Melosira indica*).

Protistan Parasites of Algae

Protists of various types can also parasitize algal cells. The saprolegnialean protist *Ectrogella perforans* (which is related to ochrophyte algae) is a virulent parasite of marine diatoms such as *Licmophora* (van Donk and Bruning, 1995). *Pirsonia diadema* is another heterotrophic flagellate related to ochrophytes that infects diatoms such as *Coscinodiscus*. It penetrates cells by piercing tubular passages (known as rimopor-

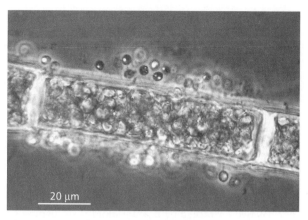

Figure 3–20 Chytrids are aquatic members of the Kingdom Fungi that, like many algae, reproduce by means of flagellate zoospores. The chytids shown here have settled upon and invaded cells of the common freshwater filamentous green alga *Oedogonium*. Algal cells that have been parasitized by chytrids can be recognized by pigment loss, evidence of cellular disorganization or senescence, and by colorless, spherical chytrid walls, which remain attached to the outer surface of the algal victim.

tulae) in the diatom frustule via pseudopodia. Part of the parasite invades and digests the algal cytoplasm, whereas part remains outside the diatom, eventually producing many infective progeny. Related freshwater forms attack a variety of algae, including *Spirogyra* (Fig. 3–21). *Stylodinium* (Fig. 3–22) and *Cystodinedria* are unusual dinoflagellates with amoeboid stages that consume cells of freshwater filamentous green algae such as *Oedogonium* and *Mougeotia*.

Bacterial Pathogens of Algae

Bacteria are often observed in association with microalgal cells and the bodies of macroalgae. Many of these associations may be mutualistic; such relationships between bacteria and algae are discussed in a later section of this chapter. Saprophytic and pathogenic bacteria also occur on or within algae. Gram-negative myxobacteria attack and lyse several types of bloom-forming cyanobacteria, causing blooms to decline (Daft, et al., 1975). *Bdellovibrio*-like bacteria lyse *Microcystis*, contributing to the decline of blooms of this cyanobacterium (Grilli Caiola and Pellegrini, 1984). One particularly noteworthy bacterial pathogen of algae is the pathogenic agent of CLOD (Coralline Lethal Orange Disease), which attacks and destroys coralline red algae that are essential in the

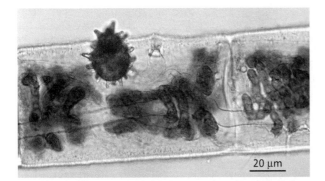

20 μm

Figure 3–21 A filament of the green alga *Spirogyra* that has been attacked by a water mold, a protistan parasite. Note the distorted plastids and parasitic hyphae ramifying throughout the host cell's cytoplasm. A spiny spore is present.

structure of coral reefs. Numerous bright orange bacilli occur within a gelatinous matrix, which is mobile because the bacteria are capable of gliding. As bacterial aggregations move across the corallines, the algal cells are completely consumed. The CLOD bacterium can infect a wide range of reef-building coralline algae, including *Porolithon onkodes*, the primary builder of algal ridges on reefs that span some 6000 km or more in the South Pacific. Marine ecologists are quite concerned about CLOD because the bacterial epidemic appears to be of recent origin and is spreading, and because it has the potential to cause very serious effects on coral reef formation and ecology (Littler and Littler, 1995).

Viral Pathogens of Algae

Viruses have been reported to infect at least some members of all major classes of freshwater and marine algae (Van Etten, et al., 1991). The viruses of cyanobacteria are known as cyanophages, whereas those of eukaryotic algae are called phycoviruses. Lytic viruses, those that result in cell lysis, are a major cause of marine phytoplankton mortality. Present in levels greater than 10^4 and less than 10^8 viruses ml^{-1}, they can result in the decline of phytoplankton blooms. It is estimated that about 3% of marine phytoplankton primary productivity is lost to viral lysis (Suttle, 1994). The cosmopolitan and abundant marine green flagellate *Micromonas* can lose 2–10% of its population each day to viral lysis (Cottrell and Suttle, 1995). In the past most reports on algal viruses consisted of accidental observations of viruslike particles within algal cells, viewed with transmission elec-

tron microscopy, but now viruses are directly observed and counted by staining them with fluorescent dyes such as DAPI (4,6-diamidino-2-phenylindole) or Yo-Pro, which bind viral DNA (Maier and Müller, 1998; Hennes and Suttle, 1995). Most viruslike particles have polyhedral capsids that are 5- or 6-sided in cross section (Fig. 3–23). The mode by which viruses enter algal cells is not well understood. It is assumed that viruses enter through breaches in algal cell walls or at locations where the cell membrane is exposed, such as the flagella. Viruses isolated from blooms of the marine ochrophyte *Aureococcus anophagefferens* in coastal areas of the North Atlantic were able to reinfect healthy cultures of the same alga. Surprisingly, the *Aureococcus* viruses had a phagelike appearance, with polygonal head and a tail (Milligan and Cosper, 1994), suggesting that viral genetic material might be injected into eukaryotic algal cells in the same way that bacteriophages infect prokaryotic cells.

Algal viruses are thought to be of potential importance in controlling algal blooms, possibly explaining sudden declines in algal populations. Viral epidemics in the marine bloom-forming haptophytes *Phaeocystis pouchetii* and *Emiliania huxleyi* are believed to the major cause of cell losses during bloom declines. A polyhedral, double-stranded DNA virus was observed to be specific for *P. pouchetii*. Lysis of cultured cells of the flagellate phase occurred within 48 hours, with release of as many as 600 viral particles per cell. Whether or not the gelatinous colonial form of *P. pouchetii* is vulnerable to the virus is not known (Jacobsen, et al., 1996).

Viruses are also known to infect seaweeds, possibly by invading swimming spores or gametes. A polyhedral virus infects the South American coastal brown seaweed *Myriotrichia clavaeformis*, causing sterility. The seaweed structures that would normally produce flagellate spores (plurilocular sporangia) are co-opted for massive production of viruses. Cell lysis results in a whitish discharge of viruses to surrounding seawater (Müller, et al., 1996).

A recent approach to detecting and identifying algal viruses is the combination of ultrafiltration to isolate and concentrate viral particles from natural waters, with comparative analysis of nucleic-acid sequences obtained from concentrated samples. Nucleotide sequences specific for the ends of genes known to generally occur in viruses are used as primers to amplify viral DNA in the polymerase chain reaction or PCR (see Chapter 5) (Chen, et al., 1996). This new approach has allowed the detection of pre-

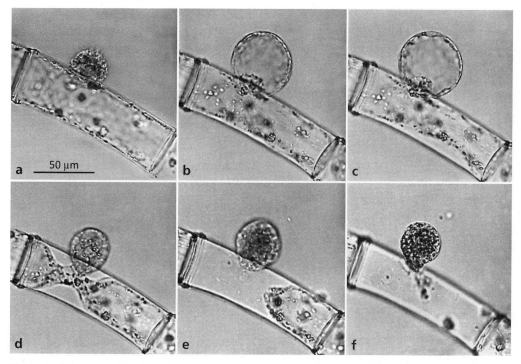

Figure 3–22 *Stylodinium* is a dinoflagellate with amoeboid stages that consume the cytoplasm of algal cells. Shown here is a sequence of an amoeba feeding upon the green alga *Oedogonium*. The entire process took only a few minutes. (Photographs by G. Wedemayer and L. Wilcox)

viously unknown phycoviruses. Such studies suggest that phycoviruses of freshwaters, coastal marine waters, and the open ocean are most likely descended from a common ancestor. As might be expected, sequences from viruses that infect the same hosts are more similar to each other than they are to sequence of viruses that infect other eukaryotic algae. The molecular sequence approach holds great promise for providing a better understanding of algal pathogens in general, because protists, fungi, bacteria, and viruses share the characteristics of being difficult to detect, isolate, and identify by conventional methods.

Algal Defenses Against Pathogens

In general, algal defenses aimed toward pathogens and parasitic attack are poorly understood. They may include structural barriers such as highly resistant sporopolleninlike polymers produced by certain dinoflagellates (Chapter 11) and green algae (Chapter 21), and algaenans in walls of other green algae (Chapter 20). Algae may also produce antibiotic compounds that deter pathogenic attack. The pro-

duction by *Spirogyra* of pentagolloyl glucose, which inhibits microbial α-glucosidase, is one such example (Cannell, et al., 1988).

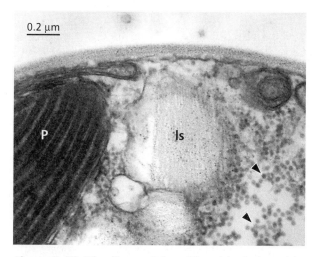

Figure 3–23 Viruslike particles with polyhedral capsids (arrowheads) within the cells of a eustigmatophycean algal cell. P=plastid, ls=lamellate storage material.

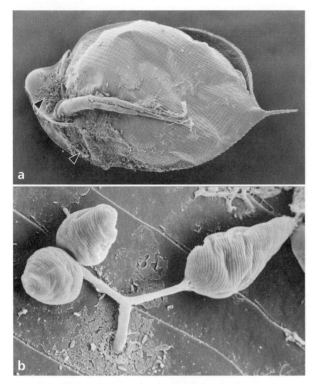

Figure 3–24 Scanning electron micrographs showing the epibiontic euglenoid *Colacium vesiculosum* growing on the cladoceran *Daphnia pulex*. (a) Low magnification view (arrowheads point to clumps of *Colacium*). (b) In higher magnification, a branched stalk, by which the algal cells are attached, is evident, as is the spiral cell covering. (Photographs courtesy R. Willey)

Algal Epibionts

Epibionts are organisms that spend most of their life cycle attached to another organism, which serves as a substrate. Numerous species of algae live on the surfaces of either protozoa, zooplankton, larger animals (most notably turtles, whales, giant sloths, and polar bears), larger algae, or plants. Mixotrophic dinoflagellates of the genus *Gyrodinium* can live on surfaces of up to 84% of the planktonic foraminifera and radiolaria at densities as great as 20,000 dinoflagellates per host cell (Spero and Angel, 1991). Some algae may be restricted to the epibiontic habit, while others can grow on nonliving substrates as well. Some of the latter are of economic concern as biofouling organisms that must be removed from the hulls of ships, surfaces of docks, or other constructions.

In many cases epibiontic surface growths are not detrimental, but at other times epibionts may be so numerous as to weigh down or shade out the organism serving as a substrate. One well-studied example is the euglenoid *Colacium vesiculosum*, which grows on the cladoceran *Daphnia* (Fig. 3–24) (Bartlett and Willey, 1998). Several hundred algal cells or colonies may be attached via carbohydrate stalks to a single animal. Heavy infestations may deleteriously affect the *Daphnia*, unless it is able to molt (shed the exoskeleton and regenerate a new one). At this point the alga produces flagellate reproductive cells that may find and attach to another animal (Al-Dhaheri and Willey, 1996). Other *Colacium* species occur in such interesting places as the rectums of damselfly larvae. Certain diatoms grow on the surfaces of marine copepods.

Algae that attach to and grow upon other algae or plants are called **epiphytes** (Fig. 3–25). It has been estimated that up to 22% of primary production can be attributed to epiphytes. Some epiphytes preferentially occupy particular algae or plants, while others are relatively indiscriminate. Coralline red algae—occupying more than 60% of the rock substrate in the intertidal zone of southern California coasts—can be epiphytized by 15–30 species of filamentous seaweeds or smaller, coralline algae. Epiphytes shade the anchor species, impede nutrient uptake, decrease growth rate, and increase drag forces, which tend to tear seaweeds from the rocks (Lobban and Harrison, 1994). Certain red algae produce halogenated organic compounds that prevent other algae from colonizing their surfaces, and brown algal phenolics serve as antifouling compounds. Some seaweeds get rid of epiphytes by shedding their surface cell layers or by copious production of surface mucilage (Round, 1992).

Algae as Parasites or Pathogens

Pathogenic microbial associations (consortia) called black band disease, which includes the dominant cyanobacterium *Phormidium corallyticum*, kill coral colonies on reefs worldwide (Richardson, 1997). Various corals (including brain and star corals) are affected, and the disease, once started at injury sites, spreads over the corals, moving a few millimeters per day (Peters, 1997). Cyanobacterial filaments facilitate the introduction of sulfide-oxidizing and sulfide-reducing bacteria into coral tissues. These create an anoxic environment and produce hydrogen sulfide, killing coral tissue, which is then degraded. It has been suggested that black band disease is correlated with water pollution in low-turbulence environ-

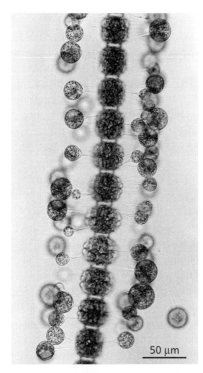

Figure 3–25 A golden-brown (ochrophyte) epiphyte (*Peroniella* sp.) growing upon the green alga *Desmidium majus*. This epiphyte does not appear to adversely affect the desmid. Related ochrophytes are often observed growing attached to this and other similar green algae.

There are a few presumably parasitic green algae, most of which are bulbous or filamentous occupants of vascular plant tissues (Chapman and Good, 1983). Examples include the bright red *Rhodochytrium*, colorless *Synchytrium*, and green pigmented *Phyllosiphon*. Although most of these endophytes seem to cause little harm to their host plants, *Phyllosiphon* infection results in yellowing of host leaves. *Cephaleuros* (Fig. 3–26) is an obligate leaf endophyte of hundreds of host plant species, including coffee, tea, and other crops, but can easily be controlled by spraying plants with a copper-sulfate solution (Chapman and Waters, 1992).

Prototheca wickerhamii and *P. zopfii* are colorless members of the green algae (identifiable as such by presence in cells of starch-bearing plastids and recent molecular data) that can cause infections in humans, cattle, and some other animals (Fig. 3–27). They are common osmotrophs in soil, sewage, and water samples. In humans they sometimes cause skin infections and occasionally bursitis and peritonitis, but these can be cured by antibiotic treatment (Gibb, et al., 1991; Sands, et al., 1991). In cattle these algae may cause a form of mastitis that is highly contagious. Infected herds are usually destroyed. *Prototheca* is associated with growth inhibition in anuran tadpoles (Wong and Beebee, 1994)

ments, as well as with other stresses to the corals. (Peters, 1997).

The human parasites that cause malaria and toxoplasmosis, classified as apicomplexans, are now known to be closely related to dinoflagellates. Their cells contain organelles that appear to be degenerate chloroplasts (Chapter 7). The pseudofungal oomycetes, many of which are pathogens, are now known to be most closely related to ochrophyte algae than to fungi (Chapter 1).

Rhodophytes are unusual among the algae in that nearly 15% of all species occur as parasites of other red algae. Most of the parasitic forms are closely related to their hosts, and in some cases it is clear that the parasite has evolved from the host. Parasitic red algae establish cellular connections with host cells and transfer parasite nuclei into host cells, transforming them. Reproductive cells that are subsequently produced may carry the parasitic genome, allowing the relationship to persist into the next generation (Goff, et al., 1996) (Chapter 16).

Figure 3–26 Parasitic growths of the green alga *Cephaleuros* on leaves of the flowering plant *Magnolia*. (Photograph courtesy R. Chapman and C. Henk)

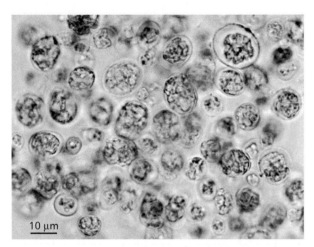

Figure 3–27 Colorless cells of the green algal relative *Prototheca*.

Algal Symbioses

Symbiosis can be defined as two or more organisms of different species that live in close physical contact. Most definitions of symbiosis imply no assumption of mutual harm or benefit, appropriate because the nature of organismal interactions can be difficult to elucidate or define. **Mutualism** is a type of symbiotic association for which there is at least presumptive evidence that all organisms involved benefit in some way. Mutualisms can help to insure the survival of the partners by providing conditions suitable for growth and reproduction that are superior to those available when growing alone. In some cases mutualistic associations maintain stable microscale environmental conditions necessary for survival that are not present outside the association (Paerl, 1992). Often we think of mutualistic symbioses as involving two partners, for example reef-building corals and their intracellular algal symbionts, zooxanthellae (described in Chapter 7). However the common occurrence and ecological significance of **multispecies consortia** are increasingly being recognized.

Algae present in mutualistic relationships often provide essential fixed carbon and oxygen to one or more heterotrophs. In the case of cyanobacteria, the benefit provided is often fixed nitrogen. Proof of mutualistic association is usually obtained by tracking the movement of radioisotope-labeled molecules from one organism to another. For example, $^{14}CO_2$ provided to the algal components of lichens can later be detected as carbohydrates or derivative molecules in cells of the fungal partner.

Aquatic algal mutualisms are common in well-lighted clear waters of low-productivity habitats, such as coral reef communities or oligotrophic and dystrophic (organically rich, brown-colored) lakes. In these environments prey availability may be limiting to heterotrophs, such that the presence of algal symbionts is beneficial. Tests of the degree of mutualistic benefit are typically done by cultivating the partners separately in the laboratory, then comparing their growth and/or reproduction to that observed in the mutualism. Often efforts are made to achieve laboratory reassociation of separated partners and to understand the nature of chemical signals passed between partners, which seem essential to mutualistic function.

Here we limit our discussion of algal symbioses to those that do not involve endosymbioses—algae living within cells and tissues of other organisms or the cells of other organisms existing within those of algae. Such intracellular mutualisms, including those between involving dinoflagellates and marine invertebrates, green algae within cells of freshwater invertebrates, or bacterial cells living within algal cells, will be covered in Chapter 7. Mutualistic relationships covered here include algal associations with heterotrophic bacteria, marine fungi, terrestrial fungi (lichens), and land plants.

Bacterial-Algal Associations

Known bacterial-algal relationships fall into three broad categories: (1) close association of more or less planktonic algae with bacterial cells, (2) macroalgal-bacterial partnerships, and (3) algae as components of highly structured benthic microbial mats.

Cyanobacteria and eukaryotic phytoplankton, including diatoms, chrysophyceans, and green algae, excrete a variety of organic compounds, including amino acids, peptides, carbohydrates, and lipopolysaccharides. These can serve as sources of fixed carbon and nitrogen for adherent bacteria. Bacteria may, in turn, provide growth factors, vitamins, chelators, or remineralized inorganic nutrients (Fe, CO_2, NH_4^+, NO_3, or PO_4^{3-}). Heterotrophic nitrogen-fixing bacteria may excrete needed fixed nitrogen to eukaryotic algae growing in N-limited environments. Aerobic heterotrophic bacteria may also help to lower O_2 concentrations, facilitating O_2-sensitive reactions such as nitrogen fixation. Bacterial associations with harmful algal bloom species are regarded as possible sources of toxigenic effects (Doucette, et al., 1998).

The association of various kinds of bacterial cells with desmid green algae in low-nutrient bog waters, anecdotally observed by several workers, has been documented more thoroughly by Fisher and Wilcox (1996). They found that bacteria typically occurred in small pockets within the mucilaginous sheath material of healthy desmids such as *Desmidium majus, D. grevillii* (Fig. 3–28), and *Hyalotheca dissiliens*. The bacteria are hypothesized to assist in providing mineral nutrients to the desmids, while desmids may provide organic nutrients to their bacterial epiphytes. Bacteria also coat healthy swimming cells of the euglenoid flagellate *Trachelomonas* (Rosowski and Langenberg, 1994), which can occur in oligotrophic freshwaters. Certain *Cryptomonas* species, characterized by a distinctive cell constriction, have been observed to harbor dense carpets of the bacterium *Caulobacter* in the constricted region. When brought into axenic (bacteria-free) culture, the algal cells lose the constriction, suggesting that presence of the constriction may not be a valid taxonomic feature (Klaveness, 1982). The surfaces of other non-senescent algae of oligotrophic waters are often colonized by large numbers of bacteria (Canter-Lund and Lund, 1995), suggesting that phytoplankton-bacterial associations may be a common phenomenon.

Species of the seaweed *Ulva* and close relatives (*Enteromorpha* and *Monostroma*) grow abnormally in bacteria-free culture but develop normal morphology in the presence of their bacterial floras (Provasoli and Pintner, 1980). Nakanishi et al. (1996) isolated hundreds of bacterial strains from the surfaces of macroalgae, and of these, about half showed morphogenetic activity when tested on *Ulva*, suggesting that bacteria produce diffusible chemical morphogens. Their work also suggests that the incidence of epiphytic bacterial effects upon macroalgal morphology may be more widespread than previously recognized. Ashen and Goff (1996) used molecular techniques (described in Chapter 5) to identify gall-causing bacteria on the surface of the marine red seaweed *Prionitis lanceolata*. DNA-based techniques offer great promise as methods for detecting and identifying symbiotic bacteria, as many are not readily culturable.

Multispecies microbial mat communities consisting of diatoms, cyanobacteria, and heterotrophic plus phototrophic bacteria, are often conspicuous features lining surface sediments of streams, hot springs, deep-sea vents, polar lakes, hypersaline lagoons, the littoral zone of hardwater lakes, coral reefs, sewage treatment plants, and estuaries. They are typically layered into

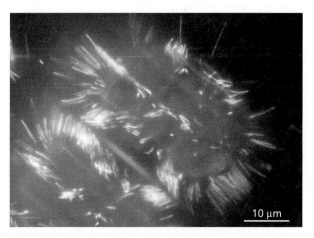

Figure 3–28 Bacterial epibionts growing within the mucilaginous sheath of the green alga *Desmidium grevillii*. The bacteria are primarily long, thin forms; they were rendered more visible through staining with the DNA-specific fluorescent dye, DAPI. (From Fisher and Wilcox, 1996 by permission of the *Journal of Phycology*)

several chemical and light microenvironments, each providing specific conditions required for major biogeochemical transformations conducted by individual mat community members (Fig. 3–29). Sunlight, trapped by the autotrophic algal members, provides the energy for mat production and chemical cycling activities. Photosynthesis by the cyanobacteria and diatoms results in release of O_2 and dissolved organic compounds, which serve as carbon sources for the heterotrophic bacteria (Paerl and Pinckney, 1996).

Lichens

Lichens are stable, self-supporting associations between fungi (known as mycobionts) and algae (green algae and/or cyanobacteria known as phycobionts or photobionts). Molecular data indicate that lichens have arisen at least five times in disparate groups of Ascomycetes and Basidiomycetes (Gargas, et al., 1995). There are estimated to be 17,000 lichen species. More than 20% of all fungi form lichens, possibly because this is a reliable way to obtain fixed carbon (from the autotrophic partner) in the form of glucose or sugar alcohols, as well as fixed nitrogen from cyanobacterial phycobionts. Some 85% of lichens contain unicellular or filamentous green algae, 10% contain cyanobacterial partners, and 4% or more contain both green and cyanobacterial algae. In the

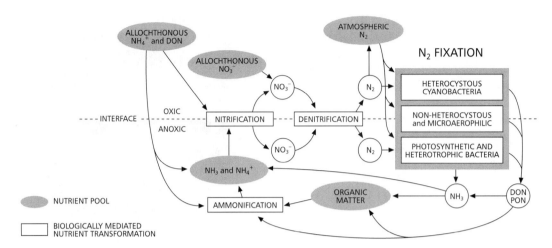

Figure 3–29 Diagram of the metabolic interactions occurring within layered microbial mats. Transformations of nitrogen-containing inorganic compounds, molecular nitrogen, particulate organic nitrogen (PON), and dissolved organic nitrogen (DON) take place in such associations. (After Paerl and Pinckney, 1996)

last case the green algae provide a source of fixed carbon, while the cyanobacterial partner provides fixed nitrogen. Although lichens have often been thought of as partnerships between one fungus and one alga, triple and greater multiple symbioses may be more common than previously recognized. Because bacteria as well as fungi and algae occur in lichens, some experts regard lichens more as microbial consortia than individuals (Honneger, 1992).

The degree of mutualism appears to vary widely among lichens. The 22% or so of lichens that are morphologically complex and internally stratified (Fig. 3–30) are considered highly mutualistic, since greater success of all partners is achieved in the association than when components are grown separately. However, more simply constructed crustose and filamentous microlichens may reflect controlled parasitism of the fungal component upon the algae. Although some lichens can produce asexual reproductive structures, which include small amounts of algae and fungi, in many cases the associations must re-form in nature by accidental contact between fungi and appropriate algae. Few lichen associations have been reestablished in the laboratory from separately cultured components (Honneger, 1992).

There is considerable evidence for selectivity in partner choice (Rambold, et al., 1998). The unicellular green algae *Trebouxia* and *Asterochloris* are by far the most common phycobionts, and in 98% of lichen species the fungal partner is an ascomycete. The production of secondary compounds by lichens (known as lichen substances, lichen acids, or lichen products)

seems to be dependent upon establishment of a stable nutritional relationship between symbionts. Evidence for this conclusion is that some laboratory-cultured fungi produce typical lichen substances when the carbohydrate normally supplied by the phycobiont is added to the growth medium (Honneger, 1992). The method by which the fungal partner may elicit excretion of algal carbohydrate or fixed nitrogen is unknown.

In deserts and grasslands lichens are typically associated with an array of soil cyanobacteria—such as *Microcoleus vaginatus*, *Nostoc* species, and *Scytonema* species, and certain mosses—to form desert-

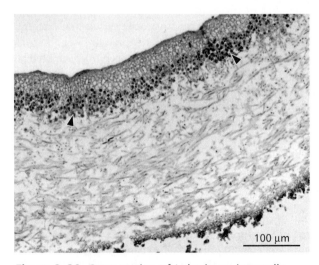

Figure 3–30 Cross section of *Lobaria*, an internally stratified lichen, showing fungal hyphae and algal cell layer (arrowheads).

Figure 3–31 Colonies of the nitrogen-fixing, filamentous cyanobacterium *Anabaena* within cavities (arrowheads) on the undersides of leaves of the aquatic fern *Azolla*. In the inset one cavity is viewed at higher magnification using fluorescence microscopy. The *Anabaena* filaments have a bright appearance due to autofluorescence of photosynthetic pigments.

crust consortia. These are regarded as important contributors to soil integrity and fertility (Garcia-Pichel and Belnap, 1996).

Cyanobacterial-Plant Associations

A number of liverwort, hornwort, moss, fern, cycad, and angiosperm species are closely associated with nitrogen-fixing cyanobacteria—typically species of *Nostoc* (Dodds, et al., 1995). The prevalence of cyanobacterial associates within ventral cavities of liverworts and hornworts suggests that early land plants may also have relied upon such associations for a supply of fixed nitrogen. Photosynthesis is primarily the job of the plant partner—the cyanobacteria mainly function in providing fixed nitrogen in exchange for a supply of fixed carbon from the plant. The plant partner typically produces enzymes that inhibit ammonium assimilation by the cyanobacteria, which facilitates excretion of larger amounts of ammonium to the plant.

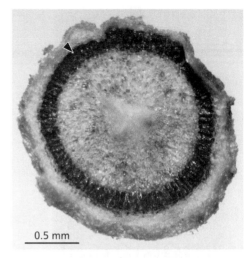

Figure 3–32 Colonies of the nitrogen-fixing, filamentous cyanobacterium *Nostoc* form a distinct dark ring (arrowhead) within the aboveground coralloid roots of the cycad *Cycas revoluta*.

The aquatic fern *Azolla*, commonly used as a biofertilizer in rice paddies, harbors nitrogen-fixing cyanobacteria (*Anabaena*) in sealed cavities on the undersides of leaves (Fig. 3–31). *Azolla* does not grow on nitrogen-free media without its algal partner, and nitrogen-fixation activity of the alga is as much as 20 times higher within *Azolla* than in free-living cyanobacteria. Cycads typically produce aboveground roots (known as coralloid roots because their branching pattern resembles that of corals) that contain *Nostoc* colonies (Fig. 3–32). Specialized root cells mediate the exchange of nutrients between algae and plant. The tropical flowering plant *Gunnera* harbors *Nostoc* colonies in special mucilage-filled glands at the bases of petioles but how the algae get into the glands is not known (Grilli Caiola, 1992).

Recommended Books

Reisser, W. (editor). 1992. *Algae and Symbiosis*. Bio-Press, Bristol, UK.

Anderson, D. M., A. D. Cembella., and G. M. Hall-egraeff. 1998. *Physiological Ecology of Harmful Algal Blooms*. Springer, New York, NY.

Chapter 4

Technological Applications of Algae

kelp-harvesting barge

Algae are involved in global biogeochemical cycles and biotic associations that provide essential ecological services. Humans reap the benefits of these algal activities in the form of atmospheric oxygen, climate modulation, and fossil fuels, as well as finfish, shrimp, and shellfish harvests, which depend upon algal primary production. In addition, humans have learned to use algae in a wide variety of technological applications. Certain algal species have become invaluable model systems for research. Algae are used as environmental monitors, both to assess the health of modern aquatic systems and to deduce environmental conditions of the past. Numerous food products are derived from algae, including food for cultivated shellfish, seaweeds eaten by humans, and protein and vitamin supplements produced from microalgae grown in pond or bioreactor cultivation systems. Some microalgae produce lipids that are potential sources of renewable fuels. Many algae manufacture compounds that have scientifically and industrially useful gelling properties. Ancient marine and lake diatom deposits (diatomite or diatomaceous earth) are mined for use in abrasives and industrial filtration. Algal phycobiliproteins can be used as fluorescent dyes in applications such as flow cytometry. Some algal compounds have potentially valuable antibiotic or antitumor activity. For example, the dinoflagellate *Amphidinium operculatum* produces potent antitumor compounds (Bauer, et al., 1995). Finally, algae have been incorporated into engineering systems utilizing algal nutrient uptake and gas-exchange properties to purify water or air.

Algae as Research Tools

The history of basic biological discoveries includes many examples in which algae played critical roles. The typical 9 doublets + 2 singlets arrangement of microtubules that is characteristic of nearly all eukaryotic cilia and flagella was first observed in algae. The existence of messenger RNA was first postulated

64

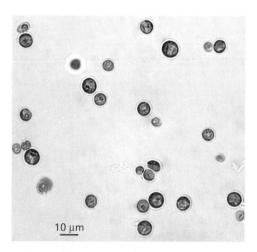

Figure 4–1 *Chlorella* is a small, unicellular green alga that has been widely used as a model system to enhance understanding of algal physiology.

from studies of algal development. The absorption spectrum for photosynthesis—the determination of the wavelengths of visible light that are captured by chlorophyll—was first demonstrated in an alga. The first products of photosynthetic carbon fixation were unknown prior to classic studies on the green alga *Chlorella* (Fig. 4–1). Currently, assessment of the role of individual amino acids in Rubisco is accomplished through study of photosynthesis-deficient algal mutants (Spreitzer, et al., 1995). Such studies may point the way toward strategies for improving carbon fixation efficiencies of Rubisco. The simple, haploid life cycles of the green flagellates *Chlamydomonas* (see Fig. 1–17)

and *Volvox* (Fig. 4–2) have been invaluable in determining the genetic basis for flagellar development and other cell processes. *Chlamydomonas* has also been useful in deducing the function of chloroplast genes (Rochaix, 1997). Other aspects of this important model system are discussed in Chapter 20. Embryos of the brown seaweed *Fucus* (Fig. 4–3) have been used to ascertain the factors that control cell polarity. Land plant relatives such as desmids (Fig. 4–4) and the related green alga *Mougeotia* (Fig. 4–5), are classic systems for determining cell differentiation and signal transduction pathways involving the light-sensing pigment phytochrome.

Algae are useful laboratory organisms because they (or their reproductive cells) are small, grow rapidly, have short generation times, and are easily cultivated under laboratory conditions. Cultures of many algal species are readily available from collections maintained by experts in several nations. An international list of culture collections is provided in Norton, et al. (1996). In the United States, the University of Texas Algal Culture Collection (UTEX) is a large and well-known repository of algae cultures. Others include the American Type Culture Collection (ATCC) and the CCMP-Provasoli-Guillard National Center for Culture of Marine Phytoplankton (Andersen, et al., 1997). Hundreds of algal cultures can be ordered and referred to in publications by their catalog numbers, which is useful to researchers who then know exactly which algal strain was used to per-

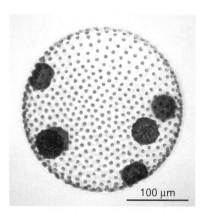

Figure 4–2 The motile colonial green alga *Volvox* is currently used as a model genetic and developmental system to enhance understanding of cellular differentiation.

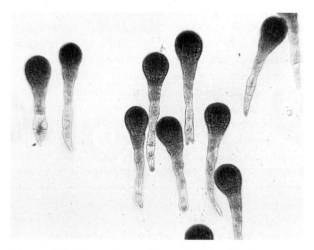

Figure 4–3 Few-celled embryos that develop by mitotic division of the zygotes of the brown seaweed *Fucus* are used as model systems for exploring cell and organismal polarity. (Photograph courtesy S. Brawley)

Figure 4–4 The desmid *Micrasterias*, a unicellular green alga that has been used in developmental studies.

Figure 4–5 Along with some of its relatives, the freshwater green alga *Mougeotia* has been one of the most important model systems for the study of phytochrome-mediated processes, such as plastid rotation.

form experiments. Researchers can choose to analyze the behavior of the same strain under different conditions or compare the behavior of various strains of the same species. Multispecies assemblages of algae have likewise been useful model systems for understanding ecological processes such as succession and responses to disturbance (Chapters 22 and 23).

Algae as Biomonitors

A bioassay is a procedure that uses organisms and their responses to estimate the effects of physical and chemical agents in the environment. Biomonitors provide early warning of possible environmental deterioration, and may provide sensitive measures of pollution. One common use of algal cultures is as biomonitors in the detection, in natural or effluent waters, of algal nutrients or substances that are toxic to algae. While daphnids and fish have traditionally been the primary biomonitor organisms in aquatic ecosystems, algae are more sensitive than animals to some pollutants, including detergents, textile-manufacturing effluents, dyes, and especially herbicides. Algal toxicity tests have become important components of aquatic safety assessments for chemicals and effluents and are required by Section 304(h) of the U. S. Federal Water Pollution Control Act, and in the registration of pesticides (Lewis, 1990). Some other countries have similar requirements.

The most widely used algal biomonitor for freshwaters is the unicellular green alga *Selenastrum capricornutum* (culture ID numbers—UTEX 1648, ATCC 22662) (Fig. 4–6). Other biomonitor algae include the freshwater green *Scenedesmus subspicatus* (UTEX 2594) (see Fig. 3–16), and the marine diatoms *Skele-*

tonema costatum and *Phaeodactylum tricornutum*. These algae were chosen for ease of laboratory cultivation and handling, and are recommended for use in tests such as the (U.S.) Environmental Protection Agency's *Algal Assay Procedure*, the similar *Algal Growth Potential Test*, and the *Algal Growth Inhibition Toxicity Test*. Such procedures are based upon the premise that cell yield is proportional to the amount of a limiting nutrient or toxin that is present in water samples. Although chemical methods can also be used to determine nutrient or toxin concentrations in water samples, these may not accurately reflect biological responses. For example, a chemical measurement of total phosphorus may overestimate the amount of this element that is biologically available to organisms (Skulberg, 1995).

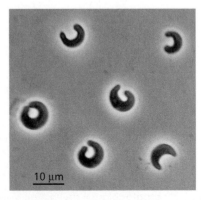

Figure 4–6 *Selenastrum capricornutum* is widely used in algal assays to assess the nutrient or toxin status of natural waters.

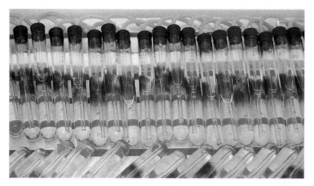

Figure 4–7 A typical algal bioassay procedure involves inoculation of multiple culture containers with different dilutions of the water sample to be tested, and uniform amounts of algae. After a growth period algal populations are assessed and compared to controls.

Algae may vary widely in their responses to nutrients and toxins, so standard procedures commonly recommend using a battery of phylogenetically distinct test species. Algal cells are dispensed into individual flasks, tubes, or wells of microtiter plates along with water samples that have been amended with varying levels of nutrients or toxins (Fig. 4–7). Initial cell density, temperature, pH, and irradiation are controlled. After a period of growth cell density is assessed by microscopic counting, electronic particle counting, or spectrophotometric methods. Use of a microtiter plate reader for rapid assessment of algal cell densities in multi-welled plates is becoming more common

(Fig. 4–8). For toxins, an "effective concentration" value (EC_x, where x reflects the percent of reduction, usually 50) and a "no observed effect concentration" (NOEC) are calculated based on growth inhibition. Some procedures also require determination of the area under the growth curve or particular statistical procedures. Because the biological relevance of EC_x and NOEC are largely undefined, the results should be interpreted only as an indication of effects on algae in the natural system and should not be over-extrapolated (Lewis, 1990). The use of algae in assessment of aquatic toxicity has been reviewed by Haglund (1997).

Use of Fossil Algae in Paleoecological Assessments

Species-specific and distinctively ornamented silica scales and walls of resting cysts (stomatocysts) of freshwater chrysophyceans (Fig. 4–9) obtained from lake sediments are widely used to deduce past lake conditions and to monitor modern environmental change such as climate warming or acidification (Smol, 1995). The decay-resistant silica walls of diatoms form layered sediments in lake and ocean sediments that are helpful in paleoclimatological studies (e.g., Williams, et al., 1997), and in learning about the history of eutrophication (e.g., Hall, et al., 1997). Diatoms are particularly helpful in correlating sedimentary layers

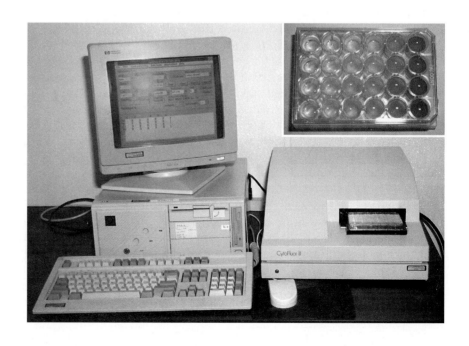

Figure 4–8 The use of microtiter plates and plate readers has greatly simplified algal bioassay procedures by reducing the labor associated with manipulation of larger growth vessels and growth assessment. Inset: a microtiter plate with various dilutions of green algae. (Photograph courtesy D. Karner and K. Schappe, Wisconsin State Laboratory of Hygiene)

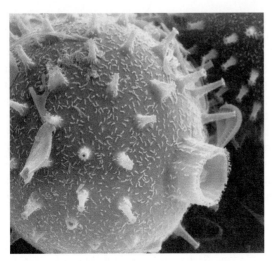

Figure 4–9 The silicified resting cysts of freshwater chrysophyceans are widely used to assess past ecological conditions affecting lakes and their biota. (Micrograph courtesy P. Siver)

from one location with those from another—a field of study known as biostratigraphy. For example, the marine diatom *Annellus californicus* lived for a period of only one million years during the early to mid-Miocene and is therefore used to identify sediments of this age in different localities (George, 1988).

Similarly, the calcified scales of coccolithophorid algae, silica skeletons of silicoflagellates, and decay-resistant cysts of dinoflagellates persist in ocean sediments for millions of years, forming records used to deduce past environmental change. Examples include abundance cycles of the coccolithophorid *Florisphaera profunda* (McIntyre and Molfino, 1996; Beaufort, et al., 1997) and dinoflagellate cyst assemblages in sediment cores from the North Atlantic (de Vernal, et al., 1996).

Microalgae in Animal Aquaculture Systems

Information gained in studies of algal herbivory (see Chapter 3) is useful in devising aquaculture systems—the cultivation of shellfish, for example. Many marine animals cannot synthesize certain essential long-chain fatty acids in quantities high enough for growth and survival and thus depend upon algal food to supply them. Nontoxic marine microalgae, including *Isochrysis*, *Pavlova*, *Nannochloropsis*, and various diatoms,

represent the primary food source for at least some stages in the life cycle of most cultivated marine animals. Algae are typically grown for food in greenhouse tanks that are often aerated from the bottom via perforated pipes or air stones. A major problem in cultivation is infestation with ciliates, amoebae, or flagellates, which can decimate the algal biomass within hours.

Growth conditions may affect algal food quality and must therefore be optimized. The diatom *Thalassiosira pseudonana* is widely used to feed a variety of molluscs, including the Pacific oyster *Crassostrea gigas* and rock scallops. In the production of microalgae as feeds for shrimp, bivalve molluscs (especially clams and oysters), and rotifers (used to feed finfish), self-shading can limit light penetration and hence productivity to less than 0.5 gram dried algal biomass per liter. Heterotrophic algae, however, can generate biomasses up to 40 g l^{-1}, and 52 of 57 microalgal strains tested were found able to utilize glucose as a carbon source (Gladue and Maxey, 1994). Sometimes larger algae may prove problematic, as when the brown seaweeds *Colpomenia peregrina* and *Sargassum muticum* thalli attach to shellfish, increasing their buoyancy and the chance that they might float away. Such seaweed infestations can reduce profits by as much as 30%.

Microalgal Mass Cultivation for Production of Food Additives, Hydrocarbons, and Other Products

Major products produced from algae and the value of their commercial markets are reviewed by Radmer and Parker (1994) and Radmer (1996). A number of microalgae are cultivated for production of human "health-foods," or food additives such as β-carotene (Ben-Amotz, et al., 1982) or protein extracts. In general, however, the high nucleic acid content of algae limits their use in human foods. Nucleic acids are converted to uric acid upon digestion. If consumed at a rate of more than 50 grams per day, uric acid may precipitate as sodium urate crystals in joint cartilage, causing gout. The cyanobacterium *Spirulina* (Fig. 4–10) has traditionally been harvested from African lakes and consumed in sauces at the rate of some 9–13 grams per meal. The protein level of *Spirulina* can be as high as nuts, grains, and soybeans, ranging from 50–70% of algal dry weight, and shows

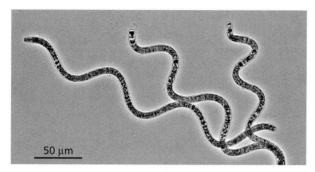

Figure 4–10 *Spirulina* is a filamentous cyanobacterium commonly grown for use in human health-food products or collected directly from natural environments for food use.

promise for use as a protein supplement for malnourished populations. Because the cell walls are composed of mucopolysaccharides (see Chapter 6) rather than cellulose, *Spirulina* is more digestible than some other microalgae. *Spirulina* is also naturally high in B vitamins and essential unsaturated fatty acids (Richmond, 1988) as well as high levels of β-carotene, which is converted to vitamin A during digestion. Vitamin A can prevent xerothalmia, a form of blindness that arises in malnourished children. *Spirulina* is commercially cultivated in shallow raceways, 500–5000 m² in area, while being stirred with paddles, or is grown in closed outdoor vessels called bioreactors (Richmond, 1988; 1990). Care must be taken to avoid contamination of the product with toxin-producing cyanobacteria. Methods are available for monitoring the toxin content of waters and algae (see Chapter 3).

Dunaliella, a green flagellate (Fig. 4–11), can manufacture 50 times the β-carotene produced by *Spirulina*. This and other microalgae are often cultivated in open-air commercial ponds that may be several thousand square meters in surface area, but less than a meter deep. Algal farms can be located on land, such as deserts, unsuitable for other crops. Problems may arise, however, if the ponds become contaminated with heavy metals, chemicals, insects, disease microbes, or weedy algae such as toxic cyanobacteria. *Dunaliella* can tolerate higher salinities than many contaminating organisms, and algal farmers use this as a means for preventing growth of contaminating organisms. Another strategy for reducing contamination is the use of enclosed bioreactors. For example, the diatom *Phaeodactylum tricornutum* produces high biomasses when grown outdoors in tubular photobioreactor vessels. It is then harvested for extraction of eicosapentaenoic acid, a long-chain polyunsaturated fatty acid used as a human food supplement (Alonso, et al., 1996). Although this fatty acid can be obtained from fish oils, there are concerns regarding heavy-metal contamination of this source. Hence cultivation of microalgae for the production of eicosapentaenoic acid is of interest. Mutation and selection programs have been successful in enhancing the eicosapentaenoic acid content of cultured *Phaeodactylum tricornutum* (Alonso, et al., 1996) and other algae (Galloway, 1990).

The green flagellate *Haematococcus* (Figs. 4–12, 4–13) is cultivated in mass quantities for extraction of **astaxanthin**, a carotenoid of value for addition to aquaculture feeds. Carotenoids are accumulated in the plastids and some aspects of their biosynthesis

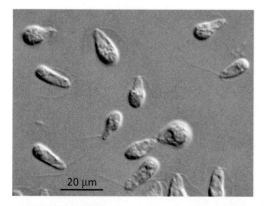

Figure 4–11 *Dunaliella* is a green algal flagellate used in the industrial production of β-carotene, an essential component of the diet of humans and other animals. Cells often appear bright orange in color because they are rich in carotene.

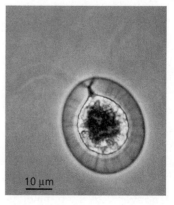

Figure 4–12 *Haematococcus* is a green algal flagellate used in industrial production of the carotenoid pigment astaxanthin.

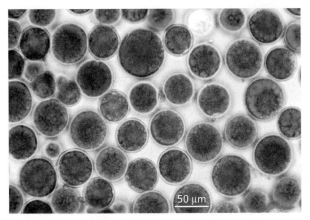

Figure 4–13 *Haematococcus* cells that have been exposed to low-nutrient conditions lose their flagella, enlarge, and produce large amounts of bright red astaxanthin. (Photograph by L. Wilcox, courtesy AgResearch International, Inc.)

have been studied (Rabbani, et al., 1998). Carotenoid accumulation is induced by decreasing the nitrogen content of the growth medium and increasing irradiance (Grünewald, et al., 1997). Although it was once thought that astaxanthin was produced in large quantities only by nonflagellate resting stages (akinetes) (see Fig. 4–13), culture experiments have shown that flagellates (see Fig. 4–12) can accumulate carotenoids as rapidly as resting cells (Lee and Ding, 1995). Biosyn-

thesis of astaxanthin by *Haematococcus lacustris* is accomplished by the same pathway as in the marine bacterium *Agrobacterium aurantiacum* (Grünewald, et al., 1997). Smaller-scale bioreactor vessels appear to work better than open-pond cultivation for growth of *Haematococcus*.

Given that at least some of our fossil fuel is derived from algae (see Chapter 2), efforts have been made to identify microalgae that produce relatively large amounts of lipids for possible mass cultivation and extraction of useful hydrocarbons. Many algal groups (excluding cyanobacteria) store substantial amounts of photosynthate as lipids. Of these, the common freshwater colonial green alga *Botryococcus braunii* (Fig. 4–14) is the best studied. Under appropriate conditions more than 30% of the dry weight of this alga consists of unique C_{17}–C_{34} unsaturated polyhydrocarbons, known as **botryococcenes**. These acyclic olefins are produced internally and excreted to the cell surface where they permeate the matrix between cells of the colony, often making it appear yellowish brown. So-called "algal coals" as well as low percentages of some crude oils consist of cooronite, derived from *Botryococcus* hydrocarbons. To date, however, cultivation of microalgae for lipid extraction has proven to be 10–20 times more expensive than production of lipids from plant crops, such as soybean and corn (Berkaloff, et al., 1984; Spencer, 1988). Aspects of environmental control of lipid production by algae and applications of these lipids in aquaculture, as lubricants and surfactants in industry, and in medicine, have been reviewed by Roessler (1990).

Human Uses of Seaweeds

It is estimated that humans have used some 500 species of seaweeds for food, fodder, or chemicals. In both ancient and modern times people have harvested wild seaweed stocks. In coastal areas of the U.S.A., "wildcrafters" collect particular seaweeds such as the brown kelp *Nereocystis* for sale. In China written records confirm that humans have harvested seaweeds there for more than 2000 years, and today the Chinese people collect 74 species of red, green, brown, or blue-green algae in 36 genera—the world record for the most diverse species list of edible seaweeds. Among the green algae eaten in China are *Ulothrix flacca*, a filamentous seaweed, the sheetlike *Monostroma nitidum* and *Ulva lactuca*, several *Enteromorpha* species, and *Caulerpa racemosa*. Brown seaweeds har-

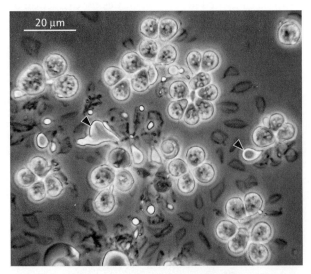

Figure 4–14 *Botryococcus* is a common planktonic freshwater green alga that is unusual in producing very large amounts of extracellular hydrocarbons (arrowheads).

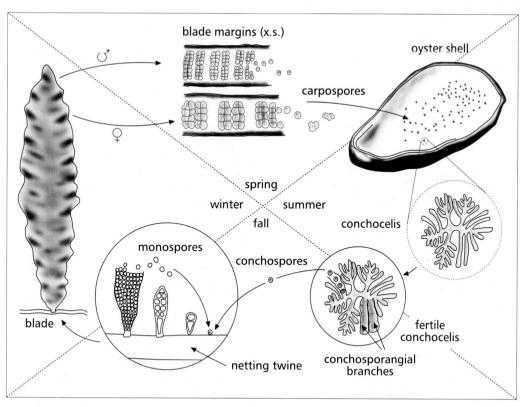

blade margins (x.s.)

oyster shell

carpospores

spring

winter   summer

fall

conchocelis

monospores

conchospores

blade

netting twine

conchosporangial
branches

fertile
conchocelis

Figure 4–15 *Porphyra* is a blade-forming red seaweed widely cultivated for use as human food. An understanding of its life history—involving both an edible blade phase and a microscopic shell-dwelling "conchocelis" phase—is essential to effective cultivation. Reproductive cells formed by blades, known as carpospores (top), colonize shells and grow into small filaments that generate conchospores. The latter can be induced to colonize twine that is suspended in the sea, and produce blades. Blades may also arise from an asexual process involving monospores. (After Mumford and Miura, 1988)

vested from the wild by the Chinese for food include *Ishige okamurai*, *Endarachne binghamiae*, *Scytosiphon lomentaria*, *Chorda filum*, and *Sargassum fissiformis*. Collected marine red algae include *Bangia fuscopurpurea*, many species of *Porphyra*, *Dermonema pulvinatum*, *Asparagopsis taxiformis*, *Grateloupia filicina*, and *Gigartina intermedia*. The cyanobacterium *Brachytrichia quoyi* is collected all along the Chinese coast and is especially abundant in Taiwan (Xia and Abbott, 1987).

A few seaweeds, mostly the reds *Porphyra*, *Eucheuma*, and *Gracilaria*, and the brown kelps *Laminaria* and *Undaria*, are cultivated in aquaculture operations for use in human foods or for extraction of gelling compounds. Dried *Porphyra* (*nori*) (see Fig. 16–2) and *Laminaria* (*kombu* or *haidai*) obtained from aquaculture operations are important crops in Asia. In China alone, hundreds of thousands of people are occupied in growth and harvest of these seaweeds, and

it is estimated that the Chinese consume well over 100 million pounds of fresh and dried seaweeds each year (Xia and Abbott, 1987). The annual *Porphyra* harvest worldwide has been estimated to be worth 2.5 billion U.S. dollars. *Porphyra* is thus the single most valuable aquacultured organism (van der Meer and Patwary, 1995). Seaweed cropping provides a valued source of employment in coastal areas and does not typically degrade the natural environment or the seascape. The economic success of these crops depends greatly upon detailed basic knowledge of the algae.

Porphyra—Food Value and Cultivation Methods

The food value of *Porphyra* lies primarily in provision of essential vitamins, such as B and C, as well as minerals, including iodine. A single sheet of high-grade *nori*—the chopped, pressed, and dried seaweed product—contains 27% of the U.S.D.A. recommended daily allowance of vitamin A, as β-carotene (Mumford and Miura, 1988). *Porphyra* also contains high proportions of digestible protein (20–25% wet weight) (Lobban and Harrison, 1994), but contributes a relatively low percentage of protein to human diets (Mumford and Miura, 1988). The distinctive taste of *Porphyra* is due to the presence of free amino acids.

Porphyra cultivation originated in Tokyo Bay about 300 years ago. Nets were strung between poles in areas where spore attachment occurred and then transported to the cultivation site. The farmers did not know where the spores came from, or that the habitat and structure of the spore-producing form of the seaweed was very different from the harvested blades, thus limiting success. The situation changed dramatically after the British phycologist Kathleen Drew Baker used culturing techniques to discover that the separately named microscopic, shell-inhabiting red algal genus *Conchocelis* was in fact the spore-producing phase in the life cycle of *Porphyra* (Fig. 4–15). Her report, published in *Nature* in 1949, transformed the industry, allowing mass cultivation of the filamentous phase in sterilized oyster shells and seeding of the nets indoors. The conchocelis phase, which serves as a survival (perennating) stage in nature, can also be maintained for long periods in laboratory cultures, reproducing vegetatively. Currently conchocelis is grown in greenhouse tanks from sexually produced carpospores, whose release from *Porphyra* blades is induced by brief drying and reimmersion in seawater for a few hours. The carpospores attach to shells or artificial substrates and are stimulated to grow by exposure to bright sunlight, abundant nutrients, and aeration. The summertime temperature increase, followed by the decrease in fall, stimulates release of conchospores, which are collected on nets—either by running nets through the indoor tanks or by placing conchocelis-bearing shells underneath nets mounted on poles or rafts in the sea. Settlement of 2–5 spores mm^{-2} is most desirable. Application of ammonium sulfate fertilizer to a final concentration of 15 μM may be required. Small *Porphyra* blades can be removed from the sea and frozen for use as much as a year later, thereby providing a method for extending the growing season or insuring the crop in case of failure. Mature blades are removed from the nets, washed to remove epiphytes, chopped, made into a slurry, spread over screens, and dried. The typical farmer buys conchocelis-occupied shells

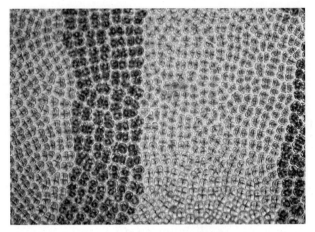

Figure 4–16 Sexually reproducing blades of *Porphyra* exhibit mosaics of male and female gamete-producing cells. The darker areas produce female gametes, and the lighter areas produce male gametes. This patterning reflects the genetic heterogeneity of the cells composing the *Porphyra* thallus.

or conchospore-"seeded" nets, and sells raw *Porphyra* to a processor (Lobban and Harrison, 1994).

Although there are about 70 species of *Porphyra* worldwide, occurring in both tropical and temperate waters, *P. yezoensis* is the most important commercial species. Because frond size is heritable, individual *nori* farmers practice selective breeding. Particularly large blades exhibiting delayed sexual maturity and rapid growth rates—on the order of 3–4 cm day^{-1}—have become the major cultivated varieties. Industry researchers are hoping to achieve vegetative propagation of the gametophytic blade phase, because elimination of conchocelis production would lower costs and reduce genetic diversity in the crop (Lobban and Harrison, 1994). Nearly all *Porphyra* blades are genetic mosaics derived from up to four different genotypes. This is reflected in the patterning of male and female gamete production commonly observed in sexually mature blades (Fig. 4–16), which results from meiotic segregation of sex-determining alleles in the germinating conchospores. The first four cells of the young gametophytes represent a linear genetic tetrad (Mitman and van der Meer, 1994). Separate regions of the resulting blade are thus clones composed of mitotic descendants from the first four cells, and hence may be genetically distinct.

Pest problems in *Porphyra* cultivation can include herbivorous fish (which are usually deterred by protective nets), weedy algal epiphytes such as the diatom

Licmophora and the green seaweeds *Monostroma* and *Enteromorpha*, and bacterial, fungal ("chytrid blight"), or oomycete infections (Lobban and Harrison, 1994).

Cultivation of Edible Kelps

Laminaria is native to the boreal arctic region and warmer temperate waters of the North Atlantic and North Pacific. Cultivation of edible *Laminaria japonica* began in China and now occurs in Japan, Korea, and other countries. An important attribute of kelps in relation to human health is their high iodine content.

The production process begins with selection of wild or cultivated sporophytes having darkly pigmented, mature **sori** on their blades. Sori are tissues bearing large numbers of meiosporangia (Fig. 4–17) (see Chapter 15). The sori are first cleaned by briefly wiping or immersing them in bleach, and then left in a cool, dark location for up to 24 hours prior to reimmersion in seawater. This treatment causes the release of numerous flagellate zoospores, which are then allowed to attach to twine in greenhouses. The strings are then suspended in the ocean where spores grow into microscopic filamentous gametophytes. Gamete production is followed by fertilization of eggs by motile sperm and the subsequent development of zygotes into young sporophytes. The time from gametogenesis to production of young blades 4–6 mm length is 45–60 days. Strings bearing young sporophytes are cut into pieces and attached to ropes that are suspended in the sea. In winter the ropes are kept about 5 m below the water surface to avoid wave damage. They are raised to a 2 m depth in summer for greater illumination. Two years are usually required to achieve sufficiently large blades. Harvest occurs in summer, with production ranging from 50 to 130 million tons ha^{-1}. Most *Laminaria* is dried for use in foods or in the alginate industry (see below). Gametophytes are cloned and crossed to generate genetically uniform populations of sporophytes. The industry is working toward cloning sporophytes, since eliminating the gametophyte stage would reduce genetic variation in addition to production time and costs (Lobban and Harrison, 1994; van der Meer and Patwary, 1995).

Undaria, also known as *wakame*, is grown and used in Asia very much like *Laminaria*. Aquaculture operations have also been established off the coast of western Europe, but efforts to expand the industry are controversial because this kelp is not native to Atlantic waters. It apparently arrived in the Mediterranean

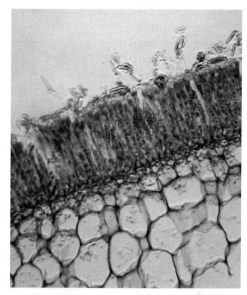

Figure 4–17 Reproducing *Laminaria* thalli typically bear patches (sori) of epidermal cells that have differentiated into unispore-producing cells. The patches often appear darker than surrounding, nonreproductive tissues and thus can be easily identified. A section through a sorus is shown here.

sometime during the 1980s on oysters imported from Japan for cultivation purposes. Some phycologists are concerned that *Undaria* might escape from cultivation and compete with native kelp populations, possibly causing negative effects on lobster fishing (Lobban and Harrison, 1994).

Production of Gelling Agents from Seaweeds

The gelling agents alginic acid (or its mineral salt, alginate), carageenan, agar, and agarose are produced from certain brown and red seaweeds. In general these products are useful because they stiffen aqueous solutions. In some cases the seaweeds are harvested from wild stocks, while in others aquaculture operations have been established.

Alginates

Alginates are structural sulfated polysaccharides that confer strength, flexibility, and toughness by forming gels and sols in the matrix between cells of brown algae. These molecules, which occur as mineral salts, help large seaweeds to cope with mechanical stresses generated by waves and currents (see Chapter 23). Alginates make up some 20–40% of the dry

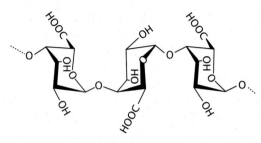

polymannuronic acid

polyguluronic acid

Figure 4–18 The chemical structures of polymannuronic acid and polyguluronic acid—polymers that constitute the alginates of brown algae. (After Stewart, 1974)

weight of brown seaweeds. They are polymers of D-mannuronic and L-guluronic acids (Fig. 4–18), with the ratios of these monomers varying among algal species. Regions of the polymer that are high in polymannuronic acid are known as M-blocks, regions rich in polyguluronic acids are G-blocks, and regions consisting of alternating monomers are M-G-blocks. Each type has different properties: M-blocks have low affinities for metals; G-blocks bind alkaline earth elements; and M-G-blocks are soluble at low pH, unlike the other two types. An M:G ratio of 1:1 gives a soft gel, whereas a stiffer gel is formed when the M:G ratio is less than 1:1. Since alginates are too complex to be synthesized chemically, the natural source is valuable, and markets are expected to increase (Jensen, 1995).

Alginates are used in the textile industry, specifically in the printing process; in manufacturing specialized wrapping papers; in production of dental creams and impression materials, shoe polish, welding electrodes, and oil drilling muds; and in food processing to thicken ice cream, jams, puddings, sauces, mayonnaise, custards, fillings, and decorations for bakery products. Future applications are expected to include immobilization of cell and tissues in biogels used in fermentation technology; entrapment of algae, bacteria, yeasts, plant cells or protoplasts, or animal cells in biomedical research applications; micropropagation systems for ornamental or crop plants; and slow-release systems for pharmaceuticals. Alginate-encapsulated islets of

Langerhans cells that are capable of producing insulin when needed, have been implanted into animals as a replacement for the endocrine functions of the pancreas (Jensen, 1995).

Commercially important sources of alginates are the brown seaweeds *Laminaria japonica*, *Laminaria digitata*, *Laminaria hyperborea*, *Ascophyllum nodosum*, *Ecklonia maxima*, *Lessonia nigresens*, some *Fucus* species, and the giant brown kelp *Macrocystis pyrifera*, which reaches lengths of 60 meters and may weigh more than 300 kg. Extensive natural *Macrocystis* beds off the coast of California are "mowed" by special barges (Fig. 4–19). This process removes only the top meter or so, leaving most of the thallus intact. Regeneration occurs within a matter of months, whereupon the seaweeds can be re-cropped. Harvested seaweeds are delivered to the extraction plant on shore, where they are treated with preservatives that prevent destabilization and discoloration of the product by polyphenolics. Dilute acid is used to extract inorganic salts and water-soluble carbohydrates and proteins, yielding insoluble alginic acid. Solubilization is accomplished by alkali treatment, after which the viscous product is dried and milled before storage (Jensen, 1995).

Carageenan, agar, and agarose

As is the case with alginates, there are a large number of industrial and scientific applications for gelling compounds derived from red algae. Similarly, there will likely always be a market for these natural products because the polymers are too complex for chemical synthesis. Agars and carageenans are linear polymers of alternating molecules of (1→3)-β-galactose and (1→4)-α-galactose. In agar, the (1→4)-bonded subunits are L-galactose, whereas in carageenans, both types of subunits are D-galactose. The polymers are extracted from dried seaweeds with cold water and then subjected to filtration, alternating with acetone, alcohol, and water washes. After grinding and extraction in hot water, alcohol precipitation yields carageenans. Agars are obtained by filtration under pressure (Cosson, et al., 1995).

Carageenans are processed from relatively few genera of red algae. *Eucheuma*, *Kappaphycus*, and *Hypnea* account for at least 75% of commercial production in tropical regions, with about half of the world's carageenans being produced in the Phillipines and Indonesia. More than 95% of the *Eucheuma* and *Kappaphycus* harvest is derived from seaweed cultivation. These two genera are particularly valued

Figure 4–19 A barge harvesting the giant kelp *Macrocystis* for alginate production. The ship is equipped with a cutting device and a conveyor belt, which transports harvested seaweeds to the storage area in the hull.

because they produce gelling carageenans in both haploid and diploid life history stages. In contrast, *Chondrus* and *Gigartina* produce strongly gelling polymers only in one life-history stage (Lobban and Harrison, 1994). In the past, *Chondrus crispus* had been intensively harvested from beds along the shores of countries bordering the North Atlantic, but the value of the wild harvest has declined in the face of competition from aquaculture production. There have been intensive efforts to improve and cultivate this species in land-based tank farms in Canada. However only the sporophytes of this species produce strongly gelling carageenans. Wild populations of *Gelidium* are harvested on a worldwide basis for extraction of high-quality agar. About 35% of the world's agar production comes from this seaweed. In addition, it is the source of low-sulfate agarose for use in biotechnology, such as in gels used to visualize DNA (van der Meer and Patwary, 1995).

Eucheuma and *Kappaphycus* aquaculture requires salinities greater than 30‰ (=3%), substantial water movement but no large waves, clear

water, and temperatures between 25° and 30° C (i.e., tropical waters). The presence of a coarse sand bottom reduces suspension of sediments and is suitable for anchoring stakes from which monofilament lines are suspended. Algal thalli are cut into small pieces, tied to the lines, and harvested in about two or three months, when thalli have grown to a mass of 1 kg or greater. About five harvests per year can be achieved, with low capital costs. Cultivation is labor intensive but is viewed as a valuable way to provide employment for coastal populations. The major problems are diseases (whose causes are not well understood) that bring about pigment loss, and undesirable hybridizations with wild seaweeds (Lobban and Harrison, 1994).

The Use of Algal Turf Systems for Removal of Mineral Nutrients in Wastewater Effluents

Decades ago, wastewater effluents such as human sewage and industrial wastes were routinely discharged to natural waters for purification. As population densities increased, effluent treatment plants were built, with a primary goal of removing particulates and pathogenic microorganisms so that water could be reused. However, even today, many sewage treatment processes still discharge waters that are relatively high in phosphate and combined nitrogen, primarily because the incorporation of procedures that more completely remove these compounds is expensive. Consequently, wastewater effluents may contribute excess algal nutrients to natural waters, causing undesirable, and sometimes toxic, algal blooms. Municipal sewage is one of the major contributors to freshwater and marine pollution and eutrophication—the conversion of high-quality, low-nutrient, high-species diversity waters into low-quality, high-nutrient, low-species diversity environments. In addition, most sewage treatment promotes heterotrophic respiration without restoring oxygen levels to natural levels. Release of heavy metals, which have harmful ecological effects in natural waters, may be another consequence of discharging incompletely treated effluents.

A few effluent treatment systems incorporate microalgal ponds or raceways, through which treated water is channeled, just prior to discharge. The algae remove inorganic nutrients and add oxygen (Nurdogan and Oswald, 1995). A disadvantage of microal-

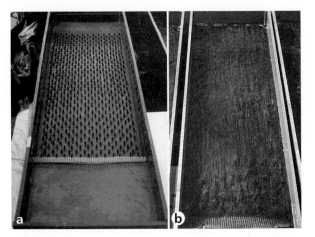

Figure 4–20 An algal turf-scrubber system developed for removal of nutrients from aquatic systems. The support bed is shown without algae in (a), and with algae in (b). (Photographs courtesy J. Hoffmann)

gal water treatment is that continual mixing is required to keep algae suspended and to maximize rates of nutrient uptake and oxygen evolution. In addition, harvesting the microalgal growths is difficult. As an alternative, nutrient-removal/oxygenation systems have been constructed that employ attached filamentous algae in sloped, artificial streams. Mats composed of natural assemblages of filamentous algae and associated biota, or algal turfs that have been selected and cultivated specifically for this purpose, are subjected to flow-through of the effluent to be treated (Fig. 4–20).

Nutrient removal/oxygenation systems employing attached algae (periphyton) have a number of advantages: they are odorless, a mechanism for suspending cells is not required, accumulated algal biomass is easily harvested for use as a soil conditioner or animal fodder, and harvesting leaves the lower portions of the algal filaments attached to substrates allowing regrowth to occur. The periphyton algae remove mineral nutrients and heavy metals in several ways: they trap particles, adsorb ions, assimilate and store nutrients, and, at high pH, precipitate layers of calcium carbonate, which may entrap pollutants. Mat and turf algae can raise oxygen levels by 100–300%. Harvesting both stimulates production by removing the shade canopy and keeps grazer populations low, so that pollutants are not transmitted to higher trophic levels and oxygen levels are not depleted (Craggs, et al., 1996).

Algal species present in mats formed in natural assemblages are of three ecological types: (1) basal mat-forming cyanobacteria such as *Oscillatoria*; (2) fil-

amentous green algae, such as *Cladophora* (in warm waters) (or *Ulothrix zonata*, *Stigeoclonium*, and *Oedogonium* in colder waters [Davis, et al., 1990a]), which grow through the cyanobacterial surface to form a canopy; and (3) epiphytic diatoms, which grow both on the canopy filaments and within the mat. In a large-scale system constructed at the wastewater treatment facility in Patterson, California, an algal turf system proved useful in reducing total phosphorus and soluble reactive phosphorus, as well as in increasing dissolved oxygen (Craggs, et al., 1996). A similar system in Florida used periphytic algae to successfully reduce phosphorus levels in effluent from fertilized sugarcane fields (Adey, et al., 1993). Algal turf systems for nutrient removal have also been designed for use in cold temperate climates (Davis, et al., 1990a, b), and turf systems that incorporate algae-grazing fish (the cichlid *Tilapia mossambica*) have also been developed (Drenner, et al., 1997).

Algae in Space Research

Algae have played a significant role in space research for several decades and will have future important applications in spacecraft life-support systems, colonization of the Moon and Mars, and eventually in terraformation of other planetary bodies. Experiments designed to study the growth of algae in the zero-gravity conditions of space have flown aboard rockets, satellites, the space shuttle, and the Salyut and Mir space stations.

The unicellular green alga *Chlorella* was one of the organisms in sealed aquaria carried aboard Cosmos 1887 and 2044. In studies aboard Salyut and Mir, chlorophyll and carotenoids levels of the *Chlorella* were found to be reduced by an average of 60% compared to ground controls. Pigment levels returned to normal after a day of cultivation on Earth (Popova, et al., 1989). Cultures of *Chlamydomonas* flown aboard Cosmos 2044 were larger in cell size and populations than ground controls (Gavrilova and Gabova, 1992). In general, algae grown in space seem to suffer few adverse affects, a fact that has engendered plans for further use of algae in life support and aquaculture systems.

Up to the present, life-support systems on spacecraft have been "open loop" systems in which air, water, and food have been stored and expended during flight, and accumulated wastes treated and stored. In partially closed systems one or more essential element is recycled, either mechanically, biologically, or both. In fully closed systems air, water, food, and wastes are all recycled. Most life-support research involving algae has focused on their use in bioreactors to recycle air by removing CO_2 and adding O_2 to the atmosphere, thus closing the air loop. Early experiments with *Chlorella* and mice often resulted in wide fluctuations in gas levels, but when light conditions were better regulated, more control over O_2 and CO_2 levels was achieved. If O_2 levels become too high, light can be reduced; if CO_2 levels become too high, raising the light level can compensate by increasing CO_2 absorption and O_2 production (Wharton, et al., 1988).

Several studies have included humans in life-support systems involving algal air-regeneration systems. Bios-3 consisted of a 315 m^3 stainless steel hull containing a 40 m^2 hydroponic garden, an algal compartment, and room for three crew members. The algae effectively processed urine wastes, and plants and algae together provided sufficient oxygen. Water and air cycles could be closed for up to six months, with food and wastes at least partially closed (see Fogg, 1995). Since 1977, NASA has operated a research program known as the Controlled Ecological Life-Support System (CELSS) to select species for both waste recycling systems and food (Schwartzkopf, 1992). The prospects for using algae directly as a food source in space systems are dim. Microalgae can be difficult to harvest, produce unpleasant tastes and odors, and can be difficult to digest. They may also produce excess nucleic acids, or toxins. However microalgae can be grown for extraction of useful chemicals, pharmaceuticals, and proteins that may be used as food supplements (Boston, 1984).

Algae figure prominently in plans for the most ambitious space projects conceivable—planetary terraforming. Mariner and Viking orbiter photos have shown that Mars—where the average surface temperature is now –60° C—once had a warmer, wetter climate. Rivers once flowed, floods scoured out channels, and lakes (and perhaps seas) were filled. Procedures to moderate the Martian climate are currently being designed. After an initial period of greenhouse-type warming—accomplished by manufacture and release of gases—experts propose seeding the planet with pioneering microorganisms, such as cyanobacteria adapted to the stresses of Antarctica or desert habitats. *Chroococcidiopsis* (see Chapter 6), for example, does not grow in mild environments but thrives under extremely arid, high- or low-temperature conditions (Friedmann and Ocampo-Friedmann,

1995). *Matteia*, which was discovered in the Negev desert, can bore into carbonate rocks—abundant on the Martian surface—and can fix nitrogen (Friedmann, et al., 1993). Growths of similar cyanobacteria are thought to have been the first terrestrial autotrophs on Earth (see Chapter 6). Because they can create and enrich soil, terrestrial algae may prove useful in paving the way for the introduction of higher plants to Mars. The plants and algae may help to modify Martian atmospheric chemistry so that it becomes more like that of Earth.

Screening of Algae for Production of Antiviral and Antifungal Compounds and Other Pharmaceuticals

Various types of algae have proved to be sources of compounds with antibiotic and antineoplastic (anti-cancer) activity (Gerwick, et al., 1994; Patterson, et al., 1994), as well as other pharmacologically useful products. The goal of screening programs is to survey cultivable organisms whose medicinal properties are unknown. Hydrophilic and lipophilic extracts of algae are initially screened for activity by testing their ability to reduce pathogenic effects on animal-cell cultures grown as monolayers in multi-well plates. For extracts showing activity, dilution studies are done to determine relative potency. Finally, efforts are made to identify the chemical structures of pharmacologically active compounds. For example, the study by Patterson, et al. (1993) found that a wide variety of cyanobacteria produce compounds, including sulfolipids, that were active against a herpes virus, a pneumonia virus, and HIV. In addition, certain cyanobacteria produce potential antitumor compounds (Patterson, et al., 1991). A potent antifungal antibiotic, tolytoxin, which has low toxicity to humans, is produced by the cyanobacteria *Tolypothrix* and *Scytonema* (Patterson and Carmeli, 1992). Tolytoxin, a macrocyclin lactone that depolymerizes actin and thereby is able to disrupt eukaryotic cell division, is produced by *Scytonema ocellatum* (Patterson, et al., 1994). A review of the antibiotics produced by cyanobacteria can be found in Patterson, et al. (1994). Once algal sources of useful medicinal compounds are identified, the algae can be cultivated on a large scale in bioreactors, harvested, and the active materials extracted, purified, and marketed.

Genetic Engineering of Algae for Improved Technological Performance

Just as crop plants are targets of genetic engineering efforts to improve productivity, broaden environmental tolerance limits, or increase pest or pathogen resistance, so too are useful algal species. DNA sequences are introduced into algal cells with the goal of modifiying biochemical pathways, either by changing expression of existing genes or by adding genes that yield new or different products. The ability to transform algal cells, i.e., to introduce and achieve desired levels of expression of foreign genes, has been made possible by a series of technical achievements. These include: (1) identification of procedures that work best for incorporating foreign DNA into algal cells; (2) development of promoter systems for linkage to the genes of interest that will allow expression after the DNA has been incorporated into the host genome; and (3) selection of an appropriate DNA marker (reporter gene) that allows the genetic engineer to determine which cells have been transformed.

DNA transfer techniques that have been used in algae include: (a) bombardment of cells with DNA-coated gold particles fired by a biolistic device (gene gun); (b) microinjection—transfer of DNA into cells via fine glass needles; (c) electroporation—the use of electrical change to temporarily open pores in the cell membrane; and (d) agitation of wall-less cells with DNA and glass beads or silicon carbide whiskers (Stevens and Purton, 1997). Recombinant viruses (Henry and Meints, 1994) and plasmid vectors represent two additional means of transferring DNA. Plasmids are relatively small circles of DNA that may exist in eukaryotic and prokaryotic cells, in addition to larger genophores or chromosomes. Either foreign or naturally occurring (endogenous) plasmids—isolated from cells of the organism to be transformed—can be modified for use in transforming algal DNA. Most cyanobacteria appear to have 2–4 plasmid types per cell, although some species may contain more. Cyanobacterial plasmids have been used to construct plasmid vectors, which have been successfully employed to stably integrate prokaryotic and eukaryotic genes into cyanobacterial cells. Endogenous plasmids that may be useful in transforming red algae have also been identified (Stevens and Purton, 1997).

In the case of diatoms and other algae, homologous promoters are used because heterologous promoters (those from another organism) do not effectively drive expression (Dunahay, et al., 1995). However, heterologous promoters were effective in transformation of the carageenan-producing red alga *Kappaphycus alvarezii* (Kurtzman and Cheney, 1991), the red *Porphyra miniata* (Kübler, et al., 1994), and the green seaweed *Ulva lactuca* (Huang, et al., 1996).

A number of reporter genes have been used in algae, but those coding for antibiotic resistance, commonly used in transforming other organisms, are avoided because several groups of algae are naturally resistant to these compounds. The gene *ARS*, which encodes the enzyme arylsulfatase and is normally expressed only under sulfur starvation, causes transformed algal cells to develop a colored product that is easily detected when it is used as a marker (Stevens and Purton, 1997).

One of the most intensive efforts to modify the genetic material of algal cells involves the design of pesticide-resistant cyanobacteria whose nitrogen fixation processes are not repressed by environmental levels of combined nitrogen (see Chapters 2 and 6 for more information about cyanobacterial nitrogen fixation). These are regarded as highly desirable properties for the cyanobacteria that inhabit and help to fertilize rice paddies. As was noted in Chapter 3, *Anabaena* (or *Nostoc*) species occurring within cavities of the water fern *Azolla* are well known to fix nitrogen in rice-cultivation ecosystems. The symbiosis helps to maintain paddy yield over time. However, modern rice cultivation frequently involves the use of herbicides and pesticides to control weeds, fungi, and other pests, and some of these chemicals, particularly those that target photosynthetic function, also inhibit cyanobacteria. In addition, nitrogenous fertilizer applications as low as 0.2 μM may result in depression of cyanobacterial nitrogen fixation (Vaishampayan, et al., 1996).

Unusual strains of *Anabaena* that fix nitrogen at higher than usual environmental levels of combined nitrogen may be useful as study organisms in discovering ways to deactivate the regulatory switch that normally inhibits expression of nitrogen-fixation genes. Efforts to achieve herbicide resistance in cyanobacteria are focused upon studies of mutations in the photosynthetic protein D-1 (encoded by the gene *psbA*), because a point mutation in amino acid "264" of the homologous protein in higher plants is associated with increased herbicide resistance. (Vaishampayan, et al., 1996). The process of identifying such mutants involves treating cyanobacterial cells with a mutagenizing agent (which can be done in a

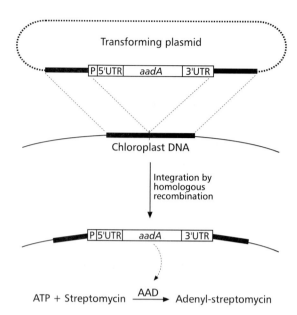

Figure 4–21 Diagram of a molecular construct that can be used to "knock out" the function of a gene (thick black line) whose sequence is known but whose function is not understood. The construct (including a promoter—P) inserts into the gene, interrupting its normal function. Cells in which homologous recombination has occurred are detected by a marker system, such as addition of a sequence encoding an enzyme (AAD) that inactivates streptomycin, allowing the cells to survive on a medium containing this antibiotic. Such cells are then examined for phenotypic changes that may correlate with loss of gene function. (Redrawn from *Trends in Plant Science* Vol. 2. Rochaix, J.-D.1997.Chloroplast reverse genetics: New insights into the function of plastid genes. pp. 419–424. ©1997, with permission from Elsevier Science)

site-directed manner) or using other methods for randomly "knocking out" gene function (Fig. 4–21). The pool of resulting mutants or potential "knockouts" is then screened for the desired phenotype. Site-directed methods for gene disruption (knockouts) utilize the phenomenon known as homologous recombination—transformation (often biolistic) of cells with a linear construct composed of a DNA sequence whose function is unknown, joined to a DNA transformation marker plus promoter. This DNA specifically binds with, and becomes incorporated into, homologous regions in the cell's original genome, disrupting them. Transformed cells are then identified and analyzed biochemically to detect changes caused by disruption of the gene of interest. This process, known as reverse genetics, is regarded as a particularly good way to

ascertain the function of unknown genes, and has been applied successfully to the plastid genome (see review by Rochaix, 1997).

Most advancements in algal genetic engineering have been achieved with cyanobacteria (Elhai, 1994; Stevens, et al., 1994; Vaishampayan, et al., 1998), because of their relatively simple genetic structure and economic importance. Transformation has been achieved in some eukaryotic algae as well. A number of reproducible methods for transforming the green algal model organisms *Chlamydomonas* and *Volvox* have been developed. In fact, methods have been developed for transforming the *Chlamydomonas* plastid and mitochondrial genomes, as well as nuclear genes (Kindle and Sodeinde, 1994; Bingham and Webber, 1994). These have allowed the identification of previously unknown genes and the study of their expression, and have been particularly useful in elucidating the molecular biology of photosynthesis (Stevens and Purton, 1997). Diatoms (including the particularly useful *Phaeodactylum tricornutum*) have also been successfully transformed (Dunahay, et al., 1995; 1996; Apt, et al., 1996). In diatoms, the major goal has been genetic manipulation of lipid metabolism for biotechnological applications, but transformation successes also pave the way for molecular dissection of photosynthesis and other aspects of diatoms that differ from green algae. The giant multicellular green alga *Acetabularia*, the multicellular green *Chara*, and the green microalgae *Dunaliella* and *Chlorella* have also been the subjects of transformation experiments, and many other algae will undoubtedly be added to this list (Stevens and Purton, 1997).

Recommended Books

Akatsuka, I. (editor). 1994. *Biology of Economic Algae.* SPB Academic Publishing, The Hague, Netherlands.

Brown, P. P. (editor). 1998. *Calcareous Nanofossil Biostratigraphy.* Chapman and Hall, London, UK.

Coleman, A. W., L. J. Goff, and J. R. Stein-Taylor. (editors). 1989. *Algae as Experimental Systems.* Alan R. Liss, New York, NY.

Lembi, C. A., and J. R. Waaland. (editors). 1988. *Algae and Human Affairs.* Cambridge University Press, Cambridge, UK.

Chapter 5

Algal Diversity and Relationships

Taxonomy, Systematics, and Phylogeny

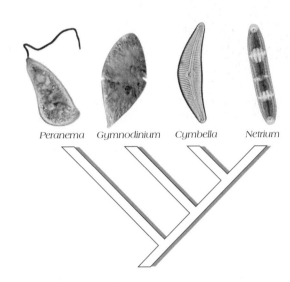

Peranema   *Gymnodinium*   *Cymbella*   *Netrium*

Algal taxonomists—phycologists who specialize in the identification, naming, and classification of algae—believe that there are from 36,000–50,000 (John and Maggs, 1997), and possibly more than 10 million (Norton, et al., 1996) species of algae. Based on pigment composition, cellular structure, and, more recently, molecular sequence or genome architecture information, the algae have been classified into several phyla (divisions). However algal systematists have long differed in their opinions regarding the number and composition of algal phyla. One of several recent classification schemes (Hoek, van den, et al., 1995) divides algal species among eleven phyla. These include Cyanophyta and Prochlorophyta, both containing oxygen-producing autotrophic prokaryotes, and nine eukaryotic phyla. However recent molecular approaches to algal systematics and phylogeny continue to revise current thinking about the systematics of a number of algal groups. For example, phylogenies based upon ribosomal RNA gene sequences suggest that the chlorophyll *a* and *b*-containing genera are distributed among cyanobacteria taxa, rather than representing a coherent group (Fig. 5–1) (Urbach, et al., 1992; Palenik and Haselkorn, 1992; Wilmotte, 1994; Palenik and Swift, 1996), which suggests that they should not be separated into a different phylum. The classification scheme used in this text includes nine phyla (see Chapter 1), for each of which exists substantial molecular (and other) evidence for monophyly (descent from a single common ancestor). These are: Cyanobacteria, Glaucophyta, Euglenophyta, Cryptophyta, Haptophyta, Dinophyta, Ochrophyta, Rhodophyta, and Chlorophyta.

In some groups there is additional molecular and other data in support of monophyletic classes, but in others, the systematics are more controversial. Therefore we have avoided organizing algae into hierarchical classifications except where these are well justified by modern findings, particularly those arising from molecular phylogenetic analyses.

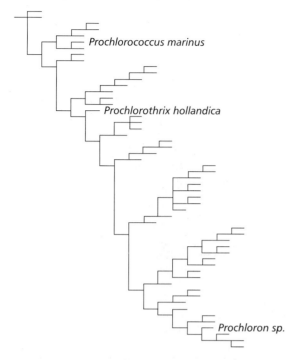

Figure 5–1 Phylogenetic tree of cyanobacteria showing the scattered placement of chlorophyll *a* and *b*-containing cyanobacteria, suggesting that they should not be taxonomically segregated from other cyanobacteria. The implication of this phylogeny is that chlorophyll *b* evolved several times. (After Wilmotte, 1994 with kind permission of Kluwer Academic Publishers)

In this chapter we will first review current knowledge of the numbers and taxonomic distribution of algal species and then consider problems in defining algal species. Examples of the importance of algal classification and species-level identification will be described. We also review modern molecular methods used to assess algal taxa and their biogeographic distribution, and discuss current approaches to the inference of algal phylogenies (patterns of evolutionary history). In view of the very substantial impact that molecular analyses have had on current views of algal evolution and relationships, an understanding of the techniques by which such data are obtained, analyzed, and interpreted is essential for all students of the algae. Finally, we will explore the utility of molecular-based phylogenetic information in attaining better understanding of the ecological roles of algae. Molecular phylogenetics is rapidly becoming an indispensible tool for algal ecologists and will play a dominant role in future approaches to algal ecology.

Numbers and Definition of Algal Species

Numbers of genera and species vary a great deal among the major algal groups. The cyanobacteria include perhaps 2000 recognized species, euglenoids 900, dinoflagellates 4000, glaucophytes 13, cryptomonads 200, red algae 6000, green algae 17,000 (many of these being desmids), haptophytes 300, chrysophyceans 1200, yellow-greens (tribophyceans) 600, eustigmatophyceans 12, raphidophyceans 27, and brown algae 1500. Diatoms are unusually diverse, with about 12,000 recognized species. Some experts have suggested that there may be as many as 10 million diatom species alone. Other algal groups that are suggested to possess many more species than are presently recognized include the dinoflagellates, red algae, green algae, haptophytes, eustigmatophyceans, and tribophyceans (Norton, et al., 1996).

Biological Species Concept

Methods for delineating and distinguishing algal species are widely discussed topics among phycologists who, because of differences in research goals, may apply different species concepts (Manhart and McCourt, 1992). Among these is the **biological species concept** described by the noted evolutionary biologist Ernst Mayr. According to the biological species concept, if two algal taxa can be demonstrated to interbreed and have viable offspring, they belong to the same species; if interbreeding does not occur, they would be regarded as separate species. Some cases of successful applications of the biological species concept to algae are described by John and Maggs (1997). In many algae, however, breeding experiments are difficult to perform. One reason is that such experiments normally require cultures, but cultures of many forms are not available, either because isolations have not been attempted or because growth requirements are not understood. Moreover, the production of gametes by algae is often strongly influenced by environmental conditions. The set of inducing conditions varies considerably, and in most cases must be experimentally determined for each organism to be tested. Worse still, sexual reproduction has not been detected in some algal groups (such as cyanobacteria and euglenoids), and sex seems to occur relatively rarely in others (such as dinoflagellates, diatoms, and cryptomonads). Furthermore, in various algae, fertile hybrids between morphologically distinguishable species (as well as intergeneric hybrids) can occur, the giant kelps providing examples of the latter (Lewis and Neushul, 1995). As a result of such problems that arise when attempting to apply the biological species concept to algae, most algal species have been defined primarily on the basis of structural (morphological) differences.

Morphological Species Concept

According to the **morphological species concept**, species are the smallest groups that can be repeatedly defined by structural characteristics that are relatively easy to distinguish. Ecologists who identify species from natural collections typically rely upon this species concept, often using pigmentation and structural information to deduce ecologically important algal functions. In many cases such procedures may be justified, but sometimes significant physiological attributes of algae (including nitrogen metabolism, toxicity, vitamin requirements, bioluminescence, and growth dependencies) can vary considerably among isolates of the same species. Frequently the physiological behaviors considered typical of algal species have been defined for only one isolate and thus cannot be reliably attributed to all members of the same morphological species. In addition, some morphological species have been observed to undergo seasonal succession at the clonal level, i.e., physiological behavior may not be consistent throughout the year. Intraspecies variation in physiological behavior is thought to be more common and ecologically significant than realized (Wood and Leatham, 1992). A review of the application of the morphological species concept to algae is provided by John and Maggs (1997).

Phylogenetic Species Concept

Phycologists interested in algal evolution have advocated use of the **phylogenetic species concept**, in which a species is the smallest group of organisms that exhibits at least one distinctive and unifying structural, biochemical, or molecular characteristic. Ideally, a phylogenetic species is also **monophyletic**, i.e., includes an ancestor and all of its descendants. Monophyletic groups are also known as **clades**. In contrast to monophyletic groups, **paraphyletic** groups do not include all of the descendants of a common ancestor. **Polyphyletic** groups include some members that are more closely related to taxa outside the group (McCourt, 1995) (Fig. 5–2). For example, the previously mentioned chlorophyll *a* and *b*-containing

cyanobacteria (Fig. 5–1) appear to be a polyphyletic group. Phylogenetic analysis is essential for identifying monophyletic, polyphyletic, or paraphyletic groups. As examples, molecular phylogenetic analysis of the common unicellular green alga *Chlorella* suggests that this is not a monophyletic genus (Huss and Sogin, 1990), and a similar analysis indicated that the green flagellate *Carteria* is paraphyletic (Buchheim and Chapman, 1992). The existence of polyphyletic or paraphyletic taxa suggests that taxonomic schemes need to be changed if a classification system reflecting phylogeny is desired. Phylogenetic analyses can also be useful in identifying structural features that are well correlated with genetic or physiological differences among species (and therefore represent the most useful features for taxonomic identification in ecological applications).

The Importance of Algal Species Identifications

Correct identification of algal species may be essential for their successful use in biotechnological applications as well as in understanding the ecology of aquatic ecosystems and global biogeochemistry. Some algal species appear to function as "keystone" taxa (Andersen, 1992). Their presence in a community is unusually influential, affecting a wide variety of other species. Examples include the endosymbiotic (dinoflagellate) *Symbiodinium* species complex in marine animals, the green alga *Chlorella vulgaris* (*C. sorokiniana*) in freshwater protists and coelenterates, and the coralline red alga *Porolithon onkodes*, the dominant coral reef algal ridge builder. In earlier chapters we noted that (1) particular algal species have been well defined for use as biomonitors; (2) changes in algal species composition of ancient aquatic communities indicate alterations in past environmental conditions; (3) certain algal species are more highly valued as foods than are close relatives; (4) particular phytoplankton species are highly toxic to zooplankton, whereas nearly indistinguishable relatives are readily consumed; (5) certain species appear to be restricted to specific epibiotic habits; and (6) related algal species can vary in carbon utilization patterns and in the extent to which the thallus becomes calcified—issues of significance in carbon cycling.

In many ecological studies species-level identifications are essential. For example, Stoermer (1978) emphasized that generic designations alone are not

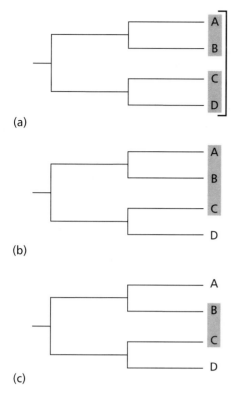

(a)

(b)

(c)

Figure 5–2 Diagrammatic representation of the differences between (a) monophyletic, (b) paraphyletic, and (c) polyphyletic groups. (Redrawn from *Trends in Ecology and Evolution*. Vol. 10. McCourt, R. M. Green algal phylogeny. Pages 159–163. ©1995, with permission from Elsevier Science)

sufficiently precise for use of diatom assemblages as indicators of environmental conditions and change. There are as many species of *Cyclotella* in the most polluted portions of the North American Great Lakes as there are in the least disturbed areas. Species determinations are thus necessary to reveal differences in water quality. As another example, two species of the highly edible cryptomonad *Rhodomonas* differ significantly in the level of light leading to optimal growth, which helps explains the spatial partitioning of the two species, which may in turn affect their availability to various herbivores. *Rhodomonas minuta* is found in lake strata that receive 50% of surface illumination, whereas *R. lens* stratifies where irradiance is 10% of surface levels (Lampert and Sommer, 1997) and where surface herbivores may not occur.

Species-level identifications are also necessary in many biotechnological applications. For example, a variety of the dinoflagellate *Amphidinium opercula-*

tum was recently found to produce compounds having potential use in the treatment of cancer. In order to apply for a patent the researchers had to carefully define the taxon, using both structural and molecular information (Maranda and Shimizu, 1996).

Problems in the Identification of Algal Species

Many biologists, particularly those engaged in ecological studies, are proficient in the identification and classification of higher plants or animals, either because this ability is essential to the conduct of their science or because they find it an enjoyable pursuit. In contrast, relatively few people are able to properly name or classify microorganisms such as algae and protozoa, even though this is an equally significant and usually agreeable activity. There are several reasons for this discrepancy. Foremost among them is the fact that many algae and related protists are of microscopic size, requiring proficiency in the use of dissecting and compound microscopes and, in many cases, scanning or transmission electron microscopy. For example, transmission electron microscopy has been necessary to unequivocally distinguish species of the common marine bloom-forming genus *Phaeocystis*. Many delicate algae do not survive collection and chemical preservation procedures, or are altered such that recognition is impossible. Loss of pigmentation, cell shrinkage, and detachment of distinctive flagella are common problems. In the case of larger seaweeds, microscopic analysis of reproductive or other structures is often necessary for identification and classification, and sometimes specialized chemical processing may also be required. As an example, identification of members of the important group of red algae known as the corallines requires removal of calcium carbonate prior to microscopic examination of reproductive cells.

Another potential barrier to facile identification of algae is the frequent occurrence of parallel evolution of very similar body (thallus) types, such as coccoid unicells or unbranched filaments, in widely divergent algal groups. For example, "little brown balls" (unicellular, coccoid, brown-colored algae) occur in at least three different lineages (ochrophytes, haptophytes, and even some green algae) (Potter, et al., 1997a). Distinguishing yellow-green (tribophycean) algae from structurally similar but unrelated forms requires pigment analysis by chromatography and/or examination of cells by electron microscopy. In a number of cases algae must be cultured to elicit expression of a particular critical taxonomic character. A classic example is the need to have information on zoospore characteristics in order to identify environmentally common coccoid green unicells. Zoospores can be observed in lab cultures but are not normally apparent in field collections. A further difficulty is that taxonomic (identification) keys for algae are typically regional, and the degree to which such keys can be applied to larger or more distant geographic areas is always uncertain. Generic taxonomic keys typically illustrate only one or a few species of a genus that may exhibit a wide array of diversity. Illustrations in keys commonly consist of line drawings devoid of color and other details that would be useful, and often essential, for correct identification. Although high-quality color images of algae are becoming increasingly more available on websites or on CD-ROM, considerable effort is required to use such resources in conjunction with microscopic observations or in the field.

Alternative Approaches to Algal Identification

In view of the difficulties involved in obtaining rapid and accurate algal identifications, scientists conducting field studies sometimes rely primarily upon measurements of chlorophyll or other pigments to estimate algal abundance and diversity. The process of using an easily measured marker to assess the ecological importance of a particular group is termed aggregation. In some cases aggregation is appropriate, but in other situations useful or even essential information is overlooked. Pigments, for example, may vary depending upon ecological conditions (e.g., Lizotte, et al., 1998) and may undergo unequal rates of destruction in natural waters (Hurley and Armstrong, 1990). Moreover, chlorophyll may be difficult to extract from many algae or their life-history stages. Taxonomic composition based on pigment signatures did not always agree with direct observations based on transmission electron microscopy in two sets of open ocean oligotrophic field samples (Andersen, et al., 1996). Thus biomass estimates made on the basis of pigments alone may be inaccurate. Because species identifications can be tedious and time consuming, aquatic ecologists need reliable procedures for determining if aggregation is justified and, if so, what approaches are most appropriate to particular studies. For ecological studies requiring species determinations, faster and more accurate identification methods

are needed. Recent advances in molecular systematics are providing assistance in both areas.

An increasing number of studies are employing molecular methods to define algal species as well as to more quickly and definitively identify members of an algal community. Medlin, et al. (1991) pioneered the use of molecular information in characterization of a new diatom species. Another example is Stiller and Waaland's (1996) use of molecular methods to compare a new species of the economically important red alga *Porphyra* with previously known species of the genus. Molecular sequence information is being used to develop class or genus-specific nucleic acid probes, which are then linked to fluorescent molecules. These tagged oligonucleotides can be hybridized to nucleic acids within whole algal cells and detected by fluorescence microscopy and/or flow cytometry (Fig. 5–3). This technique has been used to rapidly and accurately distinguish cells of domoic acid-producing species of the diatom *Pseudo-nitzschia* from similar, but nontoxic species (Scholin, et al., 1997; Miller and Scholin, 1998; Scholin, 1998). In addition, bloom-forming toxic species of the haptophyte algae *Chrysochromulina* and *Prymnesium* have been identified by the use of fluorescent nucleic acid probes (Simon, et al., 1997). The number of such probes available for algal identifications is rapidly increasing, and a list may be found in Medlin and Simon (1998). Ultimately, it may be possible to use such methods to identify most of the major species present in water samples taken from the natural environment (Lange, et al., 1996). However, since identification of the vast majority of algal fossils is (and will continue to be) based primarily upon structural data, modern taxa that are to be compared with fossils, for example, must also be well characterized structurally. It will also be necessary to define accurate structural or biochemical correlates of molecularly defined algal species for those who need to distinguish algal taxa (such as toxic versus nontoxic species), but who must rely upon microscopic examination or other relatively low-cost methods.

Phylogeny Reconstruction

The goal of phylogeny reconstruction is to understand patterns and processes of evolution, i.e., to explain the occurrence of particular characteristics in specific groups of organisms. In the absence of an extensive fossil record (the case for most algal groups) evolu-

1. phylogenetic analysis of algae

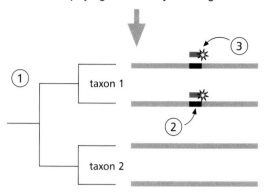

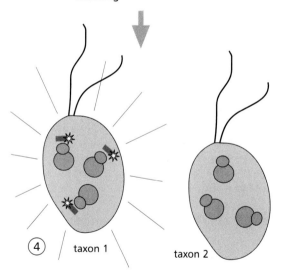

2. identification of a taxon-
   specific region

3. design of corresponding
   oligonucleotide probe and
   labelling

4. hybridization: algae + probe

5. analysis/identification using
   epifluorescence microscopy
   and/or flow cytometry

Figure 5–3 The use of fluorochrome-tagged molecular sequences as hybridization probes, combined with epifluorescence microscopy and/or flow cytometry, is an important recent advancement in the field of algal identification technology. (Based on DeLong, 1998 and Medlin and Simon, 1998 with kind permission of Kluwer Academic Publishers)

tionary history must be deduced from the characteristics of modern organisms. The characteristics used in such studies are typically structural—at the organismal, cellular, biochemical, or genome architectural levels (such as intron insertions or movement of genes from one cellular compartment to another)—or molecular sequences. There are also a variety of phylogeny inference methods, including phenetics, cladistics/parsimony, and maximum likelihood. Many algal phylogenies have been published, the results of which characteristically take the form of a dendrogram or tree diagram. Some phylogenetic trees are well supported overall, while others contain both well and poorly supported branches. A few phylogenies are almost completely unreliable. For example, the use of 5S ribosomal RNA sequences for phylogeny reconstruction has been abandoned, since there are too few informative sites and the rate of nucleotide base substitution is too high for study of ancient divergences. Phylogenies based solely upon this molecule are not regarded as reliable or robust.

It is important for anyone using phylogenetic information obtained from the literature to be familiar with the major approaches to phylogeny reconstruction (and the accompanying assumptions and pitfalls) in order to evaluate the reliability of published phylogenies. Such insight is also helpful in understanding how different workers may obtain different phylogenetic results when studying similar groups of taxa. In the subsequent discussion we will consider the utility of structural and molecular characters in phylogeny reconstruction, survey the major genes used in generating molecular sequence-based phylogenies, compare inference and validity assessment methods, and survey some examples of applications of algal phylogenetic information.

Molecular Sequences versus Structural Data in Phylogeny Reconstruction

The phylogenies of various algal (and other) groups were extensively explored with the use of structural characters prior to the development, within the last two decades, of technologies that have made the acquisition of molecular sequences relatively easy. For a time afterward, molecular sequence analyses were viewed as being highly superior to structure-based phylogenies because the former offered many more characters, which were thought to be less subject to the confounding effects of parallel or convergent evolution. Some workers suggested that the use of structural

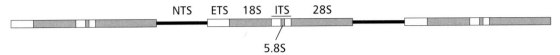

NTS   ETS   18S   ITS   28S

5.8S

Figure 5–4 Ribosomal RNA genes and internal transcribed spacer regions whose sequences are widely used in molecular systematics. NTS=non-transcribed spacer, ETS=external transcribed spacer, 18S=small subunit (18S) rRNA gene, ITS=internal transcribed spacer, 5.8S=5.8S rRNA gene, 28S=large subunit (28S) rRNA gene.

characters to deduce phylogenies, which would in turn be used to understand character evolution, represented circular logic, and should thus be avoided. It has since become clear that (1) molecular sequences can also undergo parallel or convergent evolution, (2) it is not possible to conclude that molecular trees reflect true phylogeny, and (3) sole reliance upon molecular characters excludes useful characters from analyses (Swofford, et al., 1996). While employment of molecular sequence methods has not resulted in widespread overturn of previous evolutionary hypotheses based upon structural data (Moritz and Hillis, 1996), its use often results in greater phylogenetic resolution than does use of morphological data alone.

The current practice of algal systematics reflects application of the concept of reciprocal illumination (Hennig, 1966). A first approach to phylogeny reconstruction is selected based upon appropriateness to the questions being asked and the time frame of the evolutionary events of interest (see next paragraph). Structural changes, such as differences in thallus form or cellular ultrastructure, genome architectural modifications, or alterations in amino acid sequences (typically inferred from DNA sequences) are then mapped upon the phylogeny. If they don't appear to occur in logical patterns, an alternate approach—such as a different gene sequence—is then tried. Some workers attempt to combine structural data sets with molecular sequence information into the same phylogenetic analysis.

An alternative strategy is to evaluate separate phylogenies based upon independent data sets, such as genes from different cellular organelles, or genes versus morphology, for congruence. Congruence is viewed as strong support for the validity of the phylogenetic reconstruction and evolutionary conclusions derived from it. The absence of congruence suggests that alternate approaches are required (Hillis, et al., 1996). As an example of this approach, species phylogenies for two red algal genera were inferred from sequence data obtained for two independent regions: the nuclear "internal transcribed spacer" (ITS) plus 5.8S ribosomal DNA gene (Fig. 5–4) and the plastid

Rubisco spacer region. The two phylogenies were found to be congruent (Goff, et al., 1994). Another example of congruence comes from independent molecular sequence-based studies of higher level taxa of marine red algae, one based on the gene for nuclear-encoded small subunit ribosomal RNA (Ragan, et al., 1994), and the other on the plastid gene coding for the large subunit of Rubisco (Freshwater, et al., 1994). Both studies concluded that one major group of the red algae was polyphyletic, whereas another was monophyletic, and also agreed that there were some good correlations with previous phylogenies based upon morphology (Chapter 16).

Choosing Molecular Phylogenetic Approaches

Moritz and Hillis (1996) suggest that the following questions be asked prior to collecting a data set for use in phylogeny reconstruction: Do characters exhibit an appropriate level of variation? Do characters have a genetic basis? Are characters independent? Only recently has the lack of character independence been recognized as a source of potentially confounding effects. Lack of independence is a possible pitfall not only for structural data sets, where development of one structure may influence development of another, but also for molecular data sets. Molecular function, such as the need to maintain an active site, bind cofactors, or allow for aggregation of subunits (as in the case of enzymes), or fold into three-dimensional stem and loop regions (in the case of RNA molecules), can constrain evolutionary change such that a nucleotide base alteration in one portion of a molecule influences selection for corresponding changes elsewhere (Kellogg and Juliano, 1997). Many different molecules and molecular techniques are used to study algal phylogeny, a number of which are described in the following sections. Each approach is briefly described in order to acquaint the reader with important sources of infor-

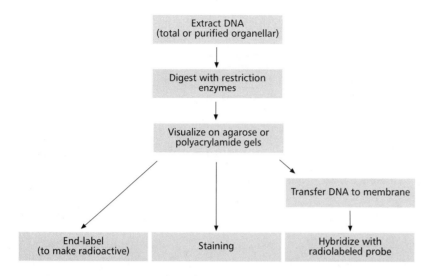

Figure 5–5 Restriction fragment length polymorphism analysis procedures. (Based on Dowling, et al., 1996)

mation regarding algal relationships and species diversity and to provide specific examples where the techniques were used with algae. Other sources of such information include reviews by Medlin and Simon (1998) and Chapman, et al. (1998).

RFLP Analysis

Restriction fragment-length polymorphisms (RFLPs) estimate DNA sequence divergence by detecting variation due to nucleotide base substitutions, deletions, or insertions that have occurred over the course of evolutionary history. DNA is extracted from the taxa under study and is subjected to digestion with a battery of restriction enzymes (endonucleases), each of which recognizes a particular sequence (the "restriction site") of four to eight (or more) nucleotides. For each taxon (and enzyme), a particular pattern results, with variously sized DNA fragments. The fragments are separated by electrophoresis on agarose gels, and stained with a dye that allows for DNA to be visualized with UV irradiation. When a restriction site is altered through nucleotide changes and/or insertions/deletions, a different pattern of bands is seen, with different fragment lengths. Patterns from the study organisms are scored relative to one another to assess the degree of variation present. The bands can also be probed with radioisotopically labeled nucleic acid probes representing cloned fragments of the genome under study (Southern hybridization) (Fig. 5–5). This process allows mapping of restriction-site mutations. RFLP analyses have been used to evaluate intergeneric,

interspecific, and intraspecific variation in algae (Manhart and McCourt, 1992). A good example of such a study is the RFLP analysis of chloroplast DNA of U.S. coastal *Codium* species and subspecies performed by Goff, et al. (1992). Disadvantages of RFLP analyses are that they are labor-intensive and require relatively large amounts of purified DNA. Further, some experts recommend that RFLP data not be used to infer phylogenies because the assumption of character independence is violated. For example, if a new restriction site were to evolve between two preexisting sites in a particular taxon, this would change RFLP patterns such that it and a related species may show no fragments in common even though they may share the original restriction sites (Swofford, et al., 1996).

RAPDs

Another approach to the study of speciation, as well as geographic variation, is the use of randomly amplified polymorphic DNAs (RAPDs). A variety of DNA amplification primers (short oligonucleotides about 10 bases in length) are used in the polymerase chain reaction (PCR) (Fig. 5–6), to synthesize many copies of anonymous (unknown coding/function) DNA regions that vary in size. When visualized on an agarose gel, these DNA products form a fingerprintlike pattern that is characteristic for each organism. Patterns can be compared with those obtained from members of other species or populations. A high degree of similarity in banding patterns indicates close relationship, whereas dissimilarity suggests more distant relationships. Advantages of using RAPDs includes the fact

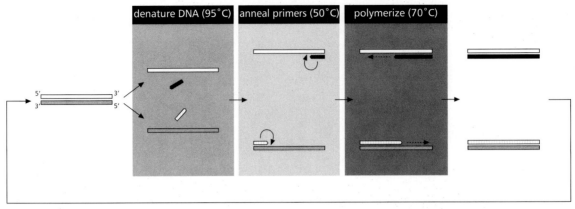

denature DNA (95°C)   **anneal primers (50°C)**   **polymerize (70°C)**

oligonucleotide amplification primer

(repeat cycle 20-40 times)

Figure 5–6 Polymerase chain reaction (PCR) procedure. Annealing temperatures, number of cycles, and times the sample spends at each temperatures varies depending upon the nature of the primers and target DNA and instrumentation.

that only small amounts of DNA are required, and procedures can also be done with relative ease, speed, and low cost. One disadvantage of the use of RAPDs is that repeatability may be difficult. Another is that band homology (degree of band relatedness) is difficult to determine without costly or time-consuming sequencing efforts. RAPDs revealed the occurrence of geographically distinct populations of North American *Fragilaria capucina* sampled across a latitudinal gradient (Lewis, et al., 1997).

Microsatellite (Minisatellite) DNA

Microsatellite DNA consists of regions of variable length having sequences of di- or trinucleotides (e.g., GAGAGAGAGA. . .) that are repeated 10–100 times per locus. Such DNA is used to study variation within species, providing a "fingerprint" for each population analyzed. Variation in the number of repeats is detected by using specific hybridization probes (Oppermann, et al., 1997). Microsatellite markers in the nuclear genome have been used to study the brown alga *Laminaria* (Billot, et al., 1998).

Nucleic Acid Sequencing

Recent technological advances have allowed for the rapid and accurate determination of the sequence of nucleotides in a nucleic acid molecule, at increasingly lower cost. Nearly all sequencing work done for phylogenetic research is conducted with DNA and relies on the Sanger dideoxy chain-termination method (Fig.

5–7). The advantage of sequencing methods is that large numbers of (presumably) independently evolving characters can be used to reconstruct phylogenies as compared to the much smaller numbers of morphological (or biochemical, etc.) characters that are available for the algae. A variety of genes have been employed in the molecular systematics of algae; the most commonly used genes are described in the following section.

Ribosomal RNA genes

Molecular study of evolutionary divergences that took place greater than 500 million years ago (which include separation of the major algal phyla) requires sequencing of very slowly evolving genes, such as nuclear-encoded genes for small (Fig. 5–8) and large subunit ribosomal RNA (SSU and LSU rDNA). Conservation of these sequences derives from their integral importance in cell processes. SSU is more highly conserved than LSU and hence is more useful for this type of analysis. However SSU rDNA also contains some regions of relatively high variability, such that the molecule can also be used at lower taxonomic levels, including species-level investigations. For example, Buchheim, et al. (1997a) used SSU rDNA sequence data to demonstrate that the green flagellate genus *Chloromonas* is not monophyletic.

One major advantage of the use of ribosomal RNA genes is that phylogenies can contain both prokaryotic and eukaryotic taxa, since homologous SSU and LSU genes occur in these groups. The Ribosomal Database Project (RDP) at the University of Illi-

5′ AAGGGTCGACACGAGGTTCGACTGATGGAT 3′
3′ TTCCCAGCTGTGCTCCAAGCTGACTACCTA 5′

Denature double-stranded DNA
to obtain single-stranded template
(or use single-stranded phage DNA)

3′ TTCCCAGCTGTGCTCCAAGCTGACTACCTA 5′

Anneal primer to template strand

primer

AAGGGTCGACACGAG
3′ TTCCCAGCTGTGCTCCAAGCTGACTACCTA 5′

Briefly extend forming strand with dNTPs,
one of which is radioactively labeled

AAGGGTCGACACGAGGTT ➤

Divide into four parts, each with a different ddNTP,
to terminate forming chains

ddATP	ddCTP	ddGTP	ddTTP
primer + GTTCGA	primer + GTTC	primer + G	primer + GT
primer + GTTCGACTGA	primer + GTTCGAC	primer + GTTCG	primer + GTT
primer + GTTCGACTGATGA		primer + GTTCGACTG	primer + GTTCGACT
		primer + GTTCGACTGATG	primer + GTTCGACTGAT
		primer + GTTCGACTGATGG	primer + GTTCGACTGATGGAT

Separate fragments by acrylamide gel electrophoresis
and visualize by autoradiography

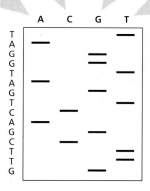

Figure 5–7 Diagram showing the steps involved in the Sanger dideoxy sequencing approach. The majority of DNA sequencing is accomplished using this method (for automated sequencing, DNA is fluorescently labeled and the bands are sensed as they run past a light detector during electrophoresis). (After Hillis et al., 1996)

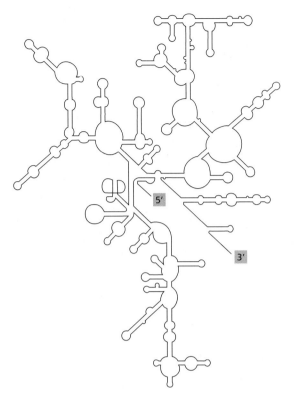

Figure 5–8 Secondary structure model for an 16S rRNA molecule (from the cyanobacterium *Chlorogloeopsis* sp. PCC 7518). There are numerous stems (based-paired regions) and loops (unpaired stretches). Different parts of the molecule evolve at different rates, allowing SSU rRNA genes to be used at a variety of taxonomic levels. (After Wilmotte, 1994 with kind permission of Kluwer Academic Publishers)

nois (Maidak, et al., 1997), contains a large array of sequences for both prokaryotes and eukaryotes. Additional rDNA sequences can be found in GenBank. The 18S (16S-like) SSU gene of eukaryotes contains approximately 1800 nucleotide positions. Variations in these positions can be treated as characters for constructing phylogenetic trees. Many investigators attempt to obtain the entire SSU sequence because this enhances phylogenetic resolution.

Ribosomal RNA genes are widely thought to exhibit **concerted evolution,** a phenomenon in which multiple copies of genes in nuclear genomes within a species (and population) undergo the same kinds of changes at nearly the same time, such that their sequences become homogeneous. The mechanism of concerted evolution is not well understood. Recently cases have been uncovered of exceptions to the concept of the homogenous nature of rDNA sequences.

Among algae, for example, multiple types of rDNA occur in the dinoflagellate *Alexandrium*. This could be problematic in phylogeny construction (Scholin and Anderson, 1996).

Internal transcribed spacer region (ITS)

For studies of species diversification events assumed to have occurred relatively recently—within the past 5 million years or so—rapidly evolving sequences are preferred. The regions of the nuclear-encoded ribosomal RNA genes known as the internal transcribed spacers (ITS) (Fig. 5–4) are particularly useful in such studies. ITS sequencing has commonly been used in intrageneric phylogenetic studies of algae. For example, Marks and Cummings (1996) used the ITS region to demonstrate low genetic diversification among freshwater isolates of the common and widespread green alga *Cladophora* obtained from a wide range of habitats and geographical locations. Peters, et al. (1997) used ITS sequences to compare species within the widespread brown algal family Desmarestiaceae.

Protein-coding genes in phylogeny reconstruction

Ancient divergences can also be illuminated by comparative sequence analysis of protein-coding genes. Frequently nuclear-encoded protein-coding genes occur in multiple versions that have arisen through gene duplication and divergence and together form a gene "family." Individual gene family members are considered to be **paralogous** genes. In addition, the genome may contain processed paralogous sequences arising from mRNA sequences that are reverse transcribed to DNA, then inserted into the genome. Such paralogs can be mistakenly incorporated into phylogenetic analyses, resulting in erroneous phylogenetic conclusions (see Bowe and dePamphilis, 1996). Extreme care must be taken to avoid comparing sequences from different organisms that are actually not the most closely related members of a multi-gene family. For this reason, single-copy protein coding genes are preferred for phylogenetic analyses. An example of a protein-coding sequence used in algal phylogeny reconstruction is the nuclear-encoded, but apparently single copy, centrin-coding region of green flagellates (Bhattacharya, et al., 1993). Other useful single copy protein-coding genes reside in organellar genomes.

Commonly used chloroplast genes are *rbcL*, which encodes the large subunit (LSU) of Rubisco,

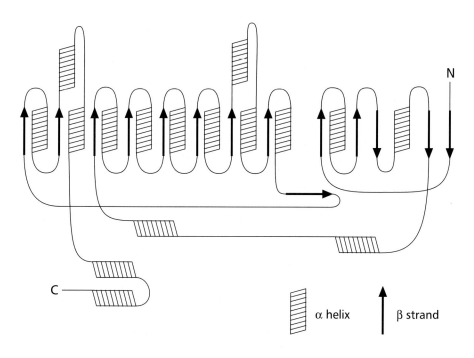

Figure 5–9 Diagrammatic representation of intramolecular associations in the large subunit of Rubisco. Functional constraints on evolutionary change limit the number of readily changeable sites that can be used in a phylogenetic analysis. (Based on Kellogg and Juliano, 1997)

ndhF (encoding a dehydrogenase subunit) and *matK* (which codes for a maturase involved in RNA splicing). Other chloroplast sequences used in phylogenetic analyses include *rpoC1* and *rpoC2* (which encode RNA polymerases), and *trnK* and *trnL* (which encode transfer RNAs).

The base substitution rate for *rbcL* is considerably higher than that of 18S SSU rDNA, and thus *rbcL* is considered less useful for resolving extremely ancient divergences, such as those of phyla, classes, and orders of algae. However, it has proven helpful in determining branching patterns within orders of the green algae (e.g., McCourt, et al., 1996a, b; Nozaki, et al., 1997a, b, c). Problems with the use of *rbcL* in phylogeny can arise from the fact that over 30% of the 476 amino acids in the Rubisco LSU are involved in intramolecular associations (Fig. 5–9). Hence the number of potentially variable (independent) genomic bases is less (around 1000) than often thought (1428). Further, it appears that nucleotide sites are more free to vary in some lineages than in others (Kellogg and Juliano, 1997). The chloroplast gene *ndhF* (approximately 2200 base pairs [bp] in length) is primarily useful in resolving relationships among families, subfamilies, and genera, whereas *matK* (about 1550 bp) is of greatest utility at the generic and species level.

Organellar genes sometimes contain mutations that are corrected in the mRNA during processing of the transcript. This correction process is known as RNA editing. It results in mRNA and protein sequences that would not have been predicted by the examination of the encoding DNA sequence. Although some workers have recommended that edited sequences not be used in phylogenetic analyses, others find no effect on phylogenetic reconstruction, as long as mRNA sequences are not used together with DNA sequences in analyses (Bowe and dePamphilis, 1996).

Genome-Level Changes in Molecular Architecture and Their Use in Phylogenetic Inference

Genomic architectural data—including gene and intron insertions and deletions, appearance of large repeated sequence regions, and transfer of genetic information from one cellular compartment to another—can also be phylogenetically useful. Although positions of some introns are known to have changed over time, in most cases intron placement seems to remain relatively constant over long time periods. Thus, although intron sequences themselves (being non-coding regions) can exhibit high rates of nucleotide change, their numbers and positions are typically conservative and useful in tracing ancient divergences. An example from the algae includes the occurrence of an intron within the vacuolar ATPase subunit A gene (which encodes the cat-

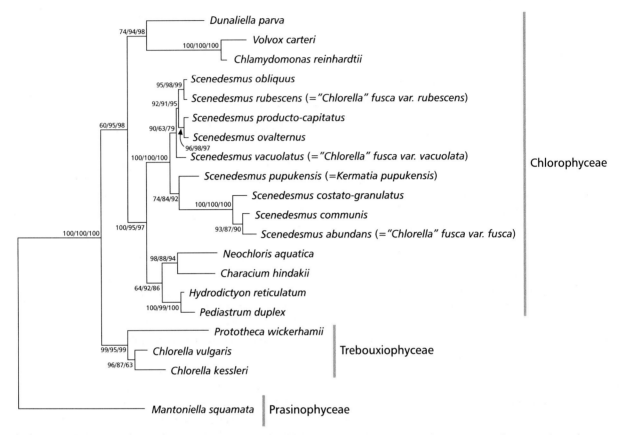

Figure 5–10 Phylogenetic relationships of some green algae, based on SSU rRNA sequence data. Separate bootstrap values are given for distance, parsimony, and maximum likelihood analyses (in that order) at each branch. (After *Botanica Acta* 110 [1997]:244-250. Kessler, E., M. Schäfer, C. Hümmer, A. Kloboucek, and V. A. R. Huss. Physiological, biochemical, and molecular characters for the taxonomy of the subgenera of *Scenedesmus* [Chlorococcales, Chlorophyta]. Georg Thieme Verlag, Stuttgart)

alytic subunit) of the green alga *Coleochaete*. This intron has apparently been inherited by vascular plants but is absent from other green algae examined and *Euglena* (Starke and Gogarten, 1993), suggesting that *Coleochaete* is closely related to the ancestry of the land plants (see Text Box 21–1). Another example is the presence of a 900 bp intron in the fucoxanthin-chlorophyll *a/c*-binding protein gene of the brown kelp *Laminaria saccharina*. Such an intron is not present in homologous genes of diatoms, even though both are members of the Ochrophyta (Caron, et al., 1996). Variations in genomic architecture are detected by a simple presence/absence test based on hybridization experiments in which specific nucleotide sequences are used as probes to reveal homologous sequences from among DNA fragments that have been separated by gel electrophoresis, then transferred onto a nitrocellulose substrate (Southern blotting). If their positions are known, introns can be amplified by employing sequences of flanking exons as primers ("EPIC"—exon-priming, intron-crossing primer pairs) in PCR and then sequenced. Intron sequences can be useful in answering species-level evolutionary questions. Another genome-level approach that has been used in algae is comparison of mitochondrial genome organization, where some surprising differences have been observed in green algae (Nedelcu, 1998) (see Chapter 17).

Generating, Identifying, and Evaluating Optimal Phylogenetic Trees

A basic understanding of how phylogenetic trees are generated and evaluated is indispensable to the consumer of such information. It allows the reader to decide whether the information is valid and, in the case of controversies, which case is best supported by molecular phylogenetic evidence. The most commonly employed tree-generating approaches are phenetic distance matrix methods, maximum parsimony (cladistic) methods, and maximum likelihood methods. Figure 5–10 illustrates variation that results when each of the three methods is applied. In distance methods the number of differences between all pairs of taxa are determined, then these numbers are used to group taxa to form a dendrogram (dichotomously branched tree diagram), which attempts to accommodate all of the pairwise distances. Parsimony methods operate under the assumption (which is not always correct) that evolution operates in the most efficient (parsimonious) manner, i.e., the most accurate phylogeny is the one requiring the fewest number of changes. One drawback of parsimony methods is that they do not detect multiple superimposed changes that have occurred in long, unbranched lineages (which are particularly common among archaic protists). Bias can be introduced if a phylogeny includes a mixture of long and short branches. Another criticism of parsimony methods is that a variety of weighting schemes can be applied to the data. Which of these is chosen may affect the structure of the resulting phylogeny and, hence, the evolutionary conclusions.

The application of maximum likelihood methods begins with construction of a model of the evolutionary process that hypothetically resulted in conversion of one sequence into another. Then the likelihood that the specified evolutionary model will yield the sequences that are actually observed is calculated as the maximum likelihood estimate. An advantage of the approach is that estimation of branch lengths is an important part of the process. In contrast to parsimony methods, maximum likelihood methods attempt to account for unobserved base substitutions. Although very demanding in terms of computational effort, maximum likelihood is gaining favor because it provides a probability estimate, i.e., a measure of the statistical significance of the phylogenetic hypothesis (Hulsenbeck and Rannala, 1997).

The number of possible phylogenies (evolutionary trees) in typical analyses can be very great, and increases with the number of taxa examined. For example, there are 2 million possible bifurcating trees linking a group of 10 taxa; for 50 taxa, the number of possible trees is 3×10^{74}. There are two basic approaches to identifying which of the possible trees is optimal. Every possible tree could be evaluated in an **exhaustive search**, but this is computationally demanding for trees involving 12 or more taxa. For studies having larger numbers of taxa, **heuristic searches** (trial and error approaches) are more efficient; these sacrifice certainty regarding the optimal tree for a reduced computing effort. Here again, there are two common approaches: (1) starting with all taxa linked (in a starlike pattern), after which a defined criterion is employed to evaluate possible neighbor linkages in a stepwise process; or (2) adding taxa in a sequential fashion to starting trees containing only three taxa. Typically authors will state whether phylogenetic trees were constructed using exhaustive or heuristic searches.

Commonly, a **consistency index** (C.I.) is calculated for phylogenetic trees; this is a measure of the amount of **homoplasy** (parallel or convergent evolution) exhibited by the characters used to construct the tree. The consistency index reflects the ability of a phylogenetic estimation method to converge upon a true value as more data accumulate. This quantity will be equal to one if there is no homoplasy; the lower the C.I., the higher the level of homoplasy. One cause of low consistency is sharply unequal rates of change in different branches of a phylogeny, causing what is known as long branch attraction. Taxa that terminate long branches (a common occurrence among archaic algae) may appear to be more closely related than they actually are. Use of maximum likelihood methods reduces the effects of long-branch attraction, yielding trees having higher consistency indices than trees generated by other methods of phylogeny construction (Hillis, et al., 1996).

Individual nodes (branch points on a phylogenetic tree) are often evaluated by calculation of a **bootstrap value** (Felsenstein, 1985; 1988). This term is derived from the colloquial phrase "to pull oneself up by the bootstrap." It simulates the collection of replicate data sets through repeated resampling of the data, including some portions and leaving out others each time. Often 100 resampling procedures are done, and the number of times a particular branch appears is determined. One hundred is thus the maximum value, and branches having this bootstrap value are regarded as very well supported. Branches having values around 50 and lower are regarded as being poorly supported. It should be noted that although bootstrap values are widely used in both phylogenetics and ecology, the statistical meaning of this procedure is unclear. Figure 5–11 shows a phylogeny for species of the red algae *Gracilaria* and *Gracilariopsis*

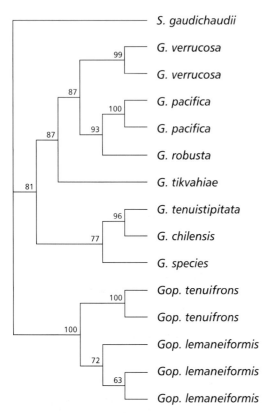

Figure 5–11 A molecular phylogeny for species of *Gracilaria* and *Gracilariopsis*, which exhibits high bootstrap values, indicating relatively strong support for most individual branches. *Sarcodiotheca gaudichaudii* was used as an outgroup taxon. (From Goff, et al., 1994 by permission of the *Journal of Phycology*)

wherein nearly every branch is supported by relatively high bootstrap values. In contrast, Fig. 5–12 shows a phylogenetic analysis aimed at uncovering phylogenetic relationships of the green flagellates known as prasinophyceans. Here the conclusion that prasinophyceans are not monophyletic is well supported by high bootstrap values, but the conclusion that charalean green algae (*Chara* and *Nitella*) are more closely related to seed plants than close relatives, is not well supported (the bootstrap value for this branch is relatively low). Recently, the "jackknife" method has gained favor as a substitute for the bootstrap as a method for evaluating the level of support for clades (Farris, 1995; 1997)

Phylogenetic trees that have been inferred by parsimony methods may sometimes include **decay values** as indices of the degree of support for particular branches (putative monophyletic groups). These values indicate the number of additional evolutionary changes that would have to occur before a monophyletic group becomes subsumed into a larger set of taxa (i.e., the branch "collapses") (Bremer, 1988). The larger the number, the more confidence is placed upon the branch. Branches having decay values of one or two are suspect. This test has been criticized for being sensitive to aspects of data set structure that are not related to the hypotheses of monophyly being evaluated. Frequently authors will provide both bootstrap (or jackknife) and decay values for all tree branches. This greatly aids the reader in evaluating the conclusions. No conclusions can be made from any phylogenetic tree that lacks such estimates of significance.

The Application of Phylogeny

As previously noted, phylogenetic approaches can be used to define monophyletic groups (clades) of algal species with some confidence. When structural, ecological, physiological, or biochemical traits are mapped onto phylogenies, clade-specific attributes are often revealed. In other cases phylogenetic analysis reveals instances in which structure, ecology, physiology, or biochemistry is not linked to a specific clade. Such information can be very useful in making decisions whether or not to use easily determined traits as markers for taxa that are to be aggregated in an ecological study. More generally, phylogenetic information can form a robust basis for predicting the extent to which physiologies or ecological behavior can be extrapolated from taxa that have been studied to related forms that have not. Traits of evolutionary, biogeochemical, ecological, or economic importance that could be mapped onto phylogenies with the goal of obtaining greater insight into important roles of algae include: loss of phagotrophic abilities, occurrence of sexual reproduction and life history types, occurrence of the highly resistant wall polymer sporopollenin, cytoplasmic starch synthesis, loss of pyrenoids, calcification ability, loss of capacity to produce flagellate reproductive cells, and toxin production.

Evolutionary Utility

Molecular phylogenetic analysis has revealed that the dinoflagellates found as endosymbionts in reef-forming corals (Chapter 7) are much more genetically diverse than previously suspected (Rowan, 1998).

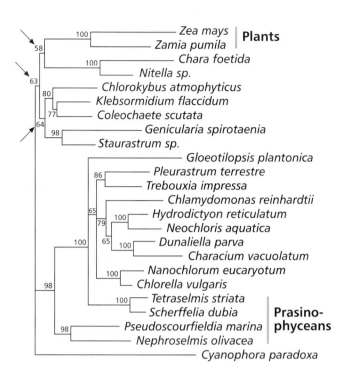

Figure 5–12 A molecular phylogenetic study of several prasinophycean and other green alga taxa. Some branches are supported by high bootstrap values, whereas others (arrows) are less well supported. This needs to be taken into account when making phylogenetic conclusions based on a phylogenetic analysis. (From Steinkötter, et al., 1994 by permission of the *Journal of Phycology*)

Other studies showed that various dinoflagellate-coral combinations are differently adapted to varying light regimes, a fact that is important in devising strategies for reef community conservation. Such an adaptation would not likely have been suspected in the absence of molecular information (Rowan and Knowlton, 1995). As another example of the utility of a phylogenetic approach, Nozaki, et al. (1997a, b, c) used *rbcL* gene sequences to study the relationships of flagellate green algae, and to reject the previously held view that anisogamous sexual reproduction (see Chapter 1) gave rise to isogamous sexual reproduction in these groups. An *rbcL* sequence analysis of Zygnematales, a group of green algae that includes both single-celled and unbranched, filamentous genera, revealed that the unicellular forms did not form a monophyletic group, but rather, chloroplast type was a better indicator of relationships (McCourt, et al.,

1995). Another molecular sequence study, of the brown seaweed group Desmarestiaceae, revealed the occurrence of a clade-specific, one-time evolution of sulfuric acid-containing vacuoles (Fig. 5–13a). In contrast, the same study showed that dioecy (production

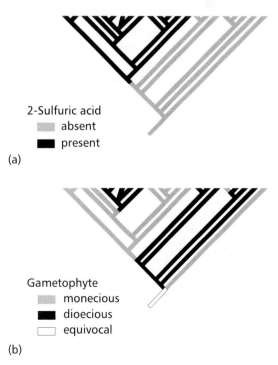

Figure 5–13 A molecular phylogeny for species of the brown seaweed family Desmarestiaceae, onto which has been mapped the occurrence of sulfuric acid (a), revealing that this character evolved only once within this group. In contrast, when dioecy (production of separate male and female thalli) is mapped onto the phylogeny (b), it appears that this trait has been gained at least twice and lost at least twice. (From Peters, et al., 1997 by permission of the *Journal of Phycology*)

of male and female gametes on separate thalli) had evolved at least twice in this group, with losses of this trait also occurring more than once (Fig. 5–13b). Plastid RFLP data were used to link heteromorphic phases of the red alga *Gymnogongrus* (Parsons, et al., 1990).

Ecological Utility

Molecular sequences obtained from DNA or RNA in natural environments are increasingly being used to detect the presence of algae and other microorganisms, such as algal-associated viruses (Chen, et al., 1996). Such "environmental sequences" can be compared to those for known taxa available in sequence databases. This technique was used by Giovannoni, et al. (1990) to detect the existence of previously unknown types of cyanobacteria in the ocean. SSU rDNA sequences obtained from perennially frozen Antarctic lake ice revealed the identities of several cyanobacterial inhabitants of this forbidding site, including a previously unknown form (Priscu, et al., 1998).

Molecular phylogenetic information can be used to evaluate differential gene expression in natural environments. For example, Xu and Tabita (1996) isolated total RNA and DNA from several sample sites in Lake Erie, and probed it with *rbcL* sequences based upon a cyanobacterium and a diatom. They observed that diatom gene expression per gene dose appeared to decrease from nearshore to open-water sites, but that a similar pattern was not observed for cyanobacterial *rbcL*. They also found that the types of *rbcL* being expressed varied with water depth, as has also been observed in marine waters (Palenik and Wood, 1998). Nucleic acid probes are also being used to evaluate changes in carbon fixation rates throughout the day and how these vary from one group of algae to another (Paul and Pichard, 1998). Yet another example of ways in which ecological information can be obtained using molecular phylogenetic approaches was the demonstration that the bloom-forming marine cyanobacterium *Trichodesmium* is itself capable of fixing nitrogen. It had previously been suspected that N_2-fixing activity associated with this alga might have been due to adherent heterotrophic bacteria. Sequencing and phylogeny studies revealed that the *nifH* gene (see Chapter 6) in *Trichodesmium* aggregates was more closely related to cyanobacterial *nifH* genes than to those of heterotrophic bacteria (Zehr and McReynolds, 1989). Other ways in which molecular systematics is being used to assess environmental N_2-fixation are described by Zehr and Paerl (1998). It is very likely that similar approaches will become more common in the future.

Recommended Books

Hillis, D. M., C. Moritz, and B. K. Mable (editors). 1996. *Molecular Systematics*. Sinauer, Sunderland, MA.

Cooksey, K. E. (editor). 1998. *Molecular Approaches to the Study of the Sea*. Chapman and Hall, London, UK.

Tomas, C. R. (editor). 1997. *Identifying Marine Phytoplankton*. Academic Press, San Diego, CA.

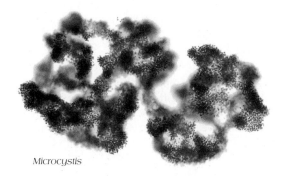

Chapter 6

Microcystis

Cyanobacteria
(Chloroxybacteria)

Cyanobacteria, also known as chloroxybacteria, blue-green algae, or cyanophytes, are significant for many reasons. Cyanobacteria were the dominant forms of life on Earth for more than 1.5 billion years. They were the most ancient oxygen-producing photosynthesizers; the first to produce chlorophylls *a* and *b* as well as a variety of accessory photosynthetic pigments; producers of massive carbonate formations in shallow waters during the Precambrian period; and the earliest (Precambrian) terrestrial autotrophs (Horodyski and Knauth, 1994). They are also the simplest organisms known to have circadian clock genes (Ishiura, et al., 1998). The chloroplasts of eukaryotic algae and plants are descended from cyanobacteria (Chapter 7). Modern cyanobacteria are recognized for their ability to occupy extreme habitats and valued for their ability to fix atmospheric nitrogen, bind and enrich soils, and produce medicinally useful compounds (Chapter 4). Cyanobacteria are of concern when they form nuisance blooms, especially when these produce toxins (Chapter 3). Cyanobacterial roles in biogeochemical cycles have been covered in Chapter 2, and various aspects of cyanobacterial ecology are discussed in Chapters 22 and 23. This chapter provides an overview of cyanobacterial paleobiology, taxonomy, phylogeny, physiology and cell biology, extreme habitat ecology, and morphological and reproductive diversity.

The Paleobiology of Cyanobacteria

The oldest fossils attributed to cyanobacteria are 3.5 billion-year-old remains from the Apex Basalt, a geological deposit in Western Australia (Schopf, 1993). The Apex fossils include a variety of types of multicellular filaments (Fig. 6–1) that resemble certain modern cyanobacteria such as *Oscillatoria* as well as some modern non-photosynthetic bacteria. Statistical comparisons of the cell sizes of the fossils with those of modern filamentous bacteria (which tend to be narrower) and cyanobacteria (which

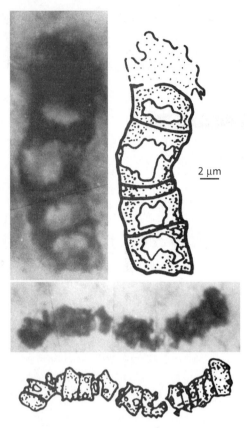

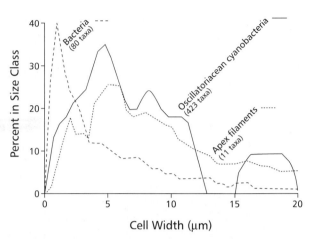

Figure 6–2 A comparison of cell widths for the filaments found in the Apex Basalt versus modern bacteria and oscillatoriacean cyanobacteria, indicating that the Apex fossils likely represent ancient cyanobacteria. (After Schopf, J. W., Microfossils of the Early Archean Apex chert: New evidence of the antiquity of life. *Science* 260:640-646. ©1993 American Association for the Advancement of Science)

Figure 6–1 Filaments discovered in the 3.5 billion-year-old Apex Basalt in Western Australia, which are thought to represent cyanobacteria. (Reprinted with permission from Schopf, J. W., Microfossils of the Early Archean Apex chert: New evidence of the antiquity of life. *Science* 260:640-646. ©1993 American Association for the Advancement of Science)

typically have wider cells) (Fig. 6–2) suggest that several fossil types were most likely cyanobacteria. The fossils *Primaevifilum laticellulosum*, *Archaeoscillatoriopsis grandis*, and *A. maxima* differ in cell width from the vast majority of non-photosynthetic bacteria, but are nearly indistinguishable from both later cyanobacterial fossils such as *Oscillatoriopsis* and modern *Oscillatoria*. In view of these morphological similarities, it has been suggested that the Apex fossils might represent the first oxygen producers.

Other evidence for ancient oxygenic photosynthesis comes from 2.7 billion-year-old layered, calcareous, mound-shaped structures known as stromatolites, found in the Tumbiana Formation of Western Australia (Buick, 1992). Stromatolites are very abundant in Precambrian deposits throughout the world and are commonly thought to have been pro-

duced by light-loving microorganisms such as cyanobacteria that trapped sediments and deposited carbonates in layers (Fig. 6–3), much as some modern cyanobacteria do today. Although there is evidence that at least some ancient stromatolites might have arisen abiotically (i.e., without the action of cyanobacteria or any other biological agent), filaments resembling cyanobacteria have been observed in the Tumbiana stromatolites, suggesting that they were biotically formed. The very low sulfate content of these deposits suggests that they arose in ancient lakes and that anaerobic photosynthetic processes were unlikely to have generated the stromatolites. Hence, Buick (1992) concluded that these structures represent early evidence for oxygenic photosynthesis. In Precambrian times stromatolites apparently grew in a variety of habitats, including shallow nearshore as well as deeper waters, forming domes or branched columns (see Fig. 1–4). Some of these types have been given generic and species names and are used as indicators of particular geological strata. The number of stromatolite types rose to a maximum some 700–800 million years ago, then declined precipitously. Some experts suggest that this decrease was related to the evolution of numerous types of herbivorous gastropods, which grazed on microbial mats, thus preventing the buildup of stromatolites. Supporting this hypothesis is the fact that modern stromatolites develop best in areas devoid

1 cm

Figure 6–3 A stromatolite (seen actual size) that has been split open, revealing the layering typical of these formations.

of gastropods. For example, the occurrence of abundant growths of stromatolites in Shark Bay, Australia (see Fig. 1–4), may be related to hypersaline conditions inimical to gastropods. Other hypotheses that might explain the Cambrian stromatolite decline include the rise of burrowing organisms that may have disrupted growth, or the appearance of red and green calcareous seaweeds (as well as metazoans) that competed with cyanobacteria for available substrate. Nevertheless, cyanobacterial stromatolites have persisted throughout the intervening time to the present (Fig. 6–4) (Tucker and Wright, 1990). Only about 20 modern habitats are known where well-developed stromatolitic algal mats can be found. Most of these are marine, hypersaline, low-latitude environments, but sediment-trapping algal mats dominated by cyanobacteria are also found in permanently ice-covered lakes in Antarctica (Parker and Wharton, 1985).

There is evidence in the fossil record for increased reproductive and structural complexity of cyanobacteria over time. The oldest microfossils whose morphology is diagnostic for cyanobacteria are 2 billion-year-old remains of *Eoentophysalis belcherensis* from the Gunflint Formation in Canada. This organism was very similar to the modern cyanobacterium *Entophysalis*, a colonial form that lacks specialized cells (Knoll, 1996). Fossils 1 to 1.6 billion years old, known as *Archeoel-*

lipsoides, are regarded as the akinetes (resting cells) of a *Nostoc*-like cyanobacterium. These fossils represent the earliest undisputed occurrence of specialized cells in cyanobacteria (Knoll, 1996). By 700 million years ago, baeocyte- (endospore-) producing cyanobacteria (Fig. 6–5a) were present, similar to certain modern

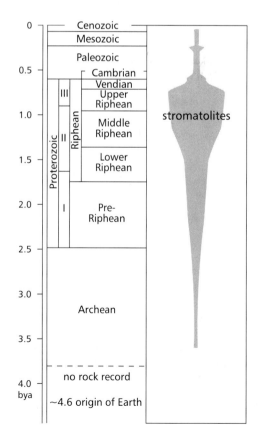

Figure 6–4 The number of stromatolites gradually increased throughout the Precambrian with a maximum number occurring about 800 million years ago, after which they rapidly declined to modern-day numbers. (Modified from Tucker and Wright, 1990)

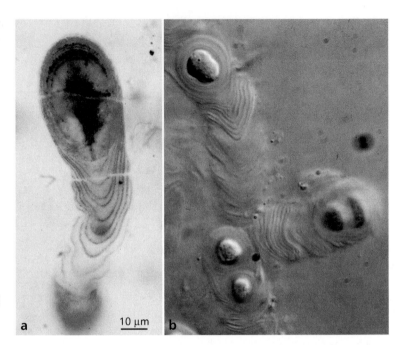

Figure 6–5 (a) *Polybessurus bipartitus*, a 700–800 million-year-old fossil cyanobacterium. (b) An unnamed modern equivalent, whose existence was predicted based on discovery of the fossil form. (a: From Knoll, 1996; b: Courtesy A. Knoll)

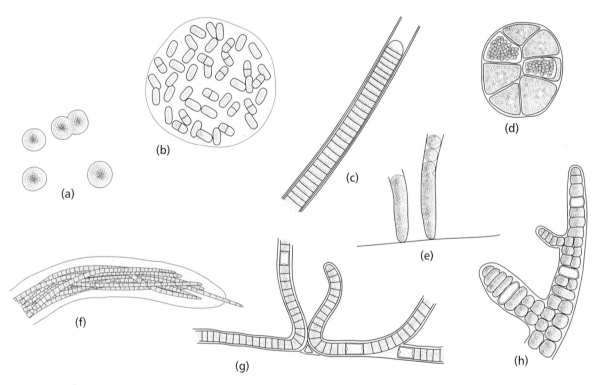

Figure 6–6 Some of the morphological types present in modern-day cyanobacteria include (a) unicells such as *Synechocystis*, (b) colonies of individual cells such as *Aphanothece*, (c) unbranched filaments including *Lyngbya*, (d) baeocyte- (endospore-) forming taxa such as *Myxosarcina*, (e) exopore-forming species, represented by *Chamaesiphon*, (f) aggregations of multiple trichomes in a common sheath as in *Microcoleus*, (g) false-branched forms including *Scytonema*, and (h) true-branched forms such as *Stigonema*. (f: After Smith, G. M. *Fresh-Water Algae of the United States.* ©1950 McGraw-Hill Companies. Reproduced with permission of the McGraw-Hill Companies)

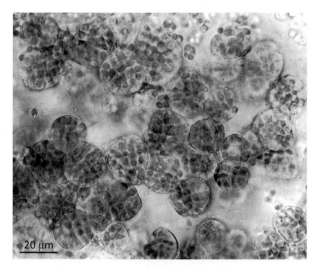

Figure 6–7 Baeocytes (endospores) in *Dermocarpa*.

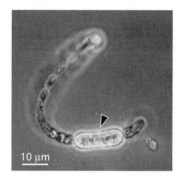

Figure 6–8 An akinete (arrowhead) in *Anabaena*.

forms (Fig. 6–5b) that grow in shallow marine waters (Green, et al., 1987). The relatively early appearance of the major modern cyanobacterial morphologies suggests that cyanobacteria have exhibited very slow evolutionary rates (Schopf, 1996).

Cyanobacterial Taxonomy and Systematics

A major controversy concerning cyanobacteria is whether they should be treated taxonomically like bacteria (as argued by Stanier, et al., 1978) or eukaryotic algae (as preferred by Lewin, 1976). Molecular biologists often tend toward the former, whereas ecologists find it easier to classify cyanobacteria on the basis of structural attributes, which is commonly done for eukaryotic algae. It is no easier to delimit cyanobacterial species than it is to define most algal species. The biological species concept (see Chapter 5) cannot be used with cyanobacteria because sexual fusion of gametes is completely absent from these prokaryotes (although other mechanisms of DNA exchange that occur in other bacteria may also operate in cyanobacteria). Therefore taxonomists have historically used morphological features to define cyanobacterial taxa, including variations in cyanobacterial thallus structure, which include occurrence as unicells, colonies, unbranched filaments, or branched filaments (Fig. 6–6). Two types of branches occur in the cyanobacteria: **false branches** (Fig. 6–6g), which are

outgrowths of filaments that occur adjacent to a dead cell or a specialized cell, and **true branches** (Fig. 6–6h), whose development involves continued division by a cell at right angles to that of the main filament axis. Cyanobacterial taxa can be further defined by the presence or absence of specialized reproductive cells, such as **exospores**, which bud off from the apex (Fig. 6–6e), or **baeocytes** (**endospores**) produced by subdivision of a cell into multiple cytoplasmic units (Figs. 6–6d, 6–7), or **akinetes** (Fig. 6–8). The distinctive cells associated with nitrogen-fixation known as **heterocysts** (Fig. 6–9) are found in some but not all cyanobacteria. Heterocysts may or may not be formed, depending upon the

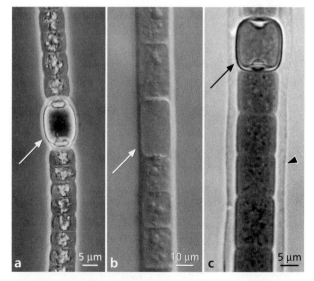

Figure 6–9 Heterocysts (arrows) in (a) *Anabaena*, (b) *Scytonema*, and (c) *Aulosira*, illustrating some of the variation in heterocyst appearance that occurs among taxa. Note the conspicuous mucilaginous sheath (arrowhead) in (c).

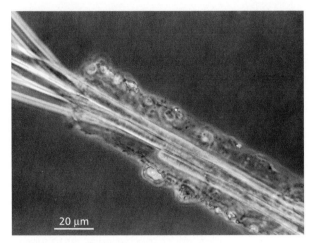

Figure 6–10 *Microcoleus.* Note the multiple filaments contained within a common sheath.

nitrogen status of the algae. Cyanobacterial taxa may also be distinguished by presence or absence of a mucilaginous sheath (Figs. 6–9, 6–10) composed of polysaccharides, which can also vary to some extent with environmental conditions. Some filamentous cyanobacteria occur singly within a sheath, but in other taxa, multiple filaments are enclosed by a single sheath (Fig. 6–10). The term **trichome** (meaning "hair") is used as a synonym for an ensheathed individual fila-

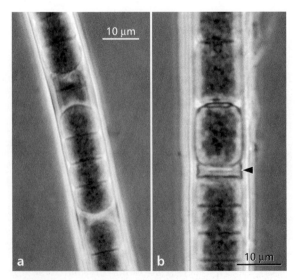

Figure 6–11 *Tolypothrix.* (a) Hormogonia. (b) Separation disk (arrowhead) adjacent to a heterocyst; this is where false branches usually occur in this organism.

ment. The term **hormogonium** is applied to a short filament (Figs. 1–21h, 6–11a) that results from breakup of longer filaments and serves as a means of vegetative reproduction. **Separation disks** or **necridia** (Fig. 6–11b) are dead and collapsed cells, whose death is probably genetically programmed (a process known as apoptosis) for production of hormogonia or false branches. Some cyanobacterial taxa are characterized by aggregates of structures visible with the light microscope (Fig. 6–12) known as gas vesicles, which are associated with buoyancy regulation (see below).

Increasingly, the principles of phylogenetic systematics are being applied to the definition of cyanobacterial taxa. Such studies have revealed that in a number of cases similar-appearing organisms, once grouped into the same genus, may not be as closely related as previously thought. Cryptic (i.e., structurally unrecognizable but ecologically significant) variants that are distinguishable only at the molecular level probably occur among cyanobacteria, as well as in other groups of algae. In view of the uncertainties regarding species delimitations, experimental work is usually conducted with well-defined strains that have been assigned a culture-collection strain designation. For example, the unicellular *Synechococcus* PCC6301 is a strain commonly used in laboratory investigations.

Ultimately, species delineation in cyanobacteria, as in other bacteria and increasingly in eukaryotic algal groups, may come to depend upon unique and defining DNA sequences; these could then be used to develop species-specific oligonucleotide probes for identification of natural samples as described in Chapter 5. However, in a review of species concepts in cyanobacteria, Castenholz (1992) cautioned that polyploidy (the occurrence of multiple genomic DNA copies per cell—as many as ten have been observed in cyanobacteria) coupled with possible variation among these multiple genomes, as well as the occurrence of horizontal DNA transfer (the uptake and genomic incorporation of foreign DNA, which results in formation of genetic chimaeras), could complicate attempts to characterize cyanobacterial species by nucleotide signatures.

Cyanobacterial Phylogeny and Evolution

The advent of molecular systematics has allowed comparison of cyanobacterial DNA sequences with

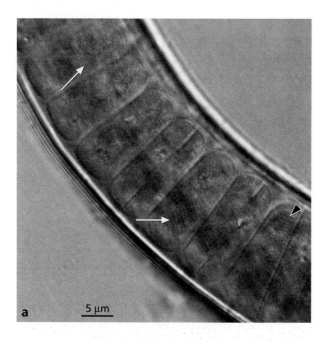

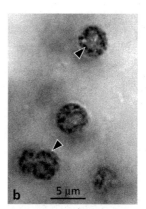

Figure 6–12 Gas vesicles in two cyanobacteria. (a) *Scytonema*, where groups of long, cylindrical gas vesicles can be seen in both side (arrows) and end-on views (arrowhead) and (b) *Microcystis*, where aggregations of gas vesicles appear as black dots (arrowheads), which give the cells a dark appearance when viewed with the light microscope.

those of other prokaryotes and eukaryotes to infer phylogeny. Analyses based upon a wide array of gene sequences show clearly that cyanobacteria constitute one of the 11 major eubacterial clades (phyla) (Fig. 6–13). This phylogenetic placement explains the many similarities in cellular structure and physiology that cyanobacteria share with other bacteria.

These similarities include prokaryotic cell construction, including absence of organelles such as nuclei, chloroplasts, mitochondria, and Golgi bodies, absence of 9+2 flagella (alternately known as cilia or undulipodia), similar cell-wall biochemistry (peptidoglycan covered by a layer of lipopolysaccharide), common occurrence of mucilaginous sheaths,

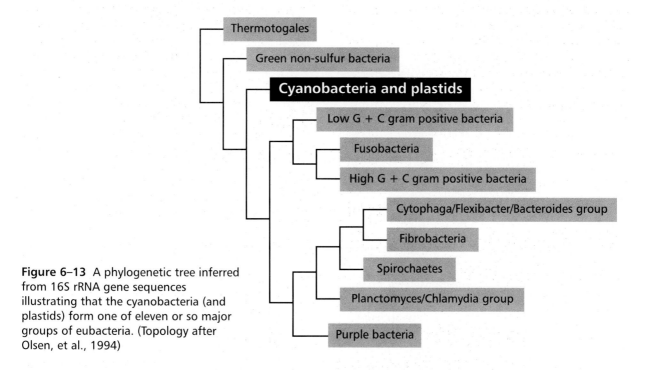

Figure 6–13 A phylogenetic tree inferred from 16S rRNA gene sequences illustrating that the cyanobacteria (and plastids) form one of eleven or so major groups of eubacteria. (Topology after Olsen, et al., 1994)

- Thermotogales
- Green non-sulfur bacteria
- **Cyanobacteria and plastids**
- Low G + C gram positive bacteria
- Fusobacteria
- High G + C gram positive bacteria
- Cytophaga/Flexibacter/Bacteroides group
- Fibrobacteria
- Spirochaetes
- Planctomyces/Chlamydia group
- Purple bacteria

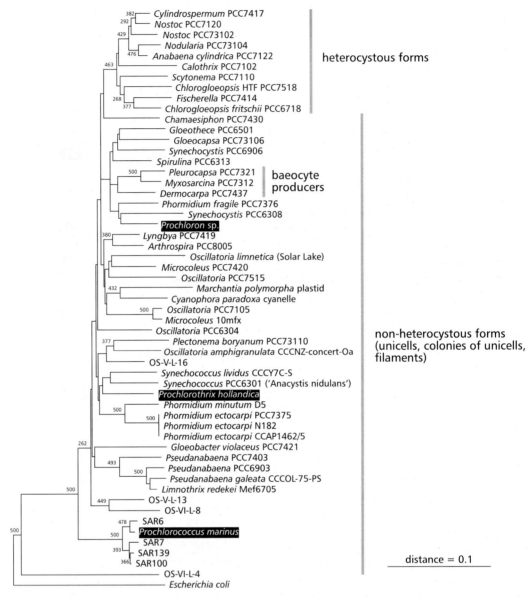

Figure 6–14 A phylogenetic tree of the cyanobacteria (and plastids, represented by the liverwort *Marchantia* and the glaucophyte *Cyanophora*) inferred from 16S rDNA sequences. Of note is the fact that the heterocyst-forming species, and possibly the baeocyte-producers, form monophyletic groups, whereas other types of cyanobacterial morphologies appear to have arisen on multiple occasions. Also note the polyphyletic nature of the chlorophyll *b*-containing forms, represented by *Prochloron*, *Prochlorothrix*, and *Prochlorococcus* (highlighted). Organisms listed as a series of letters and numbers are as yet unnamed forms known only from sequences obtained from environmental samples. The numbers on the branches represent bootstrap values based on 500 resamplings of the data (see Chapter 5). (After Wilmotte, 1994 with kind permission of Kluwer Academic Publishers)

absence of the histone proteins associated with eukaryotic DNA, and small (70S) ribosomes composed of 16S and 23S subunits. Cell division in cyanobacteria resembles that of other bacteria, with daughter cells being separated via cell-wall ingrowth (binary fission), without mitotic spindles and spindle organizing structures.

Molecular systematics has also been used to gain an understanding of phylogenetic divergence within the cyanobacteria. Most such studies have been performed using 16S rDNA (SSU) sequences. A recent tree of this type produced by Wilmotte (1994) is shown in Fig. 6–14. As noted earlier by Giovannoni, et al. (1988), unicellular (or colonial) and unbranched filaments appear to be polyphyletic, i.e., there are not separate clades composed only of unicellular (or colonial) or filamentous forms. Some unicells are more closely related to filamentous taxa than to other unicells, and vice versa. As Wilmotte (1994) suggests, the molecular changes required to maintain the integrity of filaments, as opposed to allowing separation of daughter cells after division, may be relatively slight. This might explain the occurrence of lineages containing both unicellular and filamentous taxa.

Baeocyte-forming cyanobacteria (such as *Chroococcidiopsis* and *Myxosarcina*) appear to form a clade, but the grouping of all such taxa is not strongly supported (having a bootstrap value less than 50%). In contrast, all cyanobacteria that are capable of producing heterocysts (and akinetes) form a well-supported clade, indicating common ancestry. This phylogeny also suggests that heterocyst-forming cyanobacteria, which include true-branching forms, appeared relatively late in cyanobacterial diversification, a conclusion that is supported by the fossil evidence described earlier. Figure 6–14 also suggests that chlorophyll (*a* + *b*)-containing prochlorophyte taxa (*Prochlorothrix hollandica*, *Prochloron* sp., and *Prochlorococcus marina*) are not closely related (Palenik and Haselkorn, 1992; Urbach, et al., 1992). This conclusion is supported by other sequence data: a DNA-dependent RNA polymerase (*rpoC*) gene tree also indicates polyphyletic prochlorophytes (Fig. 6–15) (Palenik and Swift, 1996). These results have important implications for the construction of evolutionarily meaningful systematic schemes and cyanobacterial taxonomy.

Most classification schemes found in textbooks on algae combine unicellular and colonial cyanobacteria within the order Chroococcales and group unbranched filamentous forms lacking specialized cells within the Oscillatoriales (e.g., Bold and Wynne, 1985; Hoek, van den, et al., 1995). The phylogenetic evidence provided by Giovannoni, et al. (1988) and Wilmotte (1994) suggests that such groupings are probably not appropriate. Limited molecular phylogenetic evidence does, however, weakly support the concept of Pleurocapsales, an order of baeocyte-producing genera advocated by Hoek, van den, et al. (1995). There is robust molecular evidence for inclusion of all heterocyst-producing cyanobacteria within

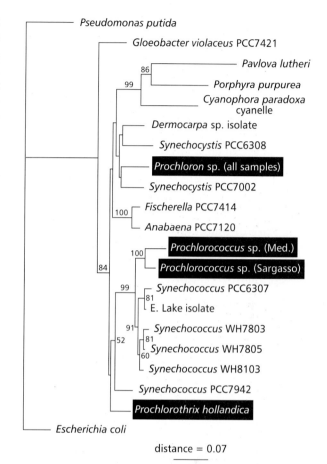

distance = 0.07

Figure 6–15 A gene tree based on sequences determined for a DNA-dependent RNA polymerase (*rpoC*). As with 16S rDNA sequence data, the prochlorophytes (shaded boxes) are revealed to be an unnatural group. (After Palenik and Swift, 1996 by permission of the *Journal of Phycology*)

the same order, supporting the concept of the order Nostocales (Hoek, van den, et al., 1995). However, relatively few sequences are available with which to evaluate the phylogenetic relevance of the true-branching Stigonematales. Additional molecular studies are needed to better clarify relationships among the cyanobacteria so that evolutionarily meaningful classification schemes can be constructed. In view of this fluid state of systematic affairs, we have organized our discussion of cyanobacterial morphological and reproductive diversity rather informally. On the basis of the strong molecular evidence cited above, chlorophyll (*a* + *b*)-producing cyanobacteria (the prochlorophytes) will not be segregated from the rest of the cyanobacteria.

DNA Exchange in Cyanobacteria

In eubacteria, DNA may move from one cell to another via one of three well-characterized processes: relatively casual transfer through the environment (transformation), transfer by infectious plasmids or viral vectors (transduction), or a highly regulated cell-cell conjugation (mating) process. The first two are regarded as examples of horizontal transfer of DNA, which can also occur in eukaryotes. Horizontal transfer represents the basis of genetic transformation (genetic engineering). Conjugation has not been observed to occur in cyanobacteria. Although natural occurrence of transformation has not been reported for cyanobacteria, possibly because thick sheaths present a barrier, lab-grown cyanobacteria can be transformed. Short pieces of DNA have been shown to move from cell to cell in rapidly growing cultures of unicellular cyanobacteria. Filamentous cyanobacteria can also be transformed with the use of vectors (see Chapter 4). In addition, the occurrence of a variety of cyanophages (cyanobacterial viruses—see Chapter 3) suggests that transduction might occur in nature, although this has not been experimentally demonstrated (Castenholz, 1992).

Photosynthetic Light Harvesting

The absorption of light energy by cyanobacteria is based upon the occurrence of one or two forms of chlorophyll, together with carotenes and phycobilins. The chemical structures of all three pigment types are characterized by rings and double bonds, which confer the ability to interact with light and to receive and transfer energy from molecule to molecule by resonance, much as a struck tuning fork causes neighboring forks to also sound. Chlorophyll *a* is the pivotal photosynthetic pigment. Accessory pigments confer extended ability to harvest light for photosynthesis and, in some cases, protection from UV and other light-induced cell damage. In most cyanobacteria, phycobilins are found in hemispherical structures on the surfaces of thylakoids.

Chlorophylls

Possession of chlorophyll *a* is a characteristic that unites all of the photosynthetic algae and plants, and is one of the characters that distinguishes cyanobacteria from other photosynthetic bacteria. Bacteriochlorophyll and chlorobium chlorophyll—found in different groups of non-oxygenic bacteria—are slightly different in chemical structure from chlorophyll *a*. Insight into evolutionary changes in chlorophyll structure that are relevant to cyanobacteria and other algae is provided by a consideration of chlorophyll biosynthesis (von Wettstein, et al., 1995) (Fig. 6–16). The first chlorophyll-specific step involves addition by the enzyme Mg^{2+}-chelatase of a magnesium ion to a four-ring, nitrogen-containing protoporphyrin. This reaction converts a red pigment into a green one. A fifth ring is then added, as is the hydrophobic phytyl tail, which anchors chlorophyll to thylakoid membranes (von Wettstein, et al., 1995).

Chlorophyll *b* occurs in at least three cyanobacterial genera (*Prochloron*, *Prochlorococcus*, and *Prochlorothrix*) as well as plastids of green algae, plants, and those (derived from green algae) of many euglenoids and the unusual protist *Chlorarachnion* and its relatives. Chlorophyll *b* can be easily formed from a chlorophyll *a* precursor molecule by a single step that involves an O_2-requiring enzymatic conversion of a methyl to a formyl group (Fig. 6–16). The relative simplicity of chlorophyll *b* synthesis (which is based on a preexisting chlorophyll *a* biosynthetic pathway) helps to explain the apparent independent evolution of chlorophyll *b* at least four times within the cyanobacterial lineage (Palenik and Haselkorn, 1992; Urbach, et al., 1992) (three prochlorophyte lineages plus the ancestor of green algal plastids, represented here by the plastids of the liverwort *Marchantia*) (Fig.

Figure 6–16 Chlorophyll biosynthesis. Four points that can be made with the aid of this illustration are that (1) at the Mg-chelatase step, the molecule switches in color from red to green; (2) oxygen is required for the formation of the fifth ring structure—an organism capable of oxygenic photosynthesis would be less likely to have this step be rate-limited by oxygen availability; (3) the phytyl tail attached by the enzyme chlorophyll synthetase serves to anchor the molecule in the thylakoid membrane; and (4) it is a relatively minor step to convert chlorophyll *a* to chlorophyll *b*, thus explaining the apparent origin of this photosynthetic pigment on multiple occasions. Also, since conversion of chlorophyll *a* to chlorophyll *b* requires oxygen, the evolution of this molecule would likely have been favored by a more oxygen-rich atmosphere. The curved arrows indicate the portions of the molecule being acted upon by the listed enzymes. (Based on von Wettstein, et al., 1995)

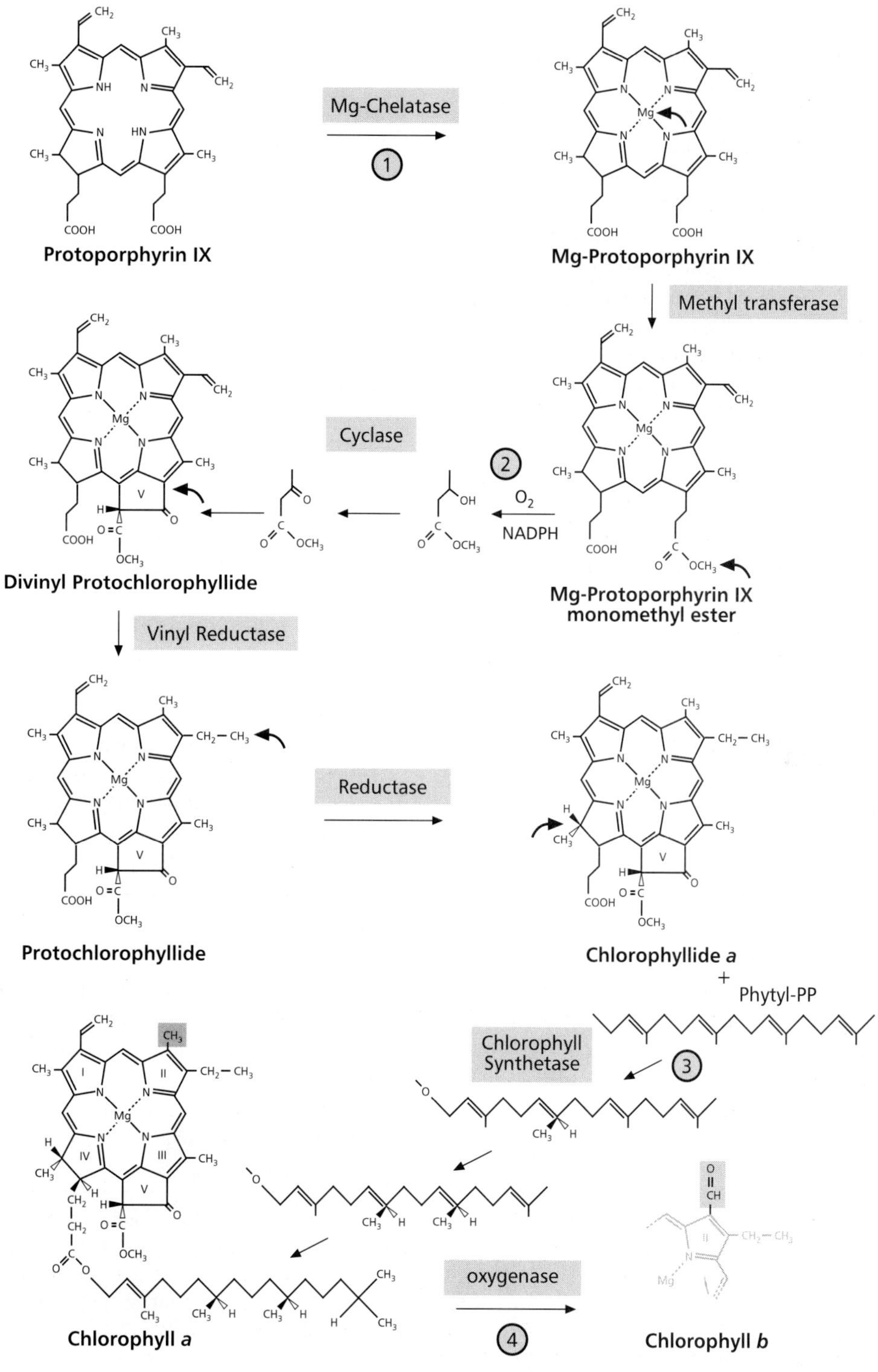

Protoporphyrin IX

Mg-Chelatase
①

Mg-Protoporphyrin IX

Methyl transferase

Divinyl Protochlorophyllide

Cyclase

②
O_2
NADPH

Mg-Protoporphyrin IX monomethyl ester

Vinyl Reductase

Protochlorophyllide

Reductase

Chlorophyllide *a*

+
Phytyl-PP

Chlorophyll Synthetase
③

Chlorophyll *a*

oxygenase
④

Chlorophyll *b*

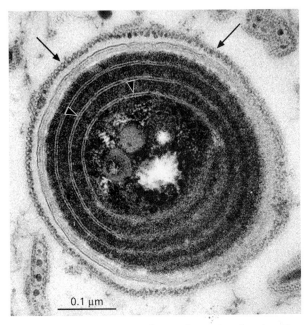

Figure 6–17 A TEM view of *Synechococcus* showing the thylakoids, which occur singly. Also note the thickened surface layer (arrows), upon which carbonates are deposited. (From Thompson, et al., 1997)

6–14). A chlorophyll *c*-like pigment has been found in *Prochlorococcus* (Goericke and Repeta, 1992) and *Prochloron* (Larkum, et al., 1994); since these are not sister taxa, an independent origin of this pigment is suggested. Chlorophyll *d*—known otherwise to occur in minor amounts in various red algae—is the major photosynthetic pigment in *Acaryochloris marina*, a cyanobacterium that has very little chlorophyll *a* and lacks phycobilisomes and phycobilin pigments (Miyashita, et al., 1996).

Cyanobacterial chlorophylls are associated with membranous thylakoids similar to those of plants and other algae. Thylakoids of most cyanobacteria (those producing only chlorophyll *a*) do not occur in stacks, but rather lie singly in the cytoplasm, typically arrayed in concentric rings at the cell periphery (Fig. 6–17). Exceptions to this generality include *Gloeobacter violaceus* PCC 7421, which contains chlorophyll *a* but lacks thylakoids, and the chlorophyll *b*-containing cyanobacteria, whose thylakoids occur in stacks of two or sometimes more (Fig. 6–18). The absence of thylakoids in *Gloeobacter* suggests that this taxon might be relatively primitive; this is also suggested by molecular evidence for a relatively early time of divergence (Figs. 6–14, 6–15). Although some experts have proposed that thylakoid stacking

depends on the presence of chlorophyll *b*, a higher plant mutant that is unable to produce chlorophyll *b* is still able to form thylakoid stacks (von Wettstein, et al., 1995). Also, thylakoid stacking has been observed in a genetically altered *Synechocystis* species lacking phycobilisomes (Swift and Palenik, 1993), suggesting that thylakoids will tend to stack when there is no spatial interference (i.e., no phycobilisomes). Chlorophyll *b* is regarded as an accessory pigment because it broadens the range of light that can be used in photosynthesis (Fig. 6–19), and transfers absorbed light energy to chlorophyll *a*.

Carotenoid Pigments

Carotenoids found in cyanobacteria include a variety of **xanthophylls**, which include oxygen in their molecular structure, and β-**carotene**, which does not. These pigments are located in thylakoids, alongside chlorophyll *a*, and like chlorophyll *b*, increase the ability of cyanobacteria to harvest blue wavelengths of light that are not directly absorbed by chlorophyll *a* (Fig. 6–19). In addition, carotenoids also provide protection from harmful photooxidation. Their association with chlorophyll prevents the formation of highly

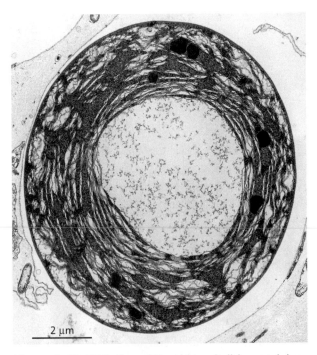

Figure 6–18 TEM view of the chlorophyll *b*-containing *Prochloron*, whose thylakoids tend to occur in stacks. (Micrograph courtesy T. Pugh and E. Newcomb)

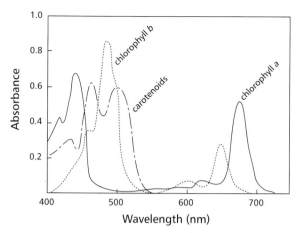

Figure 6–19 Absorption spectra for chlorophyll *a*, carotenoids, and chlorophyll *b*. The latter two pigment types act as "accessory" pigments, absorbing additional wavelengths of light and passing the energy along to chlorophyll *a*.

reactive singlet oxygen radicals that could otherwise cause irreparable damage to lipids, proteins, and other molecules (Bartley and Scolnick, 1995).

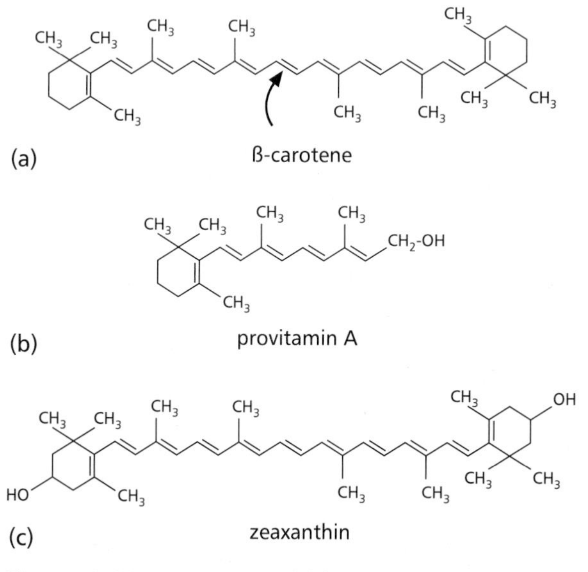

Figure 6–20 Structures of (a) β-carotene, (b) provitamin A, and (c) zeaxanthin. Provitamin A is formed by splitting β-carotene into two equal parts (at arrow). Zeaxanthin is a common xanthophyll, and is also synthesized from β-carotene. (Redrawn from BIOLOGY OF PLANTS by Raven, Evert, and Eichorn ©1971, 1976, 1981, 1986, 1992, 1999 by W. H. Freeman and Company/Worth Publishers. Used with permission.)

β-carotene (Fig. 6–20a) occurs in nearly all photosynthetic algae, and is particularly important as the main source of provitamin A (Fig. 6–20b), required by animals for synthesis of the visual pigment rhodopsin and in the regulation of genes involved in limb and skin development. β-carotene is synthesized from lycopene, which does not possess terminal rings, by the enzyme lycopene cyclase (LYC). The cyanobacterial gene for LYC, *crtL*, has been cloned and sequenced, and this information has been essential to the molecular analysis of carotenoid synthesis in higher plants. An example of a typical cyanobacterial xanthophyll is zeaxanthin (Fig. 6–20c), which is synthesized from β-carotene by hydroxylation and epi-oxidation reactions (Bartley and Scolnick, 1995). The genes and enzymes involved in carotenoid biosynthesis are reviewed by Cunningham and Gantt (1998).

Phycobilin Pigments

Cyanobacteria that do not produce chlorophyll *b* typically possess water-soluble phycobilin pigments that extend the range of wavelengths of light that can be harvested for photosynthesis (Fig. 6–21), and transfer captured energy to chlorophyll *a*. (Cyanobacteria having chlorophyll *b* do not produce phycobilin pigments.) Cyanobacterial phycobilins include the open-chain tetrapyrroles **phycoerythrobilin** and **phycocyanobilin** (Fig. 6–22), which are bound to proteins, and as such are known as phycobiliproteins. Phycocyanobilin shares the same biosynthetic pathway as chlorophyll *a* up to the point where a cyclic arrangement of tetrapyrroles is formed, after which a heme oxidase cleaves the tetrapyrrole to an open-chain arrangement. Three types of phycobiliproteins are produced by various cyanobacteria: **phycocyanin** and **allophycocyanin** (which contain phycocyanobilin), and **phycoerythrin** (containing phycoerythrobilin). However, not all taxa that contain phycobiliproteins produce all three types. Phycoerythrin-rich plankton—such as the extremely abundant open-ocean *Synechococcus* (Chisholm, et al., 1988)—can be recognized by a characteristic bright orange fluorescence. Phycoerythrin fluorescence increases with depth; very deep *Synechococcus* cells located at the lower level of the illuminated zone can fluoresce some 100 times brighter than surface cells (Olson, et al., 1990).

Absorption characteristics of the phycobiliproteins are shown in Figure 6–21. Note that they fill a gap between available light energy and the absorbance capability of chlorophyll and carotenoids (Gantt,

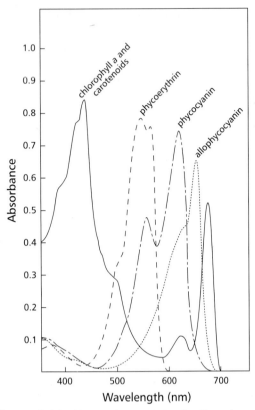

Figure 6–21 Absorption spectra for the phycobilin pigments found in non-chlorophyll *b*-containing cyanobacteria. These accessory pigments extend the range of wavelengths that can be used for photosynthesis, covering wavelengths where neither chlorophyll nor carotenoids show strong absorbances. (Redrawn from Gantt, E. *BioScience* 25:781–788. 1975. ©1975 American Institute of Biological Sciences)

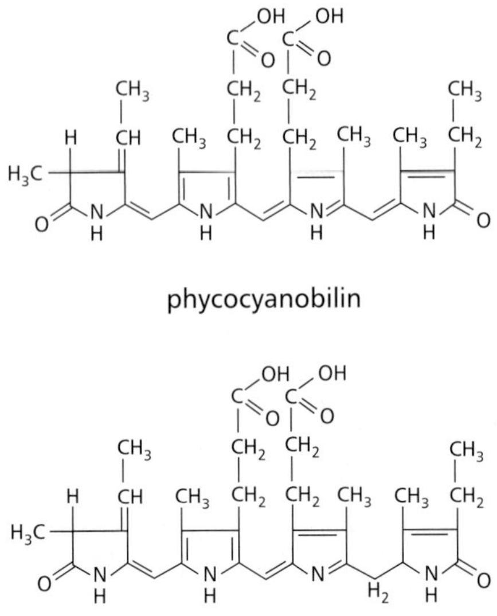

phycocyanobilin

phycoerythrobilin

Figure 6–22 Phycocyanobilin and phycoerythrobilin. Note the open-chain tetrapyrrole structure (compare with chlorophyll *a* in Fig. 6–16).

1975). Blue-green colored cyanobacteria are rich in the blue-green phycocyanin, whereas reddish cyanobacteria, such as the marine *Trichodesmium* or the freshwater *Planktothrix rubescens*, contain a higher proportion of phycoerythrin. Cyanobacteria may exhibit other color variations reflecting the presence of large amounts of UV-protective carotenoids, sheath pigments, or intracellular gas vesicles.

Phycobiliproteins occur in highly organized disk-shaped or hemispherical **phycobilisomes**, arrayed on the surfaces of thylakoids (Fig. 6–23). An exception is *Gloeobacter*, in which phycobilisomes occur on the plasma membrane because thylakoids are lacking (Ripka, et al., 1974). Phycobilisomes are nearly uniform in size with a diameter of about 40 nm, about twice the width of a ribosome. Phycobilisomes can account for as much as one quarter of the dry weight of cyanobacterial cells and make up about 40% of total soluble protein. Phycocyanin and allophycocyanin appear to be essential for the formation of phycobilisomes; phycoerythrin may be present but is not considered to be essential. The components of phycobilisomes are arranged in such a way as to maximize energy transfer to chlorophyll *a*. The protein components are responsible for aggregation of phycobilin pigments and the organization of phycobilisomes. Energy transfer efficiency, which may approach 90%, requires close association of phycobiliproteins and an arrangement that facilitates energy transfer only in one direction. On the basis of absorbance characteristics, energy transfer is expected to occur from phycoerythrin (when present) to phycocyanin to allophycocyanin to chlorophyll *a* in photosystem II (Fig. 6–24).

Models of the structure of a phycobilisome (in which all three phycobiliproteins are represented) show an outer surface consisting of fingerlike stacks (rods) composed of several phycoerythrin molecules, an inner layer of phycocyanin, and a core of allophycocyanin molecules (Fig. 6–25). This model is supported by the results of correlative laboratory studies of phycobilisome dissociation, electron microscopic imaging, and spectroscopic study of fluorescence changes in wild type cyanobacteria versus mutants

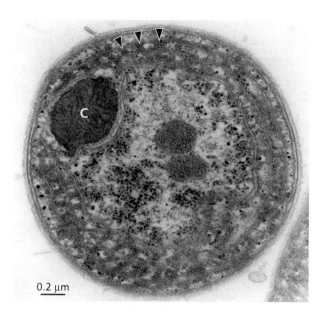

Figure 6–23 Phycobilisomes (arrowheads) in *Pseudanabaena galeata* as seen with TEM. Also note cyanophycin granule (C). (From Romo and Pérez-Martínez, 1997 by permission of the *Journal of Phycology*)

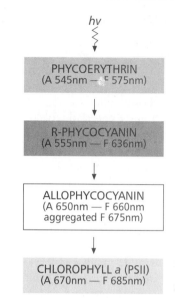

Figure 6–24 In phycobilisomes, light energy is transferred from phycoerythrin to phycocyanin to allophycocyanin and finally to chlorophyll *a* in photosystem II. (Redrawn from Gantt, E. *BioScience* 25:781–788. 1975. ©1975 American Institute of Biological Sciences)

lacking various components. When phycobilisomes are dissociated under low phosphate conditions, the three pigments are released sequentially, and when intermediate stages in the dissociation series (or mutants lacking particular components) are observed with the electron microscope, phycobilisomes appear smaller, as would be expected (Gantt, 1975; Glazer, et al., 1985). Additional details regarding light absorption and structure of cyanobacterial phycobilisomes and the genes encoding the cyanobacterial light harvesting system are found in Grossman, et al. (1993;

1995), Larkum and Howe (1997), and Falkowski and Raven (1997).

Chromatic Adaptation

Some cyanobacteria can adjust their pigment composition in response to changes in light quality. Exposure to red light increases synthesis of the blue-colored

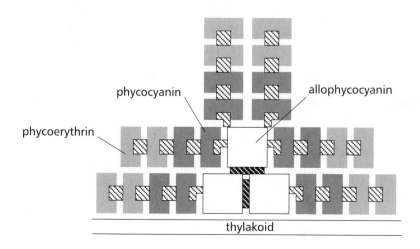

Figure 6–25 A model of the phycobilisome as inferred from several lines of evidence. Note the positions of the three types of phycobilin pigment molecules relative to the energy transfer indicated in Fig. 6–24. The pigments are held together by linker proteins (striped boxes). (Redrawn with permission from Glazer, A. N., C. Chan, R. C. Williams, S. W. Yeh, and J. H. Clark. Kinetics of energy flow in the phycobilisome core. *Science* 230:1051–1053. ©1985 American Association for the Advancement of Science

phycocyanin, whereas exposure to green light increases synthesis of phycoerythrin (Diakoff and Scheibe, 1973). Thus some cyanobacteria can undergo a color change to red under green light, and change to blue-green pigmentation in red light. Such alteration in pigment composition, known as **chromatic adaptation**, provides an adaptive advantage to cyanobacteria whose light environment may change over their lifetime. Expression of the genes that encode production of phycocyanin and phycoerythrin and their linker proteins is regulated by light quality. Cyanobacterial mutants unable to respond to light-quality changes are used to identify genes that regulate chromatic adaptation. Some of these genes encode proteins that are similar to bacterial response regulators, portions of the higher plant light-sensing pigment phytochrome, and portions of higher plant ethylene receptor proteins. When a black-colored mutant of the filamentous cyanobacterium *Fremyella diplosiphon*, which has the same pigmentation in red and green light, was transformed with such gene sequences, its ability to respond to light quality changes by altering pigment composition was restored. This evidence supports the hypothesis that the identified genes are part of the signal transduction chain that is involved in chromatic adaptation. It has been hypothesized that one of the identified regulatory proteins may be involved in sensing light-quality changes by transferring a phosphate to a particular amino acid in another regulatory protein in the series (i.e., it has kinase activity) (Kehoe and Grossman, 1996). Genes encoding proteins with similar light-sensing and kinase properties have also been identified in *Synechococcus* PCC 6803 (Wilde, et al., 1997). One of these (*cph1*) encodes a molecule described as a cyanobacterial phytochrome (Yeh, et al., 1997). This suggests that cyanobacterial chromatic adaptation is controlled by sensor kinases; such proteins are of widespread occurrence and tremendous biological significance.

Anoxygenic Photosynthesis

As noted earlier, bacteria lacking photosystem II perform photosynthesis without liberating oxygen. They use reduced compounds such as hydrogen sulfide (H_2S), hydrogen gas (H_2), or organic compounds as electron donors in photosynthesis. A number of cyanobacteria also undergo anoxygenic photosynthesis in anaerobic environments rich in hydrogen sulfide if light is available. An example is *Oscillatoria limnetica* isolated from anaerobic H_2S-rich bottom waters

of the hypersaline Solar Lake, in Israel. As a result of the reaction:

$$2H_2S + CO_2 \rightarrow CH_2O + 2S° + H_2O$$

elemental sulfur (S°) is excreted from cells and forms conspicuous granules on the cyanobacterial filament surfaces (Cohen, et al., 1975). Anoxygenic photosynthesis is characteristic of some cyanobacteria that inhabit benthic microbial mats characterized by steep gradients in sulfide, light, and oxygen (see review of cyanobacteria in microbial mats by Stal, 1995).

Carbon Fixation and Storage

In photosystem I, electrons obtained from photolysis of water (oxygenic photosynthesis) associated with photosystem II or other donors (anoxygenic photosynthesis) are further energized and used to create chemical reducing power (NADPH) and ATP. In the Calvin cycle, NADPH and ATP fuel incorporation of CO_2 into sugar. In cyanobacteria, the affinity of Rubisco for CO_2 is lower than in eukaryotic autotrophs, and hence inorganic carbon concentration systems (CCMs) occur in cyanobacteria (see Chapter 2).

The most important photosynthetic storage product of cyanobacteria is **cyanophycean starch** (also known as cyanophytan starch or glycogen), a α-1,4-linked polyglucan similar to the linear amylose portion of higher plant starch. Particles of cyanophycean starch can be visualized with the electron microscope but do not react with an iodine-iodide solution to produce a readily visible blue-black color reaction, as is typical of the branched and helically coiled amylopectin portion of green algal and plant starch. Larger **cyanophycin particles** (Fig. 6–26), often recognizable by distinctive internal structure, are composed of the amino acids arginine and asparagine. These bodies represent a store of combined nitrogen that can be mobilized under nitrogen-limiting conditions. Lipid is stored in cytoplasmic droplets, and in addition, relatively small (0.5 μm in diameter) granules containing polymers of β-hydroxybutyrate may be abundant in some cyanobacteria. Stores of phosphate are maintained as polyphosphate granules, which appear as relatively large, darkly stained, rounded particles when seen in electron micrographs (Fig. 6–26). The relative C, N, and P status of cyanobacterial cells can be estimated by assessing numbers of cyanophycean starch particles, lipid droplets, hydroxybutyrate, cyano-

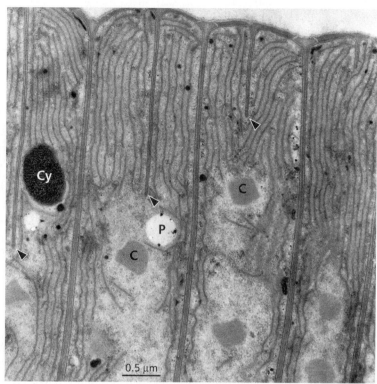

Figure 6–26 Common storage products of cyanobacteria viewed with TEM include cyanophycin particles (Cy), which store nitrogen in the form of amino acids, and polyphosphate granules (P). Arrowheads indicate cell walls in different stages of development following cell division. C=carboxysome. (From Lang and Whitton, 1973, in Carr, N. G., and B. A. Whitton [editors]. *The Biology of Blue-green Algae.* ©University of California Press)

phycin, and polyphosphate particles in electron micrographs. These inclusions and others are reviewed and illustrated by Jensen (1984).

Respiration and Heterotrophic Capabilities

Cyanophycean starch is metabolized by cellular respiration when oxygen is present, and several cyanobacteria, including *Microcystis* PCC 7806 are able to ferment stored carbohydrate when O_2 is absent (Moezelaar and Stal, 1997). A number of cyanobacteria are capable of heterotrophic growth in the dark by using organic compounds. For example, several strains of *Nostoc* are known to grow in the dark when provided with the sugars glucose, fructose, ribose, or sucrose (Dodds, et al., 1995), and some filamentous cyanobacteria also exhibit heterotrophy (Khoja and Whitton, 1975). However, heterotrophic growth of cyanobacteria is typically slower than photoautotrophic growth. This is probably related to the fact that the cyanobacterial tricarboxylic acid cycle is incomplete. Succinyl-CoA synthetase and succinoyl-CoA dehydrogenase are absent (Smith, 1973). *Cyanothece* ATCC 51142 is an

exception to the rule; it can grow rapidly in darkness by metabolizing glycerol (Schneegurt, et al., 1997). The ability to grow heterotrophically in the laboratory (under optimal conditions and in the absence of competitors) does not necessarily mean that algae normally utilize organic compounds in nature. Demonstration of naturally occurring heterotrophy should include an analysis of environmental levels of organic substrates, concurrent analysis of environmental factors that relate to carbon utilization such as levels of irradiance and dissolved inorganic carbon, demonstration that the algae have uptake systems having sufficient affinity to allow competition with other bacteria for exogenous organic compounds, demonstration that the organic compounds are absorbed by cells, and a physiological analysis of the fate of the absorbed organics (Tuchman, 1996).

Cyanobacterial Motility and Buoyancy

Cyanobacteria can change their positions within their environment in a variety of ways. Advantages conferred by mobility include achieving optimal light conditions for photosynthesis, avoiding damage caused by excess light, and obtaining inorganic nutrients. There are three basic modes of positional change: buoyancy regulation within the water column, exhibited by phytoplanktonic forms; gliding motility of cyanobacteria in contact with a substrate; and swimming through liquid without the involvement of flagella. Swimming (by an unknown mechanism) has been observed in several strains of the unicellular *Synechococcus* isolated from a number of open-ocean locations in the Sargasso Sea and temperate south Atlantic. The cells were able to move at a rate of 25 μm s^{-1} and did not change direction in response to variations in the light environment (Waterbury, et al., 1985). Further study of the swimming *Synechococcus* revealed that at least one surface protein is involved; the gene (*swmA*) encoding this protein has been cloned and sequenced. This protein is not present on surfaces of non-motile strains of *Synechococcus*, and insertional inactivation (interruption of the coding sequence) of *swmA* in strain WH8102 resulted in loss of ability to change cell position. However, some mutants can still rotate about an attachment point, suggesting the presence of additional motility factors (Brahamsha, 1996).

Various unicellular, colonial, and filamentous cyanobacteria (as well as other bacteria) have been observed to glide forward and backward at a rate up to 11 μm s^{-1} (Halfen and Castenholz, 1970; Halfen, 1973). Gliding movements may change in direction or speed in response to the light environment. Such responses can be demonstrated by spreading motile cyanobacteria evenly upon a moist surface, such as agar in a petri plate, then darkening all but a small area of the plate, which is provided with optimal irradiance conditions. After a period of time, gliding cyanobacteria tend to accumulate in the well-lighted area. Two mechanisms for gliding motility have been proposed: (1) extrusion of mucilage through small pores in the cell wall, which provide a gliding surface as well as motive force; and (2) generation of motive force by very fine, helically arranged protein fibrils extending from the cell wall. In addition to gliding, *Oscillatoria* filaments oscillate or rotate around the filament's axis. It is possible that such motions are due to a surface protein similar to that coded for by *Synechococcus swmA*. Mobility of cyanobacteria has been reviewed by Häder and Hoiczyk (1992).

Buoyancy regulation by cyanobacterial phytoplankton may involve production of intracellular

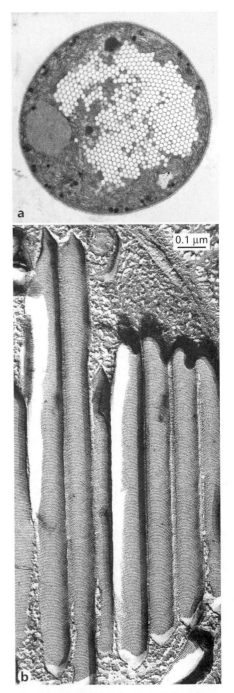

Figure 6–27 TEM views of gas vesicles. (a) Those of *Nostoc coeruleum* are seen in cross section. (Courtesy T. Jenson) (b) Those of *Nostoc muscorum*, in longitudinal view. (Reprinted with permission from Waaland, J. R. and D. Branton, Gas vacuole development in a blue-green alga. *Science* 163:1339–1341. ©1969 American Association for the Advancement of Science)

structures known as **gas vesicles** or **gas vacuoles**. Both terms are somewhat misleading—these structures are not delimited by membranes, as is suggested by the terms "vesicle" or "vacuole." Rather, they are assemblies of hollow, pointed cylinders (Fig. 6–27) whose 2 nm-thick walls are constructed of gas-permeable aggregations of a protein having a molecular weight of 20.6 kD. Gas vesicles vary in length from 62–110 μm and in width from 0.3–1 μm. While length doesn't affect strength, narrower gas vesicles are stronger than wider ones (Oliver, 1994). The concentration of gas within the buoyancy devices of cyanobacteria is no higher than that of surrounding cytoplasm. The cylinders do, however, exclude water and heavier cellular constituents. Their density (120 kg m^{-3}) is much less than that of other cell constituents (such as proteins—ca. 1300 kg m^{-3}) (Oliver, 1994). Gas-vesicle production is induced by low-light conditions. When cells have accumulated large numbers of these protein cylinders, the cytoplasm as a whole becomes less dense, and the organisms tend to float. Upward movement in the water column may increase exposure to nutrients and certainly increases light availability. If photosynthesis increases as a result, accumulation of osmotically active sugars, as well as uptake of ions, creates higher cellular turgor pressure, which may collapse the gas vesicles. This, as well as addition of ballast in the form of storage particles, cause the algae to sink in the water column. As osmotically active molecules are removed by respiration or other processes in cells located in deeper, darker waters, gas vesicles can reform, apparently by self-assembly of protein subunits, and the buoyancy cycle repeats (Fig. 6–28). In thick surface blooms, cyanobacteria may be unable to obtain sufficient resources to manufacture sugars, allowing gas vesicles to remain intact. In other cases gas vesicles are too strong to be collapsed by turgor pressure. Retention of cyanobacteria in surface waters is also favored by calm conditions. The absence of wind reduces vertical circulation of surface waters and the accompanying downward transport of cyanobacteria. Exposed to high light levels for extended periods, the algae may then succumb to photodestruction of the photosynthetic apparatus, die, and decay, with undesirable effects on water appearance and oxygen concentrations. If toxins were produced by the cells, these may be released into the water.

Aggregations of gas vesicles can refract light, causing cyanobacterial cells to appear black, or to be filled with black granules, when viewed by light microscopy.

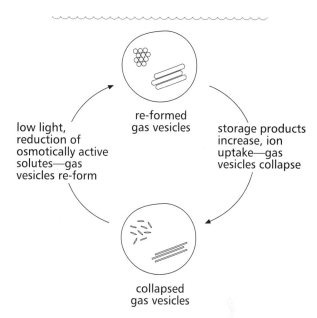

Figure 6–28 The buoyancy cycle of cyanobacteria having gas vesicles. Under low-light conditions, gas vesicles form, causing the cells/colonies to rise upward in the water column. At the surface, enhanced photosynthetic production may give rise to higher turgor pressures and accumulation of dense storage particles, causing the cells to sink. As nutrients are utilized by cells at lower depths, gas vesicles may re-form, reinitiating the cycle.

Slight pressure on the cover slip will collapse the gas vesicles, resulting in cell coloration that more accurately reflects pigment composition. Certain non-cyanobacterial prokaryotes also produce gas vesicles.

It should also be noted that some planktonic cyanobacteria appear to regulate buoyancy primarily by production of carbohydrate ballast and/or controlled reduction in the number of gas vesicles, rather than collapse of these structures. Thus, the buoyancy responses of one cyanobacterium cannot often be extrapolated to another (see review by Oliver, 1994). In some cases downward movement through the water column may occur as cyanobacteria are aggregated into iron-rich colloids.

Nitrogen-Fixation

Cyanobacteria are the only algae known to be capable of transforming molecular nitrogen gas into ammonia, which can then be assimilated into amino acids, proteins, and other nitrogen-containing cellular

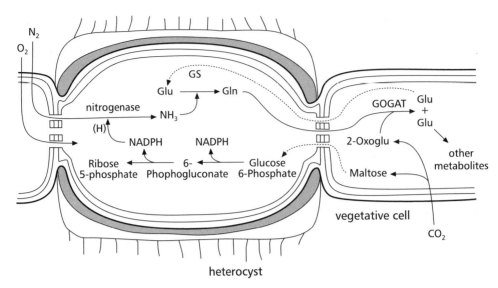

Figure 6–29 Diagram of a heterocyst and adjacent vegetative cells, illustrating the relationships between photosynthetic and nitrogen-fixation metabolism. Thick walls of heterocysts reduce diffusion of O_2 (and N_2) directly into heterocysts. Hence, entry of these gases occurs via vegetative cells and microplasmodesmata (channels) in the cell walls connecting heterocysts and vegtative cells. Such pores also allow movement of fixed nitrogen (Gln=glutamine) from heterocysts into vegetative cells. The enzyme GOGAT regenerates glutamate, which diffuses into heterocysts. Addition of NH_3 to glutamate is accomplished by glutamine synthetase (GS). The NH_3 originates from the activity of nitrogenase, using reducing equivalents (NADPH) generated from carbohydrates such as maltose or glucose-6-phosphate. These originated from photosynthetic carbon fixation in vegetative cells. (After Haselkorn, 1978 with permission from the *Annual Review of Plant Physiology and Plant Molecular Biology* Volume 29. ©1978 by Annual Reviews www.annualreviews.org)

constituents. Cyanobacteria, and a variety of other bacteria possessing this capacity, can thus avoid nitrogen-limitation of growth. The biogeochemical significance of cyanobacterial nitrogen-fixation was introduced in Chapter 2, and numerous associations of nitrogen-fixing cyanobacteria with organisms that require fixed nitrogen were described in Chapter 3. Here we will focus upon the cellular and biochemical aspects of nitrogen-fixation.

Nitrogen-fixation is an inducible process, i.e., it is regulated by environmental levels of ammonium or (to a lesser extent) nitrate. This energy-demanding process therefore does not occur if environmental levels of fixed nitrogen are sufficiently high, but is triggered by low levels. It usually occurs more rapidly in the daytime—when photosynthesis can help to generate the necessary ATP—than at night. Nitrogen-fixation is further constrained by the level of environmental oxygen, which is often high enough to inhibit activity of the nitrogenase enzyme complex. Cellular production of the principal form of this complex requires sufficient environmental levels of sulfur, molybdenum, and iron, which occur in cofactor com-

plexes. A second type of nitrogenase, which requires a vanadium cofactor, has been identified in cyanobacteria and other nitrogen-fixing bacteria. The *vnf* genes encoding this second form are repressed by molybdenum. Magnesium and cobalt ions are also required for nitrogen-fixation.

The nitrogenase complex includes two major components: (1) the Mo-Fe protein, which is a tetramer formed of two types of proteins (encoded by *nif*D and *nif*K genes), several iron-sulfur clusters, and two copies of a Mo, Fe, and S cofactor, which is the active site for nitrogen reduction, and (2) the Fe protein, which contains two copies of the *nif*H gene product along with an Fe-S cluster. Such proteins are the targets of antibodies useful in revealing the location of nitrogenase within cyanobacterial cells (Currin, et al., 1990). In the heterocyst-producing cyanobacteria, transcription of the *nif*D, *nif*K, and *nif*H genes is preceded by excision of an 11 kb nucleotide sequence with the result that these three genes are clustered and transcribed as a unit. However the 11 kb sequence is not typical of cyanobacteria lacking heterocysts; in these forms (and anoxygenic nitrogen-fixing bacteria) these three *nif* genes are always contiguous. *Anabaena variabilis* produces two different Mo-dependent nitrogenases, one that functions only in heterocysts under either anaerobic or aerobic conditions, and another that operates in both heterocysts and vegetative cells, but only under anaerobic conditions. It has been proposed that the latter is regulated by environmental conditions, and the former is developmentally regulated (Thiel, et al., 1995). These data suggest that the former is a derived form of the more ancient latter enzyme complex. Several other *nif* genes occur in cyanobacteria, some of which appear necessary for cyanobacterial nitrogen-fixation, while others do not (Lyons and Thiel, 1995). In the fully functional complex, electrons are first transferred from photosystem I (via ferredoxin) to the Fe protein, then to the Mo-Fe protein. Three successive transfers of electron pairs to Mo-bound N_2 are required to produce two ammonia molecules. The genes involved in nitrogen fixation and its regulation have been reviewed by Böhme (1998).

Cyanobacteria exhibit considerable variation in nitrogen-fixation capacities. The acetylene reduction test is used to evaluate the ability of cyanobacteria (and other bacteria) to fix nitrogen under various conditions (Stewart, et al., 1967; 1969). Some cyanobacteria appear to be incapable of nitrogen-fixation; one example is the swimming *Synechococcus* described earlier (Waterbury, et al., 1985). As previously noted, one clade of cyanobacteria is characterized by the ability to produce heterocysts. These specialized cells allow nitrogen-fixation to occur in aerobic environments by reducing oxygen toxicity effects on nitrogenase and are discussed in more detail below. Various non-heterocystous cyanobacteria are able to conduct nitrogen-fixation only under anaerobic conditions, or only at night (when oxygen is not being evolved in photosynthesis). The unicellular *Cyanothece* down-regulates photosystem II during N_2-fixation so that little O_2 is produced for a few hours of every 24-hour period, even in continuous light (Meunier, et al., 1998). The marine cyanobacterium *Trichodesmium* (formerly known as an *Oscillatoria* species), which occurs as linear tufts or spherical puffs of adherent filaments (Fig. 6–44), is unique among non-heterocystous forms in that it can conduct photosynthesis and nitrogen-fixation simultaneously. Immunolocalization was used to establish that nitrogenase occurs in some (10–40%) of the filaments of *Trichodesmium* and that there was little

nitrogenase activity by associated bacteria (Bergman and Carpenter, 1991). Microautoradiography established that carbon fixation is also of patchy distribution among trichomes (Paerl, 1994). These results suggest that there is a "division of labor" among individual cells or filaments within an aggregate or among aggregates within a population, such that photosynthetic oxygen production does not interfere with nitrogen-fixation (Paerl, 1994). Further, there appears to be a daily cycle of nitrogenase activity in *Trichodesmium*, such that it occurs only during daylight hours. This cycle is based upon an endogenous rhythm that may continue to run for up to six cycles in cultures that are subjected to continuous illumination (Chen, et al., 1996). A high requirement for iron (necessary for the synthesis of nitrogenase) is related to the production by N_2-fixing cyanobacteria of iron-binding compounds (siderochromes) (Neilands, 1967; Estep, et al., 1975; Murphy, et al., 1976; Simpson and Neilands, 1976).

Heterocyst Structure-Function Relationships

Heterocysts exhibit a number of structural and biochemical modifications that facilitate nitrogen-fixation in aerobic environments (Haselkorn, 1978) (Fig. 6–29). Many cyanobacteria produce thick mucilage, which may enclose heterocysts and function as a barrier to oxygen diffusion into cells. This was demonstrated in studies of *Nostoc cordubensis* cultured from rapidly flowing streams in Argentina where oxygen levels are high. The growth conditions of this alga can be modified such that heterocysts lacking or possessing mucilage are produced. Heterocysts coated with mucilage were able to fix nitrogen at significantly higher oxygen concentrations than were those lacking mucilage (Prosperi, 1994). Similarly, the thick walls that normally characterize heterocysts help to reduce oxygen levels within cells. Bacterial cells associated with the surfaces of heterocysts—commonly found at the junctions between heterocysts and adjacent cells— may consume oxygen, thus preventing its entry into heterocysts. A further nitrogen-fixation adaptation exhibited by heterocysts is the absence of photosystem II expression, and thus, oxygen is not generated from the splitting of water molecules in heterocysts. As a consequence, carbon fixation cannot occur within heterocysts; these cells must import carbohydrates from neighboring photosynthetic cells for respirational production of necessary ATP and reducing equivalents (Fig. 6–29). Intracellular respiration consumes oxygen, as does the oxyhydrogen reaction (reaction of hydrogen gas with oxygen to generate water). Hydrogen gas is produced by nitrogenase, which also functions as a hydrogenase.

The location of heterocysts has been widely used as a taxonomic character. In filamentous cyanobacteria such as *Anabaena*, every tenth cell or so may differentiate into a heterocyst, whereas heterocysts regularly occur basally or adjacent to branches or akinetes in other genera. The spacing of heterocysts in *Anabaena* (PCC 7120) is controlled by a gene (*PatS*) whose product is a diffusible peptide that blocks development of heterocysts from vegetative cells within the sphere of its influence (Yoon and Golden, 1998). Heterocysts are easily distinguished from vegetative cells by their pale green or nearly colorless appearance, resulting from repression of photosystem II, and by the relative absence of storage granules. Heterocysts may also differ in shape from nearby vegetative cells (Fig. 6–9).

Akinetes

The specialized cells known as akinetes are thought to function as resting cells that allow cyanobacteria to survive adverse conditions. Akinetes are produced only by the cyanobacteria that are also capable of producing heterocysts. These two types of specialized cells share a number of features that distinguish them from vegetative cells. Particular glycolipids and polysaccharides occur in the cell walls of both heterocysts and akinetes, but not vegetative cells. Also, photosystem II is inactivated in the akinetes of at least some cyanobacteria, as is typically the case for heterocysts. Finally, levels of the enzyme superoxide dismutase appear to be lower in akinetes and heterocysts than in vegetative cells of the same species. These observations suggest that early stages in the differentiation of akinetes and heterocysts may be under the control of similar genetic elements. Akinetes are typically distinguished from heterocysts, however, by absence of specializations associated with nitrogen-fixation and, often, by larger size (Figs. 6–8, 6–30). In many cases the akinete wall is distinctively ornamented and may be darkly pigmented (Fig. 6–52).

Development and germination of akinetes of *Aphanizomenon flos-aquae* were studied at the ultrastructural level by Wildman, et al. (1975). Changes that occur during development include disappearance of

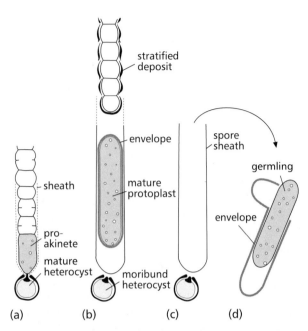

Figure 6–30 Development of the akinete of *Gloeotrichia echinulata*. (a) Differentiation begins near a heterocyst. (b) Akinetes enlarge, develop thick walls, and (c) separate from subtending vegetative filaments and heterocysts, leaving an empty sheath. (d) Germination of akinetes. (After Cmiech, et al., 1984)

gas vesicles and proliferation of storage granules (cyanophycin and glycogen). The resulting increase in cell density contributes to rapid settling to the sediments. During germination cyanophycin and glycogen are metabolized and gas vesicles reappear, permitting germlings to ascend into the plankton. The environmental stimulus for akinete formation in cyanobacteria has been controversial. Various experts have correlated different environmental factors with akinete induction, and experimental efforts designed to elucidate relationships have sometimes been confounded. Van Dok and Hart (1997) cite several lines of experimental evidence in support of the hypothesis that low cellular-phosphorus levels trigger akinete production in *Anabaena circinalis*, and probably other cyanobacteria as well. They found that akinete differentiation began when the cell phosphorus quota (QP) (see Chapter 22) fell to a critical level of 0.3–0.45 pg of phosphorus per cell. They also observed that polyphosphate granules were present in the cells of filaments lacking akinetes, but that such phosphate storage bodies were absent from filaments with akinetes, indicating phosphate depletion. In contrast, temperature was the most important factor triggering akinete formation in seven strains of phytoplanktonic *Anabaena*. There was also a positive correlation between presence of heterocysts and akinete production (Li, et al., 1997).

Cyanobacteria of Extreme Habitats

Organisms that occupy extreme environments are known as extremophiles, and for the most part are bacteria. A number of species of cyanobacteria can tolerate temperatures as high as 72° C in thermal pools of Yellowstone National Park (U.S.A.) (Fig. 6–31) and elsewhere in the world. Cyanobacteria also occur in the cold deserts of Antarctica as well as hot deserts such as the Negev (Israel). In the realm of autotrophs, cyanobacteria have unparalleled abilities to survive long periods of desiccation. For example, a dried herbarium specimen of *Nostoc* was able to resume growth more than 100 years after its collection. Various cyanobacteria produce external sheath pigmentation such as scytonemin (named for the genus *Scytonema*) and mycosporinelike amino acids, which function as UV screens, facilitating survival of cyanobacteria in high UV habitats (Dodds, et al., 1995). The yellow-brown scytonemin is produced by cyanobacteria that are exposed to intense solar radiation but whose growth is too slow to allow rapid DNA repair, such as those in Antarctic ponds (Stal, 1995). This pigment has been identified in at least 30 species. UVA (320–400 nm) is very effective in inducing scytonemin synthesis. This pigment is lipid-soluble and has an absorption maximum in the near ultraviolet, suggesting that it shields cells from UV and blue light (Garcia-Pichel and Castenholz, 1991). Scytonemin is also very resistant to microbial breakdown (Stal, 1995), suggesting that its presence in sheaths may have contributed to the formation of fossils.

The dry valleys of Antarctic's Victoria Land consist of extensive areas of rock and soil lacking snow or ice cover, possibly representing the most extreme environment on earth. Cyanobacteria occur in the mostly frozen lakes and melt streams that form in the austral summer. Here filamentous cyanobacteria such as *Anabaena*, *Lyngbya*, *Nostoc*, and *Phormidium* form benthic mats in the shoreline zone, and sometimes in the water column as well (Spaulding, et al., 1994). Cyanobacteria, including previously unknown forms,

Figure 6–31 Yellowstone thermal pools. A community of filamentous algae is evident in the lower left portion of the photograph. (Photograph courtesy D. Derouen)

have also been detected in this region by molecular sequencing methods. Such communities are regarded as possible models for life in other parts of our solar system that are also very cold (namely Mars and Europa), and for which the occurrence of seasonal meltwater is postulated (Priscu, et al., 1998). Cyanobacteria also are found at higher altitudes, where summer high temperatures reach only about 0° C, and winter temperatures drop to –60° C. Cyanobacteria can live in rock fissures and cracks and in cavities occurring in porous transparent rocks such as sandstones, granite, and marble, but not dense, dark volcanic rocks. Organisms that occupy such habitats are called cryptoendoliths (crypto=hidden, endo=inside, lith=rocks). Cryptoendolithic cyanobacteria such as *Chroococcidiopsis* live beneath rock surfaces together with cryptoendolithic lichens, fungi, and bacteria. The assemblages appear as a pigmented, subsurface layer extending horizontally. Melting snow and fog are the main sources of water (Friedmann and Ocampo, 1976; Friedmann, 1982).

Cyanobacteria are often the dominant organisms in polar lakes, ponds, and streams. Forms similar to *Oscillatoria* typically form mats or films across the bottom. Tang, et al. (1997) cultured isolates of such algae from a wide range of Arctic and Antarctic freshwater habitats and tested their ability to grow at a range of temperatures. Optimal temperatures for growth of most polar isolates ranged from 15°–35° C, suggesting that the algae were not specifically adapted for growth at low temperature but rather could tolerate suboptimal temperature conditions. Slow growth rates and tolerance to desiccation, freeze-thaw cycles, and high solar irradiation also contribute to the survival of cyanobacteria in polar aquatic systems.

Cyanobacteria also form part of the soil surface communities known as **cryptogamic crusts**, which occur over large areas of arid and semiarid lands around the world. Cyanobacteria are more common in desert crusts—on soils of pH 8–9—than they are in certain grasslands, where soil pH may be more acidic. Desert cryptogamic crusts are dominated by filamentous cyanobacteria including *Microcoleus*, *Phormidium*, *Plectonema*, *Schizothrix*, *Nostoc*, *Tolypothrix*, and *Scytonema*. These communities, which also include fungi, lichens, and mosses, are thought to enrich (via nitrogen-fixation and contribution of organic carbon) and stabilize otherwise highly erodible soils, and enhance establishment of plant seedlings. Cyanobacterial sheath mucilage aggregates soil particles and absorbs moisture. Because of these attributes, cyanobacteria are commonly included in microbial mixtures used as soil conditioners. Some people consider the concept that cryptogamic crusts are valuable in achieving long-term rangeland sustainability to be controversial. The extent to which grazing practices that destroy the crusts should be regulated merits additional study (Johansen, 1993).

Hot-desert cyanobacteria typically occupy the subsurface spaces within porous crystalline sandstone and limestone rocks, but not dark or dense rocks. Surveys conducted within arid regions of the U.S. indicate very widespread occurrence—in one study, more than 90% of sampled sandstone sites were occupied by cryptoendolithic algae. The proximity of the algal band to the surface and the thickness of the algal band varies from site to site, depending in large degree upon the color and porosity of the rock substrate. Algal biomass is greatest in the lightest colored and most porous rocks. *Chroococcidiopsis*, *Gloeocapsa*, *Gloeothece*, *Synechococcus*, *Anabaena*, *Lyngbya*, and *Phormidium* are the most common cyanobacterial cryptoendoliths in hot deserts. While some green algae can also occupy these habitats, they are less common and abundant in the most extreme conditions. At most locations, *Chroococcidiopsis* is the major form present, often occurring alone. This genus—which can survive extreme cold, heat, and aridity—may be the single autotrophic organism most tolerant of environmental extremes. Field measurements suggest that algal photosynthesis occurs only when water is present, and that when it is not, cellular metabolism becomes static. Algae such as *Chroococcidiopsis* and *Chroococcus* are not particularly well adapted to growth and photosynthesis under drought conditions but, rather, are unusually tolerant of drying (Bell, 1993).

Cyanobacteria are among the few organisms that can occupy high-temperature aquatic environments, including hot springs and thermal pools. In alkaline and neutral hot springs and streams flowing from them, cyanobacteria can form thick, colorful mats that exhibit banding patterns representing the distribution of species with different temperature tolerances (Darley, 1982). Such forms are known as thermophiles. Thermophilic cyanobacteria are defined by a temperature optimum for growth that is 45° C or greater. Some common thermophiles include *Mastigocladus* (*Fischerella*) *laminosus* and a red-brown-pigmented *Oscillatoria*, both having a worldwide distribution. *Mastigocladus laminosus* occurs in warm waters and in almost all hot springs with an upper temperature limit of 58° C. Its large, thick-walled akinetes facilitate dispersal (see review by Castenholz, 1996). *Mastigocladus laminosus*, *Phormidium fragile*, and a species of *Lyngbya* occur on hot (up to 60° C) ground in the vicinity of the active volcano Mount Erebus, Ross Island, Antarctica, where the summer air temperature is maximally –20° C (Broady, 1984). The upper temperature limit for cyanobacteria is 70°–72° C, considerably higher than for eukaryotes, but not as high as for certain thermophilic archaebacteria, which can tolerate temperatures approaching 100° C. The unicellular *Synechococcus lividus* can not only tolerate temperatures of 70° C, but appears to be adapted to such conditions, since optimal temperatures for several field populations have been determined to be close to that of the environment, and the alga will not grow at temperatures below 54° C. Most cyanobacteria are intolerant of low pH waters, disappearing as the pH approaches 5. The cyanobacteria of hot-springs environments tend not to produce the UV-protective pigment scytonemin; rapid rates of DNA repair are thought to be possible under warm, moist conditions (Stal, 1995). Cyanobacteria that are able to use sulfide as an electron donor in anoxygenic photosynthesis can tolerate the high sulfide levels present in some thermal environments, but others cannot.

Another type of extreme habitat tolerated by several cyanobacteria is hypersaline waters—those having salinities considerably higher than seawater. These include saline lakes, hypersaline marine lagoons, and solar evaporation ponds (salterns). *Aphanothece halophytica* is a unicellular, rod-shaped halophile (salt-lover) that often occurs in such habitats (Yopp, et al., 1978). Amazingly, *Synechococcus*-like forms can survive and remain metabolically active within crystalline salt deposits for as long as ten months (Rothschild, et al., 1994).

Morphological and Reproductive Diversity

Cyanobacteria exhibit an array of thallus types that is unparalleled among the prokaryotes and which approximates the morphological variation occurring in eukaryotic algal groups. A number of cyanobacterial genera are so similar in appearance to green algae, in particular, that misidentification can result. Green algae can be distinguished from similar-appearing cyanobacteria by the presence in green algae of definite green-pigmented regions (chloroplasts) and by the presence of starch, which reacts with an iodine-iodide solution to produce a blue-black product visible with the light microscope. In contrast, pigmentation of cyanobacterial cells is typically distributed homogeneously throughout the cell, and blue-black starch-iodine complexes do not form. In the generic descriptions that follow, similar eukaryotic genera will be noted.

Recent taxonomic treatments relying upon morphological characters include Anagnostidis and Komárek (1985; 1988; 1990). In view of the results of molecular studies showing that unicells, colonies, and unbranched filaments do not form distinct, monophyletic groups, we have organized the following diversity survey quite informally.

Unicellular and Colonial Forms Lacking Specialized Cells or Reproduction

SYNECHOCOCCUS (Gr. *synechos*, in succession + Gr. *kokkos*, berry) (Figs. 6–17, 6–32) is a tiny (1 μm in diameter) cylindrical unicell that is an important primary producer in the plankton of fresh and marine waters, and has also been collected from surfaces of algae and plants, films of algae on sandy beaches, outflow channels of hot springs and thermal pools, and endolithic habitats. Cell lengths are typically 2–3 times their width. There is no mucilaginous sheath. Cells may exhibit swimming motility. Both marine and freshwater forms of *Synechococcus* have been associated with "whiting" events, the production of suspended fine-grained carbonates that eventually contribute to sedimentary carbonate deposition (Hodell, et al., 1998). Such late-summer calcite precipitation and cell sedimentation also increase the flux of organic carbon to the sediments of lakes (Hodell and Schelske, 1998). A specialized cell surface layer functions as a template for carbonate deposition and is shed after it has become coated with calcium carbonate (Fig. 6–17). It is then replaced by a new surface layer and the process is repeated (Thompson, et al., 1997).

PROCHLORON (Gr. *pro*, before + Gr. *chloros*, green), is a unicellular form (Fig. 6–18) that is extensively described in a book edited by Lewin and Cheng (1989). It lives primarily in association with marine didemnid ascidians, colonial tunicates also known as sea squirts (Fig. 6–33). Ascidians are small flattened animals that attach themselves to coral rubble, mangrove roots, sea-grass blades, and other shallow

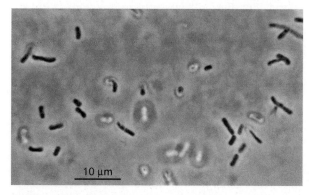

Figure 6–32 The small cylindrical cells of *Synechococcus*. (A TEM view of *Synechococcus* is shown in Fig. 6–17.)

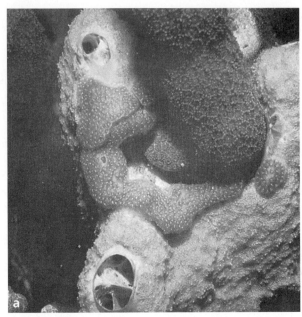
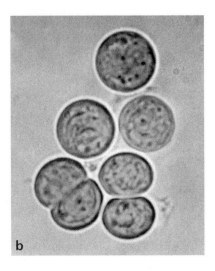

Figure 6–33 (a) An ascidian (sea squirt) containing *Prochloron*. (b) *Prochloron* cells. (a: From Lewin and Cheng, 1989; b: Photograph by R. Hoshaw from Phycological Society of America slide collection)

marine substrates in tropical regions throughout the world. *Prochloron* occurs only in coastal waters within the temperature range of 21°–31° C, and ceases photosynthesis below 20° C. The alga inhabits the animals' cloacal cavities and upper surfaces and sometimes is also found intracellularly. Although the algae release glycolate, a photosynthetic product that contributes to the animals' nutrition, the didemnids are probably not dependent upon the algae, and they require additional foods. In contrast, *Prochloron* probably does require its host in ways that are not yet understood; by itself the alga has been difficult to maintain in the laboratory. Some forms of *Prochloron* can fix nitrogen in the light, but only in association with their didemnid hosts (Paerl, 1984).

Prochloron is unusual, but not unique, among cyanobacteria in that it possesses chlorophyll *b* in addition to chlorophyll *a*, and lacks phycobiliproteins. It thus appears grass green, a color not generally found among cyanobacteria. Green pigmentation is believed to be adaptive in low-N tropical coastal waters. Production of phycobiliproteins requires about three times as much combined nitrogen as does production of the chlorophyll *a/b* protein complex. The green pigmentation may also allow enhanced absorption of the orange-red wavelengths of light present in shallow waters (Alberte, 1989). Pigments occur in stacked thylakoids, unusual for cyanobacteria. The

central portion of *Prochloron* cells is occupied by a large central space, which is also unusual for cyanobacteria (Fig. 6–18). In other ways, including cell wall and storage biochemistry, and absence of organelles, *Prochloron* resembles other cyanobacteria. At one time *Prochloron* was thought to represent the ancestor of green algal and higher plant chloroplasts, but molecular sequencing studies have since shown this to be unlikely. Recently it has been demonstrated that the chlorophyll *a/b* binding proteins of *Prochloron*, *Prochlorococcus*, and *Prochlorothrix* are closely related to cyanobacterial proteins, and not to eukaryotic chlorophyll *a/b* or *a/c* light-harvesting proteins, suggesting independent evolutionary origin of these organisms (LaRoche, et al., 1996).

PROCHLOROCOCCUS (Gr. *pro*, before + Gr. *chloros*, green + Gr. *kokkos*, berry) (not illustrated) is one of the most numerous components of open-ocean plankton and was discovered using flow-cytometric methods (Chisholm, et al., 1992). Superficially it is similar to *Synechococcus* (Fig. 6–32), but lacks phycobiliproteins and contains chlorophyll *b* in addition to chlorophyll *a*, much like *Prochloron*. However molecular sequencing studies indicate that there is not a particularly close relationship between these two chlorophyll *b*-containing taxa. In the case of *Prochlorococcus*, green pigmentation may be adaptive in harvesting the blue light that penetrates relatively deeply into ocean waters. Flow cytometry and SSU rDNA sequencing were used to establish that high- and low-irradiance ecotypes of *Prochlorococcus* occur in the North Atlantic Ocean. Although only 2% different in SSU rDNA sequence, one form grows maximally at light levels that are completely photoinhibiting to the other (Moore, et al., 1998). This work suggests that physiological diversity of the picophytoplankton may be considerably greater than previously realized.

CHROOCOCCUS (Gr. *chroa*, color of the skin + Gr. *kokkos*, berry) occupies freshwaters (particularly soft waters, including bogs), marine waters, and sometimes moist terrestrial locations. *Chroococcus* occurs as single cells or colonies of 2, 4, 16, or, less frequently, 32 hemispherical cells resulting from the adherence of daughter cells after division (Fig. 6–34). The algae may be free-floating, mixed with tangles of filamentous algae, or adherent to substrates. Many species are deep blue-green in color, but a central, paler region is often distinguishable in cells. A gelatinous sheath may or may not be noticeable. The cells of one species, *C. giganteus*, are unusually large for cyanobacteria, about 50–60 μm in diameter.

GLOEOCAPSA (Gr. *gloia*, glue + L. *capsa*, box) lives in freshwater lakes or on moist soil and other terrestrial surfaces. It can form black bands on high intertidal seacoast rocks, and live beneath rock surfaces in arid regions. It is a phycobiont in certain lichens (Rambold, et al., 1998). *Gloeocapsa* forms

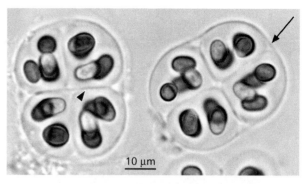

Figure 6–35 *Gloeocapsa*. Note cells occur inside of concentric sheaths (arrow and arrowhead), reflecting the pattern of cell division.

colonies having more cells than *Chroococcus*. The cells are oval or ellipsoidal, with rounded ends (Fig. 6–35). Each cell has a distinct mucilaginous sheath that is surrounded by older sheath material. These sheath layers can be used to deduce the order of cell divisions that have given rise to the colonies. Sometimes the sheath is colored yellow or brown but is most often colorless.

APHANOCAPSA (Gr. *aphanes*, invisible + L. *capsa*, box) is a free-floating member of the freshwater plankton, particularly in bog waters. Very large, usually irregularly shaped colonies consist of dozens of spherical cells evenly distributed within a colorless, or sometimes yellowish, gelatinous sheath (Fig. 6–36).

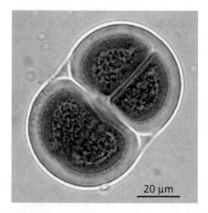

Figure 6–34 *Chroococcus turgidis*, a common bog-dwelling cyanobacterium. Following cell division, a few cells tend to remain attached to one another within the firm sheath.

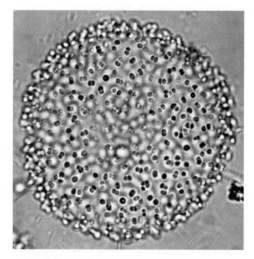

Figure 6–36 *Aphanocapsa*. Cells of this colonial cyanobacterium are spherical and without individual sheaths.

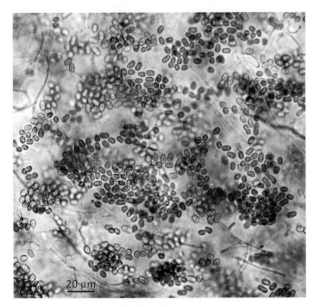

Figure 6–37 A portion of a large (ca. 2 cm in diameter), mucilaginous colony of *Aphanothece*. Note the cylindrical shape of the cells.

The individual cells are quite small, one to several micrometers in diameter, and can be pale to bright blue-green in color.

APHANOTHECE (Gr. *aphanes*, invisible + Gr. *theke*, sheath or box) occurs in the plankton of hard or soft freshwaters, or on muddy bottoms of fresh and marine waters, and sometimes in moist terrestrial sites. Very large, even macroscopic colonies contain many cylindrical cells embedded in transparent mucilage (Fig. 6–37). The mucilage of individual cells is usually indistinct. One species grows within the mucilage of other colonial cyanobacteria.

MERISMOPEDIA (Gr. *merismos*, division + Gr. *pedion*, plain) is found floating or sedentary in fresh and marine waters, including sand flats. It is a typical constituent of bog-water communities. The ovoid or spherical cells are compactly arranged in orderly rows in flat, rectangular colonies one cell thick (Fig. 6–38). The colony is held together by a mucilaginous matrix, but this sheath does not extend far beyond the cell limits, and the sheaths of individual cells are indistinct.

MICROCYSTIS (Gr. *mikros*, small + Gr. *kystis*, bladder) is a member of freshwater phytoplankton communities, but may sometimes occur as granular masses on lake bottoms. It consists of colonies of small cells (a few micrometers in diameter) that are evenly distributed throughout a gelatinous matrix. When young, the colonies are spherical, but older colonies are typically irregular in shape and frequently perforated (Fig. 6–39). In contrast to *Aphanocapsa*, whose colonies are similar, there are numerous gas vesicles in the cells of *Microcystis*, giving them a black appearance. Low numbers of *Microcystis* may occur in soft or hard oligotrophic waters. *Microcystis aeruginosa* and *M. flos-aquae* can be very common in nuisance blooms in eutrophic lakes. Differences between these two species in culture were elucidated by Doers and Parker (1988). The highest photosynthetic rates for *Microcystis* occur at surface irradiance levels, and it is resistant to photoinhibition (Paerl, et al., 1985). Thus *Microcystis* may be particularly well adapted to grow at the air-water interface where it may be able to escape carbon limitation. Phosphorus limitation does not reduce the level of gas vacuolation as much in *Microcystis* as it does in *Aphanizomenon*. The gas vacuoles of *Microcystis* are often too strong to be collapsed by turgor pressure

Figure 6–38
Merismopedia, whose cells occur in a flat sheet, often rectangular in outline (a and b). In (b), a colony is shown adjacent to a two-celled *Chroococcus*, illustrating the considerable size variation that occurs among cyanobacterial cells.

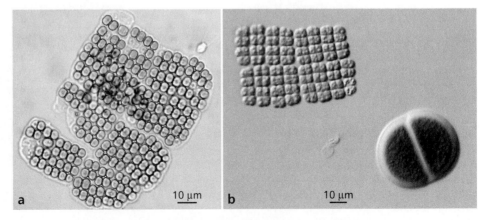

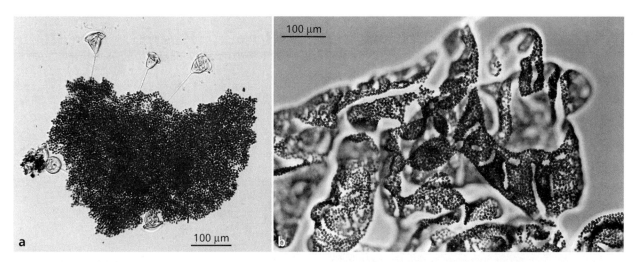

Figure 6–39 *Microcystis*, which has numerous spherical cells that form variously shaped colonies. Several vorticellid ciliates are attached to the relatively small colony shown in (a). In (b), a tortuous, perforated colony having a firm sheath is shown in an india ink preparation. Note the dark appearance of the cells, which contain numerous gas vesicles.

(see review by Oliver, 1994). These factors help to explain this organism's frequent tendency to form thick surface growths. Potassium chloride, applied at a concentration of 4–8 mM, inhibits growth of *M. flos-aquae* and may represent a potential control strategy (Parker, et al., 1997).

COELOSPHAERIUM (Gr. *koilos*, hollow + Gr. *sphairion*, small ball) (Fig. 6–40) occurs in the plankton of freshwaters, often co-occurring with *Microcystis*. *Coelosphaerium* colonies are spherical or irregular in shape, and like *Microcystis*, the cells may be so full of gas vesicles as to give them a black appearance. Unlike *Microcystis*, the cells of *Coelosphaerium* are distributed in a single layer, forming a hollow balloonlike structure. Some species occur in soft waters or bogs; these do not typically produce gas vesicles. *Coelosphaerium naegelianum* is a common constituent of nuisance surface blooms in eutrophic waters.

GOMPHOSPHAERIA (Gr. *gomphos*, wedge-shaped nail + Gr. *sphaira*, ball) occurs in hard- or soft-water lakes as a colony of spherical or heart-shaped blue-green cells mounted at the ends of gelatinous branching stalks, the whole enveloped by colorless mucilage (Fig. 6–41). The heart-shaped appearance of dividing cells results from inward growth of the outer wall, while ingrowth of the wall closest to the stalk is delayed. *Gomphosphaeria* can be confused with very similar colonies of the green algae *Dictyosphaerium* and *Dimorphococcus*, which, however, lack a mucilaginous envelope.

Filamentous Cyanobacteria Lacking Spores, Heterocysts, or Akinetes

As noted earlier, these forms do not form a clade separate from the unicellular and colonial cyanobacteria described above, but are grouped here as a matter of convenience for the reader.

PROCHLOROTHRIX (Gr. *pro*, before + Gr. *chloros*, green + Gr. *thrix*, hair) (Fig. 6–42) has been collected from shallow lakes arising from peat extrac-

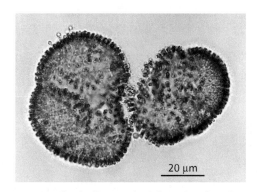

Figure 6–40 The hollow, colonial *Coelosphaerium*. (Photograph courtesy C. Taylor)

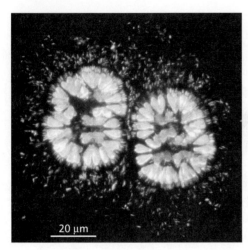

Figure 6–41 Two *Gomphosphaeria* colonies, each with a halo of bacteria. Note the heart-shaped cells.

tion in the Netherlands. Filaments are composed of elongated cells. *Prochlorothrix* is notable for having chlorophyll *b* in addition to chlorophyll *a*, and for the absence of phycobiliproteins, as is also the case for *Prochloron* and *Prochlorococcus* (Turner, et al., 1989). However *Prochlorothrix* does not appear to be closely related to either. It can readily be cultured in the laboratory (Burger-Wiersma, et al., 1986; Matthijs, et al., 1989).

OSCILLATORIA (L. *oscillator*, something that swings) has been found in a wide variety of environments, including hot springs, marine habitats, and lakes of temperate, tropical, and polar regions, as well as moist terrestrial substrates. The name reflects this organism's ability to rotate, or oscillate. In addition,

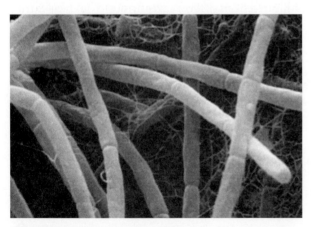

Figure 6–42 *Prochlorothrix*. (Photograph by S. Seufer, from Matthijs, et al., 1989)

gliding motility occurs. Disk-shaped cells that are typically wider than long are attached end to end to form filaments (Fig. 6–43). The end cell may be rounded or distinctive in ways that are used to define some of the species. Breakage of filaments between separation disks releases shorter hormogonia during vegetative reproduction. Filaments may sometimes adjoin in parallel to form thin films, but a readily identifiable sheath is not present, in contrast to two very similar genera, *Lyngbya* and *Phormidium*, which possess identifiable sheaths. *Phormidium* forms mats that do not readily dissociate as do those of *Oscillatoria*, and sheaths of *Phormidium* tend to be looser than the relatively rigid, distinct sheaths of *Lyngbya*, which can form thick nuisance growths in warm, high-pH lakes. However the extent to which these three genera should be separated is controversial. The planktonic species, *O. rubescens* and *O. agardhii* var. *isothrix* were changed to *Planktothrix rubescens* and *P. mougeotii*, respectively. Some other species of *Oscillatoria* have been transferred to *Limnothrix*, *Tychonema*, and *Trichodesmium*.

TRICHODESMIUM (Gr. *thrix*, hair + Gr. *desmos*, bond) is common in tropical open-ocean waters, sometimes forming visible blooms. Colonies consisting of aggregated filaments (Fig. 6–44) may be

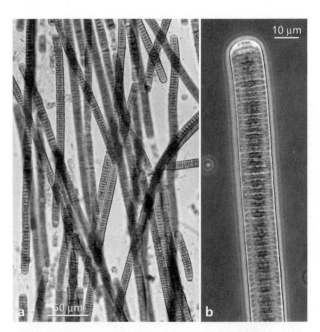

Figure 6–43 (a) A number of *Oscillatoria* filaments. (b) The tip of a filament at higher magnification. The cells are much wider than long.

Figure 6–44 *Trichodesmium* can occur as (a) tufts formed by parallel arrays of filaments, or (b) puffs—radial arrays of filaments. (Photographs courtesy S. Janson)

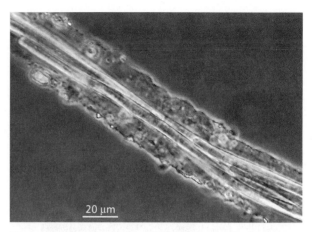

Figure 6–46 *Microcoleus*. Note the multiple filaments within the wide sheath.

visible with the naked eye, from 0.5–3 mm in diameter. It is an important phytoplankter in the tropical North Atlantic Ocean in terms of standing crop and productivity. Because it fixes significant amounts of nitrogen (30 mg m^{-2} day^{-1}), it is a major source of combined nitrogen in these oligotrophic waters. Biomass can be even higher in other oceans; the mean annual density in the Indian Ocean near Madagascar is over 16,000 colonies m^{-3}. The presence of abundant gas vesicles reduces algal density, and most of the biomass of this alga occurs in the upper 50 meters of water. Some copepods and mesograzers fish feed on *Trichodesmium* (Carpenter and Romans, 1991). The biology of *Trichodesmium* has been reviewed by Capone, et al. (1997).

SPIRULINA (Gr. *speira*, coil) is found in marine and fresh waters, including lakes high in sodium carbonate (natron lakes) and bog lakes, and on moist mud. *Spirulina* (Fig. 6–45) looks like a highly coiled version of *Oscillatoria* and exhibits similar oscillatory

and gliding motility. The genus *Arthrospira* should probably be included in *Spirulina*. The use of *Spirulina* as a human food supplement is described in Chapter 4.

MICROCOLEUS (Gr. *mikros*, small + Gr. *koleos*, sheath) occurs on the surface of silty sand in marine salt marshes and lagoons, in freshwater lakes and on their muddy banks, and in terrestrial cryptogamic crust communities. Several–many trichomes occupy a single, relatively stiff communal sheath (Figs. 6–10, 6–46). *Microcoleus chthonoplastes* (identified by its SSU rDNA sequence signature) occurs in marine microbial mats around the world (Stal, 1995).

Exospore-Producing Cyanobacteria

CHAMAESIPHON (Gr. *chamai*, dwarf + Gr. *siphon*, tube) (Fig. 6–47) is a common inhabitant of the surfaces of aquatic plants, algae, and nonliving

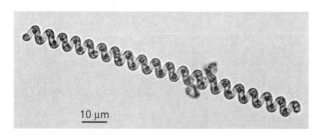

Figure 6–45 *Spirulina*.

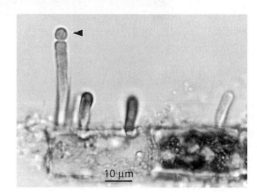

Figure 6–47 A single exospore (arrowhead) has formed on the largest of these *Chamaesiphon* individuals.

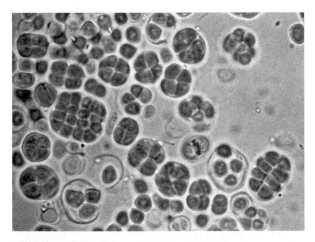

Figure 6–48 *Chroococcidiopsis*. Note the presence of baeocytes. (Photograph courtesy E. I. Friedmann)

substrates such as stones or shells in freshwater lakes and streams. The cells are cylindrical, with an open, widened sheath at the top, from which are budded off a sequence of rounded spores. The pigmentation is often grayish green.

Baeocyte- (Endospore-) Producing Cyanobacteria

CHROOCOCCIDIOPSIS (*Chroococcidium*, a genus of cyanobacteria [Gr. *chroa*, color of the skin +

Gr. *kokkos*, berry] + Gr. *opsis*, likeness) occurs in extremely arid, cold, or hot terrestrial environments as unicells (Fig. 6–48). Reproduction is by subdivision of the cytoplasm into numerous small baeocytes. These escape by breakage of the parental cell wall and each can develop into a mature vegetative cell. A number of other unicellular and filamentous baeocyte-producing genera have been defined but are not described or illustrated here.

Heterocyst and Akinete-Producing Cyanobacteria

NOSTOC (name given by Paracelsus, Swiss physician) primarily occurs in terrestrial habitats, frequently in association with fungi in lichens and with bryophytes and vascular plants of various kinds (see Chapter 3). It occurs on moist rocks and cliffs, on alkaline soils, and in wet meadows and at the edges of shallow lakes. *Nostoc* is frequently abundant in flooded rice paddies, where it contributes to the fertilization of some two million hectares. Characteristically bent or kinked filaments of round cells are held in a firm mucilaginous matrix to form colonies that can reach 50 cm in diameter, but are more commonly marble size or smaller (Fig. 6–49). The sheaths can be colored yellow, brown, or black. Transiently motile hormogonia are produced; these typically have heterocysts at the ends. Akinetes are characteristically produced midway between heterocysts, with production

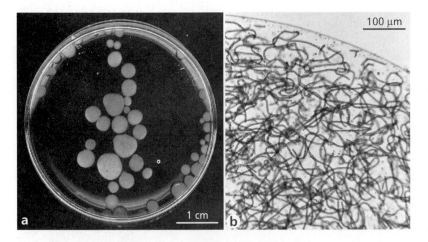
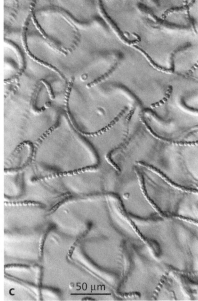

Figure 6–49 (a) *Nostoc* often forms macroscopic balls; the periphery of one is shown at higher magnification in (b). (c) The typical "kinked" appearance of the filaments is evident at greater magnification.

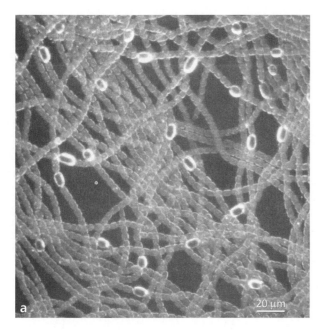

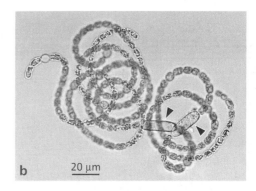

Figure 6–50 *Anabaena*. In (a), filaments are seen under dark-field illumination, where they often fluoresce a bright red color; the clear heterocysts stand out against this background. In (b), two elongate akinetes (arrowheads) are visible in a specimen taken from a freshwater bloom of the three notorious bloom-forming taxa, *Anabaena*, *Aphanizomenon*, and *Microcystis*.

of additional akinetes proceeding centrifugally. These features help to distinguish *Nostoc* from *Anabaena*, whose akinetes are produced from cells adjacent to heterocysts. The ecology of *Nostoc* was reviewed by Dodds, et al. (1995).

ANABAENA (Gr. *anabaino*, to rise), sometimes spelled *Anabaina*, is primarily planktonic in freshwaters and marine waters such as the Baltic Sea, where some forms can be nuisance bloom-formers, and toxin-producers. Like *Nostoc*, *Anabaena* consists of filaments of spherical cells resembling closely strung beads embedded in a mucilaginous matrix (Fig. 6–50). A number of features are used to distinguish the two genera: looser, more indistinct sheaths in *Anabaena*, with less-constricted filaments; habitat differences;

and differences in the location of akinetes. In addition, hormogonia of *Anabaena* are said to be more consistently motile than those of *Nostoc*, but this characteristic may best observed in cultures. *Anabaena flos-aquae* is one of the more common species, often found in bloom-causing numbers. Several species, such as the common *A. spiroides* var. *crassa*, are distinctively twisted or helically coiled.

APHANIZOMENON (Gr. *aphanizomenon*, that which makes itself invisible) (Fig. 6–51) consists of aggregations of filaments that can be so large as to be

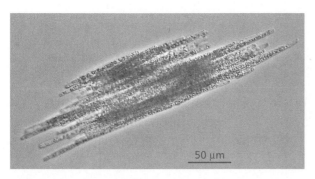

Figure 6–51 A small colony of *Aphanizomenon*, whose filaments are straight and tend to lie in parallel.

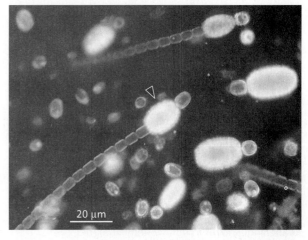

Figure 6–52 A large, ornamented akinete (arrowhead) lies adjacent a basal heterocyst in *Cylindrospermum*.

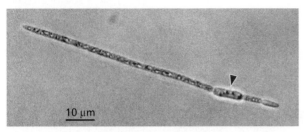

Figure 6–53 *Cylindrospermopsis*. Note the somewhat more rectangular cells than *Cylindrospermum*, and the akinete (arrowhead) one cell removed from the basal heterocyst. (From Chapman and Schelske, 1997 by permission of the *Journal of Phycology*)

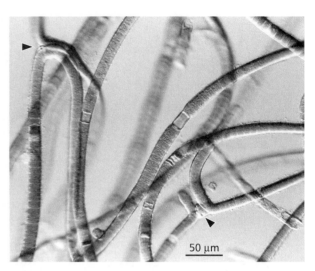

Figure 6–55 *Scytonema*. Note two occurrences of double false branching (arrowheads).

visible with the naked eye. Large blooms may form in the Baltic Sea and in eutrophic fresh waters.

CYLINDROSPERMUM (Gr. *kylindros*, cylinder + Gr. *sperma*, seed) occurs in freshwaters including soft, acid lakes, often forming dark green patches on submerged vegetation and on moist soil. Filaments of vegetative cells, whose profiles are square or rectangular, are terminated by a basal heterocyst, and enclosed in loose mucilage. A large, highly ornamented akinete may develop from the lowermost vegetative cell (Fig. 6–52).

CYLINDROSPERMOPSIS (*Cylindrospermum* + Gr. *opsis*, likeness) is an important nitrogen-fixing,

planktonic bloom-former in eutrophic temperate and tropical freshwaters around the world. It is often unrecognized because it does not appear in major taxonomic keys. Rectangular cells occur in linear or coiled filaments having a basal heterocyst. Akinetes form basally, but not immediately adjacent to the heterocyst (Fig. 6–53), as in *Cylindrospermum* (Fig. 6–52). In some locales *Cylindrospermopsis raciborskii* has in recent years replaced preexisting bloom-forming cyanobacteria, forming prodigious growths with concentrations reaching nearly 200,000 filaments ml^{-1} (Chapman and Schelske, 1997).

TOLYPOTHRIX (Gr. *tolype*, ball of yarn + Gr. *thrix*, hair) is planktonic or found entangled among submergent vegetation in freshwater lakes, including soft or acidic waters. Filaments are enclosed in a sheath of variable consistency and are highly false-branched, usually at a heterocyst (Fig. 6–54). Single false branches are formed by the continued growth of the filament on one side of the heterocyst but not the other. Sometimes double false branches occur when both ends of an interrupted filament continue to grow. The frequency of false branching gives these algae a wooly appearance, hence the name. Some forms are difficult to distinguish from *Scytonema* (below). Whether or not *Tolypothrix* and *Scytonema* should be considered separate genera is controversial.

SCYTONEMA (Gr. *skytos*, leather + Gr. *nema*, thread) forms dark tufted mats in masses of other

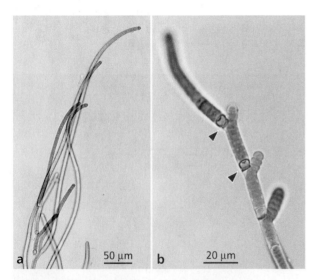

Figure 6–54 *Tolypothrix*, which exhibits single false branching, shown in (a) low and (b) higher magnification views. Branches tend to occur adjacent to heterocysts (arrowheads). (see also Fig. 6–11)

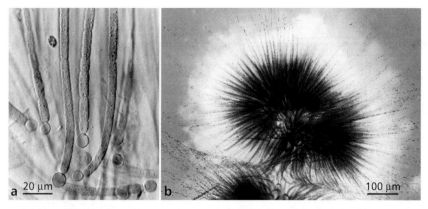

Figure 6–56 *Gloeotrichia*. In (a), the basal heterocysts are evident. The specimen in (b) was stained with india ink to show the extensive mucilage surrounding the colony.

algae or vegetation in lakes of various types or on terrestrial surfaces, including stones, wood, and soil. It is a phycobiont in several orders of lichens (Rambold, et al., 1998). Often, but not always, double false branches occur. This results from the outgrowth from both ends of a filament that has been interrupted by death of a cell or, less commonly, by heterocyst differentiation (Fig. 6–55). Heterocyst walls may be darkly pigmented. A tough and sometimes clearly layered, brown or orange-colored sheath is common.

GLOEOTRICHIA (Gr. *gloia*, glue + Gr. *thrix*, hair) occurs in freshwater habitats, attached to submerged substrates. Akinetes occur adjacent to basal heterocysts (sometimes appearing in chains) (Fig. 6–56). *Gloeotrichia* colonies often detach from substrates, becoming planktonic. Growths can reach bloom proportions.

True-branched Cyanobacteria

STIGONEMA (Gr. *stizo*, to tattoo + Gr. *nema*, thread) most commonly inhabits moist rocks and soil, but some species are aquatic, attached to submerged wood or entangled among other algae. It may form brown tufted mats or cushions on submerged portions of lake macrophytes. *Stigonema* is the primary phycobiont in certain lichens (Rambold, et al., 1998). The filaments are more than one cell in width (multiseriate) and are true-branched, i.e., branches arise by division of cells in a direction perpendicular to that of the main filament axis (Fig. 6–57). The branches may also be pluriseriate. *Fischerella* occurs in the same habitats as *Stigonema*, and these two taxa are very difficult to distinguish. One difference is that the branches of *Fischerella* are not pluriseriate.

Recommended Books

Bryant, D. A. (editor). 1994. *The Molecular Biology of Cyanobacteria*. Kluwer Academic Publishers, Dordecht, Netherlands.

Whitton, B. A. and M. Potts (editors). 1999. *The Ecology of Cyanobacteria*. Kluwer Academic Publishers, Dordecht, Netherlands.

Stal, L. J. and P. Caumette. 1994. *Microbial Mats: Structure, Development, and Environmental Significance*. NATO, Springer-Verlag, Heidelberg, Germany.

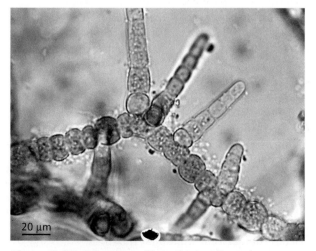

Figure 6–57 *Stigonema*. Note the true branches.

Chapter 7

Endosymbiosis and the Origin of Eukaryotic Algae

With a Focus on Glaucophytes, Chlorarachniophytes, and Apicomplexans

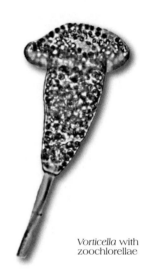

Vorticella with zoochlorellae

For more than a billion years the only autotrophic life on earth consisted of various groups of prokaryotes. An extraterrestrial visitor during this time would likely have observed landscapes covered only by thin films of terrestrial cyanobacteria (Horodyski and Knauth, 1994) in addition to somewhat more dramatic, but no more evolutionarily advanced, underwater stromatolitic mounds and columns. Then, perhaps two billion years ago, the earliest eukaryotes appeared—this point in time suggested by fossil and molecular evidence (reviewed by Knoll, 1992). A subsequent rapid radiation of more complex forms—first in the oceans, significantly later on land—radically transformed the environment, creating biospheres more like those of modern times. Eukaryotic algae played important roles in these ecological shifts; of particular significance was the origin of a terrestrial flora of diverse land plants, which were monophyletically derived from a green algal ancestor (Chapter 21).

Eukaryotes have achieved greater organismal complexity than is possible in prokaryotes because their cells are more complex. They possess metabolic compartmentation, which was conferred by the acquisition of organelles. Other features of eukaryotes that are lacking in prokaryotes include a cytoskeleton and highly regular division process (mitosis), robust motility systems (9+2 flagella—also known as cilia or undulipodia), the endomembrane system (nuclear envelope, endoplasmic reticulum, and Golgi bodies), and in many cases, an organized method for accomplishing genetic interchange (sexual reproduction). Molecular evidence suggests that the eukaryotic nucleoplasm—including its transcription and translation equipment—arose from within the Archaea (Archaebacteria) (Brown and Doolittle, 1995) (Fig. 7–1). However early eukaryotic cells soon acquired one or more eubacterial endosymbionts—formerly free-living bacteria that became permanent residents within eukaryotic cells. Molecular and structural evidence indicates that these bacteria underwent a gradual modification, becoming the organelles known as mitochondria and chloroplasts (plastids) (Fig. 7–2).

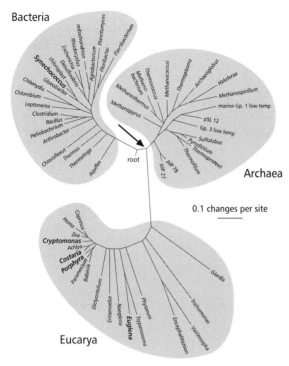

Figure 7–1 Relationships of the three major groups of life on earth as inferred from SSU rRNA gene sequences. Evidence suggests that the eukaryotes (Eucarya) and Archaea share a common ancestor (arrow) and are more closely related to each other than either is to the Bacteria. Organisms traditionally considered algae are in bold. (Redrawn with permission from Pace, N. R. A molecular view of microbial diversity and the biosphere. *Science* 276:734–739. ©1997 American Association for the Advancement of Science)

The process of incorporation and integration of bacterial endosymbionts within host cells to form eukaryotic, organelle-containing entities is called **primary endosymbiosis**. Over time, much of the genetic material of the bacterial endosymbionts became integrated into the host's nuclear genome. Complex eukaryotes that exist today thus represent genetic chimaeras; the genomes of more than one type of organism are combined in the derivative organism.

There are many modern-day examples of the occurrence of endosymbiotic bacteria within eukaryotic hosts that serve as models for understanding the adaptive advantages and early stages in establishment of primary endosymbiosis. Examples of endosymbiotic bacteria (including cyanobacteria) in the cells of algae and a wide variety of other eukaryotes are provided in this chapter. Although the mitochondria of various

algal groups are discussed, this chapter primarily focuses on the origin of algal plastids. Particular attention is paid to the glaucophyte algae (Glaucophyta)—freshwater eukaryotes whose plastids are strikingly similar to cyanobacteria; it is virtually certain that glaucophyte plastids were obtained via primary endosymbiosis. Other algal groups thought to have arisen by primary endosymbiosis are the red algae (Chapter 16) and the green algae (Chapters 17–21) (Table 7–1).

During the evolutionary radiation of the protists (Fig. 7–3), several (eukaryotic) algal lineages arose through acquisition by heterotrophic protists of photosynthetic endosymbionts that were themselves eukaryotic, further increasing cell structural and genomic complexity. The process by which eukaryotic cells are taken up and integrated into host cells is known as **secondary** or **tertiary endosymbiosis** (Figs. 7–4, 7–26). There are many examples of eukaryotic, endosymbiotic algae that can serve as models for understanding secondary and tertiary endosymbiosis, several of which will be described in this chapter. Algal groups whose plastids were acquired by secondary endosymbiosis include euglenoid flagellates (Chapter 8), cryptomonads (Chapter 9), haptophytes (Chapter 10), and ochrophytes (Chapters 12–15) (Table 7–1). In this chapter particular attention is paid to the chlorarachniophytes, a group of marine amoebae with plastids obtained by secondary endosymbiosis of green algal cells, and the apicomplexans, disease-causing protistan parasites that likewise possess plastidlike structures derived from red or green algae.

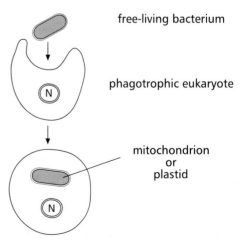

Figure 7–2 A diagrammatic representation of the process of primary endosymbiosis, in which a free-living bacterium is incorporated into a phagotrophic eukaryotic cell and eventually transformed into an organelle.

Table 7-1   Characteristics of the mitochondria and endosymbiotic origin of plastids of the major eukaryotic algal groups

group	mitochondrial cristae	plastid origin(s)
Glaucophytes	flattened	primary
Cryptomonads	flattened	secondary (red)
Red algae	flattened	primary
Green algae	flattened	primary
Euglenoids	disk-shaped	secondary (green)
Chlorarachniophytes	tubular	secondary (green)
Haptophytes	tubular	secondary (red)
Dinoflagellates	tubular	mainly tertiary (various sources)
Apicomplexans	tubular	secondary (green or red)
Ochrophytes	tubular	secondary (red)

Many dinoflagellates (Chapter 11) lack plastids altogether. Among dinoflagellates that have plastids, at least some are thought to have obtained them through tertiary endosymbiosis involving ochrophytes (chrysophyceans or diatoms) or cryptomonads (Table 7–1). Several intriguing examples are discussed in this chapter. We will also consider examples of **kleptoplastids**—plastids that have been harvested by heterotrophic protists or animals for temporary use but which are not stably integrated into host cells. All modern algal groups that possess stably integrated plastids inherited them from ancestors that first acquired them either by primary, secondary, or tertiary endosymbiosis. Endosymbiosis has thus been a common theme in the evolutionary origin of the various groups of algae and their chloroplasts.

Some experts have advocated use of the terms endocytobiont and endocytobiosis in place of endosymbiont and endosymbiosis, particularly in cases where the occurrence of mutualisic or other interactions commonly considered to reflect symbiosis have not been demonstrated. Here we have followed Reisser (1992a) in accepting de Bary's (1879) original definition of symbiosis, "living together of dissimilar organisms," and a more modern restatement, "living

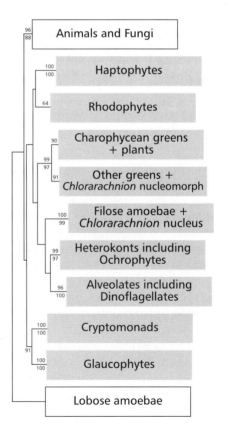

Figure 7–3 Phylogeny of eukaryotes inferred (by the maximum likelihood method) from nuclear-encoded SSU rDNA sequences. These gene sequences reflect relationships of the host nucleocytoplasm rather than relationships among plastids. Bootstrap values (see Chapter 5) above the nodes are results from parsimony analysis, whereas those below the nodes are from distance analysis of the same data. Conclusions mentioned in the text that are made from this phylogeny include the following: (1) green, red, and glaucophyte algae do not appear to be sister groups, (2) the nucleomorph of *Chlorarachnion* groups with the green algae, (3) *Chlorarachnion* is closely related to filose amoebae (Euglypha) that lack plastids, and (4) the glaucophytes and cryptomonads together form a fairly well-supported clade. Algal groups are in shaded boxes. (After Medlin and Simon, 1998 with kind permission of Kluwer Academic Publishers)

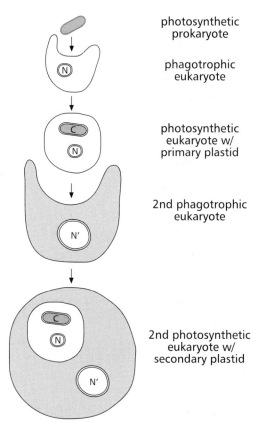

photosynthetic
prokaryote

phagotrophic
eukaryote

photosynthetic
eukaryote w/
primary plastid

2nd phagotrophic
eukaryote

2nd photosynthetic
eukaryote w/
secondary plastid

Figure 7–4 Diagram of secondary endosymbiosis, whereby a eukaryote that had earlier acquired a plastid via primary endosymbiosis is itself taken up by a second eukaryote. (After McFadden, 1993)

together in physical contact of organisms of different species," as well as extending the definition to intracellular associations (Reisser, 1992b). Our use of the term endosymbiosis does not imply mutual benefit or harm to the associated species and allows us to consider interesting and potentially significant intracellular associations for which metabolic or other interactions have not been investigated.

Phagotrophy (the ingestion of particulate food—see Chapter 1) is regarded as the major mechanism by which algal plastids were acquired, and it has been proposed that the earliest photosynthetic eukaryotes in all of the major algal lineages were phagotrophic. The selective pressure most strongly associated with endosymbiont acquisition is suggested to be the need for organic carbon and energy provided by intact endosymbionts rather than the inorganic nutrients that could be obtained by digesting ingested algae (Raven, 1997a). The engulfment and utilization of particulate food requires three important cellular features that are

absent from bacteria, but present in many protists: (1) a cell membrane that has been rendered flexible by incorporation of sterols, (2) a specialized cytoskeletal apparatus (often involving microtubules) for capturing prey, and (3) a process (endocytosis) by which particles are engulfed by invaginated host plasmalemma. Because sterol biosynthesis requires molecular oxygen, neither phagotrophy nor endosymbiosis is likely to have occurred widely prior to the buildup of O_2 in the earth's atmosphere. Furthermore, oxygen is required for aerobic respiration in most eukaryotes, and they also require the UV protection conferred by ozone, a product of atmospheric oxygen (Chapter 2). Thus the first occurrence of eukaryotes is correlated with the buildup of oxygen in the atmosphere and was hence dependent on preexisting oxygen producers, the cyanobacteria. The fossil evidence for the earliest occurrences of eukaryotic algae and earliest multicellular eukaryotic algae will next be considered.

Precambrian Fossils Attributed to Eukaryotic Algae

The oldest remains claimed to represent eukaryotic algae are numerous coiled specimens as large as 0.5 m long and 2 mm in diameter, known as *Grypania spiralis* (Fig. 7–5) occurring in 2.1 billion-year-old iron forma-

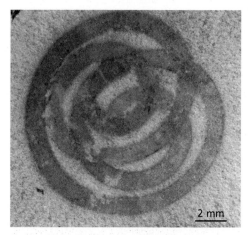

2 mm

Figure 7–5 The fossil known as *Grypania* from the 1.4 billion-year-old Geoyuzhuan Formation in China, a putative early eukaryote. (Reprinted with permission from Knoll, A. H., The early evolution of eukaryotes: A geological perspective. *Science* 256:622–627. ©1992 American Association for the Advancement of Science)

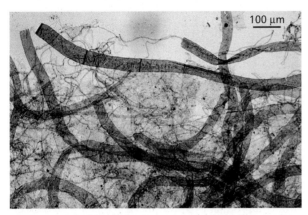

Figure 7–6 Fossils closely resembling the modern tribophycean *Vaucheria* from nearly one billion-year-old deposits in eastern Siberia. The smaller filaments are thought to have grown within the sediments, whereas the larger filaments are hypothesized to have extended above the surface. (Reprinted with permission from Knoll, A. H., The early evolution of eukaryotes: A geological perspective. *Science* 256:622–627. ©1992 American Association for the Advancement of Science)

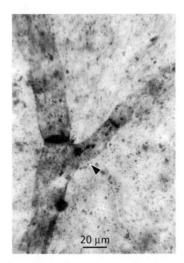

Figure 7–7 Fossils resembling the modern branched green alga *Cladophora* from 750 million-year-old deposits from Spitsbergen. Arrowhead points to branch. (Reprinted with permission from *Nature*. Butterfield, N. J., A. H. Knoll, and K. Swett. Exceptional preservation of fossils in an Upper Proterozoic shale. 334:424–427. ©1988 Macmillan Magazines Ltd.)

tions at the Empire Mine, in northern Michigan, U.S.A. (Han and Runnegar, 1992). Similar, but younger, specimens have been found in other regions of the world. *Grypania*'s discoverers suggested that it was likely photosynthetic, though there is no direct evidence of this, and eukaryotic, based on its relatively large size. Some other very ancient fossils thought to be related to modern eukaryotic algae include spherical structures 40–200 μm in diameter dated uncertainly at 1.8–1.9 billion years ago, and 300–2700 μm ornamented spheres known as **acritarchs** (Knoll, 1992). Some of the acritarchs may be remains of ancient unicellular green algae (see Chapter 17 for more information), but the relationships of many others are uncertain. Identification of Precambrian fossils as eukaryotes or as algae is often highly controversial, because intracellular structures such as nuclei, mitochondria, and plastids are not preserved, and because the fossils do not always exhibit features that place them unequivocally within known eukaryotic groups. However there are several examples of recent fossil finds that have been convincingly related to modern groups. Biochemical markers indicate that the ancestry of dinoflagellates extends back at least 800 million years (see Chapter 11).

The oldest multicellular fossils that can be confidently assigned to a modern algal group are attached, unbranched filaments similar to the modern red alga *Bangia* (Chapter 16). These fossils were found in silici-

fied carbonate rocks from Somerset Island, Canada, and are indirectly dated to between 0.54–1.26 billion years of age (Butterfield, et al., 1990). Fossil remains that are much like modern *Vaucheria*, a coenocytic tribophycean (Chapter 15), are known from 900 million to 1 billion-year-old deposits of eastern Siberia (Fig. 7–6) (Knoll, 1992; reinterpreted in Xiao, et al., 1998). Fossils similar to the modern branched green alga *Cladophora* (Fig. 7–7) have been found in 750–700 million-year-old strata at Spitsbergen (an island off the coast of Norway) (Knoll, 1992). Remains that are very similar to the modern sheetlike red alga *Porphyra* (Fig. 7–8) as well as parenchymatous fucalean brown algae occur in nearly 600 million-year-old phosphorite deposits in southern China (Xiao, et al., 1998). These new findings have revealed that the major multicellular algal lineages—red algae, ochrophytes, and green algae—are considerably more ancient than was previously thought.

Molecular and Biochemical Evidence for the Origin of the First Mitochondria and Plastids

Mitochondria

Mitochondria have long been known to possess eubacterial-like DNA and transcription and translation sys-

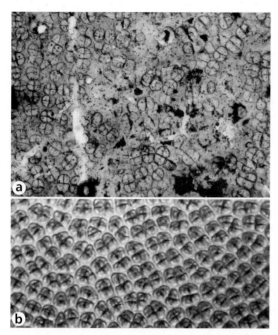

Figure 7–8 (a) Fossils that closely resemble the modern multicellular red alga *Porphyra* from the 570 million-year-old Doushantuo Formation in southern China. The phosphorite matrix has preserved cellular arrangements extremely well. These sediments are also the source of the earliest known fossils of animal embryos.(b) The extant *Porphyra*. (a: Reprinted with permission from *Nature*. Xiao, S., Y. Zhang, and A. H. Knoll. Three-dimensional preservation of algae and animal embryos in a Neoproterozoic phosphorite. 391:553–558. ©1988 Macmillan Magazines Ltd.)

tems, including similarly sized ribosomes. Also, infoldings of the cell membrane similar to the cristae of mitochondria characterize certain alpha proteobacteria (purple bacteria), where they function in photosynthesis. These similarities, together with comparative molecular sequence evidence, particularly for 16S rRNA genes, strongly support the hypothesis that mitochondria were once free-living photosynthetic proteobacteria that became endosymbiotically incorporated into host cells, lost photosynthetic capacity, and assumed the specialized function of aerobic respiration. However, genomes of modern mitochondria vary greatly and are characterized by a number of unusual features that make it difficult to trace the evolutionary history of this organelle. Eubacterial operons, for example, which are readily identifiable in plastids (where they provide useful information about the process of endosymbiosis) are nearly absent from eukaryotic mitochondrial DNA

(Gray, 1992). Among the various algal groups, three types of mitochondria have been noted; they are distinguished by morphology of the cristae (Table 7–1). Some algal groups (ochrophytes, dinoflagellates, and chlorarachniophytes) have mitochondria with tubular cristae, whereas others (glaucophytes, cryptomonads, red algae, and green algae) have flattened, platelike cristae, while euglenoids have discoid cristae. It has been proposed that the mitochondrion arose once in a common ancestor of all extant eukaryotes, possibly at the same time as the nucleus (Gray, et al., 1999).

Bacterial endosymbionts occur widely within cells of protists, including algae. For example, putatively symbiotic bacteria were detected by transmission electron microscopy in cells of the green alga *Pleodorina japonica* (Nozaki, et al., 1989) and in cells of the dinoflagellate *Peridinium balticum* (Chesnick and Cox, 1986). The latter authors provide a survey of observations of bacterial cells within algae. In some cases these may be symbionts, but in others the bacteria may be pathogenic invaders or prey cells ingested by phagotrophic algal cells.

Some recent evidence suggests that losses of the mitochondrial genome and respiration abilities were involved in the evolutionary origin of **hydrogenosomes** in various protists (such as *Giardia*), which were earlier thought to be primitive, based on lack of mitochondria (Palmer, 1997). Hydrogenosomes are like mitochondria in having a double-membrane envelope, but they lack a genome. They do, however, import nuclear-encoded, cytoplasmically produced proteins that function in pyruvate degradation, as do mitochondria. As far as is currently known, all protists that possess plastids (and are thus classified as algae) possess mitochondria, but not hydrogenosomes. However, genome extinction—hypothesized to have occurred during hydrogenosome evolution—is also associated with secondary and tertiary plastid acquisition in algae (Palmer, 1997) and will thus will be referred to later in this chapter.

Plastids

A great deal more is known about the processes that gave rise to plastids of protists and plants as compared to mitochondria. The DNA of an algal plastid may be arranged as nucleoids of various sizes and shapes that are scattered throughout the plastid (red algae, dinoflagellates, green algae, cryptomonads, haptophytes, and most eustigmatophyceans) or as a ring-shaped nucleoid lying beneath the outermost (girdle) photosynthetic lamellae of many ochrophytes (including diatoms and brown algae) (Coleman, 1985). Such DNA can be visualized using a DNA-specific fluorochrome such as DAPI. The plastid transcription and translation system is very similar to that of eubacteria, as are ribosomal sizes and the spectrum of antibiotic sensitivities. Similarities in plastid and eubacterial gene order, as well as 16S rDNA and various protein-coding sequences, indicate that all modern plastids are of cyanobacterial origin (Gray, 1992; Morden, et al., 1992). Further cementing the relationship is the presence of a self-splicing Group I intron in a leucine transfer RNA gene of the cyanobacterium *Anabaena* that is similar in sequence and position to an intron that characterizes plastids (Xu, et al., 1990). Introns are not of common occurrence in prokaryotic genomes; in this case the intron was apparently inserted into the leucine tRNA gene prior to divergence of plastids from their cyanobacterial ancestors.

Comparative studies of plastid genome structure among algae has shed light upon the mechanism by which eubacterial endosymbionts became dependent upon their hosts, and incapable of free-living existence. Plastid genomes are much smaller than the average eubacterial genome, and the majority of genes specifying photosynthetic structure and function are encoded in the nucleus. There is considerable evidence that gene transfer from plastids to the nucleus has occurred throughout the history of algae and plants. In cyanobacteria and in plastids of red, golden, and brown-pigmented algae, the genes encoding the large and small subunits of Rubisco (ribulose-1,5-bisphosphate carboxylase/oxygenase, the holoenzyme that incorporates CO_2 into organic compounds) occur in tandem, as the *rbcLS* operon. In contrast, this operon has been broken up in green algae (and plants) such that while the *rbcL* gene has persisted in the plastid, the rbcS gene has been transferred to the nuclear genome. As a result, the *rbcS* protein product is made in the cytoplasm and must be imported into plastids. There are other documented examples of transfer of plastid genes to the nucleus. The gene *tufA*, which encodes a chloroplast-specific protein synthesis elongation factor (Tu), was transferred to the nucleus in the charophycean lineage of green algae (and subsequently inherited by plants), but remains plastid-encoded in other algae (Baldauf and Palmer, 1990). Other instances include the presence in the nucleus of a gene encoding one of the subunits of plastid RNA polymerase in *Cyanidium caldarium* RK-1, a unicellular eukaryote often linked to the red algae (Tanaka, et al., 1996), and in the glaucophyte *Cyanophora*, nuclear location of the gene for ferredoxin-NADP$^+$-reductase, whose amino acid sequence is much more similar to that of cyanobacteria than to corresponding enzymes of higher plants (Schenk, et al., 1992).

In these and other cases of gene transfer from plastids (or mitochondria) to the nucleus, proteins synthesized in the cytoplasm must be imported into plastids (or mitochondria). This process requires the presence of special leader (targeting) amino-acid sequences to ensure that the protein is imported into the correct organelle. Thus the process of gene transfer from organellar genomes to the nuclear genome has required the evolutionary addition of DNA coding for targeting regions of proteins that must be imported by organelles and the modification of organellar import systems. Parts of the protein import machinery of plastids are known to be derived from ancient cell membrane transport systems in cyanobacteria—a review of higher plant plastid import systems and their evolution and models of the import process are provided by Heins, et al. (1998). A current hypothesis for the evolutionary mechanism by which such gene transfer might occur is reverse transcription of an RNA intermediate back to a DNA sequence, which is then incorporated into the nuclear genome (Gray, 1992). Gene transfer from organelles to the nucleus was discussed by Martin and Herrmann (1998).

A number of protists that have lost photosynthetic ability still possess plastids and portions of the plastid genome, suggesting the occurrence of essential non-photosynthetic genes. An example is the chloroplast DNA of the colorless euglenoid *Astasia* which retains 7×10^4 bp of DNA (Siemeister and Hactel, 1990). Other examples include plastidlike structures (also known as cyanobacterialike structures) (Fig. 7–9) in the cells of non-photosynthetic, intracellular parasites of the phylum Apicomplexa, such as *Toxoplasma gondii*, *Eimeria tenella*, and the malaria parasite *Plasmodium falciparum* (McFadden, et al., 1996; Köhler, et al., 1997), as well as the coccidian pathogen *Cyclospora* (Chiodini, 1994). ATPase gene cluster analysis suggests that apicomplexan plastids were derived from red algae (Leitsch, et al., 1999), whereas other gene sequence data suggest an origin from a green alga (Köhler, et al., 1997). Apicomplexans are included in the group of protists known as alveolates, along with ciliates and dinoflagellates (Fig. 7–3), but are thought to have acquired plastids independently of dinoflagellates (Köhler, et al., 1997). Some time after plastid acquisition, apicom-

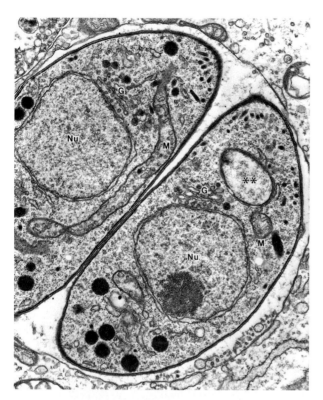

Figure 7–9 The apicomplexan *Toxoplasma gondii* apparently once possessed photosynthetic ability, based upon the presence in each cell of a small plastidlike body (asterisks) bounded by four membranes. G=Golgi body, M=mitochondrion, Nu=nucleus (Micrograph courtesy L. G. Tilney and D. S. Roos; reprinted with permission from Vogel, G., *Parasites shed light on cellular evolution. Science* 275:1422. ©1997 American Association for the Advancement of Science)

plexan plastids apparently underwent reduction and are no longer capable of photosynthesis. It has been suggested that once acquired, these organelles were not lost even when their photosynthetic capacities were no longer needed, because their genes are needed for other essential cellular processes, such as amino acid synthesis or lipid breakdown (Palmer, 1997).

Cyanobacterial Endosymbionts

Cyanobacteria occur within the cells of a variety of organisms. Some of these function like plastids to provide fixed carbon and perhaps other metabolites. An example is *Aphanocapsa*(?) *feldmanni*, a cigar-shaped unicell having a distinctive spiral thylakoid, which occurs in cells of sponges in all warm-water regions. Although relatively cryptic, *Aphanocapsa*-sponge endosymbioses may be among the most prevalent intracellular associations (Wilkinson, 1992).

Other cyanobacterial associations are based upon the nitrogen-fixation capacity of the endosymbiont. Some plastid-containing diatoms belonging to the genera *Epithemia*, *Rhopalodia*, and *Denticula* also contain 1–10 intracellular structures that are believed to have originated from endosymbiotic cyanobacteria. These spherical bodies apparently do not produce much, if any, photosynthetic pigments or phycobilisomes and are thus not highly pigmented. However phycobilisome anchor proteins have been detected, suggesting a relationship with cyanobacteria. A few thylakoids are generated by the plasma membrane, and there is a wall similar in appearance to that of gram negative bacteria. The bodies appear to divide by binary fission, and DNA has been identified by staining with DAPI (4,6-diamidino-2-phenylindole, a DNA-specific fluorescent dye) (Kies, 1992). A clue to their function is provided by the typically low-nitrogen habitats of the diatom hosts; not unexpectedly the endosymbionts have been discovered to be capable of nitrogen fixation. In an experimental study, the number of endosymbionts per cell increased as the combined nitrogen level decreased (DeYoe, et al., 1992).

A similar relationship occurs between the common marine planktonic diatoms *Rhizosolenia* and *Hemiaulus*, and an endosymbiotic, filamentous, nitrogen-fixing cyanobacterium, *Richelia*. Typically, up to 98% of the *Rhizosolenia* cells sampled from the central north Pacific gyre and the Caribbean area contain at least two *Richelia* filaments of about 4–15 cells each. The symbiosis occurs most commonly in ultra-oligotrophic, nitrogen-depleted open-ocean waters. Although *Rhizosolenia* grows well without *Richelia* in nitrogen-rich coastal waters, *Richelia* is apparently dependent upon its host (Paerl, 1992). The role of intracellular *Richelia* and other intracellular residents of diatoms is apparently the same as that of cyanobacterial associates of lichens and land plants (described in Chapter 3)—providers of combined nitrogen.

Another example of the intracellular occurrence of cyanobacterial cells is the nitrogen-fixing *Nostoc punctiforme* within the cytoplasm of the soil fungus *Geosiphon*. It represents the only known occurrence of an intracellular photosynthetic symbiont in fungi. Both partners can be grown separately. The association is interesting because it typically occurs together with a bryophyte whose culture filtrate can support growth of the fungus (Mollenhauer, et al., 1996).

Glaucophytes and Primary Endosymbiosis

The glaucophytes, also known as glaucocystophytes, are a group of freshwater microalgae that characteristically contain blue-green plastids often referred to as **cyanelles**. Glaucophyte plastids are similar to cyanobacteria, and different from plastids of other algae, in having a thin peptidoglycan wall (Fig. 7–10). The walls of both cyanobacteria and glaucophyte plastids can be destroyed with lysozyme, and their assembly prevented by penicillin treatment. In addition, ultrastructural study of freeze-fractured plastids of the glaucophyte *Cyanophora* revealed that they are surrounded by a layer very similar to the lipopolysaccharide external coat of gram negative eubacteria, and that the inner envelope membrane of the cyanelles is more similar to the plasma membrane of cyanobacteria than to plastid envelope membranes (Giddings, et al., 1983). The presence of these wall layers, together with pigment composition and other attributes, demonstrates conclusively that glaucophyte plastids originated by primary endosymbiosis of a cyanobacterium (Sitte, 1993).

Glaucophyte plastid DNA encodes the small subunit of Rubisco (*rbcS*), as do cyanobacteria and plastids of non-green algae, exhibits a number of gene clusters typical of cyanobacteria, and encodes a number of proteins not encoded by most algal plastids. Like the plastids of other algae, those of glaucophytes cannot be cultured separately from host cells. About 90% of the soluble proteins of *Cyanophora* cyanelles are encoded by the host nucleus, DNA having apparently been transferred there during the evolutionary transformation of endosymbiotic cyanobacteria to plastids. Plastid-encoded SSU rDNA sequence evidence indicates that glaucophyte plastids are monophyletic and form a clade distinct from cyanobacteria and plastids of other eukaryotic algae (Medlin and Simon, 1998) (Fig. 7–11).

Glaucophyte plastids contain chlorophyll *a* and the phycobiliproteins phycocyanin and allophycocyanin, as well as β-carotene. Phycobiliproteins occur in typical phycobilisomes on the surfaces of thylakoids, which are not stacked (Fig. 7–12). Phycoerythrin appears to be absent from glaucophytes, as are the typical cyanobacterial xanthophylls. These plastids contain polyphosphate granules and conspicuous central carboxysomes (Fig. 7–10) similar to cyanobacterial carboxysomes and pyrenoids found in plastids of other algae. The main photosynthetic product of the

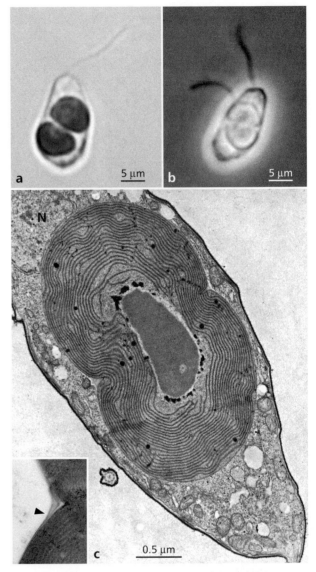

Figure 7–10 *Cyanophora*, a glaucophyte flagellate. (a) In bright-field microscopy, the cyanelle is obvious. (b) The two flagella are evident in phase-contrast microscopy. (c) With TEM, the plastid and its thin surrounding wall (arrowhead) are visible as are concentric thylakoids bearing phycobilisomes and a central carboxysome. (c: Micrograph by R. Brown, in Bold and Wynne, 1985)

plastids is glucose. Starch is not produced inside plastids, but rather by the host cytoplasm, where numerous starch granules can be observed ultrastructurally (Fig. 7–13). *Cyanophora* is not able to grow in the dark when supplied with glucose, since, lacking cytochrome c oxidase, it is unable to respire. The

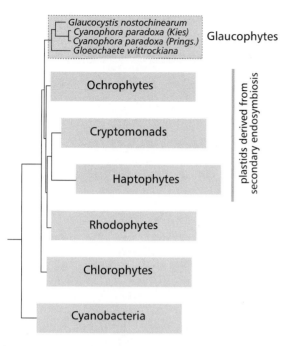

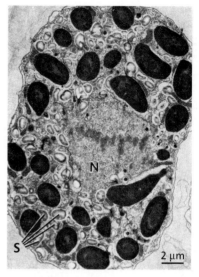

Figure 7–13 A dividing cell of the glaucophyte *Gloeochaete wittrockiana* showing a central nucleus (N) and chromosomes surrounded by darkly stained plastids. Numerous starch granules (S) occur in the cytoplasm. (From Kies, 1976)

Figure 7–11 A phylogeny inferred from plastid 16S rDNA sequences illustrating relationships between the plastids of eukaryotic algae and cyanobacteria. Monophyly of the glaucophyte plastids is indicated (top). (After Medlin and Simon, 1998 with kind permission of Kluwer Academic Publishers)

plastids of glaucophytes, like those of other algae, are unable to fix nitrogen.

Nuclear-encoded SSU rDNA data suggest that the glaucophyte nucleocytoplasm is related to that of cryptomonads (Medlin and Simon, 1998) (Fig. 7–3). Such a relationship is also supported by nuclear-encoded stress 70 protein (also known as heat shock protein) gene sequence data (Rensing, et al., 1997). Glaucophyte motile cells typically exhibit a series of flattened vesicles lying beneath the cell membrane that have con-

tents varying with the genus examined; some contain a scalelike structure, others possess a loose fibrillar material (Fig. 7–14), and yet others appear to be empty. Similar structures occur at the periphery of cryptomonads (Chapter 9). Glaucophyte flagellate cells are also characterized by distinctive multilayered structures near flagellar bases (Kies and Kremer, 1990). Similar structures are present in flagellate cells of some other algal groups. There are no known fossil representatives of the glaucophytes.

Diversity of Glaucophytes

Glaucophytes include unicellular flagellates, planktonic colonies, and attached colonies, which inhabit freshwaters, particularly soft waters such as bogs or acid swamps. With the light microscope, they can be readily distinguished from other eukaryotic algae by the bright blue-green pigmentation of the plastids, and distinguished from cyanobacteria by the presence of unpigmented cytoplasm. Nonflagellate forms may produce flagellate asexual reproductive cells (zoospores), but sexual reproduction is unknown. The three most common genera are *Cyanophora*, *Glaucocystis*, and *Gloeochaete*, which are described below. Molecular data suggest that *Gloeochaete* diverged earlier than *Cyanophora* or *Glaucocystis* (Medlin and

Figure 7–12 Portion of a plastid of the glaucophyte *Cyanophora paradoxa*, viewed by transmission electron microscopy, showing phycobilisomes on thylakoids.

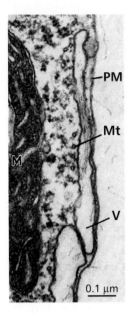

Figure 7–14 Flattened vesicles (V) occur at the periphery of glaucophyte cells beneath the plasma membrane (PM), as in *Gloeochaete wittrockiana*. Fibrillar contents of the vesicles and underlying microtubules (Mt) are distinctive. M=mitochondrion. (From Kies, 1976)

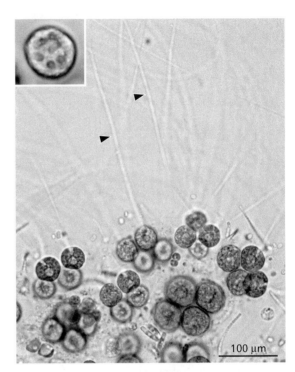

Figure 7–16 The glaucophyte *Gloeochaete*, with long pseudoflagella (arrowheads) extending from each of the cells. The inset is a higher magnification view of a single *Gloeochaete* cell showing several distinct plastids.

Simon, 1998). Several additional genera have been described; these include *Archeopsis*, *Glaucystopsis*, *Peliania*, *Strobilomonas*, *Cyanoptyche*, and *Chalarodora* (Seckbach, 1994). *Glaucosphaera*, regarded at one time as a member of the glaucophytes, has more recently been linked with red algae, on the basis of molecular evidence.

A freshwater silica-shelled amoeba, *Paulinella chromatophora*, which is unrelated to the glaucophytes (Bhattacharya, et al., 1995), contains two blue-green structures resembling the plastids of glaucophytes. This amoeba is an inhabitant of muddy bottoms of lighted regions of eutrophic lakes and ditches. Division of the amoeba is accompanied by distribution of a cyanelle into each of two daughter amoebae; the cyanelles then divide by binary fission. A related amoeba, *P. ovalis*, lacks such pigmented structures, but readily ingests and digests cyanobacteria (Johnson, et al., 1988).

CYANOPHORA (Gr. *kyanos*, blue + Gr. *phoras*, bearing) possesses two unequal flagella, one extending forward from the cell apex, and the other emerging from the same subapical depression, but extending toward the cell posterior. Typically two rounded plastids are present per cell (Fig. 7–10). Reproduction is by longitudinal division into two daughter cells. *Cyanophora* is able to swim away from very bright light; its plastids are implicated as the photoreceptors.

GLAUCOCYSTIS (Gr. *glaukos*, bluish-green or gray + Gr. *kystis*, bladder) is a colony formed by retention of ovoid daughter cells within the persistent

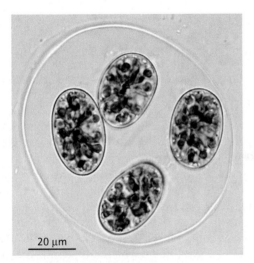

Figure 7–15 The glaucophyte *Glaucocystis* has ovoid cells containing several deeply lobed blue-green plastids.

Figure 7–17 TEM of the apical region of a *Gloeochaete wittrockiana* zoospore, showing two of the four multilayered flagellar roots (GW1 and GW2—from the German *Geisselwurzel*, meaning flagellar root). Such multilayered structures are characteristic of glaucophytes. A flagellar base occurs at the center. (From Kies, 1976)

parental cell wall, which is cellulosic. Within each pigmented cell are two starlike aggregations of several long, thin cyanelles (Fig. 7–15). Cell division results in subdivision of the cytoplasm into four autospores (nonflagellate, smaller versions of the parental cell).

GLOEOCHAETE (Gr. *gloia*, glue + Gr. *chaite*, long hair) occurs as a unicell or colony of 2–4 spherical cells, each having two distinctive long (20 times the cell diameter), thin gelatinous hairs (pseudoflagella) (Fig. 7–16). These hairs have internal microtubules similar to those of flagella but lack the two central microtubules. At one time, the presence of such hairs was thought to indicate relationship to the green alga *Tetraspora*, which has similar hairs. But now this similarity is regarded as an example of homoplasy, the result of parallel or convergent evolution. *Gloeochaete* cells are enclosed in mucilage and attached to walls of filamentous algae, leaves of aquatic mosses, or submerged macrophytes. The delicate cell wall is not composed of cellulose. The numerous plastids tend to form a cup-shaped assemblage in the basal portion of the host cell. Asexual reproduction is by flagellate zoospores, each having four distinctive multilayered structures (MLSs) as flagellar roots (Kies, 1976) (Fig. 7–17). Immediately upon zoospore settling, the germlings begin secretion of the hairs, during which time the protoplast rotates.

Eukaryotic Algal Endosymbionts

Endosymbiotic eukaryotic algae commonly occur in marine and freshwater protozoa, sponges, coelenterates, flatworms, molluscs, and other animals. In freshwaters the most common eukaryotic algal endosymbiont is the unicellular green *Chlorella*; in marine waters dinoflagellates belonging to the *Sym-*

biodinium species complex occur widely as endosymbionts. However various cryptomonads, diatoms, ochrophytes, and green algae other than *Chlorella* can also occur inside the cells of eukaryotes. The ubiquity of such endosymbioses demonstrates the adaptive value of such relationships, and these modern associations provide models for understanding the more ancient secondary and tertiary endosymbiotic associations that were pivotal in the evolution of several important algal lineages (Table 7–1).

Foraminifera are rhizopodial sarcodines (amoebae) that have calcareous shells. They are among the most successful of marine groups, having been in existence for at least 500 million years and having at least 4000 species. Many of the modern foraminifera are large enough to see with the naked eye; all of these larger forms host endosymbiotic algae. In addition, most of the planktonic foraminifera that occupy well-lighted waters have algal endosymbionts. The shells of foraminifera are constructed with complex chambers, which may represent an adaptation that fosters symbiotic relationships. Most foraminifera are selective in symbiont type, but some may have more than one kind of algal endosymbiont. Symbionts may include unicellular red algae such as *Porphyridium*, unicellu-

Figure 7–18 A green sponge containing zoochlorellae. This specimen was collected from an oligotrophic lake in northern Wisconsin.

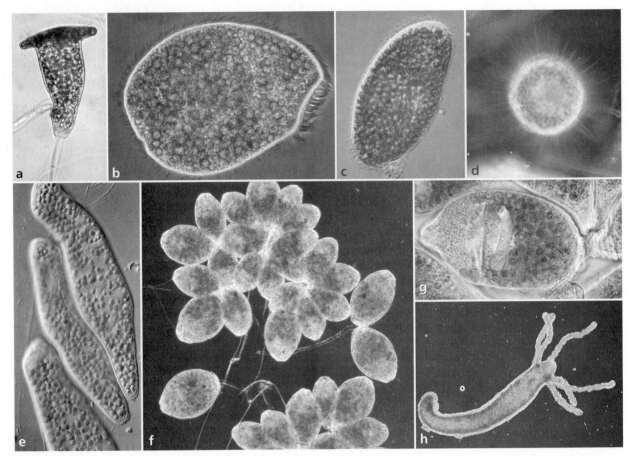

Figures 7–19 Freshwater organisms containing endosymbiotic cells of the green alga *Chlorella*. (a) *Vorticella*, an attached ciliate with an unbranched contractile stalk. (b) *Climacostomum*, a free-swimming ciliate. (c) *Paramecium bursaria*, another free-swimming form. (d) *Acanthocystis*, a heliozoan ("sun animal") that has many radiating axopodia—thin cytoplasmic extensions made rigid by internal columns of microtubules. (e) *Ophridium*, a macroscopic colonial ciliate whose numerous cells are embedded in massive amounts of gelatinous material. (f) *Carchesium*, an attached stalked, colonial ciliate; individual cells can contract within the colony without causing the whole colony to contract. (g) *Opercularia*, a noncontractile colonial ciliate that attaches to substrates via a branched stalk. (h) *Hydra*, a coelenterate.

lar green algae such as *Chlamydomonas*, more than 20 species of diatoms, or dinoflagellates (Lee, 1992a, b).

Radiolarians are marine planktonic amoebae that may be single-celled or occur in colonies; they build shells of crystalline strontium sulfate. Radiolarians occur in all of the major oceans, and there are hundreds of living and fossil species. There may be 1000 to 100,000 algal cells per radiolarian, often occurring within a halo of surface cytoplasm. Radiolarians reproduce via a flagellate stage that lacks symbionts; the mechanism by which algal symbionts are regained is not understood. Symbionts may include dinoflagellates, prasinophycean green algae, haptophytes, or ochrophytes (Anderson, 1992).

The cells of marine sponges, especially those occurring in coral reef communities, may (in addition to cyanobacterial endosymbionts, described earlier) contain diatoms such as *Nitzschia*, cryptomonads, or dinoflagellates. Sponges that contain endosymbiotic algae are better able to compete with corals and even overgrow them. After a hurricane destroyed a sponge population, sponges bearing algal endosymbionts grew back more rapidly than those lacking endosymbionts (Wilkinson, 1992). In freshwaters, *Chlorella*-containing green sponges such as *Spongilla lacustris* may be conspicuous components of the benthic community in well-lighted oligotrophic lakes (Fig. 7–18). The freshwater sponge *Corvomyenia* contains a previously

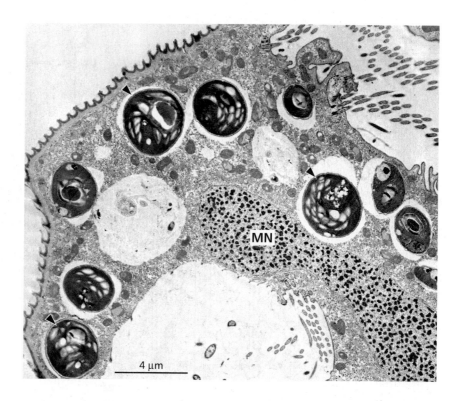

Figure 7–20 TEM of a green *Vorticella* showing numerous *Chlorella* cells within perialgal vacuoles (arrowheads). The host macronucleus (MN) is visible, as are many of its numerous cilia.

unknown eustigmatophycean endosymbiont (Frost, et al., 1997). These sponges derive as much as 80% of their growth from autotrophic processes (Frost and Williamson, 1980).

Chlorella, primarily the *Chlorella vulgaris* group (*C. vulgaris, C. sorokiniana, C. saccharophila*, and *C. lobophora*) occurs widely in freshwater protozoa (Fig. 7–19a–g). It is also encountered in the coelenterate *Hydra* (Fig. 7–19h), as well as other invertebrates. Algae are taken up only by hosts capable of phagotrophy. Ingestion involves enclosure of particles by a phagocytic (food) vacuole. If a particle is to be digested, lysosomes (membrane-bound vesicles that contain degradative enzymes) fuse with the phagocytic vacuole. In contrast to particles to be digested, endosymbiotic algal cells typically occur in perialgal vacuoles (also known as symbiosomes) (Fig. 7–20), which protect the enclosed algal cells against attack of host enzymes. The formation of a perialgal vacuole around a newly acquired green algal symbiont is initiated by interactions between specific algal cell wall carbohydrates and lectinlike proteins bound to the phagocytic vacuole. Apparently only the few strains of *Chlorella* listed above have the proper wall chemistry to elicit incorporation. The low pH (from 4–5) of host perialgal vacuoles induces release of the sugar maltose from *Chlorella* cells. Maltose released by

Chlorella cells then inhibits lysosomal fusion with the phagocytic vacuole, preventing digestion of the algal cell. Some 40–60% of newly fixed carbon can be released as maltose. The excretion of large amounts of fixed carbon limits the division rate of endosymbiotic *Chlorella* cells, probably by decreasing assimilation of inorganic nitrogen (Dorling, et al., 1997). This helps to hold algal population growth in check

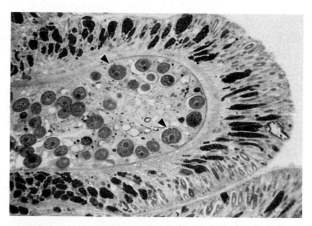

Figure 7–21 Zooxanthellae (arrowheads) in a sectioned and stained coral polyp. Zooxanthellae are readily recognized by their darkly stained spherical nuclei. (Photograph courtesy E. Newcomb)

Figure 7–22 A green sea anemone in which zooxanthellae are endosymbionts.

(Reisser, 1992a), but does not affect the photosynthetic capacity of cells. Symbiotic photosynthesis produces sufficient carbon to support the measured growth rate, respiration, and production of an extensive gelatinous matrix in the macroscopic, *Chlorella*-containing, colonial ciliate *Ophridium versatile* (Sand-Jensen, et al., 1997).

A similar relationship occurs between some dinoflagellates, known as **zooxanthellae** (Fig. 7–21), and the cells of all hermatypic (reef-forming) scleractinian corals, as well as many other marine invertebrates, such as the common green sea anemone *Anthopleura xanthogrammica* (Fig. 7–22). In these associations, and particularly those of well-studied reef-building corals, the dinoflagellate endosymbionts are fundamental to host biology, the carbon and nitrogen metabolisms of the associates being tightly linked (Allemand, et al., 1998). One square millimeter of coral tissue can contain 1–2 million dinoflagellate cells (Muller-Parker and D'Elia, 1997). The global net photosynthetic productivity of zooxanthellae is estimated to be more than 4.6×10^8 metric tons of carbon per year, much of which is made available to host corals, contributing to high rates of reef accumulation. Corals lacking zooxanthellae rarely make substantial contributions to formation of the reef framework. In anthozoans, zooxanthellae are located within cells of the tentacles, oral disk, body wall, and the tissue that connects individual polyps with each other. In contrast, in most coelenterates the algal endosymbionts are located within cells of the gastrodermis.

In contrast to free-living dinoflagellates, coral zooxanthellae release a substantial fraction of daily organic carbon production in the form of glycerol, glucose, and alanine, which are used by the host cells in respiration and growth. In the case of the Hawaiian coral *Pocillopora damicornis*, release of amino acids by host cells stimulates photosynthetic production by algae, as well as the release of photosynthates to host tissues (Gates, et al., 1995). The non-protein amino acid taurine is released by cells of the sea anemone *Aiptasia pulchella* and stimulates release of photosynthates (organic acids) from symbiotic dinoflagellates (Wang and Douglas, 1997). Such signalling processes may operate more widely. For most endosymbioses the stimulus for release of materials from endosymbionts, known as a host factor, is unknown.

Recent studies have revealed that more than one dinoflagellate type may occupy a particular coral (Rowan and Knowlton, 1995). In the symbiotic state, wall production by zooxanthellae is highly reduced. Since characteristics of the cell wall are commonly used to distinguish dinoflagellates (Chapter 11), recognition of different types require molecular sequence comparisons (Rowan, 1998). The proportions of different zooxanthellae can vary within a coral and with depth (reflecting different light environments). There are probably habitat-specific associations. Most corals do not transmit symbionts through sexual reproduction; rather, juvenile corals select zooxanthellae from among the locally available types, optimizing selection for habitat-specific variables. Such options expand the habitat versatility of corals, contributing to their abundance. Coral bleaching—loss of zooxanthellae resulting in pale to white appearance of corals—is a phenomenon of recent concern in the Caribbean and coral reef ecosystems throughout the world's oceans. Coral bleaching has been suggested to result from water temperature increases arising from global warming (Rowan and Knowlton, 1995).

Although zooxanthellae are as a rule rather rare in sponges, they occur in ectodermal cells of the most common and competitive "boring" sponges, such as the tropical *Anthosigmella varians*. The zooxanthellae play a role in the decalcification activities of boring sponges, contributing significantly to their reef destruction effects (Hill, 1996). Zooxanthellae therefore play substantial roles not only in the building of coral reefs, but also in their breakdown.

Although not of the same ecological and economic importance as reef-building corals, marine flatworms (turbellarians) may also harbor eukaryotic algal endosymbionts. *Convoluta roscoffensis* cells contain the green prasinophycean genus *Tetraselmis, C. con-*

voluta contains the diatom *Licmophora*, and *Amphiscolops* retains cells of the dinoflagellate *Amphidinium*. The eggs of these animals arc free of algae; thus, juvenile animals must acquire symbionts from free-living communities by phagocytosis.

Acquisition of Plastids by Secondary Endosymbiosis: A Focus on Chlorarachniophytes

The numerous examples of incorporation of eukaryotic algae by phagotrophs outlined above provide compelling evidence that such acquisitions are adaptively advantageous and suggest general mechanisms by which incorporation and stable metabolic integration may be accomplished. They are not only important from an ecological point of view but also provide insight into the mechanisms that may have led to secondary and tertiary acquisition of plastids in various algal groups. Groups whose plastids are widely thought to have arisen via secondary endosymbiosis—incorporation of a eukaryotic endosymbiont having a plastid derived by primary endosymbiosis—include euglenoids, cryptomonads, chlorarachniophytes, haptophytes (prymnesiophytes), the diverse heterokont

algae (ochrophytes), and some pigmented dinoflagellates, as well as their relatives, the apicomplexan parasites mentioned earlier.

The existence of secondary endosymbiosis was first indicated by the occurrence of more than two envelope membranes around the plastids of the algae listed above. The outermost of these extra membranes is usually interpreted as a remnant of secondary host phagocytotic vesicles (food vacuoles). While plastids of cryptomonads, chlorarachniophytes, haptophytes, and ochrophytes possess four bounding membranes, those of euglenoids and dinoflagellates have only three. The possibility that red and green algal plastids also arose by secondary endosymbiosis followed by loss of two membranes has been raised by Stiller and Hall (1997). Although no mitochondria are known to possess more than two envelope membranes, and are thus not usually thought to have arisen by secondary endosymbiosis, these authors suggest that the possibility should be seriously considered.

Other compelling evidence of secondary endosymbiotic origin of certain plastids was provided by electron microscopists, who first noted structures that looked like very small nuclei and eukaryotic-sized ribosomes located between the two pairs of membranes surrounding plastids of cryptomonads (Fig. 7–23) and those of chlorarachniophytes, green

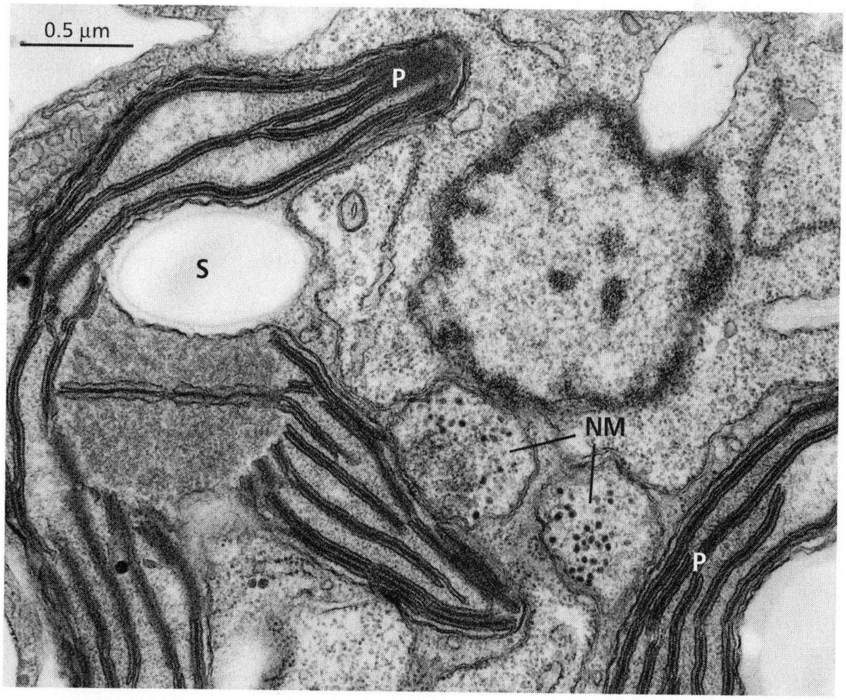

Figure 7–23 TEM of cryptomonad found within cells of the dinoflagellate *Gymnodinium acidotum*. There are two nucleomorph (NM) profiles in the periplastidal compartment—the space between the outer and inner pairs of membranes bounding the plastid (P). Also occurring in this space are eukaryotic-sized ribosomes and starch (S). (From Wilcox and Wedemayer, 1984 by permission of the *Journal of Phycology*)

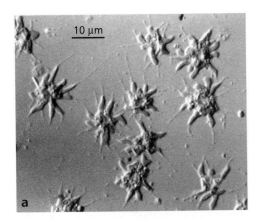

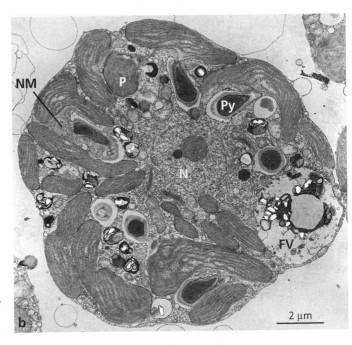

Figure 7–24 The chlorarachniophyte *Gymnochlora stellata* viewed with (a) light and (b) electron microscopy. The amoeba's central nucleus (N) and numerous plastids with nucleomorphs (NM) and pyrenoids (Py) are visible in (b). A food vacuole (FV) with degraded cellular material is also shown. (From Ishida, et al., 1996)

amoebae such as *Gymnochlora* (Fig. 7–24) and *Chlorarachnion* (Fig. 7–25). These nucleuslike structures, known as **nucleomorphs**, are bounded by an envelope with pores and contain a nucleoluslike substructure. Nucleomorphs divide by pinching into two equal halves. The nucleomorph is interpreted as the eukaryotic endosymbiont's highly reduced nucleus (Fig. 7–26). Ultrastructural data suggested that in both cryptomonads and chlorarachniophytes, most of the rest of the endosymbiont's cytoplasm was lost during transformation to a plastid, with just the nucleomorph and a few other features remaining as clues to past evolutionary history.

These dramatic ultrastructural observations stimulated further research on secondary symbioses, using molecular approaches. This work has shown that the nucleomorphs of both cryptomonads and *Chlorarachnion* plastids contain DNA, arranged in three tiny chromosomes, which principally encodes genes necessary for maintenance of the nucleomorph itself. Several genes encoded by the 380 kb *Chlorarachnion* nucleomorph genome, including those for small subunit RNA, some ribosomal proteins, a spliceosomal protein (which helps to process messenger RNA), and a protease, are known to be expressed. The protein-coding genes and their messenger RNAs show typical eukaryotic features, further supporting the hypothesis that the nucleomorph is a highly reduced eukaryotic nucleus (Gilson and McFadden, 1996). Other genes

necessary for plastid function have been transferred to the host cell's nucleus. Thus the cytoplasmically synthesized gene products destined for the plastid have to be transported across the four plastid membranes. Comparisons of DNA sequences of the nucleomorphs, host nuclei, and plastids (in addition to pigment composition) suggest independent origins of chlorarachniophytes, via capture of a eukaryotic green alga (van de Peer, et al., 1996) by a phagotrophic amoeba, and the cryptomonads, in which a eukaryotic red alga was taken in by a flagellate similar to the modern plastidless *Goniomonas* (Bhattacharya and Medlin, 1998) (Figs. 7–3, 7–12). Sorting out these relationships has relied upon special techniques for dealing with a large degree of variation in the rates of change among

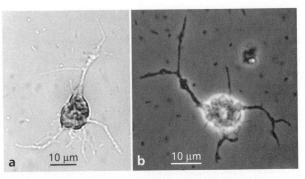

Figure 7–25 *Chlorarachnion* as seen with (a) bright-field and (b) phase-contrast light microscopy.

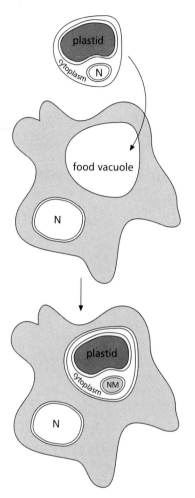

Figure 7–26 (a) Diagram of the acquisition of a green algal endosymbiont by *Chlorarachnion*, with the nucleomorph representing the nucleus of the incorporated algal cell. (Redrawn from *Trends in Ecology and Evolution* Vol. 10. McFadden, G., and P. Gilson. Something borrowed, something green: Lateral transfer of chloroplasts by secondary endosymbiosis. pp. 12–17. ©1995, with permission from Elsevier Science)

organismal genomes and nucleotide sites within nucleic acids, which is to be expected when dealing with organisms whose lineages diverged hundreds of millions to a billion or more years ago. Interesting accounts of the discovery of secondary endosymbiosis and the significance of cryptomonads and *Chlorarachnion* can be found in McFadden (1993), Cavalier-Smith, et al. (1994), McFadden and Gilson (1995), and Palmer and Delwiche (1996).

The lessons provided by cryptomonads and chlorarachniophytes suggest that similar events took place in the history of other algae having plastids with more than two membrane envelopes, but that in these cases, all vestiges of the eukaryotic endosymbiont's nucleo-cytoplasm has been lost. Apicomplexans, the non-photosynthetic but plastid-containing protists, provide an extreme example (see above). In this and most other cases of secondary endosymbiosis, the endosymbiont's nuclear genome has become extinct, as is also hypothesized to have occurred during evolution of hydrogenosomes, described earlier. Molecular evidence indicates that plastids of pigmented euglenoids are also derived from an endosymbiotic green alga, and plastids of ochrophytes and haptophytes, like those of cryptomonads, are derived from red algae by secondary endosymbiosis (Fig. 7–11) (Table 7–1).

Diversity of Chlorarachniophytes

Chlorarachniophytes are green-pigmented, unicellular protists related to the filose amoebae (Fig. 7–3). They occur in temperate and tropical marine waters. Some experts have recommended establishment of an algal phylum, Chlorarachniophyta, to include these green amoebae (Hibberd and Norris, 1984). Intriguingly, a predatory filose amoeboid organism, *Chrysarachnion insidians*, appears to have golden-brown plastids and was thus classified with chrysophyceans by Hollande (1952c).

The plastids of the chlorarachniophytes are similar to those of green algae in SSU rDNA (van de Peer, et al., 1996) and elongation factor-Tu amino acid (Ishida, et al., 1997) sequence data, and in containing chlorophylls *a* and *b*, but differ in the presence of four envelope membranes. Furthermore, there is a periplastidal compartment (the region between the inner and outer pairs of chloroplast envelope membranes) that contains a nucleomorph, as previously noted. Green algal plastids do not exhibit periplastidal compartments or nucleomorphs. The fact that no other protists are known to contain green plastids with nucleomorphs suggests that those of chlorarachniophytes originated *in situ* and that plastids have been stably integrated into cells. Multiple bilobed plastids occur in each chlorarachniophyte cell, each plastid having a large pyrenoid.

Reproduction of chlorarachniophytes often involves production of flagellate zoospores; these differ from those of most green algae in having only a single flagellum. Sexual reproduction has been observed in some. There are four known genera: *Chlorarachnion* (with two species, *C. reptans* and *C. globosum*), *Cryptochlora perforans*, *Gymnochlora stel-*

lata (Fig. 7–24) and *Lotharella*, a coccoid form. These are distinguished primarily on the basis of the ultrastructure of their pyrenoids (Ishida, et al., 1996).

GYMNOCHLORA (Gr. *gymnos*, naked + Gr. *chloros*, green) is a green star-shaped amoeba with several filopodia—filamentous extensions of the cytoplasm (Fig. 7–24). The cells do not form networks as do those of *Chlorarachnion*. Cells are 10–20 μm in diameter and often attach themselves to surfaces. The pyrenoids are capped with a photosynthetic reserve material that (surprisingly) does not stain with I$_2$KI solution as does the starch of green algae. The periplastidal envelope is not connected to the nuclear envelope and there are no ribosomes on its surface (as occur in some other plastids of secondary origin). The nucleomorph is located near the pyrenoid. One or two food vacuoles, bacterialike particles, and mitochondria with tubular cristae occur in the cytoplasm. There is no cell covering. Spherical resting stages may occur, but sexual reproduction has not been observed in this genus (Ishida, et al., 1996).

Tertiary Endosymbiosis and Horizontal Gene Transfer in Dinoflagellates

While many dinoflagellates are plastid-free, others have acquired plastids from a wide variety of eukaryotic sources, including green algae, cryptomonads, diatoms, chrysophyceans, and haptophytes. Acquisition of green algal plastids, such as has occurred in *Lepidodinium viride* (Watanabe, et al., 1990), may have involved secondary endosymbiosis, as described above. In contrast, most other dinoflagellate plastids are probably the result of tertiary endosymbiosis (Palmer and Delwiche, 1996). This means that their plastids arose by incorporation of eukaryotic cells whose plastids were themselves of secondary

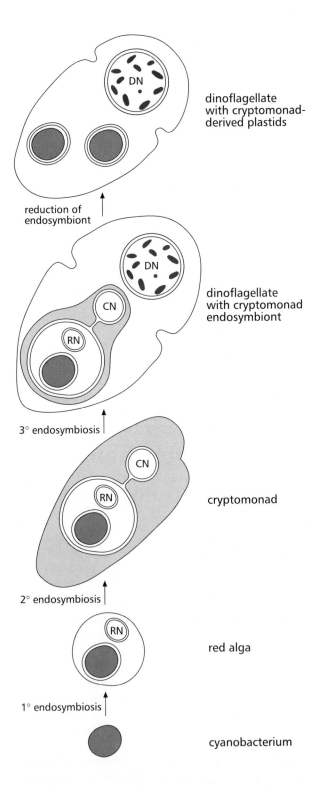

dinoflagellate with cryptomonad-derived plastids

reduction of endosymbiont

dinoflagellate with cryptomonad endosymbiont

3° endosymbiosis

cryptomonad

2° endosymbiosis

red alga

1° endosymbiosis

cyanobacterium

Figure 7–27 Diagram of the acquisition by a dinoflagellate cell of plastids (or kleptoplastids) derived from a cryptomonad by tertiary endosymbiosis. (Redrawn with permission from Wilcox, L. W., and G. J. Wedemayer. Dinoflagellate with blue-green chloroplasts derived from an endosymbiotic cryptomonad. *Science* 227:192–194. ©1985 American Association for the Advancement of Science)

endosymbiotic origin (Fig. 7–27). Such tertiary endosymbioses have occurred on multiple occasions and are exemplified here by dinoflagellates having plastids derived from cryptomonads and diatoms.

Immunolocalization of phycoerythrin in chloroplasts of two *Dinophysis* species coupled with other ultrastructural data suggest that the plastids of these marine dinoflagellates were obtained from a cryptomonad (Vesk, et al., 1996). This conclusion is supported by detection in *Dinophysis norvegica* of alloxanthin, an accessory pigment that marks cryptomonads (Meyer-Harms and Pollehne, 1998). The dinoflagellates *Peridinium balticum* and *Kryptoperidinium foliaceum* are unusual among golden-pigmented dinoflagellates in that their photosynthetic pigments are similar to those of diatoms, lacking the peridinin characteristic of most golden dinoflagellates, and they possess a second nucleus whose ultrastructure differs from the typical, distinctive dinoflagellate nucleus (Jeffrey and Vesk, 1976). Molecular analysis of the *rbcLS* operon in the dinoflagellate *Kryptoperidinium foliaceum* and nuclear SSU rDNA sequence evidence from both *Peridinium balticum* and *K. foliaceum* have now demonstrated that their plastids and additional nucleus were obtained from a benthic pennate diatom (Chesnick, et al., 1996; 1997).

At least some, and perhaps the majority of pigmented dinoflagellates, have plastids that differ dramatically from those of other algae in the form of Rubisco. In most algae (including cyanobacteria) and plants, Rubisco consists of eight large and eight small subunits (polypeptides) that are encoded by the *rbcL* and *rbcS* genes. We have previously noted differences among algae in the location of these genes; *rbcS* is encoded in the nucleus of green algae and plants, but both genes are encoded in the plastid genomes of other algal groups. Recently representatives of the peridinin-containing dinoflagellates—the largest group of pigmented dinoflagellates—were found to produce an alternate form of Rubisco, previously known only from certain anaerobic photosynthetic proteobacteria, such as the purple-pigmented *Rhodospirillum rubrum* (Morse, et al., 1995; Rowan, et al., 1996). This second type of Rubisco is known as "Form II Rubisco" to differentiate it from the more common and familiar Form I. Form II Rubisco is composed of only large subunits, which share only 25–30% amino acid sequence identity with Form I Rubisco large subunits. Further, the Form II Rubisco of peridinin-producing dinoflagellates is encoded by the nucleus, not the plastids. Plastids of these dinofla-

gellates do not produce Form I Rubisco, but they have other attributes indicating origin from oxygenic cyanobacteria. Experts have suggested that Form II genes might have undergone horizontal transfer to the dinoflagellate nucleus either directly from a proteobacterium or indirectly from mitochondria (which as noted earlier, are of proteobacterial origin), and that Form I genes were subsequently lost from the plastid, but there are other possible explanations involving horizontal gene transfer and gene loss (Palmer, 1996).

The *rbcL* genes of red, brown, and golden algae also appear to be more similar to those of certain proteobacteria than to cyanobacteria, suggesting additional instances of horizontal gene transfer, gene duplication, and gene loss. For some years molecular systematists had wondered why phylogenies of algae based upon *rbcL* gene sequences were substantially different from those obtained using other protein-coding gene sequences and SSU rDNA sequences. It appears that this discrepancy may be explainable in terms of past rampant horizontal transfer, as well as duplication and loss, of Rubisco genes in different lineages of eubacteria and algal plastids (Delwiche and Palmer, 1996). It is now known that various eukaryotic genomes contain genes transferred from endosymbionts or genomes that no longer persist; these are regarded as the result of cryptic endosymbioses (Henze, et al., 1995). Cryptic endosymbiosis is particularly likely to occur in algae because there is a long history of repeated phagotrophy, incorporation of endosymbionts, and transformation of those endosymbionts into semiautonomous organelles. Endosymbiosis should have a considerably greater potential than viral transformation for horizontal transfer of large genomic regions between unrelated nuclear lineages. The potential occurrence of migrant genes in the nuclear genomes of algae has important implications for systematics, ecology, and biotechnology.

Kleptoplastids

One way for a heterotrophic organism to obtain autotrophic and other capacities of plastids is to ingest and maintain plastids within the host cytoplasm, for at least a time. Chloroplasts are harvested from a variety of kinds of algae by a range of heterotrophs, including dinoflagellates, ciliates, and ascoglossans (Sacoglossan molluscs). Experts have argued that such relationships should not be viewed as examples of endosymbiosis, but rather should be regarded as temporary associations. The incorporated plastids are known as kleptoplastids, since they are "stolen" by their hosts. Because plastids are unable to synthesize all of the proteins needed for their maintenance, and foreign hosts lack the genes necessary to compensate for absence of the normal host encoded proteins, such associations are only temporary. Moreover, such relationships provide no benefit to the plastids or the algae from which they were harvested; thus the associations cannot be regarded as mutually advantageous. However they do serve as striking examples of the selective advantages accrued to heterotrophs when they are able to acquire plastids.

Oligotrich ciliates are protists that form an important component of the marine microplankton. They are frequently obligate or facultative mixotrophs (capable of autotrophy as well as heterotrophy). Of particular importance is *Strombidium*, some species of which occur in coastal waters and others in freshwaters. *Strombidium* harvests and maintains plastids from a variety of eukaryotic algae, including prasinophycean greens, cryptomonads, and ochrophytes. Sometimes these ciliates contain just one type of plastid, but mixtures of plastid types can also occur. Stoecker, et. al. (1987) demonstrated that algal plastids of *Strombidium* were capable of photosynthesis and found that during the height of population growth nearly 50% of *Strombidium* cells contained kleptoplastids.

Plastid harvesting and retention also occurs in the ascoglossan mollusc (sea slug) *Elysia* and its close relatives. They obtain plastids from siphonaceous algae, primarily green seaweeds belonging to the Caulerpales and the Codiales (Chapter 18), but also algae such as the coenocytic tribophycean *Vaucheria* (Chapter 15). Ascoglossans feed by piercing the algal cell wall with a specialized needlelike tooth, and then suck out large volumes of cytoplasm. Their feeding is facilitated by the large size of the algal cells and the absence of cross-wall barriers to cytoplasmic flow. Plastids are retained within cells of the animal and may continue to photosynthesize for longer than a week, providing oxygen and fixed carbon. Active transcription and translation of *Vaucheria* plastid chloroplast genes while plastids were being maintained in *Elysia* has been demonstrated (Mujer, et al., 1996). Ascoglossans are adapted structurally, physiologically, and behaviorally for plastid maintenance. For example, structural features may facilitate gas exchange and light exposure, and the animals may actively change position to provide optimal irradiance conditions for plastid photosynthesis. The animals produce a host factor that elicits leakage of photosynthates such as glycolate, glucose, and amino acids from plastids. Studies have shown that the animals may depend upon photosynthesis performed by retained plastids but not completely; other foods are also required. They must feed upon algae frequently to replace senescent plastids (Clark, 1992).

A freshwater example of acquisition of kleptoplastids is the dinoflagellate *Gymnodinium acidotum* (Fields and Rhoades, 1991). Cells of this species often contain blue-green bodies that can be identified ultrastructurally as having been derived from cryptomonads—the distinctive plastids with nucleomorphs are present as well as various components of the cryptomonad cytoplasm (Wilcox and Wedemayer, 1984) (Fig. 7–23). *Gymnodinium acidotum* has been reported to ingest cryptomonads, keep their plastids for at least ten days, then digest them (Fields and Rhodes, (1991). *Gymnodinium aeruginosum* ingests cryptomonads, then selectively digests the nucleus and nucleomorph, keeping plastids and other cell constituents (Schnepf, et al., 1989). In other cases, the occupation of *Gymnodinium* cells by plastids derived from ingested cryptomonads appears to be more stable (Wilcox and Wedemayer, 1984). It is possible that these bodies represent early stages in the formation of plastids by tertiary endosymbiosis, but it is often difficult to distinguish between such early stages and the occurrence of temporary associations.

Unresolved Issues Surrounding Endosymbiotic Origin of Eukaryotic Algae

Did primary plastids arise just once or more than once? Existing evidence strongly supports the hypothesis that various algal groups have independently acquired plastids by secondary endosymbiosis (Bhattacharya and Medlin, 1995). The evidence also indicates that various types of eukaryotic algae can be captured and stably transformed into plastids by other eukaryotes. In other words, secondary and tertiary plastids have arisen on multiple occasions. There appears to be no intrinsic barrier to transfer of genes from secondary and tertiary endosymbionts to host nuclear genomes. Repeated evolution of transit (organellar targeting) peptides and modification of import systems appears to have readily occurred when necessary. Further, the widespread occurrence of endosymbiotic algal cells and kleptoplastids within the cytoplasm of many kinds of organisms underscores the tremendous adaptive value to host cells that is conferred by acquiring photosynthetic, nitrogen fixation, and other metabolic attributes.

The occurrence of a few related *Chlorella* species in many types of freshwater invertebrates and several related species of *Symbiodinium* in numerous species of corals and other marine invertebrates suggests that some algae are more or less preadapted for uptake and use as beneficial intracellular symbionts. This phenomenon suggests that more than one member of a genetically similar group of cyanobacteria might have undergone parallel incorporation into primary endosymbioses and transformations into plastids. Further, analysis of sequences of *rpo* (RNA polymerase) genes indicates that red and green algae are not sister taxa, suggesting that their plastids were acquired by separate primary endosymbiotic events (Stiller and Hall, 1997). This result is congruent with that obtained by use of SSU rDNA sequences. Thus, it is possible that primary endosymbiosis might not necessarily have been a unique event. On the other hand, the same molecular evidence for monophyly of plastids (which is widely accepted) has also been used in support of the more controversial thesis that there was only a single primary endosymbiotic event (Morden, et al., 1992; Bhattacharya and Medlin, 1995; Palmer and Delwiche, 1996). This hypothesis implies that either (1) glaucophyte, red, and green algae are closely related, supported by the fact that they all possess similar mitochondria having flattened cristae (Table 7–1) but not by the most recent SSU rDNA sequence evidence (Medlin and Simon, 1998), or (2) that red and green plastids arose secondarily (Stiller and Hall, 1997). Analysis of the ATPase gene cluster of chloroplasts suggests that all extant algal lineages and land plants are derived from a single ancestral photosynthetic eukaryote (Leitsch, et al., 1999).

How did other components of eukaryotic cells arise? Also unresolved are the evolutionary origins of nucleo-cytoplasmic features of eukaryotes, including: 9+2 microtubule-containing flagella (cilia or undulipodia); the cytoskeletal and mitotic apparatus, comprising microfilaments composed of actin, micro-

tubules composed of tubulin proteins, and microtubule organizing centers; and secretory systems, the nuclear/endoplasmic reticulum membrane system, Golgi apparatus and other components, all of which are absent from prokaryotes, including putative archaeal ancestral types. It should be noted, however, that evidence for the origin of the eukaryotic endomembrane system by invagination of the plasma membrane of an ancestral archaeal cell is provided by similarities in sequences of genes that encode archaebacterial ATPases and vacuolar (V)-ATPases that mark eukaryotic endomembrane systems. This evidence suggests that the eukaryotic endomembrane system did not arise via endosymbiosis (Kibak, et al., 1992). Becker and Melkonian (1996) have proposed a model for evolution of a primitive endomembrane system.

Evidence related to the origin of eukaryotic cytoskeletal and mitotic systems is provided by similarities in the sequences of archaebacterial genes that encode cell division proteins, *FtsZ* and *FtsA*, to gene sequences for eukaryotic tubulins and actin, respectively (Baumann and Jackson, 1996). *FtsZ* is a GTPase that has been localized to the site of cell division in both eubacteria and archaebacteria, and while the encoding genes of both groups resemble those encoding eukaryotic tubulins, the archaeal sequences are closer to eukaryotic genes than are eubacterial sequences. Further comparative and correlative molecular, biochemical, and ultrastructural studies of archaebacteria and early divergent protists may help to explain the bases for differences in motility systems, phagotrophic ability, and extracellular matrix (wall) production that occur among eukaryotic algal groups, which are described in subsequent chapters.

Recommended Books

Bhattacharya, D. (editor). 1997. *Origins of Algae and Their Plastids*. Springer-Verlag, Wien, Germany.

Douglas, A. E. 1994. *Symbiotic Interactions*. Oxford University Press, Oxford, UK.

Reisser, W. (editor). 1992. *Algae and Symbiosis: Plants, Animals, Fungi, Viruses, Interactions Explored*. Biopress, Bristol, UK.

Chapter 8

Euglenoids

Phacus

A collecting trip to almost any wetland area, including marshes, swamps, fens, and bogs or mires will almost certainly yield a wealth of various types of the distinctive, primarily unicellular algae known as euglenoids. Indeed, euglenoids are generally found in environments where there is an abundance of decaying organic matter. Such habitats may also include nearshore marine or brackish sand and mud flats characterized by decaying seaweeds or organic contamination, farm ponds, dipteran larvae hindguts, and the rectums of tadpoles. Sometimes euglenoids form alarming blood-red surface blooms, that, so far as is known, are not harmful. However, in nearshore marine waters large populations of euglenoids have been observed to occur among blooms of potentially toxic algal species.

Because of their association with increased levels of dissolved organics, euglenoids have been used as environmental indicators of such conditions. Environments high in decaying organic materials suit the needs of euglenoids because they all require vitamins B_1 and B_{12}, which are released by the activity of associated microbial floras. Euglenoids have been used to bioassay these vitamins. In addition, many euglenoids utilize short-chain fatty acids, sugars, and other simple organic compounds produced by decay microorganisms. Euglenoids are also well adapted to life in wetlands and other habitats by virtue of flexibility in cell shape and versatility in their motility systems. They can swim across surfaces via one or more flagella, or they can ooze their way through mud or sand by a process known as euglenoid motion or **metaboly**, which is made possible by their unusual cell surface. The cell covering of euglenoids, known as a **pellicle**, and other cellular features that are of importance in their ecological behavior (as well as in their classification) will be discussed in this chapter.

The euglenoids (also known as euglenids, euglenophytes, or euglenoid flagellates) are probably the earliest divergent (most ancient) group of eukaryotic algae (Fig. 8–1). Only about one third of the known genera possess green-pigmented chloroplasts. Of the rest, many contain colorless plastids, and others lack plastids altogether (Whatley, 1993). Even the latter, however, contain typical eukaryotic mitochondria, cytoskeletal systems, 9+2 flagella, and endomembrane systems, including Golgi bod-

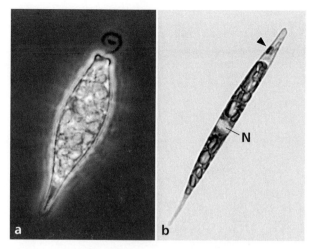

Figure 8–1 Two euglenoids, the colorless *Astasia* (a) and the photosynthetic *Euglena* (b). Paramylon granules are visible in both. The nucleus (N) and eyespot (arrowhead) of *Euglena* are also apparent.

ies. Thus the euglenoids may illustrate cellular attributes that were likely present in the earliest eukaryotes and inherited by more recently appearing eukaryotic algal groups.

A number of euglenoid genera are phagotrophic (i.e., they feed upon organic particles) and consequently possess cellular organelles that are specialized for capture and ingestion of prey, including bacteria and small algal cells. Some euglenoid predators are indiscriminate feeders, whereas others specialize, feeding only upon selected diatoms, for instance. Although euglenoid ecology has not been extensively studied, their heterotrophic (and mixotrophic) tendencies suggest that they could be important components of the microbial loop (Chapter 3) and food-web structure in general, particularly in freshwater wetlands and nearshore marine systems. In addition to the ecological significance of euglenoid phagotrophy, the ability to ingest particles as large as small algae has allowed some euglenoids to acquire plastids via secondary endosymbiosis, as discussed in Chapter 7. Further, ultrastructural examination has revealed that many euglenoids contain prokaryotic cells; some of these are tentatively regarded as endosymbionts. Euglenoids are therefore potentially important systems for research directed toward understanding the selective pressures and cellular features that facilitate capture of prokaryotic and eukaryotic cells and the transformation of endosymbionts into organelles. In this chapter we will focus upon euglenoid cell biology for the

insight it provides into the ecological and evolutionary roles of this group.

Evolution and Cell Biology of Euglenoids

Fossils that can be unequivocally designated as euglenoids are relatively rare; this is ascribed to the lack of calcified structures or cell walls that are resistant to decay processes, such as are found in some other algal groups that have, as a result, left more fossil remains. The oldest fossils attributed to euglenoids are ridged unicells known as *Moyeria*, which come from nonmarine, nearshore environments in the mid-Ordovician to Silurian, about 410–460 million years ago (Gray and Boucot, 1989). Fossils that are quite similar to the modern genera *Phacus* and *Trachelomonas* are known from the Tertiary, beginning about 60 million years ago (Taylor and Taylor, 1993). Most of what we know about the evolution of euglenoids is derived from comparative ultrastructural and molecular studies of modern representatives.

Ultrastructural surveys of cell structure and mitosis of euglenoids (Leedale, 1967; Willey, et al., 1988; Walne and Kivic, 1990; Triemer and Farmer, 1991; Walne and Dawson, 1993; Dawson and Walne, 1994) revealed close relationship to a group of flagellate protozoa known as the kinetoplastids. Subsequent molecular analysis (18S rRNA gene sequences) corroborated this relationship and strongly suggested that a kinetoplastid/euglenoid clade originated quite early within eukaryotes (Sogin, et al., 1986). The kinetoplastids are a tremendously important group because they include the parasitic trypanosomes. Examples are the human "blood pathogens" *Trypanosoma*, which is transmitted by insects and causes sleeping sickness in Africa and Chagas' disease in Latin America, and *Leishmannia*. Other trypanosomes cause serious diseases of wild and domesticated animals and of crop plants. The kinetoplastids also include the mostly free-living bodonids, such as the common genus *Bodo*, an organism that may be found in nonaxenic algal cultures and is a frequent contaminant of algal pond cultivation systems, such as are described in Chapter 4.

Kinetoplastids are characterized by a large, stainable, and easily detected mass of DNA in their single mitochondrion; this structure is not present in euglenoids. Another distinguishing characteristic of kinetoplastids is the compartmentation of glycolytic enzymes within an organelle known as a glycosome (Vickerman, 1990). In contrast, euglenoid glycolysis occurs within the cytoplasm, as it does in most eukaryotes. Like kinetoplastids, euglenoids are mostly unicellular and uninucleate, and their mitosis occurs within the confines of the nuclear envelope, i.e., there is a "closed" mitosis and an intranuclear spindle (Fig. 8–2). However, these features also characterize algae belonging to other groups.

Features unique to the kinetoplastid/euglenoid clade include disk-shaped mitochondrial cristae and the usual occurrence of two flagella that emerge from the cytoplasm into a pocket at the cell anterior. In euglenoids this pocket tends to extend more deeply into the cell than in kinetoplastids, and the euglenoid pocket is more constricted at the top to form a narrow canal through which flagella extend (O'Kelly, 1993) (Fig. 8–3). (Some authors refer to the pocket as the **reservoir**, and others refer to the reservoir and neck as the **ampulla**.) Comparative ultrastructural studies of euglenoids and related protists suggest that the euglenoid pocket evolved by coalescence of separate flagellar and cytostome ("cell mouth") openings present in ancestral phagotrophic ancestors. Typically, a contractile vacuole (Fig. 8–3) is present that collects excess water from the cytoplasm and discharges it into the flagellar pocket; the contractile vacuole is then reformed by the coalescence of smaller vesicles thought to arise from the Golgi apparatus. Another commonality is the usual occurrence of a rodlike structure that lies parallel to the flagellar axoneme, making euglenoid (and kinetoplastid) flagella thicker and easier to see than flagella of other eukaryotes (Fig. 8–18). This structure, known as a paraflagellar rod, is composed of proteins and is essential for the motility of trypanosomes (Bastin, et al., 1998) and presumably also euglenoids. At the ultrastructural level, both kinetoplastid and euglenoid flagella can be seen to bear a single row of long, thin hairs; these are not tubular as are the flagellar hairs—known as mastigonemes—that characterize the large algal group known as the ochrophytes (Chapters 12–15). In addition, shorter hairs coat the flagellar surface. Euglenoid (and kinetoplastid) flagellar hairs are thought to function to increase flagellar hydrodynamic resistance and thrust (Goldstein, 1992).

The cytoskeletal systems of early divergent kinetoplastids (bodonids) and euglenoids are also similar, consisting of three asymmetrically placed **flagellar roots** (Fig. 8–4) (Triemer and Farmer, 1991). Gener-

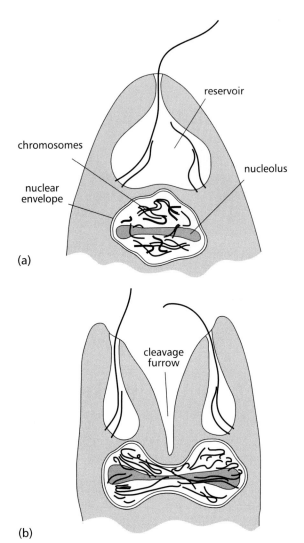

(a)

(b)

Figure 8–2 (a, b) Two stages of mitosis in *Euglena*. Mitosis is "closed," i.e., the nuclear envelope remains intact. The spindle is intranuclear. The cleavage furrow shown in (b) splits the cell longitudinally (see Fig. 8–12). (After Hollande, 1952a)

ally, flagellar roots are bands of microtubules and/or striated structures resembling miniature muscles that are biochemically similar to muscles. They extend from the **flagellar bases** (known as **basal bodies**) into the cytoplasm, usually along the cell periphery, but in some algae also toward the nucleus. Flagellar roots are believed to play important roles in maintaining cell shape. In some algae they are involved in particle capture during phagocytosis. In many algae, including euglenoids, there is a musclelike **striated connective** (Fig. 8–4) between the two flagellar bases that may function in coordinating flagellar motion. Together,

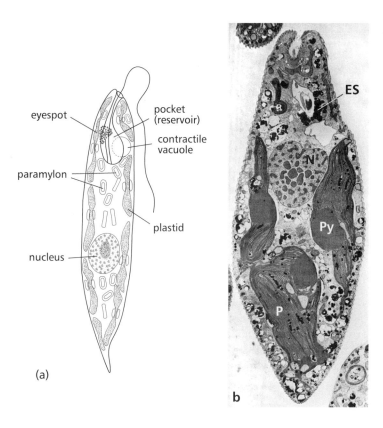

eyespot

pocket (reservoir)

contractile vacuole

paramylon

plastid

nucleus

(a)

ES

N

Py

P

b

Figure 8–3 (a) Diagram illustrating typical features of *Euglena*, which include the flask-shaped pocket (reservoir), from which one of the two flagella emerges. Adjacent the reservoir are the eyespot and a contractile vacuole. Also typically visible with the light microscope are paramylon granules, plastids, and the nucleus with nucleolus and relatively large chromosomes. (b) TEM view of longitudinally sectioned *Euglena* cell. Note eyespot (ES), plastids (P) with pyrenoids (Py) and nucleus (N). (a: After Gojdics, 1953; b: From Walne and Arnott, 1967)

Figure 8–4 Diagrams of the flagellar apparatus of euglenoids. (a) Three-dimensional view. (b) Cross section of the cell at the level of the cytostome. (c) Section at the level of the eyespot. C=cytostome, D=dorsal flagellar basal body, DR=dorsal root, ES=eyespot, IR=intermediate root, P=plastid, PR=paraflagellar rod, V=ventral flagellar basal body, VR=ventral root. (Redrawn with permission from Inouye, 1993, in *Ultrastructure of Microalgae*. ©CRC Press, Boca Raton, Florida)

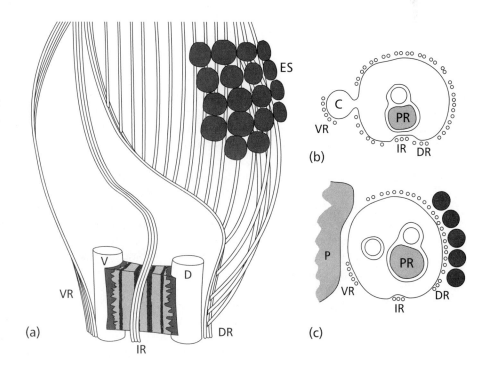

ES

V

D

VR

IR

DR

(a)

C

VR

PR

IR   DR

(b)

P

VR

PR

IR

DR

(c)

Figure 8–5 Anterior end of a *Euglena* cell stained with the vital dye neutral red. Posterior to the eyespot, which is just out of focus, lie a number of brightly stained mucilage bodies (arrow). The shape and distribution of these bodies, as visualized by neutral-red staining, have been used as taxonomic characters. Also note the two paramylon granules (arrowheads).

the flagellar roots, connectives (if present), flagella, and other nearby structures are known as the **flagellar apparatus**. Comparative architecture of the flagellar apparatus has been an important means of tracing algal relationships and phylogeny. One of the features of the flagellar apparatus that links kinetoplastids and euglenoids, and distinguishes them from other protists, is that the **transition region**—the zone linking the flagellar bases with the portion of the flagellum that extends from the cell—is rather featureless. In contrast, in other algal groups this transition region may contain complex structures that vary in detail from group to group.

Bodonids and some euglenoids also share a similar feeding apparatus, consisting of a simple tubular cytostome (cell mouth) opening into the microtubular-lined flagellar canal. A vestigial digestion system has been detected in some autotrophic (green plastid-containing) euglenoids that no longer feed on particles (Willey and Wibel, 1985). This evidence suggests that the phagotrophic mode of nutrition was the ancestral type for the kinetoplastid/euglenoid clade, and that phagotrophy was lost after acquisition of plastids in at least one major lineage of euglenoids.

Euglenoids are distinguished from kinetoplastids by two principal features. The first is the production of reserve storage granules known as **paramylon**, a β-1,3-linked glucan that does not stain blue-black with iodine-iodide solution and is found in the cytoplasm of even colorless forms (e.g., Fig. 8–5). The second distinguishing feature is a surficial pellicle composed of ribbonlike, interlocking proteinaceous strips that wind helically around cells just beneath the plasma membrane, giving cells a striated appearance (Figs. 8–6, 8–17b). Kinetoplastids do not store reserve materials, and while they may produce a glycoproteinaceous extracellular matrix, a pellicle like that of euglenoids is absent. The euglenoid genus *Petalomonas* may represent an ancestral type among euglenoids. It has a feeding apparatus, relatively few non-spiral pellicular strips, and no paramylon or chloroplasts. This hypothesis is supported by phylogenetic analysis of euglenoids based on both morphological and SSU rDNA molecular data (Montegut-Felkner and Triemer, 1997) (Fig.

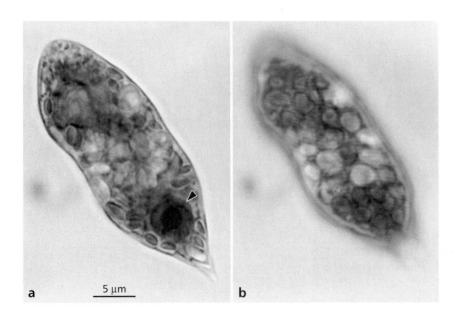

Figure 8–6 A through-focus pair of a *Euglena* cell stained with acetocarmine, which stains DNA. In (a), the nucleus with nucleolus are clearly visible (arrowhead), while in (b), the spirally arranged pellicular strips are evident.

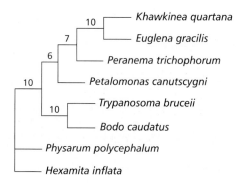

Figure 8–7 A phylogenetic tree for the euglenoids and their kinetoplastid relatives. This tree was generated using maximum likelihood and the numbers represent bootstrap values (ten resamplings). (From Montegut-Felkner and Triemer, 1997 by permission of the *Journal of Phycology*)

8–7). Results suggest that the hypothetical ancestor was phagotrophic and had two flagella (both containing a rod) joined by a striated connective. The flagellar canal then appeared, along with a longitudinally striated pellicle, as illustrated by the early divergent euglenoid *Petalomonas*. Then a clade of euglenoids having a helically organized pellicle and paramylon appeared (as illustrated by *Peranema*), followed by gain in one lineage of chloroplasts and a light-sensing system, coupled with loss of phagotrophic capacity (illustrated by *Euglena*). Finally, some members of the photosynthetic lineage secondarily lost autotrophic capability (though not their plastids), becoming reliant upon osmotrophy (Montegut-Felkner and Triemer, 1997).

The origin of plastids in euglenoids was most likely a single endosymbiotic event—capture of a eukaryotic green alga—because euglenoid plastids are typified by three envelope membranes, and the green plastids possess photosynthetic pigments similar to those of green algae. According to this hypothesis, the green algal endosymbiont's nucleocytoplasm has been permanently lost, and the outer plastid membrane represents the remains of either the host cell's phagocytotic vesicle (food vacuole), or the endosymbiont's plasma membrane (Fig. 8–8). The euglenoid outer chloroplast envelope membrane is not lined with ribosomes, nor is it usually connected to the nuclear envelope as is the case with the plastid endoplasmic reticulum (PER) of some other algae that have acquired plastids by secondary endosymbiosis. At least one cytoplasmically synthesized protein is transported to the Golgi apparatus where it is encased in a membrane prior to import by plastids (Sulli and Schwartzbach, 1995).

Euglenoid Motility and the Role of the Pellicle

Euglenoids can swim by means of one or more flagella, and flagellar features visible at the light microscopic level are commonly used as taxonomic criteria. In some forms, such as the common and familiar *Euglena*, only one flagellum emerges from the flagellar pocket; the other remains very short and is typically not visible. Other euglenoids have two emergent flagella; one extending forward from the cell anterior, the other extending laterally or backward over the cell, and therefore difficult to detect. A few euglenoids have three to seven emergent flagella. The anterior flagellum of some genera is distinctive in that it extends rigidly from the cell and is capable of moving only at the very tip (Walne and Kivic, 1990).

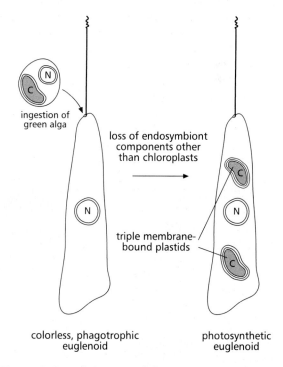

Figure 8–8 A diagram modeling acquisition of plastids by euglenoids through secondary endosymbiosis. There is uncertainty as to whether the outer membrane of the plastid is of host or endosymbiont origin.

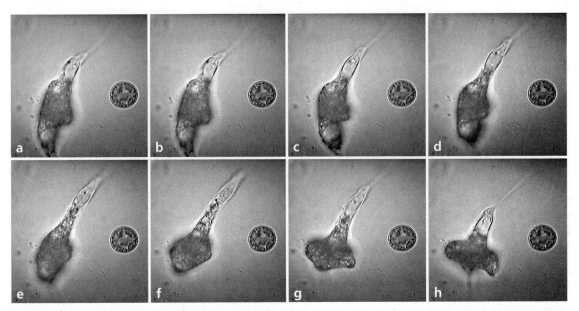

Figure 8–9 A series of video images (every fifth frame shown) of a colorless euglenoid moving metabolically. Note the considerable shape changes that occurred while the cell moved slowly upward.

In addition to swimming, euglenoids can move by cellular undulations that are known as metaboly, peristaltic movement, or euglenoid movement. These movements nearly defy description, but once observed, are readily recognizable and unique to euglenoids. Cells that are swimming are uniform in shape, whereas during metaboly cells exhibit rapid shape changes (Fig. 8–9). The cytoplasm appears to accumulate at one end of the cell and then is rapidly redistributed, resulting in forward or backward cell movement. In addition, cells can flex from side to side. The ability to move in these ways is based upon the highly flexible and plastic pellicle. Pellicles are 70–80% protein and also contain lipids. There are at least ten different polypeptides, and of these, three are major constituents of the pellicle (Leadbeater and Green, 1993). Pellicular strips are very long ribbons that typically extend helically from the cell apex to the posterior. The ribbons are curved at both edges, but in different directions, one

Figure 8–10 (a) Three-dimensional diagram of the interlocking strips that make up the euglenoid pellicle. (b) An individual pellicular strip (in cross section) and associated structures. ER=endoplasmic reticulum tubule, M=microtubules, PM=plasma membrane, PP=periodic projections of upturned part of strip, PS=pellicular strip, TF=traversing filaments (From Suzaki and Williamson, 1986)

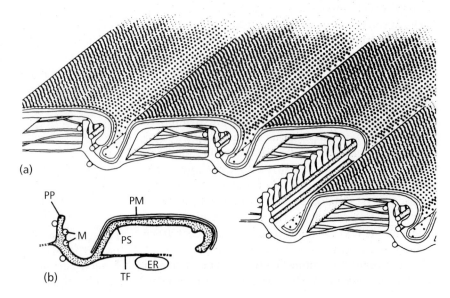

forming a ridge, and the other a groove. The ridged edge of one pellicular strip fits into the grooved edge of the adjacent strip, forming a bendable hinge, thereby conferring flexibility to the cell covering (Fig. 8–10). Moreover, one edge of each strip is associated with four microtubules, two on each side of the strip, forming rails that facilitate lateral sliding of pellicular strips past each other (Murata and Suzaki, 1998); this has been observed by video microscopy. Computer simulations based on this euglenoid motility model can account for all of the observed types of metabolic movement (Häder and Hoiczyk, 1992). Prior to cell division, the number of pellicular strips doubles, with new strips being produced between older ones, in an alternating fashion. Each daughter cell thus receives an equal number of old and new strips (Leadbeater and Green, 1993). Some euglenoids have relatively rigid pellicles and do not exhibit metaboly.

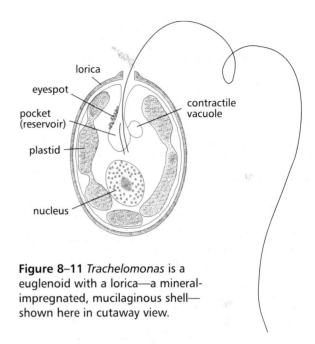

Figure 8–11 *Trachelomonas* is a euglenoid with a lorica—a mineral-impregnated, mucilaginous shell—shown here in cutaway view.

Mucilage Production and the Formation of Palmelloid Colonies or Cell Envelopes

Euglenoids can secrete polysaccharide and/or glycoproteins (Cogburn and Schiff, 1984) from mucilage bodies known as mucocysts that lie just beneath pellicular strips (Hilenski and Walne, 1983). Each is connected to the exterior via a canal that opens into the groove between adjacent pellicular strips. The mucilage bodies are thought to originate from the Golgi apparatus. They are transparent and generally not visible at the light microscopic level until stained with a neutral red (0.1% solution) (Fig. 8–5). This dye allows patterns of mucilage body shape, size, and distribution, which are used to distinguish species (Gojdics, 1953), to be visualized. Mucilage can cover the entire cell in layers of varying thickness. Mobile cells usually have only a thin mucilage layer, whereas immobile cells lacking flagella may be embedded in relatively thick layers of jelly to form scum on water or other surfaces that is one cell layer thick. Such mucilage-embedded immobile stages of cells—that under other conditions are mobile—occur in a number of algal groups and are known as **palmella** or palmelloid stages, after the green alga *Palmella* (Gr. *palmos*, quivering), which commonly exists in such a form.

The euglenoid genera *Trachelomonas*, *Ascoglena*, and *Strombomonas* are enclosed by a rigid envelope of mineral-impregnated mucilage, known as a **lorica** (Figs. 8–11, 8–23). The minerals are ferric hydroxide or manganese compounds; they confer an orange to black coloration to the lorica, the color depending upon mineral concentration. Loricas are not considered to be cell walls because they are are well separated from the cell membrane, but both loricas and cell walls are considered to be extracellular matrix. Loricas can occur in other groups of algae as well. Both genetic and environmental factors influence mineral deposition within the loricas of euglenoids. In *Trachelomonas*, for example, there are species-specific patterns of deposition that result in easily observed variations in lorica ornamentation. However the extent of mineralization can be affected by environmental levels of iron and manganese as well as redox conditions (reflective of oxygen concentrations). When oxygen concentrations are high, iron and manganese are typically precipitated into low-solubility compounds. Under the low-oxygen conditions that may occur in the sediment-water interfaces often inhabited by euglenoids, these minerals are solubilized. It has been hypothesized that oxygen production by photosynthetic euglenoids helps to precipitate iron and manganese minerals into the lorica. Some *Euglena* species that inhabit iron-rich waters bear regularly arranged, distinctive pellicular warts that contain orange-colored ferric hydroxide (Dawson, et al., 1988) (Fig. 8–19).

Euglenoid Reproduction

Sexual reproduction does not occur in euglenoids with regularity, if at all. Asexual reproduction is by longitudinal division, proceeding from apex to base, such that euglenoids in the process of cytokinesis appear to be "two-headed" (Fig. 8–12). Chromosomes of euglenoids are unusual in that they are permanently condensed, i.e., they do not undergo cell-cycle changes in DNA coiling as do those of most eukaryotes. Under favorable conditions the condensed chromosomes can be visualized by light microscopy (Fig. 8–13). Prior to mitosis, the nucleus migrates to the region just below the cell pocket (reservoir); the nuclear envelope does not break down during mitosis, as it does in some other protists, animals, and land plants. Often (but not always), a pair of basal bodies (that have replicated prior to nuclear division) forms each of the spindle poles. The spindle develops within the confines of the nuclear envelope, lying at right angles to the long axis of the cell. Following chromosomal separation, daughter nuclei form by central constriction of the parental nucleus (Fig. 8–2).

In response to changing environmental conditions, euglenoids may form resting cysts. Cyst formation involves loss of flagella, increase in the number of paramylon granules, swelling and rounding of the cells, increase in the number of mucilage bodies, and deposition of a layered mucilaginous wall, consisting primarily of polysaccharides. It has been suggested, based on laboratory and field observations, that cyst production is triggered by low nutrient levels, or by low N/P ratios (Triemer, 1980; Olli, 1996). Recognition of cysts as being euglenoid may depend upon detection of paramylon granules. Although not present in all euglenoids, the unusually large red-orange eyespots, and distinctive plastids of green forms, can also be helpful in deducing the euglenoid origin of spherical cyst stages (Fig. 8–14). In addition, euglenoids may in some cases continue to undergo metaboly, rotating within the confines of the cyst wall.

Plastids and Light-Sensing Systems

Euglenoid plastids can be shaped like plates, shallow cups, or ribbons, these sometimes being arranged in star-shaped aggregations. The edges of plastids are in

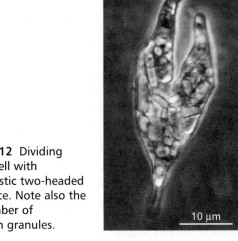

Figure 8–12 Dividing *Euglena* cell with characteristic two-headed appearance. Note also the large number of paramylon granules.

10 µm

some cases dissected into lobes. There are typically many plastids per cell; under normal circumstances plastid division is coordinated with cytokinesis. As in green algae, photosynthetic pigments include chlorophylls *a* and *b*, as well as β-carotene. However, some xanthophylls not typical of green algae may be present in euglenoids, such as diadinoxanthin. Plastid thylakoids typically occur in stacks of three; although some larger stacks can be observed, the grana typical of land plants and certain green algae are not present. Although some euglenoid plastids do not contain pyrenoids (those of *Phacus* for example), most do. In some species pyrenoids are embedded within the plastids, but in others the pyrenoid occurs at the end or edge of plastids or extends from the main body of the plastid on a stalk. There may be a shell of paramylon granules covering the portions of the pyrenoids that are in contact with the cytoplasm (Whatley, 1993). However this shell cannot be rendered visible at the light microscopic level by treatment with an iodine-iodide solution, as can the starch shell that typically forms around pyrenoids of many green algae. A number of euglenoids are colorless but possess plastids that lack chlorophyll. Treatment of euglenoids with heat, extended periods of darkness, or certain antibiotics results in plastid-bleaching, i.e., loss of pigmentation. The plastids dedifferentiate into a proplastid state in which thylakoids are few or absent; proplastids continue to divide. When exposed to light, the bleached cells may regain thylakoids and chlorophyll pigmentation. Changes in plastid proteins that are associated with heat-induced bleaching were studied by Ortiz and Wilson (1988).

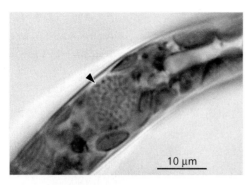

Figure 8–13 The nucleus (arrowhead) of a *Euglena* cell as seen with oil-immersion bright-field microscopy. The permanently condensed chromosomes are visible.

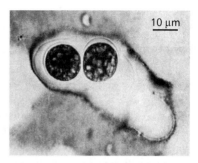

Figure 8–14 Euglenoid cysts. (From Bold and MacEntee, 1973 by permission of the *Journal of Phycology*)

The euglenoids, whether green-pigmented or colorless, frequently (but not always) possess a light-sensing system. The light-sensing system of euglenoids consists of two structures. The first of these is the paraflagellar body (PFB), a swelling at the base of at least one of the emergent flagella (Fig. 8–15) that contains blue light-sensitive flavins, that appear green when viewed by fluorescence microscopy with blue or blue-violet excitation. The second component of the light-sensing system is the eyespot (stigma) that is located in the cytoplasm adjacent to the flagellar pocket (reservoir) and opposite the basal PFB. The eyespot typically appears bright orange-red. Although eyespots occur in other algae, they are particularly conspicuous in euglenoids, being as large as 8 μm in diameter and therefore quite noticeable at the light microscopic level (Fig. 8–1). Ultrastructural examination reveals that euglenoid eyespots consist of from 50–60 globules arranged in a single layer, often bound by a membrane (Fig. 8–15). These globules contain the carotenoids astaxanthin and/or echinenone, which give the eyespot its orange-red coloration.

Euglenoid Ecology

At least one expert has suggested that there are probably no truly planktonic euglenoid species (Lackey, 1968), and others suggest that euglenoids are fundamentally occupants of interfaces, such as the air-water and sediment-water boundaries (Walne and Kivic, 1990). In such habitats euglenoids can be infected by chytrids and consumed by herbivores including other euglenoids, such as the predacious *Peranema* (*Pseudoperanema*). The euglenoid storage product, paramylon, is comparatively indigestible;

paramylon granules have been observed to pass through the gut of herbivores unharmed. In order to digest euglenoid storage products, herbivores require a gut enzyme, laminarase, that can degrade paramylon. Marine populations of *Eutreptiella gymnastica* are readily grazed by mesozooplankton in the coastal Baltic Sea (Olli, et al., 1996).

Certain euglenoids are known for tolerating extreme conditions. Some seem able to migrate into soils and persist there for long periods in a quiescent

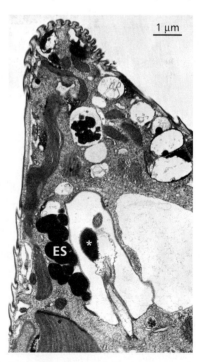

Figure 8–15 Anterior end of *Euglena granulata* cell as seen with TEM. Note eyespot granules (ES) lying opposite the paraflagellar body (asterisk). (From Walne and Arnott, 1967)

state. Though comparatively rare, *Euglena gracilis* has been recovered from cryptogamic crusts of semi-arid and arid lands of North America (Johansen, 1993). *Euglena mutabilis* is able to grow in extremely low pH waters, such as streams draining coal mines and the acidic, metal-contaminated ponds of the Smoking Hills region of the Canadian Arctic. The optimal pH for growth of this species is 3.0 but pH values lower than 1.0 can be tolerated. Euglenoids are also reported to be able to adapt to salinity increases more quickly than can other algae (see review by Walne and Kivic, 1990).

Euglenoid Diversity

It is estimated that there are more than 40 genera of euglenoids, and 800–1000 species. Species estimates, particularly for the genera *Euglena* and *Trachelomonas*, may be inflated in that minor variants observed in nature may reflect polymorphic variation rather than distinct taxa. Few species-comparison studies have been done with laboratory-grown cultures. Most genera are free-living and exist primarily in the mobile state (Walne and Kivic, 1990). Several classification systems exist, including those of Simpson (1997), Leedale (1967) with six orders or five suborders (Leedale, 1985), based on number and length of flagella and other morphological and nutritional characters, and Farmer and Triemer's (1988) system. The

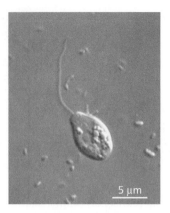

Figure 8–16 The colorless euglenoid *Petalomonas*. (Photograph courtesy R. Triemer)

last, which has seven orders, is based on ultrastructural features such as feeding apparatus, flagella, and pellicle structure. Within orders, genera are differentiated by the number of flagella.

PETALOMONAS (Gr. *petalon*, leaf + Gr. *monas*, unit) is a colorless, osmotrophic, and/or phagotrophic, relatively rigid cell (Fig. 8–16). The single emergent flagellum is held forward and moves only at the tip. Cells possess no plastids, eyespot, or paraflagellar body (swelling). It has been suggested that this euglenoid may represent the ancestral type for the group, possibly having diverged prior to acquisition (by a sep-

Figure 8–17 *Peranema*, a colorless, phagotrophic euglenoid that has very plastic cells. In (a), the thick, swimming flagellum is particularly obvious, as are the nucleus (N) and ingestion rods (arrowheads). In (b), the ornamented pellicle is evident.

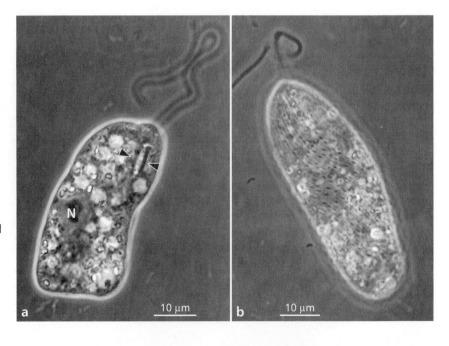

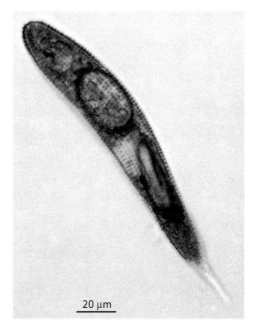

Figure 8–19 A *Euglena* species with warty protrusions on its pellicle.

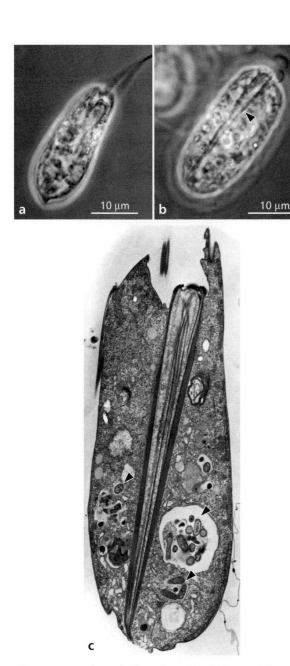

Figure 8–18 The colorless phagotrophic *Entosiphon*. In (a), the scalloped appearance of the anterior end of the cell is evident. The funnel-shaped ingestion apparatus (arrowhead) is shown with light microscopy in (b) and TEM in (c) where ingested bacteria are pointed out by arrowheads. (c: Micrograph courtesy P. Walne)

arate euglenoid lineage) of plastids and light-sensing systems.

PERANEMA (Gr. *pera*, pouch + Gr. *nema*, thread) is another colorless phagotroph (Fig. 8–17)

that, like *Petalomonas*, has a rigid, anteriorly directed flagellum that moves only at the tip. A second flagellum is directed backward and is appressed to the cell body. Ultrastructure of the flagellar root system and flagellar hairs of *Peranema* was studied by Hilenski and Walne (1985a, b). *Peranema* also exhibits metaboly. Eyespots and PFBs are lacking, but there is a specialized set of ingestion rods at the cell anterior that are involved in prey capture and ingestion (Fig. 8–17). *Peranema* is a voracious consumer of euglenoids, cryptomonads, yeasts, green algae, and other particles. The use of video microscopy and the scanning electron microscope (SEM) reveals that *Peranema* has two feeding modes. Particles can be ingested entirely or rasped open by grating action of the feeding apparatus, after which the contents are sucked into the cell and deposited in a food vacuole. The latter process can be accomplished in less than ten minutes. Prey are apparently selected by probing with the anterior flagellum (Triemer, 1997).

It has been suggested that the generic name *Peranema* may be invalid, depending upon which nomenclature system is applied, because application of the name *Peranema* to a fern has precedence according to the *Botanical Code of Nomenclature*. The name *Pseudoperanema* has been proposed as a substitute if euglenoids continue to be covered under the *Botanical*

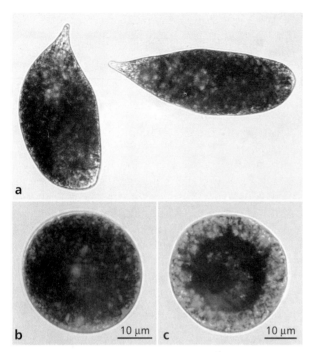

Figure 8–20 *Euglena sanguinea*, a species that can form bright red blooms. Motile cells are shown in (a). In (b) and (c), two rounded-up cells are pictured. The dark areas represent the red pigment, while green areas appear as a lighter gray. The cells are able to control the distribution of plastids and pigment globules, such that they may appear more red or green, depending upon which bodies are found closest to the cell surface. (Photographs courtesy C. Taylor)

cross section. Most *Euglena* species are elongated, with a rounded anterior and the posterior tapered to a point (Fig. 8–19). A dozen or so species produce red granules in numbers sufficient to give cells a bright or brick-red appearance (Fig. 8–20), resulting from large amounts of the carotenoid astaxanthin. When large populations of cells are present, they may form dramatic, blood-red surface scums on ponds or other water bodies. The formation of such scums is favored by the presence of high levels of dissolved organic compounds and high temperatures. Although rain can break up the scum, it can re-form when the disturbance ends. Cells of some species appear red most of the time, but those of several species can change from green to red within 5–10 minutes in response to increased light intensity such as at sunrise, then return to green coloration at the appearance of a cloud or at sunset. This color change involves differential positioning of the red globules and green plastids at the center and periphery of cells. When the red globules are at the cell periphery, cells appear red; when the red globules occupy a central position, surrounded by green plastids, cells appear green. The mechanism underlying positional changes of plastids and red globules is not understood. *Euglena sanguinea* is said to be

Code (see Walne and Kivic, 1990, for a discussion of this taxonomic issue).

ENTOSIPHON (Gr. *entos*, inside + Gr. *siphon*, tube) is a colorless phagotrophic euglenoid with a rigid cell that is longitudinally furrowed such that the anterior end has a scalloped appearance (Fig. 8–18a). It has a large, funnel-shaped ingestion apparatus that extends the length of the cell (Fig. 8–18b, c). Two flagella are present, the shorter extending forward and involved in motility (bending only near the tip) and the longer, a trailing flagellum.

EUGLENA (Gr. *eu*, good, true, or primitive + Gr. *glene*, eye), the best-known euglenoid genus, with more than 150 species having been described (Gojdics, 1953), is characterized by the presence of a single emergent flagellum. Plastid shape is variable but often discoidal, and cells are cylindrical (not flattened) in

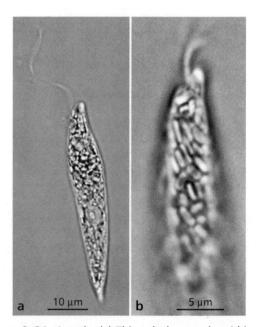

Figure 8–21 *Astasia*. (a) This colorless euglenoid is quite similar to the photosynthetic *Euglena* in overall shape and its ability to undergo metaboly. In (b), some of the numerous paramylon granules are shown in the anterior end of the cell.

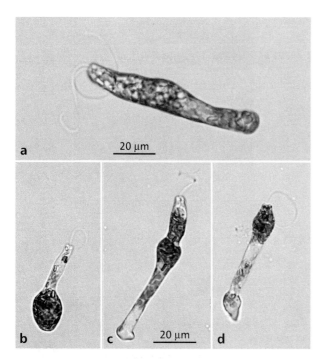

Figure 8–22 Live cells of *Eutreptia* undergoing metaboly. In (a) the two flagella are visible. Some of the different shapes assumed by cells are shown in (b)–(d).

the commonest of the red species (see review by Walne and Kivic, 1990).

ASTASIA (Gr. *astatos*, unstable or restless) is very much like *Euglena* but lacks plastids and a light-sensing system. The single emergent flagellum and multiple peripheral paramylon granules are distinctive (Fig. 8–21). Among other habitats, it lives within the guts of flatworms, nematodes, rotifers, and copepods.

EUTREPTIA (Gr. *eu*, good, true, or primitive + Gr. *treptos*, turned) is found in fresh and marine waters, where it sometimes forms blooms. The ribbonlike plastids are distinctively arranged in stellate groups of 25–30 radiating from a central pyrenoid. There are two emergent flagella (Fig. 8–22), and an eyespot and adjacent paraflagellar body are present. The genus is not known to exhibit phagotrophy. It is known for active metaboly (euglenoid movement). The ultrastructure of *E. pertyi* was studied by Dawson and Walne (1991a, b). A similar form, *Tetraeutreptia*, described from eastern Canadian marine waters, is also green but has four flagella, two of which are long and two short (McLachlan, et al., 1994).

TRACHELOMONAS (Gr. *trachelos*, neck + Gr. *monas*, unit) is a widespread genus whose green-pigmented cells are encased within a rigid, mineralized, sometimes highly ornamented lorica, which may vary in color from colorless to black-brown (Fig. 8–23). When the lorica is faintly tinted, it is easy to see the green protoplast within, but when the lorica is darkly colored, it may be impossible to detect any green coloration. The cells may be highly metabolic within the lorica, but this is detectable only when the lorica is not highly colored. A single emergent flagellum protrudes through the open neck region of the lorica. Reproduction is accomplished by emergence of one or both daughter cells from the lorica, followed by the development of a new lorica. The species are differentiated by ornamentation of the lorica. One species (*T. grandis*) does not have a spiny lorica, but appears spiny because of the attachment of numerous bacteria to the lorica surface (Rosowski and Langenberg, 1994).

PHACUS (Gr. *phakos*, lentil) cells are oval to nearly circular and are highly flattened. They have a pellicle sufficiently rigid that metaboly is never observed (Fig. 8–24). Plastids are numerous. The fairly

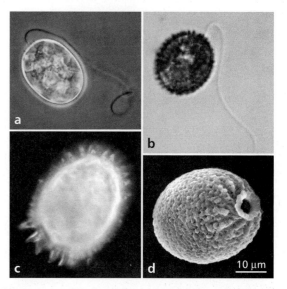

Figure 8–23 The loricate euglenoid *Trachelomonas*. In (a), the lorica is thin and translucent, while in (b), it is dark and somewhat warty in appearance. The species shown in (c) possesses characteristic spines on its lorica (the cell's anterior is toward the upper right). A lorica is seen with SEM in (d). (d: From Dunlap, et al., 1983)

Figure 8–25 *Colacium*, a stalked euglenoid, growing epibiotically on *Daphnia*. (Micrograph courtesy R. Willey)

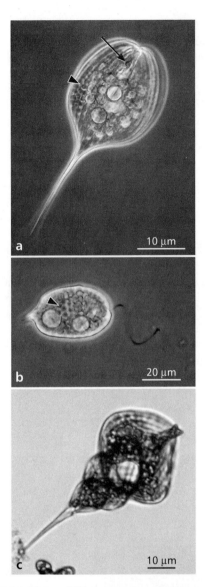

Figure 8–24 Three different species of *Phacus* are shown. The small spherical plastids (arrowheads) are evident in (a) and (b). Note the eyespot (arrow) in (a), flagellum in (b), and the twisted cell body of *P. helikoides* (c).

common *Phacus helikoides* (Fig. 8–24c) is distinctive in being twisted throughout; *P. tortus* is noticeably twisted just at the posterior. There is just a single emergent flagellum, as in *Euglena*.

COLACIUM (Gr. *kolax*, flatterer) is an attached (sessile) euglenoid of widespread occurrence on aquatic substrates, most noticeably on planktonic zooplankton (Figs. 3–24, 8–25). It forms colonies whose cells occur at the ends of branched, gelatinous strands. In the sessile stage, flagella are non-emergent, but individual cells may produce an emergent flagellum and swim away to generate a new colony elsewhere. The cells attach by the anterior end, and secrete a stalk composed of carbohydrate; subsequent cell division yields new branched colonies. The characteristic stalk is formed of carbohydrate extruded from the cell anterior in the form of Golgi-generated mucocysts. More than one hundred mucocysts may accumulate within the anterior of each cell prior to their excretion (Willey, 1984). A discussion of *Colacium*'s role as an epibiont on surfaces of *Daphnia* and other aquatic animals can be found in Chapter 3.

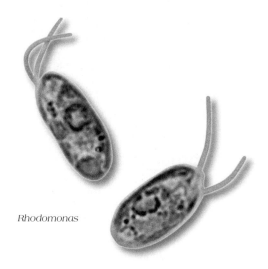

Chapter 9

Rhodomonas

Cryptomonads

Cryptomonads, their name literally meaning "hidden single cells," are among the most inconspicuous of the algae. There are several reasons for this: Cryptomonads are relatively small—3–50 μm in length—members of the phytoplankton; they are often most abundant in cold or deep waters; they are readily eaten by a wide variety of planktonic herbivores; and natural collections are not easily preserved, the cells tending to burst readily when subjected to environmental shock. Cryptomonads are probably most appreciated by plankton ecologists who recognize their high quality as food for zooplankton (discussed in Chapter 3) and algal evolutionary biologists who note the significance of cryptomonad ultrastructure and molecular biology in the study of secondary endosymbiosis (discussed in Chapter 7). This being the case, the present chapter will provide additional information on the ecology and cell biology of cryptomonads, followed by a survey of their diversity. We begin with a brief comparison of cryptomonads to the euglenoid flagellates that were discussed in the previous chapter as a useful way to continue an introduction to algal flagellates. This comparison will also aid in distinguishing these flagellates from each other and from similar algae encountered in mixed field conditions.

Cryptomonads and euglenoids share a number of characteristics. Both groups occur in a variety of aquatic environments, fresh and marine, and at least one B vitamin is required by all members of both groups. Both are fundamentally biflagellate, with flagella emerging from an apical depression (Fig. 9–1); they are primarily unicells that can also occur as nonmotile, mucilage-embedded palmelloid (palmella) stages (Fig. 9–2), and most forms of both groups are essentially naked, with the rigid portion of the cell covering (usually) occurring inside the plasma membrane. Both cryptomonads and euglenoids can produce thick-walled cysts that are able to survive adverse conditions, but neither group has a good fossil record. Putatively primitive plastidless forms as well as pigmented forms and genera having colorless plastids occur in both groups. The plastids of both groups arose by secondary

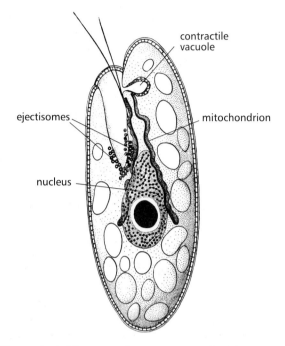

Figure 9–1 Diagram of a cryptomonad with two flagella emerging from a depression. (From Hollande, 1952b)

endosymbiosis. Photosynthetic storage of both groups occurs as granules in the cytoplasm. Such similarities do not indicate that cryptomonads and euglenoids are closely related. As noted in Chapter 7, strong ultra-structural and molecular evidence links the cryptomonads with the glaucophytes.

Differences between cryptomonads and euglenoids include the fact that cryptomonads do not

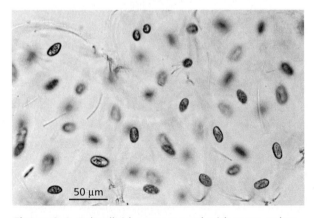

Figure 9–2 Palmelloid cryptomonad with more or less equally spaced cells in a gelatinous mass.

exhibit metabolic movements as do many euglenoids, and photosynthetic storage material in cryptomonads is starch, which, like that of plants and green algae (but unlike euglenoid paramylon), stains blue-black with an iodide-iodine solution. Cryptomonad flagellar ornamentation, roots, and transition region are distinct from those of euglenoids, and mitosis also differs in several ways. The plastid pigments of cryptomonads resemble those of red algae, in contrast to the green algallike pigments of euglenoids. In contrast to euglenoids, which tend to occupy mud, sand, or water surfaces and are less commonly planktonic, cryptomonads are widespread and important members of the phytoplankton.

Cryptomonad Ecology

Cryptomonads are especially prominent in oligotrophic, temperate, and high-latitude waters of both lakes and oceans. They seem to be more important in colder waters, typically becoming abundant in winter and early spring when they can begin growth under the ice. For example, cryptomonads may dominate the spring phytoplankton bloom in the North Sea where they are believed to make significant contributions to net primary productivity. Localized blooms of cryptomonads also occur in Antarctic waters; the blooms are correlated with the influx of water from melting glaciers (Lizotte, et al., 1998). In perennially ice-covered Antarctic lakes, *Chroomonas lacustris* or a *Cryptomonas* species may dominate the algal flora during the austral summer, contributing more than 70% of the total phytoplankton biomass (Spaulding, et al., 1994). The maximum growth rate for many cryptomonads is one division per day and occurs at a temperature of about 20° C, with growth declining rapidly at higher temperatures. Cryptomonads seem to occur only rarely in ocean waters at temperatures of 22° or higher, and they are absent from hot springs and hypersaline waters. Marine (but not freshwater) cryptomonads seem to be unusually tolerant to rapid salinity changes (Klaveness, 1988).

In oligotrophic freshwater lakes, cryptomonads typically form large populations in deep waters (15–23 m) at the junction of surface oxic (oxygen-rich) and bottom anoxic (oxygen-poor) zones, where light levels are much lower than in surface waters. Deep-water accumulations of photosynthetic organisms form what are generally known as deep-chlorophyll maxima (Chapter 22). In eutrophic lakes deep-chlorophyll maxima tend to be populated by red filamentous cyanobacteria, and in other aquatic systems purple bacteria may constitute such growths.

Ecologists have wondered what characteristics enable cryptomonads to thrive in these deep waters, suspecting both highly efficient light harvesting abilities and heterotrophic capacities. In order to test these hypothesis, Gervais (1997) isolated and grew cultures of several species of *Cryptomonas* from a deep-chlorophyll maximum layer for comparison to cryptomonad isolates from surface waters of the same lake. Two of the deep isolates, *C. phaseolus* and *C. undulata*, grew best under light-limiting conditions and were able to survive long periods of complete darkness. Neither species was able to ingest fluorescently labeled beads or bacteria, suggesting that they were probably incapable of phagotrophy. Uptake of radioactive tracer-labeled glucose was small relative to total cell carbon, suggesting that osmotrophy was probably not being used by these cryptomonads as a survival mechanism in low light environments. However, it has been suggested that uptake of organic compounds might facilitate survival during extended periods of darkness, especially when temperatures are in the range of 3°–5° C. Photosystem adaptation to low light levels, together with absence of predation, access to nutrients regenerated by benthic decomposition processes, and tolerance of sulfide, probably explain the occurrence of cryptomonad species in deep waters. Cryptomonads are also present in deep-chlorophyll maxima in marine waters (see references in Klaveness, 1988).

A bloom of the cryptomonad *Cyanomonas* is reported to have caused a massive kill of catfish in a Texas pond, but no toxin was identified. *Chilomonas paramecium* produces an ichthyotoxin (fish-killing) toxin similar to that of some chrysophyceans, but only in very small amounts, and this cryptomonad has not been associated with any fish kills (see references in Gillott, 1990).

Most cryptomonads require vitamin B_{12} and thiamine, and some also require biotin. Cryptomonads can utilize ammonium and organic sources of nitrogen, but marine forms in particular seem less able than other algae to utilize nitrate or nitrite. Production of cryptomonad resting stages seems to be induced by high light levels coupled with nitrogen deficiency. Organic compounds stimulate growth of a variety of cryptomonad species, and phagotrophy has been documented in the pigmented *Cryptomonas ovata* as well as in various colorless cryptomonads (Gillott, 1990).

Ultrastructural evidence suggests that the blue-green cryptomonad *Chroomonas pochmanni* is mixotrophic. This species possesses a specialized vacuole used for capturing and retaining bacterial cells. Bacteria are drawn into the vacuole through a small pore formed in the vestibulum, where periplast plates are absent. When it is full of bacteria, this vacuole can be seen by light microscopy, appearing as a transparent bulge at the anterior end of the cell. The bacteria are morphologically similar, suggesting selectivity on the part of the cryptomonad. Bacterial cells appear to be digested within smaller vesicles in the cryptomonad cells. This organism provides evidence that early cryptomonads were phagotrophic and that most have subsequently lost this capability (Kugrens and Lee, 1990). Cryptomonad cells may contain putatively endosymbiotic bacteria (McFadden, 1993), but the physiological impacts of such associations have not been elucidated.

A significant ecological aspect of cryptomonads is their incorporation within cells of the mixotrophic ciliate *Myrionecta rubra* (formerly *Mesodinium rubrum* or *Cyclotytrichum meunieri*). This protozoan can form dramatic, nontoxic, red-colored blooms in waters off the coast of Peru and Baja, Cal-

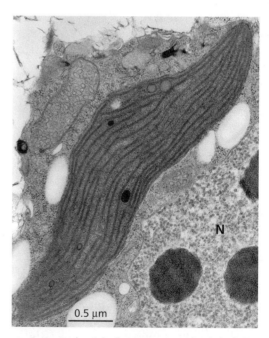

Figure 9–3 A plastid of cryptomonad origin lying adjacent the nucleus in the dinoflagellate *Amphidinium wigrense*. (Photograph by L. Wilcox and G. Wedemayer)

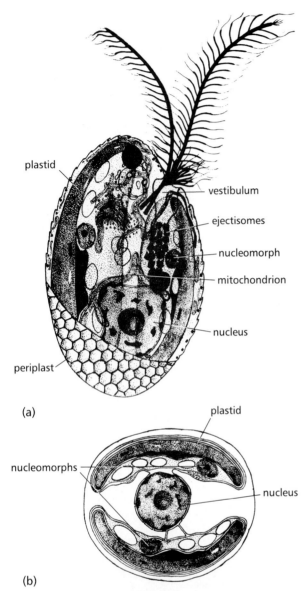

(a)

(b)

Figure 9–4 Diagram of a *Cryptomonas ovata* cell in (a) longitudinal and (b) cross-sectional view. The cell is covered by a periplast of polygonal plates. The two flagella emerge from the vestibulum. The two plastids are each associated with a nucleomorph. Also shown are ejectisomes, mitochondria, and the nucleus. (From Santore, 1985)

ifornia, and other locations, typically in upwelling conditions that bring additional nutrients to surface waters. The photosynthetic pigments of the cryptomonad endosymbionts impart a red coloration to the ciliate host. Each ciliate may contain numerous endosymbionts, each separated from host cytoplasm by two membranes. Symbionts contain a nucleus, mitochondria, plastids, and some other structures typical of cryptomonad cells, but other cryptomonad cellular features are absent. Sometimes the endosymbiont is subdivided into separate membrane-bound units, with plastids, mitochondria, and a bit of cytoplasm occurring in one compartment, and the cryptomonad nucleus and more mitochondria in another (McFadden, 1993). The plastids and mitochondria have ultrastructural features (described later) that identify them as having been derived from cryptomonads and distinguish them from organelles of the ciliate host.

Certain dinoflagellates, including *Gymnodinium acidotum* and *Amphidinium wigrense*, also regularly contain portions of cryptomonad cells, particularly plastids (Fig. 9–3). These dinoflagellates possess phagotrophic capabilities (see Chapter 11) and can thus harvest most or parts of cryptomonads or other cells. In some cases the cryptomonads possess nuclei, but in others, only the plastids persist within the dinoflagellate cytoplasm.

Cell Biology of Cryptomonads

As they swim, cryptomonad cells rotate such that the flattened, asymmetrical cell shape can be readily observed with the light microscope (LM). There are two slightly unequal flagella, each about the same length as the cell. One cryptomonad flagellum is stiff, while the other propels the cell. Flagella emerge near an anterior depression that defines the ventral (front or belly) cell surface. This depression is known as a **vestibulum**; it forms the anterior end of the cell gullet or furrow (Fig. 9–4). A prominent contractile vacuole empties into the vestibulum. Furrows occur when the depression is open along its length, whereas a gullet is formed when the depression is open only at the anterior end, so that a tube is formed (Gillott, 1990). These differences in the shape of the anterior depression may distinguish different cryptomonads. The vestibulum is normally open but can rapidly close when cells are perturbed. Fibrils of the contractile protein centrin are probably involved in this movement. Centrin contracts by supercoiling in the presence of high levels of cellular Ca^{2+}, and adenosine triphosphate (ATP) is not involved. Reextension of the protein occurs when calcium levels are lowered and ATP is present (Melkonian, et al., 1992).

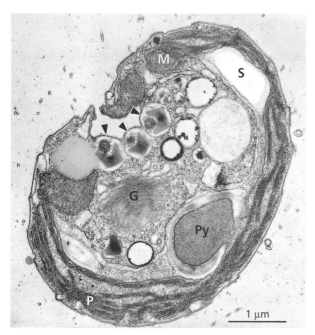

Figure 9–5 Cross section through an unidentified cryptomonad, viewed with TEM. Note the peripheral plastid (P), pyrenoid (Py) and starch (S), Golgi body (G), mitochondria (M) with flattened cristae, and three ejectisomes (arrowheads) lying next to the furrow.

Ejectisomes

The occurrence of the anterior depression is easily recognized by the nearby presence of relatively large ejectile organelles known as **ejectisomes** (ejectosomes). These can be violently discharged from cells into the vestibulum, probably as a defensive response to herbivore disturbance. The presence of an open vestibulum (furrow), rather than a tubular gullet, is an advantage in avoiding cell injury by ejectisome discharge (Santore, 1985). Cryptomonad genera differ in the number of gullet-area ejectisomes. They appear vaguely square at the LM level, whereas at the transmission electron microscopic (TEM) level, sectioned ejectisomes appear butterfly-shaped (Fig. 9–5). This shape arises from the fact that ejectisomes have a rolled ribbonlike construction, with the ribbon width decreasing toward the ends and center of the roll. Each ejectisome consists of two coiled ribbons, one smaller than the other; when coiled within the cell, the smaller ribbon lies in the depression formed by the narrow portions of the larger ribbon. When discharged, both ribbons unfurl, forming a narrow kinked barb (Fig. 9–6).

The smaller ribbon forms the shorter arm of the released ejectisome; the longer ribbon forms the longer arm. The transformation of ejectisome ribbons into a long, thin projectile resembles the common toy that consists of a stiff paper ribbon coiled onto a handled spool, which when flung outward, generates a rigid, pointed lance. Many smaller ejectisomes line the cell periphery, but these are difficult to observe with the light microscope.

When cryptomonads are more seriously irritated or experience a sudden environmental shock, such as a change in pH, osmotic conditions, or temperature, massive discharge of ejectisomes may lead to rapid cell disintegration. This is one reason why natural collections of cryptomonads are more difficult than other algae to preserve in chemical solutions for transport back to the laboratory. The use of a low concentration (1–3%) of buffered glutaraldehyde (Wetzel and Likens, 1991) favors the preservation of cryptomonads and other delicate algal cells so that they can be identified and counted using LM. Glutaraldehyde is able to quickly penetrate cells and cross-link proteins before destructive cellular reactions can occur. This is also the primary method by which cells, including those of cryptomonads, are chemically preserved for observation by SEM and

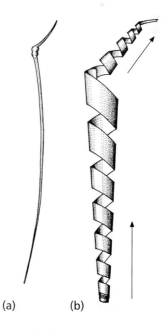

(a)   (b)

Figure 9–6 (a) A discharged ejectisome and (b) a model for the discharge of this coiled, ribbonlike structure. (From Mignot, et al., 1968)

Figure 9–7 Freeze-fracture TEM view of a cryptomonad cell (*Storeatula*) showing the periplast plates. (Micrograph courtesy P. Kugrens)

The Periplast

Beneath the plasma membrane of many cryptomonad cells lie numerous individual hexagonal, rectangular, oval, or round organic plates. These constitute the **periplast**. The pattern of these plates can be observed on surfaces of cells viewed with SEM and freeze-fracture TEM (Fig. 9–7), but is not obvious in LM view. In some cases the periplast may form a continuous layer, and in others components of the periplast may occur on the cell surface. Variations in the periplast are used to distinguish among genera and species of cryptomonads, but the most useful details are revealed only by application of freeze-fracture/etch techniques and TEM. Ultrastructurally, it can be seen that the plates are linked to the plasma membrane by small particles (Kugrens, et al., 1986). Periplast plates do not extend into the gullet or furrow region, which is instead lined by banded structures (except in the vicinity of ejectisomes) under the cell membrane. Sometimes layers of scales or other extracellular matrix material coat the cryptomonad surface.

TEM (Fig. 9–7). Cryptomonads are also unusually sensitive to the critical-point drying process widely used to prepare specimens for SEM. Freeze-drying methods are reported to work better for cryptomonads (Kugrens, et al., 1986).

Flagellar Apparatus

Visualization of small structural details characteristic of flagellar surfaces and intracellular components of the flagellar apparatus require the use of transmission electron microscopy. Both flagella are decorated with 1.5 μm long hairs and sometimes also finer hairs and small organic scales shaped like rosettes. The longer

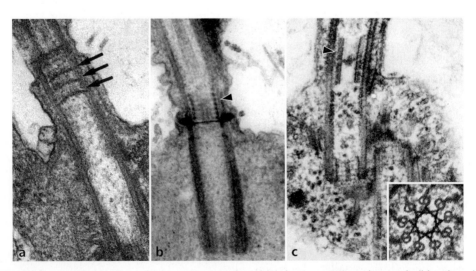

Figure 9–8 Flagellar transition region of (a) cryptomonads, which have partitions (arrows); (b) ochrophytes, which have a helical structure (arrowhead); and (c) green algae, which have a stellate structure (arrowhead). Inset in (c) shows stellate structure in cross section. (a: From Gillott, 1990; b: From Graham, et al., 1993 by permission of the *Journal of Phycology*; c: From Floyd, et al., 1985)

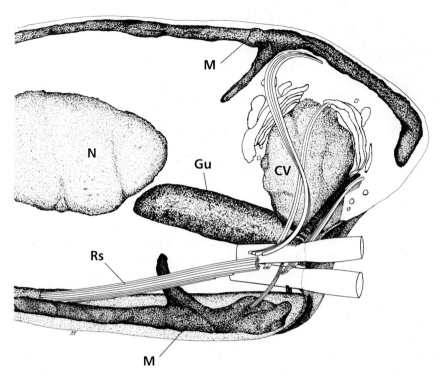

Figure 9–9 Cryptomonad flagellar apparatus components shown in diagrammatic form. CV=contractile vacuole, Gu=gullet, M=branched mitochondrion, N=nucleus, Rs=rhizostyle. (From Roberts, 1984 by permission of the *Journal of Phycology*)

flagellar hairs are properly called **mastigonemes** because they are tubular, but cryptomonad mastigonemes are two-parted, whereas the characteristic mastigonemes of the ochrophytes (heterokont or stramenopile algae) are three-parted. Another difference is that mastigonemes occur on both cryptomonad flagella (see review by Gillott, 1990), whereas they occur on only one of the two flagella of other mastigoneme-bearing algae. Presumably the mastigonemes help to increase flagellar efficiency, as is also suggested to be the role of euglenoid flagellar hairs, but the function of finer hairs and scales on cryptomonad flagella is not understood.

The link between the emergent portion of the flagellum and the intracellular basal body (centriole), known as the **transition region**, is distinctive in cryptomonads. It contains two or more platelike partitions just below the point where the two central axonemal microtubules appear (Fig. 9–8a). In contrast, this region contains a helix in many golden-brown (ochrophyte) algae (Fig. 9–8b), and a star-shaped structure in green algal flagella (Fig. 9–8c); the transitional region of euglenoids is comparatively featureless. Such highly conserved attributes are useful in deducing relationships between colorless and pigmented members of flagellate algal lineages, and may also reflect features of the ancestors that gave rise to each of these lineages.

Extending from the base of the flagella into the cytoplasm are elements of the flagellar root system (Fig. 9–9). Here too the cryptomonads show similarities to each other and differences with other algal lineages that are useful in tracing evolutionary relationships. The most prominent feature of the cryptomonad flagellar root system is the **rhizostyle**, a cluster of posteriorly directed microtubules that extends from the flagellar base (Fig. 9–9). Other flagellar roots are known as **compound roots** because they consist of both microtubules and striated, musclelike structures. A separate, striated band links the flagellar bases of cryptomonads (see Gillott, 1990). The contractile protein centrin, noted above to occur along the vestibulum, is also associated with the basal bodies, rhizostyle, and striated band of cryptomonads (Melkonian, et al., 1992). A similar striated band occurs in euglenoids and various other algae. Components of the flagellar root system are believed to contribute to cell development and maintenance of cell shape, and portions of it may also be involved in cell division.

Plastids and Photosynthetic Pigments

The plastids of cryptomonads, like those of euglenoids, have more than two envelope membranes, signifying their origin by secondary endosymbiosis. In

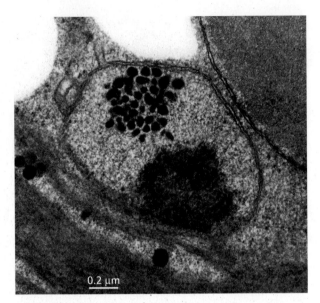

Figure 9–10 A cryptomonad nucleomorph as seen with TEM. (From Gillott and Gibbs, 1980 by permission of the *Journal of Phycology*)

contrast to plastids of euglenoids, however, those of cryptomonads are surrounded by four envelope membranes, the outer membrane bearing ribosomes on its surface, hence resembling rough endoplasmic reticulum. As a result, the outer two plastid membranes of cryptomonads (and those of some other algae) have been described as periplastidal endoplasmic reticulum (PER). In the view of some experts, only the outermost membrane is actually derived from the host's endomembrane system, while the membrane just internal to it is interpreted as the remains of the endosymbiont's plasmalemma (see references in Whatley, 1993). The outer and inner pairs of cryptomonad plastid membranes are separated by a space known as the **periplastidal compartment**. This compartment is characterized by a number of features characteristic of eukaryotic cytoplasm (particularly that of red algae), including 80S ribosomes with eukaryotic type ribosomal RNAs, starch grains, and a highly reduced nucleus, known as a **nucleomorph** (Fig. 9–10). The nucleomorph exhibits four features in common with typical eukaryotic nuclei: (1) a double membrane envelope with pores, (2) DNA, (3) self-replication, i.e., nuclear division, and (4) a nucleolus where ribosomal RNA (rRNA) genes are transcribed. In phylogenetic analyses, the 16S-like rRNA genes present in the nucleomorph group with those of red algal nuclei (see Fig. 7–3). Cryptomonad nucleo-

morph DNA is only 660 kb in size and is contained in three small chromosomes; it encodes genes required for its own maintenance. As described in Chapter 7, the evidence provided by the cryptomonad plastid envelopes, periplasmic space, and nucleomorph, strongly supports a hypothesis that cryptomonad plastids originated by secondary endosymbiosis, whereby a colorless phagotrophic flagellate ingested a eukaryotic red alga, which was then transformed into a plastid. Further evidence for this evolutionary scenario is provided by plastid pigment composition.

Photosynthetic plastids of cryptomonads contain a unique spectrum of pigments: chlorophyll *a*, α- and β-carotenes, phycoerythrin or phycocyanin—held in common with red algae (and cyanobacteria); and chlorophyll c_2—in common with various golden algae. Chlorophyll c_2 transfers energy from phycobilins to chlorophyll *a* (Falkowski and Raven, 1997). Cryptomonads also produce some unique xanthophylls, such as alloxanthin, which can be used to detect the presence of cryptomonads within phytoplankton mixtures. Cryptomonads produce either phycocyanin or phycoerythrin, but not both, in contrast to many red algae and cyanobacteria. As a result of considerable potential for variation in pigment expression, photosynthetic cryptomonads may exhibit a wide range of colors, including blue, red, olive, and brown. At one time, pigmentation was more widely used as a taxonomic criterion, but it is now understood that pigment composition may vary with environmental conditions

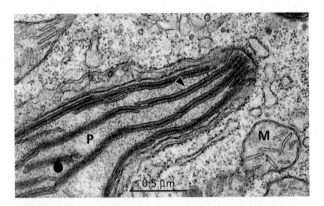

Figure 9–11 Section of a cryptomonad plastid (P) in the blue-green colored dinoflagellate *Gymnodinium acidotum*. Note the paired thylakoids (arrowhead) and their electron-opaque contents. Also note the mitochondria (M) with flattened cristae. (Micrograph by L. Wilcox and G. Wedemayer)

and is therefore not always a reliable index of relationship. The biliproteins of cryptomonads were reviewed by Hill and Rowan (1989).

The phycobiliproteins of cryptomonads are unique in that they are not located in phycobilisomes, as in cyanobacteria and red algae. Rather, they occur within the lumen of plastid thylakoids, giving this region a distinctive electron-opaque appearance (Fig. 9–11). Conclusive demonstration that the phycobiliproteins of cryptomonads are located within the thylakoid lumen and that some of the pigment-protein complex is associated with the thylakoid membrane was provided by immunolocalization studies (Spear-Bernstein and Miller, 1989). It has been hypothesized that phycobilisome linker proteins (see Fig. 6–25), once encoded in the red algal endosymbiont's nucleus, were lost during transformation of the endosymbiont into a plastid. The absence of phycobilisomes on thylakoid surfaces allows for thylakoid stacking, and indeed, cryptomonad thylakoids typically occur in stacks of two, in contrast to those of most cyanobacteria and red algae, which are unstacked.

Reproduction

Cryptomonad reproduction primarily occurs by asexual mitosis and cytokinesis, during which the cells continue to swim. Cells that are about to divide can be recognized by their more rounded shape and the presence of four flagella, in two pairs (Oakley and Dodge, 1976). Following basal body replication, division of the plastid nucleomorph occurs in preprophase by a simple pinching in two. No microtubules are involved in nucleomorph division, but there is a fibrous structure that appears only during division. Daughter nucleomorphs migrate to opposite ends of the plastid so that during subsequent plastid division, each daughter plastid receives one. Plastid division is followed by mitotic division of the host cell nucleus. Early in the mitotic process partial breakdown of the nuclear envelope occurs (in contrast to euglenoids), and there is a barrel-shaped spindle without centrioles at the poles (reviewed by Gillott, 1990; McFadden, 1993; Sluiman, 1993). Cytokinesis occurs by means of a cleavage furrow (i.e., an invagination of the cell membrane) that typically begins development at the posterior end of the cell. Thus, in contrast to dividing euglenoids, dividing cryptomonads do not appear "two-headed."

Cryptomonads may produce resting cysts that are rounded and have a thick extracellular matrix.

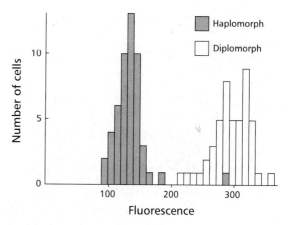

Figure 9–12 A study of the cryptomonad *Proteomonas* demonstrated the presence of two distinctive life cycle stages, one of which (the diplomorph) contains twice the amount of DNA as the other (the haplomorph). This provides evidence for some form of sexual reproduction in this organism. (After Hill and Wetherbee, 1986)

They can often be recognized by a pinkish coloration and the presence of large numbers of extraplastidic starch grains that stain blue-black with an iodine-iodide solution. In contrast, euglenoid cysts would not produce stainable starch, and the starch of coccoid green algae is located within the plastids. Although dinoflagellate cysts produce cytoplasmic starch, their cysts are typically shaped or ornamented in distinctive ways and therefore unlikely to be confused with cryptomonad cysts.

There is some intriguing evidence that at least some cryptomonads may engage in a sexual reproductive process. *Proteomonas sulcata* has a life history involving two morphologically distinct forms, one of which has twice the DNA level of the other, as determined by measurements of nuclear fluorescence in the presence of a DNA-specific fluorescent dye (Fig. 9–12). The two forms also differ in size and periplast structure, as well as in the configuration of the flagellar apparatus, but fertilization and meiosis were not observed (Hill and Wetherbee, 1986). Fertilization was observed in cultures of another cryptomonad that was similar to *Chroomonas acuta* (Kugrens and Lee, 1988). Gamete fusion is unusual in that the flagella do not seem to be involved in mating, as they often are in other algal groups. TEM analysis demonstrated that nuclear fusion occurred after fusion of isogametes (gametes that are similar in size and structure).

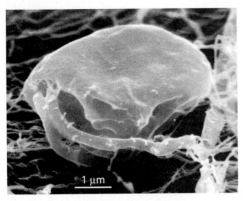

Figure 9–13 *Goniomonas* (*Cyathomonas*) seen with the SEM. (Micrograph courtesy P. Kugrens)

Diversity of Cryptomonads

Estimates of the number of cryptomonad genera range from as few as 12 (Hoek, van den, et al., 1995), to as many as 13–23 (Gillott, 1990). There are approximately 100 known freshwater species and about 100 known marine species. The molecular phylogeny of cryptomonads has been studied by Cavalier-Smith, et al. (1996), Fraunholz, et al. (1997), and Marin, et al. (1998). Novarino and Lucas (1993; 1995) recommended classification systems for the cryptomonads.

GONIOMONAS (*Cyathomonas*) (Gr. *gonia*, angle + Gr. *monas*, unit) (Fig. 9–13) lacks plastids and is regarded as a model of the kind of heterotrophic

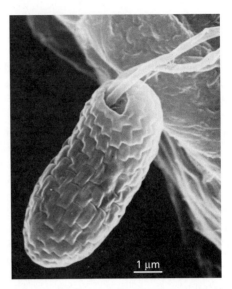

Figure 9–15 SEM view of *Chroomonas*. Note the periplast plates. (From Kugrens, et al., 1986 by permission of the *Journal of Phycology*)

host that gave rise to plastid-containing cryptomonads by secondary endosymbiosis (McFadden, et al., 1994). This hypothesis is supported by molecular evidence that *Goniomonas* is a relatively early divergent cryptomonad (Medlin and Simon, 1998) (see Fig. 7–3).

RHODOMONAS (Gr. *rhodon*, rose + Gr. *monas*, unit) (Fig. 9–14) has a single boat-shaped, red-colored plastid having a pyrenoid. There may be individual rectangular periplast plates in some species and a continuous periplast in others. Several (as many as 20 in *R. lacustris*) large ejectisomes line the vertically oriented furrow. *Rhodomonas* sometimes forms noticeable red blooms in freshwater lakes in early spring. Although these growths may appear alarming, they are not known to be harmful. The ultrastructure of *R. lacustris* was studied by Klaveness (1981). Though there has been some concern about the monophyly of the genus *Rhodomonas*, a reappraisal by Hill and Wetherbee (1989) suggests that the genus should be retained.

CHROOMONAS (Gr. *chroa*, color of the skin + Gr. *monas*, unit) (Fig. 9–15) has a single blue-green, H-shaped plastid with a pyrenoid on the bridge. Periplast plates are rectangular, and there are two large ejectisomes in the vicinity of the shallow vestibular depression. A typical gullet or furrow is absent (Kugrens, et al., 1986). Typically, the anterior end is

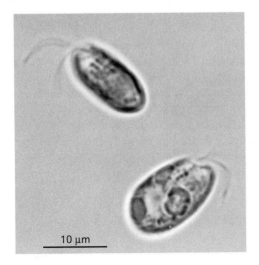

Figure 9–14 *Rhodomonas*.

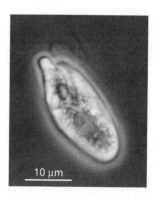

Figure 9–17 *Chilomonas*, a colorless cryptomonad.

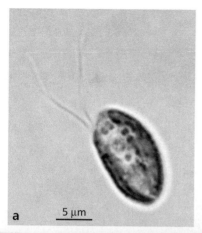

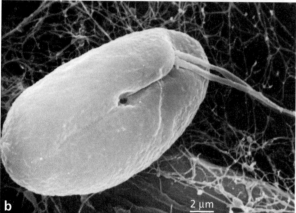

Figure 9–16 *Cryptomonas*. (a) Light microscopic view. (b) SEM view. (b: From Kugrens, et al., 1986 by permission of the *Journal of Phycology*)

somewhat broader than the posterior. The ultrastructure of *Chroomonas* and some other blue-green cryptomonads was studied by Hill (1991b).

CRYPTOMONAS (Gr. *kryptos*, hidden + Gr. *monas*, unit) (Fig. 9–16) exhibits two plastids, each having a pyrenoid. Some species have hexagonal periplast plates, whereas in others, such as *C. cryophila*, the periplast consists of a inner sheet of material that is not in direct contact with the cell membrane (Brett

and Wetherbee, 1986). The furrow is vertical in orientation, and several large ejectisomes line it. *Cryptomonas* is a very common and widespread genus, with species distinguished by such characters as length of the furrow and cell length. Optimal temperature and irradiance was found to vary among strains of freshwater *Cryptomonas* (Ojala, 1993). A revision of the species of *Cryptomonas*, based on Australian strains, was published by Hill (1991a). *Cryptomonas* is not a natural (monophyletic) genus (Santore, 1985) and probably should be divided into three genera (Kugrens, et al., 1986). A putatively related genus found in Brazil, *Pseudocryptomonas*, is distinctive in having numerous discoid plastids (Bicudo and Tell, 1988). The species formerly known as *Cryptomonas* ϕ has been renamed *Hanusia phi* (Deane, et al., 1998).

CHILOMONAS (Gr. *chilomos*, pale + Gr. *monas*, unit) (Fig. 9–17) possesses a leucoplast, a plastid that lacks pigmentation. There is no pyrenoid, but a nucleomorph is present. The genus is considered to have evolved from a pigmented form. There is typically a continuous periplast plate, i.e., the periplast does not consist of many individual units as in other cryptomonads. There are several noticeable ejectisomes in the vicinity of the vertical vestibulum. As might be expected, this cryptomonad typically occurs in waters rich in organic compounds released from decay processes.

Chapter 10

Haptophytes

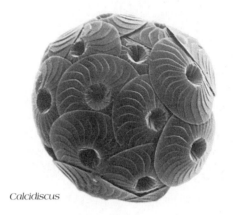

Calcidiscus

(Photograph courtesy A. Winter and P. Friedinger)

Haptophyte algae (also known as prymnesiophytes) are primarily marine unicellular biflagellates that have had a major impact on global biogeochemistry for at least 150 million years. They are probably the single modern algal group that has the greatest long-term impact on carbon and sulfur cycling, and hence, global climate. Most haptophyte species produce external body scales composed primarily of calcium carbonate that are known as **coccoliths** (Fig. 10–1). Sedimented coccoliths are the major contributors to ocean floor limestone accumulation, and represent the largest long-term sink of inorganic carbon on earth. Deep-sea carbonate deposits cover about one half of the world's seafloor, an area that represents one third of the earth's surface. Coccoliths contribute about 25% of the total annual vertical transport of carbon to the deep ocean. In addition, the coccolith-producing *Emiliania huxleyi*, and *Phaeocystis pouchetii* (Fig. 10–2) (which lacks coccoliths) are known for their formation of extensive ocean blooms with concomitant production of large amounts of dimethylsulfide (DMS), a volatile sulfur-containing molecule that increases acid rain (Fig. 10–3). Coccoliths, which readily reflect light, and DMS, which enhances cloud formation, contribute to increased albedo (reflectance of the earth's surface) and thus have a cooling influence on the climate.

Haptophytes are also important in terms of their biotic associations. Most haptophytes contain golden or brown plastids, and are thus photosynthetic primary producers. However, many are also osmotrophic or phagotrophic; thus mixotrophy is common. Phagotrophy is particularly prominent among forms that lack a cell covering formed of coccoliths, but which possess a **haptonema**, a thread-like extension from the cell that is involved in prey capture, among other functions (Fig. 10–4). A haptonema occurs in many haptophytes, hence the name of the group, although a number of taxa appear to have lost this structure. The mixotroph *Chrysochromulina* (Fig. 10–5)—which boasts an impressively long haptonema, and which forms dense coastal blooms, can readily feed on bacteria

Figure 10–1 SEM view of *Emiliania huxleyi*, showing the coccosphere made up of interlocking coccoliths. (Micrograph by A. Kleijne, in Winter and Siesser, 1994)

and small algal cells. The very widespread species *C. polylepis* (Fig. 10–5) can produce toxic offshore marine blooms that cause death of fish and invertebrates, while the related *Prymnesium parvum* causes similarly toxic blooms in brackish waters. In 1989, a *P. parvum* bloom along the Norwegian coast caused a five-million-dollar (U.S.) loss of salmon (Nicholls, 1995). The number of occurrences and the intensity of such blooms has increased within the past three decades; this is generally attributed to higher rates of pollution of ocean waters by mineral nutrients such as combined nitrogen and phosphate (Chapter 2). The physiology and bloom dynamics of *Prymnesium* and *Chrysochromulina* were reviewed by Edvardsen and Paasche (1998).

With the exception of the toxic bloom-formers mentioned above, haptophyte algae, because of their small size, fast growth rates, digestibility, and nutritional content, are considered to be high-quality foods for marine zooplankton. *Isochrysis* and *Pavlova* are widely used in the aquaculture industry (Chapter 4). Haptophytes (though substantially less important than dinoflagellates and diatoms) can be significant primary producers in polar, subpolar, temperate, and tropical waters. Haptophytes actually reach their highest species diversity in extremely low-nutrient, subtropi-

cal open-ocean waters, where a number of strange and beautiful forms mysteriously occur in nearly dark ocean waters more than 200 m deep.

Fossil Record

The haptophyte algae have one of the best fossil records among the algae, because the often round or oval calcite coccoliths are readily preserved in sediments. Coccoliths (Gr. *kokkos*, berry + *lithos*, rock) were named in 1857 by T. H. Huxley, who observed them in samples of deep-ocean sediments and was reminded of the rounded cells of the green alga *Protococcus*. Coccoliths first appear in the fossil record either as early as the Carboniferous (Siesser, 1994), or in the Late Triassic (about 220 million years ago) (Young, et al., 1994), continuing to the present time. The rise of coccolithophorids followed the most dramatic worldwide extinction episode in earth's history—the 250-million-year-ago end-Permian event. A richly diverse marine community experienced the loss of 85% of its species. The cause is thought to have been extensive volcanism with the release of large amounts of CO_2, resulting in acid rain, and cooling

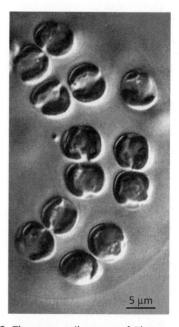

Figure 10–2 The nonmotile stage of *Phaeocystis*. The spherical cells are embedded in mucilage to form an often very large colony, the edge of which is evident at the bottom of the micrograph. (From Marchant and Thomsen, 1994 by permission of Oxford University Press)

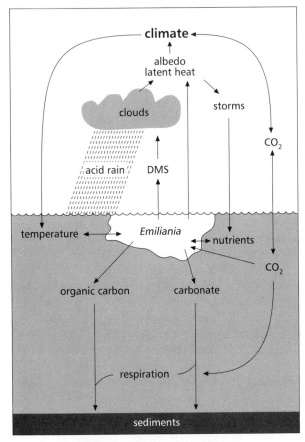

Figure 10–3 Blooms of *Emiliania huxleyi* can have important effects on the earth's climate in a variety of ways, summarized here in diagram form. (After Westbroek, et al., 1994 by permission of Oxford University Press)

caused by atmospheric ash, though other causes are also possible (Bowring, et al., 1998). This event may have created conditions suitable for the evolution of coccolithophorids.

The abundance of coccolith fossils peaked during the Late Cretaceous (63–95 million years ago), when very extensive chalk deposits were laid down across much of northern Europe and other sites around the world (Fig. 10–6). In fact, the term Cretaceous refers to this chalk. Some blackboard chalks that are derived from such deposits contain coccolith remains. Although the Cretaceous deposits were formed in relatively shallow waters over continental shelves, similar deposits are being formed today beneath large areas of the deep ocean.

The impact of a massive asteroid or comet off the Yucatan coast (the famous "K/T event"), which is associated with the demise of dinosaurs and ammonites,

also apparently caused extinction of 80% of the coccolithophorid species that had been present in the Cretaceous. In contrast, dinoflagellates and diatoms (fossils of the latter having first appeared during the Cretaceous) seem to have escaped similar drastic extinction effects. If Cretaceous coccolithophorids were like modern forms in exhibiting highest diversity in warm ocean waters, it is possible that their post-K/T decline was caused by sudden climatic cooling, an hypothesized effect of the impact event. From the surviving coccolithophorids, new forms radiated—only to experience another decline in diversity and subsequent recovery about 30–50 million years ago (Fig. 10–7).

Because coccoliths are common, small, and exhibit low endemism (restriction of certain species to particular locales), they are widely used as stratigraphic indicators to match rocks of equivalent ages from different locales. About 1000 species of fossil coccolithophorids (coccolith types) are widely used as bioindicators in the oil industry (Young, et al., 1994).

Phylogeny

The most primitive haptophytes are thought to be biflagellate unicells having a haptonema. Derived forms are considered to include flagellates with a highly reduced haptonema or none at all, as well as

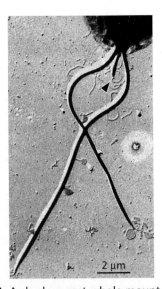

Figure 10–4 A shadow-cast whole mount (TEM view) of a motile *Phaeocystis* cell illustrating the two flagella and short haptonema (arrowhead) between them. (From Marchant and Thomsen, 1994 by permission of Oxford University Press)

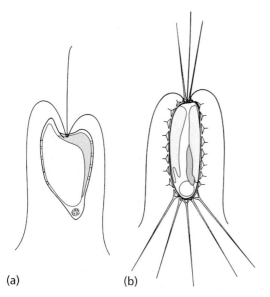

(a)   (b)

Figure 10–5 Diagrammatic view of two species of *Chrysochromulina*. (a) *C. polylepis*, (b) *C. pringsheimii*. (From Throndsen, 1997)

Figure 10–6 Chalk cliffs on the Isle of Wight, which consist largely of coccoliths deposited during the Late Cretaceous. (From Young, et al., 1994 by permission of Oxford University Press)

nonflagellate amoeboid, coccoid, palmelloid, colonial, or filamentous forms that produce biflagellate reproductive cells. Production of non-mineralized, organic fibrillar scales also seems to be a plesiomorphic (evolutionarily primitive) character for the group, with these occurring in putatively primitive flagellates lacking coccoliths, as well as in some coccolithophorids. Coccolith production is regarded as a derived feature (Cavalier-Smith, 1994). Thus haptophytes probably originated much earlier than is suggested by the first appearance of coccoliths in the fossil record, but did not leave remains because earliest forms lacked distinctive fossilizable parts.

Haptophytes possess photosynthetic pigments similar to those of the heterokont algae (ochrophytes) (Chapters 12–15), and were grouped with them until ultrastructural and molecular evidence revealed that haptophytes form a distinct group. Haptophyte plastids are similar to those of ochrophytes, cryptomonads, and chlorarachniophytes in having four envelope membranes, the outer two forming a periplastididal endoplasmic reticulum (PER), also known as chloroplast endoplasmic reticulum (CER) (Fig. 10–8). These characteristics suggest that haptophyte and ochrophyte plastids arose via secondary endosymbiosis, as have those of cryptomonads and chlorarachniophytes (Chapter 7). Cavalier-Smith (1994) has suggested that these four algal groups arose from a single common ancestor that acquired a pigmented eukaryotic endosymbiont about 600 million years ago, and that these four groups should be aggregated to form the Kingdom Chromista (which excludes dinoflagellates). However, ribosomal RNA sequence analyses indicate that the alveolate group (containing dinoflagellates, ciliates, and apicomplexans) is sister to the stramenopiles, a monophyletic group that includes oomycetes, some other heterotrophic protists, and

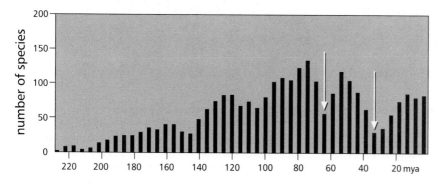

Figure 10–7 Graph showing the number of coccolithophorid species present over time. The rise of coccolithophorids followed the worst extinction event in the earth's history, 250 million years ago. Decreased numbers (arrows) are associated with major extinction events. (Based on Bown, et al., 1991)

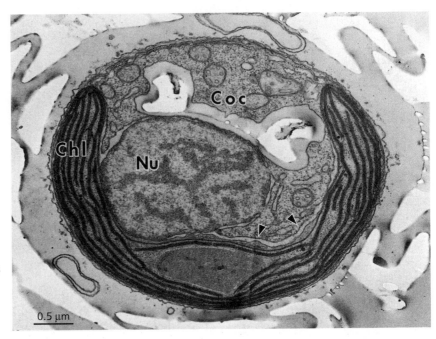

Figure 10–8 TEM of an *Emiliania huxleyi* cell. Note the periplastidic endoplasmic reticulum (PER) (arrowheads), which is continuous with the nuclear envelope. Plastid (Chl) and a developing coccolith (Coc) are also shown. (From Pienaar, 1994)

ochrophytes (heterokont algae) (Saunders, et al., 1997a; Vanderauwera and Dewachter, 1997). This phylogenetic branching pattern suggests that hapto-phytes and ochrophytes independently acquired their similar plastids (Daugbjerg and Andersen, 1997a; Medlin, et al., 1997). The haptophytes are regarded as a monophyletic group that originated with the evolution of the haptonema, possibly by duplication and modification of one of the preexisting flagellar roots (Cavalier-Smith, 1994).

Although the plastids of haptophyte algae are otherwise clearly descended from cyanobacteria, they contain a form of Rubisco that is more closely related to that of β-proteobacteria than to the Rubisco of cyanobacteria. This situation has been explained as having resulted from horizontal gene transfer (Delwiche and Palmer, 1996). These and other hypotheses regarding the ancestry and interrelationships of haptophyte taxa are currently being evaluated with the tools of molecular systematics (Medlin, et al., 1994).

Cell Biology and Reproduction

Flagella and the Haptonema

Flagellate cells usually have two smooth and approximately equal flagella, but in one early divergent group, the Pavlovales, cells have two unequal flagella,

bearing fine hairs. Haptophyte flagella are different in several ways from those of the heterokont algae (ochrophytes) (Chapters 12–15) and Oomycetes. The two flagella of ochrophytes and oomycetes differ from one another in appearance, and possess tripartite, tubular mastigonemes and a flagellar swelling associated with light sensing; all of these features are lacking from haptophyte flagella.

The haptonema, if present, emerges from the cell apex, between the flagella. The structure of the flagellar and haptonemal basal bodies and their associated roots is shown in Fig. 10–9. The haptonemal shaft includes 6–7 singlet microtubules arranged in a ring or crescent, surrounded by a ring of endoplasmic reticulum, which also extends as a flap into the cell body, the whole covered by an extension of the cell membrane (Fig. 10–10). The haptonema is about the same thickness as a flagellum and was mistaken for a flagellum until its structure was elucidated with transmission electron microscopy (Parke et al., 1955). The haptonema may be quite short or very long: *Chrysochromulina comella*'s haptonema is an astounding 160 μm long. The haptonema cannot beat (as can a flagellum), but it can bend and coil. The length and bending behavior of the haptonema seem to be correlated with phagotrophic behavior; species having a very long haptonema are inclined toward phagotrophy, whereas those having a very short or no haptonema are not. High-speed video methods

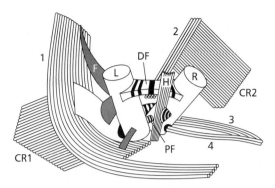

Figure 10–9 A diagram of the flagellar apparatus components of *Pleurochrysis*. There are four main microtubular roots (labeled 1 through 4). A fibrous root (F) is associated with root 1. Roots 1 and 2 have "compound components" (CR1 and CR2), which represent bundles of additional microtubules that branch off perpendicularly from the root microtubules. Basal bodies (L and R) are connected by a distal fiber (DF) and proximal fiber (PF). An intermediate fiber (shown, but not labeled) is also present. The microtubules indicated by the H represent the haptonemal base. (After Inouye and Pienaar, 1985)

Figure 10–10 A haptonema (of *Syracosphaera pulchra*) seen in cross section with the TEM. Note the microtubules surrounded by ER (arrowhead), which is in turn bound by the cell membrane. (from Pienaar, 1994)

0.125 μm

revealed the role of the haptonema in phagotrophy. As the cell swims forward, prey particles—such as bacterial cells—attach to the forward-projecting haptonema. Adhesive properties are attributed to presence of sugar groups at the haptonemal surface. Particles are moved downward to a point about 2 μm distal to the haptonemal base, known as the particle aggregating center (PAC). Once formed, the PAC moves up to the haptonemal tip, at which point the flagella stop beating while the haptonema bends toward the posterior surface, where the particle

aggregation is ingested into a food vacuole (Inouye and Kawachi, 1994) (Fig. 10–11). Whereas a long haptonema would be able to accomplish this feat, a very short one would not, hence the presumption that haptonemal length is correlated with presence or absence of phagotrophic feeding.

The coiling behavior of the haptonema appears to be related to collision avoidance responses by the cell. Upon contact with a larger obstacle, the haptonema coils very quickly (within 1/60 to 1/100 s), then the flagella change the direction of beat, propelling the cell quickly backward. Experiments have shown that coiling is induced by the rapid influx of Ca^{2+} into the cell from the environment; the threshold for activity is between 10^{-7} and 10^{-6} M. The endoplasmic reticulum present in the haptonema is thought to be involved in the calcium-induced response. In both the feeding process and collision avoidance, flagellar and haptonema behavior seem to be highly coordinated. A third function of the haptonema is attachment of haptophyte cells to substrates, observed to occur in some taxa (Inouye and Kawachi, 1994).

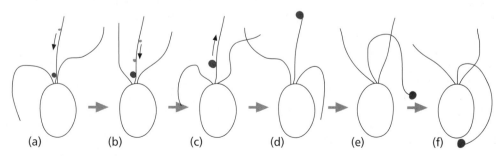

(a)   (b)   (c)   (d)   (e)   (f)

Figure 10–11 A diagram of the feeding process of *Chrysochromulina*. In (a) and (b), particles adhere to the haptonema and are translocated downward to a particle aggregating center (PAC). In (c) and (d), the aggregated, captured particles are moved to the tip of the haptonema. In (e) and (f), the aggregate is delivered to the cell surface (at the posterior end of the cell) through bending of the haptonema. Here it is taken into the cell. (Based on photomicrographs in Inouye and Kawachi, 1994)

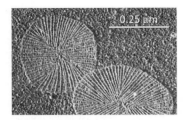

Figure 10–12 Organic fibrillar scales found on the haptophyte *Pleurochrysis* as seen in a heavy-metal–shadowed TEM preparation. This organism also produces coccoliths. (From Pienaar, 1994)

Scale Formation and Deposition

The plesiomorphic scale type for haptophytes is a thin discoid or elliptical cellulose-containing plate that has a fibrillar structure when viewed with electron microscopy. Such scales occur on flagellates that lack coccoliths, including *Chrysochromulina* and *Prymnesium*. Several coccolithophorids, such as *Pleurochry-sis*, produce organic scales (Fig. 10–12) as well as coccoliths. Although other coccolithophorids do not appear to produce organic scales, organic material is typically involved in coccolith development (Leadbeater, 1994). Both the organic scales and many types of coccoliths originate within the cisternae of a single, very large Golgi body that lies near the flagellar basal bodies, with its forming face toward the cell anterior (Fig. 10–13). This unusual Golgi apparatus is characteristic of the haptophytes.

The coccolith-forming haptophytes can be separated into two types based on coccolith structure and where coccoliths are produced: those with "heterococcoliths" formed internally (by the Golgi apparatus), and those having "holococcoliths," which are formed extracellularly (Fig. 10–14). Holococcoliths are composed of regularly packed rhombohedral or hexagonal crystals, whereas heterococcoliths possess rhombohedral crystals whose shape has been modified such that crystal faces and angles are partly or completely suppressed (Heimdal, 1997). Heterococcoliths are more robust than the smaller, more delicate holococcoliths. Some life cycles include both heterococcolith and holococcolith-producing forms. *Coccolithus* is an example of a haptophyte that produces both a nonmotile, heterococcolith-bearing stage, as well as a motile holococcolith-covered form. In the latter case an organic scale produced by the Golgi body (within the cell) is secreted to the outside, then calcite is deposited on the scale within a space covered by an organic envelope (deVrind-deJong, et al., 1994).

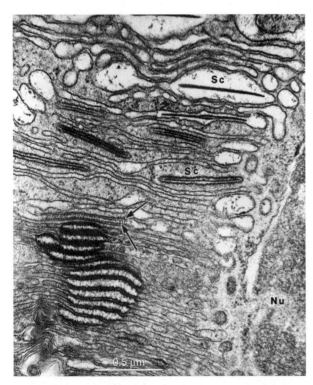

Figure 10–13 A TEM view of the large haptophyte Golgi body, shown here in *Hymenomonas lacuna*. Scales are apparent in different stages of formation. Arrows point to earliest stage of scale formation. Nu=nucleus, Sc=forming scale. (From Pienaar, 1994)

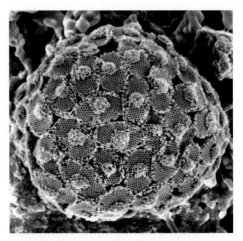

Figure 10–14 *Calyptrolithophora*, with holococcoliths. Holococcoliths are made up of smaller, similarly sized crystals (versus heterococcoliths). (Micrograph courtesy S. Nishida)

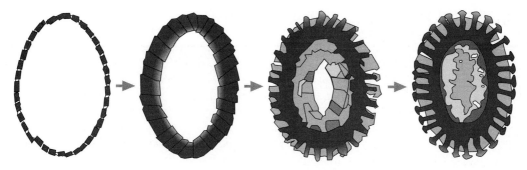

Figure 10–15 A diagram of coccolith formation in *Emiliania huxleyi*. A rim of calcite crystals is formed first. These crystals grow and interlock and the complex shape of the mature coccolith gradually becomes apparent, with the knobby, spokelike outer (distal) shield elements being illustrated with the dark shades of gray (see Fig. 10–1). (Based on electron micrographs in DeVrind-DeJong, et al., 1994)

During heterococcolith development Golgi vesicles fuse to form a large, flat compartment that becomes adherent to the nuclear envelope. A thin organic plate is then deposited within the vesicle. A "reticular body" composed of a network of tubules—which is thought to provide precursors for coccolith development—then attaches to the nonnuclear face of the vesicle. Calcification begins at the rim of the organic base plate, then calcite crystals are added to complete the structure (Fig. 10–15). Acidic polysaccharides present within the coccolith vesicle at this stage are thought to be responsible for pattern formation by preventing continued calcite crystal development at specified points. By the time coccolith development is complete, the reticulate body is no longer present. The vesicle, which previously fit closely around the developing coccolith, loses its close fit and detaches from the nuclear surface. The newly formed coccolith is then extruded onto the cell surface near the flagella, where it joins other coccoliths to form the **coccosphere**—the outer covering of the cell (deVrind-deJong, et al., 1994) (Fig. 10–16).

Calcification in coccolith-forming haptophytes is known to be highly dependent upon photosynthesis. Photosynthesis provides energy-rich molecules (ATP and NADPH) needed for transport processes at the cell and Golgi membranes. Photosynthesis also acts as a sink for carbon dioxide, thus driving the net reaction for calcification:

$$2HCO_3^- + Ca^{2+} \rightarrow CaCO_3 + CO_2 + H_2O$$

In fact, the production of CO_2—which can then be used in photosynthesis—is viewed as a major adaptive advantage of coccolith production. Physiological comparison of high- and low-calcifying strains has provided insight into the role of calcification as a source of CO_2 for photosynthesis. Recall that the concentration of dissolved CO_2 in waters of pH 8 or higher is very low; at such a pH, most of the dissolved inorganic carbon is in the form of bicarbonate (Chapter 2). At high irradiance and a pH of 8.3, photosynthesis in high-calcifying cells increases as pH rises, whereas that of low-calcifying cells is severely limited above pH 7. These results suggest that calcification provides CO_2 that can be used in photosynthesis (Brownlee, et al., 1994) (Fig. 10–17). In one hour a high-calcifying cell can produce one coccolith, containing $3.5–5.5 \times 10^{-13}$ g of calcium (Brownlee, et al., 1994). In experiments with *Emiliania huxleyi*, which is typically covered by about 15 coccoliths, the coccosphere can be dissolved from the surface of cells by growing them in low-pH medium, and then, within 12 hours after return to high pH conditions, the entire coccosphere will have been replaced (Pienaar, 1994).

There is a very great variety of coccolith types, and coccolith morphology may be correlated with ecological niche, with certain types primarily occurring on coastal taxa and others more characteristic of deep-growing oceanic forms. The simplest type of coccolith consists of a simple basal disk upon which walls of calcite are built around the edges to form a cylinder; such coccoliths are produced by *Syracosphaera* (*Coronosphaera*) (Fig. 10–18). More complex coccoliths, such as are produced by the common bloom-formers *Emiliania huxleyi* and *Gephryocapsa oceanica*, as well as other species, consist of two shields joined by a short column. This allows adjacent

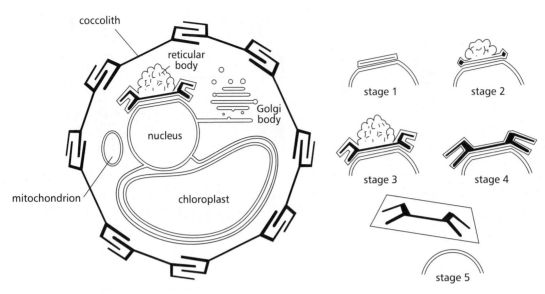

Figure 10–16 A diagram of an *Emiliania huxleyi* cell and the five discernable stages of coccolith formation that occur within the coccolith production compartment adjacent the nucleus. Stage 1: the Golgi-derived vesicle within which a coccolith will form adheres to the nuclear envelope. Stage 2: the reticular body attaches to the nonnuclear vesicle face. Stage 3: calcification becomes more extensive. Stage 4: at coccolith maturity, the reticulate body is no longer attached to the vesicle. Stage 5: the vesicle becomes less adherent to the coccolith, detaches from the nuclear surface, and moves toward the cell surface, where the coccolith is extruded. The reticular body is thought to supply precursors to the developing coccolith. The coccoliths are shown edge-on in this illustration. (After DeVrind-DeJong, et al., 1994 by permission of Oxford University Press)

coccoliths within the coccosphere to interlock (Fig. 10–19), increasing the cohesion of the cell covering. In other forms the coccoliths may overlap but do not interlock (Siesser and Winter, 1994). The evolution of crystalline structure of coccoliths was studied by Young, et al. (1992).

Coccoliths probably serve a number of functions, including restriction of access to cells by pathogenic bacteria and viruses, protection from predation by protozoa (although not larger zooplankton such as copepods, which readily consume coccolithophorids), and buoyancy regulation (by regulated production/loss of heavy coccoliths). Coccoliths have also been suggested to focus light into cells and to serve in nutrient uptake (Brand, 1994).

Plastids and Pigments

The one or two plastids in cells of haptophytes often take the distinctive form of a butterfly when viewed with epifluorescence microscopy. Plastids of all haptophytes contain chlorophylls a, c_1, and c_2, as well as β-carotene, diatoxanthin, and diadinoxanthin. Chlorophyll c, in most (but not all) haptophytes, is

unlike chlorophyll a and b in not having a phytol tail (Chapter 6), and in having a free acrylic acid group on ring V. Chlorophyll c also originates by a different biosynthetic pathway than do chlorophylls a and b

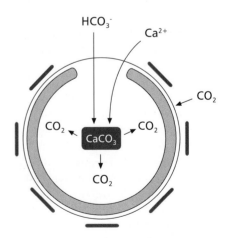

Figure 10–17 A diagram of an *Emiliania huxleyi* cell, illustrating that bicarbonate is taken into the cell and used in calcification, which then provides additional CO_2 for photosynthesis. (After Westbroek, et al., 1994 by permission of Oxford University Press)

Figure 10–18 The relatively simple coccoliths found on *Syracosphaera* (*Coronosphaera*), with a single disk having a wall at its periphery. (From Siesser and Winter, 1994)

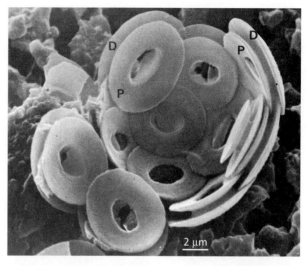

Figure 10–19 Interlocking coccoliths of *Gephyrocapsa*. D=distal shield of coccolith, P=proximal shield. (From Siesser and Winter, 1994)

(Falkowski and Raven, 1997). Chlorophyll *c* is believed to be an intermediate in energy transfer between carotenoids and chlorophyll *a* within the carotenoid/chlorophyll antenna complex.

In addition to the above-mentioned pigments, at least five variations in additional pigment content occur among haptophytes. Some contain fucoxanthin, a second group produces chlorophyll c_3 plus fucoxanthin, others have chlorophyll c_3 plus 19'-hexanoyloxyfucoxanthin, whereas chlorophyll c_3 plus 19'-butanoyloxyfucoxanthin, 19'-hexanoyloxyfucoxanthin, and fucoxanthin are found in a fourth group (Jeffrey and Wright, 1994). Yet other haptophytes, such as *Isochrysis*, are characterized by chlorophyll *c* that has a phytol tail (Zapata and Garrido, 1997). These pigment categories are not completely congruent with current taxonomic groupings, which are based on other cellular features. The fact that there is no single pigment signature that characterizes all haptophytes presents a technical problem for oceanographers who wish to characterize phytoplankton communities on the basis of pigments. Although 19'-hexanoyloxyfucoxanthin is sometimes used as a marker pigment for haptophytes, the absence of this pigment from some forms reduces its utility in assessing presence of haptophytes as a group (Jeffrey and Wright, 1994).

Plastids occur within the periplastidal endoplasmic reticulum, mentioned previously as being of phylogenetic significance. The PER may or may not be physically linked to the nuclear envelope, depending upon the species (Fig. 10–8). The occurrence of four membranes between the plastid thylakoids and stroma on the one hand and the cytoplasm on the other has been viewed as a barrier to the efficient entry into plastids of cytoplasmically encoded proteins. How protein import occurs under such structurally complex conditions is not understood. Chloroplasts usually possess pyrenoids that either bulge into the cytoplasm or, more commonly, are immersed within the chloroplast matrix (Fig. 10–20).

Anterior eyespots consisting of reddish globules contained within the plastid are present in some but not all haptophytes. The principal photosynthetic storages are cytoplasmic lipid and a polysaccharide known as **chrysolaminarin**, which occurs within cytoplasmic vacuoles. These storage products are also of widespread occurrence among ochrophytes.

Cell Division

Cell division in haptophyte algae has been reviewed by Hori and Green (1994). It commonly occurs at night, presumably because such timing results in the least disruption to photosynthesis. The first indicator of incipient cell division is elongation of the pyrenoid along the long axis of the plastid and the appearance of a

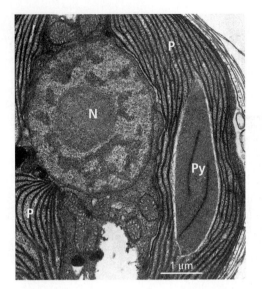

Figure 10–20 A haptophyte pyrenoid (of *Calyptrosphaera sphaeroidea*) embedded within the chloroplast stroma. N=nucleus/nucleolus, P=plastid, Py=pyrenoid. (From Pienaar, 1994)

slight depression on the plastid surface in the area of the pyrenoid midpoint. Next both the PER and the chloroplast envelope become invaginated at their midpoints. Both the pyrenoid and the plastid then complete their division. Division of the single large Golgi also begins prior to nuclear division and has been completed by metaphase. If the nuclear envelope was confluent with that of the PER, it separates from the PER prior to onset of nuclear division. Also preceding nuclear division is replication of flagellar and haptonemal basal bodies. Cells that are still in interphase or prophase may contain four flagellar and two haptonemal bases.

In the early divergent Pavlovales, the nuclear envelope remains intact throughout mitosis. The flagellar basal bodies remain aggregated until relatively late in the division process, whereupon they become located at the spindle poles. A fibrous flagellar root acts as a spindle microtubular organizing center. The spindle has an unusual V-shaped orientation. These mitotic features may be plesiomorphic for haptophytes.

In contrast, in most haptophytes the nuclear envelope partially or more completely breaks down by late prophase. Pairs of flagellar basal bodies, together with an associated haptonemal base, then move to the vicinity of the developing spindle poles, but they do not act as spindle poles. In the relatively

derived haptophyte *Phaeocystis globosa*, the spindle microtubular organizer is located on the mitochondrial surface (Hori and Green, 1994) (Fig. 10–21). These interesting observations suggest that the material that organizes the spindle has undergone evolutionary change in its location; similar evolutionary change in location of the spindle organizer has occurred during the evolution of plants (reviewed by Graham, 1993). Although microtubule organizers are clearly important in establishing the timing and polarity of nuclear divisions, little is known about their biochemical nature and behavior in algae. At the conclusion of mitosis, nuclear envelopes and continuity with the PER are reestablished, then cytokinesis occurs by invagination of scale-free cell membrane, or by fusion of vacuoles in the interzonal region. Each daughter cell receives one half of the body scales; the other half is newly generated.

Life Histories

The life histories of haptophytes may include morphologically distinct forms. For example, the coastal genera *Pleurochrysis*, *Hymenomonas*, and *Ochrosphaera* produce both coccolith-bearing cells that are thought to be diploid, as well as organic-scaled haploid forms. Other coastal forms produce planktonic, diploid flagellate stages that alternate with benthic or intertidal haploid filamentous stages. An increasing number of open-ocean forms are being discovered to have life-history stages that include both holococcolith-producing forms as well as heterococcolith-bearing forms. In spite of such compelling circumstantial evidence for sexual reproduction and alternation of generations, to date syngamy and meiosis have not been clearly documented for any haptophyte (Billard, 1994).

Diversity, Ecology, and Biogeography of Living and Fossil Haptophytes

The haptophyte algae include 11 genera (with about 80 species) that do not possess mineralized scales and about 40 genera (with more than 200 species) of coccolithophorids. Of the coccolithophorids, about 65 produce mainly holococcoliths, and the rest produce mainly heterococcoliths. A classification system for the haptophyte algae is presented by Jordan and

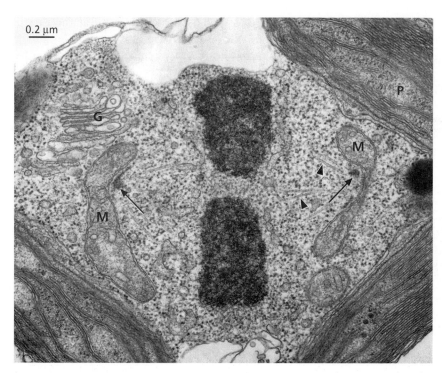

Figure 10–21 TEM view of mitosis (metaphase) in *Phaeocystis*. The microtubule organizing centers at the spindle poles (arrows) are located on the surface of mitochondria (M). Arrowheads point to microtubules. G=Golgi body, P=plastid. (Micrograph courtesy T. Hori)

Kleijne (1994). An excellent SEM atlas of the living coccolithophorids can be found in Winter and Siesser (1994). A checklist of extant Haptophyta of the world is provided by Jordan and Green (1994).

Ecological interest in haptophytes has primarily been focused upon the roles of a relatively few key bloom-forming taxa that strongly influence climate and food web organization by production of organic carbon, carbonates, DMS, and toxic compounds. Hence our discussion of haptophyte ecology and diversity will emphasize bloom-forming genera, but some other forms that are of evolutionary or economic interest are also included. A brief survey of the principal methods used to collect, identify, and enumerate haptophytes, and to evaluate their biogeochemical impact, is a useful starting point.

Sampling, Enumeration, and Preservation Methods for Living Haptophytes

Water samples of a liter or more are collected from known depths by use of sampling devices that can be remotely opened and closed. For relatively shallow sampling, water can be pumped through plastic tubing that has been lowered to the desired depth. Water samples are then concentrated by centrifugation or (more commonly) filtration through 0.22-μm pore filters. Following a rinse with alkaline water (because low pH water may dissolve coccoliths), concentrated algal samples can be preserved in 1% glutaraldehyde. If only coccolithophorids are to be examined, filters can be air dried and stored for later microscopic examination. Coccolithophorids on dried filters can be examined using a polarizing microscope; the filters are first cleared by placing a drop of immersion oil on them. Scanning electron microscopy (SEM) is generally required for identification, but it is easy to accomplish: Dried filters are cut into pieces and attached to SEM stubs with double-sided sticky tape, then gold-coated prior to examination (Winter, et al., 1994).

Glutaraldehyde-preserved samples may be enumerated with the use of a fluorescence microscope. The characteristic butterfly shape of the plastid allows discrimination of haptophytes from other algae, but species determinations cannot be made by this method. Single drops of preserved haptophytes can be placed on grids that have been coated with a thin film of plastic (Formvar), then dried for examination by transmission electron microscopy. The latter process is necessary for determination of species-specific differences in the thin fibrillar organic scales produced by non-coccolithophorids such as *Chrysochromulina*.

Sampling and Enumeration of Fossil Coccoliths

Sampling of sediments is commonly performed to survey fossil coccoliths or to determine the rate at which coccoliths transport carbon to depths in the ocean and thereby contribute to formation of sedimentary carbonates. Coring devices are used to obtain fossils. Sediment traps (Fig. 10–22), lowered to desired depths for collection, are used to capture particles settling in the water column or at the forming layers of sediments. Settling traps can be deployed for time periods varying from hours or days to months, with longer time periods more desirable because they integrate the effects of natural coccolith dissolution. Most coccoliths reach the sediments as aggregates in fecal pellets, which descend at the rate of about 200 m d^{-1}, or in marine-snow particles (amorphous agglomerations of living and dead organisms, fecal pellets, and inorganic minerals bound with mucus), that descend at about half the rate of fecal pellets. Both types of aggregates increase sedimentation far beyond that possible for single coccoliths (only about 13.8 cm d^{-1}) and protect coccoliths from dissolution that would tend to occur during passage through water that is undersaturated in $CaCO_3$. Light microscopy and a counting chamber such as a hemacytometer are employed to enumerate coccoliths in filtered samples. Coccolith flux rates are determined from the number of coccoliths counted per unit volume, the surface area of the trap opening, and the length of time that the trap was deployed. For determination of carbonate flux values, the coccolith flux values are then multiplied by an average coccolith mass; values ranging from 1.3×10^{-7} (*Coccolithus pelagicus*) to 3×10^{-9} mg (*Emiliania huxleyi*) are used by various investigators (Steinmetz, 1994a, b). Efforts to accurately calculate the rate at which coccoliths give rise to sedimentary carbonates are confounded by the lack of information about relative dissolution rates of different coccolith types in different environments.

Fossil sediments obtained from core sampling are examined with light microscopy by spreading a very small amount of sediment onto a cover-slip surface with a flat toothpick, drying on a hot plate, then mounting the cover slip onto a slide. Sediment samples similarly spread onto cover slips and dried can be gold-coated for SEM examination, necessary for species determinations (Roth, 1994). Keys for identification of modern coccolithophorids are provided by Heimdal (1997).

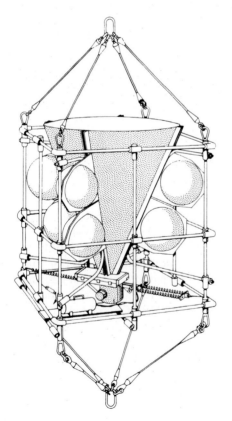

Figure 10–22 A sediment trap used to trap particles settling through the water column. Such traps are used to capture coccoliths for study. (From Honjo, et al., 1980)

Some Non-Coccolithophorid Haptophyte Taxa

PAVLOVA (named for Anna Pavlova, a famous Russian ballerina) primarily occurs as a unicellular flagellate (Fig. 10–23), but also as a palmelloid stage. The motile stage of some species is noted for active metaboly; the cell shape changes constantly in a way that is reminiscent of the behavior of various euglenoids. *Pavlova* can be found in freshwater lakes, but is primarily marine, mostly occurring in brackish environments. This genus and close relatives are placed with haptophytes because they possess a short haptonema, but differ from other haptophytes in many ways: The flagella are unequal, the long flagellum's

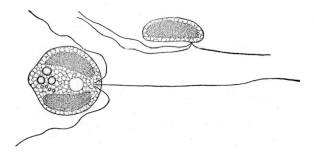

Figure 10–24 *Chrysochromulina parva*, a freshwater form. Two marine species are illustrated in Fig. 10–5. (From Smith, G. M. *Fresh-Water Algae of the United States*. ©1950 McGraw-Hill Companies. Reproduced with permission of the McGraw-Hill Companies)

Figure 10–23 *Pavlova*. (From Throndsen, 1997)

surface is decorated with particles or fine hairs; there is a gulletlike region adjacent to the flagella; the body surface-covering consists of knob- or mushroom-shaped structures, rather than fibrillar scales; an eye-spot may be present; the two parietal plastids are lemon yellow rather than brownish; and mitosis is quite distinctive. The order Pavlovales (which includes *Pavlova* and genera such as *Diacronema*) is characterized by a distinctive class of sterols, termed pavlovols. These are helpful in discerning the sources of organic matter in marine sediments (Volkman, et al., 1997).

CHRYSOCHROMULINA (Gr. *chrysos*, gold + *Chromulina*, a genus of chrysophyceans) includes 50 or more species of primarily marine flagellates (Figs. 10–5, 10–24), though several freshwater forms are known, and some produce amoeboid stages. The genus occurs worldwide, in polar as well as warmer waters. The flagella are equal, as is the case for most haptophyte flagellate cells, and there is typically a very long haptonema that functions in prey capture. The flagellar apparatus of *Chrysochromulina* was eluci-dated by Birkhead and Pienaar (1995). Although there are 2–4 parietal, golden-brown plastids, phagotrophy is common (Hansen, 1998). Experimental studies have shown that low cell-phosphate content is correlated with bactivory, suggesting that phagotrophy may be a method of acquiring phospholipids from bacteria as a source of needed phosphate (Jones, et al., 1994). Many species are also photoheterotrophic, that is, able to take up dissolved organic carbon in the pres-ence of light. This is thought to allow survival in low-light polar environments. Species are differentiated on

the basis of the structure of elaborate fibrillar organic scales, several types of which may occur on the same cell, in more than one layer. Mucilage vesicles occur beneath the cell surface. Asexual division is by longi-tudinal fission of the motile stage; production of four flagellate daughter cells from an amoeboid phase has been reported. Some marine species produce galac-tolipid toxins that are known to kill fish. The fresh-water forms are associated with odor production and tadpole deaths. Because of the cosmopolitan occur-rence of *Chrysochromulina*, some experts believe that there is the potential for it to form toxic blooms almost anywhere in the world.

PRYMNESIUM (Gr. *prymnesion*, stern-cable) is a single-celled flagellate (Fig. 10–25) noted for tolerance of an extremely broad range of salinities, and pro-duction of galactolipid toxins that cause fish kills. It is closely related to *Chrysochromulina*. There is a

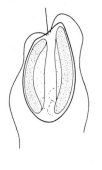
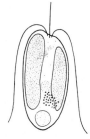

Figure 10–25 *Prymnesium*. (From Throndsen, 1997)

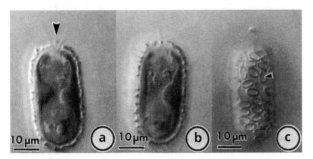

Figure 10–26 Through-focus series of a motile cell of *Pleurochrysis*, whose scale covering is evident. Arrowhead in (a) points to haptonema, small arrowhead in (c) to coccolith. (From Pienaar, 1994)

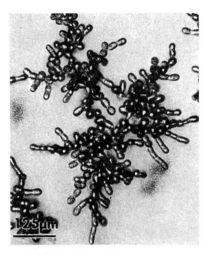

Figure 10–27 The benthic, filamentous (apistonema) stage of *Pleurochrysis*. (From Pienaar, 1994)

short haptonema that does not coil. Species are distinguished by small ultrastructural differences in their organic body scales. *Prymnesium parvum* (a known toxin-producer) and *P. patelliferum* (a suspected toxin-producer), which can be distinguished by such scale criteria, were found to be very closely related by application of molecular techniques. They are suspected to be alternating generations of the same species (Larsen and Medlin, 1997). The ultrastructure of the flagellar apparatus of a *Prymnesium* species was elucidated by Green and Hori (1990).

Coccolithophorids

The coccolithophorids are exclusively marine; some have a well-developed haptonema, but these are vestigial or absent from many forms. The presence of calcite coccoliths is unique among the algae and is regarded as an adaptive response to the need for a protective cell layer by cells occupying environments that are supersaturated with calcium carbonate. Coccolithophorids have higher sinking rates than most other phytoplankton of the same size, presumably because of the coccoliths. This enables them to harvest nutrients from deeper waters, likely to be more nutrient-rich than surface waters. Coccolithophorids can use urea, nitrite, or nitrite as sources of combined nitrogen, but ammonium is preferred. Cells can harvest phosphate at the cell surface by production of extracellular phosphatases that cleave phosphate from organic molecules. Some species require vitamins, thiamine (B_1) being a common requirement. Under optimal conditions some rapidly growing forms can divide as frequently as 2.5 times per day, but others have a maximum division rate of only once per day. Inter-

estingly, many coccolithophorids—particularly those obtained from open-ocean tropical waters—are killed by continuous light. A regular day-night cycle seems to be required by these forms. Coastal, polar, and temperate forms appear to be less sensitive to absence of a photoperiod.

PLEUROCHRYSIS (Gr. *pleuron*, rib + Gr. *chrysos*, gold) occurs as a marine flagellate (Fig. 10–26). One species is known to also have a branched, filamentous, benthic stage (known as an apistonema stage) that does not possess coccoliths (Fig. 10–27).

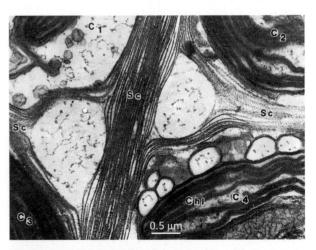

Figure 10–28 TEM view of the junction of four adjacent cells (C_1 through C_4) of *Pleurochrysis*. Note the multiple layers of scales (Sc) that surround each cell. Chl=plastid. (From Pienaar, 1994)

The motile cells are spherical to ovoid, 5–10 μm in length, with two equal flagella. There are two brown plastids, each with a pyrenoid. There are several layers of organic scales between the cell membrane and an outer layer of coccoliths (Fig. 10–28). It is readily eaten by copepods.

EMILIANIA (named for Cesare Emiliani, an Italian-American micropaleontologist and physicist) (Fig. 10–1) is ubiquitous, but most abundant in nutrient-rich temperate and subpolar waters. *Emiliania huxleyi* is the only phytoplankton species that is found in nearly every sample of ocean water and sediments from the Late Quaternary to the present. The organism has a very wide temperature range, 1°–30° C. Although no single isolate can tolerate this range, there appear to be temperature ecotypes (genetically discrete populations that are physiologically adapted to local conditions). *Emiliania* also grows in a wide range of nutrient levels, ranging from eutrophic waters to very low-nutrient, subtropical open-ocean habitats. *Emiliania* will tolerate high salinities, up to 41 ppt, characteristic of the Red Sea. It grows throughout the top 200 m of the water column and can survive at irradiance levels less than 1% that of the surface. Fossil evidence suggests that this coccolithophorid originated approximately 278,000 years ago, acquired the ability to form blooms 85,000 years ago, and has been a dominant member of the phytoplankton for the past 73,000 years (Thierstein, et al., 1977). *Emiliania huxleyi* is widely known for bloom formation today, attaining 60–80% abundance in phytoplankton communities. Such blooms may have a density of 1×10^8 cells l^{-1} and extend over areas greater than 50,000 km^2. *Emiliania* blooms are readily detected by satellite remote sensing. Although the cells themselves are pigmented yellow or golden brown, blooms have a milky-green appearance due to reflectance of light from coccoliths. Fortunately, *Emiliania huxleyi* is not known to produce toxins. Agglutination of cells, viral and bacterial infections, and grazing are the major factors that can terminate blooms. There are three cell types, all about 5–7 μm in diameter: coccolith-bearing cells that are non-motile; naked, non-motile cells; and motile, flagellate cells that have organic body scales but no coccoliths. All of these life stages reproduce asexually by cell constriction. Motile cells lack a haptonema, suggesting that phagotrophy is unlikely. *Gephryocapsa oceanica* (Fig. 10–19) is a related form that can dominate

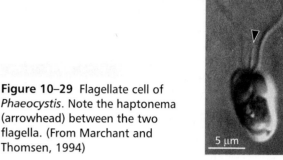

Figure 10–29 Flagellate cell of *Phaeocystis*. Note the haptonema (arrowhead) between the two flagella. (From Marchant and Thomsen, 1994)

eutrophic, tropical coastal waters, forming blooms with a density of 2.3×10^6 cells l^{-1}.

PHAEOCYSTIS (Gr. *phaios*, dusky + Gr. *kystis*, bladder) typically occurs as large gelatinous colonies (up to 8 mm in diameter) (Fig. 10–2) or as unicellular flagellates (Fig. 10–29) that have a short, non-coiling haptonema and two types of organic body scales. The biflagellate unicells possess large vesicles

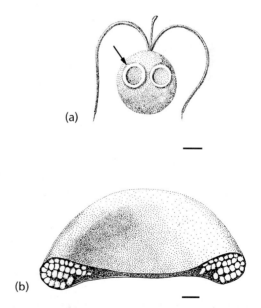

Figure 10–30 The (a) motile cell of *Phaeocystis* with internal vesicles (arrow) is illustrated. Scale bar=1 μm. (b) Diagram of red-blood-cell–shaped vesicle in cross section; these vesicles contain a ring of chitin fibrils at their periphery. Scale bar=0.1 μm. (From Chrétiennot-Dinet, et al., 1997 by permission of the *Journal of Phycology*)

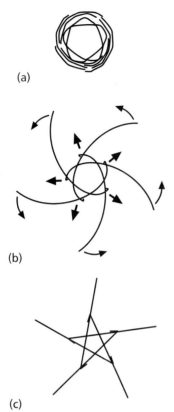

(a)

(b)

(c)

Figure 10–31 The ring of fibrils (a) uncoils (b) upon release from the vesicle and cell to form a five-rayed star (c). The biological function of these stars is unknown. (From Chrétiennot-Dinet, et al., 1997 by permission of the *Journal of Phycology*)

within which are five chitin filaments that can be discharged from the cells (Fig. 10–30). As they are released, the filaments uncoil and extend to form a 5-rayed star (Fig. 10–31). Although their function is unknown (Chrétiennot-Dinet, et al., 1997), it is possible they are analogous to the ejectisomes of cryptomonads or trichocysts of dinoflagellates. Cells of the colonial stage lack scales and a haptonema. The colonial stage develops from single motile cells that have settled on a floating substrate (such as spines extending from diatom cells). The hollow colonies consist of a single surficial layer of cells embedded in polysaccharide mucilage that does not fill the colony center (van Rijssel, et al., 1997). The mucilage is considered to serve as an energy reserve, which can be metabolized by cells when they are light-limited, and as a reservoir for phosphate and trace minerals such as iron and manganese. The mucilage also reduces density, contributing to buoyancy. These adaptations contribute to the ability of *Phaeocystis* to form blooms in the North Sea and other temperate ocean waters, as well as to dominate phytoplankton communities of marginal polar ice zones. Although considered lower in nutritional value than diatoms, copepods and krill will graze on *Phaeocystis*, and it contributes dissolved organic carbon that can be used by microbial community members (Marchant and Thomsen, 1994). In the North Sea, *Phaeocystis* is readily consumed by the ciliate *Strombidium sulcatum* and heterotrophic dinoflagellates. These graz-

Figure 10–32 Thick accumulation of *Phaeocystis* slime on a beach near Scheveningen, Netherlands. Note the two dogs on the beach, one of which is nearly covered by the foam. (Photograph courtesy P. Morehead)

ers can account for more than 90% of *Phaeocystis* losses at the ends of blooms (Peperzak, et al., 1998). *Phaeocystis* is most important for DMS production; it has been estimated to generate 10% of the total global flux of DMS to the atmosphere. Acrylic acid is produced and has antimicrobial properties, but no known toxins are produced. Fish are, however, repulsed by *Phaeocystis* blooms, possibly by the smell of DMS. Massive growths of *Phaeocystis* produced in New Zealand are known as "Tasman Bay slime." These growths clog fishing nets and wash up onto beaches, sometimes producing meter-thick foam that deters recreational use (Fig. 10–32) (Marchant and Thomsen, 1994). The ecology of *Phaeocystis* has been reviewed by Lancelot, et al. (1998).

Recommended Books

Green, J. C., and B. S. C. Leadbeater (editors). 1994. *The Haptophyte Algae*. Clarendon Press, Oxford, UK.

Winter, A., and W. G. Siesser (editors). 1994. *Coccolithophores*. Cambridge University Press, New York, NY.

Chapter 11

Dinoflagellates

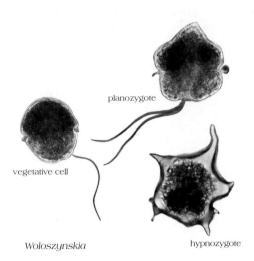

planozygote

vegetative cell

Woloszynskia

hypnozygote

Dinoflagellates probably rank first among the eukaryotic algae in terms of the current and potential future significance of their biotic associations, which may have large impacts on carbon cycling and coastal fisheries production. Dinoflagellate endosymbionts are essential to the formation and existence of coral reef ecosystems; they exhibit an amazing diversity of nutritional types, including autotrophs, mixotrophs, osmotrophs, phagotrophs, and parasites; and some are notorious for the production of toxic red tides (harmful algal blooms). Dinoflagellates are second only to diatoms as eukaryotic primary producers in coastal marine waters. Though most are too large (2–2000 μm) to be consumed by filter feeders, dinoflagellates are readily eaten by large protozoa, rotifers, and planktivorous fish, for which they can be (if not toxic) high-quality food. Common in marine and fresh waters, dinoflagellates are found in both pelagic (open-water) and benthic habitats such as sand, but also may occur in snow and inside cells or bodies of various kinds of protists and animals. They exhibit high levels of living and fossil biodiversity (more than 550 genera and 4000 species); and the internal complexity of their cells rivals that of ciliate and sarcodine protozoa. Molecular analyses reveal close relationship to the protozoan ciliates and apicomplexans (which include malarial parasites, among other human pathogens).

Most of the dinoflagellates are unicellular flagellates having two distinctive flagella that confer characteristic rotatory swimming motions. In fact, the term dinoflagellate originates from the Greek word *dineo*, meaning "to whirl." However, there are several nonflagellate amoeboid, coccoid, palmelloid, or filamentous forms whose relationships to other dinoflagellates are revealed by the characteristic structure of their flagellate reproductive cells known as **dinospores** (zoospores).

Dinoflagellates are further characterized by cell-covering components that lie beneath the cell membrane (Fig. 11–1), like the periplast of cryptomonad cells and pellicle of euglenoids. This contrasts with the cell walls, scales, or extracellular matrices of other algae, which lie outside the cell membrane.

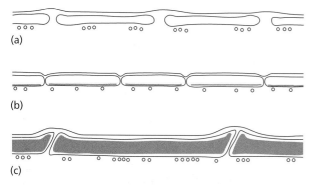

(a)

(b)

(c)

Figure 11–1 The cell covering of dinoflagellates consists of a series of vesicles beneath the plasma membrane. These vesicles can contain little or no cell wall material (a, b) or a thick thecal plate (c). (After Dodge and Crawford, 1970a)

Figure 11–3 SEM of an unidentified armored dinoflagellate. Note the highly ornamented thecal plates. (Micrograph by G. Wedemayer and L. Wilcox)

Together with the cell membrane, the dinoflagellate cell covering consists of a single layer of several to many closely adjacent, flattened amphiesmal (thecal) vesicles; the entire array is known as the **amphiesma**. In many species the thecal vesicles each contain a **thecal plate** composed of cellulose; such species are said to be armored or thecate (Figs. 11–2, 11–3). In other species the amphiesmal vesicles are devoid or nearly devoid of contents (Figs. 11–4, 11–5), such that the cells appear to be naked and are referred to as unarmored or non-thecate. Some dinoflagellates (e.g., dinospores of the invertebrate symbiont *Symbiodinium*) are regarded as intermediate between the armored and unarmored condition.

About half of the known species lack plastids and are therefore obligately heterotrophic (Fig. 11–6). The large size of many free-living dinoflagellates is thought to preclude survival by osmotrophy alone, explaining the widespread occurrence of phagotrophy. Phagotrophic dinoflagellates exhibit a fascinating array of particle capture and feeding mechanisms, some of which are described later in this chapter. Particle-feeding is regarded as a major process contributing to the occurrence of various types of endosymbionts within dinoflagellate cells and to the genesis of cellular internal complexity. An amazing array of parasitic forms also occur within the cells or tissues of fish, invertebrates, and filamentous algae. At least some of these species presumably live osmotrophically, implying the occurrence of high-affinity membrane-based solute transporter molecules.

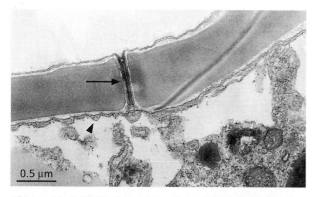

Figure 11–2 TEM view of the theca of *Diplopsalis acuta*, a freshwater colorless dinoflagellate. Note the relatively thick thecal plates within thecal vesicles. Also present is a trichocyst pore (arrow) and a few microtubules (arrowhead) lying beneath the thecal vesicles. (Micrograph by G. Wedemayer and L. Wilcox)

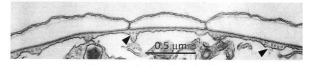

Figure 11–4 A cross section through the amphiesma of *Gymnodinium acidotum* showing the nearly empty amphiesmal vesicles, which are overlain by the cell membrane. Cytoskeletal microtubules (arrowheads) are evident beneath the vesicles. (From Wilcox and Wedemayer, 1984 by permission of the *Journal of Phycology*)

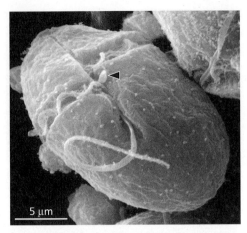

Figure 11–5 SEM of *Amphidinium cryophilum* showing the lack of thecal plates. Note the coiled transverse flagellum in the cingulum and longitudinal flagellum in the sulcus. The peduncle (arrowhead) emerges from the cell at the junction of the cingulum and sulcus. (From Wedemayer, et al., 1982 by permission of the *Journal of Phycology*)

Most of the plastid-containing dinoflagellates are similar in pigment composition: chlorophyll *a* and c_2, the unique accessory xanthophyll peridinin, β-carotene, and other carotenoids such as gyroxanthin diester, used as a pigment marker for the toxic red-tide dinoflagellate, *Gymnodinium breve* (Mille, et al., 1997). This pigment array gives photosynthetic cells a characteristic golden-brown color similar to that of ochrophytes, whose color is, however, based on presence of a different major carotenoid—fucoxanthin. A few dinoflagellates have pigments similar to ochrophytes, green, or cryptomonad algae; these represent examples of secondary or tertiary endosymbiosis (Chapter 7).

About 60 dinoflagellate species are known to produce water- or lipid-soluble toxins; these may be cytolytic, hepatotoxic, or neurotoxic (Chapter 3). The majority of toxin-producing dinoflagellates are photosynthetic, estuarine or coastal shallow-water forms that are capable of producing benthic resting cysts, and which tend to form monospecific populations. Although certain freshwater dinoflagellates may form blooms under specific conditions (Chapter 22), freshwater species are not known to produce toxins. The basis for this habitat-specific difference in toxin production is not understood.

Fossils and Evolutionary History

The fossil record is limited by preservation of only the most decay and chemically resistant structures, because nonresistant structures are usually quickly destroyed after the death of the organisms. Procedures used to remove fossil algae from rocks include treatment with strong acids, such as sulfuric and hydrofluoric acids, and strong bases. Only the most resistant of biological materials can survive such treatment. About 40% of living dinoflagellate species produce a shell of organic material just beneath the outer wall (theca) of cells, often of resting cysts or zygotes. This organic shell has been called a pellicle, but should not be confused with the quite different pellicle of euglenoids (Chapter 8). The dinoflagellate pellicle is composed of cellulose and, frequently, a complex aromatic polymer that is resistant to both strong acids and bases, as is the fossilizable land plant polymer sporopollenin. Because the dinoflagellate polymer is chemically distinct (Kokinos, et al., 1998), responds differently to dyes and stains, and its autofluorescence differs from that of sporopollenin, it is known as **dinosporin** (consult review by Lewis and Hallett, 1997). The presence of dinosporin renders dinoflagellate cells or cysts fossilizable (Fig. 11–7); in general, dinoflagellate cells which lack dinosporin are not considered to be fossilizable (Fensome, et al., 1993).

The earliest fossil that shows considerable structural similarity to modern dinoflagellates is *Arpylorus antiquus* from 400 million-year-old (Silurian) sediments. Many undoubted dinoflagellate fossil cysts

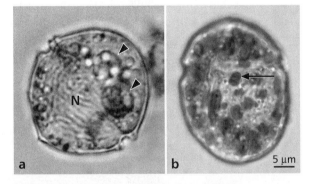

Figure 11–6 (a) The colorless phagotrophic dinoflagellate *Peridiniopsis berolinensis*. This freshwater organism has a relatively thin theca. The nucleus (N) occupies the majority of the hypotheca, with food vacuoles (arrowheads) in the epitheca. (b) The freshwater mixotrophic dinoflagellate *Amphidinium cryophilum*. This species has numerous discrete plastids (arrow).

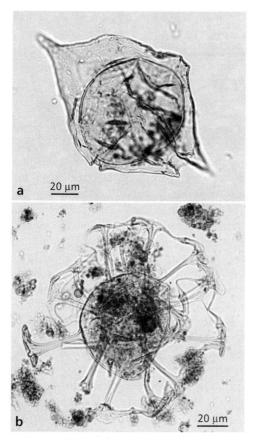

a   20 μm

b   20 μm

Figure 11–7 Fossil dinoflagellates from Late Cretaceous Arctic sediments. (a) *Alterbidinium*, (b) *Adnatosphaeridinium*. (Specimens provided by D. Clark)

occur in deposits of Late Triassic (200 million years ago) and Early Jurassic ages (Taylor, 1990); these represent a major adaptive radiation for dinoflagellates (Fensome, et al., 1993). Many of these fossils, known as **hystrichospheres** (Figs. 11–7, 11–8), are recognizable as dinoflagellate cysts because they bear characteristic spiny ornamentations that also occur on the cysts of some modern dinoflagellates. For example, the spiny cysts of certain modern *Gonyaulax* forms are quite similar to 150 million-year-old hystrichospheres (Wall, et al., 1967). The resting spore stage of the modern dinoflagellate *Pyrodinium bahamense* is identical to the hystrichosphere *Hemicystodinium zoharyi*, whose history can be traced back to the Eocene (40–50 million years ago) (Wall and Dale, 1969). There are many types of less highly ornamented, single-celled fossils extending back to Precambrian times (more than 600 million years ago); such organic-walled microfossils of uncertain relationships are known as acritarchs. While it has been suggested that some of these ancient acritarchs may be the remains of dinoflagellate cysts, morphological features that unequivocally link them with dinoflagellates are not present. However, dinoflagellate-specific molecules (dinosteranes and 4-α-methyl-24-ethyl cholestane) detected in ancient rocks serve as biomarkers linking Early Cambrian (520 million-year-old) acritarchs to the ancestry of dinoflagellates (Moldowan and Talyzina, 1998). Such acritarchs are thus now regarded as close relatives of dinoflagellates and suggest that dinoflagellates had originated by 800 million years ago.

It is widely accepted that the dinoflagellates, like euglenoids and cryptomonads, originated from heterotrophic ancestors, a hypothesis that explains the frequent occurrence of heterotrophy among modern dinoflagellates. However some experts (Hoek, van den, et al., 1995) have suggested that the dinoflagellate lineage acquired peridinin-type plastids very early

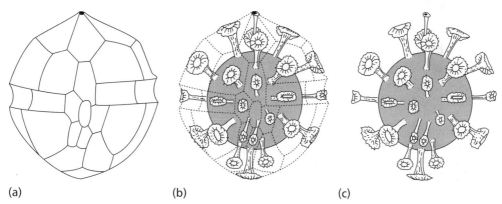

(a)   (b)   (c)

Figure 11–8 The development of the fossil cyst species *Hystrichosphaeridium tubiferum* (c), from a hypothetical parent theca (a). An intermediate stage in the developmental series is represented by (b). (After Fensome, et al., 1993)

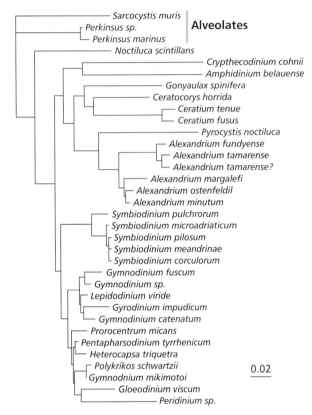

Figure 11–9 Maximum-likelihood tree inferred from complete nuclear small subunit rDNA sequences from selected dinoflagellates. Major features are congruent with distance and parsimony phylogenies. (After Saunders, et al., 1997a ©Springer-Verlag)

in evolution, and that the 50% of modern taxa that lack them subsequently lost ancestral autotrophic abilities. Another phylogenetic hypothesis is that armored dinoflagellates are more primitive than non-thecate forms (Loeblich, 1976). Yet another controversy regarding dinoflagellate phylogeny involves the "many thecal plates ancestral, few plates derived" versus "few plates ancestral, many plates derived" hypotheses (described by Steidinger and Tangen, 1997). Two unusual organisms, *Oxyrrhis* and *Noctiluca*, exhibit some but not all of the major features of most dinoflagellates. A nuclear small subunit rRNA-based phylogeny suggests *Noctiluca* to be early divergent (Saunders, et al., 1997a) (Fig. 11–9).

For the most part dinoflagellates are characterized by distinctive cellular morphologies and flagella such that they are not readily confused with other types of organisms. Moreover, dinoflagellate cell division and fertilization processes are unique. In addition, dinofla-

gellate cells contain a number of cellular organelles that are not found in other algal groups. These cellular features endow dinoflagellates with properties that explain much about their ecological behavior and therefore will be examined in some detail.

Dinoflagellate Cell Biology

There are two general dinoflagellate cell types, differentiated by the position of the flagella. A relatively few species possess **desmokont** cells that are characterized by the emergence of two dissimilar flagella from the cell apex. (By convention, the part of the cell that is directed forward during swimming is the anterior or apical pole, whereas the opposite pole is posterior or antapical.) In contrast, the vast majority of dinoflagellate species possess **dinokont** cells having dissimilar flagella that emerge from a region closer to the midpoint of one side of the cell. Dinokont cells are typically divided into two parts, an apical **epicone** (or **epitheca**) and a posterior **hypocone** (or **hypotheca**), separated by a groove that encircles the cell, known as the **cingulum** (meaning "girdle"). A smaller groove, known as the **sulcus**, extends posteriorly (in the hypocone) from the cingulum. At the intersection of the cingulum and sulcus is a pore from which the two flagella emerge. The transverse flagellum lies within the cingulum and the longitudinal flagellum, in the sulcus. The region of the cell from which the flagella emerge is defined as the ventral side.

The Dinoflagellate Cell Surface

Thecal plates of armored dinokont cells usually fit very closely together, overlapping slightly to form a continuous surface (Fig. 11–10). The regions between adjacent plates are known as **sutures**. Cell growth occurs by addition of material along the margins of the plates, forming regions known as **intercalary bands,** or growth bands (Fig. 11–11). In some dinoflagellates growth occurs along only one plate margin, but in others growth occurs along all margins. Dinoflagellate genera and species vary predictably in the numbers, sizes, and shapes of thecal plates, so these features have been widely used in taxonomy. The major types of thecal organization are shown in Figure 11–12. There is a characteristic plate formula for each species; this consists of a tabulation of plates in a specific series, from the apex toward the posterior. The plate

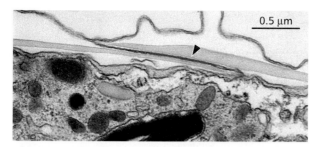

Figure 11–10 A section through the junction (suture) (arrowhead) between two overlapping thecal plates of *Peridiniopsis berolinensis*. (Micrograph by G. Wedemayer)

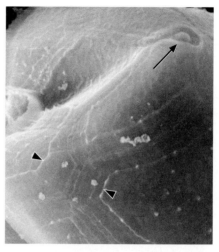

Figure 11–11 SEM view of the epitheca of an unidentified armored dinoflagellate. Note the broad intercalary (growth) bands (arrowheads) between thecal plates and the apical pore (arrow). (Micrograph by G. Wedemayer and L. Wilcox)

terminology used in compiling tabulations is shown in Figure 11–13. Tabulation of the amphiesmal vesicles can also be accomplished for unarmored dinoflagellates, even though thecal plates may be thin or absent. Dinoflagellate identification often requires breaking cells open and spreading the plates in order to tabulate them.

Desmokont cells have two large plates that cover the major portion of the cell; these two plates are known as **valves** (Figs. 11–12a). Growth occurs at the suture between the valves. In addition, there are smaller periflagellar plates at the cell apex. Identification of unarmored dinoflagellates is based upon cell shape and size, cingulum position and shape, sulcus position, and

whether or not thecal ridges, apical grooves, or feeding structures are present.

Surface Scales and Internal Skeletons

Some dinoflagellates bear organic body scales closely resembling those of prasinophycean green algae, hap-

Figure 11–12 The major types of dinoflagllate thecal organization. (a) Prorocentroid, *Prorocentrum* is an example; (b) Dinophysoid (e.g., *Dinophysis*); (c) Gonyaulacoid (e.g., *Gonyaulax*); (d) Peridinioid, exemplified by *Peridinium*; and (e) Gymnodinoid, of which *Gymnodinium* is an example. (Redrawn from *BioSystems* 13, Taylor, F. J. R. On dinoflagellate evolution, ©1980, pages 65–108, with permission from Elsevier Science)

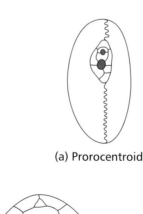

(a) Prorocentroid

(b) Dinophysoid

(c) Gonyaulacoid

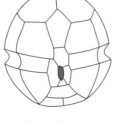

(d) Peridinioid

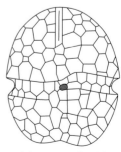

(e) Gymnodinoid

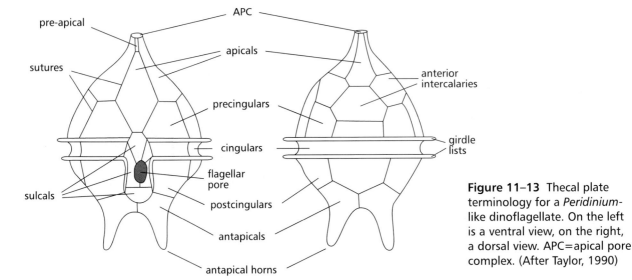

Figure 11–13 Thecal plate terminology for a *Peridinium*-like dinoflagellate. On the left is a ventral view, on the right, a dorsal view. APC=apical pore complex. (After Taylor, 1990)

tophytes (prymnesiophytes), or chrysophyceans. For example, scales similar to those frequently appearing upon the bodies of prasinophycean green flagellates also occur on the surface of the dinoflagellate *Lepidodinium viride*, consistent with the occurrence of green plastids and other organelles derived from a prasinophycean endosymbiont (Watanabe, et al., 1990). The genetic information necessary for prasinophycean scale production and transport to the cell surface has apparently been retained in this green dinoflagellate. In contrast, scales are not produced by another green dinoflagellate, *Gymnodinium chlorophorum*, which also has a prasinophycean endosymbiont (Elbrächter and Schnepf, 1996). A few dinoflagellates contain an internal siliceous skeleton. An example is the phagotrophic, unarmored dinoflagellate *Actiniscus pentasterias*, which contains two five-armed silica stars (pentasters) located near the nucleus. Fossil remains of very similar pentasters have been found (Hansen, 1993).

Dinoflagellate Motility and Flagella

Dinoflagellates are considered to be the champion swimmers among flagellate algae, achieving rates of 200–500 μm s^{-1}. In the desmokont dinoflagellates the longitudinal flagellum extends apically, while the other is coiled and lies perpendicular to the first. The coiled flagellum is attached to the cell body except at the tip; the free tip beats with a whiplash motion, while the attached portion of the flagellum undulates. The longitudinal flagellum beats with an anterior-to-posterior whipping action. The wave action is generated from

tip to base and is similar to flagellar motion in some non-algal protists, but different from that of most algal flagella, where waves are generated in a base-to-tip mode (Goldstein, 1992).

In dinokont dinoflagellates the two flagella emerge close together from the ventral side of the cell. The ribbon-shaped transverse flagellum lies in the cingulum, curving around the cell in a counterclockwise direction (as seen by a viewer looking down on the anterior end of the cell), and is usually difficult to visualize with light microscopy. If it becomes dislodged from the cingulum when the cell is disturbed, the coiled appearance of this flagellum can be seen (Fig. 11–14). Following treatment with appropriate chemical fixatives (see below), dinoflagellate flagella can be best viewed with SEM (Fig. 11–5). In addition to the usual 9 doublet + 2 singlet array of axonemal microtubules, the transverse flagellum con-

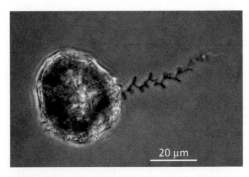

Figure 11–14 The transverse flagellum has been dislodged from the cingulum of this *Peridinium* cell. Its coiled nature is evident.

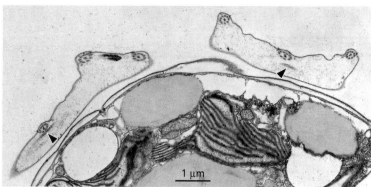

Figure 11–15 In this cross section through an *Amphidinium cryophilum* cell, two profiles of the transverse flagellum are seen lying in the cingulum. Note the outer, axonemal (9 + 2) strand and the finer striated strand to its inside (arrowheads). (From Wilcox, et al., 1982 by permission of the *Journal of Phycology*)

tains a striated, contractile strand composed of the protein centrin (Fig. 11–15). The flagellum is helically coiled in such a way that each turn appears to be stretched where it lies closest to the cingulum, and compressed where it is most exposed. The outermost edge of the flagellum projects out of the cingulum, where the ribbon edge appears to undulate. A single row of fine hairs occurs on the transverse flagellum. The waveform is generated from base to tip and propels the cell forward, accounting for about half the forward speed, as well as rotating it at a rate of about 0.7 s per turn. Cells that lose the longitudinal flagellum may continue to rotate and move forward. *Noctiluca* lacks a transverse flagellum and does not rotate while swimming (Goldstein, 1992). In aggregate, these observations support the conclusion that the transverse flagellum is responsible for rotation of swimming dinoflagellate cells.

The longitudinal flagellum of dinokont dinoflagellates (Figs. 11–5, 11–46), which may extend more than 100 μm beyond the cell body, bears two rows of fine hairs, accounts for perhaps half of the forward swimming speed, and apparently also has a steering function. In *Peridinium*, the plane of the waveform that results from beating of that flagellum is perpendicular to the ventral and dorsal cell body surfaces, but in *Ceratium*, the waveform is parallel to the cell body. Dinoflagellates can stop or reverse their swimming direction by changing the position of the longitudinal flagellum. The flagellum stops beating, points in a different direction by bending at the site where it exits from the sulcus, then resumes beating at the new attitude. This steering ability may be related to the presence in it of a fine filament (Fig. 11–16) that may be composed of the contractile protein centrin (Melkonian, et al., 1992). The intracellular portion of the flagellar apparatus of dinoflagellates is fairly consistent within the dinoflagellates (Fig. 11–17). A regular feature is a multimembered longitudinal microtubular root that extends posteriorly along the left side and beneath the sulcus. A second microtubular root is associated with the transverse flagellar basal body; it consists of a "flag" of microtubules that are attached to a "pole" of one or two microtubules. In addition to striated roots that are associated with the basal bodies, a fibrous connective links the two flagellar bases; this connective contains centrin as do fibrous connectives of other algae.

Phototaxis and Eyespots

Dinoflagellates can swim toward light (phototaxis) by repositioning the longitudinal flagellum (Goldstein, 1992). However, the light-sensing system of dinoflagellates is not well understood; it has been suggested to be a protein-bonded carotenoid, based on the results of phototactic action spectra (surveys of the wavelengths that elicit the response). Dinoflagellates lack a localized, autofluorescent, flavin-based light

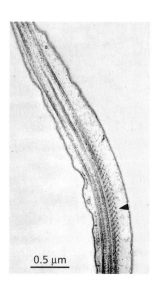

Figure 11–16 A section through a longitudinal flagellum. Note the crosshatched appearance of the fine (centrin-containing?) strand (arrowhead) that lies adjacent the axonemal microtubules. (From Wilcox, et al., 1982 by permission of the *Journal of Phycology*)

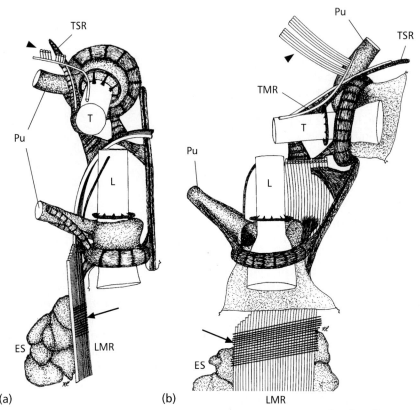

(a)   (b)

Figure 11–17 A diagrammatic representation of the major components found in the flagellar apparatus of dinoflagellates. (a) View from the cell's right (the ventral surface of the cell is toward the right). (b) Ventral view. Each of the flagella is surrounded by a striated collar. The substantial longitudinal microtubular root (LMR) originates near the longitudinal basal body (L) and extends posteriorly beneath the sulcus where it is overlain by subthecal fibers (arrows). An eyespot (ES) often underlies the LMR. The prominent transverse striated root (TSR) and transverse microtubular root (TMR) extend from the transverse basal body (T). Microtubules (arrowheads) extend from the TMR like a flag from a pole. A variety of striated and fibrous connectives are typically present, interconnecting the various components. Pusules (Pu) typically connect to the outside of the cell near the point of emergence of the flagella. (From Roberts, et al., 1995 by permission of the *Journal of Phycology*)

detector such as is associated with the flagella of euglenoids, brown algae, and some other algal groups (Kawai and Inouye, 1989). Most dinoflagellates lack eyespots (suggesting that these are not essential for phototaxis), but several types of eyespots occur among some dinoflagellates. Eyespots consisting of a single or double layer of carotenoid-containing lipid droplets are located between the three-membrane plastid envelope and the outermost layer of thylakoids of *Glenodinium*, *Gymnodinium*, and *Woloszynskia* (Fig. 11–18). Other dinoflagellates have eyespots consisting of globules that are found in the cytoplasm, rather than within plastids (Fig. 11–19). These types of eyespots underlie the sulcus, often near the point of emergence of the longitudinal flagellum; such positioning may be related to phototactic signal transduc-

tion. In *Kryptoperidinium* (formerly *Peridinium* and *Glenodinium*) *foliaceum* and *Peridinium balticum*, whose photosynthetically active plastids were acquired from a diatom endosymbiont (Chesnick, et al., 1996), the eyespots occur as carotene droplets surrounded by three membranes and are thus considered to represent the remains of highly reduced plastids of the more typical dinoflagellate type.

The complex eyespots, or **ocelli** (pl.), of a group of dinoflagellates known as the Warnowiaceae (for example, the genus *Erythropsidinium*—the "red-eyed dinoflagellate") (Fig. 11–20) bear an extraordinary resemblance to metazoan eyes. At the subcellular level there is a lenslike refractile **hyalosome** (meaning "clear body") that is constructed within the endoplasmic reticulum and which becomes surrounded by mito-

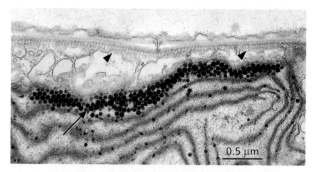

Figure 11–18 The eyespot (arrow) within the plastid of *Woloszynskia pascheri*. The eyespot is located beneath the sulcus. Numerous microtubules are seen beneath the thecal vesicles. These are part of the longitudinal microtubular root, which extends posteriorly from the flagellar apparatus.

chondria and constricting fibers that can move, changing the shape of the lens. There is also a subtending "ocular chamber," backed by a darkly pigmented, cup-shaped **retinoid** consisting of membranes backed by a layer of reddish black droplets (Fig. 11–21) (Greuet, 1968). Presumably, this assembly allows the formation of images upon the retinoid (Francis, 1967), but the mechanism of sensory transfer, with resulting changes in organismal behavior, if any, is not understood. It has been proposed that this unusually highly developed visual system has evolved in response to selective pressures of the phagotrophic habit, allowing predacious dinoflagellates to "see" their prey.

Plastids and Photosynthesis

As noted in Chapter 7, the 50% or so of dinoflagellates that possess plastids acquired them from a variety of

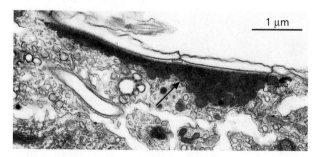

Figure 11–19 The eyespot (arrow) of the colorless dinoflagellate *Katodinium campylops*. The granules making up this eyespot are found free in the cytoplasm rather than within a plastid or surrounded by any membranes.

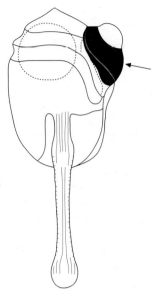

Figure 11–20 A drawing of the ocellus-containing *Erythropsidinium*, showing the large size of the ocellus and dark pigment cup (arrow) in relation to the rest of the cell. The long structure extending posteriorly is the "piston" or "tentacle" thought to be involved in phagotrophy. (After Kofoid and Swezy, 1921)

photosynthetic eukaryotes, including green algae, cryptomonads, and diatoms. However most plastid-containing dinoflagellates have golden-brown plastids with a unique accessory pigment, peridinin. The ancestry of peridinin-containing plastids is as yet unclear. Plastids of the majority of peridinin-containing dinoflagellates have thylakoids stacked in threes and are bound by an envelope of three (occasionally two) membranes that are not connected to the nuclear envelope or the endoplasmic reticulum (Fig. 11–22).

Studies of energy transmission through the light harvesting system (accomplished by measuring changes in fluorescence) have revealed that the peridinin-containing dinoflagellates are amazingly efficient in light capture and energy transmission. The reason for this high efficiency is the occurrence within thylakoid lumens of a water-soluble peridinin-chlorophyll-protein light-harvesting complex (PCP LHC) in addition to a more typical membrane-bound LHC similar to that of land plants. Peridinin (Fig. 11–23) captures light energy in the blue-green range of 470–550 nm, which is present in aquatic habitats and inaccessible to chlorophyll alone. X-ray crystallographic studies have shown that the soluble LHC protein forms a boat-shaped structure with hydrophobic cavities that are filled by

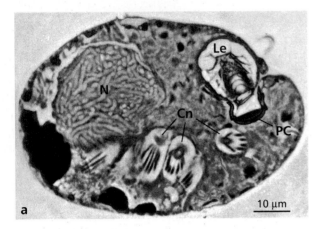

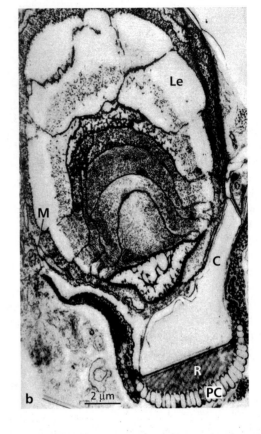

Figure 11–21 Longitudinal views of the ocellus of *Nematodinium*. (a) Low magnification view of the entire cell. Several cnidocysts (Cn) are evident as is the ocellus, with its clear lens (Le) and pigment cup (PC). N=nucleus. (b) The outer, hyaline lens (Le) is at the top and the pigment cup (PC) and retinoid (R) toward the bottom of this TEM image. M=mitochondrion, C=canal open to outside of cell. (From Mornin and Francis, 1967)

two lipid, eight peridinin, and two chlorophyll *a* molecules. The chlorophylls are completely buried within the hydrophobic environment, with half their surface area covered by peridinins, and the remainder covered by protein and fatty acid chains of the lipids. The main function of the protein is to provide a hydrophobic environment for the pigments (which would otherwise be insoluble in the watery lumen environment), but another function is to hold the peridinins at the most appropriate distance for effective energy transfer to chlorophyll. Peridinin must be located particularly close (0.33–0.38 nm) to chlorophyll *a* because the half-life of the excited

carotenoid is very short. The PCP complex allows an energy transfer efficiency of nearly 100% (Hofmann, et al., 1996). This highly organized spatial array of pigments recalls that of the phycobilisomes (Chapter 6) present in cyanobacteria, glaucophytes, and red algae.

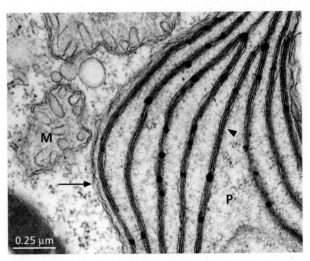

Figure 11–22 A section through the plastid (P) of *Woloszynskia pascheri*, showing the three membranes (two of which are often closely appressed to one another) bounding the plastid (arrow); the thylakoids (arrowhead) occur in stacks of three. The dark globules found on the thylakoids result from the particular preservation technique used in preparing this specimen for TEM. Note also the tubular cristae in the adjacent mitochondria (M), a feature typical of the dinoflagellates and other protist groups (see Chapter 7).

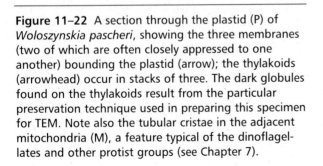

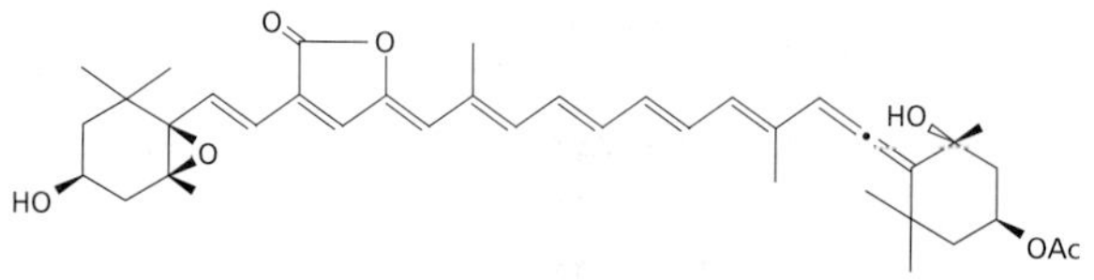

Figure 11–23 The structure of peridinin, a common accessory pigment in dinoflagellates. (Redrawn with permission from Hofmann, E., P. M. Wrench, F. P. Sharples, R. G. Hiller, W. Welte, and K. Diederichs. Structural basis of light harvesting by carotenoids: Peridinin–chlorophyll–protein from *Amphidinium carterae. Science* 272:1788–1791. ©1996 American Association for the Advancement of Science)

Peridinin-containing dinoflagellates are also notable for the occurrence within their plastids of Form II Rubisco, an unusual CO_2-fixing enzyme that is otherwise known only from certain bacteria (see Chapter 7 for more information). Dinoflagellates store photosynthate as cytoplasmic starch grains that stain blue-black with iodine-iodide solution, and lipid droplets containing C_{14}-C_{22} unsaturated fatty acids. In addition, dinoflagellates may contain sterols; more than 35 types are known. Some dinoflagellate sterols exhibit specific chemical features that link them to certain oil deposits (Robinson, et al., 1984). Thus it is thought that past dinoflagellate blooms contributed to the formation of oil deposits.

The Pusule

Dinoflagellate cells contain an unusual organelle known as a **pusule**. It is composed of an array of highly branched membranous sacs or tubules derived by invagination of the cell membrane (Fig. 11–24). Pusules do not contract as do the smaller, rounded contractile vacuoles that occur in many other algal cells. Rather, they open to the cell surface in the vicinity of the flagella. The function of the dinoflagellate pusule is unknown, but because pusules are most highly developed in non-photosynthetic marine species, it has been proposed that they serve in excretion or uptake processes.

Trichocysts, "Nematocysts," and Mucocysts

Dinoflagellate cells may contain a variety of structures that can be discharged from the cell body into the environment. **Trichocysts** (also known as extrusomes) are ejectile rods that occur almost universally at the periphery of dinoflagellate cells (Fig. 11–25). They are very similar to the trichocysts of ciliate protozoa, except that the latter are capped with a spine, which is absent from dinoflagellate trichocysts. The trichocysts lie within the amphiesma, generally oriented perpendicularly to the cell surface; in armored species, trichocysts lie directly beneath pores in the thecal plates (Fig. 11–26). Almost all of the pores found in the dinoflagellate theca are preformed sites for trichocyst discharge. Trichocysts develop within the Golgi apparatus and are produced within a sac. They consist of a paracrystalline protein rod that is a few micrometers long and rectangular in cross section. The distal end consists of twisted fibers. When dinoflagellates are irritated by temperature changes, turbulence, or other disturbance, the sac ruptures. This allows the

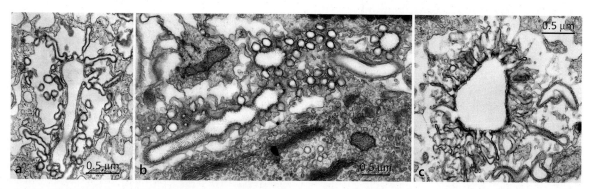

Figure 11–24 Examples of pusules from three freshwater dinoflagellates. Note the variety of sacs and tubules. The membrane lining the pusule tubules is continuous with the cell membrane. (a) *Amphidinium cryophilum*, (b) *Katodinium campylops*, and (c) *Gymnodinium acidotum*. (a: From Wilcox, et al., 1982; c: From Wilcox and Wedemayer, 1984 by permission of the *Journal of Phycology*)

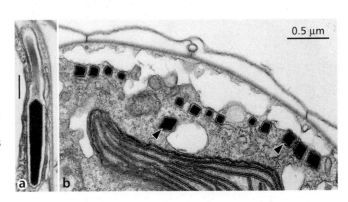

Figure 11–25 (a) Dinoflagellate trichocysts seen in longitudinal view in *Amphidinium cryophilum.* Note that there is a dense basal region and a fibrillar portion (scale bar=0.25 μm). (b) Trichocysts in cross section in *Gymnodinium acidotum.* (a: From Wilcox, et al., 1982; b: From Wilcox and Wedemayer, 1984 by permission of the *Journal of Phycology*)

entrance of water, which causes a change in the conformation of the protein, resulting in elongation of the trichocyst by eight times, and explosive release of the trichocyst from the cell. Some experts hypothesize that trichocyst release causes a jet-propulsive response that is useful in escaping from predators. Upon discharge, trichocysts become much longer and thinner and exhibit a crossbanded structure when viewed with TEM (Fig. 11–27). **Nematocysts** (Figs. 11–21, 11–28), produced by a few dinoflagellate genera (including *Polykrikos* and *Nematodinium*), are larger (up to 20 μm long) and even more elaborate than trichocysts. Some experts have compared these ejectile structures to the stinging cells (cnidocysts) of coelenterates, but others suggest that they are not comparable (Steidinger and Tangen, 1997). **Mucocysts** are relatively simple sacs that release mucilage to the cell exterior, often in the form of rather thick rod-shaped bodies, such as those extruded by *Gymnodinium fuscum* when it is perturbed (Fig. 11–29). Similar structures occur in euglenoid cells (Chapter 8).

Scintillons and Bioluminescence

Bioluminescence occurs in approximately 30 photosynthetic dinoflagellates, including *Gonyaulax*, *Protogonyaulax*, *Pyrodinium*, *Pyrocystis*, and *Ceratium*, and some non-photosynthetic marine forms, such as *Noctiluca* and *Protoperidinium* (Taylor, 1990). There are bioluminescent and nonluminescent strains of the same dinoflagellate species. As discussed in Chapter 3, bioluminescence is regarded as an adaptation that reduces attack on dinoflagellates by their predators (see review by Lewis and Hallett, 1997).

Bioluminescent dinoflagellates possess spherical intracellular structures known as **scintillons** or **microsources**. These are about 0.5 μm in diameter and are arrayed at the cell periphery. In transmission elec-

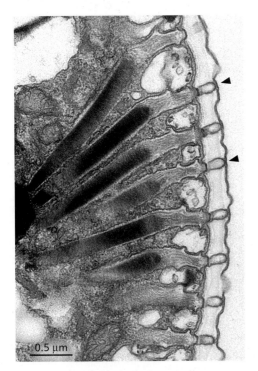

Figure 11–26 A row of trichocysts found along the cingulum of *Peridiniopsis berolinensis.* Note that each lies beneath a pore (arrowheads) in the theca. (From Wedemayer and Wilcox, 1984)

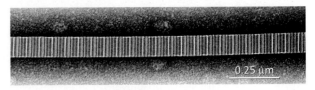

Figure 11–27 A "negatively stained" TEM preparation of a discharged trichocyst showing the fine crossbanding that is present.

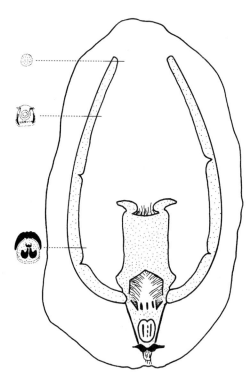

Figure 11–28 A drawing of the more elaborate nematocyst, found among various ocellus-containing dinoflagellates. (From Mornin and Francis, 1967)

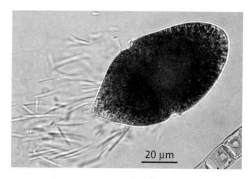

Figure 11–29 A disturbed *Gymnodinium fuscum* cell that has ejected a large number of mucocysts. Treatment of such cells with NiSO$_4$ (to stop swimming) can cause this to occur.

tron micrographs, scintillons appear as dark rodlike structures surrounded by a membrane. They are derived from the Golgi apparatus and contain luciferin, luciferase, and in some cases, also a luciferin-binding protein (LBP). Immunogold labeling was used to establish the presence in scintillons of the enzyme luciferase. In *Gonyaulax polyedra* (*Lingulodinium polyedrum*), luciferin is bound to LBP at pH 8, but is released at pH 6, becoming available for reaction with luciferase. Luciferase oxidizes the luciferin with molecular O$_2$, causing a 0.1-s flash of blue light. This reaction is thought to be triggered by environmental stimuli that cause an influx of protons across the scintillon membrane, thus lowering pH to the critical level. The number of scintillons in *G. polyedra* decreases from 540 per cell in the night phase to just 46 in day-phase cells, and the amount of bioluminescence is two orders of magnitude greater in night-phase cells. There is a daily (circadian) rhythm in synthesis and destruction of scintillons, luciferin, and luciferase. This is viewed as an adaptation that conserves energy, as bioluminescence would not be visible in the daytime (Fritz, et al., 1990). In contrast, *Pyrocystis*, which can emit about

1000 times more light than *Gonyaulax* (Swift, et al., 1973), does not undergo a daily cycle of destruction and resynthesis of scintillons, and lacks LBP (Knaust, et al., 1998). However there is a diurnal change in the positions of *Pyrocystis* scintillons—during the day they occur as a spherical mass of vesicles near the nucleus, and at night they disperse to the cell periphery, returning to the nucleus at daybreak (Fig. 11–30). The plastids exhibit the opposite pattern of movement (Sweeney, 1982).

The *Gonyaulax polyedra* genes for luciferin-binding protein and luciferase have been cloned and sequenced (Lee, et al., 1993; Bae and Hastings, 1994). They are atypical for eukaryotic genes in lacking introns and the usual 3' polyadenylation signal in addition to some other features. Although evolution of bioluminescence in dinoflagellates is thought to have occurred independently of its origin in bacteria and other organisms (Knaust, et al., 1998), the gene attributes just described suggest the possibility that this ability was acquired through cryptic endosymbiosis (Chapter 7). Bioluminescence and many other dinoflagellate processes are known to involve endogenous circadian rhythms, studied intensively by Sweeney (1987) and recently reviewed by Lewis and Hallett (1997).

Dinoflagellate Reproduction

Dinoflagellates that spend most of their life cycle as unicellular flagellates undergo population increases by mitosis, and several have been documented to undergo sexual reproduction as well. Cysts (unicellular, nonflagellate stages) of various types are produced by dinoflagellates, and in some cases these are known to

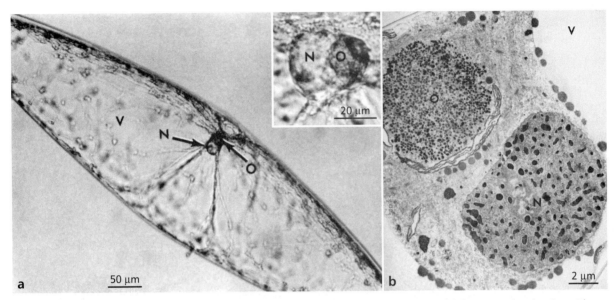

Figure 11–30 (a) Scintillons of *Pyrocystis* form an orange body (O) near the nucleus (N) during the daytime. They are seen at higher magnification in the inset. (b) The orange body seen adjacent the nucleus with TEM. During the night scintillons disperse throughout the highly vacuolate (V) cell. (From Sweeney, 1982 by permission of the *Journal of Phycology*)

result from sexual reproduction. A number of nonflagellate unicellular or multicellular dinoflagellates undergo asexual reproduction by development of flagellate zoospores (dinospores).

Mitosis

Dinoflagellate chromosomes and mitosis are quite unusual. The chromosomes of most dinoflagellates lack the histone proteins that characterize other eukaryotes and are permanently condensed (Fig. 11–31), uncoiling for only a brief time to allow DNA replication to occur. The dinoflagellate nuclear envelope remains intact throughout mitosis (as in a variety of other algae), but, uniquely in dinoflagellates, the microtubular spindle is entirely extranuclear. Bundles of spindle microtubules pass through tunnels in the mitotic nucleus (Fig. 11–32), and each chromosome attaches to microtubules via contact with the nuclear envelope at the same site (Oakley and Dodge, 1974) (Fig. 11–33). Extension of the nuclear envelope and

extension of interzonal spindle microtubules are involved in chromosomal separation and movement to the spindle poles, after which the nucleus is pinched into two at the midpoint. Cytokinesis in unarmored cells may involve a simple pinching in two along an oblique division line, with continuous synthesis of

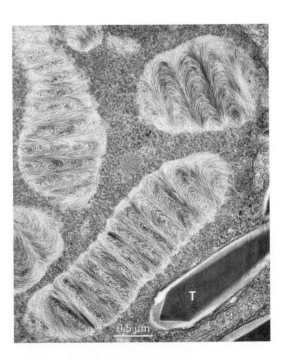

Figure 11–31 The permanently condensed chromosomes of the dinoflagellate *Prorocentrum micans*, showing the swirled appearance of the "naked" (histone-lacking) DNA. Such an arrangement of DNA is exhibited by dinoflagellates as well as the nucleoids of certain bacteria. Also note portions of trichocysts (T).

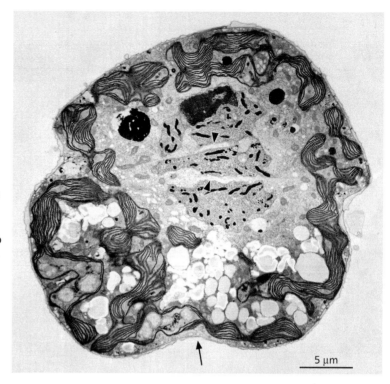

Figure 11–32 A mitotic cell of the thin-walled, photosynthetic freshwater dinoflagellate *Woloszynskia pascheri*. This species remains motile throughout mitosis and cytokinesis (which occur concurrently in this species). Note the cytoplasmic channels (arrowheads) passing from left to right through the nucleus, each of which contains spindle microtubules. Note also that a number of the chromosomes appear to be under tension—particularly near the lower cytoplasmic channel—and are seemingly being pulled toward the right. The cleavage furrow (arrow) proceeds from the posterior toward anterior end of the cell.

new amphiesma parts. Cells may remain motile, as in *Amphidinium cryophilum* (Fig. 11–34). In contrast, armored cells may shed the cell covering prior to mitosis, then each daughter cell resynthesizes a new theca. Mitosis in such forms generally occurs in non-motile cells (Fig. 11–35). In other armored forms, such as *Ceratium*, oblique fission of the theca occurs along a predetermined sequence of suture lines. Each daughter cell receives half of the parental wall components, and regenerates the missing portion (Fig. 11–36). In natural populations of *Peridinium*, division of a portion of the cells occurs every night, with dividing cells making up 1–40% of the total population. In cultures of *Peridinium*, the maximum division rate is about one per day, the estimated maximum division rate for both photosynthetic and heterotrophic dinoflagellates in nature (Naustvoll, 1998).

Sexual Reproduction

Although sexual reproduction has been studied in relatively few dinoflagellates (reviewed by Pfiester and Anderson, 1987; Pfiester, 1988), it is thought to occur much more widely than actually observed. Reasons that dinoflagellate sexual reproduction might be cryptic include the facts that gametes resemble vegetative

cells, gamete fusion is so slow that it is difficult to distinguish it from cell division, gamete fusion seems to occur at night in the photosynthetic species, and, while

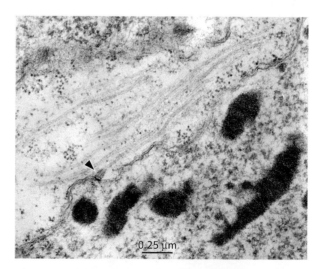

Figure 11–33 *Woloszynskia pascheri* mitotic cell. A close-up view of a cytoplasmic channel through which the extranuclear spindle microtubules pass. Note the attachment of a microtubule to a kinetochorelike structure (arrowhead) on the nuclear envelope, at a point where a chromosome is presumably also attached.

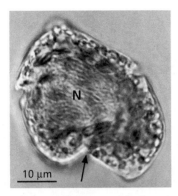

Figure 11–34 *Amphidinium cryophilum* is an unarmored dinoflagellate that undergoes mitosis and cell division while remaining motile. The cleavage furrow (arrow) proceeds from the antapex toward the apex of the dividing cell. The numerous chromosomes are visible in the large central nucleus (N). (From Wedemayer, et al., 1982 by permission of the *Journal of Phycology*)

early stages of zygote development can be recognized by presence of more than two flagella, it can difficult to recognize later zygote stages. Flagellate zygotes (**planozygotes**) (Fig. 11–37) of some species undergo meiosis rather quickly, whereas those of other dinoflagellates transform into nonflagellate **hypnozygotes** ("sleeping" zygotes, also known as dinocysts) (Fig.

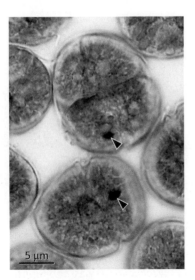

Figure 11–35 A dinoflagellate (*Woloszynskia* sp.) that divides while immobile (the top cell has recently divided). Each daughter cell makes a new theca following cell division. A large bright red eyespot (arrowheads) is found in this species.

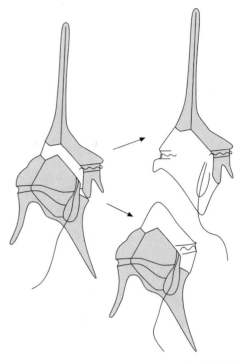

Figure 11–36 A diagrammatic view of cell division in *Ceratium hirundinella*, a common armored dinoflagellate. In this organism the original theca (shaded gray) is split among the two daughter cells, each of which must synthesize the remaining portion following division. (After Fensome, et al., 1993)

11–38), and resemble resting cysts, discussed later. It is clear that some resting cysts arise from zygotes, but whether or not this is common is an open question.

Sexual reproduction can be induced in the laboratory by reducing the nitrogen concentration of the growth medium or by changing the temperature. It is assumed that these factors are also related to induction of sexual reproduction in nature. In some dinoflagellates sexual reproduction involves isogametes—gametes that closely resemble each other in size and shape. In *Scrippsiella* (Xiaoping, et al., 1989), for example, isogametes may be about the same size as vegetative cells or noticeably smaller. In other cases (e.g., *Ceratium*), gametes are anisogamous, that is, one gamete of each fusing pair is much smaller than the other, the larger being about the same size as a vegetative cell (von Stosch, 1964). Sometimes sexual reproduction is **homothallic**, meaning that gametes produced within a clonal population will fuse, and sometimes **heterothallic**, meaning that syngamy can only be achieved by mating individuals from genetically different clones.

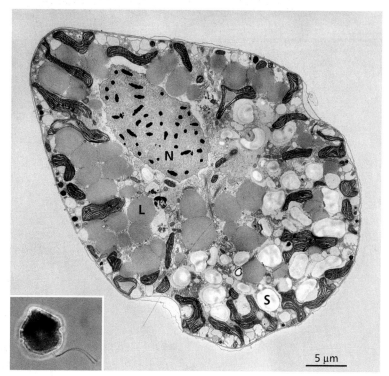

Figure 11–37 Planozygotes of the freshwater dinoflagellate *Woloszynskia pascheri*, as seen with light (inset) and transmission electron microscopy. Planozygotes are often identifiable due to the presence of two "ski-track" longitudinal flagella and a more angular outline than is seen in vegetative cells. There are also large stores of lipids (L) and starch (S). In this case, the starch is found primarily in the hypocone and lipids, in the epicone, along with the nucleus (N).

Commonly the mating process begins with movement of gametes around each other before they come into contact. In *Scrippsiella* (and some other dinoflagellates), the transverse flagellum of one gamete moves out of the cingulum, grasps the longitudinal flagellum of the other gamete, then returns to the cingulum. Fusion then begins at the sulcal region at the point where flagella emerge, and fusion of the hypocones is completed before fusion of the epicones (Xiaoping, et al., 1989) (Fig. 11–39). In other dinoflagellates the

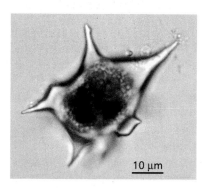

Figure 11–38 A hypnozygote of *Woloszynskia pascheri*, with an extremely thick wall bearing a number of projections.

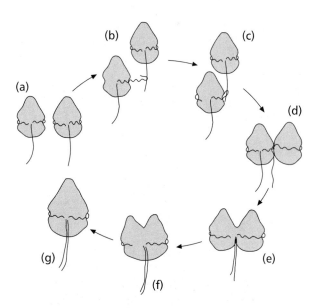

Figure 11–39 Diagram of mating in *Scrippsiella*. Using its transverse flagellum, one gamete lassos the longitudinal flagellum of the other (b–d), pulling it close, after which fusion occurs (e, f). The resulting planozygote has two longitudinal flagella. (see Fig. 11–37a) (After Xiaoping, et al., 1989)

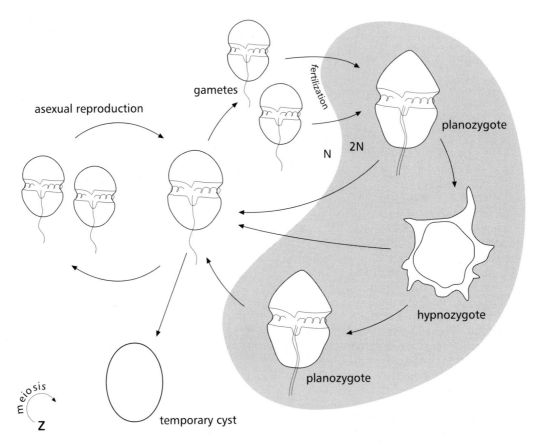

asexual reproduction

gametes

fertilization

N   2N

planozygote

hypnozygote

planozygote

meiosis

Z

temporary cyst

Figure 11–40 A diagram illustrating the life cycle of a "typical" dinoflagellate whose vegetative cells are haploid. Diploid, motile zygotes (planozygotes) may be present following gamete fusion and again following germination of nonmotile hypnozygotes. Temporary asexual cysts are formed in some species. The vegetative stage of some dinoflagellates may be nonmotile and/or strikingly different in appearance from the typical dinoflagellate motile cell morphology. However such organisms are recognizable as dinoflagellates due to the presence of dinospores (asexual or sexual motile cells with a dinoflagellate morphology) at some point in their life cycle.

epicone of one gamete attaches to the sulcal region of the other, such that the cells are oriented perpendicular to one another. In the case of the binucleate dinoflagellate *Peridinium balticum*, whose gametes behave in the latter mode, the host cells and their nuclei fuse first, then the diatom-derived endosymbionts and their nuclei fuse (Chesnick and Cox, 1987; 1989). Sexual reproduction in the nonmotile *Gloeodinium montanum* involves production of 2–4 biflagellate isogametes or anisogametes per vegetative cell. These fuse in pairs, forming a large, nonmotile zygote. After a dormancy period that may last from two months to more than a year, four nonmotile vegetative cells are produced by zygote germination (Kelley and Pfiester, 1990).

Dinoflagellates generally have a zygotic life cycle. Meiosis occurs at zygote germination, so (in most cases) the vegetative cells are considered to be haploid (Fig. 11–40). The onset of meiosis can be predicted by rotation of the diploid zygote nucleus. Most often nuclear division is followed by two cytokineses, but sometimes cytokinesis is delayed until after the second meiotic division (Beam and Himes, 1974). The enigmatic genus *Noctiluca* is exceptional among dinoflagellates in that meiosis appears to be gametic; vegetative cells are regarded as diploid in this case. A meiotic division giving rise to four daughter nuclei is followed by repeated mitoses, ultimately yielding over 1000 uniflagellate, flattened isogametes. These may fuse, with the flat side of one gamete joining the narrow edge of the other, to form zygotes that apparently differentiate into vegetative cells. However, events that follow gamete fusion in *Noctiluca* have not been completely documented (Zingmark, 1970).

The life histories of many dinoflagellates involve multiple stages, and some include amoeboid and other forms unusual in algal life cycles (see Pfiester and Popovsky, 1979; Pfiester and Lynch, 1980; Popovsky and Pfiester, 1982). The most extreme case known is *Pfiesteria piscicida*, which has some two dozen life-history stages (Burkholder and Glasgow, 1997).

Cysts

Nonmotile dinoflagellate unicells are considered to be cysts, but these may arise from a variety of processes. Sexual reproduction can, but does not always, generate cyst products. Dormant resting cysts are often found in nature, but it is not known whether these typically represent hypnozygotes or not. In *Gonyaulax polyedra*, cyst formation typically follows sexual reproduction (planozygote formation) in the autumn (Fig. 11–41). Cyst development begins with formation of a colorless peripheral region of the cytoplasm. The planozygote stops swimming and sheds its flagella, after which the thecal plates separate and pull away from the cell surface. The cyst wall, including spiny outgrowths, then forms. There is a necessary dormancy period lasting several months during which this species cannot excyst. Cyst germination is also inhibited by darkness and low O_2. Excystment in the spring

occurs through an aperture (archeopyle) in the cyst wall, and within a few hours the resulting cells have developed a normal theca. *Gonyaulax* cysts can survive in storage for as long as 12 years (see Lewis and Hallett, 1997). Like the vegetative cells, they are toxic and may be ingested by shellfish (Dale, et al., 1978).

Formation of resting cysts has been attributed to changes in nutrients, irradiance, photoperiod, or temperature. They allow survival during adverse conditions, which may include temperatures that are too high or too low for growth. High numbers of cysts have been found in sediments below coastal waters that previously supported dinoflagellate blooms. Cysts may be transported via water currents or ship ballast water (Hallegraeff and Boalch, 1991, 1992; Carlton and Geller, 1993) to new geographical locations, where they may serve as inocula for bloom formation.

Walls of resting cysts contain cellulose and dinosporin as well as mucilage, arranged in several layers. Photosynthetic pigments may be reduced and storage products increased in resting cysts. The presence of a conspicuous red-pigmented structure commonly occurs within dinoflagellate cysts (Fig. 11–42); this feature can be used to distinguish the cyst forms of dinoflagellates from similar-appearing cysts or vegetative cells belonging to other algal groups. The vast majority of fossil dinoflagellates appear to be resting

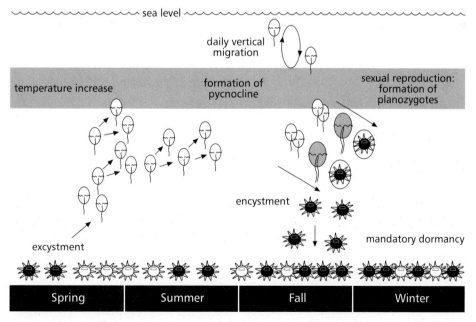

Figure 11–41 Diagram illustrating the cycle of cyst production and excystment in the bloom-forming *Gonyaulax polyedra* (*Lingulodinium polyedrum*). Dormant cysts are shown in black, excysted hypnozygotes, in white. A pycnocline is a zone separating water of different temperatures. (After Lewis and Hallett, 1997)

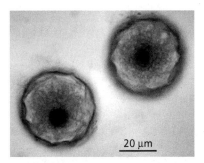

Figure 11–42 Thick-walled cysts with bright red bodies (seen here as dark areas), a common feature of dinoflagellate cysts.

cysts, and archeopyles can often be identified in fossil remains of dinoflagellate cysts. The structure of the archeopyle varies in number of plates involved, their shape, and their position. In some cases archeopyle plates detach completely, whereas in others, plates form hinges that open, releasing the cell contents (Taylor, 1990).

The piscivorous (fish-eating) dinoflagellate *Pfiesteria* (Fig. 3–6) can exist as benthic resting cysts until the presence of live finfish or their fresh excreta induces germination into flagellate stages within minutes to days, depending upon the age of cysts. One or more potent neurotoxins produced by the flagellate stage kills fish such as striped bass, southern flounder, and menhaden, then the dinoflagellate cells consume particles of flesh that slough off the dead fish. While killing the fish, the dinoflagellates complete their sexual cycle, forming poorly pigmented, anisogamous gametes. Gamete fusion results in planozygotes that generate new vegetative cells directly if live fish are present, but if fish are not present, instead produce thick-walled benthic cysts having a surface layer of bristly scales. At least some *Pfiesteria* cysts can survive treatment with concentrated sulfuric acid or ammonium hydroxide (suggesting the occurrence of a dinosporin wall layer), 35 days of desiccation, or two years of dormancy. Cyst formation is an essential component of the successful "ambush predator" strategy practiced by this dinoflagellate (see Burkholder and Glasgow, 1997 and the references cited therein).

Temporary cysts can be formed by dinoflagellates that undergo ecdysis (shedding of the theca and external membranes) in response to stress. However only species that produce a resistant, dinosporin-containing "pellicle" can survive ecdysis to form temporary cysts. Some species form a digestion cyst after feeding. As an example, the armored marine mixotroph *Fragilidium*

subglobosum undergoes ecdysis and subsequent cyst formation after feeding upon large cells of the fellow dinoflagellate *Ceratium*. In order to ingest such large particles, the thecal plates of *Fragilidium* undergo wide separation, perhaps irreparably compromising the integrity of the cell covering. Immediately following digestion, *Fragilidium* cells become immobile, lose their flagella, and the protoplast emerges through the sulcus. The empty cell covering, consisting of the former cell membrane and underlying thecal vesicles, then falls apart, while the protoplast develops into a rounded cyst (Skovgaard, 1996).

Ecology

The ecological niches of dinoflagellates are framed by the resource needs of these relatively large and fundamentally heterotrophic protists. There is a nearly universal requirement among dinoflagellates for vitamin B_{12}, and in addition, the many completely heterotrophic species must obtain sufficient organic carbon to support swimming, division, and reproduction. Possession of plastids does not seem to completely provide the energetic needs of many species, as mixotrophy is common. Photosynthetic dinoflagellates should probably be generally regarded as at least potentially capable of osmotrophy or phagotrophy or both. In addition to organic carbon, dinoflagellates must obtain inorganic nutrients such as phosphorus and combined nitrogen. Phagotrophic forms can obtain these in particulate form, whereas species that do not feed on organic particles must take up inorganic nutrients at the cell surface. Here, because of their relatively large sizes and often compact shapes, dinoflagellates are at a competitive disadvantage in comparison to nanophytoplankton, which have higher surface area to volume ratios (SA/V), and thus relatively high nutrient uptake capacities. In terms of organic carbon produced per unit of organic carbon biomass, dinoflagellates are ten or more times less efficient than are nanoplankton, primarily because of this difference in SA/V. As a result, dinoflagellate population growth rates are generally lower than those typical for nanoplanktonic algae (Pollingher, 1988). Low SA/V is also regarded as a reason why osmotrophy alone is unlikely to provide sufficient organic carbon to support growth of heterotrophic dinoflagellates (Gaines and Elbrächter, 1987).

Dinoflagellates in general are also much more vulnerable than many other algae, particularly nanophytoplankton, to high levels of turbulence. Though turbulence is beneficial in the dispersal and resuspension of benthic cysts, flagellate vegetative cells do not require turbulence to remain suspended and are rather easily damaged by it. As a consequence, researchers who must grow dinoflagellate cultures in the lab are careful not to shake or stir them vigorously. Experimental studies have shown that moderate levels of turbulence impede cell division, but when turbulence is reduced, dinoflagellate cells can resume cell division (reviewed by Lewis and Hallett, 1997). In nature, in both marine and freshwater habitats, conspicuous populations of dinoflagellates are more likely to occur in calm, windless conditions. Storm events (massive turbulence) can be associated with destruction of large numbers of dinoflagellate cells (Pollingher, 1988).

On the other hand, dinoflagellates possess a number of adaptive features that allow them not only to survive and reproduce but, under some conditions, form bloom-level populations. Efficient flagellar motion coupled with phototactic capacity provides dinoflagellates with much greater ability to migrate vertically through great depths in the water column than most other planktonic algae. Dinoflagellates can thus harvest organic particles and/or inorganic nutrients from much of the water column. Photosynthetic dinoflagellates can often be observed to ascend to surface waters during the day where they harvest photons, then descend or disperse at night, a behavior that contributes to both nutrient acquisition and predator avoidance. Swimming also confers the ability to avoid excessively high light levels.

In addition, dinoflagellates can store relatively large amounts of phosphorus, when this is readily available, for later use when ambient levels are lower. Dinoflagellate blooms can therefore occur at times when dissolved nutrient levels would seem too low to support large algal populations. They can also scavenge phosphorus from dissolved organic sources, such as ATP. By virtue of size, flagellation, and sometimes hornlike cell-wall projections, dinoflagellates can avoid predation by all but the mesoplanktonic and planktivorous fish grazers (Pollingher, 1988). With some exceptions, such as heavy chytrid infestations in some freshwaters, many dinoflagellates and their cysts seem to be relatively immune to pathogenic attack; protection may be conferred by resistant dinosporin wall layers when these are present. These constraints and adaptive responses explain many aspects of dinoflagellate distributions and behavior in nature.

Taylor (1990) has noted that while many observers have the impression that diatoms prevail in colder waters, whereas dinoflagellates are more prevalent in warmer waters, this generalization is not completely accurate. Rather, diatoms tend to prevail in coastal waters during the early portion of the growing season (also the most productive), and in high latitude (arctic and subarctic) open waters. Later in the season, as temperate coastal and open-ocean diatoms deplete what is for them an essential nutrient, silicate (see Chapters 12 and 22), subsequently declining in population numbers, dinoflagellates can flourish by virtue of nutrient harvesting abilities conferred by vertical migration and heterotrophic capacities. Here, in calm waters or along boundary fronts formed at the junction of stratified open-ocean waters and the coastal mixed zone, they achieve their greatest levels of population growth, 10^7–10^8 cells per liter. Most bloom-forming dinoflagellates are photosynthetic forms such as *Gonyaulax*, but seasonally high populations of the heterotrophs *Protoperidinium depressum* and *Oxyrrhis marina* can occur in estuaries or other nearshore waters. Bloom dynamics and physiology of the toxic, bloom-forming *Gymnodinium breve* were reviewed by Steidinger, et al. (1998). Benthic heterotrophic species can be abundant, but their ecology is not well studied (Steidinger and Tangen, 1997).

Nutrient acquisition adaptations also allow dinoflagellates to occupy notoriously nutrient-poor tropical and subtropical open-ocean waters. They may form deep chlorophyll maxima some 75–150 m below the surface in such clear nutrient-poor regions. In upwelling regions along the coast of southwest Africa, bloom-level populations may occur (Hoek, van den, et al., 1995). A number of open-ocean-dwelling dinoflagellates are characterized by elaborate horns (e.g., *Ceratium* and *Protoperidinium*) or outgrowths of precingular plates in the form of wings or sails (e.g., *Ornithocercus* and *Dinophysis*) that increase SA/V, likely providing sinking resistance; they may also retard predation. Although generally depauperate in high-latitude open-ocean waters, dinoflagellates may occur in Antarctic ice communities (Lizotte, et al., 1998).

Both non-photosynthetic (e.g., *Amphidiniopsis* and *Roscoffia*) and pigmented (e.g., *Spiniferodinium*, *Amphidinium*, and *Prorocentrum*) dinoflagellates occur within sand; the latter may noticeably color sand flats. *Spiniferodinium* has a nonmotile vegetative phase with a transparent, rigid, spiny, helmet-shaped shell. It reproduces by motile zoospores that regenerate the shell after settling (Horiguchi and Chihara, 1987). Sand-dwelling dinoflagellates are found by shoveling into the beach to the point where seawater seeps into the hole, then the seep water is collected (Horiguchi and Kubo, 1997).

Marine, but not freshwater, dinoflagellates produce toxins that affect many types of animals, including humans (see Chapter 3 for more information about the toxins and their effects). An uncharacterized toxin is produced by *Pfiesteria piscicida*, while some species of *Dinophysis* and *Prorocentrum* generate okadaic acid (see review by Turner and Tester, 1997). Saxitoxins are produced by various species of *Alexandrium* (formerly known by several other names), *Pyrodinium bahamense*, *Gonyaulax polyedra* (*Lingulodinium polyedrum*), and *Gymnodinium catenatum* (reviewed by Steidinger, 1993). *Gymnodinium breve* is the source of brevetoxins. Ciguatoxins, some 100 times more toxic than brevetoxins, are produced by the dinoflagellates *Ostreopsis siamensis*, *Prorocentrum lima*, and *Gambierdiscus toxicus*. These cause the tropical fish poisoning phenomenon known as ciguatera. *Gambierdiscus* lives epiphytically on a variety of tropical red, green, and brown macroalgae (seaweeds) that grow on coral reefs or on sea grasses (Bomber, et al., 1989). There is some evidence that the seaweeds may promote *Gambierdiscus* growth. The dinoflagellate cells occur within a mucilaginous matrix on the seaweed/sea-grass surface or may be attached by a short thread. When there is little water motion, the *G. toxicus* cells may detach and swim around their host, but when water motions increase, they reattach themselves (Nakahara, et al., 1996).

Freshwater dinoflagellates, with few exceptions, are intolerant of high salinities. There are only about 220 freshwater species, and the most prominent are large-celled species of the photosynthetic genera *Peridinium* and *Ceratium*, which, along with *Peridiniopsis*, are the most common bloom-formers. These genera are nearly cosmopolitan in waters characterized by high calcium-ion concentrations (hard waters); particular *Peridinium* and *Ceratium* species also characterize low nutrient, low pH waters. Cysts are essential to the success of dinoflagellates in freshwaters. Freshwater dinoflagellates overwinter (or oversummer, in some cases) as benthic cysts. The first vegetative cells to arise from cyst germination, located either in benthic sediments or in recently mixed waters, are in a position to take up and store newly available phosphorus. These stores are passed on to subsequent cell generations, enabling persistence or even dominance of dinoflagel-

late populations in late-season, nutrient-depleted waters. The level of recruitment from benthic cysts can determine maximal population levels.

Although a number of heterotrophic dinoflagellates occur in freshwaters, they tend to be considerably more cryptic than are pigmented forms. A number of these may grow primarily in the low-temperature season, occurring under the ice, in irradiance conditions only 1% of surface light levels. Some mixotrophic forms take advantage of their phagotrophic capabilities to supplement their nutrition under such conditions (see below). Some freshwater dinoflagellates possess a stage in their life cycle that is amoeboid and actively feeds upon filamentous algae (Fig. 3–22).

Phagotrophy and Feeding Mechanisms

Phagotrophy occurs in both marine and freshwater dinoflagellates, including photosynthetic and plastidless dinoflagellates, as well as in both armored and unarmored types. Emerging evidence indicates that phagotrophy is essential for obtaining inorganic nutrients, such as nitrogen. Dinoflagellate prey can include other dinoflagellates, members of other algal and groups, large ciliates, nematodes, polychaete larvae, and fish. Reflecting the diversity of prey organisms is a diversity of feeding structures and methods. There are three main feeding mechanisms—engulfment of whole cells, use of a feeding tube (such as a **peduncle** or **phagopod**), or extension of a feeding veil (**pallium**). Unarmored forms may engulf intact cells, digesting them inside a food vacuole. The enigmatic heterotrophic organism *Noctiluca* has a pedunclelike tentacle (Fig. 11–67) with which it snares diatoms and other planktonic particles; these are then conveyed to a cell mouth similar in function to the cytostome of ciliates. *Noctiluca* also feeds on copepods and fish eggs. *Erythropsidinium* and related forms (Fig. 11–20) have a thick "piston" that can be extended from the cell and retracted back into the cell within a fraction of a second (Greuet, 1976). This structure is regarded as a feeding device.

Armored dinoflagellates can also ingest whole cells, such as other dinoflagellates; as described earlier, *Fragilidium subglobosum*, a predator of *Ceratium*, is one such example. About 15 minutes is required for *Fragilidium* to engulf a large *Ceratium* species having volumes of over $1 \times 10^5 \ \mu m^3$, but smaller *Ceratium* species can be ingested within five minutes. *Fragilidium* breaks down its prey's theca very quickly, allowing the feeding cell to pack the large prey into a

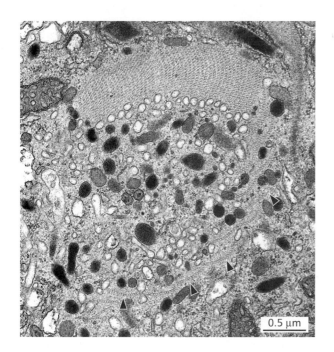

Figure 11–43 The internal portion of the peduncle of *Peridiniopsis berolinensis*, a common freshwater heterotrophic dinoflagellate. Numerous overlapping rows (arrowheads) of microtubules splay off from a dense cluster of microtubules at the top of the micrograph. When first described in dinoflagellates, this structure was termed a "microtubular basket," a term that may still be encountered in the literature. These microtubules originate near the cingulum on the dorsal side of the cell, curve along the periphery of the cell, passing near the cell's apex, and form the "backbone" of a membrane-ensheathed external peduncle that exits the cell on the ventral surface, near the cingular-sulcal intersection (see Fig. 11–46). (From Wedemayer and Wilcox, 1984)

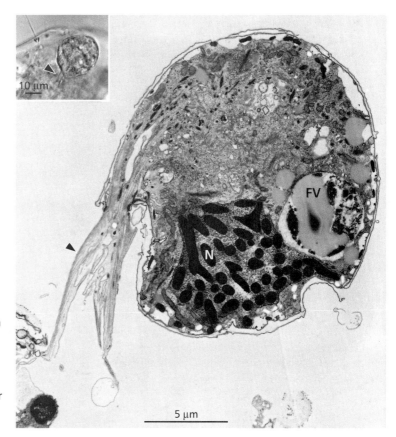

Figure 11–44 Longitudinal section of *Peridiniopsis* feeding upon an insect larva, as seen with TEM. (A feeding cell is shown with light microscopy in the inset.) Note the large peduncle (arrowhead) through which food material is ingested. The cell's nucleus (N) is also evident as is a food vacuole (FV), probably formed in an earlier feeding event. (Micrographs by L. Wilcox and G. Wedemayer)

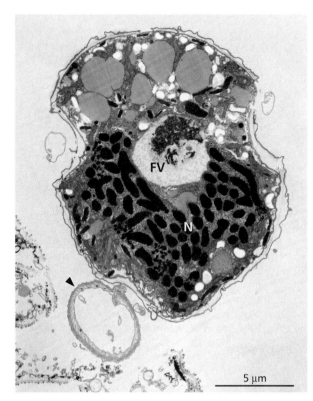

Figure 11–45 A feeding *Peridiniopsis* cell with an extended peduncle (arrowhead) seen in near cross section. Note the numerous short overlapping rows of microtubules in the walls of the hollow peduncle. FV=food vacuole, N=nucleus. (Micrographs by L. Wilcox and G. Wedemayer)

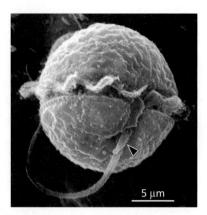

Figure 11–46 SEM view of a non-feeding *Peridiniopsis berolinensis* cell whose peduncle (arrowhead) is unextended. (From Wedemayer and Wilcox, 1984)

ysis and the freshwater *Peridiniopsis berolinensis* (Wedemayer and Wilcox, 1984; Calado and Moestrup, 1997). The latter species first uses a fine cytoplasmic filament (a "capture" or "tow" line) to establish contact with potential food items, then ingests the prey whole or its contents (Figs. 11–44, 11–45). Figure 11–46 shows the unextended peduncle in a non-feeding cell. An exception to the use of a peduncle in feeding is the cold-water mixotrophic *Amphidinium cryophilum*, an unarmored freshwater dinoflagellate that employs a specialized feeding tube or "phagopod"

manageable food vacuole; a portion of the cellulose remains is discharged as a fecal pellet. It is thought that armored dinoflagellates that have many small plates, like *Fragilidium*, can more easily expand around whole prey cells than those having few but large plates (Skovgaard, 1996).

Many dinoflagellates feed through an extensible tubelike structure termed the peduncle, which is bound by the cell membrane and contains tens to hundreds of microtubules (Fig. 11–43). This structure emanates from the ventral side of the cell in the cingular-sulcal region. In addition to its role in feeding, the peduncle may also be used to attach to surfaces. Dinoflagellates feeding through peduncles may engulf their prey whole or take in food in the form of smaller particles or liquefied, digested prey material. *Pfiesteria*, the piscivorous "ambush predator" is an example of a dinoflagellate having thin thecal plates that feeds through a peduncle (Fig. 3–6). Peduncles are also produced by armored species such as the marine *Dinoph-*

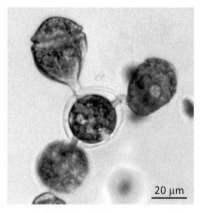

Figure 11–47 Three *Amphidinium cryophilum* cells feeding upon a prey cell whose protoplast is shrinking as its cytoplasm is being drawn into the *Amphidinium* cells. Food material passes through a feeding tube (phagopod), which extends from the posterior end of the *A. cryophilum* cell and attaches to the amphiesmal membranes of the prey cell. (From Wilcox and Wedemayer, 1991 by permission of the *Journal of Phycology*)

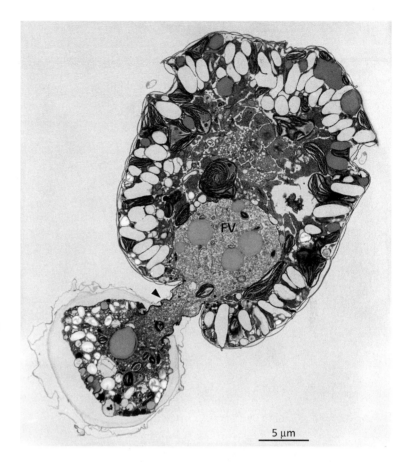

Figure 11–48 Feeding *Amphidinium* cell shown with TEM. The prey cell cytoplasm is in continuity with a forming food vacuole (FV) in the feeding cell. The maintenance of intact prey cell amphiesmal membranes appears necessary for feeding to occur in this system. Arrowhead points to phagopod. (From Wilcox and Wedemayer, 1991 by permission of the *Journal of Phycology*)

that extends from the posterior pole of the cell and contains no microtubules (Wilcox and Wedemayer, 1991) (Figs. 11–47, 11–48). This feeding structure is apparently formed anew each time the organism feeds and is left behind after feeding (Fig. 11–49).

Two dozen species of the armored dinoflagellate *Protoperidinium*, along with several *Diplopsalis* species, feed by the extension of a pallium or "feeding veil" (see Fig. 3–5) (Jacobson and Anderson, 1986; Naustvoll, 1998). This is a sheetlike extension of the cytoplasm that develops from a pedunclelike tube that emerges from the cingular-sulcal region of the cell. The pallium encloses the prey, which are first captured with a filament. The prey protoplasm is then digested by enzymes produced by the pallium, and the products of digestion are transported back into the feeding dinoflagellate. *Protoperidinium* species feed mainly on diatoms and dinoflagellates and their populations are typically most abundant following diatom blooms occurring in coastal temperate and even polar waters. *Diplopsalis* feeds on a wider variety of prey, including haptophytes, green algae, and cryptomon-

ads as well as diatoms and dinoflagellates. At 31 μm in diameter, *D. lenticula* consumes cells between 3 and 78 μm in diameter, including motile cells. Individually, bacteria are too small for capture with a

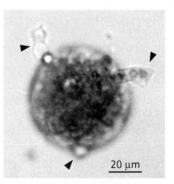

Figure 11–49 Several abandoned phagopods (arrowheads) are seen here on the remains of a dinoflagellate cell that was partially fed upon by several *Amphidinium* cells. (from Wilcox and Wedemayer, 1991 by permission of the *Journal of Phycology*)

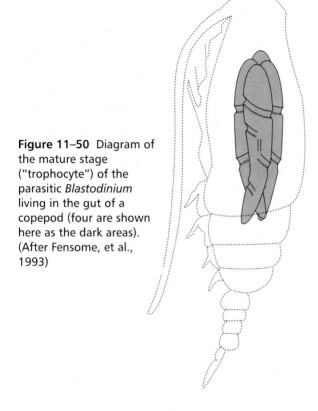

Figure 11–50 Diagram of the mature stage ("trophocyte") of the parasitic *Blastodinium* living in the gut of a copepod (four are shown here as the dark areas). (After Fensome, et al., 1993)

and species, the reader is referred to Steidinger and Tangen (1997) and Fensome, et al. (1993). The former focuses primarily upon the modern marine dinoflagellates, and the latter is an attempt to combine the taxonomy and systematics of living and fossil dinoflagellate species. A perusal of these or other taxonomic references for dinoflagellates will reveal the need for the use of specialized light-microscopic (LM) techniques and scanning electron microscopy (SEM) in order to observe species-specific details. Sometimes examination with transmission electron microscopy (TEM) is also required. Proper preservation of cells is important.

If collections are dominated by large, robust armored species, formalin or Lugol's (iodine/iodide) solution will preserve specimens sufficiently well. However, starch-staining by Lugol's solution may obscure details needed for LM identification. Such preserved cells can be individually mounted onto SEM stubs using double sticky tape, washed 10–20 times with water, air-dried, gold-coated to prevent uneven buildup of electrical charges on the specimen, and then examined with the SEM. This procedure is relatively quick and easy and readily provides essential

pallium, but if aggregated in detritus (marine snow), they can be ingested by pallium-feeding dinoflagellates (Naustvoll, 1998).

A number of nonmotile dinoflagellates, including *Protoodinium*, *Crepidoodinium*, *Piscinodinium*, and *Blastodinium*, contain plastids, but also exist as parasites, feeding on zooplankton or fish hosts. They reproduce by dinospores that use pedunclelike structures to attach to food sources. Once attached, dinospores germinate into the mature, nonmotile form (Fig. 11–50). This mode of feeding resembles that of chytrids, which also produce zoospores capable of locating specific types of living cells, attach to their surface, and form a nonmotile stage that utilizes the invaded cell's contents for food.

Techniques Used in the Identification and Classification of Dinoflagellates

For recent treatments of dinoflagellate classifications that include detailed descriptions of higher level taxa

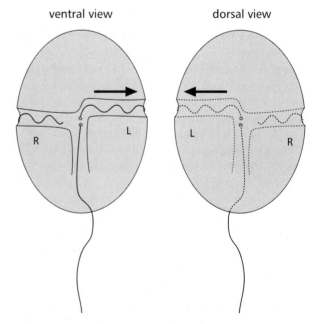

Figure 11–51 A diagram comparing ventral and dorsal views of a dinokont dinoflagellate. In ventral view, the cell's right is to the observer's left. This is reversed in dorsal view. The arrows indicate the direction in which the transverse flagellum extends (base to tip).

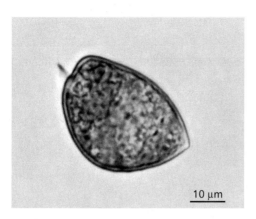

Figure 11–52 Light micrograph of *Prorocentrum micans*, a desmokont dinoflagellate.

details needed for identification. Unarmored dinoflagellates tend to be more delicate, frequently bursting or collapsing unless additional processing steps are taken. These include preservation in a buffered 2% glutaraldehyde solution (also recommended for preservation of delicate cryptomonad cells, see Chapter 9), followed by dehydration in a series of alcohol or acetone solutions of increasing concentration, to 100%, followed by "critical-point drying" prior to gold-coating and SEM study. The cells shown in Figures 11–3 and 11–5, for example, were prepared according to these procedures.

When using LM to identify dinoflagellates that are mounted on glass slides, it is important to identify the right and left sides of the cell. Thus the observer must determine whether the ventral surface (from which flagella emerge) or the dorsal surface is uppermost. If the ventral surface is not uppermost, the viewer must through-focus to it. By comparison to an image of an uppermost ventral surface, the through-focused image will be reversed because the surface has been viewed from the inside, rather than the outside of the cell. If the ventral surface of a cell is uppermost (and the apex pointing toward 12 o'clock), the right side of the cell is on the observer's left. But if the ventral surface is lowermost, then the right side of the cell is on the viewer's right (Fig. 11–51). (When observing an SEM view of the ventral side of a dinoflagellate, the right side of the cell is on the viewer's left.)

The fluorochrome dye Calcofluor White, which binds specifically to cellulose, is useful for detecting delicate thecal plates or for further defining features of more robust plates, but a fluorescence microscope

is required. Identifying armored dinoflagellates, particularly those having plastids, with LM typically requires that cells be gently crushed to spread the thecal plates and to remove obscuring pigmentation. First, excess water is removed from a specimen sandwiched between a glass slide and cover slip, after which a drop of household bleach is added, and then gentle pressure is applied with a toothpick or probe and the cover slip gently moved back and forth (Steidinger and Tangen, 1997).

Recently new techniques involving flow cytometry and specific antibodies (Costas and Lopez-Rodas, 1996) or fluorescent DNA probes have been developed for making rapid, accurate identifications of dinoflagellates, particularly toxic forms. For example, fluorescent DNA probes complementary to rRNA ITS (see Chapter 5) have been constructed for identifying toxic species of *Alexandrium* (Adachi, et al., 1996).

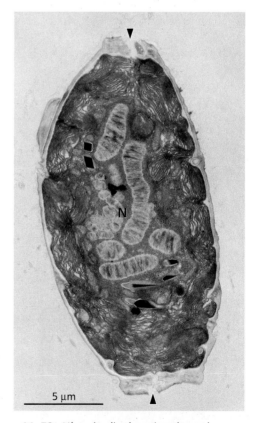

Figure 11–53 A longitudinal section through *Prorocentrum micans*. The suture (arrowheads) between the two large thecal plates (valves) is evident at the top and bottom of the cell. Note also the large chromosomes in the nucleus (N) and dark trichocyst profiles.

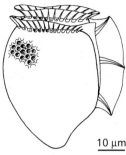

Figure 11–54 The marine *Dinophysis norvegica*. Note the large cingular lists at the top of the cell and the left sulcal list extending longitudinally. (From Steidinger and Tangen, 1997)

10 μm

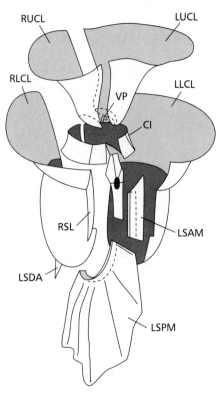

Figure 11–56 An "exploded" diagram of *Ornithocercus*, showing the various components of the elaborate theca. CI=cingulum, LLCL=left lower cingular list, LUCL=left upper cingular list, LSAM=left sulcal list, anterior moiety, LSDA=left sulcal list, posterior moiety, LSPM=left sulcal list, posterior moiety, RLCL=right lower cingular list, RUCL=right upper cingular list, RSL=right sulcal list, VP=ventral pore. The flagellar pore is indicated by the black oval. (After Taylor, 1971 by permission of the *Journal of Phycology*)

Some Examples of Dinoflagellate Diversity

The dinoflagellate genera described and illustrated here were chosen because they are common or they exhibit interesting or unusual traits. We have arranged these genera rather informally; this arrangement of taxa is not intended to reflect an a priori concept of dinoflagellate phylogeny.

PROROCENTRUM (Gr. *prora*, prow + Gr. *kentron*, center). This desmokont dinoflagellate has a theca composed of two halves (valves), both more or less flattened, so that cells appear narrow in edge view at the sutures and spherical or oval in valve view (Fig. 11–52). Additional small plates and sometimes a

10 μm

Figure 11–55 *Ornithocercus* is an impressive marine dinophysoid dinoflagellate with large cingular lists and a pronounced left sulcal list. (Photograph courtesy of C. Taylor)

spinelike process occur at the apex. Two flagella occur apically, one coiled and encircling the other, which extends outward from the cell. Most, if not all species have one or more chloroplasts per cell. The genus is of widespread occurrence in fresh, brackish, or marine waters. Some species are so abundant on tidal sand flats that they color the sand brown. *Prorocentrum micans*, a TEM view of which is shown in Figure 11–53, is a known toxic bloom-former.

DINOPHYSIS (Gr. *dineo*, to whirl + Gr. *physis*, inborn nature). The cingulum of these dinokont cells is located very close to the apex, and thecal plates in the vicinity of the cingulum and the sulcus extend outward from the cell, forming distinctive funnel-shaped,

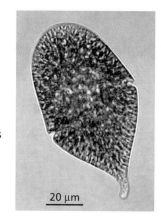

Figure 11–57
Gymnodinium caudatum is an unarmored dinoflagellate with a distinct "tail" that is common to acidic freshwater habitats.

20 μm

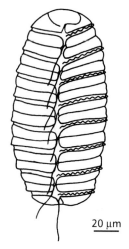

Figure 11–59 Diagram of a pseudocolony of *Polykrikos*, a marine "naked" dinoflagellate with multiple nuclei and flagella. (From Steidinger and Tangen, 1997)

20 μm

sail-like processes, known as lists (Fig. 11–54). Some species contain plastids, and all are marine. Some forms possess plastids that are apparently of cryptomonad origin (see Chapter 7). Phagotrophy is known for at least some representatives of both colorless and plastid-containing species (Jacobson and Andersen, 1994).

ORNITHOCERCUS (Gr. *ornis*, bird + Gr. *kerkos*, tail) (Fig. 11–55) is similar to and related to *Dinophysis*, but is distinguished by an extremely well-developed anterior cingular list. An "exploded" diagram of an *Ornithocercus* cell is shown in Figure 11–56. Species of this genus do not have plastids, but some harbor photosynthetic symbionts in a chamber or pouch located in the cingular area. The genus is widespread in warm marine waters.

GYMNODINIUM (Gr. *gymnos*, naked + Gr. *dineo*, to whirl) has only very thin thecal plates, and thus appears unarmored or naked. The cingulum occurs at the midpoint of the cell, so that the epicone and hypocone are approximately the same size. The cells may be dorsiventrally compressed. Not all species contain chloroplasts. Most of the species are marine; many of these have variously colored pusules. There are several important freshwater forms, including *G. fuscum* and the closely related *G. caudatum*, which occur in ponds and acid bogs (Fig. 11–57). The related genus *Woloszynskia* has thecal plates of thickness sufficient to be observable with the light microscope (Fig. 11–58). *Polykrikos* is very similar to *Gymnodinium* except that 2–16 flagellate cells, sharing a common sulcus, are united to form a linear pseudocolony (Fig. 11–59). A single nucleus serves two cells. Plastids are absent, and phagotrophy is the rule.

AMPHIDINIUM (Gr. *amphi*, on both sides + Gr. *dineo*, to whirl) is like *Gymnodinium* in being unarmored, but differs in having its cingulum much closer

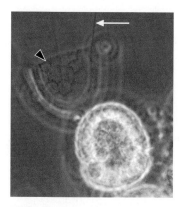

Figure 11–58 *Woloszynskia* sp. with a shed theca (arrowhead) showing a number of thin plates. These are not normally visible on living cells. A discharged trichocyst (arrow) is evident here in phase-contrast microscopy.

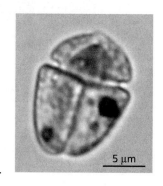

Figure 11–60
Amphidinium is a genus with a markedly smaller epicone than hypocone. Shown here is the freshwater species, *A. wigrense*, which contains cryptomonad-derived plastids (or kleptoplastids).

5 μm

Figure 11–61 *Peridinium limbatum*, a common bog-dwelling dinoflagellate with prominent antapical horns. A single cell is seen in dorsal view in (a). In (b), a number of cells are shown, illustrating the overall shape of the cells, which have a convex dorsal and concave ventral surface.

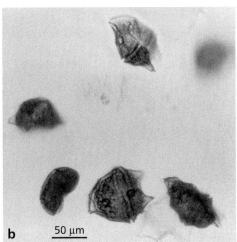

to the apical end, so that the epicone is small. Most species are photosynthetic, some containing kleptochloroplasts, and the genus can be found in freshwaters, brackish waters, and oceans. *Amphidinium carterae* is known to form extensive blooms in marine waters, and several species occur abundantly on sand flats. *Amphidinium cryophilum* (Figs. 11–4, 11–6) occurs beneath the ice of temperate lakes in winter. *Amphidinium wigrense* (Fig. 11–60) contains cryptomonad-derived plastids (Wilcox and Wedemayer, 1985). The microtubular cytoskeleton of *A. rhynchocephalum* was studied by Roberts, et al. (1988). Ultrastructure and feeding of the freshwater *Amphidinium lacustre* were studied by Calado, et al. (1998).

SYMBIODINIUM (Gr. *syn*, together + Gr. *bios*, life + Gr. *dineo*, to whirl) cells are considered unarmored because the numerous thecal plates are very thin. It exists primarily in a coccoid form (see Fig. 7–21). *Symbiodinium* is not known to occur in the free-living form, but molecular evidence suggests close relationship to the common free-living genus *Gymnodinium* (which can also occur as a symbiont in marine animals). *Symbiodinium* is found in a wide range of marine hosts, including all reef-building corals, sponges, and giant clams. Hosts may contain more than one form of *Symbiodinium*. There are at least three species groups, but it is unclear how many species exist (Rowan, 1998). Species differ in content of photoprotective carotenoid pigments and protective mycosporinelike amino acids (reviewed by Dunlap and Shick, 1998), and thus exhibit different susceptibilities to photodamage. Failure of endosymbiont photosynthesis or

expulsion of zooxanthellae cells is the cause of coral bleaching (Rowan, 1998).

PERIDINIUM (Gr. *peridineo*, to whirl around) is heavily armored with thick thecal plates and conspicuous sutures (Fig. 11–61). The cingulum is more or less median. Cells tend to be flattened, and some species, such as *P. limbatum*, illustrated here, have conspicuous horns. Most species occur in freshwater or brackish waters. *Peridinium cinctum*, also illustrated here, occurs on a worldwide basis in freshwa-

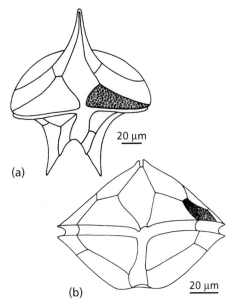

Figure 11–62 *Protoperidinium*, a common marine dinoflagellate genus. (a) *P. grande*, (b) *P. punctulatum*. (From Steidinger and Tangen, 1997)

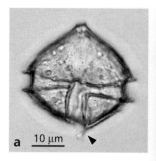

Figure 11–63 *Diplopsalis* is a dinoflagellate genus common in marine habitats, but is represented in freshwater by *D. acuta*, shown here with LM (a), and SEM (b). A prominent list occurs along the left side of the sulcus (arrowheads). The flagella were lost during the preparation of these specimens for microscopic observation. (SEM by G. Wedemayer and L. Wilcox)

ters. *Peridinium balticum* is well known for occurrence of a diatom endosymbiont. *Peridinium gatunense* is a large form that blooms February through June each year in Lake Kinneret (Israel). It forms 90% of the total phytoplankton biomass during the bloom, and has been intensively studied (see Berman-Frank, et al., 1995, for example).

PROTOPERIDINIUM (Gr. *protos*, first + *Peridinium*) (Fig. 11–62) is quite similar in external morphology to *Peridinium*, but differs in being phagotrophic and in being restricted to marine habitats. Most species lack chloroplasts. Many species are ornamented with horns or spines.

DIPLOPSALIS (Gr. *diploos*, double + Gr. *psalis*, scissors) is another armored dinoflagellate lacking plastids that occurs in both marine and freshwaters (Fig. 11–63). It has a prominent sulcal list and apical pore.

GONYAULAX (Gr. *gony*, knee + Gr. *aulax*, furrow) (Fig. 11–64) is armored and characterized by a longitudinal furrow that extends from the anterior to the posterior of the cell. The plastid-containing cells may be solitary or remain together following cell division, forming short chains. The genus is widespread in warm and temperate marine waters. Spiny benthic cysts are produced. Species of the closely related *Protogonyaulax* can form toxic blooms. *Gonyaulax polyedra* is now referred to as *Lingulodinium polyedrum* by some authorities, but others have not accepted the change. The taxonomic history of this organism is discussed in a review by Lewis and Hallett (1997). Because it is easily cultured, it has been the subject of

many ultrastructural, physiological, and molecular studies that are summarized in the review just cited.

CERATIUM (Gr. *keration*, small horn) has armored cells with three or four elongate horns, one anterior, and two or three posterior. The sutures are comparatively narrow, and the surface is ornamented with fine polygonal markings. Most species are photosynthetic, but the presence of phagocytic-type food vacuoles suggests that phagotrophy also occurs. Most of the species are marine or occur in brackish waters, but there are several important freshwater forms. *Ceratium hirundinella* (see Fig. 3–18) can occur abundantly in hard water or eutrophic lakes and may be 100 μm long and wide. This species has an unusual

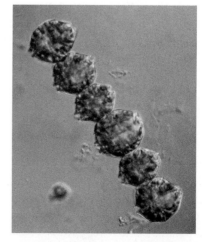

Figure 11–64 *Gonyaulax* is a dinoflagellate of freshwater and marine habitats. Shown here is the chain-forming, *G. catenella*. (Photograph courtesy C. Taylor)

Figure 11–65 *Ceratium* is a dinoflagellate common to both marine and freshwaters. The bog-dwelling *C. carolinianum* is shown here.

sulcal aperture lined by a single membrane, and contains food vacuoles with material apparently undergoing digestion, suggesting phagotrophy (Dodge and Crawford, 1970b). *Ceratium carolinianum* (Fig. 11–65) is characteristic of soft bog waters. The horns of some species can be very elaborate.

CYSTODINIUM (Gr. *kystis*, bladder + Gr. *dineo*, to whirl) (Fig. 11–66) is a member of a group of dinoflagellates that are nonmotile except for a dinospore phase. This genus is encountered quite often in bog habitats.

NOCTILUCA (Latin *nox*, night + Latin *lux*, light) has very large (up to 2-mm diameter) cells that are spherical (Fig. 11–67). Although the cells are unarmored, empty thecal vesicles are present (TEM is required to visualize this feature). A single, relatively short longitudinal-type flagellum and a thicker **tentacle** emerge from a ventral groove. The cytoplasm

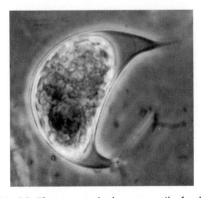

Figure 11–66 The vegetatively nonmotile freshwater dinoflagellate *Cystodinium*.

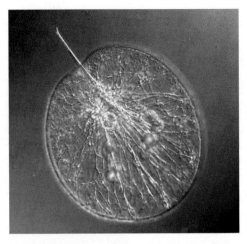

Figure 11–67 The unusual dinoflagellate *Noctiluca*, seen with differential-interference contrast optics. The cells possess a tentacle used to transfer food materials to a cytostome. (Photograph courtesy C. Taylor)

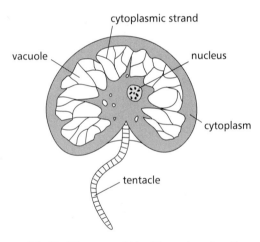

Figure 11–68 Diagram of *Noctiluca* showing the highly vacuolate nature of these large dinoflagellate cells. (After Fensome, et al., 1993)

includes a large vacuole that likely serves to increase buoyancy (Fig. 11–68). Typical plastids are absent, but photosynthetic endosymbionts or vestigial plastids have been reported. The cells are phagotrophic, often consuming other dinoflagellates; prey are captured with the tentacle and ingested through a cell mouth or cytostome. Asexual reproduction occurs by binary fission, and sexual reproduction is reported to occur by meiotic formation of large numbers of *Gymnodinium*-like gametes. Although vegetative cells have a nucleus much like that of other eukaryotes in that chromosomes are not permanently condensed, nuclei of the

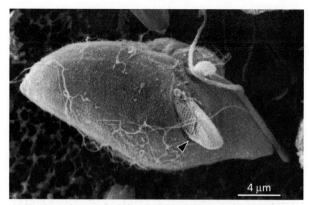

Figure 11–69 The unusual phagotrophic dinoflagellate *Oxyrrhis marina*. This cell was fixed while it was ingesting a diatom (arrowhead0. (Micrograph courtesy R. Triemer)

gametes are more like those of other dinoflagellates in containing condensed chromosomes. *Noctiluca scintillans* is widespread in coastal marine waters; it is bioluminescent and sometimes forms red tides. Population dynamics of *Noctiluca* were recently reviewed by Elbrächter and Qi (1998).

OXYRRHIS (Gr. *oxys*, sharp + Gr. *rhis*, nose) (Fig. 11–69) has some dinoflagellatelike features,

including amphiesmal vesicles, trichocysts, and lateral insertion of flagella, but the flagella are alike, and there is no intranuclear spindle as in typical dinoflagellates. It is unarmored, exhibits no cingulum or sulcus, and possesses scales on the cell body and flagella. No plastids are present. It is of widespread occurrence in temperate to tropical estuaries, marshes, and rock pools, where it forms pink blooms.

Recommended Books

Fensome, R. A., F. J. R. Taylor, G. Norris, W. A. S. Sargeant, D. I. Wharton, and G. L. Williams. 1993. *A Classification of Living and Fossil Dinoflagellates. Micropaleontology, Special Publication Number 7.* Sheridan Press, Hanover, PA.

Taylor, F. J. R. (editor). 1987. *The Biology of Dinoflagellates.* Blackwell Science, Oxford, UK.

Tomas, C. R. (editor). 1997. *Identifying Marine Phytoplankton.* Academic Press, San Diego, CA.

Spector, D. (editor). 1984. *Dinoflagellates.* Academic Press, New York, NY.

Chapter **12**

Ochrophytes I
Introduction to the Ochrophytes and a Focus on Diatoms

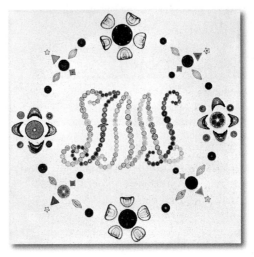

A diatom arrangement done ca. 1898 by amateur members of the San Francisco Microscopical Society. Because of their fine structural details, diatoms have long been used to check the resolving power of microscopes. (Photograph courtesy J. West)

PART 1—INTRODUCTION TO THE OCHROPHYTES

In contrast to the leafy green environment experienced by humans, many other organisms occupy a predominantly brown or golden-brown world. A brown canopy of giant kelps shelters sea otters, urchins, many kinds of fish, and other inhabitants of kelp forests. The dinoflagellate *Gambierdiscus* thrives on one of its favorite substrates, bushy thickets of brown *Turbinaria* seaweed growing on tropical reefs. Extensive areas of marine shoreline are covered by turfs of tube-dwelling golden-brown diatoms that support a myriad of tiny creatures, as do feltlike, brown diatom growths in freshwater streams and along lake edges. These brown seaweeds and golden-brown diatoms, as well as an incredibly diverse array of other algae, belong to a group that we have chosen to call the Ochrophyta, or more informally, the ochrophytes. This name reflects the ocher (golden-brown) color of many algae in this group.

In the past the algae covered in this and the following three chapters had been grouped with haptophytes (Chapter 10) in the Division Chrysophyta (Bold and Wynne, 1985) or aggregated with cryptomonads, haptophytes, and dinoflagellates to form the Chromophyta. Alternatively, some authors have elevated separate groups of ochrophytes to divisional (phylum) status—for example, the Bacillariophyta, encompassing the diatoms. However ultrastructural studies revealed that ochrophytes possess a number of features—particularly flagellar characters—that define and unite them, and exclude cryptomonads, haptophytes, and dinoflagellates (whose flagella are different). Moreover, the ultrastructural data also show that ochrophytes are actually more closely related to various plastidless protists, such as tiny colorless flagellates and oomycetes (formerly considered fungi) than to other algal groups. Ochrophytes, oomycetes, and related protists are known as the **heterokonts** or **stramenopiles**, these terms reflecting flagellar features held in common. Today this ultrastructural information, as well as a growing body of molecular sequence data, strongly support monophyly of ochrophyte algae and monophyly of the larger heterokont (stramenopile) group. We advocate use of the newer phylum concept Ochrophyta because it is defined in terms of flagellar ultrastructure and molecular data (Cavalier-Smith and Chao, 1996), and thus more closely reflects modern concepts of evolutionary relationships than do older taxonomic concepts. It avoids confusion that might ensue from the use of modified older terms such as "Chromophyta, *sensu lato*" or Heterokontophyta, which might be interpreted to include all stramenopiles or just pigmented forms.

The term heterokont literally means "different flagella." Although other groups of algae possess flagella that are also distinctively different from each other (dinoflagellates, for example), the organisms known as heterokonts typically have two flagella (or have reproductive cells with such flagella) that differ in unique ways: a long, forward-directed flagellum bears two rows of stiff, three-parted hairs and a shorter, smooth flagellum that often (but not always) bears a basal granule that functions in light sensing (Fig. 12–1).

Heterokonts include an amazing variety of organismal types, from the colorless flagellate *Cafeteria*, the parasite *Labyrinthula*, and oomycetes, to plastid-containing groups, including single-celled diatoms, the coenocytic tribophycean *Vaucheria*, and giant kelps, whose thalli are parenchymatous (Fig. 12–2). The plastid-containing heterokont groups (ochrophytes) include the diatoms, raphidophyceans, chrysophyceans, synurophyceans, eustigmatophyceans, pelagophyceans, silicoflagellates, pedinellids, sarcinochrysidaleans, phaeothamniophyceans, tribophyceans (xanthophyceans), phaeophyceans (brown algae), and some smaller groups.

Heterokonts are important in a number of ways. *Labyrinthula* has destroyed large sea-grass populations, and the thraustochytrid *Agglutinatum* digests the contents of diatom cells, after first gluing them together. The oomycete *Pizzonia* also feeds on diatoms, especially *Coscinodiscus* (Kühn, 1998). Another oomycete, *Phytophthora* ("plant destroyer"), causes late blight of potato, which contributed to the history-making Irish potato famine. Even today, *Phytophthora* is a significant pest of many crops. Other oomycetes infect fish and fish eggs. All oomycetes spread via zoospores bearing the hallmark heterokont

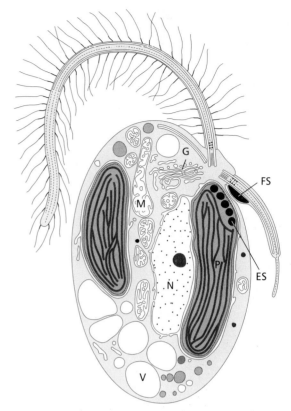

Figure 12–1 Heterokont protists possess flagella like those of the tribophycean zoospore illustrated here. The longer anterior flagellum bears two rows of stiff hairs, but the shorter posterior flagellum is smooth and often bears a swelling (FS) that is part of the light-sensing system. ES=eyespot, G=Golgi body, M=mitochondrion, N=nucleus, P=plastid.

flagella. The primarily autotrophic diatoms are ubiquitous, highly diverse, and abundant in almost all aquatic habitats as well as moist terrestrial sites. Huge fossil deposits known as **diatomite** are mined for many types of industrial applications, some diatoms are widely used as fodder in mariculture/aquaculture operations (Chapter 4), and others produce the neurotoxin domoic acid, causing amnesiac shellfish poisoning (Chapter 3). Sedimentary siliceous remains of diatoms, silicoflagellates, chrysophyceans, and synurophyceans are widely used as indicators of past environmental changes affecting aquatic habitats, or as indices of relative age in geological deposits. Past growths of sarcinochrysidalean algae have been linked to significant oil deposits, while the pelagophycean *Aureococcus* and the raphidophyceans *Chattonella* and *Fibrocapsa* cause massive marine blooms in modern waters. The brown algae dominate rocky shores of

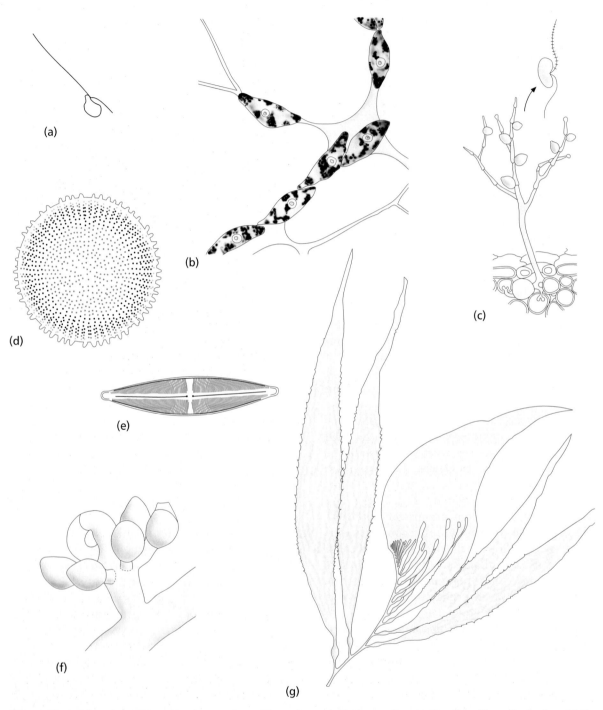

Figure 12–2 Diversity of heterokont organisms. Heterokonts include small, colorless flagellates, including (a) the bicoecid *Cafeteria* and the somewhat more complex but also plastidless pseudofungi, such as (b) *Labyrinthula* and (c) the oomycete *Phytophthora*. The heterokont algae include (d) radially symmetrical and (e) bilaterally symmetrical diatoms, (f) yellow-green algae (tribophyceans) such as *Vaucheria*, and (g) enormous kelps, including *Macrocystis*, as well as numerous other forms. (a: After Throndsen, 1997; b: Drawn from micrograph in Grell, 1973; c: After Alexopoulos, C. J., C. W. Mims, and M. Blackwell. 1996. *Introductory Mycology*. John Wiley and Sons, Inc. Reproduced with permission of The McGraw-Hill Companies)

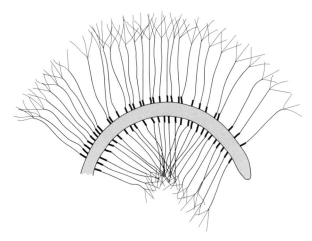

Figure 12–3 The flagella of heterokonts bear glycoproteinaceous tripartite hairs, having a basal attachment region, a long tubular shaft, and terminal fibrils.

temperate, boreal, and polar waters, forming extensive beds that shelter marine animal life. Some giant kelps are harvested for production of industrially useful carbohydrates (Chapter 4).

The following detailed examination of the features of the distinctive flagella of heterokont protists and the plastids of ochrophytes forms a useful basis for subsequent discussion of the morphological and molecular evidence bearing upon relationships among ochrophytes, as well as more detailed focus upon individual ochrophyte groups.

Flagella and Flagellar Root Systems

Although ochrophyte cells are typically heterokont, portions of the flagellar apparatus have been highly modified or lost in several groups. For example, many ochrophytes and related heterokont protists have flagellar basal bodies with a transition region that is often characterized by a helical structure lying above a basal plate. This feature distinguishes heterokonts from other protist groups (see Fig. 9–8). However this **transitional helix** has apparently been independently lost from flagella of diatoms, raphidophyceans, and phaeophyceans.

The long anterior flagellum typically bears two rows of stiff hairs that project at right angles to the long axis of the flagellum. These hairs consist of a

basal section, a longer tubular shaft, and one or more terminal fibrils (Fig. 12–3), and thus are known as "tripartite" hairs. Tripartite flagellar hairs occur only in the ochrophytes and related heterokont protists. Tripartite hair-producing protists (heterokonts) are also known collectively as stramenopiles, or stramenopila, terms derived from the Greek *stramen* meaning "straw" and *pila* meaning "hairs." In the literature, the term **mastigoneme** is sometimes restricted to use as a synonym for the tripartite flagellar hairs of stramenopiles, but has also been applied to the bipartite tubular flagellar hairs of cryptomonads and is sometimes also used to describe other types of flagellar hairs.

Tripartite flagellar hairs of ochrophytes are composed of several types of glycoprotein and are produced intracellularly (Fig. 12–4) within the **periplastidal endoplasmic reticulum** (PER) or adjacent nuclear envelope, depending on the ochrophyte group, then discharged from cells and deposited onto flagellar surfaces, most likely at their bases. The hydrodynamic effect of the tripartite hairs is due to their stiffness and orientation perpendicular to the long axis of the flagellum. In this orientation, these cylindrical structures produce twice as much drag on the

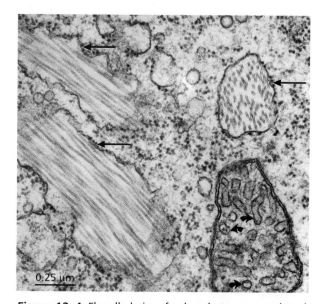

Figure 12–4 Flagella hairs of ochrophytes are produced within the periplasmic endoplasmic reticulum (arrows) or the nuclear envelope. In this transmission electron micrograph of the raphidophycean *Vacuolaria*, flagellar hairs are seen in both cross and longitudinal section. Curved arrows point to the tubular cristae in the mitochondria. (From Heywood, 1983)

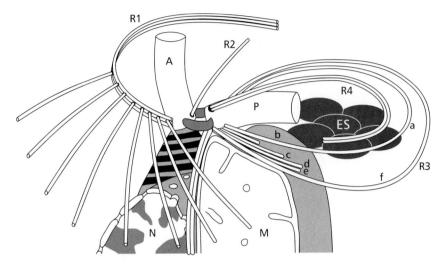

Figure 12–5 Diagram of the flagellar root system of the chrysophycean *Ochromonas*, regarded as the primitive type for ochrophytes (heterokont algae). Flagellar apparatuses that have a sliding, looplike root (R3) confer the ability to ingest particulate prey; such roots are typical for mixotrophic ochrophytes. a–f, microtubules of R3 root, A=anterior flagellar basal body, ES=eyespot, M=mitochondrion, N=nucleus, P=posterior flagellar basal body, R1–R4=flagellar roots. (Redrawn with permission from Inouye, 1993 in *Ultrastructure of Microalgae*, edited by T. Berner. ©CRC Press, Boca Raton, Florida)

surrounding fluid than they would if oriented parallel to the flagellum axis. Analysis of flagellar motion by high-speed cinematography suggests that the stiff tubular hairs act like paddles that move laterally during propagation of a wave from the tip to the base of the flagellum. Because they are stiff and numerous, the hairs generate net thrust in the opposite direction to that produced by the flagellum and smooth flagella. This is because the hairs' axes are perpendicular to the direction of movement when the flagellar axis is parallel to the direction of movement and vice versa. The long flagellum thus pulls the cell through the water. Hydrodynamic theory predicts that tripartite hairs must be held within ten degrees of perpendicular in order for thrust reversal to occur. A comprehensive review of the hydrodynamic effects of mastigonemes can be found in a review by Goldstein (1992).

Monoclonal antibodies generated against the mastigonemes of *Phytophthora* zoospores bind to the hairs, causing them to clump or detach, with the result that swimming speed is greatly reduced and forward thrust eliminated. These experimental results support the hypothesis that tripartite flagellar hairs allow the long flagellum to pull cells through the water (Cahill, et al., 1996). This finding, coupled with the fact that oomycete-caused diseases spread via flagellate zoospores, suggests that treatments designed to interfere with the synthesis, deposition, or function

of flagellar hairs might prove useful in stopping disease spread in terrestrial crops or aquaria. However, care would have to be taken to avoid impact on natural aquatic ochrophyte populations.

The short, smooth flagellum of some ochrophyte groups and both flagella of synurophyceans, have a basal swelling that is believed to be involved in light-sensing. The flagellar swelling often fits into a concave depression over the eyespot, which is typically located within the plastid. However, the eyespot of eustigmatophyceans occurs in the cytoplasm, while some ochrophytes lack an eyespot altogether. Green autofluorescence, interpreted as arising from light-sensitive flavin compounds, may occur throughout one or both flagella or in a localized region (Kawai and Inouye, 1989).

The flagellar apparatus and flagellar root system of the chrysophycean *Ochromonas* is often regarded as representative of the basal condition for ochrophytes (Fig. 12–5) (Inouye, 1993). The long anterior flagellum and the posteriorly directed, short flagellum diverge perpendicularly from each other. Within the cell two fibrous bands connect flagellar basal bodies, and a banded rhizoplast connects basal bodies to the nuclear envelope. Each basal body is associated with two microtubular roots. The four microtubular roots are designated R1, R2, R3, and R4. Of these, R1 and R2 are associated with the anterior flagellar basal

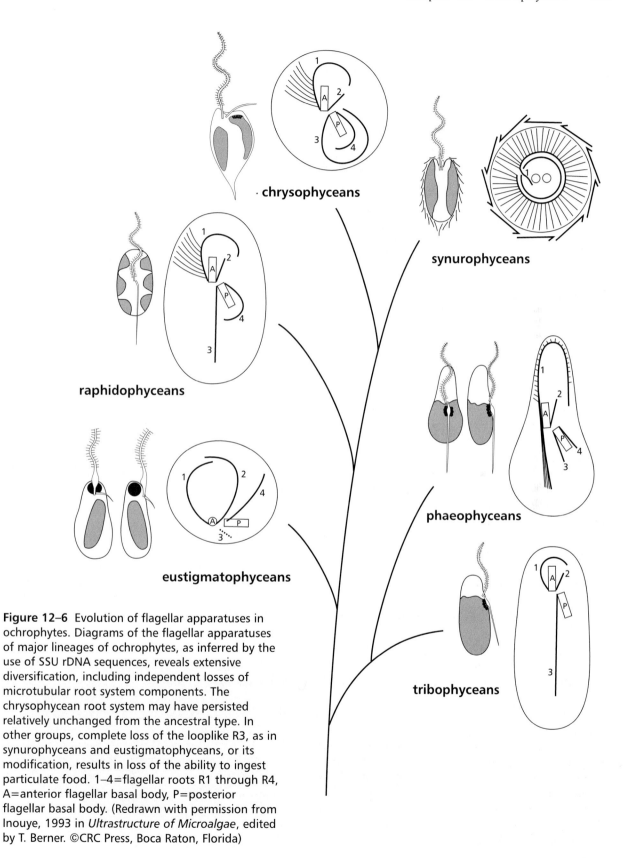

Figure 12–6 Evolution of flagellar apparatuses in ochrophytes. Diagrams of the flagellar apparatuses of major lineages of ochrophytes, as inferred by the use of SSU rDNA sequences, reveals extensive diversification, including independent losses of microtubular root system components. The chrysophycean root system may have persisted relatively unchanged from the ancestral type. In other groups, complete loss of the looplike R3, as in synurophyceans and eustigmatophyceans, or its modification, results in loss of the ability to ingest particulate food. 1–4=flagellar roots R1 through R4, A=anterior flagellar basal body, P=posterior flagellar basal body. (Redrawn with permission from Inouye, 1993 in *Ultrastructure of Microalgae*, edited by T. Berner. ©CRC Press, Boca Raton, Florida)

chrysophyceans

synurophyceans

raphidophyceans

phaeophyceans

eustigmatophyceans

tribophyceans

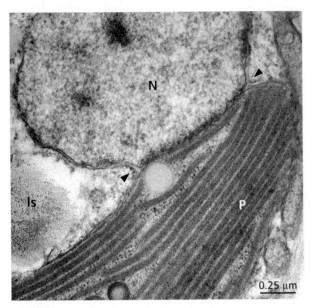

Figure 12–7 The periplastidal endoplasmic reticulum (PER), characteristic of ochrophytes, links the plastid (P) to the nucleus (N) in some taxa (arrowhead), such as the eustigmatophycean shown here. ls=lamellate storage. (From Frost, et al., 1997 ©Blackwell Science, Inc.)

body, and R1 appears to organize the microtubular cytoskeleton, since numerous microtubules emerge from it. R3 and R4 are associated with the posterior flagellar basal body. R3 is a looplike root that is associated with the ability to capture and ingest particulate prey. Prey are captured with a "feeding cup" that is formed when they contact the *Ochromonas* cell membrane. The microtubules labeled a and f (Fig. 12–5) form the rim of the cup. Prey are engulfed into a food vacuole that forms by the sliding of the f tubule. Beside *Ochromonas*, feeding behavior of this type occurs in a number of mixotrophic chryso-

phyceans (Chapter 13). Diagrammatic representations (after Inouye, 1993) of typical flagellate cells and the flagellar apparatus of other ochrophyte groups are shown in Fig. 12–6, arranged here according to current phylogenetic views (based upon SSU rDNA sequences), in order to illustrate significant evolutionary changes. For example, loss of the looplike R3 root from the synurophyceans appears to be correlated with absence of phagotrophic ability. Although synurophyceans have an encircling root, this is thought to be homologous to R1 of *Ochromonas* because it generates cytoskeletal microtubules. Silicoflagellates, pedinellids, the gametes of some diatoms, and some other ochrophytes completely lack microtubular roots associated with flagellar basal bodies, presumably as the result of evolutionary loss. Inouye (1993) has provided a review of flagellar apparatus structure in ochrophytes and other algae.

Plastids

Ochrophyte plastids are located within periplastidal endoplasmic reticulum. The PER is often, but not always, confluent with the nuclear envelope (Fig. 12–7). Targeting of cytoplasmically synthesized proteins to plastids involves transport through the PER (Bhaya and Grossman, 1991). The thylakoids typically occur in stacks (**lamellae**) of three, and in most groups there is a **girdle lamella**, which runs around the periphery of the plastid just beneath the innermost plastid membrane (Fig. 12–8). Chlorophylls *a* and usually chlorophyll *c*, in one or more of three possible forms—c_1, c_2, and c_3—are present.

Chlorophyll *c* is not considered by some experts (Falkowski and Raven, 1997) to be a true chlorophyll because its biosynthetic origin is different from that of

Figure 12–8 The plastids of many groups of ochrophytes are characterized by a girdle lamella (arrowhead) that lines the plastid periphery, just beneath the envelope, as in the raphidophycean *Vacuolaria*. (From Heywood, 1983)

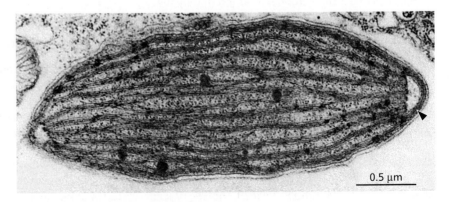

Chlorophyll a

Chlorophyll c

Figure 12–9 Chemical differences (arrows) between chlorophyll a (top) and chlorophyll c (bottom) include modifications in ring IV and its substituents, and the absence of a phytol tail from chlorophyll c. (After Falkowski and Raven, 1997. Redrawn by permission of Blackwell Science, Inc.)

haptophytes (Chapter 10), there is no single pigment signature that characterizes ochrophytes as a group. Several lines of molecular sequence evidence strongly indicate that the plastids of ochrophytes are derived from those of red algae; presumably a eukaryotic red alga was taken up as an endosymbiont by one or more ochrophyte host cells in the process known as secondary endosymbiosis (see Chapter 7).

Chlorophyll *c*, β-carotene, and other carotenoids present in the light-harvesting complexes of ochrophytes extend the ability of these algae to collect light beyond that possible with chlorophyll *a* alone. Chlorophyll *c* and fucoxanthin are believed to occur in an antenna complex. The antenna complex, also known as a pigment bed, is a group of pigment molecules that can share and transfer energy from higher to lower excitement states (Falkowski and Raven, 1997). Xanthophylls function to protect the photosystem from the deleterious effects of high-intensity light. Although a considerable amount is known regarding the gene sequences encoding proteins that bind fucoxanthin and chlorophyll in some ochrophytes (Grossman, et al., 1990) (reviewed by Larkum and Howe, 1997), the structure of their light-harvesting complexes is not understood as well as are those of cyanobacteria and red algae, dinoflagellates, and green algae (see references in Rhiel, et al., 1997).

Falkowski and Raven (1997) have noted that the Fe-containing photosynthetic electron carrier cytochrome c6 is present in ochrophytes in place of the Cu-containing plastocyanin found in other algae. This may have conferred an advantage at the end of the Permian period (about 250 million years ago) when anoxic conditions were widespread in the oceans. Under such conditions, the solubility of Cu in the upper ocean waters would be expected to decline, whereas the solubility of Fe should have increased.

A number of ochrophytes have lost photosynthetic pigments and are thus incapable of photosynthesis. Pedinellids include some species that seem to have originated from plastid-containing forms, but then completely lost their plastids, a situation uncommon and perhaps unique among algae. In other groups having plastidless representatives (cryptomonads, euglenoids, and dinoflagellates), such forms are commonly thought to have diverged prior to plastid acquisition. The most prevalent carbohydrate storage product in ochrophyte cells is **chrysolaminaran** (also known as **laminaran** or leucosin). It occurs within vacuoles often located in the posterior portion of the cell (Fig. 12–1). Cytoplasmic lipid droplets are also common.

the true chlorophylls, *a* and *b*, which are derived from chlorins, whereas chlorophyll *c* is derived from porphyrins. In addition, unlike chlorophylls *a* and *b*, chlorophyll *c* usually lacks a phytol tail, and as a result is not embedded in membranes, and is water soluble. There are additional chemical differences (Fig. 12–9) that confer different absorption properties.

The major yellow accessory pigment in ochrophytes is β-carotene. The major brown pigment is fucoxanthin in diatoms, chrysophyceans, phaeophyceans, and some others, while vaucheriaxanthin dominates in raphidophyceans, eustigmatophyceans, and tribophyceans (xanthophyceans). As is the case for

Phylogeny of Ochrophytes

Several sequencing studies of nuclear SSU rDNA indicate that ochrophytes are a monophyletic group (Potter, et al., 1997b; Vanderauwera and Dewachter, 1997), a conclusion also supported by *rbcL* sequence analyses (Daugbjerg and Andersen, 1997b). Molecular data, as well as similarities in the flagellar apparatus, indicate that their closest relatives are the oomycetes, labyrinthulids, thraustochytrids, and bicoecids (Fig. 12–10). The stramenopiles (heterokont protists) comprise the oomycetes, thraustochytrids, labyrinthulids, bicoecids, and ochrophytes. Bicoecids (also known as bicosoecids) include about eight genera of free-living marine and freshwater phagotrophs with two flagella, one with stiff hairs that is used to collect food particles. The other bicoecid flagellum attaches cells to a vaselike chitinous lorica (Dyer, 1990). Similarities in the flagellar apparatus of bicoecids and early divergent ochrophytes has been cited as evidence for close phylogenetic relationship (Moestrup, 1995).

The stramenopiles form a monophyletic group whose sister group is the alveolate cluster, which includes the dinoflagellates (Chapter 11), apicomplexans (Chapter 7), and ciliates. Their next closest algal relatives are the haptophytes (Chapter 10) (Fig. 12–10). Taxonomic schemes that group haptophytes with ochrophytes on the basis of plastid similarities, are not congruent with nuclear SSU rDNA data (Potter, et al., 1997b; Saunders, et al., 1997b). These instead suggest that haptophytes and ochrophytes arose independently by incorporation of similar eukaryotic endosymbionts into distinctive host cells. The occurrence of phagotrophic, early divergent members of both haptophytes (*Pavlova* and *Chrysochromulina*) and stramenopiles (bicoecids such as *Cafeteria* and chrysophyceans such as *Ochromonas*) increase the likelihood of such an evolutionary scenario. It has been suggested that slopalinids, a group of protozoa that includes proteromonads and opalinids, might be related to the ancestry of stramenopiles because some exhibit transitional helices, and some produce tripartite hairs on the cell body (though not on flagella) (O'Kelly, 1993).

Patterns of early branching within the ochrophyte clade are currently somewhat controversial. SSU rDNA sequences suggest that the diatoms were a very early divergent group (Cavalier-Smith and Chao, 1996), but *rbcL* data suggest that eustigmatophyceans diverged earlier (Daugbjerg and Andersen, 1997b). Our general preference has been for the use of data from the nucleocytoplasmic component of algal cells (rather than plastid-encoded genes) as the primary source of information used in systematic organization. Thus we have chosen to begin discussion of ochrophytes with diatoms. Other ochrophyte groups will then be described in an order consistent with emerging phylogenetic data. SSU rDNA sequence analyses suggest that the sister clade to diatoms can be separated into three component lineages—one composed of primarily freshwater raphidophyceans, chrysophyceans, synurophyceans, and eustigmatophyceans; a second that includes marine pelagophyceans, silicoflagellates, pedinellids, and sarcinochrysidalean forms; and a third composed of phaeothamniophyceans, tribophyceans (xanthophyceans), chrysomeridalean algae, and brown algae (phaeophyceans) (Cavalier-Smith and Chao, 1996; Potter, et al., 1997b; Saunders, et al., 1997b) (Fig. 12–11). The fact that both diatoms and the pelagophycean-silicoflagellate-pedinellid clade are distinguished by loss or great reduction of their flagellar apparatus, suggested to Saunders, et al. (1995b) that a single flagellar system reduction-event preceded the divergence of these two clades. In contrast, a SSU rDNA analysis by Cavalier-Smith and Chao (1996)

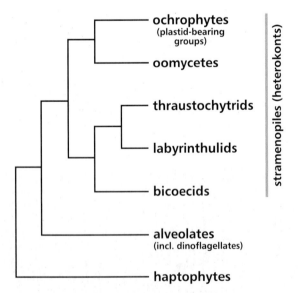

Figure 12–10 Relationships of ochrophytes (heterokont algae) to oomycetes, thraustochrytrids, labyrinthulids, the colorless flagellates known as bicoecids, alveolates (the clade that includes dinoflagellates and ciliates), and haptophytes, as revealed by molecular systematic methods. (Based on Cavalier-Smith and Chao, 1996)

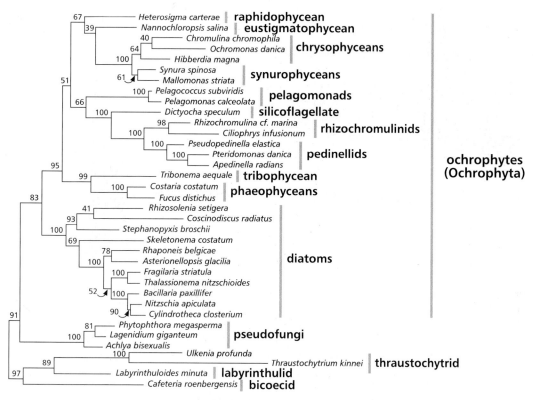

Figure 12–11 A phylogenetic tree showing relationships of the major groups of the ochrophytes, based on nuclear SSU rDNA sequences. The numbers above nodes are bootstrap values, a confidence estimate described in Chapter 5. Bootstrap values of 95–100 indicate a high level of confidence in the branch; lesser values indicate lower confidence. Monophyly of the diatoms, for example, is well supported. (After Cavalier-Smith and Chao, 1996. Redrawn by permission of Blackwell Science, Inc.)

(Fig. 12–11) does not show the diatoms and the pelagophycean-silicoflagellate-pedinellid clade to be sister taxa, suggesting that reduction of the flagellar apparatus occurred independently in these two lineages. This exemplifies the tenuous state of our understanding of ochrophyte evolutionary radiation and emphasizes the importance of using as many conservative ultrastructural and molecular characters as possible in efforts to infer relationships.

The time of origin of the ochrophyte clade is unknown. Although ochrophytes are generally regarded as having appeared substantially later in the fossil record than red or green algae, recently discovered, well-preserved remains from the 570 million-year-old Doushantuo shales in southern China suggest that highly derived brown algae had originated prior to this time (Xiao, et al., 1998). In addition, certain 900–1000 million-year-old Siberian fossils (Knoll, 1996) strongly resemble the modern ochrophyte *Vaucheria* (Xiao, et al., 1998).

PART 2—DIATOMS

In terms of evolutionary diversification, the diatoms have been wildly successful. Though occurring only as single cells or chains of cells, with 285 genera (according to Round, et al., 1990) encompassing 10,000–12,000 recognized species (Norton, et al., 1996), diatom diversity is rivaled among the algae only by the green algae. Some experts believe that many diatom species remain to be described and that diatom species may actually number in the millions (Norton, et al., 1996). In addition, diatoms are exceedingly abundant and are probably the most numerous of eukaryotic aquatic organisms. In terms of contributions to global primary productivity, diatoms are among the most important aquatic photosynthesizers. They dominate the phytoplankton of cold, nutrient-rich waters, such as upwelling areas of the oceans, and recently circulated lake waters. This abundance, coupled with

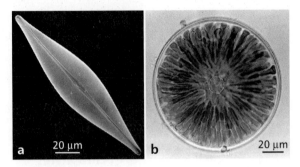

Figure 12–12 A comparison of (a) pennate diatoms, typified by bilateral symmetry, with (b) centric diatoms, which have radial symmetry. (a: SEM by O. Cholewa)

dissolution-resistant silica walls, has resulted in massive sedimentary accumulations (see, for example, Kemp and Baldauf, 1993), some of which are now mined for diatomite, useful in many industrial applications. Diatoms are ubiquitous, occurring in marine and freshwaters, where they may be planktonic, benthic, periphytic (growing on plant or seaweed surfaces), epizoic (on animals, from crustaceans to whales), or endozoic (within foraminifera, for example). In coastal regions where beaches are exposed to extremely heavy wave action, conspicuous growths of diatoms color the surf water brown. These are known as surf diatoms. Along the Oregon and Washington coasts their presence and abundance are positively correlated with rainfall and river discharge, but not temperature, salinity, day length, or nitrate/ammonium levels (Jijina and Lewin, 1984). Surf diatoms also occur on sandy South African beaches characterized by wide surf zones, medium- to high-energy surf, and well-developed rip currents that receive nutrient-rich waters from coastal dune areas (Campbell and Bate, 1997). Rich communities of diatoms occur within polar ice formations (Lizotte, et al., 1998) (see Text Box 12–1). They also occur in moist terrestrial habitats and have been isolated from air samples. Some species of the diatom *Pseudo-nitzschia* produce the toxin domoic acid, which causes amnesiac shellfish poisoning in humans and affects marine animals (Chapter 3). In addition, a few species of *Chaetoceros* are associated with fish kills but are not known to be harmful to humans (Horner, et al., 1997).

Diatoms are included within the ochrophytes based on the common presence of fucoxanthin and chlorophyll *a* + *c*, plastids within periplasmic endoplasmic reticulum, plastid girdle lamellae, chrysolaminarin and lipid reserves, and tripartite tubular

hairs on flagella of sperm of some diatoms. However, diatoms share several unique features (autapomorphies) that readily distinguish them from related algae. The most obvious of these is enclosure of the protoplast within a rigid lidded box—the **frustule**—composed of amorphous opaline silica with organic coatings. The diatom frustule is highly ornamented, with variations in this ornamentation reflecting taxonomic diversity. Only a few other groups of ochrophytes have cell coverings composed of silica. These include tiny nonflagellate forms known as the Parmales, whose walls are composed of polygonal plates, and various chrysophyceans that are covered with numerous, overlapping silica scales. A second major difference between diatoms and most other ochrophytes regards their life cycle—meiosis is gametic in diatoms. Although certain phaeophyceans also have a gametic life cycle, this has clearly evolved independently. Finally, as previously mentioned, the flagellar apparatus of diatoms has been highly reduced, and aspects of diatom mitosis and cytokinesis also distinguish them from other ochrophytes.

Diatoms are commonly grouped into two or three major categories, primarily on the basis of frustule features that can be readily observed in living cells as well as fossils. The **centric** diatoms typically have discoid or cylindrical cells having radial symmetry in face or "valve" view. A **valve** is the top or bottom of the silica frustule. In contrast, valves of **pennate** (referring to "feathery" patterns of ornamentation on the frustule) diatoms have more or less bilateral symmetry (Fig. 12–12). Some pennate diatoms possess slits in the frustule—the **raphe system**—that are associated with the ability to accomplish rapid gliding motility (see below). Such diatoms are known as raphid pennates. Other pennate diatoms lacking raphes are termed araphid pennates. On the basis of such structural features, living and fossil diatoms have been divided into three categories: centrics, araphid pennates, and raphid pennates (Round, et al., 1990). Centric diatoms typically possess many discoid plastids, whereas pennate diatoms typically contain two large, platelike plastids. Pyrenoids occur in both plastid types. Centric and pennate diatoms can also be distinguished by differences in sexual reproduction discussed below. However, as will be noted below, molecular sequence data indicate that centric diatoms and araphid pennate diatoms are paraphyletic. Thus classical taxonomic schemes for diatoms are not completely consistent with relationships deduced by the use of molecular information.

Text Box 12–1 Antarctic Sea-Ice Diatoms

An enormous band of ice formed by freezing of seawater occurs around the edges of the Antarctic continent, especially in winter, when such sea ice occupies an area of 19 million km^2. Large populations of microalgae—primarily pennate diatoms—associated with this sea ice form visible blooms that color the ice brown in spring and summer. They contribute about 20% of the primary productivity of the Southern Ocean, some 60–120 g C m^{-2} y^{-1}. This fixed carbon may be exploited directly by krill (crustaceans), which are consumed by whales, seals, birds (such as penguins), and fish, or indirectly via the microbial loop (see Chapter 3) (Nicol and Allison, 1997). In addition, sea-ice algae generate important quantities of DMSP, an osmotic- and cryo-protectant that is converted to DMS, which influences the global sulfur cycle (discussed in Chapter 2) (DiTullio, et al., 1998). Thus, sea-ice diatoms are considered to be of global ecological and climate significance and are intensively studied by oceanographers and paleoclimatologists.

(Photographs courtesy F. Taylor and A. Leventer)

Sea-ice algae are commonly found in coastal polynyas—areas of open water or thin ice that occur within the thicker pack ice—that can be tens of thousands of km^2 in area. Polynyas are regarded as oases of biological activity. The algae can occupy surface pools of melted water, brine channels that penetrate the ice, or the submerged lower ice surface. They may occur as bands that form by addition of new lower ice layers, or may be more diffusely distributed in the ice (Nicol and Allison, 1997). Sea-ice algal communities are sampled by taking ice cores with an ice auger, among other methods. As estimated by measurements of chlorophyll *a* concentrations, algal biomass in sea ice varies from 0.07–1300 μg l^{-1}, as compared with 0.01–0.26 μg l^{-1} in the underlying seawater. Analysis of accessory pigments indicates dominance of sea-ice algal communties by diatoms, which make up 65% of the community (19% is green algae, 11% haptophytes, and 5% dinoflagellates) (Lizotte, et al., 1998). Hence, sea ice appears to be a particularly favorable environment for growth of certain diatoms seemingly adapted for the sea-ice environment.

Adaptations include the DMSP mentioned earlier; an unusually high ratio of fucoxanthin to chlorophyll *a* (Lizotte, et al., 1998), interpreted as adaptation to low-light conditions; production of extracellular polymers that facilitate adhesion to ice crystals; the physiological capacity to maintain high growth rates throughout a range of irradiance, temperature, and salinity conditions that occur as the pack ice goes through melting and reformation cycles; and life histories that are tuned to the annual cycle of ice formation (Gleitz, et al., 1998). Spring/summer bloom formation is attributed to favorable environmental conditions for growth, including increased irradiance and nonlimiting levels of fixed nitrogen (Priscu and Sullivan, 1998), and low herbivore populations. Although more than 100 diatom species have been found in Antarctic sea ice, fewer than 20 contribute significantly to the blooms. Diatom species most frequently observed as significant contributors to bloom biomass are *Fragiliariopsis cylindrus*, *F. curta*, *F. kerguelensis*, *Chaetoceros dichaeta*, *Pseudonitzschia prolongatoides*, and *Cylindrotheca closterium* (Gleitz, et al., 1998).

Seafloor remains (fossils) of diatoms are especially valued by climatologists because they provide the most sensitive known indicators of change in the Southern Ocean; they serve as proxies for sea-ice distribution patterns in the past. Particular diatom species that are abundant and whose frustules are resistant to dissolution (*Fragilariopsis curta* and *F. cylindrus*) are known to be associated with pack ice, whereas other species, such as *F. kerguelensis* and *Thalassiosira antarctica*, are dominant in the open ocean. Hence seafloor patterns of these diatoms can reveal the boundaries between the pack ice and open ocean in the past (Leventer, 1998).

Global change processes that could affect sea-ice diatoms include temperature increases, changes in snowfall patterns, and increases in ultraviolet radiation arising from the Antarctic ozone hole. This annual springtime phenomenon, resulting from anthropogenic increases in atmospheric pollutants (man-made chlorofluorocarbons and other halogenated compounds), can cause short-term loss of carbon fixation in the waters of the Southern Ocean. UVA radiation (320–400 nm) interferes with light-induced increases in pigment levels, and UVB radiation (280–320 nm) is known to cause DNA damage in ice diatoms. The latter determination was made by probing DNA isolated from algae with an antibody against dipyrimidine DNA, a product of UVB damage. Such damage occurs despite the occurrence in the sea-ice algae of mycosporinelike amino acids that are widely thought to serve as UV screens (Prézelin, et al., 1998).

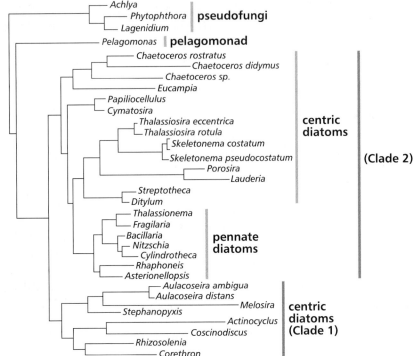

Figure 12–13 A phylogeny of the diatoms based on molecular sequence evidence suggests the existence of two major clades, one containing centric diatoms that have a tubelike process known as a rimoportula, and a second consisting of centric diatoms, featuring a different type of process (the fultoportula) together with the pennate diatoms. (After Medlin, et al., 1997)

Fossil History

The silica wall of diatoms is soluble in strong alkaline solutions or hydrofluoric acid, but resistant to dissolution in most natural waters. Sedimented diatom frustules have accumulated during the past 150 million years, since the Early Cretaceous. Thus diatoms are thought to have originated prior to the Cretaceous. The earliest fossil diatoms were centrics, while the earliest pennate diatoms were araphid forms of late Cretaceous age, about 70 million years ago. Raphid pennate diatom fossils are of somewhat more recent origin (Simonsen, 1979). According to the fossil record, freshwater pennate diatoms first appeared around 60 million years ago, and diatoms assumed a position of prominence in phytoplankton communities during the Miocene (some 24 million years ago).

Diatom Phylogeny

Molecular sequence information (SSU rDNA and other) suggests that there may be two major diatom lineages or clades (Fig. 12–13) (Medlin, et al., 1996; 1997). Clade 1 consists of centric diatoms, many of which may possess peripheral rings of tubelike structure known as **labiate processes** or **rimoportulae**. Clade 2 includes centric diatoms having a central labiate process, centric diatoms with a **central strutted process** or **fultoportula** (Fig. 12–14), and pennate diatoms, some with and others without a raphe system. As noted previously, the earlier appearance of centric diatoms in the fossil record strongly suggests that centrics are evolutionarily older than pennate forms. Molecular based phylogenies support this hypothesis and suggest that the raphe system may have arisen from a central tube-like process observed in Lower Cretaceous fossils (Medlin, et al., 1996). Molecular data also indicate that raphid diatoms are monophyletic. Flagella occur only in centric diatoms (only on sperm cells), and have presumably been lost completely from the pennate lineages.

The evolutionary origin of the diatoms and their distinctive silica frustule has long puzzled diatom experts. At one time it was thought that the ancestors of diatoms might have been silica-scaled chrysophyceans, but modern phylogenetic data suggest that this is unlikely. Another hypothesis suggests origin of diatoms from parmalean algae—small (2.0–5.5 μm in greatest dimension) tetrahedrally symmetrical ochrophytes that lack flagella but have chloroplasts and a cell covering composed of polygonal silica plates

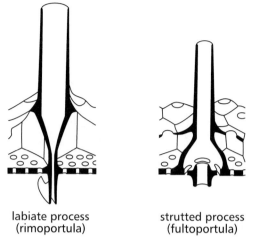

labiate process
(rimoportula)

strutted process
(fultoportula)

Figure 12–14 Diagrams of longitudinal cutaway views of two major types of tubular processes that pass through the valves of centric diatoms. Labiate processes (rimoportulae) are so-called because the internal portion of the tube is compressed to form liplike structures. Strutted processes (fultoportulae) are not compressed internally, and there are other structural differences. (After Hasle and Syvertsen, 1997)

(Mann and Marchant, 1989) (Fig. 12–15). Parmales were first discovered from Antarctic waters (Marchant and McEldowney, 1986; Booth and Marchant, 1987). More recent evidence suggests probable origin of diatoms from a newly discovered group of marine flagellates (Guillou, et al., 1999). These flagellates, described as a new algal class—the Bolidophyceae—are small (1.2 μm in diameter) and naked, lacking walls or siliceous coverings. Features linking them with ochrophytes include typical heterokont flagella (a long one with tubular hairs and a shorter, smooth one), a plastid with girdle lamellae and fucoxanthin, and SSU rDNA sequences. Molecular data suggest that bolidophyceans are sister to diatoms and that diatoms originated from unicellular flagellates (Guillou, et al., 1999).

Diatom Cell Biology

The single most striking aspect of diatom cells is their cell wall, the silica frustule. The evolutionary development of a coherent siliceous covering is thought to have provided adaptive advantages. Silica is inert to enzymatic attack, and thus diatoms may be less vulnerable to microbes or digestion within the guts of herbivores than are algae having no walls or walls

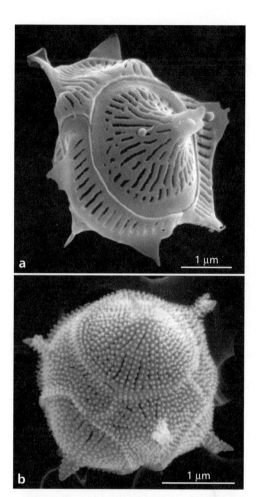

Figure 12–15 Scanning electron microscopic images of parmalean algae, showing their distinctive polygonal silica plates. (From Marchant and McEldowney, 1986 ©Springer-Verlag)

composed of polysaccharide (Pickett-Heaps, et al., 1990). In addition, silica is often plentiful in natural waters and is an energetically inexpensive source of wall material (Falkowski and Raven, 1997). Frustule shape and ornamentation influence many aspects of ecological behavior, as well as providing a convenient means for algal taxonomists to distinguish taxa. The discussion below will therefore focus upon frustule structure, variation, and development.

Frustule Structure

Diatom frustules consist of two overlapping components; the **epitheca** overlaps the **hypotheca** (Fig. 12–16). These terms are also used to describe the anterior and posterior portions of the dinoflagellate cell

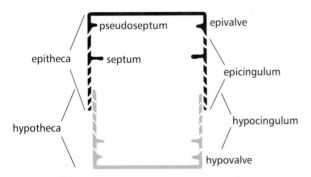

Figure 12–16 Diagrammatic representation of the diatom frustule, showing major features viewed in cross section. (After Hasle and Syvertsen, 1997)

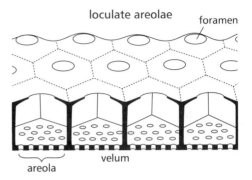

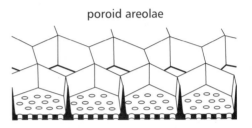

Figure 12–18 Diagrammatic representations of loculate and poroid areolae, features of the diatom frustule. (After Hasle and Syvertsen, 1997)

wall (Chapter 11), which is not homologous to that of diatoms. However, like the epitheca and hypotheca of dinoflagellates, those of diatoms are composed of more than one piece, with the pieces linked by a suture. The epitheca of diatoms consists of an **epivalve** and an **epicingulum**; these develop separately and are later joined. Similarly, the diatom hypotheca is composed of a **hypovalve** and **hypocingulum** (Fig. 12–16). The diatom frustule is analogous to a round or elongate box, where the top of the box is equivalent to the epivalve, the bottom to the hypovalve, and the sides to the two **cingula** (also known as the **girdle** regions). Diatoms differ in the construction of the cingula. In some cases these consist of complete circular bands, but in others the bands are split rings, or consist of several sections. Some cingular bands have a protrusion known as a **ligula** opposite the ring split that fits into the split of the other band (Fig. 12–17). Diatom cell expansion can only occur at the cingulum. Diatoms seen from the top or bottom are in valve view, whereas diatoms seen from their sides are in girdle view.

Both the valves and cingula are penetrated by pores of various sizes and types; these are thought to function as passageways for entry into or exit from cells of gases, nutrients, or other materials. Such pores

are called **areolae** (Fig. 12–18), and are usually complex, often having a thinly silicified layer (known as a **velum** or **rica**) penetrated by smaller pores or slits stretched over them. The form of this layer is of taxonomic value, but visualization of its details usually requires electron microscopy (Fig. 12–19). Areolae of centric diatoms may occur as small hexagonal chambers (**loculae**) where the velum stretches across one end of the chamber at the inner valve surface and the other

Figure 12–19 High-resolution low-voltage scanning electron microscopic view of poroid areolae in *Mastogloia angulata*. (From Navarro, 1993)

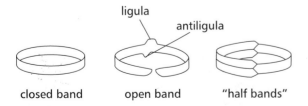

Figure 12–17 Diagrams showing variation in cingulum structure. (After Hasle and Syvertsen, 1997)

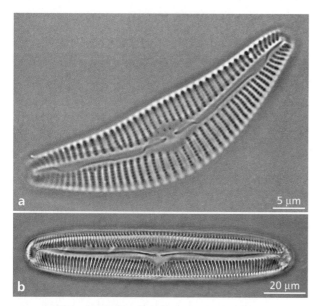

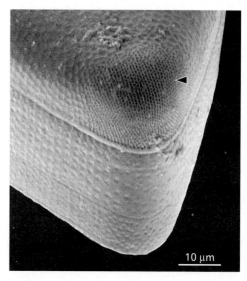

Figure 12–22 *Triceratium.* SEM view of an ocellus (arrowhead)—a rimmed, elevated plate of perforated silica. (Micrograph by O. Cholewa)

Figure 12–20 (a) Striae (rows of pores) on the valve face of an acid-cleaned specimen of *Cymbella*, viewed with light microscopy and digitally processed. (b) Costae (parallel ribs) on the valve face of an acid-cleaned specimen of *Pinnularia*, viewed with phase-contrast light microscopy.

surface is partially constricted, and hence are known as **loculate areolae**; areolae lacking this outer constriction are referred to as **poroid areolae** (Fig. 12–18). Areolae may occur in rows (**striae**) that may appear as fine parallel lines when viewed at low magnification (Fig. 12–20a). Some pennate diatoms exhibit parallel, riblike **costae**, which are hollow chambers (Fig. 12–20b).

Many centric diatoms (and pennate diatoms without raphe systems) possess a rimoportula, a tubular structure that passes through the valve, ending on the inside in a slit, as if the tube were laterally compressed (Figs. 12–14, 12–21). This structure is also known as a labiate (lipped) process and is associated with polysaccharide mucilage excretion (Medlin, et al., 1986). In some centric diatoms the rimoportula extends externally as a long tube that can sometimes be mistaken for a spine or seta. Sometimes the rimoportulae occupy extremely eccentric positions; other taxa may bear hundreds of rimoportulae on their valves (Pickett-Heaps, et al., 1990). Certain centric diatoms exhibit another type of mucilage secretion structure, the **ocellus**. This is an elevated plate of silica that is perforated by pores and surrounded by a rim (Fig. 12–22). Mucilage secreted through the ocellus is used for attachment to substrates or other cells of a colony. Members of a particular group of centric diatoms (Thalassiosirales) exhibit another type of complex tubular structure—known as a fultoportula or strutted process (Fig. 12–14)—which is associated with secretion of chitin fibrils (Fig. 12–23). Such fibrils may play roles in colony formation, herbivore deterrence, or buoyancy (Medlin, et al., 1996).

Many pennate diatoms have longitudinal slits that form a raphe system (Fig. 12–24). In some diatoms, raphes run the length of the valve, while in others they are interrupted by a **central nodule** (Fig. 12–24). Raphes extend all the way though the silica wall but

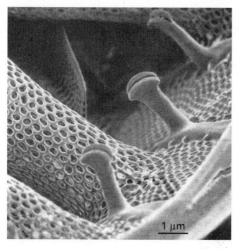

Figure 12–21 The internal opening of a rimoportula (labiate process) of *Isthmia enervis.* (From Navarro, 1993)

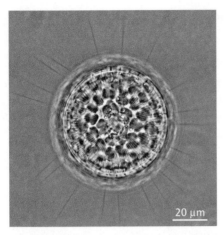

Figure 12–23 Chitin fibrils extending from a living cell of *Stephanodiscus*.

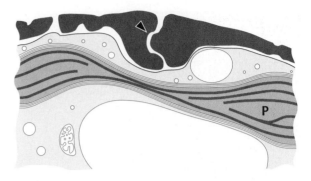

Figure 12–25 Diagrammatic and generalized representation of a cross-sectional view through the raphe region of a raphid diatom. Note the zigzag pattern taken by the slit (arrowhead).

are internally constructed such that there is an outer fissure and an inner fissure (Fig. 12–25). In other words, viewed in cross section with the TEM, the raphe appears to zigzag through the frustule, rather than taking a straight path. This is believed to add necessary structural stability, reducing the chance that the frustule might split longitudinally along the raphe (Pickett-Heaps, et al., 1990). In some pennate diatoms raphes occur on both valves, but in others a raphe may occur only on one valve. An example is *Cocconeis* (Fig. 12–55), which grows with the raphe-bearing valve in contact with solid surfaces. Raphes usually occur along the midpoint of the valve, but in some

diatoms they occur at the edge (Fig. 12–26), sometimes within a raised (keeled) structure. Raphes are associated with the ability of diatoms to move readily; diatoms having raphes can exhibit rapid motility, whereas those without raphes generally cannot.

Diatom Motility and Mucilage Secretion

A few centric diatoms can move slowly by secretion of mucilage from the rimoportula (Medlin, et al., 1986). An example is *Odontella*, which undergoes "shuffling" movements as a result of such mucilage extrusion (Pickett-Heaps, et al., 1986). However, diatoms that have raphes can move rapidly along the surfaces of substrates, usually with a rather jerky motion and frequent stops. Their speed range is about 0.2–25 μm s^{-1} (Häder and Hoiczyk, 1992). The average speeds (in μm s^{-1}) of some common diatoms are: *Stauroneis*, 4.6; *Pinnularia*, 5.3; *Nitszchia*, 10.4; and *Craticula*, 10. Diatom motility is regarded as a means by which they may avoid local nutrient limitation and shading (Cohn and Weitzell, 1996). The environmental factors that influence diatom motility were investigated by Cohn and Disparti (1994).

A number of hypothetical mechanisms have been proposed to explain diatom motility. Edgar and Pickett-Heaps (1984a) suggested that vesicles having fibrillar polysaccharide contents are liberated from the cell by exocytosis at the raphe fissure, at which point the fibrils become hydrated, transforming into rods that project toward the substrate. Although it was first thought that intracellular bundles of actin microfibrils located beneath the cell membrane provided the motive force

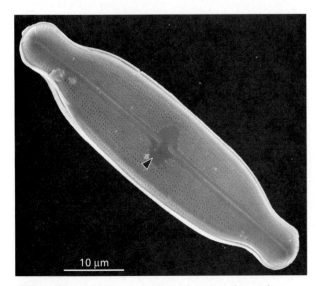

Figure 12–24 SEM of *Neidium* showing the raphe slits, which curve sharply at the central nodule (arrowhead). (Micrograph by O. Cholewa)

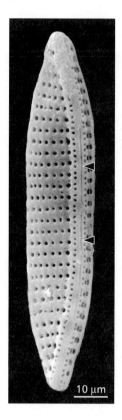

Figure 12–26 SEM of *Hantzschia*, showing the peripheral location of the raphe (arrowheads). (Micrograph by O. Cholewa)

1996). Diatom mucilages are also recognized for their importance in stabilizing sediments. The mucilage forms connecting strands between sediment particles giving rise to a gel-mud. Such sediment stabilization is thought to provide adaptive advantage to benthic diatoms and is ecologically important in reducing erosion. Studies have shown that diatom pigments and mucilages can serve as indicators of sediment stability (Sutherland, et al., 1998). A review of diatom mucilage production and its relationship to motility is provided by Wetherbee, et al. (1998).

Cell Division and Frustule Development

Production of diatom valves is always associated with cell division, and cell division always occurs in the plane of the valve. This close linkage between valve production and mitosis severely limits the extent of diatom morphological variation to single cells or cells joined by their valves into linear chains of cells that do not share a common wall, and hence are pseudofilaments. Vegetative cell division does not occur in any

for motility, a more recent hypothesis suggests that the actin bundles actually function in anchoring the cell membrane to the frustule. Raphe-associated microtubules and attached motor proteins such as kinesin and/or dynein are currently regarded as a more likely source of the motive force that drags the polysaccharide rods along, resulting in diatom motility (Schmid, 1997). The latter hypothesis more adequately explains the ability of diatoms to stop and reverse the direction of their motion, as well as the fact that motility ceases during mitosis, when microtubule proteins are recruited for spindle formation.

In the case of some raphid pennate diatoms, such as *Achnanthes*, mucilage secreted from the raphe can also form a stalk that is used to form a more stable attachment of cells to substrates (Fig. 12–27) (Wustman, et al., 1997; 1998). Application of certain polysaccharide inhibitors to *Achnanthes longipes* results in both loss of motility and loss of ability to adhere to surfaces (Wang, et al., 1997). As previously mentioned, other stalked diatoms generate attachment mucilage from polar pores or from their rimoportula. Mucilage stalks are thought to confer the ability to overcome shading and possibly also allow greater access to water-column nutrients (Cohn and Weitzell,

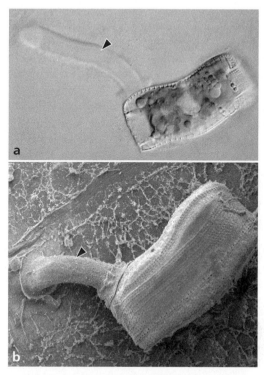

Figure 12–27 Stalk (arrowheads) production by *Achnanthes longipes*. (a) LM view, (b) SEM view. (Photographs courtesy Y. Wang and M. Gretz)

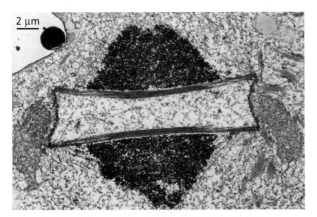

2 μm

Figure 12–28 The mitotic spindle of *Surirella robusta*. Chromosomes are grouped around the hollow spindle, composed of two interdigitating half spindles. (From Pickett-Heaps, et al., 1984. Reproduced from the **Journal of Cell Biology** 1984, Vol.99:137–143, by copyright permission of The Rockefeller University Press)

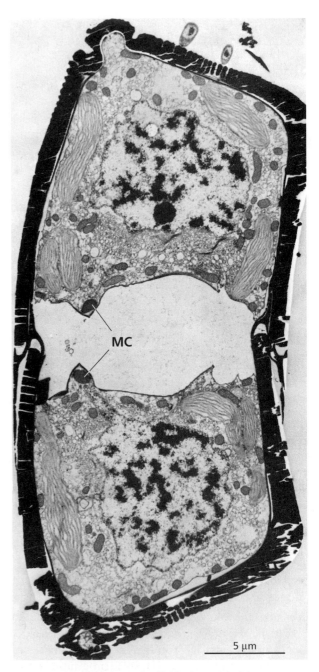

MC

5 μm

Figure 12–29 A recently divided *Hantzschia amphioxys*, showing expansion of the silica deposition vesicle across the cleavage plane. Microtubule centers (MC) occur in each daughter cell. (From Pickett-Heaps and Kowalski, 1981)

other planes and cell walls are not shared. So, there are no truly multicellular diatoms, nor can they form more complex thalli. Some diatoms form filamentous colonies by linking adjacent frustules with a specialized siliceous connector, but diatoms completely lack intercellular connections that provide for cytoplasmic continuity, such as plasmodesmata exhibited by some multicellular green algae and, most likely, all brown (phaeophycean) algae (Cook and Graham, 1999). Though evolutionarily successful in terms of speciation, in terms of the potential for thallus diversification, and intra-thallus cellular diversification, diatoms have literally "boxed themselves in."

The findings of a number of studies on diatom mitosis have been reviewed, summarized, and compared to mitosis in other algae by Pickett-Heaps, et al. (1990) and Sluiman (1993). Centrioles are absent from spindle poles; instead, an electron-opaque body—thought to function as a microtubule organizing center (MTOC)—replicates just prior to mitosis, with one copy found at each spindle pole. The spindle begins its development outside the nucleus, then enters the nucleus via a breach in the nuclear envelope. The nuclear envelope eventually disappears, so that mitosis is described as being "open." The early spindle consists mostly of microtubules that run continuously from pole to pole. By prometaphase, however, the spindle consists of two half spindles whose microtubules interdigitate in the region of overlap (Fig. 12–28). Kinetochores of diatom chromosomes appear to attach to the polar microtubules. Early in anaphase, chromosomes move along these microtubules toward the poles. Later in anaphase, the spindle elongates, reducing the amount of overlap between the half spindles. The two half spindles continue to

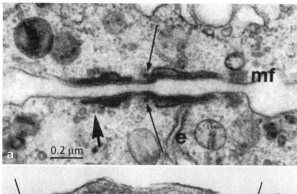

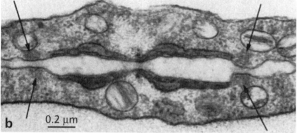

Figure 12–30 (a) Raphe fibers (thin arrows) cut transversely, associated with the developing raphe of *Navicula* (*Craticula*) *cuspidata*. thick arrow=vesicles associated with one of the two ribs forming the raphe, e=endoplasmic reticulum, mf=microfilaments. (b) Later stage, where ribs have expanded. arrows=microfilaments. (From Edgar and Pickett-Heaps, 1984b by permission of the *Journal of Phycology*)

slide past each other, eventually separating completely. By this time, cytoplasmic cleavage has begun by ingrowth of a furrow extending from the cell membrane. Nuclear envelopes then re-form around nuclei, and cytokinesis is completed.

When mitosis and cytokinesis are complete, the MTOCs migrate to a position between the nuclei and a silica deposition vesicle (SDV), which develops by fusion of Golgi vesicles (Fig. 12–29). The MTOC and associated microtubules and microfilaments appear to be involved in the process by which the two new valves will be produced through controlled silica deposition within the SDV. However, the process by which the intricate patterns of areolae and costae, as well as the raphe, ocellus, fultoportula, or rimoportula are generated is not well understood. It has been suggested that an organic template provides the pattern for silica deposition (Pickett-Heaps, et al., 1979), but other hypotheses have been advanced to explain diatom frustule development (Gordon and Drum, 1994). An electron-opaque "raphe fiber" may be involved in raphe development (Fig. 12–30). *Pinnu-*

laria is an example of a diatom in which a raphe fiber occurs at the site of the forming raphe (Pickett-Heaps, et al., 1990). However, the mechanism by which such a fiber might be associated with raphe formation is not understood.

Usually the production of the two new valves by sibling cells is identical. However, as mentioned previously, in a number of diatoms one new valve develops a raphe while the other produces a **pseudoraphe** in which the fissure is filled in with silica (Fig. 12–31). As the result of an ultrastructural analysis of valve development in *Achnanthes coarctata*, Boyle, et al. (1984) were able to explain how such differences occur. Up to a point, development in the two cells is alike. After cytokinesis both daughter nuclei and their associated MTOCs move closer to one side of the cell. Nearby, the two new tubular SDVs each form a central rib running the length of the cells. Then both SDVs generate another longitudinal rib next to the first. The gap between the two ribs becomes the future raphe fissure. Development of the two new valves now differs. In the cell that will form a raphid valve the SDV, flanked by microtubules, expands laterally in both directions. In the other cell the MTOC and its microtubules withdraw from the SDV, and the SDV

Figure 12–31 *Achnanthes coarctata*, an example of a diatom having one araphid valve (av) and one valve with a raphe (rv). (From Boyle, et al., 1984 by permission of the *Journal of Phycology*)

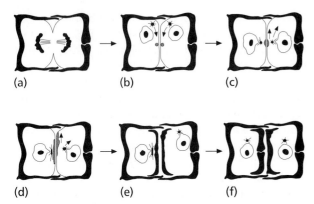

(a)   (b)   (c)

(d)   (e)   (f)

Figure 12–32 Diagram of the process by which a raphe and pseudoraphe develop in only one of the two new valves produced after a cell division in *Achnanthes coarctata*. (After Boyle, et al., 1984 by permission of the *Journal of Phycology*)

moves laterally until the double rib is at the corner of the cell. Silica is then secreted into the expanding SDV, and the raphe fissure fills in (Fig. 12–32). This explains the occurrence of a pseudoraphe at the edge of one valve of this diatom.

In many pennate diatoms the valve begins development from an axial area that runs the length of the valve, from the center outward. In raphid pennate diatoms silicification starts along the site of the raphe. In some diatoms silicification occurs from the margins inward, and uniformly across the valve in yet other taxa. Silicification in centric diatoms often starts very close to the point of emergence of a rimoportula, whose development at an earlier point in time is associated with a specialized organelle. When the valve is mature, fusion of Golgi vesicles generates a separate SDV where girdle bands are produced.

A sufficient supply of dissolved silica (orthosilic acid—$Si(OH)_4$) must be present in the surrounding medium or production of the new frustules will not occur (Darley and Volcani, 1969). Diatom walls are described as "opaline" silica because they have a granular appearance when treated with hydrofluoric acid, as does opal. With the exception of some endosymbiotic wall-less forms, diatoms do not generally appear able to live without their silica wall and therefore cannot persist and reproduce when the silicate concentration falls below critical levels. Although silicon is one of the most abundant elements in the earth's crust, spring diatom blooms can deplete levels of dissolved silica in illuminated waters, and as silicate is not readily regenerated within surface waters, diatom growth can be limited.

Silica loading from short-term seasonal input of ground water is a major source of silica for diatoms in oligotrophic lakes (Hurley, et al., 1985). Anthropomorphic changes in the silica content of natural waters may affect the continued persistence of some diatoms. For example, while examining the sediments of Lake Ontario, Stoermer, et al. (1989) found evidence that *Stephanodiscus niagarae* became unable to undergo sexual reproduction after 1947. This was attributed to changes in the seasonal supply of silica resulting from land use changes in the surrounding watershed.

The parental epitheca can be recycled several times, but seems to have a determinate life span. In one form of *Stephanodiscus*, the parental valve was observed to last only 6–8 generations, ultimately disappearing from the population (Jewson, 1992). Particularly significant to diatom biology is the fact that each new valve that is formed is always a hypotheca; the older hypotheca becomes the epitheca of one of the two new daughter cells resulting from cell division. This cell will thus be slightly smaller than its sibling cell (which is the same size as the parent). In a clonal population of diatoms, half of the cells resulting from any division will be smaller than their parents. Thus the mean cell size in the population will decline over time. This size reduction resulting from continued cell division in diatoms is known as the McDonald/Pfitzer rule after its two discoverers, who independently reported this phenomenon in 1869 (Fig. 12–33). Some diatoms, however, do not exhibit size reduction; they may have more elastic cingula that are able to expand more greatly after cell division than those of

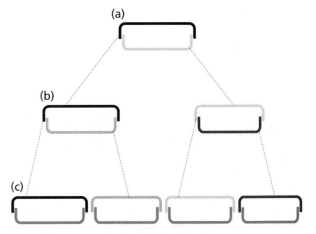

Figure 12–33 Diagrammatic view of three generations of a diatom, illustrating reduction in the mean cell size as the population increases through time.

most diatoms. Environmental factors, such as presence of grazers or parasites, are known to affect the rate of diatom size decrease (Edlund and Stoermer, 1997). When diatoms reach a critical size (about one third their maximal size) they become capable of undergoing a size regeneration process that is generally thought to involve sexual reproduction. Size regeneration is regarded as a major role played by sexual reproduction in diatoms. Sexual reproduction is also dependent on the occurrence of enabling environmental conditions (described below); if these are not present, neither sexual reproduction nor size regeneration will occur (Edlund and Stoermer, 1997). In nature, diatoms that are smaller than one third the maximal size for the species are usually unable to regenerate larger size and eventually die (Mizuno and Okuda, 1985). Such small-sized diatoms can persist in laboratory cultures, and it has been pointed out that many physiological and ecological experiments have probably been done with atypically small diatoms (Edlund and Stoermer, 1997).

Sexual Reproduction

Vegetative cells of diatoms are diploid, and gamete production involves meiosis. This was first established by Klebahn in 1896 working with *Rhopalodia*, and later by Karsten (1912), in *Surirella*. Thus diatoms, like members of a few other algal groups and animals, have a gametic life cycle. Diploidy, and the potential for heterozygosity, is believed to confer enhanced genetic and evolutionary flexibility.

Relatively little is known regarding the environmental cues that induce diatom sexual reproduction. Increases in temperature or irradiance or changes in nutrient availability are known triggers for some taxa. In addition, particular cell cycle stages and/or pheromones (hormones that induce mating) may be required, but the nature of such postulated pheromones remains to be defined. Diatom sexual reproduction is probably more common than generally believed on the basis of relatively sporadic observations in nature. Edlund and Stoermer (1997) have defined four major strategies for sexual timing in diatoms—some of these may explain the relative rarity of observation of diatom sex. First, synchronous sexuality occurs under favorable conditions for growth—large numbers of cells may be observed to engage in sexual reproduction within a relatively short period of time. Synchronous sexuality may occur among freshwater epiphytic or epilithic diatoms such as *Cymbella* and

Gomphonema in the springtime, when increasing irradiance and temperature serve as cues. Fifty percent of the competent cells in populations of the freshwater planktonic diatom *Stephanodiscus* may undergo sexual reproduction when nitrate levels in the water rise above 10 μM. Another such example is mass occurrence of sexuality in Antarctic oceanic blooms of the diatom *Corethron criophilum* (Crawford, 1995). The three remaining timing strategies defined by Edlund and Stoermer (1997) include (1) synchronous sexuality under conditions that do not favor vegetative growth (such as nitrogen or light limitation); (2) asynchronous sexuality in conditions favorable for growth (where relative few members of a population undergo sex over an extended time period); and (3) asynchronous sexuality in poor conditions, a strategy that might prove advantageous in unpredictable or fluctuating environments.

Centric diatoms are oogamous, producing one or two egg cells per parental cell (two or three of the four meiotic products die), and from 4–128 sperm per parent cell as the result of mitotic divisions following meiosis. Production of a larger number of sperm is regarded as a strategy for increasing the efficiency of fertilization in open water environments, where centric diatoms commonly live. Centric diatom sperm are flagellate, and the flagellar apparatus is highly reduced. There is only a single flagellum, having two rows of tripartite hairs, but it is unusual in that there are no central singlet microtubules, as occur in the vast majority of other flagella/cilia. The flagellar basal bodies are also unusual in that they consist of microtubule doublets, rather than triplets as in other basal bodies and centrioles. The valves surrounding sperm separate, allowing the flagellate cells to escape. The silica wall surrounding eggs also opens, exposing the protoplast. Gametes possess plastids, but not silica frustules. Gamete fusion results in formation of a zygote that develops into a large **auxospore** (Fig. 12–34). Auxospores can be completely free of the parental frustule, attached to one of the parental thecae, or enclosed by both parental thecae. A few diatoms having frustule bilateral symmetry (and thus considered to be pennate diatoms), were ultimately recognized as having closer relationship to centrics on the basis of sexual reproduction characteristics (Hasle, et al., 1983).

Pennate diatoms are almost always isogamous—the two gametes are similar in size and neither is flagellate. An exception is *Rhabdonema*, an araphid diatom whose sexual reproduction is described as "modified oogamy" because oogonia and small, non-

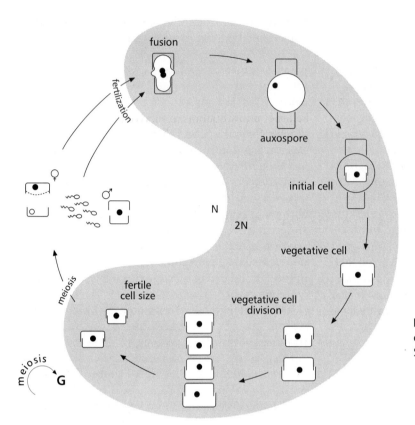

Figure 12–34 The life cycle of a typical centric diatom. (After Hasle and Syvertsen, 1997)

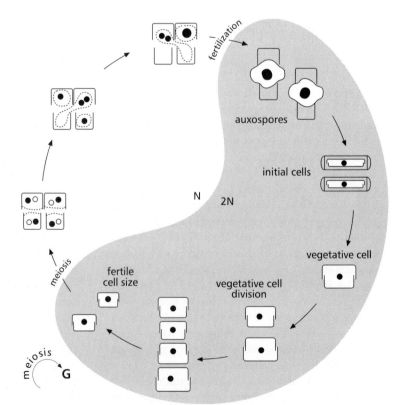

Figure 12–35 The life cycle of a typical pennate diatom. (After Hasle and Syvertsen, 1997)

flagellate amoeboid sperm are produced. Sexual reproduction in *Rhabdonema* has been suggested as a possible intermediate condition in the origin of isogamy in pennate diatoms from oogamous centric ancestors (Edlund and Stoermer, 1997), but this hypothesis requires testing. Sexual reproduction in most pennate diatoms (Fig. 12–35) begins with pairing of parental cells within a common mucilage. An unequal cytokinesis follows the first meiotic division, and the smaller of the two cells dies. The frustules then gape open, and the protoplasts begin to emerge prior to completion of the second meiotic division. Within each of the two large protoplasts one haploid nucleus resulting from the second meiotic division dies, and the other survives. Post-meiotic protoplasts thus contain a single haploid nucleus. The protoplasts then fuse, forming a binucleate cell that swells and then forms a continuous silica wall known as a **perizonium**. Nuclear fusion then occurs, restoring the diploid condition. In pennate diatoms that produce two eggs per parent cell, two auxospores are formed (Pickett-Heaps, et al., 1990). Pennate diatom mating has been described as **gametangiogamy** (mating initiated by the cells that will eventually produce gametes); a similar strategy is observed among zygnematalean green algae (Chapter 21), possibly in response to similar selective pressures (Nakahara and Ichimura, 1992).

Fossil and molecular sequence evidence indicating that centric diatoms arose prior to pennate forms suggests that in diatoms, oogamy evolutionarily preceded isogamy. This contrasts with the generally accepted view that oogamy is the more derived condition in green and brown algae. The first step in the evolution of isogamy in diatoms may have been the pairing of parental cells prior to gamete production as an adaptation that increased the frequency of fertilization (Edlund and Stoermer, 1997).

Maturing auxospores of both centric and pennate diatoms eventually produce a new two-part silica wall. Each of the two frustule components must be formed in association with a separate mitotic division. The initial epitheca is generated following a mitotic division that results in the death of one of the two daughter nuclei. A second mitosis with nuclear degeneration then results in production of the first hypotheca. Although auxospores are primarily thought to form as a result of sexual reproduction, they have also been known to arise by **autogamy**, whereby two haploid nuclei within a single cell fuse, or by **apogamy**, in which no gamete fusion occurs. Under favorable conditions, auxospores germinate by a series of divisions that ultimately generate cells having the typical frustule morphology and the maximal size for that species.

Diatom Spores and Resting Cells

Spores and resting cells serve a perennation function, i.e., they allow diatoms to survive periods that are not suitable for growth, and then germinate when conditions improve. Such periods include times of ice cover, nutrient depletion, and stratification of the water column. The development of diatom resting cells and spores does not usually involve sexual reproduction. Both resting cells and spores are characteristically rich in storage materials that supply the metabolic needs of germination. For example, the freshwater diatoms *Stephanodiscus*, *Fragilaria*, *Asterionella*, *Tabellaria*, *Diatoma*, and *Aulacoseira* produce resting cells by condensation of the cytoplasm into a dark brown mass containing many large lipid droplets and polyphosphate granules (Sicko-Goad, et al., 1986). Furnished with such provisions, resting cells of these diatoms can survive in sediments for years and perhaps decades. Some resting cells are known to require a period of dormancy before they will germinate.

Resting cells (Fig. 12–36) remain morphologically similar to vegetative cells, whereas the frustules

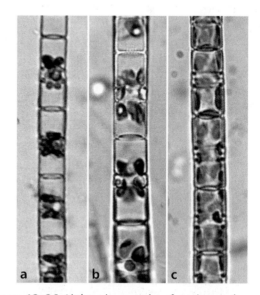

Figure 12–36 Light micrographs of resting and rejuvenating cells of *Aulacoseira granulata*. (a) Resting cells with cytoplasm aggregated in midregion. (b) Rejuvenating resting cells showing elongating plastids. (c) Fully rejuvenated cells having expanded plastids. (Micrographs courtesy L. Goad and M. Julius)

of spores (Fig. 12–37) become very thick and may assume a rounder shape and exhibit less elaborate ornamentation than vegetative cells of the same species. Hence resting cells can form under conditions of low dissolved silicate, unlike spores, which require silicate for frustule thickening. Freshwater diatoms and pennate taxa tend to produce resting cells, whereas spore production is more common by coastal marine centric diatoms. Nitrogen depletion is thought to be the inducing factor for spore development. Spores are believed to be capable of surviving for decades in benthic sediments.

If resuspended to surface waters from the sediments, resting cells and spores can germinate given sufficient light and nutrients (including vitamins). Two successive mitotic divisions are necessary to generate the normal form of the epitheca and hypotheca. Spores with a girdle germinate into two new vegetative cells that retain the spore thecae as epithecae; *Thalassiosira* is an example. Spores without a girdle (e.g., *Chaetoceros*) typically shed the spore valves during germination.

The spores of marine diatoms often occur in mucoid aggregations (marine snow), and are important in transporting organic carbon and silica to the sediments. If they are not resuspended, such stages become part of the fossil record. Spores of the common marine diatom *Chaetoceros* can account for more than 50% of the sediments in coastal regions and the northern Pacific Ocean (McQuoid and Hobson, 1996).

Aspects of Diatom Physiology and Ecology

Diatom physiology and ecology are of particular interest in the study of past and present global carbon cycling and the climatological effects of C-cycle changes (Chapter 2), aquatic food-web interactions, seafood mariculture operations, and toxin production (Chapter 3), as well as in industrial/biotechnological applications such as production of hydrocarbons (Chapter 4). The roles of diatoms in phytoplankton ecology, including light, temperature, and silicate availability effects on seasonal diatom production and successional patterns; diatom adaptations that reduce sedimentation rates; differential effects of turbulence and grazing on diatoms as compared to other algae; interspecific competition for silicate and phosphate; and the utility of diatoms in modeling phytoplankton

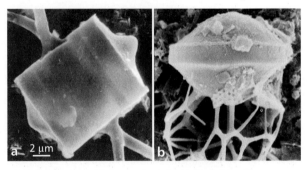

Figure 12–37 Diatom spores from Saanich Inlet, British Columbia, showing thickened frustules. (a) *Chaetoceros didymus*, (b) *C. diadema*. (Photographs courtesy M. McQuoid)

growth dynamics are discussed in Chapter 22. The role of diatoms in the structure of periphytic turf communities, and the occurrence of different diatom types in different benthic habitats are described in Chapter 23. Earlier in this chapter we have considered the adaptive advantages of diatom motility and mucilage excretions, the role of diatom mucilage in stabilizing sediments, the impact of silica concentrations on the ability of diatoms to grow and persist in aquatic habitats, and diatom reproductive ecology. Here we will focus more intensively on two aspects of diatom physiology and ecology—the relationship between diatom nutritional requirements and differences in carbon-cycle phenomena among oceans, and diatom heterotrophy, which is also relevant to carbon cycling.

Silica and Carbonate Oceans

At the present time the global carbon flux to deep ocean sediments is estimated at one billion tons per year, with a somewhat greater proportion being in the form of calcium carbonate rather than organic carbon (Honjo, 1997). Biogeochemists are interested in the rate of carbon sedimentation and differences among the world's oceans in deposition of these two forms of long-term carbon storage because they affect climate. The world's oceans can be described as either silica oceans or carbonate oceans. Silica oceans occupy only about 20% of ocean area, and include the subarctic Pacific north of 45° (Honjo, 1997) and the Southern Ocean (Falkowski and Raven, 1997). Silica oceans are characterized by sedimentation of more silica than carbonate, whereas the opposite is true for carbonate oceans. Silica deposition is based on abundant growth and sedimentation of diatom frustules, whereas carbonate deposition in open ocean regions primarily results from deposition of coccolithophorid remains (coccoliths) (Chapter 10) (although foraminiferan protists can also be important). Organic carbon is often transported together with these heavier cell surface components. Although both silica frustules and coccoliths can dissolve during transport through the water column, when vast numbers are produced, the rain of particles is so great that some survive transport to generate sediments. Further, inclusion of diatom frustules within fecal pellets and larger detrital aggregations (marine snow) protect them from dissolution, as is also the case for coccoliths (discussed in Chapters 2 and 10). Diatom frustules and coccoliths provide the ballast that is instrumental in transport of associated organic carbon to deep sediments. Determination of the composition of deep-ocean sediments (whether primarily silica frustules or carbonate coccoliths) is accomplished by microscopic examination of the contents of sediment traps (also described in Chapter 10).

Oceanographers and climatologists are interested in diatom photosynthesis and sedimentation because these processes together result in the transport of fixed CO_2 from the upper ocean waters to the deep ocean in a matter of weeks, resulting in a relatively rapid carbon dioxide drawdown that can affect atmospheric chemistry and climate on a short-term basis. Carbon dioxide drawdown is considered to occur at a slower rate in carbonate oceans, though deep carbonates are still important on a long-term basis as reservoirs of carbon that will influence the climate in future centuries (Honjo, 1997). Whether a particular ocean is a diatom-dominated silica ocean or a coccolithophorid-dominated carbonate ocean depends on nutrient availability. For example, the subarctic North Pacific Ocean is an order of magnitude richer in silicate than is the northern North Atlantic Ocean because in the former, longer-persistent upwelling phenomena recycle silica to surface waters. Thus diatoms can persist longer than in the North Atlantic, where spring diatom blooms exhaust silica resuspended only by spring circulation events, and coccolithophorids subsequently dominate.

The availability of other nutrients can also influence relative levels of diatom growth in different ocean regions. For example, diatom growth in the equatorial Pacific Ocean, the subarctic Pacific and Southern Oceans, and upwelling areas off the coast of California, is limited not by nitrate availability (as is commonly assumed for marine systems in general) but by iron. Such regions are known as high-nitrate, low-chlorophyll (HNLC) areas. This was determined experimentally by addition of iron to HNLC areas from ships or from study of naturally iron-enriched waters (described in Chapter 2). Additions of iron to ocean waters increased the abundance of large chain-forming diatoms such as *Chaetoceros* and *Coscinodiscus*-like centrics and the pennate *Pseudo-nitzschia* by six times over controls. In comparison, dinoflagellates did not respond to iron additions. Further, such iron addition experiments have revealed that iron-limited diatoms are more heavily silicified than diatoms of the same species that are not iron-limited. Thus, diatoms growing in iron-limited HNLC ocean waters are heavier, sink more readily, and hence are likely to transport more organic carbon to the sediments (Hutchins and Bruland, 1998).

Zinc is another nutrient that probably plays an important role in the abundance of diatoms (and other phytoplankton) because it is an essential cofactor in the enzyme carbonic anhydrase (Suzuki, et al., 1994). This enzyme allows many algae, including diatoms, that grow in alkaline waters to utilize bicarbonate as a source of inorganic C in photosynthesis (see Chapter 2 for more information). When zinc occurs at limiting levels, algae must rely on dissolved carbon dioxide, which occurs in only low amounts in ocean waters. Molecular sequencing studies have shown that diatoms produce a carbonic anhydrase that is distinct from that of all other algae (Roberts, et al., 1997), but the ecological implications of this finding, if any, are as yet unclear.

In ocean waters lower in nitrate than those of HNLC regions, nitrate concentrations can limit diatom growth. In the warmer waters of all oceans, multiple species of the diatom *Rhizosolenia* occur in mats that sink below the lighted zone to exploit subsurface nitrate pools, then nitrate-replete cells rise to the surface where light levels facilitate photosynthesis. Such buoyancy regulation is postulated to occur via changes in ionic balance. A comparison of chemical composition of sinking and rising mats showed that sinking mats were more nitrogen-stressed than those rising to the surface. The maximum ascent rate was 6.4 m hr^{-1}, and a complete migration cycle required 3.6–5.4 days. These diatoms provide a previously unrecognized mechanism for the transport of nitrate from deep to surface waters. It has been estimated that *Rhizosolenia* mats could convey to the surface as much as 40 μmol N m^{-2}day^{-1} (Villareal, et al., 1996). Diatoms of oligotrophic freshwaters (which are commonly limiting in phosphate) are also able to use sink-

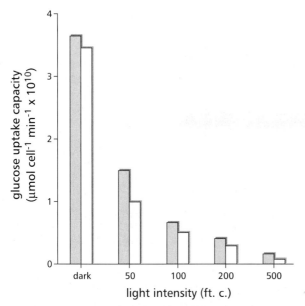

Figure 12–38 Glucose uptake by *Cyclotella cryptica*. Low illumination stimulates glucose uptake, whereas this process is reduced in the light. Pairs of bars represent replicate experiments. (From Hellebust, 1971 by permission of the *Journal of Phycology*)

ing through the water column as a mechanism for acquiring nutrients. In general, diatoms are regarded as being competitively superior to other algae for non-silicate mineral nutrients. Sinking can be regarded as an adaptation that reduces the zone of mineral nutrient depletion around phytoplankton cells in general (Sommer, 1988b), and diatoms possess the advantage of their weighted frustule.

Diatom Heterotrophy

Diatoms often occur in environments that are low in irradiance and relatively high in dissolved organic compounds. Such habitats include polar sea-ice communities, periphyton mats in shaded streams, and sediments or sand that can shift, covering the algal inhabitants. Reduced irradiance is known to induce osmotrophy (uptake of organic carbon) in a number of diatoms found in these habitats, and laboratory experiments have demonstrated their ability to grow in the dark when organic compounds are supplied (a list of 35 such diatoms and references to experimental work is provided by Tuchman, 1996). Among all algae that have been demonstrated to exhibit heterotrophy in the dark (see Tuchman, 1996), diatoms are represented more than any other group. Though

this bias may reflect the interests and prolific activity of a few researchers, it also suggests that heterotrophy may be a plesiomorphic (evolutionarily fundamental) feature of diatoms.

Many diatoms can grow as rapidly heterotrophically as autotrophically, and some species grow faster. A survey of the heterotrophic capabilities of eight benthic diatoms revealed that all were able to metabolize 94 different organic compounds, suggesting the presence of multiple types of cell membrane transport systems (Tuchman, 1996). Many marine benthic pennate diatoms migrate back and forth between the sediment-water interface where light is available and the dark, nutrient-rich sediments; such diatoms rely on heterotrophy (Lewin and Lewin, 1960). A few marine pennate diatoms, such as *Nitzschia alba*, are colorless obligate heterotrophs. Although they possess chloroplasts, photosynthetic pigments are not expressed, indicating that their obligate heterotrophy is evolutionarily derived. Such diatoms are analogous to various colorless but plastid-containing members of other protist groups, including some euglenoids, certain cryptomonads, apicomplexans (relatives of dinoflagellates), and a few green algae.

Several centric diatoms are also able to grow in the dark, using dissolved organic compounds (DOC) such as glucose. One example is *Cyclotella meneghiniana*, which grows heterotrophically in the dark if glucose concentrations between 5 mg l^{-1} and 10 g l^{-1} is provided. Light above a certain low irradiance level probably retards glucose uptake by this diatom, and 12–14 hours of darkness are required to induce glucose uptake (Lylis and Trainor, 1973). Similarly, the glucose transport system is down-regulated in highlight conditions and up-regulated in low light and darkness in the centric diatom *Cyclotella cryptica* (Hellebust, 1971) (Fig. 12–38).

Even when light is not limiting to photosynthesis, the light-independent reactions of photosynthesis (the so-called "dark reactions") are shut off and carbon-fixation ceases when sufficiently high levels of DOC such as glucose are available. This allows photosynthetic cells to use the products of the light reactions (ATP and NADPH) for energy-dependent uptake of organic carbon and for other cellular processes. The uptake of DOC in the light is known as photoorganotrophy or photoheterotrophy, and occurs in diatoms (and some other algae) occupying illuminated, DOC-rich environments such as the surfaces of dead or living animals, aquatic plants or other algae. Organic compounds are readily available as feces, decaying organic matter, or organic

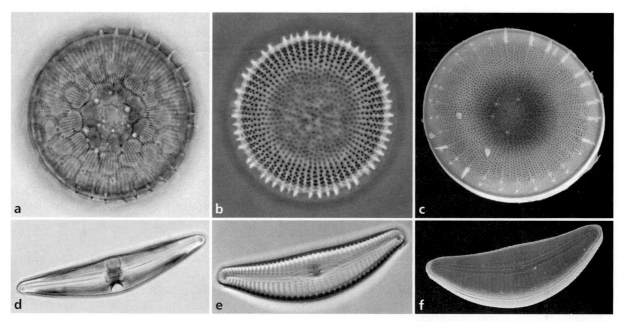

Figure 12–39 The centric diatom *Stephanodiscus* (top panel) and the pennate diatom *Cymbella* (bottom panel), viewed in three ways: (a, d) the living condition, viewed with bright-field microscopy. Note chloroplasts and lipid droplets; (b, e) after having been subjected to a cleaning process, viewed with phase or bright-field optics. Note absence of plastids and other cytoplasmic constituents; (c, f) after having been chemically cleaned and gold-coated for SEM. Each visualization process yields useful information and has advantages and disadvantages. A disadvantage of extensive processing is loss of easily damaged parts such as peripheral spines of *Stephanodiscus* (c), unless great care is taken in handling frustules. (b: Courtesy C. Taylor; f: Courtesy O. Cholewa)

exudates, the latter known to be produced by various living algae and aquatic plants. Diatoms are very common inhabitants of such surfaces, but the extent to which they use algal and plant organic exudates has not been determined (Tuchman, 1996).

Diatom Collection, Identification, and Diversity

Diatoms can be collected using plankton nets (typically with mesh size of 25 μm), water collection devices that can be remotely opened and closed, or pumps. Collections made by the latter two methods must be concentrated by allowing the diatoms to settle or by membrane filtration. They can be preserved in formaldehyde or Lugol's solution prepared as described by Hasle and Syvertsen (1997). Preservation at pH < 7 helps prevent dissolution during long-term storage. Species of such common marine phytoplankters as *Chaetoceros* and *Rhizosolenia* can be identified by gross morphology, shape of valves, and processes,

but when details of the frustule are needed, organic constituents must be removed because these interfere with visualization of fine features (Fig. 12–39a, d). The cleaning process generally separates the valves, which is also helpful in making identifications. Cleaning methods can include treatments with strong acids, solutions of potassium permanganate and acid, UV, hydrogen peroxide, or heating in a muffle furnace. The refractive index of the silica frustule is similar to that of water and glass slides and cover slips, making it difficult to see details. Use of phase-contrast, Nomarski differential interference contrast, or darkfield microscopy can be helpful. Cleaned diatoms (Fig. 12–39b, e) can be permanently mounted on glass slides by use of a mounting medium whose refractive index is higher than that of silica. Transmission or scanning electron microscopy is often required to sufficiently visualize frustule details in order to make species determinations. A drop of cleaned diatoms can be placed on a formvar-coated grid, then air-dried for TEM examination. For SEM, cleaned diatoms are dried onto a cover slip, then gold-coated (Fig. 12–39c, f). Round, et al. (1990) provide an atlas of

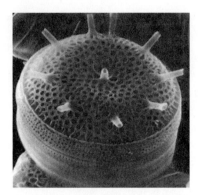

Figure 12–40 *Thalassiosira*. (Micrograph by G. Fryxell, in Bold and Wynne, 1985)

SEM images of freshwater and marine diatoms. An excellent source of information regarding identification of marine planktonic diatoms can be found in Hasle and Syvertsen (1997). Diatoms of the U.S. were covered by Patrick and Reimer (1966, 1975)

The use of molecular techniques such as RAPDs (see Chapter 5) has revealed the occurrence of geographical variants within recognized species (Lewis, et al., 1997). Such findings, together with the fact that the initial cell products following sexual reproduction and resting stages may be difficult to classify correctly, suggest that new approaches to defining diatom species may be needed. Molecular sequences offer the opportunity for precise definition of diatoms and the ability to detect rare or previously unrecognized variants in natural populations. For example, Rappe, et al. (1995) successfully used primers for SSU rDNA to amplify diatom genes from picoplankton samples taken from waters of the continental shelf off North Carolina (U.S.A.), revealing the presence of diverse taxa. These investigators were not able to identify the source of their clones because the ribosomal RNA database presently contains relatively few diatom sequences. As the database increases in size, it may be possible to circumvent many current problems in diatom identification by the use of molecular sequences or specific molecular markers. Some recent efforts to use molecular information to identify diatoms include the work of Miller and Scholin (1996, 1998), Scholin, et al. (1997), and Scholin (1998).

Currently, in systematic schemes that raise diatoms to the phylum level, centric diatoms are classified in the class Coscinophyceae, the araphid pennate diatoms into the Fragilariophyceae, and the raphid pennate diatoms into the Bacillariophyceae. In view of emerging molecular data that reveal paraphyly of the first two of these groups, we have not applied this classification system. Although there is mounting evidence for monophyly of some groups such as Thalassiosirales (Medlin, et al., 1996), until additional molecular information becomes available (allowing the erection of natural groups), we have taken a relatively informal, habitat-focused approach to the following survey of diatom diversity.

Some Marine Centric Diatoms

THALASSIOSIRA (Gr. *thalassa*, sea + Gr. *seira*, chain) occurs as cells in chains or embedded in mucilage (Fig. 12–40). Cells in chains are linked by threads extending from marginal strutted processes. Cells are about 3–5 μm or larger in diameter, and circular in valve view. Valves are ornamented with numerous areolae arranged in arcs. Areolae open externally by a large hole (**foramen**) and internally by several tiny pores. Valves bear marginal spines. Each cell has one rimoportula at the rim. There are more than 100 species.

CHAETOCEROS (Gr. *chaite*, long hair + Gr. *keras*, horn) is found as chains of cells attached by intercalary setae extending from adjacent cells and touching one another near their point of origin (Fig. 12–41). **Setae** are hollow outgrowths of the valve that project beyond the valve margin. Internally, they have a structure different from that of the valve. There are two setae per valve, formed after cell division. Their growth rates and ultrastructure were studied by Rogerson, et al., 1986. Terminal cells of the chain typically have distinctive setae. The cells are rectangular

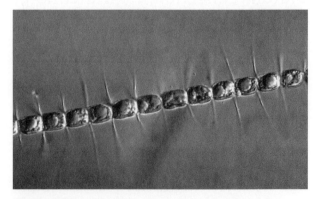

Figure 12–41 *Chaetoceros* sp. Note the distinctive elongate spinelike setae that help link adjacent cells into linear colonies. (Micrograph courtesy C. Taylor)

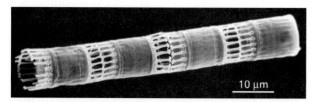

Figure 12–42 *Skeletonema* sp. Note association of cells into a linear array by means of marginal rings of tubular processes. (Micrograph by O. Cholewa)

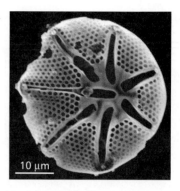

Figure 12–44 *Asteromphalus*. Note the conspicuous and distinctive raised rays; rimoportulae are located at the marginal end of each ray. (Micrograph by O. Cholewa)

in girdle view and elliptical or nearly circular in valve view. Some species of *Chaetoceros* have cells connected by silica bridges that arise following incomplete cytokinesis (Pickett-Heaps, et al., 1990). There are hundreds of described species; Rines and Hargraves (1988) is a good source of additional information.

SKELETONEMA (Gr. *skeleton*, skeleton + Gr. *nema*, thread) cells occur in chains formed by strutted tubular processes arranged in a marginal ring (Fig. 12–42). One rimoportula is found inside the ring of strutted processes or close to the valve center. Areolae are arranged radially.

TRICERATIUM (Gr. *trias*, in threes + Gr. *keras*, horn) (Fig. 12–43) is usually triangular (or sometimes quadrangular) in valve view. There are ocelli and sometimes also rimoportulae at each edge. More than 400 species have been described, mostly from coastal waters. Not all triangular diatoms are *Triceratium*, however.

ASTEROMPHALUS (Gr. *aster*, star + Gr. *omphalos*, navel) (Fig. 12–44) cells have hollow rays that are open to the valve interior as slits, and to the exterior through holes at the ends of rays. Rimoportulae occur at the marginal ends of rays. The surface of the discoid cells undulates because of the raised rays. There are more than ten species.

RHIZOSOLENIA (Gr. *rhiza*, root + Gr. *solen*, pipe) exists as single cells or chains (Fig. 12–45). Valves are small and ornamented with a spine that is open at the tip and which extends into the interior as

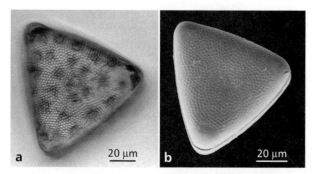

Figure 12–43 *Triceratium* sp. viewed with (a) light and (b) scanning electron microscopy. Note that ocelli at the corners are somewhat more conspicuous in the SEM view (see also Fig. 12–23). (b: Micrograph by O. Cholewa)

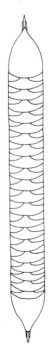

Figure 12–45 *Rhizosolenia*. (From Bold and Wynne, 1985)

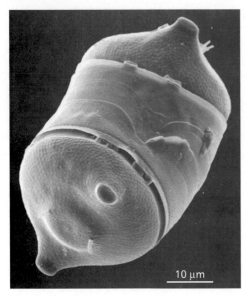

Figure 12–46 *Bidduphia*. Note the blunt processes on the valve faces. (Micrograph by O. Cholewa)

a labiate process. There is an extremely long girdle composed of multiple scalelike cingular sections. Nitrogen fixation by cyanobacterial symbionts contributes a significant amount of new nitrogen to oligotrophic waters (Martínez, et al., 1983) (see also Chapters 2 and 3). Freshwater forms have been transferred to the genus *Ursolenia*.

BIDDULPHIA (named for Susanna Biddulph, a British botanist) (Fig. 12–46), shown here as an indi-

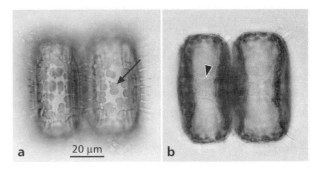

Figure 12–48 Recently divided cells of *Stephanodiscus* sp. in girdle views at two focal planes. Note the marginal spines, multiple plastids (arrow), and nucleus suspended in cytoplasm (arrowhead).

vidual cell, often occurs in zigzag chains attached to substrates such as seaweeds, but also occurring in the nearshore phytoplankton. Cells appear rectangular in girdle view, particularly when observed with the light microscope. Valves are elliptical to somewhat circular, and there are prominent blunt processes bearing pseudocelli (having areolae that decrease in diameter from the periphery toward the center). One to several rimoportulae occur near the valve center. There are internal pseudosepta (Round, et al., 1990). Some authorities cite the occasional occurrence of *Biddulphia* in freshwaters (Smith, 1950).

Figure 12–47 *Cyclotella*. (Micrograph courtesy C. Cook)

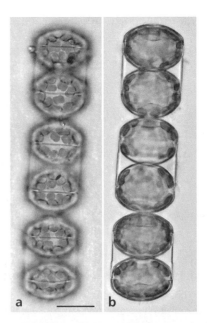

Figure 12–49 A short filament (girdle views) of *Melosira* sp. whose cells are linked by short spines. Scale bar=20 μm.

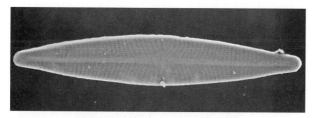

Figure 12–50 *Navicula* sp. as viewed by SEM. (Micrograph courtesy L. Goad and M. Julius)

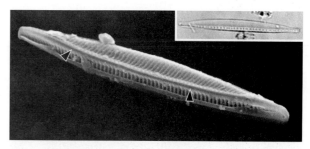

Figure 12–51 A freshwater *Nitzschia* species viewed by SEM, showing eccentric keel with raphe (arrowheads). Inset: LM view. (Micrographs courtesy L. Goad and M. Julius)

Centric Diatoms that Primarily Occur in Freshwaters

CYCLOTELLA (Gr. *kyklos*, circle) occurs as single cells, 3–5 μm in diameter (Fig. 12–47). Chitin fibrils over 150 μm long may be produced that may function in buoyancy or herbivory resistance. Rows of areolae are grouped and radiate toward the outer margin. Fultoportulae occur in a ring at the valve margin and sometimes also scattered on the valve surface. There are approximately 100 species. *C. meneghiniana* is a commonly encountered form. *Cyclotella* is unusual in its tendency to perennate in the deep chlorophyll layer of freshwater lakes (see Chapter 22 for information about such layers), rather than in benthic sediments.

STEPHANODISCUS (Gr. *stephanos*, crown + Gr. *diskos*, disk) is another single-celled diatom that produces long chitin fibrils (Figs. 12–23, 12–48). Cells are characterized by a ring of marginal spines. Fultoportulae occur at the margin, beneath some of the spines.

MELOSIRA (Gr. *melon*, apple + Gr. *seira*, chain) occurs as chains of cylindrical cells joined valve to valve (Fig. 12–49). Most of the freshwater species have been transferred to *Aulacoseira*. This includes *M. granulata*, which is a very common freshwater species in the plankton of lakes, reservoirs, and large rivers throughout the world. It has two valve types: one with long, tapering spines and straight rows of pores and a second with small spines and curved rows of pores. The short spines are important in uniting cells into long chains; the valves having them are called linking valves. In contrast, valves—known as separation valves—with long spines are involved in controlled breakage of the filaments into shorter pieces. The proportion of long-spined valves to short-spined valves determines the length of filaments in a population; filament length may be correlated with ecological variables (Davey and Crawford, 1986).

Some Freshwater or Marine Pennate Diatoms

NAVICULA (L. *navicula*, small ship) exists as single cells or ribbons of cells (Fig. 12–50). The valves are boat-shaped and each bears a raphe. *Navicula* is probably the most species-rich of all diatom genera, with nearly 2000 widely accepted species, most being bottom-dwelling forms. Some species are common on sea ice. *Navicula thallodes* is a blade-forming diatom that occurs in the Bering Sea. The blades can reach 50 cm in length, the greatest length known for colonial diatoms (Kociolek and Wynne, 1988).

NITZSCHIA (named for Christian Ludwig Nitzsch, a German naturalist) cells are linear in both valve and girdle view (Fig. 12–51). The valve face is

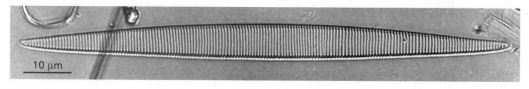

Figure 12–52 *Pseudo-nitzschia*, a toxin-producing marine diatom. (From Hasle, 1994 by permission of the *Journal of Phycology*)

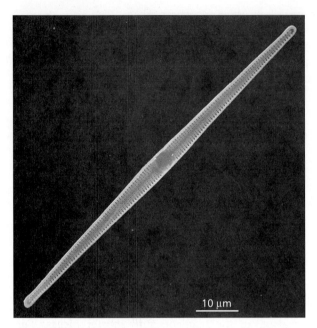

Figure 12–53 *Synedra*. (Micrograph by O. Cholewa)

marked by parallel linear striae composed of small pores. Marginal markings reveal the location of the **canal raphe.** Raphes occur on both valves, at the margin, and on the same side of the frustule. The related diatom *Hantzschia* differs in that canal raphes occur on the opposite side of the frustule.

PSEUDO-NITZSCHIA (Gr. *pseudes*, false + *Nitzschia*) was recognized as separate from *Nitszchia* by Hasle (1994). It is a marine planktonic form that occurs in colonies whose cells are overlapped at their ends. It often is weakly silicified and has an extremely eccentric raphe (Fig. 12–52). Some *Pseudo-nitzschia* species produce the toxin domoic acid (Chapter 3).

SYNEDRA (Gr. *synedria*, a sitting together) is found as single, long needle-shaped cells (Fig. 12–53),

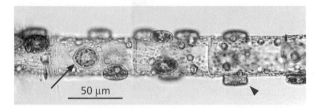

Figure 12–54 Numerous cells of *Cocconeis* attached to the green alga *Oedogonium*, in both valve (arrow) and girdle (arrowhead) views.

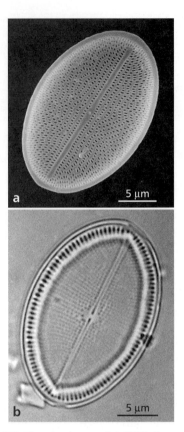

Figure 12–55 *Cocconeis* spp. (a) SEM view of the rapheless valve; (b) LM view of a cleaned raphe-bearing valve. (a: Micrograph by O. Cholewa)

or arrays of cells clustered at one pole. A marine form is known to glide, without the use of a raphe or a rimoportula, by secretion of mucilage through marginal

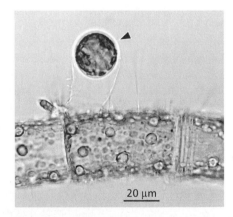

Figure 12–56 An auxospore of *Cocconeis* (arrowhead) that is adherent to a filament of the green alga *Oedogonium* by means of mucilage (the "copulation sheath") extending from the frustules of the fused cells.

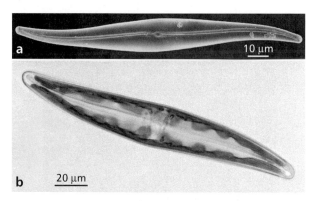

Figure 12–57 *Gyrosigma* spp. (a) Viewed by SEM and (b) in the living condition. Note that the raphe system is S-shaped, as is the frustule. (a: Micrograph by O. Cholewa)

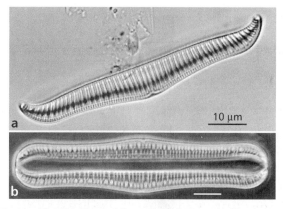

Figure 12–58 Cleaned frustules of *Rhopalodia* sp. in (a) valve and (b) girdle views (scale bar=10 μm).

grooves, in twin trails from either end of the cell. (Pickett-Heaps, et al., 1990).

COCCONEIS (Gr. *kokkos*, berry + Gr. *neos*, ship) cells are often attached to the surfaces of submerged plants and other substrata (Fig. 12–54) in marine or freshwaters. Cells are oval in valve view (Fig. 12–55). A raphe occurs only on one valve; this valve is the one appressed to substrates. Auxospores may be observed near substrates on which vegetative cells grow (Fig. 12–56).

GYROSIGMA (Gr. *gyros*, circle + Gr. *sigma*, the letter s) (Fig. 12–57) cells occur singly or within mucilage tubes. The valves are curved into an S-shape (sigmoid). There is a raphe system, also S-shaped. Two lobed plastids are present at the cell periphery. Most species are brackish in habitat, with some occurring in marine environs and some in freshwaters.

RHOPALODIA (Gr. *rhopalo*, club) (Fig. 12–58) valves are linear or curved into a strongly asymmetrical shape. A raphe system is present in an eccentric position, often raised in a keel and bordered by flanges. Cells contain a single plate-like plastid and endosymbiotic cyanobacteria.

Marine Pennate Diatom

LICMOPHORA (Gr. *likmos*, winnowing fan + Gr. *phoras*, bearing) is wedge-shaped in girdle view and oar-shaped in valve view (Fig. 12–59). The cells are typically attached to substrates such as seaweeds, marine plants, shells, stones, or animals, by branched mucilage stalks or pads (Fig. 12–59a). Mucilage is extruded through a row of slits at the frustule base. A single rimoportula occurs at one cell pole or the other. *Licmophora* is cosmopolitan in coastal areas of the world's oceans.

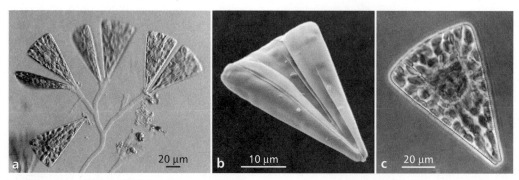

Figure 12–59 (a) A colony of *Licmophora* showing the branched mucilage stalks by which this diatom is attached to substrates. (b) SEM of *Licmophora* sp. in girdle view. (c) Live cell seen with phase-contrast light microscopy. (b: Micrograph by O. Cholewa)

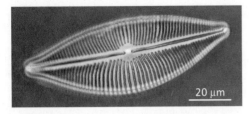

Figure 12–60 A cleaned frustule of *Cymbella* sp., in valve view. Note the bilaterally asymmetric valve and curved raphe of this form.

Some Common, Primarily Freshwater, Pennate Diatoms

CYMBELLA (Gr. *kymbe*, cup) (Figs. 12–20a; 12–39d, e, f; 12–60) often occurs at the ends of branched gelatinous stalks that are attached to submerged substrata. In stalked species, the stalk material is secreted through apical pores. A raphe is present; this is curved in species having bilaterally asymmetric valves. There is a single, H-shaped plastid, consisting of two plates joined by a bridge in which the pyrenoid is located (Fig. 12–39d). A gelatin tube-dwelling form is classified as *Encyonema*.

EPITHEMIA (Gr. *epithema*, cover) is another common epiphyte. A distinctive V-shaped raphe occurs on both valves (Fig. 12–61, 12–62). There is a single, large platelike plastid with lobed margins. A few small endosymbiotic cyanobacteria per cell is typical (Round, et al., 1990).

Figure 12–62 SEM images of *Epithemia* sp. showing primarily valve view (top) and girdle view (bottom). (Micrographs by O. Cholewa)

PINNULARIA (L. *pinnula*, small feather) cells are often quite large (Figs. 12–20b, 12–63). Frustules are elongate-oval in valve view and rectangular in girdle view. Valves are ornamented with costae that take the form of chambers within walls, with internal openings. There is a linear raphe. There are two platelike plastids or a single H-shaped plastid composed of two plates linked by a bridge. Most species are freshwater, but the genus also occurs rarely in marine habitats.

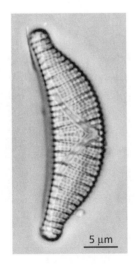

Figure 12–61 *Epithemia* sp. Valve view of a cleaned frustule showing the distinctive V-shaped raphe.

Figure 12–63 A living *Pinnularia* cell in girdle view. Scale bar=20 μm. (see Fig. 12–20b for valve view)

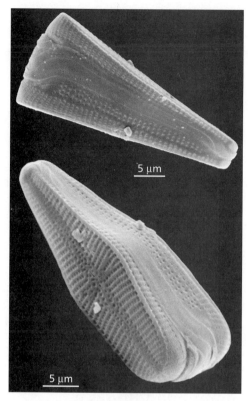

Figure 12–64 *Gomphonema* sp. viewed by SEM. (Micrographs by O. Cholewa)

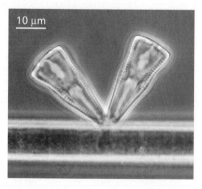

Figure 12–65 *Gomphonema* in the living condition (girdle view), attached to a substrate, as is typical.

GOMPHONEMA (Gr. *gomphos*, nail + Gr. *nema*, thread) frustules are bottle-shaped in valve view and bluntly triangular in girdle view (Fig. 12–64). Cells may be attached directly to substrates, including algae (Fig. 12–65) or via gelatinous stalks. There is a single plastid that is indented longitudinally beneath the raphe of both valves and the girdle mid-line. A single pyrenoid is present.

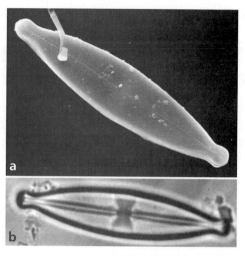

Figure 12–66 *Stauroneis* sp. (a) Valve view in SEM. (b) Valve view in LM, showing conspicuous "star," for which the genus is named. (Photographs courtesy L. Goad and M. Julius)

STAURONEIS (Gr. *stauros*, cross + Gr. *neos*, ship) (Fig. 12–66) valves are much like those of *Navicula*. This diatom occurs singly or in colonies. The pattern of valve ornamentation is interrupted by a crosspiece of non-porose silica, perpendicular to the longitudinal raphe. Together they form a cross-shaped "stauros." *Stauroneis* occurs in freshwaters, on soil, and in mosses.

CYMATOPLEURA (Gr. *kyma*, wave + Gr. *pleuron*, rib) (Fig. 12–67) has peanut-shaped valves with a raphe system that runs around the valve circumference, within a keel. Mature valve faces are undulate (wavy), and the wave pattern is bilaterally symmetri-

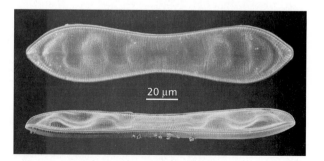

Figure 12–67 SEM images of *Cymatopleura* sp. in valve (top) and girdle (bottom) views. Note the distinctive valve surface undulations. (Micrographs by O. Cholewa)

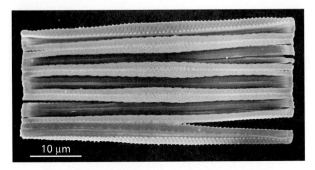

Figure 12–68 *Fragilaria* sp. SEM view of adherent cells in girdle view. (Micrograph by O. Cholewa)

Figure 12–70 *Tabellaria* sp. showing attachment of cells by means of mucilage pads (arrowheads).

cal. There is a single plastid consisting of two plates joined by a narrow isthmus.

FRAGILARIA (L. *fragilis*, easily broken) (Fig. 12–68) cells are joined to form ribbonlike colonies, presenting girdle views. There is a single rimoportula at one end of each valve. There are two platelike plastids.

TABELLARIA (L. *tabella*, small board) (Fig. 12–69) cells are joined valve to valve in short stacks that are further connected by frustule edges to form zigzag patterns. Cell-cell attachment is by mucilage pads (Fig. 12–70). There is one rimoportula per valve, located near a centrally expanded region of the valves.

ASTERIONELLA (Gr. *asterion*, a kind of spider) (Fig. 12–71) cells are elongate and joined at the ends to form star-shaped colonies. A rimoportula occurs at both ends of both valves. Short spines occur along the valve edges. Cells are held together by mucilage pads. Plastids occur as many small plates. Chytrids can often be observed attached to *Asterionella* cells.

Recommended Books

Round, F. E., R. M. Crawford, and D. G. Mann. 1990. *The Diatoms: Biology and Morphology of the Genera.* Cambridge University Press, Cambridge, UK.

Tomas, C. R. (editor). 1997. *Identifying Marine Phytoplankton.* Academic Press, San Diego, CA.

Werner, D. (editor). 1977. *The Biology of Diatoms.* University of California Press, Berkeley, CA.

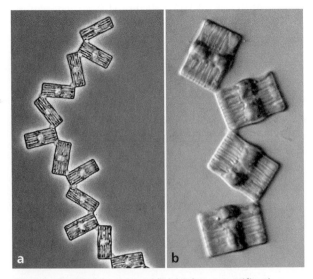

Figure 12–69 (a) Low and (b) higher magnification views of *Tabellaria* sp. colonies, with cells in girdle view. (a: Courtesy C. Taylor)

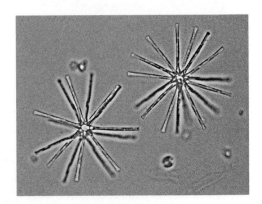

Figure 12–71 *Asterionella* sp.

Chapter 13

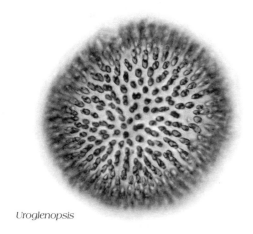

Uroglenopsis

Ochrophytes II
Raphidophyceans, Chrysophyceans, Synurophyceans, and Eustigmatophyceans

The four groups of algae covered in this chapter—raphidophyceans, chrysophyceans, synurophyceans, and eustigmatophyceans—are related to one another, as demonstrated by cellular features and SSU rDNA sequences (Cavalier-Smith and Chao, 1996). They are primarily freshwater organisms, and DNA sequence data indicate that their closest relatives are the primarily marine pelagophyceans, silicoflagellates, rhizochromulinids, and pedinellids (Cavalier-Smith and Chao, 1996), which are covered in the next chapter.

Even though most of the algae discussed here are freshwater forms, there are marine representatives of most of the component groups, and two raphidophyceans (*Chattonella* and *Fibrocapsa*) are well known for the formation of toxic marine blooms. Chrysophyceans and synurophyceans can exert significant ecological influence in oligotrophic lakes around the world. The chrysophycean *Uroglena*, for example, produces dramatic red blooms in Lake Biwa (Japan) (see review by Sandgren, 1988). Chrysophyceans and synurophyceans leave siliceous remains in lake sediments. These remains are widely used as ecological indicators in studies of past and present environmental change. The eustigmatophyceans are often regarded as a small, difficult to identify, and relatively unimportant group, but they are probably much more numerous and widespread in soils than has previously been recognized.

Many of the cellular attributes of raphidophyceans, chrysophyceans, synurophyceans, and eustigmatophyceans are shared with other ochrophytes, as discussed at the beginning of Chapter 12. Importantly, when flagellate cells are produced, they exhibit rows of tripartite hairs on one flagellum, as do other ochrophytes and heterokonts. As is more generally the case for ochrophytes, principal groups are relatively easy to distinguish on the basis of shared characteristics, but relationships of major groups to each other are more problematic. In the discussion that follows, the focus will be on (1) ways in which raphidophyceans, chrysophyceans, synurophyceans, and eustigmatophyceans are distinct from one another, (2) their evolutionary significance, and (3) their ecological roles.

269

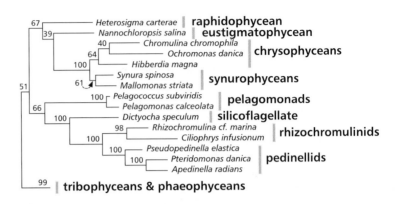

Figure 13–1 SSU rDNA sequence-based phylogeny of the primarily freshwater ochrophyte group that includes raphidophyceans, chrysophyceans, synurophyceans, and eustigmatophyceans, and shows relationships to other ochrophytes. (After Cavalier-Smith and Chao, 1996)

PART 1—RAPHIDOPHYCEANS

DNA sequence data suggest that raphidophyceans may be the earliest divergent of the algae discussed in this chapter (Cavalier-Smith and Chao, 1996) (Fig. 13–1). Commonly recognized genera of raphidophycean algae include the marine genera *Chattonella*, *Fibrocapsa*, *Heterosigma*, *Haramonas*, and *Olisthodiscus*, and the freshwater forms *Gonyostomum*, *Merotricha*, and *Vacuolaria* (Heywood, 1990; Mostaert, et al., 1998). *Haramonas* has only recently been described from mangrove swamps in Australia (Horiguchi, 1996). Raphidophyceans have also been called chloromonads, a term that is no longer in use but appears frequently in the literature. They are all relatively large (30–80 μm in their largest dimension) unicellular flagellates with a long, forwardly directed, tripartite hair-bearing flagellum and a backward-directed smooth flagellum (Fig. 13–2). Sometimes the posterior flagellum is very short and may appear to be missing altogether. As in other ochrophytes, the anterior flagellum provides most of the motive force. Raphidophycean cells "spurt" forward in a straight path or rotate. The posterior flagellum is either motionless or moves only slightly, possibly serving a steering function (Goldstein, 1992). Both flagella emerge from a shallow anterior groove (gullet) on the ventral side of the often flattened cells. In addition to flagellar-driven motility, metaboly has been observed in some forms (Horiguchi, 1996).

The cells are naked, without a cell wall or body scales (Fig. 13–3). There are no eyespots or other noticeable evidence of a photoreceptor system, such as a flagellar swelling. Several large Golgi bodies typically occur per cell. Rod-like trichocysts (also known as extrusomes) can be observed with the light microscope. These may be discharged from cells, forming long mucilaginous strands. In addition, cells contain mucocysts that can produce extracellular mucilage (Fig. 13–6). The cells occur as either motile flagellates or palmelloid groups of cells embedded in mucilage (Heywood, 1983; 1990).

The nuclei are unusually large and conspicuous (Figs. 13–3, 13–6). A fibrous flagellar root (rhizostyle or rhizoplast) contacts the nuclear envelope, as in some other ochrophytes. The periplastidal endoplasmic reticulum (PER) does not connect with the nuclear envelope. Mitosis in *Vacuolaria* begins with production of an array of microtubules from the flagellar basal body region. These microtubules extend over the nuclear envelope, and at prophase pass

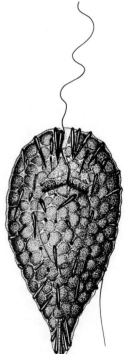
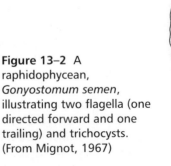

Figure 13–2 A raphidophycean, *Gonyostomum semen*, illustrating two flagella (one directed forward and one trailing) and trichocysts. (From Mignot, 1967)

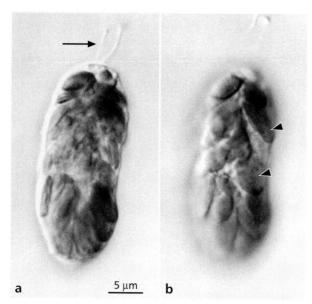

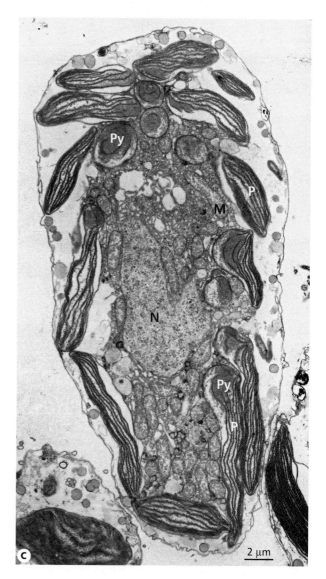

Figure 13–3 *Haramonas.* (a, b) Through-focus pair. Note two flagella (arrow) in (a) and overlapping plastids (arrowheads) in (b). (c) Transmission electron micrograph. Note that cells lack a cell wall, and that there are multiple plastids (P), each with a pyrenoid (Py). The centrally located nucleus (N) is conspicuous as are mitochondria (M). (From Horiguchi, 1996 ©Blackwell Science Asia)

through the envelope (which remains intact throughout mitosis) via polar gaps (Fig. 13–4). Cytokinesis occurs by formation of a cleavage furrow from the cell periphery (see review of algal mitosis by Sluiman, 1993). Sexual reproduction has not been reported.

Cells are photoautotrophic, typically containing multiple discoid plastids. Thylakoids occur in stacks of three as in other ochrophytes. Plastids are bright glistening green, yellow-green, or yellow-brown, and contain chlorophyll *a*, and usually both chlorophyll c_1 and c_2 (Chapman and Haxo, 1966). However, *Fibrocapsa* lacks chlorophyll c_2, and *Haramonas* does not possess chlorophyll c_1 (Mostaert, et al., 1998). The marine and freshwater forms differ in their accessory pigments. Marine species resemble chrysophyceans, eustigmatophyceans, synurophyceans, and phaeophyceans in having fucoxanthin and violaxanthin. *Fibrocapsa* and *Haramonas* contain fucoxanthin derivatives not found in the rest, and fucoxanthinol is a good pigment marker for *Fibrocapsa* (Mostaert, et al., 1998). In contrast, the freshwater forms resemble tribophyceans (xanthophyceans) in having diadinoxanthin, heteroxanthin, vaucheriaxanthin (or a deriv-

ative). Further, pyrenoids occur in the plastids of marine species but not in those of freshwater forms, and girdle lamellae are present in some, but not all, raphidophyceans. These differences in plastid structure and pigmentation have led some experts to recommend separate classification. However SSU rDNA sequence analyses unite the marine and freshwater raphidophyceans with high (100%) bootstrap values (Potter, et al., 1997b), indicating a high level of confidence (see Chapter 5). In addition, *rbcL* sequences indicate that marine and freshwater raphidophyceans are monophyletic (Daugbjerg, et al., 1997b). It is intriguing that such dramatic pigment differences occur in plastids that appear, on the basis of sequence evidence, to be closely related. Oil bodies

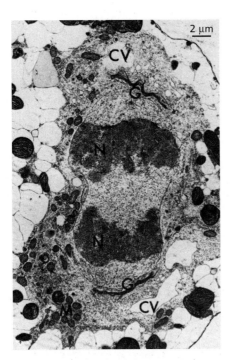

Figure 13–4 Mitosis in the raphidophycean *Vacuolaria* involves an intranuclear spindle (developed within the nuclear envelope). (From Heywood, 1983)

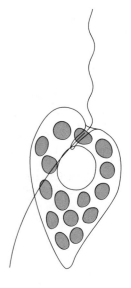

Figure 13–5 *Heterosigma akashiwo*. (After Throndsen, 1997)

represent the major photosynthetic storage product in both marine and freshwater raphidophyceans.

The marine raphidophyceans can be found in the open sea or in brackish estuaries. They usually occur infrequently but can form local blooms. *Chattonella* may bloom in organic-rich brackish waters. Red tides formed by *C. antiqua* have occurred in the Seto Inland Sea, Japan, since the 1960s, where they cause extensive losses to fisheries. Harmful blooms of this species are favored by water-column stratification when winds are weak, and by water temperature increases to 20°–22° C. This favors germination of benthic cysts (Amano, et al., 1998). *Olisthodiscus* can also form noticeable red tides. Fish mortality is associated with *Heterosigma asashiwo*, *H. carterae*, and *Fibrocapsa japonica*. Cellular features used to distinguish genera of marine raphidophyceans are summarized in Table 13–1 (from Horiguchi, 1996).

Freshwater taxa (*Gonyostomum*, *Merotricha*, and *Vacuolaria*) typically occur in waters of neutral to acidic pH where vegetation is abundant, either in the plankton or growing on mud (Heywood, 1983; 1990). *Gonyostomum* may produce local blooms. Waters impacted by acid precipitation have been reported to show an increase in numbers of *Gonyos-*

tomum cells, and bathers at beaches in Scandinavia have encountered problems with slime and skin irritation that may be associated with raphidophycean blooms (Canter-Lund and Lund, 1995).

Raphidophycean Diversity

HETEROSIGMA (Gr. *heteros*, different + Gr. *sigma*, the letter s) (Fig. 13–5) is a marine flagellate having ovoid cells with flagella inserted laterally and plastids arranged along the cell periphery. Trichocysts are not present. This marine toxic bloom-former has been the subject of physiological and molecular studies. It is an obligate photoautotroph and light changes regulate many aspects of its physiological ecology (Doran and Cattolico, 1997)

GONYOSTOMUM (Gr. *gony*, knee + Gr. *stoma*, mouth) cells are motile flagellates that are dorsiventrally compressed (Fig. 13–6). The dorsal surface is convex and the ventral surface flattened. A longitudinal ventral furrow extends from an anterior opening to a colorless, three-cornered cavity. The long flagellum projects forward, and the shorter flagellum trails posteriorly. Both flagella are about as long as the cell body (36–92 μm). The numerous discoid plastids occur at the cell periphery (Prescott, 1951). Clusters of trichocysts can occur at either end of the cells. The genus occurs in freshwater swamps, acid bogs, and other relatively low pH waters. Thiamine, biotin, and vitamin B_{12} are required for growth in culture.

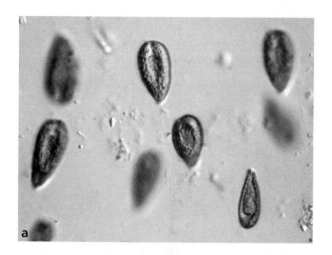

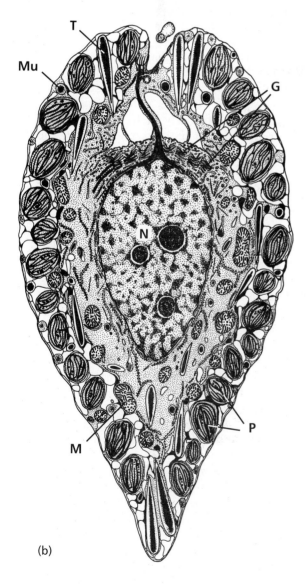

Figure 13–6 (a) *Gonyostomum* cells as seen with the light microscope. (b) Diagram of a *Gonyostomum* cell, based on TEM data. A large central nucleus (N) is present with numerous Golgi bodies (G) situated anterior to it. Peripheral cytoplasm is vacuolate with numerous plastids (P), trichocysts (T), mitochondria (M), and mucocysts (Mu). (b: From Mignot, 1967)

PART 2—CHRYSOPHYCEANS

The terms chrysophycean and chrysophyte come from the Greek word for gold, *chrysos*. Chrysophyceans are thus well named, as their plastids are golden brown, reflecting a characteristic pigment composition dominated by fucoxanthin. However other groups of algae also possess golden-brown plastids with similar pigment composition, which has led to considerable variation in the application of the more commonly used term chrysophyte. In the past some experts have applied it to haptophytes, diatoms, and other very diverse groups having golden-brown plastids that were gathered into a common taxonomic fold, the phylum Chrysophyta (Bold and Wynne, 1985). Others have included synurophyceans (synurophytes) within the chrysophytes (Sandgren, 1988). However, data obtained from ultrastructural (Andersen, 1987) pigment (Andersen and Mulkey, 1983) and molecular analyses justify a more restricted application of this term. For example, in SSU rDNA sequence-based phylogenetic analyses, the branch separating chrysophyceans (as defined below) from synurophyceans is supported by 100% bootstrap values (Cavalier-Smith and Chao, 1996). Here, we use the vernacular term "chrysophycean" (rather than chrysophyte, in order to reflect the class level status of the group) for those ochrophytes having chlorophylls *a*, c_1, and c_2 and fucoxanthin as well as a characteristic silica-walled resting stage known as a **stomatocyst** (statospore or siliceous cyst). The term stomatocyst comes from the Greek word *stoma*, meaning mouth, reflecting the occurrence of an opening that is unique to this type of resting cell.

Pigments

Chlorophylls c_1 and c_2 can be separated and identified by thin-layer chromatography. Chlorophyll c_1 is magnesium tetradehydropheoporphyrin a5 monomethyl ester, whereas chlorophyll c_2 is magnesium hexadehydropheoporphyrin a5 monomethyl ester (reviewed by

Table 13–1 Comparison of cellular features of Raphidophycean genera

	Chattonella	*Heterosigma*	*Olisthodiscus*	*Fibrocapsa*	*Haramonas*
cell shape	club-shaped, spherical	ovoid, obovate	ellipsoid	ovoid, spherical	club-shaped
flagellar insertion	apical	lateral	lateral	apical	apical
plastid arrangement	+/– radial	peripheral	peripheral	peripheral, forming superficial network	peripheral, imbricated
thylakoid arrangement	perpendicular to cell surface	parallel to cell surface	parallel to cell surface	parallel to cell surface	parallel to cell surface
thylakoids entering pyrenoid	present/absent	present	absent	present	present
cytoplasmic tubules entering pyrenoid	absent	absent	present	absent	absent
trichocysts	present/absent	absent	absent	present	absent
tubular invagination	absent	absent	absent	absent	present

Andersen and Mulkey, 1983). Important accessory and protective pigments in the chrysophyceans are fucoxanthin and violaxanthin. Although analysis of sedimentary pigments has sometimes been used in attempts to deduce presence and relative past importance of chrysophyceans and other algal groups, careful HPLC (high-performance liquid chromatography) studies of pigment transformations during sedimentation and burial have shown that this is not always a valid approach. Pigments such as fucoxanthin (and peridinin) that have 5,6-epoxide groups can be broken down to form colorless, undetectable compounds before they reach the sediments. Chrysophyceans, synurophyceans, diatoms, and other fucoxanthin-rich algae (as well as the peridinin-containing dinoflagellates) are thus severely underestimated by sedimentary pigment analysis. Other pigments are more stable, but their occurrence in the sediments does not accurately reflect phytoplankton community composition present at the time of deposition (Hurley and Armstrong, 1990). Fossil stomatocysts provide a better record of past chrysophycean occurrence, as these more closely reflect the composition of living populations (Smol, 1995).

Stomatocysts

Stomatocyst walls are heavily silicified and thus resist silica dissolution processes operating in the benthos (Fig. 13–7), particularly in freshwater lakes. The occurrence of stomatocysts in the fossil record, extending back to the Upper Cretaceous (about 80 million years ago), suggests that chrysophyceans have existed for at least this length of time (Kristiansen, 1990). Although these ancient deposits are primarily marine, modern stomatocyst deposition occurs principally in freshwater lakes. It is not known whether the marine deposits resulted from marine chrysophyceans that have since become extinct or freshwater forms whose remains were transported to the oceans.

In freshwater systems stomatocysts are widely used as bioindicators to deduce environmental conditions at the time of their deposition in the sediments (under the assumption that the sediments have been laid down in an orderly manner and have not been subsequently disturbed). Calibration data sets are first constructed; these represent statistical correlation between patterns of abundance of taxa in recent sediments and present-day physical, chemical, and biological factors. The calibration data sets are then used to deduce environments associated with the older deposits, samples of which are obtained using coring devices. The resulting data have been used to trace the onset of biotic effects resulting from acid rain and eutrophication impacts on freshwaters (Smol, 1995; Zeeb, et al., 1996).

A recent analysis of arctic peat samples, going back in age more than 7000 years, revealed 161

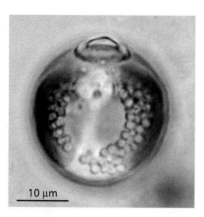

Figure 13–7 Stomatocyst from a sample collected in northern Wisconsin, U.S.A.

morphologically distinct stomatocyst types, including 52 that were previously unknown (Gilbert, et al., 1997). In modern high arctic ponds chrysophycean cysts are more commonly found attached to mosses or rocks than in surface sediments, and such periphytic chrysophycean communities are richer in diversity than are sediments derived from plankton (Wilkinson et al., 1986). A study of chrysophycean cysts from tropical lake sediments also revealed a high percentage of previously undescribed stomatocyst types (Zeeb, et al., 1996), as have analyses of very deep waters of Lake Baikal, Siberia (Sandgren, et al., 1996; Vorobyova, et al., 1996).

Ornamentation of stomatocysts is species specific, and there are a large number of stomatocyst types. Scanning electron microscopy (SEM) of stomatocysts cleaned with hydrogen peroxide is usually required to visualize these differences in adequate detail (Cronberg, 1995). Most stomatocysts are spherical, usually 2–30 μm in diameter, with a single pore whose margin is often elevated into a collar. An atlas of chrysophycean cysts (which describes more than 240 types of stomatocysts) has been prepared by Duff, et al. (1995). Unfortunately, only about 10% of the cyst types found in sedimentary deposits have been linked to well-defined taxa. Further laboratory studies are needed on the stomatocysts produced in unialgal chrysophycean cultures (Smol, 1995). Culture recommendations are provided by Kristiansen (1990).

Stomatocysts can result from asexual or sexual reproduction. For an individual species, statospores formed asexually or sexually are indistinguishable (Sandgren, 1988). Sexual reproduction involves fusion of isogametes that are structurally similar to vegetative cells. In *Dinobryon*, the female strain produces a pheromone that attracts a male flagellate cell to swim to the vicinity of a female cell. The zygote resulting from syngamy loses its flagella and forms a stomatocyst that remains attached to the colony in the place of the cell that served as the female gamete (Sandgren, 1981) (Fig. 13–8). Whether cysts arise sexually or asexually, the developmental process is the same. The cyst wall is produced inside the cell membrane, within an extensive peripheral vesicle formed by coalescence of Golgi vesicles, known as a silica deposition vesicle (SDV). Similar SDVs occur in diatoms (Chapter 12) and synurophyceans. A region at the apex remains unsilicified; this "pore" becomes plugged with polysaccharide when stomatocyst maturation is complete. Stomatocyst formation does not appear to be correlated with any particular environmental factor but rather with cell density, suggesting production of chemical inducers. Cyst germination occurs continually, with small numbers of cysts germinating at any one time (see review by Sandgren, 1988). Germination involves dissolution of the pore plug, often mitotic division of the cytoplasmic contents to form two or four initial cells, and development of flagella.

Figure 13–8 Stomatocysts of *Dinobryon* collected in February from beneath the ice of a pond in southern Wisconsin, U.S.A.

Cell Biology

In general, chrysophyceans are characterized by the following cellular features (in addition to fucoxanthin and stomatocysts): typical heterokont flagella with transitional helices, a photoreceptor system consisting of a swelling on the smooth flagellum coupled with a plastid eyespot, open mitosis (the nuclear envelope breaks down) with spindles organized by fibrillar rhizoplasts, isogamous sexual reproduction with (probably) zygotic meiosis, and water-soluble (β-1,3-linked glucan) chrysolaminarin storage in vacuoles. The cellular structure of *Ochromonas* has been particularly well studied (Bouck and Brown, 1973) (Fig. 13–9), and serves as a model for that of other chrysophyceans. Mitosis and cytokinesis have been studied in the large colonial *Hydrurus* (Vesk, et al., 1984); differences between mitosis in *Hydrurus* and *Ochromonas* (Fig. 13–10) are said to result from presence of a confining extracellular matrix in *Hydrurus*.

Nutrition

One of the most interesting features of chrysophyceans is the common occurrence of mixotrophy—utilization of dissolved organic compounds and/or particulate food by plastid-bearing algae. Some chrysophyceans may be quite nutritionally opportunistic, switching between photoautotrophy, mixotrophy, and heterotrophy depending upon cellular conditions and environmental circumstances (Sandgren, 1988). There is also wide variation in the degree to which chrysophyceans rely upon photoautotrophy, osmotrophy, or phagotrophy. Certain *Ochromonas* species and *Spumella* (a colorless form of *Ochromonas*) are strict phagotrophs. Other *Ochromonas* species and *Uroglena* are both obligately phagotrophic and photoautotrophic (i.e., particulate food cannot be ingested unless light is also present). *Dinobryon* has plastids but is facultatively phagotrophic, and *Chrysamoeba* and *Ochromonas danica* seem to be facultatively photoautotrophic and facultatively phagotrophic. *Poteri-oochromonas*, though capable of photoautotrophy, depends on photosynthesis only when the supply of particulate food is low. This species so strongly tends toward phagotrophy that the organism will engage in cannibalism rather than rely upon autotrophy (Caron, et al., 1990). The genera *Catenochrysis*, *Chromulina*, *Chrysococcus*, *Chrysosphaerella*, *Chrysostephano-*

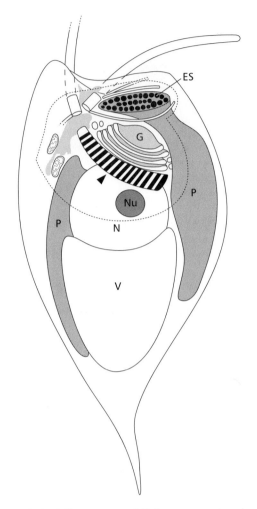

Figure 13–9 Cell structure of *Ochromonas*, showing the relative positions of the posterior vacuole (V), central nucleus (N) with nucelolus (Nu), lateral plastids (P) with apically positioned eyespot (ES), and anterior Golgi (G). The flagellar apparatus includes a robust, banded rhizoplast (arrowhead). (After Bouck and Brown, 1973. Reproduced from *The Journal of Cell Biology* (1973) 56:340–359 by copyright permission of the Rockefeller University Press.)

sphaera, and *Phaeaster* are other plastid-bearing chrysophyceans that are known phagotrophs. Phagotrophic chrysophyceans consume bacteria, yeasts, small eukaryotic algae, and nonorganismal particulate foods such as starch grains (see review by Holen and Borass, 1995). *Paraphysomonas imperforata* is omnivorous, capable of consuming algal prey such as the diatom *Phaeodactylum tricornutum* and the green flagellate *Dunaliella tertiolecta* (Caron, 1990). Members of the chrysophycean clade usually should be regarded as

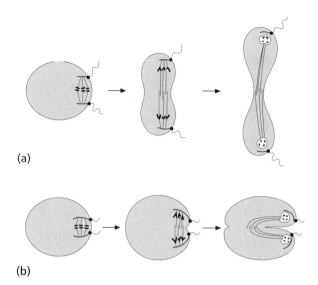

(a)

(b)

Figure 13–10 Mitosis and cytokinesis in the chrysophyceans (a) *Ochromonas* and (b) *Hydrurus*. While mitosis occurs similarly, cytokinesis appears to be more spatially constrained in *Hydrurus* than in *Ochromonas*. (After Vesk, et al., 1984 by permission of the *Journal of Phycology*)

fundamentally capable of mixotrophy under at least some conditions.

Bacteria-eating (bacterivorous) chrysophyceans are particularly important in oligotrophic lakes, where they are the primary consumers of prokaryotes. These include four species of *Dinobryon* and two species of *Uroglena*, both common colonial phytoplankters (Bird and Kalff, 1986). An average of three bacteria were consumed by each *Dinobryon* cell every five minutes—a rate approaching that of marine microflagellates that are incapable of photosynthesis. In oligotrophic lakes *Dinobryon* can remove more bacteria than crustaceans, rotifers, and ciliates combined. The maximal rate of bactivory for *Ochromonas danica* and another species of *Ochromonas* has been measured at 182–190 bacteria consumed cell⁻¹ hr⁻¹ (Holen and Borass, 1995). Populations of chrysophyceans such as *Dinobryon* can occupy lake waters as deep as seven meters, forming extensive metalimnetic growths. In the dimly lit metalimnion, they obtain about 80% of their total carbon from phagotrophic activity (Bird and Kalff, 1986). Phagotrophy provides not only fixed carbon, but also phosphorus—which is limiting to growth in oligotrophic waters—as well as vitamins. Most chrysophyceans are auxotrophic, requiring several B vitamins (Holen and Borass, 1995). Chrysophyceans also have an unusually high requirement for iron, used for syn-

thesis of an essential cytochrome involved in electron transfer between cytochromes b–f and photosystem I in photosynthesis. Phagotrophy could be useful in harvesting this nutrient as well (Raven, 1995). Rothhaupt (1996a, b) studied an *Ochromonas* species that requires high bacterial densities to reach maximal growth rates. When phagotrophy is the dominant mode of nutrition, it is a source of soluble reactive phosphate (SRP) that can be used by other algae, but when growth is primarily photoautotrophic, it is a sink for SRP. Competition experiments revealed that nutritional flexibility has a price; its phototrophic growth rates were lower than those of obligate photoautotrophic algae of comparable size, and its metabolic costs were higher than those of obligately phagotrophic flagellates.

Phagotrophic cells may attach to particulate food via anterior pseudopodia, then use a distinctive looplike microtubular root to consume prey within a feeding cup. Many phagotrophic chrysophyceans are wallless, though some (such as *Dinobryon* and *Lagynion*) may occur in organic, open vaselike loricas. The food ingestion apparatus is thus unencumbered by presence of a cell covering. Some *Chromulina* and *Phaeaster* species have platelike, organic body scales (Wujek, 1996), but the extent to which these may interfere with phagotrophic behavior is unclear. The surface of *Paraphysomonas* (some species are known phagotrophs) is covered with silica scales that apparently do not interfere with phagotrophy.

Bloom Formation

Several chrysophyceans are associated with the formation of noticeable undesirable blooms. These may have direct effects on lake biota—such as toxin poisoning or clogging of fish gills—or their effects may be indirect, involving changes in food web structure or decreases in dissolved oxygen. *Uroglena volvox* produces toxic fatty acids that affect fish. Living cells of *Uroglena* and *Dinobryon* (together with the synurophyceans *Mallomonas* and *Synura*) can excrete aldehydes and ketones (mainly n-haptanal) into the water, which can give it an unpleasant taste and odor (reviewed by Nicholls, 1995).

Habitat Preferences

Chrysophyceans typically favor slightly acid, soft (low alkalinity and conductivity) waters of moderate to low productivity. Preference for low pH waters has been

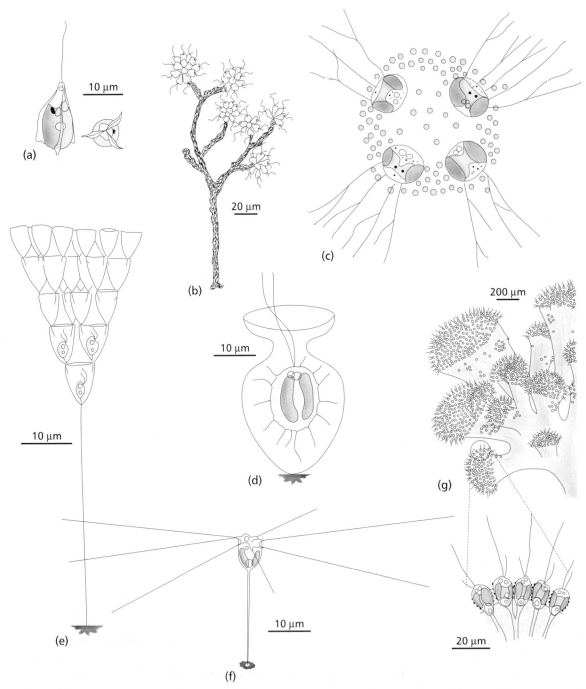

Figure 13–11 Diversity of chrysophyceans. This group includes many unique and interesting morphologies. Some examples include: (a) *Pyramidochrysis splendens*, a flagellate with distinctive winged cell covering; (b) the sessile, colonial form *Anthophysa vegetans*; (c) *Chrysostephanosphaera globulifera*, a colony of nonflagellate cells that have distinctive projections and occur in a granular matrix; (d) *Derepyxis crater*, whose cells possess what appear to be two equal flagella and are enclosed in a sessile lorica; (e) *Stylobryon*, with a colonial structure somewhat resembling that of the more familiar *Dinobryon*, but attached to substrates by a long stalk; (f) *Rhizaster crinoides*, a nonflagellate unicell that is attached by means of a stalk and possesses long cellular extensions; and (g) *Mycochrysis oligothiophila*, a large colonial aggregate of *Ochromonas*-like cells with a branched gelatinous stalk, said to have the appearance of a miniscule mushroom. (After Bourrelly, 1968)

related to their production of acid, but not alkaline, phosphatases. These enzymes are released into the water where they function to liberate phosphate from organic compounds, an ability advantageous in low nutrient waters. Chrysophycean abundance and species richness increase concomitantly with lake trophic status up to the point of slightly eutrophic conditions, and then decrease dramatically in more eutrophic waters (Elloranta, 1995). This decrease may be related to differences in herbivore presence and numbers, competition with other algae, or water chemistry. Few physiological assessments of the ability of chrysophyceans to use bicarbonate have been made. The old concept that chrysophyceans prefer cold waters (this seemed to explain their common occurrence in early spring, in deep cold layers of temperate lakes, and year-long presence in polar lakes) has been overturned by results of more extensive surveys showing that diverse chrysophycean floras can be present in warmer waters throughout the world (Siver, 1995). Water chemistry and herbivory are probably the most important factors explaining the distribution of planktonic chrysophyceans (Sandgren, 1988).

Morphology and Diversity

Chrysophyceans may occur as unicellular flagellates or colonies of flagellate cells. In addition, some taxa occur as nonmotile unicells, colonies, or rhizopodial forms that can produce flagellate zoospores. Although a number of filamentous chrysophyceans have been described, recent molecular data suggest that some (and perhaps most or all) of these are more closely related to tribophyceans and phaeophyceans (see Chapter 15). Cells of the rhizopodial species are amoeboid and generally have long, thin cytoplasmic extensions (rhizopodia). A very large array of strange and beautiful chrysophyceans have been described (Fig. 13–11), many apparently having been seen only rarely. A number of chrysophyceans closely resemble various green algae (except in color). Examples include *Chrysochaete*, a flat, circular form having gelatinous hairs that resembles the green alga *Dicranochaete*, and *Phaeoplaca*, whose sarcinoid colonies are much like those of *Chlorosarcina* (Dop, 1978). Colored iron and manganese precipitates may occur on stalks, loricas, and gelatinous extracellular matrices of chrysophyceans (Leadbeater and Barker, 1995). The generally delicate chrysophyceans require special attention in terms of collection and chemical preservation. While some occur in the phytoplankton and can thus be sampled by the use of nets, pumps, or water sampling devices, much of chrysophycean diversity is found among the periphyton—communities that occur on the surfaces of aquatic plants, rocks, or other solid substrates. Buffered glutaraldehyde in low concentration works better than some other commonly used preservatives. Drops of preserved, scaly chrysophyceans can be dried onto formvar-coated grids or cover slips for examination of the scales and flagella by transmission or scanning electron microscopy; this is often necessary for taxonomic determinations, particularly at the species level.

In the past, thallus morphology has been used to group chrysophyceans (and other ochrophytes) taxonomically, but more recent ultrastructural and molecular studies suggest that there is not always a strong correlation between thallus morphology and evolutionary relationship. For example, because the three genera *Chrysamoeba*, *Rhizochromulina*, and *Lagynion* all produce rhizopodia, they were at one time grouped together into the Chrysamoebales (Kristensen, 1990). However ultrastructural and molecular analyses showed that, although not closely related to each other, *Chrysamoeba* and *Lagynion* are allied with chrysophyceans (as they are defined in this book). *Rhizochromulina*, however, is more closely related to a second clade of ochrophytes (one that includes pedinellids, silicoflagellates, and pelagophyceans) (O'Kelly and Wujek, 1990). *Rhizochromulina* is described in Chapter 14. This is an example of parallel or convergent evolution, a phenomenon that has occurred widely among algae, generating many similar-appearing forms that are not closely related. Ultrastructural examination and molecular sequencing are necessary to sort out such relationships. The history and modern concepts of chrysophycean classification based on structural characters have been reviewed by Preisig (1995). In view of the fact that relatively few chrysophyceans have been examined by modern methods, we have taken an informal organizational approach in surveying chrysophycean diversity.

Examples of Chrysophycean Diversity

CHROMULINA (Gr. *chroma*, color) (Fig. 13–12) is a wall-less unicellular flagellate having only one emergent flagellum, of the anterior, tripartite hair-

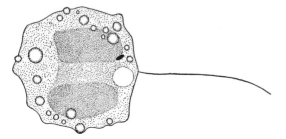

Figure 13–12 Drawing of *Chromulina*. (From Smith, G. M. *Fresh-Water Algae of the United States*. ©1950 McGraw-Hill Companies. Reproduced with permission of the McGraw-Hill Companies)

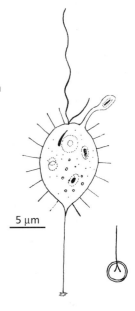

Figure 13–14 *Paraphysomonas*. Vegetative cells are covered with a layer of spined scales, one of which is shown separately. (From Bourrelly, 1968)

5 µm

bearing type. One or two golden plastids are present. There are over 100 species, some of which are marine and others freshwater.

OCHROMONAS (Gr. *ochros*, pale yellow + Gr. *monas*, unit) is a wall-less flagellate with one or two golden-brown, platelike plastids (Fig. 13–13). There are two unequal flagella of the typical heterokont type. An anterior eyespot contained within a plastid and contractile vacuoles may be visible at the anterior. The posterior portion of the cell typically comes to a point. *Ochromonas* cells may contain the remains of ingested particles such as algal cells. *Ochromonas* occurs most abundantly in oligotrophic freshwaters, but some marine forms are known (Bold and Wynne, 1985). Stomatocysts are often found. There are about

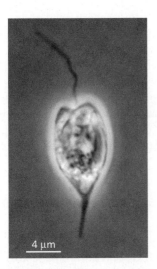

Figure 13–13 *Ochromonas*. Only one of the two flagella is visible.

80 species of *Ochromonas*. Colorless forms are known as the genus *Spumella*.

PARAPHYSOMONAS (Gr. *para*, near + *Physomonas*, a genus of flagellates [Gr. *physa*, bellows + Gr. *monas*, unit]) is a single-celled flagellate whose surface is covered with silica scales (Fig. 13–14). There are two typical heterokont flagella. Although it is not pigmented, there is a leucoplast (a colorless plastid), indicating that *Paraphysomonas* is derived from an ancestral form that was pigmented. It is osmotrophic or phagotrophic. There are about 50 species, whose identification requires electron microscopic examination of scale structure. *Paraphysomonas* is of common occurrence in tropical freshwaters (Wujek and Saha, 1995).

DINOBRYON (Gr. *dineo*, to whirl + Gr. *bryon*, moss) is a colonial organism; each cell has two heterokont flagella that contribute to the motility of the whole colony, which swims slowly. Each cell is contained in a vase-shaped organic lorica. Cells are attached to the base of their lorica by a thin cytoplasmic thread. Loricas are arranged to form a dendroid (tree-shaped) colony. This colony structure arises as the result of longitudinal mitosis and subsequent migration of one of the daughter cells to the lorica aperture, where it attaches and generates a new lorica. This process can be repeated, so that two daughter cells protrude from the parental lorica. Thus one or

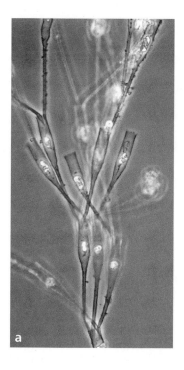

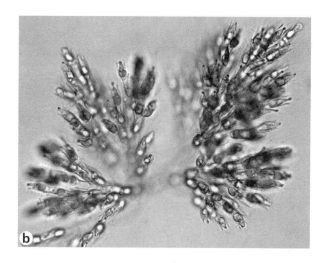

Figure 13–15 Two species of *Dinobryon*, illustrating the treelike arrangement of the flagellate cells in the colony, as viewed by light microscopy.

two loricas can be observed to extend from the mouth of the lorica just posterior in the colony (Figs. 13–15, 13–16). Loricas can be colorless or brownish in color, and the shapes of loricas vary among species. Cells contain one or two brown plastids. An eyespot and two contractile vacuoles can usually be detected at the apical end of the cell. Colonies exhibit phototaxis (Canter-Lund and Lund, 1995). A chrysolaminaran vacuole occurs in the cell posterior. *Dinobryon* is widely recognized as a bacteria-consuming mixotroph. Asexual reproduction occurs when cells swim out of their lorica and start new colonies. Sexual reproduction involves isogametes and formation of statospores that are attached to colonies (Sandgren, 1981) (Fig. 13–17). Stomatocysts of *Dinobryon* typically contain a large chrysolaminarin vacuole. *Dinobryon* may occur in the plankton of either soft or hard (high alkalinity) freshwaters. Some species grow in the mucilage of cyanobacterial colonies or on the frustules of the diatom *Tabellaria* (Prescott, 1951). *Dinobryon* can be very common in the plankton of temperate and boreal lakes that are poor in nutrients and not very alkaline. It is also common in mountainous regions and in

pools in areas where the soil is acidic (Canter-Lund and Lund, 1995). Environmental concentrations of potassium can be toxic to a number of freshwater *Dinobryon* species, limiting their seasonal and geographical occurrence. *Dinobryon sertularia* is limited to waters where temperatures are no higher than 20° C (Lehman, 1976). *Dinobryon balticum* is a dominant

Figure 13–16 Colonial habit of *Dinobryon* viewed by SEM. The cells (whose flagella have been lost) are visible through the electron-translucent loricas.

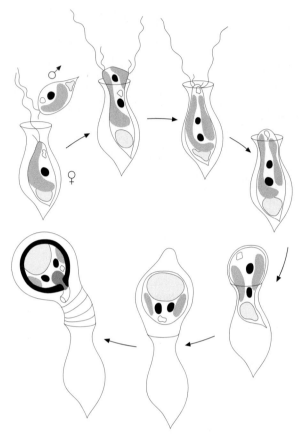

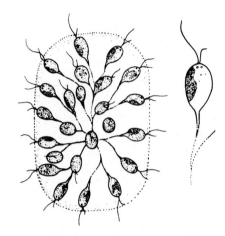

Figure 13–18 *Uroglena*. (From Prescott, G. W. 1978. *How to Know the Freshwater Algae*. Wm. C. Brown Co. reproduced by permission of The McGraw-Hill Companies)

Figure 13–17 The sexual reproductive process in *Dinobryon*. Fertilization occurs within the lorica of one gamete (termed the female) by a gamete that has left its lorica (termed the male). The resulting zygote emerges from the lorica and forms a stomatocyst at the lip. (After Sandgren, 1981 by permission of the *Journal of Phycology*)

cells arranged in a ring held together by mucilage (Fig. 13–19). The colony rotates as it swims. It occurs in freshwater ponds and pools, but is rarely observed (Canter-Lund and Lund, 1995).

UROGLENOPSIS (*Uroglena* + *Gr. opsis*, likeness) is a motile, spherical, mucilaginous colony of hundreds of ovoid cells evenly distributed along the periphery (Fig. 13–20). Colonies can reach 500 µm in diameter. It resembles the green colonial flagellate *Volvox*. Individual cells are 3–7 µm in diameter and have one or two golden-brown plastids shaped like plates or disks. Each cell has two heterokont flagella

species in cold North Atlantic coastal marine waters, for which phagotrophy has been demonstrated. It is associated with bacteria-rich fecal pellets and other detrital aggregations and is probably bacterivorous (McKenzie, et al., 1995).

UROGLENA (Gr. *oura*, tail + Gr. *glene*, eye) consists of globose motile colonies of wall-less cells held at the ends of mucilaginous threads arising from the colony center (Fig. 13–18). Cells exhibit eyespots and typical heterokont flagella, as well as golden plastids. *Uroglena* commonly occurs together with *Dinobryon* and, like *Dinobryon*, consumes bacterial cells.

CYCLONEXIS (Gr. kyklos, circle + L. *nexis*, act of swimming) is a motile colony of golden-pigmented

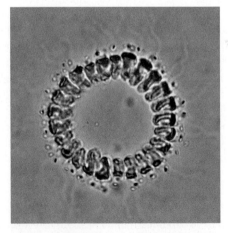

Figure 13–19 *Cyclonexis* is a relatively rare, ring-shaped swimming colony.

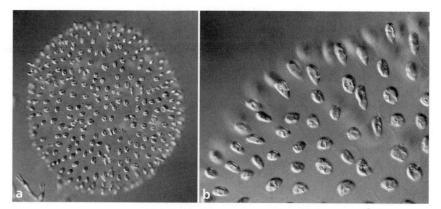

Figure 13–20 (a) *Uroglenopsis* colonies contain hundreds of cells, (b) each of which has flagella, and hence superficially resembles nonreproductive colonies of the green alga *Volvox*. (Micrographs courtesy C. Taylor)

that extend through the mucilage and contribute to colony motility. *Uroglenopsis* can be a common component of the phytoplankton in late summer, especially in lakes of fairly high alkalinity (Prescott, 1951).

CHRYSOSTEPHANOSPHAERA (Gr. *chrysos*, gold + *Stephanosphaera*, a genus of green algae) (Fig. 13–21) occurs as motile colonies of 2–16 ovoid cells. Colonies may be spherical or disk-shaped. The cells are enclosed in a distinctive mucilaginous matrix that is densely populated with particles. Cells are about 10–12 μm in diameter and have pseudopodial extensions of cytoplasm, as do those of the colonial green flagellate *Stephanosphaera*. There are two peripheral platelike plastids. Reproduction is by lon-

gitudinal division of the constituent cells and fragmentation of the colony. *Chrysostephanosphaera* occurs in shallow, low-pH waters of bogs and swamps (Prescott, 1951).

LAGYNION (L. *lagenion*, small flask) (Fig. 13–22) includes epiphytes whose cells are enclosed by a flask or bottle-shaped, transparent or, more often, brown lorica of organic composition. A rhizopodial extension of the protoplast may protrude through the narrowed lorica neck. Flagellate zoospores may be produced; when they settle onto a substrate, they first produce several robust rhizopodia, then generate the lorica (O'Kelly and Wujek, 1995). The lorica base is typically flattened against the substrate (usually filamentous algae such as the tribophycean *Tribonema* or the very similar-appearing chlorophycean *Microspora*) in soft-water lakes or bogs, especially in areas rich in organic material (Prescott, 1951).

CHRYSAMOEBA (Gr. *chrysos*, gold + Gr. *amoibe*, change) is a globose or discoid amoeboid cell with an irregular cell outline and radiating rhizopo-

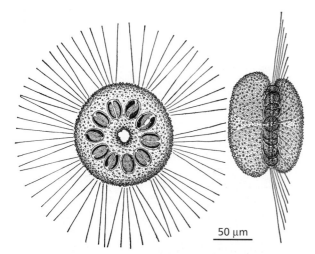

Figure 13–21 Two views of *Chrysostephanosphaera*. (From Smith, G. M. *Fresh-Water Algae of the United States.* ©1950 McGraw-Hill Companies. Reproduced with permission of the McGraw-Hill Companies)

50 μm

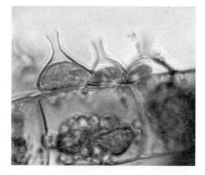

Figure 13–22 *Lagynion.*

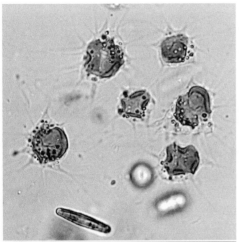

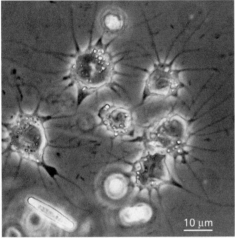

Figure 13–23 *Chrysamoeba*, viewed with (a) bright-field optics, which reveal plastids effectively, and (b) phase optics, where the colorless rhizopodia are better seen.

dia that can vary in length by active extension and retraction (Fig. 13–23). There may be a thin mucilage investment. Cells can become motile by production of a single emergent flagellum. Flagellate cells look very much like *Chromulina* (Canter-Lund and Lund, 1995). There are two platelike golden plastids per cell. Cells may occur singly or as aggregates in the plankton of freshwater lakes and in muddy areas, but is of rare occurrence (Prescott, 1951). *Chrysamoeba pyrenoidifera* (having pyrenoids that are visible with the light microscope) occurs within the hyaline cells of *Sphagnum* moss gametophores (O'Kelly and Wujek, 1995).

CHRYSONEPHELE (Gr. *chrysos*, gold + Gr. *nephele*, cloud) (Fig. 13–24) is a hollow, mucilaginous

colony including hundreds to thousands of cells, each with functional flagella. Although the flagella beat, the colony is not motile. The cells have typical chrysophycean eyespots, plastids, contractile vacuoles, and chrysolaminarin vacuoles (Pipes, et al., 1989). *Chrysonephele palustris* is known from only one freshwater swamp in central Tasmania and is thus highly endemic. Although suggested to be a link between chrysophyceans and eustigmatophyceans on the basis of ultrastructural similarities in the flagellar photoreceptor, SSU rDNA sequence data discount this possibility (Saunders, et al., 1997b).

CHRYSOCAPSA (Gr. *chrysos*, gold + L. *capsa*, box) is a planktonic colony as large as 250 μm in diameter and composed of up to 64 spherical or elliptical cells enclosed by mucilage (Fig. 13–25). The cells lack flagella but contain one or two golden-brown periph-

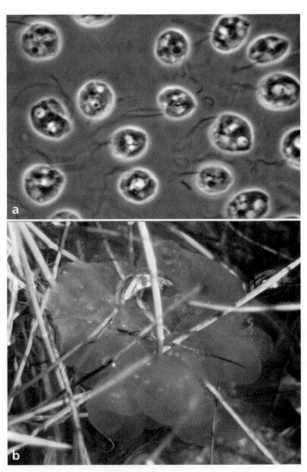

Figure 13–24 *Chrysonephele*. (a) Motile cells. (b) A large gelatinous colony. (Photographs courtesy P. Tyler)

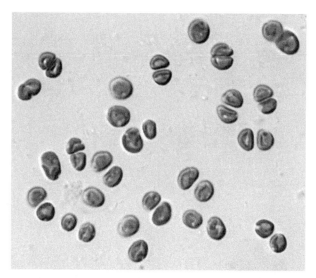

Figure 13–25 *Chrysocapsa*. (Photograph courtesy R. Andersen)

Figure 13–26 *Hydrurus* seen in low magnification in (a) and at higher magnification in (b). (Photographs courtesy P. Kugrens)

eral platelike plastids. *Chrysocapsa* is common in both hard- and soft-water lakes.

HYDRURUS (L. *hydor*, water + Gr. *oura*, tail) thalli (Fig. 13–26) can be relatively large and conspicuous to the naked eye, occurring in fast-flowing, cold (less than 10° C) streams in mountainous regions, attached to submerged rocks or other rigid substrates. *Hydrurus foetidus* emits a strong, unpleasant odor that can be detected from a distance. Spherical cells about 7–15 μm in diameter—each containing one cup-shaped golden-brown plastid—occur within a gelatinous investment that can take a pseudobranched, feathery, or leafy form. The related genus *Phaeodermatium* is also periphytic, occurring as crustose disks attached to firm substrates, but is pseudoparenchymatous, i.e., composed of closely adherent cells that form a tissuelike organization (Canter-Lund and Lund, 1995).

PART 3—SYNUROPHYCEANS

Synurophyceans are silica-scaled flagellates that have been segregated from chrysophyceans on the basis of their lack of chlorophyll c_2 and differences in their flagellar root systems correlated with loss of phagotrophic capability (Andersen, 1987). Hence the synurophyceans are regarded as primarily photoautotrophic forms (Moestrup, 1995). Molecular sequence data (SSU rDNA) and comparative study of scale cases suggest that synurophyceans are monophyletic and that *Tessellaria* is the earliest-known divergent (Lavau, et al., 1997).

Synura is infamous as a freshwater ochrophyte that can impart unpleasant tastes and odors to the water in which it grows (see below). The siliceous scales of synurophyceans may be preserved in lake sediments and can thus be used in paleolimnological reconstruction of past environments (Smol, 1995). In freshwater lakes where synurophyceans and silica scale-bearing chrysophyceans occur, these algae can have a significant impact on the local silica cycle.

Synurophyceans are regarded as important components of food webs in low-nutrient lakes.

Cell Biology

Flagella and flagellar basal bodies

A transitional helix (see Chapter 9) similar to that of other heterokont organisms is present in the flagellar bases of some synurophyceans but not others (Moestrup, 1995). Synurophyceans differ from other ochrophytes in the parallel orientation of flagella and flagellar basal bodies. The cells or colonies swim with both flagella held out in front, an unusual occurrence among heterokont flagellates. However the smooth flagellum exerts little or no motive force; the tripartite-bearing flagellum does all of the propulsive work, as in other heterokonts (Goldstein, 1992). Synurophyceans (and other algae) rotate at a rate of 0.25–1.25 turns s^{-1} during swimming. This is thought to maximize photosynthetic gain by taking advantage of variation in the light field (Raven, 1995). Plastid-based eyespots are absent from synurophycean cells, but in *Synura* and *Chrysodidymus* there is a flagellar swelling (Fig. 13–27) that autofluoresces green in blue or blue-violet irradiance, indicating presence of a flavin, as is found in several other ochrophyte groups (Kawai, 1988). Thus at least a portion of the typical ochrophyte light-sensing system is present in some taxa. *Mallomonas splendens*, with no emergent smooth flagellum, lacks a flagellar swelling (Beech and Wetherbee, 1990b).

The flagellar root system includes a cross-banded rhizoplast that extends from the flagellar bases to the nucleus, whose anterior surface is covered with a cone of rhizoplast fibrils. This musclelike structure may have a contractile function, as does the very similar banded rhizoplast of primitive green algae (Salisbury, et al., 1988). The contractile protein centrin occurs in the rhizoplast area of synurophyceans (Melkonian, et al., 1992). In addition, there is an R1 microtubular root that extends from the rhizoplast, forms a clockwise loop around the flagella, and generates numerous cytoskeletal microtubules. Some of these microtubules appear to be involved in scale formation (see below). There is also a microtubular root that is homologous with the R3 root of other ochrophytes (see review by Moestrup, 1995).

Synurophyceans exhibit a phenomenon that is widespread among algae—and possibly characteristic of

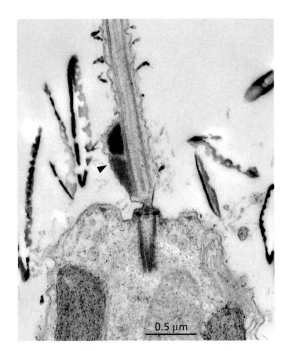

Figure 13–27 The flagellar swelling (arrowhead) of the scale-bearing synurophycean *Chrysodidymus synuroideus* consists of two parts, a more electron-opaque upper portion, and a somewhat less opaque lower region. (From Graham, et al., 1993 by permission of the *Journal of Phycology*)

all flagellate eukaryotes—known as **flagellar transformation**. In this process, complete flagellar maturity—as judged by changes in length or addition of ornamentation to the flagellar surface—requires more than one cell generation. Thus, the two or more flagella of a given cell are of different ages. Prior to cell division, synurophycean cells produce two new long flagella. In forms such as *Synura* that typically have both long and short emergent flagella, the parental long flagellum is then transformed into a new short flagellum. In other cases, including certain species of *Mallomonas*, the parental long flagellum retracts such that the basal body produces no flagellum in the next and subsequent generations (Beech and Wetherbee, 1990a, b). As a result of the discovery of flagellar transformation, a system of flagellar numbering has been developed that can be applied to all flagellates. The most mature flagellum is numbered 1, and successively younger flagella are numbered 2 and higher. This system makes it possible to compare flagella in equivalent stages of development among different taxa). In the case of *Synura*, the smooth, swelling-bearing flagellum is the oldest (number 1), whereas the younger, longer, tripartite hair-

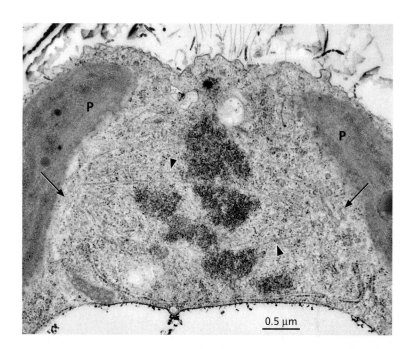

Figure 13–28 Mitosis in *Chrysodidymus*. At metaphase, a central array of chromosomes, spindle microtubules (arrowheads), and spindle poles (arrows) can be identified. (From Graham, et al., 1993 by permission of the *Journal of Phycology*)

bearing flagellum is number 2 (see references in Beech and Wetherbee, 1990b).

Mitosis

Cell division is preceded by replication of flagellar basal bodies and associated roots, including the rhizoplast. New rhizoplasts are not conspicuously striated and thus may be difficult to discern. The parental microtubular roots and rhizoplasts then disassemble, an occurrence that seems to be unusual among algae, most of which appear to recycle parental structures

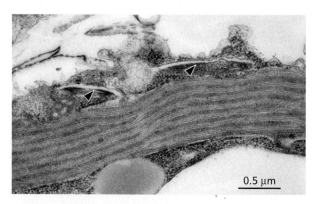

Figure 13–29 Body scales of *Chrysodidymus synuroideus* forming within silica deposition vesicles (arrowheads) located on the surface of one of the plastids.

into the next generation of cells. Although basal bodies become associated with the spindle poles in some synurophyceans, it is the new rhizoplasts that probably function as the spindle microtubular organizing centers (Beech and Wetherbee, 1990b). The nuclear envelope remains at least partially intact (Fig. 13–28), and cytokinesis proceeds via a longitudinal cleavage furrow.

Plastids

Synurophycean cells contain one large bilobed plastid or two plastids; a girdle lamella is present. Pyrenoids are rare, but present in some species. There is a periplastidic endoplasmic reticulum (PER) as in other plastid-bearing ochrophytes, but it is not connected with the nuclear envelope. Silica body scales and bristles are formed in silica deposition vesicles (SDVs) located on the surface of one of the plastids, closely adjacent to the PER (Fig. 13–29). The functional and biochemical basis of this unique association is unclear, but it has been hypothesized that the PER may serve as a template for scale shaping (Leadbeater and Barker, 1995).

Silica scale structure, development, and deployment

The surfaces of synurophycean cells or colonies are typically covered by overlapping silica scales that may also bear silica bristles. The scales are perforated

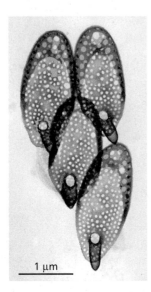

1 μm

Figure 13–30 TEM view of some of the overlapping silica scales from the surface of *Chrysodidymus synuroideus*. Note the occurrence of perforations, rims of perforated silica, and short spines. (From Graham, et al., 1993 by permission of the *Journal of Phycology*)

(Fig. 13–30), and bristles are finely perforate or very thin—this reduces their weight as well as allows access for exchange of material at the cell surface. Differences in perforation patterns of scales are used as taxonomic features in the delineation of species. The scales may have a rim that facilitates formation of a coherent covering of overlapping scales. A number of adhesive polysaccharides are involved in scale attachment: separate scale-scale, bristle-scale, and scale-cell membrane adhesives have been identified (Ludwig, et al., 1996). The pattern of *Mallomonas* scale-associated proteins appears to be species specific. Antibodies raised against the proteins of one species do not generally bind scale-associated proteins of others (Miller, et al., 1996). Large cells of *Synura petersenii* possess about 90 scales (Leadbeater and Barker, 1995). At cell division each daughter cell receives about half the scales, then generates additional scales until the surface is completely covered. However, if dissolved silicate levels are insufficient, incompletely covered or naked cells will result. Unlike diatoms, synurophyceans can continue to divide and function in the absence of an external silica covering (Sandgren, et al., 1996). Cultures of naked cells will grow indefinitely in media having undetectable levels of dissolved silicate if light and other nutrients are provided. When silicate is resup-

plied to silica-depleted cells of *Synura*, most cells will regenerate a complete cell covering within 24 hours (Leadbeater and Barker, 1995). Mature scales migrate to the cell membrane and are extruded to the surface.

Mallomonas scales typically bear long silica bristles, usually in localized positions (Fig. 13–31). Differences in the distribution of bristles are used as taxonomic characters in species differentiation. Bristles are produced in SDVs separate from those where scales are generated, but bristle SDVs are also associated with PER. In *Mallomonas splendens*, bristles destined for attachment to the cell posterior are extruded with their basal ends leading, but attached to the cell membrane by a fibrillar complex. Once free of the cell, the basal portion of the bristles is drawn back to the cell surface; this process involves a 180° reorientation of the bristle while it is outside the cell, mediated by a thin protuberance of cytoplasm (Fig. 13–32). New base-plate scales are then deposited at the cell posterior and the fibrillar complex is involved in attaching the bristle bases to the base plate scales. Anterior bristles are produced similarly, but are extruded tip first, such that no reorientation is involved (Beech, et al., 1990). A review of synurophycean scale structure and development and their relevance to synurophycean phylogeny is provided by Wee (1997).

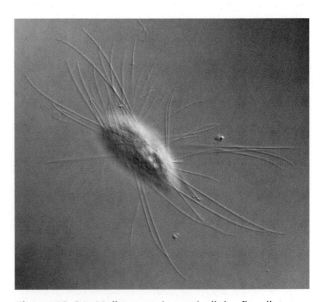

Figure 13–31 *Mallomonas* is a unicellular flagellate that bears silica scales with very long silica bristles. (Photograph courtesy C. Taylor)

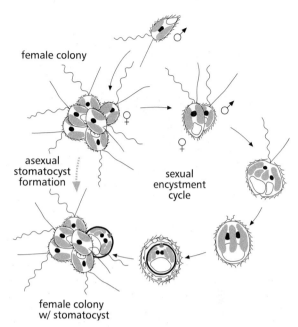

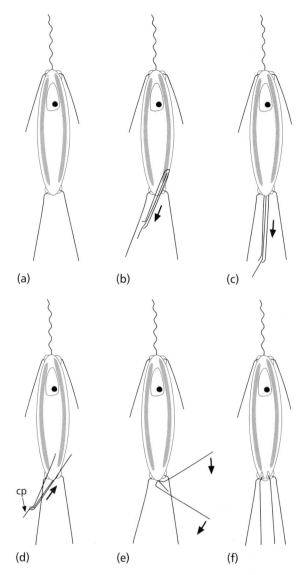

Figure 13–32 Deployment of intracellularly produced bristles to the posterior surface of *Mallomonas splendens*. Following extrusion to the outside of the cell (b, c), bristles are rotated 180°, with the involvement of a cytoplasmic protrusion (cp) (d), then attached to newly deposited base-plate scales (e, f). (After Beech, et al., 1990 by permission of the *Journal of Phycology*)

Figure 13–33 Sexual reproduction in *Synura*. One gamete, defined as the female because the cell remains associated with a colony, is fertilized by a flagellate cell that is released from its colony (and therefore designated male). The fertilized cell then develops into a stomatocyst that remains associated with the motile colony, as in *Dinobryon*. Stomatocysts may also develop asexually. (After Sandgren and Flanagin, 1986 by permission of the *Journal of Phycology*)

as the female gamete (Sandgren and Flanagin, 1986) (Fig. 13–33). Meiosis is assumed to be zygotic, but critical chromosomal or DNA level data are needed.

Synurophycean Ecology

Like their chrysophycean relatives, synurophyceans are most abundant and diverse in neutral to slightly acidic waters of low alkalinity and conductivity. As a general rule, scaly synurophyceans are indicators of low levels of pollution. However, a number of species are characteristic of eutrophic lakes or lakes of different pH. Such environmentally specific species are useful in long-term monitoring of phytoplankton aimed at detecting and understanding water-quality decline (Siver, 1995). The inverse relationship between lake phosphorus availability and the abundance of such taxa such as *Synura* and *Mallomonas* has been ascribed to the presence or absence of large herbivores such as *Daphnia*. Although small cladocerans

Sexual reproduction

Sexual reproduction in *Synura* is similar to that in *Dinobryon*. Zygotic stomatocysts can be recognized by the presence of four plastids and two nuclei. As in *Dinobryon*, the mature stomatocyst is carried along by motile colonies in the place of the cell that served

Figure 13–34 *Synura* occurs as a swimming colony of cells adjoined at their posterior ends. Note flagella at the periphery of the colony.

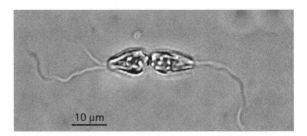

Figure 13–35 *Chrysodidymus* is a two-celled colony of *Synura*-like cells. Note the heterokont flagella.

that may be present in oligotrophic lakes are deterred from eating bristly *Mallomonas* species, *Daphnia* are not. Fortunately for synurophyceans and chrysophyceans, *Daphnia*'s food requirements are so high that it does not typically occur abundantly in cold and/or low-nutrient lakes. Thus seasonally cold periods and oligotrophic lakes provide a refuge for synurophyceans and chrysophyceans (Sandgren and Walton, 1995). *Mallomonas* and *Synura* can also escape potential predators and acquire nutrients by vertically migrating to deep waters at night and well-lighted surface waters during the day. It has also been suggested that the greater availability of dissolved iron in low pH, highly colored bog and pond waters may help explain the common occurrence of *Synura* and its relatives, which have a higher demand for iron used in photosynthetic electron transport than some other algae (Raven, 1995).

Species of *Synura* and *Mallomonas* (as well as some chrysophyceans) contribute to odor and taste problems in water supplies. Among *Synura* species, *S. petersenii* is most often associated with taste or odor problems (Nicholls, 1995).

Synurophycean Diversity

Mallomonas and *Synura* are widespread and may be locally common, whereas *Tessellaria* is rare (found only in Australia). *Chrysodidymus*, though geographically widespread, is relatively rarely encountered and is not often locally abundant. *Conradiella*, unicells having a ring of silica scales rather than being covered by scales, may or may not prove to be a member of

this group. *Tessellaria* is unique in that the surface of the whole colony is covered by scales, rather than individual cells (Pipes, et al., 1991; Pipes and Leedale, 1992); its scales are similar to those of *Synura lapponica*, suggesting these taxa may be related (Lavau, et al., 1997). The genera are easily distinguished on the basis of light microscopic characters. There are approximately 120 described species of *Mallomonas* and *Synura*. Determination of *Mallomonas* and *Synura* species typically requires electron microscopic examination of scale structure.

SYNURA (Gr. *syn*, together + Gr. *oura*, tail) is a colonial organism formed by variable numbers of cells held together at the posterior ends to form a tightly coherent spherical colony (Fig. 13–34). Colonies grow

Figure 13–36 *Mallomonas*, showing characteristic elongate silica bristles attached to flattened silica scales. (Micrograph courtesy C. Taylor)

by addition of new cells through longitudinal cell division. New colonies may form by fragmentation. Cells bear two (heterokont) flagella each and are covered with overlapping silica scales. Spines are present on anterior scales. There are two parietal golden-brown plastids per cell. *Synura ulvella* is a very common species found in the plankton of many kinds of lakes (Prescott, 1951).

CHRYSODIDYMUS (Gr. *chrysos*, golden + Gr. *didymos*, twins) is a two-celled colony of *Synura*-like cells, each bearing two heterokont flagella (Fig. 13–35). The two cells are adherent at their bases. Cells are covered with silica scales very much like those of *Synura*. *Chrysodidymus* is found in low pH *Sphagnum* bogs, as is *Synura sphagnicola*. Ultrastructural similarities suggest that *Chrysodidymus* is related to *Synura sphagnicola* (Graham, et al., 1993).

MALLOMONAS (Gr. *mallos*, wool + Gr. *monas*, unit) (Figs. 13–31, 13–36) is a single-celled flagellate with (usually) only one emergent flagellum. The cell membrane is covered with overlapping silica scales; at least some bear long silica bristles. Cells contain a single deeply divided plastid that may give the appearance of two plastids. There is a chrysolaminarin vacuole at the cell posterior. *Mallomonas* occurs in the plankton of lakes of various types.

PART 4—EUSTIGMATOPHYCEANS

The term eustigma refers to the large orange-red eyespot located (outside the plastid) in the anterior portion of flagellate cells of the eustigmatophycean algae (Fig. 13–37). This characteristic is unique among ochrophytes. As suggested by *rbcL* sequence data, eustigmatophyceans represent the most basal group of ochrophytes (Daugbjerg and Andersen, 1997b), but this conclusion is not supported by SSU rDNA sequence analyses (such as that of Cavalier-Smith and Chao, 1996). SSU rDNA sequence data from 25 strains representing five genera suggest that eustigmatophyceans are a monophyletic group (Andersen, et al., 1998b). There are no known fossils.

The eustigmatophyceans are small (2–32 μm) unicellular, coccoid algae; only some produce flagellate cells (Santos, 1996). Such cells usually have only one noticeable anterior flagellum, which bears two rows of tripartite hairs. The posterior flagellum is usually either very short or exists only as a basal body. *Ellip-*

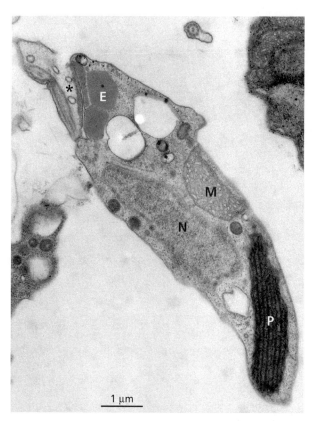

Figure 13–37 Eustigmatophyceans that produce zoospores exhibit the diagnostic features known as a eustigma (E)—an eyespot that occurs outside the plastid. This eyespot occurs at the anterior of flagellate cells, near an expanded region of the flagellum (*). M=mitochondrion, N=nucleus P=plastid. (Micrograph courtesy L. Santos)

soidion acuminatum is an exception to this general rule in having both a smooth and a hair-bearing flagellum (Hibberd and Leedale, 1970). A transitional helix is present, as in most ochrophytes. An expanded region of the flagellum is appressed to the cell membrane near the eyespot.

Eustigmatophyceans are thought to be obligate photoautotrophs. There is no evidence for their use of exogenous organic compounds (Santos, 1996). Cells have one or more yellow-green plastids that contain chlorophyll *a* but not chlorophyll *c*; this is quite unusual among ochrophytes. Violaxanthin is the major accessory pigment (Whittle and Casselton, 1975); this pigment appears to play a prominent role in light harvesting. β-carotene and vaucheriaxanthin-ester are also present, while fucoxanthin is not. Although thylakoids occur in stacks of three, as is

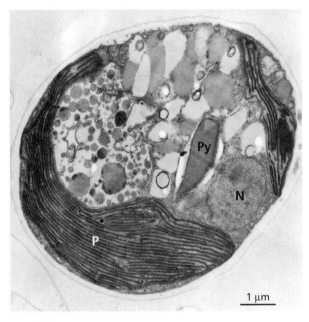

Figure 13–38 TEM view of *Eustigmatos vischeri*. Note the stalked pyrenoid (Py) extending into the cytoplasm. N=nucleus, P=plastid. (Micrograph courtesy L. Santos)

common in ochrophytes, there is no girdle lamella. Periplastidic ER is present, but is not usually connected with the nuclear envelope. Unusual stalked pyrenoids occur on the plastids of some, but not all eustigmatophyceans. Pyrenoids never occur in zoospores—they disappear during zoospore development and then regenerate after zoospores settle onto a surface (Santos, 1996). There are no plastid-based eyespots (Hibberd, 1990a). Other typical cellular features include a red-pigmented body, whose function is unknown, and storage vesicles having lamellate contents. The chemical structure of the storage product is unknown. It is not starch, as the reaction with I_2KI is negative. Moreover, the chemical nature of the eustigmatophycean cell wall is poorly understood (Santos, 1996).

Reproduction is either by production of two to four autospores (small, walled versions of parental cells), or in some cases by zoospores. Because of similarities in morphology, reproduction, cell color, and chloroplast structure, eustigmatophyceans are commonly mistaken for coccoid green algae at the light-microscopic level. Identification of an alga as a eustigmatophycean requires examination of thin sections of cells by transmission electron microscopy and/or pigment analysis by chromatography. The absence of chlorophyll *c*, coupled with presence of

violaxanthin, is definitive for the eustigmatophyceans (Whittle and Casselton, 1975).

The systematics of the Eustigmatophyceae were described by Hibberd (1981) and Santos (1996). A single order (Eustigmatales) comprises four families that are distinguished on the basis of cell shape and size, presence or absence of zoospores, and number of flagella on zoospores. The Eustigmataceae includes *Eustigmatos* and *Vischeria*, each having three species. The Pseudocharaciopsidaceae includes *Pseudocharaciopsis*, with two species and *Botryochloropsis*. Members of these groups produce zoospores. The Chlorobotryaceae (*Chlorobotrys*) and the Monodopsidaceae (*Monodopsis* and *Nannochloropsis*) do not produce zoospores. There are about seven genera and perhaps 15 species of eustigmatophyceans, most occurring in freshwater or in soil, but there are also some marine forms. A eustigmatophycean whose features do not match those of known genera occurs as a symbiont within the freshwater sponge *Corvomyenia everetti* (Frost, et al., 1997).

Eustigmatophycean Diversity

EUSTIGMATOS (Gr. *eu*, well equipped + Gr. *stigma*, eyespot) (Fig. 13–38) is a unicellular, coccoid soil alga having a single parietal plastid. Reproduction is by 2–4 autospores and uniflagellate zoospores.

PSEUDOCHARACIOPSIS (Gr. *pseudos*, false + *Characiopsis*, a genus of tribophyceans) consists of

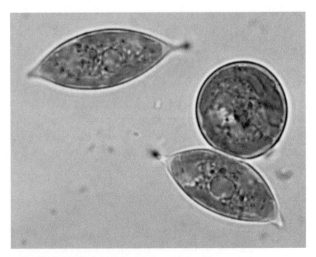

Figure 13–39 *Pseudocharaciopsis* cells are often spindle-shaped. (Photograph courtesy L. Santos)

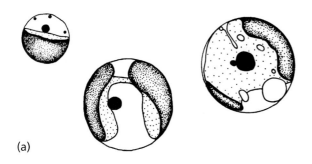

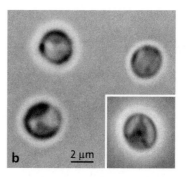

(a)

Figure 13–40 *Nannochloropsis*. Three different species are diagrammed in (a). (b) Light microscopic view of another species. Inset: Two plastid lobes are evident as is the so-called red body, which appears as a dark spot in the cell center. (a: From Andersen, et al., 1998a)

ovoid to pointed cells commonly attached to a substrate by means of a short stipe or attachment pad (Fig. 13–39). Biflagellate, naked zoospores are produced. Older cells can contain several nuclei and chloroplasts (Lee and Bold, 1973). It is very similar to, and easily confused with, the tribophycean (xanthophycean) alga *Characiopsis*, and the green alga *Characium* (and similar-appearing genera—see Chapter 20). These three organisms can be distinguished by pigment or ultrastructural analysis. *Pseudocharaciopsis* can be distinguished from *Characiopsis* and *Characium* by its possession of a conspicuous stalked pyrenoid that is visible with light microscopy.

NANNOCHLOROPSIS (*Nannochloris*, a genus of green algae [Gr. *nannos*, dwarf + Gr. *chloros*, green] + Gr. *opsis*, likeness) (Fig. 13–40) is a marine and freshwater coccoid form that resembles *Chlorella* and lacks a pyrenoid (Santos and Leedale, 1995). It produces a relatively high content of the polyunsaturated fatty acid eicosapentaenoic acid, and thus is used as a food for aquacultured (maricultured) marine animals. The fatty-acid composition can be regulated by irradiance (Sukenik, et al., 1989). A surprisingly high level of SSU rDNA sequence diversity was found in a study of 21 strains (Andersen, et al., 1998a).

Recommended Books

Sandgren, C. D., J. P. Smol, and J. Kristiansen. (editors). 1995. *Chrysophyte Algae: Ecology, Phylogeny, and Development.* Cambridge University Press, Cambridge, UK.

Duff, K. E., B. A. Zeeb, and J. P. Smol. 1995. *Atlas of Chrysophyte Cysts.* Kluwer Academic Publishers, Dordecht, Netherlands.

Chapter 14

Ochrophytes III
Pelagophyceans, Silicoflagellates, Pedinellids, and Related Forms

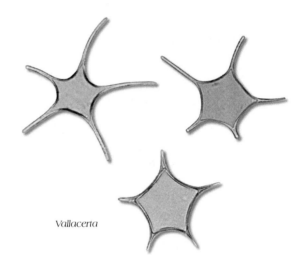

Vallacerta

Pelagophyceans (and their close relatives, the sarcinochrysidaleans), silicoflagellates, and pedinellids (and their close relatives, the rhizochromulinids), are primarily unicellular, flagellate or nonflagellate members of the phytoplankton. These groups—like those discussed in Chapters 12, 13, and 15—are members of the heterokont algae, or ochrophytes (Cavalier-Smith and Chao (1996) (see Fig. 13–1). They share many cellular features with other ochrophytes in addition to DNA sequence similarities. However pelagophyceans, silicoflagellates, and pedinellids also share some distinct characteristics, such as the presence of a paraxonemal rod within the hair-bearing anterior flagellum. In contrast to the primarily freshwater ochrophytes covered in Chapter 13, this ochrophyte group is mostly, though not exclusively, marine.

Pelagophyceans, silicoflagellates, and pedinellids are important because some representatives of each group have in the past generated marine algal blooms, or do so at present. Ancient large populations of some of these forms have been linked to the formation of important oil deposits (Moldowan, et al., 1990). The pelagophyceans *Aureococcus anophagefferens* and *Aureoumbra lagunensis* are serious marine bloom-formers. These small golden-brown cells can occur in populations as dense as 3×10^9 cells l^{-1}, forming summer "brown tides" along the New England and Texas coasts of the U.S. Although these pelagophyceans are not known to produce toxins and do not pose a direct threat to humans, the blooms have had disastrous effects on thousands of hectares of coastal shorelines. The blooms exclude light from sea grasses, causing sea-grass bed loss. They are also associated with major declines in scallop harvest. By physically interfering with filter feeding, the algal blooms cause scallops to starve to death. Fish-eating birds cannot see their prey through the dense blooms and thus leave the area. *Aureococcus* has also been responsible for an estimated several millions of dollars worth of damage to blue mussel and cultured oyster production (Nicholls, 1995). In addition, negative effects on egg production by copepods and reduced populations of ciliates are associated with dense pelagophycean blooms. The factors that stimulate pelagophycean blooms are as yet poorly defined. Although increased levels of phosphorus or nitrogen do not seem to be implicated, watershed input of iron is a possible correlate (Bricelj and Lonsdale, 1997). *Aureococcus* can utilize exogenous glutamic acid (glutamate) and the sugar glucose (Nicholls, 1995), as well as organic phosphates (Raven, 1995); these abilities may also enhance its success in waters having higher than normal levels of organic compounds.

Because of their increasing ecological importance and molecular evidence that pelagophyceans are relatively early divergent members of the clade discussed in this chapter (Cavalier-Smith and Chao, 1996), this group of algae and their close relatives the sarcinochrysidaleans will be discussed first. A focus upon silicoflagellates, pedinellids, and their relatives, the rhizochromulinids, will follow. Distinctive features of these groups and their ecological significance will be emphasized. Features that more generally characterize ochrophytes are described in Chapter 12.

Pelagophyceans and Sarcinochrysidaleans

Pelagophyceans

The pelagophyceans were formally described as a new class (Pelagophyceae) by Andersen, et al., (1993) on the basis of ultrastructural characters and SSU rDNA sequence data. Although relatively few genera have been described, there are probably additional undescribed forms. *Pelagococcus, Aureococcus,* and *Aureoumbra* (De Yoe, et al., 1997) are nonflagellate, coccoid unicells, whereas *Pelagomonas* is a unicellular flagellate. Cell structure has been most completely described for *Aureoumbra* and *Pelagomonas*; the presence of flagellar characters in the latter genus allows comparison with flagella of other ochrophyte groups (Andersen, et al., 1993).

AUREOUMBRA (L. *aureus*, golden + L. *umbra*, shadow), specifically *A. lagunensis*, is the recently named Texas Brown Tide organism (Fig. 14–1). It differs from *Aureococcus* in 18S rDNA sequence, pyrenoid structure, presence of a pair of basal bodies, and other features (De Yoe, et al., 1997).

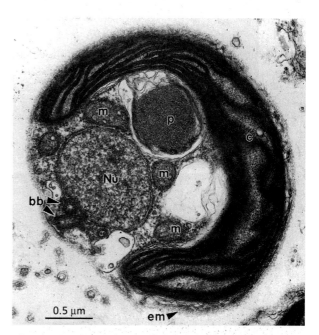

Figure 14–1 TEM view of *Aureoumbra*. bb=basal bodies, c=plastid, em=extracellular matrix, Nu=nucleus, m=mitochondrion, p=pyrenoid. (From DeYoe et al., 1997 by permission of the *Journal of Phycology*)

PELAGOMONAS (Gr. *pelagos*, sea + Gr. *monas*, unit) (Fig. 14–2) cells are very small, 1–3 μm in diameter. The flagellar apparatus is highly reduced. There is only one flagellum, and the second basal body and flagellar roots found in most other ochrophytes (Chapter 12) are absent. *Pelagomonas calceolata* is the only known eukaryote to have undergone complete loss of the mature flagellum/basal body (see discussion of flagellar maturation cycles in Chapter 13) (Heimann, et al., 1995). The flagellar basal body has a transitional helix and plates, as do other ochrophytes. However, the flagellar hairs of *Pelagomonas* are not tripartite as are those of other ochrophytes, but rather are bipartite. Although the flagellar hairs occur in two opposite rows, they are flexible rather than stiff, raising the question of whether or not their hydrodynamic function is similar to those of other heterokont organisms (see Chapter 12). The flagellum contains a rodlike structure (the paraxonemal rod) (Andersen, et al., 1993). This structure increases the apparent thickness of the flagella as do the paraxonemal rods of euglenoids (Chapter 8) and dinoflagellates (Chapter 11), but these structures are most likely not homologous, and their function(s) are not well understood. The flagellum of *Pelagomonas* does not bear a swelling that autofluoresces green as do those of some

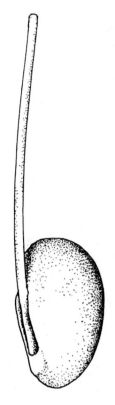

Figure 14–2 *Pelagomonas.* (From Andersen, et al., 1993 by permission of the *Journal of Phycology*)

unusually small extranuclear spindle. The spindle forms next to the nuclear envelope, then the nearby portion of the envelope breaks down, allowing the spindle to access to chromosomes. The nuclear envelope remains intact except at the spindle poles. Following chromosomal separation by spindle elongation, the nuclear envelope breaks and re-forms around each daughter nucleus. Cytokinesis occurs by cell constriction.

Sarcinochrysidalean Algae

This group of unicellular, palmelloid, and filamentous algae includes the genera *Sarcinochrysis*, *Ankylochrysis*, *Nematochrysopsis*, and *Pulvinaria* (O'Kelly, 1989). These were first thought to be closely related to brown algae (phaeophyceans) because both groups produce zoospores having heterokont flagella emerging from the side of the cell, rather than the apex. In addition, cellulose constitutes at least a portion of the cell walls of both groups. However comparative study of zoospore ultrastructure has revealed substantial differences. Unlike those of phaeophyceans, sarcinochrysidalean zoospores have a more reduced flagellar apparatus, eyespots are lacking, and there is a cell covering (theca) present. These observations suggested to O'Kelly (1989) that the similarities between sarcinochrysidaleans and brown algae were only superficial, and that the former were probably more closely related to other ochrophytes. An alliance with the pelagophyceans is indicated by the following shared features: cell coverings, similar accessory pigments, girdle lamellae in the plastids, absence of eyespots, and flagellar base structure. O'Kelly's (1989) proposals have been supported by more recent SSU rDNA sequence analyses that place sarcinochrysidaleans within the pelagophyceans (Saunders, et al., 1997b).

Sarcinochrysidalean algae are known for production of the two sterols 24-n-propylidene-cholesterol and 24-n-propylcholesterol. Such sterols are thought to have been the precursors of C_{30} steranes found in Prudhoe Bay (Alaska) oil, which is derived from organic matter deposited in the marine environment. This suggests that ancient sarcinochrysidalean algae generated at least a portion of the fossil fuel deposit. C_{30} steranes first appeared sometime between the Early Ordovician (about 500 million years ago) and the Devonian (360 million years ago), suggesting that sarcinochrysidalean algae likewise first occurred during this time period (Moldowan, et al., 1990).

other ochrophytes, and the cells of *Pelagomonas* are not known to be phototactic (Andersen, et al., 1993).

Pelagomonas cells have a very thin covering (extracellular matrix or theca) composed of organic material. Each cell possesses a peculiar, sausage-shaped ejectile structure of unknown function that cannot be visualized by light microscopy (Andersen, et al., 1993). The single plastid contains chlorophyll *a*, chlorophylls c_1, c_2, and c_3. The major accessory pigment is fucoxanthin, but other carotenoids are also present (Bidigare, 1989). The thylakoids are stacked in groups of three, and there is a girdle lamella, as in some other ochrophytes. There is no eyespot. One or more anterior storage vesicles contain material that is presumed to be chrysolaminaranlike (Andersen, et al., 1993).

Cell division in *Pelagococcus* has been reviewed by Sluiman (1993). The outer membrane of the nuclear envelope is connected to the periplastidal endoplasmic reticulum (PER) as in a number of other ochrophytes, but pelagophyceans are unusual in that this connection is retained throughout the process of mitosis. There are no flagella or basal bodies (flagellar bases or centrioles) in the interphase cells, but two centrioles make their appearance in prophase and subsequently serve as the organizing centers for the development of an

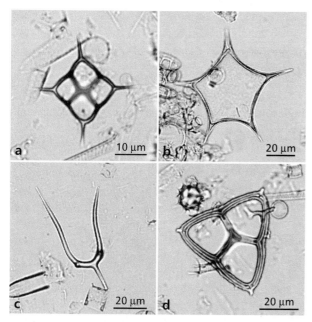

Figure 14–3 The highly perforate silica skeletons of silicoflagellates. (a) *Dictyocha*, (b) *Vallacerta*, (c) *Lyramula*, (d) *Corbisema*. (Late Cretaceous specimens provided by D. Clark)

Most of the modern sarcinochrysidalean algae occur in the benthic marine habitat, attached to substrates in the upper intertidal, but *Sarcinochrysis marina* is a phytoplankter. Some are found in brackish water.

Silicoflagellates, Pedinellids, and Rhizochromulinids

Silicoflagellates

Silicoflagellates are defined by the occurrence in at least one life cycle stage of a distinctive, one-piece external silica skeleton that is highly perforate and has been described as a basket (Fig. 14–3). Although there are only a few living taxa of silicoflagellates (represented by species of the genus *Dictyocha*), they have a substantial fossil record extending back to the Cretaceous (about 120 million years ago). Silicoflagellates reached their peak diversity during the Miocene; 100 fossil species have been described. The fossil remains of the silica exoskeletons are used as micropaleontological indicators of ancient temperature change. Modern silicoflagellates are widespread in modern oceans, but tend to be most abundant in colder waters,

sometimes growing to bloom proportions. Silicoflagellates have recently been reviewed by Prema (1996). They are currently classified as an order, Dictyochales, in the class Dictyochophyceae, established by Silva (1980) (see Preisig, 1995; Moestrup, 1995), or as a separate class, Pedinellophyceae (Cavalier-Smith, 1986; Moestrup, 1995).

DICTYOCHA (Gr. *diktyon*, net + Gr. *ochos*, holding) has three life-history stages, each being a unicellular flagellate: (1) a uninucleate, silica skeleton-producing form; (2) a uninucleate, naked stage; and (3) a multinucleate, amoeboid stage. Cell structure has been reviewed by Moestrup (1995). The cells have a distinctive, highly perforate cytoplasm that has been described as foamy or frothy (Fig. 14–4). Typical heterokont flagella are present: a long, anterior, hair-bearing flagellum and a very short, smooth flagellum. The long flagellum is extended laterally and contains a paraxonemal rod,

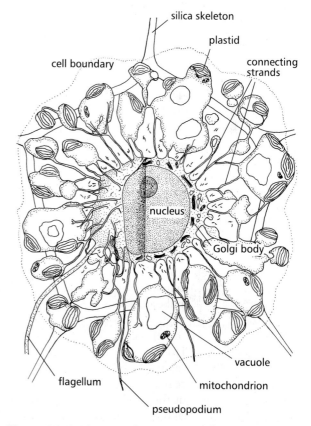

Figure 14–4 Diagram of *Dictyocha*. The cytoplasm is highly reticulate and vacuolate. (From Van Valkenberg, 1971 by permission of the *Journal of Phycology*)

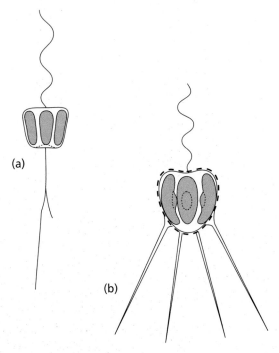

Figure 14–5 Diversity of pedinellids: (a) *Pseudopedinella* and (b) *Apedinella*, represent some of the variation known for pedinellids. Plastids (shaded) typically occur in a ring. (After Throndsen, 1997)

as does the flagellum of *Pelagomonas* (described above). However, unlike *Pelagomonas*, there is no helix in the transitional region of the flagellum, and in this characteristic *Dictyocha* more closely resembles the pedinellids (see below). There is no eyespot, no contractile vacuoles, and no nuclear envelope connection to the PER. Cells contain numerous plastids, each with a girdle lamella and one pyrenoid. Chlorophylls *a* and *c*, fucoxanthin, and other carotenoids are present. *Dictyocha* is presumed to be

photoautotrophic because mixotrophy and/or phagotrophy have not been reported.

Pedinellids

The pedinellids are unicellular flagellates having a single anterior flagellum that emerges from an anterior pit and has a paraxonemal rod located in a winglike extension. Although a transitional plate is present in the flagellar transition region (in common with other heterokonts), pedinellids lack a transitional helix (as do the flagella of silicoflagellates and *Pelagomonas*). The flagellum bears two rows of tripartite hairs, as is typical for heterokonts. Pedinellids are covered with unmineralized (organic) body scales that are produced in the Golgi apparatus (Moestrup, 1995).

Pedinellids are distinctive in being radially symmetrical along their longitudinal axis. There is usually an anterior ring of **rhizopodia**—threadlike extensions of cytoplasm supported by microtubules—also known as tentacles. The rhizopodia surround the flagellum. There are no flagellar microtubular roots, but cytoskeletal microtubules occur in characteristic groups of three (triads). These run from the nuclear surface to the cell surface, where they may extend into the tentacles. Some forms—*Pedinella, Pseudopedinella* (Fig. 14–5), *Apedinella* (Fig. 14–5), and *Mesopedinella* (Figs. 14–8, 14–9)—contain several (3–6) golden-brown plastids arranged in a circle at the cell periphery. These plastids have a girdle lamella, chlorophylls *a*, c_1 and c_2, and fucoxanthin. Other genera (*Actinomonas, Ciliophrys, Parapedinella,* and *Pteridomonas*) completely lack plastids. Cells of plastidless genera are phagotrophic, capturing bacterial cells (*Apedinella* can also ingest bacteria and is thus mixotrophic). Although the pedinellids are primarily marine, a few—such as *Palatinella* (Fig. 14–6)—occur in freshwaters.

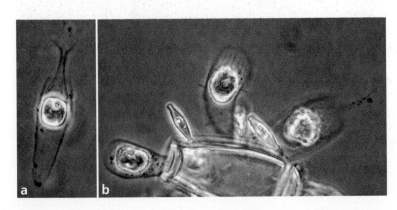

Figure 14–6 *Palatinella* is a freshwater pedinellid. (a) The single cells occur within an open-ended lorica that is usually attached to a substrate (b). A ring of tentacles emerges from the lorica.

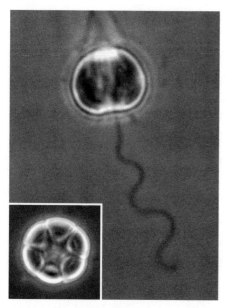

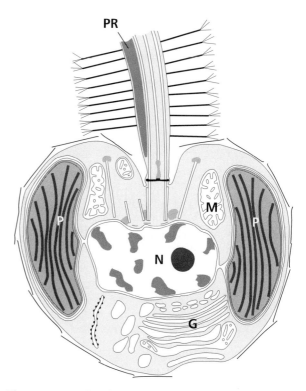

Figure 14-7 *Apedinella*. Spiny scales are at top and flagellum, at bottom. Inset: apical view of cell in which the six radially arranged plastids are evident. (Photographs courtesy A. Koutoulis)

Figure 14-8 Diagram of the internal and flagellar structure of *Mesopedinella*. Note the winglike extension of the flagellum; this contains a paraflagellar rod (PR). At the transition region there is a flagellar plate, but no helix. G=Golgi body, N=nucleus, M=mitochondrion, P=plastid. (After Daugbjerg, 1996)

APEDINELLA (Gr. *a*, not + *Pedinella*, a genus of pedinellids [Gr. *pedon*, oar]) (Fig. 14–7) cells are covered by two types of body scales that occur in overlapping layers. In addition, there are six elongate spine scales (longer than the cells that produced them). Spine scales are generated within Golgi vesicles in the posterior cytoplasm. As the spine elongates, it protrudes from the end of the cell, surrounded by both the deposition vesicle and cell membrane. The spines are deposited just below the cell apex, and they move when the cell is swimming. *Apedinella* has one of the most complex cytoskeletal apparatuses ever described for a single-celled protist; it is thought to participate in spine movement (Wetherbee, et al., 1995).

MESOPEDINELLA (Gr. *mesos*, middle + Gr. *Pedinella*, a genus of pedinellids [Gr. *pedon*, oar]) (Fig. 14–8) was recently discovered from Canadian arctic ocean waters. Apple-shaped cells possess golden plastids, lacking pyrenoids; these are arranged in a ring, as is typical of pedinellids (Fig. 14–9). *Mesopedinella* cells rotate about the longitudinal axis while swimming, similarly to *Pseudopedinella*. Tentacles are absent. A cyst stage possesses a layered wall that does not include silica. *Mesopedinella* does not survive in temperatures above

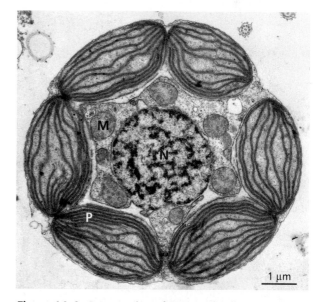

Figure 14-9 Cross section of *Mesopedinella* cell; note the ring of plastids (P) and mitochondria (M) surrounding the central nucleus (N). (From Daugbjerg, 1996)

8°–10° C, and is therefore restricted to cold waters (Daugbjerg, 1996).

Rhizochromulinids

RHIZOCHROMULINA (Gr. *rhiza*, root + *Chromulina*, a genus of chrysophyceans) (not illustrated) is an amoeboid ochrophyte that is superficially similar to the chrysophyceans *Chrysamoeba* and *Lagynion* (Chapter 13), but is actually more closely related to the pedinellids. This relationship was first detected by comparison of cell ultrastructure and later corroborated by molecular sequence analyses. The cells of *Rhizochromulina* have a dorsiventral organization with rhizopodia extending from the cell body. Rhizopodia are supported by bundles of microtubules (O'Kelly and Wujek, 1995). Flagellate zoospores may be produced. Unlike pedinellids, silicoflagellates, and pelagophyceans, their flagella lack a paraxonemal rod. Rhizochromulinids are currently classified as an order, Rhizochromulinales, within the class Dictyochophyceae (see Presig, 1995).

Recommended Book

Sandgren, C. D., J. P. Smol, and J. Kristiansen. (editors). 1995. *Chrysophyte Algae: Ecology, Phylogeny, and Development*. Cambridge University Press, Cambridge, UK.

Chapter 15

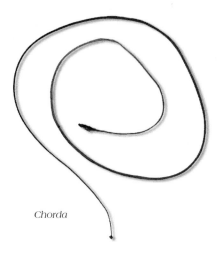

Chorda

Ochrophytes IV
Chrysomeridaleans, Phaeothamniophyceans, Tribophyceans, and Phaeophyceans

Rocky intertidal and subtidal regions of temperate, boreal, and polar marine shores are often dominated by large biomasses of macroscopic brown algae belonging to the class Phaeophyceae, also known as the phaeophytes or phaeophyceans. Many people are familiar with the rockweeds *Fucus* (Fig. 15–1) and *Ascophyllum* (Fig. 15–2), whose extensive populations may cover rocks, becoming visible at low tide, and are often used to pack fresh seafood for shipment. Kelps are economically valuable as sources of industrially useful alginates, and a few are widely cultivated for use in Asian cuisines (Chapter 4). There is some concern that large growths of brown seaweeds may contribute to ozone depletion by release of halogenated compounds to the atmosphere (Chapter 2). Brown algae are also ecologically significant because they are involved in important biotic associations and because a few may generate nuisance blooms (see Chapters 3 and 23).

Biologists interested in life-history variation regard brown algal reproductive modes—involving alternation of generations or gametic meiosis—as fascinating examples of parallel evolution of life-history types observed in other organisms, particularly land plants and metazoa (animals). Through intensive study by biochemists, the chemical signals (pheromones) involved in gamete attraction are better understood in brown algae than for any other group of photosynthetic organisms. In addition, *Fucus* has been widely used by cell and molecular biologists as a model laboratory system for studies focused on early stages in the development of multicellular thalli from unicellular zygotes. Evolutionary morphologists are interested in the mechanisms by which various brown algae, such as *Dictyota*, develop apical meristems and generate specialized tissues analogous to those of land plants.

Though not as ecologically or economically significant as phaeophyceans (and considerably less conspicuous and diverse), the classes Chrysomeridaceae (chrysomeridaleans); Phaeothamniophyceae

301

Figure 15–1 A monospecific population of *Fucus* on a rocky temperate shoreline. (Photograph courtesy C. Taylor)

(phaeothamniophyceans) and Tribophyceae (tribophyceans), the last also known as xanthophyceans or yellow-green algae, are of phylogenetic significance because they are thought to be related to the phaeophyceans. Most chrysomeridaleans and tribophyceans are of microscopic dimensions, in contrast to the usually larger phaeophyceans. While noticeable growths of tribophyceans occasionally occur in nature, these are not typically regarded as nuisance blooms. In contrast to the phaeophyceans, which are primarily marine, tribophycean diversity is greatest in freshwaters, where they exhibit intriguing examples of the evolution of thallus morphologies parallel to those of various freshwater green algae.

Fossils and Phylogeny

There are no known fossils of chrysomeridalean algae (Hibberd, 1990b), but *Vaucheria*-like remains known as *Paleovaucheria* represent tribophyceans in 900 million-year-old Russian deposits (Knoll, 1996). In addition, there are a number of fossil remains that have been attributed to phaeophyceans, including some Canadian fossils of Late Ordovician age (about 450

million years old) known as *Winnipegia* that occurred in dense beds (Fry, 1983), and *Thallocystis* from the middle Silurian (about 425 mya) (Taggart and Parker, 1976). Unfortunately, these extremely ancient remains are also similar to some modern red and green algae, so their identification as brown algae is problematic. Fossils more confidently allied to modern brown algae (because they exhibit features that occur in modern brown, but not red or green algae) include the Miocene (5–25 mya) *Zonarites* and *Limnophycus*, which resemble modern *Dictyota*; *Julescrania*, a probable kelp (Parker and Dawson, 1965); and *Cystoseirites*, *Cystoseira*, *Paleocystophora*, and *Paleohalidrys* (Fig. 15–3), which are classified in the same order as the extant genus *Fucus* (Clayton, 1990).

The phylogenetic relationships of modern brown algae and forms suspected of being their close relatives are being surveyed by the use of SSU rDNA and other molecular sequences (Fig. 15–4). Current evidence indicates that taxonomic groupings that were previously made on the basis of thallus structure do not always accurately reflect patterns of evolutionary diversification (as is true in a number of other algal groups). Parallel evolution of morphological features

Figure 15–2 *Ascophyllum*, a large brown seaweed, in the intertidal zone. (Photograph courtesy M. Cole)

Figure 15–3 *Paleocystophora* (left) and *Paleohalidrys* (right) are examples of fossil brown algae that are thought to be related to modern *Fucus* and *Ascophyllum*. (a: From Parker and Dawson, 1965; b: Courtesy B. Parker)

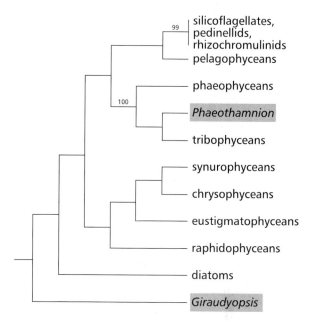

Figure 15–4 Phylogenetic tree based on SSU rDNA sequences showing relationships of tribophyceans, *Giraudyopsis*, and *Phaeothamnion* to other ochrophyte groups. (Based on Andersen, et al., 1998b by permission of the *Journal of Phycology*)

has been common. In contrast, aspects of brown algal life history and the ultrastructure of flagellate cells tend to more closely correspond to molecular sequence-based phylogenies. Molecular sequence (SSU rDNA) data strongly support an hypothesis of monophyly for the class Phaeophyceae, monophyly of the Tribophyceae, and close relationship of the tribophyceans to the phaeophyceans (Saunders, et al., 1997b; Potter, et al., 1997b). SSU rDNA sequence data also suggest that phaeothamniophyceans, ochrophytes formerly placed in the Chrysophyceae, are members of this clade (Fig. 15–4) (Andersen, et al., 1998b). *Giraudyopsis* and related algae known as Chrysomeridales have been linked to the tribophycean-phaeophycean clade by some investigators (Potter, et al., 1997b), but not by others (Andersen, et al., 1998b), on the basis of SSU rDNA evidence. While recognizing that the phylogenetic affinities of chrysomeridaleans are uncertain, we have chosen to discuss them here.

PART 1—CHRYSOMERIDALES, PHAEOTHAMNIOPHYCEAE, AND THE XANTHOPHYLL CYCLE

Giraudyopsis is the best known member of the Chrysomeridales. It resembles certain simple brown algae in having pluriseriate, branched, erect filaments produced by a multicellular (though fundamentally filamentous) disk-shaped structure that is attached to substrates (Fig. 15–5). Such morphologies are described as **heterotrichous**. Because of its similarity to certain brown algae, *Giraudyopsis* was at first thought to be a member of the Phaeophyceae, but because its cell walls lack alginates (characteristic of phaeophyceans) and for other reasons, Loiseaux (1967) proposed its separation. In addition to *Giraudyopsis*, several other genera (*Antarctosaccion*, *Chrysoderma*, *Chrysomeris*, *Chrysonephos*, *Chrysowaernella*, *Phaeosaccion*, and *Rhamnochrysis*) have tentatively been placed within the Chrysomeridales based on the absence of a zoospore wall and presence of the accessory pigment violaxanthin (O'Kelly, 1989); these provisional relationships require testing by molecular sequence analyses.

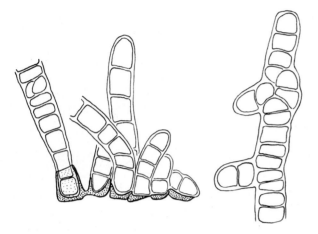

Figure 15–5 *Giraudyopsis*, a member of the Chrysomeridales, is heterotrichous, as are many simple brown algae (phaeophyceans). (From Loiseaux, 1967)

Significantly, *Giraudyopsis stellifera* exhibits light-harvesting complexes (including chlorophyll *a*, chlorophyll *c*, and fucoxanthin) as well as a suite of xanthophylls that are also present in brown algae but absent from tribophyceans and diatoms (Lichtlé, et al., 1995). These differences may prove to be of phylogenetic, physiological, and ecological significance and thus merit further description.

In many algal groups as well as land plants, xanthophylls undergo cyclic changes (through the addition or removal of oxygen), which help photosynthetic cells disperse excess energy under conditions of high illumination, thus protecting them from destructive photooxidation (Demmig-Adams and Adams, 1993). These reactions, known as epoxidation (oxygen addition to form epoxide groups) and de-epoxidation (removal of oxygen), result in the interconversion of xanthophylls located in the plastids, and constitute what is termed the xanthophyll cycle. Algae and plants can be separated into three groups based on the type of xanthophylls they contain and whether or not epoxidation cycles occur (Table 15–1). Although cyanobacteria, red algae, and cryptomonads possess the xanthophyll zeaxanthin, no epoxidation cycling seems to occur; rather, excess energy is otherwise dispersed. In contrast, reversible epoxidation of zeaxanthin to form antheraxanthin and violaxanthin is characteristic of *Giraudyopsis*, phaeophyceans, green algae, and land plants (Fig. 15–6). An epoxidation cycle involving different xanthophylls—diadinoxanthin and diatoxanthin—occurs in tribophyceans, diatoms, raphidophyceans (chloro-

monads), dinoflagellates, and euglenoids (Larkum and Howe, 1997).

The phylogenetic basis and ecological relevance of these differences in epoxidation cycles is currently unclear. Although nuclear-encoded SSU rDNA sequence analyses suggest that the nucleocytoplasmic components of tribophycean and phaeophycean cells are closely related, the xanthophyll cycle evidence suggests the possibility that plastids of chrysomeridaleans might be more closely related to those of brown algae than are those of tribophyceans.

A new class of ochrophytes, the Phaeothamniophyceae, has been established on the basis of *rbcL* sequence data, pigment composition, and cellular characters, such as lack of obvious chrysolaminaran vacuoles (Bailey, et al., 1998). Members of this class, named for *Phaeothamnion*, were formerly classified as members of the Chrysophyceae or the Tribophyceae, but are now regarded as a monophyletic group that is more closely related to the Tribophyceae and Phaeophyceae than to chrysophyceans. The new class is defined by a unique combination of pigments—fucoxanthin and heteroxanthin—in addition to chlorophylls *a* and *c* and various carotenoids. Algae included within

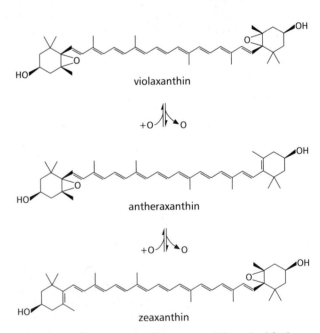

Figure 15–6 Diagram of the reversible epioxidation reactions (involving addition to or removal of oxygen from xanthophylls) that occur in *Giraudyopsis*, brown algae, and green algae (and plants). (After Falkowski and Raven, 1997 by permission of Blackwell Science, Inc.)

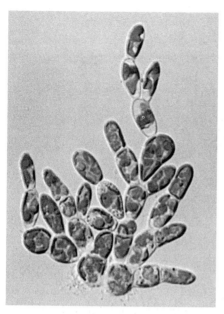

Figure 15–7 *Phaeothamnion*, a filamentous freshwater alga, previously thought to be a chrysophycean but now regarded as a close relative of tribophyceans. (Photograph courtesy R. Andersen)

the Phaeothamniophyceae are *Chrysapion, Chrysoclonium, Chrysodictyon, Phaeobotrys, Phaeogloea, Phaeoschizochlamys, Selenophaea, Sphaeridiothrix, Stichogloea, Tetrachrysis, Tetrapion, Tetrasporopsis,* and *Phaeothamnion*; only *Phaeothamnion* is further described here.

PHAEOTHAMNION (Gr. *phaios*, dusky + Gr. *thamnion*, small shrub) (Fig. 15–7) is a 1 cm-long branched filament whose cells contain one to several platelike, olive-brown plastids. It grows attached to freshwater substrates such as submerged mosses. Zoospores, released from cells through lateral pores, have two laterally inserted flagella, the longer directed toward the cell anterior, the shorter, the posterior. The long flagellum possesses tripartite tubular hairs, and there is a transitional helix. The ultrastructure of the flagellar root system resembles that of phaeophycean and tribophycean zoospores. Pigments include chlorophylls *a*, c_1, and c_2, fucoxanthin, diadinoxanthin, and diatoxanthin in addition to other carotenoids. Cysts may be produced, but these lack silica. This evidence, together with SSU rDNA sequence data, suggest closer relationship to tribophyceans than to any other group of ochrophytes (Fig. 15–4) (Andersen, et al., 1998b). This raises the possibility that other filamentous

ochrophytes presently classified as chrysophyceans may have similar affinities. The fact that *Phaeothamnion* and tribophyceans share diadinoxanthin and diatoxanthin—in contrast to zeaxanthin and violaxanthin in phaeophyceans and chrysomeridaleans—is noteworthy. If the plastids of *Phaeothamnion* are related to those of tribophyceans (as this xanthophyll evidence seems to show), presence of fucoxanthin in *Phaeothamnion* suggests the possibility of its loss in the ancestors of modern tribophyceans.

PART 2—TRIBOPHYCEAE

This class of ochrophyte algae is named for the genus *Tribonema* (and known colloquially as tribophytes or tribophyceans), but has in the past been referred to as the Xanthophyceae (colloquially, xanthophytes or xanthophyceans). The change in name has occurred in response to a modern trend of naming higher level taxa after type genera. There are about 90 genera and 600 tribophycean species, occurring primarily in freshwaters or soil; many forms are regarded as rare (Hibberd, 1990b). Cell walls are thought to be composed primarily of cellulose (with silica sometimes also present), and sometimes consist of two overlapping halves. The main photosynthetic storage product is lipid, in the form of cytoplasmic droplets. The soluble polysaccharide chrysolaminarin is probably also present, within a cytoplasmic vacuole. Importantly, starch is not present (see below). The cells usually possess several to numerous discoid plastids arranged at the periphery of the cytoplasm, and in many species, pyrenoids are present within the plastids. In addition to chlorophyll *a*, the porphyrin derivative chlorophyll *c* is also present as in most other ochrophytes, though in low amounts. Accessory pigments include β-carotene, the xanthophylls diatoxanthin and diadinoxanthin (mentioned above), and other minor pigments (see Table 1–1). Tribophyceans are distinguished by the absence of fucoxanthin, an accessory pigment otherwise widely present among ochrophytes, whose abundance in their plastids is responsible for the golden or brown color of most ochrophytes. Consequently, the chloroplasts of tribophyceans appear green or yellow-green, making tribophyceans difficult to distinguish from green algae.

Tribophyceans include a variety of morphological types, including flagellates (such as *Chloromeson*), coccoid forms (e.g., *Characiopsis* and *Ophiocytium*),

Table 15–1 Occurrence of protective xanthophylls in algal groups

Group 1 zeaxanthin	Group 2 zeaxanthin/violaxanthin	Group 3 diadinoxanthin/diatoxanthin
cyanobacteria red algae cryptomonads	chrysophyceans chrysomeridaleans phaeophyceans green algae (mosses) (ferns) (gymnosperms) (angiosperms)	euglenoids dinoflagellates raphidophyceans *Phaeothamnion* tribophyceans

filaments (e.g., *Tribonema*), and siphonaceous coenocytes (large spherical or tubular multinucleate cells) (*Botrydium* and *Vaucheria*), as well as palmelloid colonies and rhizopodial (amoebalike) forms. A number of parallel morphologies occur among the green algae. Some examples include the green algal genus *Chlorella*, which is outwardly similar to the tribophycean *Chloridella*; the green *Characium*, sometimes confused with *Characiopsis*; *Microspora*, a genus similar to *Tribonema*; and *Protosiphon*, which is much like *Botrydium*. Tribophyceans can be distinguished from green algae by use of iodine-potassium iodide solution; the intraplastidal starch of the green forms will be stained blue-black, whereas no staining will occur in cells of the starch-free tribophyceans (see Fig. 20–47).

Tribophyceans may also resemble eustigmatophycean taxa (Chapter 13) so closely that they generally cannot be distinguished without the use of chromatography to assess pigment composition, or ultrastructural comparison. Ultrastructural examination of a number of coccoid forms that were previously thought to be members of the Tribophyceae has revealed that some are more accurately placed within eustigmatophyceans (Santos, 1996). Cell division has not been widely studied among tribophyceans, but furrowing appears to be the most common form of cytokinesis (Sluiman, 1993).

Asexual reproduction can occur by means of nonflagellate cells (autospores, aplanospores, or thick-walled cysts) or by flagellate *Ochromonas*-like zoospores, depending upon the genus or environmental conditions. Autospores (which are developmentally distinct from zoospores), aplanospores (spores that develop similarly to zoospores up to the point of flagellar production), and zoospores are typically produced by division of cellular cytoplasm into one, two, or several subunits that are ultimately released from the confines of parental cell walls. Their function is to increase and disperse the population. Cysts more typically form by modification of entire vegetative cells. These favor survival of species through conditions unsuitable for growth. Cyst walls may be impregnated with silica, and consist of two overlapping parts. Sexual reproduction (isogamy or oogamy) has been observed in only a few cases. Flagellate species, zoospores, and flagellate gametes usually have two typical ochrophyte flagella—a longer anterior one with mastigonemes occurring in two opposite rows, and a shorter posterior one lacking stiff hairs. *Vaucheria* is an exception in that zoospores of this genus are multiflagellate (Goldstein, 1992).

The flagella are typically heterokont in being of different lengths. The typical ochrophyte photoreceptor system (see Chapter 12) is present in most tribophyceans. There is a photoreceptor swelling on the posterior flagellum, and it is aligned such that the swelling lies near an eyespot located within the plastid (see Fig. 12–1) (Hibberd and Leedale, 1971; Hibberd, 1990b). *Vaucheria* zoospores and sperm are exceptions in lacking eyespots.

Tribophycean Diversity

A number of genera not covered here (*Botrydiopsis, Heterococcus, Mischococcus, Xanthonema, Bumilleriopsis, Pleurochloris, Sphaerosorus, Chlorellidium*) have

Figure 15–8 *Chloridella.*

autospore formation, much like the green alga *Chlorella.*

CHARACIOPSIS (*Characium*, a genus of green algae [Gr. *charax*, pointed stake] + Gr. *opsis*, likeness) (Fig. 15–9) is a spindle-shaped unicell that is typically attached to the cell walls of larger algae, much like the green alga *Characium* (and similar-appearing forms— see Chapter 20) and the eustigmatophycean *Pseudocharaciopsis.*

OPHIOCYTIUM (Gr. *ophis*, serpent + Gr. *kytos*, cell) (Fig. 15–10) has distinctive elongate cells that may be straight, curved, or spirally twisted, sometimes with spines at the ends. Cells are attached to surfaces by a small stalk. Sometimes cells are aggregated into a colonylike arrangement. Such cell clusters result from the attachment and germination of multiple zoospores at the distal end of a parental cell. The cell wall is composed of two interlocking parts of unequal size, the apical portion being smaller. Cells can elongate by the addition of more cell wall units at the distal end. Zoospores or aplanospores are released by detachment of the lidlike apical wall piece. There are multiple discoid plastids per cell. *Ophiocytium* is widely distributed in freshwaters.

STIPITOCOCCUS (L. *stipes*, stem + Gr. *kokkos*, berry) cells are enclosed in a delicate, transparent lorica borne on a thin elongate stalk (Fig. 15–11). The cells are flagellate, and the flagella extend through small pores in the lorica. *Stipitococcus* grows attached to filamentous algae. Each cell can produce two zoospores (Hibberd, 1990b).

been demonstrated to belong to the Tribophyceae by SSU rDNA sequence analysis (Potter, et al., 1997b), ultrastructural examination (Hibberd and Leedale, 1971), or both. Here we have included descriptions of tribophycean genera that are most frequently encountered in nature, particularly forms that are often confused with morphologically similar, common green algae. Use of a starch test reagent (a solution of iodine and potassium iodide) is very helpful in distinguishing tribophyceans from green algae. Except as noted, these genera are found in freshwaters.

CHLORIDELLA (Gr. *chloros*, green) (Fig. 15–8) is a unicellular, spherical form that reproduces by

Figure 15–9 *Characiopsis.* (From Smith, G. M. *Fresh-Water Algae of the United States.* ©1950 McGraw-Hill Companies. Reproduced with permission of the McGraw-Hill Companies)

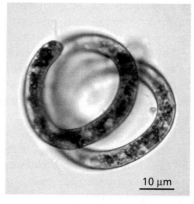

Figure 15–10 A highly elongate and coiled cell of *Ophiocytium* with many plastids.

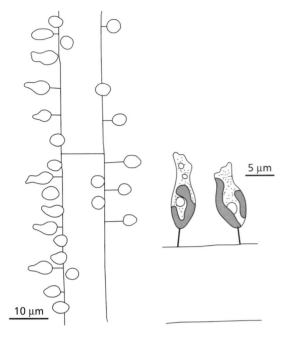

Figure 15–11 *Stipitococcus*. This epiphytic tribophycean often grows on filamentous green algae. (After Smith, G. M. *Fresh-Water Algae of the United States*. ©1950 McGraw-Hill Companies. Reproduced with permission of the McGraw-Hill Companies)

TRIBONEMA (Gr. *tribo*, to rub or wear down + Gr. *nema*, thread) is an unbranched filament having a single row of cells (Fig. 15–12). There are about 25 species. *Tribonema* may occur as bright green floating mats in still waters, including bog pools, and is more abundant in cooler seasons (spring or fall) (Hibberd, 1990b). There is a worldwide distribution in freshwaters. A distinctive feature is the cell wall, which consists of units composed of two rigid, basally attached, open-ended cylinders. Each cell is enclosed by two such cylinders that overlap at their open ends (Fig. 15–13). The closed ends of two back-to-back cylinders form the cross wall between adjacent cells. If a filament is broken, the damaged end cells will typically show what are known as H-shaped pieces (Fig. 15–12). These are views of the double cylinder wall unit described above, in optical section. A new double cylinder is constructed at each cell division, beginning with development of a ring of wall material deposited at the cleavage plane. The ring ultimately closes to generate the intermediate piece (the crossbar of the H). The cell then elongates until the lower half of the upper cell and the upper half of the lower cell occupy the new cell-wall cylinders (Sluiman, 1993).

Cell-wall development is similar in the green alga *Microspora*, which also exhibits H-shaped pieces at the light-microscopic level.

There are one to several plastids per cell, without pyrenoids. Asexual reproduction is by one or two zoospores produced per cell; these are released by separation of parental cell walls at the region of overlap. The ultrastructure of *Tribonema* zoospores and development of their flagellar hairs has been studied (Massalski and Leedale, 1969; Leedale, et al., 1970). Aplanospores and resting cysts are also known, and isogamous sexual reproduction has been reported, but the observations need to be repeated.

BOTRYDIUM (Gr. *botrydion*, small cluster of grapes) thalli are small saclike vesicles that grow in clusters on damp soil, attached by means of subterranean colorless rhizoidal filaments (Fig. 15–14). Individual vesicles may reach several millimeters in size and thus be visible to the unaided eye. *Botrydium* and *Vaucheria* (described next) are exceptions to the general case that tribophyceans are microscopic. The cytoplasm occupies a thin peripheral layer and con-

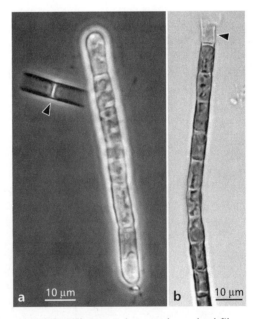

Figure 15–12 *Tribonema* is an unbranched filament characterized by cell walls that occur as two overlapping halves that look like the letter H when focused midway through the cell. The distinctive wall (arrowheads) can be detected by observing individual cell wall pieces that have separated from filaments (a) or broken ends of filaments (b).

Figure 15–13 Cutaway diagram of the cell walls of *Tribonema*, illustrating H-shaped pieces.

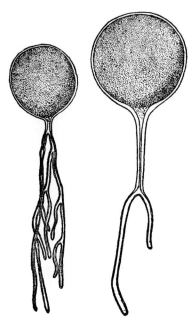

Figure 15–14 *Botrydium*, a coenocytic vesicle containing many plastids and nuclei. Two species are shown. (From Smith, G. M. *Fresh-Water Algae of the United States*. ©1950 McGraw-Hill Companies. Reproduced with permission of the McGraw-Hill Companies)

tains many plastids and nuclei. Hence the thallus morphology is described as coenocytic or siphonous (multinucleate and tubelike, without cross walls). When a water film is present, many uninucleate, biflagellate cells (zoospores or gametes) are released from the sac. When water is less abundant, aplanospores or resistant cysts may be produced. Sexual reproduction is by isogametes or anisogametes; gametes are not formed in specially differentiated structures (compare to *Vaucheria*, below). There are about eight species (Hibberd, 1990b).

VAUCHERIA (named for Jean Pierre Etienne Vaucher, a Swiss clergyman and botanist) thalli are large tubular coenocytes (Fig. 15–15) that branch, with cross-walls formed only where zoospores or gametes are produced. Hence, mitosis has been uncoupled from cytokinesis. The coenocytes are branched and have indeterminate apical tip growth, often forming extensive dark green mats on soil or sand, on brackish banks of salt marshes (Simons, 1974), or submerged in freshwaters or shallow marine waters. There are 70 or more species. The plastids of

some marine species can be harvested for temporary use by ascoglossan molluscs (Chapter 7).

For many years *Vaucheria* was regarded as a green alga, but more recent examination of its pigments and storage products, together with molecular sequence

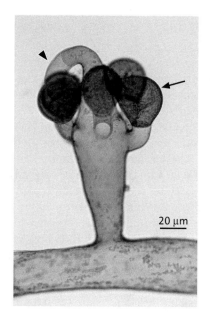

Figure 15–16 A species of *Vaucheria* having gametangia borne on a specialized portion of the coenocyte. A whorl of oogonia (arrow) surround an antheridium (arrowhead).

20 µm

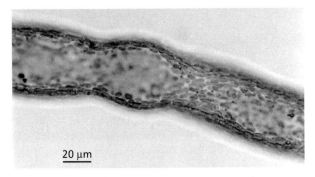

Figure 15–15 A portion of the coenocytic, tubular thallus of *Vaucheria*. The numerous plastids are located in the thin layer of cytoplasm surrounding a large central vacuole.

data, have demonstrated conclusively that *Vaucheria* is linked with other tribophyceans. However, *Vaucheria* seems to have diverged early from the tribophycean clade (Potter, et al., 1997b). The coenocytic thalli have a peripheral layer of cytoplasm (with numerous nuclei and discoid plastids) surrounding a large central vacuole. Organellar streaming is a conspicuous attribute of *Vaucheria*, and presumably functions to circulate materials within its very large cells. Chloroplasts reorient according to changes in the light environment. Ott and Brown (1972) observed closed mitosis (the nuclear envelope remains intact) in *Vaucheria*.

Vaucheria zoospores are distinctive in being very large, multinucleate, and multiflagellate. They result from incomplete cleavage and are described as "compound zoospores" or "synzoospores" (Hibberd, 1990b). Zoospore production occurs in the apical regions of vegetative filaments, where organelles accumulate, and then the tip cytoplasm is segregated from the rest of the coenocyte by cross-wall production. The segregated region functions as a zoosporangium; the process of zoospore development is known as zoosporogenesis. A pair of centrioles is associated with each of the numerous nuclei within the zoosporangium. These centrioles generate a pair of flagella that extend into internal vesicles. These vesicles fuse, resulting in the formation of a large pool of flagella that then migrate to the surface of the developing zoospore. At maturity, naked zoospores are released by rupture or disintegration of the parental wall (Ott and Brown, 1975). The presence of the cross-wall prevents loss of additional thallus cytoplasm upon zoospore release. Upon settling to a substrate, flagella retract into the cell body and a new cell wall develops. Asex-

ual reproduction can also involve large multinucleate aplanospores. Whereas asexual reproduction of *Vaucheria sessilis* was stimulated by cool temperatures (12°–15° C), sexual reproduction appears to be controlled by photoperiod, occurring more readily under long-day (18-hour days) than short-day conditions (eight-hour days), probably as a result of greater production of photosynthate under long-day conditions (League and Greulach, 1955).

Sexual reproduction is oogamous, with production of specialized antheridia (that generate large numbers of colorless, uninucleate sperm) and oogonia (each containing a single, uninucleate egg) borne either on special branches of the main thallus (Fig. 15–16) or directly upon the main thallus. Antheridial structure may be quite variable and useful in defining species. Sperm flagella are interesting in that they are atypical for most ochrophyte groups but somewhat similar to those of certain brown algae—there is a short, mastigoneme-clad anterior flagellum and a long, smooth posterior flagellum. Sperm enter an egg via a pore in the oogonial wall. A thick-walled resting zygote results that typically remains within the oogonial wall, attached to the parental thallus. Although some experts suggest that meiosis is zygotic (occurring when zygotes germinate to form new siphonous coenocytes) (Hibberd, 1990b), other workers believe meiosis to occur during gamete formation (Al-Kubaisy, et al., 1981), as in diatoms and some groups of brown algae, but not most other ochrophytes. Clearly, additional study of the timing of meiosis in *Vaucheria* is needed.

PART 3—PHAEOPHYCEANS (BROWN ALGAE)

The diversity of the brown algae—more than 250 genera and over 1500 species (Norton, et al., 1996)—is considerably greater than that of tribophyceans, and phaeophyceans also produce much greater biomass. The phaeophyceans range in structure from microscopic filaments to giant kelps many meters in length, with significant organ, tissue, and cellular specialization. Many have leaflike blades, stemlike stipes, and rootlike holdfast systems. A few possess translocatory cells bearing a striking resemblance to sieve elements of higher land plants. Large kelps can have productivity rates as high as 1 kg C m^{-2} yr^{-1}, with growth rates highest in the cool season of the year. Though some are annual, others are perennial—some living up to 15 years. Consequently, brown algae can form large biomasses in intertidal and subtidal coastal regions throughout the world, but especially in polar, boreal, and temperate latitudes. The coastal regions richest in phaeophycean species are Japan, Pacific North America, southern Australia, and the British Isles. Phaeophyceans may also produce noticeable growths in tropical and subtropical waters, including the famous Sargasso Sea. In addition, there are a few freshwater forms: *Heribaudiella* (Pueschel and Stein, 1983), some species of *Sphacelaria* (Schloesser and Blum, 1980), *Pseudobodanella*, *Lithoderma*, *Pleurocladia*, and *Porterinema* (Bold and Wynne, 1985). Although freshwater phaeophyceans tend to be inconspicuous, biogeographical and ecological findings suggested to Wehr and Stein (1985) that *Heribaudiella fluviatilis* may be a common (but often overlooked) component of river floras. *Heribaudiella* typically inhabits streams with a current velocity greater than 1 m s^{-1} and a rocky substratum, occurring in both shaded and well-illuminated stretches.

Cell Biology

Cell Walls

The cell walls of brown algae generally contain three components: cellulose, which provides structural support; alginic acid, a polymer of mannuronic and guluronic acids and their Na$^+$, K$^+$, Mg^{2+}, and Ca^{2+} salts; and sulfated polysaccharides. Cellulose is usually a minor constituent of the brown algal cell wall, making up 1–10% of thallus dry weight. Cellulose occurs as ribbonlike microfibrils generated at the cell surface by terminal complexes embedded in the cell membrane that are believed to include cellulose synthesizing enzymes. The terminal complexes of brown algae occur as linear rows of closely packed particles (Tamura, et al., 1996). Each particle consists of two subunits (Reiss, et al., 1996), as do those of the tribophycean *Vaucheria* (Mizuta and Brown, 1992). Linear terminal complexes also occur in red algae and some (but not all) green algae. Terminal complexes are deposited in the cell membrane by fusion of Golgi vesicles; cells can regulate the presence and density of cell membrane cellulose synthesizing complexes by varying the rate at which complex-bearing vesicles are produced and by targeting them to des-

ignated areas. In this way control is exerted over deposition of cellulose microfibrils and, hence, wall development. Brown algal cell membranes also contain arrays of five particles (four corner particles surrounding a central particle) that are hypothesized to produce cell-wall matrix polysaccharides (Reiss, et al., 1996).

Alginates are located primarily within the intercellular matrix, where they confer flexibility upon the thallus, help prevent desiccation, and function in ion exchange. Alginates are the primary matrix component, and in some cases constitute up to 35% of the thallus dry weight. The thalli of most brown seaweeds contain a 1:1 ratio of mannuronic and guluronic acids, but the proportions of these acids can vary with season, age and species, tissue type, and geographic location (Kraemer and Chapman, 1991). Because several investigators had proposed that alginate compositional differences might confer distinct mechanical properties on thalli, providing adaptive advantage in different environments, Kraemer and Chapman (1991) tested this hypothesis. They grew the phaeophycean *Egregia menziesii* for six to ten weeks in flowing seawater at three different current levels, and then tested thallus strength and performed correlative analysis of alginic acid composition. Although *Egregia* grown in the highest energy environments were about twice as strong and stiff as those growing in the lowest current level, there was no correlation with change in percentages of mannuronic and guluronic acids. Thus, alternate explanations for such differences in mechanical strength must be sought.

Fucans (also known as fucoidins or ascophyllans) are polymers of L-fucose and additional sugars that are sulfated. Antibodies have been used to demonstrate presence of fucose-rich fucans in the cell walls of *Fucus*. In these molecules there may be a 1:1 occurrence of fucose and sulfate, suggesting that each sugar monomer is sulfated. Sulfate linkage appears to take place in the Golgi apparatus (Callow, et al., 1978) and may occur only in specialized cells in kelps. *Ascophyllum*—a relative of *Fucus*—generates polymers whose backbone is primarily polyuronic acid with side chains of sulfated xylose and fucose (glucouronoxylofucans) (McCandless, 1981). The functions of fucans are not well understood, but they are thought to play an important role in firmly anchoring fucoid zygotes and germlings to intertidal substrates so that they are not washed off in heavy waves. It is possible that fucans serve a similar function in other brown algae.

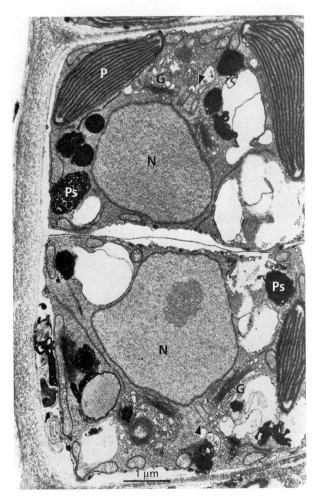

Figure 15–17 Ultrastructural features of recently divided cells of the brown alga *Pylaiella littoralis* include irregular electron-opaque structures that contain tannins, and which are known as physodes (Ps). Note centrioles (arrowheads) located near nuclei (N), and Golgi bodies (G), on each side of the centrioles. P=plastid. (From Markey and Wilce, 1975 ©Springer-Verlag)

Plastids and Photosynthetic Reserve Products

Depending upon the genus, there may be from one to many plastids per cell. Plastids contain chlorophyll *a*, two forms of chlorophyll *c*, β-carotene, violaxanthin (involved in the xanthophyll cycle), and relatively large amounts of fucoxanthin, which confers a brown color. Plastids typically possess a girdle lamella—a thylakoid stack that runs continuously beneath the plastid surface. As is more generally true for ochrophytes, there is a periplastidal endoplasmic reticulum (PER)—a

manifestation of plastid origin by secondary endosymbiosis (Chapter 7). The PER is continuous with the nuclear envelope, as in some other (but not all) ochrophytes. Pyrenoids occur in the plastids of some brown algae, but not others. The photosynthetic reserve material is **laminaran**, a β-1,3-glucan similar to the chrysolaminaran of other ochrophytes. In addition, mannitol, sucrose, and/or glycerol may also also present. Mannitol, a 6-carbon sugar alcohol, may account for 20–30% of the dry weight of brown algal thalli and may constitute up to 65% of the sap translocated in the specialized sieve elementlike cells of kelps (Schmitz, 1981). Low molecular weight compounds such as mannitol, sucrose, and glycerol may serve to lower the freezing point of the cytoplasm—an advantage in cold climates—or may function in balancing cellular osmotic pressure.

Other Cell Constituents

Brown algal cells often contain physodes—highly refractive vesicular bodies—sometimes found within the cell vacuole (Fig. 15–17). These contain tannins consisting of phloroglucinol polymers (also known as

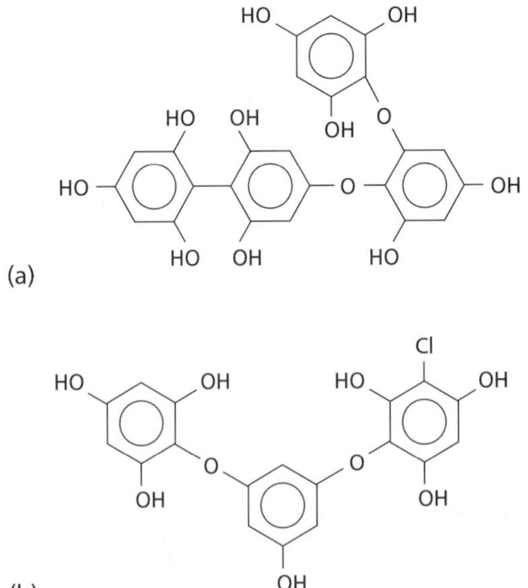

(a)

(b)

Figure 15–18 Chemical structures of some polyphenolic tannins found in brown algae. (a) A polyphloroglucinol having both biphenyl and ether linkages, and (b) a chlorinated polyphloroglucinol. (From Ragan, 1981 in Wynne, M. J. and C. S. Lobban (editors), *Biology of Seaweeds*. ©University of California Press and Blackwell Scientific, Inc.)

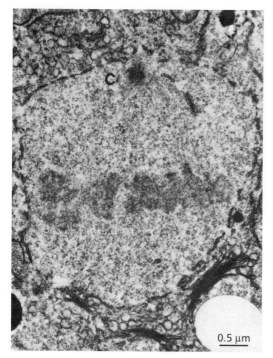

Figure 15–19 Metaphase in *Cutleria cylindrica*, showing one of the polar centrioles (at top) and the nearly intact nuclear envelope. (From La Claire, 1982b)

polyphenolics) (Fig. 15–18) and terpenes, which are thought to function in herbivore-resistance (Chapter 23) and perhaps also in strengthening cell walls by interacting with alginates. Some experimental evi-

dence suggests that phlorotannins can function as inducible UVB (280–320 nm) screens, protecting brown algae from radiation damage. (Pavia, et al., 1997). It has also been suggested that phlorotannins participate in the binding and accumulation of heavy metals (Ragan, 1981). Physodes are stainable with toluidine blue, neutral red, and osmium tetroxide, and are sometimes (but not always) autofluorescent.

The Cytoskeleton and Cell Division

Unlike those of land plants, cells of brown algae have not been observed to contain a cortical (peripheral) microtubular cytoskeleton (Reiss, et al., 1996), and microtubules are generally few in nondividing cells. Significantly, a pair of centrioles occur in interphase cells of (probably) all brown algae (Fig. 15–17), often occurring in a cleft in the nuclear envelope (Brawley and Wetherbee, 1981). The centrioles can generate flagella during development of flagellate reproductive cells, a process that appears to occur readily in all brown algae. Just prior to mitosis, centrioles are duplicated, then pairs of centrioles migrate to opposite poles of the cell early in prophase. Centriolar migration is an early event in the establishment of division polarity, and is critical in determining the direction of cytokinesis. The underlying basis for centriolar migration is unknown, but is under intense study in yeast and animal cells because it is of such fundamental importance in development.

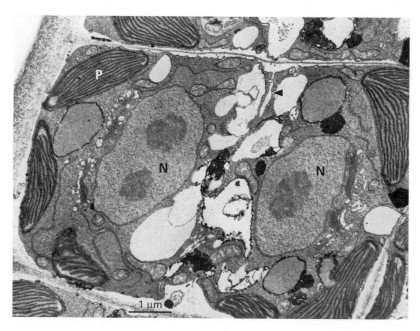

Figure 15–20 Cytokinesis in *Pylaiella littoralis* illustrating the process of furrowing (arrowhead). N=nucleus, P=plastid. (From Markey and Wilce, 1975 ©Springer-Verlag)

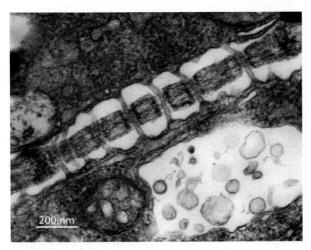

Figure 15–21 Plasmodesmata in the cross walls of a brown alga. (Courtesy J. La Claire)

The nuclear envelope of dividing brown algal cells usually remains intact (Fig. 15–19) (except for small open regions at the pores known as polar fenestrations) until anaphase, when the envelope disintegrates completely (Brawley and Wetherbee, 1981). Following the end of mitosis, cytokinesis in most brown algae occurs by infurrowing of the cell membrane (Fig. 15–20); the developing furrow is associated with microtubules. In *Fucus* and the related *Ascophyllum*, a cell plate may develop from the center of the cell toward the periphery (in a centrifugal direction), as occurs in land plants—an example of parallel evolution. Mitosis and cytokinesis in some brown algae (*Petalonia fascia* and *P. zosterifolia*) are known to peak 2–4 hours after sunset and to be completed in about two hours (Kapraun and Boone, 1987).

Plasmodesmata are generated during cytokinesis, and probably occur in the cross walls of all brown algae (Fig. 15–21), whether they divide by furrowing (La Claire, 1981) or by cell-plate formation. This is in contrast to green algae, where occurrence of plasmodesmata is tightly linked to cell-plate formation (Cook and Graham, 1999). Though the plasmodesmata of many brown algae examined to date appear to lack internal structure (desmotubules and other features characteristic of embryophytic plasmodesmata) (Bisalputra, 1966), such internal structure has been noted in a few cases (Cook and Graham, 1999). Plasmodesmata confer a high degree of intercellular continuity, and allow symplastic (cytoplasm to cytoplasm) communication. For example, modified plasmodesmata allow for the cell-to-cell transport of

photosynthates through the internal translocatory systems of kelps.

Growth Modes and Meristems

Thalli of brown algae may be basically filamentous or composed of tissues analogous to those of land plants. Some taxa are made up of individual filaments, whereas thallus formation in others results from aggregation of filaments to form a more robust body: crusts, erect and branched axes, or blades. Thalli composed of aggregated filaments (but lacking true tissues) are described as pseudoparenchymatous. In contrast, the tissues of other brown algae are described as parenchymatous, and can occur as solid axes, blades, or complex thalli with specialized blade, stipe, and holdfast regions. Parenchymatous thalli develop by cell division in various planes, yielding tissues analogous to those of higher plants. As in plants,

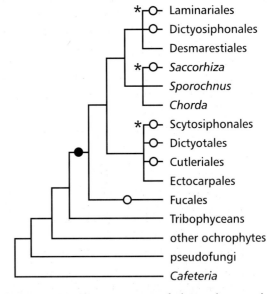

Figure 15–22 The occurrence of plasmodesmata (solid circle, indicating that this is a plesiomorphic feature of brown algae) and parenchymatous construction (open circles) are mapped onto a phylogenetic tree showing relationships of brown algae inferred from molecular sequence data by Tan and Druehl (1996). The distribution suggests that parenchyma evolved several times within brown algae, and that plasmodesmata are necessary, but not sufficient, for the evolutionary origin of tissues. Asterisks indicate clades in which branching order is as yet uncertain. (Character mapping from Cook and Graham, 1999)

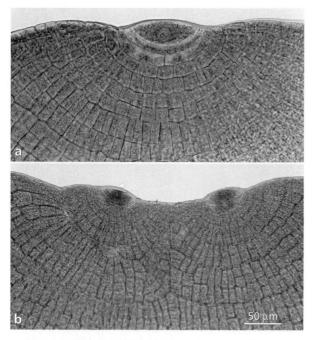

Figure 15–23 (a) A parenchyma-forming apical cell of *Dictyota*. (b) Daughter apical cells resulting from the division of the apical cell. Further production of separate tissues by each daughter cell will result in dichotomous branching of the thallus.

cell and tissue differentiation may occur. Interestingly, phylogenies based upon molecular sequences suggest that the parenchymatous growth mode arose independently in separate brown algal lineages (Fig. 15–22) (Cook and Graham, 1999).

Some early divergent brown algal lineages have what is known as diffuse growth—there is no localized meristematic region; rather, cell division occurs throughout the thallus. In contrast, among other brown algae there are a variety of well-defined types of meristems—localized regions of cell division. The

absence or presence of a localized meristem (and if present, its form) strongly influence thallus morphology. Hence, growth mode and meristem type have been used to define brown algal orders.

The simplest type of brown algal meristem has a single apical cell that repeatedly divides transversely, yielding a filament that is one cell in width (a uniseriate filament) (see Fig. 15–35). An unusual type of meristem, known only from certain brown algae, is development of a meristematic cell at the base of a hair; this is known as trichothallic growth (derived from the Greek word *thrix*, meaning hair). In other phaeophyceans there is an apical meristem composed of one or more cells that can divide in additional directions, generating three-dimensional cylindrical, ribbonlike, or thickened and branched thalli such as that of *Fucus* (Fig. 15–1). An extensive row of apical cells occurs in *Padina*, generating a fanlike thallus (Fig. 15–44), and smaller apical cell arrays occur in related algae, such as *Dictyota* (Katsaros, 1995) (Fig. 15–23). In the kelps there is typically an intercalary meristem, located between stipe and blade(s), that generates tissues in two directions, increasing the length of the thallus. Kelps also possess a surface meristematic region whose activity increases girth; this is known as the meristoderm. Meristoderm activity could be considered somewhat analogous to that of the cambia of higher land plants. Development of the intercalary meristem in *Chorda*—a member of a lineage that also includes the Laminariales (Fig. 15–24)—has been intensively studied (Fig. 15–25) (Kogame and Kawai, 1996). The growth of *Chorda* sporophytes occurs in three distinct stages: early diffuse growth, growth from a basal meristem, and finally, intercalary meristematic growth (Fig. 15–26). Such developmental transitions suggest recapitulation of possible stages in the evolution of derived types of brown algal meristems.

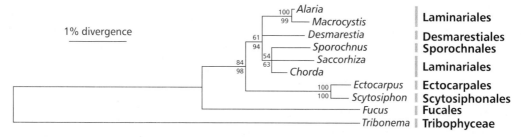

Figure 15–24 A phylogeny inferred for brown algae by comparative analysis of molecular sequences from nine phaeophycean genera; *Tribonema* has been used as the sister group (outgroup) in this analysis. (From Tan and Druehl, 1996 by permission of the *Journal of Phycology*)

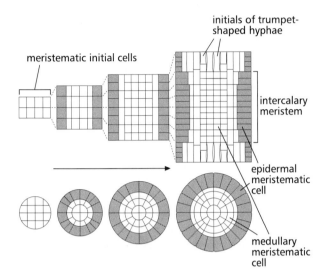

Figure 15–25 Development of the intercalary meristem of *Chorda filum*, as viewed in longitudinal section (top row) and in cross section (lower row). Meristematic cells (shaded) differentiate from unspecialized initials. (After Kogame and Kawai, 1996 ©Blackwell Science Asia)

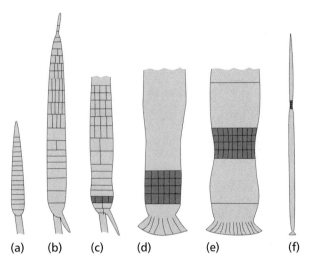

Figure 15–26 Growth stages in *Chorda filum*. Growth begins with diffuse (nonlocalized) cell divisions (a, b), is followed by a stage of basal meristematic activity (shaded region) (c, d), and culminates in development of a well-defined intercalary meristem (shaded region) (e, f). (After Kogame and Kawai, 1996 ©Blackwell Science Asia)

Reproduction

With the exception of *Fucus* and close relatives (the orders Fucales and Durvillaeales), reproductive cells and life histories of brown algae are remarkably similar. There are three types of flagellate reproductive cells—meiospores, asexual zoospores, and gametes. Often these possess typical ochrophyte heterokont flagella, but in some of the brown algae the anterior flagellum is shorter than the posterior one, and in some cases there is only a single anterior flagellum. Flagella are lacking altogether from the egg cells of some oogamous species. Phaeophycean motile cells are distinctive in that their flagella typically emerge laterally, rather than apically or subapically as in other ochrophytes (Fig. 15–27). Swimming involves beating of the anterior but not the posterior flagellum. The anterior flagellum of zoospores, meiospores, and settling gametes possess a coiled region (or knob) that is involved in attachment to the substrate (Goldstein, 1992). Flagellate reproductive cells of brown algae usually have eyespots, located within the chloroplast, but the sperm of one group lack eyespots, and both the sperm and the meiospores of certain forms lack eyespots. The ultrastructure of *Ectocarpus* male gametes was studied in detail by Maier (1997a, b).

Some of the brown algae are isogamous; the flagellate gametes are morphologically indistinguishable while motile. However, one gamete typically settles soon after it is released, whereas the other gamete actively swims for a longer period of time (see, for example, the study by Clayton [1987] of *Ascoseira mirabilis*). Thus even in the case of isogamy, the gametes can be functionally differentiated. The zygotes of isogamous species can often be recognized by the presence of two eyespots, one contributed by each gamete. Isogamous brown algae exhibit biparental inheritance of mitochondria and plastids. Other phaeophyceans are structurally anisogamous, with one flagellate gamete larger than the other. Oogamy (a nonmotile or only transiently flagellate egg cell and smaller flagellate sperm cell) also occurs in some groups. Oogamous taxa typically exhibit maternal inheritance of organelles but paternal inheritance of centrioles (Lewis, 1996b), but see Clayton, et al. (1998) for an exception.

Gametes are produced by a multicellular gametophyte generation, within specialized gametangia, variously known as plurilocular (pleurilocular) gametangia, plurisporangia, mitosporangia, or plurilocs (Fig. 15–28). It may seem quite confusing that gametes are generated within structures often

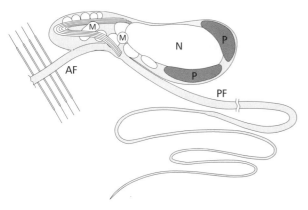

Figure 15–27 Diagrammatic representation of a laminarialean sperm cell, showing lateral emergence of heterokont flagella. AF=anterior flagellum, M=mitochondrion, N=nucleus, P=plastid, PF=posterior flagellum. (From Henry and Cole, 1982b by permission of the *Journal of Phycology*)

mon among brown algae exhibiting isogamy or anisogamy (where both gametes possess flagella) and less common in oogamous species. The absence of parthenogenesis is correlated with the lack of centrioles, which allow development of a division apparatus. Occurrence of parthenogenesis in some oogamous brown algae has been attributed to the rare presence of centrioles in eggs (Clayton, et al. 1998).

Reproductive cells are usually released from plurilocular gametangia through a single apical pore, but in some forms there is a series of peripheral openings through which zoospores or gametes escape. In the case of oogamous phaeophyceans, plurilocular gametangia are considered to have been evolutionarily reduced to single cells lacking internal compartmentation. Nevertheless, each cell generates only a single reproductive cell as in more typical plurilocs (Wynne, 1981). In many cases it is known that low molecular weight, highly volatile hydrocarbons pro-

described as "sporangia," but there is an explanation. Plurilocular gametangia (sporangia) are generated by mitotic divisions and are internally subdivided into chambers, each containing a single developing reproductive cell (Fig. 15–28). In many cases the cells produced by plurilocs are gametes that fuse in pairs during mating. However, plurilocs sometimes produce asexual zoospores that develop (without mating) into thalli similar in morphology to the parent—a situation for which use of the term sporangium is appropriate. In *Ectocarpus* and some other genera, production of zoospores (in plurilocs) can result in a series of identical generations, unpunctuated by occurrence of sexual reproduction (Papenfuss, 1935; Pedersen, 1981). Furthermore, unfused gametes of some brown algae, *Ascoseira mirabilis* (Clayton, 1987) and *Colpomenia peregrina* (Yamagishi and Kogame, 1998), for example, readily develop into multicellular thalli. In the latter case, more frequent development of larger, brighter-pigmented unfertilized female gametes (versus smaller, paler male gametes) results in populations of macroscopic gametophytic seaweeds that exhibit a female:male ratio of 19:1. The development of cells produced in plurilocs into multicellular thalli is known as **parthenogenesis** (development of gametes into parental-like thalli without mating). Brown algae that reproduce exclusively by means of plurilocular sporangia-produced zoospores are said to exhibit a direct, or monophasic, life history. This is in contrast to the more common occurrence of alternation of generations (see below). Parthenogenesis is relatively com-

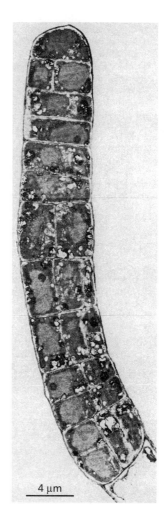

Figure 15–28 A developing plurilocular sporangium (gametangium) of *Cutleria hancockii*. (From La Claire and West, 1979 ©Springer-Verlag)

4 µm

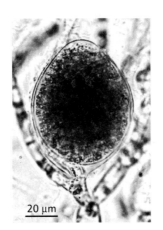

Figure 15–29 A unilocular sporangium of *Tinocladia*. (Photograph courtesy J. La Claire)

20 μm

duced by sessile (female) gametes function as pheromones, acting to attract sperm, and in some cases also inducing antheridial dehiscence. Such pheromones are known from divergent brown algal groups, including the relatively simple *Ectocarpus*, the highly specialized *Fucus*, and the kelps; all of these serve as model systems for the study of fertilization. Thus, pheromones are probably produced by most, if not all, sexually reproducing brown algae. Video observations of *Ectocarpus siliculosis* gametes show that in the absence of pheromone, flagellate gametes swim in a straight line or wide loops, but in the presence of pheromone, they swim in smaller and smaller loops (mediated by beating of the posterior flagellum) until the source of the pheromone (the sessile gamete) has been contacted (Müller, 1981).

Pheromones also form part of the basis for species isolation. The ability of two brown algal taxa to cross or hybridize depends upon successful completion of five stages in reproduction: (1) pheromone recognition, (2) gamete recognition, (3) plasmogamy (gamete fusion), (4) zygote growth into sporophytes, and (5) sporophyte meiosis. Flagellate gametes that are attracted to settled gametes make contact with the tip of the anterior, mastigoneme-covered flagellum. Antibodies and lectins—carbohydrate-binding proteins with high specificity and affinity—have been used to establish that eggs or settled gametes bear specific glycoproteins, while the motile gamete carries receptors for these glycoproteins (see review by Lewis, 1996b). Lectin-binding sites on gametes of *Ectocarpus siliculosis* can be visualized by the use of fluorescence microscopy (Maier and Schmid, 1995). Plasmogamy occurs at its maximal level about 20 minutes after gamete contact, and karyogamy about 90–120 minutes later (Brawley, 1992).

Zygotes of brown algae develop by repeated mitosis and cytokinesis into a multicellular sporophyte generation. Plurilocs may occur on the sporophytes of some (but not all) brown algae. Such plurilocs produce flagellate zoospores that either grow into sporophytes resembling the parent, or develop into smaller thalli (microthalli) with plurilocs. Sporophytes are unique in producing another type of reproductive structure, **unilocular sporangia** (also known as unisporangia, meiosporangia, or unilocs) (Fig. 15–29). Unilocs do not occur on thalli of the gametophyte generation.

The first division within uniloc initials is often assumed to be meiotic. However many cases are known where unilocs produce reproductive cells without completing meiosis. In such cases the reproductive cells have the same chromosome number (DNA level) as the parent thallus. In contrast, when meiosis occurs (evidenced in a number of cases by the presence of synaptinemal complexes—paired homologous chromosomes as viewed by transmission electron microscopy), the resulting daughter nuclei are haploid. Sex-determining chromosomes are known to occur in at least some brown algae. In *Saccorhiza polyschides*, for example, a large X chromosome and a smaller Y chromosome pair are segregated at meiosis (Evans, 1965). In some phaeophyceans only four meiospores—known as tetraspores—are produced per uniloc. In other phaeophyceans each meiotic product may continue to divide mitotically, yielding 16, 32, 64, 128, or more haploid meiospores. However, no intervening wall deposition occurs. Consequently, unilocular sporangia (Fig. 15–29) are not subdivided into chambers as are plurilocular sporangia. Environmental factors may determine whether unilocs or plurilocs are produced by sporophytes. In *Ectocarpus siliculosis*, for example, unilocular sporangia are produced when the temperature is below 13° C, whereas above 13° C only plurilocs are produced (Clayton, 1990).

In some exceptional situations, such as occurs in the widespread *Elachista fucicola*, flagellate cells produced in unilocs develop directly into thalli similar to that of the parent without undergoing life history phase change. This is another example of the so-called direct or monophasic type of life history. More generally, however, meiospores develop (by repeated mitosis and cytokinesis) into multicellular gametophytes that produce gametes in plurilocs. Such brown algae have a biphasic life cycle, commonly described as alternation of generations (Fig. 15–30). Although the sporophyte generation is commonly thought of as the diploid phase, and the game-

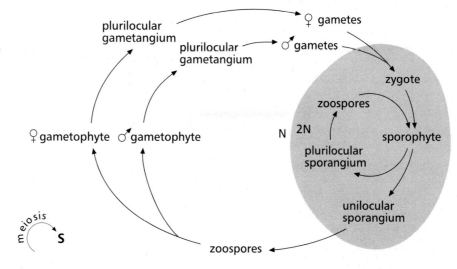

Figure 15–30 The life cycle of *Ectocarpus*.

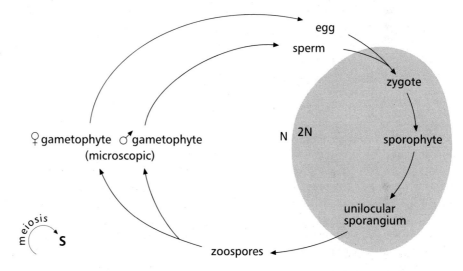

Figure 15–31 The life cycle of *Laminaria*.

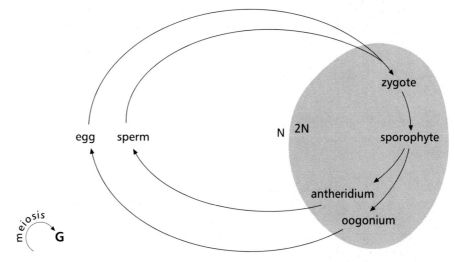

Figure 15–32 The life cycle of *Fucus*.

tophyte as the haploid phase, polyploidy and other chromosome-level variation may occur. In some cases the two alternating life-history phases are nearly indistinguishable—this is known as isomorphic alternation of generations. In other cases the two alternating phases are distinctly different (heteromorphic alternation of generations) (Fig. 15–31). Most phaeophyceans exhibiting heteromorphic alternation of generations produce sporophytes that are larger than the gametophyte. The giant kelps are notable examples of extreme size divergence, with sporophytes growing to lengths of 60 meters or so and tiny gametophytes consisting of at most a few cells. In other instances (Scytosiphonales), the sporophyte is the smaller and less conspicuous phase. Heteromorphic alternation of generations is regarded as an adaptive response to seasonal changes in selective pressures, such as herbivory, and thus is more common in temperate, boreal, and polar seaweeds than in tropical forms (John, 1994).

Alternation of thallus form (also known as phase change) is not always associated with sexual reproduction or change in chromosomal level. For example, two morphologically distinct thallus types, known by the generic names *Feldmannia* and *Acinetospora*, were discovered to actually represent different ecological manifestations of the same taxon. In the same habitat, the *Feldmannia*-form occurred only in the warm season, whereas the *Acinetospora*-form was observed only in the cold season (Pedersen, 1981). Other examples include *Petalonia* (a blade) and *Scytosiphon* (a tubular form) that alternate with microscopic crustose or filamentous stages (Wynne, 1981); in some cases, sexual reproduction is involved in alternation of these generations, and in some cases it is not. Additional examples of the production of gametophytes, sporophytes, or both, that are not correlated with change in chromosome level (ploidy) are described by Lewis (1996a). Such examples emphasize that chromosome level is not the only, or perhaps even the major, determinant of morphological and reproductive expression in phaeophyceans, and that more attention needs to be paid to interactions between environment and genome in eliciting specific structure and reproduction.

Fucus and related forms (Fucales and Durvillaeales) do not exhibit alternation of generations. Rather, meiosis followed by mitotic divisions occurs during the formation of gametes, and the multicellular, macroscopic thalli are regarded as diploid, though many cases of polyploidy occur in these and other brown algal groups (see review of brown algal chromosome numbers by Lewis, 1996a). Thus *Fucus* and relatives appear to exhibit gametic meiosis (Figs. 1–23, 15–32), as do diatoms (Chapter 12), metazoa, and most likely, various tropical green seaweeds (Chapter 18). The presence of tiny gametophytes that develop and generate gametes while confined within the tissues of the larger sporophytic thallus in Syringodermatales and Ascoseirales (Clayton, 1987) suggests that the life history of fucalean and durvillaealean algae might have arisen by reduction and retention by the sporophyte of the gametophyte phase as suggested by Jensen (1974). The evolutionary origin and adaptive advantage of gametic meiosis in brown algae is of widespread interest (e.g., Bell, 1997), but is still relatively poorly understood.

Phaeophycean Diversity and Systematics

Characteristics of thallus organization, growth mode, type of sexual reproduction (isogamy, anisogamy, or oogamy), and life history have been used to segregate phaeophycean genera into 12 or more orders (Clayton, 1990). Table 15–2 summarizes the characteristics of most of the major brown algal orders as they are classically defined. Relatively recently erected orders include Syringodermatales (Henry, 1984) and Ascoseirales (Moe and Henry, 1982; Clayton, 1987). Some of these orders are controversial, and molecular systematic studies (Fig. 15–24) suggest that some orders, such as the group including the giant kelps—the Laminariales—may not be monophyletic (Tan and Druehl, 1996). Thus some future systematic changes may be necessary. Peters and Clayton (1998) suggested the formation of a new order, Scytothamnales, based on distinctive plastids (stellate with pyrenoid, occurring singly in the center of cells) and SSU rDNA data. (In contrast, other brown algal orders have cells with single parietal plastids or multiple discoid plastids with a reduced or no pyrenoid.) These authors also note that SSU rDNA analyses may not provide a comprehensive view of phaeophycean phylogeny because of low numbers of informative sites. Hence additional types of evidence may be required to sort out phaeophycean relationships. In the remainder of this chapter, members of some of the most distinctive, and commonly encountered orders are described, but comprehensive coverage has not been attempted.

Table 15–2 A summary of the characteristics of major orders of Phaeophyceae

order	thallus type	growth mode	sexual reproduction	life-history type
Ectocarpales	uniseriate filaments, branched or unbranched	usually diffuse	morphological isogamy with functional anisogamy, or oogamy	isomorphic alternation of generations
Sphacelariales	branched multiseriate filaments	apical cell	isogamy, anisogamy, or oogamy	usually isomorphic
Dictyotales	parenchymatous	apical/marginal	oogamy	isomorphic
Scytosiphonales	small filamentous sporophyte; large parenchymatous gametophyte	sporophyte–apical; gametophyte–intercalary	anisogamy	heteromorphic
Cutleriales	parenchymatous; gametophyte larger than sporophyte	sporophyte–apical; gametophyte–trichothallic	anisogamy	heteromorphic
Dictyosiphonales	filamentous gametophyte; parenchymatous sporophyte	sporophyte–diffuse; gametophyte–apical	isogamy or anisogamy	heteromorphic
Chordariales	uniseriate filaments that may form pseudoparenchymatous disks, cushions, cords	sporophyte–trichothallic; gametophyte–apical	isogamy with functional anisogamy	heteromorphic
Sporochnales	pseudoparenchymatous; sporophyte larger than gametophyte	trichothallic	oogamy	heteromorphic
Desmarestiales	small filamentous gametophyte; larger pseudoparenchymatous sporophyte	trichothallic	oogamy	heteromorphic
Laminariales	small filamentous gametophyte; large parenchymatous sporophyte	sporophyte–intercalary; gametophyte–apical	oogamy; eggs sometimes flagellate	heteromorphic
Fucales	parenchymatous	apical cell	mostly oogamy	apparently only a single diploid generation; gametic meiosis
Durvillaeales	parenchymatous	"diffuse" at apical portions of thalli; no single apical cell	oogamy	one diploid generation; gametic meiosis
Syringodermatales	parenchymatous sporophyte; 2–4 celled gametophyte	marginal/apical	isogamy	heteromorphic; gametophytes not free-living
Ascoseirales	parenchymatous sporophyte	sporophyte–intercalary	isogamy	heteromorphic; gametophytes not free-living

Figure 15–33 The ectocarpalean genus *Elachista* often grows as tufts of filaments on the surfaces of larger seaweeds such as *Fucus* (shown here).

Ectocarpales

Some experts have recently advocated aggregation of the orders Ectocarpales, Chordariales, Dictyosiphonales, Tilopteridales, and Scytosiphonales into a single order, Ectocarpales. *rbcL* and spacer sequence evidence supports such a move but additional information is needed (Siemer, et al., 1998). Included within the classical definition of Ectocarpales are branched or unbranched filaments whose erect portions often grow from basal filamentous or crustose systems. The filaments often occur in tufts attached to substrates, such as larger seaweeds (Fig. 15–33), and some are endophytic. There may be one to many chloroplasts with pyrenoids. Growth is usually diffuse—a well-defined meristem is absent from most forms. Sexual reproduction is by isogamy or anisogamy, and involves isomorphic alternation of generations. A direct monophasic life history also occurs. Ectocarpaleans are common and widespread.

ECTOCARPUS (Gr. *ektos*, external + Gr. *karpos*, fruit) (Fig. 15–34) is widespread throughout the world, growing on rocks or larger seaweeds, and can be a significant ship-fouling organism and cause problems in mariculture. The branched filaments are uniseriate,

and cells contain several elongate, irregularly shaped chloroplasts, each having several pyrenoids (Fig. 15–35). Life history in *E. siliculosis* is biphasic and isomorphic, with both unilocs and plurilocs on the sporophyte generation, and gametophytes producing only plurilocs (Fig. 15–30). Numerous clones of the sporophyte generation result from zoospore production. After release, zoospores readily attach to substrates via adhesive material produced in the Golgi apparatus and extruded from cells (Baker and Evans, 1973a).

The gamete attractant exuded by settled gametes of *Ectocarpus siliculosus* was identified by gas chromatography, mass spectrometry, infrared spectrometry, and nuclear magnetic resonance (NMR) spectroscopy as allo-cis-1-(cycloheptadien-2',5'-yl)-butene-1 (Müller, et al., 1971). This was the first algal pheromone to have been chemically defined, and is known as sirenin, ectocarpin, or ectocarpene. Viruses are known to attack *Ectocarpus*, resulting in reduction of reproductive capacity (see review by Correa, 1997). Hybridization has been extensively studied (see review by Lewis, 1996b). Molecular systematic investigations conducted on nuclear ITS sequences (see Chapter 5) and the spacer region of the plastid-encoded Rubisco cistron from geographically diverse strains of *E. fasciculatus*, *E. siliculosus*, and the related genus *Kuckia* confirmed the validity of these taxa (Stache-Crain, et al., 1997).

STREBLONEMA (Gr. *streblos*, twisted + Gr. *nema*, thread) (Fig. 15–36) usually occurs as an endophyte. *Streblonema* occurs particularly in kelps such as *Macrocystis*, or red algae such as *Grateloupia*, and is a suspected cause of diseases in some other economically important seaweeds, such as *Undaria*. *Streblonema* lives within the intercellular spaces of its hosts, but retains typical brown coloration. Macroscopically, it appears as discolored spots on the host, with just the reproductive structures extending from the host's surface (Correa, 1997). *Laminarionema* is another ectocarpalean that occurs in sporophytes of the kelp *Laminaria*, and molecular analyses suggest that endophytism evolved more than once in the Ectocarpales (Peters and Burkhardt, 1998). Most older kelp thalli appear to be infected by ectocarpalean endophytes (Kawai and Tokuyama, 1995).

RALFSIA (named for John Ralfs, a British phycologist) occurs as radiating crustose layers composed of adherent filaments (Fig. 15–37). The thalli occur on intertidal rocks and superficially resemble lichen or

Figure 15–34 A herbarium specimen (seen life size) of *Ectocarpus* prepared in 1890 by a private collector.

fungal growths. Cells characteristically have only a single parietal plastid (Wynne, 1981). Alternation of isomorphic generations is suspected. Some species described as *Ralfsia* are actually crustose stages associated with the life cycles of other seaweeds, and thus are inaccurately placed within this genus.

Chordariales

This order is sometimes merged with the Ectocarpales, but is often segregated on the basis of occurrence of heteromorphic life histories (Bold and Wynne, 1985). The gametophyte is the smaller generation, and the sporophyte consists of heterotrichous filaments or macroscopic thalli composed of pseudotissue (pseudoparenchyma). Growth can be diffuse or apical, and reproduction is isogamous. Commonly included genera are *Elachista* (below), *Leathesia*, *Analipus*, *Myrionema*, *Chordaria*, and *Eudesme*.

Reproduction and sexuality of Chordariales were reviewed by Peters (1987).

ELACHISTA (*Elachistea*) (Gr. *elachistos*, smallest) (Fig. 15–38) grows as tufts on *Fucus* (Fig. 15–33) or within depressions in thalli of *Sargassum*. The

Figure 15–35 A light microscopic view of an *Ectocarpus* filament revealing the irregularly shaped plastids.

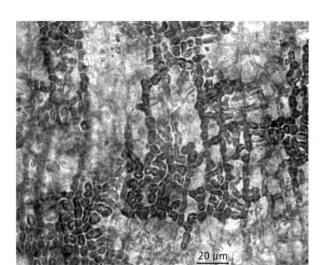

20 μm

Figure 15–36 *Streblonema* (dark filaments) creeping along the surface of a *Cutleria* gametophyte. (Photograph courtesy J. La Claire)

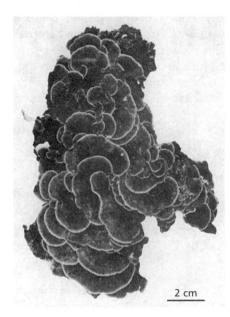

Figure 15–37 *Ralfsia*. (From Edelstein, et al., 1968 by permission of the *Journal of Phycology*)

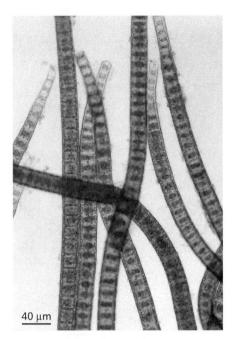

Figure 15–38 *Elachista* occurs as filaments that are commonly branched only at the bases.

macroscopic sporophyte consists of a colorless basal portion that may be embedded within the host's thallus, from which emerge uniseriate filaments, branched only at the base. There are numerous discoid plastids per cell. Culture studies suggest that it can reproduce asexually via zoospores, or produce meiospores within unilocs. Meiospores grow into a microscopic gametophytic stage.

Scytosiphonales

Members of this group have a single plastid per cell, each with a conspicuous pyrenoid. Growth is reportedly diffuse, but thalli are tissuelike tubes or small blades. Sexual reproduction is isogamous or anisogamous, and there is an alternation of generations between a macroscopic phase and a less conspicuous crustose phase (resembling *Ralfsia*). For example, gametophytic thalli of *Scytosiphon canaliculatus* alternate with the sporophytic crustose form known as *Hapterophycus canaliculatus*. The sporophytes occur in late autumn and winter, producing unilocs in response to specific temperature and photoperiod cues (15° C and short days). Subsequent development of the macroscopic, erect gametophytes occurs in early spring, and then gametophytes disappear in summer (Kogame, 1996). Alternation of phases appears to sometimes occur in the absence of sexual reproduc-

tion, influenced by changes in environmental conditions (Pedersen, 1980). The larger phase produces only plurilocs, whereas unilocs occur only on the crustose stage. SSU rDNA sequence data link Scytosiphonales with Ectocarpales at a high level of confidence (Fig. 15–24).

Figure 15–39 *Scytosiphon* thalli are tube-shaped and are constricted at intervals.

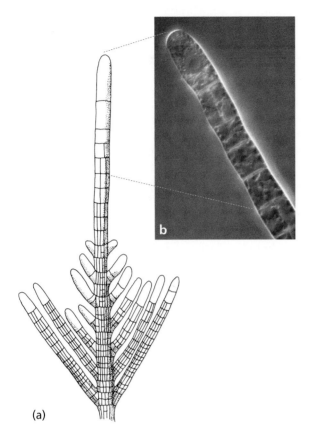

Figure 15–41 Diagram of a typical sphacelarialean asexual propagule. (After Bold and Wynne, 1985)

Figure 15–40 (a) Diagram of the apical portion of a Sphacelaria thallus. The tip of a filament is shown in (b). Note the nucleus and nucleolus in the apical cell. (a: Courtesy W. Woelkerling; b: Photograph courtesy J. La Claire)

SCYTOSIPHON (Gr. *skytos*, skin or hide + Gr. *siphon*, tube) is widespread but primarily occurs in temperate waters. The conspicuous gametophytic thalli are usually tube-shaped (though sometimes flattened) and, when mature, are noticeably constricted at regular intervals (Fig. 15–39). They are attached to substrates via a small basal disk. Uniseriate plurilocs may extend over the entire surface, interspersed with inflated hairs (paraphyses) that might be mistaken for unilocs. The sporophytic phase is an inconspicuous filamentous crust. There is evidence for day length influence on phase change in *Scytosiphon lomentaria*, mediated by blue light (Dring and Lüning, 1975). The adaptive significance of a heteromorphic life history in *S. lomentaria* was explored by Littler and Littler (1983). Although alternation of generations is not obligate, and both sporophytes and gametophytes can generate more thalli of the same type, there are adaptive advantages to each form. The crustose phase

is more resistant to herbivory, whereas the upright phase is more productive. Alternation of generations represents a kind of "bet hedging"—a way of spreading risks from different sources of mortality.

Sphacelariales

Brown algae belonging to this order are branched pluriseriate thalli that grow from a conspicuous apical cell. Cells generated from the apical cells by transverse division enlarge and then undergo additional transverse and longitudinal divisions, thus generating tissues. Cells contain many discoid plastids with no obvious pyrenoids. The genera occur as small tufts growing on rocks, other algae, or as endophytes. Alternation of generations is isomorphic, and isogamy, anisogamy, and oogamy all occur within the order.

SPHACELARIA (Gr. sphakelos, *gangrene*) (Fig. 15–40) is widespread, occurring in polar to tropical waters, but thalli are rather small, occurring as tufts or spreading mats. It is characterized by production of distinctive, often triradiate, asexual propagules (Fig. 15–41). The European species of Sphacelariales were revised by Prud'homme van Reine (1982), and the species of China described by Draisma, et al. (1998). Approximately 50 species are recognized (Prud'homme van Reine, 1982; Draisma, et al., 1998).

Dictyotales

This group resembles the Sphacelariales in having isomorphic alternation of generations, apical growth, production of tissues, and cells with many discoid, pyrenoidless plastids, but differs in that the spores are nonflagellate, and sperm are uniflagellate. The genera are most diverse in tropical and subtropical waters, but extend into temperate zones. Some taxa exhibit blue-green iridescence when submerged.

Figure 15–42 *Dictyota* growing on a coral head, San Salvador, Bahamas. (Underwater photograph courtesy Ronald J. Stephenson)

ing the cell cycle in apical cells of *Dictyota dichotoma* were studied by Katsaros and Galatis (1992).

Gametophytes are usually dioecious, with sperm-producing plurilocs borne on separate thalli from those bearing female gametangia, each of which contains one egg. *Dictyota dichotoma* is famous for its regular release of gametes at two-week intervals about an hour after dawn, following the maximum of spring tides. A sperm attractant similar to that of ectocarpen is produced by eggs of *Dictyota* (Müller, et al., 1981). Sporophytes produce unilocs containing four nonflagellate meiospores, two of which grow into male gametophytes, and two into female gametophytes. *Dictyota* primarily occurs in shallow waters but has been found as deep as 55 m (Taylor, 1972).

DICTYOTA (Gr. *diktyotos*, netlike) thalli are flattened and highly branched in a repeatedly dichotomous pattern, as in *D. dichotoma* (Fig. 15–42), or are pinnately branched. There is a single prominent apical cell (Fig. 15–23); this cuts off derivatives that contribute to development of the surface as well as the internal tissues, the medulla. Thalli are three cells thick. Division of the apical cell leads to dichotomous branching. Changes in microtubule organization dur-

ZONARIA (Gr. *zone*, belt) (Fig. 15–43) has flattened, fan-shaped blades of about eight cells in thickness, which are generated by a relatively short row of marginal apical cells (Neushul and Dahl, 1972). The banded (zoned) appearance is due to presence of parallel rows of hairs occurring at intervals. Groups of unilocs, known as sori, occur in irregular patches scattered over the thallus surface. Unilocs each produce eight nonflagellate spores. Currently there are ten recognized species of *Zonaria*; related forms are *Homoeostrichus* and the newly defined *Exallosorus* (Phillips, 1997; Phillips and Clayton, 1997).

1 cm

Figure 15–43 A pressed specimen of *Zonaria*. Note fan-shaped blades (arrowheads).

Figure 15–44 Pressed specimen of *Padina*, seen life-size.

Figure 15–45 Habit of *Padina* (upper left) growing with *Dictyota* (center) and red algae on coral rubble near San Salvador, Bahamas. (Photograph courtesy Ronald J. Stephenson)

PADINA (Gr. *pedinos*, flat) thalli, like those of *Zonaria*, are flattened and fan-shaped (Fig. 15–44), and develop from an extensive marginal row of meristematic cells. The genus is very common in tropical and subtropical waters around the world, forming large populations on rocks and coral rubble in shallow waters to depths of 14 m or so (Fig. 15–45) (Taylor, 1972). *Padina* may be lightly or heavily encrusted with calcium carbonate, one of relatively few phaeophyceans to be noticeably calcified. Heavily calcified forms may not at first be readily recognizable as brown algae because they are so pale. Because of its widespread occurrence and relatively large biomass, it is possible that calcium carbonate deposition by this seaweed may be of significance in local carbon cycling. Although alternation of isomorphic generations is the general rule, direct development of new sporophytes from the spores of *P. pavonica* has been reported (Pedersen, 1981); in this case the gametophyte generation has been bypassed.

Sporochnales

The thalli of this group are composed of pseudotissue generated by meristematic cells located at the base of a dense tuft of intensely pigmented apical hairs. Sexual reproduction is oogamous, and the life history is heteromorphic alternation of generations. Sporophytes are the larger and more conspicuous phase; gametophytes are monoecious, microscopic filaments. Most of the species of the six genera occur in the Southern Hemisphere. *Sporochnus* can occur along the southeastern and southwestern U.S. coastlines.

With the recent application of molecular systematic approaches to brown algal evolution, has come new interest in *Sporochnus* as an apparent close relative of Desmarestiales and Laminariales, occupying a key position at the base of the S-D-L clade (Tan and Druehl, 1996) (Fig. 15–24). These authors point out that a suite of shared morphological and biochemical characters also supports such an aggregation. Sporochnales, Desmarestiales, and Laminariales share pyrenoidless plastids, eyespotless sperm, longer posterior flagella (than anterior flagella) on sperm, oogamy, heteromorphic alternation of generations, similar female gametophytes and pheromones, and growth of zygotes in association with female gametophytes.

SPOROCHNUS (Gr. *spora*, seed + Gr. *chnous*, foam) (Fig. 15–46) thalli have a wiry appearance, and are highly branched. The apical tufts of hairs are quite noticeable and distinctive. There is a conspicuous fibrous holdfast that attaches thalli to rocks in shallow water to depths of approximately 100 m (Taylor, 1972). The pheromone caudoxirene is produced by members of the Sporochnales (Müller, et al., 1988).

Desmarestiales

Like their relatives the Sporochnales (Tan and Druehl, 1996), most members of the Desmarestiales are composed of pseudotissue (pseudoparenchyma), and growth is associated with meristematic cells located at the bases of hairs (trichothallic growth). When young, the entire thallus is fringed by these hairs, but they

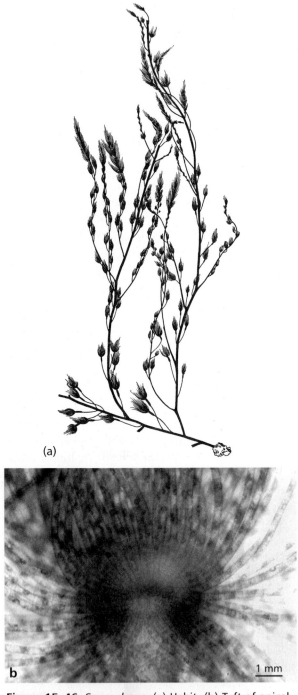

(a)

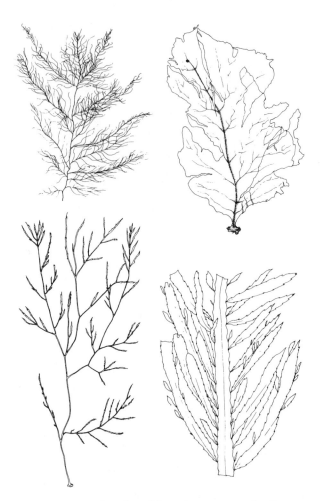

Figure 15–47 Diversity of form in *Desmarestia*. (From Bold and Wynne, 1985)

Figure 15–46 *Sporochnus*. (a) Habit. (b) Tuft of apical hairs. (a: From Taylor, 1972 ©The University of Michigan Press 1960; b: Courtesy J. La Claire)

where desmarestialeans dominate the seaweed flora. Molecular phylogenetic analysis supports the hypotheses that the Desmarestiales are monophyletic and that the family Desmarestiaceae most likely originated in the Southern Hemisphere (Peters, et al., 1997). The sporophytes are the conspicuous generation. Gametophytes are microscopic filaments, and those of at least one species grow as endophytes within red algae. The pheromone desmarestene is produced (Müller, et al., 1982). The large (10 m long and 1 m wide) thalli of the common Antarctic desmarestialean *Himantothallus* strongly resemble those of Laminariales as well as those of *Desmarestia* and *Phaeurus antarcticus* (Clayton and Wiencke, 1990). Development of *Himantothallus* from a filamentous sporeling involves the activity of trichothallic meristems whose location determines the number and position of blades in the

may be lost in more mature specimens. There is a main axis from which diverge lateral "blades" or branches (Fig. 15–47). Diversity is greatest in the Antarctic,

Figure 15–48 Pressed specimen of *Desmarestia*.

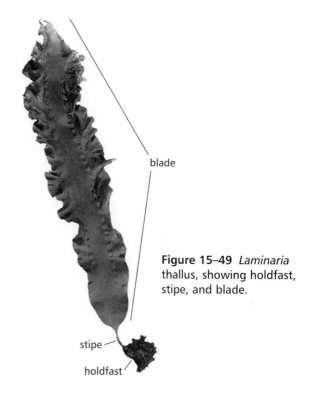

Figure 15–49 *Laminaria* thallus, showing holdfast, stipe, and blade.

Laminariales

mature sporophyte. The mature stipe and blade possess meristoderms that increase the thickness of these organs. Trumpet hyphae with perforate end walls very similar to those of some laminarialeans occur in the medulla and are postulated to function in conduction (Moe and Silva, 1981). Reproduction involves production of unilocular sporangia that occur interspersed with hair (paraphyses) in aggregations (sori) as in various kelps. The meiospores produced from unilocs develop into microscopic dioecious gametophytes, with gametes being formed under short-day conditions (Wiencke and Clayton, 1990).

DESMARESTIA (named for Anselme Gaetan Desmarest, a French naturalist) (Fig. 15–48) occurs in both high-north and south latitudes in cold waters, but also occasionally in warmer waters. One clade of species is characterized by presence of free sulfuric acid within vacuoles having a pH of 1 or lower. If collections of these forms are mixed with other seaweeds, the unavoidable breakage of cells will release acid, causing extensive disintegration of the specimens. *Desmarestia* species must be kept separate from other collections and stored under cool conditions to avoid autohydrolysis.

Laminarialean brown algae, commonly known as kelps, are characterized by large sporophytic thalli differentiated into holdfast, stipe, and blade regions (Fig. 15–49), with an intercalary meristem located at the junction of the stipe and blade. Holdfasts are branched structures formed of thickened tissues

Figure 15–50 *Macrocystis* holdfast, composed of many branched haptera.

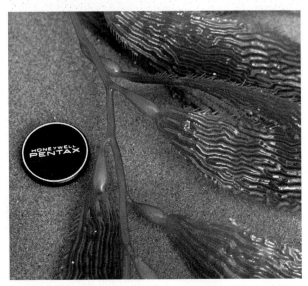

Figure 15–51 Branched stipe with blades and floats of *Macrocystis*.

known as **haptera** (Fig. 15–50). These are essential for effective anchorage of the often large and heavy kelps. Some kelps have branched stipes, others have a single unbranched stipe, and in a few the stipe is greatly reduced or missing from mature stages. In some forms the blade is entire, but in others the blade is divided by longitudinal splits. Inflated **pneumatocysts** (floats) may occur in a lateral position on the stipe, or at the base of the blade(s) (Fig. 15–51). The gas bladders of *Nereocystis luetkeana* and *Pelagophycus porra* are noteworthy in that they contain up to 10% CO (carbon monoxide gas) (Carefoot, 1977), but the significance of this is unknown. The structure, gas content, and function of pneumatocysts were reviewed by Dromgoole (1990).

Within stipes and blades, tissue specialization can occur. There is typically an outer layer of pigmented cells wherein cell divisions leading to increase in thallus circumference may occur; such a meristem is known as the **meristoderm**. Internal to the meristoderm are colorless cells of the cortex, within which may occur a central medullary area (**medulla**). The inner cortex and/or medulla contain(s) cells specialized for conduction of solutes that are referred to as sieve elements, because they are analogous to those of vascular land plants. Brown algal sieve elements have, of course, evolved independently from those of vascular plants.

Sieve elements have been reported from most of the genera in the Laminariales. They are elongate cells with perforated terminal end walls (sieve plates) (Fig. 15–52), arranged in long continuous files to form sieve tubes similar to those of vascular plants. The ends of individual sieve elements may be expanded so that they resemble trumpets, and are hence sometimes called trumpet hyphae (Fig. 15–53). The diameter of the sieve plates in *Macrocystis pyrifera* may be as large as 62 μm (Parker and Huber, 1965). The sieve tubes run throughout the thallus except for the holdfast (Fig. 15–54). Solutes translocated within sieve tubes include mannitol (about 65% of sap content), free amino acids, and inorganic ions (Parker, 1965; 1966). These solutes originate in the photosynthetic cells of the meristoderm and outer cortex. Loading into sieve elements is thought to occur by either symplastic (involving intercellular connections, i.e., plasmodesmata) or apoplastic (involving cell membrane transporter molecules) processes, or both. In order to understand the loading process, Buggeln, et al. (1985) fed radioactive bicarbonate to blades of *Macrocystis* and followed its movement within the thallus using external Geiger-Müller counters. Microscopic studies revealed a possible symplastic pathway from photosynthetic cells to sieve cells of the medulla, including connections between inner cortical cells and the thin-walled sieve cells. The latter are often sought, but infrequently observed.

Translocation occurs in both the light and the dark. Experimental efforts toward understanding this process are reviewed by Schmitz (1981). Physiological studies of translocation in laminarialean sieve tubes began with the pioneering radioisotope tracer work of Parker (1965, 1966), who found translocation rates of 65–78 cm hr^{-1} in *Laminaria*, with transport primarily toward the intercalary meristem. Translocation rates in

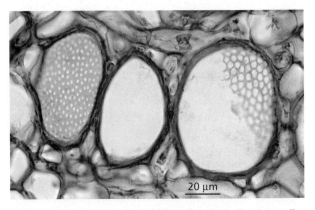

Figure 15–52 Perforated end walls of conducting cells in *Macrocystis*.

Macrocystis are about 50 cm day⁻¹. Such solute movement is of great advantage in dense *Macrocystis*
canopies where light levels may be insufficient to
allow photosynthesis to balance respiration in the
lower portions of the thallus. Isotope tracer studies
have revealed that translocation occurs from well-
illuminated surface fronds to juvenile fronds of *Macrocystis*. Immature blades may primarily import
photosynthates, but mature ones only export (Lobban,
1978a, b). Translocation to basal regions may also be
necessary for development of fertile (uniloc-bearing)
fronds at the basal region of the *Macrocystis* thallus.
Translocation also greatly benefits growth of the
annual kelp *Nereocystis*, which can grow 13 cm day⁻¹
or more and can extend 25–50 m during a single
growth season. The tremendous ecological significance of the giant kelps is largely based upon their
capacity for rapid growth, ability to achieve large size,
and translocatory competence.

Similar conducting cells occur in the central
medulla of the Antarctic *Ascoseira mirabilis* (Ascoseirales), whose sporophyte is similar to *Laminaria*.

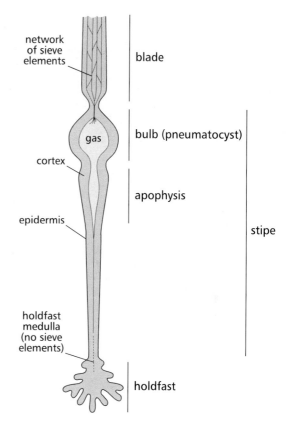

Figure 15–54 Diagrammatic representation of the
conducting system of *Nereocystis luetkeana*. (After
Nicholson, 1970 by permission of the *Journal of
Phycology*)

Sievelike walls perforated by many pits, presence of
many nuclei and plastids, and numerous physodes
characterize these putative conducting cells (Clayton
and Ashburner, 1990). Physiological analysis of possible solute transport in *Ascoseira* has yet to be
accomplished.

Reproduction in the Laminariales occurs by
means of unilocular sporangia produced on both surfaces of blades (plurilocs do not occur on the sporophytes of these taxa). The ultrastructure of the
zoospores produced in these unilocs was studied by
Henry and Cole (1982a) (Fig. 15–55). During uniloc
development, a 1:1 association between nuclei and
plastids occurs, which ensures appropriate packaging
of organelles in meiospores (Motomura, et al., 1997).
In some cases the unilocs are localized in darker
brown patches known as **sori**, and sometimes unilocs
occur on specialized reproductive blades known as
sporophylls. In the case of *Macrocystis* and *Alaria*, the
sporophylls are located close to the thallus base, facil-

Figure 15–53
Conducting cells (sieve
element) of *Laminaria*.
One linear array of
conducting cells has
been outlined in white.

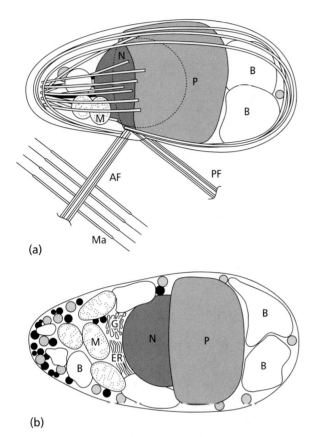

(a)

(b)

Figure 15–55 Side (a) and dorsal (from the top surface) (b) views of the zoospores of laminarialean algae, as deduced from ultrastructural studies. AF=anterior flagellum, B=vesicles, ER=endoplasmic reticulum, G=Golgi, M=mitochondrion, Ma=mastigonemes (tripartite hairs), N=nucleus, P=plastid, PF=posterior flagellum. (After Henry and Cole, 1982a by permission of the *Journal of Phycology*)

that fertilization and subsequent zygote development occur in association with the female gametophyte. The eggs of *Laminaria angustata* have two flagella that lack mastigonemes, and there are no flagellar roots. The flagella are abscised during egg liberation (Motomura and Sakai, 1988). This may be a plesiomorphic character that occurs more widely among Laminariales and related orders of brown algae; further studies are needed. The pheromone produced by kelps is lamoxirene, which is structurally distinct from the pheromones produced by other brown algal groups (Müller, et al., 1985). Culture studies have revealed that some percentage of unfertilized eggs are capable of developing into sporophytes. Intergeneric hybrids can form between members of the Laminariales (Lewis and Neushul, 1995).

The center of diversity of the Laminariales is the Pacific coast of North America. Classically, the order contains a number of families that are distinguished by structural features: These include Laminariaceae (splits in the blade do not extend into the meristematic region), Lessoniaceae (splits in the blade may extend into the meristematic region), Alariaceae (lateral blades are produced from the stipe, not from blade-splitting, and reproductive cells are produced on special sporophylls), and other groups. Recent molecular sequencing work suggests the possibility that *Chorda*, sometimes grouped with Laminariales, and *Saccorhiza* are phylogenetically isolated from *Macrocystis* (Lessoniaceae) and *Alaria* (Alariaceae), and that the former two genera are probably more closely related to the Sporochnales than previously realized (Tan and Druehl, 1996). These molecular sequence results are supported by some morphological and biochemical differences—meiospores of *Chorda* and *Saccorhiza* have eyespots as do those of Sporochnales (and Desmarestiales), and the translocatory cells and sperm attractants of *Chorda*

itating the ability of meiospores to reach suitable substrates for attachment. In the case of *Nereocystis*, large soral patches tear out of blades and readily sink to the bottom, thus depositing unilocs close to deep substrates where released meiospores can attach. Meiosis segregates sex-determining genes, such that meiospores germinate into unisexual gametophytes. Kelp gametophytes are microscopic branched filaments (Fig. 15–56) or, in some cases, single oogonial cells. Male gametophytes produce dense clusters of antheridial cells, each giving rise to a biflagellate sperm. Cells of female gametophytes may release their contents from cell walls, and these wall-less cells function as eggs. The egg cells often remain attached to their former cell wall, however, with the consequence

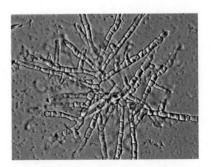

Figure 15–56 *Laminaria* gametophyte, grown in culture. (Photograph by G. McBride)

Figure 15–57 A *Laminaria* species with a highly dissected (split) blade.

Figure 15–58 *Hedophyllum* removed from its substrate and turned upside down to reveal the substantial holdfast composed of numerous haptera (arrowhead).

(multifidene) and *Saccorhiza* (ectocarpene) are different from those of other kelps.

LAMINARIA (L. *lamina*, blade) is very common worldwide on temperate and boreal rocky shorelines. There are many described species. One of the most interesting is *Laminaria solidungula*, the dominant member of an Alaskan high-Arctic community that is annually subjected to eight months of darkness (Dunton and Schell, 1986). The blade of some species is entire (Fig. 15–49), but that of others can be split (Fig. 15–57); blades do not possess a midrib. Most species are perennial. At the end of the growing season the blade may either detach or persist until it is displaced by growth of a new blade at the start of the next growing season. Unilocs may be produced at any place on the blade. An ultrastructural analysis of fertilization and zygote development in *Laminaria angustata* revealed that zygotes develop cell walls immediately after gamete fusion, possibly as a block to penetration of eggs by more than one sperm. Further, egg centrioles disappear from zygotes, such that centrioles are inherited from the sperm-contributing parent. In contrast, mitochondria are maternally inherited, since sperm mitochondria are degraded in zygotes (Motomura, 1990). *Laminaria japonica* is widely cultivated in the western Pacific region for use as food (Chapter 4).

HEDOPHYLLUM (Gr. *hedos*, seat + Gr. *phyllon*, leaf) is a sessile kelp that may be the dominant seaweed in the lower intertidal in the north Pacific region (Fig. 15–58), especially if populations of sea urchins—the primary grazers—are low (Paine and Vadas, 1969).

A stipe is not present on mature sporophytes, whose blades may become highly divided by wave action. There is a robust holdfast composed of branched haptera. There is only a single species, *H. sessile*.

COSTARIA (L. *costa*, rib) has an undivided blade that has five longitudinal ribs and a relatively thin, flattened, unbranched stipe (Fig. 15–59). Sori occur in the regions between ribs. It is perennial and common on rocks in the lower intertidal and subtidal of northern Japan as well as the eastern Pacific from Alaska to southern California (Abbott and Hollenberg, 1976).

POSTELSIA (named for Alexander Philipou Postels, an Estonian geologist and naturalist) resembles a small (to 60 cm tall) palm tree in that there is a group

Figure 15–59 Portion of the ribbed blade and unbranched stipe of *Costaria costata*.

Figure 15–60 *Postelsia palmaeformis*. Numerous palm treelike thalli are typically clustered together on wave-impacted shores of the eastern Pacific.

of terminal blades at the top of an extremely flexible, tapering stipe (Fig. 15–60). The blades have toothed margins and are ridged, with linear arrays of unilocs (sori) being produced between the ridges in late spring. *Postelsia* is well adapted for its habitat—high to mid-intertidal zones of areas exposed to very heavy wave action. Tightly adherent holdfasts and flexible stipes resist drag forces effectively, and massive production of meiospores allows rapid colonization of rocks rendered bare of other seaweeds or mussels by waves. *Postelsia* may thus dominate its habitat, frequently occurring in large populations (see Fig. 23–2). Its range is along the eastern Pacific coast from British Columbia to California. Analysis of population structure using molecular fingerprinting and RAPD analysis (Chapter 5) showed that dispersal distances are short (1–5 m), so that individuals within clusters are likely to be siblings. In addition, evidence was obtained for the occurrence of distinguishable biogeographic populations (Coyer, et al., 1997). Gametogenesis, development of the young sporophyte, and change in chromosome number during the life history

of *Postelsia palmaeformis* were studied by Lewis (1995).

NEREOCYSTIS (Gr. *Nereus*, a sea god + Gr. *kystis*, bladder) (Fig. 15–61) has a stipe that may be more than 30 m in length and is attached to rocky substrates at the bottom of subtidal habitats by a relatively massive holdfast. The apical end of the stipe is terminated by a large (about 15 cm in diameter) pneumatocyst or float, from which emerge four meristematic branches that generate as many as 100 blades per thallus. Each blade may be several meters in length. Development of the thallus was studied by Nicholson (1970) (Fig. 15–62). At reproductive maturity, blades exhibit dark brown patches (sori), which are groups of unilocs. Sori detach from blades and can be readily transported, a useful adaptation to the need for meiospores to attach to deep-water substrates. *Nereocystis* is annual, so all of its growth is amazingly attained within a single growing season. Large populations of *Nereocystis* may dominate areas within northern regions of the Pacific North American coast (Fig. 15–63).

MACROCYSTIS (Gr. *macros*, long + Gr. *kystis*, bladder) is a massive kelp that can form dense kelp forests (see Fig. 1–1). Fronds may reach lengths of 60 m and are harvested for alginate production (Chapter 4). Results of several studies of the biotic impact of kelp harvesting suggest that no problems arise as long as the integrity of the *Macrocystis* population itself was not affected (North, 1994). *Macrocystis* beds serve as important refuges and nurseries for fish

Figure 15–61 *Nereocystis* is characterized by a very long stipe that is crowned with a single large float and many blades. A group of *Postelsia* thalli at center reveals the size difference between the two kelps.

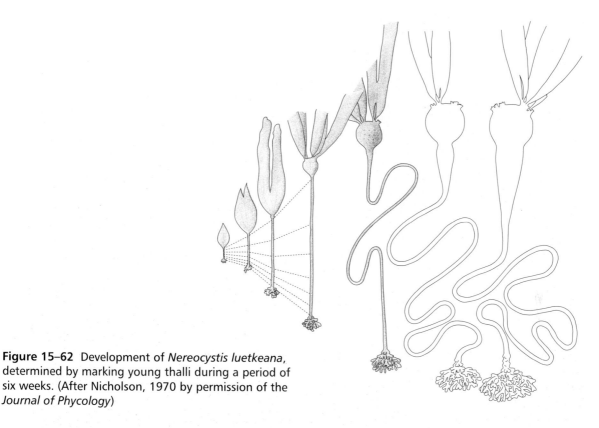

Figure 15–62 Development of *Nereocystis luetkeana*, determined by marking young thalli during a period of six weeks. (After Nicholson, 1970 by permission of the *Journal of Phycology*)

Figure 15–63 Habit of *Nereocystis luetkeana* in the Strait of San Juan de Fuca, Washington State, U.S.A. Blades and floats of numerous individuals can be seen. (Photograph courtesy C. Taylor)

Figure 15–64 Branched *Macrocystis* thalli just below the water surface. Each blade is associated with a float.

and many other ecologically and economically valuable species. They influence water motions and hence the dispersal patterns of planktonic larvae, as well as the distribution of phytoplankton. Sea otters are well known to moor themselves while sleeping by wrapping themselves with *Macrocystis* fronds. The dissolved organic carbon and particulates generated by kelp decay provide a substantial fraction of the nutrient requirements of detritivores and other organisms in kelp communities (Duggins, et al., 1989). This genus is thus regarded as a keystone taxon whose presence influences a cascade of ecological processes and numerous other organisms. The keystone properties of *Macrocystis* derive from its growth processes. These result in a proliferation of meristems that can generate the huge thallus and that confer the ability to continue growth even if some meristematic tissue is removed by herbivory.

The thallus, with branched stipe, multiple blades (each as long as 40 cm), and floats (Fig. 15–64) arises by division of a single original blade, beginning from a small hole developing near the base. The juvenile blade undergoes a longitudinal split, generating two primary fronds that continue to undergo longitudinal splitting until many blades have been formed (Fig. 15–65). The two outermost fronds serve as frond initials, and inner fronds function as the first basal meristems. Basal meristems can produce new fronds by continuing to divide by basal clefts. The innermost blade of the resulting pair continues to function as a basal meristem, and the outermost may become either a basal meristem or a frond initial. Frond initials undergo basal splitting to produce blades, and the most distal blade then becomes capable of generating

new blades (each with an individual float) and stipe tissue, thus serving as an apical meristem. Meiospores are produced in unilocs on specialized blades (sporophylls) located just above the holdfast apex; these may not possess floats. The female gametophytes are very small, consisting of just a single cell, but male gametophytes are multicellular.

There are about four species, which exhibit varying degrees of hybridization. *Macrocystis* occurs off the coasts of every major Southern Hemisphere landmass, as well as many islands, but is much less widely distributed in the Northern Hemisphere, occurring only along the northeast Pacific Ocean. Absence of this and other kelps from low latitudes is attributed to their sensitivity to water temperatures above 20°–22° C, either as a direct metabolic effect or an indirect effect of differences in occurrence or activity of grazers and microbial pathogens. DNA finger-

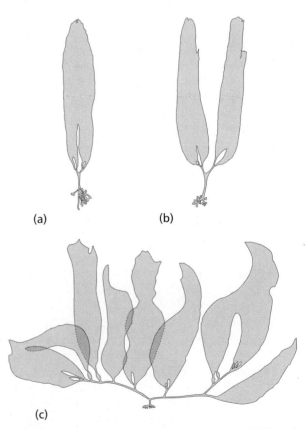

(a)   (b)

(c)

Figure 15–65 Development of the branched thallus of *Macrocystis*. A hole develops at the base of young blades (a), and this leads to longitudinal splitting of the single original blade into two (b). This process continues until multiple blades and branched stipes have been produced (c). (After Bold and Wynne, 1985)

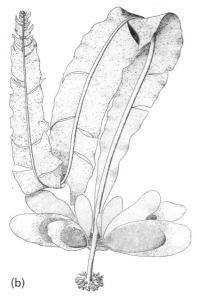

Figure 15–67 *Egregia* species having narrow blades emerging from each side of a flattened stipe.

Figure 15–66 *Alaria*. (a) Terminal portion of a blade. (b) Diagram showing the entire thallus, with the basal sporophylls. (b: From MARINE ALGAE OF THE MONTEREY PENINSULA, Second Edition, by Gilbert M. Smith, Incorporating the 1966 supplement by George J. Hollenberg and Isabella A. Abbott. Used with the permission of the publishers, Stanford University Press. ©1966, 1969 by the Board of Trustees of the Leland Stanford, Junior University.)

printing techniques are beginning to be used to assess genetic diversity within and between populations and life-history phases of *Macrocystis* (Coyer, et al., 1994). Additional information regarding the biology of *Macrocystis* can be found in an extensive review by North (1994).

ALARIA (L. *ala*, wing) has an undivided main blade with a conspicuous flat midrib (Fig. 15–66) and

numerous basal sporophylls. Blades can be as long as 25 m. It is perennial and occurs on exposed rocks from Alaska to California.

EGREGIA (L. *egregius*, remarkable) is known as the "feather boa" kelp because of the numerous lateral blades and elongate to spherical floats that emerge from a long, flattened stipe (Fig. 15–67). It is perennial.

ECKLONIA (named for Christian Friedrich Ecklon, a German botanical collector) (Fig. 15–68) typically has blades with numerous pinnately arranged outgrowths from a flattened mid-region, arising from the top of a stipe of varying length, depending on the species. Although most *Ecklonia* species produce sori of unilocs on specialized sporophylls, restriction of sori to sporophylls does not occur in all species, and one species lacks sporophylls entirely. *Ecklonia* occurs in warm temperate waters of both hemispheres and at least one tropical area (islands off the coast of Western Australia); it is considered to be the most warmth tolerant of the kelps. It forms extensive kelp forests in southern Africa, Australia, New Zealand, and China, often in association with upwelling of nutrient-rich, cool subsurface waters. Drift *Ecklonia* is collected for

Figure 15–68 *Ecklonia* in Australia. (Photograph courtesy W. Woelkerling)

alginate production. A review of *Ecklonia* biology can be found in Bolton and Anderson (1994).

Fucales

The Fucales include phaeophyceans having apical growth, thalli composed of tissues, and distinctively, a life cycle involving meiosis during gamete production. Although the Fucales have long been suggested to represent the most recently diverged group of brown algae—based on specialized life-history and relatively complex thallus construction—recent molecular systematic evidence (Tan and Druehl, 1996) suggests instead that Fucales are the earliest divergent

of the Phaeophyceae and sister to all other brown algae (Fig. 15–24). This surprising finding implies that the ancestral Fucales might have looked, and had life cycles, much like the relatively simple members of the Ectocarpales, where alternation of generations is the general rule. The molecular data also suggest that apical growth and true tissue formation evolved independently in the Fucales and other brown algae having these characteristics.

Apical growth in Fucales is thought to occur by division of an apical cell or a group of several apical initials, each having several cutting faces. Their derivatives generate a thick parenchymatous thallus, which, though not approaching the maximal sizes of the giant laminarialeans, can be a meter or more in length. Dichotomous branching results from division of the apical meristem. Radioisotope tracer experiments demonstrated long-distance transport of photosynthetic products within midribs of *Fucus serratus*. The branch tips are a strong sink, and the main translocated substance is mannitol, but amino acids and low molecular-weight organic acids also are transported (Diouris, 1989).

Gametes are generated within gametangia produced in branch termini or systems of branches known as **receptacles**. Chambers known as **conceptacles** are sunken into the surfaces of receptacles, with only an open pore (ostiole) visible from the receptacle surface (Fig. 15–69). Gamete discharge from conceptacles into seawater occurs through this pore in taxa such as *Fucus*. Depending upon the fucalean species, an individual conceptacle may con-

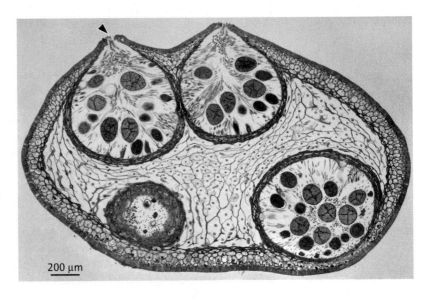

Figure 15–69 Section through a receptacle of *Fucus* showing portions of four conceptacles. This species is monecious—each conceptacle contains both oogonia and antheridia. Conceptacles open to the outside via a pore (ostiole) (arrowhead).

200 μm

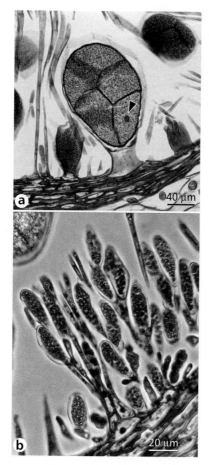

a

40 µm

b

20 µm

Figure 15–70 *Fucus* gametangia. (a) Oogonium with egg cells. A nucleus (arrowhead) is visible in one of the eggs. (b) Antheridia, containing sperm, photographed with phase-contrast optics.

tain both eggs and sperm or just one type of gamete. Some species produce both sperm- and egg-containing conceptacles on the same thallus, and others produce them on separate thalli (i.e., they are dioecious). Conceptacles are derived from single cells produced at the thallus surface close to the apical meristem. Some of the cells lining the developing conceptacle differentiate into oogonia (egg-producing gametangia) or antheridia (sperm-producing gametangia) (Fig. 15–70). Development of these gametangia is thought to be analogous to that of unilocular sporangia in other brown algae, and the gametophyte stage to consist of only a single cell (Jensen, 1974). The first divisions of the gametangia are meiotic; genera vary in the number of egg cells produced, from one to eight. In antheridia, the meiotic products continue to divide mitotically, gener-

ating 64 sperm. Fucalean sperm are distinctive in having an expanded apical region (known as a proboscis) and a posterior flagellum that is longer than the apical flagellum. Sperm possess eyespots. The wall-less gametes of *Fucus* are released when desiccated thalli are inundated with the incoming tide, and sperm are attracted to eggs as the latter release the pheromone fucoserraten (Müller, 1981). Binding between sperm and egg of *Fucus serratus* is mediated by interaction between proteins derived from the sperm that recognize sulfated glycoconjugates on the egg cell membrane; such interactions are important in activating the egg (Wright, et al., 1995a, b). Ultrastructure of the gametes and embryo of *Fucus* was studied by Brawley, et al. (1977). Fertilization of *Fucus* occurs in the water, so zygotes must rapidly attach to substrates before they are washed out to sea. After fertilization and prior to further development, zygotes of *Fucus* secrete an adhesive mucilage (Vreeland, et al., 1998).

About two hours after fertilization, *Fucus* zygotes begin to deposit a wall, and about 12 hours later the rhizoid is produced. This event reflects previous establishment of zygote polarity. Light appears to be the strongest external stimulus affecting polarity establishment. The cell wall provides information to the underlying cell membrane and cytosol that influences establishment of a polar axis and orientation of the first plane of zygote division (Quatrano and Shaw, 1997). The rhizoid forms on the shaded side of zygotes that are exposed to unidirectional illumination (Brawley and Wetherbee, 1981). Calcium signalling is involved in induction of rhizoid development, as ascertained by patch-clamp electrophysiological techniques and laser microsurgery (Taylor, et al., 1996), and by the use of inhibitors (Love, et al., 1997). Rhizoid development is dependent upon the expression of fucoidans in the cell wall (Quatrano, 1973).

Eggs of *Sargassum horneri* are extruded within a mass of mucilage, and remain attached to the conceptacle by a stalk formed from oogonial wall material. After fertilization and a few cell divisions, germlings are released from the stalk and become attached to the substrate by means of adhesive material released from rhizoids (Nanba, 1995). In genera such as *Turbinaria* and *Sargassum*, zygotes are also retained on the parental thallus. Such retention is regarded as an adaptation allowing reproduction and colonization to occur within very high-energy environments such as coral reefs (Stiger and Payri, 1997).

2 cm

Figure 15–71 A pressed specimen of *Ascophyllum*, showing branching pattern and numerous elongate floats.

FUCUS (L. *fucus* derived from Gr. *phykos*, seaweed) has flattened, dichotomously branched or pinnate fronds with a midrib (Fig. 15–1). Air bladders occur within the fronds. Receptacles occur at the frond tips and are noticeably swollen. The holdfast is discoid. A surface layer of pigmented columnar cells encloses a medulla composed of a network of colorless filaments surrounded by a mucilaginous matrix of cellulose, alginates, and fucoidan. These water-absorbing polysaccharides are thought to help mitigate desiccation effects. The surface cell layer is involved in secretion of alginates, fucoidan, and polyphenolic compounds to the outside of the thallus. *Fucus* can be very abundant on rocky shores in temperate regions of the Northern Hemisphere. Comparative studies of photosynthesis in air and seawater by *Fucus spiralis* demonstrated that this species can photosynthesize

effectively in air, revealing its tolerance to desiccation (Madsen and Maberly, 1990).

ASCOPHYLLUM (Gr. *askos*, leather bag + Gr. *phyllon*, leaf) often occurs together with or nearby *Fucus*, and both taxa are colloquially known as rockweeds. The thalli have thinner, longer axes than does *Fucus*, and a midrib is absent (Fig. 15–71). Dichotomous and lateral branching occur. Air bladders are present at intervals on the fronds.

SARGASSUM (Port. *sargaco*, name given by fishermen) thalli are highly differentiated into holdfast, cylindrical main axis, leaflike blades, and air bladders in the axils of blades (Fig. 15–72). This genus is widespread in temperate, subtropical, and tropical waters in both intertidal and subtidal zones. Some forms are free-floating, sometimes occurring in extensive rafts that harbor distinctive communities of organisms adapted to the buoyant *Sargassum* habitat. These occur in the Sargasso Sea off the western coast of Africa. During the 1940s, *Sargassum muticum* spread from Japan to the northern Pacific coast of the U.S., and by the 1970s had made its way south to California. This species of *Sargassum* has also spread to Europe, probably on oysters destined for aquaculture operations. *Sargassum* forms nuisance growths in har-

Figure 15–72 A live specimen of *Sargassum*, showing branching pattern and small round floats (air bladders).

Figure 15–73 A dense carpet of *Hormosira* in Australia. Note the distinctive beaded necklacelike appearance. (Photograph courtesy W. Woelkerling)

bors and on beaches, and it can quickly spread to new areas due to the floatation capabilities conferred by its many air bladders. Other features contributing to rapid spread include fast growth rate, fertility in the first year, and monoecious reproduction (Lüning, 1990).

HORMOSIRA (Gr. *hormos*, necklace + Gr. *seira*, chain) thalli are dichotomously branched arrays of hollow, round segments interconnected by short nodes (Fig. 15–73). It is of common occurrence on the shores of Australia and New Zealand. Parthenogenesis (development of unfertilized eggs) has been observed in *Hormosira*. By 24 hours after release from the oogonium, 74% of unfertilized eggs had secreted a cell wall, and after two days following release, up to 22% had undergone polarization. However, only a few were able to divide. These, unlike eggs that did not divide, were observed to possess a pair of centrioles (Clayton, et al., 1998).

Durvillaeales

DURVILLAEA (named for Jules Sebastien Cesar Dumont D'Urville, a French explorer), the "bull kelp," is the only genus included in this order, which is separated from the Fucales because of differences in growth mode. The four or five species are restricted to temperate to sub-Antarctic regions of the Southern Hemisphere where they may form large biomasses and become dominant components of

intertidal and subtidal communities. A cladistic, biogeographic analysis suggested that *Durvillaea* evolved in the Southern Hemisphere, and was well established prior to the dissociation of New Zealand from Gondwanaland some 80 million years ago (Cheshire, et al., 1995).

Durvillaea has a massive solid holdfast, a branched or unbranched stipe, and a large leathery blade, often ripped by wave action into several straplike segments (Fig. 15–74). One species produces honeycomblike gas-filled floats within the blade, while other species produce additional lateral blades along the stipe. The thalli are perennial and dioecious. There is a surface layer of pigmented cells, and a cortex and internal medulla formed of interwoven slender cells. Growth seems to occur by means of a peripheral meristoderm that adds new layers of cells at the beginning of new growth periods (Hay, 1994). Its mode of growth is usually described as "diffuse," but in this case the term is meant to convey the apparent absence of a distinct apical or intercalary meristem that increases thallus length, rather than the diffuse occurrence of mitosis throughout the thallus as is exhibited by ectocarpaleans. Individuals may live for 7–8 years and grow to 10 m in length, becoming fertile in the second year (Clayton, et al., 1987).

Although *Durvillaea* superficially resembles some members of the Laminariales, reproduction is more similar to that of Fucales. There is no alternation of generations, and conceptacles bear the gametangia and gametes. Conceptacles develop on the main

Figure 15–74 *Durvillaea* along the Australian coast. The blades are lacerated into straps by the action of the surf. Note the massive disklike holdfasts (arrowheads). (Photograph courtesy W. Woelkerling)

blade—usually in a subterminal position—beginning in the spring, and complete gametangial production in the autumn. Conceptacle development in *Durvillaea* was studied by Clayton, et al. (1987). There are separate male and female thalli; conceptacles of male thalli produce only sperm and those of female thalli generate only eggs. There are typically equal numbers of male and female thalli in populations. The first division leading to gamete production is meiotic. Four eggs are produced per oogonium, and 64 sperm are generated in antheridia, as in several fucaleans. Egg production can be prodigious—it is estimated that a single thallus could produce 120 million eggs in just one night (Hay, 1994).

Zygote development is quite similar to that of fucaleans. Two years' growth is required before thalli become fertile. *Durvillaea* is harvested in Chile for alginate production and for human consumption. Dried seaweed is rehydrated, diced, and boiled for about two hours before being used at a meat substitute

in casseroles or as a vegetable in salads. *Durvillaea* is also harvested for alginate production in Tasmania and New Zealand (Hay, 1994). Durvillaealeans are distinguished from Fucales in lacking an apical meristem and in having only two plastids per cell (in contrast to multiple plastids per fucalean cell).

Recommended Books

Bold, H. C., and M. J. Wynne. 1985. *Introduction to the Algae.* (2nd edition). Prentice Hall, Englewood Cliffs, NJ.

Dring, M. J. 1982. *The Biology of Marine Plants.* Arnold, London, UK.

Silva, P. C., P. W. Basson, and R. L. Moe. 1996. *Catalog of the Benthic Marine Algae of the Indian Ocean.* University of California Press, Berkeley, CA.

Chapter 16

Red Algae

Ptilota

The 5000–6000 species of red algae, classified in the division Rhodophyta, are important and distinctive in many ways. Red macroalgae (seaweeds) are common inhabits of tropical and temperate nearshore marine waters (Fig. 16–1), where they are of economic and ecological significance. *Porphyra* and a few other species are cultured for use as human food (Fig. 16–2). Cultivation or harvest of natural populations of species that are valued for sulfated polygalactans present in their extracellular matrices is an important industry. These polygalactans are extracted and purified as agar, agarose, and carrageenans, and widely used for laboratory cell-culture media, nucleic acid research, or food processing, respectively (Fig. 16–3). Such polymers have unique colloidal properties, but are too complex for industrial synthesis, and thus red algae are the only sources. The economic uses of red algae are described in greater detail in Chapter 4.

Encrusting, calcified "corallines" such as *Lithothamnion* and *Porolithon* play a major ecological role in coral reefs. These red algae occur as hard, flat sheets that consolidate and stabilize the wave-impacted crests of coral reefs, which harbor some of the world's highest species diversities. Reef crest corallines are regarded as keystone organisms, species whose decline (like the removal of the keystone from an arch) could cause collapse of larger-scale structure, namely loss of entire communities. Fossil evidence indicates that red algae have been playing this important role for the past 500 million years.

Continued growth of calcified rhodophytes on top of dead remains has, over the course of eons, built up extensive carbonate deposits in some areas of the world. Modern deposits more than 25 m in depth are known (Tucker and Wright, 1990). Rhodophyte-derived carbonate deposits can cover extensive areas of submerged coastal platforms, retaining very large amounts of carbon in long-term storage. Such carbonates are harvested in some regions of the world as soil conditioners, and some ecologists have expressed concern about the effect of their removal on global carbon cycles. Other potential ecosystem impacts of red algae include production of volatile halogenated compounds that may influence ozone levels, and emission of DMS, a volatile sulfur-containing compound that con-

Figure 16–1 Several types of red algae, including both erect thalli and expanses of encrusting corallines, in a subtidal community on the eastern shore of Newfoundland, Canada.

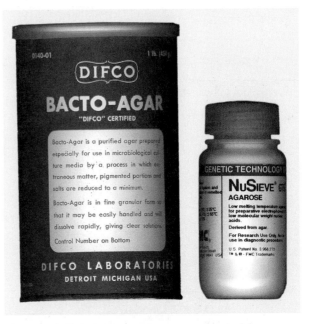

Figure 16–3 Agar and agarose are extracted from particular red algae, purified, and sold commercially for laboratory preparation of growth media for microorganisms and gels for separation of nucleic acids in molecular biology.

Figure 16–2 Commercial packaging of dried *nori* sheets derived from cultivated *Porphyra*, used in making sushi.

tributes to formation of acidic precipitation and also has climatic effects (Chapter 2).

Other ecologically interesting red algae include the deepest growing photosynthetic eukaryote, an unnamed crustlike coralline (found by the use of a submersible) living on 210 m deep seamounts near San Salvador, Bahamas (Littler, et al., 1985, 1986). Cyanidophytes are unicellular red algae that are able to grow in acidic hot springs, which are inimical to all other forms of eukaryotic life. Red algae are abundant and species-rich in warm tropical and subtropical waters. Several red algae are commonly associated with extensive mangrove forests in tropical and subtropical coastal regions around the world. Red algae are also common along temperate and boreal shorelines, while others are able to survive in Arctic and Antarctic waters where they are covered by two meters of sea ice for ten months of the year. Although most red algae are rather intolerant of salinity changes, growing best in normal seawater, a few grow best at lower salinities. There are about 150 species of freshwater red algae, far more than the number of freshwater brown algae. *Bangia* is a red alga that has invaded the Laurentian Great Lakes of North America within recent decades.

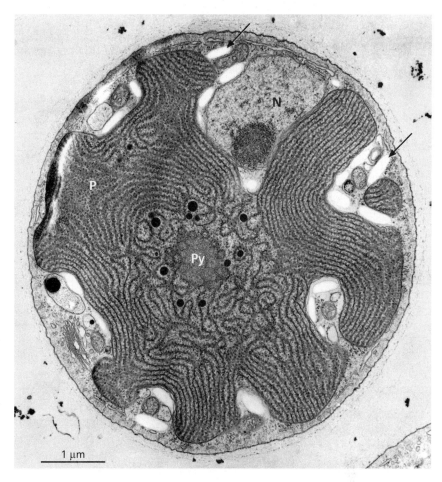

Figure 16–4 Transmission electron micrograph of *Porphyridium*, a unicellular red alga, illustrating the nucleus (N), cytoplasmic floridean starch grains (arrows), and single, lobed plastid (P) with central pyrenoid (Py) and unstacked thylakoids bearing phycobilisomes. Note the absence of a coherent cell wall and presence of a loose extracellular matrix. (From Schornstein and Scott, 1982)

Red algae are notable as obligate epiphytes on other seaweeds or marine animals. There are many species of poorly pigmented, parasitic red algae that occur on other red algal taxa. Some red algae are known to produce halogenated terpenoids and other secondary compounds that may function to inhibit herbivory or have antimicrobial properties. At least some of these compounds may prove useful as pharmaceuticals. In contrast to most other algal clades, among the Rhodophyta there are few, if any, examples of nuisance growth-formers or production of toxins harmful to humans. However, domoic acid has been identified from several red algae (Bates, et al., 1998) and paralytic shellfish-poisoning toxins have been confirmed in *Jania* sp. (Cembella, 1998). In the latter case the toxins may actually be produced by a bacterium isolated from *Jania*, *Shewanella alga* (see Doucette, et al., 1998). In other cases of supposedly toxic red algae, the culprits are probably epiphytic dinoflagellates (Tindall and Morton, 1998).

Red Algal Pigments

Most red algae are pink to deep red in color because their plastids contain large amounts of the red accessory pigment phycoerythrin, which obscures chlorophyll *a*. Phycoerythrin is extremely efficient in harvesting blue and green light in the subtidal habitats occupied by most red algae, explaining its abundance within red algal plastids. There are at least five types of phycoerythrin in red algae, differing somewhat in spectral absorption but all having an absorption peak in the green wavelengths. Within the phycobilisomes—located on unstacked thylakoids (Fig. 16–4), similar to those of cyanobacteria (Chapter 6)—captured light energy is efficiently passed to phycocyanin and allophycocyanin, and finally to chlorophyll *a*, allowing the harvesting of light energy that is otherwise inaccessible to chlorophyll.

Certain parasitic taxa may be white, cream, or yellowish in color, and some highly calcified forms

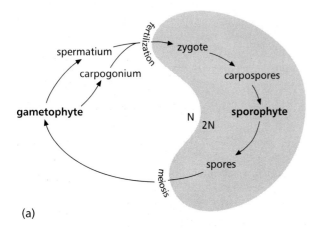

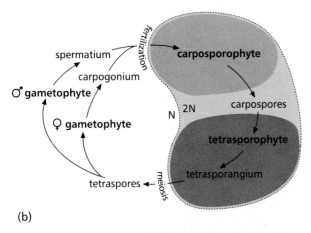

Figure 16–5 (a) Biphasic life history of early divergent red algae contrasted with (b) the triphasic life history of later divergent florideophyceans.

may appear nearly white. Freshwater rhodophytes are often colored blue-green, as their plastids contain a preponderance of red light-absorbing phycocyanin. Rhodophytes that grow in highly irradiated habitats of upper marine shorelines may be colored yellow, deep violet, brown, or black due to the presence in cells of large amounts of various photoprotective carotenoids. The carotenoids of red algae include α and β-carotene, lutein, zeaxanthin, antherixanthin, and violaxanthin. When carotenoids are abundant, rhodophytes may be difficult for the novice to distinguish from other algae, particularly brown seaweeds. A commonly used method for distinguishing red algae from similar-appearing seaweeds of other groups is to submerse the algal thallus in hot water—water-soluble phycobilins typically diffuse from red algae, revealing green chlorophyll or brown to orange

carotenoids (which are hydrophobic and remain attached to thylakoids). Pigments are not similarly leached from other macroalgae.

Other Distinctive Features of Red Algae

Primary Chloroplasts

Chloroplasts of red algae are distinctive in that they appear to have originated from cyanobacteria by primary endosymbiosis (Chapter 7), and thus lack periplastidal endoplasmic reticulum, more than two envelope membranes, or other evidence of secondary origin from eukaryotes (as is observed in ochrophytes, haptophytes, dinoflagellates, cryptomonads, and euglenoids). Molecular and ultrastructural evidence indicates that the plastids of cryptomonads, haptophytes, and ochrophytes originated by secondary endosymbiotic incorporation of red algal cells (Chapter 7). Unlike green algal plastids, those of red algae never contain starch. Rather, granules of a differently branched glucan, known as **floridean starch**, occur in the cytoplasm (Fig. 16–4). This storage product stains only slightly upon iodine treatment, in contrast to the deep purple-blue staining starch of green algae.

Absence of Centrioles and Flagella; Presence of Triphasic Life Histories

Red algae are unique among eukaryotes in lacking centrioles and flagella from their vegetative cells, spores, and gametes. In contrast, the vast majority of other eukaryotic phyla exhibit centrioles and flagella in some life stages (though parallel absence of centrioles and flagella is also characteristic of Zygomycetes, Basidiomycetes, and Ascomycetes in the Kingdom Fungi). The glaucophytes (Chapter 7)—flagellates such as *Cyanophora* and similarly pigmented nonflagellate unicells that generate flagellate zoospores—have been grouped with the red algae by some experts. The possibility of such a relationship (discussed in more detail below) suggests that centrioles and flagella may have been ancestral features that were lost early in the evolutionary radiation of red algae.

The absence of flagella is thought to have had profound effects on reproductive evolution in red algae, leading to the unique and widespread presence of life histories having three multicellular phases (Fig. 16–5). In contrast, a maximum of two phases occur in other multicellular algae and land plants. The origin of the unique third life-history stage has been regarded as an evolutionary compensation for loss of flagella, serving to enhance reproductive fecundity under this constraint (Searles, 1980). Consequently, red algae have been suggested to be useful model systems for the study of life-history evolution (Hawkes, 1990).

Pit Plugs, Cell-Cell Fusions, and Parasitic Lifestyles

Other aspects of red algal cell biology, notably the common occurrence of proteinaceous plugs in the central-most area of cross-walls (septa) (Fig. 16–6) and the widespread ability of a variety of cell types other than gametes to undergo fusion, are also very unusual features of red algae. Pit plugs are formed during cytokinesis and maintain structural linkage between adjacent cells, even when cells become well separated at maturity (Fig. 16–7). In mature filaments, pit plugs may appear as threadlike linkages between adjacent beadlike cells, and are often diagrammed in this way (Fig. 16–7). As discussed more extensively later, some other functions have been proposed for pit plugs, and variations in their structure have been very useful in red algal systematics.

Fusion of non-gamete cells, common among higher red algae, is a rather uncommon feature among other algae. Cell-cell fusion may be facilitated by the fact that red algal extracellular matrices (cell walls) are much less rigid than are those of many other algae. Cell fusion, which is under a high degree of developmental

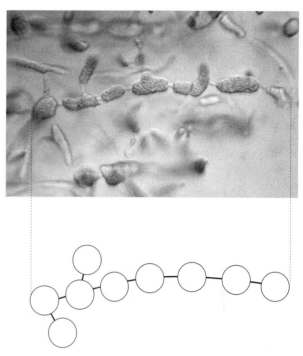

Figure 16–7 Cells in filaments of most red algae such as *Mazzaella* are separated by pit plugs that, when mature, may appear as linear connectives between sister cells. Consequently, red algal filaments are usually diagrammed as circles (representing cells) connected by lines (extended pit plugs).

control, allows red algal cells of different types to communicate directly, as nuclei and cytoplasmic information are combined. Cell fusion is an integral part of morphogenesis in various red algae, and is also an essential element in the evolution of complex interactions between cells of different life-history phases, another unusual feature that characterizes most red algae. Cell fusion is also involved in repair of wounded thalli and has been critical in the evolution of rhodophytes that are completely (or nearly so) parasitic on other red algae. Although parasitic forms occur in some other algal groups (e.g., some dinoflagellates parasitize metazoa and a few terrestrial ulvophycean green algae apparently parasitize various tropical flowering plants), in no other group of photosynthetic protists has parasitism become so widespread and successful a lifestyle as among rhodophytes.

Thallus Organization

Early divergent rhodophytes (Ragan, et al., 1994) occur as unicells, colonies, filaments, or sheets of cells

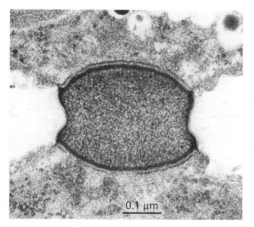

Figure 16–6 TEM view of a pit plug in the central region of a cross-wall separating adjacent sister cells of *Dumontia*. (From Pueschel and Cole, 1982)

0.1 μm

Figure 16–9 *Mazzaella* is a common example of a multiaxial, thalloid red alga. As shown in this section, the thallus is composed of aggregations of filaments (pseudoparenchyma), and there are specialized regions of pigmented surficial cortical cells that are densely packed as well as more loosely arranged, colorless filaments of the central medulla.

Figure 16–8 *Polysiphonia* is a common example of a simple, uniaxial, branched filamentous red alga.

usually lacking pit plugs and cellular differentiation in the vegetative phase. More highly derived red algae are structured as simple, though often highly branched, filaments (Fig. 16–8), or are composed of filament aggregations known as pseudoparenchyma (Fig. 16–9). In various brown algae such as giant kelps (Chapter 15), a few green algae, and most notably the land plants (derived from charophycean green algae), polyphyletic evolution of coherent tissues (parenchyma) seems to have required the earlier evolution of intercellular cytoplasmic continuity (plasmodesmata) (Cook and Graham, 1999). Red algae lack plasmodesmata, and there is as yet no direct evidence that pit plugs serve as a pathway for cytoplasmic communication. Thus parenchymatous organization with cellular integration analogous to that of some other groups of multicellular algae and plants, is absent from red algae (Stiller and Hall, 1998). Some authorities, however, maintain that multicellular early divergent forms such as *Porphyra* and *Bangia*, as well as some more advanced red algae classified in the Delesseriaceae (Ceramiales), are parenchymatous (Bold and Wynne, 1985).

Commitment to production of both male and female gametangia from apical cells relatively early in the evolution of red algae is thought by some to have been a critical event that limited growth possibilities, preventing red algae from evolving parenchyma (Hommersand and Fredericq, 1990). Although rhodophyte filaments may coalesce to form fleshy macroscopic thalli, some reaching 2 m in width (Fig. 16–10) and exhibiting thallus and cellular differentiation (Fig. 16–9), even the largest and most complex red algae do not approach the maximum sizes or internal complexity exhibited by kelps or land plants. Even so, red algae have achieved an amazing diversity of thallus types by varying the way in which their filamentous thalli develop. Among autotrophs,

Figure 16–10 One of the largest of the red algae is *Schizymenia borealis*, whose blades can be as long as two meters.

rhodophytes are the champions at effective evolutionary use of branched filaments in morphological diversification.

Fossils and Evolutionary Relationships

Fossils

The oldest convincing fossil evidence of red algae consists of Precambrian (750–1250 million-year-old) remains from Somerset Island in arctic Canada. These fossils include uni- and multiseriate filaments composed of wedge-shaped cells arranged radially, that appear to have been derived developmentally from uniseriate filaments by longitudinal divisions, similar to that of the extant rhodophyte *Bangia* (Butterfield, et al., 1990) (Fig. 16–11). Extremely well-preserved remains that have been interpreted as higher red algae were recently reported from approximately 570 million-year-old phosphorites from the Doushantuo Formation in southern China. These fossils exhibit growth forms and features that closely resemble reproductive structures of modern red algae (Xiao, et al., 1998).

Some more recent fossils believed to represent remains of red algae were preserved because they had calcified cell walls that did not readily break down and thus retained distinctive thallus structure through burial and changes associated with fossilization (diagenesis). One such group of rhodophytes, known as the Solenoporaceae, are known from the Cambrian (500–600 million years ago), reached their maximum development in the Jurassic, and became extinct in the Paleocene (60 million or so years ago) (Elliot, 1965). Solenopores were leaflike encrusting sheets or nodules, and were composed of closely packed cells in radiating filaments (Johansen, 1981). In several ways the solenopore fossils resemble fossil and modern calcified red algae known as corallines because of their resemblance to corals. They differ from corallines in cell size, however. Judging by the calcified remains of their walls, the vegetative cells of solenopores were more than 30 μm broad (sometimes reaching 60 μm), whereas those of corallines are less than 15 μm. Some authorities hypothesize that evolution of smaller cell size allowed greater carbonate production, generating a stronger, harder thallus that might be more resistant to damage (Johansen, 1981). Others suggest that demise of the solenopores occurred because their

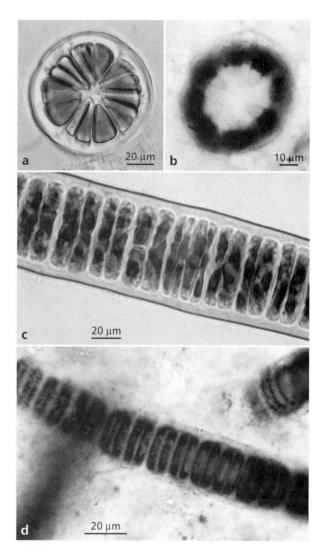

Figure 16–11 Fossil and extant *Bangia*. (a) Cross section of extant *Bangia*, showing triangular outlines of cells. (b) 750–1250 million-year-old fossil that is interpreted as a cross-sectional view of an ancient *Bangia*-like organism because the cells have a similar shape. (c) Lengthwise view of a young filament of extant *Bangia*. (d) Precambrian fossil interpreted as an ancient *Bangia*-like red algal filament because of its similarity to modern forms. (a: Courtesy A. Knoll; b and d: Reprinted with permission from Butterfield, N. J., A. H. Knoll, and K. Swett. *Science* 250:104–107 ©1990 American Association for the Advancement of Science)

thalli were less resistant to herbivores than those of corallines (Steneck, 1983).

Experts have suggested that the Solenoporaceae might have been ancestral to the coralline red algae, via a calcified group of "ancestral corallines" that were present as early as the Silurian, and perhaps mid-Ordovician (Brooke and Riding, 1998) (Fig 16–12). An example of the putative intermediate group is *Archeolithothamnion*, a fossil that differs from the modern coralline *Lithothamnion* principally in having larger cells (Wray, 1977). These ancestral forms appear to have declined in the Triassic (sometime prior to 210 million years ago). Modern corallines left definitive fossils by the Jurassic (150–200 million years ago), and have a continuous fossil record to modern times, where they reach a maximum in diversity and production. The common extant genus *Lithothamnion* (important, as noted earlier, in reef consolidation) occurred in this time period. Some uncalcified fossils thought to be red algae are known from Miocene (5–25 million years ago) deposits in California (Parker and Dawson, 1965). There are many other reports of fossil red algae (e.g., Gabrielson, et al., 1990), but the identity of such remains is often controversial when they do not possess distinctive features that would allow them to be distinguished from similar-appearing brown or green seaweeds (Taylor and Taylor, 1993).

Structural and Molecular Evidence for Red Algal Relationships

Red algae have traditionally been classified into the Phylum (Division) Rhodophyta, with either a single class, Rhodophyceae (encompassing two subclasses, Bangiophycidae and Florideophycidae) or two classes, Bangiophyceae and Florideophyceae. A modified form of the latter system has been used here, for reasons detailed below. The Bangiophyceae comprise structurally and reproductively simple forms (single cells, linear colonies, unbranched filaments such as *Bangia*, and sheets such as *Porphyra*), whereas more structurally and reproductively complex forms (including corallines) are included in the Florideophyceae. At least two structural autapomorphies (unique, derived characters) delineate the Florideophyceae: **tetrasporangia** and a **carposporophyte** composed of filamentous **gonimoblast**. No shared, derived characters define the Bangiophyceae, suggesting that this group is not monophyletic (Gabrielson, et al., 1990).

Molecular evidence from SSU rDNA and *rbcL* sequences (Ragan, et al., 1994; Freshwater, et al., 1994) has corroborated earlier inferences made on the basis of structural and reproductive characters (Gabrielson, et al., 1990) that: (1) the Rhodophyta is monophyletic, (2) bangiophytes diverged earlier than

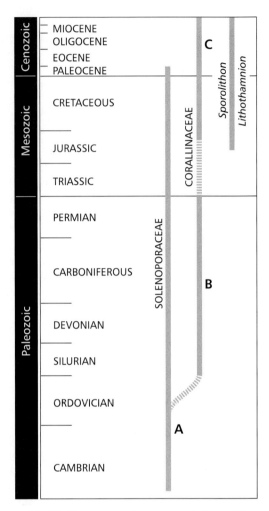

Figure 16–12 Diagrammatic representation of the hypothesized relationship between the ancient Solenoporaceae (A), a hypothetical group of ancestral corallines (B), and modern Corallinaceae (C). The occurrence of two modern coralline genera commonly associated with coral reef ridges, *Sporolithon* and *Lithothamnion* (previously known as *Lithothamnium*), are also shown. (Data from Wray, 1977; Brooke and Riding, 1998)

Florideophyceae, (3) the Florideophyceae is a monophyletic group, and (4) Bangiophyceae is a polyphyletic assemblage. The final conclusion is supported by an additional analysis of nuclear SSU rDNA and *psbA* sequences (Tsunakawa-Yokoyama, et al., 1997). Hence, class status is appropriate for the Florideophyceae, whereas the Bangiophyceae will probably be disassembled into several phylogenetically defined classes. The red algae classified into the Bangiophyceae are here referred to by the colloquial term bangiophyceans.

Bangiophyceans include the monophyletic order Bangiales (*Bangia* and *Porphyra*), which is identified as the sister group to the Florideophyceae, according to SSU rDNA data (Ragan, et al., 1994). In contrast, *rbcL* data suggest that another clade of unicellular bangiophyceans, which includes *Rhodella* and *Dixoniella*, is sister to the Florideophyceae (Freshwater, et al., 1994). Thus there is still controversy on this issue. The molecular findings correlate well with the fossil evidence that *Bangia*-like forms appeared earlier than florideophyceans such as ancient corallines.

Both molecular and ultrastructural evidence strongly suggests that acidophilic cyanidophyceans, including *Cyanidium*, are probably monophyletic, and represent a very early divergent group of rhodophytes (Tsunakawa-Yokoyama, et al., 1997). Post-endosymbiosis transfer of a chloroplast RNA polymerase subunit gene to the host nucleus, and targeting of the gene product back to the chloroplast, have been documented in *Cyanidium* (Tanaka, et al., 1996). Additional gene transfer information may aid in understanding the early evolution of red algae, as such data have been very helpful in tracing the origin of land plants from green algae (Chapter 17).

Nuclear-encoded 18S rDNA sequences (Bhattacharya and Medlin, 1995) and ultrastructural data (Broadwater and Scott, 1994) suggest that *Glaucosphaera*, which contains blue-green plastids lacking peptidoglycan walls (and was formerly classified among the glaucophytes—see Chapter 7), is sufficiently closely related to the mainstream of red algal evolution as to justify inclusion in the Rhodophyta. Glaucophytes differ in having plastids that still retain remnants of the ancestral cyanobacterial peptidoglycan wall. Some experts suggest that these organisms may also be closely related to red algae, but additional comparative molecular systematic work on nuclear-encoded genes is needed to clarify relationships between glaucophytes and red algae.

Red algae are regarded by some to be among the "crown taxa" of relatively late-diverging eukaryotes, including metazoa, fungi, green algae, and land plants (Bhattacharya and Medlin, 1998). An argument for possible sister-group relationship of red algae and green algae (and plants) was presented by Ragan and Gutell (1995). However others disagree, based on the study of RNA polymerase gene sequences (Stiller and Hall, 1997, 1998). The relationship of red algae to other major eukaryote clades is, therefore, still controversial. The placement of red algae in this book, between brown algae (Chapter 15) and marine green algae (Chapters 17 and 18) was not meant to reflect an opinion on red algal relationships, but rather to facilitate use by students in coastal regions who commonly encounter brown, red, and green seaweeds growing together.

Cell Wall Structure and Biochemical Composition

Red-algal extracellular matrices, like those of various other algae, are composed of a cellulosic microfibrillar network associated with a more amorphous matrix material, which usually includes polymers of sulfated galactans, mucilages, and cellulose. In some cases microfibrils composed of xylose substitute for cellulose and calcium carbonate occurs in the walls of several types of red algae. In contrast to the relatively rigid walls of various other algae, walls of most red algae are softer in consistency because the amorphous component typically represents a larger contribution to the wall than the fibrillar component (cellulose).

Sulphated Polygalactans

The extracellular matrix of the red algae is dominated by various types of highly hydrophilic sulfated polygalactans, polymers of β-(1→4) galactose and α-(1→3) linked 3,6 anhydrogalactose. These are the major constituents of agarocolloids (agars) and carrageenans. The presence of D-galactose and anhydro-D-galactose distinguishes the more highly sulfated carrageenans from the less highly sulfated agars (anhydro-L-galactose). Differences in galactan isomers appear to confer distinctions in sulfation patterns, which are probably related to variation in the physical properties of polygalactans. Agars are more heavily methoxylated than are carrageenans. In the presence of cations (such as K^+ or Ca^{2+}), sulfated galactans form double helices that aggregate into a three-dimensional network having colloidal properties.

Carrageenans exhibit a repeating disaccharide backbone: β-(1→4)-D-galactopyranosyl-α-(1→3)-D-galactopyranosyl. This backbone is assembled from nucleotide precursors, then sulfated within the conspicuous and highly developed Golgi apparatus (Fig. 16–13) by sulfotransferases. There are at least 17 different types of carrageenans, which occur in different combinations in different species and can also vary within a species. Carrageenan content can also differ

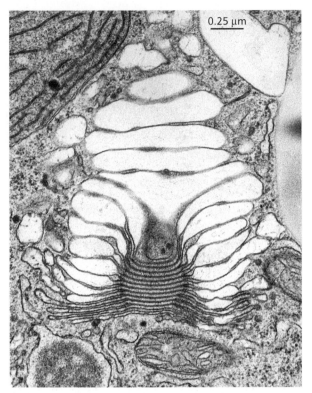

Figure 16–13 Red algal mucilages and sulfated polygalactans are produced within the large, complex Golgi apparatus, such as that of *Polysiphonia denudata*, shown here. (Micrograph by J. Scott, in Pueschel, 1990)

Conspicuous mucilage sacs are often observed within red algal reproductive cells (Fig. 16–14).

Cellulose and Other Microfibrillar Components

Freeze-fracture ultrastructural studies have revealed that red algal cell membranes are studded with protein particles and particle complexes, including linear terminal complexes (Fig. 16–15a) similar to those of certain green algae (see Fig. 21–20) that are believed to be involved in cellulose microfibril production. These protein complexes are thought to represent clusters of enzymes that convert UDP-glucose into cellulose. The form of such complexes differs among organisms that produce cellulose and determines the shape and size of the cellulose microfibrils. Microfibrils produced by linear complexes (including those of relatively primitive red algae) are typically ribbonlike (Fig. 16–15b). In contrast, at least some higher red algae generate cylindrical microfibrils (Tsekos, 1996). Cellulose synthesizing proteins are transported to the cell membrane by Golgi vesicles. A review of red algal

among red algal life-history stages. Agarocolloids consist of alternating 3-O-linked β-D-galactopyranose and 4-O-linked 3,6 anhydro-α-L-galactopyranose, but the repetition can be interrupted by blocks of repeated units of either one of the two constituents. The most complex agarocolloids are found in the most highly derived red algae (i.e., Ceramiales). Hybrid agarose/carrageenan combinations have sometimes been observed. The sulfated polygalactans of red algae have been reviewed by Craigie (1990) and Cosson, et al. (1995).

Mucilages

Mucilages are polymers of D-xylose, D-glucose, D-glucuronic acid, and galactose, the proportions differing among taxa. Sulfated mannans occur in the mucilages of certain forms. Mucilages are produced in the Golgi apparatus, and are often abundant in reproductive cells, where mucilage expansion upon hydration plays a role in reproductive cell dispersal and attachment.

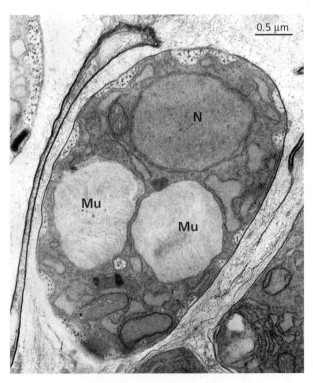

Figure 16–14 Mucilage (slime) sacs (Mu) within a reproductive cell (spermatium) of *Polysiphonia hendryi*. N=nucleus. (From Kugrens, 1980)

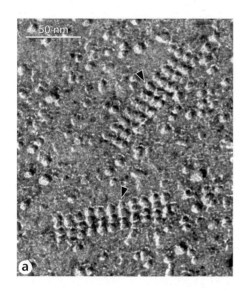

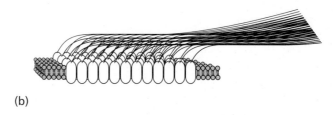

(b)

Figure 16–15 (a) Transmission electron micrograph of a freeze-fractured cell membrane of *Erythrocladia* (*Sahlingia*) *subintegra* showing rectangular terminal complexes characteristic of red algae that spin out ribbon-shaped or cylindrical cellulose microfibrils. (b) Diagrammatic representation of ribbon-shaped cellulose microfibril production by the rectangular terminal complexes of red algae. (a: From Tsekos, et al., 1996; b: After Tsekos, 1996 ©Springer-Verlag)

cellulose microfibril synthesis can be found in Tsekos (1996), and a freeze-fracture study of the development of cellulose-synthesizing linear complexes in the cell membrane of the red alga *Erythrocladia* was done by Tsekos, et al. (1996).

In gametophytes of *Porphyra*, microfibrils are composed of xylans rather than glucans. Three xylan chains are aggregated to form a macromolecular triple helix that is stabilized by hydrogen bonding interactions within and between chains. Such polymers differ from cellulose in that they are not birefringent, i.e., they do not yield a characteristic bright cross-shape when examined by polarized light. A relatively insoluble structural mannan is also present. In contrast, cellulose microfibrils are produced by the **conchocelis** life-history stage of *Porphyra* and *Bangia* (reviewed by Craigie, 1990). That cellulose occurs in the conchocelis phase of *Bangia fuscopurpurea* was discovered using X-ray diffraction techniques (Gretz, et al., 1980).

Calcification

The majority of the ecologically important coralline red algae have heavily calcified walls. In the coralline algae, calcium carbonate is deposited in a particular crystalline form known as calcite, but some non-coralline red algae (some members of the Peyssoneliaceae, as well as *Galaxaura* and *Liagora* of the Nemaliales) that also have calcified walls, deposit calcium carbonate in the form of aragonite crystals. These two forms of calcium carbonate differ in cations that substitute for calcium within the crystals—aragonite can become enriched in strontium (a larger ion than the calcium that it replaces), whereas Ca^{2+} in calcite can be replaced by smaller ions such as magnesium, iron, or zinc. Although the advantages of producing calcite versus aragonite are not well understood, Mg, Fe, and Zn are micronutrients that sometimes (at least in the case of Fe) limit algal growth. It is possible that calcite production confers a useful nutrient harvesting/storage capacity. Calcification occurs most rapidly in growing regions of red algal thalli, with the calcium carbonate being deposited onto an organic matrix. Red algal thalli are able to control deposition of calcium carbonate such that some cells become calcified, whereas others do not. The mechanism by which such control is exerted is unknown, but it is possible that regulation of the production of essential organic matrix materials is involved. Carbonate deposition is ten times faster in the light than in the dark; photosynthesis favors calcification indirectly by increasing the pH at the cell surface. Calcification by red algae has been reviewed by Craigie (1990).

Surficial "Cuticle"

The external surface of many red algal thalli is covered by a continuous layer of a densely staining, insoluble proteinaceous "cuticle." Such layers are not biochemically similar or homologous to the cuticles of land plants. The surfaces of the red alga *Mazzaella* (*Iridaea*) include as many as 17 layers of electron-opaque surface material alternating with electron-translucent layers (Fig. 16–16). This layering produces light

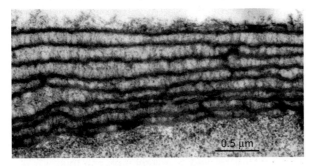

Figure 16–16 Electron micrograph of the multilayered surface "cuticle" (characteristic of *Mazzaella*), which is thought to be responsible for the iridescent appearance of submerged thalli of this genus formerly known as *Iridaea*. (Micrograph courtesy N. Lang)

interference patterns that give the seaweed an iridescent appearance when submerged. The adaptive advantage, if any, of iridescence is not understood. Little is known about the chemical structure or biosynthesis of red algal "cuticles."

Plastids and Other Cellular Constituents

Plastids occur in nearly all red algal cells, even colorless cells in the internal regions of thick thalli or parasites (Goff, 1982). Red algal plastids may be star-shaped (stellate) or discoid, and may have lobed edges (Fig. 16–4). Cells in relatively early divergent forms contain only a single plastid, but multiple plastids per cell is the general rule. In some red algae plastids undergo diurnal change in position. In *Griffithsia*, for example, plastids move away from cell poles early in the light period to form distinctive bands in the center of the cell. Later in the light period plastids return to the cell ends, remaining there throughout the night. The fact that the actin inhibitor cytochalasin B interferes with chloroplast banding indicates that actin mediates plastid movement in *Griffithsia* (Russell, et al., 1996). Plastids of some of the early divergent red algae contain a conspicuous pyrenoid (Fig. 16–4), but this structure is lacking in most red algal species.

As with many other algae, red algal plastids differ from those of green algae in that both the large and small subunits of Rubisco are encoded in the plastid genome. Nevertheless, some genes required for red algal photosynthesis are encoded in the nucleus, probably as a consequence of post-endosymbiotic transfer (Chapter 7), with the result that the cytoplasmic translation products must be targeted to the plastid. An example is the γ subunit of phycoerythrin (Apt, et al., 1993).

The cytoplasm of red algal cells typically also contains granules of floridean starch as well as protein crystals of unknown function. In addition, many red algal groups are characterized by distinctive cellular inclusions, such as the red cytoplasmic membrane-

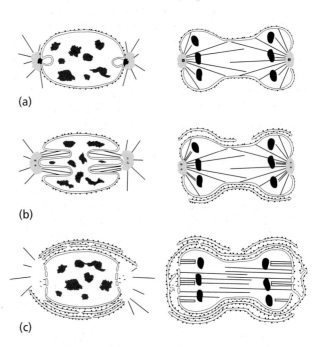

(a)

(b)

(c)

Figure 16–17 Diagrammatic representation of some of the variation in mitosis that occurs during prometaphase (at left) and anaphase (at right) in red algae. Note that the nuclear envelope remains intact except for small interruptions (fenestrations) at the poles and that centrioles are not present, but spindle microtubule organizing structures are present at the poles. (a) A putatively primitive type of mitosis illustrated by the bangiophycean unicell *Porphyridium*. (b) A somewhat more derived type (note the addition of a layer of endoplasmic reticulum) illustrated by the relatively early divergent, multicellular florideophycean *Batrachospermum*. (c) A derived type of mitosis (note the addition of more ER layers) illustrated by *Polysiphonia*. (After Scott and Broadwater, 1990)

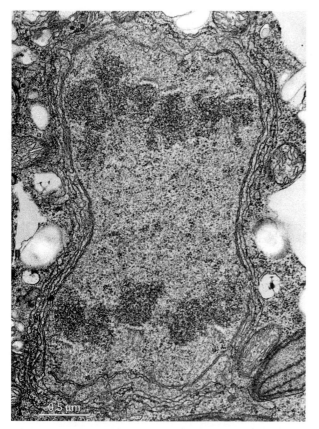

Figure 16–18 TEM view of a late-anaphase nucleus of *Dasya baillouviana*. Note the intact nuclear envelope and surrounding ER. (From Phillips and Scott, 1981 ©Springer-Verlag)

bound vesicles in certain cell types in species of the genus *Laurencia* known as corps en cerise, which are known to contain brominated secondary metabolites (Young, et al., 1980).

Cell Division, Development, and Primary Pit-Plug Formation

Among red algae, mitosis and cytokinesis vary primarily in the structure of spindle pole regions and the extent to which endoplasmic reticulum is associated with the nucleus, but are otherwise quite similar (Fig. 16–17). The nuclear envelope remains intact throughout mitosis except for gaps (fenestrations) at the poles (Fig. 16–18). As previously noted, centrioles are not present, but a distinctive organelle known as a **polar ring** or **nuclear associated organelle** (NAO) typically

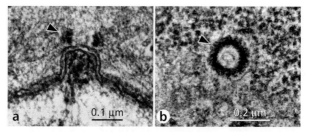

Figure 16–19 Transmission electron micrographs of the nuclear associated organelle (NAO) (arrowheads) of *Polysiphonia harveyi*. (a) Median longitudinal section through the NAO and the underlying nuclear protrusion at late prophase. (b) Transverse section of the *Polysiphonia harveyi* NAO in interphase or early prophase. (From Scott, et al., 1980 by permission of the *Journal of Phycology*)

occurs at the mitotic spindle poles (Fig. 16–19). The NAO (this term is preferred because not all polar structures are ring-shaped) often appears as a pair of short hollow cylinders (Fig. 16–20). The NAOs are about the same diameter as centrioles, but are shorter and lack the distinctive internal structure typical of centrioles. An extension of the nuclear envelope may be associated with the NAO, sometimes appearing to

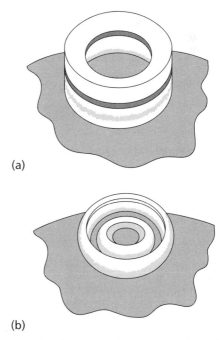

(a)

(b)

Figure 16–20 Diagrammatic representation of two of the four basic NAO types found among red algae. (a) *Polysiphonia*, (b) *Batrachospermum*. (After Scott and Broadwater, 1990)

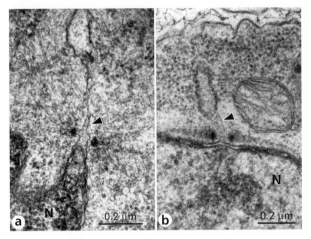

Figure 16–21 Transmission electron micrographs showing tubular extensions of the outer membrane of the nuclear envelope (arrowheads) extending through the NAO and expanding into an extranuclear cisternum. (a) *Agardhiella subulata* in late prophase. (b) *Lomentaria baileyi* in prophase. N=nucleus. (Micrographs from Scott and Broadwater, 1990)

Commonly, subapical cells divide several times radially to form whorls of branch initial cells (**periaxial cells**), which divide to form branch filaments, each with a terminal apical cell. Branches are typically determinate, meaning that only a genetically determined number of cell divisions occurs, thereby limiting the length of the branch filaments. Subapical cells also generate rhizoids—single cells or filaments at the base of the thallus that are involved in attachment (Coomans and Hommersand, 1990).

More robust and fleshy red algae are composed of multiple filamentous axes, each derived from a terminal apical cell (Fig. 16–24). These multiple axes arise during early development through the transformation of determinate branches into indeterminate axes. This means that the branch apical cells become less constrained in the number of cell divisions they can undergo. Variation in the branching patterns observed in different genera are thought to result from differences in division plane orientation (Coomans and Hommersand, 1990). Analogous multiaxial thalli are

extend through it (Fig. 16–21). Although the chemical composition and function(s) of the NAO are not known, these structures, like centrioles, are associated with microtubules. It is possible that they, or some associated substances, function as microtubule organizing centers, as does pericentriolar material in other eukaryotes (McDonald, 1972).

Daughter nuclei are kept well separated at telophase by the large plastid of monoplastidic unicells such as *Porphyridium* (Schornstein and Scott, 1982), or by formation of a large central vacuole in more advanced red algae. Cytokinesis occurs exclusively by centripetal furrowing (Scott and Broadwater, 1990). In all members of the Florideophyceae, completion of furrowing occurs only during tetraspore formation; division of all other cells is incomplete and results in pit-plug formation.

In many of the early divergent, multicellular red algae such as *Bangia* and *Porphyra*, cell division is not localized, i.e., there is no specialized meristematic region. However most red algae grow via division of an apical cell located at the terminus of a filament. Apical cells cut off derivatives at their bases in a single linear series (Fig. 16–22). Divisions of apical cells are responsible for increases in the number of cells in filaments, and together with cell enlargement, increase thallus size. Delicate red algal thalli may be uniaxial, composed of a single branched filament (Fig. 16–23).

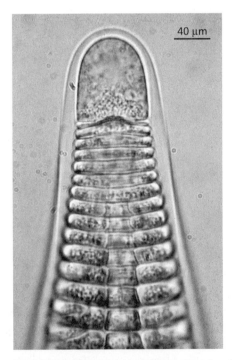

Figure 16–22 Apical meristem of the uniaxial red alga *Platysiphonia*, showing the way in which derivatives are cut off from the base of the apical cell. The subapical cells have undergone further divisions to form whorls of periaxial cells that may serve as branch initials. (Photograph by R. Norris, in Bold and Wynne, 1985)

Figure 16–23 An example of a uniaxial, branched filamentous red alga, *Aglaothamnion*, in microscopic view. (Photograph courtesy of C. Taylor)

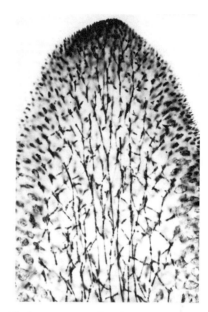

Figure 16–24 An example of a multiaxial red alga, *Agardhiella*, in microscopic view. (From Gabrielson and Hommersand, 1982 by permission of the *Journal of Phycology*)

also formed by various siphonalean green seaweeds (Chapter 18).

Development of Multinucleate Cells and Polyploid Nuclei

In many of the more highly derived red algae, nuclei continue to undergo mitosis without intervention of cytokinesis, giving rise to multinucleate cells. The uninucleate or multinucleate condition of cells can be detected by Feulgen staining, which makes nuclei visible in bright-field light microscopy or by staining with a DNA-specific fluorochrome (such as DAPI) that forms a complex detectable with fluorescence microscopy (Fig. 16–25 and Text Box 16–1) (see Goff and Coleman [1990] for additional technical details). The presence of numerous nuclei allows the maintenance of very large cells, whose cytoplasm would unlikely be efficiently served by a single nucleus, particularly since red algal cells lack cytoplasmic streaming. There is often a tight correlation between the number of nuclei in a red algal cell and the number of plastids (Goff and Coleman, 1990).

Red algal nuclei can also undergo a process known as **endoreduplication**—repeated replication of the entire nuclear genome without intervening mitosis—resulting in polyploidy. Such genome amplification is also known as nuclear polygenomy, endopolyploidy, or polyteny, and is common in red algae. Genome amplification is thought to serve as a buffer against mutation of essential genes (Goff and

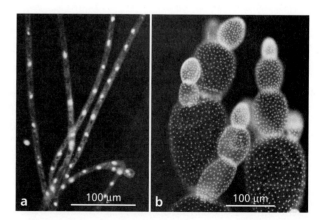

Figure 16–25 Fluorescence microscopy of DAPI-stained red algae showing uninucleate cells of *Acrochaetium pectinatum* (a) and multinucleate cells of *Griffithsia globulifera* (b). (From Goff and Coleman, 1990)

Text Box 16–1 Use of DNA Measurements in the Study of Algal Life Histories

The elucidation of algal life cycles requires an understanding of the timing of changes in nuclear DNA level, which appears often, though not always, to be associated with changes in thallus morphology or reproductive activity. The algal material being examined should be very clean and, in the case of microalgae, should be in the form of unialgal cultures to avoid confusing life-history stages of different taxa. Such studies also require knowledge of algal growth requirements, so that algae can be maintained long enough to detect life-history changes. In addition, the environmental conditions that are associated with life-history phase change must be known, so that algal cultures can be induced to undergo such changes in the laboratory.

The classical method for detecting DNA-level change is to compare chromosome counts in cells of differing life-history stages. While some phycologists have been quite successful in tracking chromosome number changes that are correlated with life-history transformations, algal chromosomes are often too small or numerous to count accurately. A more generally reliable way of measuring DNA-level change is to stain nuclei with a DNA-specific dye whose degree of binding to DNA is (1) directly proportional to the amount of DNA present and (2) can be measured easily. Feulgen staining is one such process; DNA is stained pink and the amount of the DNA-dye complex can be estimated by microspectrophotometry. An instrument known as a microspectrophotometer is required; this is fitted to a standard light microscope and allows measurement of absorption (at specified wavelengths) of individual nuclei defined by use of a correlatively sized pinhole that blocks extraneous light. Measurements are standardized by concurrent staining and measurement of nuclei whose DNA content has been spectrophotometrically determined to be constant; animal sperm or chicken red-blood cells are commonly used as standards. The major disadvantage of Feulgen microspectrophotometry is that the staining process is time-consuming and laborious—an empirical determination of optimal staining time must be performed for each cell type examined, as this parameter varies significantly among taxa.

More widely used today is DAPI microspectrophotometry. DAPI is a water-soluble compound (4,6-diamidino-2-phenylindole) that specifically binds to double-stranded DNA and is fluorescent, i.e., is a fluorochrome. An epifluorescence microscope equipped with a microspectroflurometer (or digital video imaging system) is required. The staining process is rapid and measurements are easily accomplished. Pretreatment with a fixative such as 3:1 95% ethanol: glacial acetic acid facilitates entry of the fluorochrome, but may distort cell structure. Microwave-facilitated infusion of DAPI is highly recommended as an alternative (Goff and Coleman, 1987). The pH optimum for DAPI staining is between 4.0 and 7.5. Commonly used fluorochrome concentrations range from 0.25 to 2 μg ml^{-1}, and staining time is commonly 15–30 minutes (Goff and Coleman, 1990). The wavelength range for maximum absorption is 340–380 nm, so the light source (usually a mercury vapor lamp, but see Goff and Coleman, 1990 for alternatives) must include significant UV emission for excitation of the DAPI. The wavelength range of maximum emission by excited DAPI is 460–480 nm, so the microscope must be equipped with the proper filter sets. The use of a cutoff filter that absorbs light greater than 500 nm is recommended, as it reduces the possibility of interference by autofluorescence from chlorophyll (Goff and Coleman, 1990).

DAPI

Coleman, 1990). Endoreduplication and other changes in nuclear DNA level (such as those resulting from gamete fusion or meiosis) can be detected by use of microspectrophotometry or microfluorometry. In these procedures, the extent of the Feulgen-staining reaction or DAPI binding can be measured; such measurements are directly related to the amount of DNA present (see Goff and Coleman, 1990). In some red algae the apical meristematic cell of filaments is also multinucleate (Fig. 16–26), but such cells are not typ-

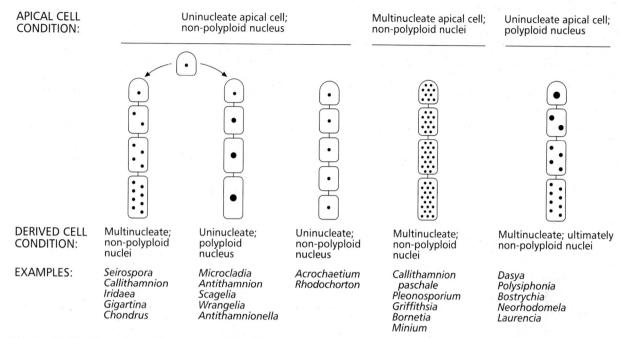

Figure 16–26 Diagrammatic representation of variation among red algae in terms of number of nuclei in apical and other cells, and in whether or not nuclei are polyploid. (After Goff and Coleman, 1990).

ically polyploid (Goff and Coleman, 1990). Examples of variation among red algae in occurrence of multinucleate and polyploid nuclei in apical and other cells of filaments is shown in Fig. 16–26).

Determination of Cytokinetic Plane

In red algae, unlike most other walled organisms, there is no obvious relationship between the position of the mitotic spindle and the plane of cytokinesis. Even in those red algae whose cells are always uninucleate, the position of the metaphase plate does not always correlate with the position of new cross-wall formation. In taxa such as *Griffithsia*, in which division of multinucleate cells occurs, nearly synchronous division of all nuclei precedes cytokinesis, but the nuclei do not position themselves along the plane of new cross-wall formation (as is the case in multinucleate green algae—see Chapter 18) (Waaland, 1990). Rings of actin are involved in determining the plane of cytokinesis (Garbary and McDonald, 1996).

Primary Pit-Plug Development

In most red algae (all of the Florideophyceae and some stages of bangialean forms), septum (cross-wall) formation is incomplete, leaving a membrane-lined pore in the central region. Tubular membranes appear in this region, then a homogeneously granular protein mass—the plug pore—is deposited around the tubules, followed by disappearance of the tubules (Fig. 16–27). In some groups, notably the relatively early divergent Bangiales (but also the more derived coralline algae), the plug core constitutes the pit plug. Pit plugs of other red algae acquire additional features, such as carbohydrate domes and cap membranes, which are continuous with the cell membrane (Fig. 16–6) (reviewed by Scott and Broadwater, 1990; Pueschel, 1990). In some groups the domes occur as two layers of material between which lies the pit membrane. Variation in pit-plug ultrastructure among red algae has proven to be very useful in deducing phylogenetic relationships among the higher rhodophytes (Pueschel and Cole, 1982; Pueschel, 1987; Gabrielson and Garbary, 1987; Saunders and Bailey, 1997).

Upon completion of pit-plug development, the two daughter cells are said to be linked by a **primary pit connection**. Pueschel (1990) has pointed out that the pit plug is not really an intercellular connection in the same way as are plasmodesmata of certain brown and green algae and land plants. Most pit plugs are covered by a membrane that is continuous with the cell membrane. The pit plug is clearly extracellular in most cases (Pueschel, 1977; 1990), hence the term

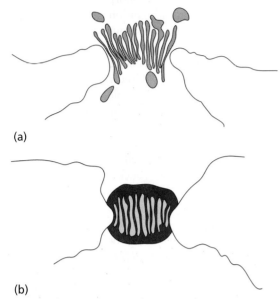

(a)

(b)

Figure 16–27 Diagrammatic representation of pit-plug formation at the ultrastructural level. At the end of cytokinesis the septal pore becomes filled with tubular membranes (a), then granular protein material is deposited around the tubules. Pore maturation (b) involves disappearance of the tubules and addition of additional pit-plug components, such as domes or cap membranes. (Based upon micrographs in Scott, et al., 1980)

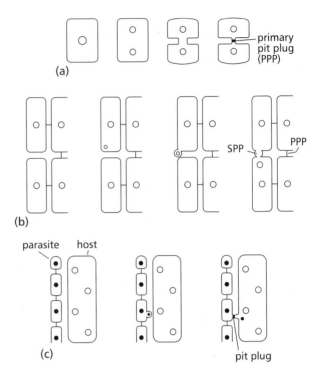

(a)

(b)

(c)

Figure 16–28 Diagrammatic representation of pit-plug formation. (a) Primary pit-plug (PPP) formation between sister cells (cells derived from a common cell division). (b) Secondary pit-plug (SPP) formation between non-sister cells involves an unequal division leading to formation of a small cell termed a conjunctor cell—fusion of the conjunctor cell with a non-sister cell results in transfer of a nucleus to the recipient cell and development of the secondary pit plug. (c) Secondary pit-plug formation also occurs between genetically unrelated cells during red algal parasitism—the parasite initiates the formation of a small conjunctor cell by asymmetric division, then this cell fuses with a host cell. (After Goff and Coleman, 1990)

"pit plug" rather than its synonym, "pit connection," is used here in order to avoid implication of function. The function of the pit plug, other than as a structural link between cells, is unknown. Although a translocatory function has been proposed on the basis of circumstantial evidence (Wetherbee, 1979), such a function has not been unequivocally demonstrated (Pueschel, 1980; 1990).

Secondary Pit-Plug Formation

Primary pit plugs are always found between related cells, but **secondary pit plugs** can also form between non-sister cells and even between the cells of red algal parasites and their hosts. Secondary pit plugs are typically formed through an unequal cell division coupled with production of a primary pit plug, followed by fusion of the smaller daughter cell with a nearby non-kindred cell (Fig. 16–28). Secondary pit plugs are thought to provide a mechanism by which cells of differing lineages can exchange information, since the cell resulting from fusion contains at least two types of nuclei and differing cytoplasms (Goff and Coleman, 1990). Secondary pit-plug formation also helps to unite the filamentous axes of multiaxial thalli, contributing greatly to structural integrity. Within a species, primary and secondary pit plugs are structurally identical. Not all red algae can produce secondary pit plugs; this capacity appears to be absent from early divergent red algae, as well as from those advanced forms that have relatively simple construction (Coomans and Hommersand, 1990).

Cell Elongation and Repair

Red algal cells can undergo a tremendous increase in volume through elongation. For example, the length

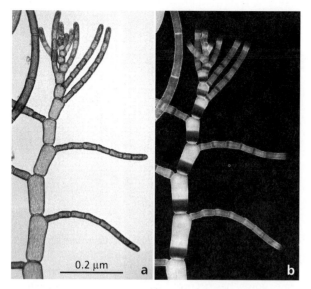

Figure 16–29 The process of cell elongation in red algae is illustrated by *Callithamnion* filaments that have been stained with Calcofluor White—a fluorescent, cellulose-specific reagent—then removed from stain and allowed to continue growth for 24 hours. (a) Thallus viewed in bright-field microscopy. (b) Thallus viewed by epifluorescence microscopy with UV excitation—new wall material added to apical cells of branches and at the bases of intercalary cells during the period following removal from stain is not fluorescent (i.e., is dark). (From Waaland and Waaland, 1975 ©Springer-Verlag)

of intercalary cells of *Ceramium echionotum* can increase by a factor of 100—from 4 μm to 420 μm—accompanied by a 14,000-fold increase in volume. Intercalary cells of the freshwater *Lemanea fluviatilis* elongate 1000 times (from 8 μm to 8000 μm), with an attendant 44,000-fold increase in volume (Waaland, 1990). Cell elongation may occur in apical cells (by growth at the tip), or in non-apical (intercalary) cells (by incorporation of wall material all along the length of the extending cell wall or in just one or more localized bands, known as zones of extension) (Fig. 16–29) (Waaland, 1990).

When a cell in the middle of a filament dies and the remains of its cell wall remain intact, the cells immediately above and below it (which normally would not divide) each divide to produce narrow-tipped, rhizoid- and apex-like cells that grow toward each other and eventually fuse, effecting a repair of the wounded filament (Fig. 16–30). In *Griffithsia pacifica*, the cell immediately above the wounded cell produces a diffusible, species-specific glycoprotein wound-repair

hormone. Known as rhodomorphin, this hormone is active at very low concentrations (10^{-13}–10^{-14} M). A signal glycoprotein analogous to rhodomorphin appears to mediate wound healing in another red alga, *Antithamnion nipponicum*, suggesting that such hormones may be common in red algae (Kim and Fritz, 1993). Tagged lectins—molecules that exhibit strong specific binding to particular carbohydrate structures—were used to demonstrate that *Antithamnion* repair hormone binds a wall glycoprotein having alpha-D-mannosyl residues (Kim, et al., 1995). Glycoproteins may mediate other occurrences of cell fusion in red algae, including fertilization, secondary pit-plug formation, and fusions between cells of different life-history stages, or genotypes (during parasitism) (Waaland, 1990). Further study of diffusible red algal glycoproteins may yield explanations for many of the unusual features of red algae.

Specialized Thallus Regions and Cell Types

Thick, fleshy red algae are commonly composed of outer highly pigmented cells that form the cortex, and inner colorless cells that make up the medulla (Fig. 16–9). Similar terms are used to describe superficial and internal tissues of parenchymatous brown algae (such as kelps) and land plants. Because the majority of red algae are filamentous, application of the term "tissue"—though often used by red algal morphologists—has been avoided here.

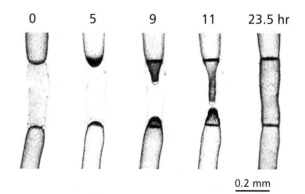

Figure 16–30 Cell repair by cell fusion in *Griffithsia pacifica* where an intercalary cell has died. A rhizoid from the cell immediately above the dead cell and a repair shoot from the cell below it grow toward each other and fuse. (From Waaland and Cleland, 1974 ©Springer-Verlag)

In addition to rhizoids and apical meristematic cells, which are present in most rhodophytes, some other cell types are present in red algae. These include thin, elongate hair cells that may extend from the thallus surface. Their function is unknown, though nutrient uptake is postulated. **Gland cells** (also known as vesicle cells) have specialized secretory functions. For example, gland cells within the inner cortex of *Botryocladia*, produce the mucilage within hollow saccate or grape-shaped structures (Fig. 16–79). In other red algae, gland cells possess crystals or halogenated (bromine or iodine-containing) compounds that are thought to have antimicrobial, anti-herbivore, or allelopathic functions. In some cases gland cells appear to lack a nucleus and plastids (Young and West, 1979).

Reproduction and Life Histories

Asexual Reproduction

Many red algae reproduce asexually by discharging unicellular **monospores** into the water (Fig. 16–31). In general, monospore production is more common in early divergent red algal groups than in more advanced forms, where monospore production may be rare or absent (Hawkes, 1990). If conditions are suitable, monospores attach to a substrate and grow by repeated mitosis into a new seaweed similar to the monospore-producing parent. Red algal spores sink relatively slowly, and a surrounding coat of mucilage aids in the initial attachment to substrates. Completion of the attachment process requires hours to days. Once attached, spores are quite resistant to shear stresses of water movement (Kain and Norton, 1990). Monospores are produced singly within cells known as monosporangia, which may be produced in clusters. The ultrastructure of *Porphyra* monospores was studied by Hawkes (1980). Production and swelling of mucilage is the mechanism for monospore release. Vegetative fragmentation, "stolon" production (analogous to asexual propagation in plants such as strawberry), and formation of specialized propagules, may also occur. A list of genera exhibiting such variations in asexual reproduction is provided by Hawkes (1990).

Sexual Reproduction

Sexual reproduction is characteristic of the vast majority of red algae. It has been well documented in ban-

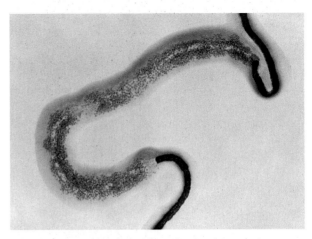

Figure 16–31 Monospore production by *Bangia* growing on an agar surface. Transformation of vegetative cells into monospores, and release of monospores, has occurred within an internal portion of the filament. (Reprinted from *Aquatic Botany* Volume 28. Graham, J. M., and L. E. Graham. Growth and reproduction of *Bangia atropurpurea* [Roth] C. Ag. [Rhodophyta] from the Laurentian Great Lakes. pp.317–331. ©1987, with permission from Elsevier Science)

giophyceans, including *Rhodochaete*, *Erythrotrichia*, *Porphyrostromium*, *Smithora*, *Bangia*, and *Porphyra* (see references in Garbary and Gabrielson, 1990), but is absent from cyanidophyceans and *Porphyridium*. It has been argued that loss of sexual reproduction has occurred in some cases, accompanying evolutionary reduction from more complex forms (Hawkes, 1990).

Sexual reproduction and life histories vary in ways that appear to correlate with evolutionary diversification of red algae. Sexual reproductive features and life-history types have therefore been widely used to classify red algae. An understanding of sexual reproduction and life history has also been critical to effective mariculture production of red algal crops such as *Porphyra* (*nori*). Laboratory cultivation research, conducted by British phycologist, Kathleen Drew Baker, linked the life history of inconspicuous red filaments, known by the generic name *Conchocelis*, with the edible blade-forming seaweed *Porphyra* (see Chapter 4). Her work formed the basis for the modern billion-dollar-per-year *nori* production industry in Japan, Korea, China, and the U.S. In recognition of the importance of her work, a memorial park was established in Drew Baker's honor in Kumamoto Prefecture, Japan, where she is annually revered with a ceremony dedicated to "the mother of the sea."

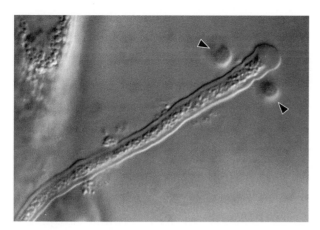

Figure 16–32 Attachment of spermatia (arrowheads) to the tip of a trichogyne (the elongate terminal portion of a carpogonium) of *Aglaothamnion*. (Photograph courtesy C. Taylor)

In all known cases, sexual reproduction of red algae is oogamous, involving fusion of a nonflagellate **spermatium** (the male gamete) with a larger, nonflagellate **carpogonium** (oogonium or female gamete) (Fig. 16–32). In most red algae, spermatia are released from **spermatangia** that are borne upon a male gametophyte thallus, while the carpogonium is produced on (and remains attached to) a separate female gametophyte thallus throughout fertilization and post-fertilization development of the zygote. However the Bangiales produce both male and female gametangia on the same thallus, within packets formed by a series of successive divisions of a precursor cell. Sexual reproduction in *Porphyra gardneri* was studied at the light and TEM levels by Hawkes (1978), and spermatial development in *Bangia* has been studied at the ultrastructural level by Cole and Sheath (1980).

In the Florideophyceae, spermatia are produced singly within spermatangia, which often occur in clusters (Fig 16–33). In these derived red algae, spermatangia originate as subapical protrusions that are cut off by oblique cell walls during division of apical initial cells. In fleshy forms, spermatangia often occur in groups (sori) at the surface of the thallus. In other cases, spermatangia may line cavities, such as the conceptacles of the coralline red algae (Fig. 16–34). *Polysiphonia*, which is widely used in classrooms to demonstrate sexual reproduction of red algae, has groups of spermatangia embedded within a common matrix (Hommersand and Fredericq, 1990). As mature spermatia are released, younger ones are produced in the same area (Kugrens, 1980). The ultra-

structure of spermatium release in *Ptilota densa* was detailed by Scott and Dixon (1973).

Red algal spermatia are generally thought to move toward female gametes passively, by means of water currents. However in *Tiffaniella*, the spermatia are released in mucilaginous strands that coalesce, extend, contract, and rotate in the water until they make contact with a female thallus (Hommersand and Fredericq, 1990). The absence of flagella from the male gametes of red algae probably reduces fecundity, because spermatia are less able than flagellate gametes (of other eukaryotes) to transport themselves to females.

Carpogonia are also produced singly from precursor cells. In most red algae, carpogonia are borne at the tips of specialized branches known as **carpogonial branches** (Figs. 16–35, 16–36), which can arise either laterally from the main axis, or terminally, by transformation of a cell at the end of a vegetative filament (Hommersand and Fredericq, 1990). Carpogonia are typically flask-shaped, with an inflated base and an elongated neck called the **trichogyne**. In general, contact between spermatia and carpogonia is brought about by water motions. It is possible that diffusible sperm attractants are released by trichogyne tips.

Fertilization is initiated by adhesion of a spermatium to the carpogonial trichogyne. Upon contact of spermatia of the appropriate type, a channel is

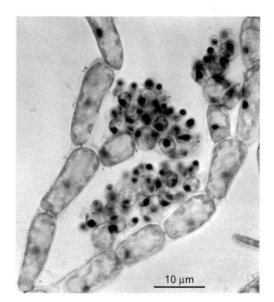

Figure 16–33 Spermatangial clusters of *Dudresnaya crassa*. Nuclei were stained with iron hematoxylin. (From Hommersand and Fredericq, 1990)

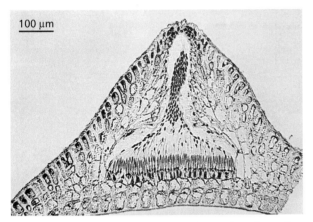

Figure 16–34 Light micrograph of a section through a spermatial conceptacle of the coralline red alga *Mastophora*. (From Woelkerling, 1988 by permission of Oxford University Press)

enzymatically digested in the carpogonial wall, allowing entry of a spermatial nucleus. Although more than one male nucleus may enter the carpogonium, only one fuses with the female nucleus. The carpogonial vacuole then contracts, which helps to pull the male nucleus into contact with the female nucleus, located at the carpogonial base. The cytoplasm then pinches closed at the top of the inflated base of the carpogonium, preventing other spermatia from entering (Broadwater and Scott, 1982). Time-lapse video has been used to document fertilization in the red alga *Bostrychia* (Pickett-Heaps and West, 1998).

Post-Fertilization Development and Life History in Bangiophyceans

In some of the sexually reproducing bangiophyceans such as *Rhodochaete* and *Smithora*, only one or two diploid spores result from zygote formation. In others, including *Porphyra* and *Bangia*, the fertilized zygote undergoes mitotic divisions to produce several diploid **carpospores**. Hence the number of spores has undergone some degree of amplification within the bangiophyceans (Hawkes, 1990).

Carpospores are released into the water, settle, and grow into an independent, diploid sporophyte filament or microthallus. Bangialian sporophytes are often referred to as the conchocelis phase, because before Kathleen Drew Baker's path-making work, such filaments were regarded as a separate red alga, *Conchocelis*. In nature, the conchocelis phase often grows within mollusc shells or the tubes of polychaetes and may live in water as deep as 78 m (Lün-

ing, 1990), far deeper than gametophytes can survive. The conchocelis phase may thus extend the range of the species beyond that occupied by the more conspicuous blade-forming life-history phase. The conchocelis phase may be perennial and proliferate by means of monospores produced in terminal cells. Under appropriate environmental conditions, it produces rows of **conchospores**, which are released into the water column, settle, and produce the more conspicuous gametophytic blade (see Fig. 4–15). Some other bangiophyceans also have heteromorphic alternation of generations; *Porphyridostromium* gametophytes are erect filaments, whereas the sporophytes are flattened (Kornmann, 1984, 1987).

Conchospores were once thought to arise through meiosis, but it is now known that reduction division does not occur until the point of conchospore germination. During development of the sheetlike thallus of *Porphyra*, conchospore meiosis first yields a four-celled linear germling, each cell of which has become haploid. The meiotic process segregates alleles for sexual determination. Thus two of the resulting cells contain genes encoding male gamete development and the other two, female development (Ma and Miura, 1984). Subsequent mitotic proliferation of each of these genotypically distinct cells may result in a

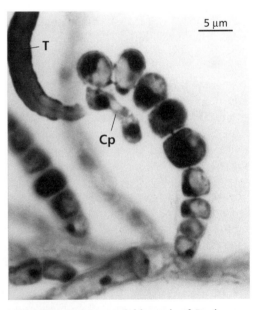

Figure 16–35 A carpogonial branch of *Dudresnaya crassa* showing division of the fertilized carpogonium (Cp) and the trichogyne (T). (From Hommersand and Fredericq, 1990)

genetically sectored thallus. In some species a patchwork nature of the thallus is revealed—some sectors produce spermatia, whereas others generate carpogonia (Fig. 4–16).

Bangialean algae such as *Porphyra* thus have a biphasic alternation of generations—a haploid gamete-producing phase (gametophyte) and a diploid spore-producing phase (sporophyte). Determination of phase change in *Porphyra* and other algae that undergo alternation of generations has been correlated with change in DNA level or chromosome number (ploidy level), typically detected by means of chromosome counts or spectrophotometric or spectrofluorometric measurements of DNA in nuclei of differing stages. Interestingly, some *Porphyra* and *Bangia* species appear to undergo alternation of gametophytic and sporophytic generations without undergoing sexual reproduction or change in chromosome level, signifying that environmental factors may have a stronger influence upon phase change than does ploidy level.

Biphasic life histories in other algal groups (e.g., phaeophyceans and ulvophycean green algae) are thought to have been evolutionarily derived from more primitive life histories in which zygotes were the only diploid cells, because early divergent members of these lineages have the latter type of life history. However, no life histories involving zygotic meiosis are known among red algae. The evolutionary precursor of the simplest type of red algal life history—alternation of biphasic generations—is unknown.

Post-Fertilization Development and Life History in Florideophycean Algae

In contrast to bangiophyceans, in most florideophyceans the fertilized carpogonium, rather than producing carpospores directly, generates one or more multicellular, diploid carposporophytes—a third type of life-history phase or generation. The carposporophyte consists of a mass of filaments that can eventually produce many carpospores. When mature, carpospores are released into the water, settle, and germinate into a second multicellular, diploid generation known as the **tetrasporophyte**, because it produces **tetraspores** (Fig. 16–5). So, alternation of generations in Florideophyceae is commonly triphasic, meaning that there is sequential alternation among three multicellular generations—gametophyte, carposporophyte, and tetrasporophyte. In a few cases carposporophytes appear to develop in the absence of fertilization (Hansen, 1977; Edelstein, et al., 1974).

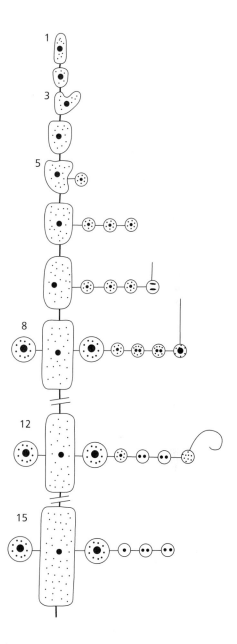

Figure 16–36 Carpogonial branch development in the female gametophyte of *Callithamnion cordatum*. For ease in illustration, stages of carpogonial branch development are shown on adjacent cells in the axis. Numbers are ages of axial cells in days. Number 1 is the apical cell, which divides once per day. Vegetative laterals are initiated in 3-day-old cells; 5-day-old cells may initiate carpogonial branches, and by 8 days, mature carpogonial branches are present. At 12 and 15 days carpogonial branches are senescent and degrading, respectively. (After O'Kelly and Baca, 1984)

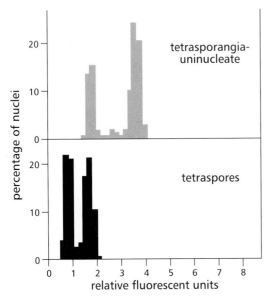

Figure 16–37 Fluorometric measurements of DAPI-stained nuclei of *Wrangelia plumosa* tetrasporangia and tetraspores provide evidence that meiosis occurs prior to tetraspore formation in at least some red algae. (From Goff and Coleman, 1990)

It has been assumed that meiosis occurs within the tetrasporangium during the formation of meiospores, and recent comparative measurements of nuclear DNA levels by DAPI microfluorometry have confirmed that this is indeed the case for some red algae, such as *Wrangelia plumosa* (Fig. 16–37). However, similar DNA-level measurements in *Scagelia pylaisaei* revealed that meiosis occurs during tetraspore germination, rather than tetraspore formation, and that tetraspores are formed mitotically. Another surprising finding of DNA-level measurement studies is that in general, the DNA level of red algal cells is determined by cell and nuclear size, not life-history phase (Goff and Coleman, 1990).

Only two red algal life-history phases, the gametophyte and tetrasporophyte, may be (though are not always) free-living. Carposporophytes always occur on the female gametophytic thallus, and probably receive nutrients from nearby female gametophytic cells (Turner and Evans, 1986). This condition is considered analogous to matrotrophy (feeding by the mother) in placental mammals and embryophytes by some workers (Graham, 1996), and is regarded as a form of parasitism by others (Hommersand and Fredericq, 1990). Evidence for a nutritional relationship between the female gametophyte and carposporophyte include various types of cellular specializations

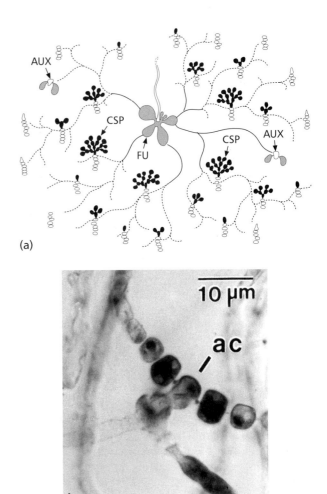

(a)

b

Figure 16–38 (a) Diagrammatic representation of the transfer of diploid nuclei (which are mitotic copies of a single zygote) from fusion cells (FU) to auxiliary cells (AUX) scattered throughout the thallus of *Hommersandia*. Successful transformation of auxiliary cells results in the formation of carposporophytes (CSP) and production of numerous carpospores. (b) Auxiliary cell (ac) in *Dudresnaya crassa*. (a: After Hansen and Lindstrom, 1984 by permission of the *Journal of Phycology*; b: From Hommersand and Fredericq, 1990)

that suggest nutritional function, such as wall and membrane proliferations in regions known as placentae. In addition, radioactive tracer work suggests that photosynthates can be translocated from parental gametophyte cells to those of developing carposporophytes (Hommersand and Fredericq, 1990).

In early divergent florideophyceans, the carposporophyte develops directly from the carpogonium or from a cell (known as a **fusion cell**) within the carpogonial branch that has fused with the carpogonium. However, in many florideophyceans, mitotically produced copies of the diploid nucleus within the zygote are transferred into other gametophyte cells, by cell-cell fusion. Cells outside the carpogonial branch that will or have received a copy of the zygote nucleus are termed **auxiliary cells.** Fusion and auxiliary cells serve as hosts and nutritional sources for repeated divisions of the adopted nucleus. Fusion cells and auxiliary cells are also thought to be sources of morphogenetic compounds that influence post-fertilization development, and to function as part of an isolating mechanism by which incompatible fertilizations can be rejected. It is emphasized that the incorporated diploid nuclei do not fuse with gametophytic cell nuclei; these retain their separate identities. Gametophytic cells that contain both their original nucleus and a nucleus derived from the zygote are heterokaryons and are regarded as having been genetically transformed. Filaments that produce the carposporangia are known as **gonimoblast.**

In many higher red algae, multiple copies of the diploid zygote nucleus are transmitted by the growth of long tubular cells, variously known as **ooblast/connecting cells,** or **ooblast/connecting filaments,** to numerous auxiliary cells throughout the thallus (Fig. 16–38). Mitotic proliferation of filaments from each successfully transformed auxiliary cell may then generate many carposporophytes, each of which at maturity generates numerous carpospores from terminal cells. Connecting filaments are of polyphyletic origin (Hommersand and Fredericq, 1990). Auxiliary cells resemble carpogonia ultrastructurally in that both have a granular cytoplasm, small vacuoles, and walls thickened with gelatinous material thought to facilitate cell-cell adherence. These similarities suggest that specific adaptations associated with red algal fertilization were preadaptive to evolution of post-fertilization cell fusion processes (Hommersand and Fredericq, 1990).

A **cystocarp** is a single carposporophyte plus the gametophyte tissues that surround or protect it (Fig. 16–39). Cystocarps can develop either from the connecting filament, adjacent to an auxiliary cell, or from the auxiliary cell itself, and typically have three layers: (1) an outermost region of photosynthetic gametophytic cells that may be modified into a distinctive **pericarp,** (2) an interior region of non-photosynthetic gametophyte cells regarded as a nutrient provision and processing center, and inside that, (3) a developing carposporophyte whose cells sometimes have sec-

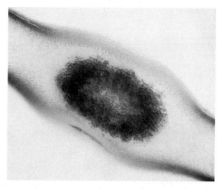

Figure 16–39 A single cystocarp, which includes masses of bright red developing carpospores, is revealed within a bladelike thallus of *Mazzaella* that has been thinly sliced for viewing by light microscopy.

ondary pit plugs in common with gametophytic cells (Fig. 16–40). Close examination of fertile red algal gametophytes, such as those of *Mazzaella*, may reveal the presence of many small, densely pigmented cystocarps (Fig. 16–41).

As the result of the nuclear transfer activities of connecting filaments, each diploid nucleus produced by fertilization may be capable of generating many cystocarps, each of which can release large numbers of carpospores into the water. Because each carpospore is potentially capable of growing into a new tetrasporophyte, given sufficient growth resources, a single fertilization event can lead to production of very

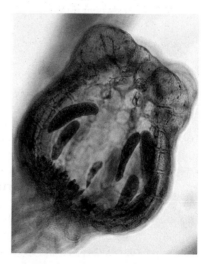

Figure 16–40 An urn-shaped pericarp, formed of gametophytic cells, surrounds developing carposporophytes of *Polysiphonia*. Photograph from a stained and mounted specimen.

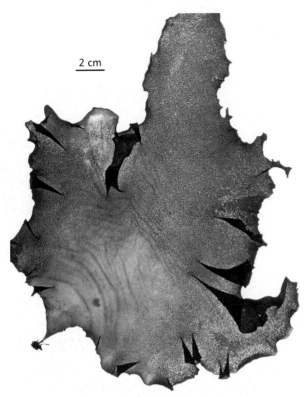

2 cm

Figure 16–41 Macroscopic view of cystocarps on a fertile thallus of *Mazzaella* (a dried specimen scanned as transparency), showing multitudinous dark spots (cystocarps) in peripheral portions of the blade.

large numbers of tetrasporophytes. Each zygote of *Schmitzia sanctae-crucis* generates some 4500 carpospores (Wilce and Searles, 1991). A single thallus of *Gelidium robustum* releases 34,000–300,000 carpospores per month, and the mean number of carpospores released by *Chondrus crispus* is estimated to be nearly 10 billion month^{-1} m^{-2} (see references in Kain and Norton, 1990).

Multicellular, and usually free-living, tetrasporophytes produce tetraspores in **tetrasporangia**, each of which develops by modification of a branch apical cell. Usually four tetraspores are produced in each tetrasporangium (Fig. 16–42). As previously mentioned, tetrasporangia are regarded as a shared, derived feature of Florideophyceae. However there are known instances where gametophytes develop directly from cells of carpospore-derived crustose thalli without the production of typical tetrasporangia (Hawkes, 1990); these may represent cases of evolutionary reduction or loss. Variation in the patterns of tetraspore production have been used taxonomically.

Prodigious numbers of tetraspores can be released by a single thallus. Individuals of *Gelidium robustum* produce 11,000–27,000 tetraspores per month, thalli of *Botryocladia pseudodichotoma*, nearly 4 million tetraspores per day, and *Chondrus crispus*, more than 200 million tetraspores month^{-1} m^{-2} (see references in Kain and Norton, 1990). After their release into the water, tetraspores attach to substrates and grow into gametophytes. When meiosis occurs during tetraspore formation, sex-determining alleles are segregated such that two members of each quartet of tetraspores give rise to male gametophytes and the others, female gametophytes.

The occurrence of the carposporophyte and tetrasporophyte in the life cycle of most red algae results in the potential for dramatic amplification of a single fertilization event. One zygote nucleus can potentially give rise to billions of gametophytes. In this way, red algae have overcome hypothesized fecundity constraints imposed by the inability of sperm to actively swim to female gametes by means of flagella.

The occurrence of carposporophytes is a defining feature of the florideophycean red algae, so there is much interest in the evolution of this distinctive structure. Some authorities suggest that the carposporophyte is of polyphyletic origin. However, substantial SSU rDNA molecular systematic evidence for monophyly of the florideophycean red algae (Fig. 16–58) (Ragan, et al., 1994) suggests that the carposporophyte evolved just once. Origin of the carposporophyte is regarded as the critical innovation that allowed extensive evolutionary radiation of the florideophycean clade, explaining why there are thousands more florideophycean species than bangialean forms. Although some researchers argue that the carposporophyte arose from a free-living, spore-producing generation that became parasitic on the female gametophyte (see references in Hawkes, 1990), others contend that the carposporophyte arose by retention of the zygote on the female gametophyte coupled with mitotic proliferation of the diploid zygotic nucleus (Searles, 1980). Some authors have recommended avoidance of the term carposporophyte, preferring the term gonimocarp because there is no evidence that this generation ever had a free-living existence (Hoek, van den, et al., 1995). We have retained the term carposporophyte because it parallels use of the term sporophyte in embryophytes. The bulk of available evidence indicates that both the red algal carposporophyte and the embryophytic sporophyte arose in the same way—by delay in zygotic meiosis and intercalation of a proliferative phase of mitotic growth.

Retention of the carposporophyte and its nourishment by the female gametophyte allows florideophycean algae to amplify carpospore production because carposporophytes have many cells (versus the unicellular carpogonium) and because the nutritional resources of gametophytic cells are made available for production of the next generation. In contrast, the number of carpospores that can be produced by bangialean zygotes is severely limited by the few resources contained within the unicellular carpogonium. A genetic analysis of reproductive success, in which hybrids of parents from geographically distinct populations of *Gracilaria verrucosa* were studied, revealed that the female parent has a strong effect on crossing success (Richerd, et al., 1993). These results may reflect powerful influence of the gametophyte-carposporophyte nutritional/developmental interaction upon fecundity in red algae.

The reproductive amplification process also contributes to greatly increased genetic diversity. Many more genetic recombinants will result from the combined meiotic activities of the millions (or even billions) of tetrasporophytes that could potentially arise from the carpospore progeny of a single carposporophyte, than could be produced from a fertilized bangialean carpogonium. Hommersand and Fredericq (1990) suggested that reproductive efficiency has also been constrained by the fact that red algal thalli are filamentous and grow only apically. These authors further hypothesized that red algal diversification has involved an evolutionary trend toward increased carposporophytic reliance upon gametophytic resources. In support of their idea is the observation that developing carposporophytes of relatively primitive florideophyceans (specifically *Nemalion*) contain mature plastids that are presumably photosynthetically functional, whereas those of the related but more derived genus *Scinaia* (and many other derived red algae) contain only proplastids, and photosynthetically active plastids do not develop until carposporangia are produced. Hommersand and Fredericq (1990) have also defined several major types of post-fertilization development of increasing specialization, reflecting evolutionary refinement of the nutritional system supporting carposporophyte development.

One group of the florideophycean red algae, which includes *Palmaria* and *Halosaccion* (Palmariales or Palmariaceae), is unusual in that the fertilized car-

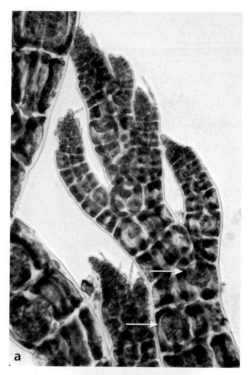

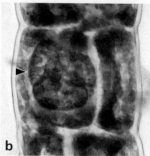

Figure 16–42 Tetraspore production in *Polysiphonia*. (a) Four tetraspores are generated terminally on one-celled lateral filaments (arrows). (b) Higher magnification view of a single tetrasporangium (arrowhead) with four tetraspores.

pogonium develops directly into the meiosporangial generation; a carposporophyte is said to be lacking (van der Meer and Todd, 1980). Some experts regard this as a primitive condition (Hawkes, 1990). However, cladistic analysis of morphological characters (Gabrielson and Garbary, 1986, 1987) and Rubisco sequence evidence (Freshwater, et al., 1994) strongly suggest that members of this group are derived from ancestral forms that possessed carposporophytes, though the factors involved in the presumed loss of carposporophytes are not understood. There are also cases (such as *Liagora tetrasporifera*) in which the tetrasporophyte generation seems to have been lost; meiosis apparently occurs during sporogenesis in the carposporophyte (see Hawkes, 1990).

Isomorphic and Heteromorphic Alternation of Generations

In most florideophycean red algae, the gametophyte and tetrasporophyte generations resemble one another closely and are therefore said to be isomorphic, as in *Polysiphonia* (Fig. 16–43). Corallines also have isomorphic life cycles, and isomorphy of the tetrasporophyte and gametophyte generations is more common than heteromorphy among red algae (John, 1994). However an increasing number of heteromorphic life cycles are being discovered, in which the tetrasporophytes either are microscopic and filamentous or consist of a thin crust that is tightly attached to the rock substrate. A list of representatives is provided by Hawkes (1990); in some cases, the alternating life phases had previously been given distinct generic names. Cultivation of the algae in the laboratory is usually necessary to establish the occurrence of such heteromorphic life histories. In some cases, specifically *Gymnogongrus* and *Phyllophora*, the tetrasporophyte lives upon the gametophyte generation, the latter being the only free-living generation in the triphasic life cycle. Some experts speculate that the selective advantages of differences in life-cycle phase appearance may include adaptation to seasonal variation in habitat and herbivory pressure, topics discussed in more detail in Chapter 23. The main basis for this conjecture is that heteromorphic alternation of generations in red algae seems to occur more commonly in seasonal temperate and polar habitats than in the tropics (John, 1994).

Red Algal Physiology and Ecology

Lobban and Harrison (1994) provide an excellent recent compilation of information regarding physiology and ecology of red algae, as well as that of other seaweeds. Seaweed community ecology, morphological types that are correlated with specific environmental selective pressures (functional forms), and biogeography are discussed in this text in Chapter 23. Here we provide a brief summary of the major physiological and ecological features that distinguish red algae.

Figure 16–43 Isomorphic alternation of generations is illustrated by *Polysiphonia*, whose (a) gametophytes and (b) tetrasporophytes are indistinguishable until they become fertile, and then only by microscopic observation (as shown here).

Carbon Metabolism

Evaluation of the literature suggested to Raven, et al. (1990) that, with the exception of rhodophyte-specific secondary compounds and some unusual osmoregulatory carbohydrates described below, carbon metabolism within red algae is similar to that of other algae. There is evidence that a number of marine red algae, including *Ceramium rubrum* and *Palmaria palmata* (Maberly, et al., 1992) can use bicarbonate, and thus have a carbon-concentration mechanism (CCM) as do many other marine and freshwater algae (Chapter 2). However, light-harvesting efficiency (determined by chlorophyll fluorescence induction methods) of *Palmaria palmata* was determined to be inversely proportional to the extent of bicarbonate use, indicating that there is a considerable physiological cost of bicarbonate acquisition (Kübler and Raven, 1995). Physiological data suggested to Colman and Gehl (1983) that the unicell *Porphyridium cruentum* can transport bicarbonate across its cell membrane. However the nature of the putative transporter system involved is unknown.

The calcification process in coralline red algae is regarded by some experts as a mechanism that facilitates inorganic carbon acquisition, as in some other algal groups (Chapter 2). An hypothesized mechanism for this process is that photosynthetic CO_2 fixation results in an increase in pH near the site of calcification, facilitating $CaCO_3$ precipitation. Coupled acidification processes are thought to be necessary (see Fig. 2–13) (McConnaughey, 1994), and a calcium-dependent ATPase that might be involved in such calcification-associated proton pumping has been isolated from the coralline *Serraticardia maxima* (Mori, et al., 1996).

Some marine subtidal red algae, such as *Delesseria sanguinea*, intertidal forms such as *Lomentaria articulata* (Maberly, et al., 1992), and the freshwater red algae of streams, appear to lack CCMs and to be dependent upon dissolved carbon dioxide. Carbon dioxide dependence of the freshwater *Lemanea mamillosa* has been extensively explored (Raven, 1997b). In the well-aerated waters of fast-flowing streams, CO_2 is not limiting to algal growth. For marine red algae that occupy deeply shaded habitats, where photosynthesis is limited by light availability, photosynthesis is saturated by ambient dissolved CO_2 levels (Raven, 1997b).

Nutritional Requirements

Many red algae require vitamin B_{12}. Some may prefer ammonium as a source of inorganic nitrogen, but most are able to use nitrate. Red algae appear capable of storing transient supplies of excess inorganic nitrogen within phycobilin accessory pigments. Hairs, which occur on various red algae, are thought to be produced in response to nutrient deficiency as a mechanism that increases absorptive surface area (Kain and Norton, 1990).

Marine Habitats

Water temperature appears to be a major factor controlling the distribution and diversity of red algae. Rhodophytes are most diverse in the tropics. Pantropical forms include the calcified genera *Galaxaura*, *Liagora*, *Jania*, and *Amphiroa*, as well as *Halymenia*, *Grateloupia*, *Laurencia*, and several agarophytes (economically important sources of agar), *Gelidium*, *Pterocladia*, and *Gracilaria* (Lüning, 1990). In contrast, red algae are relatively sparse in polar seas, though species such as *Kallymenia antarctica* and *Gymnogongrus antarctica* are among those occurring in the offshore waters of western Antarctica near the Ross Ice Shelf, and *Lithothamnion glaciale* is among the rhodophyte inhabitants of Arctic waters (Lüning, 1990). Red algae occupy a wider range of irradiance environments than any other group of photosynthetic autotrophs. Irradiance extremes include high-latitude, high-intertidal habitats subjected to long periods of full sunlight (inhabited by such forms as *Bangia atropurpurea*) as well as deepwater marine habitats where light levels of 8 nmol $m^{-2}s^{-1}$—only 0.0005% of surface irradiance—are experienced by an unnamed purple crustose coralline (Littler, et al., 1985; 1986). Day length and temperature are the main cues for seasonal growth and reproductive behavior of red algae (see Chapter 23).

Most red algae occur attached to rocky substrates by means of holdfasts, though many exceptions occur. More than one hundred species grow free-floating or entangled in other vegetation (Kain and Norton, 1990). An example of a particularly abundant free-floating form is *Phyllophora*, which may occupy areas of some 15,000 square kilometers at 10–60 m depth in the northwest Black Sea. Certain unattached corallines, such as *Lithothamnion coralloides* form marl beds 3–25 m deep, and several coralline genera form extensive beds of **rhodoliths**—rounded nodules that develop around the surfaces of stones (Lüning, 1990).

Figure 16–44 Numerous blades of *Smithora naiadum* growing on a slender leaf of the marine angiosperm *Phyllospadix*.

Some red algae grow epizootically, on the surfaces of animals. *Rhodochorton concrescens* is notable for its tendency to colonize hydroids, ectoprocts, and crustaceans having chitinous surfaces (West, 1970), as is *Audouinella membranacea* (Kain and Norton, 1990). Red algae more often occur as epiphytes, attached to other seaweeds or seagrasses as a means of support. Several cases of specific epiphytism are known, including *Porphyra nereocystis* on the kelp *Nereocystis luetkeana*, *P. subtumens* occurring obligately on species of the bull kelp *Durvillaea* (Nelson and Knight, 1996), and *Smithora naiadum* on the sea grasses *Zostera* and *Phyllospadix* (Fig. 16–44). *Microcladia californica* attaches only to the kelp *Egregia menziesii*, but a related species, *M. coulteri*, colonizes a variety of seaweeds, including the red macroalga *Prionitis* and the green *Ulva* (Gonzalez and Goff, 1989a, b). In some cases there is evidence for more than a simple epiphytic relationship. *Polysiphonia lanosa*, whose rhizoids penetrate surfaces of the fucoid brown alga *Ascophyllum* (and less frequently, *Fucus*), probably receives amino acids and mineral nutrients from *Ascophyllum*. Radiolabeled tracer experiments have revealed reciprocal exchange of about the same amount of fixed carbon between *Polysiphonia* and *Ascophyllum*, suggesting a hemiparasitic relationship (Ciciotte and Thomas, 1997).

A list of more than 100 host-restricted epiphytes has been compiled by Goff (1982); many of these are poorly pigmented or unpigmented and thus regarded as possible parasites. In addition to pigment reduction, other features used to identify red algal parasites include penetration into the host's thallus, and thallus-size reduction (Setchell, 1918). In some cases radiolabeled tracers have been used to establish net flow of photosynthate from host to parasite (Goff, 1979, 1982). Recently, microfluorometry and molecular markers have been used to detect transfer of parasite nuclei and mitochondria into host cells, a hallmark of red algal parasitism (Goff and Coleman, 1984b).

Red Algal Parasites

In contrast to red algal epiphytes (which may occur on substrates other than red algae), red algal parasites occur only on red algal hosts. This because cell-cell fusions, essential to establishment of the parasitic relationship, can only occur between cells of red algae. Red algal parasites are described as either **alloparasites** (those growing on hosts to which they are not closely related) or **adelphoparasites** (parasites that grow on

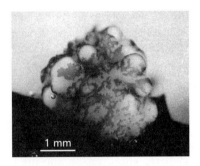

Figure 16–45 The partially pigmented red algal adelphoparasite *Gardneriella tubifera* growing on its host *Sarcodiotheca gaudichaudii* from central California, U.S.A. (From Goff and Zuccarello, 1994 by permission of the *Journal of Phycology*)

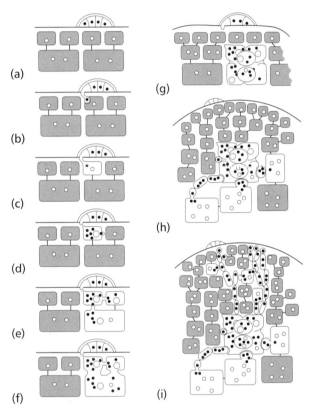

Figure 16–47 Diagram of the process of parasite (*Gardneriella tubifera*) nuclear proliferation (black circles) primarily within the host's (*Sarcodiotheca gaudichaudii*) thallus. (After Goff and Zuccarello, 1994 by permission of the *Journal of Phycology*)

closely related forms) (Goff and Zuccarello, 1994). The latter constitute some 90% of parasitic forms. An example of an alloparasitic relationship is growth of *Leachiella pacifica* on its host *Polysiphonia confusa*.

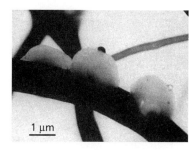

Figure 16–46 The unpigmented red algal adelphoparasite *Gracilariophila oryzoides* growing on its host *Gracilariopsis lemaneiformis* from central California, U.S.A. (From Goff and Zuccarello, 1994 by permission of the *Journal of Phycology*)

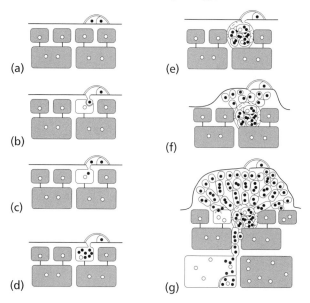

Figure 16–48 Diagrammatic representation of the process by which parasite (*Gracilariophila oryzoides*) nuclei (shown as dark circles) are able to enter host (*Gracilariopsis lemaneiformis*) cells via secondary pit-plug formation and induce them to generate a parasite thallus using host resources. (After Goff and Zuccarello, 1994 by permission of the *Journal of Phycology*)

When *Leachiella* spores germinate on the host thallus surface, a parasite nucleus, mitochondria, and ribosomes are transferred into the cytoplasm of a host cell via cell-cell fusion events. Although the parasite nucleus does not divide or undergo DNA synthesis in the host cell, the parasite nucleus is able to control aspects of host function, i.e., it has genetically transformed the host cell (Goff and Coleman, 1984b, 1985). There is no fusion of host and parasite nuclei, which are maintained together in heterokaryotic cells.

Surveys of many parasite-host relationships have established that transfer of parasite nuclei into host cells is common (see references in Goff and Zuccarello, 1994). Moreover, nuclei of adelphoparasites, in particular, undergo DNA synthesis and mitosis within host cells, thereby cloning and spreading the parasite's nuclear genome within the host. Parasitic nuclei replicate faster than those of the host cell, rapidly replacing them. A number of adelphoparasites that appear to be extremely reduced in size actually exist primarily as organelles within the host's cytoplasm, except during reproduction, when the parasite's gametes or spores are generated.

The process by which parasitic nuclei enter and control host cells has been termed "host cellular transformation" (Goff and Coleman, 1987). Two examples of this process—growth of *Gardneriella tubifera* on its only host *Sarcodiotheca gaudichaudii*, and the parasite *Gracilariophila oryzoides*, exclusive to *Gracilariopsis lemaneiformis* (Goff and Zuccarello, 1994)—are illustrated in Figs. 16–45 through 16–48. *Gardneriella* induces its host to produce an extensive surface callus that is partially pigmented (Figs. 16–45, 16–47). In contrast, *Gracilariophila* does not induce host cells to divide, but rather generates a smaller, colorless mass of cells that produces spores or gametes (Figs. 16–46, 16–48). In both cases parasite development parallels that of post-fertilization development in the host. These data, and the fact that adelphoparasites only occur on hosts that can form secondary pit plugs, suggests that aspects of parasitism were evolutionarily derived from reproductive processes involving formation of secondary pit plugs. Though often colorless when associated with a host thallus, several parasitic red algae are capable of producing pigmented cells when excised and grown separately in culture, suggesting that the genetic information for pigment production is present, but that expression is regulated by the host (Goff and Zuccarello, 1994). Reviews of red algal parasitism and its evolution are provided by Goff, et al. (1996) and Goff, et al. (1997).

Responses to Drought Stress

Many red algae occupy shoreline environments, including mangrove forests, where they are alternately submerged and exposed, or estuaries, in which they experience mixing of fresh- and seawater, creating severe osmotic stress. Following changes in salinity, ions such as sodium, potassium, and chloride, soluble carbohydrates such as digeneaside (Fig. 16–49), or

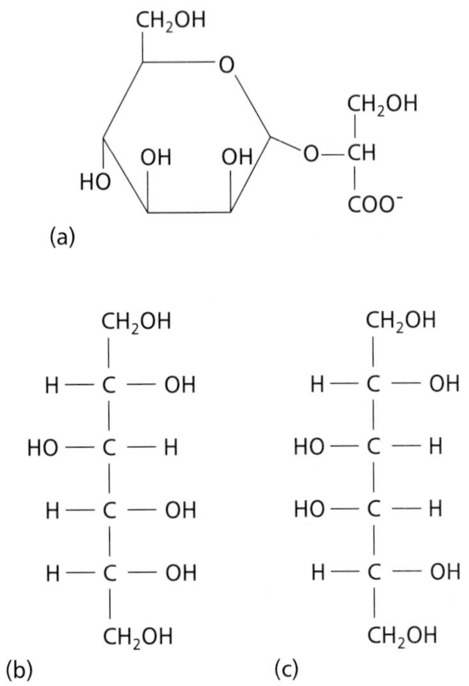

Figure 16–49 Chemical structures of low molecular weight carbohydrates that are produced and used by various red algae to regulate osmotic status of cells: (a) digeneaside, (b) D-sorbital, (c) dulcitol. (After Karsten, et al., 1995 ©Blackwell Science Asia)

amino acids are adjusted to restore normal turgor pressure to cells (Reed, 1990). *Bangia*, an inhabitant of rocky shorelines of both marine and freshwaters, can be gradually acclimated to increased or decreased salinity (Geesink, 1973; Sheath and Cole, 1980), by use of floridoside as an osmolyte (Lüning, 1990). *Bangia atropurpurea* can tolerate air drying on shore for as long as 15 consecutive days, then revive rapidly upon resubmergence (Feldmann, 1951).

Bostrychia and *Stictosiphonia*, common epiphytes of mangroves around the world, primarily use D-sorbitol (as well as dulcitol in warm waters) rather than digeneaside as osmolytes (Karsten, et al., 1995; Karsten, et al., 1996). In contrast, another common mangrove epiphyte, *Caloglossa leprieurii*, accumulates potassium and chloride ions and actively extrudes sodium to adjust osmotic conditions within cells. Mannitol is also involved, but contributes much less to salinity adjustment (Mostaert, et al., 1995). *Polysiphonia paniculata* is an intertidal form that can produce relatively large amounts of the osmolyte DMSP, which is then enzymatically degraded into

the volatile, climatically active gas, DMS (Chapter 2). The enzyme DSMP lyase has been isolated, purified, and characterized from *P. paniculata* (Nishiguchi and Goff, 1995).

Many red algae are highly sensitive to dehydration, and a number of growth habits have been interpreted as adaptations that favor maintenance of a well-hydrated state. Many reds are obligate understory species, living beneath a cover of larger, often brown seaweeds, which provides a high humidity habitat upon emersion. Corallines are particularly sensitive to drought stress, and thus tend to inhabit tide pools (Fig. 16–50) rather than nearby surfaces subject to drying during low tide.

Freshwater Red Algae

There are about 150 species and 20 genera of freshwater red algae. Most occur as small, attached filamentous individuals or tufts. These include bangiophyceans such as *Bangia atropurpurea* (Fig. 16–56) and *Compsopogon*, as well as early divergent members of the Florideophyceae such as *Audouinella violacea*, *Batrachospermum* (Figs. 16–75, 16–76), and *Hildenbrandia* (Fig. 16–61). Most occur in the winter and spring, in mildly acidic streams located in forested watersheds that have a narrow range of flow rates (29–57 cm sec^{-1}). Light is more readily available in the winter and spring, and current flow provides resupply of minerals and carbon dioxide; inability to use bicarbonate is a frequent attribute of freshwater red algae. Morphological adaptations by freshwater red algae to flow stresses were studied by Sheath and Hambrook (1988). Many species are intolerant of pollution, and few occur in large rivers.

The vast majority of freshwater red algae have a heteromorphic life history involving a perennial chantransia or audouinella stage—a small, benthic diploid filamentous phase. No tetraspores are produced. Rather, certain cells of the diploid stage undergo meiosis, giving rise to the first cell of the larger gametophyte generations (Fig. 16–51). Such a life-history type is regarded as a possible adaptation to the stream environment (Sheath and Hambrook, 1990). The need for tetraspore attachment, which might prove difficult in flowing waters, is thus avoided.

Herbivory Relationships

Freshwater red algae, particularly delicate forms, are consumed by a wide variety of animals, among which

Figure 16–50 Corallines growing in a tide pool in the intertidal zone (Washington state, U.S.A.), together with the green sea anemone *Anthopleura*. (Photograph courtesy C. Taylor)

are amphipods and caddis fly larvae. Some species of the latter also use red algae in the construction of their cases (Hambrook and Sheath, 1987).

Marine red algae are consumed by molluscs, crustaceans, sea urchins, and fish. Toughness of the outer thallus is considered to be a major factor affecting edibility. Calcified red algae and low-growing crusts appear to be more resistant to herbivory than non-calcified and erect forms (Littler, et al., 1983; Watson and Norton, 1985), but some authors have pointed out that calcified rhodophytes are not invulnerable (Padilla, 1985). It has been suggested that calcification functions primarily to prevent propagation of damage (caused by herbivores or mechanical stress) throughout an algal thallus (Padilla, 1989). When damaged, many rhodophytes are able to regenerate from lower, prostrate portions of the thallus. Herbivory resistance and environmental persistence may contribute to the abilities of some calcified or crustose red algae to live to great ages. For example, some *Petrocelis middendorfii* crusts are estimated to be 25–87 years old (Paine, et al., 1979), and individuals of the crust-forming coralline *Clathromorphum* of northwestern North America have been estimated to be more than 100 years old (Lebednik, 1977).

Herbivory is thought to have strongly affected the evolution of coralline red algae. Some of the earliest fossil corallines had relatively thin thalli with raised reproductive structures, but later fossil (as well as extant) forms are characterized by thicker thalli with sunken reproductive structures, which are thought to be more resistant to herbivores (Steneck, 1986).

In the crust-forming coralline *Clathromorphum circumscriptum*, the apical meristem and reproductive conceptacles are protected beneath a calcareous surface layer, the epithallus (Steneck, 1982), probably as a means of herbivory protection. Influences of herbivores on the growth of corallines are not always negative. Herbivores crop the turf algae that would otherwise grow on top of reef ridge corallines, thereby shading them out (Lüning 1990). Such herbivory is thought to maintain the integrity of coral reefs.

Structural and Reproductive Diversity

It is estimated that there are 500–600 genera of red algae, and estimates of the number of species range from 5000 to as many as 20,000 (Norton, et al., 1996). As previously mentioned, red algae are classically segregated into two subclasses or classes—Bangiophyceae (bangiophyceans) and Florideophyceae (florideophyceans).

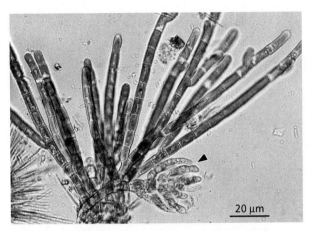

Figure 16–51 Early stage in the growth of the larger gametophyte generation (arrowhead) of *Batrachospermum* from a cell of the smaller sporophytic generation that has undergone meiosis. (Photograph courtesy R. Sheath)

Bangiophyceans

This group contains unicells, filaments, and one- or two-cell-thick membranous sheets (blades). Usually there is a single plastid per cell (though in a few cases numerous plastids are present). Sexual reproduction and pit plugs occur in some, but not all taxa. As many as five orders have been recognized: Porphyridiales, Rhodochaetales, Erythropeltidales, Compsopogonales, and Bangiales. The order Rhodochaetales contains the single species *Rhodochaete parvula*, of interest because it possesses pit plugs of a putatively primitive type, lacking both membranes and plug caps (Pueschel and Magne, 1987) and because it has an isomorphic life history and true apical cells (Gabrielson and Garbary, 1986; Garbary and Gabrielson, 1990). Of the genera described here, *Cyanidium*, *Glaucosphaera*, and *Porphyridium* are representative of the Porphyridiales; *Boldia* of the Compsopogoniales; and *Bangia* and *Porphyra*, the Bangiales (Coomans and Hommersand, 1990).

CYANIDIUM (Gr. *kyanos*, dark blue) occurs as spherical unicells, 2–6 μm in diameter, with a single blue-green, cup-shaped plastid with phycobilisome-bearing thylakoids, but no pyrenoid (Fig. 16–52). There is a distinct cell wall. Reproduction is by production of four autospores (nonflagellate unicells) per parental cell (Broadwater and Scott, 1994). Cytokinesis occurs by means of a contractile ring of actin (Takahashi, et al., 1998). It has been isolated from

acidic hot springs around the world. Usually *Cyanidium* occupies waters of pH 2–4 with a maximum temperature of 57° C. Most forms will not grow at pH greater than 5. However one isolate grows at lower temperatures and at pH greater than 5, in low-irradiance conditions (Seckbach, 1994). *Cyanidium* can be grown osmotrophically in the dark on 1% glucose, autotrophically or mixotrophically, and is used to study chloroplast and photosynthetic pigment biogenesis and regulation of chloroplast gene expression (Troxler, 1994). *Cyanidioschyzon merolae* and *Galdieria sulphuria* are related forms that also occur in acidic hot springs; all three are also capable of growing in a 100% CO_2 atmosphere (Seckbach, et al., 1992). *Cyanidioschyzon*, like *Cyanidium*, produces a storage material that is more like glycogen than floridean starch, but cells are oval and reproduce by binary fission. In contrast, *Galdieria* does possess floridean starch and produces 4–16 daughter cells during reproduction (Seckbach, et al., 1992). Although some authors classify these three taxa with Porphyridiales (Ott and Seckbach, 1994a, b), others consider their classification problematic (Broadwater and Scott, 1994).

GLAUCOSPHAERA (Gr. *glaukos*, bluish-green or gray + Gr. *sphaira*, ball) occurs as single cells, 10–20 μm in diameter, surrounded by wide mucilage sheaths (Fig. 16–53). The mucilage increases the apparent volume of cells by about seven times. TEM examination has revealed no coherent cell wall, although a layer of fibrous, densely staining material occurs just outside the cell membrane. There is a single central nucleus. Several Golgi bodies surround the

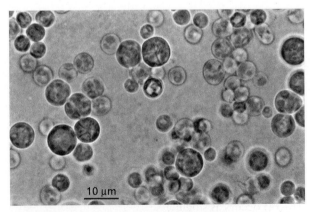

Figure 16–52 *Cyanidium* occurs as single cells having a single cup-shaped, blue-green plastid.

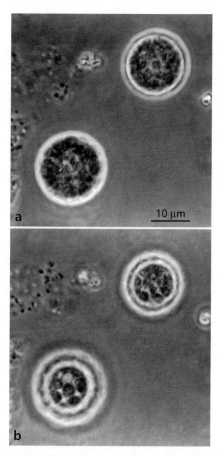

Figure 16–53 Through-focus pair of *Glaucosphaera*, showing the multi-lobed plastid.

nucleus; these are thought to manufacture the mucilage sheath. Numerous small vacuoles and floridean starch granules occur in the cytoplasm (Richardson and Brown, 1970). There is a single, highly lobed plastid that has two envelope membranes, parallel thylakoids bearing phycobilisomes, and granular regions, but a typical pyrenoid is not present. The plastid is devoid of a peptidoglycan wall such as those present around cyanelles of glaucophytes (Chapter 7). Reproduction is by cell division to form two daughter cells (Seckbach, et al., 1992). *Glaucosphaera* has recently been allied with the red algae, based on ultrastructural and molecular sequence information (Bhattacharya and Medlin, 1995; Broadwater and Scott, 1994). It is known only from freshwaters.

PORPHYRIDIUM (Gr. *porphyra*, purple) (Fig. 16–54) occurs as single cells, 5–13 μm in diameter, or as masses of such cells aggregated within a mucilaginous matrix composed of water-soluble sulfated poly-

saccharides. Mucilage is continuously produced by the cells, which lack a conventional cell wall; cells are reportedly able to move slowly by mucilage excretion. The single plastid is star-shaped (stellate), located in the cell center, and possesses a central pyrenoid. *Porphyridium aerugineum* is blue-green in color and occurs in freshwater, whereas *P. purpureum* is red-colored and inhabits areas of high salt content (Broadwater and Scott, 1994), including soils that are exposed to high salinity waters or the sides of flowerpots. Marine *Porphyridium* cells are known to be preyed upon by a dinoflagellate that can consume up to 20 algal cells at a time (Ucko, et al., 1997).

Some other unicellular red algae—*Flintiella*, *Dixoniella*, *Rhodella*, and *Rhodosorus*—were described ultrastructurally by Broadwater and Scott (1994). A molecular sequence (SSU rDNA and *psbA*) study of porphyridialean algae revealed four major lineages/groups: (1) *Cyanidium* (the earliest divergent), (2) *Flintiella* and *Porphyridium*, (3) *Rhodosorus* plus two pseudofilamentous forms—*Stylonema* and *Chroodactylon*, and (4) *Rhodella*, *Dixoniella*, and *Rhodospora*, the group closest to the Bangiales and Florideophyceae (Tsunakawa-Yokoyama, et al., 1997).

BOLDIA (named for Harold C. Bold, an American phycologist and morphologist) has a one-cell-thick sac or tubular form that is usually about 20 cm long, though it may occasionally reach 75 cm in length (Fig. 16–55). Pit plugs are not known. *Boldia* is brownish red or olive in color. It is ephemeral in freshwater streams, usually growing on snails, attached by means of a flat, discoid system of adherent fila-

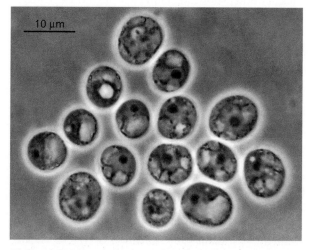

Figure 16–54 Cultured *Porphyridium purpureum* cells.

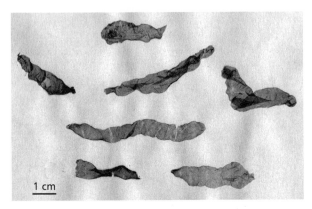

Figure 16–55 Dried and pressed tubular thalli of *Boldia* obtained from a freshwater stream.

ments. Reproduction is by means of monospores generated by unusual narrow filaments growing among other, more isodiametric cells. Following their liberation, monospores attach to substrates, then generate the prostrate systems from which, first, unicellular filaments and then subsequently the erect tubular thalli develop (Herndon, 1964). *Boldia* is considered to be heterotrichous, since it has both prostrate and upright portions.

BANGIA (named for Niels Bang, a Danish botanist) is an unbranched filament that in early developmental stages is uniseriate (Figs. 16–11c) and later becomes multiseriate (Fig. 16–56). Cells are embedded in a firm mucilaginous sheath, and there are no pit plugs. There is a single stellate plastid with a central pyrenoid per cell. *Bangia* is visible as dark red or purple strands occurring at or above the waterline on rocks or other substrates along marine shores or freshwaters of Europe and the U.S. Great Lakes (Sheath and Cole, 1980). In both marine and freshwaters asexual reproduction is by monospores (Fig. 16–31). Spermatia are produced by male gametophytes, and carpospores are produced on separate female gametophytes, following fertilization. Carpospores develop into a tiny, filamentous conchocelis phase (microthallus), which is distinctive in possessing pit plugs (lacking in the gametophyte). Conchospores liberated from the microthallus settle onto substrates and undergo divisions that begin development of the gametophyte. *Bangia* thus has a biphasic alternation of generations. The factors influencing growth and asexual reproduction by Great Lakes *Bangia atropurpurea* were studied by Graham and Graham (1987). SSU rDNA, *rbcL* and *rbcS*, and spacer region sequence analyses of the Bangiales ascertained that *Bangia* nests within *Porphyra*, implying that either *Porphyra* is a paraphyletic genus or *Bangia* is merely a filamentous form of *Porphyra* (Oliveira, et al., 1995; Brodie, et al., 1998). Gene sequence studies suggest that freshwater *Bangia* constitutes a monophyletic group and that U. S. populations arose from a European transplant (Müller, et al., 1998).

PORPHYRA (Gr. *porphyra*, purple) consists of relatively large (up to 75 cm long) cell sheets (Fig. 16–57). Thalli are either one (monostromatic) or two cells thick (distromatic). Molecular systematic studies suggest that the monostromatic condition may be ancestral and that the distromatic condition may have arisen at least twice (Oliveira, et al., 1995). Regeneration and differential display (comparisons of mRNA expression patterns) studies performed on different regions of the blade of *P. perforata* suggest that the blade is not as morphologically and physiologically simple as might appear; morphologically distinct regions are characterized by specific gene expression patterns (Polne-Fuller and Gibor, 1984; Hong, et al., 1995).

Porphyra blades are attached to substrates by means of numerous thin, colorless rhizoidal cells. Blade cells of some species possess only one plastid, but those of other species contain two. The blade (macrothallus) is the gametophyte generation, and this alternates with a microthallus—a small, branched filamentous conchocelis phase (the sporophyte generation)—similar to that of *Bangia*. Conchospores produced by the microthallus regenerate the blade-forming phase. Additional details regarding sexual reproduction, life history, and cultivation of *Porphyra*

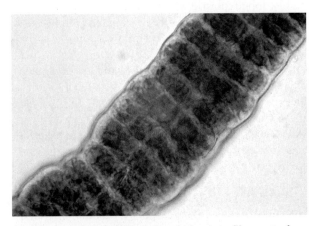

Figure 16–56 A portion of a pluriseriate filament of freshwater *Bangia atropurpurea*.

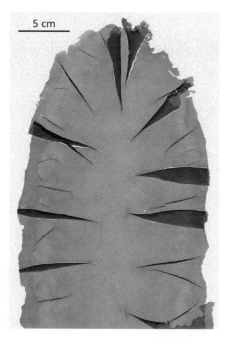

Figure 16–57 A portion of a blade of *Porphyra miniata*.

can be found in Chapter 4. *Porphyra* is common worldwide on rocky shores in the intertidal and subtidal zones. More than 70 species have been described (Kurogi, 1972).

Florideophyceans

This most diverse lineage of red algae includes 14–17 orders (Pueschel and Cole, 1982; Gabrielson, et al., 1990; Freshwater, et al., 1994), defined on the basis of reproductive and structural characters, including pit-plug features. According to SSU rDNA data, Hildenbrandiales is probably the most primitive group of florideophyceans, along with a clade including *Audouinella* (representing Acrochaetiales), *Nemalion* (Nemaliales), and *Palmaria* and *Halosaccion* (Palmariales). The Corallinales and Ahnfeltiales also diverge relatively early (Fig. 16–58) (Ragan, et al., 1994). A clade comprising the Hildenbrandiales and Corallinales is the earliest divergent in an *rbcL* tree, with a clade including *Audouinella*, *Palmaria*, and *Halosaccion* diverging next (Fig. 16–59) (Freshwater, et al., 1994). These molecular results correlate well with relationships deduced by pit-plug features (Pueschel and Cole, 1982). In contrast, phylogenies based on structural features suggest that the Acrochaetiales represent the most basal florideophyceans (Garbary and Gabrielson, 1990; Hoek, van den, et al., 1995). The

sequence of descriptions of the major florideophyceaen orders below reflects the influence of recent molecular systematic work but does not imply that a robust phylogeny yet exists. It should be recognized that molecular systematic research into relationships of red algae is an ongoing process that is likely to result in future changes in classification schemes.

Hildenbrandiales

This order was fairly recently established on the basis of distinctive pit-plug ultrastructure (Fig. 16–60) (Pueschel and Cole, 1982). It was thought to be related to the Corallinales because both groups share an unusual feature—production of secondary pit plugs without the involvement of small conjunctor cells—in contrast to most florideophyceans. Molecular phylogenetic analyses (Ragan, et al., 1994; Freshwater, et al., 1994) have supported this hypothesis. There is a single family, Hildenbrandiaceae.

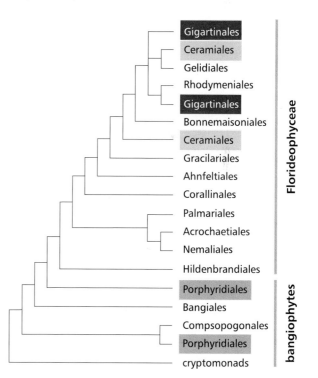

Figure 16–58 A phylogeny of the red algae inferred from comparative SSU rDNA sequences. Classical orders revealed to be polyphyletic include Porphyrdiales among the bangiophyceans, and among the Florideophyceae, Ceramiales, and Gigartinales (shaded). Hildenbrandiales emerge as the earliest divergent florideophyceans in this phylogeny. (After Ragan, et al., 1994 ©1994 National Academy of Sciences, USA)

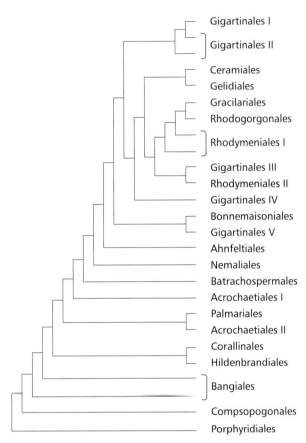

Figure 16–59 A phylogeny of red algae based on comparative *rbcL* sequences reveals Acrochaetiales and Rhodymeniales to consist of at least two clades and suggests that classically defined Gigartinales contains at least five distinct lineages. In this phylogeny a clade that includes both Hildenbrandiales and Corallinales is the earliest divergent among florideophyceans. (After Freshwater, et al., 1994 ©1994 National Academy of Sciences, USA)

HILDENBRANDIA (named for Franz Edler von Hildenbrand, a Viennese physician) (Fig. 16–61) is a pseudoparenchymatous array of filaments closely packed into a thin, flat, deep rose-colored crust growing on stones (or other substrates) in both marine and freshwater habitats; in the latter they are found in flowing streams, particularly in shady places, or in deep lake waters (Canter-Lund and Lund, 1995). The crust is several cells thick and grows by means of apical cells. The crust may be perennial, with dieback of surface cell layers during winter, a process studied ultrastructurally by Pueschel (1988). Secondary pit plugs are abundant. Reproduction appears to be exclusively asexual,

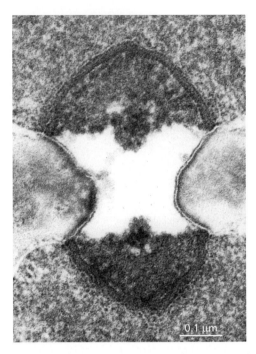

Figure 16–60 Distinctive pit plugs of Hildenbrandiales are illustrated by those of *Hildenbrandia*. (From Pueschel and Cole, 1982)

either by fragmentation, specialized propagules (gemmae), or stolons (Nichols, 1965).

Corallinales

Corallines can occur on or within macroalgae or on sea grasses, animals, coral ridges, or rocks (Figs. 16–1, 16–50). There are some parasitic corallines,

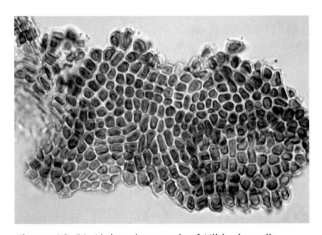

Figure 16–61 Light micrograph of *Hildenbrandia* collected from a tropical freshwater stream. (Photograph from a prepared mount made by J. Blum)

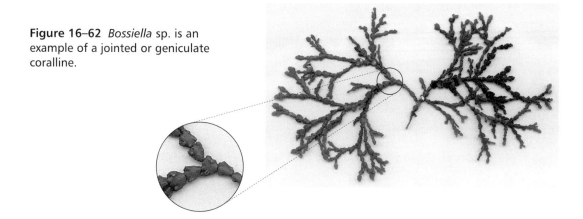

Figure 16–62 *Bossiella* sp. is an example of a jointed or geniculate coralline.

e.g., *Choreonema thuretii*, which can grow endophytically within three other corallines (Broadwater and LaPointe, 1997). Distinctive features of the Corallinales include reproductive organs produced within conceptacles (Figs. 16–33, 16–65, 16–66), and deposition of calcite in the cell walls of most forms. As noted earlier, these features are considered to be adaptations that improve herbivory resistance. Identification usually requires information on reproductive features, and gathering such data requires decalcification by pretreatment with acid, followed by the sectioning of fertile thalli. Hence, coralline taxonomy is relegated to taxonomic specialists. Some 500 extinct species and hundreds of living species have been described. The extinct species are placed into the family Solenoporaceae, and the modern species make up the Corallinaceae, which includes several subfamilies. Because of their great ecological significance and extensive fossil record, there is a considerable literature on fossil and modern corallines. Major compendia of information on coralline algae include Johansen (1981) and Woelkerling (1988).

Classically, modern corallines have been divided into two morphological groups: articulated (jointed) forms, and crustose (non-articulated forms). More recently, these two structural types have come to be known as geniculate (from the Latin word meaning knee or joint) and non-geniculate, respectively. Geniculate corallines are characterized by regions of the pseudoparenchymatous thallus that are uncalcified—these are the **genicula** or joints (Fig. 16–62). Genicula confer considerable flexibility to thalli, allowing them to reach lengths up to 30 cm without suffering physical damage from water movement. Genicula are separated by calcified regions known as intergenicula (Fig. 16–63). Geniculate corallines are typically

highly branched. Thallus growth occurs at branch apices and at intergenicular surfaces (Johansen, 1981). The presence of genicula implies that jointed corallines are able to control the occurrence and location of calcification. Similarly jointed, calcified siphonalean green algae (Chapter 18) represent an example of parallel evolution.

Non-geniculate corallines lack uncalcified regions. Although they may be many cells thick, they are usu-

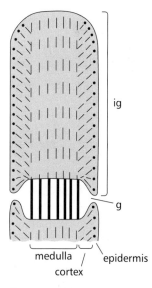

Figure 16–63 A diagrammatic representation of the flexible, non-calcified geniculate region (g) located between adjacent rigid, calcified intergeniculate regions (ig), and differentiation of intergeniculate regions of the thallus into epidermis, subsurface cortex, and central medulla. (Redrawn with permission from *Coralline Algae, A First Synthesis*, H. W. Johansen, 1981 ©CRC Press, Boca Raton, FL)

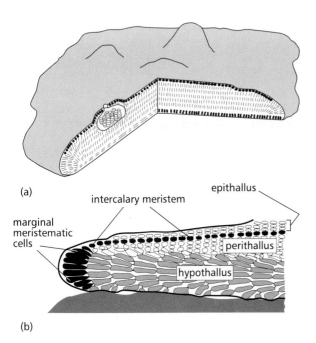

(a)

(b)

Figure 16–64 Non-geniculate corallines. (a) *Sporolithon* is an example of a crustose, or non-geniculate coralline. (b) A diagrammatic representation of thallus differentiation—marginal meristematic cells generate both the hypothallus and an intercalary meristem. The intercalary meristem cuts off cells in two directions; upper cells function as a protective epithallus, and cells cut off from the lowermost surfaces of intercalary meristematic cells form vertically oriented filaments of the perithallus. (a: Redrawn with permission from *Coralline Algae, A First Synthesis*, H. W. Johansen, 1981 ©CRC Press, Boca Raton, FL; b: After Lebednik, in Bold and Wynne, 1985)

ally low-growing crust-formers whose entire lower surface is adherent to a substrate (Fig. 16–64). The pseudoparenchymatous thalli consist of tightly packed filaments and have a dorsiventral organization. The lowermost filaments are oriented parallel to the substrate, and form the **hypothallus**. Branches arising from the upper layer of the hypothallus are oriented perpendicularly to the substrate. An apical meristem at the margin of the crust generates cells of the hypothallus. Growth of the surface cell layer occurs by division within an intercalary meristem—a region of cells lying one or more cells inward from the surface. Cells of the intercalary meristem cut off derivatives toward the surface, forming a protective **epithallus**, and toward the substrate, forming the **perithallus** (Fig. 16–64). The thickness of the perithallus generally determines the thickness of the crust. Some very thin

crusts lack a perithallus, and species that occur as thick crusts have a thick perithallus. Pigments are present in the perithallus but are typically absent from the hypothallus (Johansen, 1981).

At present, approximately 24 genera of non-geniculate and 15 geniculate corallines are recognized. A number of different classification schemes have been suggested (reviewed by Bailey and Chapman, 1996). Recently, Bailey and Chapman (1996) tested the widespread concept that geniculate and non-geniculate corallines represent evolutionarily distinct groups, by the use of nuclear-encoded SSU rDNA sequence analyses of 23 species of corallines. Two sister clades were distinguished, one including only non-geniculate species (the monophyletic subfamily Melobesiodeae) and the second including geniculate forms (monophyletic subfamilies Corallinoideae and Amphiroideae, as well as some other geniculate types) and a non-geniculate species, *Spongites yendoi*. The results of this analysis were not completely consistent with any previous classification scheme and suggested that the assumption that all natural groups will include only geniculate or non-geniculate forms is incorrect. The study also raised the possibility that various groups of geniculate corallines originated separately from different non-geniculate ancestors, but additional data are needed to test this hypothesis (Bailey and Chapman, 1996). Morphogenesis and evolution in the Amphiroideae was studied by Garbary and Johansen (1987).

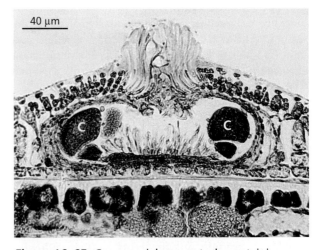

40 μm

Figure 16–65 Carpogonial conceptacle containing carpospores (C) of the crustose coralline *Pneophyllum*. Carpospores will be released via a pore in the upper thallus surface, the ostiole. (From Woelkerling, 1988 by permission of Oxford University Press)

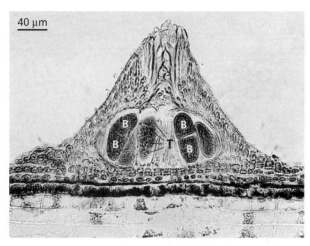

Figure 16–66 A tetrasporangial conceptacle containing tetrasporangia with four tetraspores (T) and bisporangia containing two bispores (B), in the crustose coralline *Pneophyllum*. (From Woelkerling, 1988 by permission of Oxford University Press)

The life history of corallines is triphasic, with isomorphic alternation of gametophytic and tetrasporophytic generations. Male gametophytes produce spermatangia within conceptacles, and female gametophytes generate carpogonial conceptacles (Fig. 16–65). Carpogonial branches are only two cells long, and the cells bearing them function as auxiliary cells. Similar conceptacles occur on tetrasporophytes (Fig. 16–66), where tetrasporangia and/or bisporangia are produced. Conceptacles of all types generally open to the environment via a pore known as the ostiole. The pit plugs of the Corallinales are characterized by distinctive domelike outer cap layers (Pueschel, 1989) (Fig. 16–67). Methods for collection, preservation, and preparation of non-geniculate corallines are described by Woelkerling (1988). In the following survey of coralline diversity, *Lithothamnion* is an example of the subfamily Melobesiodeae, *Corallina* represents the subfamily Corallinoideae, and the subfamily Amphiroideae is represented by *Lithothrix*.

LITHOTHAMNION (Gr. *lithos*, stone + Gr. *thamnion*, small shrub) (Fig. 16–68), a non-geniculate, whitish pink or pink-purple coralline, commonly exhibits distinctive surface protuberances and epithallial cells having flared outer cell walls (Woelkerling, 1988). Spermatangia are produced in dendroid branching filaments (Johansen, 1981). *Lithothamnion* grows throughout the world on rocks, shells, or upon other seaweeds.

CORALLINA (Gr. korallion, coral) is a purple geniculate coralline with pinnate branching, attached to substrates by a crustose base (Fig. 16–69). Branches

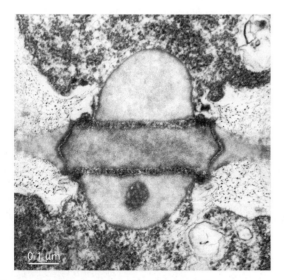

Figure 16–67 Electron micrograph of the characteristic pit plug of coralline red algae, showing dome-shaped caps. (From Pueschel and Cole, 1982)

Figure 16–68 *Lithothamnion*, a non-geniculate coralline, growing on a glass bottle.

tend to lie in the same plane, giving a flattened appearance. Conceptacles are borne singly on ends of branches (Smith, 1969). Development of the epithallial cells of *Corallina officinalis* (and *Lithophyllum*) was studied by Pueschel (1996).

LITHOTHRIX (Gr. *lithos*, stone + Gr. *thrix*, hair) is a highly branched geniculate coralline in which genicula are made up of only a single tier of very long cells (Fig 16–70). The ratio of genicular to intergenicular cell length can be as great as 40:1. Primary branching is dichotomous, and secondary branches are usually alternate. The branches do not tend to lie in one plane, which results in a bushy appearance. It is attached to substrates, usually rocks but sometimes shells, by means of a crustose base (Johansen, 1981). The reproductive and population ecology of *Lithothrix aspergillum* in southern California was analyzed by Pearson and Murray (1997).

Palmariales

This order includes seaweeds having an unusual life cycle that lacks a carposporophyte generation, probably as the result of evolutionary loss. The life history of *Palmaria* (*Rhodymenia*) *palmata* was elucidated in culture by van der Meer and Todd (1980). Male and tetrasporangial thalli are readily observed in nature, whereas the cryptic female gametophytes are not. Laboratory germination of tetraspores obtained from field-collected tetrasporophytes established that female gametophytes remain microscopic and become sexually mature when only a few days old. In contrast, male gametophytes become macroscopic and do not attain sexual maturity until they are several months old. The tiny female gametophytes are fertilized by

Figure 16–69 *Corallina*. Note the characteristic planar, pinnate-branching pattern, which gives the thallus a flattened appearance.

spermatia from males of older generations. The carpogonium has a long trichogyne, and there is no carpogonial branch or auxiliary cell. Diploid tetrasporangial thalli grow directly on the fertilized female, eventually overgrowing them. Palmariales are characterized by tetrasporangial stalk cells that can produce successive crops of tetraspores (Guiry, 1978). The male gametophyte and tetrasporophyte are isomorphic. This order includes two parasitic genera, *Neohalosacciocolax* and *Halosacciocolax*. There are two families, Rhodophysemataceae (named for *Rhodophysema*) and Palmariaceae, which includes *Palmaria*, *Devaleraea*, and *Halosaccion* (described

Figure 16–70 *Lithothrix*, seen life-size. Because branches do not lie in the same plane, the thallus has a bushy appearance.

Figure 16–71 Hollow saclike thalli of *Halosaccion* are typically filled with seawater. This specimen was collected in Newfoundland, Canada.

Figure 16–72 *Halosaccion* characteristically occurs in large populations in the littoral zone. This population was located in the intertidal region on the western coast of Vancouver Island, British Columbia, Canada.

below). The Palmariales appears to be a monophyletic group (Ragan, et al., 1994), and is thought to be closely related to the Acrochaetiales, a polyphyletic group (Freshwater, et al., 1994) which is not described here. Relationships within the Palmariales and Acrochaetiales were studied by means of nuclear-encoded SSU rDNA sequences (Saunders, et al., 1995a).

HALOSACCION (Gr. *hals*, sea + Gr. *sakkos*, sack) has yellow-green, cylindrical, hollow, saclike thalli that are normally filled with seawater (Fig. 16–71). Thalli can reach 25 cm in length and often occur in large populations in the littoral zone (Fig. 16–72). If squeezed, seawater jets out through small pores; otherwise dignified phycologists have been known to use *Halosaccion* in impromptu shoreline water fights.

Batrachospermales

The red algae grouped in the Batrachospermales (named for *Batrachospermum*) were previously classified in the Nemaliales until ultrastructural studies revealed unique features of their pit plugs that justified placing them into a separate order (Pueschel and Cole, 1982). The pit plugs of both groups are characterized by two cap layers. In contrast to the thin outer pit-plug caps of the Nemaliales, Batrachospermales have dome-shaped outer caps (Fig. 16–73). Molecular (*rbcL*) sequence studies confirm that, although closely related to Nemaliales, Batrachospermales form a distinct lineage (Freshwater, et al., 1994). Red algae in this order are freshwater, uniaxial or multiaxial, and

have determinate lateral branches. The life history involves heteromorphic alternation of a macroscopic gametophyte generation with a smaller filamentous sporophyte (also known as the chantransia phase) that lacks tetrasporangia (Fig. 16–51). Meiosis occurs in some apical cells of the sporophyte; resulting haploid cells differentiate into the gametophyte. Families include Batrachospermaceae (*Batrachospermum*) and

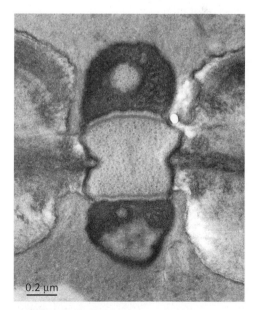

0.2 μm

Figure 16–73 TEM view of the characteristic pit-plug structure of Batrachospermales, revealing the dome-shaped outer caps. (From Pueschel and Cole, 1982)

Figure 16–74 Macroscopically, thalli of the freshwater *Batrachospermum* resemble a gelatinous mass of frog eggs.

Lemaneaceae (*Lemanea*). A description of *Lemanea*, *Paralemanea*, and *Psilosiphon* is found in Sheath, et al. (1996a), and a description of *Nothocladus lindaueri* in Sheath, et al. (1996b). SSU rDNA and *rbcL* sequences were used to evaluate the phylogeny of Batrachospermales. The order is probably monophyletic if *Rhododraparnaldia* and *Thorea violacea* are not included (Vis, et al., 1998).

BATRACHOSPERMUM (Gr. *batrachos*, frog + Gr. *sperma*, seed) has a uniaxial, branched, mucilaginous thallus that macroscopically resembles a mass of frog eggs (Fig. 16–74). Whorls of determinate lateral branches arise at intervals from 4–6 periaxial cells that extend in a circular manner from the central axis (Figs. 16–75, 16–76). Periaxial cells are initiated as apical protrusions that extend laterally and then curve apically before being cut off to form branch initials. The periaxial cells also generate descending rhizoidal filaments that may envelop the axis, making it appear more than one cell thick. Spermatia bud from terminal cells of filaments, while carpogonia (and carposporophytes) develop from a lateral branch that may be determinate or indeterminate, depending upon the species (Fig. 16–77). *Batrachospermum* occurs in cold, running streams, spring-fed ponds, bogs, or lakes around the world. Thalli may be colored blue-green, olive, gray, or deep red. Molecular sequence analyses indicate that *Batrachospermum* is probably not a monophyletic genus (Vis, et al., 1998).

Rhodymeniales

This is a large group of multiaxial forms having solid or hollow thalli, including some beautiful deep rose-colored blades. The auxiliary cell is characteris-

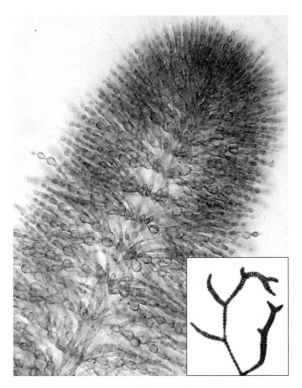

Figure 16–75 Microscopic view of the branching pattern typical of *Batrachospermum*. The alternation of branch whorls with lowermost regions of axial cells produces a beaded effect (inset) resembling strings of frog eggs.

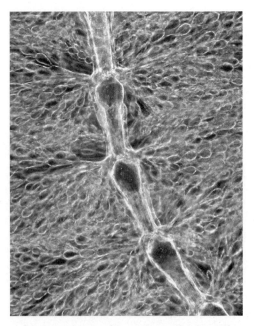

Figure 16–76 Whorls of branches arise from the apical portions of periaxial cells. This species was blue-green in color and found in a peat bog.

Figure 16–77
Carposporophytes (arrowheads) in *Batrachospermum*. (From Bold and Wynne, 1985)

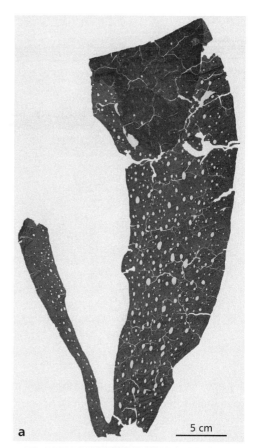

Figure 16–78 (a) *Rhodymenia pertusa* (recently renamed *Sparlingia*) has blades that are characteristically perforated, whereas other *Rhodymenia* species such as that seen life-size in (b) are not perforated.

tically located on a 2-celled filament arising from the same supporting cell as the three- or four-celled carpogonial branch. Despite this unifying reproductive characteristic, *rbcL*-based phylogenetic analysis suggests that the Rhodymeniales is probably not a monophyletic group. Representatives of the currently recognized families, Champiaceae and Rhodymeniaceae, do not form sister clades (Freshwater, et al., 1994).

RHODYMENIA (Gr. *rhodon*, rose + Gr. *hymen*, membrane) thalli are blades that may be entire or divided dichotomously or irregularly, and are attached to substrates by a holdfast that may be disc-shaped or stolonlike (Fig. 16–78). Spermatangia occur on male gametophytes in small, irregular patches (sori), while carpogonial filaments occur on the innermost cortical cells of female gametophytes. The carposporophytes grow toward the thallus surface, forming globular masses of carposporangia. Cystocarps may be distributed fairly evenly over the blades of the female gametophyte or have a more restricted location. They possess a thick pericarp with an ostiole. On the isomorphic tetrasporophyte thalli, tetrasporangia occur over the whole blade surface or only at the tips of blade segments, either singly or aggregated into groups (sori).

BOTRYOCLADIA (Gr. *botrys*, cluster of grapes + *klados*, branch) looks much like a bunch of deep red grapes (Fig. 16–79). The lowermost part of the 15 cm-tall thallus is a solid, branching cylinder attached to

substrates by a disc-shaped holdfast. Branch tips end in hollow, inflated spherical or pear-shaped vesicles that are filled with mucilage. Male gametophytes produce spermatangia in small groups (sori) on the vesicles. Carpogonial filaments are borne on vesicles of the female gametophytes. The carposporophytes grow

Figure 16–79 *Botryocladia pseudodichotoma* has a distinctive thallus that resembles a bunch of elongate grapes.

Figure 16–80 *Eucheuma* is a warm-water red alga that is commonly cultivated and harvested for production of carrageenan. (Photograph courtesy M. Hommersand)

toward the thallus surface and are enclosed in a ostiolate pericarp. Tetrasporangia occur singly on vesicles of the isomorphic tetrasporophyte thalli (Smith, 1969).

Gigartinales

This is a large order with many families that has been revealed by both SSU rDNA and *rbcL* sequencing studies to be polyphyletic (Ragan, et al., 1994; Freshwater, et al., 1994), and therefore in need of revision. Red algae that used to be classified in the Cryptonemiales have been merged into the Gigartinales (Kraft and Robins, 1985). A number of economically valuable carrageenan-producing genera (carrageenophytes) are members of this group, many occurring together in a major clade (Freshwater, et al., 1994). Some are uniaxial in construction, while others are multiaxial.

EUCHEUMA (Gr. *eu*, good, true, or primitive + Gr. *cheuma*, molten substance) is a carrageenophyte that may have bladelike or highly branched, fleshy or tough thalli (Fig. 16–80). Pericarp-enclosed carposporophytes are produced on stalks projecting from the thallus surface. Spermatia occur in groups on male

gametophytes, and female thalli generate carpogonia with reflexed trichogynes. Isomorphic tetrasporophytes produce tetrasporangia in the outer cortex.

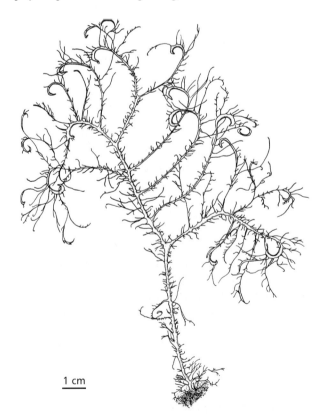

Figure 16–81 *Hypnea*. (From Taylor, 1972 ©The University of Michigan Press 1960)

Figure 16–82 *Mastocarpus*.
(Photograph courtesy J. West)

Eucheuma occurs throughout the tropics and is cultivated in the South Pacific.

HYPNEA (*Hypnum*, a genus of mosses [Gr. hypnos, sleep]) has a highly branched, wiry, or bushy multiaxial thallus that commonly displays distinctive hooklike structures (Fig. 16–81). The life history involves isomorphic gametophyte and tetrasporophyte generations, often present throughout the year. Tetrasporophytes outnumber gametophytes though, and sometimes gametophytes are not observed. It is suspected that vegetative reproduction of tetrasporophytes may be the primary mode of reproduction in some populations. Some species (e.g., *H. cornuta*) produce star-shaped asexual propagules. This desiccation-sensitive genus typically inhabits subtidal waters, but sometimes also tide pools in the intertidal zone.

Hypnea is one of the most widespread seaweeds of subtropical and tropical waters around the world, and extends into warm temperate regions including Japan, the Mediterranean, and Texas, North Carolina, and California in the U.S. In tropical regions *Hypnea* is used for food and is harvested commercially in Senegal and Brazil. It has potential application as a crop grown for production of κ carrageenan. The described species are listed by Mishigeni and Chapman (1994). Some occur as mats, others are iridescent, and several grow epiphytically on other seaweeds.

MASTOCARPUS (Gr. *masto*, breast + Gr. *karpos*, fruit) (formerly included within *Gigartina*) is a large robust blade (Fig. 16–82). The female gameto-

phytes are typically covered with distinctive rough papillae, giving it the texture of a terry-cloth towel. Carposporangial branches are produced in these papillae. Blades can be entire or divided into branches. The tetrasporophytes are dark, slippery, crustose thalli that were known as *Petrocelis* (Fig. 16–83) prior to their identification as part of the life history of *Mastocarpus*, through culture studies (West, 1972; Polanshek and West, 1975; 1977). *Petrocelis* morphology resembles that of crustose corallines in that there is a one- to several-cells-thick hypothallus of prostrate branched filaments and a perithallus of erect, sparingly branched filaments, in a gelatinous matrix. *Mastocarpus* and *Petrocelis* both occur in the lower intertidal zone of the U.S. West Coast. Heteromorphic species of *Gigartina* were transferred to *Mastocarpus* by Guiry, et al. (1984).

CHONDRUS (Gr. *chondros*, cartilage) thalli are multiaxial, bushy, dichotomously branched, flattened fronds that diverge from a tough stalk (Fig. 16–84). There are separate male and female gametophytic thalli and isomorphic tetrasporophytes. Spermatangia are colorless and occur on younger branches. Carpogonial branches develop in the inner cortex and trichogynes extend to the surface. Cystocarps do not include a pericarp. Various species occur in the lower intertidal and subtidal throughout the Northern Hemisphere. A key to the species, as well as an extensive review of the biology of this genus, is provided by Taylor and Chen (1994). *Chondrus crispus*, occurring on both sides of the North Atlantic, is commonly

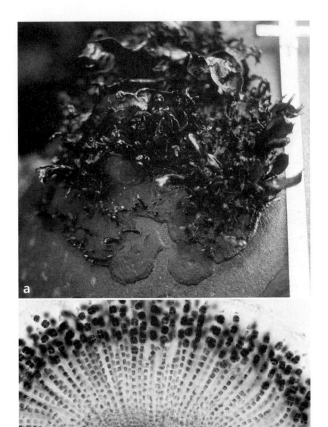

Figure 16–83 *Petrocelis*, the tetrasporophytic phase of *Mastocarpus*. (a) Habit view showing *Mastocarpus* thalli growing from the crustose *Petrocelis* thalli. (b) An outer later of tetrasporangia. (Photographs courtesy J. West)

interaction of light with a surface layer (cuticle) composed of many layers (Fig. 16–16) (Gerwick and Lang, 1977). Blades develop at the beginning of each growing season from a perennial holdfast, the principle mode of reproduction. Both gametophytes and isomorphic tetrasporophytes occur throughout the growing season, but the population of tetrasporophytes is usually larger. The effects of frond crowding in *Mazzaella cornucopiae* from British Columbia were studied by Scrosanti and DeWreede (1998).

GYMNOGONGRUS (Gr. *gymnos*, naked + *gongros*, excresence) consists of tough, branched, multiaxial thalli having thin, rounded or somewhat com-

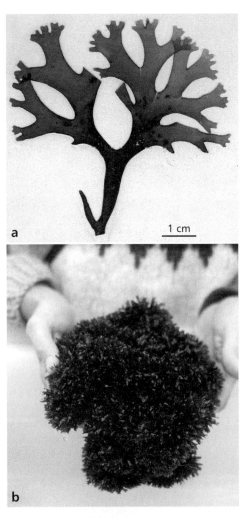

Figure 16–84 *Chondrus*. (a) A portion of the thallus, illustrating characteristic dichotomous branching of the flattened blade. (b) A bushy growth of *Chondrus crispus*. (b: Courtesy M. Cole)

known as Irish moss. The genus has long been used for food and extraction of carrageenans in western countries. In fact, the term carrageenan derives from the Celtic name for *C. crispus*, "carragheen" (Taylor and Chen, 1994). The carrageenans of the gametophytic and tetrasporophytic phases are different. ITS (internal transcribed spacer) sequences (see Chapter 5) have been used to examine species relationships (Chopin, et al., 1996).

MAZZAELLA (named for Angelo Mazza, an Italian phycologist) thalli are smooth, with entire or lobed blades that exhibit brilliant iridescence when viewed underwater (Fig. 16–85). The iridescence results from

Figure 16–85 *Mazzaella* submerged in shallow water, where blue iridescence is normally exhibited.

pressed axes (Fig. 16–86). Dense clumps of branched axes arise from a single discoid base. Some species are monoecious, with spermatangia and carpogonia both produced in the outer cortex, while others are dioecious. Carposporophytes occur as wartlike **nemathecia** (giving rise to the generic name) that were at first thought to be parasitic growths. Some species are known to have a separate crustose tetrasporophyte generation resembling *Petrocelis*, but in other species,

morphologically reduced tetrasporangia occur within the carposporangial filaments that form the nemathecia of monoecious gametophytes. *Gymnogrongrus* occurs worldwide, from Antarctica through the tropics to Alaska, in intertidal and subtidal waters. About 30 species have been described, all containing carrageenans. A list of species, information on carrageenan content, and a review of the biology of this genus are provided by Anderson (1994). Perennial *Gymnogongrus furcellatus* is commercially harvested in Chile and Peru, and ecological analyses have been used to determine the most appropriate time for harvesting (Santelices, et al., 1989).

Gracilariales

This red algal order is well known for species that are of commercial importance as sources of agar and agarose.

GRACILARIA (L. *gracilis*, slender) thalli are cylindrical to somewhat flattened and branched, with

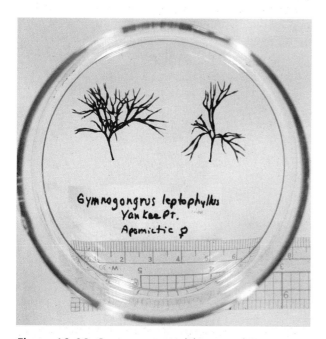

Figure 16–86 *Gymnogongrus*. (Photograph courtesy J. West)

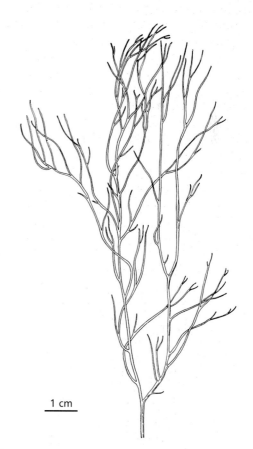

1 cm

Figure 16–87 *Gracilaria*. (From Taylor, 1972 © The University of Michigan Press 1960)

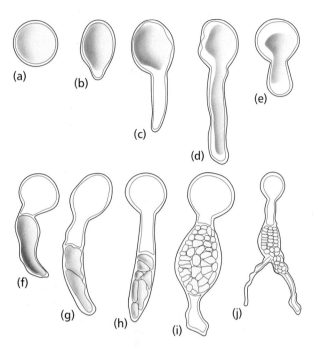

Figure 16–88 Diagrammatic representation of the characteristic tetraspore germination pattern exhibited by members of the Gelidiales. (After Ganzon-Fortes, 1994)

a cartilaginous texture (Fig. 16–87). *Gracilaria* has a uniaxial construction, growing from a single apical cell. Spermatia are produced in conceptacles. Carposporophytes occur within well-developed pericarps, the whole forming cystocarps that project from the thallus surface. The life history includes separate male and female gametophytes and isomorphic tetrasporophytes. It often grows in large clumps in shallow, tur-

bid bays or lagoons. The closely related genus *Gracelariophila* is parasitic on *Gracilaria*. *Gracilaria* and *Gracilariopsis* include more than 170 species, and species-specific plasmids are present. ITS sequences have proven useful in probing species-level, but not intergeneric relationships (Chapter 5) (Goff, et al., 1994).

Gelidiales

The Gelidiales, like the Gracilariales, includes macroscopic forms that have a uniaxial construction and are commercial sources of agar and agarose. Members of this group are characterized by a dome-shaped apical cell at the branch tips and a distinctive mode of tetraspore germination. As soon as the tetraspore attaches to the substrate, a germ tube emerges and the cytoplasmic contents of the spore flow into the tube. The empty spore is cut off by a wall, and the germ tube serves as the first cell of the next generation. A multicellular sporeling with rhizoids develops over several days, and eventually produces an apical cell (Fig. 16–88). There is typically a prostrate, branched stolon from which several erect axes develop. Stolon attachment to substrates is by unicellular or peglike rhizoids. Genera include *Acanthopeltis*, *Acropeltis*, *Beckerella*, *Gelidiella*, *Gelidium*, *Pterocladia*, *Porphryroglossum*, *Ptilophora*, *Suhria*, and *Yatabella*. Nuclear-encoded SSU rDNA and *rbcL* sequences have been used to evaluate systematics of the Gelidiales (Bailey and Freshwater, 1997). The biology and utility of *Gelidiella* were reviewed by Ganzon-Fortes (1994), and those of *Suhria* by Anderson (1994).

Figure 16–89 *Gelidium.*

Figure 16–90 *Pterocladia capillacea*, showing the characteristic bushy thallus.

GELIDIUM (Gr. *gelidus*, congealed) thalli are pinnately branched, with cylindrical or flattened axes that are stiff and cartilaginous (Fig. 16–89). Spermatangia occur in colorless aggregations at the apices of male gametophytes. Alternation of isomorphic tetrasporophytes is common. Apical growth patterns were analyzed by Vargas and Collado-Vides (1996) in an effort to understand morphological variation in this genus. Molecular phylogenetic analysis suggests that this genus may not be monophyletic (Bailey and Freshwater, 1997).

PTEROCLADIA (Gr. *pteron*, feather + Gr. *klados*, branch) (Fig. 16–90) is structurally similar to *Gelidium*, the two genera differing in cystocarp structure and occurrence of carposporangia in short chains in *Pterocladia* and singly in *Gelidium*. However some authorities have questioned the utility of these characters. The bushy thalli may be quite small, to 60 cm or so in length. Both monoecious and dioecious species occur. Colorless spermatangial sori are found on

branches at the tips of the main axes. Cystocarps occur on lateral pinnules, and tetrasporangia in sori. *Pterocladia* is found in all the warm, temperate marine waters in the world. It grows in the lower intertidal and subtidal zones. Wounding induces a regeneration reaction, and bacterial galls are reported. *Pterocladia* is harvested from natural populations for agar production in some regions, and is of interest in establishment of mariculture operations. Fertile female thalli are required for species determinations, but these are not present in some locations. A review of the biology, species, and utility of *Pterocladia* is provided by Felicini and Perrone (1994).

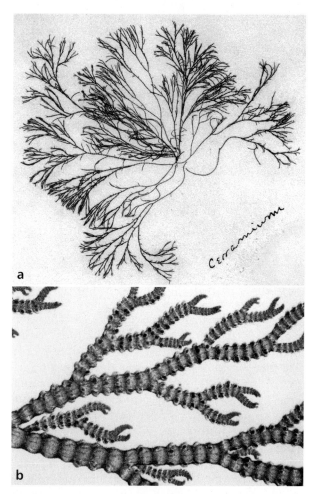

Figure 16–91 *Ceramium*. (a) Macroscopic view of a pressed specimen (life-size). (b) Microscopic view showing the typical banding pattern resulting from production of corticating cells that often cover some regions of periaxial filaments more than others. (b: Photograph courtesy M. Hommersand)

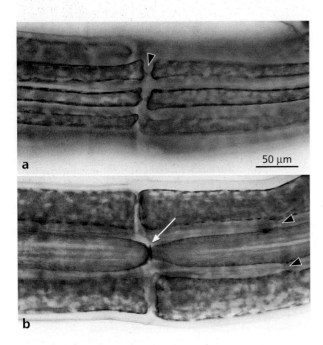

Figure 16–92 Through-focus pair of *Polysiphonia* main axis. The portions of two central cells are seen in (b), joined by a primary pit plug (arrow). Arrowhead in (a) points to a secondary pit plug between periaxial cells. Arrowheads in (b) point to secondary pit plugs between periaxial cells and a central cell. (See Figs. 16–8, 16–43 for habit views.)

Ceramiales

Ceramiales is a large order that appears to be monophyletic, according to an *rbcL* sequence analysis (Freshwater, et al., 1994). All members are uniaxial (Hoek, van den, et al., 1995). The group is also characterized by a reproductive autapomorphy—formation of auxiliary cells after fertilization, rather than before—as occurs in other Florideophyceae. The auxiliary cell is formed from the supporting cell (a periaxial cell) of the four-celled carpogonial branch. This type of female reproductive apparatus is called a **procarp**.

The life history typically involves isomorphic (and usually dioecious) gametophytic and tetrasporophytic generations, and is known as the *Polysiphonia*-type life history. There is a characteristic type of bipolar spore germination pattern—two primordia are formed, one generating a rhizoid, and the other an apical cell from which the erect frond is produced. There are many parasitic genera, e.g., *Sorellocolax stellaris*, a newly discovered species from Japan (Yoshida and Mikami, 1996). There are four families:

Ceramiaceae (illustrated by *Ceramium*); Rhodomelaceae, the largest group with about 125 genera (illustrated by *Polysiphonia*); Delesseriaceae (illustrated by *Caloglossa*); and Dasyaceae (exemplified by *Dasya*). A key to 89 genera of Delesseriaceae is provided by Wynne (1996). Preliminary molecular systematic evidence suggests that the Ceramiaceae and Rhodomelaceae, as currently delimited, are not monophyletic groups (Freshwater, et al., 1994).

CERAMIUM (Gr. *keramion*, ceramic vessel) is a delicate uniaxial filament in which junctions of the large axial cells are covered by bands of corticating cells (Fig. 16–91). Filaments consist of axial, periaxial, and cortical cells. A ring of periaxial cells develops from the upper portions of axial cells, and cortical cells are produced by division of periaxial cells. Four to ten periaxial cells are produced, the number depending on the species. In some species, spines are generated by periaxial and cortical cells, and gland cells may develop from cortical cells. Branches originate pseudodichotomously by division of the apical cell. The thallus is attached via rhizoids that develop from periaxial and corticating cells. Spermatangia occur on the cortical bands while carpogonial branches develop from pericentral cells. Tetrasporangia develop from periaxial or cortical cells. Some species produce sporangia (known as **parasporangia**) having more than four cells. *Ceramium* occurs on most marine coasts around the world. Some species serve as sources of agar. The biology of *Ceramium* was recently reviewed by Boo and Lee (1994).

Figure 16–93 *Osmundea* (*Laurencia*) *spectabilis* from California, U.S.A.

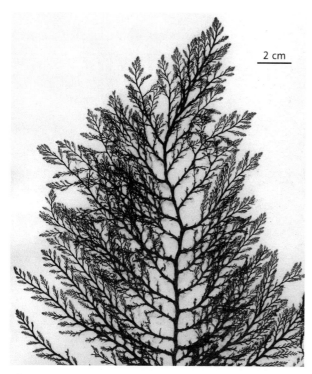

Figure 16–94 *Odonthalia.*

Figure 16–95 *Caloglossa leprieurii.* (a) Habit, showing rhizoids at the nodes. This specimen was cultured from material collected in Natal, South Africa. (b) Tetrasporic blade. (c) Tetrasporangia with cover cells. (a: From Kamiya, et al., 1998 by permission of the *Journal of Phycology*; b and c: Photographs by C. Yarish, in Bold and Wynne, 1985)

POLYSIPHONIA (Gr. *poly*, many + Gr. *siphon*, tube) gives the appearance of being composed of a cylindrical array of adherent tubes. In reality, each of the central cells of the single axis generates a whorl of periaxial cells that is aligned with periaxial whorls produced by the cells above and below in the central axis (Fig. 16–92). There may be as few as 4 or as many as 24 periaxial cells per axial cell, depending upon the species. In *Polysiphonia*, the periaxial cells are the same length as the axial cells that produced them, but in some related genera, periaxial cells undergo transverse division. Spermatia and carpogonial branches are produced on special lateral branches (known as trichoblasts) that are produced by the apical cell prior to development of periaxial cells. Carposporophytes are generated within a well-developed, urn-shaped pericarp (Fig. 16–40), the whole forming the cystocarp. Tetrasporangia are produced by periaxial cells (Fig. 16–42). More than 150 species have been described. Related genera include *Pterosiphonia*, *Osmundea* (Fig. 16–93), *Rhodomela*, and *Odonthalia* (Fig. 16–94).

CALOGLOSSA (Gr. *kalos*, beautiful + Gr. *glossa*, tongue) is a member of the Delesseriaceae, a family that includes members having flat, leaflike blades with beautiful cellular patterns (Fig. 16–95). Thalli are attached to substrates via numerous rhizoids. *Caloglossa* occurs in tropical to temperate waters, and

Figure 16–96 *Grinnellia americana*, a member of the Delesseriaceae, having a simple, translucent blade with prominent midrib.

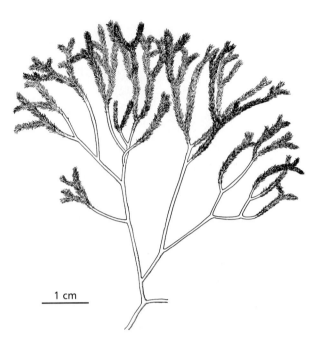

Figure 16–97 *Dasya*. (From Taylor, 1972 ©The University of Michigan Press 1960)

apices are produced by subapical regions before they become corticated, and the preceding apices generate tufts of filaments. The polysiphonous organization of the developing thallus (see description of *Polysiphonia*, above) can become obscured by growth, giving thalli a pseudoparenchymatous appearance. Tetrasporangia are produced in distinctive branches known as stichidia. A variety of species occur in the Caribbean (Taylor, 1972).

Recommended Books

Cole, K. M., and R. G. Sheath (editors). 1990. *Biology of the Red Algae*. Cambridge University Press, New York, NY.

Johansen, H. W. 1981. *Coralline Algae: A First Synthesis*. CRC Press, Boca Raton, FL.

Woelkerling, W. J. 1988. *The Coralline Red Algae: An Analysis of the Genera and Subfamilies of Nongeniculate Corallinaceae*. Oxford University Press, London, UK.

one species, *C. leprieurii*, is a common inhabitant of the mangrove seaweed flora. Related genera include *Hypoglossum* (see Wynne, 1988), *Polyneura*, *Phycodrys*, *Delesseria*, *Membranoptera* and *Grinellia* (Fig. 16–96). The Rubisco spacer region was used to evaluate relationships in *C. leprieurii* and *C. apomeiotica* (Kamiya, et al., 1998).

DASYA (Gr. *dasys*, shaggy) has a feathery thallus that can be as long as 50 cm and is branched (Fig. 16–97). The apical cell is not persistent; rather, new

Chapter 17

Halosphaera

(Photograph courtesy G. Floyd and C. O'Kelly)

Green Algae I
Introduction and Prasinophyceans

PART 1—INTRODUCTION TO THE GREEN ALGAE

The green algae are commonly known as chlorophytes because they appear bright grass green, as do most plants. This is because the chlorophylls of green algae are usually not concealed by large amounts of accessory pigments. However chlorophytes may not always have green coloration and therefore are sometimes difficult to recognize as green algae. Widely encountered examples include *Trentepohlia*, which often forms dramatic orange-red growths on cliff faces and other terrestrial substrates, the flagellate *Haematococcus*, which colors bird baths and other such structures purple-red (Fig. 17–1), and *Chlamydomonas nivalis*, which can color snow red (see Fig. 1–11, Text Box 20–1). In all three cases large amounts of carotenoid pigments obscure chlorophyll, and in the case of *C. nivalis*, at least, these red carotenoids serve a photoprotective function (Bidigare, et al., 1993).

Green algae commonly occur in nearshore marine environments, often forming conspicuous and sometimes nuisance growths. The "sea lettuce" *Ulva* (Fig. 17–2) and *Codium*, sometimes called "dead man's fingers" (Fig. 17–3), are familiar to many people, and lush growths of marine *Cladophora* and other filamentous green algae are common sights in tide pools. Tubelike thalli of the marine green seaweed *Enteromorpha* readily develop on the hulls of ships, forming macroscopic growths that must be regularly removed, to the annoyance and expense of vessel owners. In freshwaters, filamentous *Spirogyra* and colonial *Hydrodictyon* frequently form conspicuous, but harmless, springtime blooms that may cover the surface of freshwater ponds (see Fig. 1–10). Along rocky shorelines of eutrophic lakes and streams, the large, branched thalli of *Cladophora* produce summer masses of shallow water vegetation that can break off during storm events, wash up onto beaches, and generate noxious odors during decay (Fig. 17–4). Lakes that are impacted by acid rain characteristically develop massive subsurface growths of mucilaginous green algae such as the filamentous *Mougeotia* (Figs. 4–5, 21–34, 21–35), known affectionately among limnologists as "elephant snot."

397

Figure 17–1 The dark coloration in this birdbath is due to a rich, red growth of *Haematococcus* that has since dried. In warm, moist conditions, *Haematococcus* is frequently encountered in or on such concrete substrates, and is not harmful to the structure or to birds. The deep red to purple coloration is due to the pigment astaxanthin, which is used as a food colorant.

Figure 17–3 *Codium* is a seaweed whose dark green fingerlike branches are commonly observed to drape over rocks on seashores.

Green algae are also significant in some less highly visible ways. A variety of tropical green macroalgae, in particular *Halimeda*, precipitate calcium carbonate onto their bodies (Fig. 17–5). When they die, these algae contribute substantially to the production of carbonate sand, and over geological time such calcareous algae have generated important carbonate deposits. In some of these the remains of the algae can be readily identified. Some freshwater green microalgae that produce decay-resistant cell walls are frequently recovered from ancient lake sediments (Fig. 17–6). The cell walls of these algae contain fatty acid polymers, known as **algaenans**, that can withstand millions of years of burial; such algal cell walls are asso-

ciated with certain oil deposits (Gelin, et al., 1997). The microscopic green alga *Botryococcus* (Fig. 4–14) produces very large amounts of lipid and is also the source of some petroleums. This alga is a potential modern-day source of renewable energy-rich compounds. *Dunaliella* (Figs. 4–11, 20–37) and *Haematococcus* (Fig. 4–12, 4–13) are widely cultivated for production of useful organic compounds, while *Chlorella* (Figs. 4–1, 19–2) is grown for use as a human food supplement. *Selenastrum* is a single-celled green alga that is widely used in bioassays of water quality (Fig. 4–6, 20–42). These and other ways in which green and other algae are useful to humans are described further in Chapter 4.

Several green algae have served as important model systems for elucidation of eukaryotic developmental and physiological phenomena. Mutants of *Chlamydomonas* (Harris, 1989) (Figs. 1–17, 20–17) and *Volvox* (Schmitt, et al., 1992) (Figs. 4–2, 20–28) have proven useful in understanding the genetic basis for many cellular attributes. Incredibly elaborate unicellular desmids, such as *Micrasterias* (Figs. 4–4, 21–43) and the spectacular mermaid's winecup, *Acetabularia* (Mandoli, 1998) have been used to illuminate fundamental processes of cellular differentiation. The giant internodal cells of charalean algae

Figure 17–2 The "sea lettuce" *Ulva* forms blades or sheets that are very common along both rocky and sandy marine coastlines.

Figure 17–4 Shoreline covered with a dense growth of the filamentous ulvophycean *Cladophora*. Most species of this alga occur along marine shores, but one or more are found in freshwaters, where they may form nuisance growths under eutrophic conditions.

such as *Chara* (Fig. 21–90) and *Nitella* (Figs. 21–91) have been used in many electrophysiological studies (e.g., Thiel, et al., 1997) and to study the cellular basis of gravitropism (see Wayne, 1994; Kiss and Staehelin, 1993). Because they are the extant algae closest to the ancestry of land plants (embryophytes), charalean algae and *Coleochaete* (Fig. 21–56) have been used in comparative analyses aimed at tracing the evolutionary origin of fundamental cellular, biochemical, and developmental features of bryophytes and vascular plants (Graham, 1993, 1996).

The potentially serious algal pathogen *Prototheca* (Figs. 3–27, 19–5) can infect people with compromised immune systems. Infectious mastitis in cattle, caused by *Prototheca*, can result in significant economic losses. Green algae such as *Trebouxia* (Fig. 19–10), *Coccomyxa*, and *Trentepohlia* (Figs. 18–23, 18–28) are common symbionts within lichens, and *Chlorella* is a widespread endosymbiont in freshwater invertebrates (see Fig. 7–19). The cells and chloroplasts of marine green algae also widely occur as endosymbionts within the bodies of marine animals. A number of inconspicuous but biotically significant green algal endophytes grow within the matrices of corals or the thalli of marine seaweeds. These and other ways in which green algae are involved in biotic associations are described more fully in Chapters 3 and 23.

Green algae are regular components of snow algal communities and form extensive coatings on terrestrial surfaces, including mud, rocks, wood, and tree bark. Unicellular and filamentous green algae are also significant components of freshwater planktonic (Happey-Wood, 1988), and periphytic communities

Figure 17–5 Calcium carbonate-encrusted thalli of the marine seaweed *Halimeda* are important sources of carbonate sand in tropical regions. (Photograph by C. Lipke)

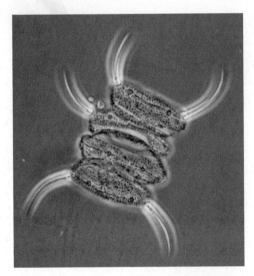

Figure 17–6 Dead remains of the freshwater green alga *Scenedesmus*. The cell walls may remain undecomposed for long periods because they contain decay-resistant compounds known as algaenans. Such sedimented remains may contribute to the formation of fossil fuel deposits.

(Stevenson, 1996). Members of the Zygnematales—including *Spirogyra*, *Mougeotia*, and desmids, are characteristic of the low pH waters of *Sphagnum* bogs—often becoming very abundant and species-rich in these biogeochemically significant habitats (Gerrath, 1993). *Chara* species can form dense, lime-encrusted lawns in shallow alkaline freshwaters and *Nitella* produces lush meadows on the bottoms of neutral to high pH lakes. Charaleans can grow at depths reaching 150 meters (Grant, 1990). Very tiny green algae such as *Ostreacoccus*—which at less than 2 μm in diameter is barely visible with the light or fluorescence microscope—are members of the extremely abundant and productive marine picoplankton (Courties, et al., 1994). Tropical nearshore waters are frequently dominated by green seaweeds having very unusual bodies composed of giant, multinucleate cells, known as **siphonalean** forms. Some of these, notably species of the genus *Caulerpa*, form very serious and extensive nuisance growths in the Mediterranean and other parts of the world (Delgado, et al., 1996). A few green algae have been collected using submersibles from very great ocean depths.

Generally regarded as a monophyletic group (i.e., derived from a single common ancestor), the green algal radiation has produced a wide array of morphological types: unicellular flagellates, nonflagellate unicells, motile colonies, nonmotile colonies, colonies of regular size and shape known as **coenobia**, unbranched filaments, branched filaments, tissuelike cellular sheets, and multinucleate coenocytes. Morphological diversity of the green algae rivals that observed among the ochrophytes (heterokont algae), which ranges from tiny flagellates to giant kelps (Chapters 12–15). Traditionally green algae have been classified according to their morphology, with unicellular flagellates grouped together, unbranched filaments grouped together, and so on (Bold and Wynne, 1985). However, evidence gathered from various lines of research—ultrastructural analyses of the flagellar apparatus, mitosis, and cytokinesis; comparative biochemistry; life-history studies; and molecular sequencing—revealed that a great deal of parallel evolution of morphological types has occurred within the green algae. In other words, very similar appearing green algae may actually be only distantly related. Therefore gross morphology (which is more subject to parallel or convergent evolution than conservative cellular, biochemical, and molecular characters) is not a reliable indicator of green algal phylogenetic relationships. Recognition of this fact has led to a dramatic reorganization of green algal classification to better reflect evolutionary history.

Green Algal Phylogeny and Evolution

Significant differences among green algae in ultrastructural details of the flagellar apparatus of motile cells, mitosis, and cytokinesis, together with the use of markedly different enzymes for fundamental biochemical functions, were among the first significant clues that the phylogeny of green algae was not well correlated with traditional morphological groupings (summarized by Pickett-Heaps and Marchant, 1972; Pickett-Heaps, 1975; Mattox and Stewart, 1984; McCourt, 1995; and Friedl, 1996). Modern molecular systematic investigations have largely corroborated the results of these earlier studies and demonstrate the usefulness of such ultrastructural and biochemical characters in assessing phylogenetic relationships. In the following section we outline the major groups of green algae, after which we discuss these phylogenetically significant features in some detail.

The Major Green Algal Lineages

Ultrastructural, biochemical, and molecular sequence evidence suggests that there are two major evolutionary lineages of green algae that contain multicellular forms (henceforth referred to as multicellular lineages), both of which include several clades (Fig. 17–7) (Mishler, et al., 1994). One of these, the Class Charophyceae or the charophyceans (named after one of its members, the "stonewort" *Chara*), includes not only stoneworts but zygnematalean algae such as the familiar *Spirogyra* and related desmids, as well as several other taxa including *Chlorokybus*, *Klebsormidium*, *Chaetosphaeridium*, and *Coleochaete* (Mattox and Stewart, 1984; Graham 1996). Sometimes this clade is referred to as the "charophytes," but this term has also been used in a more restricted fashion to include only the living and fossil stoneworts. Therefore to avoid confusion the terms "charophycean green algae" or "charophyceans" are used here. Charophyceans are characterized by **multilayered**

structures (MLSs), open mitosis, a persistent mitotic spindle, Cu/Zn superoxide dismutase, class I aldolases, and glycolate oxidase-containing peroxisomes that structurally resemble those of plants (Table 17–1). Further, their SSU rDNA, *rbcL*, and other gene sequences reveal close relationship to embryophytes (bryophytes + tracheophytes). The fact that many conservative ultrastructural, biochemical, and molecular features of charophyceans are shared with land plants, strongly suggests that embryophytes arose from charophycean ancestors (Fig. 17–7). There are, however, numerous differences that distinguish embryophytes from charophycean algae; these are discussed in Chapter 21.

The second multicellular green algal lineage (Fig. 17–7) includes the ulvophyceans (Class Ulvophyceae), trebouxiophyceans (Class Trebouxiophyceae), and the chlorophyceans (Class Chlorophyceae). Ulvophyceans primarily occupy marine waters and include the earlier mentioned seaweeds *Ulva*, *Codium*, *Enteromorpha*, *Cladophora*, *Halimeda*, *Caulerpa*, and *Acetabularia*. Trebouxiophyceans are freshwater and terrestrial algae that include such familiar forms as *Chlorella* and the common lichen phycobiont *Trebouxia*, for which the group is named (Friedl, 1995). Chlorophyceans include the familiar *Chlamydomonas* and *Volvox*, as well as many other primarily freshwater green algae.

These two multicellular lineages—the charophyceans (which we shall call the C clade) and the ulvophyceans + trebouxiophyceans + chlorophyceans (the UTC clade)—are thought to have arisen from distinct types of unicellular flagellates related to the modern prasinophyceans, a polyphyletic group that is discussed more completely later in this chapter. For example, an early divergent of the UTC clade is *Oltmannsiellopsis viridis* (Nakayama, et al., 1996b), a wall-less, scaly flagellate whose flagellar apparatus bears some resemblance to certain prasinophyceans. *Mesostigma* is a scaly flagellate formerly classified with the prasinophyceans that has been shown by SSU rDNA sequence studies to be an early divergent member of the Charophyceae (Chapter 21). Although many experts believe that the green algae originated via a single ancient acquisition of a plastid through primary endosymbiosis, this conclusion remains somewhat controversial, and the possibility of origin of green algal plastids by secondary endosymbiosis has been suggested (Stiller and Hall, 1997). Existing molecular data cited as supporting the hypothesis of monophyletic ancestry of plastids have not as yet critically distinguished between the possi-

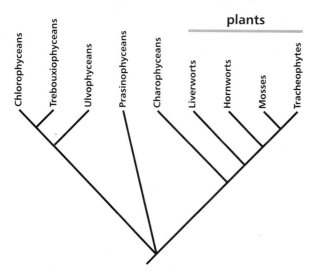

Figure 17–7 Phylogenetic analyses based upon morphological and molecular characters (and combined characters) indicate that there are four monophyletic lineages (classes) of green algae with multicellular members: Ulvophyceae, Trebouxiophyceae, Chlorophyceae, and Charophyceae. These occur in two major clades, the UTC clade and the charophyceans, each having arisen separately from unicellular flagellates related to modern prasinophyceans. Although the modern prasinophyceans are not regarded as a monophyletic group, they are often aggregated into the class Prasinophyceae for convenience. (After Mishler, et al., 1994)

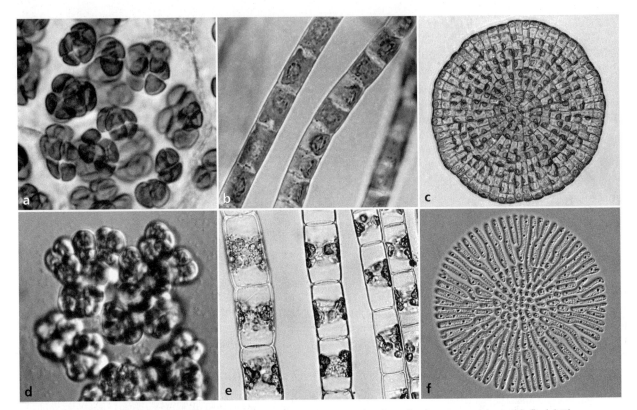

Figure 17–8 The external morphology of various charophyceans (a–c) and other green algae (d–f). (a) The sarcinoid charophycean, *Chlorokybus*, compared to (d) the sarcinoid chlorophycean, *Chlorosarcinopsis*. (b) The unbranched, filamentous charophycean, *Klebsormidium*, compared to (e) *Ulothrix*. (c) The prostrate, radially symmetrical charophycean, *Coleochaete*, compared to (f) the similar ulvophycean, *Ulvella*. These similarities are examples of parallel evolution of thallus form within the green algae; similar examples can be cited between green algae and other algal groups. (f: After Floyd and O'Kelly, 1990)

bility of a single plastid acquisition event versus multiple events involving very closely related endosymbionts and diverse hosts (Chapter 7). Colorless, phagotrophic flagellates whose flagellar apparatus ultrastructure and SSU rDNA sequences unequivocally mark them as relatives of ancient green algal hosts have not yet been identified.

A variety of different morphological types occur within each of the multicellular green algal lineages. Corollary to this is the fact that very similar-appearing forms may fall within separate lineages (i.e., they are not closely related to each other). Examples are provided by *Chlorokybus* (charophycean) and *Chlorosarcinopsis* (chlorophycean), both of which consist of packets of cells (**sarcinoid** forms); *Klebsormidium* (charophycean), *Ulothrix zonata* (ulvophycean), and *Uronema flaccidum* (chlorophycean), which are very similar unbranched filaments; and *Coleochaete* (charophycean) and *Ulvella* (ulvophycean), which can both

occur as attached, radially symmetrical flat disks formed from branched filaments that grow at the thallus margin (Fig. 17–8). It should also be noted that a number of striking parallels can be drawn between the morphology of green algae and tribophyceans (xanthophyceans or yellow-green algae—see Chapter 15). Moreover, unicellular flagellates, gelatinous colonies, and nonflagellate coccoid forms occur throughout the algae, including the green algae, and are sometimes very difficult to distinguish.

Similar body forms among unrelated algae are hypothesized to have resulted from independent adaptation to similar selection pressures. For example, environmental factors such as substrate type, light quality, temperature variation, herbivory, and competition with other algae, may be sufficiently similar in shallow water habitats to have driven the evolution of similar body types (unbranched filaments or discoid thalli) during separate radiations in fresh and marine waters. Selective pressures of desiccation, high irradiance, and other factors present in terrestrial habitats, may explain the similar appearance of unrelated sarcinoid forms. In general, hypothesized relationships between parallel morphological evolution and habitat selective pressures in green algae have been poorly explored and need to be further examined.

General Characteristics of the Green Algae

So far as is known, all forms that are included within the green algae contain at least one plastid, and most of the green algae are considered to be autotrophic. However, the green algae exhibit a surprising level of nutritional variation. Some of the prasinophyceans are known to feed on particles and therefore exhibit phagotrophy and mixotrophy. This suggests a mechanism by which the earliest green algae probably acquired their plastids (Chapter 7). A few colorless forms such as *Polytoma* and *Prototheca* are clearly related to green algae on the basis of cell structure and molecular sequence similarities. Although their cells contain a reduced plastid, they have completely lost their ability to conduct photosynthesis and are thus obligately heterotrophic. Numerous green algae are capable of supplementing photosynthesis by uptake and utilization of exogenous dissolved organic carbon, such as sugars, amino acids, and other small molecules (Neilson and Lewin, 1974; Tuchman, 1996), thus they exhibit osmotrophy and mixotrophy. Some osmotrophic green algae are **photoheterotrophic**, i.e., they utilize organic carbon only when light is present and only when their photosynthesis becomes limited by the availability of dissolved inorganic carbon (Graham, et al., 1994; Lewitus and Kana, 1994).

Features that are common to nearly all of the green algae include: flagella, commonly occurring in pairs or multiples of two, that are of approximately equal length and without tripartite, tubular hairs; stellate structures at the flagellar **transition zone** (see Fig. 9–8); chloroplasts bound by a two-membrane envelope (with no enclosing periplastidal endoplasmic reticulum) with chlorophylls *a* + *b* located in chloroplast thylakoids occurring singly or in stacks of variable numbers (Fig. 17–9); and the production and storage of starch (α-1,4-linked polyglucans) inside the chloroplasts.

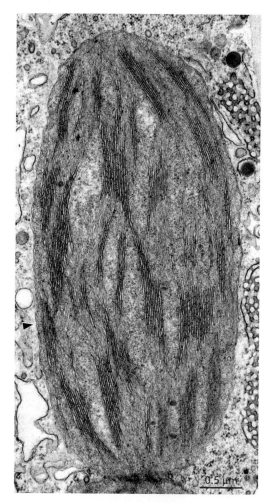

Figure 17–9 TEM view of a plastid of *Chara zeylanica*, illustrating the characteristic arrangement of thylakoids into granalike stacks and the double membrane envelope (arrowhead). Note the absence of a periplastidal endoplasmic reticulum such as that surrounding the plastids of ochrophytes and some other algae. Another characteristic of green algae is intraplastidal starch (see Fig. 19–8, for example). (Reprinted with permission from Graham, L. E., and Y. Kaneko. 1991. Subcellular structures of land plants [Embryophytes] from green algae. *CRC Critical Reviews in Plant Science* 10:323–340 ©CRC Press, Boca Raton, FL)

Production and storage of the photosynthetic reserve inside the plastid is unique to the green algae; in other eukaryotic algae the photosynthetic storage product, whether starch or some other material, is found primarily in the cytoplasm. (In the case of the cryptomonads, starch is located within the periplasti-

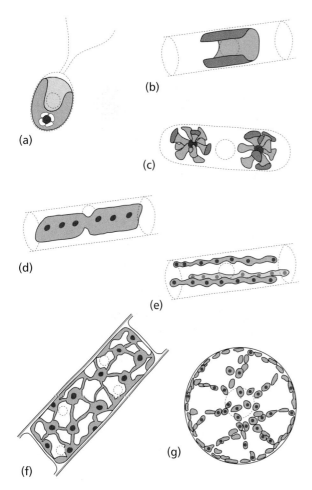

Figure 17–10 Types of plastids commonly found in green algae include: (a) cup-shaped parietal (peripheral), (b) napkin ring-shaped parietal, (c) asteroidal (star-shaped), (d) axial platelike, (e) ribbonlike, (f) reticulate (netlike), and (g) multiple discoid plastids.

dal compartment—between the periplastidal endoplasmic reticulum and the plastid proper.) Plastidal starch of green algae is reminiscent of cyanophycean glycogen storage within cyanobacterial cells (Chapter 6) thought to have been ancestral to plastids (Chapter 7). The presence and plastidal location of green algal starch can be visualized by treating cells with a solution of I_2KI, which stains starch a dark blue-black (see Fig. 20–47). Staining for starch is one of the most helpful ways to distinguish green algae from similar-appearing forms belonging to other algal groups. Chloroplasts of green algae may or may not contain eyespots and pyrenoids. If eyespots are present in green algal cells, they are always located inside the

chloroplast, never outside it as in euglenoids (Chapter 8), some dinoflagellates (Chapter 11), and eustigmatophyceans (Chapter 13). Among the green algae, chloroplasts are extremely variable in shape and number per cell (Fig. 17–10), but are typically uniform within genera. As a result, chloroplast shape and number are often useful taxonomic characters, more so than is typical for other groups of eukaryotic algae.

Green Algal Light-Harvesting Complexes

The light-harvesting systems of green algae resemble those of green plants and hence are relatively well characterized. As is the case in all algae, chlorophyll is always associated with proteins that are known as **chlorophyll-binding proteins**. There are two classes of chlorophyll *a/b* binding proteins (CAB proteins)—those of LHCI (light-harvesting complex I), which transfers energy to photosystem I, and those of LHCII, associated with photosystem II (see Chapter 6 for more details regarding photosystems I and II). All chlorophyll *a* + *b*-containing organisms have similar CAB proteins, ranging in size from 24–29 kDa (reviewed by Larkum and Howe, 1997). In addition to chlorophylls and proteins, light-harvesting complexes also include carotenoids. The major accessory photopigment of the green algae is lutein or a derivative; β-carotene is always present, and other carotenoids are also produced. Carotenoids function in light-harvesting, protection of the photosynthetic apparatus from damaging effects of excess light, dispersal of excess energy (Demming-Adams and Adams, 1992) (see Chapter 15), and are also essential for formation of the photosynthetic complex (Larkum and Howe, 1997).

LHCI and LHCII are core aggregations of proteins plus the chlorophylls that form the reaction centers of photosystems I and II, and antenna complexes of pigments and proteins that gather light energy and transfer it to the reaction centers. LHC proteins are critical to the efficient transfer of energy because they hold pigments in specific three-dimensional arrays that facilitate energy transfer.

The three dimensional structure of LHCII protein 1—the most abundant green algal and plant CAB protein—and its associations with chlorophylls and carotenoids have been determined with high resolution by electron crystallography (Kühlbrandt, et al., 1994). The 232-amino acid protein has three thylakoid membrane-spanning alpha-helices. Some of the

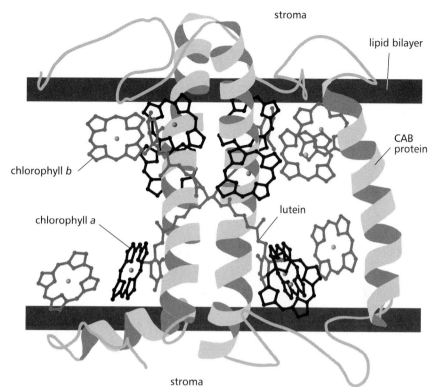

Figure 17–11 Diagrammatic view of the light-harvesting pigment complex II (LHCII). (After, with permission, *Nature*. Kühlbrandt, W., D. N. Wang, and Y. Fujiyoshi. Vol. 367:614–621. ©1994 Macmillan Magazines Limited)

protein's polar amino acids help to maintain protein shape and also attach to the magnesium atoms of at least 12 chlorophyll molecules (7 chlorophyll *a* and 5 chlorophyll *b*). By holding chlorophyll *a* in close contact with chlorophyll *b*, LHCII protein 1 helps to facilitate rapid energy transfer (Fig. 17–11). In addition, the protein binds two carotenoids, probably lutein; these prevent the formation of toxic forms of oxygen, such as singlet oxygen, a free radical that can damage DNA and other cell components.

By comparison, the three-dimensional structure of the protein complexes associated with photosystem I is not as well understood. There are two major protein subunits—A and B—each having ten thylakoid membrane-spanning regions, and a number of additional smaller subunits. Determination of the three-dimensional orientations of chlorophyll and carotenoid molecules, and their binding proteins by

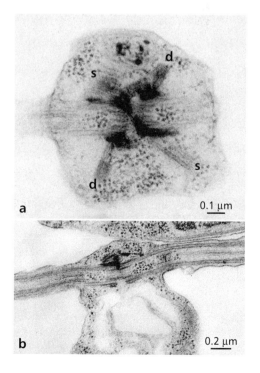

Figure 17–12 The flagellar apparatus of green algal reproductive cells reflects evolutionary relationships within the green algae. Flagellar basal bodies and associated microtubular roots of *Pediastrum* gamete viewed (a) in cross section near the cell apex and (b) in longitudinal section. (Labeling as in Fig. 17–13.) (a: From Wilcox and Floyd, 1988 by permission of the *Journal of Phycology*)

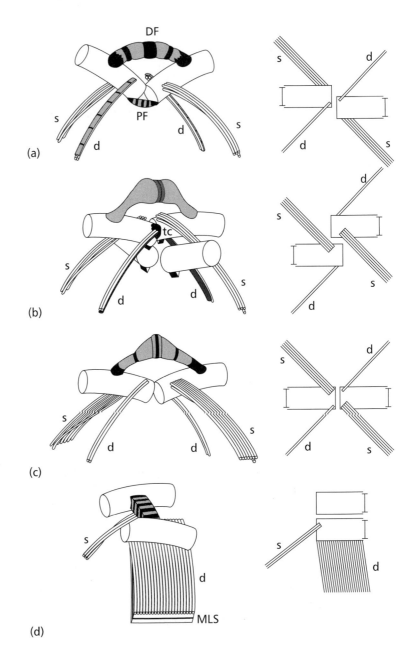

Figure 17–13 Four general types of flagellar apparatus are found among the green algae, viewed from the side (left) and from the top (right). These characterize major evolutionary lineages. The apparatuses generally include two or four basal bodies (shown here as cylinders or rectangles), microtubular roots (s or d), and distal (DF) and/or proximal (PF) connecting fibers. (a) Flagellar apparatus with cross-shaped (cruciate) roots showing clockwise displacement of the flagellar basal bodies from an imaginary line drawn parallel to and between them (when viewed from the cell's anterior). (b) Flagellar apparatus with cruciate roots and basal bodies displaced in a counterclockwise direction. (c) Flagellar apparatus with directly opposed flagellar basal bodies. (d) Flagellar apparatus with asymmetrical distribution of the flagellar roots, showing the characteristic multilayered structure (MLS). (Reprinted with permission from Inouye, I. 1993. Flagella and flagellar apparatuses of algae. Pages 99–134 in *Ultrastructure of Microalgae*, edited by T. Berner. ©CRC Press, Boca Raton, FL)

crystallographic techniques, is a very active field of research because such information illuminates fundamental processes of photosynthesis in plants and algae.

Flagellar Apparatus and Microtubular Root Systems

Flagellate cells of green algae in the UTC clade have more or less symmetrical cruciate (crosslike) root systems wherein rootlets of variable (X) numbers of microtubules alternate with rootlets composed of two microtubules to form what is known as an "X-2-X-2" arrangement (Moestrup, 1978) (Figs. 17–12, 17–13). Among the algae having cruciate root systems, there are three main variations in the orientation of the flagellar bases when cells are viewed "top-down," i.e., in an anterior-posterior direction: clockwise (CW) (Fig. 17–13a), counterclockwise (CCW) (Fig. 17–13b), and directly opposed (DO) (Fig. 17–13c) (O'Kelly and Floyd, 1984). In the first,

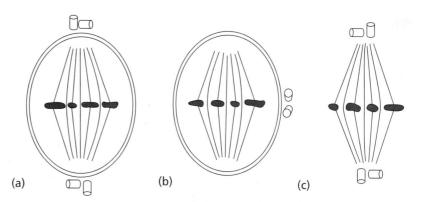

Figure 17–14 A diagrammatic comparison of (a) closed mitosis, involving an intact or nearly intact nuclear envelope, (b) metacentric mitosis, wherein centrioles are located in the plane of metaphase chromosomes rather than poles, and (c) open mitosis, characterized by dissolution of the nuclear envelope.

the basal bodies are shifted in a clockwise direction from a line drawn parallel to and between their axes. Counterclockwise flagellar apparatuses have the opposite orientation, plus in these forms, the basal bodies are often overlapping. A number of unique flagellar apparatus features are correlated with each of these three basal body arrangements.

In contrast to forms with cruciate flagellar apparatuses, the flagellar roots of the C clade (charophyceans) are highly asymmetrical (Fig. 17–13d) and are further characterized by the presence of a distinctive multilayered structure (MLS) associated with flagellar basal bodies; very similar MLSs are also characteristic of flagellate sperm produced by land plants. MLSs of charophycean green algae and plants lie adjacent to the basal bodies near the anterior end of the cell and consist of a layer of microtubules, a smaller array of parallel plates, and a layer of smaller tubules. The microtubules extend down into the cell body, serving as a cytoskeleton. MLSs have been regarded as a cellular marker for the green algal groups that share ancestry with plants. It should be noted, however, that MLSs superficially resembling those of green algae and plants occur sporadically in other algal and protist groups, for example, in glaucophytes (Kies and Kremer, 1990), the euglenoid *Eutreptiella* (Moestrup, 1978), and certain dinoflagellates (Wilcox, 1989). It is not known if the MLSs of various algae and plants are homologous. Although they appear similar, it is possible that they differ in

biochemical composition and are encoded by nonhomologous genes. Furthermore, although MLSs are believed to function primarily as microtubule organizing centers, this hypothesis has not been rigorously tested, and the function of these unusual flagellar roots needs additional study.

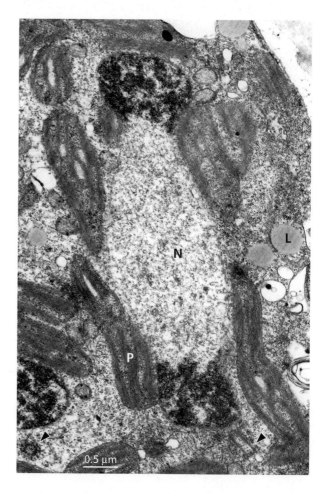

Figure 17–15 A transmission electron microscopic (TEM) view of closed mitosis (telophase in the reproductive structures of *Trentepohlia*). Arrowheads point to centrioles near bottom of image. L=lipid droplet, N=nucleus, P=plastid. (From Graham and McBride, 1978 by permission of the *Journal of Phycology*)

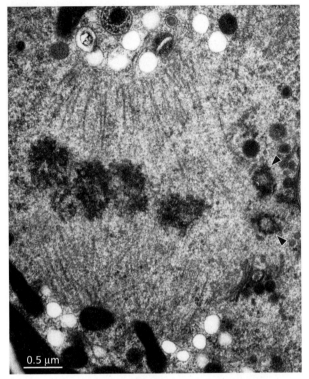

Figure 17 16 TEM of the metacentric spindle in *Friedmannia* (=*Myrmecia*; Friedl, 1995). Arrowheads point to centrioles. (Reprinted with permission from Lokhorst, G. M., P. J. Segaar, and W. Star. 1989. An ultrastructural reinvestigation of mitosis and cytokinesis in cryofixed sporangia of the coccoid green alga *Friedmannia israelensis* with special reference to septum formation and the replication cycle of basal bodies. *Cryptogamic Botany* 1:275–294)

Mitosis

In some green algae of the UTC clade the nuclear envelope persists throughout mitosis; this is known as **closed mitosis** (Figs. 17–14a, 17–15) The mitotic nuclei of such algae often appear dumbbell-shaped just prior to completion of mitosis, and after separation, the two daughter nuclei tend to drift toward each other, sometimes even flattening against each other (Pickett-Heaps, 1975). One group of green algae with closed mitosis (Trebouxiophyceans) has an unusual and distinctive **metacentric spindle**, where the centrioles are located near the metaphase plate of chromosomes (Figs. 17–14b, 17–16), rather than at the spindle poles, as is more usual. Charophyceans have an **open mitosis**, in which the nuclear envelope breaks down early in mitosis and reforms at telophase, as also occurs in plants (Figs. 17–14c, 17–17). In green algae

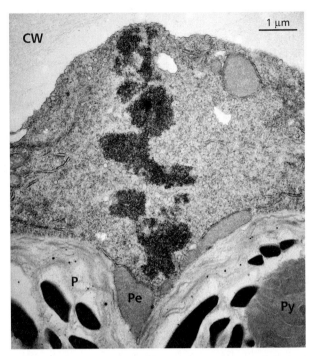

Figure 17–17 TEM of open mitosis (metaphase) in *Coleochaete*. CW=cell wall, P=plastid, Pe=peroxisome, Py=pyrenoid.

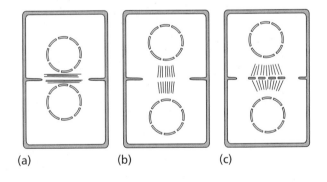

(a)   (b)   (c)

Figure 17–18 Diagrammatic comparison of specialized cytokinetic microtubular systems. Phycoplasts (a) are arrays of microtubules lying parallel to the developing cleavage furrow; these are believed to keep daughter nuclei—which may lie close together—apart during cytokinesis. Incipient phragmoplasts (b) occur together with furrowing in some green algae. Phragmoplasts (c) very similar, if not identical to those of land plants occur in a few green algae; little, if any, furrowing occurs, and cell plates develop from the center toward the cell periphery.

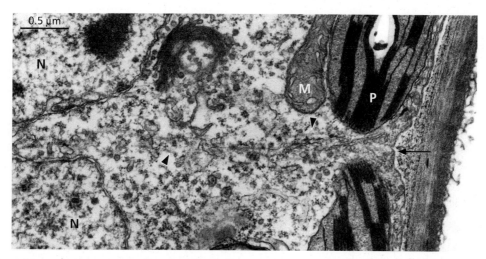

Figure 17–19 TEM view of the phycoplast in *Microspora*. Microtubules (arrowheads) lie in the same plane as the developing furrow (arrow). M=mitochondrion, N=nucleus, P=plastid. (From Pickett-Heaps, 1975)

with open mitosis, mitotic spindles tend to persist late into mitosis, holding the reforming daughter nuclei well apart from each other.

Cytokinesis

Many green algae accomplish cytokinesis by simple furrowing, as is common among protists. Members of the chlorophyceae that possess a well-developed cell wall also produce a distinctive set of microtubules—the **phycoplast**—that lies parallel to the plane of cytokinesis (Figs. 17–18a, 17–19). It has been proposed that the phycoplast microtubules help to separate daughter nuclei, which may drift close together in the region of cross-wall development, thereby ensuring that each daughter cell contains a nucleus. Phycoplast microtubules may somehow interact with the developing furrow as it grows inward from the cell periphery and eventually constricts the parental cell into two. In some chlorophyceans the phycoplast

Figure 17–20 TEM view of a phragmoplast and forming cell plate (arrows) in *Chara*. All of the structures associated with phragmoplast formation in higher land plants, such as transverse microtubules (arrowheads) and coated vesicles, also occur in *Chara*. (From Cook, et al., 1998 ©Springer-Verlag)

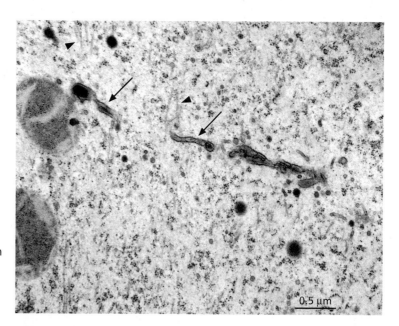

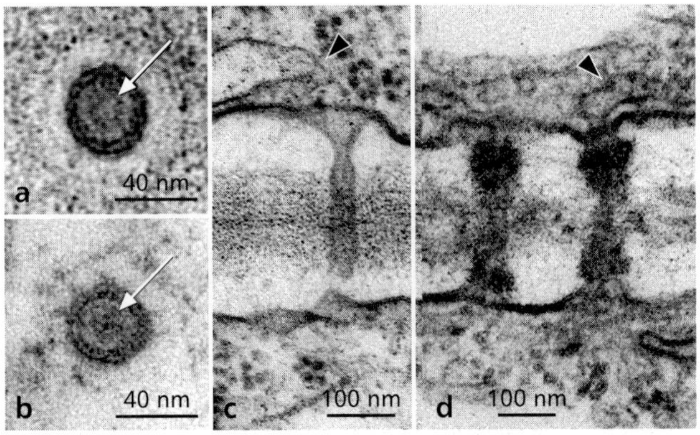

Figure 17–21 Comparison of TEM views of the plasmodesmata of *Chara* in (a) cross section and (c) longitudinal section with (b) cross-sectional and (d) longitudinal views of those found in germlings of the early divergent moss *Sphagnum*. Internal structure is visible in cross sections of both *Chara* and *Sphagnum* plasmodesmata (arrows). Spokelike structures (arrowheads) connect the central structure to the cell membrane in both *Chara* and *Sphagnum*. In (c) and (d), endoplasmic reticulum (arrowheads) can be seen to be associated with *Chara* and *Sphagnum* plasmodesmata. (From Cook, et al., 1997)

helps to organize vesicles containing cell-wall material in such a way that a **cell plate** develops outward from the center. **Plasmodesmata**—channels through the cell wall that allow for intercellular communication—may occur in the cross-walls of such algae.

Charophyceans are characterized by a persistent spindle that helps keep daughter nuclei separate until cytokinesis has been accomplished. Early divergent charophyceans undergo cytokinesis by furrowing, growth of a new wall inward (Fig. 17–18b). A few members of the charophyceae produce a **phragmoplast** (Figs. 17–18c, 17–20), cell plates (Fig. 17–20), and "primary" plasmodesmata (formed concomitantly with the cell plate) (Fig. 17–21) similar to those of plants. In contrast to phycoplasts, phragmoplasts are composed of a double set of microtubules oriented perpendicularly to the plane of cytokinesis (Fig. 17–22). Like certain phycoplasts, mentioned above, phragmoplasts help to organize the aggregation and coalescence of vesicles containing cell-wall material of which the cell plate is formed (Pickett-Heaps, 1975; Cook, et al., 1998). In algae such as *Spirogyra*, a relatively small phragmoplast occurs at the same time as a peripheral furrow, revealing an intermediate stage in the evolutionary development of the land plant phragmoplast. However, cross-walls of *Spirogyra* and its close relatives are devoid of plasmodesmata. Phragmoplasts, cell plates, and plasmodesmata that resemble those of land plants have been identified in members of the order Charales and genus *Coleochaete* and are described in more detail in Chapter 21.

Enzymes and Peroxisomes

Biochemical features, particularly differences in photorespiratory enzymes and organelles, have proven valuable in understanding green algae evolution. Green algae of the UTC clade use the enzyme glycolate dehydrogenase to break down the glycolate formed in the process of photorespiration (Fig. 17–23). Charophyceans utilize a completely different enzyme, glycolate oxidase, which is coupled to consumption of molecular oxygen, and is also present in plants (Frederick, et al., 1973) as well as members of some other algal groups. Glycolate oxidase, together with the enzyme catalase, is localized within characteristic organelles known as **peroxisomes** (see Fig. 2–8); here it operates to help recover some of the fixed carbon that would otherwise be lost to the cell through photorespiration (Fig. 17–24). Peroxisomes of charophyceans are typically relatively large, membrane-bound organelles that are very similar to plant peroxisomes. In contrast, other green algae possess smaller and enzymatically distinct **microbodies**. A review of the various types of microbodies, peroxisomes, and associated enzymes occurring throughout the algae can be found in Gross (1993).

Additional biochemical evidence relevant to green algal diversification includes differences in superoxide dismutases. Like the plants, charophycean green algae

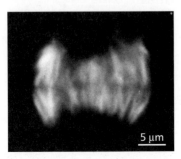

Figure 17–22 Tubulin immunofluorescence labeling demonstrates occurrence of a land plantlike phragmoplast in dividing cells of *Coleochaete*. (From Brown, et al., 1994)

Table 17–1 Characteristics of the five major green algal groups

	flagellar/cytoskeletal apparatus	photorespiratory enzymes	mitosis	cytokinesis	habitat (primary)	life history
Prasinophyceans	cruciate roots, rhizoplasts, some with MLS, flagellar & body scales common	variable	variable	furrowing	marine	zygotic meiosis
Ulvophyceae	cruciate X-2-X-2 roots, CCW orientation, +/– body & flagellar scales, rhizoplast present	glycolate dehydrogenase	closed, persistent spindle	furrowing	marine or terrestrial	zygotic meiosis or alternation of generations or gametic meiosis
Trebouxiophyceae	cruciate X-2-X-2 roots, CCW orientation, no scales, rhizoplast present	glycolate dehydrogenase	semi-closed, non-persistent spindle	furrowing	freshwater or terrestrial	zygotic meiosis
Chlorophyceae	cruciate X-2-X-2 roots, CW or DO orientation* scales occur rarely, rhizoplasts	glycolate dehydrogenase	closed, non-persistent spindle	furrowing, phycoplast, some with cell plate & plasmodesmata	freshwater or terrestrial	zygotic meiosis
Charophyceae	asymmetric roots, MLS, body & flagellar scales usually present, rhizoplast rare	glycolate oxidase & catalase in peroxisome	open, persistent spindle	furrowing, some with cell plate, phragmoplast, & plasmodesmata	freshwater or terrestrial	zygotic meiosis

* except *Hafniomonas reticulata*, whose basal bodies are CCW, but whose SSU rDNA sequences are allied with the chlorophyceans (Nakayama, et al., 1996)

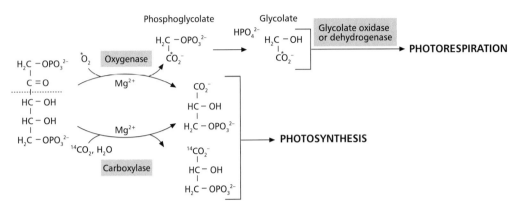

Figure 17–23 Diagrammatic comparison of photorespiration with photosynthesis. When oxygen levels are relatively high, Rubisco can function as an oxygenase, which results in the production of glycolate. In most green algae, glycolate dehydrogenase is used to oxidize the glycolate, but in a few this process is accomplished by glycolate oxidase, as in land plants. (After Falkowski and Raven, 1997. Redrawn by permission of Blackwell Science, Inc.)

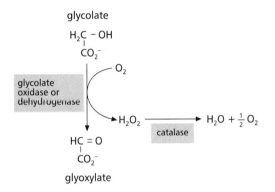

Figure 17–24 Glycolate oxidation in algae and plants. Catalase removes the reactive hydrogen peroxide (H_2O_2).

produce Cu/Zn superoxide dismutase, whereas this enzyme is absent from other green algae (deJesus, et al., 1989). The enzymes urease and urea amidolyase (Syrett and al-Houty, 1984) and class I aldolases (Jacobshagen and Schnarrenberger, 1990) have also been useful in tracing phylogenetic relationships among green algae and between green algae and land plants.

Variation in green algal flagellate cell ultrastructure, mitosis, cytokinesis, enzymes, and molecular sequences are highly correlated and are likely due to shared phylogenetic history. The green algae that possess land plantlike MLSs (i.e., charophyceans) are also characterized by open mitosis, persistent spindles or phragmoplasts, glycolate oxidase, plantlike peroxisomes, and other biochemical similarities to embryophytes. In contrast, the green algae with X-2-

X-2 roots (members of the UTC clade) typically have closed mitosis, lack glycolate oxidase, and many produce phycoplasts at cytokinesis. Table 17–1 shows the distribution of various structural and biochemical features among the green algae.

PART 2—PRASINOPHYCEANS

Prasinophyceans—a name derived from the Greek word *prasinos*, meaning green—are generally regarded as the modern representatives of the earliest green algae. This group is also known as the micromonadophytes (micromonadophyceans), but the term prasinophyceans is used here instead, as recommended by Sym and Pienaar (1993), who discuss the history of usage of the alternate names.

Prasinophyceans are primarily marine flagellates (Fig. 17–25). However several have nonmotile stages, others are nonflagellate with only brief flagellate stages, and a few, such as the coccoid *Bathycoccus* (Eikrem and Throndsen, 1990), appear to lack flagellate stages. Certain forms occur as sessile (attached) dendroid (treelike) colonies. Palmelloid (palmella) stages—mucilaginous aggregations of nonmotile cells—are also known for some species. A few inhabit freshwaters and possess contractile vacuoles, which remove excess water from cells. Although all known members of the prasinophyceans possess pigmented chloroplasts, a few (notably *Cymbomonas* and *Halosphaera*) are known phagotrophs (or more appropriately, mixotrophs) (see references in review by Sym

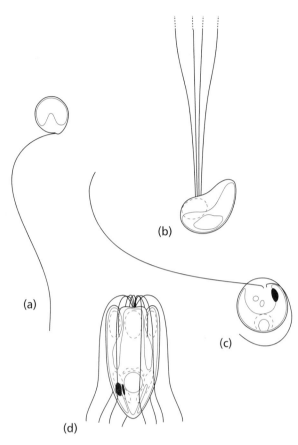

Figure 17–25 Some of the morphological variation observed within the prasinophyceans. Shown are (a) *Resultor*, (b) *Pachysphaera*, (c) *Nephroselmis*, and (d) *Pyramimonas*. (After Throndsen, 1997)

and Pienaar, 1993). The potential osmotrophic capabilities of prasinophyceans have not been intensively studied.

Within the group, the flagellar number varies from one in *Pedinomonas* to 16 in *Pyramimonas cyrtoptera*. It has been recommended that the uniflagellate, scaleless *Pedinomonas*, *Marsupiomonas*, and *Resultor* should be placed into a new class, Pedinophyceae (pedinophyceans) (Moestrup, 1991). Prasinophycean flagella typically emerge from an apical depression or pit, and the flagellar basal bodies are often very long and parallel. The cell membrane of most forms is covered with one or more layers of often extremely elaborate scales. Others are enclosed by a wall-like theca, and a few are naked. Other features that may occur, but are not ubiquitous among prasinophyceans, are a large Golgi apparatus that lies between flagellar basal bodies and the nucleus, and ejectile

structures known as **extrusomes**. A few taxa produce ejectile structures in the form of coiled ribbons that form an elongate, tapering, tubelike "trichocyst" following rapid hydration. Prasinophyceans generally also have a single plastid (though it may be highly lobed), and usually possess at least one starch-sheathed pyrenoid. Sometimes an eyespot is present; if so, it is always inside the plastid. There is typically a single, highly dissected mitochondrion with flattened cristae, like those of other green algae. A number of prasinophyceans possess a distinctive major accessory pigment, prasinoxanthin. *Pycnococcus provasolii* is a coccoid form that is difficult to recognize as a prasinophycean by the use of morphological criteria, but falls into this group based on production of prasinoxanthin as the major xanthophyll (Guillard, et al., 1991). However some other prasinophyceans lack this pigment. There are no unique and defining derived characteristics (synapomorphies) for the group, rather, prasinophyceans are characterized by the above-listed constellation of what are considered to be ancestral (plesiomorphic) features among the green algae (Sym and Pienaar, 1993).

Relatively few prasinophyceans have been studied by molecular phylogenetic analysis and few genes have been examined. Thus it is not as yet possible to construct a comprehensive phylogeny (Daugbjerg, et al., 1995). SSU rDNA studies of a few forms suggest that while prasinophyceans do diverge from other green algae relatively early, they are not monophyletic (Steinkötter, et al., 1994) (see Fig. 5–12). Chlorophylls *a* and *b* are always present, some forms possess chlorophyll variants, and there are at least six different combinations of accessory pigments. Some of these resemble suites of pigments typical of chlorophycean algae, whereas others more closely resemble certain ulvophyceans. One group of prasinophycean genera has light-harvesting pigment complexes that are very unusual among green algae (Fawley, et al., 1990b). A compendium of the pigment signatures for prasinophyceans can be found in Sym and Pienaar (1993). These data, together with variations that occur during cell division, suggest that when more taxa have been critically examined ultrastructurally, and by molecular sequence analysis, prasinophyceans will be disassembled into several monophyletic subgroups, and that some may be incorporated into other green algal classes.

Current concepts of prasinophycean relationships and classification can be found in Sym and Pienaar (1993). Because they are unicellular, prasinophycean diversity, taxonomic identification, classification, and phylogeny require an understanding of cellular features at the ultrastructural level. Therefore the next sections will focus on features of prasinophycean cell biology that are unique to the group and features that reflect evolutionary diversification within the group.

Cellular Features of Prasinophyceans

Flagella, Flagellar Apparatus, and Cytoskeleton

The flagellar transition region—typically fairly uniform within other algal groups—is unusually variable within prasinophyceans. Three major elements can occur in prasinophycean transition regions: a stellate structure (as in most green algae), a plate, and a helix (as occurs in ochrophytes). While the plate and helix may be absent or highly modified, the stellate structure is typical. This is a cylindrical structure composed of several electron-opaque filaments that connect the A-tubules of alternate peripheral doublet microtubules, forming a star-shaped pattern in cross section as in other green algae. The contractile protein, centrin, has been found in the transition region of some prasinophyceans (Melkonian, et al., 1992).

Although in most cases the flagella of an individual prasinophycean cell are all of the same length and morphology, heteromorphic flagella are known to occur in some forms. The latter also exhibit flagellar maturation, a term that describes the occurrence of flagella of different ages—or maturation states—on the same cell. As in other algae, flagellar maturation may require more than one cell generation. One prasinophycean (*Micromonas pusilla*) has flagella that are devoid of scales, but for other prasinophyceans at least one layer of flagellar scales commonly occurs. *Tetraselmis* flagella, for example, have four layers of scales, each of which is morphologically distinct. An outermost layer of long, thin hair scales is very common. A compendium of the flagellar scale types found in prasinophyceans can be found in Sym and Pienaar (1993).

The flagellar apparatus of prasinophyceans usually consists of parallel basal bodies linked by a connecting fiber that contains the contractile protein, centrin. In fact, this protein, which occurs in all groups of eukaryotes, was first discovered in studies of the flagellar apparatus of the prasinophyceans (see

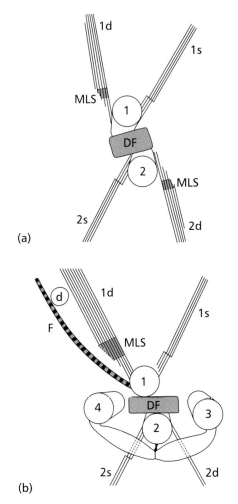

(a)

(b)

Figure 17–26 Diagrammatic views of the flagellar apparatus in (a) the charophycean flagellate *Mesostigma*, (which has two MLSs) and (b) the prasinophycean *Halosphaera* (with one MLS). Numbered circles are basal bodies (two in *Mesostigma* and four in *Halosphaera*). DF is the distal fiber, which links basal bodies number 1 and 2. Four microtubular roots are present (1d, 1s, 2d, and 2s). A fibrous root (F) and scale duct (d) are present in *Halosphaera*, but not *Mesostigma*. (After Sym and Pienaar, 1993)

Coling and Salisbury, 1992). Microtubular roots sometimes approach the characteristic X-2-X-2 pattern common among some lineages of green algae, but a high degree of variation occurs in the number of roots, their arrangement, and the number of microtubules they contain. In some cases, a non-contractile striated rootlet (known as a system I fiber) composed of a phosphoprotein is present that is thought to function in absorbing stress. **Rhizoplasts** (known as system II fibers) are striated, contractile, centrin-containing structures that link the flagellar apparatus to the anterior-most nuclear surface (e.g., Salisbury et al., 1981). These musclelike structures sometimes also form links with the chloroplast or the cell membrane. Rhizoplast contraction requires Ca^{2+} and ATP and is a mechanical adaptation to problems arising from flagellar position at the bottom of a pit (Salisbury and Floyd, 1978). Similar rhizoplasts occur in other algal groups, including various green algae, but only in prasinophyceans and ulvophycean green algae do they extend beyond the nucleus. It is typical for the rhizoplast to be associated in some way with the prasinophycean microbody, a structure that contains catalase. A summary of the flagellar root variation that occurs among prasinophyceans can be found in Sym and Pienaar (1993).

Several prasinophyceans, including *Prasinopapilla*, *Pterosperma*, and *Halosphaera*, possess a flagellar root that includes a multilayered structure (MLS) (Fig. 17–26b) very similar to MLSs observed in charophycean green algae and flagellate cells of land plants. The presence of an MLS in *Mesostigma viride* (Fig. 17–26a) led Rogers, et al. (1981) to suggest that *Mesostigma* is probably more closely related to the origin of the charophycean algae (and land plants) than many other green flagellates, and this hypothesis has been supported by molecular sequence data (Melkonian and Surek, 1995).

Cell Covering

The cells of most prasinophyceans are enclosed by one to five layers of scales attached to the cell membrane, with the scales of each layer characteristic for the species (Fig. 17–27). A compendium of these scale types and their occurrence can be found in Sym and Pienaar (1993). A continuous underlayer of small, square scales is very common, but not ubiquitous. The outer scales can be amazingly complex and three-dimensional, with lacy, basketlike shapes being common. Scales are produced within the Golgi apparatus, and all types of body and flagellar scales can be produced within the same cisterna. Scales are extruded to the cell exterior in the region of the flagellar pit, and eventually pushed to their final position on the cell surface (Sym and Pienaar, 1993). In some cases, there appears to be a permanent, specialized scale-releasing duct (Fig. 17–28), which has been hypothesized to be the vestigial remains of the cytopharyx (feeding apparatus) of ancestral phagotrophic cells.

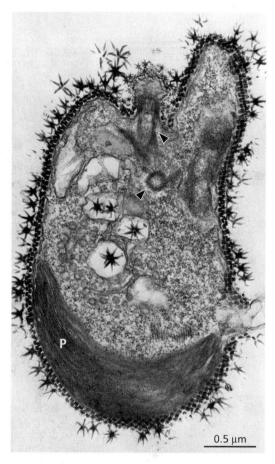

Figure 17–27 Body scales of the prasinophycean *Nephroselmis* are arranged in five layers. Note the scales within vesicles. Arrowheads point to basal bodies. P=plastid. (Micrograph courtesy S. Sym and R. Pienaar)

Some scales are known to be composed primarily of a pectinlike carbohydrate, together with a small amount of protein. Unusual 2-keto-sugar acids are also present; their occurrence in both prasinophyceans and the walls of higher plants has been suggested as evidence that the ancestry of land plants included scaly forms (Becker, et al., 1991). These acids confer a net negative charge to the scale layer that, together with calcium cations, are involved in interlocking the scales in their characteristic layers. The Golgi apparatus is also thought to give rise to mucilage vesicles that discharge their contents to the cell surface during cyst and palmella-stage formation.

In contrast to most prasinophyceans, cells of *Tetraselmis* and *Scherffelia* are enclosed by a single, coherent theca composed of two or three layers, inter-

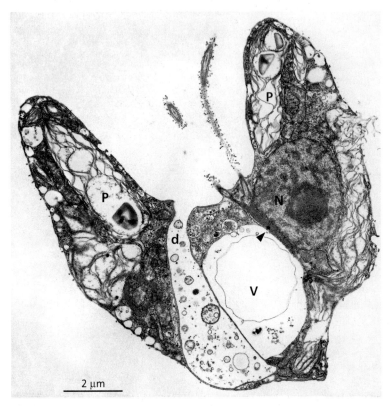

Figure 17–28 TEM of *Halosphaera* in longitudinal view, showing the specialized scale releasing duct (d), which is regarded as a vestigial cytopharynx (cell mouth or particulate feeding apparatus). Arrowhead points to rhizoplast. N=nucleus, P=plastid, V=vacuole. (Micrograph courtesy T. Hori)

rupted only by a slit at the point of flagellar emergence. Close examination reveals that the theca is composed of subunits having a size and arrangement resembling that of the scale layers of other prasinophyceans (Domozych, 1984) suggesting that the theca develops by fusion of Golgi-derived scales following their deposition on the cell surface. Such walled prasinophyceans are regarded as the most derived members of this group by Melkonian and Surek (1995), who also suggest close relationship of thecate prasinophyceans to the ulvophycean-trebouxio-phycean-chlorophycean (UTC) green algal clade.

In addition to such thecate cells, a number of prasinophyceans produce resting or other stages that are walled. *Tasmanites*, *Pterosperma*, and *Halosphaera* produce a large spherical cyst (up to 230 μm in diameter) known as a **phycoma** stage (Fig. 17–29). Phycomata develop from flagellate cells by deposition of an inner wall (formed from the contents of numerous mucilage vesicles that discharge to the outside), an increase in cytoplasmic lipids, which confer buoyancy, and the development of an outer wall. The outer phycoma wall of these taxa and the cysts of *Pyramimonas* are chemically resistant like the sporopollenin of plants. Fossil *Tasmanites* have been found in

deposits of Cambrian to Miocene age, in coaly tasmanite or oil shales. Occurrence of these and similar fossils also thought to represent prasinophycean cysts, is attributed to the presence of decay-resistant sporopolleninlike material in the cyst wall (Taylor and Taylor, 1993).

In modern forms, following a period of time—from two weeks to nearly four months, depending on

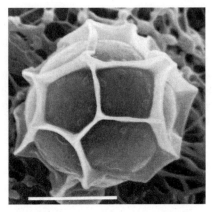

Figure 17–29 Phycoma (asexual cyst) stage of *Pterosperma*. Scale bar=10 μm. (From Inouye, et al., 1990 by permission of the *Journal of Phycology*)

Text Box 17–1 Sexual Reproduction in *Nephroselmis olivacea*

In a study of sexual reproduction in the prasinophycean *Nephroselmis olivacea*, axenic, clonal cultures were established from mud samples collected from a Japanese pond (Suda, et al., 1989). Because zygotes were never observed in clonal cultures, separate strains were combined in the wells of a multi-well dish and observed for signs of mating, which occurred. Gametes were isogamous, but of two behavioral types—one attached to the bottom of the culture dish, and the second fused with it in the region of the flagellar bases. Flagellar tip-clumping, such as is characteristic of *Chlamydomonas* (Chapter 20), was not observed. Some fused pairs engaged in swimming behavior before eventually settling, whereas others remained attached to the dish throughout the mating process. Flagella were apparently withdrawn into cells, and heart-shaped zygotes appeared. After about an hour, the two gamete nuclei had fused within the now spherical zygotes, which remained adherent to the substrate. To study subsequent events, unmated cells were washed away with sterile distilled water and the zygotes were maintained in sterile distilled water for three months in darkness at 10° C. Following this period of time, the distilled water was replaced by growth medium and the zygotes were returned to lighted culture conditions at 20° C. After one to three days, they germinated by means of meiosis. One of the products of the first meiotic division apparently aborted, and thus it was deduced that the two flagellate daughter cells were produced following meiosis II. This study represents an approach by which other prasinophyceans might be surveyed for the presence of sexual reproduction.

gamete differentiation

plasmogamy

−

+

2N

N

karyogamy

meiosis

(After Suda, et al., 1989)

meiosis

Z

environmental conditions—the phycoma/cyst nucleus and cytoplasm undergo division to form numerous flagellate cells that are released by rupture of the cyst wall. The cells produced by phycomata swim backward, which is quite unusual (Inouye, et al., 1990). Cyst/phycoma production and germination appear to represent a form of asexual reproduction. Sexual reproduction in prasinophyceans is known for *Nephroselmis olivacea* (Suda, et al., 1989), and is described in Text Box 17–1.

Cell Division

Mitosis and cytokinesis of the non-walled forms occurs while the cells are swimming. The first indications that mitosis is imminent are cell enlargement and widening of the pyrenoid. Basal bodies replicate and generate flagella, then all of the interconnections between basal bodies break down, presumably to allow for the reorganization of the flagellar apparatus that accompanies flagellar maturation cycles. The rhizoplast then divides, and the nuclear envelope develops openings at the poles. In some prasinophyceans, such as *Mantoniella*, the nuclear envelope is present throughout mitosis (Barlow and Cattolico, 1981) (as in the UTC clade of green algae), whereas in others such as *Pyramimonas amylifera* (Woods and Triemer, 1981), the nuclear envelope has almost completely broken down by mid-mitosis—closer to the condition in charophycean green algae. Cytokinesis occurs by development of a constricting furrow from the cell periphery. The interzonal spindle may persist until it is broken by the furrowing cell membrane, which suggests that the persistent spindle characteristic of charophycean green algae may be a primitive character.

Cell division of walled (thecate) prasinophyceans differs from that in scaly forms in several ways. The flagella are shed prior to mitosis, and cytokinesis occurs

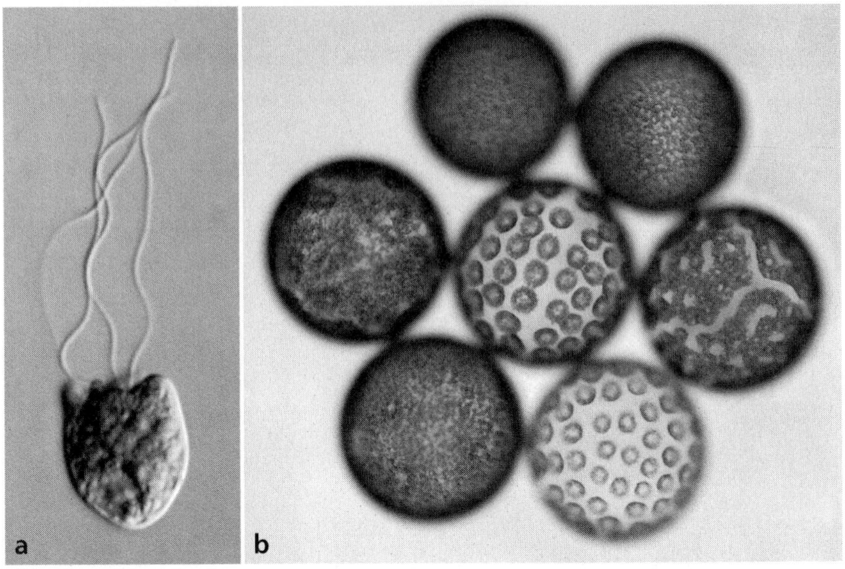

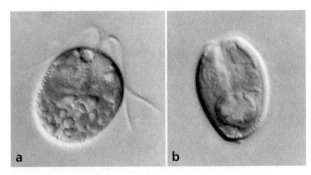

Figure 17–30 *Halosphaera*. (a) Motile cell. The four flagella emerge from an apical pit surrounded by cytoplasmic lobes. (b) Phycomata in various stages of development. (a: courtesy G. Floyd and C. O'Kelly, b: by M. Dube)

within the parental theca. The nuclear envelope never disintegrates, and the spindle is metacentric (i.e., the basal bodies [=centrioles] lie at one side of the metaphase plate) as in certain trebouxiophycean green algae. The spindle collapses quickly, allowing daughter nuclei to move close together, and a phycoplast system of microtubules develops parallel to the plane of cytokinesis, as in the UTC clade of green algae. Daughter cells then develop new thecae and flagella.

Some Examples of Prasinophycean Diversity

HALOSPHAERA (Gr. *hals*, sea + Gr. *sphaira*, ball) is a unicellular marine flagellate (Fig. 17–30). Four flagella (rarely fewer) emerge from a pit, and may be twice the length of the cell. The cells rotate as they swim, either forward or backward. The anterior end of the cell is highly lobed, as is the cup-shaped plastid, the lobes of which extend into the cytoplasmic lobes. There are two or four pyrenoids. A single eyespot occurs in the posterior portion of the cell. There are numerous mucilage vesicles arranged in longitudinal rows at the cell periphery. Their contents can be discharged as threads, rods, or spheres. A large posterior reservoir is connected to the flagellar pit by a short canal; this makes up the phagocytotic apparatus of the cell (Sym and Pienaar, 1993; Hori, et al., 1985). There is a large (250–800 μm in diameter)

planktonic phycoma/cyst stage, the outer wall of which is composed of resistant, sporopolleninlike material; this wall may have surface ornamentations. The cyst stage contains a large amount of lipid, which contributes to buoyancy. The cytoplasm eventually divides to form many small uninucleate units, which then continue dividing to form flagellate cells. These remain within the cyst wall for a time, continuing to divide, before eventual release through a slit in the wall (Sym and Pienaar, 1993).

PYRAMIMONAS (Gr. *pyramis*, pyramid + Gr. *monas*, unit) is a flagellate unicell found in marine, brackish, or freshwaters (Fig. 17–31). The 4–16 flagella can be up to five times as long as the cells, and

Figure 17–31 Two *Pyramimonas mucifera* cells. (a) Note four flagella. (b) Note pit and pyrenoid with surrounding starch. (Photographs courtesy R. Pienaar and S. Sym)

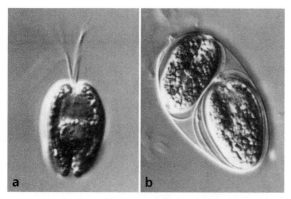

Figure 17–32 *Tetraselmis*. (a) Motile cell. (b) Daughter cells within parental theca. (Photographs courtesy R. Pienaar and S. Sym)

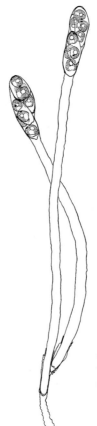

Figure 17–33 *Prasinocladus*. A diagrammatic view showing cells at the ends of elongate, branched gelatinous stalks. (From Proskauer, 1950)

emerge from a deep, narrow pit. Some species of *Pyramimonas* are benthic, associated with sand or ice; the larger number of flagella (8 or 16) of these forms is regarded as an adaptation to benthic life, facilitating attachment (Daugbjerg et al., 1994). Another such adaptation is extensive production of mucilage from muciferous vesicles as in the benthic phase of *P. mucifera* (Sym and Pienaar, 1991). The anterior is four-lobed, as is the usually single, cup-shaped chloroplast, the lobes surrounding the pit. There is usually at least one pyrenoid and one or more eyespots. Reconstructions derived from serial sections revealed that there is a single, highly branched mitochondrion in the Antarctic *Pyramimonas gelidicola*, suggesting that the same may be true for other prasinophyceans (McFadden and Wetherbee, 1982). There are several layers of body and flagellar scales. Seventy-five species have been described, but many of these have not yet been subjected to critical analysis by transmission electron microscopy. The phylogeny of 16 species of *Pyramimonas* was investigated by use of *rbcL* sequences (Daugbjerg, et al., 1994); the genus comprises at least four distinct subgeneric groups. Ultrastructural studies of the microtubular root systems suggest that *Pterosperma* and *Halosphaera* are related to *Pyramimonas*, and that all possess an X-2-X-2 root system typical of non-charophycean green algae (Inouye, et al., 1990).

TETRASELMIS (Gr. *tetras*—from *tessares*, four + Gr. *selmis*, angler's noose) (Fig. 17–32) may occur as a flagellate or a nonmotile cell attached by a gelati-

nous stalk. Flagellate cells have four flagella emerging from the pit in two pairs. Cells are covered by a distinctive theca composed of small scalelike particles in a crystalline array. Cells often stop swimming for extended periods and flagella are sometimes lost. Though usually green, some can become red by accumulation of carotenoids.

About 26 species are reported from marine and freshwaters (Sym and Pienaar, 1993). Ultrastructural studies suggest that the genus might be subdivided into as many as four subgenera (Hori, et al., 1982) on the basis of features such as pyrenoid structure. In addition, ultrastructural data (presence of a periplast of fused scales) suggest that the sessile, stalked genus *Prasinocladus* (Fig. 17–33) (Proskauer, 1950), *Platymonas*, and some other forms may actually be allied with *Tetraselmis* (Norris, et al., 1980). This hypothesis is supported by *rbcL* sequence analyses (Daugbjerg, et al., 1995).

Chapter **18**

Green Algae II
Ulvophyceans

Halimeda

Ulvophyceans include some of the largest and most conspicuous of the green algae. Casual visitors to rocky intertidal coasts in temperate regions commonly observe stringy clumps of *Enteromorpha* or limp lettucelike blades of *Ulva* that unfold into undulating filmy sheets when submerged. Under nutrient-rich conditions these weedy seaweeds can become so abundant that they are regarded as nuisances; *Enteromorpha* is of particular concern as a major hull-fouling alga on boats. Some shores are dominated by deep green, fingerlike growths of *Codium* that can wreak havoc in scallop and other shellfish beds. This heavy green seaweed can grow on the shellfish, weighing them down and preventing their escape from predators, thereby contributing to heavy mortality. Although generally innocuous, ulvophyceans include some forms (certain *Caulerpa* species) that produce neurotoxins poisonous to humans.

On a more pleasant note, snorkelers and divers commonly encounter ulvophyceans in their explorations of shallow sea-grass beds or coral reefs of warm, clear tropical waters. Expanses of jointed, pale green thalli of *Halimeda* or *Cymopolia* (Fig. 18–1), clumps of tiny winecup-shaped *Acetabularia*, and orderly rows of feathery *Caulerpa* can be as attractive a sight as schools of colorful fish. Unfortunately, in some regions of the world (including the Mediterranean Sea), an unusually large form of *Caulerpa taxifolia* has produced extensive lawns that crowd out other marine organisms (Delgado, et al., 1996).

Most ulvophycean diversity occurs in such marine habitats, but nutrient-rich freshwater lakes and streams can harbor lush green growths of the ulvophyceans *Pithophora*, *Cladophora*, and *Ulothrix*. These macroalgae, together with a diverse community of attached periphytic diatoms and other small algae, provide valuable habitat and food for numerous forms of aquatic invertebrates. *Cladophora* can form balls more than 10 cm in diameter as a result of the action of lake-bottom currents (Niyama, 1989). In the Hokkaido area of Japan such balls are the subject of folklore and are celebrated with an annual festival. However human appreciation for ulvophyceans can wain when *Pithophora* blankets

Figure 18–1 Underwater photograph of the ulvophycean *Cymopolia* growing on a coral reef off San Salvador, Bahamas. Note the white calcified segments separated by a more flexible uncalcified joint. The hazy clouds at branch tips are many thin green filaments that function in photosynthesis. (Photograph courtesy Ronald J. Stephenson)

the surfaces of ponds and interferes with recreational activities (Fig. 18–2), or beachgoers encounter masses of rotting, storm-detached *Cladophora* (see Fig. 17–4). Considerable research has been directed toward controlling or preventing undesirable growth or effects of ulvophyceans.

A number of other ulvophyceans are also of note ecologically. Among the least noticeable are microscopic filaments, such as the "boring alga" *Ostreobium*, which penetrates the tissues of marine plants,

thalli of larger algae, shells, and other carbonate substrates, including corals. A number of marine ulvophyceans precipitate calcium carbonate upon surfaces of their thalli; the carbonate contributes to the bottom sediments upon death of the algae. Such carbonate deposition has been significant in terms of global carbon cycling for hundreds of millions of years. Large carbonate deposits in some regions of the world have resulted from the action of ulvophyceans (Fig. 18–3). One group of ulvophyceans has become adapted to the terrestrial environment. Some of these forms live within and upon plant leaves, sometimes causing serious diseases of economic plants in the tropics. Another, *Trentepohlia*, forms extensive and conspicuous red-orange crusts on the surfaces of trees, wood structures, or rocks in humid regions (Fig. 18–4).

Ulvophyceans are intriguing to cell biologists because they include some forms that are coenocytes—macroscopic thalli composed of a single, large multinucleate cell. These forms are described as siphonous or **siphonaceous** (from the Greek word *siphon*, meaning "a tube"). Although multinucleate-celled forms occur in some other algal groups, notably florideophycean red algae (Chapter 16) and charalean greens (Chapter 21), as well as siphonous forms (such as *Vaucheria* and *Botrydium*) among tribophycean ochrophytes (Chapter 15) and the chlorophycean *Protosiphon* (Chapter 20), in no other algal group has the coenocytic, siphonous habit been elaborated to such a high degree as among various marine ulvophyceans. Ulvophyceans also exhibit a wide variety of other morphologies: nonflagellate unicells, branched and unbranched filaments, and tubular and bladelike thalli.

Figure 18–2 A green scum of the ulvophycean *Pithophora* covers the surface of a lake. (Photograph courtesy C. Lembi)

Figure 18–3 Calcified ulvophyceans (*Rhipocephalus*, left, and *Halimeda*, right) growing in a sea grass bed near San Salvador, Bahamas. When they die, such seaweeds contribute to the formation of carbonate sands. (Photograph courtesy Ronald J. Stephenson)

other green algal classes are distinctively different (see Chapter 17). The class Ulvophyceae is mainly defined by a cluster of plesiomorphic (ancestral) characters and by the absence of apomorphies (derived features) found in other groups. Some distinctive features of the flagellar basal bodies—such as non-striated connecting fibers and terminal caps—may prove to be apomorphic for the class Ulvophyceae (Sluiman, 1989; Floyd and O'Kelly, 1990).

Ulvophyceans are regarded as the earliest divergent (multicellular) group of the Ulvophyceae-Trebouxiophyceae-Chlorophyceae (UTC) clade (see Fig. 17–7). Evidence for this hypothesis includes the fact that flagellate reproductive cells of some ulvophyceans are covered by a layer of small diamond-shaped scales (Sluiman, 1989; Floyd and O'Kelly, 1990). Similar scales characterize many prasinophyceans (Sym and Pienaar, 1993), but no trebouxiophyceans and only a few chlorophyceans produce scale-coated cells. Close linkage to prasinophyceans is also suggested by the presence in many ulvophyceans of the xanthophyll accessory pigments siphonein or siphonoxanthin (Fawley and Lee, 1990). These are also present in some prasinophyceans (Sym and Pienaar, 1993), but not trebouxiophyceans or chlorophyceans. Additionally, the ulvophyceans and prasinophyceans are primarily marine groups, while trebouxiophyceans and

Evolution and Fossil History

Comparative analysis of reproductive cell ultrastructure and fine structural attributes of mitosis and cytokinesis suggest that ulvophyceans are a monophyletic, early divergent green algal group. They are thought to have arisen from unicellular, quadriflagellate prasinophycean ancestors (Floyd and O'Kelly, 1990), but precisely which modern prasinophyceans are most closely related to ulvophyceans is as yet unclear. Ulvophyceans are characterized by closed, centric mitosis; persistent telophase spindles and furrowing at cytokinesis; and flagellar apparatuses with cross-shaped (cruciate) X-2-X-2 roots offset in the counterclockwise (CCW) direction (Chapter 17). Telophase nuclei typically exhibit an elongated dumbbell shape (Fig. 18–5), and daughter nuclei tend to remain well separated after mitosis. A phycoplast microtubule system, such as is found in dividing cells of trebouxiophyceans or chlorophyceans, is absent, but a hooplike band of microtubules may be associated with the developing furrow. In contrast, mitosis, cytokinesis, and flagellate reproductive cells of

Figure 18–4 *Trentepohlia* is a red-pigmented, terrestrial ulvophycean that grows on wood, stones, and tree bark in humid areas. (Photograph courtesy G. McBride)

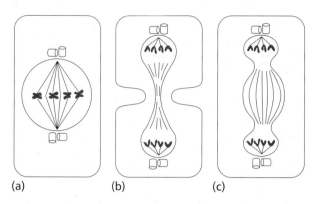

Figure 18–5 Mitosis in ulvophyceans is closed, meaning that the nuclear envelope remains intact. (a) Metaphase, showing polar position of centrioles. (b) Late telophase in uninucleate species, showing persistent spindle and cytokinesis by furrowing. (c) Telophase in multinucleate taxa, showing characteristic dumbbell shape. The absence of a cleavage furrow is a consequence of the uncoupling of mitosis and cytokinesis in multinucleate forms. (After Floyd and O'Kelly, 1990)

chlorophyceans are almost exclusively freshwater or terrestrial forms. Finally, molecular sequence analyses support the concept that ulvophyceans are an early divergent green algal group (Mishler, et al., 1994).

The earliest known fossils attributed to ulvophycean green algae are branched tubes—50–800 μm in diameter and up to 1 mm long—resembling modern *Cladophora* that were recovered from 700–800 million-year-old (Precambrian) rocks (Butterfield, et al., 1988). The next-most ancient remains are 600 million-year-old (early Cambrian) forms such as *Yakutina* that resemble a group of modern ulvophyceans—the Dasycladales (e.g., *Acetabularia* and *Cymopolia*)—in having radial symmetry and calcification. *Yakutina* is also thought to represent the oldest known calcified green alga. About 175 younger fossil dasycladalean genera are also known (Berger and Kaever, 1992). The tendency of dasycladaleans to acquire a coating of calcium carbonate is thought to be responsible for the existence of a nearly continuous record of these algae to the present time. Some prominent Ordovician dasycladaleans were *Rhabdoporella*, *Vermiporella*, and *Cyclocrinites* (Johnson and Sheehan, 1985). Diversity and density peaks occurred in the Permian (250 mya), Mid-Triassic (230 mya), Lower Cretaceous (130 mya), and the Lower Tertiary (55 mya), the peaks correlated with occurrences of high sea levels and extensive, shallow continental seas. The Cretaceous-Tertiary (KT)

extinction event (65 mya)—attributed to the impact of a large meteor or comet—strongly affected coccolithophorid diversity (Chapter 10), but did not much affect dasycladaleans. However, a number of major extinction events have occurred over time; during a Mid-Oligocene (33 mya) period lasting millions of years, dasycladaleans were apparently quite rare. The modern genus *Acetabularia* first appeared about 38 million years ago, survived the Oligocene crisis, and was joined by additional modern dasycladalean taxa only about 10,000 years ago.

The calcareous coat of dasycladalean algae—a kind of limestone cast—is often well preserved in the sediments; many dasycladalean remains do not appear to have been crushed or otherwise seriously damaged by millions of years of compressive burial, upheaval, or other alternations. Cross sections of structurally intact remains of Upper Permian *Mizzia* are quite similar to those of modern *Cymopolia* (Fig. 18–6) (Kirkland and Chapman, 1990). There are extensive carbonate deposits in the European Alps and other regions formed by dasycladalean algae. Although dasycladalean fossils are not widely useful in stratigraphic work (see Chapter 4), because their distribution tends to be provincial and their evolution very slow during some periods, they can be helpful in deducing past environmental conditions. Because most modern forms typically occur in protected shallow tropical waters and tolerance to salinity change tends to be low, fossils of their ancient relatives are thought to reflect the occurrence of normal salinity, warm temperatures, and shallow waters of low turbulence (Berger and Kaever, 1992). However Kirkland and Chapman (1990) warn that salinity tolerance and depth distribution among modern dasycladaleans can be quite broad; thus, great care should be taken when using fossil forms to interpret the paleoecological record.

Giant mounds of *Halimeda* remains—some over 50 m thick—occur in the continental shelf waters associated with the Great Barrier Reef in Australia and other regions, at depths from 12 to 100 meters. Covering hundreds of square meters, the mounds are estimated to grow at rates of nearly six meters per 1000 years. In the Bahamas, Florida Bay, and reef flats or lagoons associated with Pacific coral atolls, remains of *Halimeda* and related algae generate much of the carbonate sediments (sand and mud) upon their death and disintegration (Tucker and Wright, 1990). Fossil remains (see Fig. 2–7) are recognized by surviving calcified segments or plates characteristic of living *Halimeda* (Figs. 17–5, 18–52).

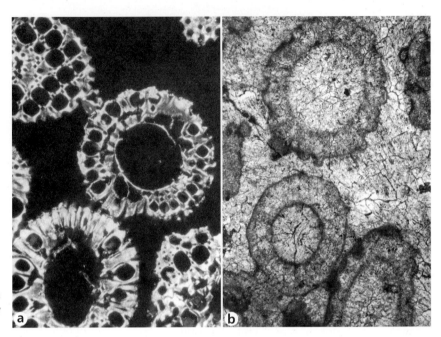

Figure 18–6 (a) Cross section of modern *Cymopolia* compared to (b) cross section of the fossil known as *Mizzia*. The fossils have persisted because the original seaweed was highly calcified, as is *Cymopolia*. (Photographs courtesy B. Kirkland)

Ulvophycean Diversity and Ecology

Approximately 100 genera and 1100 species are included in the Ulvophyceae (Floyd and O'Kelly, 1990). Ultrastructural evidence, together with emerging molecular data, suggest that the major evolutionary events in ulvophycean diversification include transition from: uninucleate to multinucleate cells; unicellular to filamentous, bladelike, or coenocytic thalli; and zygotic life history to alternation of generations (with sporic meiosis) and possibly also gametic life history (Chapter 1). Morphological, ultrastructural, and life-history differences have been used to group ulvophyceans into orders, and past authors have recommended a variety of different ways of forming and naming such groups (Sluiman, 1989; Floyd and O'Kelly, 1990).

Because there is not a substantial body of unique and defining features (autapomorphies) for the class Ulvophyceae, some experts have advocated raising subgroups to class status (Hoek, van den, et al., 1995). While recognizing the validity of this action, we have maintained the class Ulvophyceae because an integrated treatment facilitates comparative analysis of the various ways in which ulvophyceans have adapted to benthic and nearshore marine environments, and also highlights several interesting evolutionary excursions into freshwater and terrestrial habitats. We describe six orders of common ulvophycean algae; other authors include additional groups of less common forms (Sluiman, 1989; Chappell, et al., 1990). Ulotrichales include uni- or multinucleate-celled forms having a relatively simple life history involving zygotic meiosis; Ulvales and Trentepohliales possess uninucleate cells and exhibit alternation of two multicellular generations; Siphonocladales have multinucleate cells, but are not siphonous, and have alternation of generations; and Dasycladales and Caulerpales are siphonous forms (Floyd and O'Kelly, 1990). The relationships among these groups remains unclear, although some preliminary SSU rRNA results have been obtained (Zechman, et al., 1990). This classification is presented as an aid to the student, with the understanding that the phylogenetic validity of these orders needs to be critically tested by molecular systematic analysis. Mapping of cytological, morphological, ultrastructural, and reproductive characters—used to define these groups—onto robust, molecular-based phylogenies will be helpful in tracing evolutionary trends as well as the origin of innovative features that have allowed the radiation of ulvophycean clades.

Ulotrichales

This order includes a morphologically variable group of genera, including nonflagellate unicells (e.g.,

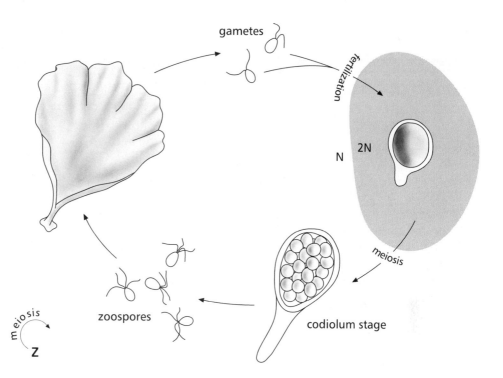

Figure 18–7 Diagram of the life history of ulotrichalean algae, such as *Monostroma*, shown here. Note the occurrence of the codiolum stage, which is a unicellular diploid zygote until just prior to meiospore production. Such a life history is regarded as ancestral within the Ulvophyceae.

Halochlorococcum), unbranched filaments of uninucleate cells such as *Ulothrix* (for which the order is named), branched filaments of uninucleate cells (e.g., *Spongomorpha*), unbranched filaments whose cells are multinucleate (for example, *Urospora*), branched filaments of multinucleate cells (*Acrosiphonia*), and blades composed of a single layer of uninucleate cells (*Monostroma*). Most ulotrichaleans occur along rocky ocean coasts, attached to stable substrates; a few occur in nearshore freshwaters. The multicellular, gamete-producing life-history stage is typically short-lived, occurring only spring or fall in temperate regions.

Common among ulotrichaleans is a distinctive, thick-walled, sac- or club-shaped unicellular life-history stage that is attached to substrates by means of a stalk (Fig. 18–7). This stage can survive a period of dormancy, typically the warm season of the year. Secure attachment is a necessary adaptive response to the turbulence often encountered in nearshore waters. In some cases this stage bores into shells or becomes endophytic within thalli of encrusting seaweeds, thereby occupying highly protected and stable microhabitats. Such reproductive stages are generally assumed, and in some cases have been demonstrated, to be diploid zygotes resulting from the fusion of gametes produced by filamentous or bladelike ulvophyceans. When conditions are favorable, the zygotes germinate, producing flagellate spores that are

assumed to be products of meiosis (and thus are known as meiospores). Following a period of motility, the meiospores settle, attach to a substrate, then grow into filaments or blades. Ulotrichalean zygotes are often described as codiolum stages because they resemble a genus known as *Codiolum*. However, most occurrences of *Codiolum* in nature may actually be the zygotic forms of multicellular ulotrichalean genera, so it is unclear whether or not the genus *Codiolum*

Figure 18–8 The dark green freshwater ulotrichalean *Ulothrix zonata* growing on emergent rocks at the edge of Lake Superior. *Ulothrix* can be found much of the year in this cool-water lake. In smaller northern temperate lakes, growths are common in the spring and sometimes fall, but are not present during the summer.

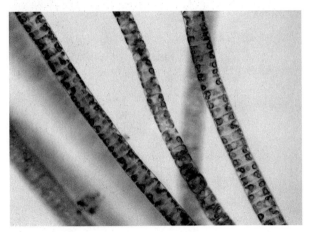

Figure 18–9 *Ulothrix zonata* is an unbranched filament. Light micrograph showing the parietal band-shaped chloroplast with numerous pyrenoids.

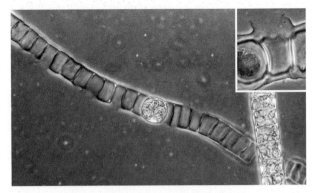

Figure 18–10 *Ulothrix zonata* filament with a single zoosporangium. The remaining cells in the filament have already discharged their flagellate reproductive cells. Inset: a pore through which flagellate reproductive cells were discharged.

should be retained (Scagel, 1966). The algae included here in the Ulotrichales, are grouped into the order Codiolales by some experts (Hoek, van den, et al., 1995). In view of the questionable status of the genus *Codiolum*, Ulotrichales seems a more appropriate name for this group.

ULOTHRIX (Gr. *oulos*, wooly + Gr. *thrix*, hair) can occur in marine or freshwaters, attached to rocks or submerged pieces of wood, and is particularly evident during cold seasons (Fig. 18–8). Cells can divide at any point along the length of the unbranched filaments. There is a single, parietal, band-shaped chloroplast per cell, and several pyrenoids are present (Fig. 18–9). Each cell can generate numerous biflagellate isogametes or quadriflagellate zoospores; these are released via a pore in the cell wall (Fig. 18–10), presumably generated by localized action of wall-dissolving enzymes. Zoospores are covered with a layer of small scales similar to those of prasinophycean flagellates. The zoospores can attach to substrates and produce new filaments, rapidly generating clonal populations. Gametes fuse in pairs to form a negatively phototactic quadriflagellate zygote that settles onto substrates and develops into a thick-walled, stalked resting cell that persists throughout the warm season. Zygote germination—by production of presumed meiospores—is induced by short days and low temperatures. In nutrient-rich lakes or running waters, *Ulothrix zonata* can develop large populations in spring, but disappears as temperatures increase above 10° C when the parental filaments disintegrate as the

result of massive zoospore production (Graham, et al., 1985). *Ulothrix zonata* is believed to have invaded freshwaters from the marine habitat, since most of its close relatives are marine.

SPONGOMORPHA (Gr. *spongos*, sponge + Gr. *morphe*, form) occurs as dense tufts of branched filaments, commonly attached to larger marine seaweeds (Fig. 18–11). It primarily occurs in the spring along temperate coastlines. The uninucleate cells are much longer than they are wide, and branches arise from the apical portions of the cells via emergence of a protru-

Figure 18–11 *Spongomorpha*. (a) Macroscopic view, showing tuft of filaments. (b) Microscopic view of an apical cell of a filament. (Photographs by C. O'Kelly)

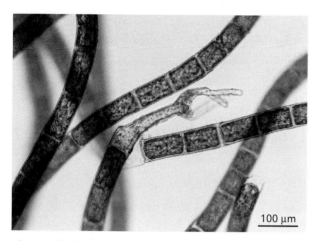

Figure 18–12 *Urospora* is an unbranched filament with multinucleate cells. A basal rhizoidal system is evident.

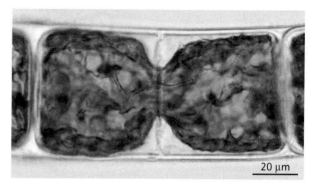

Figure 18–13 Cytokinesis in *Urospora*. The reticulate plastid is being pinched by the developing cleavage furrow.

sion that is later cut off by a cross-wall. Branches often appear curled or hooked. Growth is thought to be restricted to apical cells. Sexual reproduction occurs by means of biflagellate isogametes that are liberated from cells through a lidded pore similar to those generated by gametangia (gamete-producing cells) of many other ulvophyceans.

UROSPORA (Gr. *oura*, tail + Gr. *spora*, spore) is an unbranched filament that can attain lengths of several centimeters (Fig. 18–12). It attaches to substrates on rocky shores of cold marine waters with a holdfast consisting of basal, rhizoidlike cells. The cells are large and multinucleate. Nuclei in this and other algae can be visualized by either staining the DNA pink via the Feulgen reaction or rendering nuclei fluorescent by treatment with the DNA-specific fluorochrome DAPI (see Text Box 16–1). Mitosis in *Urospora* is uncoupled from cytokinesis. Prior to mitosis, nuclei congregate at the plane of cell division, then divide synchronously. Mitosis is closed, and cleavage is by ingrowth of a furrow, as in other ulvophyceans (Fig. 18–13). The plastid is cylindrical, parietal, and highly dissected in mature cells; there are many pyrenoids. Sexual reproduction is by anisogametes, but otherwise resembles that of *Ulothrix* and *Spongomorpha*. The flagella and flagellar basal bodies of the quadriflagellate zoospores of *U. penicilliformis* are very unusual in a number of respects; these are summarized by Sluiman (1989).

ACROSIPHONIA (Gr. *akron*, apex + Gr. *siphon*, tube) possesses a branched thallus composed of multinucleate cells (Fig. 18–14). Mitosis and cytokinesis

have been studied by immunofluorescence methods (Aruga, et al., 1996). Numerous nuclei and cortical microtubules occur in tip cells. Prior to cell division, 30–40% of these nuclei migrate downward to the region where cytokinesis will take place (Fig. 18–15), forming a nuclear ring. In this region cortical microtubules reorient, forming a transverse band, as many mitotic spindles form. Following mitosis, nuclei migrate back into the apical region, reestablishing the parallel, longitudinal arrangement of microtubules. Cytokinesis occurs by furrowing, and the transverse microtubules occupy the leading edge of the furrow. Branch development is similar to that of *Spongomorpha*. Sexual reproduction is similar to that of other ulotrichaleans. *Acrosiphonia* grows in the lower inter-

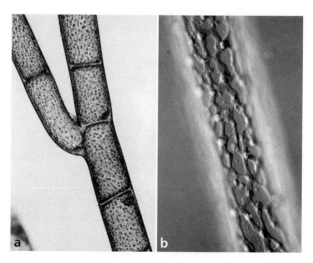

Figure 18–14 *Acrosiphonia* is a branched filament with multinucleate cells. A branch is shown in (a) and the reticulate plastid in (b). (Photographs by C. O'Kelly)

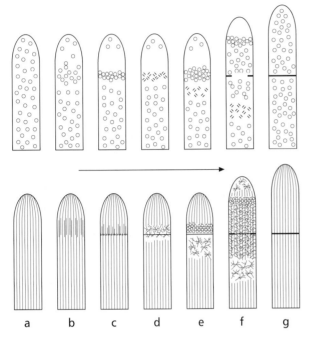

Figure 18–15 Diagrams showing behavior of nuclei (top row) and microtubules (bottom row) during cell division of *Acrosiphonia*. (a) Vegetative cells at filament tips prior to cytokinesis. (b, c) Microtubules form a band at the future site of cytokinesis and nuclei migrate to this region. (d, e, f) Stages in mitosis and onset of the formation of a septum (cross-wall). (g) Vegetative cell at the tip of a filament after septum formation has been completed. (After Aruga, et al., 1996 ©Blackwell Science Asia)

tidal zone along rocky marine coasts of temperate and polar regions; these habitats are common to most ulotrichaleans.

MONOSTROMA (Gr. *monos*, one + Gr. *stroma*, layer) is appropriately named, since it consists of a single layer of cells arranged in a sac or blade (Fig. 18–16). *Monostroma* thalli are attached to substrates via a disklike holdfast. They arise by upward protrusion of cells from an initially flat, discoid germling. Cell proliferation results in a hollow tube of cells, which in some forms persists into the adult phase. In other forms tubes open into flat blades superficially resembling *Ulva* thalli (Fig. 18–17). Normal morphological development in culture requires the presence of appropriate bacteria or medium in which bacteria-free brown or red seaweeds have been grown. This suggests that morphogenetic substances are released from bacteria and algae that could influ-

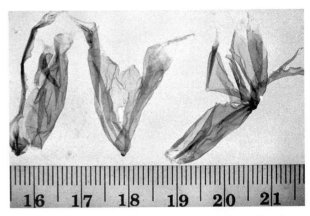

Figure 18–16 *Monostroma* occurs as a blade that is one layer thick. (Photograph by C. O'Kelly)

ence *Monostroma* development in nature (Tatewaki, et al., 1983).

Cells contain a single, parietal cup-shaped plastid with a pyrenoid. Sexual reproduction resembles that of other ulvophyceans. Anisogametes are covered with tiny prasinophycean-like scales (as is also the case for some other ulvophyceans), and zygotes are unicellular and stalked (*Codiolum* like). Zygote germination is induced by short-day and low-temperature conditions. Species possessing a life history involving alternation of generations have been transferred to the genus *Ulvaria*, in the Ulvales. *Monostroma* can be easily confused with *Ulvaria* and *Ulva*, and young stages are difficult to distinguish from the trebouxiophycean *Prasiola* (Chapter 19). *Monostroma* can be distinguished from *Prasiola* on the basis of chloroplast morphology (*Prasiola*'s plastids are axial); *Ulva* has a two-cell-layered blade; and *Ulvaria* has basal rhizoids, rather than a disk holdfast. *Monostroma* is surprisingly tolerant of variation in salinity, occurring in brackish and freshwaters (Smith, 1950), as well as marine waters of normal salinity.

Figure 18–17 *Monostroma* develops from proliferation of small filamentous germlings into hollow tubes that eventually open out into the typical blade form. (After Tatewaki, 1972)

Ulvales

Named for the common sea lettuce *Ulva*, this order includes morphologically variable forms that have uninucleate cells, a single platelike parietal chloroplast with one to several pyrenoids, and a life history involving alternation of isomorphic generations. Though a few forms may exhibit only asexual reproduction, the typical life history involves two outwardly identical, independent, multicellular generations: One produces biflagellate isogametes or anisogametes, and is thus known as the gametophyte; another produces quadriflagellate meiospores, and is thus known as the sporophyte (see Fig. 1–24). Meiosis occurs in the first division of zoospore-producing cells, known as sporangia. Sporangia can generate multiple zoospores by partitioning the cytoplasm into many small uninucleate portions. Gametangia (cells that produce gametes) typically give rise to many gametes by processes similar to zoosporogenesis. Zoospores and gametes both lack cell walls and are described as naked. Scaly coverings such as those found on flagellate cells of some ulotrichaleans have not been observed among ulvaleans. Unmated gametes are typically capable of functioning as asexual reproductive cells, attaching to substrates and growing into new multicellular gametophytes. Sporophytes contain twice the nuclear DNA level (2C) and chromosome number (2N) per cell as gametophytes (1C or 1N). Thus sporophytes are commonly described as diploid, whereas gametophytes are said to be haploid. However, many cases of polyploidy are known among algae, and such polyploids may exhibit higher DNA levels and chromosome numbers than those typically regarded as haploid or diploid for the species. For example, in a polyploid species the chromosome number of the gametophyte's cells might be 2N; in such a case, a 4N chromosome number would be predicted for sporophytic cells.

Demonstration that a change in DNA level (i.e., alternation of generations) has occurred is accomplished by making comparative chromosome counts or by spectrophotometrically measuring the amount of DNA in nuclei stained by the Feulgen procedure or DAPI (Fig. 18–18 and Text Box 16–1). These methods are useful in determining the timing and location of meiosis; this is critical in defining and comparing different kinds of life histories.

Ulvophycean alternation of generations is described as isomorphic because the two alternating generations are outwardly similar in appearance. In contrast, a variety of other seaweeds exhibit heteromorphic alternation of generations—a situation in which the alternating generations are distinguishable and often dramatically different in appearance. Such structurally distinct forms are thought to confer selective advantage in seasonally variable habitats (Chapters 15, 16, and 23). In the case of isomorphic generations, it is somewhat harder to make this argument; often there is no obvious correlation between generational change and environmental variation. Rather, a hypothesis of functional diversification within more or less stable environments may be appropriate. Diploid sporophytes, if at least partially heterozygous, may be to some degree buffered by heterozygosity from environmental selection. Sporophytes may thus be considered as repositories of a population's genetic variation. Sporophytes have the capacity through the process of meiosis to further increase genetic diversity by generating new allelic combinations by crossing-over events, and new chromosomal combinations via independent assortment. The fact that sporophytes are multicellular means that they are capable of producing many genetically variable haploid meiospores (and gametophytes that develop from them) upon which environmental selection can act directly. Favorable genotypes persist and grow into new ulvalean gametophytes, whereas unfavorable ones die at the meiospore or germling stages. Because of their greater size and energy-harvesting capacity, ulvalean sporophytes can generate much greater numbers of meiospores than can the single-

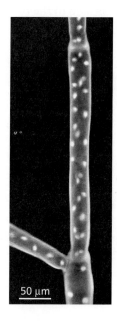

Figure 18–18 *Cladophora* nuclei stained by using the DNA-specific fluorochrome DAPI.

celled zygotes of ulotrichalean forms. Hence the multicellular sporophytes of ulvaleans confer both populational genetic diversity and increased colonization potential. Well-adapted gametophytes then perpetuate the flow of favorable alleles into the next generation. Multicellularity of gametophytes confers the ability to generate large amounts of photosynthate, and thus the capacity to produce larger numbers of gametes—and hence zygotes (potential sporophytes)—than can unicellular haploids. Alternation of multicellular generations is probably one component of a suite of characteristics favoring the widespread ecological success of ulvaleans. Their life history may also partially explain their potential to form nuisance growths.

Ulvophyceans such as *Ulva* and *Enteromorpha* are common inhabitants of coastal rocky shores around the world, and are attracting attention as fast-growing opportunists that can dominate coastal regions influenced by nutrient-rich effluents. These seaweeds are rapid colonizers of bare substrates such as jetty walls and the surfaces of seagoing vessels, where they also facilitate colonization by other seaweeds (Poole and Raven, 1997). Large growths may also occur in salt marshes. Tolerance to changes in salinity, the occurrence of high- and low temperature–adapted forms, ability to utilize bicarbonate as an inorganic carbon source, very high rates of light-saturated photosynthesis, high reproductive rates, and the ability of flagellate cells to adhere rapidly to substrates are all factors in their success. Rotting masses of these algae produce noxious H_2S, and there is some concern that continued increase in ulvalean biomass may add to the production of ozone-destroying halocarbons by marine algae (Poole and Raven, 1997).

PERCURSARIA (L. *percurso*, to run through) may be found attached to stones or other substrates in intertidal marine habitats, but is typically much less conspicuous than some other ulvophyceans (Fig. 18–19). Zoospores may grow into a flat (prostrate) adherent disk from which emerge several biseriate (two-cell-thick) ribbons, or zoospores may grow directly into biseriate ribbons that attach to substrates via basal rhizoids. Similar stages occur in the development of some other ulvalean algae. *Ulvella* forms prostrate, radially organized disks similar to those of *Percusaria* germlings, and *Blidingia* also has a prostrate disk, from which develop erect tubular thalli.

ULVARIA (From *Ulva*) occurs as small sacs or leaflike blades that can grow to 30 cm in length and

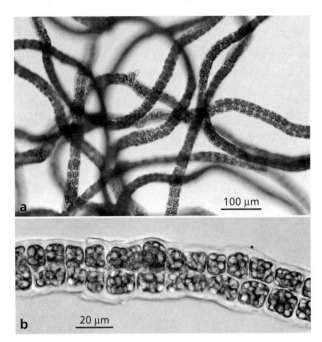

Figure 18–19 *Percursaria* occurs in the form of biseriate filaments, which are shown at two magnifications in (a) and (b).

thus resemble *Ulva* (Fig. 18–20). Thalli are attached by rhizoids in rocky marine intertidal environments similar to those inhabited by *Ulva*. Unlike *Ulva*, however, the blades are monostromatic (as are those of the ulotrichalean *Monostroma*). Development from zoospores proceeds through successive uniseriate filamentous stages, pluriseriate stages, and tubular

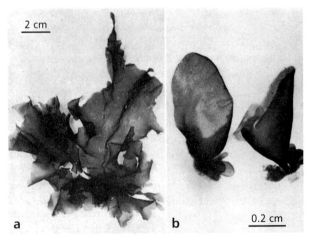

Figure 18–20 *Ulvaria* occurs as a blade (a) or sac (b) growing from an adherent disk. (From Dube, 1967 by permission of the *Journal of Phycology*)

Figure 18–21 The tubular thalli of *Enteromorpha* often occur in crowded populations attached to rocky substrates in tide pools and nearshore marine habitats.

stages, then tubes open to form blades (similarly to *Monostroma*—Fig. 18–17). Though some species appear to be entirely asexual, others are reported to exhibit isomorphic alternation of generations characteristic of ulvophyceans (Dube, 1967). Ultrastructure of biflagellate reproductive cells was studied by Hoops, et al. (1982).

ENTEROMORPHA (Gr. *enteron*, intestine + Gr. *morphe*, form) (Fig. 18–21) thalli are elongated tubes, sometimes with constrictions, that attach to substrates via rhizoidal branches that may form an attachment disk. Tubular thalli arise from phototactic zoospores, zygotes, or unmated gametes that attach to substrates (Callow, et al., 1997), then first grow into uniseriate filaments. These then form a two-cell-layered ribbon that separates in the center to generate the mature tubular thallus, similar to early stages in the development of *Monostroma* (Fig. 18–17). The tubes are only one cell thick, and may bifurcate, giving a branched appearance. The tubular construction may aid in buoyancy if gas bubbles are trapped in the central area. As in other ulvophyceans, there is a single, parietal, platelike chloroplast per cell, but interestingly, plastids appear to rotate in response to changes in the direction of light. Although zoospores are generated primarily from the distal ends of sporophytes—leaving much of the thallus intact—gametogenesis occurs throughout the thallus of gametophytes, resulting in thallus disintegration. Maximum liberation of gametes occurs a few days prior to the highest tide in a lunar cycle. There is a worldwide distribution, with some species tolerating waters as warm as 30° C, and oth-

ers occurring in Antarctic waters where the maximum temperature is 1.8° C. In addition to growing well at normal salinity, *Enteromorpha* may also occur in hypersaline lakes (at 52% salinity) or tide pools that have experienced evaporation, as well as brackish and freshwaters (Poole and Raven, 1997). Although some species appear to reproduce only asexually, others exhibit a typical alternation of isomorphic generations (Kapraun, 1970). Two of the most common species, *E. intestinalis* and *E. compressa*, are sometimes difficult to distinguish on the basis of morphology (degree of branching), but are readily distinguishable using ITS sequences and RFLPs (Blomster, et al., 1998).

ULVA (ancient Latin name for sedge) (Fig. 18–22) forms conspicuous flat blades that are composed of two cell layers. Blades can be as long as one meter. *Ulva* is attached to substrates in marine coastal waters by means of rhizoidal branches, or occurs in free-floating masses. The distromatic blade arises from zoospores via uniseriate, pluriseriate, and tubular stages of development, followed by collapse of the tube. If grown axenically (in the absence of bacteria and other organisms) in the laboratory, development is disrupted, and cushions of branched, uniseriate filaments are produced (Provasoli, 1958). However, addition of bacteria isolated from natural *Ulva* collections will restore normal development and morphology. Similar bacterial effects upon morphogenesis of *Enteromorpha* and *Monostroma* have been observed (Nakanishi, et al., 1996). Reproduction is

Figure 18–22 *Ulva* is a blade composed of two cell layers.

exclusively asexual in some forms, but alternation of isomorphic generations is expressed by others (see Fig. 1–24). Both zoospores and gametes are mainly produced in cells at the edge of the thallus. This allows reproduction to occur without causing complete disintegration of the parental thallus, thereby contributing to *Ulva*'s persistence in the environment. Sixteen gametes and sixteen zoospores are produced per parental cell of *Ulva mutabilis*, and release of flagellate cells is controlled by extracellular inhibitors (Stratmann, et al., 1996). The carbon concentrating mechanisms of three species of *Ulva* were studied by Björk, et al. (1993).

Trentepohliales

This group is named for the genus *Trentepohlia*, but is also known as the Chroolepidaceae. It consists of a few genera and about 40 species, all of which appear to be terrestrial. The algae occur as branched filaments (Fig. 18–23) on inorganic substrates such as rocks or within the leaves or upon surfaces of vascular plants, sometimes acting as economically significant parasites (Chapter 3). Some occur as phycobionts in lichens. Trentepohlialeans are often orange- or red-pigmented because of the presence in cells of lipid droplets containing β-carotenes (Fig. 18–24). Such pigments have been presumed to play a photoprotective role, but

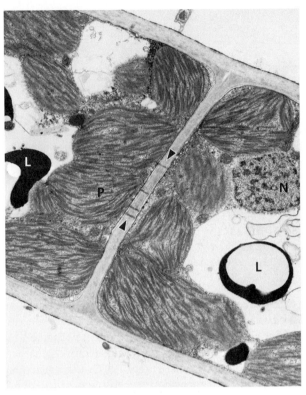

Figure 18–24 *Trentepohlia* cells are interconnected by plasmodesmata (arrowheads), as seen in this TEM view. Also shown are the many discoid plastids (P), the single nucleus (N) per cell, and carotene-containing lipid droplets (L). (From *Origin of Land Plants*. L. E. Graham. ©1993 Reprinted by permission of John Wiley & Sons, Inc.)

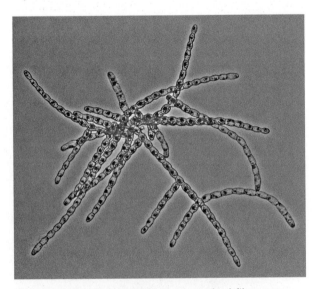

Figure 18–23 *Trentepohlia* has branched filaments composed of uninucleate cells that often have conspicuous red droplets composed of lipids with dissolved carotenoids.

experiments have revealed that nitrogen-limitation is the major stimulus for their production (Czygan and Kalb, 1966). Trentepohlialeans are thought to have arisen from marine ancestors, thus stimulating questions about the cellular adaptations needed to make such a dramatic ecological transition. Was there an intervening phase of life in freshwaters that served as a springboard to land, or is it possible that colonization of land can occur directly from marine habitats? Although a few other ulvophyceans (such as *Ulothrix zonata*) have successfully invaded freshwaters from marine habitats, the terrestrial radiation of Trentepohliales is unparalleled within the class, and is unknown from any other seaweed group. No multicellular brown or red seaweeds have colonized terrestrial habitats.

Trentepohlialeans are placed within the Ulvophyceae on the basis of molecular sequence evidence (Chapman, et al., 1995). Trentepohlialean cells, like

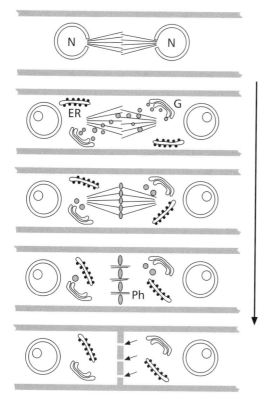

tive pressures and adaptive advantages associated with the evolution of phragmoplasts.

Production of flagellate reproductive cells—biflagellate gametes or quadriflagellate zoospores—occurs within specialized, urn-shaped gametangia or zoosporangia (Fig. 18–26). Flagellate cells are released by dissolution of a plug of material in a preformed pore (Fig. 18–27). Unmated gametes are able to grow into new thalli, as is the case with some other ulvophyceans. Flagellate cells are unusual among ulvophyceans in lacking eyespots and in possessing distinctive, layered structures at flagellar bases; the latter superficially resemble MLSs of charophyceans and land plants (Chapter 17), but detailed examination reveals some structural differences (Graham, 1984). Trentepohlialean flagella are unusual in possessing flattened, lateral "wings" or "keels" whose function is not understood. Zoosporangia may break off and be dispersed by wind, releasing zoospores when moist substrates are available. Thompson (1958) reported that at least one genus exhibits het-

Figure 18–25 Cytokinesis in *Cephaleuros virescens* involves development (top to bottom) of a system of transverse microtubules that is analogous to the phragmoplasts of some charophycean algae and land plants. Plasmodesmata are indicated by arrows. ER=endoplasmic reticulum, G=Golgi body, N=nucleus, Ph=phragmoplast. (After Chapman and Henk, 1986 by permission of the *Journal of Phycology*)

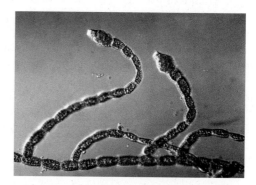

Figure 18–26 *Trentepohlia* produces specialized urnlike gametangia at the ends of branches.

Figure 18–27 *Trentepohlia* gametangium viewed by SEM, showing specialized pore through which reproductive cells are discharged.

those of ulvaleans and some ulotrichaleans, are uninucleate, but there are numerous discoid, pyrenoidless plastids per cell (as occurs in some other ulvophycean algae). Vegetative cytokinesis is quite unusual for ulvophyceans in that a phragmoplastlike array of microtubules (see Chapter 17), cell plate (Fig. 18–25), and plasmodesmata (Fig. 18–24) are present. These have been documented for *Cephaleuros* by Chapman and Henk (1986), who noted that the plasmodesmata lack internal structures such as desmotubules, which characterize the intercellular connections of land plants. Phragmoplasts are otherwise known only from certain advanced charophycean algae and their relatives, the embryophytes (land plants) (Chapters 17 and 21). It thus appears that a phragmoplastlike cytokinetic system has arisen independently at least twice within green algae; this generates questions about the selec-

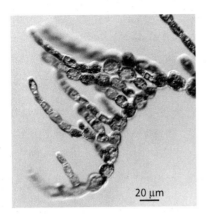

Figure 18–28 *Trentepohlia* occurs as a heterotrichous filament having a prostrate basal system of filaments from which emerge numerous erect filaments.

eromorphic alternation of generations, the sporophyte being smaller than the gametophyte generation, but the life histories of the organisms in this group require additional investigation.

Monographed by Printz (1939), the order as circumscribed here includes *Trentepohlia*, *Cephaleuros*, *Phycopeltis*, and *Stomatochroon*. The morphology, taxonomy, and ecology of the latter three taxa were summarized by Thompson and Wujek (1997). A few other algae have been suggested as members of the group, but are not well studied.

TRENTEPOHLIA (named for Johann Friedrich Trentepohl, a German clerygyman and botanist) (Fig. 18–28) is a branched, often red-orange filament hav-

ing both a prostrate portion that is adherent to the substrate and erect branches, conferring a fuzzy appearance to thalli. Reproductive structures typically occur at the ends of erect branches.

CEPHALEUROS (Gr. *kephale*, head + Gr. *euros*, breadth) (Fig. 18–29) resembles certain fungal growths, in that thalli are flattened, with branched filaments, radial in organization, with tip growth, and produce specialized reproductive structures—groups of sporangia. *Cephaleuros* is regarded as a parasite, growing beneath the cuticles of tropical and subtropical plants such as *Magnolia* (see Fig. 3–26), *Camellia*, *Citrus*, and *Rhododendron*. Erect branches bearing sporangia emerge to the surface where zoospores can be released if moisture is available, or the sporangia abscise and are wind-dispersed. Gametes are produced on the prostrate filament system; their ultrastructure was studied by Chapman (1980). The lichen *Strigula* contains *Cephaleuros* as a phycobiont (Chapman, 1976).

PHYCOPELTIS (Gr. *phykos*, seaweed + L. *pelte*, small shield) (Fig. 18–30) is an epiphyte on numerous vascular plants and some bryophytes throughout tropical and subtropical regions, but also occurs in some temperate areas (Printz, 1939; Good and Chapman, 1978). The latter authors suggested that meiosis was probably zygotic. Thalli are radially symmetrical, with growth occurring at the margins, and adjacent filaments are closely adherent. The cell walls are impregnated with highly resistant materials—identified on the basis of acid hydrolysis tests and infrared absorp-

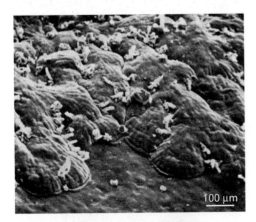

Figure 18–29 *Cephaleuros* forms radial, pseudoparenchymatous disks that produce reproductive cells similar to those of *Trentepohlia*. (From Chapman, 1976)

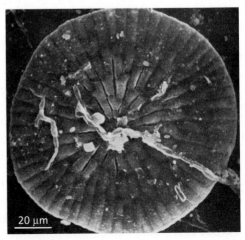

Figure 18–30 *Phycopeltis*. (From Good and Chapman, 1978)

tion spectra as sporopollenin. This polymer is believed to play a role in filament adherence and perhaps also in pathogen and desiccation resistance (Good and Chapman, 1978).

Siphonocladales

Also widely known as the Cladophorales, this group includes about 30 genera and 425 species (Floyd and O'Kelly, 1990), some of which occur as branched or unbranched filaments, whereas others form pseudo-parenchymatous blades, nets, or spherical vesicles. Vegetative cells are always multinucleate, but thalli of some species are consistently subdivided by cross-walls (septa) or undergo a form of compartmentation known as **segregative cell division**, wherein units of cytoplasm simultaneously become surrounded by cell-wall material; in other words, they are not persistently siphonous.

Wound healing in *Siphonocladus*, *Valonia*, and several other related genera that undergo segregative cell division, has been studied intensively (La Claire, 1982a, c; O'Neil and La Claire, 1984; La Claire, 1991; Goddard and La Claire, 1991). The cell contents retract from the wound site and may fragment, depending on the species. Retraction and fragmentation involve the action of actin, the associated protein calmodulin, and myosin. Each unit of the subdivided cytoplasm is viable and can grow into a new thallus. It is hypothesized that the process of segregative cell division may have originated as a wound-healing adaptation.

Some authorities advocate placing forms with septate filamentous construction into a different order (Cladophorales) than those having segregative cell division (Siphonocladales) (Bold and Wynne, 1985). However, similarities in cell-wall composition (cellulose), pyrenoid structure (bilenticular—composed of two halves), life history (alternation of isomorphic generations), and molecular evidence (Zechman, et al., 1990; Bakker, et al., 1995), support aggregating them into a single order. In the Siphonocladales, as in the multinucleate ulotrichaleans described earlier, mitosis has been uncoupled from cytokinesis. Septum (cross-wall) formation occurs by furrowing, and there are no plasmodesmata. The ultrastructure of siphono-cladalean cell membrane cellulose synthesizing complexes was studied by Mizuta and Okuda (1987). Linear plasmids have been found in several members of this group (La Claire, et al., 1997).

Siphonocladalean algae possess either highly dissected, reticulate chloroplasts or many small discoid plastids arrayed in a reticulate network. Chloroplasts usually contain pyrenoids. Siphonoxanthin occurs in some deep water forms. Genera either possess both marine and freshwater species, are exclusively freshwater, or occur only in marine waters. *Cladophora* and *Pithophora* (and to a lesser extent, *Rhizoclonium*) can form nuisance growths in freshwaters. *Cladophora* also forms conspicuous growths in tide pools and other nearshore marine habitats and *Dictyosphaeria* generates smothering growths in perturbed coral reefs and freshwaters. Other genera are much less conspicuous.

CLADOPHORA (Gr. *klados*, branch + Gr. *phoras*, bearing), which grows in both freshwater and marine habitats, is a highly branched filament of multinucleate cells (see Fig. 18–18). The chloroplast is highly reticulate and may occur as more than one piece. Many pyrenoids are present. Cells are longer than they are wide, and branches originate as cytoplasmic protrusion from the cell apex (Fig. 18–31). Branches eventually are segregated from the main axis by formation of a cross-wall (septum). Molecular phylogenetic assessment of nuclear rDNA ITS regions indicate that the genus is paraphyletic (Chapter 5), but it is viewed as impractical to undertake compensatory taxonomic alterations (Bakker, et al., 1995). *Cladophora* thalli, which can reach several meters in

Figure 18–31 Branching pattern of *Cladophora glomerata*. Note that branches (arrowheads) originate by elongation of cellular protrusions from the distal regions of the large elongate cells of the main axis.

Figure 18–32 Dense masses of *Cladophora glomerata* growing on rocks at the edge of Lake Mendota, Wisconsin, U.S.A.

length, are attached to rocks or other substrates by rhizoidal cells (Fig. 18–32). Several marine species (Bold and Wynne, 1985) and at least one freshwater species (*C. surera*) (Parodi and Cáceres, 1995) have been demonstrated to have isomorphic alternation of generations. The sporangia (of sporophytes) and gametangia (of gametophytes) look much like vegetative cells. Multiple biflagellate gametes are produced per gametangium, and numerous quadriflagellate zoospores are produced in each sporangium. Reproductive cells are primarily produced in cells of the branches, rather than in cells of the main axis; this preserves the integrity of the main axis after reproductive cell discharge has occurred (Fig. 18–33). *Cladophora* is also known to reproduce asexually via thick-walled basal cells described as akinetes; these allow survival through conditions incompatible with growth.

In marine communities that receive terrestrial nutrient inputs such as sewage effluents or agricultural runoff, large mats of *Cladophora* may form. During decomposition of such mats, anoxia may result, smothering nearshore invertebrates, including economically valuable forms. Experiments showed that growth of *C. prolifera* is controlled primarily by irradiance and nutrient availability, strongly implicating anthropomorphic sources as the agents of increased nuisance algal growth (Bach and Josselyn, 1979). *Cladophora* occurs widely in temperate and tropical seas, but is absent from polar waters.

Freshwater *Cladophora* occupies many types of aquatic habitats, from pristine streams to eutrophic lakes and estuaries, and is globally widespread. *Cladophora* has attracted the attention of freshwater biologists because it can attain high biomass that con-

tributes to the structure of benthic stream communities, and because it can generate nuisance-level populations in freshwater lakes having both rocky shorelines and high phosphate levels. Freshwater *Cladophora* occurs in a variety of morphological forms that vary in cell dimensions, branching patterns, and akinete size. Transplantation experiments have revealed that branching patterns can be influenced by environmental conditions, specifically water velocity (Bergey, et al., 1995). In order to clarify the systematics of freshwater *Cladophora*, ITS regions from the nuclear ribosomal cistron were sequenced from *Cladophora* obtained from a variety of freshwater habitats and culture collections. Results revealed that there are very few and possibly only one freshwater *Cladophora* species (Marks and Cummings, 1996). An analysis of ball-forming *Cladophora* in Japan revealed a close similarity to the ball-forming *C. aegagropila* in Europe (Niiyama, 1989).

Freshwater species of *Cladophora glomerata* commonly reproduce asexually by biflagellate zoospores (which may represent unmated gametes capable of growth, as occurs in many other ulvophyceans). Photoperiod is the primary factor involved in induction of

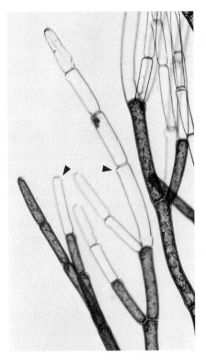

Figure 18–33 Empty terminal and subtending cells of *Cladophora glomerata*, from which reproductive cells have been released via lateral pores (arrowheads).

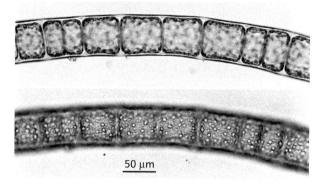

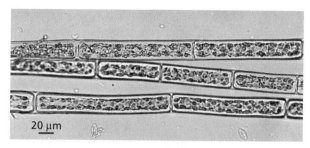

Figure 18–35 *Rhizoclonium* filaments have a coarse, wiry appearance and are sparsely branched, if at all.

Figure 18–34 Through-focus pair of *Chaetomorpha*, which consists of unbranched filaments of multinucleate cells with a highly reticulate plastid.

zoosporogenesis in freshwater *C. glomerata*; asexual reproduction is maximal in short-day conditions. Growth of freshwater *C. glomerata* is primarily influenced by temperature, although daylength, irradiance, and B-vitamin availability are also significant factors (Hoffmann and Graham, 1984; Hoffmann, 1990). Silicon occurs in the cell walls of *Cladophora glomerata* and is a required nutrient (Moore and Traquair, 1976). Other aspects of the ecology of freshwater *Cladophora*, including community associations, are described in Chapter 23. Dodds and Gudder (1992) have reviewed the ecology of *Cladophora*.

CHAETOMORPHA (Gr. *chaite*, long hair + Gr. *morphe*, form) is a marine genus, with an unbranched filament, distributed worldwide, that grows either attached to rocks or shells by holdfast cells or is free-floating (Fig. 18–34). The reticulate chloroplast has many pyrenoids, and there are multiple nuclei per cell, as is typical for the group. Isomorphic alternation of generations has been demonstrated for some species; typical quadriflagellate zoospores (meiospores) and biflagellate gametes are produced (Köhler, 1956). As appears common among ulvophyceans, unmated gametes are able to regenerate the gametophytic phase of growth, producing new haploid filaments.

RHIZOCLONIUM (Gr. *rhiza*, root + Gr. *klonion*, small twig) is an unbranched or very sparsely branched filament (Fig. 18–35) that is sufficiently coarse and wiry that individual filaments are readily visible with the unaided eye. Like *Cladophora*, it often serves as a substrate for attachment of dense coatings of periphytic cyanobacteria and diatoms. *Rhizoclonium* grows in freshwaters, brackish habitats, and

marine waters, either attached via lobed holdfast cells, or free-floating. Cells are considerably longer than broad and contain a reticulate plastid with many pyrenoids and many nuclei. Isomorphic alternation of generations has been demonstrated in some marine species (Bliding, 1957). Freshwater forms occur in hard-water lakes or streams, entangled with other vegetation, and reproduce mainly by fragmentation or the rare production of biflagellate cells.

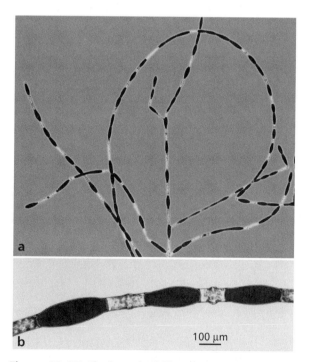

Figure 18–36 The branched filaments of *Pithophora*, seen here at two magnifications in (a) and (b), characteristically exhibit alternation between green vegetative cells and dark green or black akinetes, which serve as resting cells and as a means of asexual reproduction.

PITHOPHORA (Gr. *pithos*, earthen wine-jar + Gr. *phoras*, bearing) is a robust, branched-filamentous form that occurs only in freshwaters. Cells are multinucleate and possess a reticulate plastid with pyrenoids. The genus is common in ponds and is found year-round in warm locales and during summer in temperate regions. There are characteristic large, dark akinetes interspersed with lighter green vegetative cells. Akinetes form by the swelling of cell apices and subsequent formation of a septum (cross-wall) that separates the akinete from the subtending cell. Because of this developmental pattern, akinetes alternate with vegetative cells in both the main axis and branches (Fig. 18–36). The akinete cell walls are resistant to degradation and protect the cellular contents during periods of environmental stress. Akinetes germinate when growth conditions improve; germination was studied in culture by O'Neal and Lembi (1983). Because benthic sediments may contain very large numbers of akinetes, large nuisance-level populations may suddenly appear. These are undesirable because they form floating mats that clog boat motors and otherwise interfere with recreational use of inland waters (Lembi, et al., 1980).

BASICLADIA (Gr. *basis*, base + Gr. *klados*, branch) (Fig. 18–37) superficially appears to be an unbranched filament, but a few branches are produced at the base, just above the rhizoidal holdfast

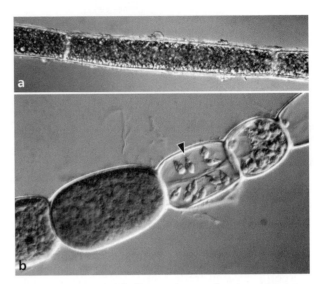

Figure 18–37 *Basicladia* may have a few branches at its base but is primarily unbranched (a). Some of the cells in (b) have formed zoospores (arrowhead). (b: Photography by D. Chappell)

cells. This genus is often found attached to the backs of snapping turtles. Because the alga will attach to inorganic substrates when grown in the laboratory, the basis for its preferred substrate in nature is unknown. The multinucleate cells are long (approximately 50 times their width) and cylindrical, with thick lamellate walls, and plastids are parietal reticulate networks.

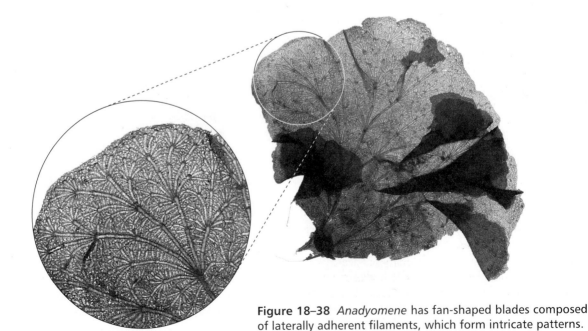

Figure 18–38 *Anadyomene* has fan-shaped blades composed of laterally adherent filaments, which form intricate patterns.

Figure 18–39 *Siphonocladus.*
(Courtesy W. Woelkerling)

Figure 18–40 *Dictyosphaeria* occurs as hollow, gray-green spheres that often break open. (a) Underwater view of numerous thalli growing on coral reef. (b) Close-up view. (Photographs courtesy S. Larned)

Sexual reproduction is reported to occur by production of isogametes in unspecialized cells at the apex, and asexual reproduction by zoospores is also known (Prescott, 1951). Like *Pithophora*, *Basicladia* is apparently restricted to freshwaters.

ANADYOMENE (Gr. mythology, name of the goddess Aphrodite when represented as arising from the sea) is one of the most beautifully formed seaweeds. The one-cell-thick fan-shaped blades are composed of adherent branched filaments wherein the cells become progressively smaller toward the thallus periphery (Fig. 18–38). Small cells also fill spaces between the large-celled filaments (known as veins). In some species the blades are perforate, but in others, blades are entire. Colorless rhizoidal filaments at the base attach the blades to substrates. *Anadyomene* can be found attached to rocks on tropical coral reefs and has been dredged from deeper tropical waters. A phylogenetic analysis based on morphological characters suggested that *Anadyomene* is a monophyletic genus (Littler and Littler, 1991).

SIPHONOCLADUS (Gr. *siphon*, tube + Gr. *klados*, branch), a resident of tropical marine waters, begins as a saccate multinucleate cell in which the cytoplasm cleaves internally and simultaneously to form individual walled units (segregative cell division). At first the subunits are spherical, but later each protrudes beyond the confines of the original sac,

forming an elongated branch. This form of branching can be repeated to form several orders of branches. The resulting thalli appear as a central axis that is densely tufted (Fig. 18–39). Thalli are attached to substrates in tropical waters.

DICTYOSPHAERIA (Gr. *dictyon*, net + Gr. *sphaira*, ball) grows attached to substrates such as coral heads in shallow tropical waters; rhizoids serve an attachment function. It begins development as small balloon-shaped cells having a large central vacuole. The peripheral, multinucleate cytoplasm cleaves into numerous small, walled cells. As these enlarge, they contact their neighbors forming a tough layer at the periphery of the spherical thalli. Older thalli may crack open, forming pebbly, pale green sheets (Fig. 18–40). *Dictyosphaeria cavernosa* competes with corals on patch and fringing reefs and can form large

Figure 18–41 *Valonia* forms bright green, balloon-shaped vesicles on coral reefs. (Photograph courtesy Ronald J. Stephenson)

growths when grazing is reduced due to overfishing and human-generated nutrient inputs (Stimson, et al., 1996).

VALONIA (Venetian vernacular name) is another balloon-shaped vesicular form that is attached to substrates in tropical waters (Fig. 18–41). Similarly to *Dictyosphaeria*, *Valonia* undergoes segregative cell division at the periphery such that many small cell units are formed just beneath the parental wall. However, these small cells are not as noticeable as those of *Dictyosphaeria*, giving the false impression that *Valonia* is composed of a single large cell. *Valonia* is typically brighter green in color than is *Dictyosphaeria*. Depending upon the species, *Valonia* can occur as single vesicles or aggregations of vesicles.

Caulerpales

Named for the beautiful genus *Caulerpa*, this order includes about 26 genera and 350 species of multinucleate, siphonous ulvophyceans (Floyd and O'Kelly, 1990). The group has also been referred to as Bryopsidaceae (Hoek, van den, et al., 1995) or Bryopsidales (Vroom, et al., 1998). Although composed of but a single cell, caulerpaleans can be surprisingly large—some extending to a meter or more in length. Such thalli are vulnerable to extensive loss of cytoplasm in

the event of cell-wall damage. Consequently, wound-healing is extremely efficient, occurring within a matter of seconds by actin-mediated contraction of the protoplast and construction of a plug of cell-wall material to seal the wound. In some taxa (or particular developmental or life-history stages), cell walls may be composed primarily of noncellulosic polysaccharides, namely mannans, xylans, or xyloglucans. The adaptive advantage of these variations in cell wall chemistry is not understood.

Chloroplasts are discoid and numerous. Pyrenoids may be either present or lacking. In forms such as *Udotea* and *Caulerpa*, there may be two different populations of plastids, some pigmented and others lacking pigments but containing starch; the latter are referred to as amyloplasts (by analogy to those of land plants). When both pigmented and unpigmented plastids are present, the algae are said to exhibit **heteroplastidy**. The accessory xanthophyll pigments siphonein and siphonoxanthin may be present.

Caulerpalean taxa are either uniaxial or multiaxial. Uniaxial forms are composed of a single branched siphon, and are consequently rather delicate. In contrast, multiaxial forms arise from extensive early dichotomization of a single siphon into multiple, branchlike, siphonous proliferations (Fig. 18–42). Each siphonous proliferation then enlarges and undergoes further dichotomization. The mature thallus is thus composed of multiple, aggregated siphons having a common developmental origin. Uncalcified multiaxial forms have a robust and spongy texture. The peripheral portions of the siphons may be inflated and aggregated to form a coherent outer surface. Such inflated areas are known as **utricles**, and reproductive structures originate from them. Multiaxial forms can be differentiated into an internal colorless region (the medulla), and a green photosynthetic surface region. In some cases chloroplasts are withdrawn from the surfaces into the medullary regions during the night, then redeployed to the surface at daybreak. This is regarded as a means of protecting chloroplasts from herbivores.

Some multiaxial forms precipitate calcium carbonate on the siphon surfaces, including those of medullary regions, as described earlier. The function of such carbonate precipitation is not known, but some researchers have suggested that it may be a method of acquiring carbon dioxide (see Chapter 2), or that calcification might retard herbivory. Among Caulerpales, *Halimeda* is distinctive in having calcified segments that alternate with uncalcified joints, thereby

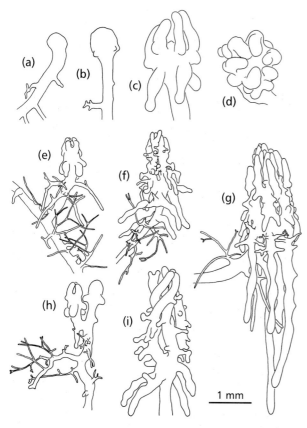

Figure 18–42 Development of a multiaxial thallus from a uniaxial germling. (From Friedmann and Roth, 1977)

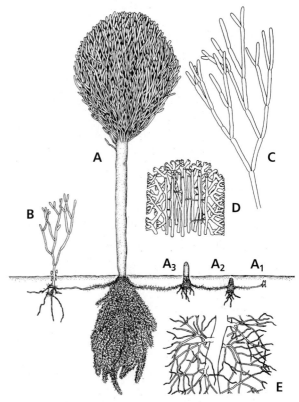

Figure 18–43 Development and anatomy of a multiaxial caulerpalean, *Penicillus*. Subsurface siphonous runners interconnect thalli of differing developmental ages (A_1–A). Young thalli are uniaxial (B, C) but rapidly undergo dichotomous branching to form numerous aerial filaments that interweave to form a thick thallus (D). The subsurface siphonous system ramifies into an extensive holdfast (E). (Diagram from Friedmann and Roth, 1977)

conferring flexibility helpful in avoiding damage from wave or current action.

Both uniaxial and multiaxial forms attach themselves to the substrate, typically sand, by colorless siphonous "branches" known as rhizoids. The rhizoid aggregates of sand-dwelling forms bind sand particles to form a massive holdfast region (Fig. 18–43).

Reproduction typically involves sexual fusion of biflagellate anisogametes; the larger gametes are usually brighter green than the smaller ones. The freshwater *Dichotomosiphon* is unusual in being oogamous. Zoospores are rare among caulerpaleans; quadriflagellate zoospores similar to those generated by other ulvophyceans are produced only by the uniaxial *Ostreobium*. This is regarded as an early divergent form for this reason (Floyd and O'Kelly, 1990). Unusual multinucleate, stephanokont (bearing a ring of flagella) zoospores are generated by *Derbesia*, *Bryopsis*, and *Bryopsidella* (Calderon-Saenz and Schnetter, 1989). In most cases the life history is very poorly understood, and little information on the timing and location of

meiosis is available. Heteromorphic alternation of generations (the only known example of the dikaryotic habit among algae) has been proposed to occur in *Derbesia* on the basis of microspectrophotometric evidence (Eckhardt and Schnetter, 1986). However an ultrastructural study of syngamy in *D. tenuissima* demonstrated that karyogamy (nuclear fusion) occurs about five minutes after plasmogamy (cell fusion) (Lee, et al., 1998). The life histories of the other genera are controversial. In the past experts had thought that most caulerpaleans had diploid vegetative thalli that underwent meiosis in gametangia, but this concept is now being challenged. Zygotes, though sometimes long-lived, do not serve as resting stages, and in some cases are thought (though not proven) to undergo meiosis prior to the formation of the macro-

coenocyte extends horizontally over substrates and gives rise to erect systems of main axes having many long, branched, featherlike divergences (known as laterals or pinnae). Gametangia develop from the laterals, and anisogametes are discharged into the water. Planozygotes resulting from fertilization attach to substrates and grow into tiny filaments that contain a single large nucleus. Nuclear division then occurs, forming many stephanokont zoospores that regenerate the vegetative thalli (Hoek, van den, et al., 1995). There is an extensive literature regarding the reproduction and life history of *Bryopsis*, linking this genus to some other uniaxial caulerpaleans (reviewed by Bold and Wynne, 1985). However, the meaning of this information is difficult to assess in the absence of detailed analysis of DNA levels of nuclei at different stages of reproduction in all of the forms and a molecular phylogeny for this genus. Microspectrophotometric measurements of nuclear DNA levels in *Bryopsis hypnoides* suggest that gametogenesis does not involve meiosis (Kapraun and Shipley, 1990). *Bryopsis* occurs in both temperate and tropical seas. At least one species has been renamed *Bryopsidella* (Calderon-Saenz and Schnetter, 1989).

DERBESIA (named for August Alphonse Derbes, a French phycologist) is a uniaxial, branched 1–10 cm coenocyte with a single plastid type. Only during development of saclike sporangia are septa formed (Fig. 18–45). A thin layer of cytoplasm surrounds a large internal vacuole. In 1938, Kornmann observed in cultures that *Derbesia* is the sporophytic stage of another marine algal genus, the small balloon-shaped *Halicystis* (Gr. *hals*, sea + Gr. *kystis*, bladder), which often grows attached to crustose red algae or other

Figure 18–44 *Bryopsis* is a uniaxial siphonous form whose erect parts appear feathery. (Courtesy W. Woelkering)

scopic vegetative thallus (Hoek, van den, et al., 1995). Caulerpalean zygotes typically have very large nuclei, reminiscent of those of the dasycladalean seaweeds, described later.

The vast majority of caulerpaleans are marine, primarily occurring in warm tropical or subtropical waters of normal salinity. They may be easily observed in shallow waters—in sea-grass beds or attached to coral heads—although some occur in much deeper waters. The widespread, and sometimes nuisance, genus *Codium* commonly extends into temperate coastal waters, and *Dichotomosiphon* occurs in sandy areas at the edges of freshwater lakes. A cladistic analysis of the order based on morphological, reproductive, and ultrastructural features, suggested that it is monophyletic, that heteroplastidy and multiaxial thalli are derived, and that *Codium* is an early divergent form (Vroom, et al., 1998).

BRYOPSIS (Gr. *bryon*, moss + Gr. *opsis*, likeness) (Fig. 18–44) is uniaxial. A prostrate portion of the

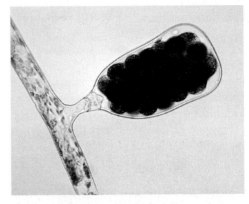

Figure 18–45 Portion of a *Derbesia* siphon bearing a sporangium with numerous round, dark spores.

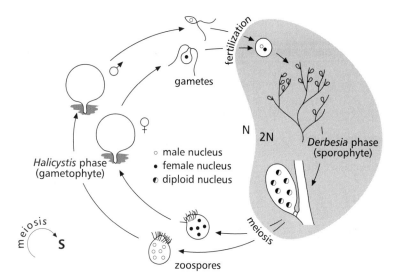

Figure 18–46 Diagram of the life history of *Derbesia-Halicystis*. The *Derbesia* phase produces stephanokont zoospores that grow into the *Halicystis* phase. Anisogametes are explosively discharged from the *Halicystis* form; their fusion leads to development of the *Derbesia* phase. (After Eckhardt, et al., 1986)

calcareous substrates. *Halicystis* thalli are also characterized by having a thin cytoplasmic layer and a large central vacuole. The life cycle for *Derbesia-Halicystis* is shown in Fig. 18–46.

Halicystis thalli are of two types: one produces small, pale gametes, and the other generates larger, dark green gametes (thalli are thus dioecious, and the organism heterothallic). *Derbesia* is the sporophytic stage in an alternation of heteromorphic generations. The stephanokont zoospores are multinucleate, but contain only one nuclear type. They are produced in and discharged from, specialized sporangia separated from the rest of the thallus by a wall (septum). Thus

the thallus persists after release of reproductive cells. Zoospores grow into the vesiculate *Halicystis* stage (the gametophyte). In addition, unmated gametes of the larger type can germinate directly into *Derbesia*-like thalli, but these are homokaryotic and haploid.

CODIUM (Gr. *kodion*, small sheepskin) includes numerous types of dark green, spongy, multiaxial thalli up to one meter in length. It grows attached to substrates, such as rocks or shells, in temperate or tropical regions. There are about 120 species, and their external morphology can be quite variable, from flat, crustlike forms to spheres and the more common

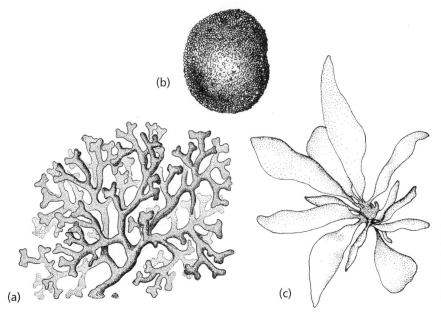

Figure 18–47 Diagrammatic views of several species of *Codium*, illustrating some of the known diversity of morphological types. (a: From Taylor, 1972 ©The University of Michigan Press 1960; b and c: From Bold and Wynne, 1985)

dichotomously branched forms (Fig. 18–47). Branched thalli are commonly known as "dead man's fingers" (see Fig. 17–3) Highly interwoven siphons are expanded at the surface to form utricles whose form is important in distinguishing the species. Numerous discoid chloroplasts lacking pyrenoids are present in utricles; there are no amyloplasts. Internal medullary portions of the thallus are colorless. Cell walls do not contain cellulose, but rather mannans or sulfated arabinogalactans.

Gametangia develop as branches from the utricles; they may be dark in color or pale, because they contain two different types of biflagellate anisogametes (as in *Derbesia-Halicystis*). Gametangia are separated from the rest of the thallus by cross-walls (septa); thus the thallus can persist after gamete discharge. Gametes are thought to arise by meiosis (Vroom, et al., 1998). Microspectrophotometric studies of DNA content in the life cycle of three *Codium* species were performed by Kapraun, et al. (1988). No stephanokont zoospores are produced. However, as in *Derbesia-Halicystis*, unmated gametes of the larger, darker type can germinate into vegetative thalli; this may be viewed as a form of asexual reproduction. On the Atlantic coast of North America an asexual, weedy form of *Codium* (*C. fragile* ssp. *tomentosoides*) has spread to nuisance proportions. Chloroplast restriction fragment mapping has been used to assess possible pathways of *Codium* invasions. For example, Goff, et al. (1992) found that the weedy Atlantic species described above did not originate from an indigenous Pacific coastal form of *C. fragile*.

CAULERPA (Gr. *kaulos*, stem + Gr. *herpo*, to creep) is a uniaxial siphonous form that occurs as a series of erect, highly differentiated photosynthetic shoots, as well as anchoring rhizoids, both of which emerge from a horizontal siphonous axis. Biologists have been impressed by *Caulerpa*'s plantlike organization since it was first described about 150 years ago; sometimes they refer to *Caulerpa*'s organs as "roots, stem, and leaves," (Jacobs, 1994), though of course the positioning of these structures is merely analogous,

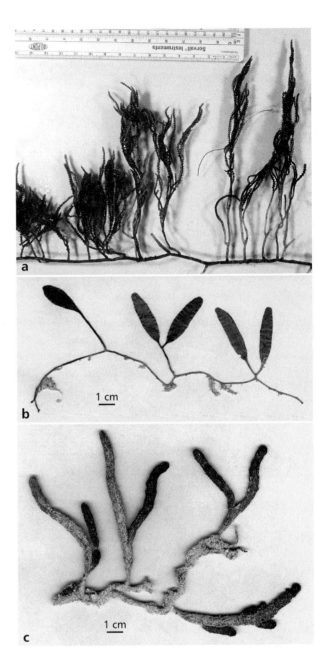

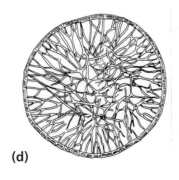

Figure 18–48 (a–c) Three species of *Caulerpa* illustrating some of the diversity of photosynthetic shoots within this genus. (d) Drawing of thallus in cross section, showing trabeculae. (a: Courtesy C. Taylor; d: Courtesy W. Woelkerling)

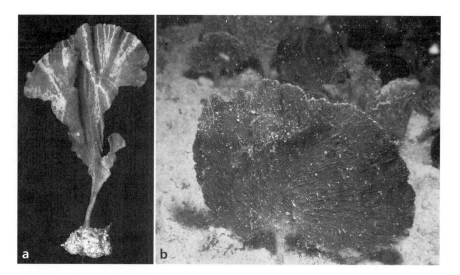

Figure 18–49 *Udotea* is a fan-shaped inhabitant of tropical seagrass beds (a, b). (b: Photograph courtesy Ronald J. Stephenson)

rather than homologous, to those of terrestrial plants. The horizontal system may extend for meters along the surface of sea-grass beds or other substrates in tropical or subtropical waters. New shoots and rhizoids are produced from the growing end of the horizontal system. If thalli are experimentally inverted, the next-produced rhizoids and green shoots will occur on the opposite sides of the thallus from their older peers, indicating the occurrence of gravity detection during morphogenesis (Jacobs, 1993).

There are about 70 species, differentiated by shoot-structure differences. Shoots can be leaflike or have radiately or bilaterally symmetrical laterals; laterals can resemble the pinnae of feathers or be spherical, resembling bunches of grapes (Fig. 18–48). Ingrowths of wall material, known as **trabeculae**, extend across the siphon lumen, probably providing structural support. The cell wall is composed of a polymer of xylose, as are those of *Halimeda* and *Udotea*. Amyloplasts as well as green discoid plastids are present. The architecture of the chloroplast genome of *C. sertularioides* has been studied by Lehman and Manhart (1997). The gene order in *Caulerpa* is unlike that of any previously characterized plastid genome. Another unusual aspect of *Caulerpa* plastids is the occurrence of DNA in the pyrenoid core (Miyamura and Hori, 1995). Reproduction occurs by liberation of the entire protoplast in the form of numerous biflagellate anisogametes, a phenomenon known as **holocarpy**. Gametes of thousands of individuals are released simultaneously, a case of "mass spawning" (see Text Box 18–1). Zygotes generate single-celled uninucleate "protospheres" from which new vegetative thalli arise.

The center of diversity is the southern Australian coast, where the genus is thought to have originated; species relationships have been studied by nuclear rDNA ITS sequences (Pillmann, et al., 1997).

UDOTEA (Gr. mythology: an ocean nymph) occurs as fan- or funnel-shaped multiaxial aggregates of siphonous branches (Fig. 18–49). There is typically a holdfast composed of aggregated colorless siphonous branches. The cell wall is composed of xylans. Calcium carbonate is deposited on the outside of the thallus and in the spaces between siphons. Cells at the periphery of the fan or funnel may be partitioned off as gametangia in some species, but in other forms the entire protoplast is converted into biflagellate anisogametes, and upon their discharge the thallus becomes white and dies (holocarpy) (Meinesz, 1980). Mass spawning has also been documented (Text Box 18–1). Zygotes develop into a nonmotile, unicellular "protosphere" stage in which there is but a single large nucleus. This is regarded as the only diploid stage in the life cycle, and meiosis is surmised to occur prior to development of the macroscopic vegetative thallus. However, this hypothesis requires testing by microspectrophotometric measurements of relative DNA levels. *Udotea* is common in sandy sea-grass beds in tropical nearshore waters and lagoons. The systematics of *Udotea* in the tropical western Atlantic Ocean were reviewed by Littler and Littler (1990). These studies suggested that *Udotea* is a monophyletic genus.

PENICILLUS (L. *penicillus*, painter's brush) is a multiaxial form whose dichotomously branched

Text Box 18–1 Mass Spawning by Caulerpalean Algae

Mass spawning—the synchronized, simultaneous release of vast numbers of male and female gametes—is an ecologically significant characteristic of the corals and gorgonians that inhabit reef systems and has been known for only a decade. Recently an impressive study by Clifton (1997) demonstrated that *Caulerpa*, *Halimeda*, *Penicillus*, *Rhipocephalus*, and *Udotea* (17 species altogether) also mass spawn. The peak occurrence of caulerpalean spawning occurs between March and July, in the predawn hours. The study involved twice daily (5:00–7:30 and 14:00–16:30) underwater observations of tagged seaweeds on a patch reef off Point San Blas, Panama, over a period of 373 days.

The onset of fertility—involving migration of the entire protoplast into terminal regions and cleavage into gametes—is easily visible with the unaided eye, as green thalli become white except for green peripheral regions. This process occurs within a 12-hour period during the night. For the dioecious genera *Halimeda*, *Penicillus*, *Udotea*, and *Rhipocephalus*, it is even possible to distinguish male and female thalli by color differences, females having a darker brownish green coloration in comparison to paler green males. Microscopic examination confirmed that the green portions contained gametes of these types. Within a species, males consistently began the process of gamete release before females, and on a given morning, typically 3–5% (though sometimes 15–20%) of the population of several hundred thalli became fertile at the same time. Gamete release in the form of green clouds or mucilaginous streams lasted for 5–15 minutes, dropping water visibility to 1 m during this time. Gametes are motile for up to 60 minutes following their release, and at least some have been observed to swim up to 100 cm within a few minutes. Zygotes rapidly become nonmotile and settle. Parental thalli—completely emptied of protoplasmic contents—often disintegrate within hours. *Caulerpa*, *Penicillus*, and *Rhipocephalus* disappear most rapidly—within 24 hours—but dead thalli of *Udotea flabellum* (inset) persist for weeks.

Unlike mass spawning of corals and gorgonians, that of caulerpalean algae was not correlated with lunar or tidal cycling. Although water temperature decreases could delay gamete release, water temperature was also not related to gamete release patterns of the algae. The environmental or biological factors that trigger mass spawning by these seaweeds is not known, but the adaptive value of synchronized release of male and female gametes is obvious. While closely related species released gametes on the same morning, there were species-specific differences in peak time of release. This is regarded as a means by which hybridization might be avoided. If you have the opportunity to visit a tropical coral reef region in spring or early summer, snorkel or dive to see the awesome mass spawning of caulerpalean algae (but you will have to get up early!).

siphons are aggregated at the base into a stipe, but occur in a free, brushlike arrangement at the apex (Fig. 18–50). Thus the thallus looks like a handled brush and is commonly known as "Neptune's shaving brush." There is a bulbous holdfast much like that of *Udotea*. The thallus is lightly encrusted with calcium carbonate. A series of *Penicillus* thalli of different ages can be generated from a horizontal "runner," as in *Caulerpa*. Sexual reproduction resembles that of *Udotea*. In at least one well-studied form, gamete production involves the entire cytoplasm, such that reproduction results in death of the vegetative thallus (Friedmann and Roth, 1977). The similar genus *Rhipocephalus* is often found with *Penicillus*, differing from it in that the filaments of the terminal tuft form small blades (Fig. 18–51).

HALIMEDA (Gr. mythology: a sea nymph) has a flattened, fan-shaped thallus composed of dichotomously branching, segmented erect portions and a large holdfast composed of aggregated rhizoidal siphons (Fig. 18–52) with extensive calcification of the

Figure 18–50 *Penicillus* has a brushlike array of photosynthetic branches emerging from a stipe having a massive holdfast.

Figure 18–51 *Rhipocephalus* shows a tiered array of bladelike branches, but has a stipe and holdfast similar to those of *Penicillus*.

shoot segments, but not joints. Sexual reproduction involves conversion of the entire biomass into anisogamous gametes whose discharge from cells at the periphery leaves the thallus empty, whereupon it dies (mass spawning) (Text Box 18–1). *Halimeda* is widespread in warm tropical or subtropical waters, growing as deep as 50 m. In part because it can produce several crops per year, *Halimeda* is a major contributor to carbonate sediments. The genus has been monographed by Hillis-Colinvaux (1980).

Dasycladales

There are eleven genera with some 19–50 species in this order. Dasycladaleans are characterized by siphonous vegetative thalli (Floyd and O'Kelly, 1990; Berger and Kaever, 1992). Some forms (e.g., *Acetabularia*) are uninucleate until reproductive development begins, but others (e.g., *Cymopolia*) appear to have multinucleate cells. Some authorities have elevated this group to class status (Dasycladophyceae) (Hoek, van den, et al., 1995). There is typically a basal

holdfast region consisting of rhizoidal branches in which the diploid or polyploid nucleus resides, an elongate erect axis, and laterals occurring in whorls

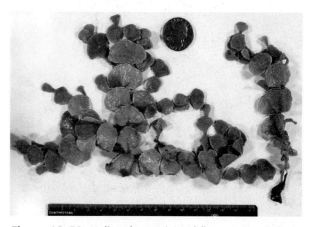

Figure 18–52 *Halimeda* consists of flattened, calcified segments interspersed with flexible joints. (Photograph courtesy J. Hackney)

Figure 18–53 The dasycladalean genus *Dascycladus* (for which the order is named) exhibits whorled branching typical of the group. These thalli are attached to carbonate sand grains by a system of rhizoids. (Photograph courtesy Ronald J. Stephenson)

(Fig. 18–53). In *Acetabularia*, adherent reproductive laterals form distinctive caps. Based on comparisons of the ultrastructure of flagellate cells, Floyd, et al. (1985) suggested that Dasycladales were closely related to siphonocladalean (cladophoralean) algae.

The single-celled thalli possess a thin layer of cytoplasm with numerous plastids, surrounding a large central vacuole. As in all green algae, starch occurs in plastids, but surprisingly, cytoplasmic polysaccharide

reserves are also reported. The cell walls are composed of a mannose polymer, but cellulose occurs in the walls of reproductive structures. Many forms are encrusted with calcium carbonate; calcification is reduced under conditions of reduced irradiance or temperature (Berger and Kaever, 1992).

Reproduction is by biflagellate gametes produced in small spherical, walled cysts formed in the reproductive laterals. Zygotes possess a single large primary nucleus and develop into vegetative thalli. Spectrophotometric measurements of nuclei in *Acetabularia* and *Batophora* indicate that the zygote undergoes meiosis sometime prior to the onset of reproductive development (Koop, 1979; Liddle, et al., 1976). Subsequent repeated mitotic divisions produce hundreds of small secondary nuclei that are incorporated into cyst cytoplasm, and eventually partitioned into gametes. Cyst formation requires the arrival into reproductive laterals of secondary nuclei. Each nucleus is surrounded by a cage of microtubules; rings of actin filaments are involved in the cleavage of cytoplasm to form gametes (Menzel, et al., 1996). Microtubules also specify the location of a circular region of the cyst wall that will become the lid or operculum. This lid opens during gamete discharge (Berger and Kaever, 1992). Gamete discharge thus does not result in the demise of the entire thallus as occurs in many caulerpaleans.

Dasycladaleans, especially the genus *Acetabularia*, are extensively used for laboratory study of cell and developmental biology (Mandoli, 1998). *Acetabularia*'s utility is based upon ease of cultivation, large cells (up to 200 mm in length), and high tolerance to surgical manipulation, thanks to rapid wound-healing

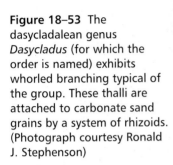

Figure 18–54 Diagrammatic representation of some of Hämmerling's famous transplantation experiments with *Acetabularia*, demonstrating the presence of species-specific nuclear and cytoplasmic (mRNA) information that determines the morphology of the cap. The nucleus occupies a basal position. (After Gibor, 1966)

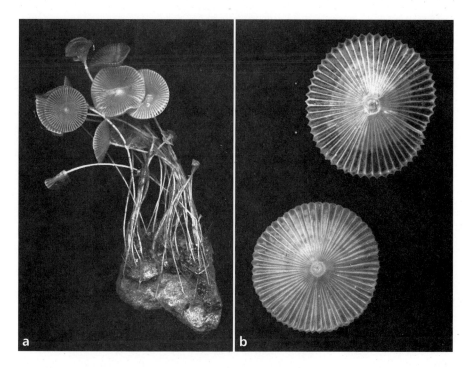

Figure 18–55 *Acetabularia* thalli often grow as clumps of individuals, attached to hard substrates such as coral rubble. (a) The unicells have a delicate holdfast, an elongate axis, and a terminal cap, wherein reproductive cells are generated. (b) Close-up view of two caps.

processes. Cells can be enucleated by removing the rhizoids with scissors. Nuclei can be isolated and then implanted into enucleated cells. Basal, nucleated rhizoidal regions from which the rest of the thallus has been removed are capable of regenerating the entire thallus. Based on the results of nuclear transplantation experiments conducted in the 1930s, Hämmerling postulated the occurrence of messenger RNA prior to its chemical characterization (Fig. 18–54). More recent experiments have demonstrated the role of localization of cell membrane calcium-ion binding sites in morphogenesis; this determines the pattern of whorl development (Berger and Kaever, 1992). An easily prepared, highly effective growth medium has been developed for *Acetabularia* (Hunt and Mandoli, 1996).

Dasycladaleans occur in shallow (to 10 m) waters of tropical or subtropical shores in protected areas. Dasycladaleans are distributed within a broad belt from 25° N and 35° S, but can occur as far north as 32° in Bermudan waters, and to 40° S on the west coast of South Africa. They can frequently be found on mangrove roots. Tolerance of most dasycladaleans to salinity variation is generally low. Thus dasycladaleans commonly occur in open lagoons, but not generally closed ones where salinities can vary because of evaporation or rainfall. However, the genus *Batophora* is more tolerant of salinity changes than are other dasycladaleans (Berger and Kaever, 1992).

ACETABULARIA (L. *acetabulum*, vinegar cup) (Fig. 18–55), commonly known as the "mermaid's wine glass," typically grows in large clumps, sometimes forming lawns. The unbranched, cylindrical

Figure 18–56 *Batophora* axes bear whorls of lateral branches.

a

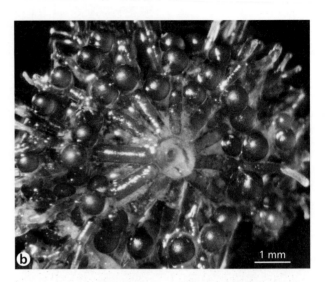

Figure 18–57 Gametophores of *Batophora*. (a) Longitudinal view of thallus. (b) Cross-sectional view showing the gametophores, which are borne on lateral branches.

stalk is anchored to substrates by a holdfast of rhizoidal branches. When thalli reach about one mm in length, the first whorls of lateral branches develop at the growing tip; these then degenerate, leaving a ring of scars on the stalk. The cap (Fig. 18–55b) is formed of adherent branches known as cap rays, which are **gametophores** (gametangia/cyst-containing structures). After development of the cap rays, protoplasm containing numerous secondary nuclei then streams into the gametophores, imparting a dark green color. A septum divides each gametophore from the rest of the thallus, and cyst walls develop around each cytoplasmic unit. When mature, cysts are released from gametophores. Later, gametes are released from cysts as they press on

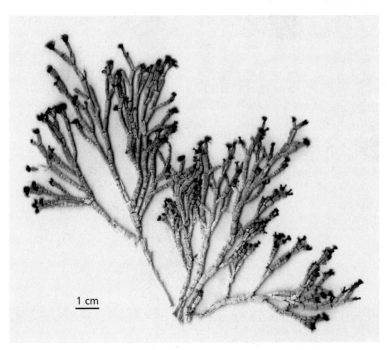

1 cm

Figure 18–58 Pressed specimen of the heavily calcified *Cymopolia*, which has repeatedly branched thalli, the branches ending in green photosynthetic tufts (see Fig. 18–1).

the operculum, forcing it open. The biflagellate gametes possess eyespots and are phototactic. Gametes can swim for up to 24 hours and mate by agglutination of flagellar tips. Zygotes are negatively phototactic and thus settle onto a substrate, lose their flagella, and generate a cylindrical structure. Rhizoids develop from one end and penetrate the substrate, whereas the other becomes tapered and elongates from the tip, beginning development of the next generation of vegetative thalli. The mechanics of the cytoskeleton during morphogenesis in *Acetabularia* are discussed by Goodwin and Briére (1994), and aspects of morphogenesis are also reviewed by Mandoli (1998) and Kratz, et al. (1998). There are about eight species (Berger and Kaever, 1992).

BATOPHORA (Gr. *batos*, bramble + Gr. *phoras*, bearing) has many whorls of repeatedly branched laterals on vertical axes that do not themselves branch (Fig. 18–56). Gametophores are produced at nodes of the laterals (Fig. 18–57). Thalli are not calcified. *Batophora* occurs in shallow waters of lagoons and has been found in waters of low salinity (Berger and Kaever, 1992).

CYMOPOLIA (Roman mythology: a daughter of Neptune) thalli have a repeatedly and dichotomously branched construction (Fig. 18–58). Branching occurs in a single plane. Calcified segments alternate with uncalcified joints, a pattern that confers flexibility to the thallus. Tufts of uncalcified laterals occur at the growing tips. Gametangia are produced individually at the tips of primary lateral branches surrounded by inflated secondary laterals in uncalcified apical regions of the thallus. Gametangia are not operculate and are produced continuously throughout the life of the thallus, in contrast to those of *Acetabularia* and *Batophora* (Berger and Kaever, 1992).

Chapter 19

Green Algae III
Trebouxiophyceans

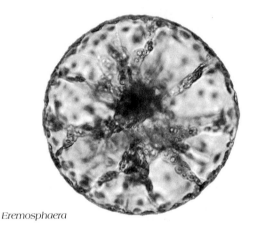

Eremosphaera

Although this green algal class will seem unfamiliar to many, it contains some familiar, widespread, and biotically significant forms. *Chlorella*—famous as the experimental system used by Melvin Calvin to perform his Nobel prizewinning research on the carbon-fixation processes of photosynthesis—is a member of this group. So is *Prototheca*, a common soil alga known to sometimes cause disease of cattle and, on occasion, humans whose immune systems have been suppressed or are in a weakened condition. Perhaps most interesting is the fact that many of the green algae in this class are lichen symbionts, known as **phycobionts**. *Trebouxia*, for which the class is named, is the single-most common lichen alga. Of the green algal phycobionts of lichens, most fall into the class Trebouxiophyceae, suggesting that there may exist some clade-specific physiological commonality that predisposes trebouxiophyceans to become lichen symbionts.

Trebouxiophyceans include a surprising variety of morphological, reproductive, and ecological types (as do the other three non-prasinophycean green algal classes). The soil or freshwater algae *Chlorella*, *Prototheca*, *Coccomyxa*, and *Stichococcus* (the latter two of which also occur in lichens) are nonflagellate unicells whose only reproductive cells are nonflagellate autospores. Other nonmotile unicellular lichen phycobionts, including *Trebouxia*, *Dictyochloropsis*, and *Myrmecia* (some species of which were formerly known as *Friedmannia*), instead generate flagellate zoospores. The giant unicell *Eremosphaera* and smaller *Golenkinia*, both freshwater forms, produce flagellate sperm during oogamous sexual reproduction. *Desmococcus* is a **sarcinoid** (cell packet-forming) inhabitant of tree bark. *Microthamnion* is a freshwater periphytic branched filament that reproduces asexually via biflagellate zoospores. *Pleurastrum terrestre* is also a branched filamentous form, but occurs on soil or in lichens. *Prasiola* is a sheetlike blade found in high mountain streams or on high intertidal rocks along marine coastlines that superficially resembles small thalli of the common ulvophycean seaweed *Ulva*.

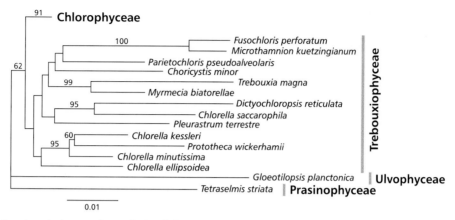

Figure 19–1 Molecular phylogenetic analysis of the Trebouxiophyceae and Chlorophyceae based on SSU rDNA sequences obtained by several workers. Within the clade formed by the Ulvophyceae, Trebouxiophyceae, and Chlorophyceae (the UTC clade—see Chapter 17), trebouxiophyceans are sister to the chlorophyceans. Many lichen symbionts fall into the trebouxiophycean clade, as does *Prototheca*, the pathogen of humans and cattle. (After Lewis, 1997 by permission of the *Journal of Phycology*)

Phylogenetic Relationships of the Trebouxiophyceans

The history of phylogenetic analyses involving these algae is interesting and instructive. The trebouxiophyceans were first recognized as a distinct group on the basis of ultrastructural studies that revealed distinctive positioning of centrioles at the sides of the spindle during mitosis (see Figs. 17–14, 17–16) (Molnar, et al., 1975). Trebouxiophyceans are thus said to have metacentric spindles. In contrast, centrioles of other green algae, as well as those of many earlier-divergent protists, are located at the spindle poles. For this reason polar location of centrioles is regarded as the plesiomorphic (primitive) condition for green algae. Consequently, metacentric spindles are thought to be a derived feature and a unique feature of the trebouxiophycean clade. This evidence was used by Mattox and Stewart (1984) to separate the algae having metacentric spindles into a new class, which they named the Pleurastrophyceae, citing a type species, *Pleurastrum insigne*. Other cell-division characteristics of the group include semiclosed mitosis, a nonpersistent telophase spindle, and cytokinesis by means of a cleavage furrow that develops from the edges in association with phycoplast microtubules. No plasmodesmata are known to occur (Melkonian, 1990b).

Further comparative ultrastructural studies of green algae next revealed that reproductive cells of

this group have a prominent rhizoplast and flagellar basal bodies arranged in a cruciate (cross-shaped) pattern. Basal bodies are displaced in the counterclockwise direction, as in ulvophyceans (Chapter 18) (Melkonian, 1990b). In contrast, the algae placed into the class Chlorophyceae (chlorophyceans) possess cruciate root systems with flagellar basal bodies either displaced in the clockwise direction or directly opposed (see Fig. 17–13). For some time it was thought that similarity in flagellar base structure was evidence of particularly close relationship between ulvophyceans and trebouxiophyceans, but now trebouxiophyceans are regarded as more closely linked to the ancestry of chlorophyceans (see below). It is still the case that the counterclockwise displaced flagellar apparatus is regarded as the ancestral condition within the UTC clade of green algae.

Later, investigators began to compare SSU rDNA sequences of green algae that had previously been difficult to classify because they lacked flagella (and basal bodies). They discovered that the trebouxiophyceans very likely represent the sister group to the chlorophyceans (Fig. 19–1) (Friedl, 1995; Lewis, 1997). Emerging knowledge regarding the architecture of green algal mitochondrial genomes (reviewed by Nedelcu, 1998) indicates that there are substantial differences between trebouxiophyceans (represented by *Prototheca*) and chlorophyceans (represented by *Chlamydomonas*). *Chlamydomonas*-like mitochondrial genomes are small, with reduced gene content and fragmented, scrambled rRNA coding regions. In con-

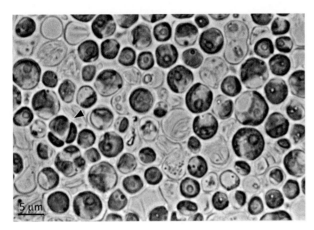

Figure 19–2 *Chlorella*. There is a single cup-shaped parietal plastid that may or may not possess a pyrenoid. Arrowhead points to a cell that has cleaved into autospores.

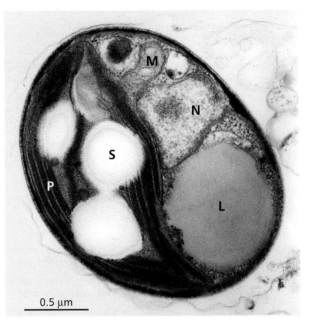

Figure 19–3 TEM of a *Chlorella* cell living within the tissues of a freshwater sponge. L=lipid droplet, N=nucleus, M=mitochondrion, P=plastid, S=starch. (From *Origin of Land Plants*. L. E. Graham. ©1993 Reprinted by permission of John Wiley & Sons, Inc.)

trast, *Prototheca*-like mitochondrial genomes are larger, and have more protein and tRNA coding genes. The name of the group was recently changed from Pleurastrophyceae (Mattox and Stewart, 1984) to Trebouxiophyceae, because the type species given for the original group—*Pleurastrum insigne*—is now known, based on molecular sequence evidence (Friedl, 1995), to be a member of the clockwise flagellar apparatus assemblage within the Chlorophyceae.

Comparing the members of the Trebouxiophyceae with those of the Ulvophyceae, Charophyceae, and Chlorophyceae reveals many examples of parallel evolution of thallus form. Unicells, sarcinoid packets, branched filaments, and bladelike thalli are found in all of these groups. The lesson is clear— *among the green algae, morphological form (unicell, colony, filament, etc.) is generally not correlated with phylogenetic relationship.* Phylogenetic divergence implies the occurrence of a physiological divergence, which is likely to influence ecological function. Filamentous trebouxiophyceans, for example, are not likely to behave ecologically in exactly the same way as would filamentous ulvophyceans, charophyceans, or chlorophyceans. Nor would coccoid unicells, or other structural variants. The power to predict algal attributes will derive from a robust understanding of evolutionary relationships in concert with identification and classification of morphological form. Once relationships are known, often subtle differences in morphology can be mapped onto clades, thereby generating greater insight into the patterns and process of algal morphological evolution.

Diversity

CHLORELLA (Gr. *chloros*, green) (Fig. 19–2) occurs as 2–12 μm spherical or ellipsoidal unicells in fresh and marine waters, soils, and as endosymbionts within the cells of freshwater invertebrates such as *Hydra viridis*, sponges, and many kinds of protozoa (see Chapter 7). The chloroplast is parietal, and there is a pyrenoid in some, but not all, species (Fig. 19–3). Asexual reproduction is by the formation of four autospores; sexual reproduction is not known. Some species have a cell wall consisting of glucose and mannose, whereas a second group of species possess a glucosamine wall (Takeda, 1991). Molecular phylogenetic analyses have shown that *Chlorella* is not monophyletic; there are are least three separate clades, one of which gave rise to *Prototheca* (see below) (Fig. 19–4) (Friedl, 1995). The molecular classification correlates well with groupings based on cell-wall composition. Ellipsoidal species having walls of a single smooth layer and lacking pyrenoids have been placed in the new genus *Watanabea* (Hanagata, et al., 1998).

PROTOTHECA (Gr. *protos*, first + Gr. *theke*, sheath or box) (Fig. 19–5) is a spherical unicell that

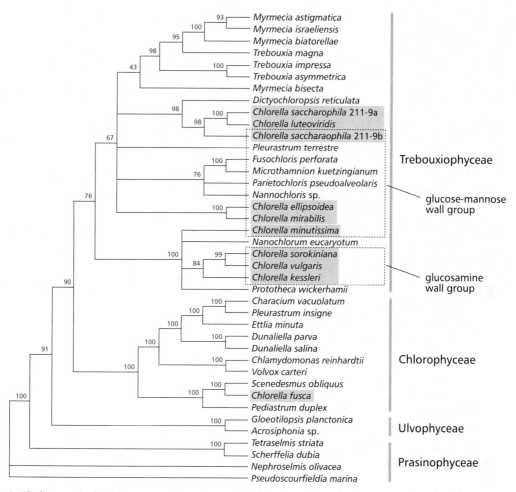

Figure 19–4 Phylogenetic (maximum parsimony) analysis of the trebouxiophyceans and related green algae based on SSU rDNA sequencing. Note that the genus *Chlorella* is polyphyletic, i.e., contains species (shown in boxes) that are not descended from a single common ancestor. One, *C. fusca*, occurs outside the Trebouxiophyceae. This analysis suggests that the generic names of several *Chlorella* species will have to be changed. Bootstrap values (1000 resamplings) are listed as percentages. Wall chemistry trait mapping is based on Takeda (1991). (After Friedl, 1995 by permission of the *Journal of Phycology*)

reproduces via autospores. Though colorless, its relationship to green algae is indicated by presence of starch-containing plastids. These can be stained by the use of an I_2KI solution, thus differentiating *Prototheca* from yeasts, with which it is often confused in medical contexts. It is an obligate heterotroph. *Prototheca* occurs in soils and freshwater environments, particularly sewage-contaminated waters. *Prototheca* is an opportunistic pathogen of animals and humans and is also associated with tree pathology (Pore, et al., 1983; Pore, 1986). It is capable of growing with adenine as the sole nitrogen source (Sarcina and Casselton, 1995). The outer walls, which have a three-layered structure, contain a hydrolysis-resistant, UV-fluores-

cent compound whose infrared spectrum differs from that of sporopollenin. Presence of this wall material is suspected to contribute to drug resistance (Puel, et al., 1987).

STICHOCOCCUS (Gr. *stichos*, row or line + Gr. *kokkos*, berry) consists of rod-shaped cells that can occur singly or be arranged end to end to form short filaments (Fig. 19–6). The cells are quite small, only 2–3 μm in diameter. The single chloroplast is parietal, with or without pyrenoids. *Stichococcus* has been isolated from soils and other terrestrial habitats (Graham, et al., 1981), freshwaters, and estuaries. Sorbitol and proline serve to maintain osmotic balance (Brown

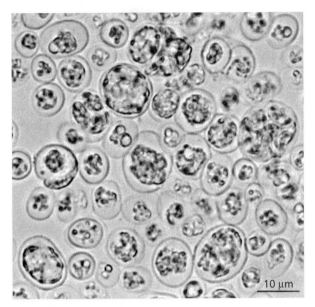

Figure 19–5 *Prototheca*. Although a plastid is present, it is not pigmented and therefore difficult to identify unless the starch contained within is stained with an I₂KI solution.

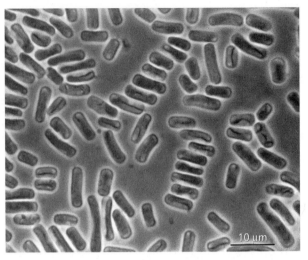

Figure 19–6 *Stichococcus* has rod-shaped cells with a single parietal plastid; this species does not have pyrenoids, but some others do.

and Hellebust, 1980). At least 11 amino acids can be taken up by *Stichococcus bacillaris*, by means of several specific carrier systems involving active transport (Carthew and Hellebust, 1982). Some lichens possess *Stichococcus* as phycobionts. Those lichens with pyrenoid-containing *Stichococcus* cells exhibited a more effective means of carbon concentration than lichens possessing pyrenoidless *Stichococcus* (Smith and Griffiths, 1996). Reproduction is by fragmentation of filaments and division of single cells into two progeny. No zoospores or sexual reproduction are known. At one time *Stichococcus* was regarded as a possible relative of the Charophyceae, but molecular sequence evidence suggests that this is not the case and that *Stichococcus* belongs with the trebouxiophyceans.

GOLENKINIA (named for Mikhail Iljitsch Golenkin, a Russian phycologist) is yet another spherical unicell, with a *Chlorella*-like parietal plastid and pyrenoid with surrounding starch shell (Figs. 19–7, 19–8), but is distinguished by the presence of several long spines per cell (Fig. 19–9). Asexual reproduction is by production of two, four, or eight autospores, and oogamous sexual reproduction has been described. Eight to sixteen biflagellate sperm are produced by some cells, whereas other vegetativelike cells function as eggs. Zygotes have spiny walls (Starr, 1963), but the

details of zygote germination and the timing of meiosis are unclear.

TREBOUXIA (named for Octave Treboux, an Estonian botanist) is a spherical unicell with an axial

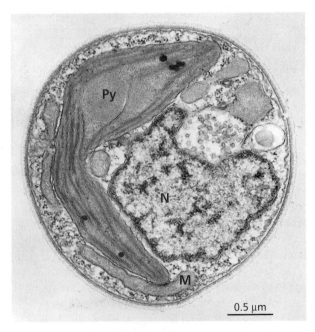

Figure 19–7 TEM of a *Golenkinia* cell showing the single parietal plastid with pyrenoid (Py), nucleus (N), and mitochondrion (M).

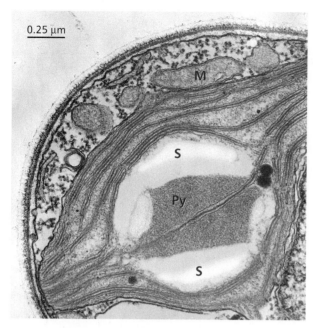

Figure 19–8 TEM of a *Golenkinia* pyrenoid (Py) that is bisected by a thylakoid and surrounded by a starch sheath (S). M=mitochondrion.

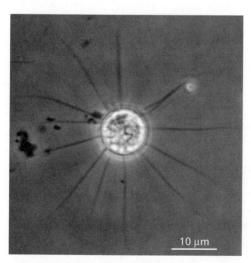

Figure 19–9 *Golenkinia* typically possess several long spines per cell. These are transparent and sometimes difficult to see with bright-field light microscopy, but show up well using phase-contrast optics, as seen here.

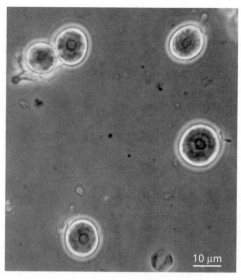

Figure 19–10 *Trebouxia* cells extracted from the lichen *Physcia*.

(suspended in the center of the cell) chloroplast containing a pyrenoid (Fig. 19–10). Wall-less zoospores are produced during asexual reproduction. It occurs either free-living on terrestrial substrates such as tree bark, or in lichens. Lichens with *Trebouxia* phycobionts possess an active carbon-concentration mecha-

nism (CCM—see Chapter 2), as determined by carbon isotope discrimination methods (Chapter 2) (Smith and Griffiths, 1996). This is in contrast to lichens whose phycobiont is a pyrenoidless green, such as *Coccomyxa*. Thus the presence of a CCM is correlated with occurrence of pyrenoids. Another coccoid, unicellular trebouxiophycean, having a pyrenoid and wall-less zoospores is *Parietochloris*. It is distinguished from *Trebouxia* by the location of the plastid—parietal (though deeply incised) in *Parietochloris*, axial in *Trebouxia* (Watanabe, et al., 1996). SSU rDNA sequences have been used to assess species relationships in *Trebouxia* (Friedl and Rokitta, 1997). Some forms previously classified as *Trebouxia* have been transferred to the genus *Asterochloris* (Tschermak-Woess, 1980).

EREMOSPHAERA (Gr. *eremos*, solitary + Gr. *sphaira*, ball) consists of spherical unicells that can reach diameters of 200 μm (Fig. 19–11). Cells have a highly layered wall and contain many small discoid chloroplasts, each with a single small pyrenoid. Chloroplasts are arranged in strands of cytoplasm radiating from the cell center to the periphery. Asexual reproduction is by production of two or four autospores (Fig. 19–11c), and the remains of parental cell walls can often be observed in the vicinity of the daughter cells. Oogamous sexual reproduction is known. Up to 64 small sperm can be produced per parental cell; eggs resemble autospores in size. Zygotes

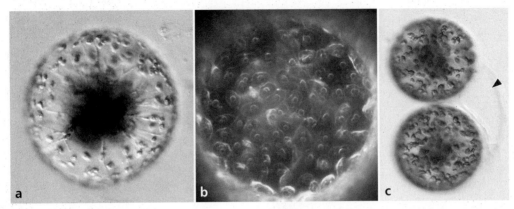

Figure 19–11 (a) *Eremosphaera* cells are unusually large and are characterized by multiple plastids and long cytoplasmic strands that are visible even at the light microscopic level. (b) The plastids contain a single pyrenoid each. (c) *Eremosphaera* reproduces asexually by means of autospores, two of which are shown along with remains of the parental cell wall (arrowhead).

produce thickened walls, but germination has not been well studied. *Eremosphaera* is a common inhabitant of soft waters and acid bogs.

MICROTHAMNION (Gr. *mikros*, small + Gr. *thamnion*, small shrub) is a highly branched filament of cells that are substantially longer than wide (Fig. 19–12). It grows on soil or in slightly acidic waters, and attaches to substrates via a specialized holdfast cell. Chloroplasts are parietal and lack pyrenoids. Biflagellate zoospores arise in pairs from the vegetative cells; their ultrastructure has been studied (Watson and Arnott, 1973; Watson, 1975).

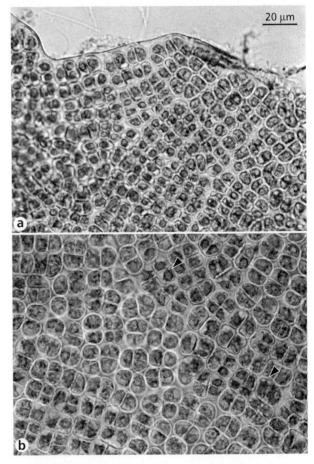

Figure 19–13 *Prasiola*. (a) Edge of bladelike thallus of an individual collected from a Colorado, U.S.A., mountain stream. Note that cells occur in packets. (b) At higher magnification each cell can be seen to contain a single stellate plastid (arrowheads).

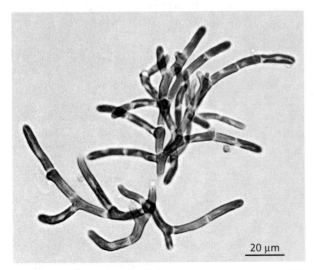

Figure 19–12 *Microthamnion* filaments are composed of thin, elongate cells.

PRASIOLA (Gr. *prason*, leek) forms small (1.5 cm), monostromatic (one-cell-thick) blades that are attached to soil or aggregations of bird droppings on rocks by hairlike rhizoids. As previously noted, some species occur only in cold, fast-flowing mountain streams (Fig. 19–13). *Prasiola* is also widespread in Antarctica on shorelines, where it is exposed to repeated freeze/thaw cycles in spring and fall, and is frozen during the winter. Amino acids such as proline are thought to serve as cryoprotectants. In addition, UV-absorbing compounds appear to help *Prasiola* adapt to high levels of solar irradiation during the austral summer (Jackson and Seppelt, 1997). Cells are conspicuously aggregated into packets, and the single chloroplast in each cell is stellate and axial. The blades are reported to be diploid, and to undergo production of nonflagellate spores or biflagellate sperm and nonmotile eggs at the blade apex. Gamete production is thought to involve meiosis (Friedmann, 1959), but needs reevaluation.

Chapter 20

Green Algae IV
Chlorophyceans

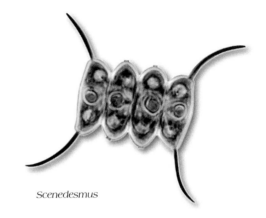

Scenedesmus

Chlorophyceans include some very familiar green algal genera. *Chlamydomonas* and *Volvox* are important laboratory model systems, *Selenastrum capricornutum* is widely recognized for its utility as a bioassay organism, and *Dunaliella* and *Botryococcus* can be valuable in production of industrially useful materials (Chapter 4). *Chlamydomonas* and *Selenastrum*, as well as other chlorophyceans, are commonly used in freshwater aquaculture systems to rear zooplankton as food for fish (Chapter 3). *Scenedesmus* and *Pediastrum*, among the most common members of the freshwater green phytoplankton, have fossil records and are implicated in formation of certain oil deposits based on their production of decay-resistant cell walls (Chapter 2). The sheer diversity of chlorophyceans in freshwaters has impressed ecologists interested in the roles of biotic diversity and functional redundancy in community and ecosystem stability.

In the past the term Chlorophyceae has been used to encompass all of the green algae—a very diverse assemblage of some 7000 described species. However, as detailed in Chapter 17, ultrastructural and molecular evidence obtained within the past few decades has demonstrated the existence of several distinct green algal lineages, including prasinophyceans (Chapter 17), ulvophyceans (Chapter 18), trebouxiophyceans (Chapter 19), and charophyceans (Chapter 21). Each lineage is characterized by specific differences in cellular features and primary habitat (marine, in the cases of prasinophyceans and ulvophyceans, and terrestrial in the case of trebouxiophyceans). Charophyceans, inhabitants of aquatic freshwaters and terrestrial sites, are closely related to plants.

The rest of the green algae, primarily freshwater or terrestrial in habitat, but not directly related to the ancestry of land plants, are currently included in the Chlorophyceae (this chapter). Some are known to belong to this lineage on the basis of ultrastructural and molecular data, whereas the membership of others is tentative, pending further study. It is possible that as new information comes to light, some of the organisms presented in this chapter may ultimately be grouped elsewhere.

Chlorophycean Thallus Types, Habitats, and Reproduction

Chlorophycean algae may occur as flagellate unicells, either as individuals or colonies; unicells that are non-flagellate in the vegetative (nonreproductive) state, but produce flagellate reproductive cells; unicells whose asexual reproductive cells are devoid of flagella; sarcinoid aggregations of nonmotile cells in which daughter cells fail to separate after division; a unique type of colony—a coenobium—in which the number and arrangement of cells can be constant (i.e., is under genetic control); unbranched filaments; branched filaments; and multinucleate, coenocytic (siphonous) forms. As discussed in Chapter 17, similar body types occur in other green algal lineages (classes), reflecting similar morphological adaptation to environmental selection pressures.

Chlorophycean cell division is also fairly uniform in that (1) mitosis is closed, (2) the telophase spindle collapses prior to cytokinesis, and (3) a system of microtubules (the phycoplast) running parallel to the cleavage plane appears at cytokinesis. In some chlorophycean genera, cleavage takes place by furrowing, the development of a centripetal ingrowth of the cell membrane from the periphery to the center of the cell. Cytokinesis by furrowing is regarded as plesiomorphic (primitive) because this mode of cytoplasmic partitioning is characteristic of early divergent protists. In other, putatively more specialized chlorophyceans, cytokinesis involves production of a cell plate that develops centrifugally by fusion of Golgi-derived vesicles. The presence of plasmodesmata—channels through the cell wall that allow for intercellular communication—is correlated with the cell-plate type of cytokinesis; plasmodesmata are absent from chlorophyceans that divide by furrowing. The presence of cell plates and correlative plasmodesmatal development in advanced chlorophyceans (Chaetophorales), certain specialized ulvophyceans (Trentepohliales), and late-diverging charophyceans (Coleochaetales and Charales) indicates that cell plates and plasmodesmata are specialized features that have evolved independently several times.

Chlorophycean flagellates inhabit fresh (or in a few cases, brackish and marine) waters. Nonflagellate forms occur in freshwaters or on soils, and may also occur on tree bark or other terrestrial substrates. Isogamous, anisogamous, or oogamous sexual reproduction gives rise to thick-walled, often spiny zygotes (also known as zygospores or hypnozygotes) (Fig. 20–1) that can persist through conditions unfavorable for vegetative growth, then germinate by meiosis when conditions allow survival of the products. Such hypnozygotes ("sleeping zygotes") thus function both as resting stages and in the generation of population-level genetic variation. In some cases decay-resistant vegetative cells seem also to be able to survive stressful environmental conditions, then resume growth when conditions improve. Dispersal can be accomplished by resistant vegetative cells and zygotes, or by asexual means. Asexual reproduction includes formation of one or more zoospores, aplanospores, or autospores within individual parental vegetative cells; the spores are subsequently released to the environment (see Fig. 1–21).

Zoospores are flagellate unicells that require liquid water for dispersal. They typically have an elongate, hydrodynamically adaptive shape and often possess eyespots, suggesting the presence of a light-sensing system. Zoospores of some genera have rigid cell walls, but those of others are naked. Scaly coverings—such as occur on many prasinophyceans and motile reproductive cells of some ulvophyceans and charophyceans—are rarely present on reproductive cells of chlorophyceans. Zoospores settle after a period of swimming, lose their flagella, become rounded (in the case of naked zoopores), then develop into the mature vegetative form. Aplanospores (nonmotile spores) are asexual reproductive cells that have some features typical of flagellate cells, such as con-

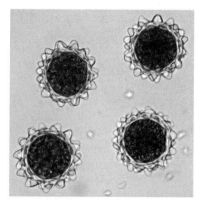

Figure 20–1 Hypnozygotes—zygotes that have thickened, often ornamented walls—allow many chlorophyceans to survive stressful environmental conditions. Hypnozygotes of the chlorophycean *Volvox* are shown here.

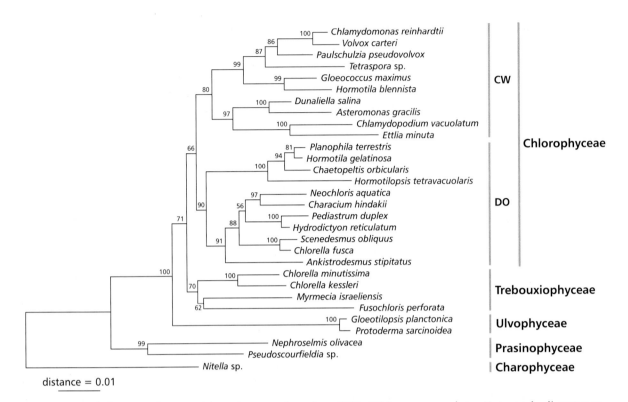

Figure 20–2 Phylogeny of some chlorophyceans based on SSU rDNA sequence data. Note early divergence into two major clades—one group with motile cells having directly opposed basal bodies (DO), and a second with clockwise-displaced basal bodies (CW). Both are thought to have been derived from ancestral forms having counterclockwise-displaced basal bodies (CCW). (After Booton, et al., 1998a by permission of the *Journal of Phycology*)

tractile vacuoles, but lack flagella. Aplanospores are regarded as cells whose further development into zoospores has been arrested. Autospores lack flagella and contractile vacuoles and appear to be small versions of vegetative cells.

Chlorophycean genera differ in the types of asexual reproductive cells they produce; some generate only zoospores or aplanospores, and others produce only autospores. Autospore-producing forms are presumed to have lost the ability to generate flagella. This is thought to be correlated with occupation of terrestrial habitats in which liquid water may not be reliably present. The kind of asexual reproductive cell produced, (whether zoospores, aplanospores, or autospores), and such features as presence of number of flagella on zoospores, occurrence of walls on zoospores, and presence of pyrenoids in reproductive cell chloroplasts are often used in identification and classification of chlorophyceans.

The flagellate cells of chlorophyceans—whether the vegetative cells of primarily flagellate genera or the zoospores and gametes of nonflagellates—have similar flagellar basal-body arrangements, reflecting close phylogenetic relationship. Flagella emerge from the cell apex (in contrast to charophyceans, where they emerge laterally), rhizoplasts typically link flagellar basal bodies with the nucleus, and there is a cruciate (cross-shaped) X-2-X-2 arrangement of microtubular roots (see Chapter 17). The number of microtubules in the X root varies among taxa. The flagellar basal bodies are displaced in a clockwise (CW) direction or are directly opposed (DO) (see Chapter 17). The reproductive cells of chlorophyceans usually possess either 2 or 4 flagella, but in a few cases, numerous flagella are present. Molecular phylogenetic analyses indicate that the arrangement of flagellar basal bodies (CW or DO) and the number of flagella reflect phylogenetic relationships in this group more accurately than do thallus structure, presence or absence of a cell wall, or occurrence of multinucleate cells (Booton, et al., 1998a; Nakayama, et al., 1996a, b).

Chlorophycean Diversity

In the past chlorophycean green algae have been segregated into orders primarily on the basis of organismal morphology. Flagellates, for example were classified into the Volvocales, coccoid unicells and nonmotile coenobial colonies were aggregated into the Chlorococcales, most unbranched filaments into the Ulotrichales, and most branched filaments into the Chaetophorales (Bold and Wynne, 1985). In the past few years the application of molecular phylogenetic methods has revealed that extensive parallel evolution of body form has occurred. It is thus no longer possible to group the chlorophycean algae into orders based only on their morphology, if classification systems are to reflect phylogeny as closely as possible. To a very large extent, the seemingly logical and convenient system by which chlorophyceans were formerly classified has collapsed, and is being replaced with a more evolutionarily accurate, but complicated, taxonomic scaffolding. In the emerging new classification orders may contain a mixture of flagellate unicells, nonflagellate cells, colonies, filaments, or other morphologies. Some taxa (such as *Chlamydomonas*) that were earlier considered to reflect ancestral forms are now revealed to be relatively derived, and others (Oedogoniales) that were previously regarded as advanced may actually prove to be relatively early divergent (Booton, et al., 1998b). The classifications provided here will reflect such changes, but must be viewed as tentative, as many chlorophycean taxa have not yet been studied by ultrastructural or molecular methods.

Most chlorophyceans fall into one of two major groups (clades)—the DO clade and the CW clade—based on flagellar ultrastructure and molecular sequence data (Fig. 20–2) (Booton, et al., 1998a). When more information is available, these clades will probably be taxonomically formalized, perhaps as subclasses, but such is currently premature. At one time it was thought that the CW clade was derived from the DO group, which was earlier derived from forms having counterclockwise-displaced flagellar basal bodies (CCW), but now the DO and CW clades are believed to have arisen independently from CCW ancestors (Booton, et al., 1998a, b). Here the DO group will be described first, followed by the CW clade, but from the standpoint of phylogenetic systematics, the reverse order would be equally appropriate. There are in addition, two major and distinctive monophyletic chlorophycean groups, Chaetophorales and Oedogoniales, that both include branched and unbranched forms, whose relationship to other members of the class Chlorophyceae is uncertain (Booton, et al., 1998b). In view of their relative structural and reproductive complexity, they will be described at the end of this chapter, but this arrangement does not reflect opinion on their evolutionary status.

The DO Clade

This chlorophycean lineage (Fig. 20–3) includes two monophyletic subgroups, the Sphaeropleales (as defined by Deason, et al., 1991, but not Bold and Wynne, 1985), and the Chaetopeltidales, which includes former members of the Tetrasporales (of older classifications) that have quadriflagellate motile cells. The Chaetopeltidales, which will not be further discussed here, is of interest because scales occur on the flagellate cells (O'Kelly, et al., 1994). The Sphaeropleales includes some taxa that produce DO motile cells and in addition, taxa lacking motile cells that are clearly related on the basis of ultrastructural and molecular sequence information. Molecular phylogenetic analysis suggests that the unicell *Ankistrodesmis stipitatus* is an early divergent member of this group. Other taxa known to belong to the Sphaeropleales include the unicells *Neochloris aquatica*, *Characium hindakii*, *Chlorella fusca*, and *Bracteacoccus*; the coenobial colonies *Pediastrum*, *Hydrodictyon*, and *Scenedesmus*; and the multinucleate filament *Sphaeroplea*.

A number of the taxa in this order have recently been disaggregated. For example, SSU rDNA and other data reveal that *Neochloris* and *Characium* contained species belonging to three different green algal classes, and *Chlorella* species occurred in two classes (Fig. 20–3) (Watanabe and Floyd, 1989; Deason, et al., 1991; Lewis, et al., 1992; Wilcox, et al., 1992). *Neochloris* species belonging to the Sphaeropleales retain this name, while others—newly recognized to be unrelated—were renamed *Ettlia* and *Parietochloris* and transferred to other groups (Watanabe and Floyd, 1989; Komárek, 1989; Deason, et al., 1991). Former *Characium* species found to be unrelated to *C. hindakii* were renamed *Chlamydopodium* and transferred to another group of chlorophyceans (Fig. 20–3).

In establishing new genera, efforts are made to identify phylogenetically significant morphological features that can be observed at the light microscopic level. For example, species retaining the generic name *Neochloris* are multinucleate, whereas *Ettlia* has uninucleate cells. *Chlamydopodium* cells produce zoospores having cell walls, whereas taxa retaining

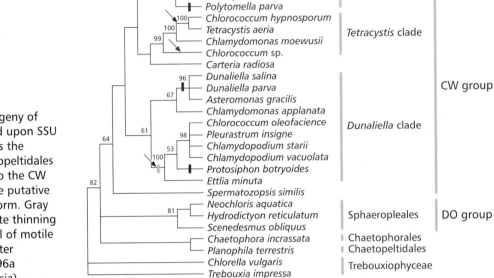

Figure 20–3 A phylogeny of chlorophyceans based upon SSU rDNA data that shows the relationship of Chaetopeltidales and Sphaeropleales to the CW clade. Arrows indicate putative gain of the coccoid form. Gray and black bars indicate thinning or loss of the cell wall of motile cells, respectively. (After Nakayama, et al., 1996a ©Blackwell Science Asia)

the name *Characium* produce wall-less (naked) zoospores. Identification of such features helps ecologists to correctly identify microalgae from mixed samples without having to resort to time-consuming and expensive ultrastructural or molecular sequence analyses.

Sphaeropleales

ANKISTRODESMUS (Gr. *ankistron*, fishhook + Gr. *desmos*, bond) occurs as long, needle- or spindle-shaped cells that may occur individually or be aggregated into groups (Fig. 20–4). Sometimes the cells are twisted around each other, but there is no mucilaginous covering. The chloroplast is parietal and a pyrenoid may be present or not. Asexual reproduction occurs by cleavage of parental cells into 1–16 autospores that are released by parental wall rupture.

Ankistrodesmus occurs in the plankton of freshwaters and in artificial ponds.

BRACTEACOCCUS (L. *bractea*, thin metal plate + Gr. *kokkos*, berry) occurs as spherical multinucleate unicells containing many discoid plastids without pyrenoids (Fig. 20–5). The biflagellate zoospores are naked. It occurs as a member of microbiotic (cryptogamic) crust communities on arid soils. A molecular phylogenetic analysis of nine strains, representing at least five species from four geographic locations,

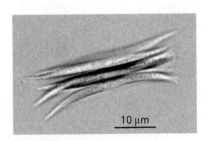

Figure 20–4 *Ankistrodesmus.*

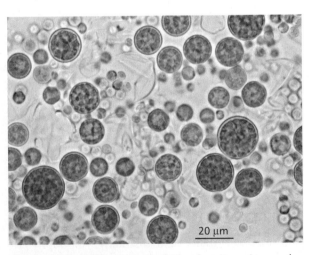

Figure 20–5 *Bracteacoccus.* Cells of various sizes and ages are evident as are remains of parental cell walls.

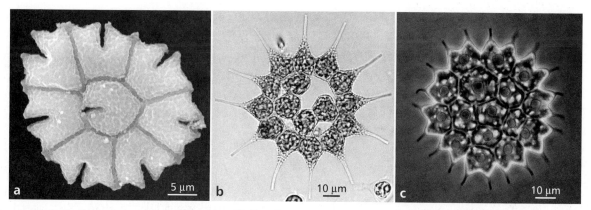

Figure 20–6 *Pediastrum* has coenobia with cells that lie in one plane. Species are readily identifiable using features visible with the light microscope. (a) SEM of *Pediastrum tetras*, showing the flattened, star-shaped pattern of cells. *Pediastrum simplex* (b) has a single horn on each cell, while *Pediastrum boryanum* (c) has two.

indicated that *Bracteacoccus* is monophyletic (Lewis, 1997).

PEDIASTRUM (Gr. *pedion*, plain + Gr. *astron*, star) is a very distinctive coenobial colony having a flattened, often starlike shape (Fig. 20–6). Each species consists of a determinate number of cells arranged in a specific pattern. The peripheral cells usually possess one or two hornlike projections, whereas internal cells may have a similar or different shape. Peripheral cells may also bear clusters of very long chitinous bristles; these are regarded as buoyancy (Gawlik and Millington, 1988) or herbivore deterrence devices (see Fig. 3–15). Cell walls contain silica and algaenans, hydrocarbon polymers that are believed to confer resistance to microbial decay and chemical hydrolysis (Gelin, et al., 1997).

Remains can be found in lake sediments, and ancient fossil remains are known. *Pediastrum* remains are also associated with certain fossil fuel deposits.

Asexual reproduction of *Pediastrum* occurs by autocolony formation. Cell protoplasts generate the same number of biflagellate zoospores as is the typical colonial cell number for the species. These zoospores are retained within a vesicle that is derived from the inner layer of the parental cell wall, and is liberated from the parental cell. After a short mobility period the zoospores aggregate in the same planar pattern that was present in the parental coenobium (Fig. 20–7). Daughter autocolonies are then liberated from the enclosing vesicle. Each colony of *Pediastrum* is thus potentially capable of generating as many autocolonies as it possesses cells. Each daughter colony will

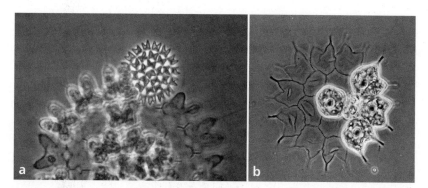

Figure 20–7 Autocolony production in *Pediastrum*. (a) The small-celled colony has just been released from one of the vegetative cells of the larger parental colony to its left. The cells of the daughter colony will expand until they reach the size characteristic for the species. In (b), autocolonies have been released from many cells of this colony through irregular openings; parental cell walls often persist for long periods because they contain decay-resistant compounds (algaenans).

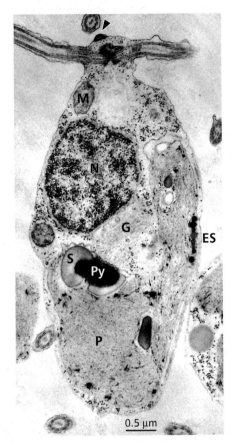

Figure 20–8 *Pediastrum* gamete. TEM view showing directly opposed flagellar bases and connecting structures, nucleus (N), and plastid (P) with eyespot (ES) (which is lacking in zoospores). An electron-opaque dome-shaped mating structure (arrowhead) is evident at the cell's apex. G=Golgi body, M=mitochondrion, Py=pyrenoid, S=starch. (From Wilcox and Floyd, 1988 by permission of the *Journal of Phycology*)

bia. Polyeders are also known to arise from asexual reproductive processes. An outstanding ultrastructural analysis of sexual and asexual reproduction in *Pediastrum* and related algae can be found in Pickett-Heaps (1975). *Pediastrum* is common in the plankton and entangled among the periphyton of many lakes, swamps, and bogs. The species can readily be distinguished by differences in colony patterning and cell ornamentation (Prescott, 1951).

HYDRODICTYON (Gr. *hydor*, water + Gr. *diktyon*, net), commonly known as the "water net," occurs widely in hard-water lakes and streams, sometimes forming nuisance growths that can blanket surfaces of freshwater ponds and small lakes, accumulate rotting masses at the edges, and clog boat engine intakes. Conspicuous growths can also be observed in waters associated with rice paddies, fish farms, and irrigation ditches. Extensive growths are usually associated with some degree of eutrophication. Its significant ecological impact can partially be explained by unusual cell and colony construction and reproductive mode.

The cells of adult colonies are very large (up to 1 cm long) and are each linked to three to eight (commonly four) other cells to form a reticulate pattern (Fig. 20–9). Cells are coenocytic, i.e., multinucleate, and contain a single, highly reticulate (netlike) chloroplast with many pyrenoids (Figs. 20–10, 20–11). The internal portion of these large cells is filled by a large vacuole; the cytoplasm is arranged around the periphery (Pickett-Heaps, 1975). The multiple nuclei arise from repeated closed mitoses without intervening cytokinesis. The number of cells in the colony is also very large, and colonies can be more than a meter long and 4–6 cm wide (Canter-Lund and Lund, 1995). It is

have the same number of cells and cell pattern as the parent. Inhibitor experiments have demonstrated that the arrangement of cytoskeletal microtubules is involved in the control of cell and colony shape (Millington, 1981; Marchant and Pickett-Heaps, 1974).

Sexual reproduction occurs by fusion of biflagellate gametes (Fig. 20–8) that are smaller than zoospores, and are liberated from their parental cell. Gametes produce eyespots, unlike vegetative cells, and have mating structures at the anterior end of the cell that are similar to those of *Chlamydomonas*. Zygote germination results in production of zoospores that quickly produce thick-walled, polyhedral unicells known as **polyeders**; these then generate new coeno-

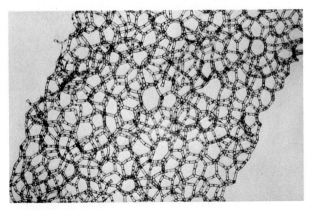

Figure 20–9 *Hydrodictyon* forms large netlike coenobial colonies.

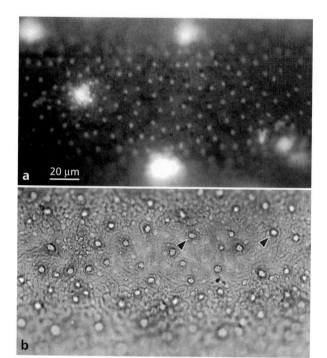

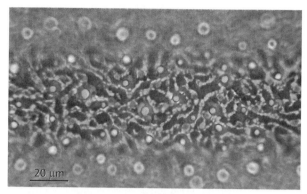

Figure 20–11 Portion of a *Hydrodictyon* cell viewed with phase-contrast optics, showing the reticulate nature of the plastid.

Figure 20–10 The same portion of a *Hydrodictyon* cell is viewed with (a) epifluorescence microscopy, showing many bright DAPI-stained nuclei (lobes of the faintly autofluorescent chloroplast are visible in the background), and (b) bright-field microscopy, showing the reticulate plastid with numerous pyrenoids (arrowheads).

parental cell wall, following a short period of motility, they adhere to other cells to form the typical net-like colony pattern. As in *Pediastrum*, microtubules are known to mediate the pattern formation process (Marchant and Pickett-Heaps, 1972). The young nets are released by disintegration of parental cell walls. Each parental net is capable of producing as many daughter nets as there are cells in the colony. As a result, if resources are plentiful, growth of *Hydrodictyon* can be explosive, explaining the occurrence of nuisance blooms.

An alternate form of asexual reproduction can occur, where zoospores are released from the parental cell and then form polygonal, thick-walled

interesting that the cell walls do not seem to accumulate significant levels of the resistant algaenans so characteristic of *Pediastrum* and *Scenedesmus* (see below). Resource allocation may have been evolutionarily diverted to rapid reproductive development rather than production of energetically expensive defensive compounds.

Asexual reproduction occurs by means of autocolony formation. Each cell's protoplast subdivides into as many as 20,000 biflagellate zoospores, and typical microtubular phycoplasts are involved in the multiple, synchronous cytokineses (Pickett-Heaps, 1975). During autocolony formation these zoospores are not released, and within the crowded confines of the

Figure 20–12 *Scenedesmus opoliensis*. Note the four adherent cells, each with a plastid and conspicuous pyrenoid (Py) and surrounding starch sheaths. Cells are ornamented with spines; these are useful in making species distinctions.

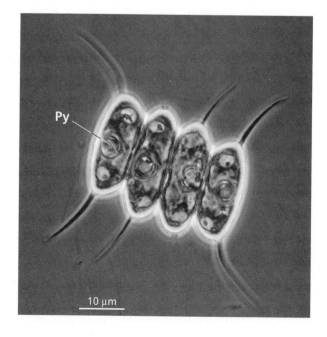

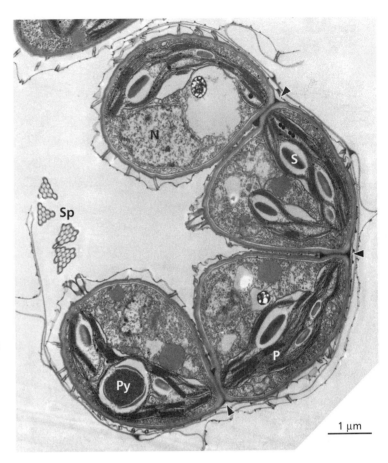

Figure 20–13 A cell of *Scenedesmus* undergoing autocolony formation. Two successive divisions of the mother cell have given rise to a linear coenobium. The cells are attached to each other by electron-opaque material (arrowheads). Note the plastid (P) with pyrenoids (Py) and intraplastidal starch (S). Cross sections of the spines (Sp) are also apparent. N=nucleus. (From Pickett-Heaps, 1975)

polyeders (similar to those of *Pediastrum*). Polyeders can germinate to produce small spherical nets by adherance of zoospores within an extruded vesicle (reminiscent of the asexual reproductive process of *Pediastrum*). Each cell of this net can then produce a new net by the process described above (Hatano and Maruyama, 1995). Sexual reproduction occurs by isogametes that are smaller than zoospores. They are released from parental cells through a pore in the wall. Gametes possess an anterior structure that is involved in the formation of a fertilization tube similar to that observed in *Chlamydomonas*. Gamete fusion produces a resting-stage hypnozygote that, after a period of dormancy, germinates (presumably by meiosis) to form four zoospores, each of which develops into a polyeder. These produce nets in a manner similar to that described above. An excellent ultrastructural survey of the events occurring in sexual and asexual reproduction of *Hydrodictyon* can be found in Pickett-Heaps (1975). The most common species is *Hydrodictyon reticulatum*, but there are several other described species. Molecular sequence analyses indi-

cate that *Hydrodictyon reticulatum* and *Pediastrum duplex* are closely related and are also related to *Scenedesmus* (see below).

SCENEDESMUS (Gr. *skene*, tent or awning + Gr. *desmos*, bond) is typically a flat coenobial colony of 4, 8, or 16 linearly arranged cells (Fig. 20–12), but sometimes one- or two-celled forms occur. Cells are cylindrical in shape and have rounded or pointed ends. Terminal cells, in particular, are often ornamented with short spines and tufts of chitinous hairs or bristles (up to 200 μm long) that are believed to confer buoyancy or to deter herbivores, or to space the algae for optimum light and nutrient availability (Trainor and Egan, 1988). In addition, cell walls may be highly ornamented with finer warts, combs, and reticulations that are more readily visible when viewed with the scanning electron microscope (see Pickett-Heaps, 1975). Cell walls contain algaenans, which are likely responsible for the occurrence of remains in lake sediments, ancient fossil deposits, and the association of *Scenedesmus*, along with

Pediastrum and *Tetraedron*, with certain fossil fuel deposits (Gelin, et al., 1997). These compounds may play an important role in adhesion of cells into colonies (Pickett-Heaps, 1975).

Cells contain a single plastid with pyrenoid and are uninucleate. Asexual reproduction is by autocolony formation, as in *Pediastrum* and *Hydrodictyon*, but with the difference that flagellate zoospores are not involved. Rather, parental cells divide to form nonflagellate cells that align themselves laterally, though often curled tightly within the confines of the parental wall (Fig. 20–13). Cleavage is typically delayed until after at least four nuclei have been produced by successive mitoses (Pickett-Heaps, 1975). Autocolonies are released by breakdown of the parental wall. Each of the more than 100 described species is characterized by distinctive cell number, cell arrangement, and wall ornamentation patterns. A single colony is capable of producing as many autocolonies as there are coenobial cells. One species has been reported to produce biflagellate isogametes under nutrient-limiting conditions (Trainor and Burg, 1965). *Scenedesmus* is a common inhabitant of the plankton of freshwaters and sometimes brackish waters, occasionally forming dense populations, but are not typically regarded as nuisance growths. As noted above, molecular sequence analyses link the ancestry of *Scenedesmus* to that of the coenobial taxa *Hydrodictyon* and *Pediastrum*. Three unicellular taxa previously regarded as species of *Chlorella* (*C. fusca* variants), are actually close relatives of *Scenedesmus* (Kessler, et al., 1997).

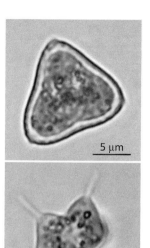

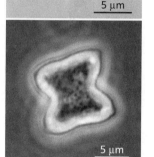

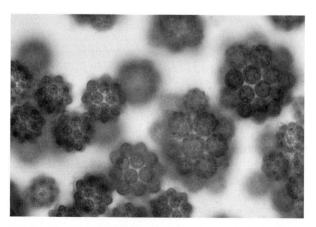

Figure 20–15 Three representatives of the genus *Tetraedron*.

COELASTRUM (Gr. *koilos*, hollow, + Gr. *astron*, star) cells are connected by blunt processes to form a hollow sphere (Fig. 20–14). The number of cells (as many as 128) and their arrangement in the colony are genetically specified, as is typical for coenobial colonies. Asexual reproduction resembles that of *Scenedesmus*, i.e., is by autocolony formation without the involvement of flagellate zoospores. *Coelastrum* and *Scenedesmus* are also similar in being phytoplankters in freshwater lakes.

TETRAEDRON (Gr. *tetra*—from *tessares*, four + Gr. *hedra*, seat or facet) (Fig. 20–15) is a single cell that may be polyhedral, pyramidal, triangular, or flat, and which often possesses distinctive spines. Dozens of species have been described from freshwater lakes and swamps. Like *Pediastrum* and *Scenedesmus*, the cell walls of *Tetraedron* contain algaenans, suggesting a close relationship among these taxa.

SPHAEROPLEA (Gr. *sphaira*, ball + Gr. *pleon*, many) is a free-floating, unbranched filament com-

Figure 20–14 *Coelastrum* cells are connected to one another by blunt processes to form hollow coenobia. Some cells have given rise to autocolonies, which can persist on the parental colony for some time.

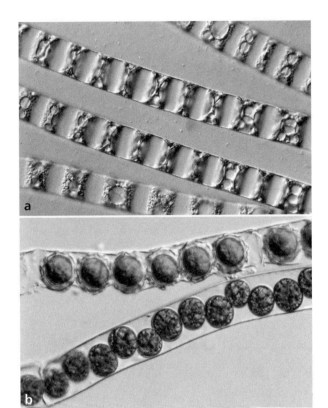

Figure 20–16 *Sphaeroplea.* (a) Vegetative cells. (b) Zygotes (top filament) and oogonia (bottom filament) are seen in this oogamous species. (Photographs courtesy Larry R. Hoffman)

posed of very elongate, multinucleate cells that are 15–60 times as long as broad (Fig. 20–16a). Cell walls are very thin, and no mucilage sheath is present. The protoplasm is segregated into ringlike units by several large, conspicuous vacuoles. Each protoplasmic ring includes several nuclei and a ring-shaped chloroplast with several pyrenoids. Mitosis and cytokinesis are uncoupled. Cross-wall development is very unusual but thought to involve a phycoplast. Sexual reproduction can be anisogamous or oogamous, and zygote (Fig. 20–16b) germination is presumed to involve meiosis; zoospores are produced that regenerate the filament. Systematic analysis of the species of *Sphaeroplea* was done by Buchheim, et al. (1990). The spindle-shaped, unicellular *Atractomorpha* is a close relative (Hoffman, 1983).

The CW Clade

This group (Fig. 20–2) includes the Volvocales, former members of the Tetrasporales that have biflagellate motile cells (Booton, et al., 1998a), the *Tetracystis* clade, and the *Dunaliella* clade (Nakayama, et al., 1996a). The Volvocales include many forms previously included in this order (Bold and Wynne, 1985), but some taxa have been removed, based on ultrastructural and molecular data. The *Tetracystis* clade was recently defined to include coccoid forms that are allied to *Chlamydomonas moewusii*. The *Dunaliella* clade includes flagellates (such as *Dunaliella*, *Haematococcus*, and *Stephanosphaera* that were formerly considered to be volvocalean), coccoid forms such as *Chlorococcum oleofaciens*, and the multicellular, siphonous *Protosiphon* (Nakayama, et al., 1996a). The new taxonomic alignment reveals that *Chlamydomonas* and *Chlorococcum* are polyphyletic, and suggests that coccoid habit and loss of the cell wall have occurred independently several times (Fig. 20–3).

Volvocales

This is an order named for the common and well-known genus *Volvox*, a colonial organism composed of *Chlamydomonas*-like cells. In this text unicellular flagellates, which will be referred to as chlamydomonads, colonial swimming forms, and colonial nonswimming forms (formerly included in Tetrasporales) are included within the Volvocales, as recommended by Hoek, van den, et al. (1995). Molecular phylogenetic analyses suggest that many of the colonial forms originated monophyletically from a particular clade of *Chlamydomonas* species (Buchheim, et al., 1994). Taxa formerly classified in this order (Bold and Wynne, 1985) that have now been moved to other groups include *Chlamydomonas moewusii*, *Chlamydomonas applanata*, *Dunaliella*, *Haematococcus*, and *Stephanosphaera* (Nakayama, et al., 1996a).

Chlamydomonads. Chlamydomonads are biflagellate unicells. The status of quadriflagellate taxa such as *Carteria* that were formerly included in this group is now uncertain (Nakayama, et al., 1996a). Non-tubular, glycoprotein flagellar hairs are common. There are usually two apical contractile vacuoles, but more may be present in some forms. There is a single plastid with pyrenoid surrounded by an I_2KI positive starch shell, and an eyespot, composed of globules of carotenoids, located within the plastid.

CHLAMYDOMONAS (Gr. *chlamys*, mantle + *monas*, unit) (Fig. 20–17), usually occurs as a biflagellate unicell having a single cuplike chloroplast and is a very common organism. There are nearly 500 described species (Ettl, 1976), separated by such features as chloroplast positioning and number of pyrenoids. It is important to recognize that *Chlamydomonas* is not a monophyletic genus, hence a number of species need to be renamed (Buchheim, et al., 1990). A list of species (including *C. reinhardtii*) determined by molecular sequence evidence to be closely related, and therefore likely to retain the name *Chlamydomonas*, can be found in Nakayama, et al. (1996a). *Chlamydomonas* occurs in a wide array of habitats including freshwaters of all types, soil and other terrestrial habitats, plus snow and ice (see Text Box 20–1). Some species grow in arctic or antarctic pools (Happey-Wood, 1988).

A few species, such as *C. reinhardtii*, have been particularly well studied (Harris, 1989), and have become major model systems for genetic (i.e., mutant-based) studies of eukaryotic cell structure and function. *Chlamydomonas* is favored for use in laboratory work, particularly in molecular genetics studies, because it is haploid, it is easy to cultivate, it grows rapidly, and sexual reproduction can readily be induced. *Chlamydomonas* is an excellent system for learning about such fundamental aspects of eukaryotes as the roles of dynein in flagellar function (Smith and Sale, 1992) and centrin in the cytoskeleton and mitosis (Salisbury, et al., 1988), genetics of the deflagellation pathway (Finst, et al., 1998), localization of GTPases within the cytoskeletal apparatus (Huber, et al., 1996), and function of contractile vacuoles (Luykx, et al., 1997a, b; Robinson, et al., 1998). It is also an excellent system for studying chloroplast genetics and maternal inheritance (e.g., Ikehara, et al., 1996) and the regulation of plastid genes (Suzuki, et al., 1997; Davies and Grossman, 1998).

The description of *Chlamydomonas* biology that follows is based almost entirely upon studies of *C. reinhardtii*; related species are assumed to be similar, but relatively few have been studied in detail. The layered cell wall is not composed of cellulose, but rather of polymers rich in the amino acid hydroxyproline that are linked to the sugars galactose, arabinase, mannose, and glucose, and hence known as glycoproteins. Flagella emerge through pores in the wall, protected by flagellar collars (which, however, are not present in all species). When growing under conditions in which liquid water is limiting, cells may occur as palmelloid stages, i.e., groups of varying numbers of nonflagellate cells held together by common mucilage. This mucilage, like the cell walls, is rich in hydroxyproline and sugars. When palmelloid aggregates are exposed to water, the cells typically transform to flagellates.

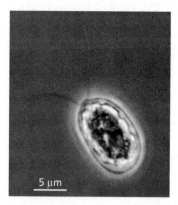

Figure 20–17 *Chlamydomonas*, which has a rigid cell wall, two equal flagella, and a parietal, cup-shaped chloroplast.

Chlamydomonas cells, like those of other flagellate chlorophycean cells, swim by breaststroking (Goldstein, 1992). During the power stroke the flagella are held in a nearly straight position, bending only at the bases. During the return stroke, a bending wave beginning at the base is propagated to the tip. Observations of changes in position of the eyespot reveal that the cells rotate while swimming. One of the flagella beats by itself about once in every 20 strokes, resulting in an overall helical path of motion. Internally, the flagella of *Chlamydomonas* are complex and resemble those of other eukaryotes. Much of what is known about eukaryotic flagella has been learned by study of *Chlamydomonas*.

A central pair of microtubules is surrounded by nine doublets, bearing arms composed of the protein dynein, a multiprotein assembly that is responsible for generation of force. The dynein arms connect adjacent doublets, and flagellar beating results from the sliding interaction of dynein with neighboring microtubules, which is coupled to ATP hydrolysis. Radial spokes link doublets with the central microtubule pair. The flagella of *Chlamydomonas* include over 200 polypeptides; 22 of these are associated with the central-pair complex, more than 20 with the dynein arms, and 17 with radial spokes (Kamiya, 1992). Analysis of mutants has demonstrated that cells can continue to swim even if flagella lack the central microtubule pair or the radial spokes.

Chlamydomonas cells are typically phototactic, swimming toward moderate light, but away from intense light. Phototactic behavior is believed to be based upon a rhodopsinlike photoreceptor located in the cell membrane just over the eyespot (Foster, et al.,

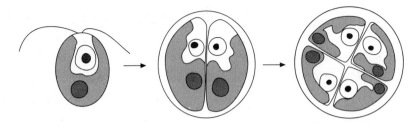

Figure 20–18 Asexual reproduction in *Chlamydomonas* usually results in the production (through two successive mitotic divisions) of four daughter cells, each of which may develop two flagella.

1984; Sineshchekov and Govorunova, 1999). Though the eyespot is probably not the primary light-sensing structure, it may function to shade the photoreceptor during rotational swimming, thus providing directional information. An instructional laboratory exercise designed to demonstrate phototactic behavior in *Chlamydomonas* is described in Harris (1989). Dutcher (1988) provides an overview of molecular analysis of flagellar basal-body function and assembly in *Chlamydomonas*. A myosinlike protein has also been identified in *Chlamydomonas*, suggesting that actin-myosin interactions occur, as in other eukaryotes (La Claire, et al., 1995).

All *Chlamydomonas* species are capable of autotrophic growth, and some, such as *C. eugametos* (Wetherell, 1958) are obligate autotrophs. However numerous species, including *C. reinhardtii*, can utilize acetate (but not glucose) as a source of exogenous dissolved organic carbon. Because *Chlamydomonas reinhardtii* can grow in the dark with acetate as a carbon source, it has become a preferred species for genetic work. Photosynthetic mutants, for example, can be rescued by growth on acetate, whereas such mutants of obligate autotrophs would be lost. Although *C. reinhardtii* does not require vitamins, some other *Chlamydomonas* species require one or more B vitamins and are thus described as auxotrophs.

Asexual reproduction of Chlamydomonas. Parental cells most often produce 2, 4, 8, or 16 progeny cells by successive mitotic divisions (Fig. 20–18). Division occurs longitudinally in cells whose flagellar apparatus has been duplicated. However, the sequence of divisions exhibited by *Chlamydomonas* is more sophisticated than the straightforward longitudinal division characteristic of many flagellate protists, involving changes in the polarity of successive divisions that have been compared to sequences of cell cleavages that occur in early animal embryos (Schmitt, et al., 1992). As in other chlorophyceans, mitosis is closed, and cytokinesis involves phycoplast microtubules. Cytokinesis occurs by furrowing. Progeny cells are released from the confines of the parental wall by production of specialized sporangial wall autolysins that digest the cell wall. These autolysins, also known as vegetative lytic enzymes (VLE), are produced only by newly formed cells; they do not occur at any other life-cycle stage, and they are not capable of digesting walls other than that of the parental cell. Such specificity is necessary to avoid digestion of the walls of the newly formed cells. The nature of changes in parental walls that enable their digestion is unknown. Comparative analysis of the VLEs obtained from cultures have been used to classify *Chlamydomonas* species, and at least one such grouping (VLE 14) is congruent with SSU rDNA sequence data (Buchheim, et al., 1997b).

Sexual reproduction in Chlamydomonas. Gamete production, syngamy (gamete fusion), and zygote germination are known best in just a few species, including *C. reinhardtii*. Some species are isogamous, some are anisogamous, and a few oogamous. Gamete development is induced by low levels of combined nitrogen. Some species are homothallic (monoecious), i.e., mating will occur in clonal cultures. However in species such as *C. reinhardtii*, cells of a single clone will not mate; genetically different mating types are required (i.e., they are heterothallic) (Fig. 20–19). In isogamous forms, mating types are commonly designated mt^+ and mt^-; for oogamous forms, the terms egg and sperm can be used. An early event in gamete development is the production of mating structures that begin as electron-opaque rings in the cell membrane at the cell apex. Filaments associated with the ring appear to be composed of actin. This region of the mt^+ cell then forms a bulge at the cell anterior and extends into a mating tube that links up with the mating ring on the mt^- cell (Goodenough, et al., 1982).

In addition, linear glycoprotein molecules (agglutinins) appear on gamete flagellar surfaces; these molecules are absent from vegetative cells. Agglutinins

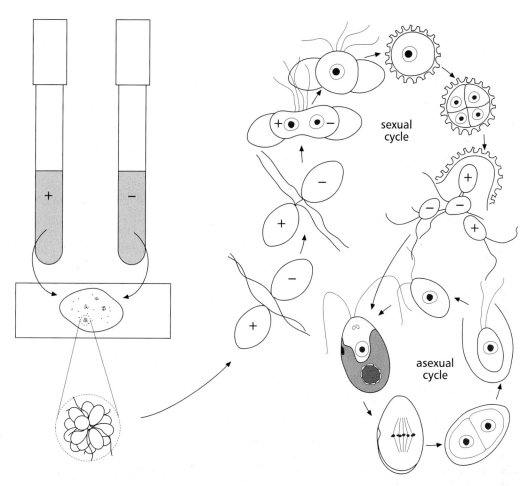

Figure 20–19 Sexual reproduction in *Chlamydomonas*. When + and − mating types are mixed on a glass microscope slide (left), the cells first clump, and then mate in pairs (right, clockwise). Contact is made first by the flagella, then a mating structure links the cells and plasmogamy begins. The quadriflagellate zygote loses its flagella and cell wall, and a new, spiny wall appears. The zygote divides by meiosis, yielding equal numbers of biflagellate daughter cells of the two mating types; these can undergo asexual reproduction repeatedly to generate large populations. (After Bold and Wynne, 1985)

promote the adhesion of flagella of cells of opposite mating types. Large groups of cells may clump together during the early stages of mating, but eventually pairs separate from the agglutinated mass. Flagellar tips adhere first, then the flagellar pairs become attached through their entire length (Fig. 20–19). As a result, the anterior regions of the mating cells are brought into close proximity, and protoplasts merge via the mating tube. The flagella then disengage and those of the mt$^+$ gamete continue to function, propelling the zygote (known as a planozygote because it is motile). Flagellar agglutination is not a necessary component of sexual reproduction for all *Chlamydomonas* species (Harris, 1989).

Autolysins of a different kind from those involved in lysing parental cell walls after asexual reproduction are expressed in mating cells; these break down the gamete cell walls. Gamete autolysins are capable of breaking down cell walls of any life-history stage of the same species, or those of closely related species. These 62 kDa proteins that include zinc, appear to act on a few cell-wall polypeptides, and are useful in preparing protoplasts of *Chlamydomonas* cells for experimental work (Harris, 1989).

Zygotes eventually become nonmotile and develop a thick, spiny wall; they are then known as hypnozygotes. Zygote walls are rich in glycoproteins, including hydroxyproline, as are vegetative cells, but

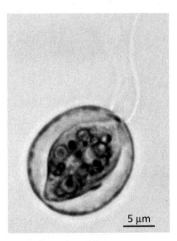

Figure 20–20 *Dysmorphococcus* has a mineralized lorica that is well separated from the cell membrane. There are two flagella that extend through pores in the lorica.

the sugars are present in different proportions. Zygotes of at least some *Chlamydomonas* species are known to include a layer of resistant material identified as sporopollenin (Van Winkel-Swift, et al., 1997). This presumbly confers some degree of resistance to microbial attack during dormancy. After a period of dormancy that can be relatively short, zygotes germinate by meiosis. Typically four but occasionally eight chlamydomonad cells are produced. These can develop flagella, if liquid water is available, but if zygotes are germinated on an agar surface, the non-flagellate meiotic products can be separated, grown into clonal populations, then isolated. This procedure is known as tetrad analysis, and is a commonly used technique in the field of haploid genetic analysis. Unmated gametes can dedifferentiate into vegetative cells. Sometimes fused gametes do not generate typical zygotes, but rather continue to exist as diploid flagellate cells. A description of the major methods used in genetic analysis of *Chlamydomonas* and a listing of mutants can be found in Harris (1989).

DYSMORPHOCOCCUS (Gr. *dys*, bad + Gr. *morphe*, form + Gr. *kokkos*, berry) consists of a *Chlamydomonas*-like protoplast within a lorica that is often colored by iron or manganese minerals (Fig. 20–20). The structure and development of the lorica was studied at the ultrastructural level by Porcella and Walne (1980). Progeny cells, formed similarly to those of *Chlamydomonas*, are liberated by rupture of the parental lorica.

Colonial Swimming Volvocaleans. This group includes about a dozen genera (60 species) of colonial flagellates. *Chlamydomonas*-like cells occur in multiples of two that are either enclosed by a common mucilaginous matrix or loosely held together by mucilage. The colonial swiming volvocaleans differ from chlamydomonads in that all sister cells produced by division of a single parental cell adhere to one another and swim as a unit after escaping from the confines of the parental cell (Schmitt, et al., 1992). Excluding *Stephanosphaera* (which has been moved to a different chlorophycean group), colonial volvocaleans so far examined by the methods of molecular systematics appear to be a monophyletic group, descended from an ancestor related to the modern *Chlamydomonas reinhardtii* (Larson, et al., 1992). There is considerable variation in size (i.e., number of cells) among colonial genera, some (*Gonium*) consisting of as few as four cells, and others (*Volvox*) containing thousands. Parallel evolution of variously sized colonial flagellates has also occurred among the freshwater ochrophytes (Chapter 13).

The flagellar apparatus of colonial swimming volvocaleans is similar to that of *Chlamydomonas*, but has undergone change in relative positioning of the basal bodies. In *Chlamydomonas* and close relatives, the flagella typically lie in a V-shape arrangement and beat in opposing directions, while in volvocalean forms the flagella are parallel and beat in the same direction. This change in orientation and beat is thought to represent a selective advantage associated with the colonial habit—when cells lie close together, beating flagella do not interfere with each other as they would if positioned as in *Chlamydomonas* (Hoops and Floyd, 1982).

Colonial swimming volvocaleans occur either in slightly curved plates, or in spherical arrays. The number of cells in the colony and colonial shape are genetically determined and specfic for each taxon. Asexual reproduction involves production of daughter colonies by successive bipartition of parental cells; these are called autocolonies (or coenobia). In some taxa all cells of the colony are capable of autocolony formation, but in others only certain cells, known as gonidia, are capable of generating daughter colonies. When the organism is a hollow sphere, as in *Volvox*, daughter colonies develop with their cell apices facing inward, and the colony must go through a programmed inversion so that flagella are in contact with the external environment (Fig. 20–21). Sexual reproduction may occur by isogamy, anisogamy, or oogamy.

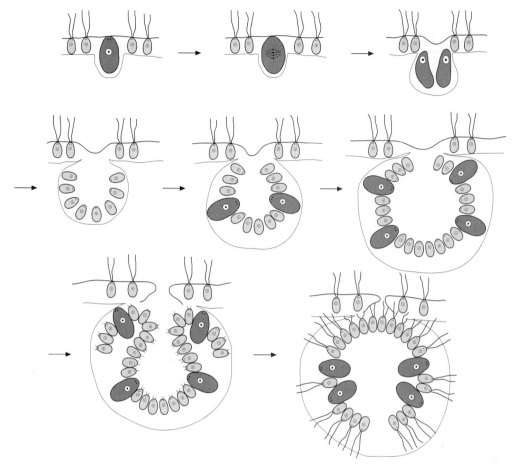

Figure 20–21 Autocolony development in *Volvox* begins with division of a specialized gonidium (darkly shaded cells) to form a spherical daughter colony, some cells of which are also gonidia. When development is complete, the colony inverts so that flagella (produced by vegetative cells, but not the gonidia) are oriented toward the outside. If colony inversion did not occur, the flagella would be produced on the inside of the hollow spheres, and the colonies would be unable to swim. Mature daughter colonies (autocolonies) eventually escape from the confines of the parental sphere. (After Pickett-Heaps, 1975)

Usually spiny, thick-walled zygotes (Fig. 20–1) are formed that germinate by meiosis.

Colonial volvocaleans commonly occur in the summer plankton of freshwater lakes, following the spring diatom bloom. Large size is regarded as adaptive in avoiding grazing pressure. Though the very large colonies of *Volvox* seem highly resistant to zooplankton grazing, young autocolonies of the smaller genus *Eudorina* are more vulnerable to predation (Happey-Wood, 1988). Colonial volvocaleans often require at least one B vitamin.

GONIUM (Gr. *gonia*, angle) is a colony of 4, 8, or 32 (the number depending upon the species) chlamydomonad cells arranged in a flat plate (Fig. 20–22). As they swim, the colonies rotate (Canter-Lund and Lund, 1995). All cells are capable of forming autocolonies. Asexual reproduction occurs by loss of motility and subsequent division of each cell to form a colony of the size typical for the species. Sexual reproduction involves dissociation of colonies into single cells, which then function as isogametes. It is likely that different mating types are generally required for syngamy to occur. Zygotes germinate by meiosis to produce a four-celled colony. *Gonium* occurs in lakes with hard waters and a high nitrogen content; it may also grow in barnyard ponds and watering troughs (Prescott, 1951). The common species *G. sociale* has been renamed as *Tetrabaena socialis* (Nozaki and Itoh, 1994).

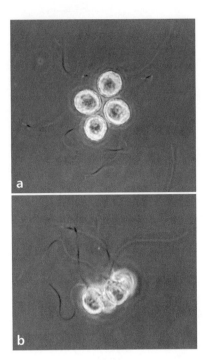

Figure 20–22 *Gonium*, typically a four- or eight-celled colony, each cell of which has two flagella, seen from the top (a), and from the side (b).

PLATYDORINA (Gr. *platys*, flat + [*Pan*]*dorina*) has flattened, slightly twisted colonies of 16–32 cells. There are characteristic lobed projections of the colonial mucilage at the posterior of the colony (Fig. 20–23). All cells are capable of autocolony formation, which begins with the formation of spherical colonies

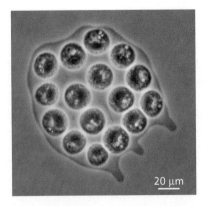

Figure 20–23 *Platydorina* is a flattened colony of 16 or 32 cells embedded in mucilage. Flagella, borne in pairs on each cell, were not on this specimen when photographed. The posterior of the colony has distinctive lobed projections.

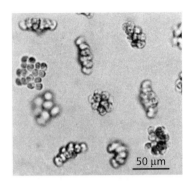

Figure 20–24 *Platydorina* colonies, like other volvocaleans, also undergo inversion during their development from hollow spheres to flat plates of cells. (Photograph by R. Starr, in Bold and Wynne, 1985)

that must turn inside out, then flatten (Fig. 20–24). Sexual reproduction is dioecious, and gametes are similar to vegetative cells. Zygote germination is by meiosis (see references in Bold and Wynne, 1985).

PANDORINA (Pandora, mythological woman) is a globular colony of 16–32 biflagellate cells that are closely adherent at their bases (Fig. 20–25). The posterior ends of cells are somewhat narrowed. Colonies swim with a rolling motion. The eyespots of cells in the anterior of the colony are larger than those of the posterior, marking the occurrence of some degree of colony polarity. All cells are capable of forming autocolonies. *Pandorina* occurs in both hard and soft waters. Molecular systematic analysis suggests that, with the exclusion of *Pandorina unicocca*, which has

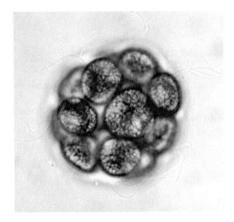

Figure 20–25 *Pandorina* is a ball of cells which are adherent at their bases and embedded in mucilage. Flagella are indistinct in this image.

been renamed as *Yamagishiella*, *Pandorina* is probably a monophyletic genus (Nozaki, et al., 1995).

EUDORINA (Gr. *eu*, good, true, or primitive + [*Pan*]*dorina*) is a globular colony of spherical biflagellate cells that are not as closely adherent as those of *Pandorina* (Fig. 20–26). The 16 or 32 cells in a colony first enlarge, then divide four or five times to form daughter colonies of 16 or 32 cells. Sometimes, however, the four most anterior cells do not enlarge or undergo divisions, eventually dying. This is viewed as an early stage in the evolutionary development of the more extensive restriction of the reproductive role to posterior cells in *Volvox*. Such division of labor within the colonies is considered to be evidence for the occurrence of true multicellularity. In contrast, the smaller species of volvocaleans (such as *Gonium*) are considered to be colonies of equivalent cells (Schmitt, et al., 1992). Sexual reproduction involves separate male and female colonies and is anisogamous. The individual cells of female colonies function as the larger female gametes, whereas cells of male colonies undergo division to form packets of adherent small sperm. These sperm packets swim as a unit to female colonies, and only upon reaching them do they dissociate into individual sperm cells. Zygotes germinate by meiosis, and variable numbers of meiotic products survive (see references in Bold and Wynne, 1985). *Eudorina* is regarded as common in the plankton of hard-water lakes (Prescott, 1951). Molecular phylogenetic analysis suggests that *Eudorina* is a polyphyletic species

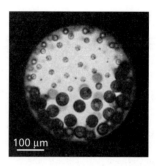

Figure 20–27 *Pleodorina* colonies are differentiated at maturity into either small anterior cells unable to produce autocolonies or larger posterior cells that can produce daughter colonies. (Photograph by R. Starr, in Bold and Wynne, 1985)

(Nozaki, et al., 1997a); some species will thus require renaming as new genera.

PLEODORINA (Gr. *pleon*, more + [*Pan*]*dorina*) differs from *Eudorina* in having both small and large cells in the same colony (Fig. 20–27). All 128 or 256 cells are initially the same size and morphology, but about two thirds of the cells in the colony posterior are able to enlarge and divide, forming autocolonies. The anterior-most cells are unable to form daughter colonies, and they ultimately die. Sexual reproduction may be dioecious (heterothallic) or monoecious (homothallic). Molecular systematic analysis suggests that this genus diverged from one clade of *Eudorina* (Nozaki, et al., 1997a).

VOLVOX (L. *volvo*, to roll) (Fig. 20–28) is a very large motile colony that is common in ponds and lakes, in warm weather sometimes forming blooms that move up and down in the water column (Smith, 1917). *Volvox* colonies contain 500 to several thousand cells arranged at the periphery of a mucilage shell. External to the mucilage there is a boundary layer that is similar to the hydroxyprotein-rich glycoprotein wall of *Chlamydomonas reinhardtii*, but innermost regions of the extracellular matrix of *Volvox* are chemically different (see references in Schmitt, et al., 1992). DNA sequence analyses suggest that Volvox is not monophyletic, i.e., that different species have arisen independently from ancestors resembling distinct modern groups of smaller colonial forms (Larson, et al., 1992; Nozaki, et al., 1995). Therefore, some species of *Volvox* will have to be renamed as new genera.

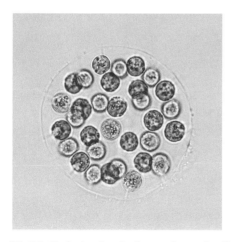

Figure 20–26 *Eudorina* is a hollow colony of cells that occupy the periphery of the mucilaginous sphere. Each cell bears two flagella.

Text Box 20–1 Red and Green Snow

Snow algae are known from every continent and occupy one of the most extreme environments on our planet. Cells may be exposed to high irradiation levels, high acidity, low temperatures, low nutrients, and extreme desiccation after snow melt (Hoham, 1980; Hoham and Duval, 1999). Red-colored

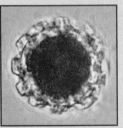

Chlamydomonas nivalis zygote—causes red-burgundy snow

snow, which is characteristic of open exposures, has been observed on snow, pack ice, and ice floes in the Arctic and the Antarctic, and on snowbanks in high-altitude mountain regions. In contrast, green- and orange-colored snows are more commonly found in high-elevation forested regions, but may also be associated with high nutrients in open-exposed bird rookeries in Antarctica. Though a wide variety of organisms can be found in red-colored snow (Marchant, 1998; Hoham and Duval, 1999), probably the most commonly encountered red snow alga is the chlorophycean *Chlamydomonas nivalis*. It is found, for example, in Yellowstone Park, Wyoming, and the Sierra Nevada, California (Weiss, 1983).

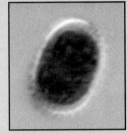

Chlamydomonas granulosa zygote—causes orange snow

Chlamydomonas nivalis commonly occurs in snow as 10–50 μm spherical resting cells that are red-pigmented; the green to red biflagellate stage is frequently overlooked. The red pigment is astaxanthin, a carotenoid that occurs as an ester in cytoplasmic lipid droplets. The ester group is thought to bind fatty acids, thus allowing accumulation of astaxanthin in the lipid droplets. The astaxanthin has been hypothesized to limit inhibition and photodamage of the photosynthetic apparatus that results from high irradiance levels in high-altitude and high-latitude habitats (Bidigare, et al., 1993). This hypothesis is supported by experiments conducted with *Haematococcus*, another chlorophycean that accumulates carotenoids in cytoplasmic droplets. Chlorophyll fluorescence measurements (see Chapter 23) were used to detect a positive correlation between higher cellular levels of carotenoids and the capacity to withstand high light (Hagen, et al., 1994).

Another interesting fact about *Chlamydomonas nivalis* is that numerous bacteria having typical gram-negative walls occupy a loose fibrous network on the surface of the cell wall. This bacterial-algal association seems to be typical of red snow populations (Weiss, 1983), but little is known about the identity of the bacteria. In addition, snow algae are often found in loose association with large populations of bacteria (Thomas and Duval, 1995), compared to control snow with low populations. The interactions between microbes in snow needs further study (Hoham and Duval, 1999).

Chloromonas (Gr. *chloros*, green + Gr. *monas*, unit) is another common snow alga whose vegetative cells or zygotes color green and orange snow (Hoham and Blinn, 1979). It is very much like *Chlamydomonas* except that pyrenoids are said to be absent from the plastid. The genus is commonly isolated from soils, snow, or cold peat. In one study of a *Chloromonas* from snow, optimal mating occurred in the blue end of the spectrum under an irradiance level of 95 μmol m^{-2} s^{-1} (Hoham, et al., 1998). A molecular phylogenetic analysis of several *Chloromonas* species indicated that this genus probably arose from *Chlamydomonas*, but is not monophyletic (Buchheim, et al., 1997a).

A number of desmids (see Chapter 21), including *Cylindrocystis*, *Mesotaenium*, and *Ancyclonema*, were found in glacial snow and ice in Nepal (Yoshimura, et al., 1997).

(Photographs courtesy R. Hoham)

Volvox colonies are so large that most forms can be seen with the naked eye. Conventional preparation of microscope slides results in crushed cells, but "hanging drop" preparations allow microscopic visualization of swimming cells without crushing them.

Volvox colonies cannot completely reverse direction while swimming, but they can change direction by stopping or slowing flagella on one side of the colony (Goldstein, 1992). They are phototactic—able to swim toward light of moderate intensity.

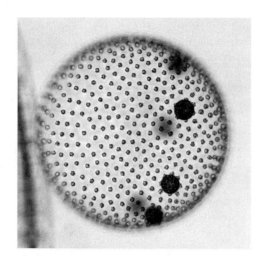

Figure 20–28 *Volvox*.

Within a *Volvox* colony each *Chamydomonas*-like cell possesses its own mucilage. Protoplasmic strands appear to interconnect adjacent cells in most species (Fig. 20–29), but such connections are lost from the mature forms of a few species. The colony has a polar organization; anterior-most cells have larger eyespots, suggesting enhanced phototactic ability, whereas daughter colonies and gametes tend to develop from one of a relatively few gonidia at the posterior. Gonidia, which arise from asymmetric divisions that occur early in autocolony formation, are significantly larger than neighboring nonreproductive cells. A number of genes have been recognized as being involved in gonidial differentiation (Schmitt, et al., 1992; Kirk, 1997). Gonidial division proceeds until a hollow ball of cells is formed. A pore appears at the eight-celled stage, beginning as a region lacking in intercellular protoplasmic connections. When the number of cells in the daughter autocolony reaches that characteristic for the species, the colony inverts through the pore (also known as a philiapore) (Schmitt, et al., 1992). Daughter colonies are released by rupture of the parental colony surface.

Sexual reproduction begins when one or a few colonies are induced to develop male gametes. The initial inducing factor is not understood; one hypothesis is that heat shock may be involved, i.e., that sexual reproduction may occur when water temperature reaches a critical high level, particularly in shallow pools or temporary ponds or puddles (Kirk, 1997). Induced male colonies produce a potent chemical sexual attractant, or pheromone. This substance is a glycoprotein that diffuses through the water, inducing many other *Volvox* colonies to become sexually competent. It has been estimated that such pheromones can be effective at concentration below 10^{-16} M. Mass sexual reproduction can occur if numerous colonies are exposed to the pheromone. Gonidia exposed to the pheromone undergo an alteration of their developmental fate from production of daughter autocolonies to production of a single egg cell or a flat packet of 16–64 elongate, pale, biflagellate sperm (Fig. 20–30). Although some species are monoecious, most are dioecious. Upon liberation of sperm packets from parental colonies, they swim to female colonies, enzymatically lyse a hole in the colony mucilage, then dissociate into individual sperm that fertilize the eggs. Eggs that have not been fertilized can develop into new colonies (Kirk, 1997). Zygotes develop thick, spiny walls (Fig. 20–1) that

Figure 20–29 Intercellular cytoplasmic strands (arrowheads) link each *Volvox* cell with its neighbors. Images (a) and (b) represent a through-focus series of a portion of a *Volvox* colony. (a) In one plane of focus the intercellular strands are visible. (b) In the other the strands are not evident, but bacterial cells (arrows) can be observed on the surface of the mucilaginous matrix.

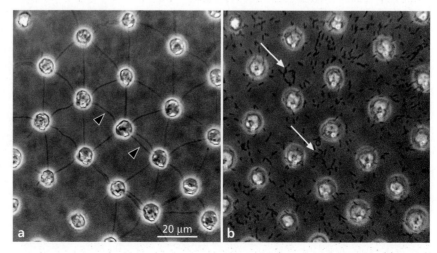

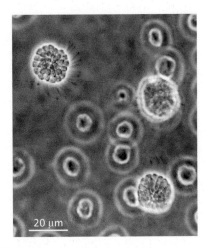

Figure 20–30 Sperm packets of *Volvox*.

serve in perennation; they may also acquire red coloration from production of carotenoid pigments. In the *Volvox* species that have been studied, only a single meiotic product—a small biflagellate cell—survives; the other three apparently die. Successive divisions of this cell regenerate the colonial form.

Nonswimming Colonial Volvocaleans. Unlike the swimming volvocaleans that are members of the freshwater phytoplankton, the nonswimming forms (formerly Tetrasporales) primarily occupy periphytic habitats, where they grow entangled among other algae or attached to plants or other substrates in shallow water. *Chlamydomonas*-like cells are embedded within mucilaginous matrices that may form stalks or flat sheets. In some cases, such as *Gloeococcus tetrasporus* (Kugrens, 1983), cells may be weakly motile within the mucilage, but in most forms the cells are nonmotile except during a reproductive phase. Cells may contain contractile vacuoles and eyespots and can easily transform into biflagellate zoospores or gametes. (Former tetrasporaleans with quadriflagellate motile cells have been transferred to the DO clade by Booton, et al. [1998a]). Many types of algae may produce stages that consist of mucilage-embedded arrays of cells. These are termed "palmelloid stages" or "palmella stages" after the genus *Palmella* (from the Greek word *palmos*, quivering), which belongs to this group.

Like the swimming colonial volvocaleans, nonswimming forms may accrue herbivory avoidance advantages by forming cell aggregates within mucilage. Absence of motility may confer energetic advantages. By attaching to shallow freshwater substrates, these algae are exposed to reasonable levels of illumination without having to swim continuously to avoid sedimentation to the benthos. Taxa that have been shown by molecular systematic analysis to belong to this group, but that are not covered here, include *Paulschultzia pseudovolvox*, *Gloeococcus maximum*, and *Hormotila blennista* (Booton, et al., 1998a) (Fig. 20–2).

TETRASPORA (Gr. *tetra*—from *tessares*, four + Gr. *spora*, spore) (Fig. 20–31) is a macroscopic tubular or saclike aggregation of chlamydomonadlike cells within a mucilaginous matrix. *Tetraspora* is usually found growing attached to substrates, often in cold, running freshwaters. Larger stringy masses or sheets may become free-floating. Cells are often arranged in groups of four that are the products of division of a parental cell, providing the basis for the generic name. Each cell has a single, parietal, cup-shaped plastid with pyrenoid, contractile vacuoles, and two pseudocilia. Pseudocilia are long, threadlike extensions from the cell to the edge of the mucilage investment. Ultrastructural examination revealed that they are internally similar to flagella, but lack the two central microtubules (Lembi and Herndon, 1966; Wujek and Chambers, 1966). Pseudocilia cannot undergo swimming motions, and their function in *Tetraspora* and its relatives is unknown. Cells can transform into zoospores, but when they do, two flagella are produced de novo, not from modification of the pseudocilia. Isogamy and zygote formation have been observed.

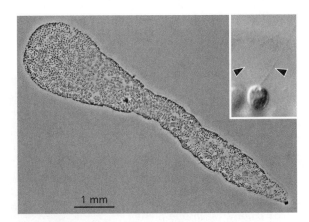

Figure 20–31 *Tetraspora* is a large nonmotile colony of *Chlamydomonas*-like cells. Rather than flagella, each cell of the vegetative colony bears two pseudocilia, which appear to have been evolutionarily derived from flagella by reduction. Inset: Higher magnification view of a single cell with pseudocilia (arrowheads)

APIOCYSTIS (Gr. *apion*, pear + Gr. *kystis*, bladder) (Fig. 20–32) is much like a miniature *Tetraspora*, except that pairs of pseudocilia extend from each cell beyond the mucilage. Young stages of *Apiocystis* and *Tetraspora* are difficult to distinguish. Biflagellate zoospores and isogametes are known. *Apiocystis* grows attached to aquatic plants and other algae in freshwaters.

The *Tetracystis* clade

This group (Fig. 20–3) was suggested by Nakayama, et al. (1996a) on the basis of SSU rDNA sequence data. It includes some unicellular, nonmotile coccoid forms that were formerly included in the order Chlorococcales (Bold and Wynne, 1985), such as *Tetracystis aeria* and *Chlorococcum hypnosporum*. Also included in this group is *Chlamydomonas moewusii* (transferred from the Volvocales). Some members of the older group Chloroccocales that are not included in this clade are *Trebouxia* and most *Chlorella* species (transferred to the Trebouxiophyceae), and *Chlorococcum oleofaciens*, *Chlamydopodium* (formerly *Characium* species), *Ettlia* (formerly *Neochloris* species), and *Protosiphon botryoides*, which are now included in the *Dunaliella* clade (Nakayama, et al., 1996a).

CHLOROCOCCUM (Gr. *chloros*, green + Gr. *kokkos*, berry) *hypnosporum*, a species commonly isolated from freshwaters and soils, occurs as single spherical or somewhat oblong cells that are sometimes aggregated into irregular masses (Fig. 20–33) (Archibald and Bold, 1970). There is a single cup-shaped, parietal chloroplast with at least one pyrenoid. Asexual reproduction occurs by distribution of parental cytoplasm amongst numerous zoospores that are released from parental cell wall. Zoospores are biflagellate and walled, with eyespots and contractile vacuoles, and thus resemble *Chlamydomonas*. If zoospore development or release is inhibited by absence of liquid water, aplanospores are produced. Both zoospores and aplanospores differentiate into mature nonmotile, spherical vegetative cells. Isogamous sexual reproduction may occur, giving rise to thick-walled zygotes or hypnospores.

Molecular sequence analyses indicate that *Chlorococcum* is a polyphyletic chlorophycean genus (Fig. 20–3) (Nakayama, et al., 1996a). There are many described species of nonmotile unicells that superficially resemble *Chlorococcum*, but are distinguished by such features as chloroplast morphology, and whether

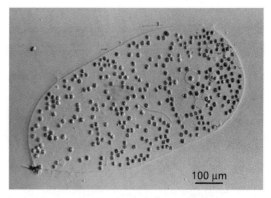

Figure 20–32 *Apiocystis*, though much like *Tetraspora*, is smaller and has pseudocilia (not visible here) that extend beyond the limits of the mucilage, unlike those of *Tetraspora*.

or not zoospores have cell walls (as judged by whether or not they retain an oblong shape after settling or become round) (Bold, 1970). Normally, observation of zoospore characters requires isolation of coccoid algae into unialgal culture. Hence, it is possible, and even likely, that identifications of *Chlorococcum* or other "little, round green things" made from mixed natural collections are often inaccurate.

TETRACYSTIS (Gr. *tetra*—from *tessares*, four + Gr. *kystis*, bladder) (Fig. 20–34) is isolated from soils. The cells resemble those of *Chlorococcum hypnospo-*

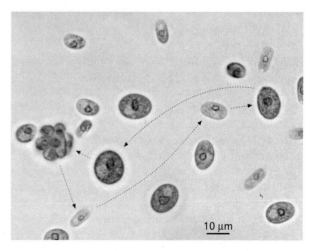

Figure 20–33 *Chlorococcum*. Arrows indicate development of elongate, recently settled zoospores into progressively larger cells that eventually cleave into zoospores or aplanospores, depending upon environmental conditions.

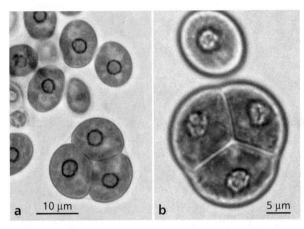

Figure 20–34 *Tetracystis* viewed at (a) low and (b) higher magnification. Cells are arranged tetrahedrally. One of the four cells is out of the plane of focus. Also note prominent pyrenoids.

rum in having a single parietal chloroplast with pyrenoids. Asexual reproduction occurs by zoospores closely resembling those of *Chlorococcum hypnosporum*. Unlike *Chlorococcum*, the cells of *Tetracystis* are aggregated into groups of four (hence the name) and even larger cell aggregations. During the successive cell divisions that generate *Tetracystis* cell aggregates, development of daughter cell walls begins in close association with parental walls; however, each daughter cell generates a complete surficial wall, and no intercellular connections (plasmodesmata) are present.

At one time the *Tetracystis* growth mode (termed **desmoschisis**), which occurs in several other green algal genera (including *Chlorosarcina* and *Chlorosarcinopsis*), was viewed as a significant evolutionary step toward origin of the tissue of land plants. Green algae having such growth were segregated into the order Chlorosarcinales (Bold and Wynne, 1985). Currently, however, such a growth pattern is viewed as an example of the parallel evolution of nonmotile cell aggregates, also known as sarcinoid forms, which is also exemplified by the charophycean alga *Chlorokybus* (Chapter 21). The phylogenetic integrity of the former group Chlorosarcinales is unclear. Identification of chlorosarcinalean genera requires zoospore characters and thus must be made from unialgal cultures.

The *Dunaliella* clade

This group, suggested by Nakayama, et al. (1996a) includes flagellates (such as *Chlamydomonas humicola* and *C. applanata*), nonmotile coccoid forms

(e.g., *Chlorococcum oleofaciens*), coenocytes (*Protosiphon*), and branched filaments (*Gongrosira*) that were formerly included in other groups of chlorophyceans. The flagellates *Chlorogonium*, *Haematococcus*, and *Stephanosphaera* were included this group because molecular sequence studies show them to be closely related to *Chlamydomonas applanata* (Buchheim, et al., 1990, 1994; Buchheim and Chapman, 1991; 1992).

HAEMATOCOCCUS (Gr. *haima*, blood + Gr. *kokkos*, berry) occurs as biflagellate unicells whose protoplast is connected to the wall by multiple thin strands of cytoplasm (Fig. 20–35). Cell structure and reproduction appear similar to those of *Chlamydomonas*. *Haematococcus lacustris* commonly occurs in granitic pools or concrete birdbaths, often forming conspicuous orange-red to deep purple growths (see Fig. 17–1). The coloration is based upon accumulation of astaxanthin (3,3'-diketo-4,4'-dihdroxy-beta carotene) during formation of nonflagellate resting cells or akinetes (see Fig. 4–13). Astaxanthin is commercially valuable as a food colorant. Some experts suggest that carotenoids help protect the cells from the deleterious effects of high light intensity (Hagen, et al., 1994), and from UV radiation in particular, but others think that the carotenoids primarily function in protection from damaging effects of highly reactive, free oxygen radicals arising from photosynthesis (Lee and Ding, 1995). The akinetes are capable of surviving complete desiccation and can be transported in wind, ostensibly germinating when favorable conditions are present. Canter-Lund and Lund (1995) note that *Haematococcus* and *Brachiomonas* (another

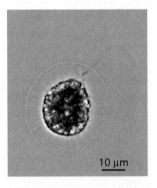

Figure 20–35 *Haematococcus* cells bear many protoplasmic extensions that traverse the space between the bulk of the cytoplasm and the cell wall. There are two flagella.

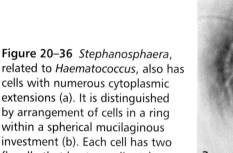
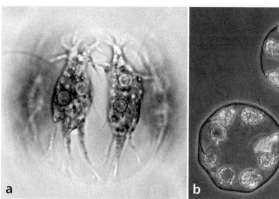
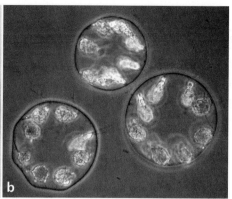

Figure 20–36 *Stephanosphaera*, related to *Haematococcus*, also has cells with numerous cytoplasmic extensions (a). It is distinguished by arrangement of cells in a ring within a spherical mucilaginous investment (b). Each cell has two flagella that beat coordinately.

chlamydomonad) can occur in brackish seashore rock pools, but that *Haematococcus* is not tolerant to high salt concentrations.

STEPHANOSPHAERA (Gr. *stephanos*, crown + Gr. *sphaira*, ball) is a motile colonial form that is phylogenetically linked with *Haematococcus*, having very similar cells with protoplastic extensions (Fig. 20–36). The cells are arranged in a ring having the appearance of a crown; the colony is enclosed in a globular mucilaginous matrix. Asexual reproduction occurs by autocolony formation, i.e., the division of each cell to

form a colony having the same number and arrangement of cells as in the parent. Such colonies can also be described as coenobia. *Stephanosphaera* occurs in granitic pools, and is reported to form reddish blooms.

DUNALIELLA (named for Michel Felix Dunal, a French botanist) has biflagellate, unicellular *Chlamydomonas*-like cells (Fig. 20–37), and occurs in extremely saline waters including the Great Salt Lake in Utah (U.S.A.) and commercial salt ponds (salterns). It uses photosynthetically produced glycerol to balance high external osmotic pressures and can tolerate

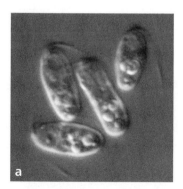
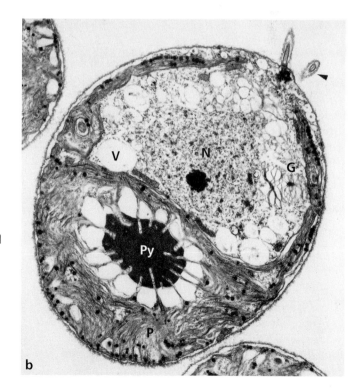

Figure 20–37 (a) *Dunaliella* cells resemble those of *Chlamydomonas* except that there is not a rigid cell wall, with the result that a variety of cell shapes can be seen. (b) *Dunaliella* cell viewed at the TEM level. Note the cup-shaped plastid (P) with pyrenoid (Py) and starch shell. The nucleus (N), Golgi body (G), and a vacuole (V) are also visible. Note also the delicate, extracytoplasmic matrix found on the cell body as well as flagella (arrowhead). (Micrograph by L. Wilcox of Chilean material cultured by O. Parra)

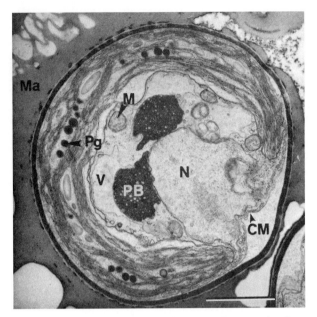

Figure 20–38 Transmission electron micrograph of *Botryococcus* showing the nucleus (N), mitochondria (M), plastoglobuli (Pg) (electron-opaque droplets within the plastids, which are common among green algae), cell membrane (CM), an electron-opaque inner cell wall layer which may be composed of decay-resistant compounds, and a lipid-rich colonial matrix (Ma). PB=polyphosphate body. Scale bar=1 μm. (From Wolf and Cox, 1981 by permission of the *Journal of Phycology*)

salt concentrations above 5M. High levels of the carotenoid β-carotene are also produced by this alga under high salinity and irradiance conditions. *Dunaliella* is grown for industrial production of glycerol and β-carotene; its ability to grow in higher salinities than can be tolerated by most potential contaminants is highly advantageous in commercial production (Chapter 4).

There is a single plastid with an eyespot, but contractile vacuoles are absent unless *Dunaliella* is grown in media of low salt concentration. Asexual reproduction is by longitudinal division, and sexual reproduction occurs by isogametes. Smooth-walled cysts can also be produced. Though often described as naked or wall-less, electron micrographs reveal that there is a fibrous extracellular matrix on the cell surface (Fig. 20–37b). Molecular phylogenetic studies suggest *Dunaliella* is probably monophyletic, that ancestral forms were walled, and that absence of substantial walls from *Dunaliella* represents evolutionary loss (Fig. 20–3) (Nakayama, et al., 1996a). A very rare coccoid green

alga (*Pachycladella umbrina*) is thought to be related to *Dunaliella* because its zoospores have a cell coat that is similar to the *Dunaliella* extracellular matrix (Friedl and Reymond, 1997).

BOTRYOCOCCUS (Gr. *botrys*, cluster of grapes + Gr. *kokkos*, berry) consists of colonies of oval or spherical cells embedded in tough, irregular, amber-colored mucilage. This mucilage contains numerous lipid globules excreted from cells (see Fig. 4–14). This lipid, and its production have been the subject of investigations focused upon development of renewable sources of energy (Tenaud, et al., 1989). *Botryococcus* occurs in at least two physiologically distinct forms: a green form that produces relatively low levels (17% dry weight) of linear olefins, and a yellow-orange form that produces up to 75% dry weight branched olefins, known as botryococcenes. Metabolic commitment to extensive lipid production is cited as the explanation for relatively slow growth rate of this alga. *Botryococcus* is believed to have contributed substantially to deposition of high-grade oil shales and coals around the world.

The lipid and mucilage of *Botryococcus* may be so highly pigmented that the green coloration of the chloroplast is obscured. Hence, it may not be immediately obvious that *Botryococcus* is a green alga; indeed, it was first thought to be an ochrophyte. The chloroplast is netlike and parietal and contains one pyrenoid and starch. Reproduction is by fragmentation

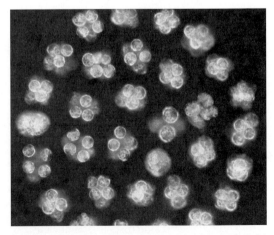

Figure 20–39 *Sphaerocystis*, viewed with dark-field microscopy. There are 4–32 vegetative cells within a mucilaginous matrix, and each is capable of producing 4–16 autospores, though often not simultaneously. In this case, most cells of the colony have done so.

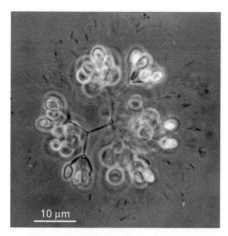

Figure 20–40 *Dictyosphaerium* cells occur at the ends of branched strands, the whole enclosed by mucilage. Note the occurrence of bacteria within the mucilage.

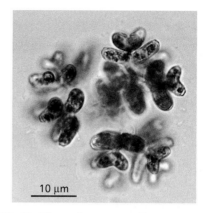

Figure 20–41 *Dimorphococcus* cells are arranged in quartets on the ends of branched strands (probably older wall material), but not enclosed within mucilage. Note that the two inner cells of each quartet are morphologically distinct from the two outer cells.

and production of autospores. *Botryococcus braunii* can be common and abundant in the plankton of moderately alkaline lakes, including eutrophic and slightly saline waters, sometimes forming surface blooms. *Botryococcus* may thus be a good candidate for development of engineering systems that are designed to link reduction of nutrients in eutrophic waterbodies with generation of useful algal products; in this case high-grade hydrocarbons for synthesis applications (Chapter 4). *Botryococcus* outer cell walls are electron-opaque and lamellate (Fig. 20–38) suggesting the occurrence of resistant biopolymers (Wolf and Cox, 1981). These are similar to the algaenans of some other green algae (Gelin, et al., 1997). Such substances may explain the occurrence of *Botryococcus* in every geological period beginning with the Carboniferous (Taylor and Taylor, 1993).

Other colonial forms that are presumed to be members of the Chlorophyceae and which commonly occur in the phytoplankton of freshwater lakes include *Sphaerocystis, Planktosphaeria, Micractinium, Dictyosphaerium, Dimorphococcus, Selenastrum,* and *Kirchneriella*. As is the case for *Botryococcus*, mucilage production is often an important element of colony formation. However molecular sequence analyses are needed to verify their relationships. Descriptions of some of these genera are located here as a matter of convenience.

SPHAEROCYSTIS (Gr. *sphaira*, ball + Gr. *kystis*, bladder) is a spherical mucilaginous aggregation of 4–32 evenly spaced, spherical cells having a single

peripheral plastid with pyrenoid (Fig. 20–39). Each cell is capable of generating 4–16 autospores, but not all cells reproduce at the same time. Colonies are quite large, up to 500 μm in diameter. *Sphaerocystis* is widespread in lakes of various types and is frequently a dominant in early summer phytoplankton communities. It is dependent upon turbulence for suspension in the water column; the mucilage envelope is thought to reduce the sinking rate (Happey-Wood, 1988).

DICTYOSPHAERIUM (Gr. *diktyon*, net + Gr. *sphaira*, ball) cells are likewise embedded in mucilage, but are distinctive in that each occurs at the end of a transparent branching structure, radiating from the center of the colony, which is believed to originate from older cell-wall material (Fig. 20–40). Cells and colony shape may be either spherical or ovoid. Cells are rather small, ranging from 3–10 μm in diameter, and have one or two parietal chloroplasts. *Dictyosphaerium* occurs in acid bogs, soft-water lakes, and hard-water lakes. Bacteria often occur in the mucilage. This alga can dominate the plankton during late spring or early summer. The name of this exclusively freshwater form should not be confused with that of the marine seaweed *Dictyosphaeria* (Chapter 18).

DIMORPHOCOCCUS (Gr. *dis*, twice + Gr. *morphe*, form + Gr. *kokkos*, berry) is a colonial form whose cells occur in groups of four (Fig. 20–41). Two of the cells in each tetrad are cylindrical, the other

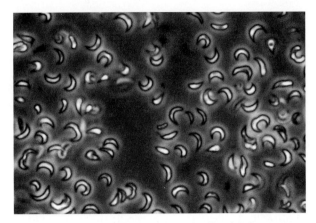

Figure 20–42 The crescent-shaped cells of *Selenastrum*.

two are kidney-shaped. The tetrads are connected to one another in an irregularly shaped colony by what appears to represent parental cell-wall material. A single plastid with pyrenoid occurs per cell. *Dimorphococcus* is a common, primarily planktonic, freshwater organism. A gelatinous envelope may surround the colony. Reproduction is by production of four autospores per cell.

SELENASTRUM (Gr. *selene*, moon + Gr. *astron*, star) is a colony of 4–16 sickle-shaped or curved cells, each having a single parietal chloroplast with pyrenoid (Fig. 20–42). A gelatinous matrix is not present. *Selenastrum capricornutum* is commonly used as a laboratory bioassay test organism (Chapter 4). It occurs in the plankton of lakes and swamps.

PROTOSIPHON (Gr. *protos*, first + Gr. *siphon*, tube) is a sac-shaped, multinucleate (coenocytic) cell that may reach lengths of 1 mm and thus is visible to the naked eye (Fig. 20–43). It is common on bare soils, often forming conspicuous green or orange patches, the color depending upon whether the soil is wet or dry, and whether the cells are metabolically active or in a resting stage that has accumulated carotenoids. The elongate, colorless, rhizoidlike basal portion of the cell extends into the soil. The chloroplast is netlike (reticulate) and contains several pyrenoids. Asexual reproduction occurs by production of many biflagellate, naked zoospores that undergo repeated mitoses without intervening cytokinesis to generate mature coenocytes. Sexual reproduction occurs by isogametes that develop by direct cleavage of parental cells. Alternatively, cytoplasm of parental sacs may become cleaved into multinucleate units that develop

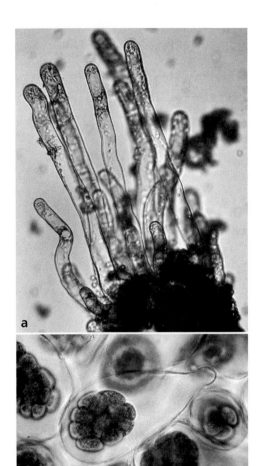

Figure 20–43 *Protosiphon*. (a) Young tubes growing from a clump of soil. (b) Coenocysts. (Photographs courtesy Larry R. Hoffman)

thick walls and are known as coenocysts (Fig. 20–43). These serve as a desiccation-resistant resting stage that can regenerate zoospores or gametes when liquid water is available.

CHARACIOSIPHON (*Characium*, a genus of green algae [Gr. *charax*, pointed stake] + Gr. *siphon*, tube) (Fig. 20–44), a coenocytic form that is even larger than *Protosiphon*, has a similar life history but differs from *Protosiphon* in having numerous discoid plastids and aquatic (freshwater) habitat; its relationships are as yet unclear.

Chaetophorales

This order, as defined by Mattox and Stewart (1984), includes chlorophyceans with branched or unbranched filaments, quadriflagellate reproductive

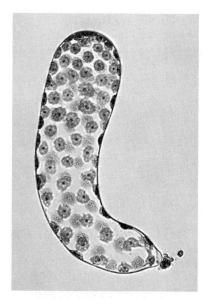

Figure 20–44 *Characiosiphon* is a multinucleate coenocyte having numerous discoid plastids. In terms of thallus construction, *Characiosiphon* parallels uniaxial members of the Caulerpales (Ulvophyceae).

cells, and plasmodesmata. *Uronema acuminata* (an unbranched form) and the branched *Fritschiella tuberosa*, *Chaetophora incrassata*, and *Stigeoclonium*

helveticum are undoubted members of this monophyletic group, based on presence of the features listed above and molecular sequence evidence (Booton, et al., 1998b). However, their relationships to other chlorophyceans are unclear at present.

At one time most unbranched green algae were grouped into the order Ulotrichales and branched forms into the Chaetophorales (Bold and Wynne, 1985; Melkonian, 1990a), but ultrastructural and molecular evidence has compelled disaggregation of these groups. The unbranched filament *Ulothrix* has been transferred to the Ulvophyceae, and a similar form, *Klebsormidium*, is charophycean. The affinities of the unbranched filaments *Geminella*, *Cylindrocapsa*, and *Microspora* are as yet unclear, but are included here as a matter of convenience. The branched green *Microthamnion* is now recognized as a member of the Trebouxiophyceae, and several branched filaments (such as members of the Trentepohliales and *Ulvella*) once considered to be chaetophoralean are now classified in the Ulvophyceae. *Coleochaete*, classified in the Chaetophorales by Bold and Wynne (1985), is now known to be in the Charophyceae. *Gongrosira* has remained within the Chlorophyceae, but has been allied with the *Dunaliella* clade.

URONEMA (G. *oura*, tail + Gr. *nema*, thread) (Fig. 20–45) filaments are typically attached to freshwater substrates by specialized disklike holdfast cells. The cells are uninucleate and have a single, parietal, band-shaped plastid with one to several pyrenoids. Cytokinesis involves production of a cell plate, and plasmodesmata are present (Floyd, et al., 1971). One or more quadriflagellate zoospores are produced per cell. These settle apical side down onto substrates, then divide to begin holdfast and filament formation

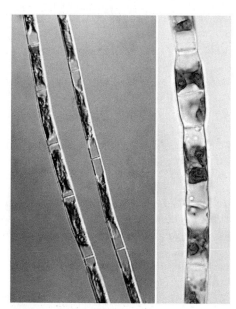

Figure 20–45 Two specimens of *Uronema*, whose filaments are unbranched, each cell having a single napkin ring-shaped chloroplast with one or more pyrenoids. There is no mucilaginous sheath.

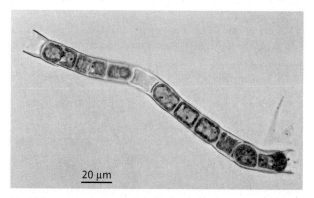

20 μm

Figure 20–46 *Microspora*.

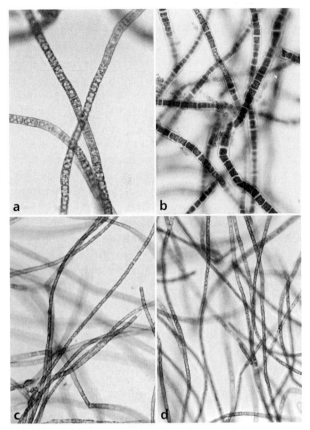

Figure 20–47 A starch test (treatment with I₂KI) can distinguish the morphological similar taxa (a) *Microsopora*, which contains starch and stains a dark purple-blue (b) and (c) *Tribonema*, a tribophycean, which lacks starch and does not stain (d).

(Bold, 1958). The zoospores of the unbranched filament formerly known as *Ulothrix belkae* (now *Uronema belkae*) are nearly identical to those of the branched filamentous forms *Stigeoclonium* and *Fritschiella* (Floyd, et al., 1980).

MICROSPORA (Gr. *mikros*, small + Gr. *spora*, spore) (Fig. 20–46) is an unbranched filament that may be terminated by a specialized holdfast cell that attaches the thallus to freshwater substrates. The cell walls occur as two open-ended cylinders firmly attached by an intermediate septum, and appear H-shaped in optical section. Each protoplast is enclosed by portions of two of these structures, with the open cylindrical portions overlapping. Similar cell walls occur in the unbranched, filamentous ochrophyte *Tribonema* (Chapter 15) but clearly evolved independently. *Tribonema* and *Microspora* can be distinguished

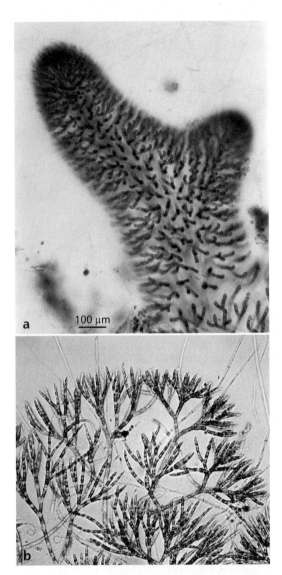

Figure 20–48 *Chaetophora* shows several levels of branching within a firm gelatinous matrix (a). At higher magnification, branches can be observed to terminate in multicellular, colorless hairs (b).

by a starch test (Fig. 20–47); *Tribonema* lacks starch, while it is present in plastids of *Microspora*. *Microspora* has a reticulate plastid, but no pyrenoids. Asexual reproduction is by fragmentation or by production of flagellate zoospores. Isogamous sexual reproduction has been observed. Placement of *Microspora* in this group is tentative, pending molecular phylogenetic analysis.

CHAETOPHORA (Gr. *chaite*, long hair + Gr. *phoras*, bearing) (Fig. 20–48) is often macroscopic, attached to submerged surfaces, including plants in freshwaters. The ends of its branches form multicellular pointed "hairs." Reproduction is by quadriflagellate zoospores and biflagellate isogametes. *Chaetophora* thalli are embedded in copious, firm mucilage.

STIGEOCLONIUM (L. *stigeus*, tattooer + Gr. *klonion*, small twig) (Fig. 20–49) occupies similar habitats as *Chaetophora*, also produces multicellular hairlike branches, and has similar reproduction, but has only a thin mucilage layer. The chloroplast is a parietal plate much like that of *Chaetophora*, and cytokinesis involves the formation of centrifugal cell plates and plasmodesmata (Floyd, et al., 1971). There is typically a prostrate system that is attached to substrates, as well as a system of erect branches. If the erect portions are eaten by herbivores, they can be regenerated from the less vulnerable, more persistent prostrate system.

FRITSCHIELLA (named for Felix Eugen Fritsch, a British phycologist) was once regarded as a model for the algal ancestors of land plants because it grows in terrestrial habitats, such as soil surfaces, and because the thallus includes tissuelike prostrate regions, as well as erect branches and colorless rhizoids (Fig. 20–50). Further, cytokinesis occurs by formation of centrifugal cell plates (McBride, 1970). However analysis of *Fritschiella*'s flagellate reproductive cells has revealed its relationship to other chlorophyceans, rather than to charophyceans (Melkonian, 1975). *Fritschiella* is now regarded as an example of parallel structural adaptation to terrestrial life.

DRAPARNALDIA (named for Jacques Philippe Raymond Draparnaud, a French naturalist) (Fig. 20–51), is commonly found attached to rocks in cold running waters, has much larger cells in the main axes than in branches, and the branch tips form multicellular "hairs." The chloroplasts are parietal bands, and thalli may be invested with a soft mucilage. Placement of *Draparnaldia* in the Chaetophorales (as defined here) is tentative, pending molecular phylogenetic analysis.

Oedogoniales

Three genera (and some 600 described species) of unbranched or branched filaments, having some very distinctive common features are placed in this order, which is inferred to be monophyletic on the basis of SSU rDNA data (Booton, et al., 1998b). However, the

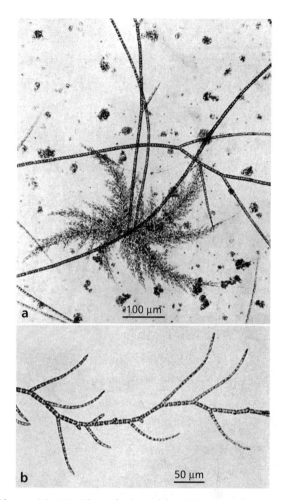

Figure 20–49 *Stigeoclonium*. (a) Habit view, showing the prostrate and erect portions. (b) Multicellular, hairlike branches. (From Cox and Bold, 1966)

relationship of this order to other chlorophyceans, in terms of molecular sequence analyses, is not well understood. Autapomorphies (shared, unique characters) of this group include an unusual form of cytokinesis, involving the precocious development of a ring of wall material; a ring of flagella on both zoospores and male gametes; and a specialized form of sexual reproduction involving dwarf male thalli. All the genera have uninucleate cells with a highly dissected, parietal, netlike chloroplast that contains numerous pyrenoids. Sexual reproduction in all genera is oogamous, and meiosis is presumed to occur at zygote germination.

OEDOGONIUM (Gr. *oidos*, swelling + Gr. *gonos*, offspring) is unbranched. Cell division generates distinctive delicate rings at the apical ends of cells.

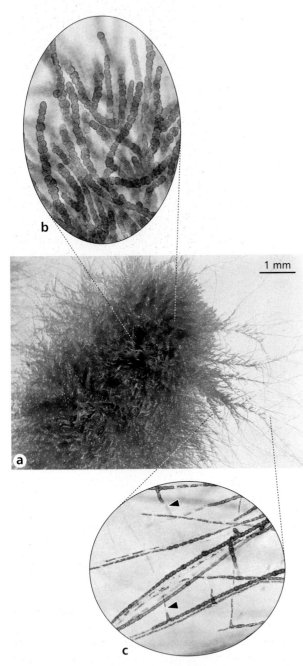

Figure 20–50 *Fritschiella*. (a) Low magnification view of thallus growing on agar. Upright (b) and prostrate (c) portions are shown at higher magnification. Arrowheads in (c) point to colorless rhizoids.

Figure 20–51 *Draparnaldia* has a main filamentous axis with relatively large cells, first order branches having smaller cells and higher order branches whose cells are smaller yet.

extensively illustrated by Pickett-Heaps (1975) (Fig. 20–53).

Asexual reproduction occurs by fragmentation and by zoospore production. The large, wall-less zoospores develop singly in parental cells. Centrioles replicate to form a ring of flagellar basal bodies, and these generate a ring of flagella that surrounds a clear apical dome, containing vesicles. Zoospores also possess an eyespot and contractile vacuoles. They emerge from the parental cell, and after a period of motility they settle onto a submerged substrate, flagellar side down. Flagella are shed, and dome vesicles release their contents, effectively gluing the germling to the attachment surface. A cell wall then develops, and cell division gives rise to a new filament (Fig. 20–54).

Sexual reproduction involves development of a single large nonflagellate egg cell within an oogonium and small multiflagellate sperm within antheridia (Fig. 20–55). Division of a vegetative cell that serves as an oogonial initial results in production of the oogonium and a subtending supporting cell. Antheridial development occurs by asymmetrical divisions that

Cells that have undergone many divisions will exhibit many such rings (Fig. 20–52). Identification of *Oedogonium* is usually based upon the presence of these rings. The ultrastructural events surrounding ring formation, mitosis, and cytokinesis are summarized and

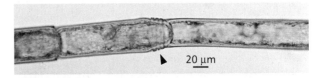

Figure 20–52 *Oedogonium* is an unbranched filament. Note the characteristic annular "rings" at the apex of the middle cell.

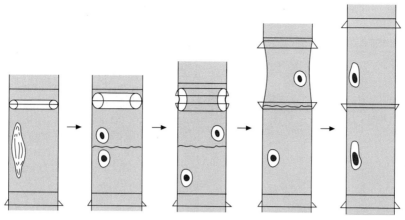

Figure 20–53 Diagram of the processes occurring in cell division in Oedogoniales. Cell that are undergoing mitosis deposit a torus of wall material at the apex, on the inside of the parental cell wall. Following cytokinesis, the parental wall breaks at a predetermined site at the same level as the torus. Expansion of torus material, now exposed to the outside, occurs as the daughter cells elongate. Eventually the new wall material contributed by the torus constitutes much of the cell wall of the uppermost of the two daughter cells. The upper point at which breakage of the parental cell wall occurred is visible as a ring at the apex of the upper cell, whereas the lower breakage site occurs just below the new cross wall, at the upper end of the lower daughter cell. If the upper cell divides again, it will acquire a second apical ring. Thus the number of times a cell has divided can be deduced by the number of apical rings it possesses. (After Pickett-Heaps, 1975)

produce very short cells and, in at least some species, may be influenced by the nearby presence of bacteria (Machlis, 1973). Antheridial cells then divide further to generate two (or sometimes more) sperm.

Some species are monoecious and presumably self-fertilization may occur in these forms. Other species are dioecious; this may promote outcrossing. Species in which the sperm develop within cells on normal-sized filaments are called macrandrous forms. In contrast, some species are nannandrous—these produce dwarf male filaments that are one or only a few cells in length and which grow on the supporting cells just below oogonia of normal-sized female filaments. The dwarf male filaments develop from androspores, cells that are intermediate in size between zoospores and sperm but likewise have numerous flagella arranged in a apical circle. The cells that produce androspores may occur on the same filament as the eggs; thus, selfing may occur in these so-called gynandrosporous species. Other species produce androspores and eggs on different filaments; this facilitates outcrossing and is known as idioandrospory. The ultrastructure of dwarf males of *O. pluviale* was studied by Leonardi, et al. (1998).

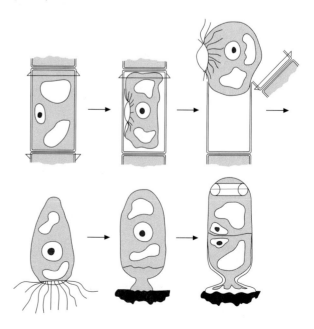

Figure 20–54 Zoosporogenesis and germling development in Oedogoniales. The entire contents of individual cells may be used in the formation of a single large zoospore having multiple flagella. Zoospores escape when the parental cell wall breaks at the anterior. Upon settling (flagella end first), attachment mucilage is exported, and mitotic divisions begin the process of filament development. (After Pickett-Heaps, 1975)

Sperm, whether produced by dwarf or normal-sized filaments, are attacted to eggs by chemical pheromones and gain access to them via a pore or fissure in the oogonial wall (Hoffman, 1973; Machlis, et al., 1974). A complex series of chemical signaling events is involved in the production of androspores and development of dwarf males, which is tantamount to a kind of mating ritual wherein each of a series of interchanges must succeed in order for mating to finally occur. Androspores are chemically attracted to oogonial initial cells of female filaments; this might be regarded as a form of mate selection. But if no female filaments are nearby, androspores can germinate into vegetative filaments, just like zoospores, a resource-saving strategy. If androspores attach and begin development into dwarf males, oogonial initials divide to produce eggs, suggesting the production of yet a second chemical signal. Such a signaling system may save resources; if sperm are not available, energy will not be diverted into oogo-

Figure 20–55 *Oedogonium* filament with numerous zygotes formed during sexual reproduction. Small, empty antheridia can be seen between zygotes.

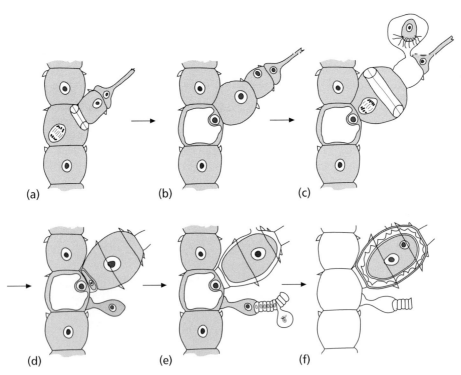

(a)   (b)   (c)

(d)   (e)   (f)

Figure 20–56 Sexual reproduction in *Bulbochaete* is similar to that in other members of the Oedogoniales. Nannandrous sexual reproduction is illustrated here. An androspore produced by an anterior cell in a branch is released as the subtending cell undergoes division preparatory to oogonial development (c). (d) The androspore settles upon a cell near the oogonium, undergoes a few mitotic divisions to form a dwarf male filament (e), then produces sperm. (f) Upon fertilization of the egg via a pore in the oogonial wall, the zygote develops a thick ornamented wall preparatory to a period of dormancy, while the dwarf male filament and the rest of the oogonial filament die. (After Pickett-Heaps, 1975)

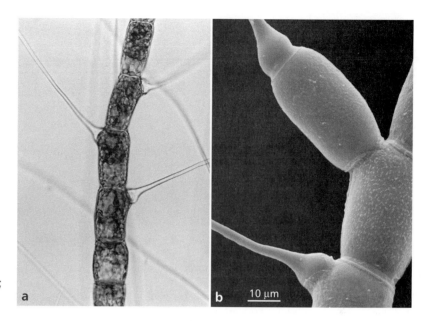

Figure 20–57 (a) *Bulbochaete* is characterized by bulbous-based hair cells that become colorless. (b) *Bulbochaete* hairs and branching viewed by SEM. (a: Courtesy C. Taylor; b: Courtesy S. Cook)

nial development. Mature oogonia produce a blanket of mucilage around themselves and the nearby male filaments, presumably limiting fertilization to sperm produced by these neighboring males, but also preventing them from wandering elsewhere. Sperm are then attracted to oogonia, suggesting occurrence of yet a third phase of chemical influence (Rawitscher-Kunkel and Machlis, 1962). After fertilization the zygotes of all species develop a thick wall and large amounts of carotenoid pigments, and enter a dormant stage, eventually germinating by meiosis to form four zoospores that regenerate the filamentous thallus. An extensive ultrastructural survey, providing many more interesting details of cell division, asexual reproduction, and sexual reproduction in *Oedogonium* can be found in Pickett-Heaps (1975).

There are about 400 described species of *Oedogonium*, varying in the size of the vegetative cells, shape and size of oogonia, ornamentation of the zygote wall, and location of antheridia (Prescott, 1951). The alga commonly occurs attached to submerged plants and other substrates in shallow fresh-waters. On occasion, *Oedogonium* can form conspicuous growths or blooms.

BULBOCHAETE (Gr. *bolbos*, bulb + Gr. *chaite*, long hair) resembles *Oedogonium* in habitat, cell structure, and details of cell division and reproduction (Fig. 20–56), but differs from it in being branched and in having distinctive, colorless bulbous-based hair cells (setae) (Fig. 20–57). Many details of the cell biology of *Bulbochaete* can be found in Pickett-Heaps (1975).

Recommended Books

Kirk, D. L. 1997. *Volvox*. Cambridge University Press, New York, NY.

Harris, E. H. 1989. *The Chlamydomonas Handbook*. Academic Press, New York, NY.

Pickett-Heaps, J. D. 1975. *Green Algae*. Sinauer Associates, Sunderland, MA.

Chapter 21

Green Algae V
Charophyceans

Micrasterias

(Photograph courtesy C. Taylor)

The single-most important thing about charophycean green algae is that they represent the lineage that is ancestral to the land plants (embryophytes=bryophytes+tracheophytes). The land plants are thought to have first appeared more than 470 million years ago—the age of the earliest fossils that are accepted as land plant remains (Gray, et al., 1982). Unfortunately, there are no known older fossils that clearly link green algae with land plants; charophyceans have been associated with the ancestry of land plants on the basis of ultrastructural, biochemical, and molecular evidence derived from the study of modern forms (Graham, 1993, 1996; McCourt, 1995; Moestrup, 1974; Pickett-Heaps and Marchant, 1972). Molecular sequence evidence (Chapman and Buchheim, 1991; Wilcox, et al., 1993; Ragan, et al., 1994; Surek, et al., 1994; Manhart, 1994; McCourt, et al., 1996a; and Huss and Kranz, 1997) has unequivocally established a close relationship of charophyceans to land-plant ancestry. Based upon earlier compelling structural and biochemical evidence for this phylogenetic linkage, Bremer (1985) suggested that charophyceans and land plants should be included in a division (Streptophyta) separate from the other green algae (Chlorophyta). This suggestion has also been advocated by other workers (Waters and Chapman, 1996). Charophycean algae have provided essential information about the evolution of several fundamental features of land plants, including the plant cytokinetic apparatus (the phragmoplast), sporopollenin-enclosed spores, and the plant life cycle (Graham, 1993, 1996).

Charophyceans include the macroscopic charalean algae such as *Chara* (for which the group is named) and *Nitella*, which can be very common in some freshwater environments and occasionally occur in brackish waters. Also known colloquially as stoneworts, these forms—referred to here as char-aleans—have a long and substantial fossil record. This record reveals that relatives of modern char-

Figure 21–1 *Chara* is a macrophytic member of the Charophyceae, the class whose name is derived from this genus. (a) Charalean algae can form dense meadows in lakes. In (b), *Chara* is seen growing in a glass tank.

aleans were the dominant form of macrophytic vegetation in freshwaters for some three hundred million years prior to the origin of flowering plants. Though seemingly much less diverse today than in some earlier time periods, modern charaleans are widespread and locally abundant. They sometimes form nuisance growths in livestock tanks and shallow waters, such as ponds (Fig. 21–1). Charaleans have long been important experimental subjects for electrophysiological studies of membrane function as well as analyses of cytoplasmic streaming and geotropism because some of their cells can be very large (up to 15 cm long), and thus easily accessible.

Charophycean algae also include microscopic desmids and the well-known *Spirogyra* and related filamentous algae, which are here referred to as zygnematealeans (named for the genus *Zygnema*) (Fig. 21–2). Although some authors have placed zygnematalean algae in a phylum of their own (Conjugaphyta) (Hoshaw, et al., 1990) and other experts treat them as a class (Zygnemaphyceae) (Hoek, van den, et al., 1995), molecular sequence analyses show clearly that zygnematealeans are closely related to (and embedded within) the charophyceans (Surek, et al., 1994). The inclusion of the zygnematealeans within the Charophyceae was proposed earlier on the basis of ultrastructural and biochemical evidence (Mattox and

Stewart, 1984). The fossil record of zygnematealeans extends back no more than about 400 million years; the oldest fossil recognized as a member of this group is *Paleoclosterium* from the Middle Devonian (Baschnagel, 1966). Modern zygnematealeans are famous for their high level of species diversity; there are more than 3,000 described species, with new forms continuing to be discovered. They are also known for the incredible beauty of their cells (Fig. 21–3), which are particularly diverse and abundant in low pH *Sphagnum* bogs (Gerrath, 1993) (Fig. 21–4).

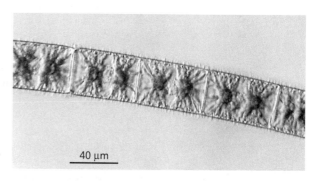

40 µm

Figure 21–2 *Zygnema*, an unbranched filament, lends its name to the order Zygnematales, a group belonging to the Class Charophyceae.

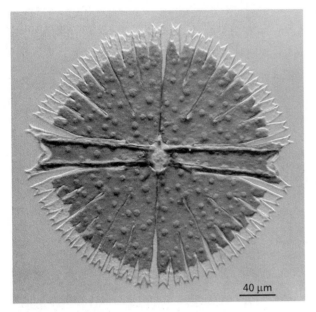

40 µm

Figure 21–3 *Micrasterias* is a well-known and extremely beautiful desmid. Desmids are members of the Zygnematales and the Charophyceae and hence related to the ancestry of land plants.

In addition, a few less conspicuous genera are included within charophyceans—*Mesostigma*, *Chlorokybus*, *Klebsormidium*, *Chaetosphaeridium*, and *Coleochaete*. These taxa are of interest because, together with charalean and zygnematalean algae, they illustrate a spectrum of morphological complexity that suggests stages in the evolution of land plants. *Mesostigma* is a unicellular flagellate formerly classified with prasinophyceans. *Chlorokybus* consists of cells grouped into rectangular arrays, described as a "sarcinoid" arrangement. *Klebsormidium* occurs as an unbranched filament, and *Chaetosphaeridium* is a highly branched filament that superficially appears to be a colonial aggregation of single cells. *Coleochaete* has a number of morphologically diverse species, most of which are branched filaments, but some of which appear to have a more tissuelike (parenchymatous) organization. The cells of all *Coleochaete* species that have been examined ultrastructurally are interconnected by plasmodesmata and so these species are regarded as truly multicellular. There are no colonial or unbranched filamentous forms of *Coleochaete*. Both *Coleochaete* and *Chaetosphaeridium* occur in shallow waters of oligotrophic lakes (Fig. 21–5) and rapidly disappear when water quality has been degraded.

Relationships of the Charophycean Algae to Land Plants

Although a small phragmoplast occurs in dividing cells of the zygnematalean *Spirogyra* (Fowke and Pick-

Figure 21–4 Zygnematalean algae are most diverse and abundant in low pH *Sphagnum* bogs, such as this one in Vilas County, Wisconsin, U.S.A.

Text Box 21–1 Molecular Architectural Characters and the Phylogeny of Charophycean Algae

Molecular architectural changes, such as intron insertion events, movement of genes from one genome to another, and other gene rearrangments have proven useful in deducing the evolutionary history of algae and plants. Such characters, whose homology can be tested, are particularly helpful when, as in charophyceans, commonly employed gene sequence data (such as SSU rDNA or Rubisco) generate conflicting results or do not resolve phylogenetic issues. Although such genomic architectural changes may undergo reversal (e.g., intron loss can occur), this is uncommon, and such changes are typically inherited by descendents of the organisms in which they first occurred.

Genomic architectural changes suggest relative divergence times for the derived charophyceans (those having sexual reproduction): Zygnematales, Charales, and Coleochaetales, and indicate which are most closely related to embryophytes.

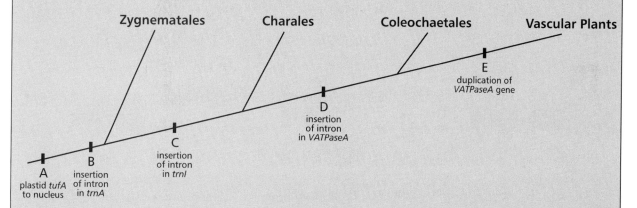

(A) The *tufA* gene (which encodes a chloroplast protein synthesis elongation factor, EF-Tu) occurs in the plastid genome of most algae, but in the nuclear genome of land plants and at least some representatives of the Zygnematales, Coleochaetales, and Charales. This suggests the occurrence of a single transfer event sometime prior to divergence of Zygnematales (Baldauf and Palmer, 1990). **(B)** In most algae there are no introns in the plastid genes encoding the tRNAs for alanine (*trnA*) and isoleucine (*trnI*), but such introns occur in both genes in land plants. An intron appears to have been inserted first in *trnA*, in a common ancestor of the zygnematalean *Spirogyra*, *Coleochaete*, Charales, and embryophytes (Manhart and Palmer, 1990). **(C)** An intron does not occur in the plastid *trnI* of *Spirogyra* but is present in *Coleochaete*, Charales, and embryophytes, suggesting that the insertion event occurred sometime after the divergence of the Zygnematales (Manhart and Palmer, 1990). **(D)** An intron having several highly conserved regions occurs in the nuclear gene encoding one of the subunits (A) of a vacuolar-type ATPase of *Coleochaete* and land plants, but is absent from *Zygnema*, and has not been reported from other algae. This suggests a single insertion event prior to the divergence of *Coleochaete* (Starke and Gogarten, 1993). **(E)** *Coleochaete* has only a single copy of the *VATPaseA* gene, whereas vascular plants have two, both having the intron. Thus a gene duplication event occurred sometime between divergence of *Coleochaete* and vascular plants (Starke and Gogarten, 1993). If future work shows evidence for nuclear *tufA* or a *VATPaseA* intron in other charophyceans, the above phylogenetic inferences would change.

continued→

ett-Heaps, 1969b; Sawitzky and Grolig, 1995), so far as is known, phragmoplasts similar to those of plants occur only in charaleans (Cook et al., 1998) and *Coleochaete* (Brown, et al., 1994). Among charo- phyceans, only charaleans and *Coleochaete* are known to possess plasmodesmata. Primary plasmod- esmata (those developing during cytokinesis) of at least one species of *Chara* have been demonstrated to

Text Box 21–1 *continued*

The introns present in *Coleochaete*'s *trnA* and *trnI* genes (top) are smaller than those of the liverwort *Marchantia* (bottom), but are similar in sequence and position within the genes, indicating homology (Manhart and Palmer, 1990).

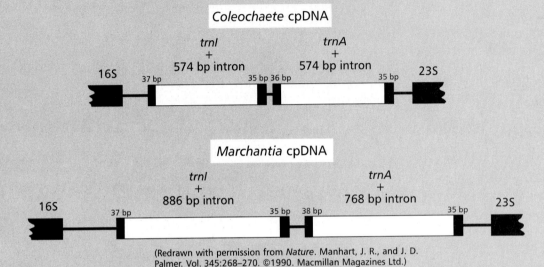

(Redrawn with permission from *Nature*. Manhart, J. R., and J. D. Palmer. Vol. 345:268–270. ©1990. Macmillan Magazines Ltd.)

Similarly, *Coleochaete*'s *VATPaseA* intron is the smallest known, but contains the same highly conserved regions also found in higher plants, indicating homology (Starke and Gogarten, 1993). Such data indicate that introns may increase in size through evolutionary time.

develop much like those of land plants and share structural features with them (Cook, et al., 1997). Thus, in terms of thallus development and morphology, charaleans and *Coleochaete* most closely approach embryophytes. Substantial molecular sequence and molecular architectural evidence, including intron insertion events and movement of genes between genomes (Fig. 21–6) (Text Box 21–1), strongly suggests that zygnemataleans diverged from the main lineage prior to *Coleochaete* and charaleans, and that the latter two taxa are more closely related to the embryophytes (plants). Since available evidence strongly indicates monophyly of the land plants (Mishler, et al., 1994; Waters and Chapman, 1996; Lewis, et al., 1997), an inescapable conclusion is that their ancestor shared some features with modern charophycean algae. Appropriate fossils have not yet been discovered, and thus comparative studies of higher charophyceans and early divergent plants are the best approaches for elucidating the earliest events in the evolutionary history of plants. In the following

discussion the evolutionary and ecological significance of charophycean groups will be described in the apparent order of their divergence.

Diversity of Charophyceans

Early Divergent Charophycean Representatives

The early divergent members of the charophycean lineage appear to include the flagellate *Mesostigma*, the sarcinoid *Chlorokybus*, and the unbranched filament *Klebsormidium*. Although the unicellular genera *Stichococcus* and *Raphidonema* were at one time considered to be members of the charophycean lineage, such a relationship is no longer accepted. So far as is known, these early divergent charophyceans do not form conspicuous growths or blooms in nature, nor are they of any economic importance. They have been studied primarily because of their evolutionary importance.

Figure 21–5 Coleochaetales are microscopic members of the Charophyceae that typically grow on rocks or macrophytes in shallow waters of oligotrophic lakes, such as Lake Tomahawk, Oneida County, Wisconsin, U.S.A. Sites having well-consolidated gravel bottoms are particularly good places to look for members of the Coleochaetales, and/or to place glass slides upon which these algae will grow within a few weeks.

MESOSTIGMA (Gr. *mesos*, middle + Gr. *stigma*, mark). The freshwater, scaly biflagellate *Mesostigma viride* (Fig. 21–7), generally classified as a member of the prasinophyceans (Chapter 17), was suggested to be the flagellate most closely related to charophyceans based on its possession of a multilayered structure (MLS) similar to MLSs in the flagellate reproductive cells of charophyceans and land plants (Rogers, et al., 1981; Melkonian, 1989) (see Chapter 17). More recently, molecular sequence analyses have confirmed close phylogenetic relationship between *Mesostigma* and charophyceans (Melkonian, et al., 1995). *Mesostigma* differs from more typical prasinophyceans in cell shape, scales, and the structure of its flagellar root system (Sym and Pienaar, 1993). The fact that *Mesostigma* is a freshwater form supports a current concept that the charophycean radiation primarily occurred in fresh, rather than marine waters (Graham, 1993; Mishler, et al., 1994, Melkonian, et al., 1995).

Mesostigma is disk-shaped and has a central or sometimes lateral flagellar pit opening that penetrates

Figure 21–6 Phylogenetic tree based upon *rbcL* sequence comparisons, showing close relationship of the clade including both Coleochaetales and Charales to the embryophytes. Zygnematales and *Klebsormidium* are more distantly related to land plants in this analysis (After McCourt, et al., 1996b)

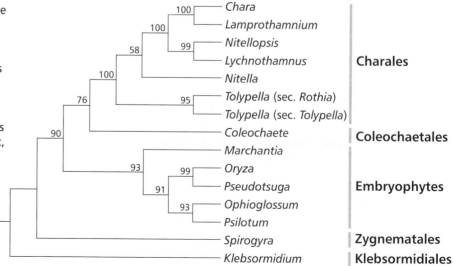

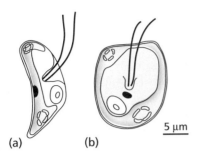

Figure 21–7 *Mesostigma* has two flagella and a single flat chloroplast that contains pyrenoids and an eyespot in the region near the flagellar basal bodies. (After Sym and Pienaar, 1993)

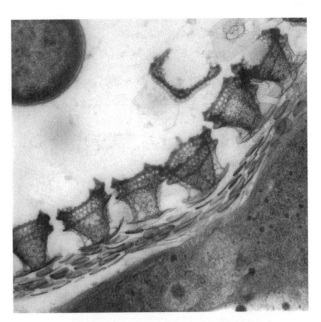

Figure 21–8 *Mesostigma* scales at the TEM level. Several layers are present. (Micrograph courtesy D. Chappell and H. Ris)

quite deeply into the cell. A layer of small, flat, polygonal scales covers the flagella, and similar scales serve as the lowermost layer of body scales. Sometimes this scale layer appears to be associated with material situated between it and the cell membrane, and it has been suggested that the cell wall of charophyceans (and land plants) evolved from substances that cemented body scales to the cell membrane (Rogers, et al., 1980). Similar small flat scales occur on the body and/or flagella of various charophyceans; it is hypothesized that these scales were inherited from a *Mesostigma*-like ancestor. *Mesostigma* has two additional layers of scales: the middle layer is composed of larger, flattened, oval scales that are ornamented with small pits; and the outer layer consists of large, distinctive basket scales (Manton and Ettl, 1965) (Fig. 21–8). These outer scales are quite different from scales occurring on the surfaces of prasinophyceans; they have not been observed to occur on reproductive cells of other charophyceans.

There is a single platelike chloroplast, which is thickened at the edges and contains several pyrenoids. Pigments are said to resemble those of the ulvophycean *Bryopsis* (Fawley and Lee, 1990). Extensions of the chloroplast appear to link with the flagellar basal bodies, which is unusual. An eyespot composed of two to three layers of pigmented globules lies within the plastid near the basal bodies. Eyespots have not been reported in charophycean flagellate cells other than *Mesostigma*. Thus if *Mesostigma* represents the ancestral flagellate, the eyespot must have been lost early in charophycean diversification. A large lobed peroxisome lies between the chloroplast and the basal bodies and is attached to the latter. This association is believed to facilitate division and distribution of the peroxisome to daughter cells at cell division.

Basal bodies are linked by fibrous connecting bands similar to those of other green algae, and each basal body is associated with a single flagellar root containing five to seven microtubules. The proximal part of each root is a multilayered structure (MLS); thus there are two MLSs per cell. Since similar MLSs occur in a variety of protozoan flagellates, such as *Karotomorpha* and *Monocercomonoides*, it has been suggested that *Mesostigma* (and other charophyceans) may be related to such protozoa (Rogers, et al., 1981). However, this hypothesis has not yet been tested by means of molecular sequence analyses. Aggregations of microtubules occupy the region of the *Mesostigma* cell beneath the flagellar pit, perhaps serving as a support system.

CHLOROKYBUS (Gr. *chloros*, green + Gr. *kybos*, cube) is a rarely encountered terrestrial or freshwater green alga (Geitler, 1955). The vegetative thallus is a packet of rounded cells surrounded by thick mucilage (Fig. 21–9). The vegetative cells are surrounded by a cell wall, and cytokinesis is followed by deposition of cell-wall material only at the new cross-wall, as in filamentous and more complex charophyceans and plants. However *Chlorokybus* is not considered to be either filamentous or parenchymatous (tissuelike) and is regarded as the simplest charophycean alga having a nonmotile vegetative stage. Some of its characteristics

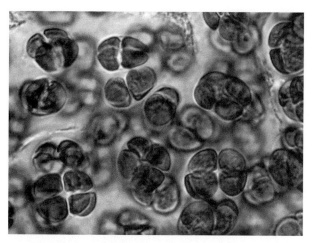

Figure 21–9 *Chlorokybus* is sarcinoid, meaning that the cells are arranged in packets of variable number, held together by mucilage.

seem to be intermediate between those of *Mesostigma* and other charophyceans.

Chlorokybus cells can be induced to produce a single biflagellate zoospore; the flagella emerge laterally and are associated with a groove, in these respects somewhat resembling *Mesostigma*. Zoospores of other charophyceans also have lateral flagella, suggesting that this is an ancient feature that can be used to distinguish charophyceans from other green algae. Zoospore release in *Chlorokybus* is by disintegration of the parental cell wall, a process that is regarded as unspecialized or primitive, compared to zoospore release mechanisms of other charophyceans (Rogers, et al., 1980). The body and flagella of the zoospores are covered with small flat scales resembling those of

Mesostigma and zoospores of some other charophyceans. The flagella also possess hairs, as do those of some other charophyceans (such hairs are lacking from *Mesostigma*). No eyespot is present in the zoospores. There is a single peroxisome that, like that of *Mesostigma* but not other charophyceans, is attached to the flagellar apparatus. The MLS and associated microtubular extension form the only known flagellar root. This MLS root contains fewer microtubules (10–11) than do those of other charophyceans (or land plants), and its microtubules do not extend as far down into the cell (Rogers, et al., 1980).

After the zoospores swim for about an hour, they become round, retract their flagella, and begin to deposit a cell wall beneath the body scales. After the cell wall becomes well developed, the scale layer is lost. In this behavior, *Chlorokybus* zoospores resemble those of *Coleochaete* and *Chaetosphaeridium*. Cells of *Chlorokybus* possess a single cup-shaped chloroplast resembling that of some other charophyceans, but atypically, there are two distinct types of pyrenoids—one embedded within the plastid containing numerous traversing thylakoids, and another located at the periphery of the plastid and lacking thylakoids. The pyrenoids of other charophyceans (and

Figure 21–10 Cell division in *Chlorokybus* illustrates a number of features characteristic of all charophyceans: Mitosis is open, and persistent spindle microtubules keep the daughter nuclei apart at cytokinesis. Other mitotic attributes of *Chlorokybus* occur in some other charophyceans, but not all: Mitosis in *Chlorokybus* is centric (there are centrioles at the spindle poles), and plastid and pyrenoid division occur prior to nuclear division. (a) Chloroplast (P) and pyrenoid (Py1, Py2) division precedes that of the nucleus (N). (b) Centriole pairs (twin circles) have moved to poles. (c) In metaphase/anaphase, the peroxisome (Pe) is closely associated with the spindle. (d) At telophase, a cleavage furrow develops alongside parallel-oriented microtubules (MT). (After Lokhorst, et al., 1988 by permission of the *Journal of Phycology*)

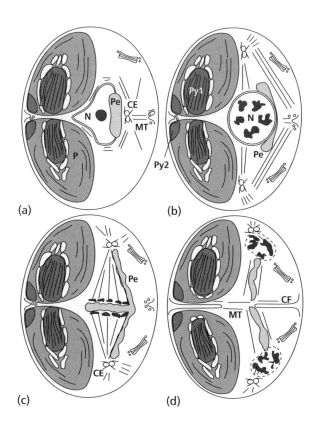

(a)　(b)　(c)　(d)

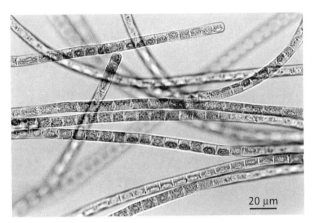

Figure 21–11 *Klebsormidium* is an unbranched filament; cells have a single, platelike, parietal plastid.

hornworts among land plants) are embedded and traversed by thylakoids; the simpler peripheral pyrenoid type is lacking. The physiological and evolutionary relevance of these differing pyrenoids in *Chlorokybus* is not known.

At the initiation of mitosis in *Chlorokybus*, centrioles are positioned at the plane of cell division, which is indicated by a precocious (early developing) cleavage furrow. Centrioles then duplicate and move apart, forming the poles of the mitotic spindle; they are associated with material that serves as a microtubule organizing center. The centrioles are also associated with an array of astral microtubules, which is typical for many eukaryotes but not plants. By metaphase, the nuclear envelope has completely broken down; mitosis is thus described as "open." The spindle persists until it is disrupted by the formation of a cross-wall (septum) as the cleavage furrow is completed. Microtubules are arrayed in the plane of the developing furrow (Lokhorst, et al., 1988) (Fig. 21–10). Sexual reproduction has not been observed.

KLEBSORMIDIUM (Klebs [Georg Albrecht Klebs, a German phycologist] + *Hormidium*, a genus of green algae [Gr. *hormidion*, small chain]) is a freshwater, unbranched filament (Fig. 21–11). Noting that the cells produced only a single zoospore, and that this zoospore lacked an eyespot and had two laterally emergent flagella (cf. *Chlorokybus* above), Marchant, et al. (1973) predicted that *Klebsormidium* zoospores would be found to possess an MLS-type flagellar root. Their subsequent ultrastructural examination confirmed the aptness of this prediction. *Klebsormidium* is regarded as more specialized than *Chlorokybus*

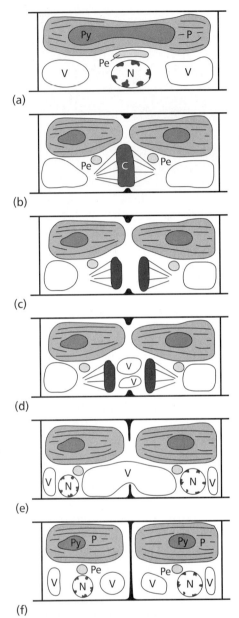

Figure 21–12 Mitosis and cytokinesis in *Klebsormidium*. (a, b) Division of the chloroplast (P), pyrenoid (Py), and associated peroxisome (Pe) is accomplished prior to completion of nuclear division. (c, d) Mitosis is open, and a precocious cleavage furrow begins to develop prior to cytokinesis. (e, f) A large vacuole (V) is instrumental in retaining separation of daughter nuclei (N) during cytokinesis, and is cleaved during cross-wall development by furrowing. (After Floyd, et al., 1972 by permission of the *Journal of Phycology*)

because zoospores are discharged through differentiated pores in the cell wall. Production of the pore involves protrusion of the cytoplasm from a specific wall site, then deposition of presumed cell-wall lytic vesicles, rather than the generalized dissociation of the entire zoosporangial wall, as occurs in *Chlorokybus*. The ovoid zoospores are devoid of body scales, and the flagella also lack scales and hairs, which is unique among charophyceans. It is presumed that scales and hairs were lost sometime after divergence of *Klebsormidium* from the main line of charophycean evolution. After swimming for about an hour, the zoospores become round, retract their flagella, and form a cell wall without attaching to a substrate or forming a holdfast, as do some other green filamentous algae.

Klebsormidium vegetative cells and zoospores each contain a single parietal chloroplast with a pyrenoid surrounded by starch grains, but no eyespot. Cells contain a single, land plantlike peroxisome, but this organelle is not attached to the basal bodies as it is in *Chlorokybus* and *Mesostigma*. Rather, the peroxisome lies appressed to the midpoint of the chloroplast and is segregated at mitosis along with the plastid (Fig. 21–12), as in *Coleochaete* (see below). Mitosis is open. Centrioles are present at the spindle poles; these are believed to be associated with spindle microtubule organizing material. An unusual large vacuole that forms in the interzonal region about halfway through anaphase appears to help complete chromosomal separation. Cytokinesis occurs by the development of a constricting furrow from the periphery (Floyd, et al., 1972). Sexual reproduction is not known.

Klebsormidium may be difficult to distinguish at the light microscopic level from several other unbranched filamentous green algae that are not allied to the charophyceans. However, if it can be observed that only a single eyespotless zoospore is produced per cell, the filament is probably *Klebsormidium*. In contrast, green unbranched filaments such as the ulvophycean *Ulothrix zonata* (see Fig. 18–9) typically produce multiple zoospores/gametes per cell, and these flagellate cells typically possess eyespots. Comparative taxonomic studies of European species of *Klebsormidium* were done by Lokhorst (1996).

Zygnemataleans

In contrast to *Mesostigma*, *Chlorokybus*, and *Klebsormidium*, zygnematalean green algae can be more conspicuous. Ponds, ditches, sheltered nearshore regions of lakes, and slow-flowing streams may exhibit blooms of *Spirogyra* or related forms each spring, the growths sometimes assuming nuisance proportions (see Fig. 1–10) (Graham, et al., 1995). *Mougeotia* and

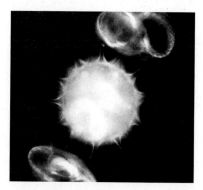

Figure 21–13 Zygote of the desmid *Cosmarium*. Zygnematalean zygotes are often thick-walled with spiny or warty ornamentation, as seen here. The function of the ornamentations is unknown. The empty cell walls lying just above and below the zygote are those of the parental cells from which gametes emerged. (Photograph by L. Wilcox, in *Origin of Land Plants*. Graham, L. E. ©1993. Reprinted by permission of John Wiley & Sons, Inc.)

some relatives can produce metaphytic (subsurface) clouds in lake waters that have been impacted by acidic precipitation (acid rain) or experimental acidification (Schindler, et al., 1985; Watras and Frost, 1989; Howell, et al., 1990; Turner, et al., 1991). Single-celled and "filamentous" desmids can generate conspicuous growths in *Sphagnum* bogs (Woelkerling, 1976). In addition to their ecological significance, filamentous and single-celled zygnematealeans are important as model research systems to investigate the plant cytoskeleton, cell-wall biosynthesis, photomorphogenesis, and other cellular and physiological features; this is justified by close phylogenetic relationship to the land plants, evidence of which has been earlier described.

Zygnematealeans differ from the early divergent charophyceans described above in having sexual reproduction of a distinctive type known as conjugation. This characteristically involves nonflagellate gametes, and in fact, production of flagella and centrioles is unknown among zygnematealeans. It is commonly assumed that their immediate ancestors had flagellate cells, but that the capacity to produce flagella was lost early in the radiation of zygnematealeans (Hoshaw, et al., 1990).

Zygotes having distinctive shapes and wall structure (Fig. 21–13) result from sexual reproduction; these typically confer the ability to survive long periods when conditions are not favorable for growth, germinating when the environment improves

Figure 21–14 Conjugation tubes link cells of two mating filaments in *Spirogyra*. Such conjugation tubes occur commonly in related forms; flagellate gametes are not produced in any of the modern zygnemataleans. Conjugation tubes allow transfer of gamete cytoplasm from one filament to the other, or to the mid-region of the tube, where zygote formation may occur. (From *Origin of Land Plants*. Graham, L. E. ©1993. Reprinted by permission of John Wiley & Sons, Inc.)

Classically, zygnematalean algae have been divided into three major groups based upon their external morphology and features of the cell wall. Unbranched filamentous forms, such as *Spirogyra* and *Mougeotia*, which generally reproduce sexually by the movement of gametes through conjugation tubes (canals) (Fig. 21–14), have been classified in the family Zygnemataceae. Unicellular zygnemataleans that have a homogeneous wall lacking pores are known informally as the "saccoderm" desmids, and have been classified in the family Mesotaeniaceae. Unicells and "pseudofilaments" (linear arrays of loosely adherent cells) having a pore-containing, two-part wall of different ages and origin that are typically separated by a narrow isthmus, have been classified as the family Desmidiaceae (the "placoderm" desmids) (Bold and Wynne, 1985). In addition, some desmids not fitting neatly into any of the above three groups have been segregated into the families Peniaceae or Closteriaceae (Gerrath, 1993).

Recent molecular analyses have shown that these classical classification schemes do not always accu-

(Hoshaw, et al., 1990). Zygote persistence has been attributed to the presence of acid hydrolysis-resistant, sporopolleninlike polymers in the cell wall (DeVries, et al., 1983). Zygotes similar to those of modern zygnematalean algae have been recovered from sediments of Carboniferous age (some 250 million years old) and younger. A compilation of records of fossil zygnemataleans can be found in Hoshaw, et al., (1990). Some desmids produce acid and microbe-resistant polyphenolic polymers in vegetative cell walls (Gunnison and Alexander, 1975a, b; Kroken, et al., 1996); similar wall compounds may be responsible for preservation of the oldest known fossil desmid vegetative cells, *Paleoclosterium leptum* from the Middle Devonian (about 380 million years ago) (Baschnagel, 1966). Sexual reproduction involving production of sporopolleninlike materials in zygote walls as wells as the inclusion of resistant polyphenolic materials in walls of vegetative cells also occurs in more derived charophyceans. Resistant cell-wall polymers may also explain the survival of desmids such as *Closterium*, *Micrasterias*, *Pleurotaenium*, and *Euastrum* in drying mud at the edges of lakes. Desmids have been found to be alive after three months of drying and at depths of 6 cm in sediments (Brook and Williamson, 1988).

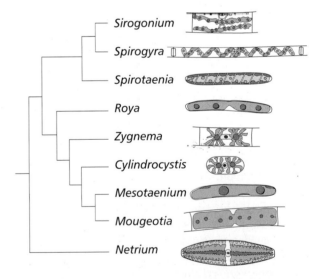

Figure 21–15 Molecular sequence analyses have revealed that zygnematalean thallus morphology reflects evolutionary pattern less accurately than does plastid shape. Genera that have ribbon-shaped plastids are closely related, those having stellate plastids are closely related, and those with plate-shaped plastids are related. In contrast, neither the filaments (*Spirogyra, Zygnema, Mougeotia*) nor the unicells (*Spirotaenia, Cylindrocystis, Mesotaenium*) form monophyletic groups. (Topology from McCourt, et al., 1995 by permission of the *Journal of Phycology*)

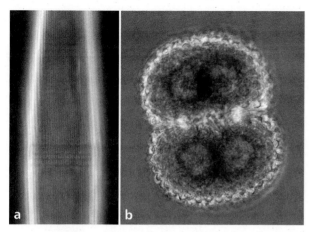

Figure 21–16 (a) *Closterium* cell walls may be marked with distinctive ridges; (b) *Cosmarium* cell walls are often ornamented with numerous warts.

rately reflect phylogenetic relationships. For example, saccoderm desmids having spiral, ribbonlike chloroplasts (*Spirotaenia*) appear more closely related to filaments having similar plastids (*Spirogyra*) than to other saccoderm desmids. Likewise, both the saccoderm desmid *Mesotaenium* and the filamentous *Mougeotia* have similar axial platelike plastids that can be rotated much like a solar panel to take maximum advantage of irradiance; molecular data show that these taxa are more closely related to each other than either is to other saccoderms or filaments (McCourt, et al., 1995) (Fig. 21–15). Such evidence suggests that additional assessment of the relationships of zygnematalean algae will be needed to evaluate older classifications. We will therefore take a fairly informal

approach to descriptions of zygnematalean diversity and general biology.

Cell biology of zygnemataleans

Cell Walls and Mucilage. The extracellular matrix of zygnematalean cells typically includes several layers. An outer mucoid layer of calcium pectate and hemicelluloses, a thin fibrillar primary wall, and a thicker, fibrillar secondary wall are commonly produced by *Spirogyra*, *Zygnema*, *Mougeotia*, and other filamentous forms, as well as by saccoderm desmids. The fibrillar portion of the wall of one *Mougeotia* species was found to consist of both noncellulosic carbohydrates (64%) and cellulose (13%), comparable to the cell walls of *Klebsormidium* (Hotchkiss, et al., 1989). In contrast, cell walls of charalean algae and land plants—although of similar biochemical constitution—are richer in cellulose. The extensive mucilage envelope that characterizes many saccoderm desmids and their filamentous relatives appears to be extruded through the cell wall (Gerrath, 1993). This mucilage sheath confers a characteristic slimy feel to masses of *Spirogyra* and its relatives. Ultra-

Figure 21–18 Mucilage prisms produced from individual pores (arrows) form a confluent sheath of mucilage on *Desmidium grevillii* (captured video image of a filament viewed with light microscopy). The small black dots (arrowheads) are end-on views of bacteria that live in the mucilaginous sheath.

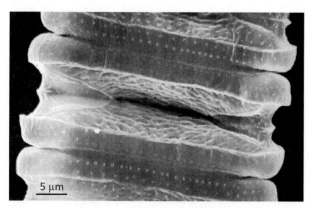

Figure 21–17 Linear arrays of mucilage pores in the cell wall of *Desmidium*, viewed by SEM. (Micrograph courtesy S. Cook)

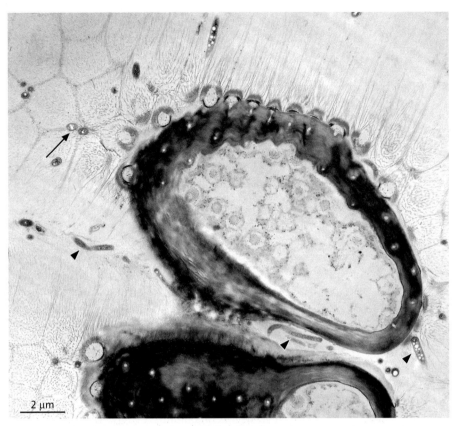

Figure 21–19 Mucilage prisms and their association with wall pores of *Desmidium grevillii*, as viewed with TEM. Note occurrence of associated bacterial cells (arrowheads) in the pocket formed at the isthmus region and at the interfaces of mucilage prisms (arrow). (From Fisher and Wilcox, 1996 by permission of the *Journal of Phycology*)

2 μm

structural and immunocytochemical analysis of *Closterium* mucilage revealed that mucilage vesicles are produced by the Golgi, then released through flask-shaped cell-wall pores. Under normal conditions, each cell can produce about 3 μg of mucilage in 30 days, but mucilage production increases three to four times when cells are grown under low phosphate or nitrate levels (Domozych, et al., 1993; Domozych and Domozych, 1993).

Cell walls of *Closterium*, *Genicularia*, and *Penium* are distinctive in that the outer layer is electron-opaque when viewed by transmission electron microscopy, and is often decorated with warts, ridges, or spines visible with light microscopy (Fig. 21–16); there are also primary and secondary walls that are presumed to contain cellulose. Placoderm desmids have a primary wall consisting of pectins and cellulosic microfibrils that is often discarded after the development of an often highly ornamented secondary wall. However, filamentous placoderm desmid cells may be held together in long chains by retained primary wall material (Krupp and Lang, 1985). The secondary wall of placoderm desmids is typically perforated with dis-

tinctive cylindrical pores (Fig. 21–17) through which pectinaceous mucilage is extruded. The mucilage produced by adjacent pores can be visualized at the light microscope (Fig. 21–18) and ultrastructural level (Fig. 21–19) as "prisms" of extruded material. Mucilage extrusion is the mechanism by which many desmids can undergo a form of motility (see below). A number of zygnematalean algae harbor putatively symbiotic bacteria within the confines of their sheaths (Figs. 3–28, 21–18, 21–19) (Gerrath, 1993; Fisher and Wilcox, 1996; Fisher, et al., 1998). In other cases the mucilage may confer a reduced sinking rate to planktonic forms or serve as an attachment mechanism for periphytic species. Additional possible functions for zygnematalean sheaths include water retention and resistance to desiccation, nutrient trapping, and absorption of harmful ultraviolet radiation. Iron is sometimes deposited in the outer wall layers of zygnematalean algae; it is rather evenly distributed in some, but more localized in others, giving cell walls a yellow or brown color (Gerrath, 1993).

Cellulose-Synthesizing Complexes. In cellulose-walled algae and land plants, cellulose microfibrils are

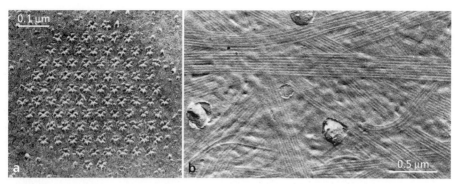

Figure 21–20 (a) Hexagonal arrays of rosettes occur in the cell membranes of desmids such as *Micrasterias*. These are believed to be enzyme complexes involved in cellulose synthesis, spinning out ribbons of microfibrils to form the secondary cell wall (b). (From Giddings, et al., 1980. Reproduced from **The Journal of Cell Biology**, 1980, 84:327–337, by copyright permission of the Rockefeller University Press)

synthesized at the cell membrane by arrays of membrane particles (Fig. 21–20) (Kiermayer and Sleytr, 1979; Hotchkiss and Brown, 1987; Giddings and Staehelin, 1991; Okuda and Brown, 1992). These particle groups are thought to include cellulose synthase, an enzyme or enzyme-complex that generates cellulose from precursor UDP-glucose molecules. Among the algae, cellulose synthesizing arrays can take the form of linear particle aggregates (characteristic of non-charophyceans), rosettes consisting of six particles (characteristic of plants and some charophyceans), and a unique structure in *Coleochaete* hav-

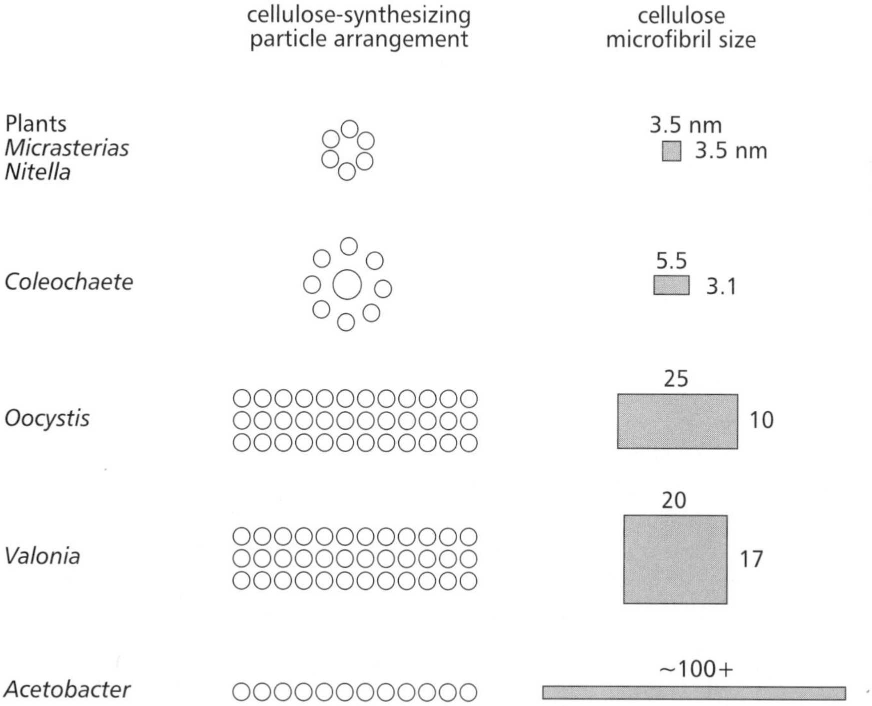

	cellulose-synthesizing particle arrangement	cellulose microfibril size
Plants *Micrasterias* *Nitella*		3.5 nm 3.5 nm
Coleochaete		5.5 3.1
Oocystis		25 10
Valonia		20 17
Acetobacter		~100+

Figure 21–21 Diversity of cellulose synthesizing complexes within the green algae. Rosettes similar to those of land plants occur in desmids and charaleans, and a modified type of rosette occurs in at least one species of *Coleochaete*; these rosettes generate relatively thin microfibrils (shaded boxes). In contrast, cellulose synthesizing complexes of chlorophyceans (*Oocystis*) and ulvophyceans (*Valonia*) are linear and are composed of three rows of particles that generate thicker microfibrils. The bacterium *Acetobacter* is regarded as having the ancestral type of cellulose synthesizing complex. (After Okuda and Mizuta, 1993)

ing a larger number of particles in rosettes (Okuda and Brown, 1992) (Fig. 21–21). Zygnemataleans are unusual in having hexagonal arrays of particle rosettes in membranes beneath developing secondary walls. Callose (β-1,3-glucan) can also be generated at the surfaces of zygnematalean algae (Dubois-Tylski, 1981) and other charophyceans and land plants, possibly by the cellulose-synthesizing array. Callose may play a role in the initial stages of cell-wall development and in reproductive-cell development. The cellulose and callose-synthesizing systems of charophycean algae and land plants are thought to have arisen from ancestral, patchy arrays of particles similar to those occurring in the cell membranes of cellulose-producing bacteria (Brown, et al., 1976).

Cell Motility. As mentioned earlier, many desmids are capable of gliding motility through mucilage extrusion. Movement is typically quite slow, only about 1 μm per second. Gliding movement is accompanied by continuous secretion of a slime trail that can be visualized at the light microscopic level by adding india ink to the preparation. *Closterium* and some other desmids secrete mucilage only from opposing cell tips, in alternate fashion, resulting in a somersaultlike movement. The placoderm desmid *Cosmarium*, among others, secretes mucilage from the older cell-wall half, while the younger wall half is lifted from the substrate at a slight angle. Subsequent swelling of the mucilage by water absorption then propels the cell forward (Häder and Hoiczyk, 1992). Phototaxis has been observed in many desmids (including the saccoderms *Cylindrocystis*, *Netrium*, and *Spirotaenia*, and the placoderms *Cosmarium* and *Micrasterias*) (Gerrath, 1993).

Plastid Movements. The filamentous *Mougeotia*, the closely related unicell *Mesotaenium* (see McCourt, et al., 1995), and the filamentous *Groenbladia* (Fig. 21–22) each possess the ability to orient their single, axial, platelike plastids to achieve optimal exposure to light, much as a motor-driven solar panel can be repositioned to follow changes in direction of solar radiation through the day. This process requires three components: one or more sensory pigments to perceive the wavelength and direction of the light signal, a transducer to convert the light signal into chemical information, and a mechanical effector to receive the message and move the chloroplast accordingly. In *Mougeotia* and *Mesotaenium*, the red/far red-light–sensitive pigment **phytochrome**, and a blue-light sensor (Gabrys, et al., 1984) are involved in signal perception. *Mesotaenium* phytochrome has been extracted, puri-

fied, and partially characterized (Kidd and Lagarias, 1990), and compared to higher plant phytochrome (Morland, et al., 1993).

Phytochrome molecules are thought to be oriented in a helical pattern at the cell surface, probably associated with the cell membrane. When red light is present, a structural change occurs in the phytochrome molecule, much as if a light switch were flipped. This is thought to effect changes in the binding of actin microfilaments at the cell periphery with myosin molecules that are linked to the chloroplast surface. The actomyosin motor then causes changes in positioning of the chloroplast. The result is that the platelike chloroplast is pulled into a position such that its broad face is directed toward the light when irradiance is low or optimal, but so the edge faces the light when irradiance is too high (Wagner and Grolig, 1992). The blue-light sensor is thought to be important in detecting excessively high light levels (Kraml and Herrmann, 1991). Calcium influx, interaction of calcium ions with the calcium-binding protein calmodulin, and the binding of calcium ions with protein kinases are also thought to be involved in the light signal transduction process in these algae. Actomyosin interactions are also involved in organelle movements that occur in

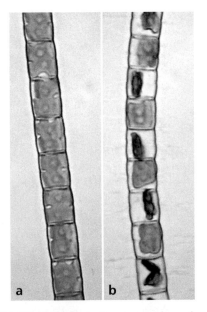

Figure 21–22 Plastid rotation occurs in various zygnemataleans, such as *Groenbladia*. Platelike plastids are seen in face view in (a), and those of alternate cells are observed in edge view within the filament shown in (b).

interphase cells of other zygnemataleans, such as *Spirogyra* (whose plastids do not rotate) (Grolig, 1990).

Mitosis and Cytokinesis of Zygnemataleans. Cell division in zygnemataleans typically occurs during the dark portion of the light-dark cycle. Most studies of division have been conducted with *Spirogyra* and other filamentous forms, in addition to the desmids *Closterium* and *Micrasterias*. Mitosis and cytokinesis in zygnemataleans resemble those of *Chlorokybus* and *Klebsormidium*, except that centrioles are absent. In *Closterium*, the first sign of impending mitosis is the appearance of indentations in the two chloroplasts at positions about two fifths of the distance from the centrally located nucleus. These represent the early stages of chloroplast division by constriction mediated by actin microfilaments (Hashimoto, 1992). The nuclear envelope breaks down in prophase, and a precocious cleavage furrow begins to appear at the same time. At telophase, as furrowing continues, the spindle abruptly disintegrates, and transverse microtubules develop near the forming cross-wall (Pickett-Heaps and Fowke, 1970), but a phragmoplastlike apparatus is absent from *Closterium*. Upon completion of cross-wall development, the chloroplast constrictions become deeper and a vacuole forms in the resulting groove. The vacuole then increases in size as the chloroplasts continue to constrict. After the daughter cells separate from each other, the chloroplasts complete their division, and the nucleus moves into a central position between them (Fig. 21–23). The cells then elongate on the side closest to the cross-wall so that cell symmetry is reestablished (Hashimoto, 1992).

In placoderm desmids, often characterized by elaborately lobed semicells, new semicells develop by expansion after mitosis has been completed. The process by which these new semicells come to look almost identical to the older ones is still not completely understood. It has primarily been studied in *Closterium*, *Cosmarium*, *Euastrum*, and *Micrasterias* (Fig. 21–24) (Waris, 1950; Gerrath, 1993; Pickett-Heaps, 1975). In highly constricted desmids the nucleus undergoes what is known as post-telophase migration after the formation of a cross-wall; this is necessary to provide a pathway through the constricted isthmus region for expansion of the chloroplast from the old semicell into the newly developing semicell. An urn-shaped cage of microtubules radiating from a microtubular organizing center often surrounds the moving nucleus, suggesting a mechanism for organellar transport—possibly microtubule-associated motor proteins. In *Euastrum oblongatum*,

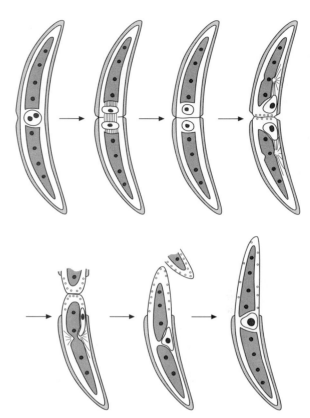

Figure 21–23 Cell division in the desmid *Closterium*. Mitosis and cytokinesis are followed by plastid division at predetermined sites, then the nuclei migrate to a position between them as a new cell half expands. (After Pickett-Heaps, 1975)

nuclear movement begins about 80 minutes after cross-wall formation (Url, et al., 1993). Usually the nucleus gets out of the way by moving into the expanding semicell, but in some desmids (*Netrium* and *Closterium*) the nucleus migrates into the older semicell. Expansion of the new semicell appears to be driven by turgor pressure because plasmolysis stops the expansion process. As the new semicell develops, the nucleus migrates back to its usual position in the isthmus, between old and new semicells. In *Euastrum oblongatum*, the nucleus begins to move back to the isthmus about 12 hours after the new semicell has been completely formed (Url, et al., 1993). A new semicell is completely formed within about 16 hours in *Micrasterias* (Kiermayer, 1981; Meindl, 1983).

Lobing of new semicells seems to occur by greater deposition of primary wall material (reviewed by Kiermayer and Meindl, 1989) at the tips of the growing lobes as compared to regions between lobes. The factors influencing this differential distribution of wall

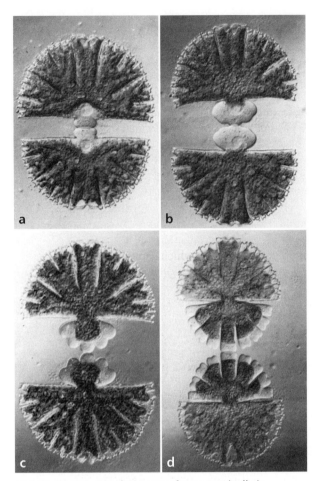

Figure 21–24 Development of new semicells in *Micrasterias*. (From Pickett-Heaps, 1975)

material are not well understood. However, it is hypothesized that at the cross-wall development stage, the cell membrane is chemically imprinted with a pattern (Kiermayer, 1981) that may later influence localized accumulation of calcium ions and targeted fusion of pectin-bearing Golgi vesicles at the lobes. Cellulose-synthesizing rosettes also appear to be directed to lobe regions, such that cellulose synthesis occurs more rapidly there. Rosettes are thought to arrive at the cell membrane as Golgi-derived vesicles, but the mechanisms by which they are directed to lobe regions are unknown. After primary wall deposition ends, development of the secondary wall of *Micrasterias denticulata* occurs for eight or so hours. Flat Golgi-derived vesicles deliver hexagonal arrays of cellulose-synthesizing rosettes to the cell membrane; these spin out bands of 2–17 adherent cellulose microfibrils (Giddings, et al., 1980; Kiermeyer and Meindl, 1984). At the beginning of secondary wall deposition the cell

membrane develops concave, circular invaginations that are about 0.2 μm in diameter. These mark the positions where pores will develop. Pore vesicles containing neutral polysaccharides then deposit these materials at the invaginations. These polysaccharides prevent cellulose microfibril deposition across the sites; subsequent disappearance of the plugs occurs when secondary wall development has been completed, leaving open pores in the walls.

In *Spirogyra* and *Mougeotia*, the nuclear envelope remains intact until metaphase, when it disintegrates, as in other charophyceans. These algae, and *Zygnema*, are also unusual among zygnemataleans in that a small phragmoplast—an array of perpendicular microtubules as well as membranous tubules and vesicles—occurs at the central region of the cell at the plane, and at the same time as furrow extension from the periphery (Fowke and Pickett-Heaps, 1969a, b; Pickett-Heaps and Wetherbee, 1987). This small phragmoplast has been suggested to represent an intermediate stage in the evolutionary origin of more highly developed, plantlike phragmoplasts of *Coleochaete* and Charales (Pickett-Heaps, 1975; Graham and Kaneko, 1991), which are described below. A study of the *Spirogyra* phragmoplast via fluorescent tagging of cytoskeletal components and video microscopy revealed several differences between phragmoplast structure and behavior as compared to those of other charophyceans and higher plants (Sawitzky and Grolig, 1995). In contrast to *Coleochaete* and charalean algae, plasmodesmata are not known to occur in the cross-walls of zygnematalean algae. Furrowing in *Spirogyra* involves actin microfilaments (Goto and Ueda, 1988; Nishino, et al., 1996); these presumably help to constrict the cytoplasm in a fashion analogous to the tightening of purse strings. Immunolocalization of microtubules has been followed throughout the cell cycle in *Mougeotia* (Galway and Hardham, 1991).

Zygnematalean reproduction

Asexual Reproduction. Zygnematalean algae, in contrast to *Chlorokybus*, *Klebsormidium*, *Coleochaete*, and its close relative *Chaetosphaeridium*, do not produce flagellate zoospores. This has been suggested to be correlated with the absence of centrioles, necessary for the generation of flagella (Pickett-Heaps, 1975). However filamentous zygnemataleans may reproduce asexually by fragmentation, and populations of single-celled zygnemataleans grow by means of mitosis and cytokinesis. Vegetative cells are dispersible by wind,

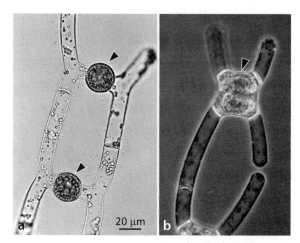

Figure 21–25 (a and b) Zygotes (arrowheads) of two species of *Mougeotia*, developed within the conjugation tube. These species both exhibit scalariform (ladderlike) conjugation.

insects, and water birds (as are desiccation-resistant zygotes) (Hoshaw, et al., 1990). Zygnemataleans may sometimes produce thick-walled resting cells known as akinetes, aplanospores, asexual spores, or parthenospores (if they originated from unpaired cells in sexual populations). Such resting cells occur in *Closterium* and the placoderm desmids *Cosmarium*, *Hyalotheca*, *Staurastrum*, *Staurodesmus*, *Tetmemorus*, and *Euastrum* (Gerrath, 1993), and probably occur more widely among zygnemataleans.

Sexual Reproduction. Because flagellate cells are not produced by this group of algae, the fusion of flagellate isogametes or fertilization of a nonmotile egg by a flagellate sperm cell (oogamy)—such as occurs in most other algae having sexual reproduction (and land plants)—does not occur in zygnemataleans. Rather, the physical pairing of filaments or single cells, their enclosure within common mucilage, and subsequent fusion of nonflagellate gametes—a process known as **conjugation**—is characteristic. Although this type of sexual reproduction is unique to zygnemataleans among the charophyceans, pennate diatoms (Chapter 12) can also accomplish sexual reproduction without flagellate gametes. It has been suggested that both zygnematalean conjugation and sexual reproduction in pennate diatoms may have been derived from oogamy, in response to selective pressures operating in shallow waters (Nakahara and Ichimura, 1992).

Both homothallic and heterothallic strains occur among zygnemataleans. In the laboratory, cultures of zygnematalean algae have been induced to undergo

conjugation by reducing the combined nitrogen concentration in their growth medium (Biebel, 1973), by increasing carbon dioxide levels (Starr and Rayburn, 1964), or increasing temperature, light levels, and/or concentrations of Ca^{2+} and Mg^{2+} (Gerrath, 1993), but little is known about the factors responsible for inducing sexual reproduction in nature.

Spirogyra, *Mougeotia*, *Zygnema*, and other filamentous forms may undergo **scalariform conjugation**, where filament pairs align themselves laterally and develop modified branches known as **conjugation tubes**, that link opposing cells (Figs. 21–14, 21–25, 21–26). The conjugation tube is composed of an outgrowth (papilla) from each opposing cell; when they meet, the wall at their interface is degraded to form an open tube through which gametes can move. Ultrastructural studies have suggested that papilla growth occurs by addition of new cell-wall material, under the influence of enlarging vacuoles in the papilla, but the mechanism by which the end walls disintegrate is unclear (Pickett-Heaps, 1975). Except in *Closterium*, the conjugation tube always forms at the narrow isthmus region of placoderm desmids, but can occur in various regions of the walls of saccoderm desmids. The placoderm *Closterium* behaves more like a saccoderm desmid in this respect, as conjugation papillae do not always develop from a central area in this organism (Pickett-Heaps, 1975).

Although gamete motion is often described as "amoeboid," there is no direct evidence that zygnematalean cells can actually move in the same way as

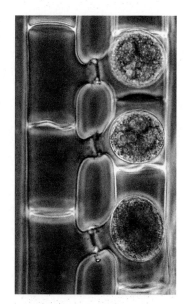

Figure 21–26
Zygotes of *Zygnema*, formed within the parental walls of one of the gametes.

true amoeboid cells. Gamete protoplasts can be observed to shrink as they lose water; increased external hydrostatic pressure or mucilage accumulation in the surrounding area may be responsible for their propulsive movement (Pickett-Heaps, 1975), but the mechanism is not well understood. Sometimes both gametes move into the conjugation tube, whereupon a zygote is formed in the center of the tube (Fig. 21–25). In other cases the cytoplasm of only one of the connected cells moves across the tube, forming a zygote within the confines of the second cell (Fig. 21–26). The timing of gamete nuclear fusion and meiosis in filamentous zygnemataleans appears to vary from organism to organism.

An alternate form of conjugation, known as **lateral conjugation**, occurs when gametes develop from adjacent cells within the same filament. In this case filament pairing does not occur. Rather, a short curved tube extends from one cell to the next in the filament. Because lateral conjugation is a form of selfing, this process would be expected to result in reduced levels of genetic variability relative to scalariform conjugation, which may involve filaments that are genetically distinct. Both result in the production of resistant-walled zygotes that serve a perennation function, and during zygote germination (involving meiosis) there exists the potential for recombination events. However, little population genetic work has been done to examine the relative roles of lateral and scalariform conjugation.

Among the single-celled zygnemataleans—the saccate and placoderm desmids—sexual reproduction involves cell aggregation, formation of gametes through mitotic division, papilla formation, release of gamete protoplasts from enclosing walls, and zygospore formation. Sexual cell divisions (those essential to gamete formation), papilla formation, and gamete release involve cell-cell interactions via chemical messengers (pheromones) (Hogetsu and Yokoyama (1979), but relatively little is known about the nature of these chemical inducers. A 20 kDa heat-labile, diffusible protein produced by one of the mating types of *Closterium ehrenbergii* induces divisions that generate gametes in the other mating type (Fukumoto, et al., 1997). Secreted glycoproteins are thought to induce the release of gamete protoplasts from enclosing cell walls (Sekimoto, et al., 1993). Although meiosis is zygotic, changes in DNA level during sexual reproduction have been shown to be somewhat more complex than expected, due to the occurrence of polyploidy in vegetative cell nuclei. Changes in nuclear DNA levels have been followed throughout the sexual reproductive process in *Netrium* and *Closterium* (Hamada, 1987) by microspectrophotometry, a technique that allows measurement of DNA levels in single cells. In *Closterium ehrenbergii*, vegetative cells were observed to possess two or four times the nuclear DNA level of gametes. There is no increase in DNA level immediately prior to the divisions that give rise to gametes, so that gametes possess half the DNA of parental vegetative cells. Desmid zygotes often include unfused gamete nuclei that fuse only shortly before or during germination, which also involves meiosis. When zygotes undergo meiosis, DNA levels are reduced, as expected, but frequently two or three of the nuclear products do not survive. In surviving meiotic nuclei, the DNA level is duplicated twice, then the cell divides, partitioning DNA such that each daughter-cell nucleus receives at least two copies of the genome, restoring the typical vegetative DNA level (Fig. 21–27).

Zygote development involves formation of a thick wall consisting of as many as six distinct layers; callose and sporopollenin (DeVries, et al., 1983) have been demonstrated to occur among the layers of zygotc walls of at least some zygnematalean species. The sporopollenin layer is said to be responsible for the resistant properties of zygotes. Mature zygotes are often highly ornamented and colored orange-brown as the result of wall formation and chlorophyll degradation. In nature, zygotes germinate in spring, or the end of a period of dry dormancy, as when a temporary pool is reformed in the wet season. Zygnematalean zygotes can withstand burial in mud for long periods (Brook and Williamson, 1988) and zygotes have been germinated after dry storage for more than 20 years (Coleman, 1983). Zygote germination can be induced in the laboratory by allowing the culture medium to slowly evaporate, storing the zygotes (also known as zygospores) in the dark (with or without refrigeration) for one to 12 months, then rewetting zygotes with fresh culture medium. A few hours following rehydration, zygotes become green due to synthesis of chlorophyll, and one to three days later the wall ruptures and a germination vesicle containing the meiotic products emerges. In placoderm desmids, the first cells produced—known as **gones** (rhymes with bones)—do not generally resemble normal vegetative cells, typically being less ornamented. As they divide, the new semicells exhibit a normal morphology. Early spring collections from nature or laboratory cultures established from recently germinated zygotes may contain some cells having both a typical semicell and a gone semicell.

Zygnematalean ecology

Zygnematalean algae are almost exclusively found in freshwater habitats, although a few have been collected from brackish waters. They are ubiquitous in freshwaters, occurring in pools, lakes, streams, rivers, marshes, and especially bogs and mildly acidic, nutrient-poor streams. In addition, zygnemataleans may be abundant in reservoirs, cattle tanks, roadside ditches, irrigation canals and other water bodies of human construction. *Spirogyra*, for example, was found at nearly one third of the more than 1000 locations sampled by McCourt, et al. (1986), and in a North American continent-wide survey, Sheath and Cole (1992) located *Spirogyra* in streams from a wide variety of biomes, including tundra, temperate and rain forests, and desert chaparral. In streams and shallow lakes *Spirogyra* is typically attached to stable substrates, but can also occur as free-floating mats that originate from benthic zygotes or filaments (Lembi, et al., 1988). As growth and photosynthesis occur, oxygen bubbles become entrapped in the mats and provide floatation; at the water surface the algal mats are exposed to high temperatures and light levels. Optimal temperature and irradiance conditions for photosynthesis for one species of *Spirogyra* were determined to be 25° C and 1500 μmol photons m^{-2}, respectively. Net photosynthesis was observed to be positive at 5° C under high irradiance conditions, explaining the widespread occurrence of surface growths in the cool waters of early spring. However the alga could not maintain positive photosynthesis at the low light levels that can result from self-shading when temperatures were high

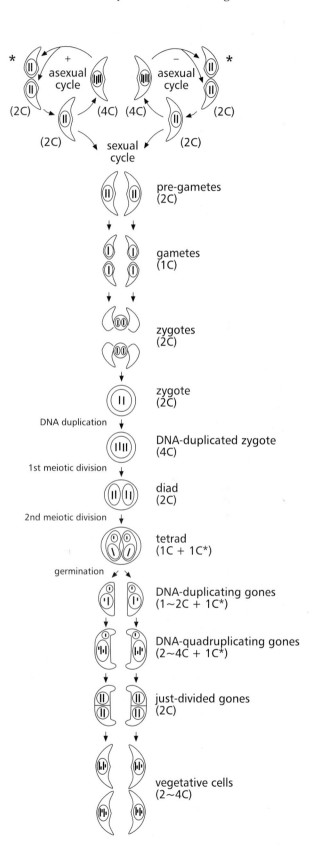

Figure 21–27 DNA levels in nuclei of sexually reproducing *Closterium*, measured by DAPI microspectrophotometry. Note that G1 interphase vegetative cells (asterisks) actually contain twice the expected levels (2C) of DNA, as do the cells that will divide to form gametes (pregametes). Gamete-forming mitoses are not preceded by DNA replication, as would be expected; thus, DNA levels are reduced to the 1C level in gametes. The 2C number is restored in zygotes. Meiotic reduction of the nuclear DNA level to 1C (as expected), gives rise to gone cells at germination. Gones undergo a series of two mitotic divisions during which normal morphology of both semicells is reestablished, and the G1 interphase DNA level returns to 2C. (After Hamada, 1987 by permission of the *Journal of Phycology*)

20 μm

Figure 21–28 *Closterium aciculare,* a desmid that may be abundant in eutrophic waters. (Reprinted from SYNOPSIS OF NORTH AMERICAN DESMIDS. PART II. DESMIDIACEAE: PLACODERMAE, SECTION I by permission of the University of Nebraska Press. ©1975 by the University of Nebraska Press.)

explain the rise to dominance of large subsurface *Mougeotia* growths in acidified lakes (Graham, et al., 1996a, b).

Desmids are more common in oligotrophic (low-nutrient) and dystrophic (highly colored) lakes and ponds, but some species can occur in mesotrophic (higher nutrient) and eutrophic (high-nutrient) water bodies. *Closterium aciculare* (Fig. 21–28) in particular, is regarded as an indicator of eutrophic conditions and is sometimes abundant, occasionally growing to bloom proportions. This species is unable to utilize nitrate as a source of combined nitrogen because nitrate reductase (see Chapter 2) is lacking (Coesel, 1991); a requirement for ammonium ion may explain occurrence of this desmid in highly eutrophic waters. Cells of the very unusual, slow-growing desmid *Oöcardium stratum* live at the tops of branched calcareous tubes in calcareous streams and waterfalls, in association with deposits of tufa and travertine. A number of desmids occur in soils and other terrestrial habitats, such as moist rocks and among bryophytes, and some can tolerate low temperatures, occasionally occurring in ice and snow (Gerrath, 1993; Yoshimura, et al., 1997).

Desmid diversity is particularly high in low pH freshwaters (Woelkerling, 1976), where hundreds of species may co-occur, despite very low levels of nutrients such as phosphate and combined nitrogen. Their success is due in part to their ability to store phosphate when pulses are available, and perhaps also to microbial associations. In nutrient-poor streams, desmids can make up some 2–10% of the community, and they are persistent residents, rather than forms that have come to be there via incidental drift. More than 200 desmid species have been observed among stream periphyton (algae attached to substrates), associated with plants such as the moss *Fontinalis*. There, desmids may achieve cell concentrations as high as 10^6 per gram of substrate (Burkholder and Sheath, 1984). In bogs and fens, desmids can also constitute a significant portion of the algal biomass and species diversity.

Diversity of zygnemataleans

Genus and species definitions within the Zygnematales have classically depended upon characters of the cell wall (including wall pores, ridges, knobs, and spines in desmids), width of filaments and chloroplast morphology (particularly for filamentous forms such as *Spirogyra*), and the shape of gametangial papillae and zygotes. Although some authors have suggested that there may be many thousands of

(30°–35° C), explaining late spring and summer declines in zygnematalean mats (Graham, et al., 1995).

Mougeotia, a filamentous zygnematalean, often forms large nuisance growths in subsurface freshwaters impacted by acid precipitation or experimental acidification. Such waters are characterized by increases in the concentration of metals such as aluminum and zinc, reduced levels of dissolved inorganic carbon, and food web changes, including reduction in numbers of herbivores (Stokes, 1983; Webster, et al., 1992; Fairchild and Sherman, 1993). The appearance of *Mougeotia* mats is widely regarded as an early indicator of environmental change (Turner, et al., 1991). The optimal light, temperature, and pH conditions for photosynthesis, as well as effects of the metals zinc and aluminum on photosynthesis, were determined for this alga so that its common and specific association with acidification could be better understood. Net photosynthesis was high (on average over 40 mg O_2 was produced per gram dry weight per hour) over a wide range of irradiances (300–2300 μmol quanta m^{-2} s^{-1}). The optimal temperature was 25° C, and the organism exhibited tolerance of a wide range of pH (3–9) and metal concentrations. These results, together with release from herbivores, help to

desmid species, a careful analysis of all of the described forms, together with addition of 10% to account for as yet undescribed forms, led Gerrath (1993) to the conclusion that there were on the order of 3000 species of desmids. Accurately identifying zygnemataleans can be difficult for a number of reasons. Among these are that zygotes are not always present, and in some cases sexual reproduction is quite rare or nonexistent. For example, Prescott (1951) found that of about 250 species of *Spirogyra* occurring in the western Great Lakes region, only 55 had been observed in the reproductive state. Moreover, analysis of a clonal culture of *Spirogyra communis* revealed the presence of three significantly different filament types, as defined by width and chromosome number; each of these, in fact, corresponded to a separate described species. Variation in filament width appeared to correlate with differences in ploidy level (Hoshaw, et al., 1985). This means that algae of very different morphology may actually be closely related. However, the use of standard identification keys would result in their classification into different species. In addition, zygnemataleans that look similar, and thus are classified as the same species, may actually not be closely related. One of the most valuable sources of taxonomic information is the series by Prescott, et al. (1975, 1977, 1981, 1982) and Croasdale, et al. (1983) on North American desmid taxa.

Saccoderm Desmids and Related Filamentous Forms

SPIROTAENIA (Gr. *speira*, coil + Gr. *taenia*, band) (Fig. 21–29) is a saccoderm desmid that is closely related to the filamentous *Spirogyra* (McCourt, et al., 1995). The cells are straight or somewhat curved with a single spirally twisted, ribbonlike chloroplast resembling that of *Spirogyra*. Sexual reproduction was studied in *S. condensata* (Hoshaw and Hilton, 1966); pairing in mucilage occurs between cells of different sizes—the smaller-sized cells undergo gametangial divisions before the larger cells. Zygotes develop an unusual honeycomblike wall, lose pigmentation, and form large cytoplasmic oil droplets. Gamete nuclear fusion occurs three days after cytoplasmic fusion, an occurrence unusual among zygnemataleans where it is more common for gamete nuclei to remain unfused for long periods after syngamy. Spontaneous germination of *Spirotaenia* zygospores begins 22 days after their formation; meiosis occurs at

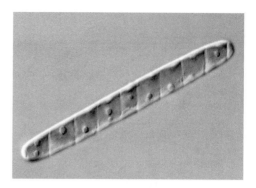

Figure 21–29 *Spirotaenia* is single-celled and has a spirally twisted, ribbonlike chloroplast.

the onset, and four germling products are released. These may be observed in the vicinity of the empty zygote wall, surrounded by mucilage. The short life cycle of this species suggests that it would be suitable for demonstrating sexual reproduction in instructional laboratories.

SPIROGYRA (Gr. *speira*, coil + Gr. *gyros*, twisted) is a filament composed of cells having 1–16 spiral, ribbon-shaped chloroplasts per cell (Fig. 21–30). The plastid edges are often beautifully sculpted, and numer-

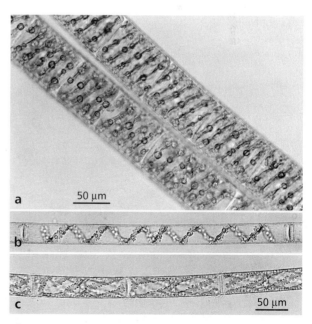

Figure 21–30 (a, b, c) Three species of *Spirogyra*—an unbranched filament of cells attached end to end. Note the helically twisted ribbon-shaped chloroplasts with multiple round pyrenoids.

Figure 21–31 Higher magnification view of spiral, lobed, ribbonlike plastid of *Spirogyra* with numerous globular pyrenoids.

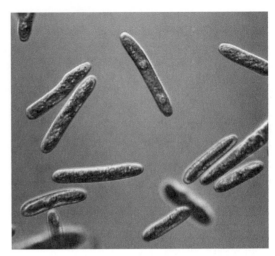

Figure 21–33 *Mesotaenium* is a unicell having just one plate-shaped axial plastid. (From *Origin of Land Plants*. Graham, L. E. ©1993. Reprinted by permission of John Wiley & Sons, Inc.)

ous pyrenoids are present in plastids (Fig. 21–31). *Spirogyra* plastid genomes contain an operon for the *str* genes (for streptomycin resistance) that is nearly identical to that of plants, yet quite distinct from that of cyanobacteria (Lew and Manhart, 1993). Sometimes rhizoidal processes occur at the basal end of the filament; these are involved in attachment to substrates. Cytoplasmic streaming—based on the action of actin microfibrils—can often be observed in the peripheral cytoplasm. The nucleus is suspended in the center of the cells (Fig. 21–32). More than 300 species have been described, all from field collections. Both scalariform and lateral conjugation have been observed. During

conjugation previously bright green filamentous masses turn noticeably brownish in color, reflecting the loss of chlorophyll pigments from zygotes and development of brown zygote walls. Zygotes germinate to form a single filament; from this it is deduced that only a single meiotic product survives.

MESOTAENIUM (Gr. *mesos*, middle + Gr. *tainia*, ribbon) cells are shaped like cylinders, each with a single platelike plastid having several pyrenoids (Fig. 21–33). Culture studies have shown that sexual reproduction in *Mesotaenium kramstei* involves the formation of a broad conjugation tube that can grow from any portion of the cell wall. Mucilage is secreted inside the wall as gametes shrink during development. Mature zygospores are mahogany brown (Biebel, 1973).

MOUGEOTIA (named for Jean Baptiste Mougeot, an Alsatian physician and botanist) consists of long, unbranched, free-floating filaments, each cell of which is characterized by a single platelike chloroplast (Fig. 21–34). The plastids are suspended in the central area of the cell (i.e., are axial in location). Orientation of the plastid to achieve optimal light conditions can be commonly observed; in the same filament some plastids may be in face view, others in edge view, and yet others twisted so that part or parts of the plastid are in face view and others in edge view (Fig. 21–35). Pyrenoids are either arranged in a single row

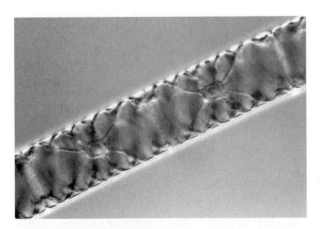

Figure 21–32 The nucleus of zygnemataleans is suspended in cytoplasm in the cell center, as illustrated here by *Spirogyra*. (Photograph courtesy M. Cook)

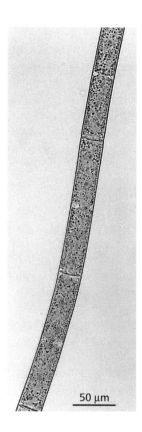

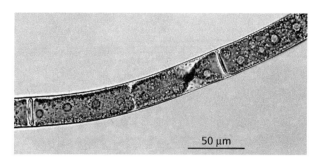

Figure 21-35 The *Mougeotia* plastid can rotate around a lengthwise axis, and portions of the plastid can turn independently, generating a twisted plastid.

Figure 21-34 *Mougeotia* is an unbranched filament having a single platelike plastid per cell.

or scattered throughout the chloroplast. (*Mougeotiopsis* is a form in which pyrenoids are said to be absent.) Conjugation in *Mougeotia* is usually scalari-

form, and zygotes usually form in the conjugation tube (Fig. 21–25). Only a single filament is produced upon zygote germination. *Mougeotia* cells contain numerous small vacuoles filled with phenolic compounds, believed to serve in protecting the cells from herbivores (Wagner and Grolig, 1992).

CYLINDROCYSTIS (Gr. *kylindros*, cylinder + Gr. *kystis*, bladder) unicells are cylindrical and contain two axial, stellate chloroplasts similar to those of *Zygnema*. Sexual reproduction of a number of species has been studied in culture. Easily grown forms that might prove useful in molecular genetics studies are heterothallic strains of *C. brebissonii* (Fig. 21–36) (UTEX 1922 and 1923) isolated by Biebel (1973). They grow well in defined culture medium (Bold's

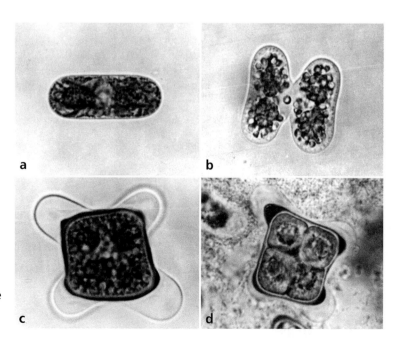

Figure 21-36 Sexual reproduction in *Cylindrocystis*. (a) Vegetative cell. (b) Mating gametes with short conjugation tube. (c) Pillow-shaped zygote with remains of parental gamete cell walls at the corners. (d) Zygote that has undergone meiosis to form four daughter products. (From Biebel, 1973)

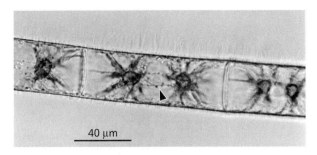

Figure 21–37 *Zygnema* is an unbranched filament having two stellate plastids per cell, each with a large conspicuous pyrenoid. Arrowhead points to cytoplasmic bridge between plastids, which contains nucleus.

Basal Medium—Stein, 1973), and when one-week-old cultures of the two mating types are mixed and placed into culture medium lacking nitrogen salts, conjugating pairs appear within four days. A high yield of zygotes results within two weeks. Conjugation begins with lengthwise pairing, and wall papillae appear at the mid-regions of the lateral walls, then they fuse to form a tube (Fig. 21–36b). The plastids remain visible during zygote dormancy. Zygotes (Fig. 21–36c) can be concentrated and placed on agar medium, dried, and kept for two months in a dormant condition, then induced to germinate by transfer to fresh agar medium and exposure to illumination. Germination (Fig. 21–36d) occurs about one month later. At the onset of germination, the plastids migrate to new positions, one pair lying above the other, and meiosis is presumed to commence. All four meiotic products (gones) survive, and could presumably be isolated for tetrad analysis.

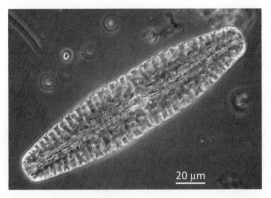

Figure 21–38 *Netrium* is a unicellular desmid that is not constricted and has two elaborately lobed chloroplasts, one in each cell half.

ZYGNEMA (Gr. *zygon*, yoke + Gr. *nema*, thread) occurs as relatively short, unbranched filaments of cylindrical cells, enclosed by a mucilage sheath. As is the case for *Spirogyra*, there may be basal rhizoidal outgrowths that are used in attachment. *Zygnema* is frequently found in nature together with *Spirogyra* and/or *Mougeotia*. There are two stellate chloroplasts per cell, each with a central pyrenoid (Fig. 21–37). Conjugation is similar to that of *Spirogyra*; both scalariform and lateral conjugation have been observed. Akinetes—asexually produced, thick-walled cells containing abundant oil and starch storage products—can be formed; these are able to survive for over one year before germinating into vegetative filaments. At least 80 species have been described. Prescott (1951) observed that in the western Great Lakes region, species growing in higher pH waters more often possessed zygotes than forms occurring in low pH habitats. Zygotes germinate to form a single filament, suggesting that only a single meiotic nucleus survives.

NETRIUM (Gr. netrion, small spindle) (Fig. 21–38) unicells are elongated and cylindrical, usually with rounded ends. Two large, elaborately lobed and ridged chloroplasts, each with a pyrenoid, occur per cell. Copious mucilage is excreted into the surrounding environment. Homothallic and heterothallic strains of *N. digitus* were studied by Biebel (1973). A broad conjugation tube is formed, gametes shrink when they contact each other, the protoplasts then fuse, and the zygote develops in the conjugation tube. The plastid pigmentation disappears, and a golden-brown wall develops on the zygote. After two months in culture medium, the zygotes germinate spontaneously. Typical plastid pigmentation reappears, and the protoplast swells, causing the zygote wall to burst. Usually only two meiotic products survive to form germlings.

Placoderm desmids

CLOSTERIUM (Gr. *klosterion*, small spindle) (Fig. 21–39) is considered to be a placoderm desmid (even though the cells are not constricted as are those of most placoderms) because mucilage secretion pores occur in the cell walls. Vegetative cells are crescent-shaped or elongate and somewhat curved. There is a single axial, ridged plastid with several pyrenoids in each semicell and a central nucleus. Conspicuous vacuoles occur at the cell tips; these contain barium sulfate crystals of unknown function (Brook, 1980) that

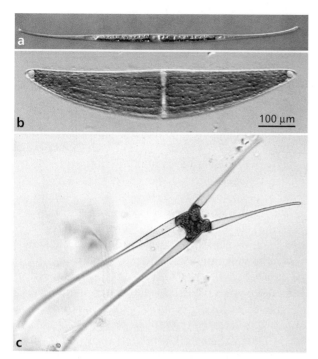

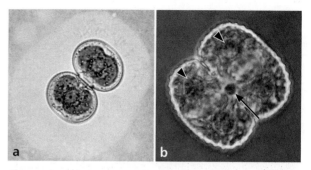

Figure 21–40 *Cosmarium*. Note the extensive mucilage investment (a). The pyrenoids (arrowheads) and nucleus (arrow) are conspicuous in this cell seen under phase-contrast microscopy (b).

Figure 21–39 *Closterium* species vary from needlelike cells (a) to broader cells (b). Note the presence of terminal vacuoles and ridged chloroplast in the latter. (c) Zygote and remains of parental walls of a species of *Closterium*.

Closterium have been described. Students should avoid confusing this algal name with the similar sounding name of a common toxin-producing bacterium, *Clostridium*.

COSMARIUM (Gr. *cosmarion*, small ornament) (Fig. 21–40) occurs as single cells that are deeply divided at the mid-region to form a short isthmus and two semicells that are rounded in front view, but flattened, oval, or elliptic in side view. Walls may be smooth or ornamented; spines are not present on vegetative cells. One or sometimes more axial or parietal chloroplast(s) with pyrenoids occurs in each semicell. Sexual reproduction in *C. turpinii* involves separate mating types, with pairs of cells becoming enclosed by mucilage. Mating cells open at the isthmus and the emerging protoplasts function as gametes. Their fusion results in production of a thick-walled, spiny zygote, which can often be found together with the empty parental cell walls, enclosed in the same

move by Brownian motion. Almost all natural populations of the *Closterium ehrenbergii* species complex have been found to be heterothallic (outbreeding); i.e., there are distinct mating types segregated in a 1:1 ratio upon zygote germination. Mating type is determined by one gene and the mt⁻ allele is dominant to mt⁺ (Kasai and Ichimura, 1990). About 140 species of

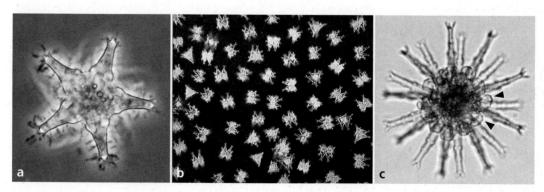

Figure 21–41 *Staurastrum*. (a) An end-on view of one of the semicells of a species having five arms. (b) A population of a species having triradiate semicells, viewed with dark-field microscopy. (c) A species having nine projections or arms on each semicell, note also the infestation of spherical chytrid cells (arrowheads).

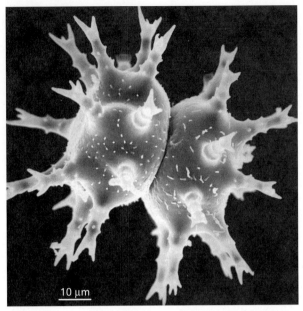

Figure 21–42 A multiple-armed *Staurastrum* species viewed with SEM. (Photograph by S. Cook, in *Origin of Land Plants*. Graham, L. E. ©1993. Reprinted by permission of John Wiley & Sons, Inc.)

enveloping mucilage. Over 1000 species have been described.

STAURASTRUM (Gr. *stauron*, cross + Gr. *astron*, star) unicells are highly constricted unicells that are radially symmetrical in end-on (polar) view. The semicells are often triradiate or hexaradiate and may be highly ornamented with spines and other protuberances (Figs. 21–41, 21–42). The walls are impregnated with polyphenolic compounds that confer decay resistance (Gunnison and Alexander, 1975 a, b). These materials explain the recovery of fossil remains of *Staurastrum* walls from lake sediment cores that are thousands of years old. Some 800 species have been described, primarily on the basis of cell-wall characters. At conjugation the gamete protoplasts escape as the semicell walls separate at the isthmus; their fusion generates spiny zygotes. Zygote germination, presumably by meiosis, produces one to four gones.

MICRASTERIAS (Gr. *mikros*, small + Gr. *aster*, star) is a flattened, often highly incised and lobed pla-

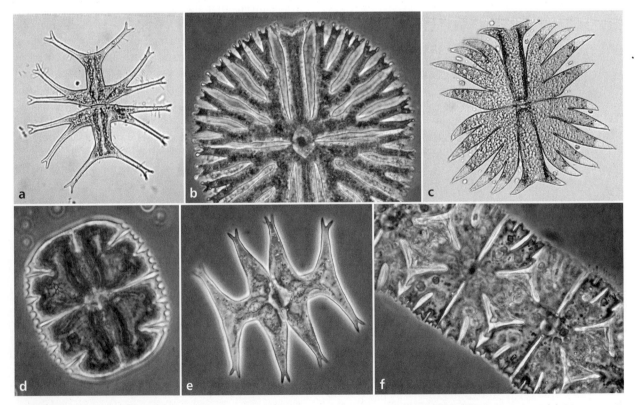

Figure 21–43 Diversity of *Micrasterias*—(a) *M. radiata*, (b) *M. radiosa*, (c) *M. torreyi*, (d) *M. truncata*, (e) *M. pinnatifida*, and (f) *M. foliacea*, which forms filaments by means of interlocking prongs on the ends of semicells. (b: Courtesy C. Taylor)

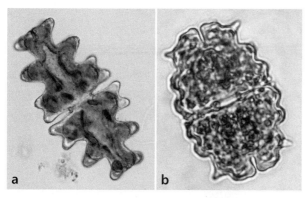

Figure 21–44 *Euastrum* is a lobed, flattened placoderm desmid. (a) *E. pinnatum*, (b) *E. bidentatum*.

coderm desmid. Some species look like flattened disks, and others are so highly dissected that they look like stars (Fig. 21–43). Each semicell has a single large lobed plastid that extends into lobes of the cell. Plastids are studded with numerous pyrenoids. The nucleus lies in the isthmus. Conjugation involves the formation of papillae that allow gamete fusion. *Micrasterias foliacea* is exceptional within the genus for its ability to form filamentlike arrays by overlapping polar lobes and interlocking apical teeth (Lorch and Engels, 1979). Zygospores are usually spherical with spines that are sometimes forked. Gones are very simple in construction.

EUASTRUM (Gr. *eu*, good, true, or primitive + Gr. *astron*, star) is a unicellular placoderm desmid that, like *Micrasterias*, has flattened cells (Fig. 21–44). There is a characteristic notch in the apices of most species. Semicells are typically lobed, and there may be

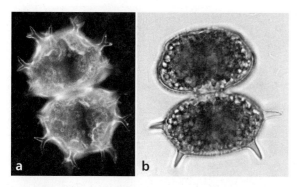

Figure 21–45 *Xanthidium* is characterized by three-dimensional protuberances on each of the semicells (a). In (b), a recently divided cell is shown, where the characteristic spines have yet to develop on the newly formed semicell.

small bumps on the surface that are visible at the light microscopic level only in side view. The incision or sinus between semicells is usually closed, as the semicell walls come close together. One or sometimes two chloroplasts with pyrenoids occur in each semicell. Chloroplasts are often conspicuously ridged. Zygotes are round and ornamented with spines or other protuberances. There are some 265 described species.

XANTHIDIUM (Gr. *xanthos*, yellow) unicells are mostly characterized by wall protuberances rising from semicell surfaces perpendicular to the flat plane of the surface (Fig. 21–45). They are visualized at the light microscopic level by through-focusing. The protuberances may be ornamented with pits or granules or may be pigmented, but these features are difficult to see at the light microscopic level. Another distinctive feature is presence of robust, often paired spines. In general the wall is less ornamented than that of *Cosmarium, Euastrum,* or *Staurastrum*. There are usually four axial or parietal plastids, each with a pyrenoid, per semicell. Zygotes have only rarely been observed; they are round and most bear spines. Some 115 species have been described.

GONATOZYGON (Gr. *gonatos*, joint or knee + Gr. *zygon*, yoke) (Fig. 21–46) occurs as a single cell or a short chain of cells adherent at their poles. Cells are not constricted but bear distinctive spines on wall surfaces. Edges of cells have a squared-off appearance, and the single chloroplast is a flat plate.

PLEUROTAENIUM (Gr. *pleuron*, rib + Gr. *tainia*, ribbon) (Fig. 21–47) is a cylindrical desmid with long, blunt-ended cells. Cells are 4–35 times longer than wide. There may be a noticeable ringlike thickening where semicells join. The chloroplast appears as parietal bands or axial with lamellae, and pyrenoids are present. Sometimes apical vacuoles with crystalline inclusions are present. Conjugation resembles that of *Closterium*.

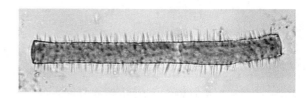

Figure 21–46 *Gonatozygon* features spines along the length of the cell wall.

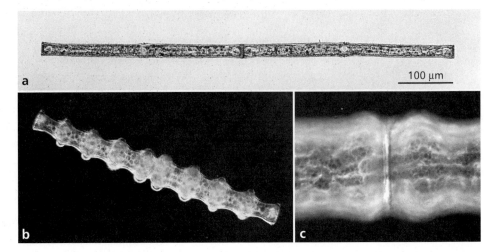

Figure 21–47
Pleurotaenium. In (a), two cells have remained connected following cell division. Chains of several cells are sometimes encountered. *Pleurotaenium nodosum* is shown in (b), while (c) is a high magnification view of the semicell junction, showing a ring that is characteristic of some species.

100 µm

TRIPLOCERAS (Gr. *triploos*, triple + Gr. *keras*, horn) cells are elongate, 8–20 times longer than they are wide, and are not noticeably constricted. The cells are more extensively ornamented than is *Pleurotaenium*; with various kinds of spines and undulating margins present (Fig. 21–48).

COSMOCLADIUM (Gr. *kosmos*, ornament + Gr. *klados*, branch) has relatively small cells that are constricted into semicells and compressed. Usually there is one axial chloroplast with pyrenoid in each semicell. Individual cells look like small *Cosmarium* cells, but they are interconnected by branching mucilaginous strands to form colonies of variable numbers of cells, the whole enclosed in mucilage (Fig. 21–49).

BAMBUSINA (*Bambusa*, a genus of bamboos [bambu, Indian vernacular name]) cells are cylindrical or barrel-shaped and arranged in linear series to form pseudofilaments (Fig. 21–50). There is only a slight constriction, and the wall region on either side of the isthmus is swollen. Fine striations may be visible at the apices. The axial chloroplasts have radiating lamellae and a single central pyrenoid. Zygospores are globular or elliptical and their walls are smooth.

SPONDYLOSIUM (Gr. *spondylos*, vertebra) is a filamentous form consisting of flattened cells that are deeply constricted and have a distinct sinus region (Fig. 21–51). Cell walls are not highly ornamented. Plastids are axial. Filaments are often twisted and have an extensive sheath. Zygospores are globose and smooth-walled or with short spines.

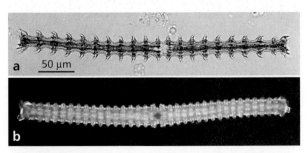

Figure 21–48 Two species of *Triploceras*, one seen with bright-field (a) and the other with dark-field optics (b).

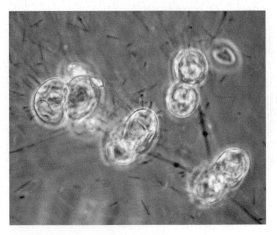

Figure 21–49 *Cosmocladium* cells occur at the ends of radiating strands.

Figure 21–50 *Bambusina*, a pseudofilamentous desmid whose semicells are not deeply incised.

HYALOTHECA (Gr. *hyalos*, glass + Gr. *theke*, sheath or box) is a pseudofilamentous member of the placoderm desmids (Fig. 21–52). Each cell is indented at the midpoint, but the indentation may be very slight. Cells contain two axial, radiately ridged chloroplasts with pyrenoids, one per semicell. The similar genus, *Groenbladia* (Fig. 21–22), has flat plastids that are very slightly indented. There is an extensive mucilaginous sheath. Filaments can break apart into fragments, a form of asexual reproduction. Conjugation involves formation of conjugation tubes; zygotes may form in the tubes or within one of the parental cell walls. Zygotes are globose and smooth-walled. Meiosis occurs during zygote germination and there are two filaments produced. Cells of filaments can round up and dissociate into thick-walled aplanospores, a form of asexual reproduction.

DESMIDIUM (Gr. *desmos*, bond) is another pseudofilamentous placoderm desmid, often having a prominent gelatinous sheath (Figs. 3–13, 21–53). Cells

Figure 21–52 *Hyalotheca*, a pseudofilamentous desmid often having a wide mucilaginous sheath; it is shown here in an india-ink preparation, which allows the sheath to be better visualized. Sessile chrysophytes (arrows) are often found associated with *Hyalotheca* and other filamentous desmids having similar sheaths. Note also the central nuclei in the cells (arrowheads).

are bi-, tri-, or quadriradiate in end view, depending upon the species. The filaments appear to be spirally twisted because the axes of adjacent cells are offset by a slight angle (Fig. 21–54). There are two axial chloroplasts with pyrenoids per cell; one plastid per semicell.

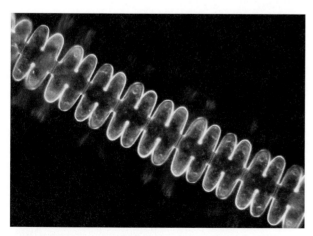

Figure 21–51 *Spondylosium*, a pseudofilamentous desmid whose semicells are deeply incised. There is a substantial mucilaginous sheath in which bacterial cells often occur. The fine details in the sheath of this specimen represent structure inherent in the sheath.

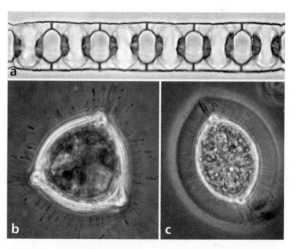

Figure 21–53 *Desmidium*, a pseudofilamentous genus. (a) A distinctive species, *Desmidium baileyi*. (b) View from the end of a semicell, showing a tri-radiate structure of another species, and numerous attached bacteria within the mucilage sheath. (c) End view of an isolated semicell from a biradiately symmetrical species showing sheath.

Figure 21–54 SEM view of *Desmidium grevillii*, a pseudofilamentous desmid. Note the twisted orientation of adjacent cells, a general feature of *Desmidium*. The mucilaginous sheath has been maintained in this preparation. The periodic indentations reflect the underlying constrictions between semicells and adjacent cells in the filament. (From Fisher and Wilcox, 1996 by permission of the *Journal of Phycology*)

Filaments can fragment. Conjugation tubes are formed between opposing cells of paired filaments, and zygotes form within the tubes or in one of the parental cell walls, depending upon the species. Zygospores are round or ellipsoidal, smooth-walled or with short projections.

ONYCHONEMA (Gr. *onyx*, claw + Gr. *nema*, thread) is a twisted filamentous arrangement of small, deeply constricted cells (Fig. 21–55). Semicells possess distinctive overlapping apical processes that can be nearly as long as the semicells themselves. There is a wide mucilage sheath. Zygospores are globose with short spines.

Coleochaetaleans

Recent molecular systematic analyses (McCourt, et al., 1995; Huss and Kranz, 1997) and molecular architectural evidence (Manhart and Palmer, 1990; Baldauf and Palmer, 1990; Starke and Gogarten, 1993) indicate that members of this group are closely related to the ancestry of land plants (Fig. 21–6, Text Box 21–1). It is not yet clear whether coleochaetaleans or charaleans are the sister group to the embryophytes; these three groups may have diverged at about the same time.

Distinguishing Features

COLEOCHAETE (Gr. *koleos*, sheath + Gr. *chaite*, long hair), including approximately 15 species (Pringsheim, 1860; Jost, 1895; Printz, 1964; Szymánska, 1989) (Fig. 21–56), and **CHAETOSPHAERIDIUM** (Gr. *chaite*, long hair + Gr. *sphaira*, ball), with only four described species (Thompson, 1969) (Fig. 21–57) are the only members of this group. Prominent among the characters they share are sheathed hairs. These extensions of the cell wall enclose a small amount of cytoplasm and are believed to serve as protection against herbivores (Marchant, 1977) (Fig. 21–58). The hairs may attain a length greater than 100 times the vegetative cell diameter. The base of the hairs is enclosed by a rigid sheath of wall material; the delicate hairs often break off at the rim of the sheath. Such sheathed hairs are not known to occur outside this group of algae. Whereas all *Chaetosphaeridium* cells bear one or more sheathed hairs, only 3–5% of the cells of *Coleochaete scutata* thalli produce hairs, and then only one per cell (McBride, 1974). Such cells, known as **seta cells**, are specialized in their structure (Marchant, 1977) (Fig. 21–59). The walls of *Coleochaete* seta cells are composed of several layers, whereas those of *Chaetosphaeridium* are of simpler construction. The chloroplasts of *Coleochaete* seta cells typically have a

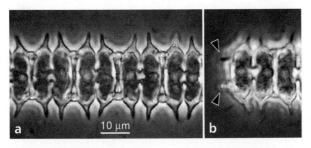

Figure 21–55 *Onychonema*, a pseudofilamentous desmid. (a) Arrangement of cells in the filament. (b) Terminal cell with projecting prongs (arrowheads) that interlock with adjacent cells.

C-shaped appearance (Fig. 21–60), and may be observed to rotate. Sometimes the hairs of *Chaetosphaeridium* are highly coiled, and occasionally this is true for *Coleochaete* setae, too, but the significance, if any, of such coiling is unknown. An 18S rDNA analysis suggested to Sluiman and Guihal (1999) that *Chaetosphaeridium* and *Coleochaete* are not closely related.

Ecology

Both *Coleochaete* and *Chaetosphaeridium* are periphytic, attached to submerged portions of higher plants such as the bulrush *Scirpus*, *Potomogeton*, and the undersides of water lily leaves (Fig. 21–60). They also grow on nonliving substrates such as pebbles near the edges of oligotrophic (low-nutrient) freshwater lakes and ponds. Here they are frequently, but unpredictably, exposed to desiccation as a result of variable wave action and changes in water levels. Such environments are thought to generate selective pressures that may have been involved in the origin of adaptations useful in colonization of land. Further, at least some forms of *Coleochaete* grow in moisture-saturated air, suggesting the capability for terrestrial existence (Graham, 1993). Therefore, the ecology of coleochaetaleans may shed light on ancient events related to the transition from charophyceans to earliest land plants. *Coleochaete nitellarum* can occur on the surfaces of deep-water charalean algae, particularly *Nitella* (described later in this chapter). Coleochaetaleans are sensitive to the effects of eutrophication; they disappear from water bodies that suffer excessive input of nutrients. They are not known to occur in brackish waters or extreme habitats. Although both *Coleochaete* and *Chaetosphaeridium* can occur in low pH *Sphagnum* bogs, most species do not appear to regularly occur in waters of pH higher than about 9.

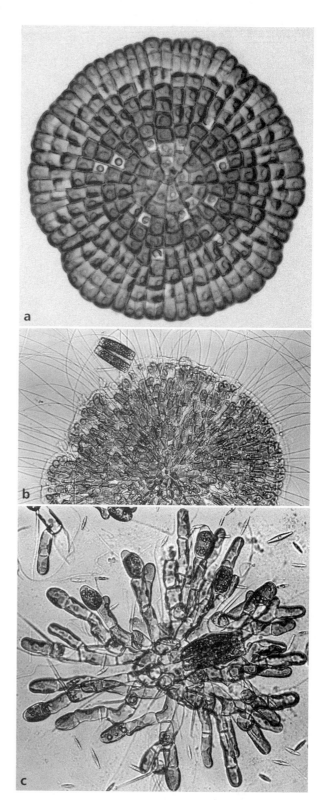

Figure 21–56 Morphological diversity of *Coleochaete*. (a) *Coleochaete orbicularis* has a thallus that is one-cell-layer thick. It grows by means of a marginal meristem. (b) *Coleochaete soluta* thalli are pseudoparenchymatous, composed of laterally adherent branched filaments enclosed within a mucilage layer. Note the extensive number of long hairs (setae), which are thought to function in protection from herbivory (a pair of diatoms lies among the setae). (c) *Coleochaete pulvinata* is a heterotrichous, branched filament; there is a prostrate and an erect system of branches, all enclosed by extensive mucilage, which often harbors small diatoms.

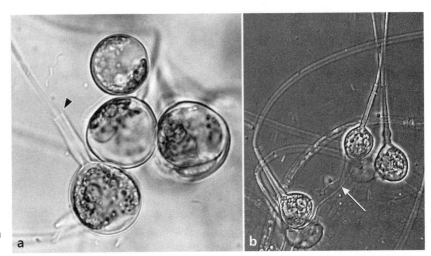

Figure 21–57 *Chaetosphaeridium.* (a) The spherical cells of *Chaetosphaeridium* are actually linked to form branched filaments, though this is often difficult to discern. Note the sheathed hairs (arrowhead), a hallmark of the order Coleochaetales. (b) Delicate colorless strands (arrow) are the remains of the cellular linkages that denote filamentous construction of *Chaetosphaeridium*. Note that each cell bears a hair.

In laboratory cultures, one *Coleochaete* species has been demonstrated to utilize exogenous dissolved organic carbon in the form of hexose sugars and sucrose (Graham, et al., 1994), as can certain zygnemataleans, such as *Mesotaenium* (Taylor and Bonner, 1967), but not the charalean, *Chara* (Forsberg, 1965). It has been suggested that this capability would be useful in low pH or other environments in which dissolved inorganic carbon levels might be too low to saturate photosynthetic requirements. Utilization of exogenous dissolved organic carbon is also regarded as a possible evolutionary preadaptation that was

inherited by land plants from their charophycean ancestors (Graham, 1996).

Structure and Development

Although *Chaetosphaeridium* appears, at first glance, to be composed of single cells held together by a gelatinous matrix, it is actually a branched filament (Thompson, 1969). The rounded portions of

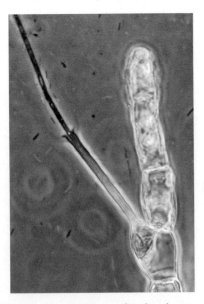

Figure 21–58 Sheathed hair of *Coleochaete pulvinata*.

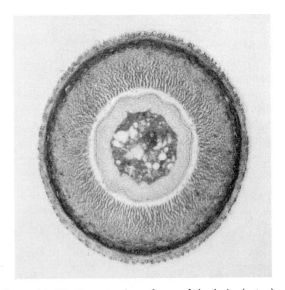

Figure 21–59 Cross section of one of the hairs (setae) of *Coleochaete pulvinata*, at the level of the sheath (collar), as viewed by TEM. Note the occurrence of multiple cell-wall layers, and the central column of cytoplasm that runs throughout the length of the hair. Breakage of hairs is thought to release chemicals inhibitory to herbivores. (From *Origin of Land Plants*. Graham, L. E. ©1993. Reprinted by permission of John Wiley & Sons, Inc.)

Figure 21–60 *Coleochaete*, like *Chaetosphaeridium*, is a member of the periphyton, growing on rocks or the leaves and stems of aquatic macrophytes. (a) Flat thallus of *Coleochaete orbicularis* and (b) hemispherical thallus of *C. pulvinata* on *Scirpus* stem. The C-shaped structures within some cells of *C. orbicularis* are the unusual plastids of seta cells (arrows); these are capable of rapid rotation. The dark structures on *C. pulvinata* are zygotes (arrowhead).

Chaetosphaeridium cells are interconnected by thin colorless branches that are difficult to discern with the light microscope. Division is regularly oblique to the plane of the substrate, giving rise to a new cell located somewhat beneath the parental cell. A bulge develops into which the new protoplast moves, then the new cell grows into a position next to the parental one. The portion of the cell lying beneath the parental cell collapses, and the new cell develops a hair (Fig. 21–61). Lateral branching may take place by the same method. Thompson (1969) did not consider this form of growth to be equivalent to the apical growth exhibited by *Coleochaete*. Thalli are covered by abundant mucilage.

Many (if not all) species of *Coleochaete* appear to possess terminal or marginal meristems (i.e., the only cells that undergo vegetative mitotic divisions are those at the tips or edges of thalli). In contrast to other algal genera, *Coleochaete* exhibits an unusual degree of thallus variability. *Coleochaete pulvinata* consists of both radially symmetrical prostrate portions whose branched filaments grow flat against the substrate and radially symmetrical, erect branching systems. In contrast, other species occur only as radially branched prostrate filaments, and yet others are prostrate branched filaments that lack radial symmetry. *Coleochaete orbicularis* is a radially symmetrical species that grows as a single layer of cells, but rather than having an overtly filamentous organization, the thallus resembles that of plant tissues, such as protonemata of the early divergent moss *Sphagnum* (Fig. 21–62). The tissuelike species tend to occur in shallower waters than filamentous forms (Graham,

1982a; Graham, 1993). Tissuelike organization may thus confer adaptive advantage in very shallow waters (Graham, 1993). It has been suggested that the thallus variants exhibited by modern *Coleochaete* illustrate evolutionary transformations involved in the origin of land plant tissue (Graham, 1982a).

Thalli of the filamentous *Coleochaete pulvinata* are covered with an abundant, clear mucilaginous material similar to that of *Chaetosphaeridium*. Mucilage is not as conspicuous on other species of *Coleochaete*. In addition, surfaces of the tissuelike forms are coated with a ridged material (Marchant and Pickett-Heaps, 1973) (Fig. 21–62b) that may represent a modified form of mucilage. It exhibits some similarities to cuticles of land plants (Graham, 1993), especially those of various bryophytes (Cook and Graham, 1998).

Chloroplasts occur singly in most cells of *Chaetosphaeridium* and *Coleochaete*, and contain one or more pyrenoids that are similar to those of other charophyceans and hornworts (Graham and Kaneko,

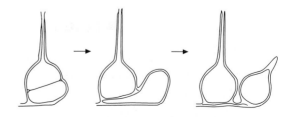

Figure 21–61 Development of *Chaetosphaeridium* filaments. (After Thompson, 1969 by permission of the *Journal of Phycology*)

Figure 21–62 *Sphagnum* moss germlings (a) are radially symmetrical and grow by means of a marginal meristem as do thalli of *Coleochaete orbicularis* (b). Note occurrence of concentric ridges of surficial cuticle-like material on *Coleochaete*. (a: Courtesy M. Cook; b: Courtesy S. Cook)

1991). The photosynthetic enzyme Rubisco has been localized in the pyrenoids of *Coleochaete* and hornworts (McKay and Gibbs, 1991) and Rubisco activase localizes to the pyrenoids of *Coleochaete* and *Chlamydomonas* (McKay, et al., 1991). The thylakoids of *Coleochaete*—organized into grana—are regarded as having a more plantlike arrangement than those of other algae (Whatley, 1993).

Cell Division

Vegetative cell division has been studied at the ultrastructural level only in the tissuelike forms *Coleochaete scutata* (Marchant and Pickett-Heaps, 1973), and *C. orbicularis* (Graham, 1993). In addition, mitosis and cytokinesis in *C. orbicularis* was investigated by immunofluorescence techniques (Brown, et al., 1994). Cell division differs, depending upon whether the marginal cells divide radially (perpendicularly to the direction of thallus growth) or circumferentially (parallel to the direction of thallus growth) (Fig. 21–63). In each case, pairs of centrioles appear at the poles of developing spindles, and at prometaphase the nuclear envelope begins to disintegrate, such that mitosis is open. A small ingrowth of the outer wall appears very early in division of cells dividing in the radial direction only; this marks the location of later cross-wall formation. Graham (1982a) suggested that this ingrowth is a manifestation of precocious furrowing (such as occurs in other charophyceans) and is an evolutionary remnant of ancestral processes associated with branching. Cytokinesis in radially dividing cells begins with the centrifugal development of a cell plate that is associated with a phragmoplast indistinguishable from that of

land plants (Brown, et al., 1994). In contrast, cytokinesis in circumferentially dividing cells occurs primarily by furrowing, though a small central phragmoplast is also present. It has been suggested that circumferential division in *Coleochaete* reflects the ancestral condition, such as is illustrated by *Spirogyra*, whereas radial cytokinesis reflects evolutionary transition toward the land-plant condition (Marchant and Pickett-Heaps, 1973; Graham, 1993).

Like land plants and charalean algae, the cross-walls of *Coleochaete* are penetrated by numerous plasmodesmata, whereas these are absent from the cross-walls of early divergent charophyceans such as *Chlorokybus*, *Klebsormidium*, and the zygnemataleans. Whether or not plasmodesmata occur in *Chaetosphaeridium* is unknown. The evolutionary origin of land plant plasmodesmata is significant because these structures are so developmentally important (Cook, et al., 1997) and seem to be a necessary prerequisite to the origin of histogenetic (tissue-producing) meristems (Cook and Graham, 1999).

In *Coleochaete*, the land plantlike peroxisome becomes closely associated with the chloroplast and divides (by invagination) at the same time, thus achieving regular partitioning to daughter cells (Fig. 21–64). In contrast, as noted earlier, some other charophyceans have different ways of partitioning their peroxisomes to daughter cells.

Asexual Reproduction

Both *Coleochaete* and *Chaetosphaeridium* produce biflagellate zoospores (lacking eyespots) that serve as asexual reproductive cells. They generate new thalli of the same type from which they were produced

Radial Division

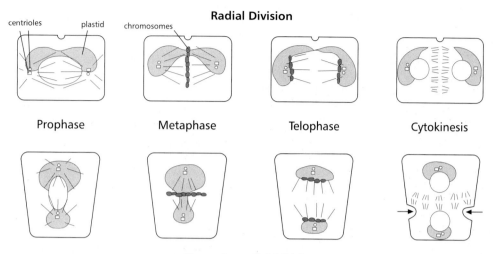

| Prophase | Metaphase | Telophase | Cytokinesis |

Circumferential Division

Figure 21–63 Diagram of radial versus circumferential division in *Coleochaete orbicularis*, as deduced from immunofluorescent localization of tubulin. Mitosis is open and centric, and associated with plastid division in both cases. However a phragmoplast and cell plate similar to that of embryophytes occurs only in radial division. A highly modified phragmoplast occurs together with more extensive furrowing (arrows) in circumferential division. The midzone of the phragmoplast does not mark the path of cytokinesis. Circumferential division in *Coleochate orbicularis* is thought to have arisen by superposition of an innovation (phragmoplast-associated cell plate development) on top of the more ancient persistent spindle and furrowing process. (After Brown, et al., 1994)

and are capable of rapidly generating a clonal population from a single parental thallus. Younger thalli can often be seen clustered around larger, parental thalli.

In *Chaetosphaeridium*, zoospores can be produced from any cell. Zoospores can arise from terminal cells in the filamentous *Coleochaete pulvinata*, but are never produced by marginal cells of the tissuelike species (*C. orbicularis* and *C. scutata*) (Fig. 21–65). Zoospore production in *Chaetosphaeridium* involves a precursor cell division, and typically the lower cell differentiates into the zoospore, which escapes by dissolution of the cell wall. Usually only a single zoospore is produced per parental cell of *Chaetosphaeridium*, but sometimes two are generated. In contrast, zoospores of *Coleochaete* are always produced singly, a precursor division is not involved, and zoospores escape through a specialized discharge pore (Fig. 21–66) that is presumed to arise by the localized action of hydrolytic enzymes on the cell wall. Temperature is more influential than irradiance or day length in inducing *Coleochaete* zoospore production (Graham, et al., 1986). Such information is useful in generating unialgal cultures for laboratory investigations. Zoospores of *Coleochaete* settle, attach to surfaces, and develop a cell wall beneath the scale layer (as is also the case in *Chlorokybus*). A transverse cell

division then occurs (reminiscent of the growth pattern in *Chaetosphaeridium*), and the upper cell terminally differentiates into a seta cell. The lower cell continues to divide, and its derivatives serve as the marginal meristem, dividing either radially or circumferentially, depending upon spatial constraints.

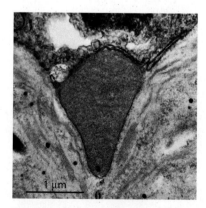

Figure 21–64 The peroxisome of *Coleochaete orbicularis* is closely associated with the dividing plastid during mitosis; this is also the case in the monoplastidic hornworts. (Reprinted from Graham, L. E., and Y. Kaneko. 1991. *Critical Reviews in Plant Science* 10:323–342. ©CRC Press, Boca Raton, Florida)

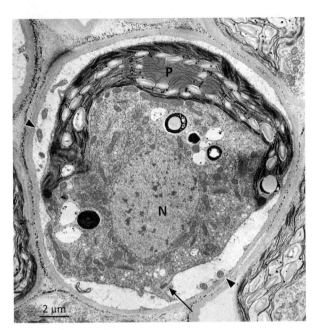

Figure 21–65 *Coleochaete* zoospore prior to release from parental thallus cell, as viewed by TEM. Note the single parietal plastid (P) with numerous starch grains, and small multilayered structure (arrow). Profiles of the two flagella, which are coiled around the cell, are visible just inside the parental cell wall (arrowheads). (From Graham and McBride, 1979)

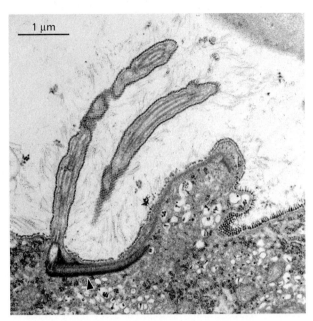

Figure 21–67 *Coleochaete* zoospore multilayered structure (arrowhead) associated with flagellar bases. Note presence of numerous scales on the cell and flagellar surface, as well as the numerous long flagellar hairs. These are produced within cytoplasmic vesicles, then transported to the surface. (From Graham and McBride, 1979)

In both *Coleochaete* (Graham, 1993) and *Chaetosphaeridium* (Moestrup, 1974), zoospore surfaces and flagella are coated with scales (Fig. 21–67) that resemble the lower layer of body scales of *Mesostigma* and various prasinophyceans. The body and flagella of *Coleochaete* gametes are also covered

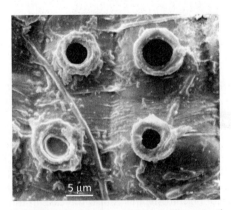

Figure 21–66 *Coleochaete scutata* wall pores through which zoospores were released. (Photograph courtesy of S. Cook)

with such scales (Fig. 21–68). The scale covering gives the surface of motile cells a frosty or granular appearance (Thompson, 1969); the function of the scales is not completely understood. Flagellate reproductive cells of *Coleochaete* and *Chaetosphaeridium* also exhibit a single MLS-containing flagellar root (Fig. 21–67), which is a marker of evolutionary relationship to land plants (Chapter 17). In addition, *Coleochaete* contains another flagellar root composed of just a few microtubules (Sluiman, 1983). A fibrous connective links the two flagellar bases in coleochaetaleans, as in other charophyceans, and the flagella emerge laterally, as is also characteristic of the larger group.

Sexual Reproduction

Observations on sexual reproduction have been made on cultured *Chaetosphaeridium globosum* (Thompson, 1969). Thalli can be homothallic or heterothallic. Sexual reproduction involves naked nonmotile egg cells that are larger than vegetative cells and biflagellate sperm cells that are released in pairs from colorless precursor cells. The eggs are expelled before fertilization and held within the thallus mucilage. The

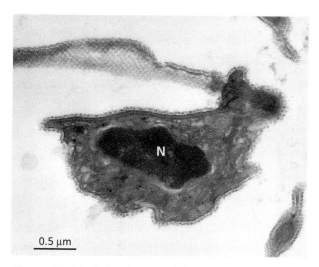

Figure 21–68 *Coleochaete pulvinata* spermatozoid as viewed by TEM. Note scales on body and flagellar surfaces and the highly condensed chromatin in the nucleus (N). (From Graham and Wedemayer, 1984 by permission of the *Journal of Phycology*)

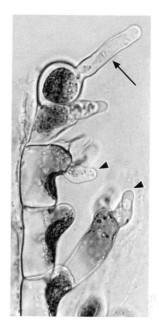

Figure 21–69 *Coleochaete pulvinata* oogonia with elongate trichogyne (arrow) and small colorless antheridial branches (arrowheads). (From *Origin of Land Plants*. Graham, L. E. ©1993. Reprinted by permission of John Wiley & Sons, Inc.)

sperm are without a wall (though ultrastructural studies have not been done to determine if scales are present on their surfaces) and also lack eyespots. Fertilized eggs develop into oval smooth-walled zygotes (zygospores) that are about the same size as egg cells. There is a delay between plasmogamy (cytoplasmic fusion) and karyogamy (nuclear fusion). The zygote wall is several layers thick, and internal layers may be yellow or deep brown in color.

In contrast to *Chaetosphaeridium*, egg cells of *Coleochaete* are not released from the thallus. Rather, they develop a cell-wall protuberance that is relatively short in *Coleochaete orbicularis*, but may be quite long in *Coleochaete pulvinata* (Fig. 21–69). The tip of the protuberance (sometimes called a **trichogyne**) disintegrates when the egg is ready for fertilization (Oltmanns, 1898) and exudes cytoplasmic contents that appear to attract flagellate sperm. In *C. pulvinata*, the sperm are produced in small colorless branches that occur in groups on precursor cells near egg cells (Fig. 21–69) (Graham and Wedemayer, 1984), while in *C. orbicularis* and *C. scutata*, they are formed in small cells occurring in packets derived from asymmetric cell divisions (Fig. 21–70). These packets can be formed from cells lying just internal to egg cells (in *C. orbicularis*), or in concentric rings that occur about midway between the thallus center and periphery (in *C. scutata*). Production of sperm is an exception to the general rule that non-peripheral cells do not divide. Sperm release is by localized wall dissolution.

Coleochaete scutata is heterothallic (i.e., sperm packets are found on only some thalli, and egg cells only occur on thalli lacking sperm), but *C. orbicularis*

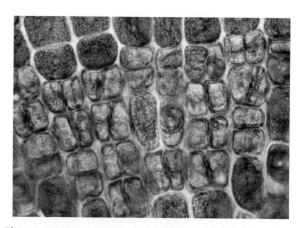

Figure 21–70 *Coleochaete scutata* antheridia occur as packets of cells formed in concentric rings between the thallus periphery and its center. (From *Origin of Land Plants*. Graham, L. E. ©1993. Reprinted by permission of John Wiley & Sons, Inc.)

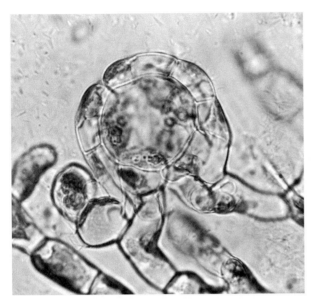

Figure 21–71 *Coleochaete pulvinata* zygotes induce neighboring cells to grow toward and cover them, forming a protective cortical layer. This is circumstantial evidence of chemically mediated interactions between generations, a process that also occurs in embryophytes. (From *Origin of Land Plants*. Graham, L. E. ©1993. Reprinted by permission of John Wiley & Sons, Inc.)

orbicularis (Fig. 21–72), these cells exhibit cellular features similar to those of placental transfer cells at the gametophyte-embryo junction of land plants. These function in the transport of nutrients such as amino acids and sugars across the maternal-embryo interface, a process that has been described as **matrotrophy** (Graham, 1996).

In both placental transfer cells of land plants and the cortical cells surrounding *Coleochaete orbicularis* zygotes, elaborate wall ingrowths develop on a localized basis (Fig. 21–72) (Graham and Wilcox, 1983). In the case of *Coleochaete*, the wall ingrowths occur only on the vegetative cell walls that are closest to zygotes, suggesting that zygotes have an inductive influence, perhaps by exuding chemical signals. The wall ingrowths vastly increase the surface area of the cell membrane, across which nutrients moving from vegetative cells to zygotes must pass. The ability of vegetative thalli of *Coleochaete* to take up and utilize sugars (Graham, et al., 1994) suggests that these algae may possess the requisite cell membrane carbohydrate transporter molecules. Concurrently, zygotes begin to enlarge through accumulation of massive storages of starch and lipid (Fig. 21–73). These storage materials are thought to arise in part from maternal contributions, though zygotes maintain green chloroplasts at

and *C. pulvinata* are homothallic, as both eggs and sperm are produced by the same thallus. The advantage of heterothallism (dioecy) is that outcrossing is mandated, and this may increase genetic variability in the population. The disadvantage is that zygotes may not be formed if individuals are too far separated to allow fertilization. In natural collections, fertile thalli of the homothallic *C. orbicularis* and *C. pulvinata* appear to be much more common than fertile *C. scutata*, perhaps because of this reproductive constraint.

In *Coleochaete*, sexual reproduction occurs in mid to late summer, with zygote maturation occurring in the fall. When several species co-occur, it appears that sexual reproduction may be temporally separated, perhaps as a species isolation mechanism. Zygotes are not released from parental thalli, a major distinction between this alga and many others, but similar to retention of zygotes within the female gametangia of early divergent land plants. Immediately after fertilization zygote cytoplasm at first shrinks, then begins massive enlargement. During enlargement, surrounding vegetative cells are induced to divide, forming a layer of cortical cells that cover zygotes partially or completely (Fig. 21–71). In *C.*

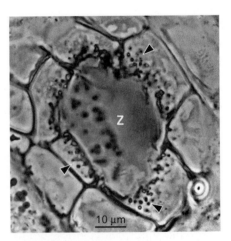

Figure 21–72 *Coleochaete orbicularis* exhibits numerous, localized wall projections (arrowheads) in the cortical cells surrounding a zygote (Z). Production of such ingrowths appears to be induced by the presence of zygotes and is additional evidence of chemical interactions between parent and progeny. Similar wall ingrowths occur at the interface of parent and progeny tissues in land plants, and the cells in which they occur are known as placental transfer cells. (From Graham and Wilcox, 1983)

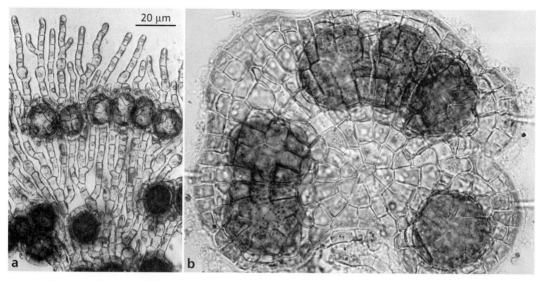

Figure 21–73 (a) *Coleochaete pulvinata* with zygotes; the most peripheral ones (top) are the youngest, and the larger, darker zygotes at bottom were formed earlier. (b) *Coleochaete orbicularis* zygotes are the large, dark spheres, which are filled with storage products. Note that in both cases, zygotes are completely covered by a layer of vegetative cells. (a: From Graham, 1985)

this stage, and therefore are presumably capable of generating photosynthates. The storage materials are needed to fuel production of an unusually large number of meiotic products when zygotes germinate in early spring. At maturity, *Coleochaete* zygote walls become lined with a thin layer of material that is similar to sporopollenin in higher plant spores and pollen walls (Delwiche, et al., 1989; Graham, 1990). Further, the walls of vegetative cells that surround zygotes, and indeed, any cells within the apparent sphere of influence of zygotes, accumulate highly resistant phenolic compounds (Delwiche, et al., 1989; Kroken, et al., 1996). These, together with the sporopollenin layer, are presumed to protect zygotes against microbial attack during the dormant period.

In north temperate latitudes, *Coleochaete* zygotes are induced to germinate, perhaps by warmer temperatures and longer day lengths. At least some of the zygotes remain in shallow nearshore waters—the same location favored by vegetative thalli—because zygotes are attached to senescent parental thalli, which often remain attached to stable substrates. Though dead, cell walls of parental thalli do not completely decay during this period because they are impregnated with phenolic materials. In at least one species of *Coleochaete*, meiosis occurs during the first divisions of the polyploid zygotes, followed by several mitotic divisions (without further DNA replication) (Hopkins and

McBride, 1976), yielding 8–32 haploid products. These cells, known as meiospores, develop two flagella, an MLS flagellar root, and a layer of surface scales, then escape from zygotes as the wall cracks open (Graham and Taylor, 1986a, b). Upon settling in nearby, well-illuminated nearshore habitats, the meiospores seed the development of vegetative populations of *Coleochaete* during the subsequent growing season, completing the life cycle.

Sperm development in *Coleochaete* involves cell divisions that are similar in some ways to those occurring at the tissue-generating apical meristems of bryophytes. Retention of the egg and zygote on the parental thallus is another example, as is the presence of wall ingrowths at the maternal-zygote/embryo interface, a manifestation of matrotrophy. Such maternal/embryo nutritional and developmental interactions occur in all land plants—from bryophytes to flowering plants—thereby defining them as embryophytes. Finally, development of highly resistant phenolic wall polymers in cells associated with *Coleochaete* zygotes is paralleled by production of similar compounds in walls at the gametophyte-sporophyte junction of bryophytes (Graham, 1996). Bryophytes have apparently extended such production of resistant cell-wall compounds to other locations (such as sporangial epidermis, rhizoids, and elaters) where decay resistance is adaptive (Kroken, et al.,

1996). These similarities are viewed as evidence that ancestral shallow water charophyceans acquired critical adaptations that later proved useful in the colonization of land (Graham, 1985; Graham, 1996). Although no fossils older than those attributed to the earliest land plants have as yet been linked directly to coleochaetaleans, Niklas (1976) suggested that Silurian-Devonian plants known as *Parka* were very similar in growth pattern and reproduction to some *Coleochaete* species.

Charaleans

Modern charalean algae are important both ecologically and evolutionarily. They are closely related to the ancestry of land plants (McCourt, et al., 1995), and thus may provide information about the features of the ancient charophycean progenitors of embryophytes. They have a long fossil history, based primarily upon calcified reproductive structures that provide useful information about the evolutionary process and patterns of extinction (Tappan, 1980; Feist and Grambast-Fessard, 1991). In addition, charaleans can form massive growths in both deep and shallow lake and pond waters, sometimes to the point of being regarded as nuisance weeds. However, charaleans are also an important food for waterfowl and provide a nursery area for fish. Some also occur in gently flowing waters such as irrigation canals (Hussain, et al., 1996). A few, including *Chara evoluta*, occur in brackish waters having salt content of 20–40 ppt, but no modern forms are marine. They occur in waters of pH range 5–10. Some form extensive meadows in fairly deep waters (Stross, 1979); *Chara contraria* has been collected from 150 meters in Lake Tahoe. Charaleans are considered to be adapted in various ways to low-irradiance, benthic habitats (Andrews, et al., 1984). High phosphate levels are regarded as deleterious to charaleans. Because many forms accumulate surface layers of calcium carbonate in the form of calcite, the group is known colloquially as stoneworts, muskgrasses, bassweeds, or brittleworts. Calcification gives some forms a white or pale green appearance (Grant, 1990). Charalean algae are now, and for the past few hundred million years have been, major carbonate sediment producers in freshwater lakes, because they may be more heavily encrusted by calcium carbonate than aquatic higher plants. Most of the charalean thallus (except zygotes) readily disintegrates in the benthos, forming marl deposits at rates that can reach several hundred g m^{-2}

Figure 21–74 *Chara zeylanica* (left) compared to *Ceratophyllum*, an aquatic flowering plant (right). These are similar in size, and have similar nodal and internodal organization, with whorled branching; thus, they are often confused. *Ceratophyllum* can be distinguished from *Chara* by its possession of bifurcating branch tips. (From *Origin of Land Plants*. Graham, L. E. ©1993. Reprinted by permission of John Wiley & Sons, Inc.)

year^{-1} (Tucker and Wright, 1990). However, the significance of past and present charalean carbonate sequestration upon local and global carbon cycling is not known.

Charaleans are the largest and most morphologically, developmentally, and reproductively complex group of charophycean green algae. Reaching lengths of one meter or more, with whorls of branches at nodes, some are regularly confused with similar-appearing aquatic flowering plants, such as *Ceratophyllum* (Fig. 21–74). In view of their apparent distinctiveness, some authors have placed them into a separate division (phylum) (Bold and Wynne, 1985) or class (Hoek, van den, et al., 1995). However considerable molecular sequence and architectural evidence closely links charaleans with the rest of the charophyceans, and *Coleochaete* in particular (Manhart and Palmer, 1990; Baldauf and Palmer, 1990; McCourt, et al., 1996a, b). Moreover, careful examination of many of the seemingly distinctive traits of charaleans reveal

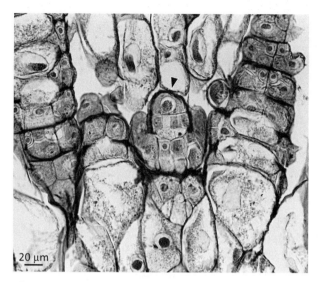

Figure 21–75 Apical meristematic cell of *Chara* (arrowhead).

regularly alternating discoidal nodal and long, cylindrical internodal cells (Fig. 21–76).

Internodal cells can reach lengths of 15 cm and contain well over a thousand nuclei, which are produced by the replication of a single original nucleus by a process that does not involve the typical mitotic apparatus. The interphase nuclei of vegetative cells in young shoots of *Chara* and *Nitella* can undergo endoreduplication—increase in the amount of DNA beyond the haploid level (Michaux-Ferrière and Soulié-Märsche, 1987). An increased number of nuclei is presumably required to balance the large increase in cell volume, which is mediated by development of a large internal vacuole. The cytoplasm nearest the central vacuole of internodal cells is an ideal site for visualizing cytoplasmic streaming, resulting from actin microfibril activity (Allen, 1974; Palevitz and Hepler, 1975; Williamson, 1979, 1992). Presumably, such streaming is necessary to achieve mixing and long-distance transport of cell constituents in long cells hav-

similarities to the features of other charophyceans (Graham, et al., 1991). Hence we regard high-level taxonomic separation of stoneworts as inappropriate. In view of solid molecular evidence for monophyly of modern stoneworts, as well as monophyly of the charalean-coleochaetalean clade (McCourt, et al., 1996a, b), at present it seems most appropriate to categorize modern stoneworts as an order, the Charales, as recommended by Mattox and Stewart (1984) on the basis of ultrastructural evidence. So, we refer to this modern group of algae as charaleans. Fossils are often referred to as charophytes.

Vegetative Structure

Charaleans are fundamentally branched filaments, though the main axis is differentiated at the apex, nodes, and basal region. The erect shoot possesses a single specialized apical meristematic cell; this cell cuts off derivatives from its lower surface only (in contrast to the apical meristematic cells of bryophytes, which have three or four cutting faces, and hence generate tissues) (Fig. 21–75). No centrioles are present at spindle poles in these or other cell divisions in charalean algae. The immediate derivative of transverse apical cell mitotic division (known as a segment cell) divides again transversely. The uppermost of the resulting two cells will continue to divide to produce a complex node with lateral branches, whereas the lower cell develops into a very long internodal cell without further division. Therefore, the main axis consists of

Figure 21–76 *Nitella* viewed by fluorescence microscopy. Note the large internodal cells, branches emerging in whorls from much smaller nodal cells, and the numerous autofluorescent, discoid plastids arranged in vertical files.

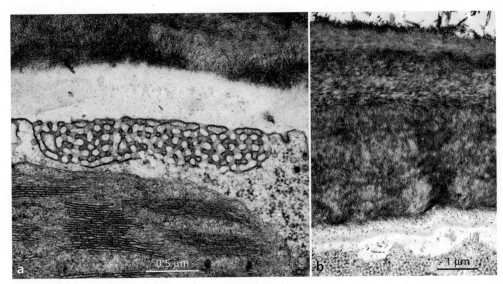

Figure 21–77 (a) *Chara zeylanica* charasome—an array of infolded cell membrane associated with proteinaceous material whose function is unknown. (b) Cell wall. Note the electron-opaque, layered appearance in TEM. (a: Reprinted from Graham, L. E., and Y. Kaneko. 1991. *Critical Reviews in Plant Science* 10:323–342. ©CRC Press, Boca Raton, Florida)

ing large cytoplasmic volumes. It has been hypothesized that such large internodal cells represent adaptation to shade, i.e., benthic habitats (Raven, et al., 1979).

Within the internodal cells a layer of nonmobile peripheral cytoplasm contains numerous discoid chloroplasts arranged in rows (Fig. 21–76) having granalike thylakoid stacks and starch grains but lacking pyrenoids (see Fig. 17–9), as well as mitochondria and peroxisomes containing conspicuous catalase crystals (see Fig. 2–8). The plastids are generated by repeated fission. Elaborate invaginations of the cell membrane, known as **charasomes**, may occur along the cell periphery (Fig. 21–77a); their function is uncertain. The presence of charasomes is apparently not correlated with the ability to use bicarbonate ion in photosynthesis (Lucas, et al., 1989). The cell wall of charaleans is composed of distinct layers, including cellulose, and can be relatively thick (Fig. 21–77b), possibly providing protection against microbial and/or herbivore attack. Because of the axial, filamentous construction of these algae, much of the rest of the thallus could be lost should an internodal cell be damaged or destroyed.

The cell plate separating nodal from internodal cells is characterized by very large pores. These appear to have originated by the coalescence of several plasmodesmata, themselves reported (in at least one species) to have been derived secondarily by the action of wall hydrolytic enzymes (Franceschi, et al., 1994; Lucas, 1995). These large pores are thought to facilitate passage of materials throughout the vertical axis of the thallus (Cook, et al., 1997). Internodal cells are so large that microelectrodes can easily be inserted for electrophysiological studies. An excellent review of electrophysiological investigations in charalean algae can be found in Wayne (1994).

Nodal initial cells first divide vertically (commonly into two halves), then each of the resulting cells undergoes a very highly controlled series of asymmetric divisions that produce the branch initials (Fig. 21–78). Divisions are synchronized in the two cells making up the node; divisions leading to formation of branch initials occur in a sequential, radial manner. The pattern of such divisions closely resembles the manner in which antheridial initials are generated in *Coleochaete* (Graham, 1996). Charalean branch initials serve as the apical cells for development of branches having the same kind of alternating nodal and internodal cells as the main axis. The branches in turn produce smaller branchlets. Unlike the apical cell of the main axis, which can continue to divide on an indeterminate basis, the branch apices cease division after a determined number of cells have been produced. However, cells of the nodal region may generate other branches that mirror the growth habit of the main axis; i.e., have indeterminate growth. In addition, in most species of *Chara*, basal nodal cells of

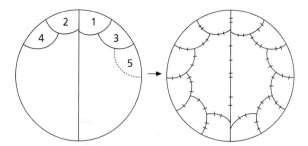

Figure 21–78 *Chara zeylanica*. Structure and development of the nodal complex. A first division gives rise to two equal nodal cells. Each of these then undergoes coordinated, directional, asymmetrical divisions, giving rise to whorls of branch initials. During early stages of development the branches are of unequal length, reflecting different times since origin. However, branch growth is determinate, and more recently produced branches eventually catch up; in the mature state, all branches of a given whorl are the same length. Short lines denote plasmodesmata. (From Cook, et al., 1998 ©Springer-Verlag)

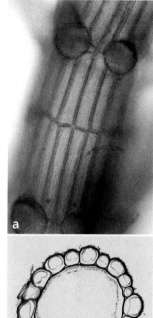

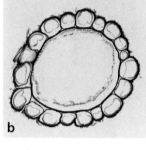

Figure 21–79 (a) *Chara zeylanica* corticating filaments grow from both nodes over the surface of internodal cells. (b) Cross-sectional view.

the branches can generate multiple rows of filaments that grow up or down over the internodal cell surfaces, forming what is known as a corticating layer. Corticating filaments originating from adjacent nodes meet in the middle of the internodal cell (Fig. 21–79).

A land plantlike phragmoplast is present during cytokinesis in charalean algae. In addition to the characteristic longitudinal microtubule array, charalean algae possess actin microfilaments, membrane tubules, coated vesicles, and fenestrated sheets such as are associated with cell plate development in higher plants (Pickett-Heaps, 1975; Cook, et al., 1997; Braun and Wasteneys, 1998). However, charalean algae that have been investigated differ in a number of ways from plants in regulation of the cytokinetic process. These include: later dissolution of the phragmoplast; absence of an organellar and ribosome exclusion zone from developing cell plates; cell-plate development that is patchy and not regularly centrifugal; co-occurrence of different cell-plate developmental stages; and an earlier development of plasmodesmata (Cook, et al., 1997).

Although some of the literature (Grant, 1990) implies that charalean algae produce pre-prophase microtubule bands homologous to those of plants, to date, such bands have not been demonstrated to occur in any charophycean alga.

Ultrastructural studies of the charalean node suggest that it can be considered to be a tissuelike, or parenchymatous structure (Pickett-Heaps, 1975;

Cook, et al., 1998). Comparative immunolocalization studies of actin and microtubules in internodal and nodal cells suggest that the latter more closely resemble higher plant meristem cells (Braun and Wasteneys, 1998). Moreover, cells in the nodal region of at least one charalean species possess primary plasmodesmata, i.e., those produced during formation of the cell plate at cytokinesis, much like plasmodesmata of land plants (Cook, et al., 1997). These observations suggest that the ability of plants to produce tissues capable of intercellular communication originated in their charophycean ancestors. Some of the attributes of charalean algae may thus reflect features of ancestral algae that were preadaptive to the evolution of distinctive features of land plant structure and development (Graham, 1996).

The basal portion of charalean algae in nature is typically attached to muddy or silty substrates by numerous colorless rhizoids. These are very long cells containing pigmentless plastids and 30–60 barium and sulfur-containing crystals believed to have a geotropic function. Rhizoids grow at the tip and are not differentiated into nodes and internodes, but do exhibit a definite polarity, with cells showing at least seven distinct zones (Fig. 21–80) (Kiss and Staehelin, 1993).

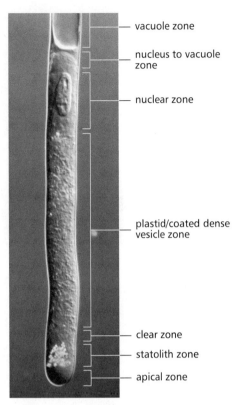

vacuole zone

nucleus to vacuole zone

nuclear zone

plastid/coated dense vesicle zone

clear zone

statolith zone

apical zone

Figure 21–80 *Chara* rhizoid, showing zonation of the cytoplasm that is related to gravitropism. (From Kiss and Staehelin, 1993)

Asexual Reproduction

Adventitous (from the shoot) development of new thalli can occur from rhizoids and the nodal complexes. Bulbils are white spherical or star-shaped structures that form on rhizoids of some species and function in dispersal and perennation.

Sexual Reproduction

The charaleans have probably the most conspicuous sexual structures of any green algae (Figs. 21–81, 21–82). These specialized structures (gametangia) are of two types—(male) antheridia, where thousands of biflagellate spermatozoids develop, and (female) oogonia, each containing a single egg cell. Both types of gametangia possess protective nonreproductive cells in addition to gametes, prompting some experts to compare them to the male and female gametangia of bryophytes (antheridia and archegonia). While developmental analysis has revealed that charalean gametangia are probably not directly ancestral to those of land plants

Figure 21–81 *Chara braunii* oogonia are evident on this isolated node and its whorl of branches. (Photograph courtesy of M. Cook)

(Pickett-Heaps, 1975), it is possible, and indeed probable, that some cellular aspects of sexual reproduction in charaleans reflect ancestral traits related to the origin of embryophyte gametangia. Antheridial development in *Chara vulgaris* was found to involve changes in the occurrence and ultrastructure of plasmodesmata (Kwiatkowska and Maszewski, 1986).

Figure 21–82 *Chara* antheridia borne on nodes of branchlets. Note petallike arrays of surface shield cells. (Photograph courtesy of M. Cook)

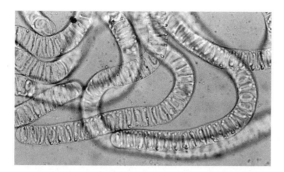

Figure 21–83 *Chara zeylanica* spermatangial filaments. Each of the equal-sized colorless cells will produce a single, elongate, spirally twisted biflagellate sperm. (From *Origin of Land Plants*. Graham, L. E. ©1993. Reprinted by permission of John Wiley & Sons, Inc.)

sperm structural development can be found in Moestrup (1970) and Pickett-Heaps (1975). When the sperm are ready for release, the shield cells separate, allowing sperm to swim away.

Oogonia arise from branchlet nodal cells as well. A primordial cell divides twice transversely, and the uppermost of the resulting cell stack becomes the egg. The cell just below the egg repeatedly divides, gener-

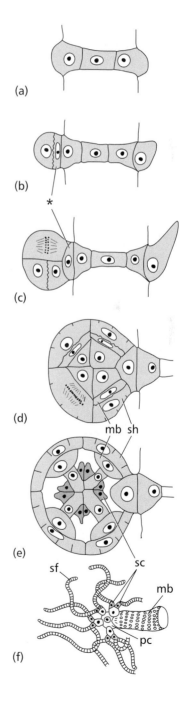

The antheridia of charalean algae are bright orange at maturity and are visible without the use of a microscope. In monoecious species, antheridia and oogonia are usually borne together at the node of a branchlet. Microscopic examination of antheridia reveals that the orange pigmentation is generated by carotenoid droplets within an outer layer of cells arranged in groups of eight, forming a flowerlike pattern (Fig. 21–82). These are known as the **shield cells**, and their form is used in classification and identification of some species. The shield cells are attached to a columnar-shaped cell known as a **manubrium**, which is also associated, at its other end, with a group of eight cells known as the **primary capitulum**. Cells derived from division of the primary capitulum generate long, unbranched filaments of small cells, each producing a single thin, helically twisted spermatozoid (Figs. 21–83, 21–84). The process of spermatogenesis involves an increase in the level of DNA methylation (Olszewski, et al., 1997). Additional details of

Figure 21–84 *Chara* antheridial development. Antheridia develop on short branches generated at the nodes (a). (b) Antheridial induction results in an unequal division, with the smaller cell (*) serving as a stalk. (c) The larger cell undergoes further division. (d) Diagonal divisions are integral to the formation of specialized regions within the developing antheridia. (e) An innermost set of columnlike manubria (mb) are attached to radiating shield cells (sh) at the surface, and also generate sperm filaments (sf) in the intervening space. (f) Detail of the manubrium and attachment of sperm filaments. pc=primary capitula, sc=secondary capitula. (After Pickett-Heaps, 1975)

ating a ring of five peripheral cells surrounding a central cell (Fig. 21–85). The five peripheral cells elongate to form **tube cells** (also known as sheath cells) (Fig. 21–86) that grow upwards along the surface of the egg, extending to keep pace with enlargement of the egg. As each tube cell elongates, it takes a counterclockwise helical path, and at its tip, a transverse division(s) gives rise to one or two **coronal (crown) cells** (the number depending on the genus) (Figs. 21–86, 21–87). This process resembles that of zygote cortication in *Coleochaete*, and it has been proposed that both are based on similar cell-cell signal-response events (Graham, 1993). As the egg enlarges, it becomes filled with storage products—usually many white starch grains, but lipid droplets are also abundant. Such storage buildup resembles that occurring in *Coleochaete* zygotes, and its purpose is similar—to support later zygote germination. A major difference is that food storage occurs following fertilization in *Coleochaete* (i.e., resources are not committed until after fertilization has been assured). At maturity, openings form between the tube cells that allow sperm to reach the egg.

After fertilization a thick, darkly pigmented zygote wall develops that contains a sporopolleninlike layer. Calcification of the concave inner walls of the spiral tube cells of *Chara* also typically occurs after fertilization. In *Tolypella*, calcium carbonate is deposited on the outside of the tube cells, and *Nitella* zygotes are not calcified. As thalli are degraded at the end of the growing season, zygotes (also known as oospores or zygospores) together with their protective tube cells and perhaps a few other vegetative remnants, fall to the sediments. The thick, resistant wall and calcified tube-cell layer (if present) contribute to zygote survival during a period of dormancy. The calcified impressions of the tube cells and enclosed structures may persist in the fossil record; these are known as

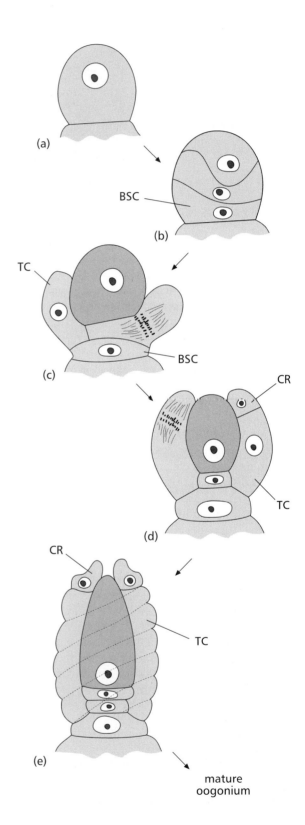

Figure 21–85 Development of oogonia in *Chara*. (a) A cell derived by division of a nodal cell undergoes further division (b) into a basal stalk cell (BSC), a middle cell that generates the five tube cells (TC) and the terminal egg (darker shaded cell). (c–e) Elongation of the tube cells forms a twisted cortical layer that covers the egg cell entirely except for a pore at the top. A final division at the tip of each of the five tube cells gives rise to the five coronal cells (CR) of *Chara*. *Nitella* undergoes two such divisions at the ends of each of the five tube cells, giving rise to a total of ten coronal cells. (After Pickett-Heaps, 1975)

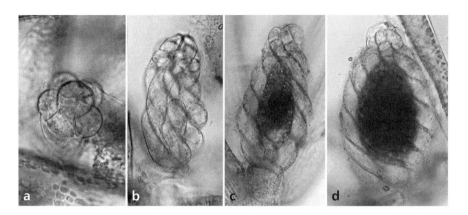

Figure 21–86 Oogonial development in *Nitella tenuissima* as viewed by light microscopy. Note the presence of two tiers of coronal cells on mature oogonium. (d: From *Origin of Land Plants*. Graham, L. E. ©1993. Reprinted by permission of John Wiley & Sons, Inc.)

gyrogonites. Fossil gyrogonites reveal that the ancient relatives of modern charaleans often had more than five tube cells; the number has apparently been reduced over time. Another interesting (and unexplained) change has occurred in charalean tube cell orientation over time. Tube cells of older taxa such as the lower Devonian *Trochiliscus* were twisted to the right (Fig. 21–88a), whereas tube cells of taxa appearing near the end of the Devonian were twisted toward the left, as are those of modern forms and fossils of intervening ages. The gyrogonite fossil *Eochara* (Fig. 21–88b) is regarded as representative of the ancestral forms of modern charalean species; as in modern forms, the number of tube cells is few, tube cells are

twisted to the left, and there is no open pore at the apex, as is often typical of more ancient fossils. A few examples of calcified or silicified remains of antheridia and vegetative parts are also known (Fig. 21–89). *Paleonitella* had non-calcified vegetative thalli with nodal organization, and was preserved in the geological deposit known as the Rhynie Chert as petrifactions (mineral-impregnated remains) (Taylor, et al., 1992). Another example is 405-million-year-old fossil evidence for the occurrence of numerous antheridia around oogonia, the oldest record of the occurrence of bisexuality among autotrophs (Feist and Feist, 1997).

Zygotes of charaleans are believed to germinate by meiosis, with only one meiotic product surviving. However this assumption is based entirely upon circumstantial evidence such as the observation of four nuclei within germinating zygotes (Hoek, van den, et al., 1995) and the fact that sperm and vegetative nuclei contain the same level of DNA (suggesting that meiosis is not gametic) (Shen, 1967). Analysis of DNA level changes or chromosome counts during zygote germination have been difficult to accomplish because the massive amounts of storage photosynthate and thick, dense zygote wall preclude easy observation of nuclear phenomena. Zygote germination occurs when a colorless filamentous "protonema" emerges from a break in the zygote wall. The protonema possesses a colorless primary rhizoid and, under the influence of blue or white light, undergoes transverse divisions to form a short filament with green chloroplasts appearing in the uppermost cells. Additional rhizoids, nodal regions, and an apical cell then form. Germination of charalean zygotes is notoriously difficult to achieve at high levels of efficiency in the laboratory, further adding to difficulties in clarifying the life history of these algae. Cold temperature and red-light treat-

Figure 21–87 Mature oogonium of *Chara*, viewed by SEM, showing five coronal cells. (Micrograph courtesy M. Cook)

100 μm

(a)

(b)

Figure 21–88 Fossil gyrogonites. (a) *Trochiliscus*, and (b) *Eochara*. Note the difference in direction of surface ornamentation twisting. (From teaching documents of the late Professor L. Grambast, by courtesy of M. Feist)

evolution of living and fossil charaleans has been discussed by Grambast (1974) and Feist and Grambast-Fessard (1991). Cytosystematics and biogeography of charaleans was reviewed by Khan and Sarma (1984) and Proctor (1980). The molecular systematics of modern charalean genera has been recently reviewed by McCourt, et al. (1996a). The *rbcL* data suggest that modern charaleans form a monophyletic group, and there is some support for a monophyletic tribe Chareae (*Chara, Lamprothamnion, Nitellopsis,* and *Lychnothamnus*), but monophyly of the tribe Nitelleae (*Nitella* and *Tolypella*) was not supported (McCourt, et al., 1996b) (Fig. 21–6). *Tolypella* was found to be monophyletic, in contrast to earlier hypotheses, but additional studies are needed to test monophyly of genera such as *Chara*. Molecular sequence data suggest that the genera *Nitella* and *Tolypella* are basal to the more derived Chareae.

ments (Takatori and Imahori, 1971) are said to increase the rate of zygote germination. Resistance to germination, and germination only after extended storage in wet conditions or treatment with fluctuating water levels, irradiance, or temperature are regarded as adaptations that increase the chances that wetland charaleans will survive one or more unfavorable seasons. Variation among species in zygote germination behavior helps to explain seasonal and geographical distribution patterns. For example, *Nitella cristata* var. *ambigua* zygotes germinated well in response to cues for the onset of winter, explaining the occurrence of this winter form (Casanova and Brock, 1996).

Diversity

According to Wood and Imahori (1965), there are six genera of living charaleans: *Chara, Lamprothamnion, Lychnothamnus, Nitellopsis, Nitella,* and *Tolypella.* Most of the 81–400 species (the number depending upon the expert) belong to *Chara* or *Nitella* (Grant, 1990). Thus *Chara* and *Nitella* will be featured here. Species determinations are based on such characters as the arrangement of the gametangia, presence or absence of cortical layers, and the number and arrangement of cortical cells. Phylogeny and

Figure 21–89 Fossilized *Chara* from Argentina. Note the dark zygotes and remnants of cortical and tube cells. (Photograph courtesy M. Cook)

Figure 21–90 *Chara braunii*, an uncorticated *Chara* species. (Photograph by M. Cook and C. Lipke)

(a)

(b)

Figure 21–91 *Nitella*. (a) *Nitella flexilis*. (b) *Nitella tenuissima*. Note characteristic dense tufts of branches from nodes that are well separated by internodal cells. (Photographs by M. Cook and C. Lipke)

CHARA (pre-Linnaean name of unknown origin) is characterized by structures known as **stipules** (or stipulodes), which are single-celled, often sharply tipped structures occurring below the branchlets. The main axes of most species are corticated, but some species, such as *Chara braunii* (Fig. 21–90), lack this corticating layer and can be mistaken for *Nitella*. There is a single layer of five oogonial coronal cells (in contrast, *Nitella* possesses two tiers of coronal cells, totalling ten). *Chara* species are often calcified and thus may have a stony, gray-green appearance. Calcification, together with cortication, gives *Chara* a generally more robust appearance than *Nitella*. Species such as *Chara vulgaris* are regarded as marl formers, because they deposit large amounts of calcium carbonate at the bottoms of water bodies. They primarily occur in relatively high-alkalinity waters. Some forms, including *C. vulgaris*, produce a foul odor which is variously described as "skunky" or "like spoiled garlic."

NITELLA (L. *nitella*, brightness or splendor) is characterized by very regular, symmetrical branching (Fig. 21–91). The branches that bear the gametangia are repeatedly forked. Thalli are uncorticated and not typically calcified. The oogonia are either solitary or occur below the antheridia. Oogonia have ten coronal cells in two tiers of five each (Fig. 21–86). The species range greatly in size from minute and delicate forms such as *N. tenuissima* (Fig. 21–91), which occurs in shallow, silty waters, to meter-long *N. flexilis*, which

has very long internodal cells and occupies waters 10–12 meters deep. In contrast to most *Chara* species, *Nitella* most commonly occurs in soft, slightly acidic waters.

Recommended Books

Graham, L. E. 1993. *The Origin of Land Plants*. John Wiley, New York, NY.

Kenrick, P., and P. R. Crane. 1997. *The Origin and Early Diversification of Land Plants*. Smithsonion Institution Press, Washington, DC.

Pickett-Heaps, J. D. 1975. *Green Algae*. Sinauer Associates, Sunderland, MA.

Chapter **22**

Phytoplankton Ecology
Dr. James M. Graham

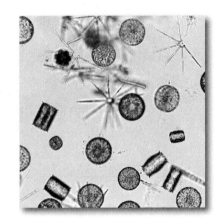

freshwater planktonic diatoms

What are phytoplankton? The prefix "phyto" comes from the Greek word for plant, *phytos*. "Plankton" derives from another Greek word meaning wanderer. Hence phytoplankton refers to organisms that wander in the surface waters of lakes, rivers, and the oceans.

The previous chapters have described many groups of algae. Among these groups, cyanobacteria, greens, diatoms, dinoflagellates, haptophytes, and chrysophyceans are especially rich in planktonic species. The dominant taxonomic groups differ between freshwaters and oceans. In freshwaters, cyanobacteria and greens are conspicuous and morphologically diverse while in the oceans these groups are primarily represented by small coccoid cyanobacteria and green microflagellates. Dinoflagellates occur in both freshwater and marine environments but are much more dominant and diverse in the oceans. Diatoms are abundant in both aquatic systems.

Phytoplankton cover a vast range of sizes and forms, both as single cells and as colonies (Fig. 22–1). Generally, cells or colonies with a maximum linear dimension less than 2 μm are considered to be picoplankton. Those greater than 30 μm in length are considered net plankton or microplankton. Cells and colonies between these two extremes are nanoplankton. Individual cells and colonies may possess flagella and be motile or lack flagella and be nonmotile. *Chlamydomonas* and *Ochromonas* are examples of small flagellate motile cells while *Chlorella* is a small nonmotile cell. *Scenedesmus* is a small nonmotile colony, and *Gonium* is a small motile one. *Staurastrum* and many other genera of desmids are large nonmotile cells. The dinoflagellate *Ceratium hirundinella* is an exceptionally large (> 100 μm), single-celled, motile alga. Large colonial phytoplankton include the nonmotile cyanobacterium *Microcystis aeruginosa* and diatom *Asterionella formosa* and the flagellate green *Volvox aureus*. Planktonic filamentous algae such as *Anabaena flos-aquae* can be thought of as long linear colonies. Why do phytoplankton vary so much in size?

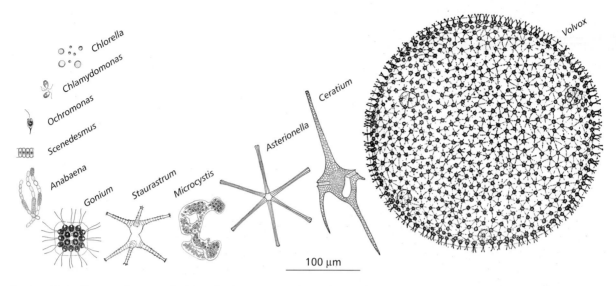

Figure 22–1 Sizes and types of phytoplankton (approximately to scale). (Drawings from Smith, G. M. 1950. *Freshwater Algae of the United States*. McGraw-Hill. Reproduced with permission of the McGraw-Hill Companies, except for *Anabaena*: originally from G. M. Smith, in Prescott, G. W. 1951. *Algae of the Western Great Lakes Area*. Cranbrook Institute of Science, a division of Cranbrook Educational Community)

Size is the most important single characteristic affecting the ecology of phytoplankton. Table 22–1 presents volumes and surface areas of a number of phytoplanktonic species (Reynolds, 1984). As phytoplankton become larger, their volumes increase as the cube of their radius, while their surface areas increase in proportion to only the square of the radius. Consequently, as algal species become larger, their surface-to-volume ratio (SA/V) becomes smaller. Note the high SA/V ratios for tiny *Synechococcus* and *Ankistrodesmus* compared to the low SA/V ratios for *Microcystis* and *Volvox*. Since growth of phytoplankton involves the exchange of materials at the cell surface, there is a close relationship between cell volume and maximum rate of reproduction (Table 22–2). Large cells and colonies generally have low rates of reproduction, although the correlation holds less well for flagellates (Sommer, 1981).

Size affects more than just phytoplankton growth rates. Small cells can respond to a pulse of nutrients with a rapid burst of growth. Large cells, however, can take up and store more nutrients such as phosphate than can small cells. These size-related phenomena result in a spectrum of competitive strategies among phytoplankton species from small, fast-growing cells to large, slow-growing cells able to store nutrients through periods of scarcity. Between these extremes lie a range of intermediate adapta-

tions. The danger of sinking out of the lighted (euphotic) zone in a body of water is a continual problem. Large cells tend to sink more rapidly through water than small cells (see below). To avoid losses due to sinking, phytoplankton have evolved a wide range of adaptations to increase buoyancy, reduce density, or increase physical-form resistance to sinking; these adaptations are more pronounced in large cells and colonies. In turn these same adaptations are responsible for much of the morphological diversity within the phytoplankton. Finally, the susceptibility of phytoplankton to losses due to grazing or predation is size-dependent. Small cells are generally rapidly consumed by grazers such as protozoa, rotifers, and crustaceans (see below). Large cells and colonies above about 50 μm in diameter are largely immune to predation by crustaceans (Burns, 1968), but are more prone to attack by parasites (see below). Intermediate cell and colony sizes may be free from predation by protozoa but susceptible to crustaceans. Many of the adaptations that reduce sinking rates also perform anti-herbivore roles.

The importance of cell size is one example of how scale factors into phytoplankton ecology. Different aspects of phytoplankton ecology occur at different scales of time and space. Nutrient regeneration and grazing of phytoplankton can take place at scales of seconds to minutes over distances of millimeters.

Table 22–1   Volumes, surface areas, and surface-to-volume ratios for some selected phytoplanktonic algae (Data from Reynolds, 1984)

species	volume (μm^3)	surface area (μm^2)	SA/V (μm^{-1})
Synechococcus sp.	18	35	1.94
Aphanizomenon flos-aquae	610	990	1.62
Anabaena circinalis	2040	2110	1.03
Oscillatoria agardhii	46,600	24,300	0.52
Microcystis aeruginosa	4.2×10^6	1.26×10^5	0.03
Cryptomonas ovata	2710	1030	0.38
Ceratium hirundinella	43,740	9600	0.22
Fragilaria crotonensis	6230	9290	1.48
Asterionella formosa	5160	6690	1.30
Tabellaria flocculosa	13,800	9800	0.71
Melosira granulata	8470	4915	0.58
Synedra ulna	7900	4100	0.52
Cyclotella meneghiniana	1600	780	0.49
Stephanodiscus astraea	5930	1980	0.33
Chrysochromulina parvula	85	113	1.33
Mallomonas caudata	4200	3490	0.83
Uroglena lindii	2.2×10^6	8.1×10^4	0.04
Ankistrodesmus falcatus	30	110	3.67
Ankyra judayi	24	60	2.50
Chlorella sp.	33	50	1.52
Pediastrum boryanum	16,000	18,200	1.14
Scenedesmus quadricauda	1000	908	0.91
Staurastrum pingue	9450	6150	0.65
Cosmarium depressum	7780	2770	0.36
Sphaerocystis schroederi	5.1×10^4	6.65×10^3	0.13
Eudorina unicocca	1.15×10^6	5.31×10^4	0.05
Volvox globator	4.77×10^7	6.36×10^5	0.01

Lehman and Scavia (1982) showed that zooplankton excrete nutrients in patches that significantly affect nutrient uptake by algae over scales of millimeters. Growth rates operate at scales of hours to days. Phytoplankton patches develop at scales of weeks and kilometers, and successions of species occur at scales of entire seasons across entire lake basins and oceans (Harris, 1986). Dynamic processes in lakes and oceans are very similar but operate on different scales. A patch of phytoplankton in a lake may extend for hundreds of meters whereas a patch in an ocean may be hundreds of kilometers across.

As an example of the importance of scale in phytoplankton ecology, consider the routine problem of determining the abundance of phytoplankton in a body of water. Counts of phytoplankton are usually done by collecting a sample of water and placing a subsample into a settling chamber. After the algae have settled overnight, they are counted and identified with an inverted microscope. The results are then reported as number of cells of each species per unit of volume (ml or l). If an alga were present at a density of one cell per liter, it would likely be missed or considered unimportant. But consider a typical lake like Lawrence Lake in southwestern Michigan (Wetzel, 1975). Lawrence Lake has a volume of 293,500 m^3, which is 293,500,000 liters! Thus even a rare alga existing at an average density of one per liter would have a total population of 293,500,000—more than the number of people in the United States! Many such rare species contribute to the high species diversity of phytoplankton in lakes and oceans.

Table 22–2 Volumes and maximum rates of reproduction for selected phytoplankton (Data from Reynolds, 1984; Sandgren, 1988)

species	volume (μm^3)	μ_{max} (day^{-1})
Synechococcus sp.	18	2.01
Aphanizomenon flos-aquae	610	0.98
Oscillatoria agardhii	46,600	0.86
Microcystis aeruginosa	4.2×10^6	0.48
Cryptomonas ovata	2710	0.83
Ceratium hirundinella	43,740	0.26
Stephanodiscus hantzschii	600	1.18
Cyclotella meneghiniana	1600	0.85
Asterionella formosa	5160	1.74
Tabellaria flocculosa	13,800	0.76
Monodus subterraneus	105	0.64
Dinobryon cylindricum	290	0.58
Synura petersenii	431	0.76
Mallomonas cratis	1516	0.55
Mallomonas caudata	10,625	0.30
Ankistrodesmus falcatus	30	1.59
Chlorella sp.	33	2.15
Scenedesmus quadricauda	1000	2.84
Eudorina unicocca	1.15×10^6	0.62

Even though a phytoplankton species may be rare in a body of water at a particular time, the importance of temporal scale in phytoplankton ecology guarantees that it may not remain so for long. Rates of reproduction in phytoplankton vary from two to three doublings per day to one doubling every week to ten days. If an alga started at a density of one per liter with a reproduction rate of two doublings per day, it would reach a density of 16,384 cells l^{-1} in one week (if we assume no losses during this time). It is therefore essential when planning a program to sample phytoplankton to be aware of the time scale at which events occur. In Figure 22–2, the upper graph shows the rapid changes in population densities of chrysophyceans as was determined by sampling every two to three days (Sandgren, 1988). Figure 22–2b shows what the same data would have looked like if these populations had been sampled at weekly intervals (which is a more common practice). The population maxima of several species would have been missed. The phenomenon of missing important phenomena by sampling at an inappropriate scale is called **aliasing**. Because of the short scales at which many events involving phytoplankton occur, aliasing is a persistent problem.

Phytoplankton ecology has been characterized by the development of a large body of theory and mathematical modeling. Theory and modeling have centered about two topics—competition theory and trophic dynamics. The original source of competition theory lies in the work of Gause (1934) on various microorganisms in test tubes and flasks. These experiments led to the development of the competitive exclusion principle and later to niche theory. The basic idea is that biological interactions—not physical and chemical external factors—are paramount in community dynamics. Species are assumed to exist close to their maximum density in the environment and to compete for scarce resources. If two species occupy the same niche, one must inevitably drive the other out through competitive displacement. These concepts entered plankton ecology when Hutchinson (1961) wrote his famous paper on the paradox of the plankton. If we assume that species are close to their

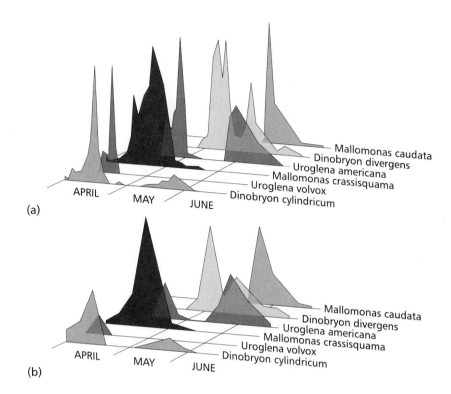

Figure 22–2 (a) Patterns in populations of some opportunistic chrysophyceans sampled every two to three days. (Data from Sandgren, 1988) (b) A plot of the same data had sampling been conducted only every seven days. Several population maxima are missed.

maximum density in aquatic systems and competitive exclusion is a general rule, how can 50 to 100 species of phytoplankton possibly coexist in only a few milliliters of lake or ocean water? The controversy concerning this paradox has fueled experimental and theoretical research to the present and has even shaped aquatic management practices.

The second focus of theory and modeling is that of trophic dynamics. The concept of top-down control of phytoplankton communities originated with Porter (1977). Top-down theories assume phytoplankton are controlled by herbivory, which directs species compositions and seasonal patterns of biomass. The opposite concept of bottom-up control maintains that phytoplankton are fundamentally controlled by nutrients rather than herbivory. These concepts have also shaped management practices. We will discuss both competition theory and trophic dynamics in later sections. We will consider phytoplankton population dynamics under two general categories: (1) growth processes, including photosynthesis and nutrient uptake, and (2) loss processes, including competition, grazing, sedimentation, parasitism, washout, and death. But first, we will consider briefly the physical and chemical environment of lakes and oceans, the arena of phytoplankton ecology.

The Physical Environment

Water as a Fluid Medium

The physical environment in which phytoplankton reside is determined by the physical properties of water as a molecule and the interaction of water with solar radiation. In a molecule of water the two hydrogen atoms and single oxygen atom are arranged as if they were at the points of an isosceles triangle with an obtuse angle of 104.5° at the oxygen atom. Consequently, the hydrogen atoms bear a weak positive charge and the oxygen atom a weak negative charge. Water molecules are capable of forming hydrogen bonds between adjacent molecules. These bonds in turn allow water to act as a liquid crystal—a property that is almost unique among compounds. If water (H_2O) behaved like structurally similar compounds such as H_2S or NH_3, it would be a gas at normal environmental temperatures.

Water is a highly effective solvent for inorganic salts, soluble hydrophilic organic compounds, and a wide array of gases including O_2, N_2, and CO_2. Therefore it is an ideal medium for a vast number of chemical interactions. The intermolecular bonding properties of water give it a high specific heat. This

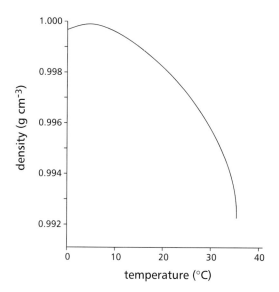

Figure 22–3 The density of freshwater changes with temperature. Maximum density occurs at 4° C. (After Goldman, C. R., and A. J. Horne. 1983. *Limnology*. McGraw-Hill. Reproduced with permission of the McGraw-Hill Companies)

water is quickly insulated from the cold air above and therefore remains liquid. Many lakes in Antarctica are covered by permanent ice caps up to 4 m thick; liquid water and phytoplankton survive under the ice. In the oceans, high salinity (on average, 35 g l⁻¹) increases density and depresses the freezing point to –1.91° C, but ice still floats and insulates the underlying waters.

Two final properties of water arise from hydrogen bonding. Surface tension is the tendency of water molecules to bond together at the air-liquid interface. This property makes it possible for a number of species of phytoplankton, including some diatoms and chrysophyceans, to hang from or sit atop the surface film. Collectively they are termed **neuston** (Wetzel, 1975). Viscosity is the tendency of water to resist flow and impose drag on organisms moving through it. Viscosity increases as temperature decreases. While water is much less viscous than maple syrup, it still has profound effects on the shapes of phytoplankton. If water

means that liquid water can store a large quantity of solar heat and that a large quantity of heat is required to raise the temperature of a body of water. Especially large inputs of heat (540 cal g⁻¹) are required for the phase transition from liquid to gas. Conversely, when a body of water cools, it gives off large amounts of heat to the overlying air and surrounding land. Aquatic environments are very stable thermal environments for phytoplankton, which are normally not subjected to sharp temperature changes over short periods of time as are terrestrial organisms.

Water shows very marked changes in density with temperature (Fig. 22–3). Density increases rapidly as temperature falls from 35° C, reaching a maximum density at 4° C. With further cooling, water becomes less dense until it freezes as ice, which floats because it is less dense than liquid water. The fact that ice floats is profoundly significant for aquatic life. If, like most substances, water had its greatest density when it became solid, ice would sink to the bottom of lakes or the oceans, which would consequently freeze from the bottom up. Bodies of water would be inhospitable places for life since few large organisms can withstand being frozen solid. Water ice furthermore has a specific heat (0.5) about half that of liquid water, and so ice forms and melts readily. The ease with which it forms means that the underlying warmer and denser

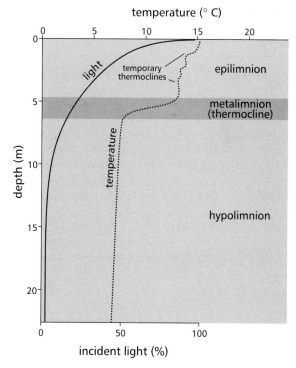

Figure 22–4 Distribution of light and heat in a summer stratified lake. Light declines exponentially with depth. The thermocline divides the lake into upper mixed epilimnion, metalimnion, and lower, colder hypolimnion. Temporary thermoclines are due to warm, calm days. (After Goldman, C. R., and A. J. Horne. 1983. *Limnology*. McGraw-Hill. Reproduced with permission of the McGraw-Hill Companies)

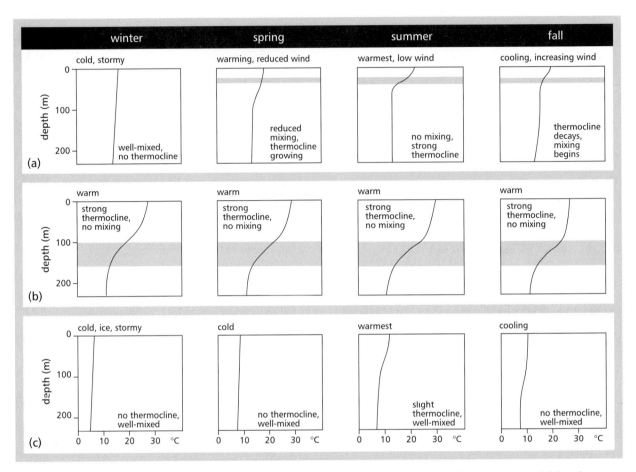

Figure 22–5 Generalized patterns of temperature and mixing in (a) temperate, (b) tropical, and (c) polar oceans as functions of depth and seasonality. Line denotes temperature as a function of depth. Shaded zone indicates thermocline, which prevents mixing in the waters beneath it. (After MARINE BIOLOGY 2nd. Edition, by J. W. Nybakken. ©1987 by HARPER COLLINS PUBLISHERS. Reprinted by permission of Addison-Wesley Educational Publishers)

were significantly less viscous than it is, phytoplankton would have difficulty remaining in suspension.

Light and Heat

Most of the solar radiation entering a lake or ocean is converted into heat. Light entering a body of water declines as a function of depth according to a simple exponential decay function.

$$I_z = I_o \, e^{-\eta z} \quad (1)$$

Here I_z is the irradiance (μmol quanta m^{-2} s^{-1}) at depth z. I_o is the irradiance at the surface (about 2000 μmol quanta m^{-2} s^{-1}), e the base of natural logarithms, and η the extinction coefficient. The value of the extinction

coefficient varies with the wavelength of light, being greater than 2.0 for red and infrared light and less than 0.01 for blue and violet light. Therefore most heat is absorbed near the surface, and only blue light penetrates to any depth. The penetration of light into a lake is shown in Figure 22–4. The depth at which light becomes too dim for photosynthesis defines the bottom of the **euphotic zone**. Temperature would follow a similar curve except that wind sets up currents in the water column that mix the heat down into the water column. This mixing process divides a lake or an ocean into distinct layers or strata and is called **stratification**. The upper warm and less dense layer is termed the **epilimnion** in lakes and the **epipelagic** in the oceans. In lakes the cooler denser lower layer is known as the **hypolimnion**. In the oceans there are four cold lower

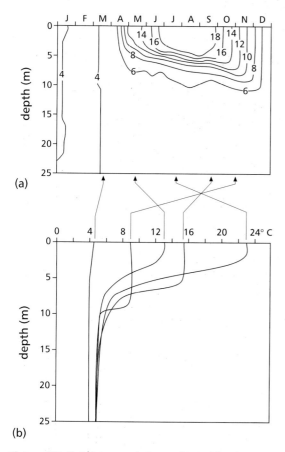

(a)

(b)

Figure 22–6 Mixing and thermal stratification in Lake Plußsee in Holstein, Germany, in 1986. (a) Temperatures are shown as isotherms—lines of equal temperature. (b) Temperature plotted as a function of depth on five dates. Lake Plußsee is an example of a dimictic lake (one that mixes twice a year, in the spring and fall). (After Lampert and Sommer, 1997 ©Georg Thieme Verlag)

strata—the **mesopelagic, bathypelagic, abyssal pelagic,** and **hadalpelagic** (see below). Between the upper and lower zones lies a region of sharp change in temperature and density (termed the **metalimnion** in lakes), which is marked by a decrease in temperature called the **thermocline** (Fig. 22–4). The depth of the epilimnion varies with the area of the lake; it may be as shallow as 2 m or greater than 20 m. Metalimnions are usually several meters in depth. A hypolimnion may be absent entirely in shallow lakes or extremely deep as in the Great Lakes or Lake Baikal in Russia. In oceans the epipelagic may be 20–100 m deep with its lower boundary marked by the thermocline. The mesopelagic extends from the epipelagic down to the 10° C isotherm at 700–1000 m depth, depending on geo-

graphic area. The bathypelagic lies between the 10° C and 4° C isotherms (2000–4000 m depth). The abyssal pelagic reaches to 6000 m, and the hadalpelagic occurs in deep trenches, from 6000–10,000 m. There are many patterns of formation and breakdown of stratification in lakes and oceans, which vary with size and latitude (Figs. 22–5, 22–6).

Stratification has important consequences for phytoplankton. The upper, warmer epilimnion is generally well mixed and well lighted. Phytoplankton will circulate within the epilimnion over 30 minutes to a few hours. In the metalimnion, however, the mixing time is on the order of weeks (Harris, 1986). Phytoplankton may occupy distinct vertical layers or patches within the epilimnion where light may be optimal for growth. Motile species may migrate up to the top of the epilimnion for photosynthesis during the day and back to the darker lower levels to acquire nutrients at night. Nonmotile species depend on vertical mixing to maintain position in the lighted water column. If the metalimnion is lighted (within the euphotic zone), species such as *Oscillatoria* may be concentrated there (Fig. 22–7). In the oceans, tropical and subtropical regions have a perennial thermocline that retards upward movement of nutrients (Fig. 22–5). Phytoplankton biomass in these regions is low year-round, except at coastal upwelling zones (Fig. 22–13) where it is enhanced and persistent. In temperate oceans the epipelagic is recharged with nutrients when the thermocline breaks down. In stratified open ocean waters phytoplankton typically show a deep chlorophyll maximum, which represents a trade-off between decreas-

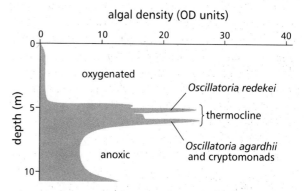

Figure 22–7 Vertical distribution of cryptomonads and *Oscillatoria* spp. in a stratified kettle lake. (Goldman, C. R., and A. J. Horne. 1983. *Limnology*. McGraw-Hill. Reproduced with permission of the McGraw-Hill Companies, based on Baker and Brook, 1971)

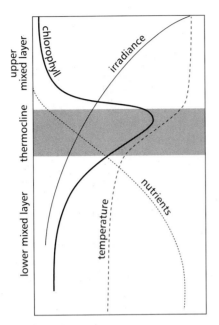

Figure 22–8 Vertical profile of chlorophyll, temperature, irradiance, and nutrients in a stratified water column of the open ocean. (After Falkowski and Raven, 1997. Reprinted by permission of Blackwell Science, Inc.)

the haptophyte *Phaeocystis antarctica* dominates in deeply mixed waters whereas diatoms dominate where waters are more highly stratified. Diatoms are less effective than *Phaeocystis* in CO_2 drawdown (Chapter 2) and transport to the deep oceans. Hence oceanographers are concerned that global warming, which is predicted to increase upper-ocean stratification, may result in a dramatic change in the biological pump (transfer of carbon to deep oceans) (Arrigo, et al., 1999).

Turbulence

Oceans and lakes are turbulent environments. The interplay of wind action and solar heating with tides and the rotation of the earth creates many different types of water motions. Texts on limnology and oceanography are a good source for detailed descriptions of these various motions. Water motions may affect populations of phytoplankton by concentrating or dispersing patches of phytoplankton. In addition, there is evidence that species of algae differ in their tolerance of turbulence (Fogg, 1991; Willen, 1991), and these differences may affect competitive interactions and the formation of blooms. Here we will focus on a few types of water motions that significantly affect the ecology of phytoplankton.

Convection cells, or Langmuir cells, are elongate, wind-driven surface rotations in lakes marked by con-

ing light and increasing nutrients in a more stable environment (Fig. 22–8). Polar oceans mix year-round and provide a consistent level of nutrients to sustain phytoplankton growth (Fig. 22–5). In the Southern Ocean

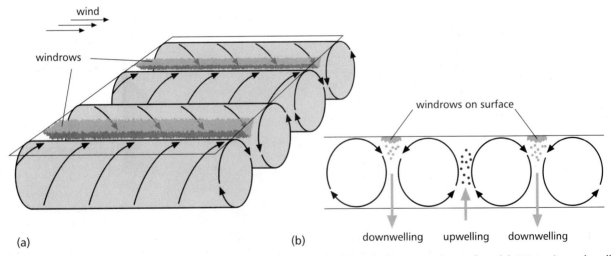

Figure 22–9 The formation of Langmuir cells in a lake is marked by windrows on the surface (a). Water in each cell follows a spiral path through the water in the direction of the wind. Adjacent cells, seen in cross section in (b), rotate in opposite directions, creating regions of downwelling (where the windrows occur) and upwelling. Foam and buoyant algae (gray circles) accumulate in the downwelling regions where currents converge. Negatively buoyant algae (black circles) accumulate in the upwelling zones beneath the surface.

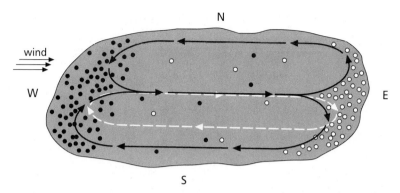

Figure 22–10 Diagrammatic representation of a lake under a steady westerly wind. Return currents flow back toward the west along the north and south shores. Internally, there is also a return flow (dashed white lines), which acts much like a conveyor belt on suspended phytoplankton. Buoyant phytoplankton (white circles) build up on the eastern (downwind) shore. Phytoplankton that avoid the surface waters (black circles) may accumulate along the western shore to which they are carried by the subsurface return flow. (Based on diagrams in Verhagen,1994; George and Edwards, 1976. Reprinted by permission of Blackwell Science, Ltd.)

spicuous lines of foam called windrows (Fig 22–9a). A minimum wind speed of about 11 km hr^{-1} is necessary for their formation. Within each Langmuir cell, water moves in the direction of the wind but in a spiral pattern. Adjacent convection cells rotate in opposite directions, creating alternating zones of upwelling and downwelling between them. The foam streaks on the surface occur in the zones of downwelling where buoyant phytoplankton and bubbles accumulate. Between the rotating cells in the zones of upwelling, negatively buoyant particles and algal cells also concentrate (Fig. 22–9b). In both cases phytoplankton are moved across the lake in the direction of the wind.

Imagine a steady wind blowing from the west across a lake for an extended period of time. Such a wind transmits about 3% of its energy to the water surface (Goldman and Horne, 1983) and generates normal surface waves. The same kind of normal waves occur on the surface of the oceans, but on a much larger scale. When that flow of water reaches the eastern shore of the lake, waves, of course, crash

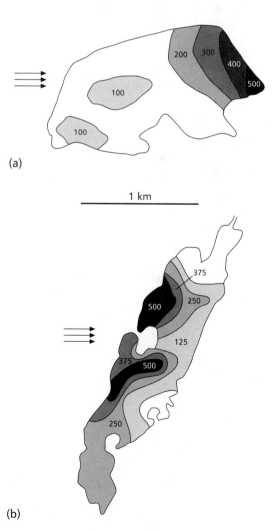

Figure 22–11 Horizontal patches of phytoplankton due to westerly winds. (a) buoyant *Microcystis aeruginosa* builds up on the downwind (eastern) shore in Eglwys Nynydd, South Wales, reservoir. Isopleths in µmol chl *a* l^{-1} (b) *Ceratium hirundinella* favors deeper strata in Esthwaite Water, Lake District, England, and is carried back toward the western shore by subsurface return currents. Isopleths are cells ml^{-1} (a: After George and Edwards, 1976. Reprinted by permission of Blackwell Science, Ltd.; b: After Heany, 1976)

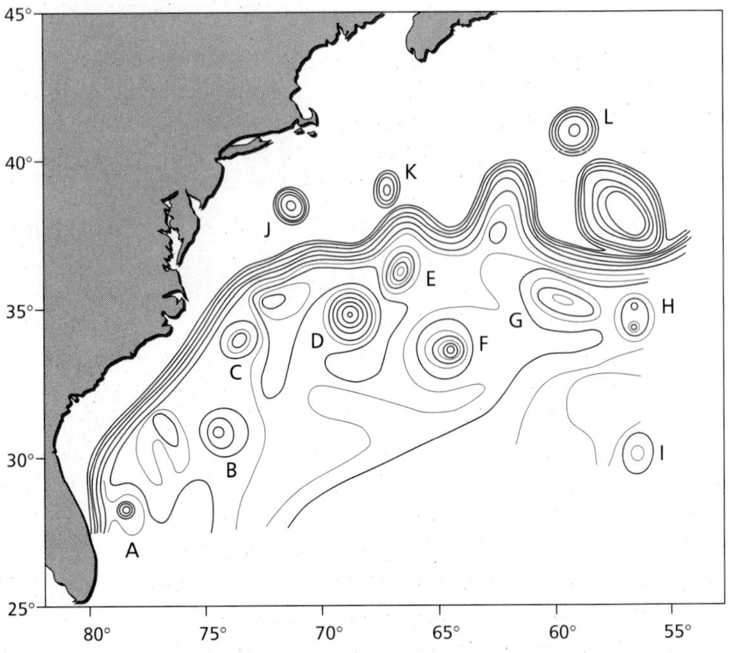

Figure 22–12 Gulf stream rings in the western North Atlantic (A–L) from bathythermograph and infrared satellite imagery. Rings J, K, L are warm core rings north of Gulf Stream. Rings A–I are cold core rings. Warm core rings are headed northeasterly, while cold core rings flow southwesterly. (After Richardson, P. L., R. E. Cheney, and L. V. Worthington. *Journal of Geophysical Research* 83:6136–6144 ©American Geophysical Union)

on the beach, but also a surface current will move north and south along the shore, returning to the west along the north and south shores (Fig. 22–10). Internally there will also be an east-to-west return flow under the surface. This return flow acts like a conveyor belt, and it can have a substantial effect on the distribution of phytoplankton. If the phytoplankton are buoyant near the surface, then the surface flow of the conveyor tends to pile them up on the downwind (eastern) shore. Figure 22–11a shows an example where *Microcystis aeruginosa*, a buoyant cyanobacterium, piled up on the downwind shore in Eglwys Nynydd, South Wales (U.K.) (George and Edwards, 1976). *Ceratium hirundinella* maintains its position at a lower level in the water column. In Esthwaite Water, English Lake District (U.K.), westerly winds caused this dinoflagellate to accumulate on the western shore of the lake due to the action of the subsurface return currents (Fig. 22–11b) (Heaney, 1976). Clearly, winds can affect the distribution of phytoplankton patches. If sampling were done at a single fixed station, these types of large-scale population movements could result in widely varying population estimates.

In the oceans water movements occur on much larger scales than in lakes and largely without the edge effects of closed basins. Patch sizes are much larger, and phytoplankton are often monitored by remote sensing (see Text Box 22–1). While phytoplankton patches may extend for hundreds of meters in large lakes, they may extend for hundreds of kilometers in oceans. Two types of currents are of special importance to phytoplankton dynamics in the seas. A major ocean current like the Gulf Stream wanders north and south as it makes its way across the North Atlantic (see Fig. 23–18). As it wanders, it forms eddies or loops that become pinched off as rings varying in diameter from 100 km to over 250 km (Fig. 22–12). These rings rotate and move across the ocean carrying whole ecosystems as discrete parcels, which maintain their structure for up to two years. Warm core rings move northward into European coastal waters, and cold core rings may push south into the Sargasso Sea (Harris, 1986).

The second important type of current is that responsible for upwelling zones (Fig. 22–13). The most famous of these currents is the Peruvian upwelling. Off the coast of Peru, cold nutrient-rich water rises up from great depth to create an abundant growth of phytoplankton. Zooplankton graze on the phytoplankton and in turn support a rich fishery. An El Niño event occurs when warm Pacific water floods over the cold water and closes it off. The phytoplankton and then the fishery collapse. Moreover, the global climate is altered as storms become more severe on the west coast of North America. Why this occurs is not yet clear, but the strength of the southeast trade winds in the central Pacific is involved.

Text Box 22–1 Remote sensing of phytoplankton

Phytoplankton pigments interact with light in ways that can be detected from space by means of satelliteborne sensors (radiometers). Sensor data can be used to construct global to local views of phytoplankton distribution patterns. Images can be used to study changes through time or to compare phytoplankton of different regions. For example, the lightest areas of the adjacent global map (a composite image, from September 1997 through July 1998) indicate regions of high chlorophyll concentrations/algal populations, most notably along the coasts of North America, northern Europe, northern Asia, Indonesia, southeastern South America, and southwestern Africa. More diffuse phytoplankton populations are indicated by bands extending

through the northern and southern oceans; these contrast with mid-ocean regions where phytoplankton populations are low. A more localized example (left) reveals the extent of a phytoplankton (coccolithophorids) bloom in the Bering Strait. The bloom—showing up as the lighter gray areas off the Alaskan coast—imparted an aqua coloration to these normally dark waters.

Until mid-1986, eight years of such records were made using the Coastal Zone Color Scanner (CZCS) operated on the Nimbus-7 environmental satellite mission. The CZCS was the first source of remote sensing data on phytoplankton, but was not continuously operational, so only partial data bases are available. The data obtained by the CZCS are maintained at the Goddard Distributed Active Archive Center (DAAC), which also collects all of the ocean color data from NASA satellite missions. These data (and those described below) are available free of charge through the World Wide Web and by other means.

Currently operating is the Sea-viewing Wide Field-of-view Sensor (SeaWiFS). It observes more than 90% of the oceans every two days, with a resolution of 4.5 km. This technology has allowed improved measurement of phytoplankton pigment concentration in both open ocean and turbid nearshore areas. The Goddard DAAC also archives SeaWiFS images.

In 1999 and 2000 a new imaging system will be orbited: this is the Moderate Resolution Imaging Spectroradiometer (MODIS). The new system will provide higher quality images as well as new capabilities. It will be possible to monitor chlorophyll fluorescence (an indicator of the physiological condition of phytoplankton), atmospheric levels of marine aerosols, detached coccolith concentrations, and other features.

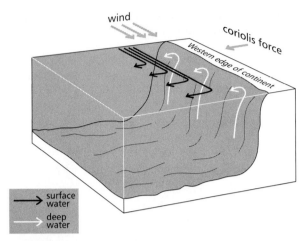

Figure 22–13 Coastal upwelling. The combined action of wind and the Coriolis force of Earth's rotation move surface waters offshore along the western margins of continents (black arrows). This water is replaced by deep water rising to the surface (white arrows), which brings nutrients for phytoplankton growth. (After MARINE BIOLOGY 2nd. Edition, by J. W. Nybakken. ©1987 by HARPER COLLINS PUBLISHERS. Reprinted by permission of Addison-Wesley Educational Publishers)

All of these large-scale water movements affect the growth and distribution of phytoplankton populations. But what effect does turbulence have on the cells of individual species of phytoplankton? It has long been known that certain species of algae, including many dinoflagellates (White, 1976, Pollingher and Zemel, 1981), will not grow if stirred or aerated in small culture flasks. Cell division is inhibited at low rates of turbulence, and at high rates, cells disintegrate. Among marine red-tide dinoflagellates, *Prorocentrum micans* was more tolerant of turbulence than either *Gonyaulax polyedra* or *Gymnodinium sanguineum* (Thomas and Gibson, 1990, 1992; Thomas, et al., 1997). Dinoflagellate red tides are associated with extended periods of relatively calm seas. Thus red tides may be disrupted by turbulence, which damages dinoflagellate cells.

Turbulence has the opposite effect on diatoms. In the oceans, blooms of *Chaetoceros armatum* and *Asterionella socialis* occur in the surf zone of beaches in Washington State and New Zealand (Lewin and Norris, 1970). In cultures, the diatoms *Skeletonema costatum* and *Asterionella glacialis* grew better when subjected to turbulence than in controls in stationary culture (Thomas, et al., 1997). These preliminary results suggest that in the marine environment turbu-

lence may act as a selective agent in determining competitive dominance. Turbulence and tolerance among freshwater forms has been discussed in reviews by Fogg (1991) and Willen (1991).

The Chemical Environment

Dominant Ions of Lakes and Oceans

Oceans contain about 34.8 g of dissolved salts per kg of seawater. Of this amount, sodium accounts for 10.77 g and chloride for 19.35 g. Oceans are dominated by sodium and chloride followed by magnesium (1.29 g kg^{-1}) and sulfate (2.71 g kg^{-1}). Lakes are dominated by calcium and bicarbonate, and their concentrations may vary greatly. Major ions in oceans vary by only a few percent from place to place, and turnover times are extremely long. Exceptions are aluminum and iron, which turn over at scales of 10^2 years. Major ions in lakes have turnover times of weeks to at most one hundred years.

Phytoplankton have little use for sodium. In saline lakes, the high levels of sodium actually restrict the diversity of algae. Chloride is used in some biochemical transformations such as photolysis of water and ATP production. Calcium is necessary in small amounts for growth and cell wall integrity. Magnesium is needed by all cells for energy metabolism in converting ATP to ADP. It is also the central metal in the chlorophyll molecule. Sulfur as sulfate (SO_4^{2-}) is important in disulfide bonds, which stabilize the structure of enzymes and other proteins. Most of the major ions in freshwaters and the oceans are only needed by phytoplankton in small quantities and therefore rarely limit growth. Because their turnover times are relatively long, they are considered conservative substances in aquatic systems. Carbonate is the exception. As the principle source of carbon in aquatic systems, it is needed in large quantities.

Carbon is rarely limiting in aquatic systems because carbon dioxide—although relatively scarce in the atmosphere—is highly soluble in water. The form that carbon takes in water depends on the pH of that water. Most lakes have a pH of 6 to 9, where pH=7 is considered neutral. Acid waters may have a pH as low as 2, while some alkaline lakes may have a pH>10. Seawater is slightly alkaline, ranging from pH 7.5 to 8.4. Below pH 6, the bulk of CO_2 dissolved in water is present as soluble carbon dioxide or undissociated carbonic acid (H_2CO_3) (Fig. 22–14). Above

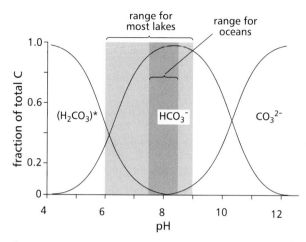

Figure 22–14 Forms of inorganic carbon as a function of pH in lakes and oceans. Most lakes and all oceans are bicarbonate (HCO_3^-) solutions. $(H_2CO_3)^* = H_2O + CO_2$. (After Goldman, C. R., and A. J. Horne. 1983. *Limnology*. McGraw-Hill. Reproduced with permission of the McGraw-Hill Companies)

pH 6, bicarbonate ion (HCO_3^-) becomes increasingly important. Note that the normal range of pH for seawater spans the maximum for bicarbonate. Above pH 10, carbonate becomes increasingly important. Most lakes and all oceans are predominantly bicarbonate solutions, and phytoplankton acquire their carbon by taking up bicarbonate or converting it to CO_2 for uptake. In acidic lakes, carbon can occasionally become limiting to growth since green algae respond positively to additions of CO_2 or bicarbonate (Fairchild and Sherman, 1993). For more detail, see Chapter 2. The carbonate–bicarbonate–CO_2 equilibrium plays a major role in the buffering capacity of lakes. If a lake has abundant calcium, it will have abundant carbonate. When acid is added to such a lake in the form of protons (H^+), the protons combine with carbonate (CO_3^{2-}) to form bicarbonate (HCO_3^-), and the pH is stable. Calcium-rich lakes are generally well buffered. Calcium-poor lakes, such as are found in Scandinavia, the northeastern United States, and eastern Canada, have little buffering capacity and are thus susceptible to acidification by nitric and sulfuric acids in acid rain.

Dynamics of phytoplankton are primarily correlated with spatial and temporal fluxes of major nutrient ions such as N, P, and Si. Redfield (1958) determined that marine phytoplankton growing at their maximum growth rate possessed a characteristic ratio of major nutrient ions of 106 C : 16 N : 1 P.

Diatoms, and perhaps other algae that require silica, should have an ion ratio of 106 C : 16 Si : 16 N : 1 P. Phytoplankton should show the Redfield ratio when growing close to their maximum growth rate, at which they are unlikely to be nutrient limited. In the remainder of this section we will consider the roles of nitrogen, phosphorus, and silicon as important limiting nutrients.

Nitrogen

Nitrogen is present in waters primarily as dissolved dinitrogen gas (N_2). Ionic forms include the ammonium ion (NH_4^+), nitrite ion (NO_2^-), and nitrate ion (NO_3^-). In oceans, 95% of nitrogen occurs as N_2. About two thirds of the remainder is nitrate ion, which turns over very rapidly and is often so scarce in ocean waters as to be undetectable, especially in tropical and subtropical waters. In tropical oceans, the filamentous, nitrogen-fixing cyanobacterium *Trichodesmium* is fairly common. For a discussion of nitrogen fixation, refer to Chapters 2 and 6. Cells of the common marine diatom *Rhizosolenia* typically contain two filaments of the nitrogen-fixing cyanobacterium *Richelia*, which effectively acts as a nitrogen-fixing organelle (see Chapter 7). Nitrate only appears in quantity in zones of coastal upwelling or pollution.

The main source of nitrate in lakes is stream and river discharge. In most temperate oligotrophic and mesotrophic freshwaters, nitrate is present in relative excess and exceeds the supply of phosphorus. Nitrogen fixation plays a relatively minor role in such low-nutrient freshwaters. But in some western U.S. lakes whose basins are predominately volcanic rock, nitrogen is the limiting nutrient (Reuter and Axler, 1992). In tropical lakes, nitrogen may be in low supply due to low levels in the surrounding soils of the watershed. In oligotrophic freshwaters nitrate is the major form of dissolved inorganic nitrogen (DIN) while in eutrophic waters, NH_4^+ and NO_2^- may be present at depth if oxygen is reduced. Nitrite is normally present in insignificant amounts, although in eutrophic lakes a narrow layer in the thermocline may contain elevated levels of nitrite (Fig. 22–15). If oxygen is present, NO_2^- is converted to NO_3^-, whereas in anoxic waters, it is reduced to ammonia. The supply of phosphorus may exceed the supply of nitrogen in eutrophic waters. Under these circumstances nitrogen-fixing cyanobacteria such as *Anabaena*, *Aphanizomenon*, and *Gloeotrichia* may generate nuisance blooms. Such

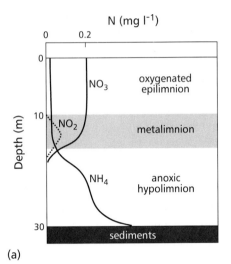

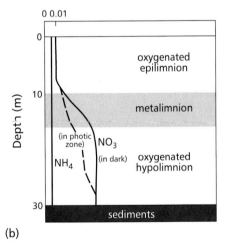

Figure 22–15 Generalized distribution of nitrate, nitrite, and ammonia with depth in a eutrophic (a) and an oligotrophic stratified lake (b) in midsummer. (After Goldman, C. R., and A. J. Horne. 1983. *Limnology*. McGraw-Hill. Reproduced with permission of the McGraw-Hill Companies)

nitrogen-fixing cyanobacteria are common in freshwaters even when not at bloom levels (Chapter 6).

Phosphorus

As indicated by the Redfield ratio, phytoplankton need phosphorus in relatively small amounts compared to carbon, nitrogen, and silicon. It is frequently growth-limiting, however, because it is often in short supply in many watersheds. Most of the total phosphorus (TP) in waters is in the form of living or dead particulates. Dissolved phosphorus occurs as either inorganic phosphates (DIP) or dissolved organic phosphorus (DOP). Most dissolved phosphorus is DOP. DIP is primarily orthophosphate (PO_4^{3-}) with much lower amounts of monophosphate (HPO_4^{2-}) and dihydrogen phosphate ($H_2PO_4^-$). Phytoplankton can only use dissolved inorganic phosphate, which is termed soluble reactive phosphate (SRP). When the supply of SRP is exhausted, phytoplankton can release alkaline phosphatases, which are extracellular enzymes capable of freeing phosphate bound to organic substances. In addition, when brief pulses of SRP do occur, many phytoplankton can take up and store excess phosphate as polyphosphate bodies within the cell. This so-called luxury consumption is an important mechanism for dealing with phosphate shortages. During nutrient pulses, an algal cell may be able to store enough phosphate to provide for as many as 20 cell divisions. Finally, many phytoplankton can take up phosphate at extremely low ambient levels, well below the level of detection (see below).

In many lakes the amount of phosphorus present in late winter determines the size of the phytoplankton populations that can develop in the summer. In winter, DIP turnover is relatively slow in all freshwaters—on the order of hours to days—as a result of low populations of algae with low growth rates. DIP represents less than 10% of TP in oligotrophic freshwaters, where, during summer, the demand for DIP is high, and the DIP pool may become unmeasurable. Phytoplankton depend on DIP recycling. Using the radioisotopes $^{32}PO_4$ and $^{33}PO_4$, Hudson and Taylor (1996) measured the rate of regeneration of phosphate (during summer) in two oligotrophic lakes. Measured rates ranged from 15 to 205 ng P l^{-1} hr^{-1}. Grazers smaller than 40 μm accounted for 77% of the measured regeneration. Many marine waters are similar to oligotrophic freshwaters with respect to phosphate dynamics. In coastal marine waters, DIP builds up during the periods of vertical mixing. If the ocean stratifies, the DIP pool is depleted and phytoplankton depend on phosphorus recycling, as in oligotrophic freshwaters. DIP may approach 100% of TP in eutrophic freshwaters. The DIP pool may exhibit slow turnover due to the fact that DIP input may exceed algal growth requirements and thus build up to high levels.

Silicon

Silicon can be a limiting nutrient for the growth of diatoms in lakes during summer stratification when

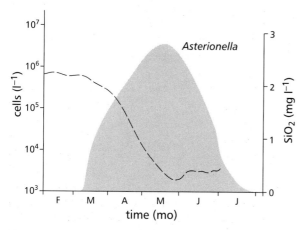

Figure 22–16 The spring bloom of the diatom *Asterionella* in the north basin of Windemere, Lake District, England, and the depletion of the limiting nutrient silica (dashed line). (After Goldman, C. R., and A. J. Horne. 1983. *Limnology*. McGraw-Hill. Reproduced with permission of the McGraw-Hill Companies, based on data in Lund, 1964)

the pool of silica (SiO_2) falls to unmeasurable levels. The decline of silica is certainly one major factor in bringing about the decline of the spring diatom bloom in many lakes (Fig. 22–16). Since polymerized silica is dense (2.6 g cm^{-3}), it adds considerably to the density of diatoms, which consequently suffer population losses due to sedimentation. Polymerized silica decomposes slowly (on the order of 50 days) into soluble orthosilicic acid (H_2SiO_4) or soluble reactive silicate. It does not dissolve during gut passage through herbivorous zooplankton. Thus there is essentially no rapid recycling of silica in the epilimnion. Dissolved silica is only restored to surface waters by external inputs and the turnover of the water column at spring and fall mixing.

Growth Processes of Phytoplankton Populations

Populations take up space, show various levels of mobility, and are distributed in a variety of patterns in time and space. Populations are characterized by the fact that they grow, that is, they change in size (numbers) and density (number per unit area or volume). The concept of growth rate is essential for describing population dynamics. In field research, growth rate usually means the net rate of change in numbers or

biomass and represents the balance between additions due to reproduction and losses due to various sources of mortality or export. In culture studies, however, phytoplankton essentially grow without losses, and so growth rate is equivalent to the rate of reproduction or the "birth rate" in field research. These ideas can be expressed mathematically as:

$$\frac{dN}{dt} = rN \qquad (2)$$

In this equation, dN/dt represents the change in numbers or biomass during a unit of time (such as one day). Sometimes dN/dt is referred to as the growth rate, when it is actually the population growth rate or the change in population size in a unit of time. This change in population size is equal to r, the net growth rate (with units of reciprocal time or, in this case, day^{-1}) multiplied by N, the number of individuals in the population or its biomass. If we isolate r on the right side of the equation by dividing both sides by N, we can see that r is the change in the population size per individual in that population N or the net per capita growth rate. Since the net growth rate is equal to the rate of reproduction minus any losses due to death, the net growth rate is equivalent to:

$$r = \mu - \lambda \qquad (3)$$

where μ is the gross growth rate, rate of reproduction, or the birth rate, and λ is the death rate or the loss rate. The net growth rate may be positive or negative. If positive, then the population will increase in numbers or biomass. If negative, the population will decrease in numbers or biomass over time. Note that a population could have a high rate of gross growth (μ) but show no net growth at all. This situation can arise anytime the gross growth rate (μ) is roughly balanced by the loss rate (λ). In that situation (r = 0), a phytoplankton population could be photosynthesizing, taking up nutrients, and dividing at a significant rate, but repeated sampling would show no net change in numbers because losses balance production. Failure to understand the difference between gross growth and net growth has led to a great deal of confusion in the literature. Lack of evidence of net growth (r) among phytoplankton does not mean there is no gross growth or that the phytoplankton are not physiologically active.

Equation (2) (above) is an example of a differential equation because of the presence of the term dN/dt.

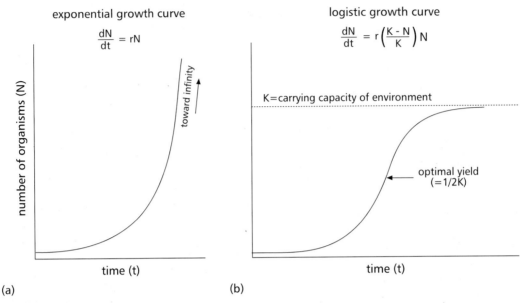

Figure 22–17 Two basic models of population growth: (a) exponential and (b) logistic. (After Wilson and Bossert, 1971)

If equation (2) is solved explicitly to remove the differential term dN/dt, the result can be expressed as:

$$N = N_o \, e^{rt} \quad (4)$$

A population of phytoplankton growing according to equation (4) will increase exponentially without an upper limit (Fig. 22–17a). If the natural logarithms of both sides of equation (4) are taken, the equation becomes that of a straight line whose slope is the net growth rate (r):

$$\ln N = rt + \ln N_o \quad (5)$$

Thus, the net growth rate can be estimated from the natural logarithms of the population sizes. Equation (5) can also be used to calculate the doubling time of the population (t_d). If N is taken as twice the value of N_o, then equation (5) can be rearranged into the form:

$$\ln N - \ln N_o = \ln \frac{N}{N_o} = rt \quad (6)$$

Now, when N=2 (N_o), ln (N/N_o) = ln 2 or 0.69, and we get t_d = ln2/r. If the net growth rate were 0.69 day^{-1}, then the doubling time would be 0.69/0.69 day^{-1} or 1.0 day. Therefore, an r of 0.69 day^{-1} corresponds to one doubling per day. If the net growth rate

were –0.69 day^{-1}, the population would decrease by 50% per day.

No population of phytoplankton can continue to grow indefinitely at an exponential rate as described by equation (4). Even a slow-growing population would soon overflow its habitat. There must be some sort of upper limit to population density due to limits in available resources such as nutrients. This upper limit is called the **carrying capacity** of a population. The first widely used expression that included a carrying capacity was the logistic equation:

$$\frac{dN}{dt} = rN \frac{(K - N)}{K} \quad (7)$$

In the logistic equation, the carrying capacity is represented by the parameter K, which has the same units as N (cells ml^{-1} or mg ml^{-1}). Initially, when N is small, (K–N)/K is close to 1, and the population grows at nearly an exponential rate. As N approaches K, (K–N)/K approaches zero, and the population ceases to grow when at its carrying capacity (Fig. 22–17b). The logistic equation makes no explicit reference to the cause of the carrying capacity, that is, there is no term in the equation for the consumption of some resource that determines the magnitude of the final population size. For some practice in working with the exponential and logistic equations, refer to Question Box 22–1.

Question Box 22–1 Working with the exponential and logistic equations

Sample calculation. The table and figure to the right present the results from a batch culture experiment in which the cyanobacterium *Anabaena flos-aquae* grew for 120 hours. Use these data to estimate the parameters r and K in the logistic growth equation (7).

time (hr)	population (cells ml^{-1})
0	2,360,742
6	3,335,056
12	5,025,320
15	5,498,578
24	9,478,620
35	13,660,240
48	17,979,420
71	23,399,540
96	23,431,136
120	22,698,608

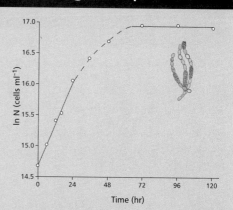

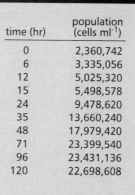

Solution. As shown in equation 5, we can determine the net growth rate (r) of *Anabaena* by taking the natural logarithms of the population sizes. We plot the natural logs against time, as shown in the graph. The first five points make a straight line. Using a calculator, you can run the linear regression of ln (N) on t for these five points. The resulting equation is Y = 0.0578 (X) + 14.68, and the correlation coefficient is 0.99. The slope of this line is the net growth rate, r = 0.0578 hr^{-1}. Multiplying this by 24 hr day^{-1}, r becomes 1.38 day^{-1}. The doubling time is given by t_d = ln (2)/r. For *Anabaena*, t_d = 0.69/1.38 day^{-1}, which equals 0.5 days. Determining the value of K is much simpler. The values of ln (N) for 71, 96, and 120 hours are roughly the same, and K can be taken as their average. Therefore K = 16.96 or 23,207,820 cells ml^{-1}.

Practice problems. Work the following problems with a pocket scientific calculator. You will find it essential to plot the data in order to decide which are to be used in making the estimates. Answers are given at the end of the chapter.

1. The following data give the results of three replicated batch cultures of *Chlamydomonas reinhardtii*. Calculate the net growth rate r for each batch culture. Take the average of the three as the net growth rate of *Chlamydomonas* in this study.

| | (cells ml^{-1}) | | |
time (hr)	culture #1	culture #2	culture #3
0	9,280	13,890	7,860
24	38,550	42,490	24,300
48	100,680	143,480	129,620
71	249,250	355,780	321,440
95	946,230	1,283,390	1,131,070

2. The euglenoid flagellate *Euglena gracilis* grows in a culture medium containing acetate. While *Euglena* can photosynthesize, it can also grow in the dark by using acetate as an energy source. The table below contains data on growth of *E. gracilis* in the light and dark. Calculate r for each growth experiment. How much did photosynthesis increase the growth rate of *Euglena* over the rate in the dark? By how much did photosynthesis reduce the doubling time?

| dark-grown | | light-grown | |
time (hr)	cells ml^{-1}	time (hr)	cells ml^{-1}
0	915	0	2,830
24	1,605	24	4,695
48	2,590	46	14,910
147	56,454	72	31,120
219	360,464		

continued→

Question Box 22–1 *continued*

3. Growth of *Chlorella pyrenoidosa* was followed in batch culture for 24 days. Use the following table of population counts (cells ml^{-1}) to estimate the logistic parameters r and K.

day	cells ml^{-1}	day	cells ml^{-1}
1	45,740	10	3,710,040
2	71,904	13	5,266,840
3	85,740	14	8,326,400
4	229,640	15	8,753,200
5	517,580	17	7,576,200
6	570,640	20	10,357,000
7	1,189,320	24	9,341,700
9	2,112,780		

An alternative way to write the logistic equation is to remove the parentheses:

$$\frac{dN}{dt} = rN - \frac{rN^2}{K} \qquad (8)$$

In this form, the logistic equation embodies much of the controversy and theory that prevails in plankton ecology. The first term to the right of the equal sign is the familiar exponential growth term. If populations of phytoplankton fundamentally show exponential growth, then some external forces such as water turbulence, washout, or high temperatures must occur frequently enough to keep the phytoplankton below their carrying capacity. Such external forces occur independently of whatever density the algal population may have reached at the time they occur. Hence they are density-independent, and the population may be said to be under density-independent control. Algal populations under density-independent control are regulated by external, abiotic (physical and chemical forces) in the environment. Such algal populations would likely show wide fluctuations in abundance because external disturbances would occur randomly.

On the other hand, if the second term on the right of the equal sign also is important, then phytoplankton are limited by density-dependent factors. As the population approaches its carrying capacity, resources are depleted, reproduction rates decrease, and/or rates of mortality increase. In either case the controlling factors are biological and density-dependent. Algal populations under density-dependent control would probably show fairly stable levels of abundance. Fur-

thermore, if many phytoplankton species were all close to their carrying capacities at the same time, then they would likely be in competition with each other for one or more resources. Thus, if phytoplankton are regulated by density-dependent factors, competition and other biotic interactions are more likely to control the structure of phytoplankton communities than would physical and chemical factors.

Are phytoplankton communities made up largely of species subject to density-independent control or are many of the planktonic algae at or close to their carrying capacities and subject to density-dependent control? The answer is more than just of academic interest because the way we view the structure of planktonic communities affects the kinds of management plans that are employed on lakes and in the oceans. Those plans in turn affect the quality and use of aquatic resources.

Yet another way to gain insight into ecological theory from the logistic equation in (8) is the concept of r and K-strategists, where r and K are the parameters in the logistic equation (MacArthur and Wilson, 1967). It is a readily demonstrated fact that phytoplankton differ with respect to their gross growth rates or rates of reproduction. There is also a good correlation between cell size and growth rate (Table 22–2). Small cells have faster growth rates than large cells and larger surface to volume ratios (Table 22–1). This has led to the idea that there are r-strategists and K-strategists in phytoplankton communities. R-strategists are small and can grow exponentially to exploit temporarily favorable conditions. Being small, they are readily eaten by grazers. They are opportunistic species. At the opposite end of the spectrum are K-strategists, which are large and grow relatively slowly. These are largely immune to predation and grow close to their carrying capacity. They are considered to be good competitors.

Today, r and K-strategists are not thought of as two opposite poles but as a continuum. Species are thought to have characteristics selected to exploit certain environmental conditions. An r-strategist can exploit a temporary pulse of nutrients washed in by a storm and rapidly build up its population. The resulting bloom can just as quickly be dispersed by turbulence in the water. A K-strategist grows slowly but can survive between pulses of nutrients by living off stored nutrients. K-strategists maintain more constant population levels. The set of characteristics that constitute an r-strategy may be best suited to the highly mixed and turbulent waters of spring and fall, while those of a K-strategy may be appropriate for the stable, stratified waters of summer. Between extreme r and K strategists are the large number of phytoplankton of intermediate size with a range of growth rates and nutrient storage capacities that may be adapted to intermediate disturbances of nutrients and turbulence. Phytoplankton communities are mixtures of species with various combinations of the characteristics of these strategies.

Growth and Light

In the previous section we described the general process of phytoplankton growth without regard to the specific resources that might be directing that growth. In the following sections we will discuss growth as a function of specific factors such as light and nutrients.

The most fundamental aspect of phytoplankton ecology is the conversion of light energy into biomass through photosynthesis. Photosynthesis is the biochemical process by which light energy is used to transform inorganic molecules into organic matter. Details of the biochemistry are beyond the scope of this chapter. Excellent recent treatments are available in Geider and Osborne (1992) and Falkowski and Raven (1997). The process of photosynthesis can be summarized by the familiar equation:

$$6CO_2 + 6H_2O \rightarrow C_6H_{12}O_6 + 6O_2 \quad (9)$$

Light energy is used to strip protons and electrons from water molecules with the resulting production of

oxygen. Those protons and electrons are used to reduce CO_2 to an organic carbon molecule such as glucose, as shown in equation (9). A significant fraction of carbon metabolism may be coupled to nitrate assimilation with the final production of amino acids and proteins (Turpin, 1991).

Plant physiologists quantify photosynthesis in two ways. Gross photosynthesis (P_g) is the light-dependent rate of electron flow from water to CO_2 in the absence of respiratory losses (Lawlor, 1993). Respiration (R_l) is the flow of electrons from organic carbon to O_2 with the production of CO_2. Photosynthesis only occurs in the light, and respiratory losses in the light reduce the level of gross photosynthesis. The difference between gross photosynthesis and respiration is called net photosynthesis (P_n). Therefore:

$$P_n = P_g - R_l \text{ or } P_g = P_n + R_l \quad (10)$$

All three terms are rates and therefore time-dependent.

Equation (9) indicates two ways to measure photosynthesis: by carbon uptake and by oxygen evolution. In the past, the most commonly used method was the uptake rate of ^{14}C (usually as $H^{14}CO_3^-$) into organic matter. The resulting rates may be expressed as mg C g^{-1} hr^{-1}, mg C cell^{-1} hr^{-1} or mg C (mg chl *a*)$^{-1}$ hr^{-1}. Their interpretation is complicated by the fact that it is not clear whether gross or net photosynthesis is being measured. If short incubation times are used, the ^{14}C uptake method should approximate gross photosynthesis. If incubation runs to equilibrium, the rates may approach net photosynthesis (Falkowski and Raven, 1997). Radiocarbon techniques provide no information about respiration.

Interpretation is more straightforward if oxygen evolution techniques are used. If photosynthesis is measured with an oxygen electrode, the data represent net photosynthesis since respiration removes oxygen. Oxygen methods allow direct measurement of respiration. If a chamber containing algae is covered immediately after net photosynthesis measurements with a light-tight bag, the consumption of oxygen gives a good approximation for R_l, the respiration rate in the light. Furthermore, if the chamber remains in the dark for several hours, the dark or basal rate of respiration can also be obtained. Both methods share the problem of requiring confined spaces. If natural phytoplankton assemblages are used, bacteria and microzooplankton may alter the observed rates through their own metabolic activities.

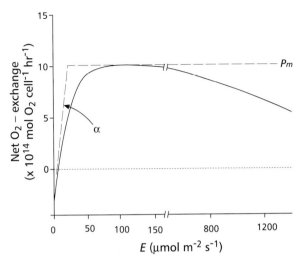

Figure 22–18 Relationship between irradiance E and photosynthesis in *Chlamydomonas reinhardtii*. P_m is the maximum rate of light-saturated photosynthesis, and α is the initial slope of the curve. The compensation point lies at the intersection of the P-curve and the zero net–O_2 exchange line (dashed). (Based on Neale, 1987, in Geider and Osborne, 1992 with kind permission of Kluwer Academic Publishers)

Phytoplankton show a characteristic response curve to increasing light intensity (Fig. 22–18). At some low light level, the rate of net photosynthesis (P_n) will just balance the rate of respiration (R). This light level is called the **compensation point.** For any light level greater than the compensation point, the alga will make a net gain in photosynthesis over losses due to respiration. If algal cells are exposed to low light levels for several hours, they will adapt physiologically by increasing their chlorophyll content. Initially, when light is limiting, there is a linear increase in photosynthesis with increasing light. The slope of this linear increase is termed α, a parameter used in modeling. As light levels continue to increase, the rate of increase in photosynthesis declines and the photosynthesis-versus-light curve bends over and levels off at a maximum value (P_m). Different phytoplankton appear to be adapted to different optimal light levels. Diatoms are especially efficient at photosynthesis under the low light levels and photoperiods that prevail in spring when the water column is mixing and the elevation of the sun is low. Thus diatoms are numerically abundant in temperate lakes and oceans in spring and fall and during most of the ice-free season in polar latitudes. The red-pigmented cyanobacterium *Planktothrix* is adapted to very low

light levels in European lakes (Mur and Bejsdorf, 1978).

At high light levels approaching that of full sunlight, many phytoplankton species show **photoinhibition,** a decline in photosynthesis with increasing light. In this case, respiration increases with light while photosynthesis remains constant such that the photosynthesis versus light curve declines with increasing light (Fig. 22–18). Phytoplankton will exhibit decreased levels of chlorophyll in their cells at these high levels. Prolonged exposure to such light levels can result in damage to the photosynthetic apparatus in the cell. Kromkamp (1990) has examined photoinhibition in *Anabaena flos-aquae.*

Surface waters are a turbulent environment, however, and vertical mixing can carry the phytoplankton from the surface down to 10 m depth and back again in as little as 30 minutes to a few hours. Thus, within as little as 30 minutes an individual algal cell could experience light levels ranging from photoinhibiting to below the compensation point and back again.

A large number of models have been proposed for gross photosynthesis as a function of light (Geider and Osborne, 1992). Two of the most widely tested and successful are the exponential function (equation 11) and the hyperbolic tangent function (equation 12):

$$P = P_m \left[1 - \exp\left(\frac{\alpha E}{P_m} \right) \right] \qquad (11)$$

$$P = P_m \tanh\left(\frac{\alpha E}{P_m} \right) \qquad (12)$$

These models were reformulated by Jassby and Platt (1976) to express them using the same two parameters. P_m is the maximum rate of gross photosynthesis, and E is the light intensity expressed as μmol quanta m^{-2} s^{-1}. The parameter α is the slope of the initial increase in gross photosynthesis with light. It can be estimated by the same process as the net growth rate in equation (5). The above models have been successful in the study of coral reefs, phytoplankton, and macroalgae.

Growth and Nutrient Uptake

As phytoplankton grow, they consume mineral resources such as C, N, P, S, and Si. Earlier it was noted that C and S are rarely limiting in natural aquatic systems. This leaves N and P (plus Si for

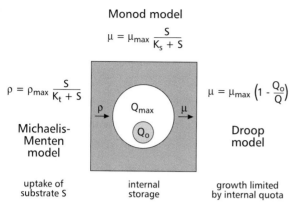

Monod model

$$\mu = \mu_{max} \frac{S}{K_s + S}$$

$$\rho = \rho_{max} \frac{S}{K_t + S}$$

$$\mu = \mu_{max} \left(1 - \frac{Q_o}{Q}\right)$$

Q_{max}

Q_o

Michaelis-Menten model

Droop model

uptake of substrate S

internal storage

growth limited by internal quota

Figure 22–19 Relationships among the various nutrient-based models of phytoplankton growth. The Michaelis-Menten model describes the process of nutrient uptake from the environment. The Droop model describes growth as a function of internal nutrient stores, while the Monod model defines growth in terms of external nutrient supply.

diatoms) as essential macronutrients that must be taken up from the environment for growth to occur. Other (micro)nutrients are essential but seldom limiting because they are needed in such small amounts.

Nutrient Uptake

The process by which nutrients are taken up from the environment and transported into the algal cell has been described by the Michaelis-Menten model, which is based on the kinetics of enzyme function. Authors differ somewhat in their choice of variables, but the following equation is widely used to describe nutrient transport into algal cells:

$$\rho = \rho_{max} \left(\frac{S}{K_t + S}\right) \qquad (13)$$

In this formulation, ρ is the velocity of the nutrient transporter or the nutrient transport rate in units of μmol of nutrient per cell per minute. If the cell is not the chosen unit for the algal population but some measure of biomass, then the units could be μmol mg^{-1} min^{-1}. Refer to the diagram of nutrient-based models of growth (Fig. 22–19). The term ρ_{max} is the maximum velocity of the nutrient transporter; ρ approaches ρ_{max} when the level of the external nutrient—the substrate concentration S—is high and the internal store of that same nutrient (Q) is low. K_t is the half-saturation constant, which equals the value of S where $\rho = 1/2 \, \rho_{max}$;

that is, where the enzyme-based nutrient transporter is half-saturated with the substrate S. S and K_t have the same units of concentration (μmol l^{-1}). Figure 22–20 illustrates the relationships between ρ, ρ_{max}, K_t and S for *Anabaena* taking up phosphorus. Note that as the concentration of phosphorus (S) increases, the measured level of nutrient uptake, ρ, first increases rapidly and essentially linearly. The curve of ρ then bends over to a plateau as ρ approaches ρ_{max}. The point where ρ reaches $1/2 \, \rho_{max}$ defines the phosphorus concentration, which is K_t. For some measured values of K_t and ρ_{max}, refer to Table 22–3, which is taken from Sandgren (1988). In this table, ρ_{max} is referred to as V_{max}. In some papers, V_{max} has a different meaning and has units of t^{-1} (min^{-1} or hr^{-1}). In this context, V_{max} is the maximum specific uptake rate and:

$$V_{max} = \frac{\rho_{max}}{Q} \qquad (14)$$

Internal Nutrient Stores

For the next steps in nutrient dynamics, refer again to the diagram (Fig. 22–19). As the nutrient transporter pumps nutrients like P or N into the cells, they go into an internal pool Q with units of μmol cell^{-1}. Q varies between two fixed levels, which are set by natural selection and vary with cell size. Q_o is the minimum cell quota or minimum stored nutrient supply, and Q_{max} is the maximum internal quota of nutrient. The alga draws off the internal storage pool as it grows.

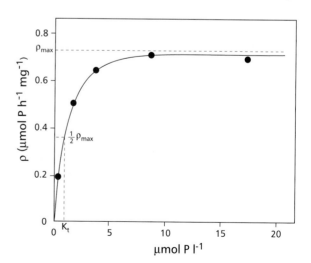

Figure 22–20 Phosphorus transport in *Anabaena*. (Based on Lampert and Sommer, 1997 ©Georg Thieme Verlag)

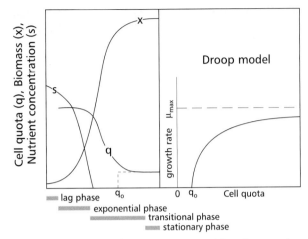

Figure 22–21 Droop model (right) and batch growth of organism X on substrate S (left). (After Lampert and Sommer, 1997 ©Georg Thieme Verlag)

Droop (1983) advanced the following equation for the relationship between growth rate and internal nutrient quota:

$$\mu = \mu_{max}\left(1 - \frac{Q_o}{Q}\right) \qquad (15)$$

Here μ is the gross growth rate or rate of reproduction with units of t^{-1} (day^{-1} or hr^{-1}). The maximum rate of reproduction is μ_{max}. Note when Q approaches Q_o, then μ is zero. When Q reaches Q_{max}, μ is approximately μ_{max}. Figure 22–21 presents graphs of the Droop model μ versus Q in the right-hand graph and in the left-hand graph changes in biomass of a hypothetical organism X as substrate S is consumed and internal cell quota Q is depleted to Q_o (or q_o). In the right-hand figure, μ does not begin until Q is greater than q_o. The value of μ then increases rapidly and levels off as Q approaches Q_{max} and μ approaches μ_{max}. In the left-hand graph, S drops away rapidly as it is entirely picked up by the nutrient transporter and joins the internal quota. Growth of X then continues as long as Q is greater than Q_o. As Q approaches Q_o, growth stops and X reaches its carrying capacity.

For actual values of Q_o and Q_{max} refer to Table 22–3 from Sandgren (1988) for chrysophycean flagellates. Measured Q_{max} values are considerably larger than Q_o. In fact, the ratio of Q_o/Q_{max} varies from about 3% to 9.5%. Larger volume cells have higher values of Q_{max} than do small cells, presumably because

they have more room inside for storage of polyphosphate. Values of μ_{max} vary over a rather narrow range. The K_s values are quite low for phosphorus among these algae.

Growth and External Nutrient Supply

Our final consideration is the process of population growth as a function of nutrient concentration in the surrounding environment. Here we must consider two types of culture: the batch culture and the continuous culture or chemostat. In the batch culture there is a fixed initial amount of nutrient or substrate S. The organism X is introduced and grows at the expense of S. In Figure 22–21, the left-hand graph shows a batch growth process for X growing on S and storing S as Q. The process of depletion of S and growth of X can be described by a Monod (1950) equation, which is similar to a Michaelis-Menten equation:

$$\mu = \mu_{max}\frac{S}{K_s + S} \qquad (16)$$

Here μ, μ_{max}, and S have been defined previously. K_s is the half-saturation constant for growth as a function of substrate S and is that value of S at which $\mu = 1/2$ μ_{max}. When S is very large relative to K_s, then $S/(K_s + S)$ is close to 1 and μ equals μ_{max}. To describe the batch growth process, we write equations for the use of S and the growth of X as follows:

$$\frac{dS}{dt} = -\mu_{max}\left(\frac{S}{K_s + S}\right)\frac{X}{Y} \qquad (17)$$

$$\frac{dX}{dt} = \mu_{max}\left(\frac{S}{K_s + S}\right)X \qquad (18)$$

In words, equation (17) says the change in S in the batch culture is equal to the consumption of S by the growth of alga X. The parameter Y is the yield coefficient for conversion of S into X. If X were expressed as cells ml^{-1} and S in μmol l^{-1}, then Y would have units of cells ml^{-1}/μmol l^{-1}. Equation (18) then says X grows as a function of S. For some practice working with the Michaelis-Menten and Monod models, refer to Question Box 22–2.

In continuous cultures, nutrients are continuously being added to a culture vessel and unconsumed nutrients and organisms continuously being washed out at a dilution rate D with units of t^{-1}. In continuous cul-

Table 22–3 Values of parameters in models of growth and phosphorus uptake for P–limited chrysophyte flagellates (from Sandgren, 1988)

species	clone	mean cell volume (μm^3)	μ_{max} (day^{-1})	K_S (μM)	V_{max} (10^{-9} μmol cell^{-1} min^{-1})	K_t (μM)	Q_{max} (10^{-9} μmol cell^{-1})	Q_o (10^{-9} μmol cell^{-1})
Dinobryon cylindricum	1	—	0.90	—	—	—	—	2.40
	5	272	0.51	0.014	—	—	18.5	1.77
	7	—	0.58	—	—	—	—	2.15
	13	290	0.75	0.021	—	—	21.0	1.87
					0.34	0.72		
Dinobryon bavaricum		80	—	—	0.10	0.11	—	—
					0.22	0.01		
Dinobryon sociale		—	—	—	2.32	0.39	—	—
Dinobryon divergens		—	—	—	—	0.10-0.27	—	—
Synura petersenii	2b	374	0.51	0.003	5.1	1.19	90.0	3.04
	7c	431	0.76	0.001	21.8	1.35	55.2	1.96
Mallomonas cratis	UW-126	1516	0.55	0.001	14.2	0.36	152.0	7.90
Mallomonas caudata	2j	10,625	0.30	—	—	—	—	—

ture, D represents a loss rate like λ, which we discussed previously with exponential growth. The equations for the same substrate and organism in continuous culture as in equations (17) and (18) are:

$$\frac{dS}{dt} = D\,(S_o - S) - \mu_{max}\left(\frac{S}{K_S + S}\right)\frac{X}{Y} \qquad (19)$$

$$\frac{dX}{dt} = \mu_{max}\left(\frac{S}{K_S + S}\right)X - DX \qquad (20)$$

In words, equation (19) says that the change in substrate concentration in the continuous culture vessel is equal to the amount delivered (DS_o), minus any residual that is carried out in the overflow ($-DS$), minus the amount consumed by growth of alga X. Equation (20) then says the change in population of X is equal to growth (μX), minus loss due to washout ($-DX$). While the equations may look formidable, they can be solved because in continuous culture, a steady state or equilibrium can be reached where growth equals losses and thus both dS/dt and $dX/dt = 0$.

If we set equation (20) equal to zero and substitute μ for $\mu_{max}\,[S/(K_s + S)]$, we get:

$$0 = \mu\,X - DX \qquad (21)$$

And consequently:

$$\mu = D \qquad (22)$$

This means that at steady state, the growth of X equals the loss rate or dilution rate, D. If we now set equation (19) equal to zero we get:

$$0 = D\,(S_o - S) - \mu\left(\frac{X}{Y}\right) \qquad (23)$$

Since $\mu = D$ at steady state, we can substitute D for μ in (23) above:

$$0 = D\,(S_o - S) - \left(\frac{DX}{Y}\right) \qquad (24)$$

And:

$$X = Y\,(S_o - S) \qquad (25)$$

which in words means the steady state concentration of X equals the amount of substrate consumed times the yield coefficient, which converts the amount of S used into units of alga X. If D is greater than μ_{max}, the population of algae cannot maintain itself against this loss rate and will wash out of the continuous culture vessel.

Continuous cultures have been a major research tool for the study of phytoplankton populations in culture because the cultures can be studied for long time periods. Single species of phytoplankton and mixtures of species in competitive and predation interactions have been examined. The equations describing species interactions in continuous culture can be

Question Box 22–2 Working with the Michaelis-Menten and Monod models of nutrient uptake and growth

Sample calculations.

1. Tilman (1976) calculated the following growth parameters for the diatoms *Asterionella formosa* and *Cyclotella meneghiniana* under both PO$_4$ limitation and SiO$_2$ limitation.

	SiO$_2$ limitation		PO$_4$ limitation	
	μ_{max} (day^{-1})	K$_s$ (µM SiO$_2$)	μ_{max} (day^{-1})	K$_s$ (µM PO$_4$)
Asterionella formosa	1.59	3.9	0.76	0.04
Cyclotella meneghiniana	1.74	1.4	0.76	0.25

Use these data in the Monod equation (16) to plot the growth rates of each alga as a function of the concentration of SiO$_2$ and separately as a function of the concentration of PO$_4$. Which alga will dominate under PO$_4$ limitation? Which will dominate under SiO$_2$ limitation? These results will be useful in a later section on nutrient competition.

Solution. We choose a convenient range of values of the nutrient concentration of 0 to 10 µM. The Monod equations are:

	SiO$_2$	PO$_4$
A. formosa	$\mu = 1.59\,[S/(3.9 + S)]$	$\mu = 0.76\,[S/(0.04 + S)]$
C. meneghiniana	$\mu = 1.74\,[S/(1.4 + S)]$	$\mu = 0.76\,[S/(0.25 + S)]$

From the graphs below, it is clear that if SiO$_2$ is limiting, *C. meneghiniana* will dominate over *A. formosa*. At every concentration of SiO$_2$ its growth rate μ is higher, and because its K$_s$ value is lower, it can reduce SiO$_2$ to levels where *A. formosa* cannot grow. Conversely, if PO$_4$ is limiting, *A. formosa* dominates *C. meneghiniana* at all levels of PO$_4$. *Asterionella formosa* can reduce the levels of PO$_4$ to exclude *C. meneghiniana* because its K$_s$ value is so much smaller. Note that as K$_s$ values become smaller, the growth curves rise more rapidly toward μ_{max} and appear more angular. These results give us an important insight into nutrient competition among phytoplankton—if two species are competing for a single nutrient such as silicate, the one with the lower K$_s$ value will always win because it can reduce that nutrient's level to the point where the other species cannot grow.

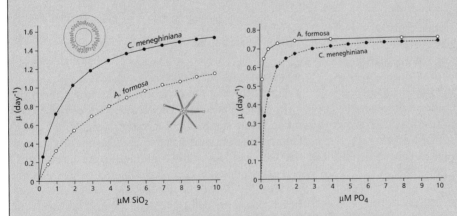

Redrawn with permission from Tilman, D. Ecological competition between algae: Experimental confirmation of resource-based competition theory. *Science* 192:463–465 ©1976 American Association for the Advancement of Science

2. The following data show the process of phosphorus uptake by *Cyclotella meneghiniana* as a function of phosphate concentration (adapted from Tilman and Kilham, 1976).

Question Box 22–2 *continued*

S (µM PO$_4$)	ρ (µM PO$_4$ cell^{-1} hr^{-1} x 10^{-9})
1.00	2.87
1.38	3.40
2.08	5.21
2.86	3.30
4.00	4.65
4.76	4.48
6.66	5.05

Determine the parameters K_t and ρ_{max} in the Michaelis-Menten equation (13) for this set of data. Values of S represent the initial phosphate concentrations. Uptake rates were measured by following the removal of PO$_4$ and the increase in cell numbers over several hours.

Solution. In research applications, the final parameters are determined by nonlinear regression techniques on a computer. You can, however, obtain good estimates of the parameters by a linear transformation of the Michaelis-Menten equation, where ρ and S are the variables and ρ_{max} and K_t are constants. We use the algebraic axiom that if x = y then it is also true that 1/x = 1/y. The first step is to invert equation 13 to yield:

$$\frac{1}{\rho} = \frac{K_t + S}{\rho_{max}\, S}$$

We can now separate the two terms on the right-hand side to give us:

$$\frac{1}{\rho} = \frac{K_t}{\rho_{max}}\left(\frac{1}{S}\right) + \frac{S}{\rho_{max}\, S}$$

Canceling out the S in the far-right term we finally derive:

$$\frac{1}{\rho} = \frac{K_t}{\rho_{max}}\left(\frac{1}{S}\right) + \frac{1}{\rho_{max}}$$

This is the equation of a straight line of the form y = mx + b, where y = 1/ρ, x = 1/S, the slope m = K_t/ρ_{max} and the intercept b = 1/ρ_{max}. To derive this equation for our example, we have only to take the inverses of the values of S and ρ, plot them on graph paper, and take the linear regression of 1/ρ on 1/S. The intercept will be the inverse of ρ_{max}, and the slope will give us K_t. The inverses are:

x = 1/S	y = 1/ρ
1.00	0.348
0.72	0.294
0.48	0.192
0.35	0.303
0.25	0.215
0.21	0.223
0.15	0.198

If we use all seven points, the resulting linear equation is Y = 0.1529 (X) + 0.1842 with a correlation coefficient of 0.78 (refer to the plot below). The maximum uptake rate ρ_{max} is the inverse of 0.1842 or 5.4 × 10^{-9} µM PO$_4$ cell^{-1} hr^{-1}. K_t = 0.1529 ρ_{max}, so K_t = 0.83 µM PO$_4$. These values are quite close to those reported by Tilman and Kilham (1976) from nonlinear regression (ρ_{max} = 5.1 × 10^{-9} and K_t = 0.8). In most cases real data are much noisier than this set, and linear regression can at most provide a starting point for a nonlinear regression routine.

continued→

Question Box 22–2 *continued*

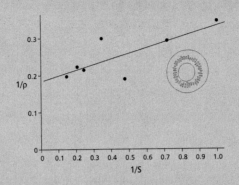

Problems.

1. Tilman and Kilham (1976) measured the uptake of SiO_2 by the diatom *Cyclotella meneghiniana*. Use the following data adapted from their paper to estimate the uptake parameters ρ_{max} and K_t. How do the dynamics of SiO_2 uptake compare to those of PO_4 calculated in the above sample problem?

SiO_2 (μM)	ρ μM cell^{-1} hr^{-1} ($\times 10^9$)
0.893	1.66
2.270	4.29
3.860	5.77
7.270	6.72
15.000	8.93
27.300	12.56

2. Tilman, et al. (1981) examined silicate and phosphate growth kinetics of four Lake Michigan diatoms to test his resource competition theory. The following parameters were obtained for *Fragilaria crotonensis* and *Synedra filiformis* under silicate-limiting conditions.

	μ_{max} (day^{-1})	K_s (μmol l^{-1})
Fragilaria	0.62	1.5
Synedra	1.11	19.7

Plot the curves of the Monod growth model (16) for each species. Based on these graphs, what SiO_2 conditions would favor growth of each species?

3. Tilman and Kilham (1976) also measured the growth of *Asterionella formosa* as a function of SiO_2 concentration. Use the following data adapted from their paper to estimate the growth parameters μ_{max} and K_s in the Monod equation (16). *Hint*: the same procedure can be used as with the Michaelis-Menten model of uptake dynamics.

SiO_2 (μM)	μ (doublings day^{-1})
3.5	0.508
3.7	0.515
6.5	0.690
7.0	0.700
10.2	0.783
13.2	0.813
18.6	0.833
22.0	0.933

Note: in the actual data set, the values of μ at low SiO_2 levels are prone to large variation. Therefore the values of $1/\mu$ are also subject to high variation and were not used in the regression. Answers are found at the end of the chapter.

solved for steady-state equilibrium values, and the solutions of the equations can be compared to observations. Unfortunately, this research also has led to the assumption that steady-state equilibrium models can be applied to natural aquatic systems. No matter how attractive the equations may be, however, this does not mean steady-state models work in the real world.

Growth and Uptake of Organic Carbon

The previous two sections have covered autotrophic processes of growth, that is, processes by which algae generate their own food through photosynthesis coupled with uptake of inorganic nutrients. But phytoplankton also exhibit two common heterotrophic processes of growth, consumption of particulate organic matter and uptake of dissolved organic carbon (DOC). Such versatile organisms are called mixotrophic because they have the capability for a mixture of feeding processes. The chrysophycean *Ochromonas* is a good example of a mixotrophic flagellate. For more details refer to Chapters 1 and 13 in the section on variations in algal nutrition. Some DOC-users have chlorophyll (*Euglena*) and others not (*Prototheca*). For more information concerning DOC use by algae, refer to Chapter 2.

Loss Processes

What are loss processes? Recall our earlier discussion of net growth rate being equal to the gross growth rate or rate of reproduction minus the death rate or the sum of all loss processes (equation 3). Loss processes are all those factors that remove or displace phytoplankton in the aquatic environment. Following the formulation of Lampert and Sommer (1997), loss processes can be described by the expression:

$$\lambda = \gamma + \sigma + \chi + \delta + \pi + \omega \quad (26)$$

Here the Greek letter λ stands for losses and the remaining Greek letters for specific types of loss processes. Greek gamma (γ) stands for grazing—the loss of algal cells to herbivores such as ciliates, rotifers, and crustaceans (see Chapter 3). The letter sigma (σ) stands for sedimentation—the sinking of nonmotile algal cells out of the euphotic zone. Losses due to com-

petition are represented by the letter chi (χ). Delta (δ) stands for death or physiological mortality, and pi (π) is parasitism. The final symbol, omega (ω), represents loss due to washout.

Perennation

We will discuss each of these processes, but first we need to mention an adaptation used by many groups of phytoplankton to reduce losses altogether, namely **perennation**. Perennation is the formation of some sort of resting stage that allows the algal population to avoid a period of adverse environmental conditions. Many cyanobacteria form akinetes, a type of asexual spore that develops from vegetative cells (Chapter 6). Akinetes may be resuspended into the

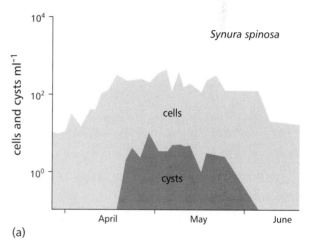

(a)

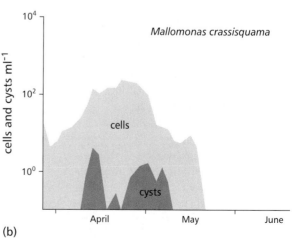

(b)

Figure 22–22 Cells and cysts per ml of two planktonic chrysophyceans from Egg Lake, Washington, U.S.A., in spring 1976. (After Sandgren, 1988)

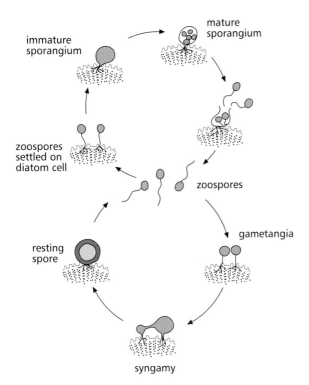

immature
sporangium

mature
sporangium

zoospores
settled on
diatom cell

zoospores

gametangia

resting
spore

syngamy

Figure 22–23 Life cycle of the chytrid *Zygorhizidium planktonicum*. (Adapted from van Donk and Ringelberg, 1983 by permission of Blackwell Science, Ltd.)

water column where they will germinate (Reynolds, 1972). One study found that akinetes could survive and germinate after more than 60 years buried in sediments (Livingstone and Jaworski, 1980). *Microcystis*, however, can survive as vegetative cells on sediment surfaces with neither light nor oxygen for several years (Reynolds, et al., 1981). Some centric diatoms such as *Melosira* and *Stephanodiscus* form a resting vegetative cell in which the protoplast contracts into a ball within the frustule; these resting cells can survive on the sediments for several months to a year (see also Chapter 12).

Green algae often produce resting zygotes (Chapters 20 and 21), and dinoflagellates form cysts (Chapter 11). The dinoflagellate *Ceratium hirundinella* forms cysts in response to declining nutrients; this is an example of control by an external factor or exogenous control. Chrysophyceans, on the other hand, produce a resting spore or stomatocyst whose formation is under endogenous (internal) control. The production of stomatocysts is directly dependent on the cell density (Fig. 22–22). Most chrysophyceans are only present in the plankton for a few weeks per year

and produce their resting cysts when the population is most abundant.

Perennation stages play important roles in both the freshwater and marine environments. In oceans, however, if perennation stages sink, they would need to do so over shallow continental shelves to have a chance for survival. A cyst sinking into the abyssal depths of the open oceans is a lost cyst.

Mortality and Washout

A few of the loss terms can be dealt with very briefly. Physiological mortality (δ) can only be determined if "bodies," such as diatom frustules can be recovered. Most algal cells are either eaten or sink out of the mixing layer. Similarly, washout (ω) is not often a significant factor for phytoplankton. Most algal populations are sufficiently numerous and dispersed that washout due to a flood event is unlikely to reduce the population significantly.

Parasitism

Parasitism (π) upon phytoplankton populations has been relatively little studied. Viruses have been recovered from cyanobacteria and have been shown to lyse them in lab cultures. Van Etten, et al. (1991) surveyed viruses and viruslike particles in eukaryotic algae, especially in symbiotic *Chlorella* species. In the marine environment, Sieburth, et al. (1988) reported viruses in the pelagophycean picoplankter *Aureococcus anophagefferens*, and Suttle, et al. (1990) showed that viruses could reduce primary productivity in marine waters.

Fungal parasites of algae are better known. Most belong to the phylum Chytridiomycota. Free-swimming uniflagellate zoospores seek out and attach to host cells (Fig. 22–23) A mycelial thread penetrates the host cell and supplies nutrients to the enlarging zoospore, which becomes a sporangium. Zoospores are released from the sporangium to complete the cycle. The host cell is killed by the infection. Van Donk (1989) followed the populations of *Asterionella formosa* and their infection with the chytrid *Zygorhizidium planktonicum* in Lake Maarsseveen (Netherlands) during the spring from 1978 to 1982 (Fig. 22–24). Heavy infection of *A. formosa* allowed the diatoms *Fragilaria crotonensis*, *Stephanodiscus astraea*, and *S. hantzschii* to become abundant. Some small protozoa are known to attack large colonial green algae as parasites causing heavy mortality (Canter, 1979). Little is known about the dynamics of most parasites acting on planktonic algae or the impact they may have on phytoplankton communities. This is one area of aquatic ecology in which it is not a cliché to say that more research is needed. For additional information, refer to Chapter 3.

Sedimentation (σ)

If an algal cell sinks below the euphotic zone, it will be lost unless it functions as a perennation stage. Many phytoplankton, however, can minimize sedimentation losses by controlling their position in the water column.

Swimming and buoyancy

Flagellates can swim at speeds sufficient to maintain their position in the water column. The average swimming speeds of flagellated phytoplankton are about ten times greater than the average sinking rate, which is on the order of 0.5 m day^{-1} (Sournia, 1982). Flagellates may perform extensive daily vertical migrations, often moving from upper waters by day to deeper waters at night. *Peridinium cinctum* migrates 8–10 m daily in Lake Kinneret, Israel (Berman and Rodhe, 1971), and *Cryptomonas ovata* carries out a daily migration of 5 to 7.5 m in Finstertaler See, Austria (Tilzer, 1973). In the oceans, dinoflagellates may migrate to depths of 10–20 m daily (Eppley, et al., 1968). The largest migration in freshwaters has been reported from Lake Cahora Bassa, Mozambique. There *Volvox* sp. migrates as much as 20 m into the deeper strata at descent rates of 1.8 to 3.6 m hr^{-1} and returns every day. *Volvox* require high levels of phosphorus for growth. Its half-saturation constant (K_s) for P-limited growth is on the order of 19 to 59 μg P l^{-1} (Senft, et al., 1981). In the euphotic zone of Lake Cahora Bassa, P-levels are undetectable (Fig. 22–25). Thus *Volvox* may be migrating down to a depth of 20 or more meters to take up phosphorus at night and then return to upper waters for photosynthesis and growth in the day (Sommer and Gliwicz, 1986).

Many cyanobacteria also control their position in the water column by adjusting the formation of intracellular gas vacuoles and regulating cellular ballast. See Chapter 6 for a discussion of buoyancy regulation in cyanobacteria. Genera that generate nuisance blooms, such as *Microcystis*, *Anabaena*, and *Aphanizomenon*, use such buoyancy mechanisms to adjust their vertical position. During wind-

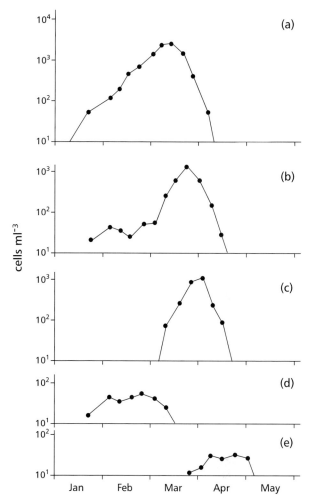

Figure 22–24 Population of *Asterionella formosa* in spring of 1982 in the surface waters of Lake Maarsseveen, Netherlands, divided into five categories. (a) uninfected *A. formosa* cells, (b) cells with zoospores of *Zygorhizidium planktonicum*, (c) *A. formosa* infected with sporangia, (d) cells infected with thick-walled spores, (e) cells with resting spores of *Z. planktonicum*. Most of the *A. formosa* population was infected by late March (After van Donk, 1989 ©Springer-Verlag)

less periods, daily rhythms of vertical migration have been observed (Reynolds, et al., 1987). Vertical migration rates can be high. In Lake George, Uganda, *Microcystis aeruginosa* migrated at speeds greater than 3 m hr^{-1} (Ganf, 1975). *Aphanizomenon flos-aquae* reached speeds of 40 cm hr^{-1} to 2.75 m hr^{-1} in the Chowan River of North Carolina (Paerl and Ustach, 1982). *Oscillatoria* also uses the same buoyancy control mechanisms to maintain stable

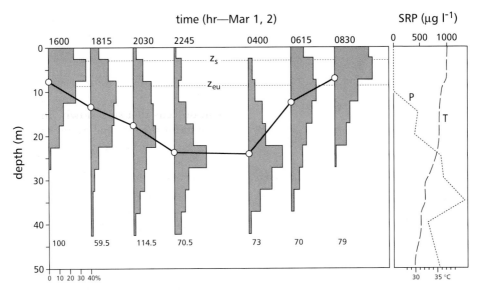

Figure 22–25 Evening descent and morning ascent of *Volvox* in Lake Cahora Bassa on March 1–2, 1983. The upper horizontal axis gives 24-hour time. The column diagrams show the depth distribution of *Volvox* in percent of total population. The numbers at the bottoms of the column diagrams give the areal population density in units of 10^3 colonies m^{-2}. The solid lines track the median depth. Z_s marks Secchi-disk transparency in m, and Z_{eu} the depth of the euphotic zone. Note soluble reactive phosphorus (P) is unmeasureable within the euphotic zone. The lake is quite hot by comparison to north temperate lakes (> 30° C). (From Sommer and Gliwicz, 1986)

populations in the metalimnion of lakes for periods of weeks (Fig. 22–7).

Buoyancy regulation and vertical positioning work best under stable water conditions. Turbulent mixing causes cyanobacteria to circulate along with other phytoplankton. Prolonged turbulent mixing can shift dominance away from cyanobacteria and toward other algae such as greens (Harris, et al., 1980).

Sinking

Phytoplankton that can neither swim by flagella nor regulate their buoyancy with gas vacuoles are subject to sinking through the water column. Most phytoplankton are only slightly more dense than water at 1.02 to 1.05 g ml^{-1}, but diatoms have densities around 1.3 g ml^{-1}. The silicate making up their frustules is very dense at about 2.6 g ml^{-1}. Only a few substances within algal cells are less dense than water—lipids have densities around 0.86 g ml^{-1} and the gas vacuoles of cyanobacteria have values around 0.12 g ml^{-1}.

The sinking velocity of a falling sphere was originally described by Stokes' law. Since phytoplankton are living rather than inert and are often shaped other than as spheres, Ostwald's modification of Stokes' law was formulated to adjust sinking rates for the form

resistance of nonspherical phytoplankton cells. The modified Stokes' law has the form:

$$v_s = \left(\frac{2}{9}\right) g \, r_s^2 \, (q' - q) \, \upsilon^{-1} \, \phi^{-1} \qquad (27)$$

Here, v_s is the sinking velocity in m s^{-1}, g is the gravitational acceleration of the earth (9.8 m s^{-2}), and r_s is the radius (in meters) of a sphere of volume equivalent to that of the algal cell. The term q' is the density of the algal cell expressed in kg m^{-3}, and q is the density of the fluid medium. For water, q has the value 1000 kg m^{-3}. Thus, $(q' - q)$ is the difference between the density of the algal cell and the surrounding water. This difference is positive for most algae, which are more dense than water, but it may be negative for cyanobacteria with gas vacuoles or an alga with a high intracellular concentration of lipids. The term υ (Greek upsilon) is the viscosity of water, which is expressed in units of kg m^{-1} s^{-1}. Water is more viscous at lower temperatures, and thus $\upsilon = 1 \times 10^{-3}$ kg m^{-1} s^{-1} at 20° C but 1.8×10^{-3} kg m^{-1} s^{-1} at 0° C (Lampert and Sommer, 1997). The final term ϕ (Greek phi) stands for form resistance and is dimensionless. Form resistance accounts for the fact that most algae have nonspherical shapes and may possess horns and spines.

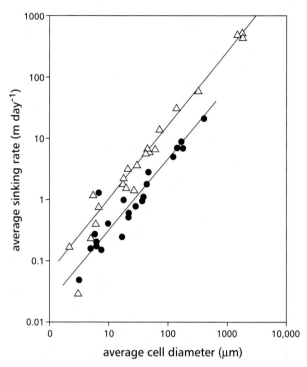

Figure 22–26 Data for sinking rates of phytoplankton. Triangles represent senescent cells, closed circles, actively growing cells. (Based on Smayda, 1970, in Harris, 1986 with kind permission of Kluwer Academic Publishers)

In words, equation (27) says that the sinking velocity of an algal cell is proportional to the gravitational force of the earth times a measure of the size of the cell (the radius squared) times the difference between the densities of the cell and its fluid medium. According to the equation, larger cells or colonies or more dense cells such as diatoms should sink more rapidly. Conversely, the more viscous the water and the greater the form resistance of the algal cell, the slower the sinking velocity. Some representative data and a sample calculation are presented in Question Box 22–3 together with a few practice problems.

Observed sinking rates of living algal cells and colonies frequently do not correlate well with the predictions of the Ostwald modification of Stokes' law. Living cells do not sink as rapidly as dead or senescent cells of the same species (Fig. 22–26). Reynolds (1984) found that the observed sinking velocity of killed cells of *Stephanodiscus astraea* was highly correlated (r = 0.98) with the radius of a sphere having a volume identical to that of the killed cells (r_s), the parameter used in the Stokes' equation. There was no correlation between r_s and the observed

sinking rates of living cells of *S. astraea*. Thus living cells do not behave in a water column as the inert spheres of Stokes' Law but dead cells do, with allowance for form resistance.

Phytoplankton that sink rapidly require more rapid vertical mixing to remain in the water column than do cells that sink slowly. In part this fact may account for the seasonal occurrence of large diatoms and desmids during periods of strong vertical mixing.

Phytoplankton have a number of adaptations to reduce sinking rates. Small cell size is the most obvious one. Large cells are not simply bigger small cells. Larger diatom species have a reduced surface-to-volume ratio compared to small diatoms. This means they have proportionally less siliceous frustule material and are thus less dense than small diatoms. Some large marine diatoms such as species of *Rhizosolenia* and *Ethmodiscus* are actually positively buoyant. Among *Rhizosolenia* spp., the rate of ascent was highly correlated with cell diameter (Fig. 22–27). The mechanism of this positive buoyancy is still unclear, but the ascent rates can be quite rapid with upper limits approaching 7–8 m hr^{-1}.

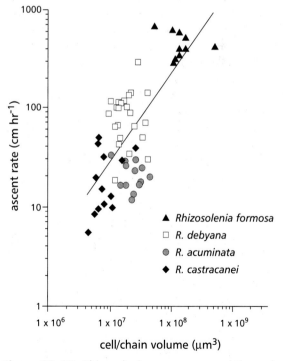

Figure 22–27 *Rhizosolenia* spp. can ascend through the water column. Larger forms rise faster than smaller forms. (After Moore and Villareal, 1996)

Question Box 22–3   Working with the Ostwald modification of Stokes' law

The table below shows some representative algal species, their cell or colony volumes, the radius of a sphere with a volume equal to that of the algal cell or colony, and the algal density.

Algal species	volume (μm^3)	radius (μm)	density (kg m^{-3})
Stephanodiscus astraea	5,930[a]	11.23	1091[a]
Asterionella formosa	5,160[a]	10.72	1130[a]
Chlorella vulgaris	30[b]	1.9	1095[c]
Microcystis aeruginosa	4.2 x 10^6 [a]	100.1	999.4[d]

[a] Reynolds, 1984
[b] Bellinger, 1974
[c] Oliver, et al., 1981
[d] Reynolds, et al., 1981

Sample calculation for Stephanodiscus astraea *at 20° C.* From the text, $g = 9.8$ m s^{-2}, $q = 1000$ kg m^{-3}, and $\upsilon^{-1} = 10^3$ m s kg^{-1}, the inverse of υ. The value of q' is given in the above table. Convert the table values of radius in μm to meters by multiplying by 10^{-6} m μm^{-1}. For the moment we do not concern ourselves with ϕ. Then:

$$v_s = \left(\frac{2}{9}\right) 9.8 \text{ m s}^{-2} (11.23 \times 10^{-6} \text{ m})^2 (1091 - 1000 \text{ kg m}^{-3}) 10^3 \text{ m s kg}^{-1} \phi^{-1}$$

Which becomes:

$$v_s = 2.177 \text{ m s}^{-2} (126.1 \times 10^{-12} \text{ m}^2) (91 \text{ kg m}^{-3}) 10^3 \text{ m s kg}^{-1} \phi^{-1}$$

And finally:

$$v_s = 24.98 \times 10^{-6} \text{ m s}^{-1} \phi^{-1}$$

If we multiply this result by 3600 s hr^{-1}, we find that *S. astraea* is predicted to sink at the rate of 0.09 m hr^{-1}. If we multiply the result above by 10^6 μm m^{-1} instead, we get $v_s = 24.98$ μm s^{-1}. Reynolds (1984) reported that the actual sinking rate of killed *S. astraea* cells was 27.62 μm s^{-1}. Thus, ϕ is 24.98/27.62 or 0.904.

Practice calculations. Work the following problems with a pocket scientific calculator. Answers are given at the end of the chapter.

1. Calculate the predicted sinking velocity of an 8-celled colony of *Asterionella formosa* according to the Ostwald modification of Stokes' law. Reynolds (1984) reports that the actual sinking velocity of this algal colony is 7.33 μm s^{-1}. What therefore must be the value of ϕ?

2. Calculate the sinking velocity of *Chlorella vulgaris* using equation (27). Assume for this spherical alga a ϕ of 1.0. What does this tell you about the effect of size on sinking rate?

3. Calculate the ascending velocity of a large buoyant colony of *Microcystis aeruginosa* from the data in the table. Assume the colonies are essentially spherical, so $\phi = 1.0$. How does this rate compare to some of the upward migration rates reported in the text for cyanobacteria?

4. In an earlier section on algae in space research (Chapter 4), it was suggested that algae may one day be introduced to Mars as part of a terraforming effort. If *Stephanodiscus astraea* were introduced to a reformed Martian lake, how fast might it sink through a Martian lake's water column? *Hint*: the gravitation force of Mars is 0.38 of that of Earth.

Many large phytoplankton cells and colonies, including the desmid *Staurastrum* and cyanobacteria such as *Coelosphaerium* and *Microcystis*, possess a mucilaginous sheath or matrix in which the cells are embedded. Since the density of mucilage is close to that of water, a mucilaginous sheath can reduce the overall cell or colony density to levels closer to that of water, but it cannot make a cell or colony buoyant. At the same time, adding mucilage can increase cell or colony diameter, which should (according to Stokes' law) increase sinking rate. Reynolds (1984) has shown that there is a trade-off of effects such that adding mucilage decreases sinking up to a point, after which further additions cause sinking rate to rise again. Mucilage may serve other purposes than decreasing sinking rates, as will be discussed later.

A further adaptation of phytoplankton to sinking is contained in the Ostwald modification of Stokes' law as the parameter ϕ for form resistance. According to the original Stokes' equation, larger cells should sink more rapidly. But if larger cells or colonies depart from a spherical shape, they acquire a form resistance, which slows their sinking rates. Thus many large desmids like *Staurastrum* have long arms extending out from the cell, and dinoflagellates such as *Ceratium* have long, straight or curved horns (Fig. 22–1). Conway and Trainor (1972) showed that spine-bearing *Scenedesmus* spp. sink less rapidly than species without spines. Walsby and Xypolyta (1977) removed chitinous fibers from the marine diatom *Thalassiosira weissflogii* with the enzyme chitinase and observed that fiberless cells sank twice as fast as cells with fibers, although the fibers were more dense than the cells. According to Stokes' law, colonies should sink faster because they have larger equivalent spherical volumes than their component cells. As cells are added to colonies, the predicted sinking rate v_s from Stokes' law should increase nearly exponentially. Reynolds (1984) showed, however, that many diatom colonies increase their form resistance by adding cells to such an extent that their actual sinking rates approach a constant rate. In *Asterionella formosa*, the sinking velocity increased to a constant at 6–8 μm s^{-1} as the number of cells reached eight to nine cells per colony. Beyond this number, additional cells filled in the colony, decreased the form resistance, and increased the sinking rate. Thus an eight to nine cell colony appears to be optimal to maximize size and minimize sinking rate (Reynolds, 1984).

Recently form resistant shapes have been interpreted as a response to grazing pressure. Spined *Scenedesmus* are less taken by certain grazers than non-spiny species of the same size. In nearshore marine waters, long-spined *Thalassiosira* are not consumed by microzooplankton including ciliates, while non-spiny *Thalassiosira* are readily eaten. The long-spined diatoms can be taken by crustacean zooplankton (Gifford, et al., 1981). Phytoplankton with mucilaginous sheaths may be rejected by grazers because they are too big to be eaten. Porter (1977) has shown that some algae with mucilaginous sheaths may pass through the guts of grazers unharmed. These species may even benefit from the gut passage by acquiring nutrients during the process. There seems to be little reason to claim that features such as horns, arms, spines, or mucilaginous sheaths are either flotation devices or defense mechanisms exclusively. They may well have arisen in diverse forms because they serve more than one useful function simultaneously.

Competition

No subject in phytoplankton ecology generates as much controversy and passionate rhetoric as the concept of competition. The idea that competition plays a central role in phytoplankton ecology derives from early ecological theory. Darwin's theory of natural selection proposed that natural populations deplete their resources and press against their carrying capacity. The research of Gause in the 1930s later showed that if two species were grown together in the laboratory, the natural outcome would be depletion of resources and competition in which one of the two species would eventually be eliminated. Gause's work then led to the competitive exclusion principle and niche theory, and their application as the dominant paradigm in the natural world. But the application of this paradigm to phytoplankton communities led to a paradox, as recognized by Hutchinson (1961). If phytoplankton species were all at or near their carrying capacity and competition were a major force in community structure, how could one explain the coexistence of 50 to 100 species in a milliliter of water under the same apparent conditions? That paradox has led to a great deal of research in an effort to circumvent the competitive exclusion principle and allow coexistence of so many species in an homogeneous environment.

Lotka-Volterra model

The first attempt to formulate mathematically the process of competition between two species was made by Volterra in 1926 and by Lotka in 1932. Their efforts have come down to us as the Lotka-Volterra competition equations. They are structurally very similar to the logistic growth equation presented earlier:

$$\frac{dN_1}{dt} = \frac{r_1 N_1 (K_1 - N_1 - \alpha N_1)}{K_1} \qquad (28)$$

$$\frac{dN_2}{dt} = \frac{r_2 N_2 (K_2 - N_2 - \beta N_2)}{K_2} \qquad (29)$$

As in the logistic equation, N_1 and N_2 are the population densities of species 1 and species 2 in units such as cells ml^{-1} or mg ml^{-1}. K_1 and K_2 are the carrying capacities of the two species expressed in the same units, and r_1 and r_2 represent the net per capita growth rates of the two species with units of days^{-1} or hr^{-1}. The only new elements are the terms αN_2 and βN_1, which represent the competitive interactions between species 1 and species 2. In the logistic equation, as species 1 grows, the increasing density of its own population slows down its population growth rate as N_1 approaches K_1. How does competition by species 2 affect species 1? Species 2 acts as if it were additional members of species 1 and thus slows down the population growth rate of species 1 faster than species 1 would alone. If α were equal to 1, species 2 would act exactly like additional members of species 1. If α is less than 1, then species 2 has less impact on species 1 than members of species 1 have on each other. If α is greater than 1, individuals of species 2 have a more severe impact on species 1 than does its own population. The same arguments apply to β and species 1 impacting species 2. The parameters α and β are the competition coefficients. Alpha (α) translates units of species 2 into equivalent units of species 1, while beta (β) translates species 1 into units of species 2. Another way to look at these equations is that species 2 acts on species 1 as a loss process; any growth of species 2 results in a reduction in growth of species 1 by the amount αN_2.

The outcome of competition between the two species is determined by the values of the carrying capacities and competition coefficients. If we set both equations equal to zero such that $dN_1/dt = dN_2/dt = 0$, the net growth rates and the carrying capacity terms in the denominators disappear, leaving:

$$N_1 = K_1 - \alpha N_2 \quad (30)$$

$$N_2 = K_2 - \beta N_1 \quad (31)$$

These equations represent two straight lines of slopes α and β. If the x-axis represents N_1 and the y-axis N_2,

then equation (33) intercepts the x-axis (when $N_2 = 0$) at $N_1 = K_1$ and the y-axis (when $N_1 = 0$) at $N_2 = K_1/\alpha$. Equation (34) intercepts the x-axis at $N_1 = K_2/\beta$ and the y-axis at $N_2 = K_2$. These lines are known as zero growth isoclines. We will have further use for zero growth isoclines in the following section on mechanistic models of competition.

Four possible arrangements of these two lines are shown in Figure 22–28. These graphs represent four possible types of competitive interaction between two species. The arrows in the figures indicate the direction in which the populations of each species will move over time. In Figure 22–28a, species N_1 wins in competition against species N_2. In the region between the two isoclines, species 1 increases but species 2 decreases in population size. The arrows point downward to the right, and they terminate when species 1 reaches K_1.

The case where species 2 displaces species 1 is shown in Figure 22–28b. Here the zero growth isocline for species 2 is above that of species 1. Thus, in the region between the two isoclines, species 2 can still increase but species 1 will decrease in population size. The arrows now point upward and to the left and will end at K_2, the carrying capacity for species 2, which is the stable equilibrium point.

The third case shown in Figure 22–28c is somewhat unusual. The two zero growth isoclines intersect, and the point of intersection is an equilibrium point. But the equilibrium is unstable. Any perturbation will drive the system away from equilibrium.

The last case in Figure 22–28d is of more interest. Here again the two zero growth isoclines intersect, but in this case the intersection is a stable equilibrium. As indicated by the directions of the arrows, any perturbation that moves the populations of both species away from the equilibrium point, causes changes in the populations so as to drive the species back to that equilibrium.

How well do these equations represent real interactions between organisms? The Russian microbiologist G. F. Gause (1934) was the first to investigate the Lotka-Volterra equations experimentally using microbial systems. The single set of experiments for which Gause is most famous are the competition experiments with the ciliate protozoans *Paramecium caudatum* and *Paramecium aurelia*, which can be easily distinguished under a microscope by the differences in their lengths. Gause grew these ciliates in 5 ml volumes of physiological salt solution with bacteria as food. Every day, he withdrew 0.5 ml for counting and replaced it with

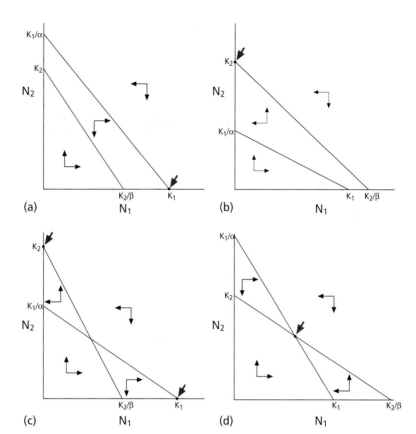

Figure 22–28 Zero growth isocline analysis of the Lotka-Volterra model of competition between two species. The outcome depends on the relative values of K_1, K_2, K_1/α, and K_2/β. Large arrows indicate final equilibrium. (a) Species 1 wins at K_1, (b) Species 2 wins at K_2, (c) unstable equilibrium where either species 1 or 2 may win, (d) stable equilibrium with species 1 and 2 coexisting.

fresh solution. The results of one such experiment, which was run for 16 days, is shown in Figure 22–29. In every experimental run, *P. aurelia* always drove down the population of *P. caudatum*. But note that *P. aurelia* had not eliminated *P. caudatum* after 16 days. These observations became the key data in the establishment of the competitive exclusion principle and niche theory, and their application to all of ecological theory. In fact, these were the only solid data for a competitive exclusion principle.

Despite the historical significance of the Lotka-Volterrra equations, they have a serious drawback. They cannot be used to make predictions about the outcome of competition. A researcher must actually perform a competition experiment, derive values for the competition coefficients and then fit the Lotka-Volterra equations to the data. The Lotka-Volterra equations say nothing about the mechanism of competition—that is, for what are the two species competing?

Tilman's mechanistic model

Tilman (1976) was the first to address this problem directly by adapting the Monod equation for nutrient-limited growth to the two species competi-

$$\frac{dS}{dt} = D\,(S_o - S) - \mu_{m1}\left(\frac{S}{K_{S1} + S}\right)\left(\frac{N_1}{Y_1}\right) - \mu_{m2}\left(\frac{S}{K_{S2} + S}\right)\left(\frac{N_2}{Y_2}\right) \qquad (32)$$

$$\frac{dN_1}{dt} = \mu_{m1}\left(\frac{S}{K_{S1} + S}\right)N_1 - DN_1 \qquad (33)$$

$$\frac{dN_2}{dt} = \mu_{m2}\left(\frac{S}{K_{S2} + S}\right)N_2 - DN_2 \qquad (34)$$

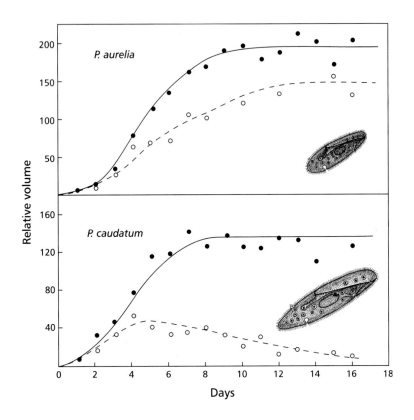

Figure 22–29 Growth of *Paramecium caudatum* and *P. aurelia* in separate and mixed culture with bacteria as food. Sample removal and replacement effectively provided a dilution rate of D = 0.1 day⁻¹. Solid lines—growth in separate culture. Dashed lines—growth in mixed culture. Cells counted in 0.5 ml samples. Relative volume calculated by assuming volume of *P. caudatum* equal to 1.0, and *P. aurelia* equal to 0.43 of volume of *P. caudatum*. (After Gause, 1934. Drawings from Jahn, T. L., E. C. Bovee, and F. F. Jahn. 1979. *How to Know the Freshwater Protozoa*. McGraw-Hill. Reproduced with permission of the McGraw-Hill Companies)

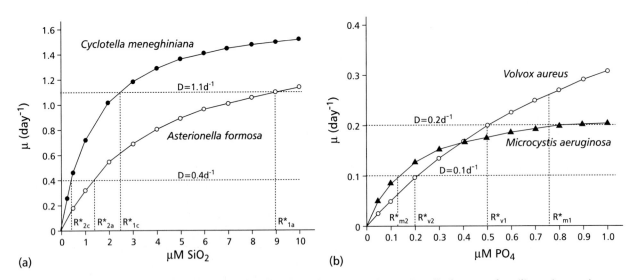

Figure 22–30 (a) Competition between *Cyclotella meneghiniana* and *Asterionella formosa* for silicate in continuous culture. *Cyclotella meneghiniana* always displaces *A. formosa* because it can reduce the level of silicate below that at which *Asterionella* can grow, at any dilution rate (R^*_c always < R^*_a). (b) Competition between *Volvox aureus* and *Microcystis aeruginosa* for phosphate in continuous culture. *Volvox* wins above the point of intersection of the curves, *Microcystis* wins below the intersection. (a: Parameters from Tilman, 1976; b: Kinetic constants from Holm and Armstrong., 1981; Senft, et al., 1981)

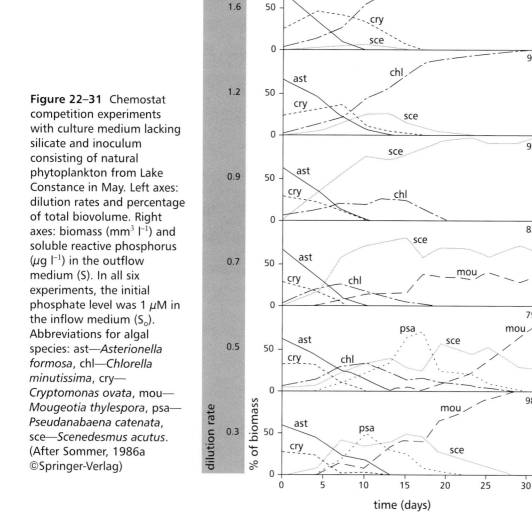

Figure 22–31 Chemostat competition experiments with culture medium lacking silicate and inoculum consisting of natural phytoplankton from Lake Constance in May. Left axes: dilution rates and percentage of total biovolume. Right axes: biomass (mm³ l⁻¹) and soluble reactive phosphorus (µg l⁻¹) in the outflow medium (S). In all six experiments, the initial phosphate level was 1 µM in the inflow medium (S₀). Abbreviations for algal species: ast—*Asterionella formosa*, chl—*Chlorella minutissima*, cry—*Cryptomonas ovata*, mou—*Mougeotia thylespora*, psa—*Pseudanabaena catenata*, sce—*Scenedesmus acutus*. (After Sommer, 1986a ©Springer-Verlag)

tion problem for algae. In the simplest case, assume that two species are competing for a single nutrient in a continuous culture. Figure 22–30a (from Tilman, 1976) shows the growth curves for the diatoms *Cyclotella meneghiniana* and *Asterionella formosa* in continuous culture under silicate-limited conditions. The equations for the growth of these two species and the consumption of the limiting nutrient (SiO_2) in chemostat culture are given in equations 32 through 34 below. These equations are simply an extension of the equations presented earlier for growth of a single species in continuous culture. In words, these equations state that the concentration of nutrient in the chemostat equals the amount flowing in (DS_o) minus the unused nutrient flowing out

($-DS$) minus the consumption of limiting nutrient by the growth of each species. The growth of each species occurs according to the Monod model of nutrient-limited growth (μN) minus the loss of cells due to washout ($-DN$). At equilibrium, $D = \mu$, and *C. meneghiniana* always wins in competition for silicate with *A. formosa* over the entire range of dilution rates because it can always reduce the nutrient level below that at which its competitor can grow (Fig. 22–30a). R^*_c is always less than R^*_a at any D. Their Monod growth curves never cross.

In Figure 22–30b, however, the growth curves for *Volvox aureus* and *Microcystis aeruginosa* in continuous culture do cross. Now if the dilution rate is high ($D = 0.2$ day⁻¹), *V. aureus* has a higher growth rate

Table 22–4 Values of R* for *A. formosa* and *C. meneghiniana*

| | nutrient required at D = 0.25 day^{-1} (R*) | |
	PO$_4$ (µM)	SiO$_2$ (µM)
A. formosa	0.01	1.9
C. meneghiniana	0.20	0.6

than *M. aeruginosa* and can reduce the level of phosphate to R*$_{v1}$. *Microcystis aeruginosa* will wash out because it needs at least R*$_{m1}$ to grow at a dilution rate of 0.2 day^{-1}. If the dilution rate is lower (D = 0.1 day^{-1}), however, *M. aeruginosa* now has a higher growth rate and can reduce the nutrient level to R*$_{m2}$, a level where *V. aureus* cannot grow and washes out of the chemostat. In each case, the species that wins in competition is the one that can grow and reduce the nutrient level to the lowest level.

Sommer (1986a) tested this idea by inoculating natural phytoplankton from Lake Constance, on the German/Swiss/Austrian border, into a chemostat with a culture medium lacking silicate. After several weeks only one species of green alga remained. At the lowest dilution rate only *Mougeotia thylespora* was present. At intermediate dilutions, *Scenedesmus acutus* dominated, and at the highest dilution rate, *Chlorella minutissima* took over the culture vessel (Fig. 22–31). Despite uncertainties in the estimation of the half-saturation constants (K$_s$), these results were in accord with the predictions of Monod kinetics. Thus, unlike the Lotka-Volterra equations, the Monod-based model of competition is predictive.

When many species compete for a single limiting nutrient under steady-state conditions, the mechanistic model predicts that only one species—the one with the lowest value of K$_s$ for that nutrient—can persist. But what happens when there is more than one nutrient limiting growth? Tilman (1976, 1977) developed a Monod-based model to predict the outcome of competition between two species of diatom for two limiting nutrients, silicate and phosphate. Because of the importance of this model in phytoplankton ecology, we will devote some space to the development of these ideas. In an earlier section on growth and nutrient uptake, we presented Tilman's results for growth of the diatoms *Asterionella formosa* and *Cyclotella meneghiniana* under both phosphate and silicate limitation (Question Box 22–2). Under phosphate limitation, *A. formosa* always wins because its K$_s$ for PO$_4$ (0.04 µM) is lower than that of *C. meneghiniana* (0.25 µM). Under silicate limitation, however, *C. meneghiniana* dominates because its K$_s$ for silicate (1.4 µM) is lower than that of *A. formosa* (3.9 µM).

The Monod growth curves for each diatom species growing separately in continuous culture limited by silicate and phosphate are shown again in Fig. 22–32. Assume each continuous culture is operating at a dilution rate of 0.25 day^{-1}. At steady state in a chemostat, D = µ. The dilution rate is represented in each graph by a straight line parallel to the x-axis (y = 0.25). The intersection of this line with the Monod growth curve gives us the minimum amount of PO$_4$ or SiO$_2$ (R*) that each diatom needs to grow at the dilution rate D = 0.25 day^{-1}. For a mathematical formula to calculate R*, refer to Question Box 22–4: Working with the Mechanistic Model of Competition. The values of R* are given in Table 22–4. Consider a resource plane in which the concentrations of PO$_4$ are plotted on the y-axis and those of SiO$_2$ on the x-axis (Tilman, 1982). *Asterionella formosa* cannot grow in continuous culture at D = 0.25 day^{-1} and washes out if [PO$_4$] < 0.01 µM or [SiO$_2$] < 1.9 µM. These conditions define two lines in the resource plane (x = 1.9 µM SiO$_2$ and y = 0.01 µM PO$_4$. In the rectangular area above these lines, *A. formosa* can grow, but below them it washes out (Fig. 22–33a). The lines are zero net growth isoclines, just as we used in the Lotka-Volterra model. On these lines, growth of *A. formosa* just balances losses due to washout so dN/dt = 0.

Similarly, *C. meneghiniana* cannot grow at D = 0.25 day^{-1} if [PO$_4$] < 0.2 µM or [SiO$_2$] < 0.6 µM. These values define the zero net growth isoclines for *C. meneghiniana* as x = 0.6 µM SiO$_2$ and y = 0.2 µM PO$_4$ (Fig. 22–33b). *Cyclotella meneghiniana* shows positive net growth in the region above these lines and washes out below them.

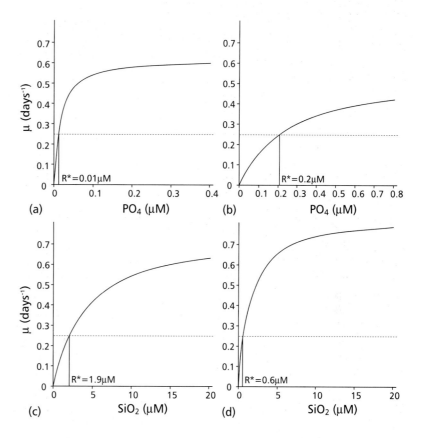

Figure 22–32 Tilman's (1977) mechanistic model of competition. Monod growth curves for *Asterionella formosa* and *Cyclotella meneghiniana* as functions of [PO₄] (a and b) and [SiO₂] (c and d). (After Lampert and Sommer, 1997 ©Georg Thieme Verlag)

If we now plot both sets of zero growth isoclines on the same resource plane, the rectangular areas intersect and define four regions (Fig. 22–33c). Both diatoms will wash out in region 1. In region 2 defined by x = 0.6 and x = 1.9 μM SiO₂ and above y = 0.2 μM PO₄, only *C. meneghiniana* can grow. *Asterionella formosa* would wash out because [SiO₂] < 1.9 μM. Region 3 is defined by x > 1.9 μM SiO₂ and y between y = 0.01 μM and 0.2 μM PO₄; *A. formosa* can grow there, but *C. meneghiniana* cannot because the level of PO₄ is too low. Finally, in region 4 where x > 1.9 uM SiO₂ and y > 0.2 uM PO₄, both diatoms will be able to grow separately.

What happens in region 4 if both diatoms are in continuous culture together? Three outcomes are possible: *C. meneghiniana* displaces *A. formosa*, both diatoms coexist in a stable equilibrium, or *A. formosa* displaces *C. meneghiniana*. What determines the areas within region 4 where these three outcomes may occur? At the intersection of the two rectangles in Fig. 22–33c, there is a point of stable coexistence where *A. formosa* is limited by SiO₂ and *C. meneghiniana* is limited by PO₄. What happens in region 4 above this point depends on the relative amounts of PO₄ and

SiO₂ delivered to the continuous culture compared to the half-saturation constants of the two diatoms for each nutrient. To make this clearer, refer to Table 22–5, where we list the values for the half-saturation constants for each nutrient and diatom and the ratio of these constants.

From the information in Table 22–5 we can then derive the information given in Table 22–6, which defines the nutrient supply ratios that determine whether PO₄ or SiO₂ is limiting growth of each diatom.

If the ratio of supply of phosphate to silicate is less than 0.0103, both diatoms are limited by PO₄. *Asterionella formosa* wins in competition by virtue of having the lower value of K_s for PO₄. If the ratio of supply is greater than 0.1786, both diatoms are limited by SiO₂, and *C. meneghiniana* will displace *A. formosa* because it has the lower K_s value for silicate and can therefore reduce silicate to a level where its competitor cannot grow. If the supply ratio is greater than 0.0103 but less than 0.1786, *A. formosa* is limited by silicate and *C. meneghiniana* is limited by phosphate. Note each diatom is limited by the nutrient that it is least effective at acquiring. Under these conditions each diatom species has more effect on itself than do

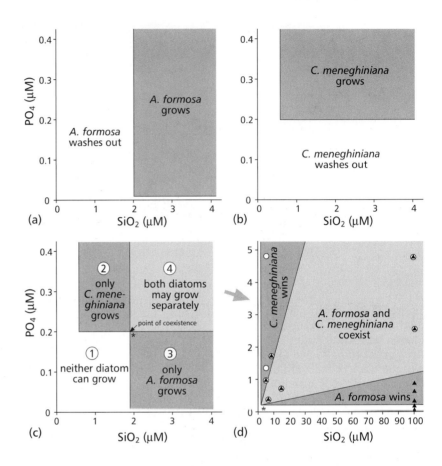

Figure 22–33 Zero growth isoclines for *A. formosa* and *C. meneghiniana* as functions of PO₄ and SiO₂ at D = 0.25 day⁻¹. (a) zero growth isoclines for *A. formosa*, (b) zero growth isoclines for *C. meneghiniana*, (c) combined zero growth isoclines for both diatoms, (d) region 4 of (c) expanded with consumption vectors added. Symbols indicate experimental results. open circle—*A. formosa* only, black triangle—*C. meneghiniana* only, circled triangles—both diatoms coexisted. (d: After Lampert and Sommer, 1997 ©Georg Thieme Verlag)

members of the other species. This is the zone of stable coexistence under steady-state chemostat conditions.

How do we put this information into our nutrient resource plane to define the zones of competitive displacement and coexistence? We treat the above ratios of nutrient supply as the slopes of consumption lines or vectors originating at the point where the two rectangles intersect, namely the point x = 1.9 μM SiO₂, y = 0.2 μM PO₄, which is a point of stable coexistence. We wish to construct two straight lines starting at x = 1.9, y = 0.2 with slopes m = 0.0103 for *A. formosa* and m = 0.1786 for *C. meneghiniana*. The following algebraic formula will generate the equations of the straight lines.

$$m = \frac{y - y_1}{x - x_1} \qquad (35)$$

where m is the required slope (the ratio of half-saturation constants) and (x_1, y_1) is the origin (1.9, 0.2). Thus, 0.0103 = (y − 0.2)/(x − 1.9) yields y = 0.0103x + 0.1805 for *A. formosa*, and 0.1786 =

(y − 0.2)/(x − 1.9) gives y = 0.1786x − 0.1393 for *C. meneghiniana*. These equations are the consumption vectors that divide region 4 into areas of competitive displacement or coexistence. We have added these lines in Figure 22–33d. Above the upper consumption vector, *C. meneghiniana* displaces *A. formosa*. Below the lower consumption vector, *A. formosa* out competes *C. meneghiniana*. The region between the two consumption vectors is the zone of coexistence where the two diatoms are limited by different nutrients. The results of Tilman's competition experiments have been placed on the figure to show the excellent agreement between prediction and observation. Refer to Question Box 22–4—Working with the Mechanistic Model of Competition—to try a similar analysis on data for *Fragilaria crotonensis* and *Synedra filiformis*.

Tilman's research on competition between *A. formosa* and *C. meneghiniana* is presented in several papers including Tilman (1976, 1977) and Tilman and Kilham (1976). His mechanistic theory of competition is developed fully in Tilman (1982) and Tilman, et al. (1982), which also contains a table of Monod kinetic constants for a variety of phytoplankton. Tilman then

Table 22–5 Values of K_s and ratios of K_s values for *A. formosa* and *C. meneghiniana*

	K_s (μM PO$_4$)	K_s (μM SiO$_2$)	K_s (PO$_4$)/K_s (SiO$_2$)
A. formosa	0.04	3.9	0.01026
C. meneghiniana	0.25	1.4	0.17857

extended his studies to four species of Lake Michigan diatoms including *A. formosa*, *Fragilaria crotonensis*, *Synedra filiformis*, and *Tabellaria flocculosa*. He found that *A. formosa* and *F. crotonensis* were competitively equal; they coexisted at all ratios of silicate to phosphate because they appear to have similar resource requirements. This result suggests a novel way species diversity may increase. Even if two phytoplankton species are morphologically distinct, they can coexist if they are physiologically very similar. It is not known how common this phenomenon might be.

Temperature was later found to be capable of altering the outcome of competitive interactions (Tilman, et al., 1981, Mechling and Kilham, 1982). Monod kinetic constants changed as a function of temperature. In one set of experiments, *A. formosa* displaced *Synedra ulna* below 20° C, but *S. ulna* displaced *A. formosa* above 20° C.

While Tilman, the Kilhams, and their students have primarily focused on freshwater diatoms, and silicate and phosphate as limiting nutrients, they have extended their observations to green algae and cyanobacteria. Cyanobacteria studied include *Oscillatoria agardhii* (Ahlgren, 1978) and *Microcystis*

aeruginosa (Holm and Armstrong, 1981). Among green algae, we have data on *Volvox aureus* and *V. globator* (Senft, et al., 1981), *Mougeotia thylespora*, *Scenedesmus acutus*, and *Chlorella minutissima* (Sommer, 1986a), and *Scenedesmus quadricauda*, *Oocystis pusilla*, and *Sphaerocystis schroeteri* (Grover, 1989). Work by Sandgren (1988) and Lehman (1976) has provided data on chrysophycean flagellates. Data on Monod kinetics for marine phytoplankton and for nitrogen as a limiting nutrient are available. In one marine study, Sommer (1986b) examined Tilman's mechanistic model of competition among five species of marine diatoms from Antarctic waters for limiting supplies of nitrate and silicate (see Question Box 22–4 problem 2). In general, the predictions of the mechanistic model and experimental observations of competition have been in excellent agreement. Despite some gaps in specific taxonomic groups, Tilman's model has been well verified in continuous cultures at steady state.

While the model successfully predicts coexistence between two species competing for two different limiting nutrients, there are a very limited number of potentially limiting nutrients. The list includes only sil-

Table 22–6 Regions of nutrient limitation defined by the ratios of K_s values

	limited by PO$_4$	limited by SiO$_2$
A. formosa	$\dfrac{[PO_4]}{[SiO_2]} < 0.01026$	$0.01026 < \dfrac{[PO_4]}{[SiO_2]}$
C. meneghiniana	$\dfrac{[PO_4]}{[SiO_2]} < 0.17857$	$0.17857 < \dfrac{[PO_4]}{[SiO_2]}$

Question Box 22–4 Working with the mechanistic model of competition

1. Tilman (1981) tested his resource competition theory on four species of Lake Michigan, U.S.A., diatoms. The Monod kinetic parameters μ_{max} (day^{-1}), K_s (μM) and cell quotient Q (μmol cell^{-1}), the amount of nutrient necessary to produce one cell of a species, were determined. Q is the inverse of the yield coefficient. The kinetic parameters under phosphate and silicate limitation for two of the diatoms, *Fragilaria crotonensis* and *Synedra filiformis* are shown in the following table.

parameters under phosphate limitation

	μ_{max} (day^{-1})	K_s (μM)	Q (μmol cell^{-1})
Fragilaria	0.80	0.011	4.7×10^{-8}
Synedra	0.65	0.003	1.1×10^{-7}

parameters under silicate limitation

	μ_{max} (day^{-1})	K_s (μM)	Q (μmol cell^{-1})
Fragilaria	0.62	1.5	9.7×10^{-7}
Synedra	1.11	19.7	5.8×10^{-5}

The diatoms were all grown in continuous culture at D = 0.25 day^{-1}, as were *A. formosa* and *C. meneghiniana* in the main text. To calculate the minimum concentration of each nutrient that just allows each diatom species to persist in continuous culture at D = 0.25 day^{-1}, we rearrange the terms in the Monod growth equation and substitute in the parameter values from the table above. Thus:

$$\mu = \mu_{max} \left(\frac{S}{K_s + S} \right)$$

At equilibrium, D = μ = 0.25 day^{-1} and S = R*, the nutrient level that allows the diatom to remain in culture with dN/dt = 0. Then substituting in the Monod equation, we get:

$$D = \frac{\mu_{max} R^*}{K_s + R^*}$$

We now solve for R* by multiplying both sides by (K_s + R*):

$$D (K_s + R^*) = \mu_{max} R^*$$

Then a few algebraic manipulations to isolate R* gives us:

$$DK_s + DR^* = \mu_{max} R^*$$

$$\mu_{max} R^* - DR^* = DK_s$$

$$(\mu_{max} - D) R^* = DK_s$$

$$R^* = \frac{DK_s}{\mu_{max} - D}$$

Question Box 22–4 *continued*

As a practice problem, graph the four Monod curves for *Fragilaria* and *Synedra*, each limited by phosphate and silicate. Use the above formula for R* to calculate the minimum levels of silicate and phosphate at which each diatom can just maintain a population in continuous culture at D = 0.25 day^{-1}. Then plot the zero growth isoclines on a resource plane where y = [PO$_4$] and x = [SiO$_2$]. (*Hint*: you will have to make the axis for PO$_4$ extend over a very small range, on the order of 0.05 μM). Tilman (1981) used the ratios of the nutrient quotients (Q$_p$/Q$_{si}$) as his slopes of the consumption vectors. If you use the ratios of the half-saturation constants, the final figure is not greatly changed. Finally, plot the consumption vectors in your resource plane.

2. Sommer (1986b) studied nitrate and silicate competition among marine diatoms from the frigid waters around Antarctica. He obtained the following values for Monod kinetic constants at 0° C for four species of marine diatoms:

	μ_{max} (day^{-1})	K_n (μM)	K_{si} (μM)
Corethron criophilum	0.39	0.3	60.1
Nitzschia kerguelensis	0.56	0.8	88.7
Thalassiosira subtilis	0.40	0.9	5.7
Nitzschia cylindrus	0.59	4.2	8.4

Sommer grew these diatoms in continuous culture at D = μ = 0.25 day^{-1}. Use the formula in problem 1 above to calculate the values of R* for each diatom and nutrient. Then use this information to plot the zero growth isoclines in a resource plane where x = μM NO$_3^-$ and y = μM SiO$_2$. You can then use the values of the half-saturation constants to calculate the slopes of the consumption vectors that extend from the intersection points of the zero growth isoclines. (The consumption vectors should be plotted on a separate graph where the axes cover a wider range than that used for the zero growth isoclines.) Answers are found in the section at the end of the chapter.

icon (for diatoms), nitrogen, phosphorus, carbon (possibly in some acidic waters), iron in oceans, and perhaps vitamins. Recent work by Huisman and Weissing (1994) has shown that light can act as a limiting nutrient. It is unlikely that many other items will be added to this list.

When more than two species compete for two limiting nutrients, the mechanistic model predicts that no more than two species can coexist stably at any given combination of those two nutrients. This result is usually presented as a model diagram such as Figure 22–34 from Lampert and Sommer (1997). Here, four species compete for two nutrients in a resource plane. Regions labeled with one number indicate dominance by that one species. Regions with two numbers indicate coexistence by those two species. For all four species to persist, the ratio at which both resources are supplied must move along a gradient through these

Figure 22–34 Tilman's model for steady-state competition for two resources among four species of algae. Zero growth isoclines and consumption vectors are shown. Single numbers indicate regions where one species dominates, double numbers where two species coexist. (After Sommer, 1989 ©Springer-Verlag)

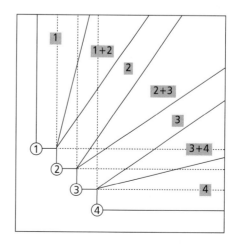

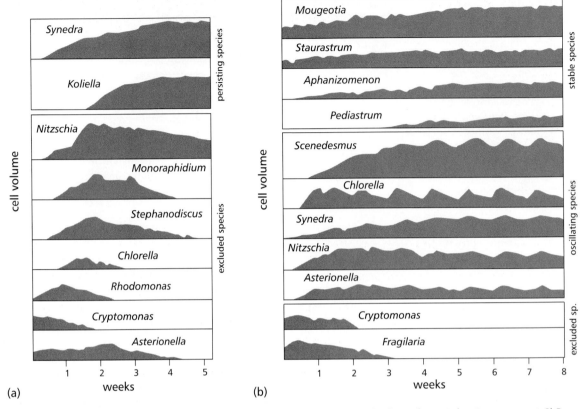

Figure 22–35 (a) Chemostat competition experiment with natural phytoplankton from Lake Constance at Si:P = 20:1. (b) Pulsed P and Si (1 week intervals) competition experiment with the same phytoplankton inoculum at Si:P = 20:1. Biomass—$\log_{10}(\mu m^3 \text{ ml}^{-1})$—of all species that represent more than 5% of total biomass. Under continuous culture conditions, only two species persisted. Under weekly pulsed P and Si conditions, nine species persisted for over eight weeks. (After Sommer, 1985)

regions. Thus at one time species 1 and 2 may be coexisting while species 3 and 4 are in decline. The ratio of resources moves from the region where species 1 and 2 coexist to the region where species 2 and 3 coexist and then to the region of coexistence of species 3 and 4 before either is completely displaced. Then the ratio has to return to the original state before species 1 and 2 are displaced. If more than two species are to coexist on two limiting nutrients, the model must invoke not steady-state conditions but incorporate a cycle of steady states. Resources must change over time in a manner determined by the growth dynamics of the competing species if additional species are to persist. For an example of four marine diatoms involved in competition for two limiting nutrients, refer to the second problem in Question Box 22–4.

The mechanistic model of competition under steady state continuous culture conditions cannot account for more species than the number of poten-

tially limiting nutrients, which is no more than about seven. Researchers have therefore turned to examining the role that perturbations in the supply of nutrients or non-steady–state conditions may have on competition and the number of coexisting species of phytoplankton. In these experiments, continuous culture results become a control to compare against the results of various severities and frequencies of disturbances.

In an early study of perturbations, Turpin and Harrison (1980) studied the marine diatoms *Chaetoceros* spp., *Skeletonema costatum*, and *Thalassiosira nordenskioldii* in natural and artificial assemblages, under both continuous and pulsed additions of ammonium. Continuous cultures were dominated by *Chaetoceros* spp. *Skeletonema costatum* dominated cultures receiving a spike of ammonium every third day. When the ammonium spike was delivered every seventh day, *T. nordenskioldii* dominated cultures. The authors noted that as nutrient addition became

Table 22–7 Species number in chemostat and pulsed experiments as a function of the Si:P molar ratio. (adapted from Sommer, 1985)

# of species	Si:P (molar) 4:1	10:1	20:1	30:1	40:1	80:1	140:1	≥200:1
chemostat	1	2	2	3	2	1	–	–
pulsed P	8	–	6	–	6	–	5	5
pulsed P + Si	8	8	9	–	10	8	7	6

less frequent, the dominant diatom had a larger average cell size. Presumably this is because larger cells can store up more nutrient to survive the depleted conditions between additions.

Sommer (1985) took natural phytoplankton communities from Lake Constance and compared competition in chemostats to competition in cultures where either phosphate or both phosphate and silicate were added in pulses at one week intervals. Si:P ratios from 0:1 to 280:1 were tested. Pulsed additions of nutrients led to a marked increase in number of persisting species from 1–2 to 9–10 (Table 22–7). Continuous nutrient additions favored diatoms, while pulsed additions promoted greens (Fig. 22–35). These results applied to a single fixed interval of pulses of nutrients.

In a later study, Gaedeke and Sommer (1986) varied the frequency of pulses from 1 to 14 days. At one-day intervals, only one species dominated cultures—the cyanobacterium *Pseudanabaena catenata* in silicate-enriched systems and the green alga *Koliella spiculiformis* in silicate-free cultures. As the interval between nutrient additions reached three days, species diversity rose and reached a maximum at seven days. At greater intervals than seven days, diversity declined slowly (Fig. 22–36). Perturbations had to occur at an interval greater than the average generation time of the algae before they affected species diversity. Interestingly, the maximum diversity occurred at seven days, an interval roughly equal to that of the passage of weather events across bodies of water. Gaedeke and Sommer (1986) believe their results support the intermediate disturbance hypothesis of Connell (1978), which states that moderate levels of disturbance promote species diversity.

Perturbation studies have established that competitive exclusion can be circumvented by intermediate disturbances in the nutrient supplies. The mechanistic models assume that if nutrients are limiting, then competition will occur. Conversely, if nutrients are not limiting, competition will not occur. How often are various nutrients limiting? Sommer (1988a) tried to determine whether silicon, phosphorus, and nitrogen were limiting in Lake Constance by adding spikes of each nutrient to water samples from the lake at weekly intervals over a year. If the phytoplankton did not respond by growing, then the nutrient was not likely to have been limiting in the sample. Silicate was limiting for diatoms throughout the stratified period, but surprisingly nitrate and phosphate were limiting to growth in an intermittent manner. When they were limiting, that limitation was weak to moderate and

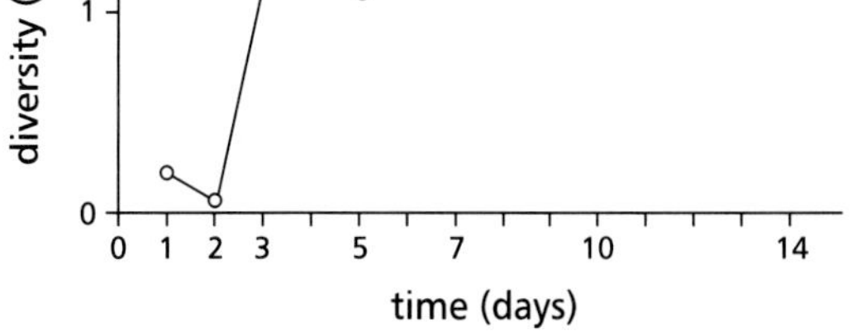

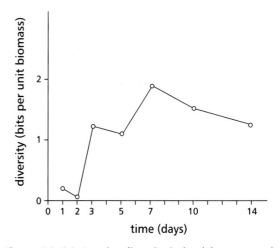

Figure 22–36 Species diversity index (Shannon and Weaver, 1949) as a function of the interval between nutrient additions in semicontinuous competition experiments. Maximum diversity (the highest number of coexisting species) was reached when the interval between nutrient additions was seven days. (After Gaedeke and Sommer, 1986 ©Springer-Verlag)

rarely severe. These results indicate that in Lake Constance, diatoms are likely to be competing for silicate over the stratified period, but other phytoplankton may be competing only during brief, widely separated intervals. If these results are generally true (more research on lakes of various trophic states would be useful), then competition is a sporadic event, and many species may be able to coexist because they are not competing all the time. The diversity of phytoplankton is only a paradox if we assume competition is continuous. This viewpoint is very different from the old idea that all species are at their carrying capacity and in continuous competition, and suggests that competition is only one loss process among many that shape phytoplankton communities.

Grazing

The most dramatic phytoplankton loss process is grazing loss. Grazing refers to predator-prey interactions where the prey are bacteria and algae, and the predators are rotifers and crustaceans. Some phytoplankton can ingest bacteria and even other algae. For examples, refer to Chapter 3. Protozoa may act both as predators on algae and bacteria as well as prey to crustaceans. Rotifers are mainly filter feeders, but a few specialists feed raptorially on large algae. Among crustaceans the lower boundary of phytoplankton cell sizes is determined by the mesh width of the filtering appendages, which range from 0.16 to 4.2 μm in cladocerans (Geller and Müller, 1981). Cladocerans with the finest mesh sizes (*Chydorus sphaericus*, *Daphnia magna*) can remove large bacteria, while those with the coarsest filters (*Sida cristallina*, *Holopedium gibberum*) miss picoplanktonic algae (0.5 to 2 μm). For copepods, which generally feed by raptorial capture, the upper limit of algal cell size is about 50 μm (Burns 1968). Thus cell or colony size in part determines the relative susceptibility of phytoplankton to grazing losses.

How important are losses due to grazing on phytoplankton? The significance of grazing losses is indicated by the existence of a clear-water phase in late spring and early summer in many mesotrophic and eutrophic lakes. The development of a clear-water phase is shown in Lake Schöhsee near Plön, Germany, in May 1983 (Fig. 22–37). As the dominant grazers (species of *Daphnia* and *Eudiaptomus*) increase in numbers, the smaller (<35 μm) algae decline, and Secchi disk transparency increases (Lampert, et al., 1986). For such a clear-water phase to occur, the loss rate due

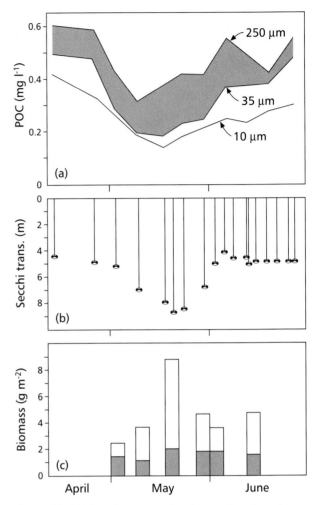

Figure 22–37 Development of a clear-water phase in Lake Schöhsee, Germany, in 1983. (a) Particulate organic carbon (POC) separated by size (< 10 μm, < 35 μm, < 250 μm). (b) Secchi disk transparency (m). (c) Biomass (g m⁻²) of *Daphnia* spp. (open bar) and *Eudiaptomus* spp. (shaded bars). (After Lampert, et al., 1986)

to grazing (γ) must exceed the phytoplankton production rate (r).

Cladocerans play a major role in the generation of these clear-water phases. *Daphnia*, in particular, has a strong impact on the structure of phytoplankton communities and on the microbial food web in general. Compare the two food webs in Figure 22–38. In Figure 22–38a, *Daphnia* dominates the system and suppresses all the protozoans and edible phytoplankton. The microbial food web consists of large, inedible, or grazing-resistant algae and small bacteria that escape the mesh size of *Daphnia* (Jürgens 1994). Phytoplankton are too large to be grazed or else survive gut

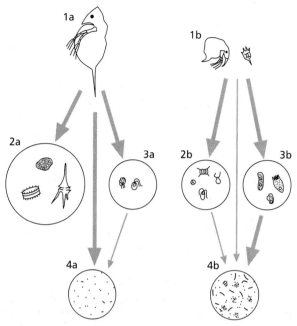

Figure 22–38 Impact of different zooplankton assemblages on the microbial food web. (a) Domination by *Daphnia* leads to grazing resistant algae, small flagellates and small bacteria. (b) Domination by *Chydorus* and rotifers produces small edible algae, diverse protozoa and grazing resistant bacteria.

passage because of protective sheaths (Porter 1977). In contrast, Figure 22–38b shows a food web dominated by rotifers and *Chydorus*. Protozoa are numerous and diverse, while phytoplankton are small and mostly edible. The protozoa graze heavily on the bacteria, leaving grazing-resistant (filamentous and clumped) forms (Jürgens and Güde, 1994). Thus, the type of grazing pressure affects the composition of phytoplankton and microbial communities.

Grazing can also alter the outcome of competition between phytoplankton. Rothhaupt (1992) added the diatom *Synedra*, *Cryptomonas*, and bacteria to a chemostat. Both *Synedra* and bacteria had low half-saturation constants for phosphorus and grew rapidly. *Cryptomonas*, however, had a higher K_s value for phosphorus and remained in culture only at low density (Fig. 22–39). *Synedra* reached a constant population level due to silica limitation. At week three, Rothhaupt introduced the heterotrophic flagellate *Spumella*, which grazed on the bacteria and released the bacterial phosphorus into the vessel. Since *Synedra* was silica limited, it could not use the added phosphorus, and *Cryptomonas* eventually became the dominant alga in the chemostat. Grazing pressure can

thus alter competitive interactions. The dominance of grazing-resistant algae after the clear-water phase is in large part due to the release of nutrients from edible algae by grazers. Large phytoplankton, or those with protective sheaths, are generally poor nutrient competitors. Grazers act to shuttle nutrients from edible to inedible algae. This capacity of grazing pressure to change algal assemblages has led to a great deal of research on trophic dynamics and to the use of biomanipulations to affect changes in aquatic systems.

Before we consider trophic dynamics and biomanipulations, we need to ask first how grazing is measured. Terminology and mathematical expressions vary among authors, and definitions should be checked carefully in references. Here we define the feeding rate or ingestion rate as the number or mass of cells ingested per individual grazer in a unit of time. Thus:

$$I = C\ \text{ind}^{-1}\ t^{-1} \quad (36)$$

C is expressed in cells or some measure of biomass such as biovolume (μm^3), mg dry weight or mg C.

The filtering rate F of a zooplankton is the volume of water filtered by that herbivore to ingest C cells in time t, assuming 100% efficiency in filtration and retention. F has units of ml ind^{-1} t^{-1}. In practice, two formulas are used in the determination of a filtration rate, depending on the method of measurement. If cell counts are used, then Gauld's (1951) equation is used:

$$F = \frac{V\ (\ln C_o - \ln C_t)}{N \times t} \quad (37)$$

The equation assumes an exponential decline in cells or biomass from C_o to C_t as a result of ingestion by N herbivores over time t. V is the volume of the container in ml. C_o and C_t are measured by direct cell counts or, more commonly now, by an electronic particle counter. Because t is necessarily fairly long (a few hours), correction must be made for any increase in number of algal cells. For additional precautions, refer to Peters (1984).

Alternatively, a radioisotope such as ^{32}P or ^{14}C may be used. Grazing animals ingest the labeled algal cells, and the activity of each is measured by liquid scintillation. The form of this equation is:

$$F = \frac{A_a\ 60}{A_s\ (N)\ t} \quad (38)$$

where A_a is the activity in counts min^{-1} of the N herbivores measured after t minutes of exposure to the

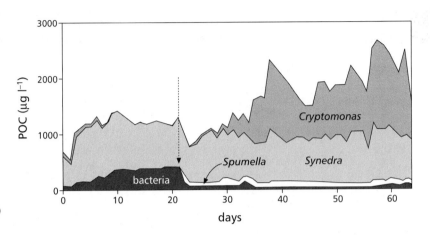

Figure 22–39 Competition between *Synedra*, *Cryptomonas*, and bacteria for phosphorus in a chemostat. The addition of the heterotrophic flagellate *Spumella* on day 21 reduces the bacterial biomass and makes phosphorus available for growth of *Cryptomonas*. (After Rothhaupt, 1992)

labeled cells. A_s is the activity of the labeled algae in counts min^{-1} ml^{-1}. The number 60 converts the minutes of exposure into the more useful units of ml ind^{-1} h^{-1}. The equation assumes a linear uptake of cells over the short incubation time. The exposure period must be short because it must be less than the gut passage time of the herbivore. Otherwise, labeled cells will be ingested but not counted because they have been expelled. Gut passage times increase with body size; most are about 15 minutes or more. Peters (1984) provides an extensive table.

Filtration rates can be calculated for individual zooplankton species and for separate life cycle stages such as copepodids—the juvenile stages of copepods between the naupliar stage and the adult. Table 22–8 gives individual filtration rates for zooplankton in Blelham Tarn, Cumbria, U.K. (Thompson, et al., 1982).

If we multiply each of these filtration rates by the number of individuals of each species or stage in a liter, we get a new parameter, the community grazing rate:

$$G = \Sigma\, F_i\, N_i \quad (39)$$

G represents the losses due to grazing by the entire zooplankton community on the algae used as a tracer and all other phytoplankton with the same relative edibility as the tracer alga. Here it is the same as the γ term in our loss equation (26). Thompson, et al. (1982) measured the community grazing rate (or index) by summing the individual filtration rates on *Chlorella* as tracer alga for all significant herbivores in Blelham Tarn enclosures multiplied by their respective population densities. G varied from 0.2% to 200%

day^{-1}, meaning that on some occasions the entire volume of water in an enclosure was filtered twice in one day! Such a rate exceeds the growth rate of many phytoplankton.

Haney (1973) pioneered the direct measurement of G by radioactivity as:

$$G\ (\text{day}^{-1}) = \frac{A_a\ 60\ (24)}{\Lambda_s\ V\ t} \quad (40)$$

V is the volume of the chamber (Haney chamber) used in the field, in ml. A_a is the radioactivity of all animals in the chamber, and A_s is that of the tracer alga introduced to the chamber. The value of t is expressed in minutes, and (60 × 24) converts readings from minutes into days. In eutrophic Heart Lake, Ontario, Canada, Haney found that the summer value of G often exceeded 100% day^{-1} with the average G from June through September being 62–80% day^{-1}. The net growth rate of the phytoplankton (r) was generally balanced by the community grazing rate G (or γ) in Heart Lake. In oligotrophic Hall's Lake, Ontario, Canada, however, r exceeded γ.

Because of the labor involved, relatively few researchers have followed the procedures of Haney (1973) or Thompson, et al. (1982). Furthermore, neither procedure takes into account the relative edibility or selectivity of different phytoplankton species. If we assume that our tracer alga is highly edible and defenseless, we can define its selectivity as w = 1.0. Relative to it, all other algae would have a selectivity w of 1.0 or less. Figure 22–40 gives the relative selectivity coefficients for *Daphnia magna* feeding on a variety of phytoplankton (Sommer, 1988b). In this

Table 22–8 Ranges and means of individual filtration rates for each category of zooplankton in enclosure A in Blelham Tarn, Cumbria, U.K., in 1978. (from Thompson, et al., 1982)	
Species and stage	**Filtration rates: range (mean) (ml ind⁻¹ day⁻¹)**
Daphnia hyalina V	42.1 – 62.6 (53.0)
Daphnia hyalina IV	14.0 – 60.0 (28.4)
Daphnia hyalina III	9.3 – 29.3 (19.8)
Daphnia hyalina II	5.7 – 19.3 (11.8)
Daphnia hyalina I	4.0 – 7.6 (6.8)
Diaptomus gracilis, adult + cop 5	0.5 – 10.2 (4.4)
Diaptomus gracilis, cops 1-4	0.5 – 6.7 (2.7)
Chydorus sphaericus	0.6 – 2.6 (1.0)

case, the loss rate due to grazing by *D. magna* on each of these algae is:

$$\gamma_i \ = \ G\,w_i \quad (41)$$

G is the grazing rate of *D. magna* on a tracer alga, and the w_i are the selectivity coefficients of each of the other algae. To derive a community grazing index for all herbivores on all phytoplankton, one would need a similar table for each zooplankton species as well as for many life-cycle stages. For obvious reasons this has not been done. Lampert, et al. (1986) developed an inter-

esting approach to this problem. They used two different tracer algae in a Haney chamber. *Scenedesmus acutus* labeled with ¹⁴C had a diameter of 10 μm and was used to measure grazing on edible nanoplankton; ³²P-labeled *Synechococcus elongatus* with a diameter of 1 μm was used to measure grazing on picoplankton. Grazing rates varied from 40% to 170% day⁻¹. In general, values of G in excess of 1.0 may persist in lakes for several months (Sterner 1989).

As grazing studies gradually accumulated, it was recognized that zooplankton could have significant impact on phytoplankton communities and even gen-

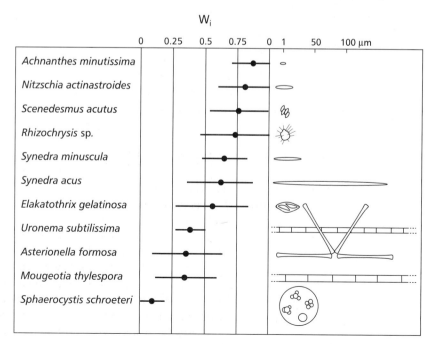

Figure 22–40 Selectivity coefficients (w_i) for various phytoplankton fed on by *Daphnia magna*. Note that large species and those with gelatinous sheaths are not grazed upon well. (After Lampert and Sommer, 1997 ©Georg Thieme Verlag)

erate clear-water phases in some lakes (Porter, 1977; Bergquist, et al., 1985; Bergquist and Carpenter, 1986; Jürgens, 1994). At the same time, evidence was gathered indicating that other trophic levels in aquatic systems could have major impact on the levels below them (Carpenter, 1989). Piscivorous fish can shape the assemblage of planktivorous fish (Tonn and Magnuson, 1982). Planktivorous fish can, in turn, determine the species composition and size structure of zooplankton (Brooks and Dodson, 1965). All these studies led to the idea of a **trophic cascade** in which effects cascade from top to bottom down a chain of linked trophic levels (Carpenter, et al., 1985). The top predators—piscivorous fish—are seen as keystone species (Paine, 1969) with effects extending down many trophic levels below their own position in the food web. The keystone species concept led to the hypothesis that, in aquatic systems, control proceeds from the top of the food web downward. Thus, if one were to manipulate the top trophic levels, one might be able to alter the phytoplankton community to achieve a clear-water lake with enhanced aesthetic and recreational value.

These trophic cascade concepts are diagrammed in Figure 22–41. The top diagram (a) shows a lake without fish, where the dominant herbivores are invertebrates like *Chaoborus*. *Chaoborus* prefers small zooplankton prey. Consequently, large *Daphnia* and copepods dominate and suppress the phytoplankton as in Figure 22–38a. Figure 22–41b shows a lake with large stocks of piscivorous fish that feed on the planktivorous fish. A mixed assemblage of various sizes of zooplankton results with strong grazing pressure on the phytoplankton. In Figure 22–41c, few piscivorous fish lead to numerous planktivorous fish and small zooplankton. The small zooplankton in turn lead to a microbial food web as in Figure 22–38b, with many small, edible algae and protozoa. Both Figures 22–41a and b represent biomanipulations.

The keystone species concept is often referred to as the "top-down" hypothesis of aquatic system control. In its simplest form, the top-down hypothesis predicts that adjacent trophic levels are negatively correlated. In other words, more piscivorous fish means fewer planktivorous fish, which leads to more and larger zooplankton and fewer algae. Conversely, more planktivorous fish would mean fewer zooplankton and more algae. The top-down hypothesis is often contrasted to the older, "bottom-up" hypothesis. The bottom-up hypothesis says that all trophic levels are positively correlated. Thus more nutrients means more phytoplankton and more zooplankton. Comparison across many lakes of different trophic status generally supports a bottom-up hypothesis. Figure 22–42 shows two significant cross-lake correlations between phosphorus and total phytoplankton chlorophyll *a* (a) and between phytoplankton and zooplankton (b). Biomanipulations within lakes and enclosures, however, often support the top-down hypothesis. Before we discuss these contrasting hypotheses further, we will present some of the top-down data.

Tests of the trophic cascade hypothesis are difficult to carry out because of the large workload. Many trophic levels must be monitored in both manipulated and control systems over a significant period of time. Biomanipulations are done as whole-lake manipulations or enclosure experiments, the latter being amenable to replication.

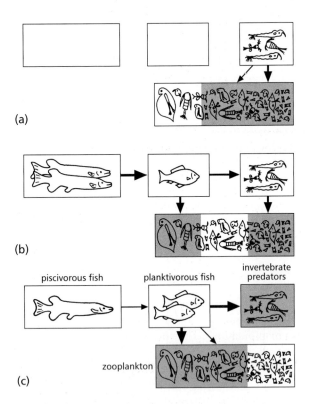

Figure 22–41 The trophic cascade in lakes. (a) With only invertebrate predators, large zooplankton result. (b) Large populations of piscivorous fish suppress the planktivorous fish and lead to medium-sized zooplankton. (c) Having few piscivorous fish leads to many planktivorous fish and small zooplankton. (After Lampert, 1987 ©Springer-Verlag)

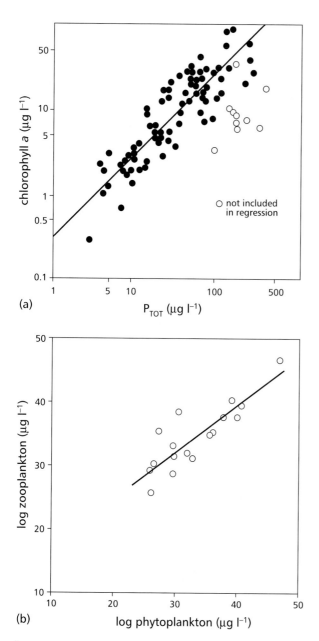

(a)

(b)

Figure 22–42 Bottom-up control data from many lakes of various trophic status. (a) Phytoplankton biomass (as chlorophyll *a*) as a function of phosphorus. (b) Log zooplankton biomass as a function of log phytoplankton biomass. (a: Based on Vollenweider, 1982 ©Springer-Verlag; b: After McCauley and Kalff, 1981)

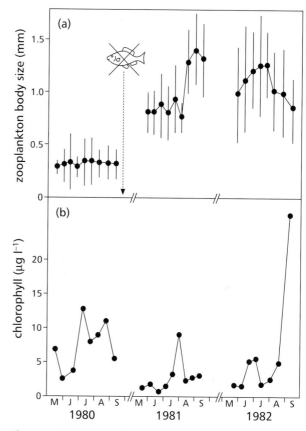

Figure 22–43 Biomanipulation of Round Lake, Minnesota. The entire fish population was eliminated with rotenone in 1980. (a) The average body size of zooplankton increased after the elimination of planktivorous fish. (b) Phytoplankton (as chlorophyll) were reduced in 1981, but a massive bloom of the cyanobacterium *Aphanizomenon flos-aquae* occurred in 1982. (Data from Shapiro and Wright, 1984; figure after Lampert and Sommer, 1997 ©Georg Thieme Verlag)

In enclosure experiments, fish may be excluded, or added at various densities in combination with nutrient treatments. Fish generally reduce zooplankton (top-down) but nutrients increase phytoplankton (bottom-up) (Vanni, 1987). Brett and Goldman (1997) collected data from 11 enclosure studies that employed both nutrient additions and planktivorous fish manipulations. They performed a meta-analysis of all these data to examine the regulatory roles of both planktivores and nutrients. They found that both top-down and bottom-up controls were significant. Zooplankton were under strong planktivore control (top-down) but were only weakly stimulated by nutrients. Phytoplankton were under strong nutrient control (bottom-up) and were moderately controlled by fish through their effects on zooplankton. Specifically, phytoplankton biomass increased, on average, 179% in nutrient treatments, but 77% in fish treatments where zooplankton were suppressed. Therefore both top-down and bottom-up control of phytoplankton occurs, but the consensus of these enclosure experiments is that bottom-up control was stronger than top-down control.

Results from whole-lake studies are more variable than results from enclosures. Whole-lake studies may involve natural experiments, where a mass mortality of fish has occurred, or artificial manipulation of fish stocks. In at least one class of lakes—small, shallow, eutrophic lakes—biomanipulations have been generally successful (Sondergaard, et al., 1990; Meijer, et al., 1990; van Donk, et al., 1990a). Initially, in summer, these lakes were dominated by blooms of cyanobacteria. Following removal of planktivorous and benthivorous fish, the zooplankton shifted from rotifers to larger *Daphnia* and a clear-water phase occurred. Submerged macrophytes reestablished themselves due to improved light conditions at the lake bottoms. The lakes remained clear because nutrients were taken up by the macrophytes. When the same manipulations were tried on larger (>100 ha) shallow eutrophic lakes, results were variable. Lake Breukeleveen, Netherlands, (180 ha) showed no improvement, due to wind-induced turbidity that prevented macrophytes from developing (van Donk, et al., 1990b), but Lake Christina (1619 ha), central Minnesota, U.S.A., improved dramatically because submerged macrophytes increased (Hanson and Butler, 1990).

Biomanipulations in other types of lakes are less predictable. Shapiro and Wright (1984) poisoned the entire fish stock of Round Lake, Minnesota (U.S.A.) with rotenone (Fig. 22–43). In 1980 the lake was dominated by planktivorous fish and small zooplankton. After fish removal, the zooplankton size increased and phytoplankton (as chlorophyll) declined. But two years after the manipulation, the grazing-resistant cyanobacterium *Aphanizomenon flos-aquae* bloomed, and chlorophyll levels were higher than previously.

In a seven year study of Lake St. George (Canada), McQueen, et al. (1989) found evidence of both top-down and bottom-up controls. In the winter of 1981–1982, low oxygen under the ice caused a winter fish kill (Fig. 22–44). Zooplankton quickly reached maximum body size and biomass, but they had no impact on the phytoplankton. The correlation coefficients show that piscivorous fish had a strong negative impact on planktivorous fish, which, in turn, had a negative impact on zooplankton. Nutrients had a strong positive impact on phytoplankton. Both top-down and bottom-up control occurred, with the

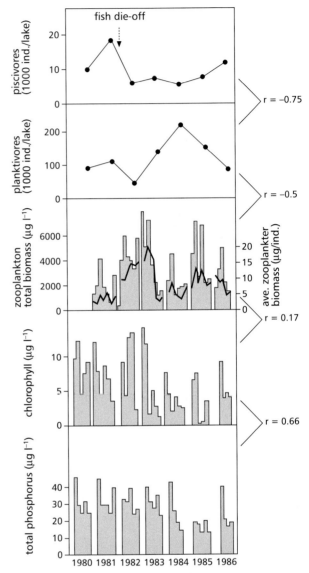

Figure 22–44 Seven years of food web data from Lake St. George, Canada. In the winter of 1981-1982, a winter kill reduced all the fish populations. Zooplankton biomass and body size (center) increased due to reduced planktivore pressure. The planktivores recovered quickly and reduced the zooplankton in 1983. Phytoplankton and phosphorus show little change. Correlation coefficients between adjacent trophic levels are shown on the right. (Data from McQueen, et al., 1989; figure after Lampert and Sommer, 1997 ©Georg Thieme Verlag)

effects diminishing as they passed up or down the food web. These results were supported by the meta-analysis of enclosure experiments discussed previously.

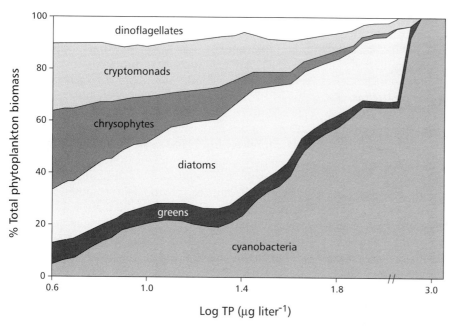

Figure 22–45 Percentage contribution of taxonomic groups of phytoplankton to total summer biomass as a function of average total phosphorus of lakes. As log TP rises, cyanobacteria become increasingly dominant. (After Watson, et al., 1997)

The concept of top-down and bottom-up interaction has become increasingly recognized in the literature. In 1990, Vanni, et al., reported on the effects of a mass mortality of planktivorous fish in Lake Mendota, Wisconsin. After the fish die-off, the larger *Daphnia pulicaria* replaced *D. galeata mendotae* and generated a longer spring clear-water phase. These authors suggested that high phosphorus levels may reduce effects of trophic cascades on phytoplankton communities. In a comparative study of manipulations in Peter, Paul, and Long lakes, Michigan (U.S.A.), Carpenter, et al. (1996) concluded that the potential effects of phosphorus input on phytoplankton were stronger than the potential for controlling phytoplankton through food-web manipulation. In other words, phosphorus enrichment could overpower the mitigating effects of a trophic cascade and affect the success of a biomanipulation. If nutrient inputs have a stronger effect on phytoplankton blooms than does food-web manipulation, then effective phytoplankton management will require the difficult (and expensive) job of controlling nutrient inputs through erosion control, wetland preservation and restoration, and effluent treatment.

Is it top-down or bottom-up? Carpenter and Kitchell (1993) reply, "A simplistic view of trophic interactions (Is it top-down or bottom-up?) is insufficient and misleading." Clearly, both controls are important and interact with each other. Two current hypotheses have been proposed to address the inter-

action between nutrients and food-web effects (Carpenter and Kitchell, 1993). The "mesotrophic maximum hypothesis" states that food-web effects are greatest in mesotrophic lakes (Elser and Goldman, 1991). One study of fish removal from an oligotrophic lake, however, showed that significant food-web effects could occur even at low nutrient levels (Henrikson, et al., 1980). The "nutrient attenuation hypothesis" states that as nutrients increase, phytoplankton escape grazer control (McQueen, 1990). Watson, et al. (1997) examined patterns of phytoplankton taxonomic composition across temperate lakes of varying trophic status (Fig. 22–45). In summer, the taxonomic composition of the phytoplankton changes drastically with total phosphorus (TP) concentration. Lakes with high TP are dominated by cyanobacteria. *Daphnia* cannot survive in dense blooms of cyanobacteria (Gliwicz 1990). Therefore, once lakes are in the upper range of phosphorus enrichment, they will be dominated by cyanobacteria and grazing may have little effect. Recent work suggests, however, that *Daphnia* may be able to reduce the frequency of cyanobacterial blooms by recycling nitrogen (ammonium) and reducing N_2 fixation by cyanobacteria (MacKay and Elser, 1998). Biomanipulations in nutrient-rich lakes may even cause lake quality to deteriorate if they result in the transfer of nutrients from edible algae to inedible algae, as may have happened in the Little Round Lake manipulation. Carpenter and Kitchell (1993) conclude that our ability to predict the outcome of trophic level interactions will improve "in proportion to our mechanistic understanding of species interactions and life-history characteristics." In their study of biomanipulations in Peter, Paul, and Long lakes, Carpenter and Kitchell's (1993) trophic dynamic models did not predict the actual phytoplankton response largely due to some unknown species interactions and life-history characteristics of *Peridinium limbatum*.

Patterns of Loss Processes

The previous sections have discussed a wide range of phytoplankton loss processes. All loss processes do not affect all species or taxonomic groups of phytoplankton to the same extent. Although the data collection is laborious, a few studies have attempted to evaluate the relative impact of various loss processes on phytoplankton populations. Jewson, et al. (1981) examined loss rates from sedimentation, parasitism, and grazing in two diatoms, *Melosira italica* and *Stephanodiscus astraea*, in Lough Neagh, Northern Ireland. Losses due to washout from the lake were estimated to lie between 0.25% and 0.5% day^{-1}. Parasitism was observed at 2% to 4% of the *M. italica* cells, and grazing by copepods was calculated at 0.05% day^{-1} early in the spring to 0.15% day^{-1} when the diatoms were declining. Once silica became limiting, calm weather led to 90% of the diatoms settling out of the water column. Although virtually the entire crop reached the sediment surface and formed resting stages, grazing by the benthic fauna removed all but 1% of the population. That 1% could provide an inoculum of 65 cells ml^{-1} for the following year.

Reynolds, et al. (1982) examined a wider range of species of phytoplankton for loss processes in their enclosures in Blelham Tarn (U.K.). Estimates were made of net growth rate (r) and losses due to sedimentation and grazing. Small algae, such as *Ankyra*, *Chromulina*, and *Cryptomonas*, were heavily grazed. The diatoms, *Asterionella* and *Fragilaria*, were grazed very little—even when the populations were in decline. Sedimentation was the major loss process. Neither sedimentation nor grazing could account for losses of *Microcystis*. Here, a mechanistic study of death and lysis of *Microcystis* could contribute significantly to our understanding of bloom dynamics.

More recently, Hansson (1996) examined grazing, sinking, and algal recruitment from the sediments in four northern Wisconsin lakes. Sinking losses were negligible because the phytoplankton chosen for study were all motile. Grazing was the main loss process. Algal recruitment from shallow sediments in the littoral zones played a significant part in determining dominant species in the phytoplankton. Algae able to occupy the sediment surface had a refuge from grazing losses and could augment their populations in the water column by 10% to 50% per day. Studies of loss processes and recruitment among different taxonomic groups of phytoplankton underscore the importance of understanding species differences in predicting the outcomes of trophic-level interactions. Thus the characteristics and differences of individual species of phytoplankton can be important even at the level of aquatic management.

Recommended Books

Harris, G. P. 1986. *Phytoplankton Ecology*. Chapman & Hall, London, UK.

Lampert, W., and U. Sommer. 1997. *Limnoecology*. Oxford University Press, New York, NY.

Reynolds, C. S. 1997. *Vegetation Processes in the Pelagic: A Model for Ecosystem Theory*. Ecology Institute, Oldendorf/Luhe, Germany.

Answers:

Box 22–1 Exponential and logistic growth

1. Culture #1: Y = 0.04692(X) + 9.2469 and r = 0.04692 hr^{-1}. Culture #2: Y = 0.04717(X) + 9.5380 and r = 0.04717 hr^{-1}. Culture #3: Y = 0.05285(X) + 8.9762. The average net growth rate of *Chlamydomonas* is 0.04898 hr^{-1} or 1.175 day^{-1}.

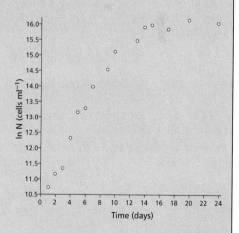

2. In the dark, Y = 0.02797(X) + 6.7091 so r = 0.02797 hr^{-1} or 0.67 day^{-1}. In the light, Y = 0.03482(X) + 7.8708 so r = 0.03482 hr^{-1} or 0.84 day^{-1}. Photosynthesis increased the net growth rate by 0.17 day^{-1}. The corresponding doubling times are 1.03 days in the dark and 0.83 days in the light. Thus, photosynthesis decreased the doubling time by 0.2 days.

3. The net growth rate can be estimated from a linear regression of the first 7 data points (or the first 9). Refer to the figure (below) for this decision. The results are:

> For n = 7: Y = 0.5612(X) + 10.0433 and r = 0.56 day^{-1}
> For n = 9: Y = 0.5013(X) + 10.2387 and r = 0.50 day^{-1}

The last 4 or 5 data points can be used to estimate the carrying capacity K:

> For n = 4, K = 8,949,870 and for n = 5, K = 8,821,480.

   Which is used is a matter of choice, but a useful rule would be to accept the highest estimates consistent with a reasonable number of data points. Thus an estimate based on only two points is not as reliable as one based on three or more.

Box 22–2 Nutrient uptake and growth

1. First take the reciprocals of the values of SiO$_2$ and ρ (see table below). Plot these values as shown in the accompanying figure. The regression is Y = 0.4676(X) + 0.06429 and the correlation coefficient is 0.993.

x = 1/S	y = 1/ρ
1.120	0.602
0.440	0.233
0.259	0.173
0.138	0.149
0.066	0.112
0.037	0.080

Now $1/\rho_{max} = 0.06429$ and so $\rho_{max} = 15.5 \times 10^{-9}$ μM cell^{-1} hr^{-1}. $K_t/\rho_{max} = 0.4676$ and $K_t = 7.3$ μM SiO$_2$. These are quite close to the values reported in Tilman and Kilham (1976), who used nonlinear regression ($\rho_{max} = 15.1 \times 10^{-9}$ and $K_t = 7.5$). The uptake rate for silicate (15.5) is much higher than that for phosphate (5.4), but the half-saturation constant for phosphate uptake is much lower (0.83) than that for silicate (7.3). This means that *C. meneghiniana* can take up phosphate at much lower external concentrations than it can silicate.

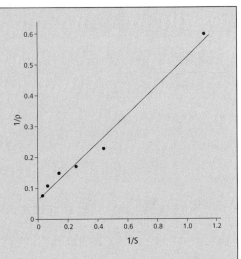

2. The equations are:

$$\frac{1}{\mu} = \frac{K_s}{\mu_{max}}\left(\frac{1}{S}\right)\frac{1}{\mu_{max}}$$

(a)

Fragilaria

$$\mu = 0.62\left(\frac{S}{1.5 + S}\right)$$

Synedra

$$\mu = 1.11\left(\frac{S}{19.7 + S}\right)$$

(b)

The plot is shown in the accompanying figure (right). According to these graphs, if the supply rate of SiO$_2$ is less than about 22 μM, *Fragilaria* should be able to outgrow *Synedra*. But if the silicate supply remains above 22 μM silicate, *Synedra* should be able to outgrow *Fragilaria*. This could lead to a replacement of *Synedra* by *Fragilaria* as silicate is depleted from well-mixed spring surface waters.

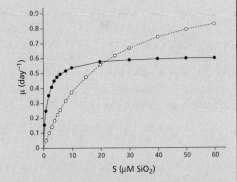

3. Take the reciprocals of the data in the table (below). Then plot the values of 1/S and 1/μ on graph paper as shown. Now take the linear regression of 1/μ on 1/S using equation a (above).

x = 1/S	y = 1/μ
0.2860	1.97
0.2700	1.94
0.1540	1.45
0.1430	1.43
0.0980	1.28
0.0758	1.23
0.0538	1.20
0.0454	1.07

The result is Y = 3.5982(X) + 0.9398 for which the correlation coefficient is 0.994. $1/\mu_{max} = 0.9398$ so $\mu_{max} = 1.06$ day^{-1}. $K_t = 3.5982$ (μ_{max}) and therefore $K_t = 3.83\mu$M SiO$_2$. The values reported by Tilman and Kilham (1976) were $\mu_{max} = 1.06$ day^{-1} and $K_S = 3.94$ μM.

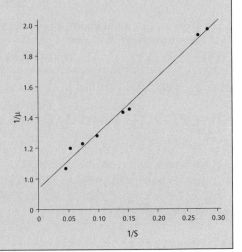

Box 22–3 Sinking rates/Stokes' law

1. $v_s = 32.51$ μm s^{-1} $\phi = 4.43$
2. $v_s = 0.75$ μm s^{-1}. Small cells sink much slower than large ones.
3. $v_s = -13.09$ μm s^{-1} or -0.047 m hr^{-1}. This ascent rate is much slower than the rates reported in the text, suggesting that cyanobacteria can achieve much lower densities through gas vacuoles than in this example.
4. Since g on Mars is (0.38) 9.8 m s^{-2} or 3.72 m s^{-2}, *S. astraea* would sink at the rate of 9.49 μm s^{-1}, or 10.5 μm s^{-1} if you used the observed value of Reynolds (1984).

Box 22–4 Tilman's mechanistic model of competition

1. The graphs showing the Monod growth curves for each diatom and limiting nutrient are given below (a through d). The calculated values of R* for *Fragilaria* are: 0.005 μM PO$_4$ and 1.0 μM SiO$_2$. R* values for *Synedra* are: 0.002 μM PO$_4$ and 5.7 μM SiO$_2$. The following figure (e and f) shows the same resource plane with the consumption vectors based on the ratio of half-saturation constants.

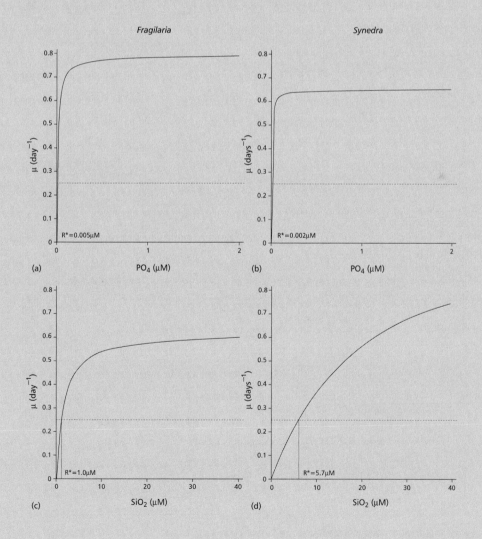

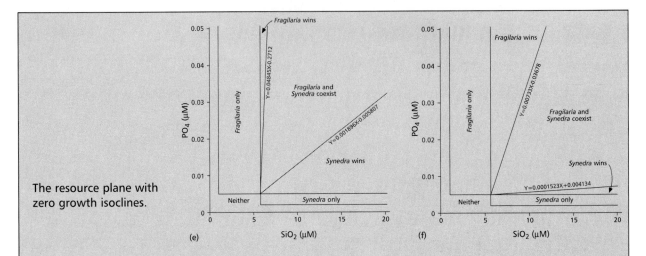

The resource plane with zero growth isoclines.

2. The values of R* for each species and nutrient are given in the table below. From these values the zero growth isoclines can be plotted as shown below:

	R* (μM NO$_3^-$)	R* (μM SiO$_2$)
Corethron criophilum	0.54	107.3
Nitzschia kerguelensis	0.64	71.5
Thalassiosira subtilis	1.50	9.5
Nitzschia cylindrus	3.10	6.2

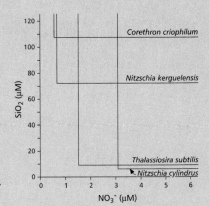

Zero growth isoclines.

There are three significant intersection points: (0.64, 107.3), (1.5, 71.5) and (3.1, 9.5) where the values are (μM NO$_3^-$, μM SiO$_2$). In addition, there are three other intersection points of the zero growth isoclines. These intersections only have meaning in two species interactions; in the four species system any population at these junctions would fall back to the isoclines further to the left.

There are two consumption vectors radiating from each of the significant intersection points listed above. The vectors have been plotted on the graph below. The equations of these lines radiating from each point of intersection are:

point of intersection	*Corethron criophilum*	*Nitzschia kerguelensis*
(0.64, 107.3)	Y = 200.3(X) − 21.9	Y = 110.8(X) + 35.8
	Nitzschia kerguelensis	*Thalassiosira subtilis*
(1.5, 71.5)	Y = 110.8(X) − 94.7	Y = 6.33(X) + 62.0
	Thalassiosira subtilis	*Nitzschia cylindrus*
(3.1, 9.5)	Y = 6.33(X) − 10.12	Y = 2(X) + 3.3

(Consumption vectors based on nutrient quotient.)

Chapter 23

Pelvetiopsis

(Photograph courtesy C. Taylor)

Macroalgal and Periphyton Ecology

Macroalgae, also known as seaweeds, and **periphyton** algae (smaller unicellular, colonial, or filamentous algae) typically grow attached to substrata, in contrast to the floating or swimming phytoplankton discussed in the previous chapter. Substrata for macroalgae and periphyton include rocks, coral, carbonate or silica sands, other algae, and aquatic animals and plants. Most seaweeds occur along ocean coastal regions (Fig. 23–1), but a few, such as *Cladophora*, *Ulothrix*, and *Bangia*, have colonized freshwaters. Periphyton, also known as microphytobenthos or *Aufwuchs*, occur in both the oceans and freshwaters. Both macroalgae and the periphyton are immensely important as they provide the community structure and primary productivity that supports a wide array of other organisms. An understanding of the composition and function of periphyton and/or macroalgal communities is essential to effective management, protection, and restoration of freshwater wetlands, lakes and streams, coastal salt marshes and mangrove forests (mangals), coral reefs (see Text Box 23–1), and rocky ocean coastlines.

On a global basis, nearshore ocean waters inhabited by macroalgae represent sites of intense primary productivity that support coastal ecosystems, including fisheries. Primary productivity in seaweed communities is equal to or greater than that of the most productive terrestrial plant communities. For example, productivity of the *Postelsia* community, which inhabits high energy, wave-exposed sites along the Pacific U.S. coast (Fig. 23–2), is estimated to be 14.6 kg m^{-2} yr^{-1}, in contrast to less than 2 kg m^{-2} yr^{-1} for rain forests (Leigh, et al., 1987). Kelp-dominated seaweed communities along the eastern U.S. coast have an estimated primary productivity of 1.75 kg m^{-2} yr^{-1}. An estimated 10% of seaweed biomass is directly consumed by herbivores, whereas 90% enters detrital food webs, thereby contributing indirectly to animal biomass (Branch and Griffiths, 1988) (Fig. 23–3).

Fossil evidence (Xiao, et al., 1998) suggests coastal marine seaweed communities (including red, brown, and green macroalgae, as well as periphyton forms much like modern *Bangia*) have been in existence for the past 500 or more million years, most likely playing a similarly important role in marine primary productivity throughout this time. Coastal marine communities have proven valuable

603

Figure 23–1 Macroalgae (seaweeds) are commonly seen along coastal areas of the world, including Rialto Beach, Washington, U.S.A.

as ecological models for determining the relative importance of physical factors, algal morphological and physiological adaptations, and biotic factors such as herbivory and interspecific competition in structuring macroalgal communities.

Freshwater periphyton communities also contribute significantly to primary productivity, particularly in streams or lakes having a relatively large proportion of nearshore substrate to open water. In many bodies of water, periphyton contribute more productivity than do the phytoplankton. For example, in Lawrence Lake (located in southwestern Michigan, U.S.A.), less than one third of the primary productivity was due to the dense growth of submerged macrophytes, and phytoplankton contributed only about 13%, but periphyton generated 71% of the primary productivity (Burkholder and Wetzel, 1989). Similar ratios are thought to be representative of other shallow lakes (Wetzel and Søndergaard, 1998). Nutrient perturbation of freshwater bodies may result in nuisance growths of periphytic algae such as *Cladophora*, which have deleterious effects on

the quality of such waters for fisheries production or human recreational uses.

As a result of human population pressures, coastal ecosystems and inland freshwaters around the world are being seriously affected by inputs of nutrients and contaminants, exposure to increased levels of ultraviolet radiation, overexploitation of resources, loss of species diversity, and colonization by nonnative species. Data are needed to more accurately estimate ecosystem health and make predictions regarding the direction and effects of change. Most studies of nearshore communities have been performed at very small scales, on local levels, with just a few ecosystem components or for short time periods. Many ecologists believe that it is necessary to undertake longer-term, regional and global scale analyses that include as many of the significant biotic and abiotic ecosystem components as possible. The U.S. National Science Foundation's Long-Term Ecological Research (LTER) Program includes several intensively studied aquatic sites, while the Global Ocean Observing System (National Research Council) is an example of an international

Figure 23–2 The annual sea palm, *Postelsia*, is characteristic of high-energy environments where heavy wave action removes sessile animals that would otherwise compete for space. *Postelsia*'s flexible stipe and extensive holdfast are regarded as adaptations to wave drag forces.

effort to increase the range of scientific information available to coastal zone planners and policy makers. Remote sensing techniques—either airborne or satellite radiation scanners or acoustic detectors—are now being used to survey seaweeds and freshwater algae; technological progress in this field has recently been reviewed by Guillaumont, et al. (1997).

Macroalgal and periphyton ecologists are interested in defining the factors that influence algal occurrence and distribution patterns. They ask questions such as "Why do particular seaweeds or periphyton algae grow where they do? How do structural or life-history characteristics of attached algae or pelagic seaweeds relate to environmental variation? Under what conditions is a particular algal community regularly replaced by another? To what extent do functions of algal communities propagate to higher ecosystem levels? What effects do environmental perturbations have upon key community functions?" Studies suggest that in both seaweed and periphyton communities, there is a hierarchy of controlling factors, with the potential limits of algal growth set primarily by the interplay between algal physiological constraints and physical environmental factors. Within these limits, biotic factors—specifically grazing by herbivores and the activities of pathogens—can modify algal distribution patterns. Because interspecific competition can only occur in a stable environment (see Chapter 22), competition is most important

as a structuring agent when disturbance and stress are at low levels (Carpenter, 1990). Commonly, the habitats of seaweeds and periphytic algae are highly variable, or have a high level of disturbance. In such conditions, interspecific competition is not regarded as being a particularly influential factor. However, several studies have revealed situations in which reduced levels of herbivory can occur, either because herbivore populations are relatively low or because the algae possess herbivory avoidance mechanisms. In such cases, competition among algal species may control patterns of succession. There is also some evidence for the occurrence of intraspecific competition (self-thinning) in dense algal growths. In view of this apparent hierarchy of controlling factors, physical factors and associated algal adaptations will be discussed first, followed by consideration of the roles of herbivory and competition.

Although there are some circumstances in which growth of freshwater periphyton (e.g., the transmigrant red *Bangia atropurpurea* and greens *Ulothrix zonata* and *Cladophora glomerata*) greatly resembles that of marine turf-forming seaweeds, the habitats of seaweeds and marine periphyton on the one hand, and freshwater periphyton on the other, typically differ substantially in degree of wave energy, tide-related exposure to desiccation and solar radiation, salinity, temperature variation, nutrient supply, and types of herbivores. As a consequence, organismal sizes and

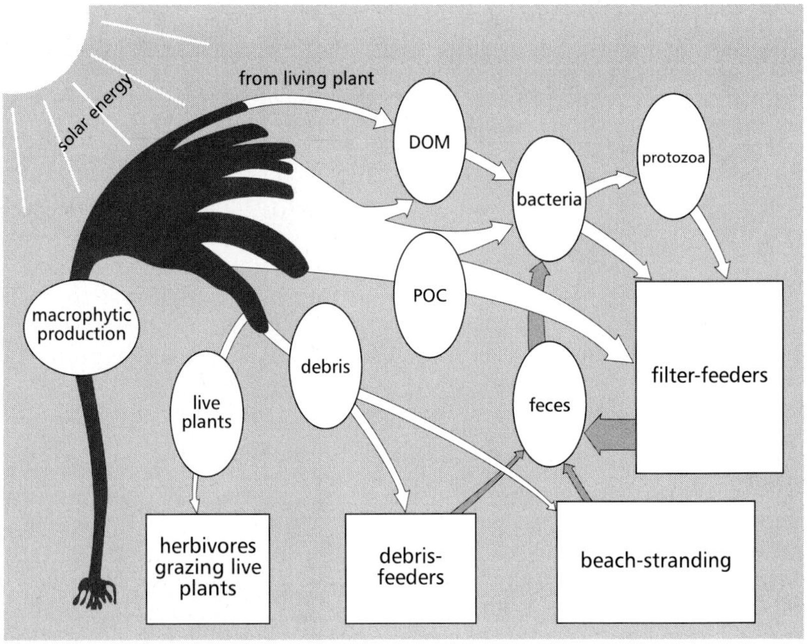

Figure 23–3 The fate of macroalgal primary production. Arrow thickness indicates relative amounts of carbon excreted from living seaweeds in the form of dissolved organic material (DOM), carbon consumed by grazers on living seaweed, and particulate organic carbon (POC) and debris resulting from macroalgal decay. (After Branch and Griffiths, 1988)

compositions of marine macroalgal and freshwater attached algal communities differ greatly. The branched, filamentous green alga *Cladophora* is the largest, most conspicuous alga occurring in the freshwater periphyton. Attached diatoms, cyanobacteria, and filamentous eukaryotic algae dominate the microscopic periphyton world of both fresh and marine waters. In contrast, extensive growths of larger red, green, and brown macroalgae, or seaweeds, occur in marine coastal regions, but not in freshwaters. The first portion of this chapter focuses upon marine macroalgae, a second section is concerned with turf-forming periphyton algae of marine waters, and freshwater periphyton algae are discussed in the last section.

PART 1—MARINE MACROALGAL ECOLOGY

Physical Factors and Macroalgal Adaptations

Coastal regions are influenced strongly by tide-related variations in water level and other water motions, including waves and currents. These, combined with weather changes and seasonal variation in solar irradiance levels, subject seaweeds exposed at low tides to frequent variation in light, temperature, nutrient avail-

ability, and level of hydration. Seaweeds that are exposed to air (emersed) for long periods lose their source of dissolved nutrients, may become desiccated, and may receive damaging levels of solar radiation. When covered by seawater (immersed) from the incoming tide a few hours later, the same seaweeds may experience much reduced irradiance levels, and be subjected to stressful mechanical forces from wave action. Water motions also influence the abundance and activities of herbivores, and the extent to which interspecific competition occurs. Hence, our discussion of the physical factors influencing seaweed distribution will begin with tides and other water motions, their effects on algal distribution patterns, and ways in which macroalgae are adapted to different degrees of water motion. This will be followed by examples of ways in which seaweeds deal with variation in light environment, salinity changes, desiccation, and nutrient availability.

Tides

Tidal changes in water level are characteristic of most marine shorelines, though the magnitude of water-level changes can vary greatly from one geographical region to another. Tides are periodic, gravitation-induced waves generated by attraction of the sun and moon. The moon has the larger effect because it is closer to the earth. Lunar tidal cycles have a period-

Text Box 23–1 Algae and Coral Reefs

Although reefs of more ancient type have occurred, the modern coral reef form originated more than 200 million years ago (Wood, 1998). The optimum temperatures for coral reef development are between 23° and 29° C. Today, coral reefs occur between the latitudes of 30° N and 30° S. Biogeographically, there are two major reef regions: slowly sinking Pacific Ocean volcanic islands, and platforms of Pleistocene age in the Caribbean Sea. Coral reefs are common throughout the tropics, except the west coast of Africa, where upwelling is thought to provide increased nutrient levels, which favor development of macroalgae. Generally, higher levels of dissolved phosphate are correlated with competitive superiority of macroalgae, whereas lower phosphate levels favor coral reef development (Lüning, 1990). Competition between reef-building corals and seaweeds for domination of open, sunlit substrates has been hypothesized to be a direct cause of the restriction of coral reefs to tropical waters (Miller, 1998).

As noted in Chapter 7, most reef-building (scleractinian) corals contain zooxanthellae, and thus have a light requirement of at least 3% of surface irradiance. Though such corals may be found at a maximal depth of 145 m, they grow optimally at depths less than 10 m. Vertical growth rates of coral reefs range from 5–10 mm yr^{-1}. Over millions of years, such growth has built structures that are hundreds of meters thick (and over 1300 m in the case of Enewetak Atoll). There are three major classes of coral reefs: atolls, barrier reefs, and fringing reefs. Atolls are ring-shaped coral islands that nearly or completely surround a central lagoon. Barrier reefs are ridges of coral, close to, and parallel to the coastline, that can be 300–500 m wide and up to 2000 km long (in the case of the Great Barrier Reef). Fringing reefs occur quite close to shore, with little or no water between. Living shallow water coral reefs presently cover about 600,000 square km.

(Reef distribution from World Conservation Monitoring Centre)

Coral reefs are important in long-term global C-cycles; an estimated 700 billion kg of carbon are deposited as calcium carbonate in coral reefs every year. In addition, reefs are a major repository of biodiversity, harboring the greatest vertebrate diversity of any community on earth. Further, they are among the most productive aquatic communities, offering a wealth of resources to humanity, but are extremely sensitive to overexploitation (Birkeland, 1997).

Algae of coral reefs include the essential endosymbiotic zooxanthellae of the *Symbiodinium* complex (Chapter 7), turf communities, siphonalean green algae (Chapter 18), and a variety of red algae, notably the reef-consolidating encrusting corallines (Chapter 16). Two thirds of coral reef productivity is attributed to the dinoflagellates, and one third to the other algae. The zooxanthellae, which normally occur at densities of over 10^6 cells cm^2 of coral surface, greatly enhance coral calcification rates (Muller-Parker and D'Elia, 1997). Several species of filamentous cyanobacteria are known as endolithic microborers because their growth within the coral rock results in its dissolution. Such coral-eroders include *Hyella*, *Plectonema*, *Mastigocoleus*, and *Entophysalis*; they occur from the splash zone to 75 m deep. *Ostreobium* is a siphonalean green algal borer that often forms a green band (the "*Ostreobium* band") a few cm beneath the surface of living coral. The relative importance of algae as bioeroders is controversial, with some experts believing that algae are significant and others, that they are not (Glynn, 1997).

Coral reefs are increasingly suffering from anthropogenic effects, including increases in seawater temperature and violent weather, which have been attributed to global climate change, increased UV exposure, oil pollution, coral mining, overfishing, sedimentation, and nutrient enrichment (Brown, 1997). Corals respond to sharp increases (3°–4° C) for a short time (several days), or more moderate temperate increases over longer periods, by undergoing bleaching—loss of zooxanthellae or loss of zooxanthellae pigments. If not too severely damaged, coral

continued→

Text Box 23–1 *continued*

can "rebrown," acquiring a new population of zooxanthellae that may be better adapted to the environmental change (Muller-Parker and D'Elia, 1997). Stressed corals are more vulnerable than healthy corals to pathogenic infestations such as black band disease. This disease is caused by the cyanobacterium *Phormidium corallyticum* and associated microorganisms, which occur as a black mat a few millimeters wide that moves across the coral surface at the rate of a few mm per day. The microbes generate anoxic conditions and hydrogen sulfide that kills coral tissues, whereupon the microbes use the organic compounds released from the dying coral. A bare coral skeleton is left behind (Peters, 1997).

Overfishing removes herbivorous fish, such as surgeonfishes, rabbitfishes, parrot fishes (Choat and Clements, 1998), and damselfishes that would otherwise prevent domination of coral reefs by algae. Intense herbivory by such fish, as well as polychaete worms, limpets, amphipods, isopods, and certain crabs is regarded as essential for maintenance of reef-building corals (Carpenter, 1997; Hixon, 1997). In the absence of herbivores, reefs can be smothered by excessive algal growths, particularly when nutrients are abundant.

icity of 24 hours and 50 minutes, whereas solar tidal influences run on a slightly shorter, 24-hour cycle. The strongest gravitational effects occur when the earth, sun, and moon are in a linear alignment, i.e., when there is a full or new moon. This results in very high and low "spring" tides. The weakest, or "neap," tides occur when the sun, moon, and earth are oriented perpendicularly, i.e., when the moon is in the first or third quarter. There is a thus a monthly cycle of neap and spring tides (reviewed by Dawes, 1998).

The frequency and amplitude of tides are affected by the morphology of the ocean basin, such that different coasts may have distinct types of tidal cycles. Uncommonly, there is a single high and low tide per day; such a tidal cycle can occur in the Gulf of Mexico. Along the open coast of the Atlantic, two high and two low tides, all of which are equal in amplitude,

occur per day. Along the Pacific and Indian ocean coasts, as well as in the Caribbean and the Gulf of St. Lawrence, there are also two high and low tides per day, but these are of unequal magnitude (Fig. 23–4). Weather and large-scale climatic features such as El Niño can also influence tides.

Tidal patterns influence the structure of algal communities by creating an intertidal region—the portion of shoreline lying between a line defined by the highest tides and another by the lowest tides (i.e., the intertidal is submerged at high tide and exposed at low tide). Higher portions of the intertidal region are exposed for significantly longer periods of time than the lower, creating environmentally distinct zones of organismal distribution. Critical tide levels are defined as those that result in significant increases or decreases in the period of submergence or emersion; they may

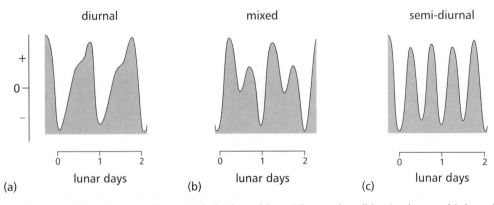

Figure 23–4 Types of tides: (a) diurnal—a single high and low tide per day; (b) mixed—two high and two low tides per day (of unequal amplitude); (c) semidiurnal—two equal high and two equal low tides per day. (Redrawn from *Journal of Experimental Marine Biology* Vol. 62. Swinbanks, D. D. 1982. Intertidal exposure zones: A way to subdivide the shore, pages 69–86. ©1982 with permission from Elsevier Science)

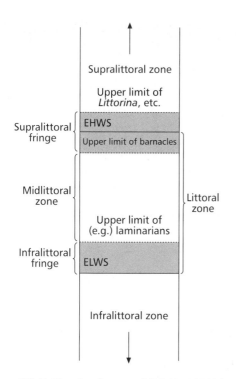

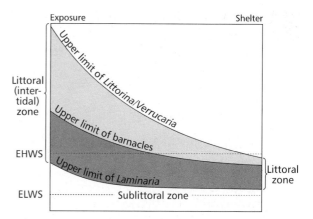

Figure 23–6 Lewis's (1964) modification of the Stephensons' scheme, showing variation in the extent and location of intertidal zones between extremely exposed and extremely sheltered sites. (Modified from Lewis, 1964)

Figure 23–5 The Stephensons' (1949, 1972) intertidal zonation system of terminology, based on both water levels—extreme high water of spring tides (EHWS) and extreme low water of spring tides (ELWS), and organismal distributions—locations of littorinid snails, barnacles, and laminarialean brown seaweeds. (Modified from Stephenson and Stephenson, 1949 ©British Ecological Society)

restrict occurrence of some seaweeds to specific zones in the intertidal region (Doty, 1946). In a worldwide survey, Stephenson and Stephenson (1949; 1972) found that such zones are defined by both water levels and the types of organisms present in them, and developed a system of terminology (Fig. 23–5). Their system is still used, modified by Lewis (1964) (Fig. 23–6), who took into account the effects of wave action on exposed versus protected shores. Some variation exists among different countries in the terms used to denote particular shoreline zones. Here, we use the terms "littoral zone" and "intertidal" regions interchangeably to mean the stretch of shoreline between the extreme high water of spring tides (EHWS) and extreme low water of spring tides (ELWS). The supralittoral region lies above the littoral, and receives seawater spray, whereas the sublittoral region (also known as the subtidal or infralittoral) is submersed

Figure 23–7 One of three universal intertidal zones, the "black zone" (arrowheads) observed in the supralittoral, is the habitat of desiccation-resistant cyanobacteria, lichen, and snails.

Figure 23–8 Barnacles (arrowheads), shown here with the brown seaweed *Hedophyllum*, are characteristic of the intermediate of the three universal intertidal zones.

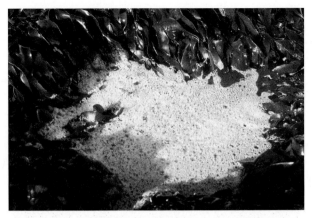

Figure 23–9 *Laminaria* or other brown algae are characteristic of the lowermost of the three universal intertidal zones. The foam shown in the center of this photo is formed by incorporation of air (via wave energy) into colloidal suspensions of polysaccharides (alginates) exuded by the kelp. This is known as kelp foam.

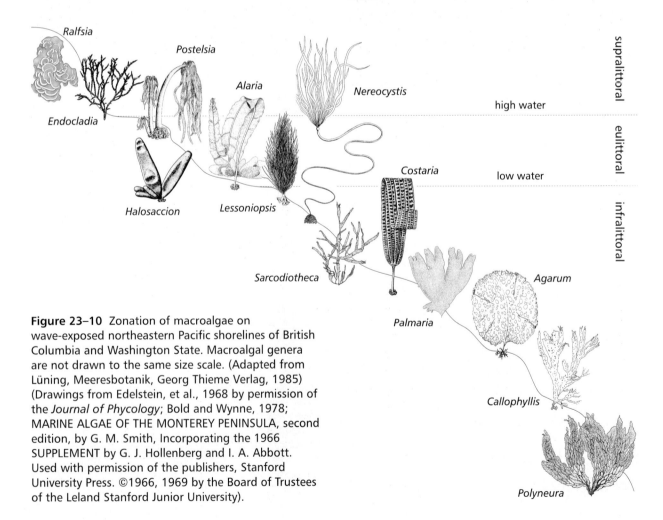

Figure 23–10 Zonation of macroalgae on wave-exposed northeastern Pacific shorelines of British Columbia and Washington State. Macroalgal genera are not drawn to the same size scale. (Adapted from Lüning, Meeresbotanik, Georg Thieme Verlag, 1985) (Drawings from Edelstein, et al., 1968 by permission of the *Journal of Phycology*; Bold and Wynne, 1978; MARINE ALGAE OF THE MONTEREY PENINSULA, second edition, by G. M. Smith, Incorporating the 1966 SUPPLEMENT by G. J. Hollenberg and I. A. Abbott. Used with permission of the publishers, Stanford University Press. ©1966, 1969 by the Board of Trustees of the Leland Stanford Junior University).

most of the time and its lower limit is defined by the deepest-growing algae. For most regions the depth of the deepest-occurring seaweeds is unknown, as survey by submersible is required to make such determinations (Lobban and Harrison, 1994).

Stephenson and Stephenson found that there are three nearly universal intertidal zones: (1) an uppermost black strip of highly desiccation-tolerant cyanobacteria, a marine lichen (*Verrucaria*), and littorinid snails (Fig. 23–7), (2) an intermediate zone of various seaweeds together with barnacles and limpets (Fig. 23–8), and (3) a lowermost zone inhabited by laminarialean brown algae (in temperate or high latitude environments) or corals (in tropical regions) (Fig. 23–9). Aside from these general features, zonation patterns may differ considerably from one biogeographical region to another. Zonation of seaweeds typical of wave-exposed shorelines of British Columbia (Canada) and Washington State (U.S.A.) is illustrated in Figure 23–10, seaweeds often found on the wave-exposed southern coast of Newfoundland are diagrammed in Figure 23–11a, and seaweed zonation on a Japanese shore (Rikuchu National Park on North Honshu Island) is shown in Figure 23–11b. Additional examples are illustrated by Lüning (1990). Community analysis is relatively easy to accomplish in the intertidal. At low tide, transect lines can be laid down the shoreline and quadrats used to defined smaller areas for detailed quantification and identification of seaweeds.

Zonation also occurs in the sublittoral (subtidal) region, as the result of light attenuation with increasing depth. Upper, middle, and lower sublittoral zones have been defined, which are characterized by distinct seaweed assemblages (Fig. 23–12). The mid-sublittoral kelp community with the canopy-forming *Macrocystis* and understory brown and red seaweeds is depicted in Figure 23–13. Characterization of sublittoral communities is considerably more logistically challenging than in the intertidal. Although some information can be obtained by study of samples obtained by means of a dredging apparatus operated from a boat, SCUBA or submersible technology is required for more detailed analysis. Several prominent phycologists have lost their lives in diving accidents while engaged in the pursuit of knowledge about sublittoral macroalgal communities (Lüning, 1990).

Waves and Currents

Waves are periodic surge events generated by wind, tides, earthquakes, and landslides; small-scale surface

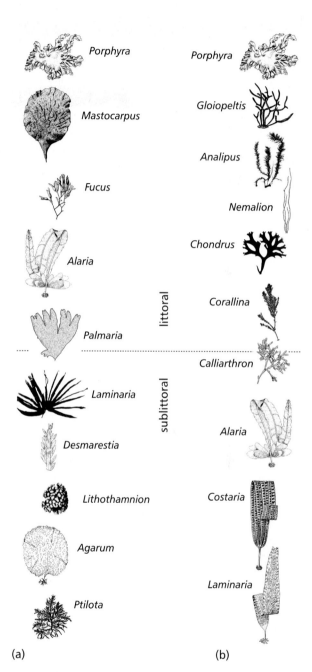

Figure 23–11 Comparison of macroalgal zonation patterns along (a) the northwestern Atlantic (Newfoundland) and (b) northwestern Pacific (Japan). Supralittoral forms are at the top, and sublittoral forms are at the bottom of the diagram. (Seaweeds are not drawn to the same size scale.) (Adapted from Lüning, *Meeresbotanik*, Georg Thieme Verlag, 1985) (Drawings from Smith, 1969; Bold and Wynne, 1978; 1985; Woelkerling, 1988)

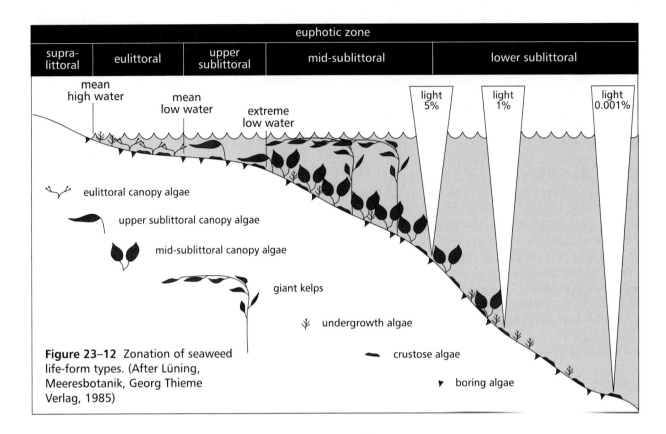

Figure 23–12 Zonation of seaweed life-form types. (After Lüning, Meeresbotanik, Georg Thieme Verlag, 1985)

currents are unidirectional motions that are strongly influenced by coastal topography. Waves and currents are a tremendous source of energy to nearshore environments, contributing some ten times more energy than does solar radiation. This extraordinary energy input explains the high productivity of coastal macroalgal communities described earlier. Thus, water motions have positive effects on seaweeds. By constantly moving the fronds of macroalgae, waves and currents reduce shading and bring a constant resupply of inorganic nutrients to the algae. They also reduce nutrient depletion by reducing the thickness of the boundary layer of unmoving water associated with seaweed surfaces. Such boundary layers act as a diffusive barrier to the movement of nutrients from the water column to the algal surface, and the thickness of the boundary layer influences the rate of diffusion. For example, the rate of nitrate uptake by *Macrocystis pyrifera* increases by 500% if the current velocity increases from 0 to 4 cm s^{-1}, allowing the photosynthetic rate to increase by 300% (see references in Lüning, 1990). When phycologists grow seaweeds in culture, aeration or other motion has the same effect—reduction of boundary

layers and increase in nutrient uptake and growth rates (Hanisak and Samuel, 1987). In addition, waves and currents can remove sessile animals (such as mussels) that compete with macroalgae for attachment space in the intertidal zone. Thus the annual seaweed *Postelsia* is dependent upon wave removal of barnacles and mussels for establishment. The tropical sea-grass bed inhabitants *Penicillus* and *Codium* reportedly require water motion for normal morphological development; in culture, the multiaxial thallus develops only if thalli are shaken (Ramus, 1972).

Seaweeds also experience negative impacts of water motions, including removal, destruction, or damage as the result of mechanical forces. Water-motion velocities can exceed 10 m s^{-1}, and because of the density of water, waves and currents can exert much stronger forces than wind. Existence in such an environment has been compared to that of humans experiencing a hurricane (Lobban and Harrison, 1994). Mathematical expressions of water motions and the resulting forces operating upon seaweeds (and attached animals) are described in Denny (1988), Lobban and Harrison (1994), and Vogel (1994), and references cited therein.

Phycologists visualize water motions at the surfaces of macroalgae using dyes and/or flakes of mica. These methods are then used to study the responses of seaweeds of different morphological type to currents of various flow rates established in seawater tanks (for an example, see Hurd and Stevens, 1997). Macroalgal surfaces slow down water, setting up fluid or surface shear stress. Shear stress is a force applied parallel to a surface, in contrast to compressive force, which operates perpendicularly to surfaces. Drag forces pulling on seaweeds are proportional to the square of the velocity of the water, and the larger the organism, the more it will experience such acceleration forces. Seaweeds are morphologically adapted in ways that allow them to cope with these forces. Many large seaweeds have stretchy, flexible thalli; such flexibility reduces the effects of drag forces, allowing them to grow to greater size than if they were stiff. The flexible stipe of the sea palm *Postelsia* helps it cope with heavy wave action on extremely exposed shorelines (Fig. 23–2). The length of *Nereocystis*'s flexible stipe appears to be tuned to resonate with wave frequencies (Denny, 1988). Midribs, such as

those in the blades of *Alaria*, may function to reduce tearing across the thallus. *Fucus* and some other seaweeds resist mechanical damage by having strong thalli. Seaweeds of the intertidal possess effective attachment adhesives (Vreeland, et al., 1998) and holdfasts that resist separation from substrates. The force required to detach the holdfast of seaweeds such as *Fucus* or *Laminaria* is about 40 kg cm^{-2} (Schwenke, 1971). Other morphological adaptations that reduce the effects of drag forces include growth as bushy forms, short turfs, or flat adherent crusts. Flat, encrusting algae and microscopic germlings avoid damage by living within the boundary layer of the substrate to which they are attached. Crustose corallines are the only seaweeds that can resist the extreme wave environments of the coral reef crest. Crustose or highly reduced stages in the heteromorphic life histories of various intertidal brown, red, and green seaweeds could help them survive water motions that are more violent at particular times of the year.

As currents flow over macroalgal surfaces, eddies are set up at the edges, contributing to turbulence, which involves vertical motions as well as laminar

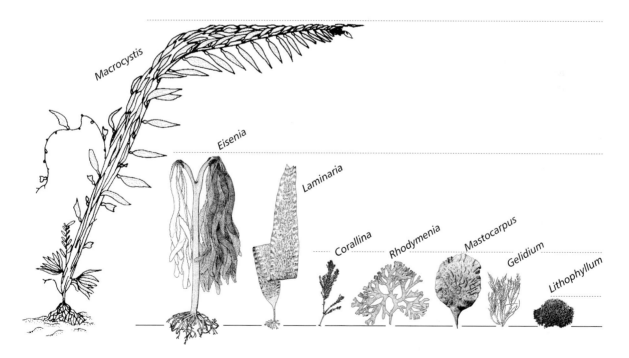

Figure 23–13 Composition of mid-sublittoral macroalgal communities dominated by the giant kelp *Macrocystis*, such as occur on the California (U.S.A.) coast. *Macrocystis* is a canopy species that modifies the light environment of smaller understory seaweeds, including the brown algae *Eisenia* and *Laminaria* and the red algae *Corallina*, *Rhodymenia*, *Gigartina*, *Gelidium*, and *Lithophyllum*. (Modified from Lüning, Meeresbotanik, Georg Thieme Verlag, 1985) (Drawings from Smith, 1969; North, 1994; Woelkerling, 1988)

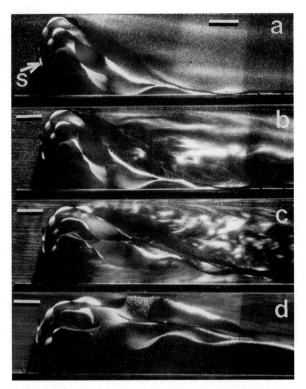

Figure 23–14 Patterns of the flow of seawater around a region of stipe (S, arrow) bearing three blades of *Macrocystis integrifolia* at four different velocities (a—0.5, b—1.5, c—3, and d—5 cm s^{-1}) in a recirculating flow tank. Shear is visualized by the use of mica flakes, which minimize drag by aligning with the shear and thus become less reflective (the seawater appears black). Where shear forces are weak, the mica particles are randomly aligned, scatter light, and thus appear white. (Scale bars=5 cm) (From Hurd and Stevens, 1997 by permission of the *Journal of Phycology*)

flow. If the seaweed surface is rough, turbulent motions will develop more rapidly than if the surface is smooth. Turbulent motions reduce the thickness of the boundary layer of slow-moving water very close to the algal surface, thus increasing the availability of nutrients. Studies of seaweed morphology have revealed correlations between particular structural features and turbulence features of the environment. For example, sublittoral *Macrocystis* blades are characterized by both a wrinkled surface and marginal spines. These features become more prominent in kelps growing in rough waters and thus are thought to reduce the damaging effects of drag forces. In low- and medium-current environments, wrinkled surfaces and spines are thought to serve to enhance nutrient uptake (Lobban and Harrison, 1994).

Nereocystis, a sublittoral kelp, has narrow, smooth blades in rapidly moving water, but wider, ruffled blades in more sheltered habitats. The latter blade type is thought to be advantageous in creating turbulence that helps reduce surface boundary layers and increase nutrient acquisition, but tends to tear when flow rates are high. Smooth blades, in contrast, tend to bundle together, forming a streamlined profile that is more resistant to damage. However, the latter condition may contribute to self-shading and reduction in ability to obtain nutrients, because individual blades are less well exposed to the water (Koehl and Alberte, 1988). The cabbagelike intertidal brown alga *Hedophyllum* likewise has narrow, flat blades in high-energy habitats, but puckered (bullate) blades in calmer sites (Armstrong, 1987). The conspicuous holes in blades of the brown seaweed *Agarum* and the red *Rhodymenia pertusa* likely represent morphological adaptations that increase turbulence on the blade surface thereby enhancing nutrient uptake.

Recently, water moving past single blades of the giant kelp *Macrocystis* and entire thalli of *Nereocystis*, *Laminaria*, *Costaria*, *Alaria*, *Egregia*, and *Fucus* was found to undergo transition from laminar flow to a turbulent boundary layer at a current velocity of around 2 cm s^{-1}, which is at the lowest end of the range of water velocities experienced by seaweeds (Hurd and Stevens, 1997) (Fig. 23–14). These results suggest that all seaweeds that are within the general size range of single *Macrocystis* blades will achieve turbulent flow across their surfaces when seawater velocity is greater than 2 cm s^{-1}, regardless of the presence of corrugations, undulations, or bulbs (floats or pneumatocysts).

There are some habitats in which seaweeds do not experience much mechanical stress, and consequently appear relatively delicate or fragile. An example is the Arctic—the stormy period occurs in winter, during which time seaweeds are covered by a protective layer of ice. The summer open water period is characterized by relatively calm conditions. Under these conditions, perennial algae may grow to very large sizes (though very slowly), as thalli do not suffer wave damage (Lüning, 1990).

Light

Intertidal seaweeds must cope with both visible light—photosynthetically active radiation (PAR) of wavelengths 400–700 nm—and ultraviolet radiation, particularly UVB (280–320 nm), which may damage

Text Box 23–2 Fluorescence Methods for Assessing Photosynthetic Competency and Nitrogen Limitation

When illuminated by a beam of light positioned at right angles to a detector, chlorophyll exhibits distinctive red fluorescence having an emission peak at 685 nm. The intensity of this fluorescence varies with that of the incident light and can be measured by means of an instrument known as a fluorometer (Geider and Osborne, 1992). The level of fluorescence is an indicator of the functional status of photosystem II, hence, measuring fluorescence is a common way of assessing the degree of photoinhibition of algal cells or tissues. Nitrogen-limitation also results in changes in levels of algal chlorophyll fluorescence. Measurements of algal fluorescence can thus also be used to assess the degree to which natural algal populations are nitrogen-limited (Falkowski and Raven, 1997).

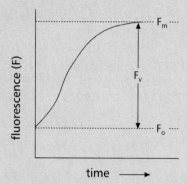

(Redrawn from Falkowski and Raven, 1997 with permission from Blackwell Science, Inc.)

Immediately upon illumination by light that is photochemically active, such as white light, the fluorescence of chlorophyll a increases from essentially zero to a low level (F_0) (see graph above). As the irradiance continues and the sample continues to absorb photons, fluorescence will increase to a maximum level (F_M). The quantity F_M–F_0 is known as the variable fluorescence (F_V). The ratio of variable fluorescence to maximum fluorescence (F_V/F_M) is the maximum quantum yield for stable charge separation at photosystem II—an estimate of the functional status of PSII. (In general, the quantum yield (ϕ) is the ratio of moles of product formed or substrate consumed to the moles of photons absorbed in a photochemical reaction.) The quantum yield for fluorescence (ϕ_f) is the ratio of light emitted as fluorescence to the light absorbed.

Under conditions of maximum quantum yield (F_V/F_M), all PSII reaction centers are considered to be open. Reaction centers are regarded as closed when they are unable to transmit energy to the secondary electron acceptor quinone B (QB). Normally QB is bound to the photosystem II protein D1, which may be damaged or destroyed under high irradiance conditions. The greater the number of closed PSII reaction centers, the greater the amount of energy that will be dissipated as fluorescence. Experiments have shown that the quantum yield of variable fluorescence is inversely related to O_2 evolution. In other words, when PSII is maximally functional, maximal O_2 evolution occurs, but minimal variable fluorescence will be detected. If PSII is not optimally functional, O_2 evolution levels will be reduced, and variable fluorescence will increase.

The quantum yield of fluorescence is an estimate of the fraction of photosynthetically competent reaction centers. Changes in the value of F_0 are associated with the onset of photodamage and signal the induction of photoprotective mechanisms (Falkowski and Raven, 1997). Fluorescence measurement of intact algal cells and tissues can be made in the laboratory, or fluorometers equipped with light sources (needed for stimulating fluorescence) can be lowered into the ocean. Differences between fluorescent quantum yields in the light versus dark can be used to make instantaneous, *in situ* estimates of the rate of photosynthetic electron transport.

DNA and cellular protein, such as the D1 protein essential to photosystem II in chloroplasts. High light conditions cause conformational changes in or destruction of D1, which can affect the efficiency of electron flow from reaction centers to plastoquinone. Thymine dimers may form in DNA that is exposed to UVB, leading to misreading of coding regions and, if repair does not occur, to heritable mutations. Some

algae produce protective UV-absorbing compounds such as β-carotene and aromatic amino acids. Algae possess DNA repair mechanisms, but at high UVB levels, repair rates may not be able to keep up with damage. Damaged proteins must be resynthesized, a process that involves transcription and translation. UVB radiation is a significant component of sunlight that is transmitted well in water. Intertidal and reef surface macroalgae are particularly vulnerable in that they regularly receive biologically effective doses of UVB radiation. In addition, destruction of atmospheric ozone by anthropogenic halocarbons has already been documented to have resulted in increased UVB transmission to the earth's surface; deleterious effects on earth's biota, including macroalgae (Franklin and Forster, 1997) are anticipated.

Several recent studies have revealed that macroalgae such as *Laminaria saccharina*, which are attached in the intertidal region, and *Sargassum natans*, a **pelagic** form occupying surface tropical waters, undergo regulated, photoprotective responses to high levels of solar radiation that involve changes in photosynthetic efficiency. Such changes can be monitored with an instrument that measures *in vivo* fluorescence changes in photosystem II (see Text Box 23–2). High light stress causes a phenomenon known as **photoinhibition**, which results in a depression of photosynthetic activity and, at its extreme, photooxidation of chlorophyll. Photoinhibition is exacerbated by UV (Sagert, et al., 1997), and has been documented in brown, red, and green seaweeds, as well as phytoplankton. Many algae display a daily pattern of photosynthesis characterized by rising rates in the morning, followed by decline at noon, then recovery in the afternoon. The noon decline reflects the effects of photoinhibition, and subsequent increase represents recovery from photoinhibition. Intertidal algae exposed at low tide and surface pelagic forms experience the highest levels of photoinhibition.

During photoinhibition, energy that has been absorbed, but which cannot be used in photochemical reactions, is dissipated in the form of harmless thermal radiation or fluorescence. In green and brown algae such energy dissipation can occur through an increase in the zeaxanthin content of photosystem II (operation of the protective xanthophyll cycle—see Chapter 21). High light induces an increase in the relative proportion of zeaxanthin to violaxanthin. These two carotenoids can be extracted from seaweeds (with 100% acetone), and relative amounts of each measured by use of HPLC (high performance liquid chromatography). Correlation between changes in these xanthophylls and the degree of photoinhibition has been demonstrated in the brown seaweeds *Dictyota dichotoma* (Uhrmacher, et al., 1995), *Laminaria saccharina* (Hanelt, et al., 1997), and *Sargassum natans* (Schofield, et al., 1998), as well as other algae (Falkowski and Raven, 1997).

Another widely employed protective scheme—dynamic photoinhibition—involves an increase in the number of inactive (closed) photosynthetic reaction centers. In this process, photosystem II becomes less efficient in the conversion of light energy into chemical energy, because light-harvesting complexes become disconnected from photochemical energy conversion processes during periods of dangerously high irradiance. When irradiance levels decline, recovery of the photosynthetic system occurs. In *Sargassum natans* the efficiency of energy conversion decreased by 50–60% from predawn levels at noon, then recovered to predawn levels three hours after sunset (Schofield, et al., 1998). An analysis of the sensitivity of *Chondrus crispus* from a range of depths (from 3.5 to 8.5 m below high tide level) revealed that deeper-growing specimens experienced a greater depression of fluorescent yield (see Text Box 23–2) and slower recovery from photoinhibition than did those from shallower water (Sagert, et al., 1997). Studies of photoinhibition and recovery after high light stress in the kelp *Laminaria saccharina* revealed differences in sensitivity among the various life-history stages. Older sporophytes and gametophytes were relatively less sensitive to deleterious effects of high irradiation levels (55 μmol m^{-2} s^{-1} for a period of two hours) than were young sporophytes, which suffered severe photodamage under these conditions (Hanelt, et al., 1997).

Subtidal seaweeds face irradiance environments depleted in both light quantity and quality. At 100–140 m depth in even very clear waters, the irradiance is only about 1% that of the surface. Ocean waters are classified into more than a dozen types based upon light transmission characteristics. In clear waters, differential absorption of short and long wavelengths results in an irradiance environment dominated by blue light (Fig. 23–15a). Turbid coastal waters often contain yellow dissolved organic compounds known as "gelbstoff" (German for gold substance). These are humic materials derived from terrestrial plants. Waters containing such compounds strongly absorb blue light, with the result that the irradiance environment is yellow (Fig. 23–15b).

Macroalgae cope with low irradiance by modifying thallus structure for optimal light absorbance, and red, brown, and green seaweeds employ efficient light-harvesting mechanisms (covered in Chapters 6, 15, 16, and 17). The accessory pigments phycoerythrin, fucoxanthin, β-carotene, and siphonoxanthin are able to absorb green light, often more available at depth,

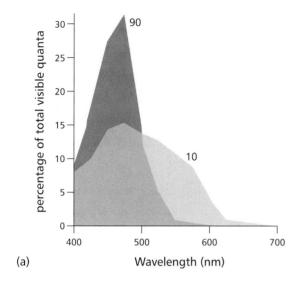

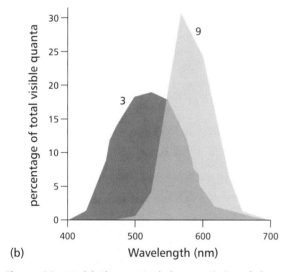

Figure 23–15 (a) The spectral characteristics of clear open-ocean waters at 10 and 90 m depth. The light environment is dominated by blue light. (b) The spectral characteristics of nearshore, coastal waters at 3 and 9 m depth. Blue light is strongly absorbed, and the irradiance environment is dominated by yellow light. (Adapted from Saffo, M. B. *BioScience* 37:654–664 ©American Institute of Biological Sciences)

and transfer this light energy to chlorophyll-based photosynthesis reaction centers. Red algae that grow in deeper waters, such as *Chondrus crispus*, show an increase in phycoerythrin content compared to shallower forms of the same species, and the efficiency of photosynthesis increases with depth (Sagert, et al., 1997). In all algal groups, β-carotene is sited close to

reaction centers of photosystem II, suggesting that efficient energy transfer probably occurs. An old idea that green algae are best adapted to the light environment of shallow waters, red algae to deepest waters, and brown algae to waters of intermediate depth has been refuted (Ramus, 1983; Saffo, 1987). It is now evident that green algae can inhabit very deep waters and that algal communities at all depths are known to be composed of mixtures of brown, red, and green algae. For example, collections made in Hawaii by dredging at 55 m depth revealed the presence of 22 species of red algae (including crustose corallines), ten species of greens (such as *Codium mamillosum*), and five species of brown seaweeds (including *Sargassum hawaiiensis*) (Lüning, 1990).

Polar seaweeds, such as the reds *Constantinea subulifera*, *Maripelta rotata*, and *Delesseria sanguinea*, and the brown kelp *Laminaria solidungula*—which grows to depths of 20 m in the Alaskan high Arctic—experience only short annual periods of illumination. At the northernmost locations at which seaweeds can be collected, there may be only 4–10 ice-free weeks during which illumination levels are sufficient for photosynthesis rates that allow formation of storage products. In fact, 80% of the yearly light supply reaching seaweeds at 10 m depth is received during these few ice-free weeks (Chapman and Lindley, 1980). During August in the Arctic, at a depth of 6–7 m, irradiance levels are 100–200 μmol m^{-2} during the day, and 20–30 μmol m^{-2} at night (Dunton and Schell, 1986). These light levels explain the occurrence of large sublittoral seaweeds at relatively substantial depths in the Arctic, and allow carbon stores to be produced in photosynthesis that can be used for growth during the dark winter.

During the nine months that *Laminaria solidungula* is covered by a turbid ice canopy, the alga loses some 30% of its carbon content. But amazingly, at the same time, it produces a new blade and undergoes 90% of its annual growth before the new season of irradiance begins. Reallocation of stored reserve materials allows growth to occur under such conditions (Dunton and Schell, 1986; Dunton, 1990). Glycolytic catabolism of stored carbohydrate yields the substrates for so-called dark carbon assimilation, a process involving enzymes such as PEPCK (phosphoenolpyruvate carboxykinase), PEPC (phosphoenol pyruvate carboxylase), and PC (pyruvate carboxylase), none of which have any oxygenase activity (in contrast to Rubisco). This process, also known as β-carboxylation (because the β-carbon of 3-carbon substrates is carboxylated) or anaplerotic carbon fixation, also occurs more generally, in many kinds of organisms. β-carboxylation is a source of compounds needed for growth that are not generated by the Calvin cycle, namely several amino acids, tetrapyrroles, pyrimidines, purines, and lipids. β-carboxylation is measured by following the incorporation of ^{14}C in the dark. Typically, β-carboxylation rates rise when ammonium is added to N-deficient algae. Consequently, β-carboxylation has been used as a test for N-limitation in natural populations (Falkowski and Raven, 1997).

Light is also an important environmental signal that controls the timing of seaweed growth and reproduction. Sensor pigments, including a phytochromelike red/far-red reversible protein (Lopez-Figueroa, et al., 1989) and a cryptochromelike blue-absorbing substance (Lopez-Figueroa and Niell, 1989) have been isolated from several seaweeds. These pigments are involved in photoperiodism (dependence of a growth or reproductive response on day length) and photomorphogenesis (dependence upon a particular portion of the spectrum). An example of photomorphogenesis is the blue light induction of oogonia and antheridia in *Laminaria* gametophytes. Short-day (8 hour) (actually long-night) responses of seaweeds include formation by *Laminaria hyperborea* of a new blade in early winter, the spring appearance of the erect thallus of *Scytosiphon* (Chapter 15), transition from the codiolum phase to the blade phase of *Monostroma* via spore production (Chapter 18), and production of conchospores by the conchocelis phase of *Porphyra* and marine *Bangia* (Chapter 4). Conchosporangial production is suppressed by red light (660 nm) and induced by a subsequent exposure to far-red light (730 nm). Long-day (16 hour) responses (actually short-night responses) of seaweeds include formation of gametangia (*Sphacelaria*) and production of the gametophyte phase (*Batrachospermum moniforme*) (Lüning, 1990).

Salinity and Desiccation

The definition of salinity is the weight of solids that is obtained from drying 1 kg of water. The ten most abundant ions in seawater are, in decreasing order: chloride, sodium, sulfate, magnesium, calcium, potassium, bicarbonate, bromide, boric acid, and carbonate (Dawes, 1998). In marine waters, salinities of 25–35‰ (parts per thousand) are most common, but

salinity may range from 10–70‰. Seawater of 35‰ has an osmolality (moles per kg of solvent) of 1050 mosmol kg^{-1} at 20° C.

Tide-pool inhabitants may experience salinity increases as a result of evaporation, and salinity decreases resulting from rainfall. Estuarine seaweeds face variation in salinity resulting from freshwater influxes.

An increase in salinity lowers the external water potential, which triggers rapid cell plasmolysis as well as stress responses by the seaweeds. Responses may include uptake of ions (e.g., K^+, Na^+, or Cl^-); water loss; and synthesis of osmotically active carbohydrates, such as sucrose in green seaweeds, mannitol in brown algae, and digeneaside in reds. Other osmotically active compounds can occur in seaweeds, including amino acids such as proline, quaternary ammonium compounds (betaines), and DMSP (β-dimethylsulfoniopropionate).

Decrease in salinity results in increased external water potential. Responses include uptake of water by the seaweeds with concomitant increase in cell volume and turgor pressure (the internal osmotic pressure against the cell wall or cell membrane), and loss of ions and organic solutes, leading to osmotic adjustment. If plasmolyzed cells are exposed to solutions having a lower ionic content that their cytoplasm, they undergo deplasmolysis; if this is too extensive or rapid, cell damage may result. Regardless of whether salinity is increased or decreased, if osmotic adjustments are not made, damage to cell membranes, organelles, and enzymes (and possibly cell death) will result. Extended exposure to higher than optimal salinity inhibits cell division and may result in stunted growth and abnormal branching patterns.

Intertidal seaweeds tolerate a wide range of salinities (10–100‰), while subtidal forms tolerate a narrower range (18–52‰). *Ulva* and *Enteromorpha* are remarkable in that they can be repeatedly plasmolyzed and deplasmolyzed without harm, a feature that explains their success in estuarine environments. Red and brown algae, in contrast, rarely penetrate as far into estuaries as do these two ulvophyceans. *Fucus* can recover after loss of as much as 25% of its water content (Lobban and Harrison, 1994). The related *Pelvetia canaliculata* apparently requires periodic emersion, since it decays if submerged for periods longer than six out of every 12 hours. This seaweed is often host to an endophytic fungus, and it is hypothesized that desiccation helps to keep the fungus from causing decay of its host (Schonbeck and Norton, 1978).

Exposure to air results in dehydration, increased cellular solute concentrations, and loss of the ability to take up dissolved nutrients. The degree of water loss is related to seaweed surface area and volume—thicker thalli are more resistant to water loss than are thinner forms. The robust thalli of *Fucus*, for example, allow this seaweed to remain well hydrated. As long as *Fucus* is well hydrated, photosynthetic rates are as high when emersed as submersed. In general, intertidal algae are regarded as desiccation-tolerators rather than desiccation-avoiders (Surif and Raven, 1990). Intertidal understory species that are able to retain moisture because of protection provided by large intertidal canopy species such as *Hedophyllum*, as well as saccate, water-storing forms such as *Halosaccion*, are exceptions to this general rule.

Nutrients

For growth to occur, seaweeds require (in addition to water and light) combined N, inorganic carbon, phosphate, iron, cobalt, manganese, and other elements. Here, our discussion is limited to N, C, and P—the major nutrients known to influence macroalgal growth. In general the mineral requirements and uptake mechanisms of seaweeds are similar to those of phytoplankton (Chapter 22). Mineral uptake by macroalgae is influenced by irradiance, temperature, water motion, desiccation, and age. Rates of macroalgal mineral uptake are studied by tracking the disappearance of a nutrient from culture medium or by monitoring the uptake and conversion of a radiolabeled substance into algal biomass.

The nutrient that most frequently limits seaweed growth is combined nitrogen. N uptake by seaweeds was reviewed by Hanisak (1983), who noted that in general, the rate of uptake of ammonium is greater that that for nitrate. However, at very high concentrations (>30–50 μM) ammonium can be toxic to macroalgae. Ammonium can be used directly in the synthesis of amino acids. Nitrate may be stored in the cell's vacuole, but before it can be used, it must first be converted to nitrite by means of the cytoplasmic enzyme nitrate reductase; nitrite must next be reduced to ammonium within the chloroplast (Chapter 2). The chloroplast electron carrier ferredoxin is the source of reductant for the latter reaction. Assays of nitrate reductase activity are sometimes used as an index of the ability of algae to use nitrate. A newer approach is molecular assessment of expression of genes that encode the components of nitrate reductase (by determining mRNA levels).

Marine macroalgae vary in their use of bicarbonate, which is more abundant in seawater than is CO_2. One way to assess this capacity is to measure the natural abundance of the stable carbon isotopes ^{13}C and ^{12}C in seaweed thalli, and calculate the ratios of these two substances (Chapter 2). In a large survey of the stable carbon isotopic ratios of seaweeds, Maberly, et al. (1992) found that six species of green seaweeds, 12 species of brown seaweeds, and eight species of red macroalgae were able to use bicarbonate, but that six other rhodophytes were obligate CO_2 users. An explanation for the occurrence of the latter is that many subtidal red algae may experience light limitation of photosynthesis to such a degree that photosynthesis is not limited by CO_2 availability.

Phosphate can limit the growth of macroalgae under some conditions, particularly during rapid growth, partly because it is required for production of DNA and ATP. At least some seaweeds are able to supplement uptake of dissolved inorganic phosphate with phosphate cleaved from organic sources, such as phosphomonesters. Phosphate is released from organic compounds extracellularly through the activity of cell surface alkaline phosphatases, the production of which is induced by phosphorus deficiency. For example, alkaline phosphatase activity was detected in tips of *Fucus spiralis* that had undergone a sharp decrease in the P:N ratio as a result of rapid meristematic growth (Hernandez, et al., 1997). Assay of alkaline phosphatase activity has been used as an indicator of the degree of P-limitation experienced by seaweeds.

Biological Factors and Macroalgal Adaptations

Herbivore Interactions

The herbivores that affect macroalgae are relatively small mesograzers such as amphipods, copepods, and polychaetes, in addition to larger macrograzers including urchins and fish. Some seaweeds respond to grazing pressure via production of secondary defensive compounds and/or structural defenses (Duffy and Hay, 1990). Presence or absence of defenses can influence the direction of seaweed community change. For example, in a mesocosm (tank-based) study, the chemically defended red alga *Hypnea spinella* dominated the attached algal community after grazers (gammarid amphipods) effectively removed non-defended filamentous algae (Brawley and Adey, 1981).

Macroalgal defensive compounds include terpenes, acetogenins, alkaloids, or phenolics, which unlike the secondary compounds of terrestrial plants, usually contain halogens, such as bromine. A seaweed can often produce more than one defensive compound, particularly when protection against several different types of herbivores is required—species of the red alga *Laurencia* (some transferred to *Osmundea*) produce more than 500 different types of terpenes (Faulkner, 1993). The Hawaiian brown alga *Dictyota mertensii* produces a sterol known as Dictyol-H, which is ineffective against amphipods but protects against herbivorous fish (Fleury, et al., 1994). Phlorotannins are very widely produced as a herbivory defense by brown seaweeds of both tropical and temperate regions. Phlorotannins not only deter feeding, but result in reduced food quality for animals that do consume defended seaweeds. The occurrence, diversity, and ecological function of phlorotannins are reviewed by Targett and Arnold (1998).

Figure 23–16 *Turbinaria*, a tropical brown alga related to *Fucus*, is characterized by a tough, spiny thallus and thus has structural defenses against herbivory. (Photograph courtesy Ronald J. Stephenson)

Defensive chemicals may be constitutive (produced continuously) or inducible (produced only when required). An example of an inducible chemical defense is the production by *Fucus* of higher levels of phenolic compounds when herbivores are present than when they are absent (Van Alstyne, 1988). Production of defense compounds results in energetic costs to seaweeds. Such compounds are not simple proteins, and therefore multiple biosynthetic steps, enzymes, and genes are required for their synthesis. Seaweeds that produce large amounts of defensive compounds may have lower growth rates than related forms that lack them (DeMott and Moxter, 1991). In view of energy costs, producing defense chemicals only when they are needed—in response to induction by herbivores—should be favored. However, if herbivory occurs at unpredictable intervals on a nonseasonal basis, the time period required by an alga to synthesize protective compounds may be too long to provide effective protection (Padilla and Adolph, 1996). In such cases constitutive production of defense compounds may be more effective. The high cost of producing defensive compounds also explains why some seaweeds (such as the greens *Ulva* and *Enteromorpha*) that employ the strategy of rapid growth to reduce the impact of herbivory often lack such compounds.

Interestingly, marine animals may also take advantage of algal secondary compounds to gain protection for themselves, analogous to terrestrial monarch butterflies, which take advantage of bad-tasting secondary compounds in their milkweed food plants to acquire protection from predatory birds. For example, the tube-building temperate Atlantic amphipod *Ampithoe* selectively lives and feeds on the chemically defended brown alga *Dictyota menstrualis*. This seaweed exudes diterpene alcohols that discourage fish from consuming both the seaweed and the epizooic amphipod (Duffy and Hay, 1994). The Caribbean amphipod *Pseudamphithoides* constructs an enclosure from the similarly defended macroalga *Dictyota bartrayesii*—the alga's secondary compounds induce the building behavior in the amphipod. Predatory fishes readily consume these amphipods if they are removed from enclosures, but not animals that remain enclosed. These and other examples of marine chemical ecology are summarized by Hay (1996) and in other review articles cited therein.

The tough, spiny thalli of the tropical brown alga *Turbinaria* (Fig. 23–16) and calcified thalli of the coralline red algae (none of which are chemically defended) represent conspicuous examples of structural defenses (Padilla, 1985; 1989). In the case of some macroalgae, such as *Halimeda*, the timing of growth, secondary defense compounds, and calcification are all involved in herbivory protection. New segments (see Chapter 18) are produced at night, when fish—which find food visually—are less active. The poorly calcified young segments contain more defensive compounds than do older, more extensively calcified regions of the thallus (Hay, 1996). Other aspects of algal structure, including life histories involving heteromorphic alternation of generations, have been interpreted as ecological responses to herbivore interactions.

Several workers have contributed to the development of form-function theory, a body of thought that seeks to provide ecological explanations of algal morphologies and life histories (Littler and Littler, 1980; Littler, et al., 1983; Hay, 1981; and Padilla, 1985). The theory allows predictions to be made regarding correlations between morphological type, grazer resistance, food quality, productivity, reproductive success, growth mode, and role in succession. Seaweeds have been classified into functional-form groups such as sheets, filamentous, coarsely branched thalli, leathery thalli, jointed calcareous forms, and crusts. Often there appears to be a tradeoff between the ability to grow rapidly and the ability to resist herbivores (Fig. 23–17). For example, fast-growing sheets of *Ulva* and *Porphyra* are vulnerable to grazing, whereas the tough, stony thalli of corallines are grazer resistant but incredibly slow-growing. Some corallines are thought to be hundreds of years old. In general, the more highly structured seaweeds are thought to gain persistence in time at the cost of high productivity levels (Littler, et al., 1983). However success of corallines is dependent upon grazer removal of epiphytic diatoms, which in the absence of grazers can damage surfaces of corallines by their overgrowth (Steneck, 1990).

Competition

Important interactions occur between physical factors, grazing, and levels of interspecific competition in macroalgae. Competition is defined as the simultaneous use by two or more individuals or species of some limiting resource, such as light, space, or nutrients. Annual seaweeds—those not persisting from one growing season to the next—are typically not herbivore-resistant or good competitors. Rather, they are well adapted, by virtue of fast growth rate, to colonize newly available substrates such as recently formed

volcanic islands, and shores scoured by waves or ice. Common annuals include *Ulva*, *Enteromorpha*, and *Porphyra*. Perennials—seaweeds that can live for more than one year—are typically able to out-compete annuals. *Macrocystis pyrifera* lives as long as seven years, *Cystoseira barbata* and *Constantinea rosamarina* 18–20 years, and *Ascophyllum nodosum* more than 30 years. Longevity of individuals is estimated by counting annually formed vesicles in *Ascophyllum* and in *Constantinea* by counting scars left by the annual loss of the peltate blade (Lüning, 1990). Perennial seaweeds are often well adapted to resist herbivores. The recent approaches and data relating to occurrence of interspecific competition have been reviewed by Olson and Lubchenco (1990), Carpenter (1990), and Paine (1990). The role of allelochemical interactions in macroalgae was reviewed by Harlin (1987).

On the New England coast the red alga *Chondrus crispus* dominates sheltered subtidal areas, out-competing the browns *Fucus* and *Ascophyllum*, but farther north, severe ice scouring in the winter gives the edge to the latter two algae. In similar areas on the Scottish coast, heavy grazing by limpets also favors fucoid algae (Lubchenco, 1980). In some cases both herbivores and potential algal competitors are responsible for structuring the community. In New England midlittoral tide pools, *Fucus* and its relative *Pelvetia* are noticeably absent unless both herbivores such as the snails *Littorina* and *Achmaea*, and the seaweeds *Chondrus crispus*, *Ulva lactuca*, *Rhizoclonium tortuosum*, *Spongomorpha spinescens*, *Polysiphonia* sp., *Dumontia incrassata*, and *Scytosiphon lomentaria* are removed (Lubchenco, 1982). In the rocky intertidal of the Washington State (U.S.A.) coast, *Hedophyllum* is often dominant in areas of moderate wave exposure, but is out-competed by *Laminaria* and *Lessoniopsis* in more exposed areas with heavy wave action. A predator in this system—the carnivorous sea urchin *Strongylocentrotus*—can influence algal communities by overexploiting its prey, molluscan herbivores. In addition, the carnivorous starfish *Pycnopodia* and the sea anemone *Anthopleura* consume *Strongylocentrotus*, clearing patches of substrate for seaweed colonization (Dayton, 1975). These studies emphasize the importance of considering as many direct and indirect physical and biotic factors as possible in attempts to understand the ecological structure of seaweed communities.

Algal epiphytes may compete with their hosts for light and nutrients. Consequently, some seaweeds possess effective methods for ridding themselves of such

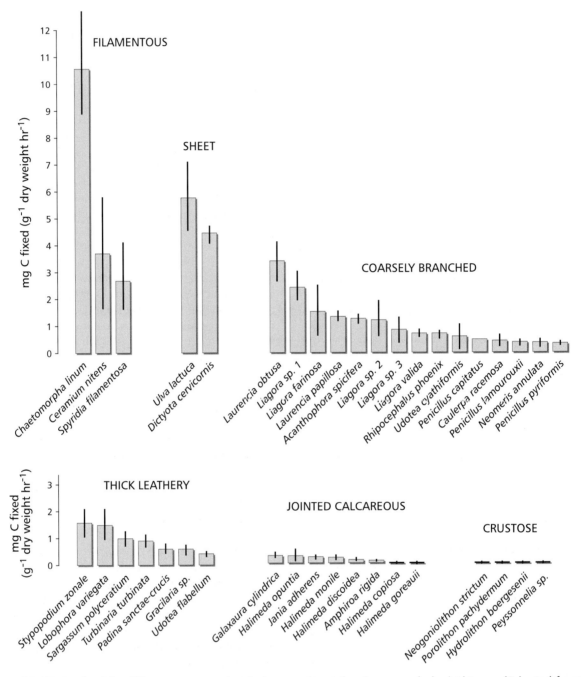

Figure 23–17 Productivity differences among tropical macroalgae of various morphological types. (Adapted from Littler, et al., 1983 by permission of the *Journal of Phycology*)

competitors. Fucalean brown algae and crustose corallines, which often appear conspicuously free of epiphytes, shed their outer cell layers or the outer portions of cell walls. Other seaweeds undergo continuous erosion of the blade tips or thallus edges, thus eliminating the epiphyte-ridden older portions. Such strategies also work against animal epiphytes, including the common bryozoan *Membranipora membranacea*, which can grow to cover blades of *Laminaria hyperborea* or *Nereocystis luetkeana* within months (Lüning, 1990). A study of epiphyte loads on crustose coralline algae in Britain showed that an overlying canopy of larger seaweeds was the most important factor in reducing epiphytes, but that herbivory and coralline surface features were also important (Figueiredo, et al., 1996).

Intraspecific competition has not been as intensively explored as interspecific competition. Intraspecific competition may result in self-thinning, a process in which biomass increase is coupled to decrease in density over time, as is commonly observed in terrestrial plant populations. In other words, better-adapted individuals survive and utilize available resources more effectively than less well-adapted individuals of the same species, which may perish. However, there has been some controversy as to whether self-thinning actually occurs in seaweeds, since there are documented advantages to growing in crowded stands. Crowded seaweeds may protect each other from desiccation, grazing, or both, particularly when young. In an investigation of populations of *Himanthalia elongata*, a brown alga that frequently occurs in dense monospecific stands on rocky north temperate shores, Creed (1995) detected evidence for self-thinning. Young individuals were initially regularly spaced, but over time, spatial patterning became more irregular, with larger individuals surviving better than smaller ones.

The Ecological Roles of Pathogens

Seaweed pathogens include viruses, bacteria, fungi, other algae, and heterotrophic protists (Chapter 3). Determination that a particular pathogen causes a particular disease of seaweeds (or any organism) requires fulfillment of Koch's postulates. These postulates are: (a) the pathogen must always be associated with the disease symptoms and no symptoms should be present in the absence of the pathogen; (b) the pathogen must be isolated and grown either in laboratory culture or in a susceptible host; (c) when inoculated onto a disease-free host, the isolated pathogen must be able to cause the symptoms associated with the disease; and (d) the pathogen must be re-isolated from the experimentally infected host and shown to be the same organism as the original isolate. Such information is helpful in designing strategies to prevent or reduce effects of the disease. New molecular methods such as *in situ* hybridization have been successfully applied to the process of identifying seaweed pathogens (Ashen and Goff, 1998).

Recent data on seaweed pathogens have been reviewed by Correa (1997), while aspects of algal attack by viruses, bacteria, protists, and fungi were discussed in this text in Chapter 3, and a survey of the major diseases that affect economically important seaweeds is found in Chapter 4. In general the ecological impacts of pathogenic attack on seaweeds is poorly documented. The role of physical factors, herbivores, and interspecific and intraspecific competition on the probability of pathogenic attack, and the role of pathogens in seaweed resistance to physical damage or competition have not been widely investigated. The possible role for pathogens in influencing seaweed evolution has also not been explored. The focus here is upon algal pathogens of seaweeds, because such pathogenesis may affect or reflect interspecific competition.

Algae (including cyanobacteria) that occur as endophytes ramifying within host thalli may have pathogenic effects (i.e., cause noticeable symptoms). Endophytes often cause spots (areas of increased pigmentation) or abnormal twisting of seaweed stipes or blades. Such symptoms may sufficiently influence seaweed appearance to be of concern to those who harvest wild or cultivated seaweeds for markets, or they may be ecologically significant if growth is seriously impacted. Algal endophytes of seaweeds include filamentous green (e.g., *Acrochaete*), red (e.g., *Adouinella*), and brown algae (e.g., *Streblonema*), which typically grow in host intercellular spaces, rather than intracellularly. It should be noted, however, that there are numerous parasitic red algae that infect rhodophycean hosts by the proliferation of parasitic nuclei within host cells; these are discussed in Chapter 16. Movement of photosynthates between the epiphytic red alga *Polysiphonia* and its host, the brown *Ascophyllum*, and from red algal hosts to rhodophycean parasites having reduced pigment content (Chapter 16) has been documented. However, in the case of most hosts and their endophytes, the extent of nutrient transfer, or whether this occurs, is not known. By associating with larger hosts, endophytes may obtain protection from desiccation, high irradiance, or herbivory.

A deformative disease of *Mazzaella* (formerly *Iridaea*) *laminarioides* is caused by infection with a species of the cyanobacterium *Pleurocapsa* (Correa and Sánchez, 1996). Green patch disease of *M. laminarioides*, which can co-occur with the *Pleurocapsa* infection, is caused by a filamentous green endophyte, the appropriately named *Endophyton*. Infections are present throughout the year, and from 20–80% of the *Mazzaella* population can be affected by *Endophyton*.

Endophyton, as well as two endophytic species of the green alga *Acrochaete*, are pathogens of *Chondrus crispus*, causing similar symptoms to those observed in *Mazzaella*. Host specificity of *Acrochaete operculata* appears to be determined by host cell-wall polysaccharides, specifically lambda carrageenans. The sporophytes of *Chondrus crispus* and *Mazzaella cordata* (which possess lambda carrageenans) become infected, but the gametophytic generations (which lack lambda carrageenans) are not susceptible (Correa and Sanchez, 1996). This suggests that attachment of propagules of the green algal pathogens may be dependent upon specific chemical attributes of host surfaces.

Other macroalgal diseases caused by endophytes include brown spot disease of cultivated *Undaria pinnatifida* and gall induction in kelps, both of which result from infection with the filamentous brown endophyte *Streblonema* (see references in Correa, 1997). *Streblonema* is a fully pigmented form that is found widely within kelps; some *Streblonema* species also occur within red algae such as *Grateloupia*. *Laminaria saccharina* often contains the brown endophyte *Laminarionema*, which causes "twisted stipe disease." There is some evidence that brown algal reproductive cells, like those of the green endophytes described above, are able to detect the presence of suitable hosts, probably through chemical means, and to then attach to them specifically. However, the biochemical mechanisms underlying specificity are very poorly understood. Green and brown algal endophytes/pathogens, as well as numerous cases of red algal parasitism (discussed in Chapter 16), would seem to offer intriguing opportunities for ecological analysis of interspecific competition at a complex and intimate scale of interaction, the costs and benefits of the endophytic habit, and pathogen-host evolutionary interactions.

Macroalgal Biogeography

Biogeography is the study of the geographical distribution patterns of organisms and the mechanisms producing these patterns. This subject is gaining the interest of phycologists as they seek to predict the effects of global change on algal biodiversity and occurrence. As we will later note, temperature is a major factor determining the distribution of seaweeds, hence global warming could influence their distributions. Changes in climate, UV levels, and nutrient content of coastal waters are also expected to affect algal biodiversity, with possible consequences for associated organisms. Human-aided migrations of seaweeds to new areas—where, in the absence of natural herbivores or pathogens, they sometimes grow to nuisance levels—has also become a major concern. These and other aspects of seaweed biogeography have been reviewed by Lüning (1990). The brief survey of seaweed biodiversity presented here will focus on the geographical sites of highest diversity and biomass of algae as places where global change could have a large impact, and on locations where endemism, and thus extinction rates, could be high.

The present distribution of seaweeds has resulted from both patterns of migration (dispersal from a center of origin) and historical speciation events related to the establishment of geographical barriers to interbreeding. Such barriers include boundaries of dramatic water temperature change or the geological opening or closing of marine basins. Ancient taxon splitting events (adaptive radiations) can be detected by the use of cladistic (phylogenetic) analyses, which infer the phylogeny of a related group of organisms. In recent years molecular approaches have increasingly been used to deduce the relationships among species (Chapter 5). Such a phylogenetic approach to biogeographical analysis is known as vicariance (meaning "split up"). Vicariance describes the situation in which closely related species occupy distinct and often widely separated geographical locations. In the past, experts argued about the relative importance of long-distance dispersal versus vicariance in explaining the present-day distributions of seaweeds (and other organisms), but today phycologists recognize that both processes have played important roles.

Seven major seaweed floristic regions are commonly recognized: Arctic, Northern Hemisphere Cold Temperate, Northern Hemisphere Warm Temperate, Tropical, Southern Hemisphere Warm Temperate, Southern Hemisphere Cold Temperate, and Antarctic. Aside from the Arctic and Antarctic, each of these regions is further subdivided into component areas that are not covered here (see Lüning, 1990). These regions are characterized by distinctive climatic features (notably seawater temperature) and by the organisms that inhabit them. Both are influenced by the present-day positioning of the continental land masses and ocean currents (Fig. 23–18). Currents that are particularly important in explaining seaweed occurrence include the California Current on the west coast of North America, the Humboldt Current (also known as the Peru Current) off the western coast of

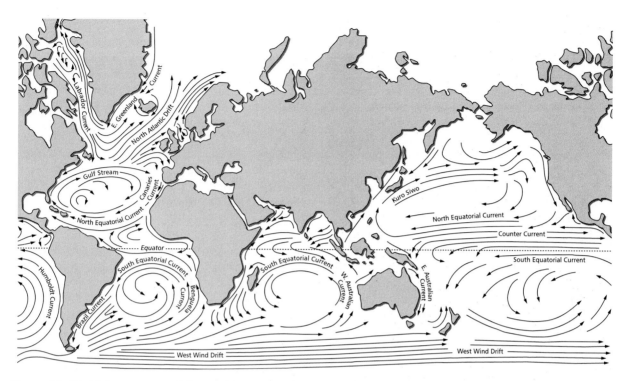

Figure 23–18 Major currents of the world's oceans. (After Lüning, Meeresbotanik, Georg Thieme Verlag, 1985)

South America, the Benguela Current, which influences the south and west coasts of Africa, and the West Wind Drift, a current that influences all of the southern hemisphere continents. The northbound, cold-water Humboldt Current is responsible for coastal upwelling, as are the California and Benguela currents. Upwelling increases the level of inorganic nutrients, and thus primary productivity of phytoplankton and seaweeds in these regions. The major ocean currents also set up temperature belts (also known as isotherms) that do not usually run parallel to the equator or Tropics of Cancer and Capricorn. Isotherms are boundaries defined by the same surface water temperature averaged over many years, for a particular month. Along any coast, the northern and southern limits for growth of a particular seaweed species are marked by the February and August isotherms (Lüning, 1990).

Every seaweed species has high and low lethal temperature limits set by the tolerance of the least hardy life-history stage. This is often a cryptic microthallus capable of persisting for long periods, reproducing only asexually, under conditions that do not permit completion of the sexual life cycle or production of more conspicuous life-history stages. If conditions change in a direction that allows sexual reproduction to occur, the conspicuous life-history stage may appear to have suddenly invaded the area.

Some seaweeds have a wide temperature tolerance (i.e., are eurythermal), whereas others (termed stenothermal) have narrow temperature tolerances. If a particular geographical region experiences environmental conditions beyond those of stenothermal seaweeds, such seaweeds will not occur there. For example, the brown seaweed *Saccorhiza polyschides*, which is common along shores of the eastern Atlantic Ocean (on European coasts), is excluded from the western Atlantic region (North America) because sporophytes die at temperatures above 24° C and below 3° C and all North American shores experience extremes beyond these temperature limits (Lüning, 1990).

Tropical Seaweeds

Many tropical marine genera have a circumglobal distribution. Examples include the green ulvophycean genera *Codium*, *Caulerpa*, and *Acetabularia*; the browns *Dictyota*, *Dictyopteris*, *Padina*, *Sargassum*, and *Turbinaria*; and the reds *Gelidium*, *Pterocladia*, *Galaxaura*, *Liagora*, *Halymenia*, *Grateloupia*, *Jania*, *Amphiroa*, and *Laurencia*. The common occurrence of pantropical seaweeds is explained in terms of the great age of the habitat. Since 180 million years ago, there has been a continuous equatorial belt of warm water that has been interrupted only relatively recently by land bridges (closure of the Mediterranean about 17 million years ago and origin of the Central American Land Bridge some 3–4 million years ago). Thus for many millions of years there was no major barrier to tropical seaweed migrations. Moreover, tropical waters did not experience much cooling in the Pleistocene. The tropical warm water belt is bordered by north and south 20° C winter isotherms, and water temperatures may rise to 30° C.

The boundaries of tropical seaweed occurrence coincide with that of hermatypic (reef-forming) corals—a habitat of very high animal-species diversity. The tropics might likewise be expected to also harbor the world's largest diversity of seaweeds, but this is not the case. Rather, the geographical regions that are presently richest in seaweed species are the warm temperate coasts of southern Australia and the Mediterranean. The 5500 kilometer-long coast of southern Australia hosts 1100 seaweed species with a high level of endemism. About 30% of the 800 species of red algae, 20% of the 231 species of browns, and 11% of green seaweed species are restricted to this region (Womersley, 1984; 1987). One explanation for this seeming paradox is that in the tropics, corals effectively compete with macroalgae for settling space. Another is that constraints on colonization of remote tropical islands may have contributed to reduced diversity. The number of immigrating species decreases along with island size and increasing distance from continental coasts. The seaweed flora of Easter Island, the most remote island of the Indo-Pacific region, consists of about 170 species, 80% of which are widely distributed in warm waters (Santelices and Abbott, 1987), whereas there are more than 300 species in the seaweed flora of the Galapagos Islands (see references in Lüning, 1990), which are closer to the South American mainland where some 400 seaweed species occur (Santelices, 1980). Compared with temperate floras, the diversity of brown algae is lower in the tropics, but the diversity of red seaweeds, particularly Ceramiales (Wynne, 1983), as well as green algae, is higher.

Cold Temperate and Polar Seaweeds

The cold temperate and polar floras of the Northern Hemisphere differ substantially from those of the Southern Hemisphere, presumably because they have been long separated (for at least 65 million years) by the tropical warm belt. Northern and Southern Hemisphere cold temperate and polar seaweeds are thus thought to have arisen in isolation, though there are a few examples of transmigrants. A few seaweeds are thought to have crossed the equator during the Pleistocene, when the mean temperature of the tropical belt was somewhat lower and the mean width somewhat less. *Macrocystis pyrifera* is thought to have crossed the equator along the Pacific Coast of Central America, moving southward against the Humboldt Current (Fig. 23–18), while *Laminaria ochroleuca* appears to have migrated southward along the west African coast. In the Southern Hemisphere the cold temperate seaweed flora is similar around the world, primarily because the southern continents have a common history (i.e., they once had common coastlines) as components of Gondwanaland, and because the West Wind current (Fig. 23–18) facilitates migrations.

In contrast, the cold temperate seaweed floras of the North Atlantic differ greatly from those of the North Pacific, the latter having much greater diversity and higher proportion of endemic species. Two thirds of the 60 or so cold-water North Pacific red algal genera are endemics, compared to only four of the 30 cool-water red algae in the Arctic North Atlantic (Lindstrom, 1987). Greater Pacific diversity and endemism is explained by the greater age of the North Pacific basin (Mesozoic, 65 million years or more) compared to the North Atlantic (no earlier than the Tertiary, 65 million years or less), and the fact that biotic exchange via the Arctic Ocean has been possible only since the late Tertiary (Lüning, 1990), and not at all since the last interglacial (Wilce, 1990). The Pacific basin is thought to have been the center of origin of the Laminariales. Only one of the ten Pacific *Alaria* species, *A. esculenta*, also occurs in the North Atlantic. Several *Laminaria* species (or closely related sister species) occur in both the North Atlantic and North Pacific, *L. saccharina* being a common and conspicuous example (Fig. 23–19). The site of origin of the non-laminarialeans that occur in both the North Pacific and North Atlantic is difficult to determine (Wilce, 1989).

The Arctic Sea itself has few or no endemic species; its comparatively depauperate flora derives from invasion of cold-adapted North Atlantic species at the end of the Pleistocene. Extremely stressful (ice-scoured) habitats, such as the heads of northern fjords, or sites lacking solid substrata for seaweed attachment, have very sparse algal communities, giving the impres-

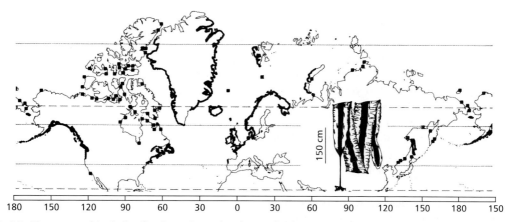

Figure 23–19 Biogeographical distribution of *Laminaria saccharina*, a common brown macroalga. This species occurs in both the North Pacific and the North Atlantic. (Adapted from Lüning, Meeresbotanik, Georg Thieme Verlag, 1985—an excellent source of many other seaweed distribution maps).

sion that the Arctic is characterized by few species and low biomass. But more favorable sites with hard substrata, such as the rocky coasts of northwest and east Greenland and much of the eastern coastline of Canada, are more productive. Examples of Arctic green seaweeds include *Ulothrix flacca, Rhizoclonium riparium, Blidingia minima,* and *Endocladia viridis.* Some Arctic red seaweeds are *Palmaria palmata, Clathromorphum compactum, Lithothamnion glaciale,* and *Polysiphonia arctica.* Arctic browns include *Sphacelaria arctica, Desmarestia aculeata, Chorda tomentosa,* and *Laminaria saccharina* (Wilce, 1990).

The eastern North Atlantic is rich in seaweed species compared to the western North Atlantic because the North Atlantic Drift current (Fig. 23–18) brings warm water, allowing migration of many warm-adapted species of the Mediterranean and the northwest African coast as far north as Ireland (Lüning, 1990). The temperature and photoperiod responses of more than 60 species of North Atlantic seaweeds have been defined in the laboratory, and serve to explain distributions quite well (Hoek, van den and Breeman, 1989).

The temperate and polar seaweed floras of the Southern Hemisphere are cold-adapted communities that have evolved in parallel to those of temperate and polar floras of the Northern Hemisphere. This reflects separation of the coasts of Gondwana and Laurasia 135 million years ago by the tropical warm water belt. Compared with the Northern Hemisphere, where migrations occur mainly along coastlines, ocean currents are more significant in the migration of seaweeds

of the Southern Hemisphere. Temperate Southern Hemisphere eulittoral waters are regarded as the site of origin of the Fucales, because the greatest diversity of fucaleans occurs there (Clayton, 1984). Fucales of tropical and northern waters are thought to have migrated there, probably aided by the air-filled floats common in this group. Approximately 100 species of seaweeds occur in Antarctic waters, about one third of which are endemics. This is a much higher level of endemism than occurs in the Arctic (about 5%). It has been attributed to the longer history of the Antarctic and the absence of coastal connections to cold temperate regions. No laminarialeans occur in the Antarctic, rather, the seaweed flora is dominated by perennial members of the Desmarestiales, which are thought to have originated in the Southern Hemisphere. Dense thickets of the 10 m-long, one m-wide *Himantothallus grandifolius* and *Desmarestia* occur from 5–35 m in depth (Fig. 23–20). The smaller thalli of the desmarestialean *Phaeurus antarctica* occur in tide pools and to depths of 10 meters (Fig. 23–20) (Clayton and Wiencke, 1990). As is the case for Arctic seaweeds, carbohydrates are stored in the high-light season for use in growth during the low-irradiance austral winter.

Seaweed Migration

Seaweeds capable of migrating and colonizing new regions are less vulnerable to extinction than those having very restricted distributions. Seaweed migrations can occur either by short distance "stepping-stone" transfers from one coast to another close by, or by long distance movements via currents. Migrations

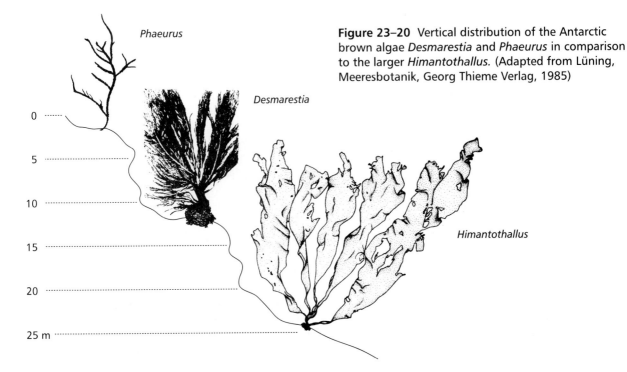

Phaeurus

Desmarestia

0

5

10

15

20

25 m

Himantothallus

Figure 23–20 Vertical distribution of the Antarctic brown algae *Desmarestia* and *Phaeurus* in comparison to the larger *Himantothallus*. (Adapted from Lüning, Meeresbotanik, Georg Thieme Verlag, 1985)

are accomplished by production of propagules such as zoospores or by floatable thalli. Zoospores of the annual colonizers *Enteromorpha*, *Blidingia*, and *Ulothrix* can remain alive in the plankton during migrations over tens of kilometers. Hence they have been able to colonize sub-sea portions of remote offshore oil platforms in the North Sea (Lüning, 1990). These seaweeds were also early colonizers of the island of Surtsey, which was formed in 1964 some 35 km from the nearest coast (Iceland), as well as on artificial substrates mounted some 35 km away from the North American coast (Amsler and Searles, 1980). Phycologists have also cultured *Enteromorpha*, *Blidingia*, and *Ulothrix* from offshore seawater (Hruby and Norton, 1979; Zechman and Mathieson, 1985). *Macrocystis pyrifera*, which is equipped with many floats (pneumatocysts), and one floatable species of *Durvillaea* (*D. antarctica*) have been able to migrate throughout the cool temperate Southern Hemisphere via the West Wind current. In some cases, these large seaweeds may have served as rafts for co-migration of epiphytic forms.

During a warming period in the first half of the twentieth century, an increase in seawater temperature (by an average of 1° C) in the Northern Hemisphere has been correlated with northward migration of several warm water macroalgal species (*Dictyota*

dichotoma, *Desmarestia ligulata*, and *Gracilaria verrucosa*) as far as Norway, the migration of some Mediterranean brown algae (*Taonia atomaia* and *Dictyopteris membranacea*) to the British Isles, and spread of *Laminaria ochroleuca* from France to Britain. Detection of such events helps phycologists predict the effects of future global warming on seaweed distribution.

More recently, marine ecologists have become greatly concerned about the sudden appearance and spread of nonnative algal species via human activities, such as the dumping of ship ballast waters. Several cases of seaweed introductions have been documented, and are reviewed by Lüning (1990) and Ribera and Boudouresque (1995). The red alga *Bonnemaisonia hamifera* was introduced to European waters from Japan early in the twentieth century, after which it spread to the western Mediterranean and Scandinavia, and still later had made its way to eastern North America by 1927. The rhodophyte *Asparagopsis armata* has been introduced from Australia to northern Europe, and the ulvophycean *Codium fragile*, a fouling seaweed on ships, has been transferred from Japan throughout both hemispheres. The ecology of invasive and noninvasive forms of *C. fragile* has been reviewed by Trowbridge (1998).

Although most introduced seaweed species appear to have little effect on native communities, there are several nuisance growth-forming migrants. The brown seaweed *Colpomenia* apparently entered European waters from Japan through the exchange of oyster cultures. *Colpomenia*'s gas-inflated thalli have readily spread and have become a nuisance problem in that thalli growing on shellfish will often make them so buoyant that the shellfish float away. Hence the name "oyster thief" is commonly applied to *Colpomenia*. *Sargassum muticum* is another floatable brown seaweed (owing to its numerous air bladders) that was probably introduced from Japan to the U.S. by oyster aquaculturists. It has since become a nuisance alga in harbors and along beaches in the U.S. Following its introduction to Europe, *S. muticum* has spread rapidly. Because it grows quickly, is monoecious, and is fertile in the first year, this seaweed has out-competed native fucaleans and laminarialeans. Since the latter are often considered to be keystone organisms, replacement by *Sargassum muticum* is expected to have serious ecological effects (Ribera and Boudouresque, 1995). *Undaria pinnatifida* was purposefully transplanted from the eastern Pacific to the northern Mediterranean, Brittany, New Zealand, and Tasmania for cultivation (Lüning, 1990). However, ecologists are becoming concerned that *Undaria* may displace native seaweeds.

In Hawaii the introduced red seaweeds *Hypnea musciformis* and *Acanthophora spicifera* occupy the same niche as the native *Hypnea cervicornis*, which has seriously declined as a result of the introduction. Recently concern has emerged over ecological consequences of the introduction or migration and subsequent spread of *Caulerpa taxifolia* in the Mediterranean (Delgado, et al., 1996). A second species of *Caulerpa*, *C. filiformis*, has become a dominant in Sydney Harbor, Australia, replacing a native species of this genus (Ribera and Boudouresque, 1995).

Pollution Effects on Macroalgae

Pollutants that affect seaweeds include chlorinated hydrocarbons, oil, herbicides, insecticides, heavy metals, chlorine and copper used to reduce ship fouling, and heated wastewater from power plants and some industries. Seaweeds are sometimes used as indicators for monitoring the occurrence and severity of pollution events (Chapter 4). Oil spills affect seaweeds by reducing photosynthetic rates or interfering with gamete or spore release. Mercury is the single- most

toxic metal; its mechanism of action is inhibition of enzyme activity. Copper and cadmium are the next-most toxic metals. Copper can interfere with cell permeability and enzyme function, while cadmium inhibits photosynthesis and protein synthesis. Seaweeds growing in water released from nuclear power plants can accumulate radioactive metals, and hence humans should avoid consuming macroalgae in the vicinity of outfalls. Nuisance growths of green seaweeds such as *Enteromorpha, Ulva, Cladophora* as well as the coral-smothering *Dictyosphaeria* and/or phytoplankton may result from input of sewage. Raffaeli, et al. (1998) reviewed the ecological aspects of such marine green algal blooms, commonly known as "green tides." It has been estimated that on a worldwide basis, more than 90% of sewage from coastal areas enters the ocean untreated (Lobban and Harrison, 1994). It is expected that expanding human populations will result in further increases in nutrient and other pollution, and that negative impacts on seaweeds will continue to occur for the foreseeable future. Nuisance macroalgal responses to increases in nutrients have recently been reviewed from a geographical perspective (Schramm and Nienhuis, 1996). Thermal pollution can also be expected to rise in the future, with serious impacts on seaweeds. The number of seaweed species can be expected to decrease by 10% with increase in temperature of 1° C (Devinny, 1980).

Global Environmental Change

Seawater temperatures have risen on average 0.5° C within the last century, as the result of human activities, particularly burning of fossil fuels and deforestation. Further temperature increases are predicted as a result of positive feedback effects (warming increases the rate of warming). Possible effects might include migration of coral reefs (and their associated algae) northward and changes in the biogeographic distribution of seaweeds. It is possible that increasing atmospheric CO_2 levels might shift the balance toward sea grasses, thereby influencing nearshore food webs (Dawes, 1998). Stratospheric ozone depletion—already demonstrated to have resulted in increased UVB radiation in polar regions—may alter depth distributions of seaweeds, such that intolerant forms may have to retreat to deeper, less well-illuminated waters. New remote sensing techniques are currently being developed that should allow global-scale changes in benthic macroalgae to be monitored by airborne spec-

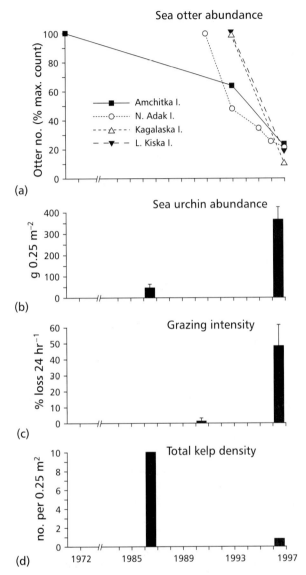

(a) Sea otter abundance

Otter no. (% max. count)

- —■— Amchitka I.
- ····○···· N. Adak I.
- --△-- Kagalaska I.
- —▼— L. Kiska I.

(b) Sea urchin abundance

g 0.25 m^{-2}

(c) Grazing intensity

% loss 24 hr^{-1}

(d) Total kelp density

no. per 0.25 m^2

1972   1985   1989   1993   1997

Figure 23–21 In a study by Estes, et al. (1998), dramatic declines in the density of kelp beds off the coasts of several islands in the Aleutian archipelago during the period between 1986 and 1997 were shown to be correlated with changes in food web structure. (a) Prior to 1990, sea otter abundance was relatively high, but had plummeted by 1997. (b) In contrast, sea urchin biomass, low in 1987, had sharply increased by 1997, as had grazing intensity, shown in (c). (d) Total kelp density, which had been high in 1986, had dropped sharply by 1997. (Redrawn with permission from Estes, J. A., M. T. Tinker, T. M. Williams, and D. F. Doak. Killer whale predation on sea otters linking oceanic and nearshore ecosystems. *Science* 282:473–475. ©1998 American Association for the Advancement of Science)

tral scanners or satellite sensors (reviewed by Guillaumont, et al., 1997).

Overfishing—depletion of fish stocks below sustainable levels—has become a worldwide problem. Recent studies have linked decreases in fish populations in the North Pacific to dramatic declines in kelp density (Estes, et al., 1998). As fish populations have declined, killer whales, which normally do not prefer to feed on sea otters, are forced to prey upon them when seals (the preferred food, which depend on fish) suffer population declines. Sea otters normally control sea urchin populations (Fig. 23–21), but when otter numbers are low, urchins flourish and graze on kelps, reducing their densities. Because kelp forests are widely regarded as essential to the maintenance of coastal ecosystems where they occur, such food web changes are of great concern.

PART 2—MARINE TURF-FORMING PERIPHYTON

Some marine macroalgae (such as the brown *Dictyota* and the green *Halimeda*) grow as short, tightly compacted patches known as turfs. Turf-forming members of the Dictyotales, for example, often colonize sublittoral regions from which kelp canopies have been removed (Kennelly, 1987). However, this kind of macroalgal turf is distinguished from turfs composed of smaller single-celled, colonial, or filamentous marine periphyton. The latter are similar in taxonomic composition (at the class level), and morphological type, to periphyton communities in freshwaters. Here, we use the term "marine periphyton turfs" to distinguish them from turf-forming macroalgae. Marine periphyton turfs are common on coral reefs throughout the world, in nutrient-poor oceans having very low, sometimes undetectable, levels of nitrate (Hackney, et al., 1989).

In terms of percent cover, cyanobacteria dominate marine periphyton turfs, usually occupying some 50% of the total area, and sometimes forming erect tufts within the size range of macroalgae. An example of a dominant turf cyanobacterium is *Hormothamnion enteromorphoides* ("looks like *Enteromorpha*"). *Hormothamnion* can form dense blooms that dominate hundreds of square meters of reef flat off Guam (Pennings, et al., 1997). Many turf community cyanobacteria (including *Hormothamnion*) are nitrogen-fixers, and thus contribute significantly to community N-resources.

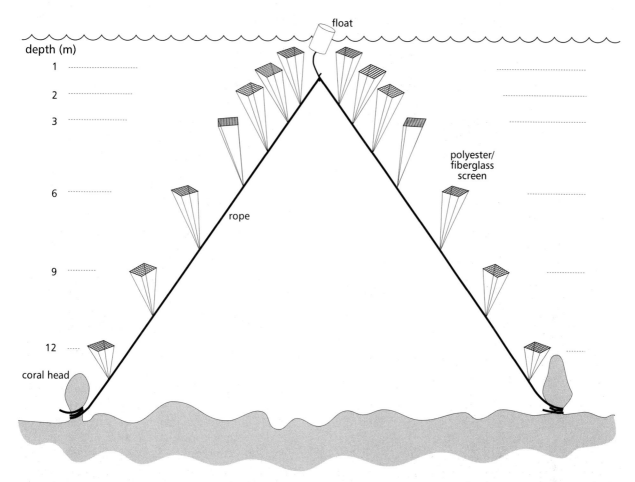

Figure 23–22 An experimental scheme for determining productivity of periphyton turfs at different depths (and different irradiance environments). The turf algae grow on screens that have been mounted on frames attached to a line anchored at both ends and suspended by a float at mid-length. (From Adey and Goertemiller, 1987)

Succession in marine periphyton turf communities begins with a film of attached diatoms, and is followed within five to seven weeks by a mixture of diatoms and cyanobacteria. Later the proportions of red and brown algae may increase. Many kinds of unicellular, colonial, or filamentous algae are found in periphyton turfs. For example, an extensive species list has been compiled for turf communities on reefs off Mayaguana Island in the southeastern Bahamas. Some representative periphyton turf species include the cyanobacteria *Oscillatoria* and *Anabaena*, the green algae *Cladophora* and *Bryopsis*, the red algae *Ceramium* and *Polysiphonia*, and the brown algae *Sphacelaria* and *Ectocarpus* (Adey and Goertemiller, 1987). In a study of the marine periphytic turfs of coral reefs in St. Croix, U.S. Virgin Islands, the single-most abundant eukaryotic, non-crust member of the turf community was *Ectocarpus rhodochortonoides*

(Hackney, et al., 1989). The specific epithet indicates that this brown alga resembles the filamentous red alga *Rhodochorton*.

In contrast to macroalgal communities—characterized by high standing crop and very high primary productivity per unit area, but low primary productivity per unit biomass—marine periphytic turfs feature low standing crop, moderate primary productivity per unit area, and very high primary productivity per unit biomass. High primary productivity is based on very high surface area-to-volume ratio of the turf algal cells and filaments. This facilitates nutrient uptake in low-nutrient waters. Turf communities are also best developed along shores facing incoming currents, which break up diffusion boundary layers at the turf surface and bring in a constant nutrient supply. In the Bahamas, for example, the North Equato-

intensity of macrograzing pressure

	low to absent	moderate to high	high	extreme
predominant benthic algal component	macroalgae	algal turfs	crusts	(bare substrates)
standing crop	very high	low	very low	
primary productivity per unit area	very high	moderate	low	
primary productivity per unit biomass	low	very high	very low	
predominant types of adaptations	maintain upright morphology	growth	resistance	

Figure 23–23 The relationship between grazing pressure, macroalgal life-form type, and effects on standing crop, primary productivity per unit area, primary productivity per unit biomass, and growth form adaptations. (From Hackney et al., 1989)

rial Current (Fig. 23–18) may provide a sufficiently high-energy environment that turf communities escape nutrient limitation, even though the nutrient levels in surrounding waters are exceedingly low. Productivity of periphyton turfs at different depths has been studied by suspending screens bearing turf communities in the ocean (Fig. 23–22) (Adey and Goertemiller, 1987).

Moderate to high grazing pressure is characteristic of reef periphytic turf communities, in contrast to relatively low herbivory pressure on macroalgae (which are often well defended) (Fig. 23–23). Important grazers include herbivorous fish (Choat and Clements, 1998). Herbivory helps to maintain the turf community by preventing domination by macroalgae. Damage to apices of branched, filamentous turf algae, resulting from either herbivory or mechanical damage, increases the extent of branching, yielding a more compact turf. Dense turfs are well adapted to survive desiccation during low tides. After severe damage the basal portions of turf algae may persist as resting stages until conditions allow regeneration of erect photosynthetic filaments (Hay, 1981). Some tropical turf cyanobacteria possess effective chemical defenses against herbivory, possibly explaining their success, despite not being calcified or having tough thalli. For

example, *Hormothamnion entermorphoides* produces a suite of cyclic peptides—principally laxaphycin A—that strongly deter feeding by parrot fish, sea urchins, and crabs. These compounds also appear to have antifungal activity (Pennings, et al., 1997).

PART 3—FRESHWATER PERIPHYTON

Freshwater periphyton (also known as benthic algae) are unicells, colonies, or filaments that grow attached to substrata in streams and in the littoral zones of lakes. Such communities are very common where light levels are sufficient for their growth. Freshwater periphyton are important because they are the primary source of fixed carbon in shallow lakes and streams, and because they are unusually rich in species. They provide an essential source of food for a wide variety of stream and lake micrograzers, mesograzers, and larger animals. The algae of periphyton communities also sequester nutrients such as N and P in wetlands, helping to prevent lake eutrophication. Though macrophytic vegetation (primarily vascular plants) are often credited with this function, they actually obtain most of their N and P from the sediments, whereas the attached algae are able to harvest N and P from the water column. The algal periphyton on macrophytes traps nutrients and transfers them to the sediments upon their death, thus reducing the rate of nutrient transfer from wetlands to associated lakes (Wetzel, 1996). Periphyton communities are also recognized for their ability to stabilize unconsolidated substrata, such as loose sediments and sand, preventing their disaggregation by water movements. For example, sticky mucilages exuded by the diatom *Nitzschia curvilineata* decrease erodibility of sediments in laboratory experiments modeling stream flow (Sutherland, et al., 1998).

Periphyton is a complex mixture of microalgae, bacteria, and fungi, often held together in a gluelike mucilaginous matrix produced by the algal or bacterial inhabitants (Fig. 23–24). Thus, the term "periphyton" cannot be used as a synonym for the attached algae alone. The matrix also typically includes dead algal cells, fecal pellets generated by micrograzers, and calcite particles, which may be produced by some algae, and bound inorganic sediment (Burkholder, 1996). Such communities may occur attached to rocks (epilithic), sand (epipsammic), sediments (epipelic), plants or larger algae (epiphytic), or animals (epizooic). In addition, periphyton communities may occur as loose tangles among the branches of larger algae or macrophytes. Attached algae may also break loose from the substrate and occur as a floating mass in the water column. Such floating algae are known as **metaphyton**. Subsurface summer clouds of the filamentous zygnematalean green algae *Spirogyra*, *Mougeotia*, *Temnogametum*, and *Zygnema* are the most common metaphytic forms in lakes—particularly those subjected to acidification—but metaphytic mats of *Cladophora*, *Chaetophora*, *Oedogonium*, and *Spirogyra* also occur widely in wetlands (Goldsborough and Robinson, 1996). Filamentous tribophyceans (xanthophyceans) may also form metaphytic clouds in lakes and ponds, particularly during spring and fall.

Freshwater periphyton algae consist largely of cyanobacteria, green algae, and diatoms, but representatives of other groups such as red algae, chrysophyceans, and tribophyceans may also be present. Examples of common periphytic cyanobacteria are *Schizothrix*, *Rivularia*, and *Tolypothrix*; common green members of the periphyton include *Ulothrix*, *Cladophora*, *Rhizoclonium*, *Stigeoclonium*, *Draparnaldia*, and *Coleochaete*; and frequently observed periphytic diatoms include *Epithemia*, *Cocconeis*,

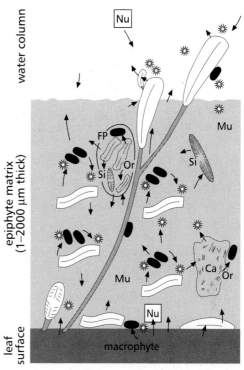

Figure 23–24 Nutrient supplies (Nu) to the periphytic algal community on a macrophyte leaf surface may come from the substrate, the water column, and the epiphytic matrix itself. The epiphytic matrix, defined by the extent of mucosaccharide (Mu) arising from the algae and associated bacteria is a site of dynamic nutrient regeneration. Fissures in the matrix (not shown) promote exchange of materials between the water column and lower portions of the matrix. Living diatoms are shown in white, whereas diatom frustule remains, including those in fecal pellets (FP), are gray. The latter are a source of silica (Si). Bacterial cells (black ovals) and free phosphatase enzymes (sunbursts) are active in recycling materials from organic matter (Or) and calcium carbonate (Ca). (After Burkholder, 1996)

Gomphonema, and *Cymbella*. *Batrachospermum*, *Lemanea*, and *Audouinella* from the red algae, and the tribophyceans *Tribonema* and *Vaucheria* are also widely encountered. The chrysophycean *Hydrurus* is a common inhabitant of cold mountain streams. Zygnematalean green algae dominate the periphyton communities of acidic *Sphagnum* bogs, with the greatest diversity associated with the higher plant *Utricularia*, which serves as a substrate (Woelkerling, 1976). In general, similar genera occur in streams and the shallow nearshore waters of lakes. Some of the larger filamentous algae, particularly *Cladophora*, may become covered by a dense community of microperiphyton, including cyanobacteria and diatoms. In contrast, zygnematalean filaments are commonly algal epiphyte-free, as they constantly generate mucilage that prevents adherence of algal epiphytes. Filamentous desmids are occasional exceptions to this generalization in that particular algal epiphytes may be present within their mucilaginous sheaths (see Fig. 3–28).

Structure and development of periphyton communities is studied by examination of natural substrates throughout the season, or by placing artificial substrates in streams or lakes and retrieving them at intervals (e.g., Sabater, et al., 1998). Glass microscope slides or ceramic tiles are often used as artificial substrates that mimic rock surfaces. They may be fastened within a holder that is anchored at the desired depth (Fig. 23–25). Algae are typically scraped from tiles for analysis of chlorophyll content and determination of species composition. Slides may be fixed in a 1% solution of glutaraldehyde to preserve cellular structure, followed by dehydration in an alcohol series, staining (fast green is recommended), and infiltration with xylenes. Mounting medium is then used to affix cover slips for preparation of permanent slides for microscopic assessment of species composition and spatial relationships (Fig. 23–26). Population density estimates are made by counting (and identifying) 50–500 cells per sample. Density is often expressed in terms of biovolume rather than cell number, because small populations of very large cells may actually prove to be more important than larger populations of very small cells. A table of volume estimates for a large number of periphyton species (primarily diatoms) and additional details of periphyton analysis methods are provided by Lowe and Pan (1996).

The Influence of Physical Factors on Periphyton

The architecture and species complexity of periphyton communities can vary greatly, depending upon the degree of disturbance, current, substrate type, temperature, irradiance, and nutrient levels. Heterogeneous (patchy) distribution of periphyton communities has been attributed to variation in irradiance and substrate characteristics, among other factors such as flow regime, nutrients, and grazing pressure (DeNicola and McIntire, 1990a, b). Disturbance frequency, in the form of flooding, wave action on lake shores, currents in streams and rivers, and

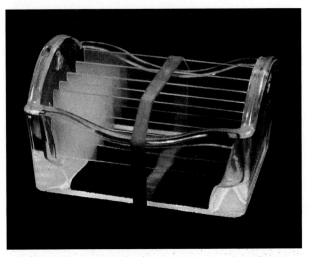

Figure 23–25 A simple and relatively inconspicuous apparatus for growth and analysis of periphyton communities. Glass slides are arranged within a glass holder designed for staining procedures, and held in place with a rubber band. The slide holder can then be anchored at various depths along transects within the littoral, and slides can be removed for processing and observation at intervals throughout the growing season.

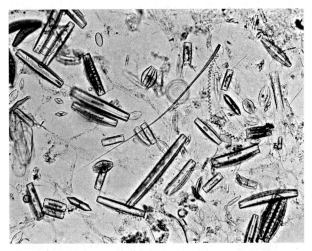

Figure 23–26 A permanent slide preparation derived from application of the method shown in Fig. 23–25 to an oligotrophic lake. After removal from the environment, the slide was rinsed to remove sand, sediment, and unattached algae, and then dehydrated and stained with fast green. Various common members of the freshwater periphyton can be observed. Quadrats can be marked on slides, and periphyton species abundance quantified in much the same way as for macrophytes.

water motions in general, is probably the strongest influence on periphyton community structure, as is the case for marine intertidal communities, where tides, waves, and currents are major structuring influences. This is in part because water motions strongly affect the balance between biomass accumulation and loss, as well as controlling nutrient supply. In conditions of frequent flooding or strong current, periphyton communities are subjected to shear stresses that can cause disturbance-sensitive species to slough off their substrates, and prevent community re-establishment. In contrast, where the current is slower, biomass can accumulate to the point that filamentous forms can generate long, streaming, conspicuous masses, if irradiance and nutrient levels are sufficient. *Cladophora* and *Lemanea* are robust enough to withstand the drag forces of waves and currents, and consequently can attain considerable biomasses. The extensively branched morphology of *Cladophora* provides an enormous surface area for attachment of large numbers of smaller periphytic forms, with positive feedback effects on community productivity. The positive effects of low levels of disturbance and relatively high nitrogen levels on periphyton chlorophyll *a* at 15 sites is shown in Figure 23–27 (Biggs, 1996).

As in marine environments, water motions are important in reducing the boundary layer of non-moving water at algal surfaces, and in replenishment of nutrients. In currents slower than 10 cm s^{-1}, growth of periphyton algae is limited because the surface boundary layer of unmixed water is not disturbed and continues to restrict nutrient diffusion (Stevenson, 1996). Hence, periphytic algal density is often higher in intermediate current velocities than in slow- or fast-moving water. Nevertheless, freshwater attached algae experience lower levels of water turbulence than do phytoplankton. One consequence of this fact is that freshwater periphyton (like marine periphyton) are enriched in the stable carbon isotope ^{13}C (see Chapter 2), as compared to phytoplankton, and a similar difference is apparent in their respective consumers. This has led to the hypothesis that littoral food webs based on attached algal productivity operate independently (are uncoupled) from pelagic food webs based on phytoplankton productivity (France, 1995).

Radiotracer evidence suggests that in oligotrophic to mesotrophic waters, nutrients such as phosphorus can be transferred from macrophyte host surfaces to periphytic algae. For example, some periphytic diatoms obtain more than 50% of their phosphorus

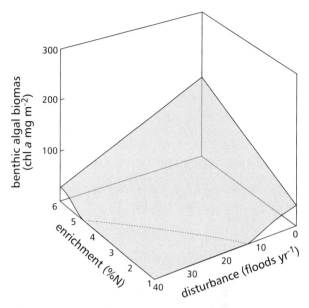

Figure 23–27 A three-dimensional graphical representation (a response surface) showing the effects of N-enrichment and disturbance (floods) on the biomass of periphytic algae estimated by measurements of chlorophyll *a*. (After Biggs, 1996)

from *Najas flexilis* tissues (Fig. 23–28) (Wetzel, 1996). Under higher nutrient conditions, many periphytic algae are thought to obtain most of their nutrients from the water column.

An analysis of data from 42 studies on nutrient effects on periphyton revealed that in 23 of these cases, nutrients are most often only secondarily limiting to algal growth, with disturbance, light availability, and grazing playing more significant roles (Borchardt, 1996). If light levels are too low to saturate photosynthesis, nutrient enrichment has no effect on periphyton algal growth rate. This weaker association between water-column nutrients and algal biomass is a major difference between periphyton algae and phytoplankton. As a result of the reduced coupling of nutrient levels in the water to algal growth rates, the biomass of periphytic algae does not always reflect water nutrient levels. Periphytic algae growing in the shade of macrophytes or streamside tree canopies may not exhibit a level of biomass that would be expected on the basis of water nutrient levels, and conversely, well-illuminated periphyton may be more abundant than predicted by instantaneous measurements of water nutrient levels (Borchardt, 1996). Light penetration also influences the depth to which periphytic algae can grow in lakes and rivers.

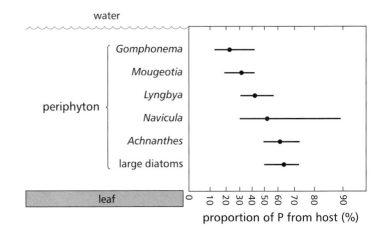

Figure 23–28 Experimental determination of the extent to which P is transferred from living leaf surfaces of the macrophyte *Najas flexilis* to periphytic algal species. (After Wetzel, 1996)

Light limitation at the base of thick periphyton assemblages may stimulate heterotrophic activity by various algae. Tuchman (1996) points out that eukaryotic algae are evolutionarily derived from heterotrophic protists and thus are likely equipped with nutrient uptake mechanisms, which provide a safety net when conditions are not suitable for photosynthesis. Uptake and utilization of 21 organic compounds, including sugars and amino acids, by many species of periphytic algae—primarily diatoms—has been documented (Tuchman, 1996). Increased uptake of the sugar glucose in the dark has been demonstrated in the diatom *Cyclotella*, for example (Hellebust, 1971). The periphytic green alga *Coleochaete* is also able to use glucose for growth, though light is required, and DOC utilization occurs only under conditions of inorganic carbon limitation (Graham, et al., 1994).

Grazing

Grazing usually results in significant declines in periphyton algal biomass. Grazers on periphyton algae include snails, caddis-fly larvae, mayfly nymphs, fish, shrimp, chironomid larvae, and tadpoles (see Kupferberg, et al., 1994; Peterson, et al., 1998). A review of the evidence for top-down versus bottom-up control suggested to Lamberti (1996) that consumers might play a pivotal role in structuring periphyton communities. Pleurocerid snails in streams and the widespread crayfish *Orconectes rusticus*, which occurs in the littoral of lakes and sometimes streams, are particularly important. The latter is one of few animals that can consume *Cladophora*, thereby making substrate available for other periphyton taxa. Many periphyton graz-

ers are thought to be omnivores that exhibit little specificity in food choice within the size range available to them (Steinman, 1996). Grazing modes include raspers and scrapers that harvest prostrate or crustose algae and gelatinous colonies, scrapers and gatherers that feed on stalked diatoms and short filaments, and gatherers, shredders, and piercers that obtain filamentous algae and their epiphytes (Fig. 23–29). A number of types of periphytic algae can survive passage through the gut of herbivorous snails and other grazers. An analysis of algae from snail feces revealed that more than 60% were able to resume growth (Underwood and Thomas, 1990).

Under conditions of intense disturbance or heavy grazing, the periphytic community may include only a few species of closely adherent diatoms, such as the disturbance and grazing-resistant *Cocconeis*. When disturbance and resource levels (light and nutrients) are at medium to low levels, periphyton communities can be dominated by filamentous cyanobacteria, including the nitrogen-fixers *Nostoc* and *Tolypothrix*, red algae such as *Audouinella*, and diatoms, such as *Epithemia*, that have N-fixing endosymbionts. In contrast, where disturbance and grazing are at a medium to low level, and resources are moderate to high, a more complex community, including stalked diatoms, filamentous greens, cyanobacteria, and adherent diatoms may be present (Biggs, 1996). In an experiment conducted in streams of the southwestern Ozark mountains, cyanobacterial mats dominated by the nitrogen fixer *Calothrix* were overgrown by benthic diatoms within four to ten days after exclusion of herbivorous fish and invertebrates. When exposed to grazers, the diatom growths were removed, and cyanobacterial mats regenerated (Power, et al., 1988).

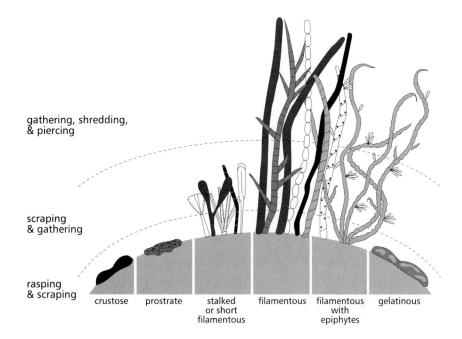

gathering, shredding, & piercing

scraping & gathering

rasping & scraping

crustose   prostrate   stalked or short filamentous   filamentous   filamentous with epiphytes   gelatinous

Figure 23–29 Grazers of various types and feeding modes and the types of freshwater periphyton upon which they feed. (After Steinman, 1996)

Community development begins with colonization of substrate by bacteria that generate an organic matrix onto which small diatoms can adhere. These are followed by erect or stalked diatoms, then filamentous greens and reds, forming a layered community. At its maximal development, a peak in biomass is achieved (Fig. 23–30). The time required for development of peak biomass varies from two to many weeks, depending upon the availability of light and nutrients and grazing intensity. Peak biomass may vary from very low levels to more than 1200 mg chlorophyll a m^{-2}. The carrying capacity of the environment is defined as the biomass present when rates of accumulation and loss are balanced (Fig. 23–31). Losses can result from grazing (see, for example, Cuker, 1983), disease (see Peterson, et al., 1993, for example), parasitism, age, or sloughing. As the community increases in size, it may become vulnerable to removal by water motions. In addition, algae close to the substrate may senesce as the result of light or nutrient limitation, leading to the process known as autogenic (self-generated) sloughing (Biggs, 1996). Dense, low-profile communities consisting of adnate diatoms and cyanobacteria-dominated mats are resistant to being dislodged by shear stress. High current velocity enhances their biomass by increasing nutrient diffusion to these encrusting periphyton. In contrast, more open growths, such as filamentous green algae, are vulnerable to sloughing in high current velocities (Biggs, et al., 1998). Though erect filaments of het-

erotrichous algae such as *Stigeoclonium* can be broken off in fast currents, the prostrate portions may be retained. *Stigeoclonium* is able to regenerate erect filaments in the post-sloughing phase. Experimental studies suggest there is a trade-off between productivity and persistence in the case of *Stigeoclonium* (Rosemond and Brawley, 1996). Grazing effects on algal succession appear to be highly specific to habitat; currently few generalizations can be made regarding presence or absence of grazer effects on algal succession (Steinman, 1996).

Temporal and Spatial Variation

There are three major temporal patterns of periphyton occurrence: (1) relatively constant low biomass under conditions of frequent disturbance, (2) cycles of accumulation and loss linked to less frequent disturbance, and (3) seasonal cycles resulting from seasonal variance in disturbance, temperature, grazer behavior, and/or irradiance (Biggs, 1996). Spatial variation within lakes is primarily attributable to depth-related decreases in water motion and irradiance. The upper littoral is characterized by wave action and high light levels, and is dominated by stalked diatoms and filamentous green algae and their epiphytes. *Cladophora*, *Ulothrix*, and *Oedogonium* are common littoral genera, with *Bangia* occurring in the high littoral in sites where there is significant influx of halides (from road

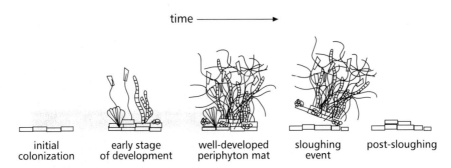

time ——————→

initial colonization   early stage of development   well-developed periphyton mat   sloughing event   post-sloughing

Figure 23–30 Development of a periphyton mat through time. Following initial colonization by diatoms (rectangles), and early development, a complex mat having maximum biomass accumulates. The larger aggregation is more vulnerable than early developmental stages to sloughing as a result of stream turbulence. (After Tuchman, 1996)

salting, for example). In contrast, the sublittoral of lakes is characterized by low levels of light and turbulence, and is dominated by cyanobacteria together with growths of epipelic and epipsammic diatoms. Though biomass is low, species richness is reportedly higher in the sublittoral than in the upper littoral; reasons for this difference are unknown (Lowe, 1996). The biomass of periphytic, littoral algae is characteristically low in oligotrophic lakes, whereas it may accumulate to nuisance levels in eutrophic lakes.

Considerable data on the effects of variations in temperature, irradiance, day length, and nutrients on photosynthesis and/or reproduction of freshwater periphytic algae have been obtained by multifactorial studies (Graham, 1982b; Graham, et al., 1982; Hoffmann and Graham, 1984; Graham, et al., 1984, 1985, 1986; Graham and Graham, 1987; Graham, et al., 1996a). Genera studied in this fashion include *Cladophora glomerata, Ulothrix zonata, Spirogyra* sp., *Coleochaete scutata,* and *Bangia atropurpurea.* Irradiance, temperature, and photoperiod explain most of the natural variation in net photosynthesis and reproduction of these periphytic algae. Periphyton occurring in the same habitat (*Cladophora, Bangia,* and *Ulothrix*) may partition the environment spatially (by growth at different levels in the littoral) or temporally (dominating at different times of the year). Reproduction of *Ulothrix zonata* is stimulated by long days, and results in the loss of most biomass before *Cladophora* becomes well established. Although *Bangia* and *Cladophora* may co-occur, *Bangia* is better able to tolerate high littoral conditions (high irradiance and sporadic desiccation) and thus *Cladophora* occupies the littoral region below *Bangia.*

In streams, spatial variation can occur between pools, runs, or riffle regions and among streams of

increasing size (distance from their headwaters). Riffle areas of streams are characterized by high shear stress and are consequently inhabited by low-growing diatoms such as *Cocconeis.* In contrast, lower velocity runs and pools harbor larger filamentous forms such as *Spirogyra, Oedogonium,* and *Cladophora. Rhizoclonium* can form conspicuous, long strands in nutrient-enriched streams (Fig. 23–32). A model of periphytic algal biomass in streams of increasing dimensions—known as the river continuum concept (Vannote, et al., 1980)—predicts that biomass should increase as streams coalesce and become wider, such that shading from streamside vegetation is reduced, and then decrease in rivers because light decreases with depth and increased turbidity. Some stream systems appear to behave according to this model, but others do not. Headwater streams are often domi-

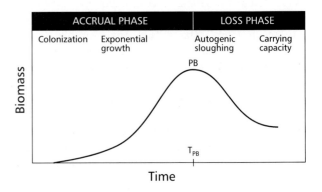

Figure 23–31 Graphical representation of phases in the development of freshwater periphyton communities. Following initial colonization and exponential growth phases, the community reaches a peak biomass (PB) (at time T_{PB}) that erodes by sloughing. (After Biggs, 1996)

Figure 23–32 *Rhizoclonium* generates very long filamentous growths in nutrient-rich streams.

nated by cyanobacteria such as *Schizothrix*, red algae including *Batrachospermum*, low-biomass green algae such as *Stigeoclonium*, and diatoms like *Cymbella*. With increasing downstream enrichment (arising from human land-use practices), nutrient-demanding forms such as *Vaucheria* and *Cladophora* become more abundant. Nuisance growths of *Ulothrix* occur in cold waters, and *Cladophora* and *Rhizoclonium* in warmer waters, of both streams and lakes. Variations among stream and river systems in biomass and species composition of the periphyton arise from differences in flood frequency and stream enrichment. High biomasses develop only under the combined conditions of low flood frequency and high nutrients. Control of nuisance stream- and lake-edge algae is a major issue in water resources management (Biggs, 1996).

Pollution Effects

The major inorganic compounds (including a variety of heavy metals) that inhibit growth of marine and freshwater algae were reviewed by Genter (1996). In general, these metals are thought to influence algal enzymes, and their effects vary among taxa—levels that inhibit one species may stimulate growth of another. Increases in zinc concentration are associated with a shift from dominance by diatoms to filamentous green algae and then to unicellular green algae. The inhibitory effects of a wide variety of organic compounds such as polycyclic aromatic hydrocarbons, polyhalogenated biphenyls, herbicides, insecticides,

surfactants, detergents, dyes, oils, solvents, and resins on periphytic algae was reviewed by Hoagland, et al. (1996). Substantial variability in sensitivity among attached taxa was the rule. Dramatic shifts in community species composition occur when herbicides reach concentrations in the μg per liter range. It is recommended that both single species- and community-bioassays be used to further evaluate the impact of organic compounds on periphytic algae.

Acidification is correlated with a decline in species richness, possibly because of decline in the macrophytes that provide support for periphyton communities. The cyanobacteria are most sensitive to lowered pH. In addition, acidification is correlated with dramatic metaphytic growths of zygnematalean filamentous algae in lakes and slower streams, resulting in increases in algal biomass as compared to unperturbed conditions. For *Mougeotia*, low pH is not optimal for growth. In fact, for one nuisance metaphytic growth-forming species, the optimum for photosynthesis was pH 8 (Graham, et al., 1996b). The reasons for dominance of such algae in acidified waters is not understood (Planas, 1996). Tolerance to heavy metals such as aluminum and zinc (which become more available as pH drops) (Graham, et al., 1996b), ability to obtain scarce dissolved inorganic carbon under low pH conditions, and release from the constraints of nutrient competition and herbivory all most likely contribute to these patterns. Some diatoms, such as *Eunotia* spp. are acid-tolerant and become dominant in fast flowing streams affected by acidification.

Recommended Books

Dawes, C. J. 1998. *Marine Botany*. John Wiley, New York, NY.

John, D. M., S. J. Hawkins, and J. H. Price. 1992. *Plant-Animal Interactions in the Marine Benthos*. Clarendon Press, Oxford, UK.

Lobban, C. S., and P. J. Harrison. 1994. *Seaweed Ecology and Physiology*. Cambridge University Press, New York, NY.

Lüning, K. 1990. *Seaweeds: Their Environment, Biogeography, and Ecophysiology*. John Wiley, New York, NY.

Schramm, W., and P. H. Nienhuis. 1996. *Marine Benthic Vegetation: Recent Changes and the Effects of Eutrophication*. Springer-Verlag, New York, NY.

Stevenson, R. J, M. L. Bothwell., and R. L. Lowe. (editors). 1996. *Algal Ecology: Freshwater Benthic Ecosystems*. Academic Press, San Diego, CA.

accessory pigment: Pigment that absorbs light energy and transfers it to a reaction center of chlorophyll *a* for use in photosynthesis.

acritarch: The general term for unicellular fossils whose relationships are uncertain; many are regarded as resistant cyst stages of planktonic algae, often prasinophyceans or dinoflagellates.

actin: Protein filaments about 8 nm in width that constitute microfilaments, a component of the cytoskeletal system often associated with contractile elements.

adelphoparasite: In red algae, a parasite that is closely related to its host.

aerobe: An oxygen-requiring organism (**aerobic:** having to do with an oxygen-rich environment).

agar: A sulfated polygalactan extracted from walls of various red algae that is used as a gelling agent.

agglutinins: Glycoproteins occurring on the surfaces (commonly flagella) of gametes; involved in the mate-recognition and gamete-adhesion process in sexual reproduction.

akinete: A thick-walled spore that functions in asexual reproduction, frequently serving as a resistant stage that undergoes a period of dormancy.

algaenans: Decay-resistant polymers of unbranched hydrocarbons produced in the cell walls of some algae.

alginic acids: Polysaccharides (a mixture of mannuronic and guluronic acids) extracted from the walls of brown algae for industrial applications; may occur in the salt form (alginates or algin).

aliasing: A sampling problem resulting from the too infrequent collection of samples.

alloparasite: In red algae, a parasite that is not closely related to its host.

allophycocyanin: A type of phycobiliprotein produced by cyanobacteria (except chlorophyll *a* + *b*-containing taxa), glaucophytes, red algae, and cryptomonads.

alternation of generations (diplohaplontic): A life history type in which there are two (or more in some red algae) multicellular stages that can be distinguished by type of reproductive cell produced and sometimes also by morphological features.

amoeboid: A type of cell organization in which a wall is absent and the protoplasm undergoes rapid shape changes.

amphiesma: The covering of dinoflagellate cells, which, in addition to an overlying plasma membrane, consists of membranous thecal vesicles that may contain little or no material or cellulosic thecal plates of varying thickness.

ampulla: (a) The flask-shaped reservoir and neck of euglenoids; (b) cortical reproductive regions in the thalli of some red algae.

anaerobe: An organism that can grow in an oxygen-free environment (**anaerobic:** having to do with an oxygen-free environment).

anisogamy: A type of sexual reproduction characterized by two types of gametes that differ in size.

anoxic: Without oxygen.

anoxygenic: Without production of oxygen; for example, use of hydrogen sulfide rather than water as a source of photosynthetic reductant by some cyanobacteria.

antenna pigments: Chlorophyll *a* and major accessory pigments that harvest light in photosynthesis.

antheridium: A cell that undergoes internal division and/or differentiation to form male gametes.

apical cell(s): The single apical cell or group of cells that occurs at the tip of a thallus, often capable of meristematic proliferation (apical growth).

apicomplexans: A phylum of protozoan parasites whose cells contain a nonpigmented plastid.

aplanospore: A nonflagellate spore that has the genetic potential to produce flagella under appropriate conditions; produced by subdivision of a parental cell.

apogamy: Generally, the development of an adult organism without the occurrence of sexual reproduction; sometimes used as a synonym for **parthenogenesis** (see).

archeopyle: In dinoflagellates, the exit pore of a germinating cyst.

areolae (pl.): Perforations in the siliceous walls of diatoms, commonly occurring in regular patterns such as rows.

astaxanthin: A red carotenoid pigment (3,3'-diketo-4,4'-dihydroxy-β-carotene) produced by some green algae that is commercially useful; also known as haematochrome.

autapomorphy: Derived character that defines a group of organisms.

autocolony: A type of asexual reproductive colony that is a miniature of the adult colony; produced by single cells of the adult.

autogamy: A type of sexual reproduction in which a zygote is formed by the fusion of two gametes from the same individual.

autolysins: Enzymes that degrade algal cell walls during release of spores from sporangia or gametes from gametangia.

autospore: A type of nonflagellate spore that lacks the genetic potential to produce flagella.

autotrophic: The capacity to produce organic compounds from inorganic compounds (i.e., to fix carbon) using the energy of light (photoautotrophs) or chemical energy (chemoautotrophs).

auxiliary cell: In the higher red algae, a cell into which the zygote nucleus or one of its mitotic progeny is deposited and which generates the carposporophyte generation by mitotic proliferation.

auxospore: A cell produced by diatoms that undergoes enlargement, compensating for the reduction in size that often occurs during population growth; commonly also the zygote of diatoms.

auxotrophy: The nutritional requirement for one or more vitamins.

axenic culture: In phycology, a laboratory-maintained single strain or algal species that is free of other algae, bacteria, or fungi.

axial (axile): A type of chloroplast that occurs in the center of the cell along its longitudinal axis.

baeocytes (endospores): Spores of cyanobacteria formed by internal division of vegetative cells.

basal body: The basal portion of a eukaryotic algal flagellum lying within the cell, consisting of a short cylinder of nine triplet microtubules; structurally equivalent to a centriole.

benthic: Having to do with the benthos—the bottom of a lake, stream, or marine system.

β-carotene: An accessory carotenoid pigment lacking oxygen that occurs in major algal groups; the source of provitamin A.

biflagellate: Having two flagella.

biological species concept: Distinction of species on the basis of breeding incompatibility.

bioluminescence: Production of light by living organisms (among the algae, by dinoflagellates).

bloom: Visible growth of planktonic algae, often associated with nutrient-enriched waters.

bootstrap value: An estimate of the validity of a branch in a phylogenetic tree that is determined by the number of times the branch appears after the data are repeatedly resampled.

botryococcenes: Unsaturated polyhydrocarbons excreted into the intercellular spaces of colonies of the planktonic green alga *Botryococcus*; associated with the formation of certain types of fossil fuel deposits.

calcification: Deposition of calcium carbonate.

callose: A β-1,3-linked polymer of glucose that is commonly associated with blockage of plasmodesmata or sieve plates in brown and green algae and land plants; also occurs in sexual reproductive cells of charophycean green algae.

Cambrian: The period from 490 to 543 million years ago, during which the oceans were inhabited by major algal and invertebrate groups.

canal raphe: A tubelike structure extending longitudinally along the valves of some pennate diatoms that opens externally via the raphe slit.

carbon fixation: Conversion of carbon dioxide into organic carbon as the result of the light-independent reactions of photosynthesis.

carbonic anhydrase: An enzyme that converts carbon dioxide to bicarbonate and vice versa.

Carboniferous: The period from 290 to 362 million years ago, during which atmospheric carbon dioxide levels underwent steep decline and atmospheric oxygen increased dramatically.

carboxysome (polyhedral body): Polygonal structure within the cells of cyanobacteria that contains the enzymes ribulose bisphosphate carboxylase/oxygenase (Rubisco) and carbonic anhydrase.

carpogonial branch: In red algae, a short branch consisting of a few vegetative cells, terminated by a female gamete, the carpogonium.

carpogonium: The female gamete of red algae.

carpospore: In red algae, the spore released from a carposporangium; usually assumed to be diploid.

carposporophyte: A multicellular phase in the life cycle of florideophycean red algae that generates carpospores via gonimoblast filaments; in all cases, attached to the female gametophytic thallus (i.e., not free-living); also known as the gonimocarp.

carrageenans: Mucilaginous sulfated polygalactans in the cell walls of red algae that are extracted for use as gelling agents in the food industry.

carrying capacity: The number of individual organisms that can be supported with available resources.

cell plate: A planar array of vesicles containing cell wall material that assembles during early cytokinesis and gives rise by centrifugal extension to new crosswalls; present in certain filamentous or parenchymatous green and brown algae, and in land plants (embryophytes).

cell wall: A more or less rigid enclosure surrounding the cell membrane; can be composed of various materials, depending on the algal group (see also **lorica** and **extracellular matrix**).

cellulose: A β-1,4-linked glucose polymer that forms microfibrils in the cell walls of various algae and all land plants.

cellulose synthesizing complexes: A group of enzymes that constitute particles in the plasmalemma and produce cellulose microfibrils outside the plasmalemma from glucose manufactured inside the cell.

Cenozoic: The period from 65 million years ago (the time of the KT extinction event) to the present.

central cells: In ceramialean red algae, the cells of the main axis that are surrounded by pericentral cells.

central nodule: A thickened region in the siliceous valves of motile pennate diatoms that separates the raphe into two longitudinal portions.

central strutted process: In certain diatoms, a tubular siliceous structure located at the midpoint of the valve (see **fultoportula**).

centric diatom: A diatom that has a frustule with radial valve symmetry (or that is phylogenetically linked to other diatoms having radially symmetrical valves).

centric mitosis: A type of mitosis in which centrioles are present at the spindle poles.

centrin: A contractile protein that is associated with the flagellar apparatus and cytoskeleton.

charasome: Latticelike, tubular array of membranes at the cytoplasmic surface of some regions of charalean cells.

chlorophyll-binding proteins: Proteins that link chlorophylls with other components of the photosynthetic apparatus.

chloroplast (plastid): A membrane-bound DNA-containing organelle (the number of bounding membranes varying among groups) that typically contains membranous thylakoids bearing chlorophyll *a* and other components of photosynthetic systems (some plastids lack thylakoids and chlorophyll, however).

chloroplast endoplasmic reticulum (CER): See **periplastidal endoplasmic reticulum**.

chromatic adaptation: The ability of algae to modify the amounts or proportions of photosynthetic pigments in relation to changes in the light environment.

chrysolaminaran: See **laminaran**.

ciliate: The informal name for a protist characterized by numerous cilia, presence of two types of nuclei (a larger macronucleus and a smaller micronucleus), and a cell mouth (cytostome).

cingulum (pl. cingula): (a) The transversely oriented girdle region of dinoflagellate cells; (b) the region of overlap of the two halves of diatom frustules.

circadian rhythm (endogenous rhythm): Cyclic changes in organisms that are endogenously controlled, often on a 24-hour cycle.

clade: A group of organisms descended from a common ancestor.

cladistics (cladistic analysis): A formal method of organizing organisms on the basis of relationships established by the presence or absence of derived characters.

Cladocera: A group (order) of planktonic crustaceans, many of which consume algae as food.

cladogram: A treelike diagram reflecting hypotheses of organismal relationships based on application of cladistic analysis.

cleavage furrow: An invagination of the cell membrane that grows from the cell surface to its interior, thereby accomplishing cytokinesis; present in many protists.

closed mitosis (intranuclear mitosis): Mitosis that occurs within an intact nuclear envelope; present in many protists.

coccoid: The morphology of unicellular algae that have cell walls and are often, but not always, spherical in shape.

coccolithophorids: Unicellular members of the haptophyte algae that are characterized by a covering of small, ornate calcium carbonate scales.

coccoliths: Ornate calcium carbonate scales, produced internally, that occur on the surfaces of coccolithophorid algae.

coccosphere: The aggregate of coccoliths forming a coalescent outer covering of coccolithophorid algal cells.

coenobium (pl. coenobia): A type of colony whose shape and cell number is genetically determined, established early in development, and does not change during the life of the organism (though cells commonly enlarge during colony development).

coenocytic: Multinucleate and without transverse walls (except, in many cases, during reproductive development); see also **siphonous**.

colony: A type of thallus consisting of a group of cells held together by mucilage or cell-wall material.

compensation point: The depth within the water column at which photosynthetic rate equals respiratory rate.

compound root: Flagellar root that consists of more than one type of element, for example, a group of microtubules plus a banded strand.

conceptacle: Cavity that contains the reproductive cells of some algae (particularly coralline red algae and fucalean brown algae).

concerted evolution: Parallel changes in each copy of a multicopy gene; usually refers to ribosomal RNA genes.

conchocelis: A filamentous, sporophytic phase in the life cycle of bangialean red algae that produces conchospores in conchosporangia; commonly occurs in shells or other calcareous materials.

conchospores: Spores produced by the sporophyte (conchocelis phase) of bangialean red algae.

conjugation: Mating of zygnematalean green algae involving nonflagellate gametes.

connecting cell (connecting filament, ooblast): In florideophycean red algae, a cell (often long and filamentlike) through which a zygote nucleus is transferred from the fertilized carpogonium to an auxiliary cell.

consistency index: An estimate of the degree of homoplasy (parallel or convergent evolution) in a phylogeny.

consortium: A close spatial association of two or more organisms, which may or may not confer mutualistic advantages.

contractile vacuole: A membrane-bound vesicle that can fill and contract, thereby expelling excess water from a cell.

convergent evolution: The evolution of similar features (from a different starting point) in unrelated organisms.

coronal cells: The five or ten cells found at the tips of the tubular cells that form an investment around oogonia of charalean green algae.

cortex: In a thallus, the layer of cells or tissues lying between the epidermis on the outside and the medulla on the inside.

costa (pl. costae): An elongate, hollow siliceous rib occurring on the frustule of some diatoms, commonly in parallel rows.

Cretaceous: The period between 65 and 145 million years ago, during which extensive chalk deposits were laid down (involving sedimentation of coccolithophorids).

cryptogamic crusts: Consortia of bacteria (including cyanobacteria), eukaryotic algae, fungi, lichens, and mosses that occur on soils or rocks, commonly in arid lands.

cyanelles (cyanellae): The blue-green plastids of glaucophyte algae; endosymbiotic cyanobacteria of certain other protists.

cyanophycean starch: Polyglycan granules (glycogen) that serve as the carbohydrate storage material of cyanobacterial cells.

cyanophycin: Granules consisting of polymers of the amino acids arginine and asparagine that serve as a proteinaceous reserve for cyanobacterial cells.

cyst: A thick-walled, dormant cell.

cystocarp: The carposporophyte (gonimocarp) and enveloping gametophytic tissues (pericarp) of florideophycean red algae.

cytokinesis: Division of the cytoplasm during cell division.

cytostome: A mouthlike opening through which food particles may be ingested; present in various phagotrophic protists.

DAPI: A DNA-specific fluorochrome (4,6-diamidino-2-phenylindole) used to visualize DNA in cell compartments and measure changes in nuclear DNA level during the cell cycle or the life history.

DIC: Dissolved inorganic carbon, including H_2CO_3 (carbonic acid) and HCO_3^- (bicarbonate).

DIN: Dissolved inorganic nitrogen, such as nitrate or ammonium.

DOC: Dissolved organic carbon, such as sugars and amino acids.

DON: Dissolved organic nitrogen, such as amino acids.

decay index: In parsimony, the number of additional steps beyond the shortest tree in which a particular branch collapses.

desmokont: In dinoflagellates, the condition in which the two flagella extend from the apical region of the cell.

determinate branch: A branch that does not increase further in length once its genetically determined maximum has been reached (compare with **indeterminate branch**).

Devonian: A period from 362 to 408 million years ago.

diatomite (diatomaceous earth): A mineral consisting of the siliceous remains of diatoms that have accumulated in sediments; mined for industrial applications.

diazotrophy: The ability to convert diatomic, gaseous N_2 gas into ammonium ion, which can be used by cells to produce amino acids.

diffuse growth: Not having a localized point of cell division.

dikaryon: The presence in cells of more than one genetic type of nucleus.

dinokont: In dinoflagellates, the condition of having one flagellum wrapped around the cell at the equatorial cingulum and the other extending posteriorly over the ventral sulcus.

dinospore: The flagellate reproductive cell of dinoflagellates (analogous to zoospores of various other algae).

dinosporin: A decay-resistant compound deposited at the surfaces of the cells of some dinoflagellates and their cyst stages that allows fossilization to occur.

dioecious: Having male or female gametes (or gametes of different mating types) produced by separate individuals.

diplobiontic: Having two free-living phases in the life history.

diplohaplontic: See **alternation of generations**.

diploid: Having two sets of homologous chromosomes.

diplontic: Having a life history in which there is only one multicellular, diploid phase, with the gametes representing the only haploid cells.

distromatic: Having two layers of cells.

domoic acid: A neurotoxin produced by the diatom *Pseudo-nitzschia multiseries* that causes amnesiac shellfish poisoning and memory loss (toxic encephalopathy) in humans.

dystrophic waters: Acidic, nutrient-poor waters that contain brown-colored humic acids.

ecdysis: Shedding of the thecal wall of dinoflagellates.

ejectisome (ejectosome): In cryptomonads, structure that is explosively discharged from the cell, presumably as a defense mechanism.

embryophytes: Bryophytes and vascular plants.

endocytosis (phagocytosis): The process by which particles are engulfed and taken into protistan cells.

endogenous rhythm: See **circadian rhythm**.

endolithic: Living within rocks.

endophytic: Living within the tissues of plants or thalli of seaweeds.

endoplasmic reticulum (ER): A system of cytoplasmic membranes occurring as flat or tubular vesicles that function in processing and transport of proteins and other cell constituents.

endoreduplication: The process by which the nuclear DNA undergoes repeated rounds of replication without intervening mitotic separation, yielding DNA levels higher than the haploid or diploid state.

endospores: See **baeocytes**.

endosymbiont: Organism that lives within the cell of another organism (the host) without causing disease or other obvious negative consequences for the host.

endosymbiosis: The condition in which one or more organisms live within the cells of a host without causing disease or other conspicuous harmful consequences.

endotoxins: Toxins produced within a cell and released through cell death and lysis.

endozooic: Living within an animal's body.

epibiont: Organism that grows on the surfaces of another organism.

epicingulum: In diatoms, the portion of the side wall that is associated with the epivalve.

epicone: In dinoflagellates, the portion of the cell anterior to the cingulum.

epilimnion: The upper layer of a stratified lake, whose waters are warmer in summer (and are typically more oxygen-rich) than bottom waters (hypolimnion).

epilithic: Living on the surfaces of rocks.

epipelagic: Upper warm ocean waters.

epipelic: Living on the surfaces of mud or sand.

epiphyte: An organism that grows on the surfaces of plants or algae.

epithallus (epithallium): In coralline red algae, the portion of the thallus arising from the intercalary meristem, and lying above it.

epitheca: (a) In dinoflagellates, the portion of the cell covering (theca) lying anterior to the cingulum; (b) in diatoms, the epivalve plus epicingulum.

epivalve: In diatoms, the usually flat (or slightly curving) covering of one end of the cell.

epizooic: Living on the surfaces on animals.

estuary: A partially enclosed body of seawater that usually receives input of freshwater.

eukaryotic: A type of cell possessing a nucleus, endomembrane and cytoskeletal systems, and (usually) organelles, including mitochondria, and in the case of most algae, plastids.

eulittoral: Living near the shore in the region that is uncovered by low tides and covered by high tides (i.e., the intertidal zone).

euryhaline: Having a broad tolerance to salinies of various levels.

eutrophic waters: Waters that contain relatively high levels of nutrients such as phosphate and/or combined nitrogen and; typically exhibit high levels of primary productivity.

eutrophication: The process by which natural waters are converted from the nutrient-poor, low-productivity, high-diversity condition to the nutrient-rich, high-productivity, low-diversity condition.

exhaustive search: In phylogenetic systematics, the finding of all possible trees that can be constructed from the data.

exospores: In cyanobacteria, spores that are cut off from one end of the parental cell (compare with **baeocytes**).

extracellular matrix (ECM): Materials generated by a cell that are secreted from, or produced on, the external surface; includes mucilage, cell walls, and loricas.

extrusome: See **trichocyst**.

euphotic zone: The (upper) portion of the water column that receives enough light for photosynthesis to occur.

eyespot (stigma): A red-colored spot, consisting of lipid droplets with carotenoids, that is involved in light perception; occurs in flagellate unicellular or colonial algae (or the reproductive cells of multicellular algae); may be found in plastid or cytoplasm, depending upon group.

facultative heterotrophy: The ability of photosynthetic organisms to also carry on heterotrophic nutrition.

false branches (false branching): In cyanobacteria, the continued growth of one or both ends of a filament that has been interrupted by cell death or cell differentiation, with retention of the pieces by an enveloping sheath.

filament: A type of algal thallus consisting of a linear array of cells in which neighboring cells share a common wall (see also **pseudofilament**).

filter feeding: A mode of food collection by herbivores that involves sieving large volumes of water for particles.

flagellar apparatus: Flagellar basal bodies together with connectives, flagellar roots, and sometimes rhizoplasts.

flagellar bases: See **basal bodies**.

flagellar root: Elongate structure consisting of groups of microtubules or banded material (or both—see **compound root**) that extends from the flagellar bases.

flagellar transformation: A maturation process by which the younger flagella in a parental cell become the older in a progeny cell.

flagellates: Unicellular or colonial protists whose cells bear one or more flagella.

flagellum (pl. flagella): In eukaryotes, a long, thin cellular projection that functions in motility; contains a peripheral ring of nine doublet microtubules and (usually) two central single microtubules.

floridean starch: In red algae, a branched α-1,4-linked glucose polymer with some α-1,6 linkages, that occurs as granules within the cytoplasm.

foramen: In diatoms, an opening or pore through the frustule.

foraminifera: A group of marine protists having calcareous shells; often possess algal endosymbionts.

fragmentation: Accidental or programmed breakage of a thallus into pieces that serve as asexual propagules.

frustule: In diatoms, the silica enclosure or wall.

fultoportula (pl. fultoportulae) (strutted process): In diatoms, a complex tubular structure extending from the frustule; often associated with chitin fibril formation.

furrowing: A form of cytokinesis in which a **cleavage furrow** grows from the cell edge toward the center; involves actin.

fusion cell: In red algae, (a) generally, a cell resulting from the coalescence of two or more non-gamete cells; (b) specifically, the cell produced by fusion of an auxiliary cell with one or more neighbors.

gametangiogamy: In desmids and diatoms, a term used to describe the process by which two nonflagellate cells associate by means of mucilage or a tube formed of cell-wall material, combine their cytoplasm and nuclei, and form a zygote.

gametangium: A container formed of one or more cells, in which gametes are produced.

gamete: A cell capable of fusing with another to form a zygote, with the two nuclei uniting to form a single zygotic nucleus.

gametic meiosis: Meiosis occurring during the production of gametes.

gametophore: A branch that bears one or more gametangia.

gametophyte: The multicellular, gamete-producing phase in the life history of organisms having alternation of generations.

gas vesicles: In cyanobacterial cells (and those of some other aquatic bacteria), cylindrical structures whose protein walls are permeable only to gases, and increase the buoyancy of the cells.

gas vacuoles: In cyanobacteria, aggregates of gas vesicles; sometimes used as a synonym for gas vesicles.

genicula: The uncalcified, flexible regions occurring as joints between calcified, non-flexible regions of the thalli of jointed coralline red algae and some ulvophycean green algae.

genophore: Ring-shaped, DNA-containing structure in plastids.

girdle: See **cingulum**.

girdle lamella: A flat sheet composed of three thylakoids that extends just under the plastid envelope in some ochrophyte algae.

gland cell (vesicle cell): In red algae, a specialized cell that serves in secretion or storage.

gliding motility: Movement of organisms (e.g., some cyanobacteria, diatoms, and desmids) that lack flagella or pseudopodia while in contact with a substrate.

glycoprotein: A protein having attached carbohydrates.

Golgi body (dictyosome): In eukaryotes, an organelle made up of stacks of flattened discoid membrane-delimited chambers (cisternae) in which materials are chemically modified, then secreted in vesicles.

gone: In desmid green algae, a cell derived from zygote germination, one or both of whose semi-cells may not possess the typical morphology of the species.

gonidium: In colonial volvocalean green algae, an enlarged, nonmotile cell that can generate new (daughter) colonies.

gonimoblast: In red algae, one or all of the filaments that bear carpospores (the filaments known in aggregate as the carposporophyte).

Gram stain: A procedure applied to bacteria for use in their classification; gram-positive bacteria have walls that retain a purple stain, whereas walls of gram-negative species do not.

grana (pl.): Stacks of discoid thylakoids that are characteristic of the plastids of land plants and certain green algae.

gullet: A deep depression at the anterior of some flagellate cells from which the flagella emerge.

gyrogonites: The calcified zygotes of charalean green algae; includes those occurring in the fossil record and those in modern sediments.

haplobiontic (haplontic): Having a life history exhibiting one haploid vegetative phase; zygotes are the only diploid cells in such a life cycle.

haploid: Having a single set of homologous chromosomes.

haptera: The attachment structures of the kelps.

haptonema: In haptophytes, a flagellumlike structure arising from the cell apex, near the flagella, that contains several microtubules (but not the 9+2 microtubular arrangement of flagella); may function in attachment, feeding, or avoidance responses.

hepatotoxins: Toxins (such as microcystin) produced by certain cyanobacteria that specifically affect liver cells; a cause of illness and death in livestock, waterfowl, dogs, and humans.

herbivory: The consumption and utilization of autotrophic cells as food.

heterococcolith: A type of coccolith in which there are crystals of more than one shape or size.

heterocyst: In some cyanobacteria, a thick-walled, weakly pigmented cell that is the site of nitrogen fixation.

heterogamy: Sexual reproduction involving gametes that are distinctive in size or behavior.

heterokaryon: A cell containing genetically distinct nuclei, or a thallus composed of such cells.

heterokont: Having flagella of unequal length or different ornamentation, position, or behavior; commonly used to describe flagellate cells of ochrophytes and related protists (though other algal groups also possess structurally different flagella on the same cell).

heteromorphic (heteromorphy): Morphologically different; in algae, usually applied to distinctive gametophyte and sporophyte phases in diplohaplontic life cycles (alternation of generations).

heteroplastidy: In certain ulvophycean green algae, having two kinds of plastids—green chloroplasts and colorless, starch-storing amyloplasts (leucoplasts).

heterothallic: Two different clones are required for sexual reproduction; self-incompatible.

heterotrichous: Having a filamentous body consisting of both an erect and a prostrate (attached) system.

heterotrophy: Nonautotrophic nutrition.

heuristic search: In systematics, a search technique that is not guaranteed to find the optimal solution, but which can greatly reduce computational time (compare to **exhaustive search**).

histones: In eukaryotes, DNA-binding proteins that are responsible for coiling and packing of chromosomes.

holdfast: A cell or multicellular structure (see **haptera**) that functions in attachment to a substrate.

holocarpy: In certain ulvophycean green seaweeds, conversion of the entire thallus cytoplasm into reproductive cells, whose release results in death of the parental alga.

holococcolith: A type of coccolith in which there is a single type of crystal.

homologous characters: Features of different species that are similar because they were inherited from a common ancestor.

homoplasy: A group of phenomena that lead to a similarity of characters in two or more taxa for reasons other than inheritance.

homothallic: Only one clone is required for sexual reproduction; self-compatible.

horizontal transfer: Movement of genetic material from one organism to another without the involvement of sexual reproduction (vertical transfer).

hormogonium: In filamentous cyanobacteria, a few-celled, usually motile filament that functions in asexual reproduction, dispersal, and in some cases, colonization of a host.

host: An organism whose cells or body serves as habitat for another.

hyalosome: In certain predaceous dinoflagellates, a colorless lenslike region in the eyelike ocellus.

hydrogenosomes: Membrane-bound organelles in certain protists; believed to represent the remains of prokaryotic endosymbionts that have lost all of their DNA.

hyphae: (a) In some kelps and fucoid brown algae, colorless filaments of elongate cells present in the central medulla; (b) colorless filaments that form the body (mycelium) of fungi.

hypnospore: A thick-walled resting structure.

hypnozygote: A thick-walled, resting zygote that may germinate in favorable circumstances following a required period of dormancy.

hypocingulum: In diatoms, the side wall of the hypotheca.

hypocone: The portion of a dinoflagellate cell posterior to the cingulum.

hypolimnion: The cold deep layers of stratified lakes.

hypothallus (hypothallium): In crustose (non-geniculate) coralline red algae, densely packed filaments forming the lowermost portion of the thallus, derived from divisions of the marginal meristem.

hypotheca: (a) In dinoflagellates, the portion of the wall (theca) lying below the cingulum; (b) in diatoms, the hypovalve plus the hypocingulum.

hypovalve: In diatoms, the end wall of the hypotheca.

hystrichosphere: In modern and fossil dinoflagellates, resting cysts commonly bearing distinctive projections and/or markings and an excystment pore (archeopyle).

indeterminate growth, branch, or lateral: Having no genetically imposed limits.

intercalary bands: In dinoflagellates, the regions of the theca that lie between the plates; in diatoms, a portion of the girdle region of the silica frustule.

intercalary growth: Growth in the middle of a thallus rather than at the tip (apical growth).

intergenicula: In articulated (geniculate) coralline red algae and certain jointed ulvophycean green algae, the nonflexible, highly calcified regions between joints (genicula).

internodal cell: A cell located between nodal cells.

intertidal: See **eulittoral**.

isogamy: Sexual reproduction involving gametes that are morphologically indistinguishable; structurally similar, but behaviorally distinguishable gametes are described as physiologically anisogamous.

isokont: Having flagella that are indistinguishable in length and structure.

isomorphic: Referring to a type of diplohaplontic life cycle (alternation of generations) in which the multicellular vegetative phases (sporophyte and gametophyte) are morphologically indistinguishable.

isthmus: In desmid green algae, the narrow region between two semicells, where the nucleus is located.

karyogamy: Fusion of gamete nuclei.

kelp: A member of the brown algal order Laminariales.

kleptoplastid: A plastid that has been harvested for temporary use by heterotrophic protists or metazoa, but is not stably integrated into the host's cell(s).

labiate process (rimoportula): In certain diatoms, a siliceous tube extending through the frustule; on the inside is laterally compressed to form a slit surrounded by liplike (labiate) structures.

lagoon: Shallow salt or brackish water that is separated from the open sea.

lamellae (adj. lamellate): Structures that occur as layers or a stack of flat plates.

lamina: Blade

laminaran: In brown algae and other ochrophytes, a soluble polysaccharide storage product composed of β-1,3-linked glucose units together with some branch-producing β-1,6-linkages.

lateral conjugation: In some filamentous zygnematalean green algae, the formation of a conjugation tube between adjacent cells of the same filament.

leucoplast: A colorless plastid that, if containing starch, is also known as an amyloplast.

ligula: In diatoms, a projection from a girdle band that extends toward the epivalve.

lipopolysaccharides (LPS): In general, polysaccharides having attached lipids; specifically, toxins produced by certain bacteria and bloom-forming cyanobacteria that can cause fever and inflammation in humans.

list: In some armored dinoflagellates, a saillike extension of the theca usually emerging from the cingulum or sulcus.

littoral: Occuring in the nearshore region (see also **eulittoral**).

loculus (loculate areola): In diatoms, a chamber within the frustule having a constricted opening on one side and a velum on the other.

lorica: A protective covering, often mineralized, that surrounds cells of various algae; a form of extracellular matrix (ECM) that is usually more distant from the cell membrane than are more typical forms of ECM.

macrandrous: Species of oedogonialean green algae in which the male thallus is the same size as the female thallus, and does not grow epiphytically on the female thallus (compare with **nannandrous**).

macrothallus: The larger of two alternating phases in a heteromorophic life history.

mannan: A polymer of the hexose sugar mannose.

mannitol: A polyhydroxyalcohol derived from the hexose sugar mannose.

manubria (s. manubrium): In charalean green algae, columnar cells that connect the pedicel to the shield cells within male gametangia.

marine snow: Particulate aggregates of algal cells or their remains, fecal pellets, bacteria, and heterotrophic protists, held together by mucilage, that are important in the transformation and transport of organic carbon to deep ocean sediments.

mastigonemes: (a) Stiff, three-parted hairs occuring in two lateral rows on one of the two flagella of heterokont protists (ochrophytes and related protists); (b) two-parted flagellar hairs of cryptomonads.

matrotrophy: The provision of nutrients by cells of the parental generation to cells of the next generation that have been retained on the maternal thallus.

medulla: Cells or tissues occurring in the center of a fleshy multicellular thallus.

meiosis: A form of nuclear division in which pairing and separation of homologous chromosomes yields four progeny nuclei having half the chromosome number and DNA level of the parental nucleus.

meiosporangium: A structure in which meiospores are produced.

meiospores: Reproductive cells that arise by the process of meiosis, and are therefore haploid.

meristem: A cell or group of cells that is capable of repeated division and thus adds to the number of cells in a thallus.

meristoderm: In fleshy brown algae, particularly Laminariales, a surface layer of cells (epidermis) that is capable of dividing (i.e., is meristematic).

mesoplankton: A size class of plankton, consisting of organisms that are between 0.2 mm and 2 mm in diameter.

mesotrophic: Waters that are intermediate in nutrient content and productivity to oligotrophic and eutrophic waters.

metaboly: In euglenoids, a form of motility that does not involve flagellar action.

metacentric spindle: In trebouxiophycean green algae, the positioning of centrioles at the midpoint of the mitotic spindle, rather than at the poles, as is more common.

microplankton: A size class of plankton, consisting of organisms that are between 20 μm and 200 μm in diameter.

microsource (scintillon): In certain marine dinoflagellates, the particulate location of bioluminescence.

microthallus: The smaller of two alternating phases in a heteromorphic life history.

microtubule organizing center: Dark-staining cloud of material in cells viewed at the ultrastructural level that is thought to be involved in microtubule assembly; in algae, often associated with centrioles (most groups) or polar rings (red algae).

microtubules: In eukaryotes, components of the cytoskeleton and flagella composed of the protein tubulin arranged to form hollow tubes 25 nm in diameter.

mitochondrion: In eukaryotes, a DNA-containing respiratory organelle derived from an endosymbiotic bacterium, characterized by a two-membrane envelope, the innermost of which is convoluted, forming cristae.

mitosis: A form of nuclear division that generates two progeny nuclei identical in chromosome level and DNA content to each other and to the parental nucleus.

mixotrophy: A form of nutrition in which both autotrophy and heterotrophy may be utilized, depending on the availability of resources (see also **heterotrophy, phagotrophy,** and **osmotrophy**).

monoccious: Producing male and female gametes on the same thallus.

monophyletic: Used to describe a group of organisms that have descended from a single common ancestor (compare to **polyphyletic**).

monosporangium: A sporangium in which only a single spore is produced.

monospore: A nonflagellate spore produced singly in a monosporangium.

monostromatic: Composed of a single layer of cells.

morphological species concept: The use of structural differences and similarities to distinguish species and classify them.

mucocyst (muciferous body): Saclike structure within cells of dinoflagellates and euglenoids from which thick, rod-shaped mucilage body can be extruded to the cell surface when the organisms are disturbed.

mucopolysaccharides: Polysaccharides containing amino sugars and uronic acids that bind water, forming mucilaginous slime.

multiaxial: A thallus form composed of many similar axial filaments arranged in parallel (also known as fountain-type structure).

multilayered structure (MLS): In the cytoskeleton of flagellate cells of various protists and sperm of nonflowering land plants, a layered structure located near the flagellar basal bodies that includes microtubules; thought to function as a microtubule organizing center.

multiseriate: Having more than two rows of cells.

multispecies consortium: Association of several species, often of disparate taxonomic groups, that are involved in metabolic interactions and which occupy the same habitat.

mutualism: Symbiotic association of two or more organisms in which all members benefit.

mycobiont: The fungal partner in a lichen.

myosin: A protein often associated with actin to form contractile structures (as in animal muscles); occurs in the cytoskeleton and is involved in cytoplasmic streaming.

nannandrous: Species of oedogonialean green algae in which the male thallus is a tiny filament attached to the larger female thallus.

nanoplankton: A size class of plankton, consisting of organisms that are between 2 μm and 20 μm in diameter.

necridia (separation disks): In filamentous cyanobacteria, dead cells whose production is associated with development of hormogonia and in some forms, false branching.

nemathecia: In some red algae, a raised wartlike structure on the thallus surface that contains reproductive structures.

nematocyst: In certain dinoflagellates, harpoonlike ejectile structure that is morphologically distinct from the more common trichocyst.

neurotoxin: Toxin (such as saxitoxin) produced by some algae; blocks neuromuscular activity in animals.

neuston: The community of organisms living in surface films.

nitrogenase: In cyanobacteria (and some other bacteria), the holoenzyme that performs nitrogen fixation—conversion of diatomic N_2 gas into ammonium ion.

node: A site on a thallus from which branches arise.

non-geniculate (non-articulate): A thallus that is not segmented into flexible joints and non-flexible regions.

nori: Sheets of chopped, pressed thalli of the red seaweed *Porphyra* that are used in Asian cuisines.

nuclear associated organelle (NAO): In red algae, a ring-shaped structure that occurs at the spindle poles during cell division.

nucleomorph: In some algae having plastids of secondary origin (cryptomonads, chlorarachniophytes, and some dinoflagellates), a plastid-based, double-membrane enclosed structure containing DNA arranged in small chromosomes and other features suggesting its origin from a eukaryotic nucleus.

ocellus (pl. ocelli): (a) In certain marine phagotrophic dinoflagellates, a complex structure consisting of a lens and retinalike cup that is thought to provide a vision system of utility in prey location; (b) in some diatoms, a rimmed perforated plate of elevated silica.

okadaic acid: A toxin produced by certain dinoflagellates that inhibits serine- and threonine-specific phosphates (which occur widely in eukaryotes); the cause of diarrhetic shellfish poisoning in humans.

oligotrophic: Waters that are low in nutrients such as phosphate and combined nitrogen, and consequently low in primary productivity and biomass, but typically high in species diversity.

ooblast: See **connecting cell**.

oogamy: Sexual reproduction involving syngamy of a small flagellate male gamete and a larger, nonflagellate (or only transiently flagellate) female gamete.

oogonium: A structure that produces one or more eggs.

open mitosis: A form of mitosis in which the nuclear envelope disintegrates.

osmoregulation: Regulation of cell water content.

osmotrophy: A form of nutrition in which dissolved organic carbon (**DOC**) is imported from the environment into cells.

ostiole: (a) In coralline red algae or fucoid brown algae, an opening in a conceptacle; (b) in red algae, an opening in the cystocarp.

oxygenic: Oxygen-producing, as in photosynthetic breakdown of water by cyanobacteria and eukaryotic autotrophs.

pallium: A sheetlike extension of cytoplasm, also known as a feeding veil, that is produced by some phagotrophic dinoflagellates for the purpose of capturing and digesting prey.

palmella: A stage produced by algae genetically capable of producing flagella consisting of nonmotile cells embedded in an amorphous mucilaginous matrix.

paraflagellar rod: In euglenoids and certain ochrophytes (in which case it may be called a paraxonemal rod), a long proteinaceous rod that runs alongside the microtubular axoneme in flagella.

parallel evolution: The independent evolution of similar characters in organisms belonging to separate lineages, but whose genetic backgrounds are similar.

paralogous genes: Individual members of gene families.

paramylon: In euglenoids, a β-1,3-glucose polymer that occurs as granules in the cytoplasm.

paraphyletic: In systematics, a group that does not include all of the descendants of a common ancestor.

paraphysis (pl. paraphyses): A nonreproductive filament that occurs among sporangia or gametangia.

parasite: An organism that lives at the expense of a host.

parasporangium: In red algae, a sporangium that produces many spores; not equivalent to a tetrasporangium.

parenchyma (adj. parenchymatous): A form of thallus in which true tissues are produced (see also **pseudoparenchyma**).

parietal plastid: A plastid that lies along the cell margin.

parthenogenesis: Production of a new individual from an unfertilized gamete.

peduncle: In dinoflagellates, a microtubule-containing tubular extension of cytoplasm that can be extended to attach to prey cells or tissues and either extract their contents or engulf them.

pelagic: Living in open ocean waters rather than nearshore coastal or inland waters.

pellicle: (a) In euglenoids, an often flexible, protein-containing surface layer consisting of interlocking, helically oriented strips that occurs just inside the cell membrane; (b) in some dinoflagellates, a cellulosic and sporopollenin-containing layer beneath the theca that remains in place when the theca is shed; (c) the surface layers of some other algae.

pennate: One of the major diatom frustule types, exhibiting bilateral valve symmetry.

perennation: Living through unfavorable conditions from one favorable period to another.

periaxial cells (pericentral cells): In uniaxial red algae, cells that are cut off from the main axis that occur in a whorl that surrounds the parental axial cell.

pericarp: In florideophycean red algae, a coherent layer of gametophytic filaments that surrounds a carposporophyte; the outermost, presumably protective, layer of a cystocarp.

periphyton: The organisms that occur on the surfaces of plants, algae, and inorganic substrates in shallow benthic or nearshore littoral habitats.

periplast: In cryptomonads, the cell covering, consisting of intersecting plates lying beneath the cell membrane.

periplastidal compartment: In cryptomonads, haptophytes, and ochrophytes, the space between the outer and inner pairs of plastid envelope membranes (the outer two of which are known as the periplastidal endoplasmic reticulum).

periplastidal endoplasmic reticulum (PER): An extension of the endoplasmic reticulum that encloses the plastids of some algal groups; sometimes also connected to the nuclear envelope.

perithallus (perithallium): In crustose (non-geniculate) coralline red algae, the layer of cells that is produced inwardly from the intercalary meristem (compare with **epithallus**).

perizonium: In pennate diatoms, the silicified outer layer of the auxospore (zygote) wall, consisting of series of bands.

peroxisome: Organelle bound by a single membrane that occurs in the cytoplasm of embryophyte cells and those of some eukaryotic algae, and contains characteristic enzymes such as glycolate oxidase and catalase; this term also sometimes used more generally as a synonym for microbodies—small single-membrane-bound structures that occur in some algae and contain catalase but not glycolate oxidase.

phagopod: In dinoflagellates, a tubular structure lacking microtubules that can be formed on cells for the purpose of attaching to and extracting the contents of prey cells.

phagotrophy (phagocytosis): A form of nutrition in which particles such as cells are ingested by protists via invagination of the cell surface.

pheromone: A substance produced by one organism that evokes a sexual response in another.

phialopore: In certain colonial volvocalean green algae, a hole in the colony surface through which the colony undergoes inversion during development.

phlorotannins: In brown algae, polymers of 1,3,5-trihydroxybenzene, which, like tannins of embryophytes, are able to bind metal ions, precipitate proteins, and have an unpleasant taste, thus probably serving as herbivore deterrents.

photoautotroph (photoautotrophy): Organism that obtains its organic nutrients by means of photosynthesis; obligate photoautotrophs are restricted to this form of nutrition.

photoheterotrophy (adj. photoheterotrophic): A form of nutrition in which light is captured and used as an energy source by a pigmented alga, which, at the same time, takes up dissolved organic compounds from the environment.

photoinhibition: The reduction of photosynthetic rates by high irradiance.

photorespiration: A process by which some of the organic carbon produced in photosynthesis is wasted through the excretion of glycolate; based on competition between O_2 and CO_2 for the active site in Rubisco; occurs more rapidly under conditions of high irradiance, low dissolved inorganic carbon levels, and high levels of oxygen.

phototaxis: Movement toward (positive phototaxis) or away (negative phototaxis) from light.

phragmoplast: In post-mitotic cells of certain green algae and embryophytes, an array of microtubules arranged perpendicularly to the plane of division that, together with coated vesicles and endoplasmic reticulum, is involved with cytokinesis by centrifugal cell plate formation (in contrast, see **furrowing** and **phycoplast**).

phycobilisome: In cyanobacteria, glaucophytes, and red algae, a hemispherical or discoidal structure on the surfaces of thylakoids that contains the phycobilin accessory pigments.

phycobiont: The algal component of a lichen.

phycocyanin: A blue-green phycobiliprotein found in cyanobacteria, glaucophytes, cryptomonads, and red algae.

phycocyanobilin: The open-chain tetrapyrrole that, when bound to proteins, constitutes phycocyanin.

phycoerythrin: A red phycobiliprotein found in cyanobacteria, cryptomonads, and red algae.

phycoerythrobilin: The open-chain tetrapyrrole that, when bound to proteins, constitutes phycoerythrin.

phycoma: A resting-cyst stage in the life history of some prasinophyceans whose wall contains a highly resistant compound resembling sporopollenin, enabling fossilization.

phycoplast: In chlorophycean green algae, an array of microtubules oriented parallel to the plane of cytokinesis, through which the developing wall grows.

phylogenetic: Pertaining to phylogeny (evolutionary relationships).

phylogenetic species concept: A species is the smallest monophyletic group of organisms that exhibits at least one distinctive, unifying characteristic.

phylogenetic tree: A treelike diagram that represents a hypothesis regarding relationships of a group of organisms.

phylogeny: A hypothesis of the way in which a group of organisms is evolutionarily related.

phytochrome: A red/far-red light-sensitive pigment that is involved in perception of light and photomorphogenesis in algae and embryophytes.

phytoferritin: Intracellular, semicrystalline arrays of iron-binding proteins.

phytoplankton: Floating or swimming microscopic algae.

picoplankton: A size class of plankton, consisting of organisms such as bacteria and certain small eukaryotic algae that are between 0.2 μm and 2 μm in diameter.

pit plug (pit connection): In red algae, a plug that develops in the wall between adjacent cells, and which, in an elongate form, can maintain structural association even when other portions of the cell walls become spatially separated.

placental cell: In red algae, a cell that results from the fusion of an auxiliary cell and nearby cells and that ultimately generates carpospores; in certain charophycean green algae and embryophytes, cells located at the junction of haploid and diploid generations that are modified for apoplastic nutrient transfer from parental cells to zygotes/sporophytes.

placoderm desmid: A member of the Zygnematales, in which the walls of semicells are of different ages and commonly contain pores through which mucilage is extruded.

plankton: Microscopic organisms that are suspended or swim in the water column.

planozygote: Motile zygote.

plasmalemma: The cell (or plasma) membrane.

plasmid: Small ring of DNA that is capable of replication; found in prokaryotic and at least some eukaryotic cells.

plasmodesmata (s. plasmodesma): In certain green and probably all brown algae and embryophytes, protoplasmic connections between adjacent cells; primary plasmodesmata are formed during cytokinesis, whereas secondary plasmodesmata are produced after a new cell wall has been formed.

plasmogamy: During sexual reproduction, fusion of the cytoplasm of gametes; may or may not be immediately followed by karyogamy.

plastid: See **chloroplast.**

ploidy: The number of complete sets of chromosomes in a nucleus.

plurilocular sporangium: In brown algae, a reproductive structure that is subdivided by cell walls into numerous small chambers (locules), each of which produces a single flagellate cell.

pneumatocyst: In some fleshy brown algae, a bulbous, gas-filled structure that functions in buoyancy.

polar nodule: (a) In some pennate diatoms, thickened regions at the ends of the valves; (b) in cyanobacteria, thickened regions of the heterocyst wall where it attaches to adjacent vegetative cells.

polar rings: In red algae, densely staining rings located at the spindle poles that are thought to play a role in microtubule organization.

polyeder: Angular, polyhedral cell produced by a flagellate meiospore that arises from zygote germination in *Hydrodictyon* and *Pediastrum*; gives rise to new coenobium.

polyhedral body: See **carboxysome.**

polyphosphate: Storage granules composed of phosphate polymers.

polyphyletic: In systematics, describes a group that contains some members that are actually more closely related to organisms outside the group.

polysaccharide: A polymer of sugar monomers, having the general formula $C(H_2O)_n$.

polysiphonous: Thallus form of some red algae, having the appearance of multiple, parallel tubes (siphons), but arising from the linear abutment of adjacent periaxial cells.

polysporangium: In red algae, a sporangium that produces many spores; homologous to a tetrasporangium.

polystromatic: Thallus composed of many cell layers.

poroid areola: An areola with little or no constriction on one end and occluded by a velum at the other (compare with **loculus [loculate areola]**).

Precambrian: The period in the earth's history extending from 543 million to 4500 million years ago, which was dominated by cyanobacteria.

primary endosymbiosis: Incorporation of a free-living prokaryote into a host eukaryotic cell, with subsequent transformation into an organelle.

primary pit plug (primary pit connection): A pit plug that develops during cytokinesis, occurring in the wall between two sibling cells.

primary productivity: The net amount of organic carbon that results from photosynthesis, usually expressed per unit volume or area and per unit time (often estimated by measuring oxygen production).

procarp: In certain florideophycean red algae, presence of an auxiliary cell in the carpogonial branch.

prokaryotic: Lacking membrane-bound DNA (a

nucleus), membrane-bound organelles (mitochondria, plastids), and endomembrane systems such as endoplasmic reticulum and a Golgi apparatus; prokaryotes also lack sexual reproduction systems and $9+2$ flagella (cilia) like those of eukaryotes.

propagule: A multicellular structure that serves in asexual reproduction.

protists: Eukaryotic organisms that are not members of the Kingdoms Fungi, Animalia, or Plantae (embryophytes); includes eukaryotic algae and many kinds of non-pigmented organisms.

protozoa: An informal term for a polyphyletic array of motile, heterotrophic protists.

pseudofilament (adj. pseudofilamentous): A linear array of cells that do not share common walls; rather, end walls of individual cells are attached.

pseudoparenchyma (adj. pseudoparenchymatous): A form of thallus composed of interwoven, continuous filaments; superficially resembles parenchyma.

pseudoraphe: A siliceous region, uninterrupted by a slit, on one or both valves of some pennate diatoms, that occupies the central valve axis and sometimes mimics a true raphe.

pusule: In dinoflagellates, a branched system of tubes, lined by cell membrane, that opens to the cell exterior.

pyrenoid: A proteinaceous region in the plastids of many types of algae; known in some cases to contain Rubisco, and commonly associated with formation of storage compounds.

quadriflagellate: Having four flagella.

radiolarians: Marine planktonic protozoans having radially symmetrical siliceous skeletons and stiff, radiating rhizopods; often possess algal symbionts.

raphe system: An elongate fissure in one or both valves of many pennate diatoms that has both an external and internal opening.

raptorial feeding: Obtaining food by seizing prey with specialized appendages.

receptacle: Fertile area of fleshy thallus in which reproductive structures develop.

refractory carbon: Forms of organic carbon that are resistant to microbial, chemical, and physical degradation.

remineralization: Transformation of elements from organic to inorganic form; e.g., the conversion of organic carbon to inorganic carbon.

reservoir: In euglenoids, the base of the flasklike apical invagination from which flagella emerge.

retinoid: In certain dinoflagellates, a portion of the eyelike ocellus that is cup-shaped and has a retinalike layer of dark red droplets.

rhizoid: A single cell or multicellular filament, growing from the basal region of a thallus, that functions in attachment.

rhizoplast: A striated, contractile strand that extends from the flagellar basal bodies into the cell, often connecting with the nuclear surface.

rhizostyle: In cryptomonads, a cluster of posteriorly directed microtubules that extends from the flagellar bases.

rhodolith: Rounded nodules of coralline red algae that develop around the surfaces of stones.

rimoportula: See **labiate process**.

Rubisco: Ribulose 1,5-bisphosphate carboxylase/oxygenase, the enzyme that catalyzes incorporation of carbon dioxide into carbohydrate.

sarcinoid: A type of thallus consisting of a three-dimensional packet of cells.

saxitoxins: Toxins produced by certain bacteria, cyanobacteria, and dinoflagellates that block cell membrane sodium channels; the cause of paralytic shellfish poisoning in humans.

scalariform conjugation: In filamentous zygnematalean green algae, parallel alignment of two filaments, followed by formation of multiple conjugation tubes between opposed cells, giving a ladderlike appearance.

scintillon: See **microsource**.

secondary endosymbiosis: The incorporation of a photosynthetic eukaryote, whose plastid was derived from a prokaryote, into a eukaryotic host cell.

secondary pit plug (secondary pit connection): Pit plug formed between a pair of non-sibling cells, by the cutting off of a small cell from one of the two cells (which involves formation of a pit plug) and subsequent fusion of the small cell with the other member of the pair.

segregative cell division: In siphonocladalean green algae, cleavage of the multinucleate protoplast into spherical units that then expand, develop walls, and function as independent regions.

semicell: In desmid green algae, one of the two halves of the cell.

separation disks: See **necridia**.

seta (pl. setae): Long, thin projections from a cell.

seta cells: In some charophycean green algae, cells that generate long, thin sheathed hairs that serve in herbivory deterrence.

shield cells: In charalean green algae, cells that form the outer walls of the multicellular male gametangia.

siderophores (siderochromes): Extracellular compounds that bind iron, making it available for import into cells.

sieve tubes: In brown algae, longitudinal series of sieve elements—cells with porous end walls, through which photoassimilates are transported.

silica deposition vesicle: A membrane-bound intracellular vesicle in which siliceous structures such as diatom frustule components are deposited.

Silurian: The period from 408 to 439 million years ago.

siphonous (siphonalean): Describes a type of thallus composed of large multinucleate coenocytic cells wherein cross-walls are rarely formed.

sorus (pl. sori): Group of reproductive structures.

spermatangia: In red algae, male reproductive structures that produce spermatia.

spermatia: The male gametes of red algae; no flagella are present.

spermatogenous threads: In charalean green algae, filaments of cells that each produce a sperm, located within a complex antheridium.

sporangium: A cell whose contents become subdivided to form spores.

spore: An asexual reproductive cell.

sporic meiosis: Meiosis that occurs during production of spores.

sporophyll: In kelps, a fertile, spore-producing blade.

sporophyte: The spore-producing phase in a life history that involves alternation of spore- and gamete-producing generations.

sporopollenin: In walls of certain green algae and spores of embryophytes, a resistant biopolymer.

starch: A polymer of glucose having α-1,4- and α-1,6-linkages.

stellate structure: In green algae, a structure in the flagellar transition region that is star-shaped in cross-sectional view; its contraction is associated with flagellar abscission.

stephanokont: In reproductive cells of certain green algae, presence of many flagella arranged in an anterior ring, like a crown.

stigma: See **eyespot**.

stipules: In charalean green algae, spinelike cells in rings beneath whorls of branches.

stomatocyst (statospore): A resting spore of chrysophycean and synurophycean algae.

stonewort: A calcified member of the charalean algae.

stramenopiles (heterokonts): Organisms that possess two flagella of distinct type, one bearing characteristic three-part flagellar hairs, namely ochrophytes (heterokont algae), oomycetes, and some other protists.

stratification: Formation of a surface layer of warm water over deeper, cold water as the result of density differences that develop during warm-season heating.

striae (pl.): In diatoms, rows of pores or areolae in the frustule.

striated connective: In flagellate cells, a banded structure that links flagella and other components of the flagellar apparatus.

stromatolite: Calcareous, layered assemblage of cyanobacteria, occurring as fossils or in modern waters in sheltered areas.

strutted process: See **fultoportula**.

sublittoral: Shoreline at depths below the lowest low tide.

sulcus: A longitudinal furrow or depression on the ventral side of a dinoflagellate cell, in which lies the longitudinal flagellum.

supralittoral: Shoreline above the uppermost limit of high tides.

sutures: In dinoflagellates, the regions between adjacent plates.

symbiosis: Two or more organisms living in close physical contact.

systematics: The scientific study of organismal diversity and the relationships among organisms.

taxon: A general term for any taxonomic category.

tentacle: A slender, contractile feeding structure found in certain dinoflagellates.

tertiary endosymbiosis: The incorporation of a photosynthetic eukaryote, whose plastid was derived from a eukaryotic endosymbiont, into a eukaryotic host cell.

tetrasporangium: In florideophycean red algae, meiosporangium containing four spores produced by meiosis.

tetraspore: In florideophycean red algae, one of the four spores produced within a tetrasporangium.

tetrasporophyte: In florideophycean red algae, the phase of the life history that produces tetraspores.

thallus (pl. thalli): The body of an alga, which is not differentiated into vascularized leaves, roots, and stems.

thecal plate: A portion of the cellulosic theca, or wall, of dinoflagellates.

thermocline: The region of greatest rate of vertical temperature change in a stratified water body.

thylakoid: A flattened, saclike membranous structure in cyanobacterial cells and plastids of eukaryotic algae and plants.

trabecula (pl. trabeculae): In some ulvophycean green algae, an extension of the cell wall into the cell lumen, providing structural support.

transition region (zone): In eukaryotic flagella, the zone between the flagellum and its basal body, at the point where the flagellum exits the cell.

transitional helix: In many ochrophytes, a coiled structure present in the transition region between the axoneme and the flagellar base.

trichocysts: In dinoflagellates and raphidophyceans, ejectile structures occurring within cytoplasmic membranous vesicles that greatly elongate when discharged, serving a defensive function.

trichogyne: In red algae and some charophycean green algae, the elongated apical portion of the female gamete (carpogonium or oogonium) that is receptive to male gametes.

trichome: In cyanobacteria, a filament exclusive of the sheath.

trichothallic growth: In some brown algae, active cell division occurring in an intercalary position, in a stack of short cells located at the base of filaments.

true branching: In filamentous cyanobacteria, divisions occurring parallel to the long axis of the filament, leading to production of branches (compare to **false branching**).

trumpet hyphae (pl.): In some kelps, colorless, elongate cells in the medulla that are expanded (like the bell of a trumpet) at the cross-walls.

tube cells: In charalean green algae, elongate helically twisted tubular cells that surround oogonia.

uniaxial construction: A thallus consisting of a single axial filament.

unilocular sporangium: In brown algae, a sporangium in which all of the spores are produced within a single compartment, usually by meiosis.

uniseriate: Having a single row of cells.

utricle: In certain ulvophycean green algae, the swollen terminal end of a siphonaceous tube.

valve: (a) In diatoms, the flattened, siliceous end wall; (b) in desmokont dinoflagellates, one of the two large thecal plates.

velum: In diatoms, a thin, perforated siliceous layer over areolae.

vestibulum: In cryptomonads, an anterior depression from which flagella emerge; the anterior end of the cell gullet or furrow.

xanthophylls: Oxygen-containing, yellow-pigmented carotenoids.

zoospore: Flagellate spore.

zooxanthellae (pl.): Unicellular golden-pigmented cells (usually dinoflagellates) that are endosymbiotic in marine animals, including reef-building corals.

zygote: The product of gamete fusion (syngamy).

zygotic meiosis: Meiosis occurring during zygote maturation or germination.

Abbott, I. A., and G. J. Hollenberg. 1976. *Marine Algae of California.* Stanford University Press, Stanford, CA.

Abrahams, M. V., and L. D. Townsend. 1993. Bioluminescence in dinoflagellates: A test of the burglar alarm hypothesis. *Ecology* 74:258–260.

Adachi, M., Y. Sako, and Y. Ishida. 1996. Identification of the toxic dinoflagellates *Alexandrium catenella* and *A. tamarense* (Dinophyceae) using DNA probes and whole-cell hybridization. *Journal of Phycology* 32:1049–1052.

Adey, W. H., and T. Goertemiller. 1987. Coral reef algal turfs: Master producers in nutrient poor seas. *Phycologia* 26:374–386.

Adey, W. H., C. Luckett, and K. Jensen. 1993. Phosphorus removal from natural waters using controlled algal production. *Restoration Ecology* 1:29–39.

Ahlgren, G. 1978. Growth of *Oscillatoria agardhii* in chemostat culture. II. Dependence of growth constants on temperature. *Internationale Vereinigung für Theoretische und Angewandte Limnologie* 21:88–102.

Al-Dhaheri, R. S., and R. L. Willey. 1996. Colonization and reproduction of the epibiotic flagellate *Colacium vesiculosum* (Euglenophyceae) on *Daphnia pulex. Journal of Phycology* 32:770–774.

Al-Kubaisy, K. H., H. O. Schwantes, and G. Seibold. 1981. Cytophotometrische Untersuchungen zum Generationswechsel autotropher und heterotropher siphonaler Organismen (*Vaucheria sessilis* und *Saprolegnia ferax*). *Nova Hedwigia* 34:301–316.

Alberte, R. S. 1989. Physiological and cellular features of *Prochloron*. Pages 31–52 in *Prochloron, A Microbial Enigma*, edited by R. A. Lewin and L. Cheng. Chapman & Hall, London, UK.

Alexopoulos, C. J., C. W. Mims, and M. Blackwell. 1996. *Introductory Mycology*. 4th edition. John Wiley Inc., New York, NY.

Allemand, D., P. Furia, and S. Benazet-Tambutte. 1998. Mechanisms of carbon acquisition for endosymbiont photosynthesis in Anthozoa. *Canadian Journal of Botany* 76:925–941.

Allen, N. S. 1974. Endoplasmic filaments generate the motive force for rotational streaming in *Nitella. Journal of Cell Biology* 63:270–287.

Alonso, D. L., C. I. S. del Castillo, E. M. Grima, and Z. Cohen. 1996. First insights into improvement of eicosapentaenoic acid content in *Phaeodactylum tricornutum* (Bacillariophyceae) by induced mutagenesis. *Journal of Phycology* 32:339–345.

Aluwihare, L., D. J. Repeta, and R. F. Chen. 1997. A major biopolymeric component to dissolved organic carbon in surface sea water. *Nature* 387:166–169.

Amano, K., M. Watanabe, K. Kohata, and S. Harada. 1998. Conditions needed for *Chattonella antiqua* red tide outbreaks. *Limnology and Oceanography* 43:117–128.

Amoroso, G., C. Weber, D. Sültemeyer, and H. Fock. 1996. Intracellular carbonic anhydrase activities in *Dunaliella tertiolecta* (Butcher) and *Chlamydomonas reinhardtii* (Dangeard) in relation to inorganic carbon concentration during growth: Further evidence for the existence of two distinct carbonic anhydrases associated with the chloroplasts. *Planta* 199:177–184.

Amsler, C. D., and R. B. Searles. 1980. Vertical distribution of seaweed spores in a water column offshore of North Carolina. *Journal of Phycology* 16:617–619.

Anagnostidis, K., and J. Komárek. 1985. Modern approach to the classification system of the cyanophytes I—Introduction. *Algological Studies* 38/39:291–302.

Anagnostidis, K., and J. Komárek. 1988. Modern approach to the classification system of the cyanophytes III—Oscillatoriales. *Algological Studies* 50–53:327–472.

Anagnostidis, K., and J. Komárek. 1990. Modern approach to the classification system of the cyanophytes IV—Stigonematales. *Algological Studies* 59:1–73.

Andersen, R. A. 1987. Synurophyceae classis nov., a new

class of algae. *American Journal of Botany* 74:337–353.

Andersen, R. A. 1992. Diversity of eukaryotic algae. *Biodiversity and Conservation* 1:267–292.

Andersen, R. A., R. R. Bidigare, M. D. Keller, and M. Latasa. 1996. A comparison of HPLC pigment signatures and electron microscopic observations for oligotrophic waters of the North Atlantic and Pacific Oceans. *Deep-Sea Research* 43:517–537.

Andersen, R. A., R. W. Brett, D. Potter, and J. P. Sexton. 1998a. Phylogeny of the Eustigmatophyceae based upon 18S rDNA, with emphasis on *Nannochloropsis. Protist* 149:61–74.

Andersen, R. A., S. L. Morton, and J. P. Sexton. 1997. CCMP-Provasoli-Guillard National Center for culture of marine phytoplankton. *Journal of Phycology* 33 (supplement).

Andersen, R. A., and T. J. Mulkey. 1983. The occurrence of chlorophylls c_1 and c_2 in the Chrysophyceae. *Journal of Phycology* 19:289–294.

Andersen, R. A., D. Potter, R. R. Bidigare, M. Latasa, K. Rowan, and C. J. O'Kelly. 1998b. Characterization and phylogenetic position of the enigmatic golden alga *Phaeothamnion confervicola*: Ultrastructure, pigment composition and partial SSU rDNA sequence. *Journal of Phycology* 34:286–298.

Andersen, R. A., G. W. Saunders, M. P. Paskind, and J. P. Sexton. 1993. Ultrastructure and 18S rRNA gene sequence for *Pelagomonas calceolata* gen. et sp. nov. and the description of a new algal class, the Pelagophyceae classis nov. *Journal of Phycology* 29:701–715.

Anderson, D. M. 1994. Red tides. *Scientific American* 271:62–68.

Anderson, O. R. 1992. Radiolarian algal symbioses. Pages 93–110 in *Algae and Symbioses*, edited by W. Reisser. BioPress, Bristol, UK.

Anderson, R. J. 1994. *Suhria* (Gelidiaceae, Rhodophyta). Pages 227–244 in *Biology of Economic Algae*, edited by I. Akatsuka. SPB Academic Publishing, The Hague, Netherlands.

Andrews, M., R. Box, S. McInroy, and J. A. Raven. 1984. Growth of *Chara hispida*. II. Shade adaptation. *Journal of Ecology* 72:885–895.

Apt, K. E., N. E. Hoffmann, and A. R. Grossman. 1993. The γ-subunit of R-phycoerythrin and its possible mode of transport into the plastid of red algae. *Journal of Biological Chemistry* 268:16208–16215.

Apt, K. E., P. G. Korth-Pancic, and A. R. Grossman. 1996. Stable transformation of the diatom *Phaeodactylum tricornutum. Molecular & General Genetics* 252:572–579.

Archibald, P., and H. C. Bold. 1970. Phycological studies XI. the genus *Chlorococcum* Meneghini. *University of Texas Publication* 7015. 115 pp.

Armstrong, S. L. 1987. Mechanical properties of tissues of the brown alga *Hedophyllum sessile* (C. Ag.) Setchell: Variability with habitat. *Journal of Experimental Marine Biology and Ecology*. 114:143–151.

Arrigo, K. R., D. H. Robinson, D. L. Worthen, R. B. Dunbar, G. R. DiTullio, M. Van Woert, and M. P. Lizotte. 1999. Phytoplankton community structure and the drawdown of nutrients and CO_2 in the Southern Ocean. *Science* 283:365–367.

Aruga, H., T. Motomura, and T. Ichimura. 1996. Immunofluorescence study of mitosis and cytokinesis in *Acrosiphonia duriuscula* (Acrosiphoniales, Chlorophyta). *Phycological Research* 44:203–213.

Ashen, J. B., and L. J. Goff. 1996. Bacteria in galls of the marine red alga *Prionitis lanceolata. Journal of Phycology* 32:286–296.

Ashen, J. B., and L. J. Goff. 1998. Galls on the marine red alga *Prionitis lanceolata* (Halymeniaceae): Specific induction and subsequent development of an algal bacterial symbiosis. *American Journal of Botany* 85:1710–1721.

Bach, S. D., and M. N. Josselyn. 1979. Production and biomass of *Cladophora prolifera* (Chlorophyta, Clado-

phorales) in Bermuda. *Botanica Marina* 22:163–168.

Bae, Y. M., and J. W. Hastings. 1994. Cloning, sequencing and expression of dinoflagellate luciferase DNA from a marine alga *Gonyaulax polyedra*. *Biochimica Biophysica Acta* 1219:449–456.

Baden, D. G., K. S. Rein., and R. E. Gawley. 1998. Marine toxins: How they are studied and what they can tell us. Pages 487–514 in *Molecular Approaches to the Study of the Ocean*, edited by K. E Cooksey. Chapman & Hall, London, UK.

Badger, M. R., T. J., Andrews, S. M. Whitney, M. Ludwig, D. C. Yellowlees, W. Leggat, and G. D. Price. 1998. The diversity and co-evolution of Rubisco, pyrenoids, and chloroplast-based CO_2-concentration mechanisms in algae. *Canadian Journal of Botany* 76:1052–1071.

Badger, M. R., and G. D. Price. 1994. The role of carbonic anhydrase in photosynthesis. *Annual Review of Plant Physiology and Plant Molecular Biology* 45:369–392.

Bailey, J. C., R. R. Bidigare, S. J. Christensen, and R. A. Andersen. 1998. Phaeothamniophyceae classis nova: A new lineage of chromophytes based upon photosynthetic pigments, *rbcL* sequence analysis and ultrastructure. *Protist* 149:245–263.

Bailey, J. C., and R. L. Chapman. 1996. Evolutionary relationships among coralline red algae (Corallinaceae, Rhodophyta) inferred from 18S rRNA gene sequence analyses. Pages 363–376 in *Cytology, Genetics, and Molecular Biology of Algae*, edited by B. R. Chaudhary and S. B. Agrawal. SPB Academic Publishers, Amsterdam, Netherlands.

Bailey, J. C., and D. W. Freshwater. 1997. Molecular systematics of the Gelidiales: Inferences from separate and combined analyses of plastid *rbcL* and nuclear SSU gene sequences. *European Journal of Phycology* 32:343–352.

Baker, A. L., and A. J. Brook. 1971. Optical density profiles as an aid to the study of microstratified phytoplankton populations in lakes. *Archiv für Hydrobiologie* 69:214–233.

Baker, J. R. J., and L. V. Evans. 1973a. The ship-fouling alga *Ectocarpus*. I. Ultrastructure and cytochemistry of plurilocular reproductive stages. *Protoplasma* 77:1–13.

Baker, J. R. J., and L. V. Evans. 1973b. The ship-fouling alga *Ectocarpus*. II. Ultrastructure of the unilocular reproductive stages. *Protoplasma* 77:181–189.

Bakker, F. T., J. L. Olsen, and W. T. Stam. 1995. Evolution of nuclear rDNA ITS sequences in the *Cladophora albida/sericea* clade (Chlorophyta). *Journal of Molecular Evolution* 40:640–651.

Baldauf, S. L., and J. D. Palmer. 1990. Evolutionary transfer of the chloroplast *tufA* gene to the nucleus. *Nature* 344:262–265.

Ban, S., C. Burns, J. Castel, Y. Chaudron, E. Christou, R. Escribano, S. F. Umani, S. Gasparini, F. G. Ruiz, M. Hoffmeyer, A. Ianora, H.-K. Kang, M. Laabir, A. Lacoste, A. Miralto, X. Ning, S. Poulet, V. Rodriguez, J. Runge, J. Shi, M. Starr, S. Uye, and Y. Wang. 1997. The paradox of diatom-copepod interactions. *Marine Ecology Progress Series* 157:287–293.

Barlow, S. B., and R. A. Cattolico. 1981. Mitosis and cytokinesis in the Prasinophyceae. I. *Mantoniella squamata* (Manton and Parke) Desikachary. *American Journal of Botany* 68:606–615.

Bartlett, R., and R. Willey. 1998. Epibiosis of *Colacium* on *Daphnia. Symbiosis* 25:291–299.

Bartley, G. E., and P. A. Scolnick. 1995. Plant carotenoids: Pigments for photoprotection, visual attraction, and human health. *The Plant Cell* 7:1027–1038.

Baschnagel, R. A. 1966. New fossil algae from the Middle Devonian of New York. *Transactions of the American Microbiological Society* 85:297–302.

Bastin, P., T. Sherwin, and K. Gull. 1998. Paraflagellar rod is vital for trypanosome motility. *Nature* 391:548.

Bates, S. S., D. L. Garrison, and R. A. Horner. 1998. Bloom dynamics and physiology of domoic-acid producing *Pseudo-nitzschia* species. Pages 267–292 in *Physiological Ecology of Harmful Algal Blooms*, edited by D. M. Anderson, A. D. Cembella, and G. M. Hallegraeff. Springer-Verlag, Berlin, Germany.

Bauer, I., L. Maranda, K. A. Young, Y. Shimizu, C. Fairchild, L. Cornell, J. Macbeth, and S. Huang. 1995. Isolation and structure of caribenolide I, a highly potent antitumor amcrolide from a cultured free-swimming Caribbean dinoflagellate, *Amphidinium* sp S1-36-5. *Journal of Organic Chemistry* 60:1084–1086.

Baumann, P., and S. P. Jackson. 1996. An archaebacterial homologue of the essential eubacterial cell division protein FtsZ. *Proceedings of the National Academy of Science* 93:6726–6730.

Beam, C. A., and M. Himes. 1974. Evidence for sexual fusion and recombination in the dinoflagellate *Crypthecodinium* (*Gyrodinium*) *cohnii*. *Nature* 250:435–436.

Beaufort, L., Y. Lancelot, P. Camberlin, O. Cayre, E. Vincent, F. Bassinot, and L. Labeyrie. 1997. Insolation cycles as a major control of equatorial Indian Ocean primary production. *Science* 278:1451–1454.

Becker, B., and M. Melkonian. 1996. The secretory pathway of protists: Spatial and functional organisation and evolution. *Microbiological Reviews* 60:697–721.

Becker, B., D. Becker, J. P. Kamerling, and M. Melkonian. 1991. 2-keto-sugar acids in green flagellates: A chemical marker for prasinophycean scales. *Journal of Phycology* 27:498–504.

Beech, P. L., and R. Wetherbee. 1990a. Direct observations on flagellar transformation in *Mallomonas splendens* (Synurophyceae). *Journal of Phycology* 26:90–95.

Beech, P. L., and R. Wetherbee. 1990b. The flagellar apparatus of *Mallomonas splendens* (Synruophyceae) at interphase and its development during the cell cycle. *Journal of Phycology* 26:95–111.

Beech, P. L., R. L. Wetherbee, and J. D. Pickett-Heaps. 1990. Secretion and deployment of bristles in *Mallomonas splendens*. *Journal of Phycology* 26:112–122.

Beer, S. 1994. Mechanisms of inorganic carbon acquisition in marine macroalgae (with special reference to the Chlorophyta). *Progress in Phycological Research* 10:179.

Bell, G. 1997. The evolution of the life cycle of brown seaweeds. *Biological Journal of the Linnean Society* 60:21–38.

Bell, R. A. 1993. Cryptoendolithic algae of hot semiarid lands and deserts. *Journal of Phycology* 29:133–139.

Ben-Amotz, A., A. Katz, and M. Avron. 1982. Accumulation of β-carotene in halotolerant algae: Purification and characterization of β-carotene-rich globules from *Dunaliella bardawil* (Chlorophyceae). *Journal of Phycology* 18:529–537.

Berger, S., and M. J. Kaever. 1992. *Dasycladales: An Illustrated Monograph of a Fascinating Algal Order*. Thieme, Stuttgart, Germany.

Berges, J. A. 1997. Minireview: algal nitrate reductases. *European Journal of Phycology* 32:3–8.

Bergey, E. A., C. A. Boettiger, and V. H. Resh. 1995. Effects of water velocity on the architecture and epiphytes of *Cladophora glomerata* (Chlorophyta). *Journal of Phycology* 31:264–271.

Bergman, B., and E. J. Carpenter. 1991. Nitrogenase confined to randomly distributed trichomes in the marine cyanobacterium *Trichodesmium thiebautii*. *Journal of Phycology* 27:158–165.

Bergquist, A. M., and S. R. Carpenter. 1986. Limnetic herbivory effects on phytoplankton populations and primary production. *Ecology* 67:1351–1360.

Bergquist, A. M., S. R. Carpenter, and J. C. Latino. 1985. Shifts in phytoplankton size structure and community composition during grazing by contrasting zooplankton assemblages. *Limnology and Oceanography* 30:1037–1045.

Berkaloff, C., R. Rosseau, A. Couté, E. Casadevall, P. Metzger, and C. Chirac. 1984. Variability of cell wall structure and hydrocarbon type in different strains of *Botryococcus braunii*. *Journal of Phycology* 20:377–389.

Berman, T., and W. Rodhe. 1971. Distribution and migration of *Peridinium* in Lake Kinneret. *Internationale Vereinigung für Theoretische und Angewandte Limnologie* 19:266–276.

Berman-Frank, I., and J. Erez. 1996. Inorganic carbon pools in the bloom-forming dinoflagellate *Peridinium gatunense*. *Limnology and Oceanography* 41:1780–1789.

Berman-Frank, I., A. Kaplan, T. Zohary, and Z. Dubinsky. 1995. Carbonic anhydrase activity in the bloom-forming dinoflagellate *Peridinium gatunense*. *Journal of Phycology* 31:906–913.

Berner, R. A. 1997. The rise of land plants and their effect on weathering and CO_2. *Science* 276:544–546.

Bhattacharya, D., T. Helchen, C. Bebeau, and M. Melkonian. 1995. Comparisons of nuclear-encoded small-subunit ribosomal RNAs reveal the evolutionary position of the Glaucocystophyta. *Molecular Biology and Evolution* 12:415–420.

Bhattacharya, D., and L. Medlin. 1995. The phylogeny of plastids: A review based on comparisons of small-subunit ribosomal RNA coding regions. *Journal of Phycology* 31:489–498.

Bhattacharya, D., and L. Medlin. 1998. Algal phylogeny and the origin of land plants. *Plant Physiology* 116:9–15.

Bhattacharya, D., J. Steinkötter, and M. Melkonian. 1993. Molecular cloning and evolutionary analysis of the calcium-modulated contractile protein, centrin, in green algae and land plants. *Plant Molecular Biology* 23:1243–1254.

Bhaya, D., and A. R. Grossman. 1991. Targeting proteins to diatom plastids involves transport through an endoplasmic reticulum. *Molecular and General Genetics* 229:400–404.

Bicudo, C. E. deM., and G. Tell. 1988. *Pseudocryptomonas*, a new genus of Cryptophyceae from southern Brazil. *Nova Hedwigia* 46:407–411.

Bidigare, R. R. 1989. Photosynthetic pigment composition of the brown tide alga: Unique chlorophyll and carotenoid derivatives. Pages 57–75 in *Novel Phytoplankton Blooms*, edited by E. M. Cosper, V. M. Bricelj, and E. J. Carpenter. *Estuarine and Coastal Studies* 35. Springer-Verlag, Berlin, Germany.

Bidigare, R. R., M. E. Ondrusek, M. C. Kennicutt, R. Iturriaga, H. R. Harvey, R. W. Hoham, and S. A. Macko. 1993. Evidence for a photoprotective function for secondary carotenoids of snow algae. *Journal of Phycology* 29:427–434.

Biebel, P. 1973. Morphology and life cycles of saccoderm desmids in culture. *Nova Hedwigia* 42:39–57.

Biggs, B. J. F. 1996. Patterns in benthic algae of streams. Pages 31–56 in *Algal Ecology. Freshwater Benthic Ecosystems*, edited by R. L. Stevenson, M. L. Bothwell, and R. L. Lowe. Academic Press, San Diego, CA.

Biggs, B. J. F., D. G. Goring, and V. I. Nikora. 1998. Subsidy and stress responses of stream periphyton to gradients in water velocity as a function of community growth form. *Journal of Phycology* 34:598–607.

Billard, C. 1994. Life cycles. Pages 167–186 in *The Haptophyte Algae*, edited by J. C. Green and B. S. C. Leadbeater. Clarendon Press, Oxford, UK.

Billot, C., S. Rousvoal, A. Estoup, J. T. Epplen, P. Saumitou-Laprade, M. Valero, and B. Kloareg. 1998. Isolation and characterization of microsatellite markers in the nuclear genome of the brown alga *Laminaria digitata* (Phaeophyceae). *Molecular Ecology* 7:1179–1180.

Bingham, S. E., and A. N. Webber. 1994. Maintenance and expression of heterologous genes in chloroplast of *Chlamydomonas reinhardtii*. *Journal of Applied Phycology* 6:239–246.

Bird, D. F., and J. Kalff. 1986. Bacterial grazing by planktonic lake algae. *Science* 231:493–495.

Birkeland, C. 1997. Introduction. Pages 1–12 in *Life and Death of Coral Reefs*, edited by C. Birgeland. Chapman & Hall, New York, NY.

Birkhead, M., and R. N. Pienaar. 1995. The flagellar apparatus of *Chrysochromulina* sp. (Prymnesiophyceae). *Journal of Phycology* 31:96–108.

Bisalputra, T. 1966. Electron microscopic study of the protoplasmic continuity in certain brown algae. III. Cytokinesis and the multicellular embryo. *Canadian Journal of Botany* 44:89–93.

Björk, M., K. Haglund, Z. Ramazanov, and M. Pedersén.

1993. Inducible mechanisms for HCO_3^- utilization and repression of photorespiration in protoplasts and thalli of three species of *Ulva* (Chlorophyta). *Journal of Phycology* 29:166–173.

Bliding, C. 1957. Studies in *Rhizoclonium*. I. Life history of two species. *Bot. Notiser* 110:271–275.

Blomster, J., C. A. Maggs, and M. J. Stanhope. 1998. Molecular and morphological analysis of *Enteromorpha intestinalis* and *E. compressa* (Chlorophyta) in the British Isles. *Journal of Phycology* 34:319–340.

Böhme, H. 1988. Regulation of nitrogen fixation in heterocyst-forming cyanobacteria. *Trends in Plant Science* 3:346–351.

Bold, H. C. 1958. Three new chlorophycean algae. *American Journal of Botany* 45:737–743.

Bold, H. C. 1970. Some aspects of the taxonomy of soil algae. *Annals of the New York Academy of Science* 175:601–616.

Bold, H. C., and F. J. MacEntee. 1973. Phycological notes. II. *Euglena myxocylindracea* sp. nov. *Journal of Phycology* 9:152–156.

Bold, H. C., and M. J. Wynne. 1978, 1985. *Introduction to Phycology*, 1st and 2nd editions. Prentice Hall, Englewood Cliffs, NJ.

Bolton, J. J., and R. J., Anderson. 1994. *Ecklonia*. Pages 385–406 in *Biology of Economic Algae*, edited by I . Akatsuka. SPB Academic Publishing, The Hague, Netherlands.

Bomber, J. W., M. G. Rubio, and D. R. Norris. 1989. Epiphytism of dinoflagellates associated with the disease ciguatera: Substrate specificity and nutrition. *Phycologia* 28:360–368.

Boo, S. M., and I. K. Lee. 1994. *Ceramium* and *Campylaephora* (Ceramiaceae, Rhodophyta). Pages 1–33 in *Biology of Economic Algae*, edited by I. Akatsuka. SPB Academic Publishing, The Hague, Netherlands.

Booth, B. C., and H. J. Marchant. 1987. Parmales, a new order of marine chrysophytes, with descriptions of three new genera and seven new species. *Journal of Phycology* 23:245–260.

Booton, G. C., G. L. Floyd, and P. A. Fuerst. 1998a. Polyphyly of tetrasporalean green algae inferred from nuclear small-subunit ribosomal data. *Journal of Phycology* 34:306–311.

Booton, G. C., G. L. Floyd, and P. A. Fuerst. 1998b. Origins and affinities of the filamentous green algal orders Chaetophorales and Oedogoniales based on 18S rRNA gene sequences. *Journal of Phycology* 34:312–318.

Borchardt, M. A. 1996. Nutrients. Pages 184–228 in *Algal Ecology. Freshwater Benthic Ecosystems*, edited by R. J. Stevenson, M. L. Bothwell, and R. L. Lowe. Academic Press, New York, NY.

Boston, P. J. 1984. Critical life issues for a Mars base, in *The Case for Mars*, edited by P. J. Boston. *American Astronautical Society Science and Technology Series* 57, Univelt, San Diego, CA.

Bouck, G. B., and D. L. Brown. 1973. Microtubule biogenesis and cell shape in *Ochromonas*. *Journal of Cell Biology* 56:340–359.

Bourne, D. G., G. L. Jones, R. L. Blakeley, A. Jones, A. P. Negri, and P. Riddles. 1996. Enzymatic pathway for the bacterial degradation of the cyanobacterial cyclic peptide toxin microcystin LR. *Applied and Environmental Microbiology* 62:4086–4094.

Bourrelly, P. 1968. *Les Algues d'eau douce. Initiation à la Systématique. Tome II: Les Algues jaunes et brunes, Chrysophycées, Phéophycées, Xanthophycées et Diatomées* [freshwater chrysophytes, phaeophytes, xanthophytes=tribophytes and diatoms]. Boubée & Cie, Paris, France.

Bowe, L. M., and C. W. DePamphilis. 1996. Effects of RNA editing and gene processing on phylogenetic reconstruction. *Molecular Biology and Evolution* 13:1159–1166.

Bown, P. R., J. A. Burnett, and L. Gallagher. 1991. Critical events in the evolutionary history of calcareous nannoplankton. *Historical Biology* 5:279–290.

Bowring, S. A., D. H. Erwin, Y. G. Jin, M. W. Martin, K. Davidek, and W. Wang. 1998. U/Pb geochronology and tempo of the end-Permian mass extinction. *Science* 280:1039–1045.

Boyle, J. A., J. D. Pickett-Heaps, and D. B. Czarnecki. 1984. Valve morphogenesis in the pennate diatom *Achnanthes coarctata*. *Journal of Phycology* 20:563–573.

Brahamsha, B. 1996. An abundant cell-surface polypeptide is required for swimming by the nonflagellated marine cyanobacterium *Synechococcus*. *Proceedings of the National Academy of Science* 93:6504–6509.

Branch, G. M., and Griffiths, C. L. 1988. The Benguela ecosystem. Part V. The coastal zone. *Oceanography and Marine Biology Annual Review* 26:395–486.

Brand, L. E. 1994. Physiological ecology of marine coccolithophores. Pages 39–50 in *Coccolithophores*, edited by A. Winter and W. G. Siesser. Cambridge University Press, Cambridge, UK.

Braun, M., and G. O. Wasteneys. 1998. Reorganization of the actin and microtubule cytoskeleton throughout blue-light-induced differentiation of characean protonemata into multicellular thalli. *Protoplasma* 202:38–53.

Brawley, S. H. 1992. Fertilization in natural populations of the dioecious brown alga *Fucus ceranoides* and the importance of the polyspermy block. *Marine Biology* 113:145–157.

Brawley, S. H., and W. H. Adey. 1981. The effect of micrograzers on algal community structure in a coral reef microcosm. *Marine Biology* 61:167–177.

Brawley, S. H., R. S. Quatrano, and R. Wetherbee. 1977. Fine-structural studies of the gametes and embryo of *Fucus vesiculosus* L. (Phaeophyta). *Journal of Cell Science* 24:275–294.

Brawley, S. H., and R. Wetherbee. 1981. Cytology and ultrastructure. Pages 248–299 in *The Biology of Seaweeds*, edited by C. S. Lobban and M. J. Wynne. Blackwell Scientific, Oxford, UK.

Bremer, K. 1985. Summary of green plant phylogeny and classification. *Cladistics* 1:369–385.

Bremer, K. 1988. The limits of amino acid sequence data in angiosperm phylogeny reconstruction. *Evolution* 42:795–803.

Brett, M. T., and C. R. Goldman. 1997. Consumer versus resource control in freshwater pelagic food webs. *Science* 275:384–386.

Brett, S. J., and R. Wetherbee. 1986. A comparative study of periplast structure in *Cryptomonas cryophila* and *C. ovata* (Cryptophyceae). *Protoplasma* 131:23–31.

Bricelj, V. M., and D. L. Lonsdale. 1997. *Aureococcus anophagefferens*: Causes and ecological consequences of brown tides in U. S. mid-Atlantic coastal waters. *Limnology and Oceanography* 42:1023–1038.

Broadwater, S. T., and E. A. LaPointe. 1997. Parasitic interactions and vegetative ultrastructure of *Choreonema thuretii* (Corallinales, Rhodophyta). *Journal of Phycology* 33:396–407.

Broadwater, S. T., and J. Scott. 1982. Ultrastructure of early development in the female reproductive system of *Polysiphonia harveyi* Bailey (Ceramiales, Rhodophyta). *Journal of Phycology* 18:427–441.

Broadwater, S. T., and J. L. Scott. 1994. Ultrastructure of unicellular red algae. Pages 215–230 in *Evolutionary Pathways and Enigmatic Algae:* Cyanidium caldarium *(Rhodophyta) and Related Cells*, edited by J. Sechbach. Kluwer Academic Publishers, Boston, MA.

Broady, P. A. 1984. Taxonomic and ecological investigations of algae on steam-warmed soil on Mt. Erebus, Ross Island, Antarctica. *Phycologia* 23:257–271.

Brodie, J., P. K. Hayes, G. L. Barker, L. M. Irvine, and I. Bartsch. 1998. A reappraisal of *Porphyra* and *Bangia* (Bangiophycidae, Rhodophyta) in the northeast Atlantic based on the *rbcL-rbcS* intergenic spacer. *Journal of Phycology* 34:1069–1074.

Brook, A. J. 1980. Barium accumulation by desmids of the genus *Closterium* (Zygnemaphyceae). *British Phycological Journal* 15:261–264.

Brook, A. J., and D. B. Williamson. 1988. The survival of desmids on the drying mud of a small lake. Pages 185–196 in *Algae and the Aquatic Environment*, edited by F. E. Round. BioPress, Bristol, UK.

Brooke, C., and R. Riding. 1998. Ordovician and Silurian coralline red alga. *Lethaia* 31:185–195.

Brooks, J. L., and S. I. Dodson. 1965. Predation, body size and composition of plankton. *Science* 150:28–35.

Brown, B. E. 1997. Disturbances to reefs in recent times. Pages 354–379 in *Life and Death of Coral Reefs*, edited by C. Birkeland. Chapman & Hall, New York, NY.

Brown, J. R., and W. F. Doolittle. 1995. Root of the universal tree of life based on ancient aminoacyl-tRNA synthetase gene duplications. *Proceedings of the National Academy of Science* 92:2441–2445.

Brown, L. M., and J. A. Hellebust. 1980. The contribution of organic solutes to osmotic balance in some green and eustigmatophyte algae. *Journal of Phycology* 16:265–270.

Brown, R. C., B. E. Lemmon, and L. E. Graham. 1994. Morphogenetic plastid migration and microtubule arrays in mitosis and cytokinesis in the green alga *Coleochaete orbicularis*. *American Journal of Botany* 81:127–133.

Brown, R. M., J. H. M. Willison, and C. L. Richardson. 1976. Cellulose biosynthesis in *Acetobacter xylinum*: visualization of the site of synthesis and direct measurement of the *in vivo* process. *Proceedings of the National Academy of Science* 73:4565–4569.

Brownlee, C., N. Nimer, L. F. Dong, and M. J. Merrett. 1994. Cellular regulation during calcification in *Emiliania huxleyi*. Pages 133–148 in *The Haptophyte Algae*, edited by J. C. Green and B. S. C. Leadbeater. Clarendon Press, Oxford, UK.

Buchheim, M. A., J. A. Buchheim and R. L. Chapman. 1997a. Phylogeny of *Chloromonas* (Chlorophyceae): A study of 18S ribosomal RNA gene sequences. *Journal of Phycology* 33:286–293.

Buchheim, M. A., J. A. Buchheim, and R. L. Chapman. 1997b. Phylogeny of the VLE-14 *Chlamydomonas* (Chlorophyceae) group: A study of 18S rRNA gene sequences. *Journal of Phycology* 33:1024–1030.

Buchheim, M. A., and R. L. Chapman. 1991. Phylogeny of the colonial green flagellates: A study of 18S and 26S RNA sequence data. *BioSystems* 25:85–100.

Buchheim, M. A., and R. L. Chapman. 1992. Phylogeny of the genus *Carteria* (Chlorophyta) inferred from organismal and molecular evidence. *Journal of Phycology* 28:362–374.

Buchheim, M. A., M. A. McAuley, E. A. Zimmer, E. C. Theriot, and R. L. Chapman. 1994. Multiple origins of colonial green flagellates from unicells: Evidence from molecular and organismal characters. *Molecular Phylogeny and Evolution* 3:322–343.

Buchheim, M. A., D. L. Nickrent, and L. R. Hoffman. 1990. Systematic analysis of *Sphaeroplea* (Chlorophyceae). *Journal of Phycology* 26:173–181.

Buggeln, R. G., D. S. Fensom, and C. J. Emerson. 1985. Translocation of ¹¹C-photoassimilate in the blade of *Macrocystis pyrifera* (Phaeophyceae). *Journal of Phycology* 21:35–40.

Buick, R. 1992. The antiquity of oxygenic photosynthesis: Evidence from stromatolites in sulphate-deficient Archean lakes. *Science* 255:74–77.

Bunt, J. S. 1969. Observations on photoheterotrophy in a marine diatom. *Journal of Phycology* 5:37–42.

Burger-Wiersma, T., M. Veenhuis, H. J. Korthals, C. C. M. VandeWiel, and L. R. Mar. 1986. A new prokaryote containing chlorophylls *a* and *b*. *Nature* 320:262–264.

Burkholder, J. M. 1996. Interactions of benthic algae with their substrata. Pages 253–298 in *Algal Ecology. Freshwater Benthic Ecosystems*, edited by R. J. Stevenson, M. L. Bothwell, and R. L. Lowe. Academic Press, New York, NY.

Burkholder, J. M., and H. B. Glasgow. 1997. *Pfiesteria piscicida* and other *Pfiesteria*-like dinoflagellates: Behavior, impacts, and environmental controls. *Limnology and Oceanography* 42:1052–1075.

Burkholder, J. M., H. B. Glasgow, and A. J. Lewitus. 1998. Physiological ecology of *Pfiesteria piscicida* with general comments on "ambush-predator" dinoflagellates. Pages 175–191 in *Physiological Ecology of Harmful Algae Blooms*, edited by D. M. Anderson, A. D. Cembella, and G. M. Hallegraeff. Springer-Verlag, Berlin, Germany.

Burkholder, J. M., E. J. Noga, C. H. Hobbs, and H. B. Glasgow. 1992. New 'phantom' dinoflagellate is the causative agent of major estuarine fish kills. *Nature* 358:407–410.

Burkholder, J. M., and R. G. Sheath. 1984. The seasonal distribution, abundance and diversity of desmids (Chlorophyta) in a softwater, north temperate stream. *Journal of Phycology* 20:159–172.

Burkholder, J. M., and R. G. Wetzel. 1989. Epiphytic microalgae on a natural substratum in a phosphorous-limited hardwater lake: Seasonal dynamics of community structure biomass and ATP content. *Archiv für Hydrobiologie* 83:1–56.

Burns, C. W. 1968. The relationship between body size of filter-feeding cladocera and the maximum size of particles ingested. *Limnology and Oceanography* 13:675–678.

Butterfield, N. J., A. H. Knoll, and K. Swett. 1988. Exceptional preservation of fossils in an Upper Proterozoic shale. *Nature* 334:424–427.

Butterfield, N. J., A. H. Knoll, and K. Swett. 1990. A bangiophyte red alga from the Proterozoic of Arctic Canada. *Science* 250:104–107.

Cahill, D. M., M. Cope, and A. R. Hardham. 1996. Thrust reversal by tubular mastigonemes: Immunological evidence for a role of mastigonemes in forward motion of zoospores of *Phytophthora cinnamomi*. *Protoplasma* 194:18–28.

Calado, A. J., S. C. Craviero, and Ø. Moestrup. 1998. Taxonomy and ultrastructure of a freshwater, heterotrophic *Amphidinium* (Dinophyceae) that feeds on unicellular protists. *Journal of Phycology* 34:536–554.

Calado, A. J., and Ø. Moestrup. 1997. Feeding in *Peridiniopsis berolinensis* (Dinophyceae): New observations on tube feeding by an omnivorous, heterotrophic dinoflagellate. *Phycologia* 36:47–59.

Calderon-Saenz, E., and R. Schnetter. 1989. Life cycle and morphology of *Bryopsidella ostreobiformis* (spec. nov.) (Bryopsidaceae, Chlorophyta) from the Mediterranean under culture conditions, with comments on the phylogeny of the *Bryopsis/Derbesia* complex. *Botanica Acta* 102:249–260.

Callow, M. E., J. A. Callow, J. D. Pickett-Heaps, and R. Wetherbee. 1997. Primary adhesion of *Enteromorpha* (Chlorophyta, Ulvales) propagules: Quantitative settlement studies and video microscopy. *Journal of Phycology* 33:938–947.

Callow, M. E., S. J. Coughlan, and L. V. Evans. 1978. The role of Golgi bodies in polysaccharide sulphation in *Fucus* zygotes. *Journal of Cell Science* 32:337–356.

Campbell, E. E., and G. C. Bate. 1997. Coastal features associated with diatom discoloration of surf-zones. *Botanica Marina* 40:179–185.

Cannell, R. J. P., P. Farmer, and J. M. Walker. 1988. Purification and characterization of pentagalloylglucose, a alpha-glucosidase inhibitor/antibiotic from a freshwater green alga *Spirogyra varians*. *Biochemical Journal* 255:937–941.

Canter, H. M. 1979. Fungal and protozoan parasites and their importance in the ecology of the phytoplankton. *Freshwater Biological Association Report* 47:43–50.

Canter, H. M., and J. W. G. Lund. 1968. The importance of protozoa in controlling the abundance of planktonic algae in lakes. *Proceedings of the Linnean Society of London* 179:203–219.

Canter-Lund, H., and J. W. G. Lund. 1995. *Freshwater Algae, Their Microscopic World Explored*. BioPress, Bristol, UK.

Capone, D. G., J. P. Zehr, H. W. Paerl, B. Bergman, and E. J. Carpenter. 1997. *Trichodesmium*, a globally significant marine cyanobacterium. *Science* 276:1221–1229.

Carefoot, T. H. 1977. *Pacific Seashores. A Guide to Intertidal Ecology*. J. J. Douglas, Vancouver, BC, Canada.

Carlton J. T., and J. B. Geller. 1993. Ecological roulette: The global transport of nonindigenous marine organisms. *Science* 261:78–82.

Carmichael, W. W. 1997. The cyanotoxins. *Advances in Botanical Research* 27:211–255.

Caron, D. 1990. Carbon utilization by the omnivorous flagellate *Paraphysomonas inperforata*. *Limnology and Oceanography* 35:192–201.

Caron, D. A., K. G. Porter, and R. W. Sanders. 1990. Carbon, nitrogen, and phosphorus budgets for the mixotrophic phytoflagellate *Poterioochromonas malhamensis* (Chrysophyceae) during bacterial ingestion. *Limnology and Oceanography* . 35:433–443

Caron, L., D. Douady, M. Quinet-Szely, S. de Goer, and C. Berkaloff. 1996. Gene structure of a chlorophyll *a/c* binding protein from a brown alga: presence of an intron and phylogenetic implications. *Journal of Molecular Evolution* 43:270–280.

Carpenter, E. J., and K. Romans. 1991. Major role of the cyanobacterium *Trichodesmium* in nutrient cycling in the North Atlantic Ocean. *Science* 254:1356–1358.

Carpenter, R. C. 1990. Competition among marine macroalgae: A physiological perspective. *Journal of Phycology* 26:6–12.

Carpenter, R. C. 1997. Invertebrate predators. Pages 198–229 in *Life and Death of Coral Reefs*, edited by C. Birkeland. Chapman & Hall, New York, NY.

Carpenter, S. R. 1989. Temporal variance in lake communities: Blue-green algae and the trophic cascade. *Landscape Ecology* 3:175–184.

Carpenter, S. R., and J. F. Kitchell. 1993. *The Trophic Cascade in Lakes*. Cambridge University Press, Cambridge, UK.

Carpenter, S. R., J. F. Kitchell, K. L. Cottingham, D. E. Schindler, D. L. Christensen, D. M. Post, and N. Voichick. 1996. Chlorophyll variability, nutrient input, and grazing: evidence from whole-lake experiments. *Ecology* 77:725–735.

Carpenter, S. R., J. F. Kitchell, and J. R. Hodgson. 1985. Cascading trophic interactions and lake productivity. *BioScience* 35:634–639.

Carrias, J.-F., C. Amblard, and G. Bourdier. 1996. Protistan bacterivory in an oligomesotrophic lake: Importance of attached ciliates and flagellates. *Microbial Ecology* 31:249–268.

Carthew, R. W., and J. A. Hellebust. 1982. Transport of amino acids by the soil alga *Stichococcus bacillaris*. *Journal of Phycology* 18:441–446.

Carthew, R. W., and J. A. Hellebust. 1983. Regulation of a glucose transport system in *Stichococcus bacillaris*. *Journal of Phycology* 19:467–473.

Casanova, M. T., and M. A. Brock. 1996. Can oospore germination patterns explain charophyte distribution in permanent and temporary wetlands? *Aquatic Botany* 54:297–312.

Castenholz, R. W. 1992. Species usage, concept, and evolution in the cyanobacteria (blue-green algae). *Journal of Phycology* 28:737–745.

Castenholz, R. W. 1996. Endemism and biodiversity of thermophilic cyanobacteria. *Nova Hedwigia* 112:33–47.

Cavalier-Smith, T. 1986. The Kingdom Chromista: Origin and systematics. *Progress in Phycological Research* 4:309–347.

Cavalier-Smith, T. 1994. Origin and relationships of Haptophyta. Pages 413–436 in *The Haptophyte Algae*, edited by J. C. Green and B. S. C. Leadbeater. Clarendon Press, Oxford, UK.

Cavalier-Smith, T., M. T. E. P. Allsopp, and E. E. Chao. 1994. Chimeric conundra: Are nucleomorphs and chromists monophyletic or polyphyletic? *Proceedings of the National Academy of Science* 91:11368–11372.

Cavalier-Smith, T., and E. E. Chao. 1996. 18S rRNA sequence of *Heterosigma carterae* (Raphidophyceae), and the phylogeny of heterokont algae (Ochrophyta). *Phycologia* 35:500–510.

Cavalier-Smith, T., J. A. Couch, K. E. Thorsteinsen, P. Gilson, J. A. Deane, D. R. A. Hill, and G. I. McFadden. 1996. Cryptomonad nuclear and nucleomorph 18S rRNA phylogeny. *European Journal of Phycology* 31:315–328.

Cembella, A. D. 1998. Ecophysiology and metabolism of paralytic shellfish toxins in marine microalgae. Pages 381–404 in *Physiological Ecology of Harmful Algae Blooms*, edited by D. M. Anderson, A. D. Cembella, and G. M. Hallegraeff. Springer-Verlag, Berlin, Germany.

Chapman, A. D., and C. L. Schelske. 1997. Recent appearance of *Cylindrospermopsis* (Cyanobacteria) in five hypereutrophic Florida lakes, *Journal of Phycology* 33:191–195.

Chapman, A. R. O., and J. E. Lindley. 1980. Seasonal growth of *Laminaria solidungula* in the Canadian High Arctic in relation to irradiance and dissolved nutrient concentrations. *Marine Biology* 57:1–5.

Chapman, D. J., and F. T. Haxo. 1966. Chloroplast pigments of Chloromonadophyceae. *Journal of Phycology* 2:89–91.

Chapman, R. L. 1976. Ultrastructure of *Cephaleuros virescens* (Chroolepidaceae:Chlorophyta). I. Scanning electron microscopy of zoosporangia. *American Journal of Botany* 63:1060–1070.

Chapman, R. L. 1980. Ultrastructure of *Cephaleuros virescens* (Chroolepidaceae: Chlorophyta). II. Gametes. *American Journal of Botany* 67:10–17.

Chapman, R. L., J. C. Bailey, and D. A. Waters. 1998. Macroalgal phylogeny. Pages 389–408 in *Molecular Approaches to the Study of the Oceans*, edited by K. E. Cooksey. Chapman & Hall, London, UK.

Chapman, R. L., and M. A. Buchheim. 1991. Ribosomal RNA gene sequences: Analysis and significance in the phylogeny and taxonomy of green algae. *Critical Reviews in Plant Science* 10:343–368.

Chapman, R. L., and B. H. Good. 1983. Subaerial symbiotic green algae: Interactions with vascular plant hosts in *Algal Symbiosis: A Continuum of Interaction Strategies*, edited by L. J. Goff. Cambridge University Press, New York, NY.

Chapman, R. L., and M. C. Henk. 1986. Phragmoplasts in cytokinesis of *Cephaleuros parasiticus* (Chlorophyta) vegetative cells. *Journal of Phycology* 22:83–88.

Chapman, R. L., and D. A. Waters. 1992. Epi- and endobiotic chlorophytes. Pages 619–640 in *Algae and Symbioses*, edited by W. Reisser. BioPress, Bristol, UK.

Chapman, R. L., D. A. Waters, and J. M. Lopez-Bautista. 1995. Phylogenetic affinities of the Trentepohliales (Chlorophyta) inferred from small subunit rRNA gene sequences. *Journal of Phycology* 31:7 (abstract).

Chappell, D. F., C. J. O'Kelly, L. W. Wilcox, and G. L. Floyd. 1990. Zoospore flagellar apparatus architecture and the taxonomic position of *Phaeophila dendroides* (Ulvophyceae, Chlorophyta). *Phycologia* 29:515–523.

Chen, F., C. A. Suttle, and S. M. Short. 1996. Genetic diversity in marine algal virus communities as revealed by sequence analysis of DNA polymerase genes. *Applied and Environmental Biology* 62:2869–2874.

Cheshire, A. C., J. G. Conran, and N. D. Hallam. 1995. A cladistic analysis of the evolution and biogeography of *Durvillaea* (Phaeophyta). *Journal of Phycology* 31:644–655.

Chesnick, J. M., and E. R. Cox. 1986. Specialization of endoplasmic reticulum architecture in response to a bacterial symbiosis in *Peridinium balticum* (Pyrrhophyta). *Journal of Phycology* 22:291–298.

Chesnick, J. M., and E. R. Cox. 1987. Synchronized sexuality of an algal symbiont and its dinoflagellate host, *Peridinium balticum* (Levander) Lemmermann. *BioSystems* 21:69–78.

Chesnick, J. M., and E. R. Cox. 1989. Fertilization and zygote development in the binucleate dinoflagellate *Peridinium balticum* (Pyrrhophyta). *American Journal of Botany* 76:1060–1072.

Chesnick, J. M., W. H. C. F. Kooistra, U. Wellbrock, and L. K. Medlin. 1997. Ribosomal RNA analysis indicates a benthic pennate diatom ancestry for the endosymbionts of the dinoflagellates *Peridinium foliaceum* and *Peridinium balticum* (Pyrrhophyta). *Journal of Eukaryotic Microbiology* 44:314–320.

Chesnick, J. M., C. W. Morden, and A. M. Schmeig. 1996. Identity of the endosymbiont of *Peridinium foliaceum* (Pyrrophyta): Analysis of the *rbcLS* operon. *Journal of Phycology* 32:850–857.

Chiodini, P. L. 1994. A "new" parasite: Human infection with *Cyclospora cayetanensis*. *Transactions of the Royal Society of Tropical Medicine and Hygiene* 88:369–371.

Chisholm, J. R. M., C. Dauga, E. Ageron, P. A. D. Grimont, and L. M. Jaubert. 1996. "Roots" in mixotrophic algae. *Nature* 381:382.

Chisholm, S. W., S. I. Frankel, R. Goericke, R. J. Olson, J. B. Waterbury, L. West-Johnson, and E. R. Settler. 1992. *Prochlorococcus marinus* nov. gen. et nov. sp.: An oxy-phototrophic marine prokaryote containing divinyl chlorophyll *b*. *Archiv für Microbiologie* 157:297–300.

Chisholm, S. W., R. J. Olson, E. R. Zettler, R. Goericke, J. B. Waterbury, and N. A. Welschmeyer. 1988. A novel free-living prochlorophyte abundant in the oceanic euphotic zone. *Nature* 334:340–343.

Choat, J. H., and K. D. Clements. 1998. Vertebrate herbivores in marine and terrestrial environments: A nutritional ecology perspective. *Annual Review of Ecology and Systematics* 29:375–403.

Chopin, T., C. J. Bird, C. A. Murphy, J. A. Osborne, M. U. Patwary, and J.-Y. Floc'h. 1996. A molecular investigation of polymorphism in the North Atlantic red alga *Chondrus crispus* (Gigartinales). *Phycological Research* 44:69–80.

Chrétiennot-Dinet, M.-J., M.-M. Giraud-Guille, D. Vaulot, J.-L. Putaux, Y. Saito, and H. Chanzy. 1997. The chitinous nature of filaments ejected by *Phaeocystis* (Prymnesiophyceae). *Journal of Phycology* 33:666–672.

Ciciotte, S. L., and R. J. Thomas. 1997. Carbon exchange between *Polysiphonia lanosa* (Rhodophyceae) and its brown algal host. *American Journal of Botany* 84:1614–1616.

Clark, K. B. 1992. Plant-like animals and animal-like plants: Symbiotic co-evolution of ascoglossan (=sacoglossan) molluscs, their algal prey and algal plastids. Pages 515–530 in *Algae and Symbiosis*, edited by W. Reisser. BioPress, Bristol, UK.

Clayton, M. N. 1984. Evolution of the Phaeophyta with particular reference to the Fucales. *Progress in Phycological Research* 3:11–46.

Clayton, M. N. 1987. Isogamy and a fucalean type of life cycle in the Antarctic brown alga *Ascoseira mirabilis* (Ascoseirales, Phaeophyta). *Botanica Marina* 30:447–454.

Clayton, M. N. 1990. Phaeophyta. Pages 698–714 in *Handbook of the Protoctista*, edited by L. Margulis, C. O. Corliss, M. Melkonian, and D. J. Chapman. Jones and Bartlett Publishers, Boston, MA.

Clayton, M. N., and C. M. Ashburner. 1990. The anatomy and ultrastructure of "conducting channels" in *Ascoseira mirabilis* (Ascoseirales, Phaeophyceae). *Botanica Marina* 33:63–70.

Clayton, M. N., N. D. Hallam, and C. M. Shankly. 1987. The seasonal pattern of conceptacle development and gamete maturation in *Durvillaea potatorum* (Durvillaeales, Phaeophyta). *Phycologia* 26:35–45.

Clayton, M. N., K. Kevekordes, M. E. A. Schoenwaelder, C. E. Schmid, and C. M. Ashburner. 1998. Parthogenesis in *Hormosira banksii*. *Botanica Marina* 41:23–30.

Clayton, M. N., and C. Wiencke. 1990. The anatomy, life history and development of the Antarctic brown alga *Phaeurus antarcticus* (Desmarestiales, Phaeophyceae). *Phycologia* 29:303–315.

Clifton, K. E. 1997. Mass spawning by green algae on coral reefs. *Science* 275:1116–1118.

Cmiech, H. A., G. F. Leedale, and C. S. Reynolds. 1984. Morphological and ultrastructural variability of planktonic cyanophyceae in relation to seasonal periodicity. I. *Gloeotrichia echinulata*: Vegetative cells, polarity, heterocysts, akinetes. *British Phycological Journal* 19:259–275.

Codd, G. A., C. Edwards, K. A. Beattie, L. A. Lawton, D. L. Campbell, and S. G. Bell. 1995. Toxins from cyanobacteria (blue-green algae). Pages 1–17 in *Algae, Environment and Human Affairs*, edited by W. Wiessner, E. Schnepf, and R. Starr. BioPress, Bristol, UK.

Coesel, P. F. M. 1991. Ammonium dependency in *Closterium aciculare* T. West, a planktonic desmid from alkaline, eutrophic waters. *Journal of Plankton Research* 13:913–922.

Cogburn, J. N., and J. A. Schiff. 1984. Purification and properties of the mucus of *Euglena gracilis* (Euglenophyceae). *Journal of Phycology* 20:533–544.

Cohen, Y., B. B. Jorgensen, E. Padan, and M. Shilo. 1975. Sulphide-dependent anoxygenic photosynthesis in the cyanobacterium *Oscillatoria limnetica*. *Nature* 257:489–492.

Cohn, S., and R. E. Weitzell. 1996. Ecological characterization of diatom cell motility. 1. Characterization of motility and adhesion in four diatom species. *Journal of Phycology* 32:928–939.

Cohn, S. A., and N. C. Disparti. 1994. Environmental factors influencing diatom cell motility. *Journal of Phycology* 30:818–828.

Cole, G. A. 1994. *Textbook of Limnology*. Waveland Press, Prospect Heights, IL.

Cole, K., and R. G. Sheath. 1980. Ultrastructural changes in major organelles during spermatial differentiation in *Bangia* (Rhodophyta). *Protoplasma* 102:253–279.

Coleman, A. W. 1983. The roles of resting spores and akinetes in chlorophyte survival. Pages 1–21 in *Survival Strategies of the Algae*, edited by G. A. Fryxell. Cambridge University Press, Cambridge, UK.

Coleman, A. W. 1985. Diversity of plastid DNA configuration among classes of eukaryote algae. *Journal of Phycology* 21:1–16.

Coling, C. E., and J. L. Salisbury. 1992. Characterization of the calcium-binding contractile protein centrin from *Tetraselmis striata* (Pleurastrophyceae). *Journal of Protozoology* 39:385–391.

Colman, B., and K. A. Gehl. 1983. Physiological characteristics of photosynthesis in *Porphyridium cruentum*: Evidence for bicarbonate transport in a unicellular red alga. *Journal of Phycology* 19:216–219.

Colman, B., and C. Rotatore. 1995. Photosynthetic inorganic carbon uptake and accumulation in two marine diatoms. *Plant, Cell and Environment* 18:919–924.

Connell, J. 1978. Diversity in tropical rainforests and coral reefs. *Science* 199:1304–1310.

Conway, K., and F. R. Trainor. 1972. *Scenedesmus* morphology and flotation. *Journal of Phycology* 8:138–143.

Cook, M. E., and L. E. Graham. 1998. Structural similarities between surface layers of selected charophycean algae and bryophytes, and the cuticles of vascular plants. *International Journal of Plant Science* 159:780–787.

Cook, M. E., and L. E. Graham. 1999. Evolution of plasmodesmata. In *Plasmodesmata: Nanochannels with Megatasks*, edited by A. van Bel and C. Kesteren. Springer-Verlag, Berlin, Germany.

Cook, M. E., L. E. Graham, C. E. J. Botha, and C. A. Lavin. 1997. Comparative ultrastructure of plasmodesmata of *Chara* and selected bryophytes: Toward an elucidation of the evolutionary origin of plant plasmodesmata. *American Journal of Botany* 84:1169–1178.

Cook, M. E., L. E. Graham, and C. A Lavin. 1998. Cytokinesis and nodal anatomy in the charophycean green alga *Chara zeylanica*. *Protoplasma* 203:65–74.

Coomans, R. J., and M. H. Hommersand. 1990. Vegetative growth and organization. Pages 275–304 in *Biology of the Red Algae*, edited by K. M. Cole and R. G. Sheath. Cambridge University Press, Cambridge, UK.

Correa, J. A. 1997. Infectious diseases of marine algae: Current knowledge and approaches. *Progress in Phycological Research* 12:149–180.

Correa, J. A., and P. Sánchez. 1996. Ecological aspects of algal infectious diseases. *Hydrobiologia* 326/327:89–95.

Cosson, J., E. Deslandes, M. Zinoun, and A. Mouradi-Givernaud. 1995. Carrageenans and agars, red algal polysaccharides. *Progress in Phycological Research* 11:269–324.

Costas, E., and V. Lopez-Rodas. 1996. Enumeration and separation of the toxic dinoflagellate *Alexandrium minutum* from natural samples using immunological procedures with blocking antibodies. *Journal of Experimental Marine Biology and Ecology* 198:81–87.

Cottrell, M. T., and C. A. Suttle. 1995. Dynamics of a lytic virus infecting the photosynthetic marine picoflagellate *Micromonas pusilla*. *Limnology and Oceanography* 40:730–739.

Courties, C., A. Vaquer, M. Troussellier, J. Lautier, M. J. Chrétiennot-Dinet, J. Neveux, C. Machado, and H. Claustre. 1994. Smallest eukaryotic organism. *Nature* 370:255.

Cox, E. R., and H. C. Bold. 1966. Phycological Studies. VII. Taxonomic Investigations of *Stigeoclonium*. *The University of Texas Publication* No. 6618.

Coyer, J. A., J. L. Olsen, W. T. Stam. 1997. Genetic variability and spatial separtion in the sea palm kelp *Postelsia palmaeformis* (Phaeophyceae) as assessed with M13 fingerprints and RAPDs. *Journal of Phycology* 33:561–568.

Coyer, J. A., D. L. Robertson, and R. S. Alberte. 1994. Genetic variability within a population and between diploid/haploid tissue of *Macrocystis pyrifera* (Phaeophyceae). *Journal of Phycology* 30:545–552.

Craggs, R. J., W. H. Adey, K. R. Jensen, M. S. St. John, F. B. Green, and W. J. Oswald. 1996. Phosphorus removal from waste water using an algal turf scrubber. *Water Science and Technology* 33:191–198.

Craigie, J. S. 1990. Cell walls. Pages 221–258 in *Biology of the Red Algae*, edited by K. M. Cole and R. G. Sheath. Cambridge University Press, Cambridge, UK.

Crawford, R. M. 1995. The role of sex in the sedimentation of a marine diatom bloom. *Limnology and Oceanography* 40:200–204.

Creed, J. C. 1995. Spatial dynamics of a *Himanthalia elongata* (Fucales, Phaeophyta) population. *Journal of Phycology* 31:851–859.

Croasdale, H., C. E. deM. Bicudo, and G. W. Prescott. 1983. *A Synopsis of North American Desmids. Part II. Desmidiaceae: Placodermae, Section 5*, University of Nebraska Press, Lincoln, NE.

Cronberg, G. 1995. *Mallomonas variabilis*, sp. nov. (Synurophyceae) with stomatocysts found in Lake Konneresi, Finland. Pages 333–344 in *Chrysophyte Algae. Ecology, Phylogeny and development*, edited by C. D. Sandgren, J. P. Smol, and J. Kristiansen. Cambridge University Press, New York, NY.

Cuker, B. E. 1983. Grazing and nutrient interactions in controlling the activity and composition of the epilithic algal community of an arctic lake. *Limnology and Oceanography* 28:133–141.

Cullen, J. J. 1995. Status of the iron hypothesis after the open-ocean enrichment experiment. *Limnology and Oceanography* 40:1336–1343.

Cunningham, F. X., and E. Gantt. 1998. Genes and enzymes of carotenoid biosynthesis in plants. *Annual Review of Plant Physiology and Plant Molecular Biology* 49:557–583.

Currin, C. A., H. W. Paerl, G. K. Suba, and R. S. Alberte. 1990. Immunofluorescence detection and characterization of N_2-fixing microorganisms from aquatic environments. *Limnology Oceanography* 35:59–71.

Czygan, F. C., and K. Kalb. 1966. Untersuchungen zur Biogenese der Carotinoide in *Trentepohlia aurea*. *Zeitschrift Pflanzenphysiologie* 55:59–64.

Daft, M. J., S. B. McCord, and W. D. P. Stewart. 1975. Ecological studies on algal-lysing bacteria in fresh waters. *Freshwater Biology* 5:577–596.

Dale, B., C. M. Yentsch, and J. W. Hurst. 1978. Toxicity in resting cysts of the red-tide dinoflagellate *Gonyaulax excavata* from deeper water coastal sediments. *Science* 201:1223–1225.

Darley, W. M. 1982. *Algal Biology: A Physiological Approach*. Blackwell Scientific, Oxford, UK.

Darley, W. M., and B. E. Volcani. 1969. Role of silicon in diatom metabolism. A silicon requirement for deoxyribonucleic acid synthesis in the diatom *Cylindrotheca fusiformis* Reimann and Lewin. *Experimental Cell Research* 58:334–342.

Daugbjerg, N. 1996. *Mesopedinella arctica* gen. et sp. nov. (Pedinellales, Dictyochophyceae) I: Fine structure of a new marine phytoflagellate from Arctic Canada. *Phycologia* 35:435–445.

Daugbjerg, N., and R. A. Andersen. 1997a. Phylogenetic analyses of the *rbcL* sequences from haptophytes and heterokont algae suggest their chloroplasts are unrelated. *Molecular Biology and Evolution* 14:1242–1251.

Daugbjerg, N., and R. A. Andersen. 1997b. A molecular phylogeny of the heterokont algae based on analysis of chloroplast-encoded *rbcL* sequence data. *Journal of Phycology* 33:1031–1041.

Daugbjerg, N. , Ø. Moestrup, and P. Arctander. 1994. Phylogeny of the genus *Pyramimonas* (Prasinophyceae, Chlorophyta) inferred from the *rbcL* gene. *Journal of Phycology* 30:991–999.

Daugbjerg, N., Ø. Moestrup, and P. Arctander. 1995. Phylogeny of genera of Prasinophyceae and Pedinophyceae (Chlorophyta) deduced from molecular analysis of the *rbcL* gene. *Phycological Research* 43:203–213.

Davey, M. C., and R. M. Crawford. 1986. Filament formation in the diatom *Melosire granulata*. *Journal of Phycology* 22:144–150.

Davies, J. P., and A. R. Grossman. 1998. The use of *Chlamydomonas* (Chlorophyta:Volvocales) as a model algal system for genome studies and the elucidation of photosynthetic processes. *Journal of Phycology* 34:907–917.

Davis, L. S., J. P. Hoffmann, and P. W. Cook. 1990a. Seasonal succession of algal periphyton from a wastewater treatment facility. *Journal of Phycology* 26:611–617.

Davis, L. S., J. P. Hoffmann, and P. W. Cook. 1990b. Production and nutrient accumulation by periphyton in a wastewater treatment facility. *Journal of Phycology* 26:617–623.

Dawes, C. J. 1998. *Marine Botany*. John Wiley, New York, NY.

Dawson, N. S., J. R. Dunlap, and P. A. Walne. 1988. Structure and elemental composition of pellicular warts of *Euglena spirogyra* (Euglenophyceae). *British Phycological Journal* 23:61–69.

Dawson, N. S., and P. L. Walne. 1991a. Structural characterization of *Eutreptia pertyi* (Euglenophyta). I. General description. *Phycologia* 30:287–302.

Dawson, N. S., and P. L. Walne. 1991b. Structural characterization of *Eutreptia* (Euglenophyta). III. Flagellar structure and possible function of the paraxial rods. *Phycologia* 30:415–437.

Dawson, N. S., and P. L. Walne. 1994. Evolutionary trends in euglenoids. *Archiv für Protistenkunde* 144:221–225.

Dayton, P. K. 1975. Experimental evaluation of ecological dominance in a rocky interidal algal community. *Ecological Monographs* 45:137–159.

Deane, J. A., D. R. A. Hill, S. J. Brett, and G. I. McFadden. 1998. *Hanusia phi* gen. et sp. nov. (Cryptophyceae): Characterization of "*Cryptomonas phi*." *European Journal of Phycology* 33:149–154.

Deason, T. R., P. C. Silva, S. Watanabe, and G. L. Floyd. 1991. Taxonomic status of the green algal genus *Neochloris*. *Plant Systematics and Evolution* 177:213–219.

DeBaar, H. J. W., J. T. M. deJong, D. C. E. Bakker, B. M. Löscher, C. Veth, U. Bathman, and V. Smetacek. 1995. Importance of iron for plankton blooms and carbon dioxide drawdown in the Southern Ocean. *Nature* 373:412–415.

DeJesus, M. D., F. Tabatabai, and D. J. Chapman. 1989. Taxonomic distribution of copper-zinc superoxide dismutase in green algae and its phylogenetic importance. *Journal of Phycology* 25:767–772.

DeLong, E. F. 1998. Molecular phylogenetics: New perspectives on the ecology, evolution and biodiversity of marine organisms. Pages 1–28 in *Molecular Approaches to the Study of the Ocean*, edited by K. E. Cooksey. Chapman & Hall, London, UK.

Delgado, O., C. Rodriguez-Prieto, E. Gacia, and E. Ballesteros. 1996. Lack of severe nutrient limitation in *Caulerpa taxifolia* (Vahl) C. Agardh, an introduced seaweed spreading over the oligotrophic northeastern Mediterranean. *Botanica Marina* 39:61–67.

Delwiche, C. F., L. E. Graham, N. Thomson. 1989. Lignin-like compounds and sporopollenin in *Coleochaete*, an algal model for land plant ancestry. *Science* 245:399–401.

Delwiche, C. F., and J. D. Palmer. 1996. Rampant horizontal transfer and duplication of Rubisco genes in eubacteria and plastids. *Molecular Biology and Evolution* 13:873–882.

Demmig-Adams, B., and W. W. Adams. 1992. Photoprotection and other responses of plants to high light stress. *Annual Review of Plant Physiology and Plant Molecular Biology* 43:599–626.

DeMott, W., and F. Moxter. 1991. Foraging on cyanobacteria by copepods: Responses to chemical defenses and resource abundance. *Ecology* 72:1820–1834.

DeNicola, D. M., and C. D. McIntire. 1990a. Effects of substrate relief on the distribution of periphyton in laboratory streams. I. Hydrology. *Journal of Phycology* 26:624–633.

DeNicola, D. M., and C. D. McIntire. 1990b. Effects of

substrate relief on the distribution of periphyton in laboratory streams. II. Interactions with irradiance. *Journal of Phycology* 26:634–641.

Denny, M. W. 1988. *Biology and the Mechanics of the Wave-Swept Environment*. Princeton Unversity Press, Princeton, NJ.

DeVernal, A., C. Hillaire-Marcel, and G. Bilodeau. 1996. Reduced meltwater outflow from the Laurentide ice margin during the Younger Dryas. *Nature* 381:774–777.

Devinny, J. S. 1980. Effects of thermal effluents on communities of benthic marine macroalgae. *Journal of Environmental Management* 11:225–242.

DeVries, P. J. R., J. Simons, and A. P. VanBeem. 1983. Sporopollenin in the spore wall of *Spirogyra* (Zygnemataceae, Chlorophyceae). *Acta Botanica Neerlandica* 32:25–28.

DeVrind-deJong, E. W., P. R. van Emburg, and J. P. M. de Vrind. 1994. Mechanisms of calcification: *Emiliania huxleyi* as a model system. Pages 149–166 in *The Haptophyte Algae*, edited by J. C. Green and B. S. C. Leadbeater. Clarendon Press, Oxford, UK.

DeYoe, H., D. A. Stockwell, R. R. Bidigare, M. Latasa, P. W. Johnson, P. E Hargraves, and C. A. Suttle. 1997. Description and characterization of the algal species *Aureoumbria lagunensis* gen. et sp. nov. and referral of *Aureoumbria* and *Aureococcus* to the Pelagophyceae. *Journal of Phycology* 33:1042–1048.

DeYoe, H. R., R. L. Lowe, and J. C. Marks. 1992. Effects of nitrogen and phosphorus on the endosymbiont load of *Rhopalodia gibba* and *Epithemia turgida* (Bacillariophyceae). *Journal of Phycology* 28:773–777.

Diakoff, S., and J. Scheibe. 1973. Action spectra for chromatic adaptation in *Tolypothrix tenuis*. *Plant Physiology* 51:382–385.

Diouris, M. 1989. Long-distance transport of ^{14}C-labelled assimilates in the Fucales: Nature of translocated substances in *Fucus serratus*. *Phycologia* 28:504–511.

DiTullio, G. R., D. L. Garrison., and S. Mathot. 1998. Dimethylsulfonioproprionate in sea ice algae from the Ross Sea polynya. Pages 139–146 in *Antarctic Sea Ice Biological Processes, Interactions, and Variability*, edited by M. P. Lizotte and K. R. Arrigo. Antarctic Research Series 73, American Geophysical Union, Washington, DC.

Dodds, W. K., and D. A. Gudder. 1992. The ecology of *Cladophora*. *Journal of Phycology* 28:415–427.

Dodds, W. K., D. A. Gudder, and D. Mollenhauer. 1995. The ecology of *Nostoc*. *Journal of Phycology* 31:2–18.

Dodge, J. D., and R. M. Crawford. 1970a. A survey of thecal fine structure in the Dinophyceae. *Botanical Journal of the Linnean Society* 63:53–67.

Dodge, J. D., and R. M. Crawford. 1970b. The morphology and fine structure of *Ceratium hirundinella* (Dinophyceae). *Journal of Phycology* 6:137–149.

Doers, M. P., and D. L. Parker. 1988. Properties of *Microcystis aeruginosa* and *M. flos-aquae* (Cyanophyta) in culture: Taxonomic implications. *Journal of Phycology* 24:502–508.

Domozych, C. R., K. Plante, P. Blais, L. Paliulis, and D. S. Domozych. 1993. Mucilage processing and secretion in the green alga *Closterium*. I. Cytology and biochemistry. *Journal of Phycology* 29:650–659.

Domozych, D. S. 1984. The crystalline cell wall of *Tetraselmis convolutae* (Chlorophyta): a freeze fracture analysis. *Journal of Phycology* 20:415–418.

Domozych, D. S., and C. R. Domozych. 1993. Mucilage processing and secretion in the green alga *Closterium*. II. Ultrastructure and immunocytochemistry. *Journal of Phycology* 29:659–667.

Dop, A. J. 1978. Systematics and morphology of *Chrysochaete brittannica* (Godward) Rosenberg and *Phaeoplaca thallosa* chodat (Chrysophyceae). *Acta Botanica Neerlandica* 27:35–60.

Doran, E., and R. A. Cattolico. 1997. Photoregulation of chloroplast gene transcription in the chromophytic alga *Heterosigma carterae*. *Plant Physiology* 115:773–781.

Dorling, M., P. J. McAuley, and H. Hodge. 1997. Effect of pH on growth and carbon metabolism of maltose-releasing *Chlorella* (Chlorophyta). *European Journal of Phycology* 32:19–24.

Doty, M. S. 1946. Critical tide factors that are correlated with the vertical distribution of marine algae and other organisms along the Pacific Coast. *Ecology* 27:315–328.

Doucette, G. J., M. Kodama, S. Franca, and S. Gallacher. 1998. Bacterial interactions with harmful algal bloom species: Bloom ecology, toxigenesis, and cytology. Pages 619–647 in *Physiological Ecology of Harmful Algal Blooms*, edited by D. M. Anderson, A. D. Cembella, and G. M. Hallegraeff. Springer-Verlag, Berlin, Germany.

Dowling, T. E., C. Moritz, J. D. Palmer, and L. H. Rieseberg. 1996. Nucleic acids III: Analysis of fragments and restriction sites. Pages 249–320 in *Molecular Systematics*, edited by D. M. Hillis, C. Moritz, and B. K. Mable. Sinauer, Sunderland, MA.

Draisma, S. G. A., Y.-S. Keum, W. F. Prud'homme van Reine, and G. M. Lokhorst. 1998. The species of *Sphacelaria* (Sphacelariales, Phaeophyceae) in China with a description of a new species. *Botanica Marina* 41:181–190.

Drenner, R. W., D. J. Day, S. J. Basham, J. D. Smith, and S. I. Jensen. 1997. Ecological water treatment system for removal of phosphorus and nitrogen from polluted water. *Ecological Applications* 7:381–390.

Drew, K. M. 1949. *Conchocelis*-phase in the life history of *Porphyra umbilicalis* (L.) Kuetz. *Nature* 164:748–749.

Dring, M. J., and K. Lüning. 1975. A photoperiodic response mediated by blue light in the brown alga *Scytosiphon lomentaria*. *Planta* 125:25–32.

Dromgoole, F. I. 1990. Gas-filled structure, buoyancy and support in marine macro-algae. *Progress in Phycological Research* 7:169–211.

Droop, M. R. 1983. 25 years of algal growth kinetics. *Botanica Marina* 26:99–112.

Duarte, C. M., and J. Cebrián. 1996. The fate of marine autotrophic production. *Limnology and Oceanography* 41:1/58–1766.

Dube, M. A. 1967. On the life history of *Monostroma fuscum* (Postels et Ruprecht) Wittrock. *Journal of Phycology* 3:64–73.

DuBois-Tylski, T. 1981. Utilisation de fluorochromes pour l'observation des parois cellulaires chez troi especes de *Closterium* (Desmidiales) au cours de leur reproduction en sexuée. *Cryptogamie:Algologie* II 4:277–287.

Duff, K. E., B. A. Zeeb, and J. P. Smol. 1995. *Atlas of Chrysophycean Cysts*. Kluwer Academic Publishers, Dordrecht, Netherlands.

Duffy, J. E., and M. E. Hay. 1990. Seaweed adaptations to herbivory. *BioScience* 40:368–375.

Duffy, J. E., and M. E. Hay. 1994. Herbivore resistance to seaweed chemical defense: The roles of mobility and predation risk. *Ecology* 75:1304–1319.

Duggins, D. O, C. A. Simenstad., and J. A. Estes. 1989. Magnification of secondary production by kelp detritus in coastal marine ecosystems. *Science* 245:170–173.

Dunahay, T. G., E. E. Jarvis, S. S. Dais, and P. G. Roessler. 1996. Manipulation of microalgal lipid production using genetic engineering. *Applied Biochemistry and Biotechnology* 58:223–231.

Dunahay, T. G., E. E. Jarvis, and P. E. Roessler. 1995. Genetic transformation of the diatoms *Cyclotella cryptica* and *Navicula saprophila*. *Journal of Phycology* 31:1004–1012.

Dunlap, J. R., P. L. Walne, and J. Bentley. 1983. Microarchitecture and elemental spatial segregation of envelopes of *Trachelomonas leferrei* (Euglenophyceae). *Protoplasma* 117:97–106.

Dunlap, W. C., and J. M. Schik. 1998. Ultraviolet radiation-absorbing mycosporine-like amino acids in coral reef organisms: A biochemical and environmental perspective. *Journal of Phycology* 34:418–430.

Dunton, K. H. 1990. Growth and production in *Laminaria solidungula*: Relation to continuous underwater light levels in the Alaskan High Arctic. *Marine Biology* 106:297–304.

Dunton, K. H., and D. M. Schell. 1986. Seasonal carbon budget and growth of *Laminaria solidungula* in the Alaskan High Arctic. *Marine Ecology Progress Series* 31:57–66.

Dutcher, S. K. 1988. Linkage group XIX in *Chlamydomonas reinhardtii* (Chlorophyceae): Genetic analysis of basal body function and assembly. Pages 39–53 in *Algae as Experimental Systems*, edited by A. W. Coleman, L. J. Goff, and J. R. Stein-Taylor. Alan R. Liss, New York, N.Y.

Dyer, B. D. 1990. Bicoecids. Pages 191–193 in *Handbook of Protoctista*, edited by L. Margulis, J. O. Corliss, M. Melkonian, and D. J. Chapman. Jones and Bartlett, Publishers, Boston, MA.

Eckhardt, R., R. Schnetter, and G. Siebold. 1986. Nuclear behavior during the life cycle of *Derbesia* (Chlorophyceae). *British Phycological Journal* 21:287–295.

Edelstein, T., L. Chen, and J. McLachlan. 1968. Sporangia of *Ralfsia fungiformis* (Gunn) Setchell and Gardner. *Journal of Phycology* 4:157–160.

Edelstein, T., L. Chen, and J. McLachlan. 1974. The reproductive structures of *Gigartina stellata* (Stackh.) Batt. (Gigartinales, Rhodophyceae) in nature and culture. *Phycologia* 13:99–108.

Edgar, L. A., and J. D. Pickett-Heaps. 1984a. Diatom locomotion. *Progress in Phycological Research* 3:47–88.

Edgar, L. A., and J. D. Pickett-Heaps. 1984b. Valve morphogenesis in the pennate diatom *Navicula cuspidata*. *Journal of Phycology* 20:47–61.

Edlund, M. B., and E. F. Stoermer. 1997. Ecological, evolutionary, and systematic significance of diatom life histories. *Journal of Phycology* 33:897–918.

Edvardsen, B., and E. Paasche. 1998. Bloom dynamics and physiology of *Prymnesium* and *Chrysochromulina*. Pages 193–208 in *Physiological Ecology of Harmful Algal Blooms*, edited by D. M. Anderson, A. D. Cembella, and G. M. Hallegraeff. Springer-Verlag, Berlin, Germany.

Eikrem, W., and J. Throndsen. 1990. The ultrastructure of *Bathycoccus* gen. nov. and *B. prasinos* sp. nov., a non-motile picoplanktonic alga (Chlorophyta, Prasinophyceae) from the Mediterranean and Atlantic. *Phycologia* 29:344–350.

Elbrächter, M., and Y.-Z. Qi. 1998. Aspects of *Noctiluca* (Dinophyceae) population dynamics. Pages 315–335 in *Physiological Ecology of Harmful Algal Blooms*, edited by D. M. Anderson, A. D. Cembella, and G. M. Hallegraeff. Springer-Verlag, Berlin, Germany.

Elbrächter, M., and E. Schnepf. 1996. *Gymnodinium chlorophorum*, a new, green, bloom-forming dinoflagellate (Gymnodiniales, Dinophyceae) with a vestigial prasinophyte endosymbiont. *Phycologia* 35:381–393.

Elhai, J. 1994. Genetic techniques appropriate for the biotechnological exploitation of cyanobacteria. *Journal of Applied Phycology* 6:177–186.

Elliot, G. F. 1965. Tertiary solenoporacean algae and the reproductive structures of the Soleonoporaceae. *Palaeontology* 7:695–702.

Elloranta, P. 1995. Biogeography of chrysophytes in Finnish lakes. Pages 214–231 in *Chrysophyte Algae. Ecology, Phylogeny and development*, edited by C. D. Sandgren, J. P. Smol, and J. Kristiansen. Cambridge University Press, Cambridge, UK.

Elser, J. J., and C. R. Goldman. 1991. Zooplankton effects on phytoplankton in lakes of contrasting trophic status. *Limnology and Oceanography* 36:64–90.

Eppley, R. W., O. Holm-Hansen, and J. D. Strickland. 1968. Some observations on the vertical migrations of dinoflagellates. *Journal of Phycology* 4:333–340.

Estep, M., J. E. Armstrong, and C. Van Baalen. 1975. Evidence for the occurrence of specific iron (III)-binding compounds in near-shore marine ecosystems. *Applied Microbiology* 30:186–188.

Estes, J. A., M. T. Tinker, T. M. Williams, and D. F. Doak. 1998. Killer whale predation on sea otters linking oceanic and nearshore ecosystems. *Science* 282:473–475.

Ettl, H. 1976. Die Gattung *Chlamydomonas* Ehrenberg. *Nova Hedwigia* 49:1–1122.

Evans, L. V. 1965. Cytological studies in the Laminariales.

Annals of Botany 29:541–562.

Fairchild, G. W., and J. W. Sherman. 1993. Algal periphyton response to acidity and nutrients in softwater lakes: lake comparison vs. nutrient enrichment approaches. *Journal of the North American Benthological Society* 12:157–167.

Falkowski, P. 1997. Evolution of the nitrogen cycle and its influence on the biological sequestration of CO_2 in the ocean. *Nature* 387:272–275.

Falkowski, P. G., and J. A. Raven. 1997. *Aquatic Photosynthesis*. Blackwell Science, Malden, MA.

Farmer, M. A., and R. E. Triemer. 1988. Flagellar systems in the euglenoid flagellates. *BioSystems* 21:283–291.

Farris, J. S. 1995. JAC: The Parsimony Jacknife Program v. 4.2. Museum of Natural History, Laboratory of Molecular Systematics, Stockhom, Sweden.

Farris, J. S. 1997. PAX: The Parsimony Jacknife for Amino Acids v. 1.0. Museum of Natural History, Laboratory of Molecular Systematics, Stockhom, Sweden.

Faulkner, D. J. 1993. Marine natural products. *Natural Products Reports* 10:497–539.

Faust, M., S. L. Morton, and J. P. Quod. 1996. Further SEM study of marine dinoflagellates: The genus *Ostreopsis* (Dinophyceae). *Journal of Phycology* 32:1053–1065.

Fawley, M. W., and M. A. Buchheim. 1995. Loroxanthin, a phylogenetically useful character in *Chlamydomonas* and other chlorophycean flagellates. *Journal of Phycology* 31:664–667.

Fawley, M. W., C. A. Douglas, K. D. Stewart, and K. R. Mattox. 1990a. Light-harvesting pigment-protein complexes of the Ulvophyceae (Chlorophyta): Characterization and phylogenetic significance. *Journal of Phycology* 26:186–195.

Fawley, M. W., and C. M. Lee. 1990. Pigment composition of the scaly green flagellate *Mesostigma viride* (Micromonadophyceae) is similar to that of the siphonous green alga *Bryopsis plumosa* (Ulvophyceae). *Journal of Phycology* 26:666–670.

Fawley, M. W., N. Osterbauer, C. M. Lee, and S. Jiao. 1990b. The light-harvesting complex of *Mamiella filva*: a character linking scaly and naked members of the Micromonadophyceae (Chlorophyta). *Phycologia* 29:511–514.

Feist, M., and R. Feist. 1997. Oldest record of a bisexual plant. *Nature* 385:401.

Feist, M., and N. Grambast-Fessard. 1991. The genus concept in Charophyta: Evidence from Palaeozoic to Recent. Pages 189–203 in *Calcareous Algae and Stromatolites*, edited by R. Riding. Springer-Verlag, Berlin, Germany.

Feldmann, J. 1951. Ecology of marine algae. Pages 313–334 in *Manual of Phycology*, edited by G. M. Smith. Waltham: Chronica Botanica.

Felicini, G. P., and C. Perrone. 1994. *Pterocladia*. Pages 283–344 in *Biology of Economic Algae*, edited by I. Akatsuka. SPB Academic Publishing, The Hague, Netherlands.

Felsenstein, J. 1985. Confidence limits in phylogenies: An approach using the bootstrap. *Evolution* 39:783–791.

Felsenstein, J. 1988. Phylogenies from molecular sequences: Inference and reliability. *Annual Review of Genetics* 22:521–565.

Fensome, R. A., F. J. R. Taylor, G. Norris, W. A. S. Sargeant, D. I. Wharton, and G. L. Williams. 1993. *A Classification of Living and Fossil Dinoflagellates*. American Museum of Natural History, *Micropaleontology Special Publication #7*, Micropaleontology Press, Hanover, MA.

Fields, S. D., and R. G. Rhodes. 1991. Ingestion and retention of *Chroomonas* spp. (Cryptophyceae) by *Gymnodinium acidotum* (Dinophyceae). *Journal of Phycology* 27:525–529.

Figueiredo, M. A. deO., J. M. Kain (Jones), and T. A. Norton. 1996. Biotic interactions in the colonization of crustose coralline algae by epiphytes. *Journal of Experimental Marine Biology and Ecology* 199:303–318.

Finst, R. J., P. J. Kim, and L. M. Quarmby. 1998. Genetics of the deflagellation pathway in *Chlamydomonas*.

Genetics 149:927–936.

Fisher, M. M., and L. W. Wilcox. 1996. Desmid-bacterial associations in *Sphagnum*-dominated Wisconsin peatlands. *Journal of Phycology* 32:543–549.

Fisher, M. M., L. W. Wilcox, and L. E. Graham. 1998. Molecular characterization of epiphytic bacterial communities on charophycean green algae. *Applied and Environmental Microbiology* 64:4384–4389.

Fleury, B. G., A. Kelecom, R. C. Pereira, and V. L. Teixeira. 1994. Polyphenols, terpenes and sterols in Brazilian Dictyotales and Fucales (Phaeophyta). *Botanica Marina* 37:457–462.

Floyd, G. L., H. J. Hoops, and J. A. Swanson. 1980. Fine structure of the zoospore of *Ulothrix belkae* with emphasis on the flagellar apparatus. *Protoplasma* 104:17–31.

Floyd, G. L., and C. J. O'Kelly. 1990. Ulvophyceae. Pages 617–635 in *Handbook of Protoctista*, edited by L. Margulis, J. O. Corliss, M. Melkonian, and D. J. Chapman. Jones & Bartlett Publishers, Boston, MA.

Floyd, G. L., C. J. O'Kelly, and D. F. Chappell. 1985. Absolute configuration analysis of the flagellar apparatus in *Cladophora* and *Chaetomorpha* motile cells, with an assessment of the phylogenetic position of the Cladophoraceae (Ulvophyceae, Chlorophyta). *American Journal of Botany* 72:615–625.

Floyd, G. L., K. D. Stewart, and K. R. Mattox. 1971. Cytokinesis and plasmodesmata in *Ulothrix*. *Journal of Phycology* 7:306–309.

Floyd, G. L., K. D. Stewart, and K. R. Mattox. 1972. Cellular organization, mitosis, and cytokinesis in the ulotrichalean alga, *Klebsormidium*. *Journal of Phycology* 8:176–184.

Fogg, G. E. 1991. The phytoplanktonic ways of life. Tansley Review No. 30. *New Phytologist* 118:191–232.

Fogg, M. J. 1995. *Terraforming. Engineering Plant Environments*. Society of Automotive Engineers, Warrendale, PA.

Forsberg, C. 1965. Nutritional studies of *Chara* in axenic cultures. *Physiologia Plantarum* 18:275–290.

Foster, K. W., J. Saranak, N. Patel, G. Zarilli, M. Okabe, T. Kline, and K. Nakanishi. 1984. A rhodopsin is the functional photoreceptor for phototaxis in the unicellular eukaryote *Chlamydomonas*. *Nature* 311:756–759.

Fowke, L. C., and J. D. Pickett-Heaps. 1969a. Cell division in *Spirogyra*. I. Mitosis. *Journal of Phycology* 5:240–259.

Fowke, L. C., and J. D. Pickett-Heaps. 1969b. Cell division in *Spirogyra*. II. Cytokinesis. *Journal of Phycology* 5:273–281.

France, R. L. 1995. Differentiation between littoral and pelagic food webs in lakes using stable carbon isotopes. *Limnology and Oceanography* 40:1310–1313.

Franceschi, V. R., P. Ding, and W. J. Lucas. 1994. Mechanism of plasmodesmata formation in characean algae in relation to evolution of intercellular communication in higher plants. *Planta* 192:347–358.

Francis, D. 1967. On the eyespot of the dinoflagellate *Nematodinium*. *Journal of Experimental Biology* 47:495–502.

Franklin, L. A., and R. M. Forster. 1997. The changing irradiance environment: Consequences for marine macrophyte physiology, productivity and ecology. *European Journal of Phycology* 32:207–232.

Fraunholz, M. J., J. Wastl, S. Zauner, S. A. Rensing, M. M. Scherzinger, and U.-G. Maier. 1997. the evolution of cryptophytes. *Plant Systematics and Evolution* Supplement 11:163–174.

Frederick, S. E., P. J. Gruber, and N. E. Tolbert. 1973. The occurrence of glycolate dehydrogenase and glycolate oxidase in green plants. *Plant Physiology* 52:318–323.

Freshwater, D. W., S. Fredericq, B. S. Butler, and M. H. Hommersand. 1994. A gene phylogeny of the red algae (Rhodophyta) based on plastid *rbcL*. *Proceedings of the National Academy of Science* 91:7281–7285.

Friedl, T. 1995. Inferring taxonomic positions and testing genus level assignments in coccoid green lichen algae: A phylogenetic analysis of 18S ribosomal RNA sequences from *Dictyochloropsis reticulata* and from

members of the genus *Myrmecia* (Chlorophyta, Trebouxiophyceae cl. nov.). *Journal of Phycology* 31:632–639.

Friedl, T. 1996. Evolution of the polyphyletic genus *Pleurastrum* (Chlorophyta): inferences from nuclear-encoded ribosomal DNA sequences and motile cell ultrastructure. *Phycologia* 35:456–469.

Friedl, T., and O. L. Reymond. 1997. Zoospore ultrastructure of the rare coccoid green alga *Pachycladella umbrina* (Chlorophyta, Chlorophyceae). *Nova Hedwigia* 65:377–384.

Friedl, T., and C. Rokitta. 1997. Species relationships in the lichen alga *Trebouxia* (Chlorophyta, Trebouxiophyceae): Molecular phylogenetic analyses of nuclear-encoded large subunit rRNA gene sequences. *Symbiosis* 23:125–148.

Friedmann, E. I. 1982. Endolithic microorganisms in the Antarctic cold desert. *Science* 215:1045–1053.

Friedmann, E. I., and R. Ocampo. 1976. Endolithic blue-green algae in the dry valleys: Primary producers in the Antarctic desert ecosystem. *Science* 24:1247–1249.

Friedmann, E. I., and W. C. Roth. 1977. Development of the siphonous green alga *Penicillus* and the *Espera* state. *Botanical Journal of the Linnean Society* 74:189–214.

Friedmann, I. 1959. Structure, life-history and sex determination of *Prasiola stipitata* Suhr. *Annals of Botany* 23:571–594.

Friedmann, I., M. Hua, and R. Ocampo-Friedmann. 1993. Terraforming Mars: Dissolution of carbonate rocks by cyanobacteria. *Journal of the British Interplanetary Society* 46:291–292.

Friedmann, I., and R. Ocampo-Friedmann. 1995. A primitive cyanobacterium as pioneer microorganism for terraforming Mars. *Advances in Space Research* 15:243–246.

Fritz, L., D. Mores, and J. W. Hastings. 1990. The circadian bioluminescence rhythm of *Gonyaulax* is related to daily variation in the number of light emitting organelles. *Journal of Cell Science* 95:321–328.

Fritz, L., M. A. Quilliam, J. L. C. Wright, A. M. Beale, and T. M. Work. 1992. An outbreak of domoic acid poisoning attributed to the pennate diatom *Pseudonitzschia australis*. *Journal of Phycology* 28:439–442.

Frost, T. M., L. E. Graham, J. E. Elias, M. J. Haase, D. W. Kretchmer, and J. A. Kranzfelder. 1997. A yellow-green algal symbiont in the freshwater sponge, *Corvomyenia everetti*: Convergent evolution of symbiotic associations. *Freshwater Biology* 38:395–399.

Frost, T. M., and C. E. Williamson. 1980. *In situ* determination of the effect of symbiotic algae on the growth of the freshwater sponge *Spongilla lacustris*. *Ecology* 61:1361–1370.

Fry, B., C. S. Hopkinson, and A. Nolin. 1996. Long-term decomposition of DOC from experimental diatom blooms. *Limnology and Oceanography* 41:1344–1347.

Fry, W. L. 1983. An algal flora from the upper Ordovician of the Lake Winnipeg region, Manitoba, Canada. *Review of Palaeobotany and Palynology* 39:313–341.

Fukumoto, R.-H., T. Fuji, and H. Sekimoto. 1997. Detection and evaluation of a novel sexual pheromone that induces sexual cell division of *Closterium ehrenbergii* (Chlorophyta). *Journal of Phycology* 33:441–445.

Gabrielson, P. W., and D. J. Garbary. 1986. Systematics of red algae (Rhodophyta). *CRC Critical Reviews in Plant Science* 3:325–366.

Gabrielson, P. W., and D. J. Garbary. 1987. A cladistic analysis of Rhodophyta:Florideophycidean orders. *British Phycological Journal* 22:125–138.

Gabrielson, P. W., D. J. Garbary, M. R. Sommerfeld, R. A. Townsend, and P. L. Tyler. 1990. Rhodophyta. Pages 102–118 in *Handbook of Protoctista*, edited by L. Margulis, J. O. Corliss, M. Melkonian, and D. J. Chapman. Jones & Bartlett Publishers, Boston, MA.

Gabrielson, P. W., and M. H. Hommersand. 1982. The morphology of *Agardhiella subulata* representing the Agardhiellaeae, a new tribe in the Solieriaceae (Gigartinales, Rhodophyta). *Journal of Phycology* 18:46–58.

Gabrys, H., T. Walczak, and W. Haupt. 1984. Blue-light-induced chloroplast orientation in *Mougeotia*. Evidence for a separate sensor pigment besides phytochrome. *Planta* 160:21–24.

Gaedeke, A., and U. Sommer. 1986. The influence of the frequency of periodic disturbances on the maintenance of phytoplankton diversity. *Oecologia* 71:25–28.

Gaines, G, and M. Elbrächter. 1987. Heterotrophic nutrition. Pages 224–268 in *The Biology of Dinoflagellates*, edited by F. J. R. Taylor. Blackwell Scientific Publications, Oxford, UK.

Galloway, R. E. 1990. Selective conditions and isolation of mutants in salt-tolerant, lipid-producing microalgae. *Journal of Phycology* 26:752–760.

Galway, M. E., and A. R. Hardham. 1991. Immunofluorescent localization of microtubules throughout the cell cycle in the green alga *Mougeotia* (Zygnemataceae). *American Journal of Botany* 78:451–461.

Ganf, G. G. 1975. Photosynthetic production and irradiance-photosynthesis relationships of the phytoplankton from a shallow equatorial lake (L. George, Uganda). *Oecologia* 18:165–183.

Gantt, E. 1975. Phycobilisomes: Light-harvesting pigment complexes. *BioScience* 25:781–788.

Ganzon-Fortes, E. T. 1994. *Gelidiella*. Pages 149–184 in *Biology of Economic Algae*, edited by I. Akatsuka. SPB Academic Publishing, The Hague, Netherlands.

Garbary, D. J., and P. W. Gabrielson. 1990. Taxonomy and evolution. Pages 477–498 in *Biology of the Red Algae*, edited by K. M. Cole and R. G. Sheath. Cambridge University Press, Cambridge, UK.

Garbary, D., and H. W. Johansen. 1987. Morphogenesis and evolution in the Amphiroideae (Rhodophyta, Corallinaceae). *British Phycological Journal* 22:1–10.

Garbary, D. J., and A. R. McDonald. 1996. Fluorescent labelling of the cytoskeleton in *Ceramium strictum* (Rhodophyta). *Journal of Phycology* 32:85–93.

Garcia-Pichel, F., and J. Belnap. 1996. Microenvironments and microscale productivity of cyanobacterial desert crusts. *Journal of Phycology* 32:774–782.

Garcia-Pichel, F., and R. W. Castenholz. 1991. Characterization and biological implications of scytonemin, a cyanobacterial sheath pigment. *Journal of Phycology* 27:395–409.

Gargas, A., P. T. DePriest, M. Grube, and A. Tehler. 1995. Multiple origins of lichen symbioses in fungi suggested by SSU rDNA phylogeny. *Science* 268:1492–1495.

Gates, R. D., O. Hoegh-Guldberg, M. J. McFall-Ngai, and K. Y. Bil. 1995. Free amino acids exhibit anthozoan "host factor" activity: They induce the release of photosynthate from symbiotic dinoflagellates *in vitro*. *Proceedings of the National Academy of Science* 92:7430–7434.

Gauld, D. T. 1951. The grazing rate of planktonic copepods. *Journal of the Marine Biological Association of the UK* 29:695–706.

Gause, G. F. 1934. *The Struggle for Existence*. Williams & Wilkins. (Dover ed. 1971), Dover Publications, Minneola, NY.

Gavrilova, O. V., and A. V. Gabova. 1992. Experiment "*Chlamydomonas*" aboard Biosatellite Cosmos-2044. *Physiologist* 35, supplement S 212.

Gawlik, S. R., and W. F. Millington. 1988. Structure and function of the bristles of *Pediastrum boryanum* (Chlorophyta). *Journal of Phycology* 24:474–482.

Geesink, R. 1973. Experimental investigations on marine and freshwater *Bangia* (Rhodophyta) from The Netherlands. *Journal of Experimental Marine Biology and Ecology* 11:239–247.

Geider, R. J., and B. A. Osborne. 1992. *Algal Photosynthesis*. Chapman & Hall, New York, NY.

Geitler, L. 1955. Über die cytologisch bemerkenswerte Chlorophycee *Chlorokybus atmophyticus*. *Osterreichische Botanische Zeitschrift* 102:20–29.

Gelin, F., I. Boogers, A. A. M. Noordeloos, J. S. Sinninghe Damsté, R. Riegman, and J. W. de Leeuw. 1997. Resistant biomacromolecules in marine microalgae of the classes Eustigmatophyceae and Chlorophyceae: Geochemical implications. *Organic Geochemistry* 26:659–675.

Geller, W., and H. Müller. 1981. The filtration apparatus of cladocera: filter mesh sizes and their implications on food selectivity. *Oecologia* 49:316–321.

Genter, R. B. 1996. Ecotoxicology of inorganic chemical stress to algae. Pages 404–468 in *Algal Ecology: Freshwater Benthic Ecosystems*, edited by R. J. Stevenson, M. L. Bothwell, and R. L. Lowe. Academic Press, New York, NY.

George, D. G., and R. W. Edwards. 1976. The effect of wind on the distribution of chlorophyll *a* and crustacean plankton in a shallow eutrophic reservoir. *Journal of Applied Ecology* 13:667–690.

George, R. W. 1988. Products from fossil algae. Pages 305–334 in *Algae and Human Affairs*, edited by C. A. Lembi and J. R. Waaland. Cambridge University Press, Cambridge. UK.

Gerrath, J. F. 1993. The biology of desmids: a decade of progress. *Progress in Phycological Research* 9:79–192.

Gervais, F. 1997. Light-dependent growth, dark survival, and glucose uptake by cryptophytes isolated from a freshwater chemocline. *Journal of Phycology* 33:18–25.

Gerwick, W. H., and N. J. Lang. 1977. Structural, chemical and ecological studies on iridescence in *Iridaea*. *Journal of Phycology* 13:121–127.

Gerwick, W. H., M. A. Roberts, P. J. Proteau, and J.-L. Chen. 1994. Screening cultured marine algae for anticancer-type activity. *Journal of Applied Phycology* 6:143–149.

Gibb, A. P., R. Aggarwal, and C. P. Swainson. 1991. Successful treatment of *Prototheca* peritonitis complicating continuous ambulatory peritoneal dialysis. *Journal of Infection* 22:183–185.

Gibor, A. 1966. *Acetabularia*: A useful giant cell. *Scientific American* 215:118–124.

Giddings, T. H., D. L. Brower, and L. A. Staehelin. 1980. Visualization of particle complexes in the plasma membrane of *Micrasterias denticulata* associated with the formation of cellulose fibrils in primary and secondary cell walls. *Journal of Cell Biology* 84:327–339.

Giddings, T. H., and L. A. Staehelin. 1991. Microtubule-mediated control of microfibril deposition: A re-examination of the hypothesis, in *The Cytoskeletal Basis of Plant Growth and Form*, edited by C. W. Lloyd. Academic Press, London, UK.

Giddings, T. H., C. Wasmann, and L. A. Staehelin. 1983. Structure of the thylakoids and envelope membranes of the cyanelles of *Cyanophora paradoxa*. *Plant Physiology* 71:409–419.

Gifford, D. J., R. N. Bohrer, and C. M. Boyd. 1981. Spines on diatoms: Do copepods care? *Limnology and Oceanography* 26:1057–1061.

Gilbert, S., B. A. Zeeb, and J. P. Smol. 1997. Chrysophyte stomatocyst flora from a forest peat core in the Lena River Region, northeastern Siberia. *Nova Hedwigia* 64:311–352.

Gillott, M. 1990. Cryptophyta (Cryptomonads). Pages 139–151 in *Handbook of Protoctista*, edited by L. Margulis, J. O. Corliss, M. Melkonian, and D. J. Chapman. Jones and Bartlett Publishers, Boston, MA.

Gillott, M. A., and S. P. Gibbs. 1980. The cryptomonad nucleomorph: Its ultrastructure and evolutionary significance. *Journal of Phycology* 16:558–568.

Gilson, P. R., and G. I. McFadden. 1996. The miniaturized nuclear genome of a eukaryotic endosymbiont contains genes that overlap, genes that are cotranscribed, and the smallest known spliceosomal introns. *Proceedings of the National Academy of Science* 93:7737–7742.

Giovannoni, S. J., T. B. Britschi, C. L. Moyer, and K. G. Field. 1990. Genetic diversity in Sargasso Sea bacterioplankton. *Nature* 345:60–63.

Giovannoni, S. J., S. Turner, G. J. Olsen, S. Barns, D. J. Lane, and N. R. Pace. 1988. Evolutionary relationships among cyanobacteria and green chloroplasts. *Journal of Bacteriology* 170:3584–3592.

Gladue, R. M., and J. E. Maxey. 1994. Microalgal feeds for aquaculture. *Journal of Applied Phycology* 6:131–142.

Glazer, A. N., C. Chan, R. C. Williams, S. W. Yeh, and J. H. Clark. 1985. Kinetics of energy flow in the phycobilisome core. *Science* 230:1051–1053.

Gleitz, M., A. Bartsch, G. S. Dieckmann, and H. Eichen. 1998. Composition and succession of sea ice diatom assemblages in the eastern and southern Weddell Sea, Antarctica. Pages 107–120 in *Antarctic Sea Ice Biological Processes, Interactions, and Variability*, edited by M. P. Lizotte and K. R. Arrigo. Antarctic Research Series 73, American Geophysical Union, Washington, DC.

Gliwicz, Z. M. 1990. *Daphnia* growth at different concentrations of blue-green filaments. *Archiv für Hydrobiologie* 120:51–65.

Glynn, P. W. 1997. Bioerosion and coral reef growth: A dynamic balance. Pages 68–95 in *Life and Death of Coral Reefs*, edited by C. Birkeland. Chapman & Hall, New York, NY.

Goddard, R. H., and J. W. La Claire. 1991. Calmodulin and wound healing in the coenocytic green alga *Ernodesmis verticillata* (Kützing) Borgesen. *Planta* 183:281–293.

Goericke, R., and D. J. Repeta. 1992. The pigments of *Prochlorococcus marina*: The presence of divinylchlorophyll *a* and *b* in a marine procaryote. *Limnology and Oceanography* 37:425–433.

Goff, L. J. 1979. The biology of *Harveyella mirabilis* (Cryptonemiales, Rhodophyceae). VII. Structure and proposed function of host-penetrating cells. *Journal of Phycology* 15:87–106.

Goff, L. J. 1982. The biology of parasitic red algae. *Progress in Phycological Research* 1:289–370.

Goff, L., J. Ashen, and D. Moon. 1997. The evolution of parasites from their hosts: A case study in the parasitic red algae. *Evolution* 51:1068–1078.

Goff, L. J., and A. W. Coleman. 1984a. Elucidation of fertilization and development in a red alga by quantitative DNA microspectrofluorometry. *Developmental Biology* 102:173–194.

Goff, L. J., and A. W. Coleman. 1984b. Transfer of nuclei from a parasite to its host. *Proceedings of the National Academy of Science* 81:5420–5424.

Goff, L. J., and A. W. Coleman. 1985. The role of secondary pit connections in red algal parasitism. *Journal of Phycology* 21:483–508.

Goff, L. J., and A. W. Coleman. 1990. DNA. Microspectrophotometric studies. Pages 43–72 in *Biology of the Red Algae*, edited by K. M. Cole and R. G. Sheath. Cambridge University Press, Cambridge, UK.

Goff, L. J., L. Liddle, P. C. Silva, M. Voytek, and A. W. Coleman. 1992. Tracing species invasion in *Codium*, a siphonous green alga, using molecular tools. *American Journal of Botany* 79:1279–1285.

Goff, L. J., D. A. Moon, and A. W. Coleman. 1994. Molecular delineation of species and species relationships in the red algal agarophytes *Gracilariopsis* and *Gracilaria* (Gracilariales). *Journal of Phycology* 30:521–537.

Goff, L. J., D. A. Moon, P. Nyvall, B. Stache, K. Mangin, and G. Zuccarello. 1996. The evolution of parasitism in the red algae; molecular comparisons of adelphoparasites and their hosts. *Journal of Phycology* 32:297–312.

Goff, L. J., and G. Zuccarello. 1994. The evolution of parasitism in red algae: Cellular interactions of adelphoparasites and their hosts. *Journal of Phycology* 30:695–720.

Gojdics, M. 1953. *The Genus Euglena*. University of Wisconsin Press, Madison, WI.

Goldman, C. R., and A. J. Horne. 1983. *Limnology*. McGraw-Hill, New York, NY.

Goldsborough, L. G., and G. G. C. Robinson. 1996. Pattern in wetlands. Pages 78–120 in *Algal Biology. Freshwater Benthic Ecosystems*, edited by R. T. Stevenson, M. L. Bothwell, and R. L. Lowe. Academic Press, New York, NY.

Goldstein, S. F. 1992. Flagellar beat patterns in algae. Pages 99–154 in *Algal Cell Motility*, edited by M. Melkonian. Chapman & Hall, London, UK.

Gonzalez, M. A., and L. J. Goff. 1989a. The red algal epiphytes *Microcladia coulteri* and *M. californica* (Rhodophyeae, Ceramiaceae). I. Taxonomy, life history and phenology. *Journal of Phycology* 25:545–558.

Gonzalez, M. A., and L. J. Goff. 1989b. The red algal epiphytes *Microcladia coulteri* and *M. californica* (Rhodophyceae, Ceramiaceae). II. Basiphyte specificity. *Journal of Phycology* 15:558–567.

Good, B. H., and R. L. Chapman. 1978. The ultrastructure of *Phycopeltis* (Chroolepidaceae: Chlorophyta). I. Sporopollenin in the cell walls. *American Journal of Botany* 65:27–33.

Goodenough, U., P. A. Detmas, and C. Hwang. 1982. Activation of cell fusion in *Chlamydomonas*: Analysis of wild-type gametes and nonfusing mutants. *Journal of Cell Biology* 92:378–386.

Goodwin, B. C., and C. Briére. 1994. Mechanics of the cytoskeleton and morphogenesis of *Acetabularia*. *International Review of Cytology* 150:225–242

Gordon, R., and R.W. Drum. 1994. The chemical basis for diatom morphogenesis. *International Review of Cytology* 150:243–372.

Goto, Y., and K. Ueda. 1988. Microfilament bundles of F-actin in *Spirogyra* observed by fluorescence microscopy. *Planta* 173:442–446.

Graham, J. M. 1991. Symposium introductory remarks: A brief history of aquatic microbial ecology. *Journal of Protozoology* 38:66–69.

Graham, J. M., P. Arancibia-Avila, and L. E. Graham. 1996a. Physiological ecology of a species of the filamentous green alga *Mougeotia* under acidic conditions: Light and temperature effects on photosynthesis and respiration. *Limnology and Oceanography* 41:253–262.

Graham, J. M., P. Arancibia-Avila, and L. E. Graham. 1996b. Effects of pH and selected metals on growth of the filamentous green alga *Mougeotia* under acidic conditions. *Limnology and Oceanography* 41:263–270.

Graham, J. M., M. T. Auer, R. P. Canale, and J. P. Hoffmann. 1982. Ecological studies and mathematical modeling of *Cladophora* in Lake Huron: 4. Photosynthesis and respiration as functions of light and temperature. *Journal of Great Lakes Research* 8:100–111.

Graham, J. M., and L. E. Graham. 1987. Growth and reproduction of *Bangia atropurpurea* (Roth) C. Ag. (Rhodophyta) from the Laurentian Great Lakes. *Aquatic Botany* 28:317–331.

Graham, J. M., L. E. Graham, and J. A. Kranzfelder. 1985. Light, temperature and photoperiod as factors controlling reproduction in *Ulothrix zonata* (Ulvophyceae). *Journal of Phycology* 21:235–239.

Graham, J. M., C. A. Lembi, H. L. Adrian, and D. F. Spencer. 1995. Physiological responses to temperature and irradiance in *Spirogyra* (Zygnematales, Charophyceae). *Journal of Phycology* 31:531–540.

Graham, L. E. 1982a. The occurrence, evolution, and phylogenetic significance of parenchyma in *Coleochaete* Bréb. (Chlorophyta). *American Journal of Botany* 69:447–454.

Graham, L. E. 1982b. Cytology, ultrastructure, taxonomy, and phylogenetic relationships of Great Lakes filamentous algae. *Journal of Great Lakes Research* 8:3–9.

Graham, L. E. 1984. An ultrastructure re-examination of putative multilayered structures in *Trentepohlia aurea*. *Protoplasma* 123:1–7.

Graham, L. E. 1985. The origin of the life cycle of land plants. *American Scientist* 73:178–186.

Graham, L. E. 1990. Meiospore formation in charophycean algae. Pages 43–54 in *Microspores: Evolution and Ontogeny*, edited by Blackmore & Knox. Academic Press, London, UK.

Graham, L. E.. 1993. *The Origin of Land Plants*. John Wiley, New York, NY.

Graham, L. E. 1996. Green algae to land plants: An evolutionary transition. *Journal of Plant Research* 109:241–251.

Graham, L. E., C. F. Delwiche, and B. D. Mishler. 1991. Phylogenetic connections between the "green algae" and the "bryophytes." *Advances in Bryology* 4:213–244.

Graham, L. E., J. M. Graham, and J. A. Kranzfelder. 1986. Irradiance, daylength and temperature effects on zoosporogenesis in *Coleochaete scutata* (Charophyceae). *Journal of Phycology* 22:35–39.

Graham, L. E., J. M. Graham, W. A. Russin, and J. M. Chesnick. 1994. Occurrence and phylogenetic significance of glucose utilization by charophycean algae: Glucose enhancement of growth in *Coleochaete orbicularis*. *American Journal of Botany* 81:423–432.

Graham, L. E., J. M. Graham, and D. E. Wujek. 1993. Ultrastructure of *Chrysodidymus sunuroideus* (Synurophyceae). *Journal of Phycology* 29:330–341.

Graham, L. E., and Y. Kaneko. 1991. Subcellular structures of relevance to the origin of land plants (Embryophytes) from green algae. *CRC Critical Reviews in Plant Sciences* 10:323–342.

Graham, L. E., F. J. Macentee, and H. C. Bold. 1981. An investigation of some subaerial green algae. *Texas Journal of Science* 23:13–16.

Graham, L. E., and G. E. McBride. 1978. Mitosis and cytokinesis in sessile sporangia of *Tretepohlia aurea* (Chlorophyceae). *Journal of Phycology* 14:132–137.

Graham, L. E., and G. E. McBride. 1979. The occurrence and phylogenetic significance of a multilayered structure in *Coleochaete scutata* spermatozoids. *American Journal of Botany* 66:887–894.

Graham, L. E., and C. Taylor. 1986a. The ultrastructure of meiospores of *Coleochaete pulvinata* (Charophyceae). *Journal of Phycology* 22:299–307.

Graham, L. E., and C. Taylor. 1986b. Occurrence and phylogenetic significance of "special walls" at meiosporogenesis in *Coleochaete*. *American Journal of Botany* 73:597–601.

Graham, L. E., and G. J. Wedemayer. 1984. Spermatogenesis in *Coleochaete pulvinata* (Charophyceae): Sperm maturation. *Journal of Phycology* 20:302–309.

Graham, L. E., and L. W. Wilcox. 1983. The occurrence and phylogenetic significance of putative placental transfer cells in the green alga *Coleochaete*. *American Journal of Botany* 70:113–120.

Grambast, L. J. 1974. Phylogeny of the charophytes. *Taxon* 23:463–481.

Grant, M. C. 1990. Charophyceae (Order Charales). Pages 641–648 in *Handbook of Protoctista*, edited by L. Margulis, J. O. Corliss, M. Melkonian, and D. J. Chapman. Jones and Bartlett Publishers, Boston, MA.

Gray, J., and A. J. Boucot. 1989. Is *Moyeria* a euglenoid? *Lethaia* 22:447–416.

Gray, J., D. Massa, and A. J. Boucot. 1982. Caradocian land plant microfossils from Libya. *Geology* 10:197–201.

Gray, M. W. 1992. The endosymbiont hypothesis revisited. *International Review of Cytology* 141:233–357.

Gray, M. W., G. Burger, and B. F. Lang. 1999. Mitochondrial evolution. *Science* 283:1476–1481.

Green, J. C., and T. Hori. 1990. The architecture of the flagellar apparatus of *Prymnesium patellifera* (Prymnesiophyceae). *The Botanical Magazine, Tokyo* 103:191–207.

Green, J. C., and B. S. C. Leadbeater (editors). 1994. *The Haptophyte Algae*, Clarendon Press, Oxford, UK.

Green, J. W., A. H. Knoll, S. Golubíc, and K. Swett. 1987. Paleobiology of distinctive benthic microfossils from the Upper Proterozoic limestone-dolomite "series," central East Greenland. *American Journal of Botany* 74:928–940.

Grell, K. G. 1973. *Protozoology*. Springer-Verlag, Berlin, Germany.

Gretz, M. R., J. M. Aronson, and M. R. Sommerfeld. 1980. Cellulose in the cell walls of the Bangiophyceae (Rhodophyta). *Science* 207:779–781.

Greuet, C. 1968. Organization ultrastructurale de l'ocelle de deux Péridiniens Warnowiidae, *Erythropsis pavillardi* Kofoid et Swezy et *Warnowia pulchra* Schiller. *Protistologica* 4:202–230.

Greuet, C. 1976. Organisation ultrastructurale du tentacule D' *Erythropsis pavallardi* Kofoid et Swezy péridinien *Warnowiidae* Lindemann. *Protostologica* 3:335–345.

Griffith, J. K., M. E. Baker, D. A. Rouch, M. G. P. Page, R. A. Skurray, I. T. Paulsen, K. F. Chater, S. A. Baldwin, and P. J. F. Henderson. 1992. Membrane transport proteins: Implications of sequence comparisons. *Current Opinion in Cell Biology* 4:684–695.

Grilli Caiola, M. 1992. Cyanobacterian symbiosis with bryophytes and tracheophytes. Pages 231–254 in *Algae and Symbioses*, edited by W. Reisser. BioPress, Bristol, UK.

Grilli Caiola, M., and S. Pellegrini. 1984. Lysis of *Microcystis aeruginosa* by *Bdellovibrio*-like bacteria. *Journal of Phycology* 20:471–475.

Grolig, F. 1990. Actin-based organelle movements in interphase *Spirogyra*. *Protoplasma* 155:29–42.

Gross, W. 1993. Peroxisomes in algae: their distribution, biochemical function and phylogenic [phylogenetic] importance. *Progress in Phycological Research* 9:47–78.

Grossman, A. R., D. Bhaya, K. E. Apt, and D. M. Kehoe. 1995. Light harvesting complexes in oxygenic photosynthesis: Diversity, control and evolution. *Annual Review of Genetics* 29:231–288.

Grossman, A. R., Manodori, A., and D. Snyder. 1990. Light-harvesting proteins of diatoms: Their relationship to the chlorophyll *a/b* binding proteins of higher plants and their mode of transport into plastids. *Molecular and General Genetics* 224:91–100.

Grossman, A. R., M. R. Schaefer, G. G. Chiang, and J. L. Collier. 1993. The phycobilisome, a light harvesting complex responsive to environmental conditions. *Microbial Reviews* 57:725–749.

Grover, J. P. 1989. Phosphorus-dependent growth kinetics of 11 species of freshwater algae. *Limnology and Oceanography* 34:341–348.

Grünewald, K., C. Hagen, and W. Braune. 1997. Secondary carotenoid accumulation in flagellates of the green alga *Haematococcus lacustris*. *European Journal of Phycology* 32:387–392.

Guillard, R. R. L., M. D. Keller, C. J. O'Kelly, and G. L. Floyd. 1991. *Pycnococcus provasoli* gen. et sp. nov., a coccoid prasinoxanthin-containing phytoplankter from the western North Atlantic and Gulf of Mexico. *Journal of Phycology* 27:39–47.

Guillaumont, B., T. Bajjouk, and P. Talec. 1997. Seaweed remote sensing: A critical review of sensors and data processing. *Progress in Phycology Research* 12:213–282.

Guillou, L., M.-J. Chrétinnot-Dinet, L. K. Medlin, H. Claustre, S. Loiseaux-de Goër, and D. Vaulot. 1999. *Bolidomonas*: A new genus with two species belonging to a new algal class, the Bolidophyceae (Heterokonta). *Journal of Phycology* 35:368–381.

Guiry, M. D. 1978. The importance of sporangia in the classification of the Florideophyceae. Pages 111–114 in *Modern Approaches to the Taxonomy of Red and Brown Algae*, edited by D. E. G. Irvine and J. H. Price. Systematics Association Special Volume 10. Academic Press, London, UK.

Guiry, M. D., D.-H. Kim, and M. Masuda. 1984. Reinstatement of the genus *Mastocarpus* Kützing (Rhodophyta). *Taxon* 33:53–63.

Gunnison, D., and M. Alexander. 1975a. Basis for the resistance of several algae to microbial decomposition. *Applied Microbiology* 29:729–738.

Gunnison, D., and M. Alexander. 1975b. Resistance and susceptibility of algae to decomposition by natural microbial communities. *Limnology and Oceanography* 20:64–70.

Hackney, J. M., R. C. Carpenter, and W. H. Adey. 1989. Characteristic adaptations to grazing among algal turfs on a Caribbean coral reef. *Phycologia* 28:109–119

Häder, D.-P., and E. Hoiczyk. 1992. Gliding motility. Pages 1–38 in *Algal Cell Motility*, edited by M. Melkonian. Chapman & Hall, London, UK.

Hagen, C., W. Braune, and L. O. Björn. 1994. Functional aspects of secondary carotenoids in *Haematococcus lacustris* (Volvocales). III. Action as a "sunshade." *Journal of Phycology* 30:241–248.

Haglund, K. 1997. The use of algae in aquatic toxicity assessment. *Progress in Phycological Research* 12:182–212.

Halfen, L. N. 1973. Gliding motility of *Oscillatoria*: Ultrastructural and chemical characterization of the fibrillar layer. *Journal of Phycology* 9:248–253.

Halfen, L. N., and R. N. Castenholz. 1970. Gliding in a blue-green alga: A possible mechanism. *Nature* 225:1163–1165.

Hall, R. I., P. R. Leavitt, J. P. Smol, and N. Zirnhelts. 1997. Comparison of diatoms, fossil pigments and historical records as measures of lake eutrophication. *Freshwater Biology* 38:401–417.

Hallegraeff, G. M., and C. J. Bolch. 1991. Transport of toxic dinoflagellate cysts via ships' ballast water. *Marine Pollution Bulletin* 32:27–30.

Hallegraeff, G. M., and C. J. Boalch. 1992. Transport of diatom and dinoflagellate resting spores in ships' ballast water: Implications for plankton biogeography and aquaculture. *Journal of Plant Research* 14:1067–1084.

Hamada, J. 1987. Diploidy in DNA content in vegetative cells of *Closterium ehrenbergii* (Chlorophyta). *Journal of Phycology* 23:541–546.

Hambrook, J. A., and R. G. Sheath. 1987. Grazing of freshwater Rhodophyta. *Journal of Phycology* 23:656–662.

Han, T.-M., and B. Runnegar. 1992. Megascopic eukaryotic algae from the 2.1-billion-year-old Negaunee iron-formation, Michigan. *Science* 257:232–235.

Hanagata, N., I. Karube, M. Chihara, and P. C. Silva. 1998. Reconsideration of the taxonomy of ellipsoidal species of *Chlorella* (Trebouxiophyceae, Chlorophyta) with establishment of *Watanabea* gen. nov. *Phycological Research* 46:221–229.

Hanelt, D., C. Wiencke, U. Karsten, and W. Nultsch. 1997. Photoinhibition and recovery after high light stress in different developmental and life-history stages of *Laminaria saccharina* (Phaeophyta). *Journal of Phycology* 33:387–395.

Haney, J. F. 1973. An *in situ* examination of the grazing activities of natural zooplankton communities. *Archiv für Hydrobiologie* 72:87–132.

Hanisak, M. D. 1983. The nitrogen relationships of marine macroalgae. Pages 699–730 in *Nitrogen in the Marine Environment*, edited by E. J. Carpenter and D. G. Capone. Academic Press, New York, NY.

Hanisak, M. D., and M. A. Samuel. 1987. Growth rates in culture of several species of *Sargassum* from Florida, U.S.A. *Hydrobiologia* 151/152:399–404.

Hansen, G. I. 1977. *Cirrulicarpus carolinensis*, a new species in the Kallymeniaceae (Rhodophyta). *Occasional Papers of the Farlow Herbarium* 12:1–22.

Hansen, G. 1993. Light and electron microscopical observations of the dinoflagellate *Actiniscus pentasterias* (Dinophyceae). *Journal of Phycology* 29:486–499.

Hansen, G. I., and S. C. Lindstrom. 1984. A morphological study of *Hommersandia maximicarpa* gen. et sp. nov. (Kallymeniaceae, Rhodophyta) from the North Pacific. *Journal of Phycology* 20:476–488.

Hansen, P. 1998. Phagotrophic mechanisms and prey selection in mixotrophic phytoflagellates. Pages 525–537 in *Physiological Ecology of Harmful Algal Blooms*, edited by D. M. Anderson, A. D. Cembella, and G. M. Hallegraeff. Springer-Verlag, Berlin. Germany

Hanson, M. A., and M. G. Butler. 1990. Early responses of plankton and turbidity to biomanipulation in a shallow prairie lake. *Hydrobiologia* 200/201:317–327.

Hansson, L.-A. 1996. Algal recruitment from lake sediments in relation to grazing, sinking, and dominance patterns in the phytoplankton community. *Limnology and Oceanography* 41:1312–1323.

Happey-Wood, C. E. 1988. Ecology of freshwater planktonic green algae. Pages 175–226 in *Growth and Reproductive Strategies of Freshwater Phytoplankton*, edited by C. D. Sandgren. Cambridge University Press, Cambridge, UK.

Harlin, M. M. 1987. Allelochemistry in marine macroalgae. *CRC Critical Reviews in Plant Science* 5:237–249.

Harris, E. 1989. *The Chlamydomonas Handbook*, Academic Press, New York, NY.

Harris, G. P. 1986. *Phytoplankton Ecology*. Chapman & Hall, London, UK.

Harris, G. P., G. D. Haffner, and B. B. Piccinin. 1980. Physical variability and phytoplankton communities II. primary productivity by phytoplankton in a physically variable environment. *Archiv für Hydrobiologie* 88:393–425.

Haselkorn, R. 1978. Heterocysts. *Annual Review of Plant Physiology* 29:319–344.

Hashimoto, H. 1992. Involvement of actin filaments in chloroplast division of the alga *Closterium ehrenbergii*. *Protoplasma* 167:88–96.

Hasle, G. R. 1994. *Pseudo-nitzschia* as a genus distinct from *Nitzschia*. *Journal of Phycology* 30:1036–1039.

Hasle, G. R., H. A. von Stosch, and E. E. Syvertsen. 1983. Cymatosiraceae, a new diatom family. *Bacillaria*

6:9–156.

Hasle, G. R., and E. E. Syvertsen. 1997. Marine Diatoms. Pages 5–385 in *Identifying Marine Phytoplankton*, edited by C. R. Tomas. Academic Press, New York, NY.

Hatano, K., and K. Maruyama. 1995. Growth pattern of isolated zoospores in *Hydrodictyon reticulatum* (Chlorococcales, Chlorophyceae). *Phycological Research* 43:105–110.

Hawkes, M. W. 1978. Sexual reproduction in *Porphyra gardneri* (Smith et Hollenberg) Hawkes (Bangiales, Rhodophyta). *Phycologia* 17:329–353.

Hawkes, M. W. 1980. Ultrastructure characteristics of monospore formation in *Porphyra gardneri* (Rhodophyta). *Journal of Phycology* 16:192–196.

Hawkes, M. W. 1990. Reproductive strategies. Pages 455–476 in *Biology of the Red Algae*, edited by K. M. Cole, and R. G. Sheath. Cambridge University Press, Cambridge, UK.

Hay, C. H. 1994. *Durvillaea*. Pages 353–384 in *Biology of Economic Algae*, edited by I. Akatsuka. SPB Academic Publishing, The Hague, Netherlands.

Hay, M. E. 1981. The functional morphology of turf-forming seaweeds: Persistence in stressful marine habitats. *Ecology* 63:739–750.

Hay, M. E. 1996. Marine chemical ecology: What's known and what's next? *Journal of Experimental Marine Biology and Ecology* 200:103–134.

Heaney, S. I. 1976. Temporal and spatial distribution of the dinoflagellate *Ceratium hirundinella* O. F. Müller within a small productive lake. *Freshwater Biology* 6:531–542.

Heimann, K., R. A. Andersen, and R. Wetherbee. 1995. The flagellar development cycle of the uniflagellate *Pelagomonas calceolata* (Pelagophyceae). *Journal of Phycology* 31:577–583.

Heimdal, B. R. 1997. Modern coccolithophorids. Pages 731–831 in *Identififying Marine Phytoplankton*, edited by C. R. Tomas. Academic Press, San Diego, CA.

Hein, M. 1997. Inorganic carbon limitation of photosynthesis in lake phytoplankton. *Freshwater Biology* 37:545–552.

Heins, L., I. Collinson, and J. Soll. 1998. The protein translocation apparatus of chloroplast envelopes. *Trends in Plant Science* 3:56–61.

Hellebust, J. A. 1971. Glucose uptake by *Cyclotella cryptica*: Dark induction and light inactivation of transport system. *Journal of Phycology* 7:345–349.

Hennes, K. P., and C. A. Suttle. 1995. Direct counts of viruses in natural waters and laboratory cultures by epifluorescence microscopy. *Limnology and Oceanography* 40:1050–1055.

Hennig, W. 1966. *Phylogenetic Systematics*, University of Illinois Press, Urbana, IL.

Henrikson, L., H. G. Nyman, H. G. Oscarson, and J. A. E. Stenson. 1980. Trophic changes without changes in the external nutrient loading. *Hydrobiologia* 68:257–263.

Henry, E. C. 1984. Syringodermatales ord. nov. and *Syringoderma floridana* sp. nov. (Phaeophyceae). *Phycologia* 23:419–426.

Henry, E. C., and K. M. Cole. 1982a. Ultrastructure of swarmers in the Laminariales (Phaeophyceae). I. Zoospores. *Journal of Phycology* 18:550–569.

Henry, E. C., and K. M. Cole. 1982b. Ultrastructure of swarmers in the Laminariales (Phaeophyceae). II. Sperm. *Journal of Phycology* 18:570–579.

Henry, E. C., and R. H. Meints. 1994. Recombinant viruses as transformation vectors of marine algae. *Journal of Applied Phycology* 6:247–255.

Henze, K., A. Badr, M. Wettern, R. Derff, and W. Martin. 1995. A nuclear gene of eubacterial origin in *Euglena* gracilis reflects cryptic endosymbioses during protist evolution. *Proceedings of the National Academy of Science* 92:9122–9126.

Hernández, I., M. Christmas, J. M. Yelloly, and B. A. Whitton. 1997. Factors affecting surface alkaline phosphatase activity in the brown alga *Fucus spiralis* at a North Sea intertidal site (Tyne Sands, Scotland). *Journal of Phycology* 33:569–575.

Herndon, W. R. 1964. *Boldia*: A new rhodophycean genus. *American Journal of Botany* 51:575–581.

Hessen, D. O., and E. Van Donk. 1993. Morphological

changes in *Scenedesmus* induced by substances released from *Daphnia*. *Archiv für Hydrobiologie* 127:129–140.

Heywood, P. 1983. The genus *Vacuolaria* (Raphidophyceae). *Progress in Phycological Research* 2:53–86.

Heywood, P. 1990. Raphidophyta. Pages 318–325 in *Handbook of Protoctista*, edited by L. Margulis, J. O. Corliss, M. Melkonian, and D. J. Chapman. Jones & Bartlett Publishers, Boston, MA.

Hibberd, D. J. 1981. Notes on the taxonomy and nomenclature of the algal classes Eustigmatophyceae and Tribophyceae (synonym Xanthophyceae). *Botanical Journal of the Linnean Society* 82:93–119.

Hibberd, D. J. 1990a. Eustigmatophyta. Pages 326–333 in *Handbook of Protoctista*. edited by L. Margulis, J. O. Corliss, M. Melkonian, and D. J. Chapman. Jones & Bartlett Publishers, Boston, MA.

Hibberd, D. J. 1990b. Xanthophyta. Pages 686–697 in *Handbook of the Protoctista*, edited by L. Margulis, C. O. Corliss, M. Melkonian, and D. J. Chapman. Jones and Bartlett Publishers, Boston, MA.

Hibberd, D. J., and G. F. Leedale. 1970. Eustigmatophyceae—a new algal class with unique organization of the motile cell. *Nature* 225:758–760.

Hibberd, D. J., and G. F. Leedale. 1971. Cytology and ultrastructure of the Xanthophyceae. II. The zoospore and vegetative cell of coccoid forms, with special reference to *Ophiocytium majus* Naegeli. *British Phycological Journal* 6:1–23.

Hibberd, D. J., and R. E. Norris. 1984. Cytology and ultrastructure of *Chlorarachnion reptans* (Chlorarachniophyta divisio nova, Chlorarachniophyceae, classis nova). *Journal of Phycology* 20:310–330.

Hilenski, L. L., and P. L. Walne. 1983. Ultrastructure of mucocysts in *Peranema trichophorum* (Euglenophyceae). *Journal of Protozoology* 30:491–496.

Hilenski, L. L., and P. L. Walne. 1985a. Ultrastructure of the flagella of the colorless phagotroph *Peranema trichophorum* (Euglenophyceae). I. Flagellar mastigonemes. *Journal of Phycology* 21:114–125.

Hilenski, L. L., and P. L. Walne. 1985b. Ultrastructure of the flagella of the colorless phagotroph *Peranema trichophorum* (Euglenophyceae). II. Flagellar roots. *Journal of Phycology* 21:125–134.

Hill, D. R. A. 1991a. A revised circumscription of *Cryptomonas* (Cryptophyceae) based on examination of Australian strains. *Phycologia* 30:170–188.

Hill, D. R. A. 1991b. *Chroomonas* and other blue-green cryptomonads. *Journal of Phycology* 27:133–145.

Hill, D. R. A., and K. S. Rowan. 1989. The biliproteins of the Cryptophyceae. *Phycologia* 28:455–463.

Hill, D. R. A., and R. Wetherbee. 1986. *Proteomonas sulcata* gen. et sp. nov. (Cryptophyceae), a cryptomonad with two morphologically distinct and alternating forms. *Phycologia* 25:521–543.

Hill, D. R. A., and R. Wetherbee. 1989. A reappraisal of the genus *Rhodomonas* (Cryptophyceae). *Phycologia* 28:143–158.

Hill, M. S. 1996. Symbiotic zooxanthellae enhance boring and growth rates of the tropical sponge *Anthosignella varians* forma *varians*. *Marine Biology* 125:649–654.

Hillis, D. M., C. Moritz, and B. K. Mable. 1996. *Molecular Systematics*, 2nd ed., Sinauer Associates, Sunderland, MA.

Hillis-Colinvaux. L. 1980. Ecology and taxonomy of *Halimeda*: Primary producer of coral reefs. *Advances in Marine Biology* 17:1–327.

Hixon, M. A. 1997. Effects of reef fishes on corals and algae in *Life and Death of Coral Reefs*, edited by C. Birkeland. Chapman & Hall, New York, NY.

Ho, T. S.-S., and M. Alexander. 1974. The feeding of amebae on algae in culture. *Journal of Phycology* 10:95–100.

Hoagland, K. D., J. P. Carder, and R. L. Spawn. 1996. Effects of organic toxic substances. Pages 469–496 in *Algal Ecology: Freshwater Benthic Ecosystems*, edited by R. J. Stevenson, M. L. Bothwell, and R. L. Lowe. Academic Press, New York, NY.

Hodell, D. A., and C. L. Schelske. 1998. Production, sedimentation, and isotopic composition of organic matter in Lake Ontario. *Limnology and Oceanography*

43:200–214.

Hodell, D. A., C. L. Schelske, G. L. Fahnenstiel, and L. L. Robbins. 1998. Biologically induced calcite and its isotopic composition in Lake Ontario. *Limnology and Oceanography* 43:187–199.

Hoek, van den C., and A. M. Breeman. 1990. Seaweed Biogeography of the North Atlantic: Where are we now? In *Evolutionary Biogeography of the Marine Algae of the North Atlantic*, edited by D. J. Garbary and G. R. South. Springer-Verlag, Berlin, Germany.

Hoek, van den C., D. G. Mann, and H. M. Jahns. 1995. *Algae: An Introduction to Phycology*. Cambridge University Press, Cambridge, UK.

Hoffman, L. R. 1973. Fertilization in *Oedogonium*. II. Polyspermy. *Journal of Phycology* 9:296–301.

Hoffman, L. R. 1983. *Atractomorpha echinata* gen. et sp. nov., a new anisogamous member of the Sphaeropleaceae (Chlorophyceae). *Journal of Phycology* 19:76–86.

Hoffmann, J. P. 1990. Dependence of photosynthesis and vitamin B_{12} uptake on cellular vitamin B_{12} concentration in the multicellular alga *Cladophora glomerata* (Chlorophyta). *Limnology and Oceanography* 35:100–108.

Hoffmann, J. P., and L. E. Graham. 1984. Effects of selected physicochemical factors on growth and zoosporogenesis of *Cladophora glomerata*. *Journal of Phycology* 20:1–7.

Hofmann, E., P. M. Wrench, F. P. Sharples, R. G. Hiller, W. Welte, and K. Diederichs. 1996. Structural basis of light harvesting by carotenoids: Peridinin-chlorophyll-protein from *Amphidinium carterae*. *Science* 272:1788–1791.

Hogetsu, T., and M. Yokoyama. 1979. Cell expansion and microfibril deposition in *Closterium ehrenbergii*. *The Botanical Magazine, Tokyo* 92:299–303.

Hoham, R. W. 1980. Unicellular chlorophytes-snow algae. Pages 61–84 in *Phytoflagellates*, edited by E. R. Cox. Elsevier/North Holland, Amsterdam, Netherlands.

Hoham, R. W., and D. W. Blinn. 1979. Distribution of cryophilic algae in an arid region, the American Southwest. *Phycologia* 14:133–145.

Hoham, R. W., and B. Duval. 1999. Microbial ecology of snow and fresh-water ice with emphasis on snow algae. In *Snow Ecology*, edited by H. G. Jones, J. W. Pomeroy, D. A. Walker, and R. W. Hoham. Cambridge University Press, Cambridge, UK. (in press)

Hoham, R. W., E. M. Schlag, J. Y. Kang, A. J. Hasselwander, A. F. Behrstock, I. R. Blackburn, R. C. Johnson, and S. C. Roemer. 1998. The effects of irradiance levels and spectral composition on mating strategies in the snow alga, *Chloromonas* sp. -D, from the Tughill Plateau, New York State. *Hydrological Processes* 12:1627–1639.

Holen, D. A., and M. E. Borass. 1995. Mixotrophy in chrysophytes. Pages 119–140 in *Chrysophyte Algae. Ecology, Phylogeny and development*, edited by C. D. Sandgren, J. P. Smol, and J. Kristiansen. Cambridge University Press, Cambridge, UK.

Hollande, A. 1952a. Classe des Eugleniens. *Traité de Zoologie* 1:238–284.

Hollande, A. 1952b. Classe des Cryptomonadines. *Traité de Zoologie* 1:285–308.

Hollande, A. 1952c. Classe de Chrysomonadines. *Traité de Zoologie* 1:471–570.

Holm, N. P., and D. E. Armstrong. 1981. Role of nutrient limitation and competition in controlling the populations of *Asterionella formosa* and *Microcystis aeruginosa* in semicontinuous culture. *Limnology and Oceanography* 26:622–634.

Hommersand, M. H., and S. Fredericq. 1990. Sexual reproduction and cystocarp development. Pages 305–346 in *Biology of the Red Algae*, edited by K. M. Cole, and R. G. Sheath. Cambridge University Press, Cambridge, UK.

Hommersand, M. H., M. D. Guiry, S. Fredericq, and G. L. Leister. 1993. New perspectives on the taxonomy of the Gigartinaceae. *Hydrobiologia* 260/261:105–120.

Honegger, R. 1992. Lichens: Mycobiont-photobiont relationships. Pages 255–276 in *Algae in Symbioses*, edited by W. Reisse. BioPress, Bristol, UK.

Hong, Y.-K., C. H. Sohn, M. Polne-Fuller, and A. Gibor. 1995. Differential display of tissue-specific messenger RNAs in *Porphyra perforata* (Rhodophyta) thallus. *Journal of Phycology* 31:640–643.

Honjo, S. 1997. The rain of ocean particles and earth's carbon cycle. *Oceanus* 40:4–7.

Honjo, S., J. F. Connell, and P. L. Sachs. 1980. Deep-ocean sediment trap:design and function of PARFLUX Mark II. *Deep-Sea Research* 27:745–753.

Hoops, H. J., and G. L. Floyd. 1982. Ultrastructure of the flagellar apparatus of *Pyrobotrys* (Chlorophyceae). *Journal of Phycology* 18:455–462.

Hoops, H. J., G. L., Floyd, and J. A. Swanson. 1982. Ultrastructure of the biflagellate motile cells of *Ulvaria oxysperma* (Kütz.) Bliding and phylogenetic relationships among ulvaphycean algae. *American Journal of Botany* 69:150–159.

Hopkins, A. W., and G. E. McBride. 1976. The life history of *Coleochaete scutata* (Chlorophyceae) studied by a Feulgen microspectrophotometric analysis. *Journal of Phycology* 12:29–35.

Hori, T., and J. C. Green. 1994. Mitosis and cell division. Pages 91–110 in *The Haptophyte Algae*, edited by J. C. Green and B. S. C. Leadbeater. Clarendon Press, Oxford, UK.

Hori, T. , I. Inouye, T. Horiguchi, and G. T. Boalch. 1985. Observations on the motile stage of *Halosphaera minor* (Ostenfeld) (Prasinophyceae) with special reference to the cell structure. *Botanica Marina* 28:529–537.

Hori, T., R. E. Norris, and M. Chihara. 1982. Studies on the ultrastructure and taxonomy of the genus *Tetraselmis* (Prasinophyceae). I. Subgenus *Tetraselmis*. *The Botanical Magazine, Tokyo* 95:49–61.

Horiguchi, T. 1996. *Haramonas dimorpha* gen. et sp. nov. (Raphidophyceae), a new marine raphidophyte from Australian mangrove. *Phycological Research* 44:143–150.

Horiguchi, T., and M. Chihara. 1987. *Spiniferodinium galeiforme*, a new genus and species of benthic dinoflagellates (Phytodiniales, Pyrrhophyta) from Japan. *Phycologia* 26:478–487.

Horiguchi, T., and F. Kubo. 1997. *Roscoffia minor* sp. nov. (Peridiniales, Dinophyceae): A new, sand-dwelling, armored dinoflagellate from Hokkaido, Japan. *Phycological Research* 45:65–69.

Horner, R. A., D. L. Garrison, and F. G. Plumley. 1997. Harmful algal blooms and red tide problems on the U.S. west coast. *Limnology and Oceanography* 42:1076–1088.

Horodyski, R. J., and L. P. Knauth. 1994. Life on land in the Precambrian. *Science* 263:494–498.

Hoshaw, R. W., and R. L. Hilton. 1966. Observations on the sexual cycle of the saccoderm desmid *Spirotaenia condensata*. *Journal of the Arizona Academy of Science* 4:88–92.

Hoshaw, R. W., R. M. McCourt, and J.-C. Wang. 1990. Conjugaphyta. Pages 119–131 in *Handbook of Protoctista*, edited by L. Margulis, C. O. Corliss, M. Melkonian, and D. L. Chapman. Bartlett & Jones Publishers, Boston, MA.

Hoshaw, R. W., J.-C. Wang, R. M. McCourt, and H. M. Hull. 1985. Ploidal changes in clonal cultures of *Spirogyra communis* and implications for species definitions. *American Journal of Botany* 72:1005–1011.

Hotchkiss, A. T., and R. M. Brown. 1987. The association of rosette and globule terminal complexes with cellulose microfibril assembly in *Nitella translucens* var. *axillaris* (Charophyceae). *Journal of Phycology* 23:229–237.

Hotchkiss, A. T., M. R. Gretz, K. C. Hicks, and R. M. Brown. 1989. The composition and phylogenetic significance of the *Mougeotia* (Charophyceae) cell wall. *Journal of Phycology* 25:646–654.

Howell, E. T., M. A. Turner, R. L. France, M. B. Jackson, and P. M. Stokes. 1990. Comparison of zygnematacean (Chlorophyta) algae in the metaphyton of two acidic lakes. *Canadian Journal of Fisheries and Aquatic Science* 47:1085–1092.

Hruby, T., and T. A. Norton. 1979. Algal recolonization on rocky shores in the Firth of Clyde. *Journal of Ecology* 67:65–77.

Huang, X., J. C. Weber, T. K. Hinson, A. C. Mathieson, and S. C. Minocha. 1996. Transient expression of the GUS reporter gene in the protoplast and partially digested cells of *Ulva lactuca* L (Chlorophyta). *Botanica Marina* 39:467–474.

Huber, H., K. Beyser, and S. Fabry. 1996. Small G proteins of two green algae are localized to exocytic compartments and to flagella. *Plant Molecular Biology* 31:279–293.

Hudson, J. J., and W. D. Taylor. 1996. Measuring regeneration of dissolved phosphorus in planktonic communities. *Limnology and Oceanography* 41:1560–1565.

Huisman, J., and F. J. Weissing. 1994. Light-limited growth and competition for light in well-mixed aquatic environments: an elementary model. *Ecology* 75:507–520.

Huelsenbeck, J. P., and B. Rannala. 1997. Phylogenetic methods come of age: Testing hypothesis in an evolutionary context. *Science* 276:227–232.

Hunt, B. E., and D. F. Mandoli. 1996. A new, artificial seawater that facilitates growth of large numbers of cells of *Acetabularia acetabulum* (Chlorophyta) and reduces the labor inherent in cell culture. *Journal of Phycology* 32:483–495.

Hurd, C. L., and C. L. Stevens. 1997. Flow visualization around single- and multiple-bladed seaweeds with various morphologies. *Journal of Phycology* 33:360–367.

Hurley, J. P., and D. E. Armstrong. 1990. Fluxes and transformations of aquatic pigments in Lake Mendota, Wisconsin. *Limnology and Oceanography* 35:384–398.

Hurley, J. P., D. E. Armstrong, G. J. Kenoyer, and C. J. Bowser. 1985. Ground water as a silica source from diatom production in a precipitation-dominated lake. *Science* 227:1576–1578.

Huss, V. A. R., and H. D. Kranz. 1997. Charophyte evolution and the origin of land plants. *Plant Systematics and Evolution Supplement* 11:103–114.

Huss, V. A. R., and M. L. Sogin. 1990. Phylogenetic position of some *Chlorella* species with the Chlorococcales based upon complete small-subunit ribosomal RNA sequences. *Journal of Molecular Evolution* 31:432–442.

Hussain, M. I., T. M. Jhoja, and M. Guerlesquin. 1996. Chemistry, ecology and seasonal succession of charophytes in the Al-Kharj irrigation canal, Saudi Arabia. *Hydrobiologia* 333:129–137.

Hutchins, D. A. 1995. Iron and the marine phytoplankton community. *Progress in Phycological Research* 11:1–49.

Hutchins, D. A., and K. W. Bruland. 1998. Iron-limited diatom growth and SI:N uptake ratios in a coastal upwelling regime. *Nature* 393:561–564.

Hutchinson, G. E. 1961. The paradox of the plankton. *American Naturalist* 95:137–145.

Huxley, T. H. 1858. Appendix A in *Deep-sea soundings in the North Atlantic Ocean between Ireland and Newfoundland* (J. Dayman), pp. 63–68. Her Majesty's Stationery Office, London.

Ianora, A., S. A. Poulet, A. Miralto, and R. Grottoli. 1996. The diatom *Thalassiosira rotula* affects reproductive success in the copepod *Arcartia clausi*. *Marine Biology* 125:279–286.

Iglesias-Rodriguez, M. D., and M. J. Merrett. 1997. Dissolved inorganic carbon utilization and the development of extracellular carbonic anhydrase by the marine diatom *Phaeodactylum tricornutum*. *New Phytologist* 135:163–168.

Ikehara, T., H. Uchida, L. Suzuki, and S. Nakamura. 1996. Chloroplast nucleoids in large number and large DNA amount with regard to maternal inheritance in *Chlamydomonas reinhardtii*. *Protoplasma* 194:11–17.

Inouye, I. 1993. Flagella and flagellar apparatuses of algae. Pages 99–134 in *Ultrastructure of Microalgae*, edited by T. Berner. CRC Press, Boca Raton, FL.

Inouye, I., T. Hori, and M. Chihara. 1990. Absolute configuration analysis of the flagellar apparatus of *Pterosperma cristatum* (Prasinophyceae) and consideration of its phylogenetic position. *Journal of Phycology* 26:329–344.

Inouye, I., and M. Kawachi. 1994. The haptonema. Pages 73–90 in *The Haptophyte Algae*, edited by J. C. Green and B. S. C. Leadbeater. Clarendon Press, Oxford, UK.

Inouye, I, and R. N. Pienaar. 1985. Ultrastructure of the flagellar apparatus in a species of *Pleurochrysis* (class Prymnesiophyceae). *Protoplasma* 125:24–35.

Ishida, K.-I., Y. Cao, M. Hasegawa, N. Okuda, and Y. Hara. 1997. The origin of chlorarachniophyte plastids, as inferred from phylogenetic comparisons of amino acid sequences of EF-Tu. *Journal of Molecular Evolution* 45:682–687.

Ishida, K.-I., T. Nakayama, and Y. Hara. 1996. Taxonomic studies on the Chlorarachniophyta. II. Generic delimitation of the chlorarachniophytes and description of *Gymnochlora stellata* gen. et sp. nov. and *Lotharella* gen. nov. *Phycological Research* 44:37–45.

Ishiura, M., S. Katsuna, S. Aoki, H. Iwasaki, C. R. Andersson, A. Tanabe, S. S. Golden, C. H. Johnson, and T. Kondo. 1998. Expression of a gene cluster *kaiABC* as a circadian feedback process in cyanobacteria. *Science* 281:1519–1523.

Jackson, A. E., and R. D. Seppelt. 1997. Physiological adaptations to freezing and UV radiation exposure in *Prasiola crispa*, an Antarctic terrestrial alga. Pages 226–233 in *Antarctic Communities: Species, Structure, and Survival*, edited by B. Battaglia, J. Valencia, and D. W. H. Walton. Cambridge University Press, Cambridge, UK.

Jacobs, W. P. 1993. Rhizome gravitropism precedes gravimorphogenesis after inversion of the green algal coenocyte *Caulerpa prolifera* (Caulerpales). *American Journal of Botany* 80:1273–1275.

Jacobs, W. P. 1994. *Caulerpa. Scientific American* 271:100–105.

Jacobsen, A., G. Bratbak, and M. Heidal. 1996. Isolation and characterization of a virus infecting *Phaeocystis pouchetii* (Prymnesiophyceae). *Journal of Phycology* 32:923–927.

Jacobshagen, S., and C. Schnarrenberger. 1990. Two class I aldolases in *Klebsormidium flaccidum* (Charophyceae): An evolutionary link from chlorophytes to higher plants. *Journal of Phycology* 26:312–317.

Jacobson, D. M., and R. A. Andersen. 1994. The discovery of mixotrophy in photosynthetic species of *Dinophysis* (Dinophyceae): Light and electron microscopical observations of food vacuoles in *Dinophysis acuminata*, *D. norvegica*, and two heterotrophic dinophysoid dinoflagellates. *Phycologia* 33:97–110.

Jacobson, D. M., and D. M. Anderson. 1986. Thecate heterotrophic dinoflagellates: Feeding behavior and mechanisms. *Journal of Phycology* 22:249–258.

Jacobson, D. M., and D. M. Anderson. 1996. Widespread phagocytosis of ciliates and other protists by marine mixotrophs and heterotrophic thecate dinoflagellates. *Journal of Phycology* 32:279–285.

Jahn, T. L., E. C. Bovee, and F. F. Jahn. 1979. *How to Know the Freshwater Protozoa*. Wm. C. Brown, Dubuque, IA.

Jassby, A. D., and T. Platt. 1976. Mathematical formulation of the relationship between photosynthesis and light for phytoplankton. *Limnology and Oceanography* 21:540–547.

Jeffrey, S. W., and M. Vesk. 1976. Further evidence for a membrane-bound endosymbiont within the dinoflagellate *Peridinium foliaceum*. *Journal of Phycology* 12:450–456.

Jeffrey, S. W., and S. W. Wright. 1994. Photosynthetic pigments in the Haptophyta. Pages 111–132 in *The Haptophyte Algae*, edited by J. C. Green and B. S. C. Leadbeater. Clarendon Press, Oxford, UK.

Jensen, A. 1995. Production of alginates. Pages 79–92 in *Algae, Environment and Human Affairs*, edited by W. Wiessner, E. Schnepf, and R. C. Starr. BioPress, Bristol, UK.

Jensen, J. B. 1974. Morphological studies on Cystoseiraceae and Sargassaceae (Phaeophyceae) with special reference to apical organization. *University of California Publications Botany* 68:1–61.

Jensen, T. E. 1984. Cyanobacterial cell inclusions of irregular occurrence: Systematic and evolutionary implications. *Cytobios* 39:35–62.

Jewson, D. H. 1992. Life cycle of a *Stephanodiscus* sp. (Bacillariophyta). *Journal of Phycology* 28:856–866.

Jewson, D. H., B. H. Rippey, and W. K. Gilmore. 1981. Loss rates from sedimentation, parasitism, and grazing during the growth, nutrient limitation, and dormancy of a diatom crop. *Limnology and Oceanography* 26:1045–1056.

Jijina, J. G., and J. Lewin. 1984. Persistent blooms of surf diatoms (Bacillariophyceae) along the Pacific coast, USA. III. Relationships between diatom populations and environmental variables along Oregon and Washington beaches (1977 and 1978). *Phycologia* 23:471–483.

Johansen, H. W. 1981. *Coralline Algae, A First Synthesis*. CRC Press, Boca Raton, FL.

Johansen, J. R. 1993. Cryptogamic crusts of semiarid and arid lands of North America. *Journal of Phycology* 29:140–147.

John, D. M. 1994. Alternation of generations in algae: Its complexity, maintenance and evolution. *Biological Review* 69:275–291.

John, D. M., and C. A. Maggs. 1997. Species problems in eukaryotic algae: A modern perspective. Pages 83–105 in *Species: The Units of Biodiversity*, edited by M. F. Claridge, H. A. Dawah, and M. R. Wilson. Chapman & Hall, London, UK.

John-McKay, M. E., and B. Colman. 1997. Variations in the occurrence of external carbonic anhydrase among strains of the marine diatom *Phaeodactylum tricornutum* (Bacillariophyceae). *Journal of Phycology* 33:988–990.

Johnson, P. W., P. E. Hargraves, and J. Mc. N. Sieburth. 1988. Ultrastructure and ecology of *Calycomonas ovalis* Wulff, 1919 (Chrysophyceae) and its redescription as a testate rhizopod, *Paulinella ovalis*. N. comb. (Filosea:Euglyphina). *Journal of Protozoology*. 35:618–626.

Johnson, R F., and R. M. Sheehan. 1985. Late Ordovician dasyclad algae of the eastern Great Basin. Pages 79–84 in *Paleoalgology: Contemporary Research and Applications*, edited by D. F. Toomey and M. H. Nitecki. Springer-Verlag, Berlin, Germany.

Jónasdóttir, S. H., and T. Kiorboe. 1996. Copepod recruitment and food composition: do diatoms affect hatching success? *Marine Biology* 125:743–750.

Jones, H. J. 1997. A classification of mixotrophic protists based on their behavior. *Freshwater Biology* 37:35–43.

Jones, H. L. J., B. S. C. Leadbeater, and J. C. Green. 1994. Mixotrophy in haptophytes. Pages 247–264 in *The Haptophyte Algae*, edited by J. C. Green, and B. S. C. Leadbeater. Clarendon Press, Oxford, UK.

Jordan, R. W., and J. C. Green. 1994. A check list of the extant Haptophyta of the world. *Journal of the Marine Biological Association of the UK* 74:149–174.

Jordan, R., and A. Kleijne. 1994. A classification system for living coccolithophores. Pages 83–106 in *Coccolithophores*, edited by A. Winter and W. G. Siesser. Cambridge University Press, Cambridge, UK.

Jost, L. 1895. Beiträge zur Kenntniss der Coleochaeteen. *Ber d. Deutsch Botanische Gesellschaft* 13:433–452.

Jürgens, K. 1994. Impact of *Daphnia* on planktonic microbial food webs—a review. *Marine Microbial Food Webs* 8:295–324.

Jürgens, K., and H. Güde. 1994. The potential importance of grazing-resistant bacteria in planktonic systems. *Marine Ecology Progress Series* 112:169–188.

Kain, J. M., and T. A. Norton. 1990. Marine ecology. Pages 377–422 in *Biology of the Red Algae*, edited by K. M. Cole and R. G. Sheath. Cambridge University Press, Cambridge, UK.

Kamiya, M., J. A. West, R. J. King, G. C. Zuccarello, J. Tanaka, and Y. Hara. 1998. Evolutionary divergence in the red algae *Caloglossa leprieurii* and *C. apomeiotica*. *Journal of Phycology* 34:361–370.

Kamiya, R. 1992. Molecular mechanisms of flagellar movement. Pages 155–198 in *Algal Cell Motility*, edited by M. Melkonian. Chapman & Hall, London, UK.

Kaplan, A., M. Ronen-Iavazi, H. Zer, R. Schwarz, D. Tchernov, D. J. Bonfel, D. Schatz, A. Vardi, M. Hassidim, and L. Reinhold. 1998. The inorganic carbon concentrating mechanisms in cyanobacteria: Induction and ecological significance. *Candian Journal of Botany* 76:917–924.

Kapraun, D. F. 1970. Field and cultural studies of *Ulva* and *Enteromorpha* in the vicinity of Port Aransas, Texas. *Contributions in Marine Science* 15:205–283.

Kapraun, D. F., and P. W. Boone. 1987. Karyological studies of three species of Scytosiphonaceae (Phaeophyta) from coastal North Carolina. *Journal of Phycology* 23:318–322.

Kapraun, D. F., M. G. Gargiulo, and G. Tripodi. 1988. Nuclear DNA and karyotype variation in species of *Codium* (Codiales, Chlorophyta) from the North Atlantic. *Phycologia* 27:273–282.

Kapraun, D. F., and M. J. Shipley. 1990. Karyology and nuclear DNA quantification in *Bryopsis* (Chlorophyta) from North Carolina, USA. *Phycologia* 29:443–453.

Karsten, G. 1912. Über die Reduktionstellung bei der Auxosporenbildung von *Surirella saxonica*. *Zeitschrift Botanische* 4:417–426.

Karsten, U., C. Bock, and J. A. West. 1995. ^{13}C-NMR spectroscopy as a tool to study organic osmolytes in the mangrove red algal genera *Bostrychia* and *Stictosiphonia* (Ceramiales). *Phycological Research* 43:241–247.

Karsten, U., A. S. Mostaert, R. J. King, M. Kamiya, and Y. Hara. 1996. Osmoprotectors in some species of Japanese mangrove macroalgae. *Phycological Research* 44:109–112.

Kasai, F., and T. Ichimura. 1990. A sex determining mechanism in the *Closterium ehrenbergii* (Chlorophyta) species complex. *Journal of Phycology* 26:195–201.

Katsaros, C. I. 1995. Apical cells of brown algae with particular reference to Sphacelariales, Dictyotales and Fucales. *Phycological Research* 43:43–59.

Katsaros, C., and B. Galatis. 1992. Immunofluorescence and electron microscopic studies of microtubule organization during the cell cycle of *Dictyota dichotoma* (Phaeophyta, Dictyotales). *Protoplasma* 169:75–84.

Kawai, H. 1988. A flavin-like autofluorescent substance in the posterior flagellum of golden and brown algae. *Journal of Phycology* 24:114–117.

Kawai, H., and I. Inouye. 1989. Flagellar autofluorescence in forty-four chlorophyll *c*-containing algae. *Phycologia* 28:222–227.

Kawai, H., and M. Tokuyama. 1995. *Laminarionema elsbetiae* gen. et sp. nov., new endophyte of *Laminaria japonica*. *Phycological Research* 43:185–190.

Kehoe, D. M., and A. R. Grossman. 1996. Similarity of a chromatic adaptation sensor to phytochrome and ethylene receptors. *Science* 273:1409–1412.

Kelley, I., and L. A. Pfiester. 1990. Sexual reproduction in the freshwater dinoflagellate *Gloeodinium montanum*. *Journal of Phycology* 26:167–173.

Kellogg, E. A., and N. D. Juliano. 1997. The structure and function of RuBisCO and their implications for systematic studies. *American Journal of Botany* 84:413–428.

Kemp, A. E. S., and J. G. Baldauf. 1993. Vast Neogene laminated diatom mat deposits from the eastern equatorial Pacific Ocean. *Nature* 362:141–143.

Kennelly, S. J. 1987. Inhibition of kelp recruitment by turfing algae and consequences for an Australian kelp community. *Journal of Experimental Marine Biology and Ecology* 112:49–60.

Kessler, E., M. Schäfer, C. Hümmer, A. Kloboucek, and V. A. R. Huss. 1997. Physiological, biochemical, and molecular characters for the taxonomy of the subgenera of *Scenedesmus* (Chlorococcales, Chlorophyta). *Botanica Acta* 110:244–250.

Khan, M., and Y. S. R. K. Sarma. 1984. Cytogeography and Cytosystematics of Charophyta. Pages 303–330 in *Systematics of the Green Algae*, edited by D. E. G. Irvine and D. M. John. Academic Press, New York, NY.

Khoja, T. M., and B. A. Whitton. 1975. Heterotrophic growth of filamentous blue-green algae. *British Phycological Journal* 10:129–148.

Kibak, H., L. Taiz, T. Starke, P. Bernasconi, and J. P. Gogarten. 1992. Evolution of structure and function of V-ATPases. *Journal of Bioenergetics and Biomembranes* 24:415–424.

Kidd, D. F., and J. C. Lagarias. 1990. Phytochrome from the green alga *Mesotaenium caldariorum*. Purification and preliminary characterization. *Journal of Biological Chemistry* 265:7029–7035.

Kiermayer, O. 1981. Cytoplasmic basis of morphogenesis in *Micrasterias*. Pages 147–189 in *Cytomorphogenesis in Plants*, edited by O. Kiermayer. Springer-Verlag, New York, NY.

Kiermayer, O., and U. Meindl. 1984. Interactions of the Golgi apparatus and the plasmalemma in the cytomorphogenesis of *Micrasterias*. Pages 175–182 in *Compartments in Algal Cells and Their Interaction*, edited by W. Wiessner, D. Robinson, and R. C. Starr. Springer-Verlag, Berlin, Germany

Kiermayer, O., and U. Meindl. 1989. Cellular morphogenesis: The desmid (Chlorophyceae) system. Pages 149–167 in *Algae as Experimental Systems*, edited by A. W. Coleman, L. J. Goff, and J. R. Stein-Taylor. Liss, New York, NY.

Kiermayer, O., and U. B. Sleytr. 1979. Hexagonally ordered "rosettes" of particles in the plasma membrane of *Micrasterias denticulata* Bréb. and their significance for microfibril formation and orientation. *Protoplasma* 101:133–138.

Kies, L. 1976. Untersuchungen zur Feinstructure und taxonomischen Einordnung von *Gloeochaete wittrockiana*, einer apoplastidalean capsalen alga mit blaugrünen Endosymbionten (Cyanellen). *Protoplasma* 87:419–446.

Kies, L. 1992. Glaucocystophyceae and other protists harboring prokaryotic endocytobionts. Pages 353–379 in *Algae and Symbioses*, edited by W. Reisser. BioPress, Bristol, UK.

Kies, L. 1995. Algal snow and the contribution of algae to suspended particulate matter in the Elbe estuary. Pages 92–121 in *Algae, Environment and Human Affairs*, edited by W. Wiessner, E. Schnepf, and R. C. Starr. BioPress, Bristol, UK.

Kies, L., and B. P. Kremer. 1990. Glaucocystophyta. Pages 152–166 in *Handbook of Protoctista*, edited by L. Margulis, J. O. Corliss, M. Melkonian, and D. J. Chapman. Jones & Bartlett Publishers, Boston, MA.

Kim, G. H., and L. Fritz. 1993. Signal glycoprotein with α-D-mannosyl residues is involved in wound-healing processes of *Antithamnion sparsum*. *Journal of Phycology* 29:89–90.

Kim, G. H., I. K. Lee, and L. Fritz. 1995. The wound-healing responses of *Antithamnion nipponicum* and *Griffithsia pacifica* (Ceramiales, Rhodophyta) monitored by lectins. *Phycological Research* 43:161–166.

Kindle, K. E., and O. A. Sodeinde. 1994. Nuclear and chloroplast transformation in *Chlamydomonas reinhardtii*: Strategies for genetic manipulation and gene expression. *Journal of Applied Phycology* 6:231–238.

Kirk, D. L. 1997. *Volvox. Molecular Genetic Origins of Multicellularity and Cellular Differentiation*. Cambridge University Press, Cambridge, UK.

Kirkland, B. L., and R. L. Chapman. 1990. The fossil green alga *Mizzia* (Dasycladaceae): A tool for interpretation of paleoenvironment in the upper Permian Capitan Reef complex, southeastern New Mexico. *Journal of Phycology* 26:569–576.

Kiss, J. Z., and L. A. Staehelin. 1993. Structural polarity in the *Chara* rhizoid: A reevaluation. *American Journal of Botany* 80:273–282.

Klaveness, D. 1981. *Rhodomonas lacustris* (Pascher & Ruttner) Javornicky (Cryptomonadida): Ultrastructure of the vegetative cell. *Journal of Protozoology* 28:83–90.

Klaveness, D. 1982. The *Cryptomonas-Caulobacter* consortium: Facultative ectocommensalism with possible taxonomic consequences? *Nordic Journal of Botany* 2:183–188.

Klaveness, D. 1988. Ecology of the cryptomonadida: A first review. Pages 105–133 in *Growth and Reproductive Strategies of Freshwater Phytoplankton*, edited by C. D. Sandgren. Cambridge University Press, Cambridge, UK.

Klebahn, H. 1896. Beiträge zur Kenntnis der Auxosporenbildung. I. *Rhopalodia gibba* (Ehrenb.) O. Müller. *Jahrbuch wissenschaftliche Botanik* 29:595–654.

Kleijne, A. 1992. Extant Rhabdosphaeraceae (coccolithophorids, class Prymnesiophyceae) from the Indian Ocean, Red Sea, Mediterranean Sea, and North Atlantic Ocean. *Scripta Geologica* 100:1–63.

Knaust, R., T. Urbig, L. Li, W. Taylor, and J. W. Hastings. 1998. The circadian rhythm of bioluminescence in *Pyrocystis* is not due to differences in the amount of luciferase: A comparative study of three bioluminescent marine dinoflagellates. *Journal of Phycology* 34:167–172.

Knoll, A. H. 1992. The early evolution of eukaryotes: A geological perspective. *Science* 256:622–627.

Knoll, A. 1996. Archean and Proterozoic paleontology. Pages 51–80 in *Palynology: Principles and Applications*, edited by J. Jansonius, and D. C. McGregor. American Association of Stratigraphic Palynologists Foundation.

Kociolek, J. P., and M. J. Wynne. 1988. Observations on *Navicula thallodes* (Bacillariophyceae), a blade-forming diatom from the Bering Sea. *Journal of Phycology* 24:439–441.

Koehl, M. A. R., and R. S. Alberte. 1988. Flow, flapping, and photosynthesis of *Nereocystis luetkeana*: A functional comparison of undulate and flat blade morphologies. *Marine Biology* 99:435–444.

Kofoid, C. A., and O. Swezy. 1921. The free-living unarmored Dinoflagellata. *Memoirs of the University of California* 5:1-562.

Kogame, K. 1996. Morphology and life history of *Scytosiphon canaliculatus* com. nov. (Scytosiphonales, Phaeophyceae) from Japan. *Phycological Research* 44:85–94.

Kogame, Y., and H. Kawai. 1996. Development of the intercalary meristem in *Chorda filum* (Laminariales, Phaeophyceae) and other primitive Laminariales. *Phycological Research* 44:247–260.

Köhler, K. 1956. Entwicklungsgeschichte, Geschlectsbestimmung und Befruchtung bei *Chaetomorpha*. *Archiv für Protistenkunde* 101:223–268.

Köhler, S., C. F. Delwiche, P. W. Denny, L. G. Tilney, P. Webster, R. J. M. Wilson, J. D. Palmer, and D. S. Roos. 1997. A plastid of probable green algal origin in apicomplexan parasites. *Science* 275:1485–1489.

Kokinos, J. P., T. I. Eglinton, M. A. Goni, J. Boon, P. A. Martoglios, and D. M. Anderson. 1998. Characterization of a highly resistant biomacromolecular material in the cell wall of a marine dinoflagellate resting cyst. *Organic Geochemistry* 28:265–288.

Komárek, J. 1989. Polynuclearity of vegetative cells in coccal green algae from the family Neochloridaceae. *Archiv für Protistenkunde* 137:255–273.

Koop, H.-U. 1979. The life cycle of *Acetabularia* (Dasycladales, Chlorophyceae): A compilation of evidence for meiosis in the primary nucleus. *Protoplasma* 100:353–366.

Korb, R. E., P. J. Saville, A. M. Johnston, and J. A. Raven. 1997. Sources of inorganic carbon for photosynthesis by three species of marine diatom. *Journal of Phycology* 33:433–440.

Kornmann, P. 1938. Zur Entwicklungsgeschichte von *Derbesia* und *Halicystis*. *Planta* 28:464–470.

Kornmann, P. 1984. *Erythrotrichopeltis*, eine neue Gattung der Erythropeltidaceae (Bangiophyceae, Rhodophyta) *Helgolander Meeresuntersuchungen* 38:207–224.

Kornmann, P. 1987. Der Lebenszyklus von *Porphyrostromium obscurum* (Bangiophyceae, Rhodophyta). *Helgolander Meeresuntersuchungen* 41:127–137.

Kraft, G. T., and P. Robins. 1985. Is the order Cryptonemiales defensible? *Phycologia* 24:67–77.

Kramer, G. P., and D. J. Chapman. 1991. Biomechanics and alginic acid composition during hydrodynamic adaptation by *Egregia menziesii* (Phaeophyta) juveniles. *Journal of Phycology* 27:47–53.

Kraml, M., and H. Herrmann. 1991. Red-blue-interaction in *Mesotaenium* chloroplast movement: Blue seems to stabilize the transient memory of the phytochrome signal. *Photochemistry and Photobiology* 53:255–259.

Kratz, R. F., P. A. Young, and D. F. Mandoli. 1998. Timing and light regulation of apical morphogenesis during reproductive development in wild-type populations

of *Acetabularia acetabulum* (Chlorophyceae). *Journal of Phycology* 34:138–146.

Kristiansen, J. 1990. Chrysophyta. Pages 438–454 in *Handbook of Protoctista*, edited by L. Margulis, J. O. Corliss, M. Melkonian, and D. J. Chapman. Jones & Bartlett Publishers, Boston, MA.

Kroken, S. B., L. E. Graham, and M. E. Cook. 1996. Occurrence and evolutionary significance of resistant cell walls in charophytes and bryophytes. *American Journal of Botany* 83:1241–1254.

Kromkamp, J. 1990. The kinetics of photoinhibition in the cyanobacterium *Anabaena flos-aquae*. *British Phycological Journal* 25:91.

Krupp, J. M., and N. J. Lang. 1985. Cell division and filament formation in the desmid *Bambusina brebissonii* (Chlorophyta). *Journal of Phycology* 21:16–25.

Kübler, J. E., S. C. Minocha, and A. C. Mathieson. 1994. Transient expression of the GUS reporter gene in protoplasts of *Porphyra miniata* (Rhodophyta). *Journal of Marine Biotechnology* 1:165–169.

Kübler, J. E., and J. A. Raven. 1995. The interaction between inorganic carbon acquisition and light supply in *Palmaria palmata* (Rhodophyta). *Journal of Phycology* 31:369–375.

Kugrens, P. 1980. Electron microscopic observations on the differentiation and release of spermatia in the marine red alga *Polysiphonia hendryi* (Ceramiales, Rhodomelaceae). *American Journal of Botany* 67:519–528.

Kugrens, P. 1983. *Gloeococcus tetrasporus* sp. nov. (Tetrasporales, Palmellaceae), from Colorado mountain lakes. *Journal of Phycology* 19:511–516.

Kugrens, P., and R. E. Lee. 1988. Ultrastructure of fertilization in a cryptomonad. *Journal of Phycology* 24:385–393.

Kugrens, P., and R. E. Lee. 1990. Ultrastructural evidence for bacterial incorporation and myxotrophy in the photosynthetic cryptomonad *Chroomonas pochmanni* Huber-Pestalozzi (Cryptomonadida). *Journal of Protozoology* 37:263–267.

Kugrens, P., R. E. Lee, and R. A. Andersen. 1986. Cell form and surface patterns in *Chroomonas* and *Cryptomonas* cells (Cryptophyta) as revealed by scanning electron microscopy. *Journal of Phycology* 22:512–522.

Kühlbrandt, W., D. N. Wang, and Y. Fujiyoshi. 1994. Atomic model of plant light-harvesting complex by electron crystallography. *Nature* 367:614–621.

Kühn, S. F. 1998. Infection of *Coscinodiscus* spp. by the parasitoid nanoflagellate *Pirsonia diadema*: II Selective infection behaviour for host species and individual host cells. *Journal of Plankton Research* 20:443–454.

Kupferberg, S. J., J. C. Marks, and R. E. Power. 1994. Effects of variation in natural algal and detrital diets on larval anuran (*Hyla regilla*) life-history traits. *Copeia* 1994:446–457.

Kurogi, M. 1972. Systematics of *Porphyra* in Japan. Pages 167–191 in *Contributions to the Systematics of Benthic Marine Algae of the North Pacific*, edited by I. A. Abbott and M. Kurogi. Japanese Society of Phycology, Kobe, Japan

Kurtzman, A. L., and D. P. Cheney. 1991. Direct gene transfer and transient gene expression in a marine red alga using the biolistic method. *Journal of Phycology* 27(Supplement):42 (abstract).

Kwiatkowska, M., and J. Maszewski. 1986. Changes in the occurrence and ultrastructure of plasmodesmata in antheridia of *Chara vulgaris* L. during different stages of spermatogenesis. *Protoplasma* 132:179–188.

Lackey, J. B. 1968. Ecology of *Euglena*. Pages 28-44 in *The Biology of Euglena*, Vol. I, edited by D. E. Buetow. Academic Press, New York, NY.

La Claire, J. W. 1981. Occurrence of plasmodesmata during infurrowing in a brown alga. *Biology of the Cell* 40:139–142.

La Claire, J. W. 1982a. Cytomorphological aspects of wound healing in selected Siphonocladales (Chlorophyceae). *Journal of Phycology* 18:379–384.

La Claire, J. W. 1982b. Light and electron microscopic studies of growth and reproduction in *Cutleria* (Phaeophyta). III. Nuclear division in the trichothallic meristem of *C. cylindrica*. *Phycologia* 21:273–287.

La Claire, J. W. 1991. Immunolocalization of myosin in intact and wounded cells of the green alga *Ernodesmis verticillata* (Kützing) Borgesen. *Planta* 184:209–217.

La Claire, J. W., R. Chen, and D. L. Herrin. 1995. Identification of a myosin-like protein in *Chlamydomonas reinhardtii* (Chlorophyta). *Journal of Phycology* 31:302–306.

La Claire, J. W., and J. A. West. 1979. Light- and electron-microscopic studies of growth and reproduction in *Cutleria* (Phaeophyta). II. Gametogenesis in the male plant of *C. hancockii*. *Protoplasma* 101:247–267.

La Claire, J. W., G. C. Zuccarello, and S. Tong. 1997. Abundant plasmid-like DNA in various members of the orders Siphonocladales and Cladophorales (Chlorophyta). *Journal of Phycology* 33:830–837.

Lamberti, G. A. 1996. The role of periphyton in benthic food webs. Pages 533–572 in *Algal Ecology: Freshwater Benthic Ecosystems*, edited by R. J. Stevenson, M. L. Bothwell, and R. L. Lowe. Academic Press, San Diego, CA.

Lampert, W. 1987. Predictability in lake ecosystems: The role of biotic interactions. Pages 333–346 in *Potentials and Limitations of Ecosystem Analysis*, edited by E. D. Schulze and H. Zwolfer. *Ecological Studies* 61, Springer-Verlag, Berlin, Germany.

Lampert, W., W. Fleckner, H. Rai, and B. E. Taylor. 1986. Phytoplankton control by grazing zooplankton: A study on the spring clear-water phase. *Limnology and Oceanography* 31:478–490.

Lampert, W., and U. Sommer. 1997. *Limnoecology*. Oxford University Press, New York, NY.

Lancelot, C., M. D. Keller, V. Rosseau, W. O. Smith, and S. Mathot. 1998. Autecology of the marine haptophyte *Phaeocystis* sp. Pages 209–224 in *Physiological Ecology of Harmful Algal Blooms*, edited by D. M. Anderson, A. D. Cembella, and G. M. Hallegraeff. Springer-Verlag, Berlin, Germany.

Lang, N. J., and B. A. Whitton. 1973. Arrangement and structure of thylakoids. Pages 66–79 in *The Biology of Blue-Green Algae*, edited by N. G. Carr and B. A. Whitton. University of California Press, Berkeley, CA.

Lange, M., L. Guillou, D. Vaulot, N. Simon, R. I. Amamm, W. Ludwig, and L. K. Medlin. 1996. Identification of the class Prymnesiophyceae and the genus *Phaeocystis* with ribosomal RNA-targeted nucleic acid probes detected by flow cytometry. *Journal of Phycology* 32:858–868.

Larkum, A. W. D., C. Scaramuzzi, G. C. Cox, R. G. Hiller, and A. G. Turner. 1994. Light-harvesting chlorophyll *c*-like pigment in *Prochloron*. *Proceedings of the National Academy of Science* 91:679–683.

Larkum, T., and C. J. Howe. 1997. Molecular aspects of light-harvesting processes in algae. *Advances in Botanical Research* 27:258–330.

LaRoche, J., G. W. M. van der Staay, P. Partensky, A. Ducret, R. Aebersold, R. Li, S. S. Golden, R. G. Hiller, P. M. Wrench, A. W. D. Larkum, and B. R. Green. 1996. Independent evolution of the prochlorophyte and green plant chlorophyll *a/b* light-harvesting proteins. *Proceedings of the National Academy of Science* 93:15244–15248.

Larsen, A., and L. K. Medlin. 1997. Inter- and intraspecific genetic variation in twelve *Prymnesium* (Haptophyceae) clones. *Journal of Phycology* 33:1007–1015.

Larson, A., M. M. Kirk, and D. L. Kirk. 1992. Molecular phylogeny of the volvocine flagellates. *Molecular Biology and Evolution* 9:83–105.

Lavau, S., G. W. Saunders, and R. Wetherbee. 1997. A phylogenetic analysis of the Synurophyceae using molecular data and scale case morphology. *Journal of Phycology* 33:135–151.

Lawlor, D. W. 1993. *Photosynthesis: Molecular, physiological, and environmental processes*. Longman Scientific, London, UK.

Laws, E. A., R. R. Bidigare, and B. N. Popp. 1997. Effect of growth rate and CO_2 concentration on carbon isotopic fractionation by the marine diatom *Phaeodactylum tricornutum*. *Limnology and Oceanography* 42:1552–1560.

Leadbeater, B. S. C. 1994. Cell coverings. Pages 23–46 in *The Haptophyte Algae*, edited by J. C. Green and B.

S. C. Leadbeater. Clarendon Press, Oxford, UK.

Leadbeater, B. S. C., and D. A. N. Barker. 1995. Biomineralization and scale production in the Chrysophyta. Pages 141–164 in *Chrysophyte Algae: Ecology, Phylogeny and Development*, edited by C. D. Sandgren, J. P. Smol, and J. Kristiansen. Cambridge University Press, Cambridge, UK.

Leadbeater, B. S. C., and J. Green. 1993. *Ultrastructure of Microalgae*, edited by T. Berner. CRC Press, Boca Raton, FL.

League, E. A., and V. A. Greulach. 1955. Effects of daylength and temperature on the reproduction of *Vaucheria sessilis*. *The Botanical Gazette* 117:45–51.

Lebednik, P. A. 1977. Postfertilization development in *Clathromorphum, Melobesia, and Mesophyllum* with comments on the evolution of the Corallinaceae and the Cryptonemiales (Rhodophyta). *Phycologia* 16:379–406.

Lee, D.-H., M. Mittag, S. Sczekan, D. Morse, and J. W. Hastings. 1993. Molecular cloning and genomic organization of a gene for luciferin-binding protein from the dinoflagellate *Gonyaulax polyedra*. *Journal of Biological Chemistry* 268:8842–8850.

Lee, J. J. 1992a. Symbiosis in Foraminifera. Pages 63–78 in *Algae and Symbioses*, edited by W. Reisser. BioPress, Bristol, UK.

Lee, J. J. 1992b. Taxonomy of algae symbiotic in Foraminifera. Pages 79–92 in *Algae and Symbioses*, edited by W. Reisser. BioPress, Bristol, UK.

Lee, K. W., and H. C. Bold. 1973. *Pseudocharaciopsis texensis* gen. et sp. nov., a new member of the Eustigmatophyceae. *British Phycological Journal* 8:31–37.

Lee, S. H., T. Motomura, and T. Ichimura. 1998. Karyogamy follows plasmogamy in the life cycles of *Derbesia tenuissima* (Chlorophyta). *Phycologia* 37:330–333.

Lee, Y.-K., and S.-Y. Ding. 1995. Effect of dissolved oxygen partial pressure on the accumulation of astaxanthin in chemostat cultures of *Haematococcus lacustris* (Chlorophyta). *Journal of Phycology* 31:922–924.

Leedale, G. F. 1967. *Euglenoid Flagellates*, Prentice Hall, Englewood Cliffs, NJ.

Leedale, G. F. 1985. Euglenida Butschli 1884. Pages 41–50 in *An Illustrated Guide to the Protozoa*, edited by J. J. Lee, S. H. Hutner, and E. C. Bovee. Society of Protozoologists, Lawrence, KS.

Leedale, G. F., B. S. C. Leadbeater, and A. Massalski. 1970. The intracellular origin of flagellar hairs in the Chrysophyceae and Xanthophyceae. *Journal of Cell Science* 6:701–719.

Lehman, J. T. 1976. Ecological and nutritional studies on *Dinobryon* Ehrenb.: Seasonal periodicity and the phosphate toxicity problem. *Limnology and Oceanography* 21:646–658.

Lehman, J. T., and D. Scavia. 1982. Microscale patchiness of nutrients in plankton communities. *Science* 216:729–730.

Lehman, R. L., and J. R. Manhart. 1997. A preliminary comparison of restriction fragment patterns in the genus *Caulerpa* (Chlorophyta) and the unique structure of the chloroplast genome of *Caulerpa sertularioides*. *Journal of Phycology* 33:1055–1062.

Leigh, E. G., R. T. Paine, J. T. Quinn, and T. H. Suchanek. 1987. Wave energy and intertidal productivity. *Proceedings of the National Academy of Science* 84:1314–1318.

Leitsch, C. E.W., K. W. Kowallik, and S. Douglas. 1999. The *atpA* gene cluster of *Guillardia theta* (Cryptophyta): A piece in the puzzle of chloroplast genome evolution. *Journal of Phycology* 35:128–135.

Lembi, C. A., and W. R. Herndon. 1966. Fine structure of the pseudocilia of *Tetraspora*. *Canadian Journal of Botany* 44:710–712.

Lembi, C. A., S. W. O'Neal, and D. F. Spencer. 1988. Algae as weeds: Economic impact, ecology, and management alternatives. Pages 455–481 in *Algae and Human Affairs*, edited by C. A. Lembi and J. R. Waaland. Cambridge University Press, Cambridge, UK.

Lembi, C. A., N. L. Perlmutter, and D. F. Spencer. 1980. *Life Cycle, Ecology, and Management Considerations of the Green Filamentous Alga, Pithophora*. Technical Report 130. Purdue University Water Resources Center, W. Lafayette, IN.

Leonardi, P. I., E. J. Cáceres, and C. G.Vélez. 1998. Fine structure of dwarf males in *Oedogonium pluviale* (Chlorophyceae). *Journal of Phycology* 34:250–256.

Leventer, A. 1998. The fate of Antarctic "sea ice diatoms" and their use as paleoenvironmental indicators. Pages 121–127 in *Antarctic Sea Ice Biological Processes, Interactions, and Variability*, edited by M. P. Lizotte and K. R. Arrigo. Antarctic Research Series 73, American Geophysical Union, Washington, DC.

Levin, E. D., D. E. Schmechel, J. M. Burkholder, H. M. Glasgow, N. J. Deamer-Melia, V. C. Moser, and G. J. Harry. 1997. Persisting learning deficits in rats after exposure to *Pfiesteria piscicida*. *Environmental Health Perspectives* 105:1320–1325.

Lew, K. A., and J R Manhart. 1993. The *rps12* gene in *Spirogyra maxima* (Chlorophyta) and its evolutionary significance. *Journal of Phycology* 29:500–505.

Lewin, J. C. 1953. Heterotrophy in diatoms. *Journal of General Microbiology* 9:305–313.

Lewin, J. C., and R. A. Lewin. 1960. Auxotrophy and heterotrophy in marine littoral diatoms. *Canadian Journal of Microbiology* 6:127–134.

Lewin, J. C., and R. E. Norris. 1970. Surf-zone diatoms of the coasts of Washington and New Zealand (*Chaetoceros armatum* T. West and *Asterionella* spp). *Phycologia* 9:142–149.

Lewin, R. A. 1976. Naming the blue-greens. *Nature* 259:360.

Lewin, R. A., and L. Cheng. 1989. Collection and handling of *Prochloron* and its hosts in *Prochloron, a Microbial Enigma*, edited by R. A. Lewin and L. Cheng. Chapman & Hall, London, UK.

Lewis, J., and R. Hallett. 1997. *Lingulodinium polyedrum* (*Gonyaulax polyedra*) a blooming dinoflagellate. *Oceanography and Marine Biology: An Annual Review* 35:97–161.

Lewis, J. R. 1964. *The Ecology of Rocky Shores*. English Universities Press, London, UK.

Lewis, L. A. 1997. Diversity and phylogenetic placement of *Bracteacoccus tereg* (Chlorophyceae, Chlorophyta) based on 18S ribosomal RNA gene sequence data. *Journal of Phycology* 33:279–285.

Lewis, L. A., B. D. Mishler, and R. Vilgalys. 1997. Phylogenetic relationships of the liverworts (Hepaticae), a basal embryophyte lineage, inferred from nucleotide sequence data of the chloroplast gene *rbcL*. *Molecular Phylogenetics and Evolution* 7:377–393.

Lewis, L. A., L. W. Wilcox, P. A. Fuerst, and G. L. Floyd. 1992. Concordance of molecular and ultrastructural data in the study of zoosporic chlorococcalean green algae. *Journal of Phycology* 28:375–380.

Lewis, M. A. 1990. Are laboratory-derived toxicity data for freshwater algae worth the effort? *Environmental Toxicology and Chemistry* 9:1279–1284.

Lewis, R. J. 1995. Gametogenesis and chromosome number in *Postelsia palmaeformis* (Laminariales, Phaeophyceae). *Phycological Research* 43:61–64.

Lewis, R. J., S. I. Jensen, D. M. DeNicola, V. I. Miller, K. D. Hoagland, and S. G. Ernst. 1997. Genetic variation in the diatom *Fragilaria capucina* (Fragilariaceae) along a latitudinal gradient across North America. *Plant Systematics and Evolution* 204:99–108.

Lewis, R. J., and M. Neushul. 1995. Intergeneric hybridization among five genera of the family Lessoniaceae (Phaeophyceae) and evidence for polyploidy in a fertile *Pelagophycus × Macrocystis* hybrid. *Journal of Phycology* 31:1012–1017.

Lewis, R. J. 1996a. Chromosomes of the brown algae. *Phycologia* 35:19–40.

Lewis, R. J. 1996b. Hybridization of brown algae: Compatibility and speciation. Pages 275–289 in *Cytology, Genetics and Molecular Biology of Algae*, edited by B. R. Chaudhary and S. B. Agrawal. SPB Academic Publishing, Amsterdam, Netherlands.

Lewitus, A. J., H. B. Glasgow, and J. M. Burkholder. 1999. Kleptoplastidy in the toxic dinoflagellate *Pfiesteria piscicida* (Dinophyceae). *Journal of Phycology* 35:303–312.

Lewitus, A. J., and T. M. Kana. 1994. Responses of estuarine phytoplankton to exogenous glucose: Stimulation versus inhibition of photosynthesis and respiration. *Limnology and Oceanography*

39:182–189.

Li, R., M. Watanabe, and M. M. Watanabe. 1997. Akinete formation in plankton *Anabaena* spp. (Cyanobacteria) by treatment with low temperature. *Journal of Phycology* 33:576–584.

Lichtlé, C., W. Arselane, J. C. Duval, and C. Passaquet. 1995. Characterization of the light harvesting complex of *Giraudyopsis stellifer* (Chrysophyceae) and effects of light stress. *Journal of Phycology* 31:380–387.

Liddle, L. B., S. Berger, and H. Schweiger. 1976. Ultrastructure during development of the nucleus of *Batophora oerstedtii* (Chlorophyta, Dasycladaceae). *Journal of Phycology* 12:261–272.

Lindstrom, S. 1987. Possible sister groups and phylogenetic relationships among selected North Pacific and North Atlantic Rhodophyta. *Helgolander Meeresunters* 41:245–260.

Littler, D. S., and M. M. Littler. 1990. Systematics of *Udotea* species (Bryopsidales, Chlorophyta) in the tropical western Atlantic. *Phycologia* 29:206–252.

Littler, D. S., and M. M. Littler. 1991. Systematics of *Anadyomene* species (Anadyomenaceae, Chlorophyta) in the tropical western Atlantic. *Journal of Phycology* 27:101–118.

Littler, M. M., and D. S. Littler. 1980. The evolution of thallus form and survival strategies in benthic marine macroalgae: Field and laboratory tests of a functional form model. *American Naturalist* 116:25–43.

Littler, M. M., and D. S. Littler. 1983. Heteromorphic life-history strategies in the brown alga *Scytosiphon lomentaria* (Lyngb.) Link. *Journal of Phycology* 19:425–431.

Littler, M. M., and D. S. Littler. 1995. Impact of CLOD pathogen on Pacific coral reefs. *Science* 267:1356–1360.

Littler, M. M., D. S. Littler, S. M. Blair, and J. N. Norris. 1985. Deepest known plant life discovered on an uncharted seamount. *Science* 227:57–59.

Littler, M. M., D. S. Littler, S. M. Blair, and J. N. Norris. 1986. Deep-water plant communities from an uncharted seamount off San Salvador Island, Bahamas: Distribution, abundance, and primary productivity. *Deep-Sea Research* 33:881–92.

Littler, M. M., D. S. Littler, and P. R. Taylor. 1983. Evolutionary strategies in a tropical barrier reef system: Functional form groups of marine macroalgae. *Journal of Phycology* 19:229–237.

Livingstone, D., and G. H. M. Jaworski. 1980. The viability of akinetes of blue-green algae recovered from the sediments of Rostherne Mere. *British Phycological Journal* 15:357–64.

Lizotte, M. P., D. H. Robinson, and C. W. Sullivan. 1998. Algal pigment signatures in Antarctic sea ice. Pages 93–105 in *Antarctic Sea Ice Biological Processes, Interactions, and Variability*, edited by M. P. Lizotte, and K. R. Arrigo. Antarctic Research Series 73, American Geophysical Union, Washington, DC.

Lobban, C. S. 1978a. The growth and death of the *Macrocystis* sporophyte (Phaeophyceae, Laminariales). *Phycologia* 17:196–212.

Lobban, C. S. 1978b. Translocation of ^{14}C in *Macrocystis* (giant kelp). *Plant Physiology* 61:585–589.

Lobban, C. S., and P. J. Harrison. 1994. *Seaweed Ecology and Physiology*. Cambridge University Press, Cambridge, UK.

Loeblich, A. R. 1976. Dinoflagellate evolution: Speculation and evidence. *Journal of Protozoology* 23:13–28.

Loiseaux, S. 1967. Sur la position systématique du genre *Giraudyopsis* P. Dangeard. *Revue Génerale Botanique* 74:389–396.

Lokhorst, G. M. 1996. Comparative taxonomic studies on the genus *Klebsormidium* (Charophyceae) in Europe. *Cryptogamic Studies* 5:1–132.

Lokhorst, G. M., P. J. Segaar, and W. Star. 1989. An ultrastructural reinvestigation of mitosis and cytokinesis in cryofixed sporangia of the coccoid green alga *Friedmannia israelensis* with special reference to septum formation and the replication cycle of basal bodies. *Cryptogamic Botany* 1:275–294.

Lokhorst, G. M., P. J. Sluiman, and W. Star. 1988. The ultrastructure of mitosis and cytokinesis in the sarcinoid *Chlorokybus atmophyticus* (Chlorophyta,

Charophyceae) revealed by rapid freeze fixation and freeze substitution. *Journal of Phycology* 24:237–248.

Lopez-Figueroa, F., P. Lindemann, S. E. Braslavsky, K. Schaffaer, H. A. W. Schneider-Poetsch, and W. Rudiger. 1989. Detection of a phytochrome-like protein in macroalgae. *Botanica Acta* 102:178–180.

Lopez-Figueroa, F., and F. X. Niell. 1989. Red light and blue light photoreceptors controlling chlorophyll *a* synthesis in the red alga *Porphyra umbilicalis* and in the green alga *Ulva rigida*. *Physiologia Plantarum* 76:391–397.

Lorch, D. W., and M. Engels. 1979. Observations on filament formation in *Micrasterias foliacea* (Desmidiaceae, Chlorophyta). *Journal of Phycology* 15:322–325.

Lotka, A. J. 1932. The growth of mixed populations: two species competing for a common food supply. *Journal of the Washington Academy of Science* 22:461.

Love, J., C. Brownlee, and A. J. Trewavas. 1997. Ca^{2+} and calmodulin dynamics during photopolarization in *Fucus serratus* zygotes. *Plant Physiology* 115:249–261.

Lowe, R. L. 1996. Periphyton patterns in lakes. Pages 57–76 in *Algal Ecology. Freshwater Benthic Ecosystems*, edited by R. J. Stevenson, M. L. Bothwell, and R. L. Lowe. Academic Press, New York, NY.

Lowe, R. L., and Y. Pan. 1996. Benthic algal communities as biological monitors. Pages 705–739 in *Algal Ecology. Freshwater Benthic Ecosystems*, edited by R. J. Stevenson, M. L. Bothwell, and R. L. Lowe. Academic Press, New York, NY.

Lu, M., and G. C. Stephens. 1984. Demonstration of net influx of free amino acids in *Phaeodactylum tricornutum* using high performance liquid chromatography. *Journal of Phycology* 20:584–589.

Lubchenco, J. 1980. Algal zonation in the New England rocky intertidal community: An experimental analysis. *Ecology* 61:333–334.

Lubchenco, J. 1982. Effects of grazers and algal competitors of fucoid colonization in tide pools. *Journal of Phycology* 18:544–550.

Lucas, W. J. 1995. Plasmodesmata: Intercellular channels for macromolecular transport in plants. *Current Opinion in Cell Biology* 7:673–680.

Lucas, W. J., F. Brechnignac, T. Mimura, and J. W. Oross. 1989. Charasomes are not essential for photosynthetic utilization of exogenous HCO_3^- in *Chara corallina*. *Protoplasma* 151:106–114.

Ludwig, M., J. L. Lind, E. A. Miller, and R. Wetherbee. 1996. High molecular mass glycoproteins associated with the siliceous scales and bristles of *Mallomonas splendens* (Synurophyceae) may be involved in cell surface development and maintenance. *Planta* 199:219–228.

Lund, J. W. G. 1964. Primary productivity and periodicity of phytoplankton. *Internationale Vereinigung für Theoretische und Angewandte Limnologie* 15:37–56.

Lüning, K. 1990. *Seaweeds: Their Environment, Biogeography and Ecophysiology*, John Wiley, New York, NY.

Luykx, P., M. Hoppenrath, and D. G. Robinson. 1997a. Structure and behavior of contractile vacuoles in *Chlamydomonas reinhardtii*. *Protoplasma* 198:73–84.

Luykx, P., M. Hoppenrath, and D. G. Robinson. 1997b. Osmoregulatory mutants that affect the function of the contractile vacuole in *Chlamydomonas reinhardtii*. *Protoplasma* 200:99–111.

Lylis, J. C., and F. R. Trainor. 1973. The heterotrophic capabilities of *Cyclotella meneghiniana*. *Journal of Phycology* 9:365–369.

Lyons, E. M., and T. Thiel. 1995. Characterization of *nifB*, *nifS*, and *nifU* genes in the cyanobacterium *Anabaena variabilis*: *nifB* is required for the vanadium-dependent nitrogenase. *Journal of Bacteriology* 177:1570–1575.

Ma, J., and A. Miura. 1984. Observations on the nuclear division in conchospores and their germlings in *Porphyra yezoensis* Ueda. *Japanese Journal of Phycology* 32:373–378.

Maberly, S. C., J. A. Raven, and A. M. Johnston. 1992. Discrimination between ^{12}C and ^{13}C by marine plants. *Oecologia* 91:481–492.

MacArthur, R. H., and E. O. Wilson. 1967. *The Theory of Island Biogeography*. Princeton University Press, Princeton, NJ.

MacKay, N. A., and J. J. Elser. 1998. Nutrient recycling by *Daphnia* reduces N_2 fixation by cyanobacteria. *Limnology and Oceanography* 43:347–354.

Machlis, L. 1973. The effects of bacteria on the growth and reproduction of *Oedogonium cardiacum*. *Journal of Phycology* 9:342–344.

Machlis, L., G. C. Hill, K. E. Steinback, and W. Reed. 1974. Some characteristics of the sperm attractant in *Oedogonium cardiacum*. *Journal of Phycology* 10:199–204.

Madsen, T. V., and S. C. Maberly. 1990. A comparison of air and water as environments for photosynthesis by the intertidal alga *Fucus spiralis* (Phaeophyta). *Journal of Phycology* 26:24–30.

Maidek, B. L., G. J. Olson, N. Larsen, R. Overbeck, M. J. McCaughey, and C. R. Woese. 1997. Ribosomal database project. *Nucleic Acids Research* 25:109–111.

Maier, I. 1997a. The fine structure of the male gamete of *Ectocarpus siliculosus* (Ectocarpales, Phaeophyceae). I. General structure of the cell. *European Journal of Phycology* 32:241–253.

Maier, I. 1997b. The fine structure of the male gamete of *Ectocarpus siliculosus* (Ectocarpales, Phaeophyceae). II. The flagellar apparatus. *European Journal of Phycology* 32:255–266

Maier, I., and D. G. Müller. 1998. Virus binding to brown algal cells as visualized by DAPI fluorescence microscopy. *Phycologia* 37:60–63.

Maier, I., and C. E. Schmid. 1995. An immunofluorescence study on lectin binding sites in gametes of *Ectocarpus siliculosus* (Ectocarpales, Phaeophyceae), *Phycological Research* 43:33–42.

Malin, G., and G. O. Kirst. 1997. Algal production of dimethyl sulfide and its atmospheric role. *Journal of Phycology* 33:889–896.

Malin, G., S. M. Turner, and P. S. Liss. 1992. Sulfur: The plankton/climate connection. *Journal of Phycology* 28:590–597.

Mandoli, D. F. 1998. Elaboration of body plan and phase change during development of *Acetabularia*: How is the complex architecture of a giant unicell built? *Annual Review of Plant Physiology and Plant Molecular Biology* 49:173–198.

Manhart, J. 1994. Phylogenetic analysis of green plant *rbcL* sequences. *Molecular Phylogenetic Evolution* 13:114–127.

Manhart, J. R., and R. M. McCourt. 1992. Molecular data and species concepts in the algae. *Journal of Phycology* 28:730–737.

Manhart, J. R., and J. D. Palmer. 1990. The gain of two chloroplast tRNA introns marks the green algal ancestors of land plants. *Nature* 345:268–270.

Mann, D. G., and H. J. Marchant. 1989. The origins of the diatom and its life cycle. Pages 307–323 in *The Chromophyte Algae: Problems and Perspectives*, edited by J. C. Green, B. S. C. Leadbeater, and W. L. Diver. *Systematics Association Special Volume* 38, Clarendon Press, Oxford, UK.

Manton, I., and H. Ettl. 1965. Observations on the fine structure of *Mesostigma viride* Lauterborn. *Botanical Journal of the Linnean Society* 59:175–184.

Maranda, L., and Y. Shimizu. 1996. *Amphidinium operculatum* var. nov. *gibbosum* (Dinophyceae), a free-swimming marine species producing cytotoxic metabolites. *Journal of Phycology* 32:873–879.

Marchant, H. J. 1977. Ultrastructure, development and cytoplasmic rotation of seta-bearing cells of *Coleochaete scutata* (Chlorophyceae). *Journal of Phycology* 13:28–36.

Marchant, H. J. 1998. Life in the snow: Algae and other microorganisms. Pages 83–97 in *Snow (A Natural History; an Uncertain Future)*, edited by K. Green. Australian Alps Liaison Committee, Canberra, Australia.

Marchant, H. J., and A. McEldowney. 1986. Nanoplanktonic siliceous cysts from Antarctica are algae. *Marine Biology* 92:53–57.

Marchant, H. J., and J. D. Pickett-Heaps. 1972. Ultrastructure and differentiation of *Hydrodictyon reticulatum*. III. Formation of the vegetative daughter net.

Australian Journal of Biological Science 25:265–278.

Marchant, H. J., and J. D. Pickett-Heaps. 1973. Mitosis and cytokinesis in *Coleochaete scutata*. *Journal of Phycology* 9:461–471.

Marchant, H. J., and J. D. Pickett-Heaps. 1974. The effect of colchicine on colony formation in the algae *Hydrodictyon*, *Pediastrum*, and *Sorastrum*. *Planta* 116:291–300.

Marchant, H. J., J. D. Pickett-Heaps, and K. Jacobs. 1973. An ultrastructural study of zoosporogenesis and the mature zoospore of *Klebsormidium flaccidum*. *Cytobios* 8:95–107.

Marchant, H. J., and H. A. Thomsen. 1994. Haptophytes in polar waters. Pages 209–228 in *The Haptophyte Algae*, edited by J. C. Green and B. S. C. Leadbeater. Clarendon Press, Oxford, UK.

Margulis, L., and K. V. Schwartz. 1988. *Five Kingdoms*: *An Illustrated Guide to the Phyla of Life on Earth*. W. H. Freeman and Company, New York, NY.

Marin, B., M. Klingberg, and M. Melkonian. 1998. Phylogenetic relationsips among the Cryptophyta: Analysis of nuclear-encoded SSU rDNA sequences support the monophyly of extant plastid-containing lineages. *Protist* 149:265–276.

Markey, D. R., and R. T. Wilce. 1975. The ultrastructure of reproduction in the brown alga *Pylaiella littoralis*. I. Mitosis and cytokinesis in the plurilocular gametangia. *Protoplasma* 85:219–241.

Marks, J. C., and M. P. Cummings. 1996. DNA sequence variation in the ribosomal internal transcribed spacer region of freshwater *Cladophora* species (Chlorophyta). *Journal of Phycology* 32:1035–1042.

Martin, J., and S. E. Fitzwater. 1988. Iron deficiency limits phytoplankton growth in the north-east Pacific subarctic. *Nature* 331:341–343.

Martin, J. H., K. H. Coale, K. S. Johnson, S. E. Fitzwater, R. M. Gordon, S. J. Tanner, C. N. Hunter, V. A. Elrod, J. L. Nowicki, T. L. Coley, R. T. Barber, S. Lindley, A. J. Watson, K. Van Scoy, C. S. Law, M. I. Liddlcoat, R. Ling, T. Stanton, J. Stockel, C. Collins, A. Anderson, R. Bidigare, M. Ondrusek, M. Latasa, F. J. Millero, K. Lee, W. Yao, J. Z. Zhang, G. Friederich, C. Sakamoto, F. Chavez, K. Buck, Z. Kolber, R. Greene, P. Falkowski, S. W. Chisholm, F. Hoge, R. Swift, J. Yungel, S. Turner, P. Nightingale, A. Hatton, P. Liss, and N. W. Tindale. 1994. Testing the iron hypothesis in ecosystems of the equatorial Pacific Ocean. *Nature* 371:123–129.

Martin, W., and R. G. Herrmann. 1998. Gene transfer from organelles to the nucleus: How much, what happens, and why? *Plant Physiology* 118:9–17.

Martínez, L., M. W. Silver, J. M. King, and A. L. Alldredge. 1983. Nitrogen fixation by floating diatom mats: A source of new nitrogen to oligotrophic ocean waters. *Science* 221:152–154.

Massalski, A., and G. F. Leedale. 1969. Cytology and ultrastructure of the Xanthophyceae. I. Comparative morphology of the zoospores of *Bumilleria sicula* Borzi and *Tribonema vulgare* Pascher. *British Phycological Journal* 4:159–180.

Matthijs, H. C. P., T. Burger-Wiersma, and L. R. Mur. 1989. A status report on *Prochlorothrix hollandica*, a free-living prochlorophyte. Pages 83–87 in *Prochloron*, edited by R. A. Lewin and L. Cheng. Chapman & Hall, London, UK.

Mattox, K. R., and K. D. Stewart. 1984. A classification of the green algae: A concept based on comparative cytology. Pages 29–72 in *Systematics of the Green Algae*, edited by D. E. G. Irvine, and D. M. John. Academic Press, London, UK.

McBride, G. E. 1970. Cytokinesis in the green alga *Fritschiella*. *Nature* 216:939.

McBride, G. E. 1974. The seta-bearing cells of *Coleochaete scutata* (Chlorophyceae, Chaetophorales). *Phycologia* 13:271–285.

McCandless, E. 1981. Polysaccharides of the seaweeds. Pages 559–588 in *The Biology of Seaweeds*, edited by C. S. Lobban and M. J. Wynne. Blackwell Scientific, Oxford, UK.

McCauley, E., and J. Kalff. 1981. Empirical relationships between phytoplankton and zooplankton biomass in lakes. *Canadian Journal of Fisheries and Aquatic Science* 38:458–463.

McConnaughey, T. A. 1994. Calcification, photosynthesis, and global carbon cycles. *Bulletin de l'Institut Océanographique, Monaco* 13:137–161.

McConnaughey, T. A. 1998. Acid secretion, calcification, and photosynthetic carbon concentrating mechanisms. *Canadian Journal of Botany* 76:1119–1126.

McConnaughey, T. A., and J. F. Whelan. 1996. Calcification generates protons for nutrient and bicarbonate uptake. *Earth Science Reviews* 42:95–118.

McCourt, R. M. 1995. Green algal phylogeny. *Trends in Ecology and Evolution* 10:159–163.

McCourt, R. M., R. W. Hoshaw, and J.-C. Wang. 1986. Distribution, morphological diversity and evidence for polyploidy in North American Zygnemataceae (Chlorophyta). *Journal of Phycology* 22:307–313.

McCourt, R. M., K. G. Karol, S. Kaplan, and R. W. Hoshaw. 1995. Using *rbcL* sequences to test hypotheses of chloroplast and thallus evolution in conjugating green algae (Zygnematales, Charophyceae). *Journal of Phycology* 31:989–995.

McCourt, R. M., K. G. Karol, M. Guerlesquin, and M. Feist. 1996a. Phylogeny of extant genera in the family Characeae (Charales, Charophyceae) based on *rbcL* sequences and morphology. *American Journal of Botany* 83:125–131

McCourt, R. M., S. T. Meiers, K. G. Karol, and R. L. Chapman. 1996b. Molecular systematics of the Charales. Pages 323–336 in *Cytology, Genetic and Molecular Biology of Algae*, edited by B. R. Chaudhary, and S. B. Agrawal. SPB Academic Publishing, Amsterdam, Netherlands.

McDonald, K. 1972. The ultrastructure of mitosis in the marine red alga *Membranoptera platyphylla*. *Journal of Phycology* 8:156–166.

McFadden, G. I. 1993. Second-hand chloroplasts: Evolution of cryptomonad algae. *Advances in Botanical Research* 19:190–230.

McFadden, G., and P. Gilson. 1995. Something borrowed, something green: Lateral transfer of chloroplasts by secondary endosymbiosis. *Trends in Ecology and Evolution* 10:12–17.

McFadden, G. I., P. R. Gilson, and D. R. A. Hill. 1994. *Goniomonas* rRNA sequences indicate that this phagotrophic flagellate is a close relative of the host component of cryptomonads. *European Journal of Phycology* 29:29–32.

McFadden, G. I., M. E. Reith, J. Mulholland, and N. Lang-Unnasch. 1996. Plastid in human parasites. *Nature* 381:482.

McFadden, G. I., and R. Wetherbee 1982. Serial reconstruction of the mitochondrial reticulum in the Antarctic flagellate, *Pyramimonas gelidicola* (Prasinophyceae, Chlorophyta). *Protoplasma* 111:79–82.

McIntyre, A., and B. Molfino. 1996. Forcing of Atlantic equatorial and subpolar millenial cycles by precession. *Science* 274:1867–1870.

McKenzie, C. H., D. Deibel, M. A. Paranjape, and R. J Thompson. 1995. The marine mixotroph *Dinobryon balticum* (Chrysophyceae): Phagotrophy and survival in a cold ocean. *Journal of Phycology* 31:19–24.

McLachlan, J. L., M. R. Seguel, and L. Fritz. 1994. *Tetreutreptia pomquetensis* gen. et sp. nov. (Euglenophyceae). A quadriflagellate, phototrophic marine eugenoid. *Journal of Phycology* 30:538–544.

McQueen, D. J. 1990. Manipulating lake community structure: where do we go from here? *Freshwater Biology* 23:613–620.

McQueen, D. J., M. R. S. Johannes, J. R. Post, T. J. Stewart, and D. R. S. Lean. 1989. Bottom-up and top-down impacts on freshwater pelagic community structure. *Ecological Monographs* 59:289–309.

McQuoid, M. R., and L. A. Hobson. 1996. Diatom resting stages. *Journal of Phycology* 32:889–902.

Mechling, J. A., and S. S. Kilham. 1982. Temperature effects on silicon limited growth of the Lake Michigan diatom *Stephanodiscus minutus* (Bacillariophyceae). *Journal of Phycology* 18:199–205.

Medlin, L. K., G. L. A. Barker, M. Baumann, P. K. Hayes, and M. Lange. 1994. Molecular biology and systematics. Pages 393–412 in *The Haptophyte Algae*, edited by J. C. Green and B. S. C. Leadbeater. Clarendon Press, Oxford, UK.

Medlin, L. K., R. M. Crawford, and R. A. Andersen. 1986. Histochemical and ultrastructureal evidence of the labiate process in the movement of centric diatoms. *British Phycological Journal* 21:297–301.

Medlin, L. K., H. J. Elwood, S. Stickel, and M. L. Sogin. 1991. Morphological and genetic variation within the diatom *Skeletonema costatum* (Bacillariophyceae): Evidence for a new species, *Skeletonema pseudo-costatum*. *Journal of Phycology* 27:514–524.

Medlin, L. K., W. H. C. K. Kooistra, R. Gersonde, and U. Wellbrock. 1996. Evolution of the diatoms (Bacillariophyta): III. Molecular evidence for the origin of the Thalassiosirales. *Nova Hedwigia* 112:221–234.

Medlin, L., and N. Simon. 1998. Phylogenetic analysis of marine phytoplankton. Pages 161–186 in *Molecular Approaches to the Study of the Oceans*, edited by K. E. Cooksey. Chapman & Hall, London, UK.

Medlin, L. K., H. C. Wiebe, C. F. Kooistra, D. Potter, G. W. Saunders, and R. A. Andersen. 1997. Phylogenetic relationship of the 'golden algae' haptophytes, heterokont chromophytes and their plastids. *Plant Systematics and Evolution* 11:187–219.

Meijer, M.-L., M. W. deHaan, A. W. Breukelaar, and H. Buiteveld. 1990. Is reduction of the benthivorous fish an important cause of high transparency following biomanipulation in shallow lakes? *Hydrobiologia* 200/201:303–315.

Meindl, U. 1983. Cytoskeletal contol of nuclear migration and anchoring in developing cells of *Micrasterias denticulata* and the change caused by the anti-microtubular herbicide amiprophos-methyl (APM). *Protoplasma* 118:75–90.

Meinesz, A. 1980. Connaisances actuelles et contribution á l'étude de la reproduction et du cycle des Udoteacées (Caulerpales, Chlorophytes). *Phycologia* 19:110–138.

Melkonian, M. 1975. The fine structure of the zoospores of *Fritschiella tuberosa* Iyeng. (Chaetophorineae, Chlorophyceae). *Protoplasma* 86:391–404.

Melkonian, M. 1989. Flagellar apparatus ultrastructure in *Mesostigma viride* (Prasinophyceae). *Plant Systematics and Evolution* 164:93–122.

Melkonian, M. 1990a. Chlorophyceae. Pages 608–616 in *Handbook of Protoctista*, edited by L. Margulis, J. O. Corliss, M. Melkonian, and D. J. Chapman. Jones & Bartlett Publishers, Boston, MA.

Melkonian, M. 1990b. Microthamniales. Pages 652–654 in *Handbook of Protoctista*, edited by L. Margulis, J. O. Corliss, M. Melkonian, and D. J. Chapman. Jones & Bartlett Publishers, Boston, MA.

Melkonian, M., P. L. Beech, C. Katsaros, and D. Schulze. 1992. Centrin-mediated cell motility in algae. Pages 179–221 in *Algal Cell Motility*, edited by M. Melkonian. Chapman & Hall, London, UK.

Melkonian, M., B. Marin, and B. Surek. 1995. Phylogeny and evolution of the algae. Pages 153–176 in *Biodiversity and Evolution*, edited by R. Arai, M. Kato, and Y. Doi. *The National Science Museum Foundation*, Tokyo, Japan.

Melkonian, M., and B. Surek. 1995. Phylogeny of the Chlorophyta: Congruence between ultrastructural and molecular evidence. *Bulletin de la Société zoologique de France* 120:191–208.

Menzel, D., H. Jonitz, and C. Elsner-Menzel. 1996. The perinuclear microtubule system in the green alga *Acetabularia*: anchor or motility device? *Protoplasma* 193:63–76.

Mercado, J. M., F. X. Niell, and F. L. Figueroa. 1997. Regulation of the mechanism for HCO_3^- use by the inorganic carbon level in *Porphyra leucosticta* Thur. in Le Jolis (Rhodophyta). *Planta* 201:319–325.

Meunier, P. C., M. S. Colón-López, and L. A. Sherman. 1998. Photosystem II cyclic heterogeneity and photoactivation in the diazotrophic, unicellular cyanobacterium *Cyanothece* species ATCC 51142. *Plant*

Physiology 116:1551–1562.

Meyer-Harms, B., and F. Pollehne. 1998. Alloxanthin in *Dinophysis norvegica* (Dinophysiales, Dinophyceae) from the Baltic Sea. *Journal of Phycology* 34:280–285.

Michaux-Ferrier, N., and I. Soulie-Märsche. 1987. The quantities of DNA in the vegetative nuclei of *Chara vulgaris* and *Tolypella glomerata* (Charophyta). *Phycologia* 26:435–442.

Mignot, J.-P. 1967. Structure et ultrastructure de quelques Chloromonadines. *Protistologica* 3:5–23.

Mignot, J.-P., L. Joyon, and E. G. Pringsheim. 1968. Complements a l'etude cytologique des Crypto-monadines. *Protistologica* 4:493–506.

Mille, D. F., O. M. Schofield, G. J. Kirkpatrick, G. Johnsen, P. A. Tester, and B. T. Vinyard. 1997. Detection of harmful algal blooms using photopigments and absorption signatures: A case study of the Florida red tide dinoflagellate, *Gymnodinium breve*. *Limnology and Oceanography* 42:1240–1251.

Miller, E. A., J. L. Lind, and R. Wetherbee. 1996. The scale-associated molecules of *Mallomonas* spp. (Synurophyceae) are immunologically distinct. *Nova Hedwigia* 114:45–55.

Miller, M. W. 1998. Coral/seaweed competition and the control of reef community structure within and between latitudes. *Oceanography and Marine Biology. An Annual Review* 36:65–96.

Miller, P. E., and C. A. Scholin. 1996. Identification of cultured *Pseudo-nitzschia* (Bacillariophyceae) using species-specific LSU rRNA-targeted fluorescent probes. *Journal of Phycology* 32:646–655.

Miller, P. E., and C. A. Scholin. 1998. Identification and enumeration of cultured and wild *Pseudo-nitzschia* (Bacillariophyceae) using species-specific LSU rRNA-targeted fluorescent probes and filter-based whole cell hybridization. *Journal of Phycology* 34:371–382.

Milligan, K. L., and E. M. Cosper. 1994. Isolation of virus capable of lysing the brown tide microalga, *Aureococcus anophagefferens*. *Science* 266:805807.

Millington, W. F. 1981. Form and pattern in *Pediastrum*. *Cell Biology Monographs* 8:94–118.

Mimura, T., R. Müller, W. M. Kaiser, T. Shimmen, and K.-J. Dietz. 1993. ATP-dependent carbon transport in perfused *Chara* cells. *Plant, Cell and Environment* 16:653–661.

Mishigeni, K. E., and D. J. Chapman. 1994. *Hypnea*, Gigartinales, Rhodophyta. Pages 245–281 in *Biology of Economic Algae*, edited by I. Akatsuka. SPB Academic Publishing, The Hague, Netherlands.

Mishler, B. D., L. A. Lewis, M. A. Buchheim, K. A. Renzaglia, D. J. Garbary, C. F. Delwiche, F. W. Zechman, T. S. Kantz, and R. L. Chapman. 1994. Phylogenetic relationships of the "green algae," and the "bryophytes." *Annals of the Missouri Botanical Garden* 81:451–483.

Mitman, G. G., and J. P. van der Meer. 1994. Meiosis, blade development, and sex determination in *Porphyra purpurea* (Rhodophyta). *Journal of Phycology* 30:147–159.

Miyamura, S., and T. Hori. 1995. Further confirmation of the presence of DNA in the pyrenoid core of the siphonous green algal genus *Caulerpa* (Caulerpales, Ulvophyceae, Chlorophyta). *Phycological Research* 43:101–104.

Miyashita, H., H. Ikemoto, N. Kurano, K. Adachi, M. Chihara, and S. Miyachi. 1996. Chlorophyll *d* as a major pigment. *Nature* 383:402–403.

Mizuno, M., and K. Okuda. 1985. Seasonal change in the distribution of cell size of *Cocconeis scutellum* var. *ornata* (Bacillariophyceae) in relation to growth and sexual reproduction. *Journal of Phycology* 21:547–553.

Mizuta, S., and R. M. Brown. 1992. High resolution analysis of the formation of cellulose synthesizing complexes in *Vaucheria hamata*. *Protoplasma* 166:187–199.

Mizuta, S., and K. Okuda. 1987. A comparative study of cellulose synthesizing complexes in certain cladophoralean and siphonocladalean algae. *Botanica Marina* 30:205–215.

Moe, R. L., and E. C. Henry. 1982. Reproduction and early development of *Ascoseira mirabilis* Skottsberg

(Phaeophyta) with notes on Ascoseirales Petrov. *Phycologia* 21:55–66.

Moe, R. L., and P. C. Silva. 1981. Morphology and taxonomy of *Himantothallus* (including *Phaeoglossum* and *Phyllogigas*), an Antarctic member of the Desmarestiales (Phaeophyceae). *Journal of Phycology* 17:15–29.

Moestrup, Ø. 1970. The fine structure of mature spermatozoids of *Chara corallina*, with special reference to microtubules and scales. *Planta* 93:295–308.

Moestrup, Ø. 1974. Ultrastructure of the scale-covered zoospores of the green alga *Chaetosphaeridium*, a possible ancestor of the higher plants and bryophytes. *Biological Journal of the Linnean Society* 6:111–125.

Moestrup, Ø. 1978. On the phylogenetic validity of the flagellar apparatus in green algae and other chlorophyll *a* and *b* containing plants. *BioSystems* 10:117–144.

Moestrup, Ø. 1991. Further studies of presumedly primitive green algae, including the description of Pedinophyceae class. nov. and *Resultor* gen. nov. *Journal of Phycology* 27:119–133.

Moestrup, Ø. 1995. Current status of chrysophyte 'splinter groups': Synurophytes, pedinellids, silicoflagellates. Pages 75–91 in *Chrysophyte Algae. Ecology, Phylogeny and development*, edited by C. D. Sandgren, J. P. Smol, and J. Kristiansen. Cambridge University Press, Cambridge, UK.

Moezelaar, R., and L. J. Stal. 1997. A comparison of fermentation in the cyanobacterium *Microcystis* PCC7806 grown under a light/dark cycle and continuous light. *European Journal of Phycology* 32:373–378.

Moldowan, J. M., F. J. Fago, C. Y. Lee, S. R. Jacobson, D. S. Watt, N.-E. Slougui, A. Jeganathan, and D. C. Young. 1990. Sedimentary 24-n-propylcholestanes, molecular fossils diagnostic of marine algae. *Science* 247:309–312.

Moldowan, J. M., and N. M. Talyzina. 1998. Biogeochemical evidence for dinoflagellate ancestors in the Early Cambrian. *Science* 281:1168–1170.

Mollenhauer, D., R. Mollenhauer, and M. Kluge. 1996. Studies on initiation and development of the partner association in *Geosiphon pyriforme* (Kütz.) v Wettstein, a unique endocytobiotic system of a fungus (Glomales) and the cyanobacterium *Nostoc punctiforme* (Kütz.) Hariot. *Protoplasma* 193:3–9.

Molnar, K. E., K. D. Stewart, and K. R. Mattox. 1975. Cell division in the filamentous *Pleurastrum* and its comparison with the unicellular *Platymonas* (Chlorophyceae). *Journal of Phycology* 3:287–296.

Monod, J. 1950. La technique de culture continue: Theorie et applications. *Annales de l'Institut Pasteur Lille* 79:390–410.

Montegut-Felkner, A. E., and R. E. Triemer. 1997. Phylogenetic relationships of selected euglenoid genera based on morphological and molecular data. *Journal of Phycology* 33:512–519.

Moore, J. K., and T. A. Villareal. 1996. Size-ascent rate relationships in positively buoyant marine diatoms. *Limnology and Oceanography* 41:1514–1520.

Moore, J. F., and J. A. Traquair. 1976. Silicon, a required nutrient for *Cladophora glomerata* (L.) Kütz. *Planta* 128:179–182.

Moore, L. R., G. Rocap, and S. W. Chisholm. 1998. Physiology and molecular phylogeny of coexisting *Prochlorococcus* ecotypes. *Nature* 393:464–467.

Morden, C. W., C. F. Delwiche, M. Kuhsel, and J. D. Palmer. 1992. Gene phylogenies and the endosymbiotic origin of plastids. *BioSystems* 28:75–90.

Mori, I. C., G. Sato, and M. Okazaki. 1996. Ca^{2+}-dependent ATPase associated with plasma membrane from a calcarous alga, *Serraticardia maxima* (Corallinaceae, Rhodophyta). *Phycological Research* 44:193–202.

Moritz, C., and D. M. Hillis. 1996. Molecular systematics: Context and controversies. Pages 1–13 in *Molecular Systematics*, edited by D. M. Hillis, C. Moritz, and B. K. Mable. Sinauer Associates, Sunderland, MA.

Morland, Z., D. G. Kidd, and J. C. Lagarias. 1993. Phytochrome levels in the green alga *Mesotaenium caldariorum* are light-regulated. *Plant Physiology* 101:97–103.

Mornin, L., and D. Francis. 1967. The fine structure of

Nematodinium armatum, a naked dinoflagellate. *Journal Microscopie* (Paris) 6:759–772.

Moroney, J. V., and Z.-Y. Chen. 1998. The role of the chloroplast in inorganic carbon uptake by eukaryotic algae. *Canadian Journal of Botany* 76:1025–1034.

Moroney, J. V., and A. Somanchi. 1998. How do algae concentrate CO_2 to increase the efficiency of photosynthetic carbon fixation? *Plant Physiology* 119:9–16.

Morse, D. M., P. Salois, P. Markovic, and J. W. Hastings. 1995. A nuclear-encoded form II RuBisCO in dinoflagellates. *Science* 268:1622–1624.

Mostaert, A. S., U. Karsten, Y. Hara, and M. M. Watanabe. 1998. Pigments and fatty acids of marine raphidophytes: A chemotaxonomic re-evaluation. *Phycological Research* 46:213–220.

Mostaert, A. S., U. Karsten, and R. J. King. 1995. Physiological responses of *Caloglossa leprieurii* (Ceramiales, Rhodophyta) to salinity stress. *Phycological Research* 43:215–222.

Motomura, T. 1990. Ultrastructure of fertilization in *Laminaria angustata* (Phaeophyta, Laminariales) with emphasis on the behavior of centrioles, mitochondria and the chloroplasts of the sperm. *Journal of Phycology* 26:80–89.

Motomura, T., T. Ichimura, and M. Melkonian. 1997. Coordinative nuclear and chloroplast division in unilocular sporangia of *Laminaria angustata* (Laminariales, Phaeophyceae). *Journal of Phycology* 33:266–271.

Motomura, T., and Y. Sakai. 1988. The occurrence of flagellated eggs in *Laminaria angustata* (Phaeophyta, Laminariales). *Journal of Phycology* 24:282–285.

Mujer, C. V., D. L. Andrews, J. R. Manhart, S. K. Peirce, and M. E. Rumpho. 1996. Chloroplast genes are expressed during intracellular symbiotic association of *Vaucheria litorea* plastids with the sea slug *Elysia chlorotica*. *Proceedings of the National Academy of Science* 93:12333–12338.

Müller, D. G. 1981. Sexuality and sex attraction. Pages 661–674 in *Biology of Seaweeds*, edited by C. S. Lobban and M. J. Wynne. Blackwell Scientific, Oxford, UK.

Müller, D. G., W. Boland, U. Becker, and T. Wahl. 1988. Caudoxirene, the spermatozoid-releasing and attracting factor in the marine brown alga *Perithalia caudata* (Sporochnales, Phaeophyta). *Phycologia* 24:467–473.

Müller, D. G., G. Gassman, W. Boland, F. Marner, and L. Jaenicke. 1981. *Dictyota dichotoma* (Phaeophyceae): Identification of the sperm attractant. *Science* 212:1040–1041.

Müller, D. G., I. Maier, and G. Gassman. 1985. Survey on sexual hormone specificity in Laminariales (Phaeophyceae). *Phycologia* 24:475–477.

Müller, D. G., L. Jaenicke, M. Donike, and A. Akintobi. 1971. Sex attractant in a brown alga: Chemical structure. *Science* 171:815–816.

Müller, D. G., A. Peters, G. Gassman, W. Boland, F.-J. Marner, and L. Jaenicke. 1982. Identification of a sexual hormone and related substances in the marine brown alga *Desmarestia*. *Naturwissenschaften* 69:290–292.

Müller, D. G., S. Wolf, and E. R. Parodi. 1996. A virus infection in *Myriotrichia clavaeformis* (Dictyosiphonales, Phaeophyceae) from Argentina. *Protoplasma* 193:58–62.

Müller, K. M., R. G. Sheath, M. L. Vis, T. J. Crease, and K. M. Cole. 1998. Biogeography and systematics of *Bangia* (Bangiales, Rhodophyta) based on the Rubisco spacer, *rbcL* gene and 18S rRNA gene sequences and morphometric analysis. I. North America. *Phycologia* 37:195–207.

Muller-Parker, G., and C. F. D'Elia. 1997. Interactions between corals and their symbiotic algae. Pages 96–113 in *Life and Death of Coral Reefs*, edited by C. Birkeland. Chapman & Hall, New York, NY.

Mumford, T. F., and A. Miura. 1988. *Porphyra* as food: Cultivation and economics. Pages 87–118 in *Algae and Human Affairs*, edited by C. A. Lembi and J. R. Waaland. Cambridge University Press, Cambridge, UK.

Mur, L., and R. O. Bejsdorf. 1978. A model of the succession from green to blue-green algae based on light limitation. *Internationale Vereiningung für Theoretische und Angewandte Limnologie* 20:2314–2321.

Murata, K., and T. Suzaki. 1998. High-salt solutions prevent reactivation of euglenoid movement in detergent-treated cell models of *Euglena gracilis*. *Protoplasma* 203:125–129.

Murphy, T. P., D. R. S. Lean, and C. Nalewajko. 1976. Blue-green algae: Their excretion of iron-selective chelators enables them to dominate other algae. *Science* 192:394–395.

Nakahara, H., and T. Ichimura. 1992. Convergent evolution of gametangiogamy both in the zygnematalean green algae and in the pennate diatoms. *Japanese Journal of Phycology* 40:161–166.

Nakahara, H., T. Sakami, M. Chinain, and Y. Ishida. 1996. The role of macroalgae in epiphytism of the toxic dinoflagellate *Gambierdiscus toxicus* (Dinophyceae). *Phycological Research* 44:113–117.

Nakanishi, K., M. Nishijima, M. Nishimura, K. Kuwano, and N. Saga. 1996. Bacteria that induce morphogenesis in *Ulva pertusa* (Chlorophyta) grown under axenic conditions. *Journal of Phycology* 32:479–482.

Nakayama, T., S. Watanabe, and I. Inouye. 1996b. Phylogeny of wall-less green flagellates inferred from 18SrDNA sequence data. *Phycological Research* 44:151–161.

Nakayama, T., S. Watanabe, K. Mitsui, H. Uchida, and I. Inouye. 1996a. The phylogenetic relationship between the Chlamydomonadales and Chlorococcales inferred from 18S rDNA data. *Phycological Research* 44:47–55.

Nanba, N. 1995. Egg release and germling development in *Sargassum horneri* (Fucales, Phaeophyceae). *Phycological Research* 43:121–125.

Naustvoll, L.-J. 1998. Growth and grazing by the thecate heterotrophic dinoflagellate *Diplopsalis lenticula* (Diplopsalidaceae, Dinophyceae). *Phycologia* 37:1–9.

Navarro, J. N. 1993. Three-dimensional imaging of diatom ultrastructure with high resolution low-voltage SEM. *Phycologia* 32:151–156.

Neale, P. J. 1987. Algal photoinhibition and photosynthesis in the aquatic environment. Pages 39–65 in *Photoinhibition*, edited by D. J. Kyle, C. B. Osmond, and C. J. Aratzen. Elsevier, New York, NY.

Nedelcu, A. M. 1998. Contrasting mitochondrial genome organizations and sequence affiliations among green algae: Potential factors, mechanisms, and evolutionary scenarios. *Journal of Phycology* 34:16–28.

Neilands, J. B. 1967. Hydroxamic acids in nature. *Science* 156:1443–1447.

Neilson, A. H., and R. A. Lewin. 1974. The uptake and utilization of organic carbon by algae: An essay in comparative biochemistry. *Phycologia* 13:227–264.

Nelson, W. A., and G. A. Knight. 1996. Life history in culture of the obligate epiphyte *Porphyra subtumens* (Bangiales, Rhodophyta) endemic to New Zealand. *Phycological Research* 44:19–25.

Neushul, M., and A. L. Dahl. 1972. Zonation in the apical cell of *Zonaria*. *American Journal of Botany* 59:393–400.

Newman, S. M., and R. A. Cattolico. 1990. Ribulose bisphosphate carboxylase in algae: Synthesis, enzymology and evolution. *Photosynthesis Research* 26:69–85.

Nichols, H. W. 1965. Culture and development of *Hildenbrandia rivularis* from Denmark and North America. *American Journal of Botany* 52:9–15.

Nicholls, K. H. 1995. Chrysophyte blooms in the plankton and neuston of marine and freshwater systems. Pages 181–213 in *Chrysophyte Algae: Ecology, Phylogeny and Development*, edited by C. D. Sandgren, J. P. Smol, and J. Kristiansen. Cambridge University Press, Cambridge, UK.

Nicholson, N. L. 1970. Field studies on the giant kelp *Nereocystis*. *Journal of Phycology* 6:177–182.

Nicol, S., and I. Allison. 1997. The frozen skin of the Southern Ocean. *American Scientist* 85:426–439.

Niiyama, Y. 1989. Morphology and classification of *Cladophora aegagropila* (L.) Rabenhorst (Cladophorales, Chlorophyta) in Japanese lakes. *Phycologia* 28:70–76.

Niklas, K. J. 1976. Morphological and ontogenetic reconstruction of *Parka decipiens* Fleming and *Pachytheca hooker* from the Lower Old Red Sandstone, Scotland. *Transactions of the Royal Society of Edinburgh* 69:483–499.

Nimer, N. A., M. D. Iglesias-Rodriguez, and M. J. Merrett. 1997. Bicarbonate utilization by marine phytoplankton species. *Journal of Phycology* 33:625–631.

Nishiguchi, M. K., and L. J. Goff. 1995. Isolation, purification, and characterization of DMSP lyase (dimethylpropiothetin dethiomethylase [4.4.1.3]) from the red alga *Polysiphonia paniculata*. *Journal of Phycology* 31:567–574.

Nishino, T., S. Asakawa, and S. Ogawa. 1996. Fine filaments observed within the cytoplasm surrounding the leading edge of the septum in telophase cells of *Spirogyra* (Zygnematales, Chlorophyta). *Phycological Research* 44:163–166.

Norris, R. E., T. Hori, and M. Chihara. 1980. Revision of the genus *Tetraselmis* (Class Prasinophyceae). *The Botanical Magazine, Tokyo* 93:317–339.

North, W. J. 1994. Review of *Macrocystis* biology. Pages 447–528 in *Biology of Economic Algae*, edited by I. Akatsura. SPB Academic Publishing, The Hague, Netherlands.

Norton, T. A., M. Melkonian, and R. A. Andersen. 1996. Algal biodiversity. *Phycologia* 35:308–326.

Novarino, G., and I. A. N. Lucas. 1993. Some proposals for a new classification system of the Cryptophyceae. *Botanical Journal of the Linnean Society* 111:3–21.

Novarino, G., and I. A. N. Lucas. 1995. A zoological classification system of cryptomonads. *Acta Protozoologica* 34:173–180.

Nozaki, H., and M. Ito. 1994. Phylogenetic relationships within the colonial Volvocales (Chlorophyta) inferred from cladistic analysis based on morphological data. *Journal of Phycology* 30:353–365.

Nozaki, H., M. Ito, R. Sano, H. Uchida, M. M. Watanabe, T. Kuroiwa. 1995. Phylogenetic relationships within the colonial Volvocales (Chlorophyta) inferred from *rbcL* gene sequence data. *Journal of Phycology* 31:970–979.

Nozaki, H., M. Ito, R. Sano, H. Uchida, M. M. Watanabe, H. Takahashi, and T. Kuroiwa. 1997c. Phylogenetic analysis of *Yamagishiella* and *Platydorina* (Volvocaceae, Chlorophyta) based on *rbcL* gene sequences. *Journal of Phycology* 33:272–278.

Nozaki, H., M. Ito, D. Uchida, M. M. Watanabe, and T. Kuroiwa. 1997a. Phylogenetic analysis of *Eudorina* species (Volvocaceae, Chlorophyta) based on *rbcL* gene sequences. *Journal of Phycology* 33:859–863.

Nozaki, H., M. Ito, M. M. Watanabe, H. Takano, and T. Kuroiwa. 1997b. Phylogenetic analysis of morphological species of *Carteria* (Volvocales, Chlorophyta) based on *rbcL* gene sequences. *Journal of Phycology* 33:864–867.

Nozaki, H., H. Kuroiwa, T. Mita, and T. Kuroiwa. 1989. *Pleodorina japonica* sp. nov. (Volvocales, Chlorophyta) with bacteria-like endosymbionts. *Phycologia* 28:252–267.

Nurdogan, Y., and W. J. Oswald. 1995. Enhanced nutrient removal in high-rate ponds. *Water Science and Technology*. 31:33–43.

Nybakken, J. W. 1982. *Marine Biology*. Harper & Row, New York, NY.

O'Kelly, C. J. 1989. The evolutionary origin of brown algae: Information from studies of motile cell ultrastructure. Pages 255–278 in *The Chromophyte Algae: Problems and Perspectives*, edited by J. C. Green, B. S. C. Leadbeater, and W. L. Diver. Clarendon Press, Oxford, UK.

O'Kelly, C. J. 1993. Relationships of eukaryotic algal groups to other protists. Pages 269–294 in *Ultrastructure of Microalgae*, edited by T. Berner. CRC Press, Boca Raton, FL.

O'Kelly, C. J., and B. J. Baca. 1984. The time course of carpogonial branch and carposporophyte development in *Callithamnion cordatum* (Rhodophyta, Ceramiales). *Phycologia* 23:407–417.

O'Kelly, C. J., and G. L. Floyd. 1984. Flagellar apparatus absolute orientations and the phylogeny of the green algae. *BioSystems* 16:227–251.

O'Kelly, C. J., S. Watanabe, and G. L. Floyd. 1994. Ultrastructure and phylogenetic relationships of Chaetopeltidales ord. nov. (Chlorophyta, Chlorophyceae). *Journal of Phycology* 30:118–128.

O'Kelly, C. J., and D. E. Wujek. 1995. Status of the Chrysamoebales (Chrysophyceae): Observations on *Chrysamoeba pyrenoidifera*, *Rhizochromulina marina* and *Lagynion delicatulum*. Pages 361–372 in *Chrysophyte Algae: Ecology, Phylogeny and Development*, edited by C. D. Sandgren, J. P. Smol, and J. Kristiansen. Cambridge University Press, Cambridge, UK.

O'Neal, S. W., and C. A. Lembi. 1983. Physiological changes during germination of *Pithophora oedogonia* (Chlorophyceae) akinetes. *Journal of Phycology* 19:193–199.

O'Neil, R. M., and J. W. La Claire. 1984. Mechanical wounding induces the formation of extensive coated membranes in giant algal cells. *Science* 225:331–333.

Oakley, B. R, and J. D. Dodge. 1974. Kinetochores associated with the nuclear envelope in the mitosis of a dinoflagellate. *Journal of Cell Biology* 63:322–325.

Oakley, B. R., and J. D. Dodge. 1976. The ultrastructure of mitosis in *Chroomonas salina*. *Protoplasma* 88:241–254.

Ojala, A. 1993. Effects of temperature and irradiance on the growth of two freshwater photosynthetic cryptophytes. *Journal of Phycology* 29:278–284.

Okuda, K., and R. M. Brown. 1992. A new putative cellulose-synthesizing complex of *Coleochaete scutata*. *Protoplasma* 168:51–63.

Okuda, K., and S. Mizuta. 1993. Diversity and evolution of putative cellulose-synthesizing enzyme complexes in green plants. *Japanese Journal of Phycology* 41:151–173.

Oliveira, M. C., J. Kurniawan, C. J. Bird, E. L. Rice, C. A. Murphy, R. K. Singh, R. G. Gutell, and M. A. Ragan. 1995. A preliminary investigation of the order Bangiales (Bangiophycidae, Rhodophyta) based on sequences of nuclear small-subunit ribosomal RNA genes. *Phycological Research* 42:71–79

Oliver, R. L. 1994. Floating and sinking in gas-vacuolate cyanobacteria. *Journal of Phycology* 30:161–173.

Olli, K. 1996. Resting cyst formation of *Eutreptiella gymnastica* (Euglenophyceae) in the northern coastal Baltic Sea. *Journal of Phycology* 32:535–542.

Olli, K., A.-S. Heiskanen, and J. Seppälä. 1996. Development and fate of *Eutreptiella gymnastica* bloom in nutrient-enriched enclosures in the coastal Baltic Sea. *Journal of Plankton Research* 18:1587–1604.

Olsen, G. J., C. R. Woese, and R. Overbeek. 1994. The winds of (evolutionary) change: Breathing new life into microbiology. *Journal of Bacteriology* 176:1–6.

Olson, A. M., and J. Lubchenco. 1990. Competition in seaweeds: Linking plant traits to competitive outcomes. *Journal of Phycology* 26:1–6.

Olson, R. J., S. W. Chisholm, E. R. Zettler, and E. V. Armbrust. 1990. Pigments, size and distribution of *Synechococcus* in the North Atlantic and Pacific Oceans. *Limnology and Oceanography* 35:45–58.

Olszewski, M. J., D. Gernand, M. Godlewski, and A. Kunachowicz. 1997. DNA methylation during antheridial filament development and spermiogenesis in *Chara vulgaris* (Charophyceae) analysed by *in situ* nick-translation driven by methylation-sensitive restriction enzymes. *Phycological Research* 44:163–166.

Oltmanns, F. 1898. Die Entwicklung der Sexualorgane bei [development of sexual organs in] *Coleochaete pulvinata*. *Flora* 85:1–14.

Oppermann, B., P. Karlovsky, and W. Reisser. 1997. M13 DNA fingerprinting in unicellular and filamentous green algae. *European Journal of Phycology* 32:103–110.

Ortiz, W., and C. J. Wilson. 1988. Induced changes in chloroplast protein accumulation during heat bleaching in *Euglena gracilis*. *Plant Physiology* 86:554–561.

Ott, D. W., and R. M. Brown. 1972. Light and electron microscopical observations on mitosis in *Vaucheria litorea* Hofman ex C Agardh. *British Phycological Journal* 7:361–374.

Ott, D. W., and R. M. Brown. 1975. Developmental cytology of the genus *Vaucheria*. III. Emergence, settlement and germination of the mature zoospore of *V. fontanalis* (L.) Christianson. *British Phycological*

Journal 10:49–50.

Ott, F. D., and J Seckbach. 1994a. A review on the taxonomic position of the algal genus *Cyanidium* Geitler 1933 and its ecological cohorts *Galdieria* Merola in Merola, et al., 1981 and *Cyanidoschyzon* De Luca, Taddei and Varano 1978. Pages 113–132 in *Evolutionary Pathways and Enigmatic Algae: Cyanidium caldarium (Rhodophyta) and Related Cells*, edited by J. Seckbach. Kluwer Academic Publishers, Boston, MA.

Ott, F. D., and J Seckbach. 1994b. New classification for the genus *Cyanidium* Geitler 1933. Pages 145–152 in *Evolutionary Pathways and Enigmatic Algae: Cyanidium caldarium (Rhodophyta) and Related Cells*, edited by J. Seckbach. Kluwer Academic Publishers, Boston, MA.

Pace, N. R. 1997. A molecular view of microbial diversity and the biosphere. *Science* 276:734–739.

Padilla, D. K. 1985. Structural resistance of algae to herbivores: A biomechanical approach. *Marine Biology* 90:103–199.

Padilla, D. K. 1989. Algal structural defenses: Form and calcification in resistance to tropical limpets. *Ecology* 70:835–842.

Padilla, D. K., and S. C. Adolph. 1996. Plastic inducible morphologies are not always adaptive: The importance of time delays in a stochastic environment. *Evolutionary Ecology* 10:105–117.

Paerl, H. W. 1984. N$_2$-fixation (nitrogenase activity) attributable to a specific *Prochloron* (Prochlorophyta)-ascidian association in Palau, Micronesia. *Marine Biology* 81:251–254.

Paerl, H. W. 1992. Epi- and endobiotic interactions of cyanobacteria. Pages 537–565 in *Algae and Symbioses*, edited by W. Reisser. BioPress, Bristol, UK.

Paerl, H. W. 1994. Spatial segregation of CO$_2$ fixation in *Trichodesmium* spp: Linkage to N$_2$ fixation potential. *Journal of Phycology* 30:790–799.

Paerl, H. W., P. T. Bland, N. D. Bowles, and M. E. Haibach. 1985. Adaptation to high-intensity, low-wavelength light among surface blooms of the cyanobacterium *Microcystis aeruginosa. Applied and Environmental Microbiology* 49:1046–1052.

Paerl, H. W., and J. L. Pinckney. 1996. A mini-review of microbial consortia: Their roles in aquatic production and biogeochemical cycling. *Microbial Ecology* 31:225–247.

Paerl, H. W., R. C. Richards, R. C. Leonord, and C. R. Goldman. 1975. Seasonal nitrate cycling as evidence for complete vertical mixing in Lake Tahoe, California-Nevada. *Limnology and Oceanography* 20:1–8.

Paerl, H. W., and J. F. Ustach. 1982. Blue-green algal scums: An explanation for their occurrence during freshwater blooms. *Limnology and Oceanography* 27:212–217.

Paine, R. T. 1969. A note on trophic complexity and community stability. *American Naturalist* 103:91–93.

Paine, R. T. 1990. Benthic macroalgal competition: Complications and consequences. *Journal of Phycology* 26:12–17.

Paine, R. T., C. J. Slocum, and D. O. Duggins. 1979. Growth and longevity in the crustose red alga *Petrocelis middendorfii. Marine Biology* 51:185–192.

Paine, R. T., and R. L. Vadas. 1969. The effects of grazing by sea urchins, *Strongylocentrotus* spp., on benthic algal populations. *Limnology and Oceanography* 14:710–719.

Palenik, B., and R. Haselkorn. 1992. Multiple evolutionary origins of prochlorophytes, the chlorophyll *b*-containing prokaryotes. *Nature* 355:265–267.

Palenik, B., and S. E. Henson. 1997. The use of amides and other organic nitrogen sources by the phytoplankton *Emiliania huxleyi. Limnology and Oceanography* 42:1544–1551.

Palenik, B., and H. Swift. 1996. Cyanobacterial evolution and prochlorophyte diversity as seen in DNA-dependent RNA polymerase gene sequences. *Journal of Phycology* 32:638–646.

Palenik, B., and A. M. Wood. 1998. Molecular markers of phytoplankton physiological status and their application at the level of individual cells. Pages 187–207 in *Molecular Approaches to the Study of the Oceans*,

edited by K. E. Cooksey. Chapman & Hall, London, UK.

Palevitz, B. A., and P. K. Hepler. 1975. Identification of actin *in situ* at the ectoplasm-endoplasm interface of *Nitella. Journal of Cell Biology* 65:29–38.

Palmer, J. D. 1996. Rubisco surprises in dinoflagellates. *The Plant Cell* 8:343–345.

Palmer, J. D. 1997. Organellar genomes: Going, going, gone! *Science* 275:790–791.

Palmer, J. D., and C. F. Delwiche. 1996. Second-hand chloroplasts and the case of the disappearing nucleus. *Proceedings of the National Academy of Science* 93:7432–7435.

Pan, Y., D. V. S. Rao, and K. H. Mann. 1996. Changes in domoic acid production and cellular chemical composition of the toxigenic diatom *Pseudo-nitzschia multiseries* under phosphate limitation. *Journal of Phycology* 32:371–381.

Papenfuss, G. F. 1935. Alternation of generations in *Ectocarpus siliculosus. Botanical Gazette* 96:421–446.

Parke, M., I. Manton, and B. Clarke. 1955. Studies on marine flagellates. II. Three new species of *Chrysochromulina. Journal of the Marine Biology Association of the UK* 34:579–609.

Parker, B. C. 1965. Translocation in the giant kelp *Macrocystis*. I. Rates, direction, quantity of ^{14}C-labeled products and fluorescein. *Journal of Phycology* 1:41–46.

Parker, B. C. 1966. Translocation in *Macrocystis*. III. Composition of sieve tube exudate and identification of the major ^{14}C label and products. *Journal of Phycology* 2:38–41.

Parker, B. C., and E. Y. Dawson. 1965. Non-calcareous marine algae from California Miocene deposits. *Nova Hedwigia* 10:273–295.

Parker, B. C., and J. Huber. 1965. Translocation in *Macrocystis*. II. Fine structure of the sieve tubes. *Journal of Phycology* 1:172–179.

Parker, B. C., and W. A. Wharton. 1985. Physiological ecology of blue-green algal mats (modern stromatolites) in Antarctic oasis lakes. *Archiv für Hydrobiologie* 71:331–348.

Parker, D. L., H. D. Kumar, L. C. Rai, and J. B. Singh. 1997. Potassium salts inhibit growth of the cyanobacteria *Microcystis* spp. in pond water and defined media: Implication for control of microcystin-producing aquatic blooms. *Applied and Environmental Microbiology* 63:2324–2329.

Parodi, E. R., and E. J. Cáceres. 1995. Life history of *Cladophora surera* sp. nov. (Cladophorales, Ulvophyceae). *Phycological Research* 43:223–231.

Parsons, T. J., C. A. Maggs, and S. E. Douglas. 1990. Plastid DNA restriction analysis links the heteromorphic phases of an apomictic red algal life history. *Journal of Phycology* 26:495–500.

Patrick, R., and C. W. Reimer. 1966. *The Diatoms of the U.S. exclusive of Alaska and Hawaii, Vol I—Fragilariaceae, Eunotiaceae, Achnanthaceae, Naviculaceae.* Academy of Natural Sciences, Philadelphia, Monographs 13.

Patrick, R., and C. W. Reimer. 1975. *The Diatoms of the U.S. exclusive of Alaska and Hawaii, Vol II—Part 1, Entomoneidaceae, Cymbellaceae, Gomphonemaceae, Epithemiaceae.* Academy of Natural Sciences, Philadelphia, Monographs 13.

Patterson, G. M. L., K. K. Baker, C. L. Baldwin, C. M. Bolis, F. R. Caplan, L. K. Larsen, I. A. Levine, R. E. Moore, C. S. Nelson, K. D. Tschappat, G. D. Tuang, M. R.. Boyd, J. H. Cardellina, R. P. Collins, K. R. Gustafson, K. M. Snader, O. S. Weislow, and R. A. Lewin. 1993. Antiviral activity of cultured blue-green algae (Cyanophyta). *Journal of Phycology* 29:125–130.

Patterson, G. M. L., C. L. Baldwin, C. M. Bolis, F. R. Caplan, H. Karuso, L. K. Larsen, I. A. Levine, R. E. Moore, C. S. Nelson, K. D. Tschappat, G. D. Tuang, E. Furusawa, S. Furusawa, T. R. Norton, and R. B. Raybourne. 1991. Antineoplastic activity of cultured blue-green algae (Cyanophyta). *Journal of Phycology* 27:530–536.

Patterson, G. M. L., and S. Carmeli. 1992. Biological effects of tolytoxin (6-hydro-7-O-methyl scytophycin b), a potent bioreactive metabolite from cyanobacteria. *Archives of Microbiology* 157:406–410.

Patterson, G. M. L., L. K. Larsen, and R. E. Moore. 1994. Bioactive natural products from blue-green algae. *Journal of Applied Phycology* 6:151–157.

Paul, J. H., and S. L. Pichard. 1998. Phytoplankton activity through the measurement of ribulose bisphosphate carboxylase gene expression (RuBisCO). Pages 207–226 in *Molecular Approaches to the Study of the Oceans*, edited by K. E. Cooksey. Chapman & Hall, London, UK.

Pavia, H. G. Cervin, A. Lindgren, and P. Åberg. 1997. Effects of UV-B radiation and simulated herbivory on phlorotannins in the brown alga *Ascophyllum nodosum. Marine Ecology Progress Series* 157:139–146.

Pearson, E. A., and S. N. Murray. 1997. Patterns of reproduction, genetic diversity, and genetic differentiation in California populations of the geniculate coralline alga *Lithothrix aspergillum* (Rhodophyta). *Journal of Phycology* 33:753–763.

Pedersen, P. M. 1980. Culture studies on complanate and cylindrical *Scytosiphon* (Fucophyceae, Scytosiphonales) from Greenland. *British Phycological Journal* 15:391–398.

Pedersen, P. M. 1981. Phaeophyta, life histories. Pages 194–217 in *The Biology of Seaweeds*, edited by C. S. Lobban and M. J. Wynne. Blackwell Scientific, Oxford, UK.

Pennings, S. C., S. R. Pablo, and V. J. Paul. 1997. Chemical defenses of the tropical, benthic marine cyanobacterium *Hormothamnion enteromorphoides*: Diverse consumers and synergisms. *Limnology and Oceanography* 42:911–917.

Peperzak, L., F. Colijn, W. W. C. Gieskes, and J. C. H. Peeters. 1998. Development of the diatom-*Phaeocystis* spring bloom in the Dutch coastal zone of the North Sea: The silicon depletion versus the daily irradiance threshold hypothesis. *Journal of Plankton Research* 20:517–537.

Perl, T. M., L. Bedard, J. C. Hockin, E. C. D. Todd, and R. S. Remis. 1990. An oubreak of toxic encephalopathy caused by eating mussels contaminated with domoic acid. *New England Journal of Medicine* 322:1775–1780.

Peters, A. 1987. Reproduction and sexuality in the Chordariales (Phaeophyceae): A review of culture studies. *Progress in Phycological Research* 5:223–263.

Peters, A., M. J. H. van Oppen, C. Wiencke, W. T. Stam, and J. L. Olsen. 1997. Phylogeny and historical ecology of the Desmarestiaceae (Phaeophyceae) support a Southern Hemisphere origin. *Journal of Phycology* 33:294–309.

Peters, A. F., and E. Burkhardt. 1998. Systematic position of the kelp endophyte *Laminarionema elsbetiae* (Ectocarpales *sense lato*, Phaeophyceae) inferred from nuclear ribosomal DNA sequences. *Phycologia* 37:114–120.

Peters, A. F., and M. N. Clayton. 1998. Molecular and morphological investigations of three brown algal genera with stellate plastids: Evidence for Scytothamnales ord. nov. (Phaeophyceae). *Journal of Phycology* 37:106–113.

Peters, E. C. 1997. Diseases of coral-reef organisms. Pages 114–139 in *Life and Death of Coral Reefs*, edited by C. Birkeland. Chapman & Hall, New York, NY.

Peters, R. 1984. Methods for the study of feeding, grazing and assimilation by zooplankton. Pages 336–412 in *A Manual on Methods for the Assessment of Secondary Productivity in Fresh Waters*, edited by J. A. Downing and F. H. Rigler. Blackwell Scientific, Oxford, UK.

Peterson, C. G., T. L. Dudley, K. D. Hoagland, and L. M. Johnson. 1993. Infection, growth, and community-level consequences of a diatom pathogen in a Sonoran desert stream. *Journal of Phycology* 29:442–452.

Peterson, C. G., K. A. Vormittag, and H. M. Valett. 1998. Ingestion and digestion of epilithic algae by larval insects in a heavily grazed montane stream. *Freshwater Biology* 40:607–623.

Pfiester, L. A. 1988. Dinoflagellate sexuality. *International Review of Cytology* 114:249–272.

Pfiester, L. A., and D. M. Anderson. 1987. Dinoflagellate reproduction. Pages 611–648 in *The Biology of Dinoflagellates*, edited by F. J. R. Taylor. Blackwell,

Oxford, UK.

Pfiester, L. A., and R. Lynch. 1980. Amoeboid stages and sexual reproduction of *Cystodinium bataviense* and its similarity to *Dinococcus* (Dinophyceae). *Phycologia* 19:178–183.

Pfiester, L. A., and J. Popovsky. 1979. Parasitic, amoeboid dinoflagellates. *Nature* 24:421–424.

Phillips, D., and J. Scott. 1981. Ultrastructure of cell division and reproductive differentiation of male plants in the Florideophyceae (Rhodophyta). Mitosis in *Dasya baillouviana*. *Protoplasma* 106:329–341.

Phillips, J. A. 1997. Genus and species concepts in *Zonaria* and *Homoeostrichus* (Dictyotales, Phaeophyceae), including the description of *Exallosorus*, gen. nov. *European Journal of Phycology* 32:303–311.

Phillips, J. A., and M. N. Clayton. 1997. Comparative studies on gametangial distribution and structure in species of *Zonaria* and *Homoeostrichus* (Dictyotales, Phaeophyceae) from Australia. *European Journal of Phycology* 32:25–34.

Pickett-Heaps, J. D. 1975. *Green Algae. Structure, Reproduction and Evolution of Selected Genera.* Sinauer Associates, Sunderland, MA.

Pickett-Heaps, J. D., and L. C. Fowke. 1970. Mitosis, cytokinesis, and cell elongation in the desmid, *Closterium littorale*. *Journal of Phycology* 6:189–215.

Pickett-Heaps, J. D., D. R. A. Hill, and R. Wetherbee. 1986. Cellular movement in the centric diatom *Odontella sinensis*. *Journal of Phycology* 22:334–339.

Pickett-Heaps, J. D., and S. E. Kowalski. 1981. Valve morphogenesis and the microtubule center of the diatom *Hantzschia amphioxys*. *European Journal of Cell Biology* 25:150–170.

Pickett-Heaps, J. D., and H. J. Marchant. 1972. The phylogeny of the green algae: A new proposal. *Cytobios* 6:255–264.

Pickett-Heaps, J., A.-M. M. Schmid, and L. A. Edgar. 1990. The cell biology of diatom valve formation. *Progress in Phycological Research* 7:1–157.

Pickett-Heaps, J. D., T. Spurck, and D. H. Tippit. 1984. Chromosome motion and the spindle matrix. *Journal of Cell Biology* 99:137–143.

Pickett-Heaps, J. D., D. H. Tippet, and J. A. Andreozzi. 1979. Cell division in the pennate diatom *Pinnularia* L. IV. Valve morphogenesis. *Biology of the Cell* 35:199–203.

Pickett-Heaps, J. D., and J. A. West. 1998. Time-lapse video observations on sexual plasmogamy in the red alga *Bostrychia*. *European Journal of Phycology* 33:43–56.

Pickett-Heaps, J. D., and R. Wetherbee. 1987. Spindle formation in the green alga *Mougeotia*: Absence of anaphase A correlates with postmitotic nuclear migration. *Cell Motility and Cytoskeleton* 7:68–77.

Pienaar, R. N. 1994. Ultrastructure and calcification of coccolithophores. Pages 13–38 in *Coccolithophores*, edited by A. Winter and W. G. Siesser. Cambridge University Press, Cambridge, UK.

Pillmann, A., G. W. Woolcott, J. L. Olsen, W. T. Stam, and R. J. King. 1997. Inter- and intraspecific genetic variation in *Caulerpa* (Chlorophyta) based on nuclear rDNA ITS sequences. *European Journal of Phycology* 32:379–386.

Pipes, L. D., and G. F. Leedale. 1992. Scale formation in *Tesellaria volvocina* (Synurophyceae). *British Phycological Journal* 27:1–19.

Pipes, L. D., G. F. Leedale, and P. A. Tyler. 1991. Ultrastructure of *Tesellaria volvocina* (Sunurophyceae). *British Phycological Journal* 26:259–278.

Pipes, L. D., P. A. Tyler, and G. F. Leedale. 1989. *Chrysonephele palustris* gen. et sp. nov. (Chrysophyceae), a new colonial chrysophyte from Tasmania. *Nova Hedwigia* 95:81–97.

Planas, D., 1996. Acidification effects. Pages 497–532 in *Algal Ecology: Freshwater Benthic Ecosystems*, edited by R. J. Stephenson, M. L. Bothwell, and R. L. Lowe. Academic Press, New York, NY.

Plumley, F. G. 1997. Marine algal toxins: Biochemistry, genetics, and molecular biology. *Limnology and Oceanography* 42:1252–1264.

Polanshek, A. R., and J. A. West. 1975. Culture and hybridization studies on *Petrocelis* from Alaska and

California. *Journal of Phycology* 11:434–439.

Polanshek, A. R., and J. A. West. 1977. Culture and hybridization studies on *Gigartina papillata* (Rhodophyta). *Journal of Phycology* 13:141–149.

Pollingher, U. 1988. Freshwater armored dinoflagellates: Growth, reproduction strategies, and population dynamics. Pages 134–174 in *Growth and Reproductive Strategies of Freshwater Phytoplankton*, edited by C. D. Sandgren. Cambridge University Press, Cambridge, UK.

Pollingher, U., and E. Zemel. 1981. *In situ* and experimental evidence of the influence of turbulence on cell division processes of *Peridinium cinctum* forma Westii (Lemm.) Lefevre. *British Phycological Journal* 16:281–287.

Polne-Fuller, M., and A. Gibor. 1984. Developmental studies in *Porphyra*. I. Blade differentiation in *Porphyra perforata* as expressed by morphology, enzymatic digestion, and protoplast regeneration. *Journal of Phycology* 20:609–616.

Poole, L. J., and J. A. Raven. 1997. The biology of *Enteromorpha*. *Progress in Phycological Research* 12:1–148.

Popova, A. F., K. M. Sytnik, E. L. Kordyum, G. I. Meleshko, V. N. Sychev, and M. A. Levinskykh. 1989. Ultrastructure and growth indices of *Chlorella* culture multicomponent aquatic systems under space flight conditions. *Advances in Space Research* 9:79.

Popovsky, J., and L. A. Pfiester. 1982. The life histories of *Stylodinium sphaera* Pascher and *Cystodinedria inermis* (Geitler) Pasher (Dinophyceae), two freshwater facultative predator-autotrophs. *Archiv für Protistenkunde* 125:115–127.

Porcella, R. A., and P. A. Walne. 1980. Microarchitecture and envelope development on *Dysmorphococcus globusus* (Phacotaceae, Chlorophyceae). *Journal of Phycology* 16:280–290.

Pore, R. S. 1986. The association of *Prototheca* spp. with slime flux in *Ulmus* and other trees. *Mycopathologia* 94:67–73.

Pore, R. S., E. A. Barnett, W. C. Barnes, and J. D. Walker. 1983. *Prototheca* ecology. Fungi, ubiquitous inhabitants of a raw and treated rural and municipal sewage, contamination of aquatic systems and food, which is consumed by humans and animals and then excreted, infection of both after traumatic inoculation. *Mycopathologia* 81:49–62.

Porter, K. G. 1977. The plant-animal interface in freshwater ecosystems. *American Scientist* 65:159–170.

Potter, D., T. C. Lajuenesse, G. W. Saunders, and R. A. Andersen. 1997a. Convergent evolution masks extensive biodiversity among marine coccoid picoplankton. *BioDiversity and Conservation* 6:99–107.

Potter, D., G. W. Saunders, and R. A. Andersen. 1997b. Phylogenetic relationships of the Raphidophyceae and Xanthophyceae as inferred from nucleotide sequences of the 18S ribosomal RNA gene. *American Journal of Botany* 84:966–972.

Power, M. E., A. J. Stewart, and W. J. Matthews. 1988. Grazer control of algae in an Ozark mountain stream: Effects of short-term exclusion. *Ecology* 69:1894–1898.

Preisig, H. R. 1995. A modern concept of chrysophyte classification. Pages 46–74 in *Chrysophyte Algae: Ecology, Phylogeny and Development*, edited by C. D. Sandgren, J. P. Smol, and J. Kristiansen. Cambridge University Press, Cambridge, UK.

Prema, P. 1996. Taxonomy and phylogeny of silicoflagellates. *Nova Hedwigia* 112:379–389.

Prescott, G. W. 1951. *Algae of the Western Great Lakes Area*, Wm. C. Brown, Dubuque, IA.

Prescott, G. W. 1978. *How to Know the Freshwater Algae*. Wm. C. Brown, Dubuque, IA.

Prescott, G. W., C. E. deM. Bicudo, and W. C. Vinyard. 1982. *A Synopsis of North American Desmids, Part II. Desmidiaceae: Placodermae, Section 4*. University of Nebraska Press, Lincoln, NE.

Prescott, G. W., H. T. Croasdale, and W. C. Vinyard. 1975. *A Synopsis of North American Desmids, Part II. Desmidiaceae: Placodermae, Section 1*. University of Nebraska Press, Lincoln, NE.

Prescott, G. W., H. T. Croasdale, and W. C. Vinyard.

1977. *A Synopsis of North American Desmids, Part II. Desmidiaceae: Placodermae, Section 2*. University of Nebraska Press, Lincoln, NE.

Prescott, G. W., H. T. Croasdale, W. C. Vinyard, and C. E. deM. Bicudo. 1981. *A Synopsis of North American Desmids, Part II. Desmidiaceae: Placodermae Section 3*. University of Nebraska Press, Lincoln, NE.

Prézelin, B. B., M. A. Moline, and A. Matlick. 1998. Icecolors '93: Spectral UV radiation effects on Antarctic frazil ice algae. Pages 45–83 in *Antarctic Sea Ice Biological Processes, Interactions, and Variability*, edited by M. P. Lizotte and K. R. Arrigo. Antarctic Research Series 73, American Geophysical Union, Washington, DC.

Price, G. D., D. Sültemeyer, B. Klughammer, M. Ludwig, and M. R. Badger. 1998. The functioning of the CO_2 concentrating mechanism in several cyanobacterial strains: A review of general physiological characteristics, genes, proteins, and recent advances. *Canadian Journal of Botany* 76:973–1002.

Pringsheim, N. 1860. Beiträge zur Morphologie und Systematik der Algen III. Die Coleochaeteen. *Jahrbuch fuer wissenschaftliche Botanik* 2:1–38.

Printz, H. 1939. Vorarbeiten zu einer Monographie der Trentepohliaceen. *Nytt Magasin f Naturv* 80:137–210.

Printz, H. 1964. Die Chaetophoralean der Binnengewässer. Eine systematische Ubersicht. *Hydrobiologia* 24:1–376.

Priscu, J. C., C. H. Fritsen, E. A. Adams, S. J. Giovannoni, H. W. Paerl, C. P. McKay, P. T. Doran, D. A. Gordon, B. D. Lanoil, and J. L.Pinckney. 1998. Perennial Antarctic lake ice: An oasis for life in a polar desert. *Science* 280:2095–2098.

Priscu, J. C., and C. W. Sullivan. 1998. Nitrogen metabolism in Antarctic fast-ice microalgal assemblages. Pages 147–160 in *Antarctice Sea Ice: Biological Processes, Interactions, and Variability*, edited by M. D. Lizotte and K. R. Arrigo. American Geophysical Union. *Antarctic Research Series* 73, American Geophysical Union, Washington, DC.

Proctor, V. W. 1980. Historical biogeography of *Chara* (Charophyta): An appraisal of the Braun-Wood classification plus a falsifiable alternative for future consideration. *Journal of Phycology* 16:218–233.

Proskauer, J. 1950. On *Prasinocladus*. *American Journal of Botany* 37:59–66.

Prosperi, C. H. 1994. A cyanophyte capable of fixing nitrogen under high levels of oxygen. *Journal of Phycology* 30:222–224.

Provasoli, L. 1958. Nutrition and ecology of protozoa and algae. *Annual Review of Microbiology* 12:279–308.

Provasoli, L., and I. J. Pintner. 1980. Bacteria induced polymorphism in an axenic laboratory strain of *Ulva lactuca* (Chlorophyceae). *Journal of Phycology* 16:196–201.

Prud'homme van Reine, W. F. 1982. *A taxonomic revision of the European Sphacelariaceae (Sphacelariales, Phaeophyceae)*. *Leiden Botanical Series* 6:1–293.

Puel, F., C. Largeau, and G. Giraud. 1987. Occurrence of a resistant biopolymer in the outer walls of the parasitic alga *Prototheca wickerhamii* (Chlorococcales): Ultrastructural and chemical studies. *Journal of Phycology* 23:649–656.

Pueschel, C. M. 1977. A freeze-etch study of the ultrastructure of red algal pit plugs. *Protoplasma* 91:15–30.

Pueschel, C. M. 1980. Pit connections and translocation in red algae. *Science* 209:422–423.

Pueschel, C. M. 1987. Absence of cap membranes as a characteristic of pit plugs of some red algal orders. *Journal of Phycology* 23:150–153.

Pueschel, C. M. 1988. Cell sloughing and chloroplast inclusions in *Hildenbrandia rubra* (Rhodophyta, Hildenbrandiales). *British Phycological Journal* 23:17–23.

Pueschel, C. M. 1989. An expanded survey of the ultrastructure of red algal pit plugs. *Journal of Phycology* 25:625–636.

Pueschel, C. M. 1990. Cell structure. Pages 7–42 in *Biology of the Red Algae*, edited by K. M. Cole and R. G. Sheath. Cambridge University Press, Cambridge, UK.

Pueschel, C. M. 1996. Development of epithallial cells in

Corallina officinalis and *Lithophyllum impressum* (Corallinales, Rhodophyta). *Phycologia* 35:161–169.

Pueschel, C. M., and K. M. Cole. 1982. Rhodophycean pit plugs: An ultrastructural survey with taxonomic implications. *American Journal of Botany* 69:703–720.

Pueschel, C. M., and F. Magne. 1987. Pit plugs and other ultrastructural features of systematic value in *Rhodochaete parvula* (Rhodophyta, Rhodochaetales). *Cryptogamie Algologie* 8:201–209.

Pueschel, C. M., and J. R. Stein. 1983. Ultrastructure of a freshwater brown alga from western Canada. *Journal of Phycology* 19:209–215.

Quatrano, R. S. 1973. Separation of processes associated with differentiation of two-celled *Fucus* embryos. *Developmental Biology* 30:209–213.

Quatrano, R. S., and S. L. Shaw. 1997. Role of the cell wall in the determination of cell polarity and the plane of cell division in *Fucus* embryos. *Trends in Plant Science* 2:15–21.

Rabbani, S., P. Beyer, J. V. Lintig, P. Hugueney, and H. Kleinig. 1998. Induced β-carotene synthesis driven by triacylglycerol deposition in the unicellular alga *Dunaliella bardawil*. *Plant Physiology* 116:1239–1248.

Radmer, R. J. 1996. Algal diversity and commercial algal products. *BioScience* 46:263–270.

Radmer, R. J., and B. C. Parker. 1994. Commercial applications of algae: Opportunities and constraints. *Journal of Applied Phycology* 6:93–98.

Raffaeli, D. G., J. A. Raven, and L. G. Poole. 1998. Ecological impact of green macroalgal blooms. *Oceanography and Marine Biology. An Annual Reveiw* 36:97–126.

Ragan, M. A. 1981. Chemical constituents of seaweeds. Pages 598–626 in *The Biology of Seaweeds*, edited by C. S. Lobban and M. J. Wynne. Blackwell Scientific, Oxford, UK.

Ragan, M. A., C. J. Bird, E. L. Rice, R. S. Gutell, C. A. Murphy, and R. K. Singh. 1994. A molecular phylogeny of the marine red algae (Rhodophyta) based on the nuclear small-subunit rRNA gene. *Proceedings of the National Academy of Science* 91:7276–7280.

Ragan, M. A., and R. R.Gutell. 1995. Are red algae plants? *Botanical Journal of the Linnean Society* 118:81–105.

Rambold, G., T. Friedl, and A. Beck. 1998. Photobionts in lichens: Possible indicators of phylogenetic relationships. *The Bryologist* 101:392–397.

Ramus, J. 1972. Differentiation of the green alga *Codium fragile*. *American Journal of Botany* 59:478–482.

Ramus, J. 1983. A physiological test of the theory of complementary chromatic adaptation. II. Brown, green and red seaweeds. *Journal of Phycology* 19:173–178.

Rappe, M. S., P. F. Kemp, and S. J. Giovannoni. 1995. Chromophyte plastid 16S ribosomal RNA genes found in a clone library from Atlantic Ocean seawater. *Journal of Phycology* 31:979–988.

Rassoulzadegan, F., M. Laval-Peuto, and R. W. Sheldon. 1988. Partitioning of the food ration of marine ciliates between pico- and nanoplankton. *Hydrobiologia* 159:75–88.

Raven, J. A. 1995. Comparative aspects of chrysophyte nutrition with emphasis on carbon, phosphorus and nitrogen. Pages 95–118 in *Chrysophyte Algae: Ecology, Phylogeny and Development*, edited by C. D. Sandgren, J. P. Smol, and J. Kristiansen. Cambridge University Press, Cambridge, UK.

Raven, J. A. 1997a. Phagotrophy in phototrophs. *Limnology and Oceanography* 42:198–205.

Raven, J. A. 1997b. Inorganic carbon acquisition by marine autotrophs. *Advances in Botanical Research* 27:86–209.

Raven, J. A. 1997c. Putting the C in phycology. *European Journal of Phycology* 32:319–333.

Raven, J. A., A. M. Johnston, and J. J. MacFarlane. 1990. Carbon metabolism. Pages 171–202 in *Biology of the Red Algae*, edited by K. M. Cole and R. G. Sheath. Cambridge University Press, Cambridge, UK.

Raven, J. A., F. A. Smith, and S. M. Glidewell. 1979. Photosynthetic capacities and biological strategies of giant-celled and small-celled macroalgae. *The New Phytologist* 83:249–309.

Raven, P. H., R. F. Evert, and S. E. Eichhorn. 1999. *Biology of Plants*, 6th ed. W. H. Freeman, Worth Publishers, New York, NY.

Rawitscher-Kunkel, E., and L. Machlis. 1962. The hormonal integration of sexual reproduction in *Oedogonium*. *American Journal of Botany* 49:177–183.

Redfield, A. C. 1958. The biological control of chemical factors in the environment. *American Scientist* 46:205–221.

Reed, R. H. 1990. Solute accumulation and osmotic adjustment. Pages 147–170 in *Biology of the Red Algae*, edited by K. M. Cole and R. G. Sheath. Cambridge University Press, Cambridge, UK.

Reimers, C. E. 1998. Carbon cycle feedbacks from the sea floor. *Nature* 391:536–537.

Reiss, H. D., C. Katsaros, and B. Galatis. 1996. Freeze-fracture studies in the brown alga *Asteronema rhodochortonoides*. *Protoplasma* 193:46–57.

Reisser, W. (editor). 1992a. *Algae and Symbioses*. BioPress, Bristol, UK.

Reisser, W. 1992b. Basic mechanisms of signal exchange, recognition, specificity, and regulation in endosymbiotic systems. Pages 657–674 in *Algae and Symbioses*, edited by W. Reisser. BioPress, Bristol, UK.

Reisser, W. 1992c. Endosymbiotic associations of algae with freshwater protozoa and invertebrates. Pages 1–20 in *Algae and Symbioses*, edited by W. Reisser. BioPress, Bristol, UK.

Rensing, S. A., P. Obrdlik, N. Rober-Kleber, S. B. Müller, C. J. B. Hofmann, Y. Van de Peer, and U.-G. Maier. 1997. Molecular phylogeny of the stress-70 protein family with reference to algal relationships. *European Journal of Phycology* 32:279–285.

Reuter, J. E., and R. P. Axler. 1992. Physiological characteristics of inorganic nitrogen uptake by spatially separate algal communities in a nitrogen-deficient lake. *Freshwater Biology* 27:227–236.

Reynolds, C. S. 1972. Growth, gas vacuolation and buoyancy in a natural population of a planktonic blue-green alga. *Freshwater Biology* 2:87–106.

Reynolds, C. S. 1984. *The Ecology of Freshwater Phytoplankton*. Cambridge University Press, Cambridge, UK.

Reynolds, C. S., G. H. M.. Jaworski, H. A. Cmiech, and G. F. Leedale. 1981. On the annual cycle of the blue-green alga *Microcystis aeruginosa* Kütz emend Elenkin. *Philosophical Transactions of the Royal Society of London*, B293:419–477.

Reynolds, C. S., J. M. Thompson, A. J. D. Ferguson, and S. W. Wiseman. 1982. Loss processes in the population dynamics of phytoplankton maintained in closed systems. *Journal of Plankton Research* 4:561–600.

Reynolds, C. S., A. E. Walsby, and R. L. Oliver. 1987. The role of buoyancy in the distribution of *Anabaena* s. in Lake Rotaongaio. *New Zealand Journal of Marine and Freshwater Research* 21:525–26.

Rhiel, E., J. Marquardt, M. Eppart, E., Mörschel, and W. E. Krumbein. 1997. The light harvesting system of the diatom *Cyclotella cryptica*. Isolation and characterization of the main light harvesting complex and evidence for the existence of minor pigment proteins. *Botanica Acta* 110:109–117.

Ribera, M. A., and C. F. Boudouresque. 1995. Introduced marine plants, with special reference to macroalgae: Mechanisms and impact. *Progress in Phycological Research* 11:188–217.

Richardson, F. L., and R. E. Brown. 1970. *Glaucosphaera vacuolata*, its ultrastructure and physiology. *Journal of Phycology* 6:165–171.

Richardson, L. 1997. Occurrence of the black band disease cyanobacterium on healthy corals of the Flordida Keys. *Bulletin of Marine Science* 61:485–490.

Richardson, P. L., R. E. Cheney, and L. V. Worthington. 1978. A census of Gulf Stream Rings, Spring 1975. *Journal of Geophysical Research* 83:6136–6144.

Richerd, S., C. Destombe, J. Cuguen, and M. Valero. 1993. Variation of reproductive success in a haplo-diploid red alga, *Gracilaria verrucosa*: Effects of parental identities and crossing distance. *American Journal of Botany* 80:1379–1391.

Richmond, A. 1988. *Spirulina*. Pages 85–121 in *Micro-Algal Biotechnology*, edited by M. A. Borowitzka and L. J. Borowitzka. Cambridge University Press, Cambridge, UK.

Richmond, A. 1990. Large scale microalgal culture and applications. *Progress in Phycological Research* 7:269–330.

Riebesell, U., D. A. Wolf-Gladrow, and V. Smetacek. 1993. Carbon dioxide limitation of marine phytoplankton growth rates. *Nature* 361:249–251.

Rinehart, K. L., M. Namikoshi, and B. W. Choi. 1994. Structure and biosynthesis of toxins from blue-green algae (cyanobacteria). *Journal of Applied Phycology* 6:159–176.

Rines, J. E. B., and P. E. Hargraves. 1988. The *Chaetoceros* Ehrenberg (Bacillariophyceae) flora of Narragansett Bay, Rhode Island, U.S.A. *Bibliotheca Phycologica* 79:1–196.

Ripka, R., J. Waterbury, and G. Cohen-Bazire. 1974. A cyanobacterium which lacks thylakoids. *Archiv für Mikrobiologie* 100:419–436.

Roberts, K. R. 1984. Structure and significance of the cryptomonad flagellar apparatus. I. *Cryptomonas ovata* (Cryptophyta). *Journal of Phycology* 20:590–599.

Roberts, K. R., M. A. Farmer, R. M. Schneider, and J. E. Lemoine. 1988. The microtubular cytoskeleton of *Amphidinium rhynchocephalum*. *Journal of Phycology* 24:544–553.

Roberts, K. R., G. Hansen, and F. J. R. Taylor. 1995. General ultrastructure and flagellar apparatus architecture of *Woloszynskia limnetica* (Dinophyceae). *Journal of Phycology* 31:948–957.

Roberts, S. B., T. W. Lane, and F. M. M. Morel. 1997. Carbonic anhydrase in the marine diatom *Thalassiosira weissflogii* (Bacillariophyceae). *Journal of Phycology* 33:845–850.

Robinson, D. G., M. Hoppenrath, K. Oberbeck, P. Luykx, and R. Ratajczak. 1998. Localization of pyrophosphatase and V-ATPase in *Chlamydomonas reinhardtii*. *Botanica Acta* 111:108–122.

Robinson, N., G. Eglinton, S. C. Brassell, and P. A. Cranwell. 1984. Dinoflagellate origin for sedimentary 4α-methylsteroids and 5α(H)-stanols. *Nature* 308:439–442.

Rochaix, J.-D. 1997. Chloroplast reverse genetics: New insights into the function of plastid genes. *Trends in Plant Science* 2:419–424.

Roessler, P. G. 1990. Environmental control of glycerolipid metabolism in microalgae: Commercial implications and future research directions. *Journal of Phycology* 26:393–399.

Rogers, C. E., D. S. Domozych, K. D. Stewart, and K. R. Mattox. 1981. The flagellar apparatus of *Mesostigma viride* (Prasinophyceae): Multilayered structures in a scaly green flagellate. *Plant Systematics and Evolution* 138:247–258.

Rogers, C. E., K. R. Mattox, and K. D. Stewart. 1980. The zoospore of *Chlorokybus atmophyticus*, a charophyte with sarcinoid growth habit. *American Journal of Botany* 67:774–783.

Rogerson, A., A. S. W. DeFreitas, and A. G. McInnes. 1986. Growth rates and ultrastructure of siliceous setae of *Chaetoceros gracilis* (Bacillariophyceae). *Journal of Phycology* 22:56–62.

Romo, S., and C. Perez-Martinez. 1997. The use of immobilization in alginate beads for long-term storage of *Pseudanabaena galeata* (cyanobacteria) in the laboratory. *Journal of Phycology* 33:1074–1076.

Rosemond, A. D., and S. H. Brawley. 1996. Species-specific characteristics explain the persistence of *Stigeoclonium tenue* (Chlorophyta) in a woodland stream. *Journal of Phycology* 32:54–63.

Rosowski, J. R., and W. G. Langenberg. 1994. The near-spineless *Trachelomonas grandis* (Euglenophyceae) superficially appears spiny by attracting bacteria to its surface. *Journal of Phycology* 30:1012–1022.

Roth, P. H. 1994. Distribution of coccoliths in oceanic sediments. Pages 199–218 in *Coccolithophores*, edited by A. Winter and W. G. Siesser. Cambridge University Press, Cambridge, UK.

Rothhaupt, K. O. 1992. Stimulation of phosphorus limited phytoplankton by bacterivorous flagellates in laboratory experiments. *Limnology and Oceanography* 37:750–759.

Rothhaupt, K. O. 1995. Algal nutrient limitation affects rotifer growth rate but not ingestion rate. *Limnology and Oceanography* 40:1201–1208.

Rothhaupt, K. O. 1996a. Laboratory experiments with a mixotrophic chrysophyte and obligately phagotrophic and phototrophic competitors. *Ecology* 77:716–724.

Rothhaupt, K. O. 1996b. Utilization of substitutable carbon and phosphorus sources by the mixotrophic chrysophyte *Ochromonas* sp. *Ecology* 77:706–715.

Rothschild, L. J., L. J. Giver, M. R. White, and R. L. Mancinelli. 1994. Metabolic activity of microorganisms in evaporites. *Journal of Phycology* 30:431–438.

Round, F. E. 1992. Epibiotic and endobiotic associations between chromophyte algae and their hosts. Pages 593–618 in *Algae and Symbioses*, edited by W. Reisser. BioPress, Bristol, UK.

Round, F. E., R. M. Crawford, and D. G. Mann. 1990. *The Diatoms—Biology and Morphology of the Genera*. Cambridge University Press, Cambridge, UK.

Rowan, R. 1998. Diversity and ecology of zooxanthellae on coral reefs. *Journal of Phycology* 34:407–417.

Rowan, R., and N. Knowlton. 1995. Intraspecific diversity and ecological zonation in coral-algal symbiosis. *Proceedings of the National Academy of Science* 92:2850–2853.

Rowan, R., S. M. Whitney, A. Fowler, and D. Yellowlees. 1996. Rubisco in marine symbiotic dinoflagellates: Form II enzymes in eukaryotic oxygenic phototrophs encoded by a nuclear multigene family. *The Plant Cell* 8:539–553.

Russell, C. A., M. D. Guiry, A. R. McDonald, and D. J. Garbary. 1996. Actin-mediated chloroplast movement in *Griffithsia pacifica* (Ceramiales, Rhodophyta). *Phycological Research* 44:57–61.

Sabater, S., S. V. Gregory, and J. R. Sedell. 1998. Community dynamics and metabolism of benthic algae colonizing wood and rock substrata in a forest stream. *Journal of Phycology* 34:561–567.

Saffo, M. B. 1987. New light on seaweeds. *BioScience* 37:654–664.

Sagert, S., R. M. Forster, P. Feuerpfiel, and H. Schubert. 1997. Daily course of photosynthesis and photoinhibition in *Chondrus crispus* (Rhodophyta) from different shore levels. *European Journal of Phycology* 32:363–371.

Salisbury, J. L., A. T. Baron, and M. A. Sanders. 1988. The centrin-based cytoskeleton of *Chlamydomonas reinhardtii*: Distribution in interphase and mitotic cells. *Journal of Cell Biology* 107:635–641.

Salisbury, J. L., and G. L. Floyd. 1978. Calcium-induced contraction of the rhizoplast of a quadriflagellate green alga. *Science* 202:975–977.

Salisbury, J. L., J. A. Swanson, G. L. Floyd, R. Hall, and N. J. Maihle. 1981. Ultrastructure of the flagellar apparatus of the green alga *Tetraselmis subcordiformis* with special consideration given to the function of the rhizoplast and rhizanchora. *Protoplasma* 107:1–11.

Sanders, R. W., C. E. Williamson, P. L. Stutzman, R. E. Moeller, C. E. Goulden, and T. Aoki-Goldsmith. 1996. Reproductive success of "herbivorous" zooplankton fed algal and nonalgal food resources. *Limnology and Oceanography* 41:1295–1305.

Sandgren, C. D. 1981. Characteristics of sexual and asexual resting cyst (statospore) formation in *Dinobryon cylindricum* Imhof (Chrysophyta). *Journal of Phycology* 17:199–210.

Sandgren, C. D. (editor). 1988. *Growth and Reproductive Strategies of Freshwater Phytoplankton*. Cambridge University Press, Cambridge, UK.

Sandgren, C. D., and J. Flanagin. 1986. Heterothallic sexuality and density dependent encystment in the chrysophycean alga *Synura petersenii* Korsh. *Journal of Phycology* 22:206–216.

Sandgren, C. D., S. A. Hall, and S. B. Barlow. 1996. Siliceous scale production in chrysophyte and synurophyte algae. I. Effects of silica-limited growth on cell silica content, scale morphology, and the construction of the scale layer of *Synura petersenii*. *Journal of Phycology* 32:675–692.

Sandgren, C. D., and W. E. Walton. 1995. The influence of zooplankton herbivory on the biogeography of chrysophyte algae. Pages 269–302 in *Chrysophyte Algae: Ecology, Phylogeny and Development*, edited by C. D. Sandgren, J. P. Smol, and J. Kristiansen. Cambridge University Press, Cambridge, UK.

Sand-Jensen, K., O. Pedersen, and O. Geertz-Hansen. 1997. Regulation and role of photosynthesis in the colonial symbiotic ciliate *Ophrydium versatile*. *Limnology and Oceanography* 42:866–873.

Sands, M., D. Poppel, and R. Brown. 1991. Peritonitis due to *Prototheca wickerhamii* in a patient undergoing chronic ambulatory peritoneal dialysis. *Reviews of Infectious Diseases* 13:376–378.

Santelices, B. 1980. Phytogeographic characterization of the temperate coast of Pacific South America. *Phycologia* 19:1–12.

Santelices, B., and I. A. Abbott. 1987. Geographic and marine isolation: An assessment of the marine algae of Easter Island. *Pacific Science* 41:1–20.

Santelices, B., P. Camus, and A. J. Hoffmann. 1989. Ecological studies for harvesting and culturing *Gymnogongrus furcellatus* (Rhodophyta, Gigartinales) in Central Chile. *Journal of Applied Phycology* 1:171–181.

Santore, U. J. 1985. A cytological survey of the genus *Cryptomonas* (Cryptophyceae) with comments on its taxonomy. *Archiv für Protistenkunde* 130:1–52.

Santos, L. M. A. 1996. The Eustigmatophyceae: Actual knowledge and research perspectives. *Nova Hedwigia* 112:391–405.

Santos, L. M. A., and G. F. Leedale. 1995. Some notes on the ultrastructure of small azoosporic members of the algal class Eustigmatophyceae. *Nova Hedwigia* 60:219–225.

Sarcina, M., and P. J. Casselton. 1995. Degradation of adenine by *Prototheca zopfii* (Chlorophyta). *Journal of Phycology* 31:575–576.

Sauer, N., and W. Tanner. 1993. Molecular biology of sugar transporters in plants. *Botanica Acta* 106:277–286.

Saunders, G. W., and J. C. Bailey. 1997. Phylogenesis of pit-plug-associated features in the Rhodophyta: Inference from molecular systematic data. *Canadian Journal of Botany* 75:1436–1447.

Saunders, G. W., C. J. Bird, M. A. Ragan, and E. L. Rice. 1995a. Phylogenetic relationships of species of uncertain taxonomic position within the Acrochaetiales-Palmariales complex (Rhodophyta): Inferences from phenotypic and 18S rDNA sequence data. *Journal of Phycology* 31:601–611.

Saunders, G. W., D. R. A. Hill, J. P. Sexton, and R. A. Andersen. 1997a. Small-subunit ribosomal RNA sequences from selected dinoflagellates: Testing classical evolutionary hypotheses with molecular systematic methods. *Plant Systematics and Evolution* 11:237–259.

Saunders, G. W., D. Potter, and R. A. Andersen. 1997b. Phylogenetic affinities of the Sarcinochrysidales and Chrysomeridales (Heterokonta) based on analyses of molecular and combined data. *Journal of Phycology* 33:310–318.

Saunders, G. W., D. Potter, M. P. Paskind, and R. A. Andersen. 1995b. Cladistic analyses of combined traditional and molecular data sets reveal an algal lineage. *Proceedings of the National Academy of Science* 92:244–248.

Sawitzky, H., and F. Grolig. 1995. Phragmoplast of the green alga *Spirogyra* is functionally distinct from the higher plant phragmoplast. *Journal of Cell Biology* 130:1359–1371.

Scagel, R. F. 1966. *Marine Algae of British Columbia and Northern Washington. Part I: Chlorophyceae (Green Algae)*. National Museum of Canada Bulletin 207.

Schenk, H. E. A., M. G. Bayer, T. L. Maier, A. Luettke, U. G. Gebhart, and S. Stevanovic. 1992. Ferredoxin-NADP oxidoreductase of *Cyanophora paradoxa* nucleus encoded, but cyanobacterial: gene transfer from symbiont to host, an evolutionary mechanism originating new species. *Zeitschrift für Naturforschung: Section C Biosciences* 47:387–393.

Schindler, D. E., S. R. Carpenter, J. L. Cole, J. F. Kitchell, and M. L. Pace. 1997. Influence of food web structure on carbon exchange between lakes and the atmosphere. *Science* 277:248–251.

Schindler, D. W., K. H. Mills, D. F. Malley, D. L. Findlay, J. A., Shearer, I. J. Davies, M. A. Turner, G. A. Linsey, and D. R. Cruikshank. 1995. Long-term ecosystem stress: The effects of years of experimental acidification on a small lake. *Science* 228:1395–1401.

Schlegel, M. 1994. Molecular phylogeny of eukaryotes. *Trends in Ecology and Evolution* 9:330–335.

Schloesser, R. E., and J. L. Blum. 1980. *Sphacelaria lacustris* sp. nov., a freshwater brown alga from Lake Michigan. *Journal of Phycology* 16:201–207.

Schmid, A.-M. M. 1997. Putative main function of actin bundles in raphid diatoms: Necessity for a new locomotion model. *Phycologia* 36:99 (abstract).

Schmitt, R., S. Fabry, and D. L. Kirk. 1992. In search of molecular origins of cellular differentiation in *Volvox* and its relatives. *International Review of Cytology* 139:189–265.

Schmitz, K. 1981. Translocation. Pages 534–558 in *The Biology of Seaweeds*, edited by C. S. Lobban and M. J. Wynne. Blackwell Scientific, Oxford, UK.

Schneegurt, M. A., D. M. Sherman, and L. A. Sherman. 1997. Growth, physiology, and ultrastructure of a diazotrophic cyanobacterium *Cyanothece* sp strain ATCC 51142, in mixotrophic and chemoheterotrophic cultures. *Journal of Phycology* 33:632–642.

Schnepf, E. S. Winter, and D. Mollenhauer. 1989. *Gymnodinium aeruginosum* (Dinophyta): A blue-green dinoflagellate with a vestigial, anucleate, cryptophycean endosymbiont. *Plant Systematics and Evolution*. 164:75–92.

Schofield, O., T. J. Evans, and D. F. Mille. 1998. Photosystem II quantum yields and xanthophyll-cycle pigments of the macroalga *Sargassum natans* (Phaeophyceae) responses under natural sunlight. *Journal of Phycology* 34:104–112.

Scholin, C., and D. M. Anderson. 1996. LSU rDNA-based RFLP assays for discriminating species and strains of *Alexandrium* (Dinophyceae). *Journal of Phycology* 32:1022–1035.

Scholin, C., P. Miller, K. Buck, F. Chavez, P. Harris, P. Haydock, J. Howard, and G. Cangelosi. 1997. Detection and quantification of *Pseudo-nitzschia australis* in cultured and natural populations using LSU rRNA-targeted probes. *Limnology and Oceanography* 42:1265–1272.

Scholin, C. A. 1998. Development of nucleic acid probe-based diagnostics for identifying and enumerating harmful algal bloom species. Pages 337–350 in *The Physiological Ecology of Harmful Algal Blooms*, edited by D. M. Anderson, A. D. Cembella, and G. M. Hallegraeff. Springer-Verlag, Berlin, Germany.

Schonbeck, M, and T. A. Norton. 1978. Factors controlling the upper limits of fucoid algae on the shore. *Journal of Experimental Marine Biology and Ecology* 31:303–313.

Schopf, J. W. 1993. Microfossils of the Early Archean Apex chert: New evidence of the antiquity of life. *Science* 260:640–646.

Schopf, J. W. 1996. Cyanobacteria: Pioneers of the early earth. *Nova Hedwigia* 112:13–32.

Schornstein, K. L., and J. Scott. 1982. Ultrastructure of cell division in the unicellular red alga *Porphyridium purpureum*. *Canadian Journal of Botany* 60:85–97.

Schramm, W., and P. H. Nienhuis. 1996. *Marine benthic vegetation: Recent changes and the effects of eutrophication*. Springer Verlag, Berlin, Germany.

Schwartzkopf, S. M. 1992. Design of a controlled ecological life support system. *BioScience* 42:526–535.

Schwenke, H. 1971. Water movement. Pages 1091–1121 in *Marine Ecology*, edited by O. Kinne. Wiley-Interscience, London, UK.

Scott, J., C. Bosco, K. Schornstein, and J. Thomas. 1980. Ultrastructure of cell division and reproductive differentiation of male plants in the Florideophyceae (Rhodophyta): Cell division in *Polysiphonia*. *Journal of Phycology* 16:507–524.

Scott, J., and S. Broadwater. 1990. Cell division. Pages 123–146 in *Biology of the Red Algae*, edited by K. M. Cole, and R. G. Sheath. Cambridge University Press,

Cambridge, UK.

Scott, J. L, and P. S. Dixon. 1973. Ultrastructure of spermatium liberation in the marine red alga *Ptilota densa*. *Journal of Phycology* 9:85–91.

Scrosati, R., and R. E. DeWreede. 1998. The impact of frond crowding on frond bleaching in the clonal intertidal alga *Mazzaella cornucopiae* (Rhodophyta, Gigartinaceae) from British Columbia, Canada. *Journal of Phycology* 34:228–232.

Searles, R. B. 1980. The strategy of the red algal life history. *American Naturalist* 115:113–120.

Seckbach, J. 1994. The natural history of *Cyanidium* (Geitler 1933: past and present perspectives). Pages 99–112 in *Evolutionary Pathways and Enigmatic Algae: Cyanidium caldarium (Rhodophyta) and Related Cells*, edited by J. Seckbach. Kluwer Academic Publishers, Boston, MA.

Seckbach, J., E. González, and I. M. Wainright. 1992. Peroxisomal function in the Cyanidiophyceae (Rhodophyta): A discussion of phylogenetic relationships and the evolution of microbodies (peroxisomes). *Nova Hedwigia* 55:99–109

Sekimoto, H., S. Satoh, and T. Fujii. 1993. Analysis of binding of biotinylated protoplast-release-inducing protein that induces release of gametic protoplasts in the *Closterium peracerosum-strigosum-littorale* complex. Planta 189:468–474.

Senft, W. H., R. A. Hunchberger, and K. E. Roberts. 1981. Temperature dependence of growth and phosphorus uptake in two species of *Volvox* (Volvocales, Chlorophyta). *Journal of Phycology* 17:323–329.

Setchell, W. A. 1918. Parasitism among the red algae. *Proceedings of the American Philosophical Society* 57:155–172.

Shannon, C. E., and Weaver, W. 1949. The mathematical theory of communication. University of Illinois Press, Urbana, IL.

Shapiro, J., and D. I. Wright. 1984. Lake restoration by biomanipulation: Round Lake, Minnesota, the first two years. *Freshwater Biology* 14:371–83.

Sheath, R. G., and K. M. Cole. 1980. Distribution and salinity adaptations of *Bangia atropurpurea* (Rhodophyta), a putative migrant into the Laurentian Great Lakes. *Journal of Phycology* 16:412–420.

Sheath, R. G., and K. M. Cole. 1992. Biogeography of stream macroalgae in North America. *Journal of Phycology* 28:448–460.

Sheath, R. G., and J. A. Hambrook. 1988. Mechanical adaptations to flow in freshwater red algae. *Journal of Phycology* 24:107–111.

Sheath, R. G., and J. A. Hambrook. 1990. Freshwater ecology. Pages 423–454 in *Biology of the Red Algae*, edited by K. M. Cole and R. G. Sheath. Cambridge University Press, Cambridge, UK.

Sheath, R. G., K. M. Müller, M. L. Vis, and T. J. Entwhistle. 1996a. A re-examination of the morphology, ultrastructure and classification of genera in the Lemaneaceae (Batrachospermales, Rhodophyta). *Phycological Research* 44:233–246.

Sheath, R. G., M. K. Müller, A. Whittick, and T. J. Entwhistle. 1996b. A re-examination of the morphology and reproduction of *Nothocladus lindauerii* (Batrachospermales, Rhodophyta). *Phycological Research* 44:1–10.

Shen, E. Y. F. 1967. Microspectrophotometric analysis of nuclear DNA in *Chara zeylanica*. *Journal of Cell Biology* 35:377–384.

Shiraiwa, Y., and Y. Umino. 1991. Effect of glucose on the induction of the carbonic anhydrase and the change in $K_{1/2}(CO_2)$ of photosynthesis in *Chlorella vulgaris* 11h. *Plant Cell Physiology* 32:311–314.

Sicko-Goad, L., E. F. Stoermer, and G. Fahnenstiel. 1986. Rejuvenation of *Melosira granulata* (Bacillariophyceae) resting cells from the anoxic sediments of Douglas Lake, Michigan. I. Light microscopy and ^{14}C uptake. *Journal of Phycology* 22:22–28.

Sieburth, J. M., P. W. Johnson, and P. W. Hargraves. 1988. Ultrastructure and ecology of *Aureococcus anophagefferens* gen. et sp. nov. (Chrysophyceae): The dominant phytoplanter during a bloom in Narragansett Bay, Rhode Island, Summer 1985. *Journal of Phycology* 24:416–425.

Siemeister, G., and W. Hactel. 1990. Structure and expression of a gene encoding the large subunit of ribulose-1,5-bisphosphate carboxylase (*rbcL*) in the colourless flagellate *Astasia longa*. *Plant Molecular Biology* 14:825–833.

Siemer, B. L., W. T. Stam, J. L. Olsen, and P. M. Pedersen. 1998. Phylogenetic relationships of the brown algal orders Ectocarpales, Chordariales, Dictyosiphonales, and Tilopteridales (Phaeophyceae) based on Rubisco large subunit and spacer sequences. *Journal of Phycology* 34:1038–1048.

Siesser, W. G., 1994. Historical background of coccolithophore studies. Pages 1–11 in *Coccolithophores*, edited by A. Winter and W. G. Siesser. Cambridge University Press, New York, NY.

Siesser, W. G., and A. Winter. 1994. Composition and morphology of coccolithophore skeletons. Pages 51–62 in *Coccolithophores*, edited by A. Winter and W. G. Siesser. Cambridge University Press, Cambridge, UK.

Silva, P. C. 1980. Names of classes and families of living algae. *Regnum Vegetabile* 103:1–156.

Simek, K., M. Macek, J. Pernthaler, V. Straskrabová, and R. Psenner. 1996. Can freshwater planktonic ciliates survive on a diet of picoplankton? *Journal of Plankton Research* 18:597–613.

Simon, N., J. Brenner, B. Edvardsen, and L. K. Medlin. 1997. The identification of *Chrysochromulina* and *Prymnesium* species (Haptophyta, Prymnesiophyceae) using fluorescent or chemiluminescent oligonucleotide probes: a means for improving studies on toxic algae. *European Journal of Phycology* 32:393–401.

Simons, J. 1974. *Vaucheria compacta*: A euryhaline estuarine algal species. *Acta Botanica Neerlandica* 23:613–626.

Simonsen, R. 1979. The diatom system: Ideas on phylogeny. *Bacillaria* 2:9–71.

Simpson, A. G. B. 1997. The identity and composition of the Euglenozoa. *Archiv für Protistenkunde* 148:318–328.

Simpson, F. B., and J. B. Neilands. 1976. Siderochromes in cyanophyceae: Isolation and characterization of schizokinen from *Anabaena* sp. *Journal of Phycology* 12:44–48.

Sineshchekov, D. A., and E. G. Govorunova. 1999. Rhodopsin-mediated photosensing in green flagellated algae. *Trends in Plant Science* 4:58–63.

Sitte, P. 1993. Symbiogenetic evolution of complex cells and complex plastids. *European Journal of Protistology* 29:131–143.

Siver, P. A. 1995. The distribution of chrysophytes along environmental gradients: Their use as biological indicators. Pages 232–268 in *Chrysophyte Algae: Ecology, Phylogeny and Development*, edited by C. D. Sandgren, J. P. Smol, and J. Kristiansen. Cambridge University Press, Cambridge, UK.

Sivonen, K. 1996. Cyanobacterial toxins and toxin production. *Phycologia* 35:12–24.

Skovgaard, A. 1996. Engulfment of *Ceratium* spp. (Dinophyceae) by the thecate photosynthetic dinoflagellate *Fragilidium subglobosum*. *Phycologia* 35:490–499.

Skulberg, O. M. 1995. Use of algae for testing water quality. Pages 181–199 in *Algae, Environment and Human Affairs*, edited by W. Wiessner, E. Schnepf, and R. C. Starr. BioPress, Bristol, UK.

Sluiman, H. 1993. Nucleus, nuclear division, and cell division. Pages 221–268 in *Ultrastructure of Microalgae*, edited by T. Berner. CRC Press, Boca Raton, FL.

Sluiman, H. J. 1983. The flagellar apparatus of the zoospore of the filamentous green alga *Coleochaete pulvinata*: Absolute configuration and phylogenetic significance. *Protoplasma* 115:160–175.

Sluiman, H. J. 1989. The green algal class Ulvophyceae. An ultrastructural survey and classification. *Cryptogamic Botany* 1:83–94.

Sluiman, H. J., and C. Guihal. 1999. Phylogenetic position of *Chaetosphaeridium* (Chlorophyta), a basal lineage in the Charophyceae inferred from 18S rDNA sequences. *Journal of Phycology* 35:395–402.

Smayda, T. J. 1970. The suspension and sinking of phytoplankton in the sea. *Oceanography and Marine Biology Annual Review* 8:353–414.

Smith, A. J. 1973. Synthesis of metabolic intermediates. Pages 1–38 in *The Biology of Blue-green Algae*, edited by N. G. Carr and B. A. Whitton. University of California Press, Berkeley, CA.

Smith, E. C., and H. Griffiths. 1996. The occurrence of the chloroplast pyrenoid is correlated with the activity of a CO_2-concentrating mechanism and carbon isotope discrimination in lichens and bryophytes. *Planta* 198:6–16.

Smith, E. F., and W. S. Sale. 1992. Regulation of dynein-driven microtubule sliding by the radial spokes in flagella. *Science* 257:1557–1558.

Smith, G. M. 1917. The vertical distribution of *Volvox* in the plankton of Lake Mendota. *American Journal of Botany* 5:78–185.

Smith, G. M. 1950. *The Fresh-Water Algae of the United States*, McGraw-Hill, New York, NY.

Smith, G. M. 1969. *Marine Algae of the Monterey Peninsula, CA* (incorporating 1966 Supplement by G. J. Hollenberg and I. A. Abbott), Stanford University Press, Stanford, CA.

Smol, J. P. 1995. Application of chrysophytes to problems in paleoecology. Pages 303–329 in *Chrysophyte Algae: Ecology, Phylogeny and Development*, edited by C. D. Sandgren, J. P. Smol, and J. Kristiansen. Cambridge University Press, Cambridge, UK.

Sœmundsdóttir, S., and P. A. Matrai. 1998. Biological production of methyl bromide by cultures of marine phytoplankton. *Limnology and Oceanography* 43:81–87.

Sogin, M. L., H. J. Elwood, and J. H. Gunderson. 1986. Evolutionary diversity of eukaryotic small-subunit rRNA genes. *Proceedings of the National Academy of Science* 83:1383–1387.

Sommer, U. 1981. The role of r- and K-selection in the succession of phytoplankton in Lake Constance. *Acta Oecologica/Oecologia Generalis* 2:327–342.

Sommer, U. 1985. Comparison between steady state and non-steady state competition: Experiments with natural phytoplankton. *Limnology and Oceanography* 30:335–346.

Sommer, U. 1986a. Phytoplankton competition along a gradient of dilution rates. *Oecologia* 68:503–506.

Sommer, U. 1986b. Nitrate- and silicate-competition among antarctic phytoplankton. *Marine Biology* 91:345–351.

Sommer, U. 1988a. Does nutrient competition among phytoplankton occur *in situ*? *Internationale Vereinigung für Theoretische und Angewandte Limnologie* 23:707–712.

Sommer, U. 1988b. Growth and survival strategies of planktonic diatoms. Pages 227–260 in *Growth and Reproductive Strategies of Freshwater Phytoplankton*, edited by C. D. Sandgren. Cambridge University Press, Cambridge, UK.

Sommer, U. 1989. The role of competition for resources in phytoplankton succession. Pages 57–106 in *Plankton Ecology*, edited by U. Sommer. Springer-Verlag, Berlin, Germany.

Sommer, U., and Z. M. Gliwicz. 1986. Long range vertical migration of *Volvox* in tropical Lake Cahora Bassa (Mozambique). *Limnology and Oceanography* 31:650–653.

Sondergaard, M., E. Jeppesen, E. Mortensen, E. Dall, P. Kristensen, and O. Sortkjaer. 1990. Phytoplankton biomass reduction after planktivorous fish reduction in a shallow, eutrophic lake: a combined effect of reduced internal P-loading and increased zooplankton grazing. *Hydrobiologia* 200/201:229–240.

Sournia, A. 1982. Form and function in marine phytoplankton. *Biological Review* 57:347–394.

Spaulding, S. A., D. M. McKnight, R. L. Smith, and R. Dufford. 1994. Phytoplankton population dynamics in perennially ice-covered Lake Fryxell, Antarctica. *Journal of Plankton Research* 16:527–541.

Spear-Bernstein, L., and K. R. Miller. 1989. Unique location of the phycobiliprotein light-harvesting pigment in the cryptophyceae. *Journal of Phycology* 25:412–419.

Spencer, K. G. 1988. Lipids and polyols from microalgae. Pages 237–253 in *Algae and Human Affairs*, edited by C. A. Lembi and J. R. Waaland. Cambridge University Press, Cambridge, UK.

Spero, H. J., and D. L. Angel. 1991. Planktonic sarcodines: Microhabitat for oceanic dinoflagellates. *Journal of Phycology* 27:187–195.

Spreitzer, R. J., G. Thow, and G. Zhu. 1995. Pseudoreversion substitution at large-subunit residue 54 influences the CO_2/O_2 specificity of chloroplast ribulose-bisphosphate carboxylase/oxygenase. *Plant Physiology* 109:681–685.

Stache-Crain, B., D. G. Müller, and L. J. Goff. 1997. Molecular systematics of *Ectocarpus* and *Kuckuckia* (Ectocarpales, Phaeophyceae) inferred from phylogenetic analysis of nuclear- and plastid-encoded DNA sequences. *Journal of Phycology* 33:152–168.

Stadler, R., K. Wolf, C. Hilgarth, W. Tanner, and N. Sauer. 1995. Subcellular localization of the inducible Chlorella HUP1 monosaccharide-H$^+$ symporter and cloning of a co-induced galactose-H$^+$ symporter. *Plant Physiology* 107:33–41.

Stal, L. J. 1995. Physiological ecology of cyanobacteria in microbial mats and other communities. *New Phytologist* 131:1–32.

Stanier, R. V., W. R. Sistrom, T. A. Hansen, B. A. Whitton, R. W. Castenholz, N. Pfennig, V. N. Gorlenko, E. N. Kondratieva, K. E. Eimhjellen, R. Whittenbury, R. I. Gherna, and H. G. Trüper. 1978. Proposal to place the nomenclature of the cyanobacteria (blue-green algae) under the rules of the International Code of Nomenclature of Bacteria. *International Journal of Systematic Bacteriology* 28:335–336.

Starke, T., and J. P. Gogarten. 1993. A conserved intron in the V-ATPase A subunit genes of plants and algae. *Federation of European Biological Societies* 315:252–258.

Starr, R. C. 1963. Homothallism in *Golenkinia minutissima*. Pages 3–6 in *Studies in Microalgae and Bacteria*. Japanese Society of Plant Physiology. University of Tokyo Press, Tokyo, Japan.

Starr, R. C., and W. R. Rayburn. 1964. Sexual reproduction in *Mesotaenium kramstai*. *Phycologia* 4:22–26.

Steidinger, K. A. 1993. Some taxonomc and biologic aspects of toxic dinoflagellates. Pages 1–28 in *Algal Toxins in Seafood and Drinking Water*, edited by I. R. Falconer. Academic Press, London, UK.

Steidinger, K. A., J. M. Burkholder, H. B. Glasgow, C. W. Hobbs, J. K. Garrett, E. W. Truby, E. J. Noga, and S. A. Smith. 1996. *Pfiesteria piscicida* gen. et sp. nov. (Pfiesteriaceae fam. nov.), a new toxic dinoflagellate with a complex life cycle and behavior. *Journal of Phycology* 32:157–164.

Steidinger, K. A., and K. Tangen. 1997. Dinoflagellates. Pages 387–584 in *Identifying Marine Phytoplankton*, edited by C. R. Tomas. Academic Press, San Diego, CA.

Steidinger, K. A., G. A. Vargo, P. A. Tester, and C. A. Tomas. 1998. Bloom dynamics and physiology of *Gymnodinium breve* with emphasis on the Gulf of Mexico. Pages 133–153 in *Physiological Ecology of harmful Algal Blooms*, edited by D. M. Anderson, A. D. Cembella, and G. M. Hallegraeff. Springer-Verlag, Berlin. Germany.

Stein, J. R. (editor). 1973. *Handbook of Phycological Methods. Culture Methods and Growth Measurements*. Cambridge University Press, Cambridge, UK.

Steinkötter, J., D. Bhattacharya, I. Semmelroth, C. Bibeau, and M. Melkonian. 1994. Prasinophytes form independent lineages within the Chlorophyta: Evidence from ribosomal RNA sequence comparisons. *Journal of Phycology* 30:340–345.

Steinman, A. D. 1996. Effects of grazers on freshwater benthic algae. pages 341–374 in *Algal Ecology: Freshwater Benthic Ecosystems*, edited by R. J. Stevenson, M. L. Bothwell, and R. L. Lowe. Academic Press, New York, NY.

Steinmetz, J. C. 1994. Sedimentation of coccolithophores. Pages 179–198 in *Coccolithophores*, edited by A. Winter and W. G. Siesser. Cambridge University Press, Cambridge, UK.

Steinmetz, J. C. 1994. Stable isotopes in modern coccolithophores. Pages 219–230 in *Coccolithophores*, edited by A. Winter and W. G. Siesser. Cambridge University Press, Cambridge, UK.

Stemler, A. J. 1997. The case for chloroplast thylakoid carbonic anhydrase. *Physiologica Plantarum* 99:348–353.

Steneck, R. S. 1982. A limpet-coralline algal association: Adaptation and defences between a selective herbivore and its prey. *Ecology* 63:507–522.

Steneck, R. S. 1983. Escalating herbivory and resulting adaptive trends in calcareous algal crusts. *Paleobiology* 9:44–61.

Steneck, R. S. 1986. The ecology of coralline algal crusts: Convergent patterns and adaptive strategies. *Annual Review of Ecology and Systematics* 17:273–303.

Steneck, R. S. 1990. Herbivory and the evolution of nongeniculate coralline algae (Rhodophyta, Corallinales) in the North Atlantic and North Pacific. Pages 107–129 in *Evolutionary Biogeography of the Marine Algae of the North Atlantic*, edited by D. J. Garbary and G. R. South. NATO *Advanced Research Workshop Series G*, Volume 22, Springer-Verlag, Berlin, Germany.

Stephenson, T. A., and A. Stephenson. 1949. The universal features of zonation between tide-marks on rocky coasts. *Journal of Ecology* 38:289–305.

Stephenson, T. A., and A. Stephenson. 1972. *Life Between Tidemarks on Rocky Shores*. W. H. Freeman, San Francisco, CA.

Sterner, R. W. 1989. The role of grazers in phytoplankton succession. Pages 107–170 in *Plankton Ecology*, edited by U. Sommer. Springer-Verlag, Berlin, Germany.

Sterner, R. W., J. J. Elser, and D. O. Hessen. 1992. Stoichiometric relationships among producers and consumers in food webs. *Biogeochemistry* 17:49–67.

Stevens, D. R., and S. Purton. 1997. Genetic engineering of eukaryotic algae: Progress and prospects. *Journal of Phycology* 33:713–722.

Stevens, S. E., J. R. C. Murphy, W. J. Lamoreaux, and L. B. Coons. 1994. A genetically engineered mosquitocidal cyanobacterium. *Journal of Applied Phycology* 6:187–198.

Stevenson, R. J. 1996. An introduction to algal ecology in freshwater benthic habitats. Pages 3–30 in *Algal Ecology. Freshwater Benthic Ecosystems*, edited by R. J. Stevenson, M. L. Bothwell, and R. L. Lowe. Academic Press, New York, NY.

Stewart, W. D. P. (editor). 1974. *Algal Physiology and Biochemistry*, University of California Press, Berkeley, CA.

Stewart, W. D. P., G. P. Fitzgerald, and R. H. Burris. 1967. *In situ* studies on N_2 fixation using the acetylene reduction technique. *Proceedings of the National Academy of Science* 58:2071–2078.

Stewart, W. D. P., A. Haystead, and H. W. Pearson. 1969. Nitrogenase activity in heterocysts of blue-green algae. *Nature* 224:226–228.

Stiger, V., and C. Payri. 1997. Strategies of reef invasion by two brown algae in Tahiti (French Polynesia): Reproduction, dispersion, competition. *Phycologia* 36 (supplement):108 (abstract).

Stiller, J. W., and B. D. Hall. 1997. The origin of red algae: Implications for plastid evolution. *Proceedings of the National Academy of Science* 94:4520–4525.

Stiller, J. W., and B. D. Hall. 1998. Sequences of the largest subunit of RNA polymerase II from two red algae and their implications for rhodophyte evolution. *Journal of Phycology* 34:857–864.

Stiller, J. W., and J. R.Waaland. 1996. *Porphyra rediviva* sp. nov. (Rhodophyta): a new species from northeast Pacific salt marshes. *Journal of Phycology* 32:323–332.

Stimson, J., S. Larned, and K. McDermid. 1996. Seasonal growth of the coral reef macroalga *Dictyosphaeria cavernosa* (Forskål) Borgensen and the effects of nutrient availability, temperature and herbivory on growth rate. *Journal of Experimental Marine Biology and Ecology* 196:53–77.

Stoecker, D. K., A. E. Michaels, and L. H. Davis. 1987. Large proportion of marine planktonic ciliates found to contain functional chloroplasts. *Nature* 326:790–792.

Stoermer, E. F. 1978. Phytoplankton as indicators of water quality in the Laurentian Great Lakes. *Transactions of the American Microscopical Society*. 97:2–16.

Stoermer, E. F., G. Emmert, and C. L. Schelske. 1989. Morphological variation of *Stephanodiscus niagarae* Ehrenb. (Bacillariophyta) in a Lake Ontario sediment core. *Journal of Paleolimnology* 2:227–236.

Stokes, P. M. 1983. Responses of freshwater algae to metals. *Progress in Phycological Research* 2:87–112.

Stratmann, J., G. Paputsogle, and W. Oertel. 1996. Differentiation of *Ulva mutabilis* (Chlorophyta) gametangia and gamete release are controlled by extracellular inhibitors. *Journal of Phycology* 32:1009–1021.

Stross, R. G. 1979. Density and boundary regulation of the *Nitella* meadow in Lake George, NY. *Aquatic Botany* 6:285–300.

Suda, S., M. M. Watanabe, and I. Inouye. 1989. Evidence for sexual reproduction in the primitive green alga *Nephoselmis olivacea* (Prasinophyceae). *Journal of Phycology* 25:596–600.

Sukenik, A., Y. Carmeli, and T. Berner. 1989. Regulation of fatty acid composition by irradiance level in the eustigmatophyte *Nannochloropsis* sp. *Journal of Phycology* 25:686–692.

Sukenik, A., D. Tchernov, A. Kaplan, E. Huertas, L. M. Lubian, and A. Livne. 1997. Uptake, efflux, and photosynthetic utilization of inorganic carbon by the marine eustigmatophyte *Nannochloropsis* sp. *Journal of Phycology* 33:969–974.

Sulli, C., and S. D. Schwartzbach. 1995. A soluble protein is imported into *Euglena* chloroplasts as a membrane-bound precursor. *Plant Cell* 8:43–53.

Sültemeyer, D., B. Klughammer, M. R. Badger, and G. D. Price. 1998. Protein phosphorylation and its possible involvement in the induction of high affinity CO_2 concentrating mechanism in cyanobacteria. *Canadian Journal of Botany* 76:954–961.

Sültemeyer, D., C. Schmidt, and H. P. Fock. 1993. Carbonic anhydrases in higher plants and aquatic microorganisms. *Physiologia Plantarum* 88:179–180.

Surek, B., U. Beemelmanns, M. Melkonian, and D. Bhattacharya. 1994. Ribosomal RNA sequence comparisons demonstrate an evolutionary relationship between Zygnematales and charophytes. *Plant Systematics and Evolution* 191:171–181.

Surif, M. B., and J. A. Raven. 1990. Photosynthetic gas exchange under emersed conditions in eulittoral and normally submersed members of the Fucales and the Laminariales: Interpretation in relation to C isotope ratio and N and water use efficiency. *Oecologia* 82:68–80.

Sutherland, T. F., C. L. Amos, and J. Grant. 1998. The effect of buoyant biofilms on the erodibility of sublittoral sediments of a temperate microtidal estuary. *Limnology and Oceanography* 43:225–235.

Suttle, C. A. 1994. The significance of viruses to mortality in aquatic microbial communities. *Microbial Ecology* 28:237–243.

Suttle, C. A., A. M. Chan, and M. T. Cottrell. 1990. Infection of phytoplankton by viruses and reduction of primary production. *Nature* 347:467–469.

Suzaki, T., and R. E. Williamson. 1986. Ultrastructure and sliding of pellicular structures during eugenoid movement in *Astasia longa* Pringsheim (Sarcomastigophora, Euglenida). *Journal of Protozoology* 33:179–184.

Suzuki, E., Y. Shiraiwa, and S. Miyachi. 1994. The cellular and molecular aspects of carbonic anhydrase in photosynthetic microorganisms. *Progress in Phycological Research* 10:2–54.

Suzuki, H., J. Ingersoll, D. B. Stern, and K. L. Kindle. 1997. Generation and maintenance of tandemly repeated extrachromosomal plasmid DNA in *Chlamydomonas* chloroplasts. *The Plant Journal* 11:635–648.

Suzuki, K., K. Iwamoto, S. Yokoyama, and T. Ikawa. 1991. Glycolate-oxidizing enzymes in algae. *Journal of Phycology* 27:492–498.

Sweeney, B. M. 1982. Microsources of bioluminescence in *Pyrocystis fusiformis* (Pyrrhophyta). *Journal of Phycology* 18:412–416.

Sweeney, B. M. 1987. *Rhythmic phenomena in plants*, 2nd ed. Academic Press, San Diego, CA.

Swift, E., W. H. Biggley, and H. H. Seliger. 1973. Species of oceanic dinoflagellates in the genera *Dissodinium* and *Pyrocystis*: Interclonal and interspecific comparisons of the color and photon yield of bioluminescence. *Journal of Phycology* 9:420–426.

Swift, H., and B. Palenik. 1993. Prochlorophyte evolution

and the origin of chloroplasts: Morphological and molecular evidence. Pages 123–139 in *Origins of Plastids*, edited by R. A. Lewin. Chapman & Hall, New York, NY.

Swinbanks, D. D. 1982. Intertidal exposure zones: A way to subdivide the shore. *Journal of Experimental Marine Biology* 62:69–86.

Swofford, D. L., G. L. Olsen, P. J. Waddell, and D. M. Hillis. 1996. Phylogenetic inference. Pages 407–514 in *Molecular Systematics* (2nd. ed), edited by D. M. Hillis, C. Moritz, and B. K. Mable. Sinauer Associates, Sunderland, MA.

Sym, S. D., and R. N. Pienaar. 1991. Light and electron microscopy of a punctate species of *Pyramimonas, P. mucifera* sp. nov. (Prasinophyceae). *Journal of Phycology* 27:277–290.

Sym, S. D, and R. N. Pienaar. 1993. The class Prasinophyceae. *Progress in Phycological Research* 9:281.

Syrett, P. J., and F. A. A. Al-Houty. 1984. The phylogenetic significance of the occurrence of urease/urea amidolyase and glycollate oxidase/glucollate dehydrogenase in green algae. *British Phycological Journal* 19:11–21.

Szymánska, H. 1989. Three new *Coleochaete* species (Chlorophyta) from Poland. *Nova Hedwigia* 49:435–446.

Taggart, R. E., and L. R. Parker. 1976. A new fossil alga from the Silurian of Michigan. *American Journal of Botany* 63:1390–1392.

Takahashi, H., H. Takano, H. Kuroiwa, R. Itoh, K. Toda, S. Kawano, and T. Kuroiwa. 1998. A possible role for actin dots in the formation of the contractile ring in the ultramicroalga *Cyanidium caldarium* RK-1. *Protoplasma* 202:91–104.

Takatori, S., and K. Imahori. 1971. Light reactions in the control of oospore germination of *Chara delicatula*. *Phycologia* 10:221–228.

Takeda, H. 1991. Sugar composition of the cell wall and the taxonomy of *Chlorella* (Chlorophyceae). *Journal of Phycology* 27:224–232.

Tamura, H., I. Mine, and K. Okuda. 1996. Cellulose-synthesizing terminal complexes and microfibril structure in the brown alga *Sphacelaria rigidula* (Sphacelariales, Phaeophyceae). *Phycological Research* 44:63–68.

Tan, I. H., and L. D. Druehl. 1996. A ribosomal DNA phylogeny supports the close evolutionary relationships abong the Sporochnales, Desmarestiales, and Laminariales (Phaeophyceae). *Journal of Phycology* 32:112–118.

Tanaka, K., K. Oikawa, N. Ohta, H. Kuroiwa, T. Kuroiwa, and H. Takahashi. 1996. Nuclear encoding of a chloroplast RNA polymerase sigma subunit in a red alga. *Science* 272:1932–1935.

Tang, E. P. Y. 1996. Why do dinoflagellates have lower growth rates? *Journal of Phycology* 32:80–84.

Tang, E. P. Y., R. Tremblay, and W. F. Vincent. 1997. Cyanobacterial dominance of polar freshwater ecosystems: Are high-latitude mat-formers adapted to low temperature? *Journal of Phycology* 33:171–181.

Tappan, H. 1980. *The Paleobiology of Plant Protists*, W. H. Freeman, New York, NY.

Targett, N. M., and T. M. Arnold. 1998. Predicting the effects of brown algal phlorotannins on marine herbivores in tropical and temperate oceans. *Journal of Phycology* 34:195–205.

Tatewaki, M. 1972. Life history and systematics in *Monostroma*. Pages 1–15 in *Contributions to the Systematics of Benthic Marine Algae of the North Pacific*, edited by I. A. Abbott and M. Kurogi. Japanese Society of Phycology, Kobe, Japan.

Tatewaki, M., L. Provasoli, and I. J. Pintner. 1983. Morphogenesis of *Monostroma oxyspermum* (Kütz.) Doty (Chlorophyceae) in axenic culture, expecially in bialgal culture. *Journal of Phycology* 19:409–416.

Taylor, A. O., and B. A. Bonner. 1967. Isolation of phytochrome from the alga *Mesotaenium* and the liverwort *Sphaerocarpos*. *Plant Physiology* 42:726–766.

Taylor, A. R., N. F. H. Manison, C. Fernandez, J. Wood, and C. Brownlee. 1996. Spatial organization of calcium signaling involved in cell volume control in the *Fucus* rhizoid. *The Plant Cell* 8:2015–2031.

Taylor, A. R. A, and L. C.-M. Chen. 1994. *Chondrus*. Pages 35–76 in *Biology of Economic Algae*, edited by I. Akatsuka. SPB Academic Publishing, The Hague, Netherlands.

Taylor, F. J. R. 1971. Scanning electron microscopy of thecae of the dinoflagellate genus *Ornithocercus*. *Journal of Phycology* 7:249–258.

Taylor, F. J. R. 1980. On dinoflagellate evolution. *BioSystems* 13:65–108.

Taylor, F. J. R. 1990. Dinoflagellata (Dinomastigota). Pages 419–437 in *Handbook of Protoctista*, edited by L. Margulis, C. O. Corliss, M. Melkonian, and D. J. Chapman. Jones & Bartlett Publishers, Boston, MA.

Taylor, T. N., H. Hass, and W. Remy. 1992. Devonian fungi: Interactions with the green alga *Palaeonitella*. *Mycologia* 84:901–910.

Taylor, T. N., and E. L. Taylor. 1993. *The Biology and Evolution of Fossil Plants*. Prentice Hall, Englewood Cliffs, NJ.

Taylor, W. R. 1972. *Marine Algae of the Eastern Tropical and Subtropical Coasts of the Americas*. University of Michigan Press, Ann Arbor, MI.

Tenaud, M., M. Ohmori, and S. Miyachi. 1989. Inorganic carbon and acetate assimilation in *Botryococcus braunii* (Chlorophyta). *Journal of Phycology* 25:662–667.

Thiel, G., U. Homann, and C. Plieth. 1997. Ion channel activity during the action potential in *Chara*: New insights with new techniques. *Journal of Experimental Botany* 48:609–622.

Thiel, T., E. M. Lyons, J. C. Erker, and A. Ernst. 1995. A second nitrogenase in vegetative cells of a heterocyst-forming cyanobacterium. *Proceedings of the National Academy of Science* 92:9358–9362.

Thierstein, H. R., K. R. Geitzenauer, B. Molfino, and N. J. Shackleton. 1977. Global synchroncity of late Quarternary coccolith datum levels: Validation by oxygen isotopes. *Geology* 5:400–404.

Thomas, W. H., and B. Duval. 1995. Sierra Nevada, California, U.S.A., snow algae: Snow albedo changes, algal-bacterial interrelationships, and ultraviolet radiation effects. *Arctic and Alpine Research* 27:389–399.

Thomas, W. H., and C. H. Gibson. 1990. Quantified small-scale turbulence inhibits a red tide dinoflagellate, *Gonyaulax polyedra* Stein. *Deep-Sea Research* 37:1583–1593.

Thomas, W. H., and C. H. Gibson. 1992. Effects of quantified small-scale turbulence on the dinoflagellate, *Gymnodinium sanguineum* (*splendens*): Contrast with *Gonyaulax* (*Lingulodinium*) *polyedra* and fishery implication. *Deep-Sea Research* 39:1429–1437.

Thomas, W. H., C. T. Tynan, and C. H. Gibson. 1997. Turbulence-phytoplankton interrelationships. *Progress in Phycological Research* 12:284–324.

Thompson, J. B., S. Schultze-Lam, T. J. Beveridge, and D. J. DesMarais. 1997. Whiting events: Biogenic origin due to the photosynthetic activity of cyanobacterial picoplankton. *Limnology and Oceanography* 42:133–141.

Thompson, J. M., A. J. D. Ferguson, and C. S. Reynolds. 1982. Natural filtration rates of zooplankton in a closed system: the derivation of a community grazing index. *Journal of Plankton Research* 4:545–560.

Thompson, R. H. 1958. The life cycles of *Cephaleuros* and *Stomatochroon*. Proceedings of the 9th International Botanical Congress (Montreal) 2:397.

Thompson, R. H. 1969. Sexual reproduction in *Chaetosphaeridium globosum* (Nordst.) Klebahn (Chlorophyceae) and description of a species new to science. *Journal of Phycology* 5:285–290.

Thompson, R. H., and D. E. Wujek. 1997. *Trentepohliales. Cephaleuros, Phycopeltis, and Stomatochroon: Morphology, Taxonomy and Ecology*. Science Publishers, Enfield, NH.

Throndsen, J. 1997. The planktonic marine flagellates. Pages 591–730 in *Identifying Marine Phytoplankton*, edited by C. R. Tomas. Academic Press, New York, NY.

Tilman, D. 1976. Ecological competition between algae: Experimental confirmation of resource-based competition theory. *Science* 192:463–465.

Tilman, D. 1977. Resource competition between planktonic algae: an experimental and theoretical approach.

Ecology 58:338–348.

Tilman, D. 1981. Tests of resource competition theory using four species of Lake Michigan algae. *Ecology* 62:802–815.

Tilman, D. 1982. *Resource Competition and Community Structure*. Princeton University Press, Princeton, NJ.

Tilman, D., and S. S. Kilham. 1976. Phosphate and silicate growth and uptake kinetics of the diatoms *Asterionella formosa* and *Cyclotella meneghiniana* in batch and semicontinuous culture. *Journal of Phycology* 12:375–383.

Tilman, D., S. S. Kilham, and P. Kilham. 1982. Phytoplankton community ecology: the role of limiting nutrients. *Annual Review of Ecology and Systematics* 13:349–372.

Tilman, D., M. Mattson, and S. Langer. 1981. Competition and nutrient kinetics along a temperature gradient: an experimental test of a mechanistic approach to niche theory. *Limnology and Oceanography* 26:1020–1033.

Tilzer, M. M. 1973. Diurnal periodicity in the phytoplankton assemblage of a high mountain lake. *Limnology and Oceanography* 18:15–30.

Timpano, P., and L. A. Pfiester. 1986. Observations on *Vampyrella pendula-Stylodinium sphaera* and the ultrastructure of the reproductive cyst. *American Journal of Botany* 73:1341–1350.

Tindall, D. R., and S. L. Morton. 1998. Community dynamics and physiology of epiphytic/benthic dinoflagellates associated with ciguatera. Pages 293–313 in *Physiological Ecology of Harmful Algal Blooms*, edited by D. M. Anderson, A. D. Cembella, and G. M. Hallegraeff. Springer-Verlag, New York, NY.

Tonn, W. M., and J. J. Magnuson. 1982. Patterns in the species composition and richness of fish assemblages in northern Wisconsin lakes. *Ecology* 63:1149–1166.

Trainor, F. R., and C. A. Burg. 1965. *Scenedesmus obliquus* sexuality. *Science* 148:1094–1095.

Trainor, F. R., and P. F. Egan. 1988. The role of bristles in the distribution of a *Scenedesmus*. *British Phycological Journal* 23:135–141.

Triemer, R. E. 1980. Role of Golgi apparatus in mucilage production and cyst formation in *Euglena gracilis* (Euglenophyceae). *Journal of Phycology* 16:46–52.

Triemer, R. E. 1997. Feeding in *Peranema trichophorum* revisited (Euglenophyta). *Journal of Phycology* 33:649–654.

Triemer, R. E., and Farmer, M. A. 1991. An ultrastructural comparison of the mitotic apparatus, feeding apparatus, flagellar apparatus, and cyskeleton in euglenoids and kinetoplastids. *Protoplasma* 164:91–104.

Trowbridge, C. D. 1998. Ecology of the green macroalga *Codium fragile* (Suringar) Hariot 1889: Invasive and non-invasive subspecies. *Oceanography and Marine Biology: An Annual Review* 36:1–64.

Troxler, R. F. 1994. The molecular aspects of pigments and photosynthesis in *Cyanidium caldarium*. Pages 263–282 in *Evolutionary Pathways and Enigmatic Algae: Cyanidium caldarium (Rhodophyta) and Related Cells*, edited by J. Sechbach. Kluwer Academic Publishers, Boston, MA.

Tschermak-Woess, E. 1980. *Asterochloris phycobiontica* gen. et spec. nov., der phycobiont der Flechte *Varicellaria carneonivea*. *Plant Systematics and Evolution* 135:279–294.

Tsekos, I. 1996. The supramolecular organization of red algal cell membranes and their participation in the biosynthesis and secretion of extracellular polysaccharides: A review. *Protoplasma* 193:10–32.

Tsekos, I., K. Okuda, and R. M. Brown. 1996. The formation and development of cellulose-synthesizing linear terminal complexes (TCs) in the plasma membrane of the marine red alga *Erythrocladia subintegra* Rosenv. *Protoplasma* 193:33–45.

Tsuji, K., T. Watanuki, F. Kondo, M. F. Watanabe, H. Nakazawa, M. Suzuki, H. Uchida, and K.-I. Harada. 1997. Stability of microcystins from cyanobacteria—IV. Effect of chlorination on decomposition. *Toxicon* 35:1033–1041.

Tsunakawa-Yokoyama, A., J. Yokoyama, H. Ohashi, and Y. Hara. 1997. Phylogenetic relationship of Porphyridiales (Rhodophyta) based on DNA sequences of

18SrRNA and *psbA* genes. *Phycologia* 36 (supplement), p. 13 (abstract).

Tuchman, N. C. 1996. The role of heterotrophy in algae. Pages 299–319 in *Algal Ecology: Freshwater Benthic Ecosystems*, edited by R. J. Stevenson, M. L. Bothwell, and R. L. Lowe. Academic Press, San Diego, CA.

Tucker, M. E., and V. P. Wright. 1990. *Carbonate Sedimentology*. Blackwell Scientific, Oxford, UK.

Turner, C. H. C., and L. V. Evans. 1986. Translocation of photoassimilated ^{14}C in the red alga *Polysiphonia lanosa*. *British Phycological Journal* 13:51–55.

Turner, J. T., and P. A. Tester. 1997. Toxic marine phytoplankton, zooplankton grazers, and pelagic food webs. *Limnology and Oceanography* 42:1203–1214.

Turner, M. A., E. T. Howell, M. Summerby, R. H. Hesslein, D. L. Findlay, and M. B. Jackson. 1991. Changes in epilithon and epiphyton associated with experimental acidification of a lake to pH 5. *Limnology and Oceanography* 36:1390–1405

Turner, S., T. Berger-Wiersma, S. J. Giovannoni, L. R. Mur and N. R. Pace. 1989. The relationship of a prochlorophyte *Prochlorothrix hollandica* to green chloroplasts. *Nature* 337:380–385.

Turpin, D. H. 1991. Effects of inorganic N availability on algal photosynthesis and carbon metabolism. *Journal of Phycology* 27:14–20.

Turpin, D. H., and P. J. Harrison. 1980. Cell size manipulation in natural marine, planktonic, diatom communities. *Canadian Journal of Fisheries and Aquatic Science* 37:1193–1195.

Ucko, M., M. Elbrächter, and E. Schnepf. 1997. A *Crypthecodinium cohnii*-like dinoflagellate feeding myzocytotically on the unicellular red alga *Porphyridium* sp. *European Journal of Phycology* 32:133–140.

Uhrmacher, S., D. Hanelt, and W. Nultsch. 1995. Zeaxanthin content and the photoinhibitory degree of photosynthesis are linearly correlated in the brown alga *Dictyota dichotoma. Marine Biology* 123:159–165.

Underwood, G. J. C., and J. D. Thomas. 1990. Grazing interactions between pulmonate snails and epiphytic algae and bacteria. *Freshwater Biology* 23:505–522.

Urbach, E., D. L. Robertson, and S. W. Chisholm. 1992. Multiple evolutionary origins of prochlorophytes within the cyanobacterial radiation. *Nature* 355:267–270.

Url, T., M. Höftburger, and U. Meindl. 1993. Cytochalasin B influences dictyosomal vesicle production and morphogenesis in the desmid *Euastrum*. *Journal of Phycology* 29:667–674.

Vaishampayan, A., T. Dey, and A. K. Awasthi. 1996. Genetic improvement of nitrogen-fixing cyanobacteria in response to modern rice agriculture. Pages 395–420 in *Cytology, Genetics and Molecular Biology of Algae*, edited by B. R. Chaudhary and S. B. Agrawal. SPB Academic Publishing, Amsterdam, Netherlands.

Vaishampayan, A., R. P. Sinha, and D.-P. Häder. 1998. Use of genetically improved nitrogen-fixing cyanobacteria in rice paddy field: Prospects as a source material for engineering herbicide sensitivity and resistance in plants. *Botanica Acta* 111:176–190.

Van Alstyne, K. L. 1988. Herbivore grazing increases polyphenolic defenses in the intertidal brown alga *Fucus*. *Ecology* 69:655–663

Van de Peer, Y., S. A. Rensing, U.-G. Maier, and R. De Wachter. 1996. Substitution rate calibration of small subunit ribosomal RNA identified chlorarachniophyte endosymbionts as remnants of green algae. *Proceedings of the National Academy of Science* 93:7732–7736.

Vanderauwera, G., and Dewachter. 1997. Complete large subunit ribosomal RNA sequence from the heterokont algae *Ochromonas danica, Nannochloropsis salina*, and *Tribonema aequale* and phylogenetic analysis. *Journal of Molecular Evolution* 45:84–90.

van der Meer, J. P., and M. U. Patwary. 1995. Genetic studies on marine macroalgae: A status report. Pages 235–258 in *Algae, Environment and Human Affairs*, edited by W. Wiessner, E. Schnepf, and R. C. Starr. Bio-Press, Bristol, UK.

van der Meer, J. P., and E. R. Todd. 1980. The life history of *Palmaria palmata* in culture. A new type for the Rhodophyta. *Canadian Journal of Botany* 58:1250–1256.

Van Dok, W., and B. T. Hart. 1997. Akinete germination in *Anabaena circinalis* (Cyanophyta). *Journal of Phycology* 33:12–17.

van Donk, E. 1989. The role of fungal parasites in phytoplankton succession. Pages 171–194 in *Plankton Ecology*, edited by U. Sommer. Springer-Verlag, Berlin, Germany.

van Donk, E., and K. Bruning. 1995. Effects of fungal parasites on planktonic algae and the role of environmental factors in the fungus-alga relationship. Pages 223–234 in *Algae, Environment and Human Affairs*, edited by W. Weissner, E. Schnepf, and R. Starr. Bio-Press, Bristol, UK.

van Donk, E., M. P. Grimm, R. D. Gulati, and J. P. G. Klein Breteler. 1990a. Whole-lake food-web manipulation as a means to study community interactions in a small ecosystem. *Hydrobiologia* 200/201:275–289.

van Donk, E., M. P. Grimm, R. D. Gulati, P. G. M. Heuts, W. A. de Kloet, and L. van Liere. 1990b. First attempt to apply whole-lake food-web manipulation on a large scale in The Netherlands. *Hydrobiologia* 200/201:291–301.

van Donk, E., M. Lürling, and W. Lampert. 1999. Consumer induced changes in phytoplankton: Inducibility, costs, benefits, and the impact on grazers. Pages 89–103 in *The Ecology and Evolution of Inducible Defenses*, edited by R. Tollrian and C. D. Harvell. Princeton University Press, Princeton, NJ.

van Donk, E., M. Lürling, D. O. Hessen, and G. M. Lokhorst. 1997. Altered cell wall morphology in nutrient-deficient phytoplankton and its impact on grazers. *Limnology and Oceanography* 42:357–364.

van Donk, E., and J. Ringelberg. 1983. The effect of fungal parasitism on the succession of diatoms in Lake Maarsseveen (The Netherlands). *Freshwater Biology* 28:241–251.

Van Etten, J. L., L. C. Lane, and R. H. Meints. 1991. Viruses and viruslike particles of eukaryotic algae. *Microbiological Reviews* 55:586–620.

van Gool, E., and J. Ringelberg. 1996. Daphnids respond to algae-associated odours. *Journal of Plankton Research* 18:197–202.

van Rijssel, M., C. E. Hamm, and W. W. C. Gieskes. 1997. *Phaeocystis globosa* (Prymnesiophyceae) colonies: Hollow structures built with small amounts of polysaccharides. *European Journal of Phycology* 32:185–192.

Vanni, M. J. 1987. Effects of food availability and fish predation on a zooplankton community. *Ecological Monographs* 57:61–88.

Vanni, M. J., C. Luecke, J. F. Kitchell, and J. J. Magnuson. 1990. Effects of planktivorous fish mass mortality on the plankton community of Lake Mendota, Wisconsin: Implications for biomanipulation. *Hydrobiologia* 200/201:329–336.

Vannote, R. L., G. W. Minshall, K. W. Cummins, J. R. Sedell, and C. E. Cushing. 1980. The river continuum concept. *Canadian Journal of Fisheries and Aquatic Science* 37:130–137.

Van Valkenburg, S. D. 1971. Observations on the fine structure of *Dictyocha fibula* Ehrenberg. II. The protoplast. *Journal of Phycology* 7:118–132.

Van Winkle-Swift, K. P., and W. L. Rickoll. 1997. The zygospore wall of *Chlamydomonas monoica* (Chlorophyta): Morphogenesis and evidence for the presence of sporopollenin. *Journal of Phycology* 33:655–665.

Vargas, D. R., and L. Collado-Vides. 1996. Architectural models for apical patterns in *Gelidium* (Gelidiales, Rhodophyta): Hypothesis of growth. *Phycological Research* 44:95–100.

Verhagen, J. H. G. 1994. Modeling phytoplankton patchiness under the influence of wind-driven currents in lakes. *Limnology and Oceanography* 39:1551–1565.

Verity, P. G. 1985. Grazing, respiration, excretion, and growth rates of tintinnids. *Limnology and Oceanography* 30:1268–1282.

Vesk, M., T. P. Dibbayawan, and P. A. Vesk. 1996. Immunogold localization of phycoerythrin in chloroplasts of *Dinophysis acuminata* and *D. fortii* (Dinophysiales, Dinophyta). *Phycologia* 35:234–238.

Vesk, M., L. R. Hoffman, and J. D. Pickett-Heaps. 1984. Mitosis and cell division in *Hydrurus foetidus* (Chrysophyceae). *Journal of Phycology* 20:461–470.

Vickerman, K. 1990. Kinetoplastida. Pages 215–238 in *Handbook of Protoctista*, edited by L. Margulis, J. O. Corliss, M. Melkonian, and D. J. Chapman. Jones & Bartlett Publishers, Boston, MA.

Villareal, T. A., S. Woods, J. K. Moore, and K. Culver-Rymsza. 1996. Vertical migration of *Rhizosolenia* mats and their significance to NO_3^- fluxes in the central North Pacific gyre. *Journal of Plankton Research* 18:1103–1121.

Villarejo, A., M. I. Orús, and F. Martinez. 1997. Regulation of the CO_2-concentrating mechanism in *Chlorella vulgaris* UAM 101 by glucose. *Physiologia Plantarum* 99:293–301.

Vis, M. L., G. W. Saunders, R. G. Sheath, K. Dunse, and T. J. Entwhistle. 1998. Phylogeny of the Batrachospermales (Rhodophyta) inferred from *rbcL* and 18S ribosomal DNA gene sequences. *Journal of Phycology* 34:341–350.

Vitousek, P. M., H. A. Mooney, J. Lubchenco, and J. M. Melilo. 1997a. Human domination of earth's ecosystems. *Science* 277:494–499.

Vitousek, P. M., J. D. Aber, R. W. Howarth, G. E. Likens, P. A. Matson, D. W. Schindler, W. H. Schlesinger, and D. G. Tilman. 1997b. Human alteration of the global nitrogen cycle: Sources and consequences. *Ecological Applications* 71:737–750.

Vogel, G., 1997. Parasites shed light on cellular evolution. *Science* 275:1422

Vogel, S. 1994. *Life in Moving Fluids*. 2nd edition. Princeton University Press, Princeton, NJ.

Volkman, J. K., C. L. Farmer, S. M. Barrett, and E. L. Sikes. 1997. Unusual dihydroxysterols as chemotaxonomic markers for microalgae from the order Pavlovales (Haptophyceae). *Journal of Phycology* 33:1016–1023.

Vollenweider, R. 1982. Eutrophication of waters, monitoring, assessment and control. OECD, Paris, France.

Volterra, V. 1926. Variazioni e fluttuazioni del numero d'individui in specie animali conviventi. *Memorie. Reale Accademia Nazionale dei Lincei. Classe di Scienze Fisiche, Matematiche, e Naturali*. Ser. VI, Vol. 2.

von Stosch, H. A. 1964. Zum problem der sexuellen Fortpflanzung in der Peridineengattung *Ceratium. Helgolander wissenschaftliche Meeresuntersuchungen* 10:140–153.

von Wettstein, D., S. Gough, and C. G. Kannangara. 1995. Chlorophyll biosynthesis. *The Plant Cell* 7:1039–1057.

Vorobyova, S. S., G. V. Pomazkina, E. Y. Baranova, Y. V. Likhoshway, and C. D. Sandgren. 1996. Chrysophycean cysts (stomatocysts) from Lake Baikal and Irkutsk Reservoir, Siberia. *Journal of Paleolimnology* 15:271–277.

Vreeland, V., J. H. Waite, and L. Epstein. 1998. Polyphenols and oxidases in substratum adhesion by marine algae and mussels. *Journal of Phycology* 34:1–8.

Vroom, P. S., C. M. Smith, and S. C. Keeley. 1998. Cladistics of the Bryopsidales: A preliminary analysis. *Journal of Phycology* 34:351–360.

Waaland, J. R., and D. Branton. 1969. Gas vacuole development in a blue-green alga. *Science* 163:1339–1341.

Waaland, S. D. 1990. Development. Pages 259–274 in *Biology of the Red Algae*, edited by K. M. Cole and R. G. Sheath. Cambridge University Press, Cambridge, UK.

Waaland, S. D., and R. E. Cleland. 1974. Cell repair through cell fusion in the red alga *Griffithsia pacifica*. *Protoplasma* 79:185–196.

Wagner, G., and F. Grolig. 1992. Algal chloroplast movements. Pages 39–72 in *Algal Cell Motility*, edited by M. Melkonian. Chapman & Hall, London, UK.

Wall, D., and B. Dale. 1969. The "hystrichosphaerid" resting spore of the dinoflagellate *Pyrodinium bahamense*, Plate, 1906. *Journal of Phycology* 5:140–149.

Wall, D., R. R. L. Guillard, and B. Dale. 1967. Marine

dinoflagellate cultures from resting spores. *Phycologia* 6:83–86.

Walne, P. L., and H. J. Arnott. 1967. The comparative ultrastructure and possible function of eyespots: *Euglena granulata* and *Chlamydomonas eugametos*. *Planta* 77:325–353.

Walne, P. L., and N. S. Dawson. 1993. A comparison of paraxial rods in the flagella of euglenoids and kinetoplastids. *Archiv für Protistenkunde* 143:177–194.

Walne, P. L., and P. A. Kivic. 1990. Euglenida. Pages 270–287 in *Handbook of Protoctista*, edited by L. Margulis, J. O. Corliss, M. Melkonian, and D. J. Chapman. Jones & Bartlett Publishers, Boston, MA.

Walsby, A. E., and A. Xypolyta. 1977. The form resistance of chitan fibres attached to the cells of *Thalassiosira fluviatilis* Hustedt. *British Phycological Journal* 12:215–223.

Wang, J. T., and A. E. Douglas. 1997. Nutrients, signals, and photosynthetic release by symbiotic algae. The impact of taurine on the dinoflagellate alga *Symbiodinium* from the sea anemone *Aiptasia pulchella*. *Plant Physiology* 114:631–636.

Wang, Y., J. Lu, J.-C. Mollet, M. R. Gretz, and K. D. Hoagland. 1997. Extracellular matrix assembly in diatoms (Bacillariophyceae). II. 2,6-dichlorobenzonitrile inhibition of motility and stalk production in the marine diatom *Achnanthes longipes*. *Plant Physiology* 113:1071–1080.

Waris, H. 1950. Cytophysiological studies on *Micrasterias* I. Nuclear and cell division. *Physiologia Plantarum* 3:1–16.

Watanabe, S., and G. L. Floyd. 1989. Ultrastructure of the zoospore of the coenocytic algae *Ascochloris* and *Urnella* (Chlorophyceae) with emphasis on the flagellar apparatus. *British Phycological Journal* 24:143–152.

Watanabe, S., S. Hirabayashi, S. Boussiba, Z. Cohen, A. Vonshak, and A. Richmond. 1996. *Parietochloris incisa* comb. nov. (Trebouxiophyceae, Chlorophyta). *Phycological Research* 44:107–108.

Watanabe, M. M., S. Suda, I. Inouye, T. Sawaguchi, and M. Chihara. 1990. *Lepidodinium viride* gen. et sp. nov. (Gymnodiniales, Dinophyta), a green dinoflagellate with a chlorophyll *a*- and *b*-containing endosymbiont. *Journal of Phycology* 26:741–751.

Waterbury, J. B., J. M. Willey, D. G. Franks, F. W. Valois, and S. W. Watson. 1985. A cyanobacterium capable of swimming motility. *Science* 230:74–76.

Waters, D. A., and R. L. Chapman. 1996. Molecular phylogenetics and the evolution of green algae and land plants. Pages 337–349 in *Cytology, Genetics and Molecular Biology of Algae*, edited by B. R. Chaudhary and S. B. Agrawal. SPB Publishing, Amsterdam, Netherlands.

Watras, C. J., and T. M. Frost. 1989. Little Rock Lake (Wisconsin): Perspectives on an experimental ecosystem approach to seepage lake acidification. *Archives of Environmental Contamination* 18:157–165.

Watson, D. C., and T. A. Norton. 1985. The physical characteristics of seaweed thalli as deterrents to littorine grazers. *Botanica Marina* 28:383–387.

Watson, M. W. 1975. Flagellar apparatus, eyespot and behavior of *Microthamnion kuetzingianum* (Chlorophyceae) zoospores. *Journal of Phycology* 11:439–448.

Watson, M. W., and H. J. Arnott. 1973. Ultrastructural morphology of *Microthamnion* zoospores. *Journal of Phycology* 9:15–29.

Watson, S. B., E. McCauley, and J. A. Downing. 1997. Patterns in phytoplankton taxonomic composition across temperate lakes of different nutrient status. *Limnology and Oceanography* 42:487–495.

Wayne, R. 1994. The excitability of plant cells: With a special emphasis on characean internodal cells. *Botanical Review* 60:265–367.

Webster, K. E., T. M. Frost, C. J. Watras, W. A. Swenson, M. Gonzalez, and P. J. Garrison. 1992. Complex biological responses to the experimental acidification of Little Rock Lake, Wisconsin, USA. *Environmental Pollution* 78:73–78.

Wedemayer, G. J, and L. W. Wilcox. 1984. The ultrastructure of the freshwater, colorless dinoflagellate *Peridiniopsis berolinense* (Lemm.) Bourrelly

(Mastigophora, Dinoflagellida). *Journal of Protozoology* 31:444–453.

Wedemayer, G. J., L. W. Wilcox, and L. E. Graham. 1982. *Amphidinium cryophilum* sp. nov. (Dinophyceae), a new freshwater dinoflagellate. I. Species description using light and scanning electron microscopy. *Journal of Phycology* 18:13–16.

Wee, J. L. 1997. Scale biogenesis in synurophycean protists: Phylogenetic implications. *CRC Critical Reviews in Plant Science* 16:497–534.

Weers, P. M. M., and R. D. Gulati. 1997. Growth and reproduction of *Daphnia galeata* in response to changes in fatty acids, phosphorus, and nitrogen in *Chlamydomonas reinhardtii*. *Limnology and Oceanography* 42:1584–1589.

Wehr, J. D., and J. D. Stein. 1985. Studies on the biogeography and ecology of the freshwater phaeophycean alga *Heribaudiella fluviatilis*. *Journal of Phycology* 21:81–93.

Weiss, R. L. 1983. Fine structure of the snow alga (*Chlamydomonas nivalis*) and associated bacteria. *Journal of Phycology* 19:200–204.

Wells, M. L. 1998. Marine colloids, a neglected dimension. *Nature* 391:530–531.

West, J. A. 1970. The life history of *Rhodochorton concrescens* in culture. *British Phycological Journal* 5:179–186.

West, J. A. 1972. The life history of *Petrocelis francisciana*. *British Phycological Journal* 7:299–308.

Westbroek, P., J. E. Van Hinte, G. J. Brummer, M. Veldhuis, C. Brownlee, J. C. Green, R. Harris, and B. R. Heimdal. 1994. Pages 321–334 in *The Haptophyte Algae*, edited by J. C. Green and B. S. C. Leadbeater. *The Systematics Association Special Volume* 51, Oxford University Press, Oxford, UK.

Wetherbee, R. 1979. "Transfer connections": Specialized pathways for nutrient translocation in a red alga? *Science* 204:858–859.

Wetherbee, R., J. L. Lind, J. Burke, and R. S. Quatrano. 1998. The first kiss: Establishment and control of initial adhesion by raphid diatoms. *Journal of Phycology* 34:9–15.

Wetherbee, R., M. Ludwig, and A. Koutoulis. 1995. Immunological and ultrastructural studies of scale development and deployment in *Mallomonas* and *Apedinella*,. Pages 165–178 in *Chrysophyte Algae: Ecology, Phylogeny and Development*, edited by C. D. Sandgren, J. P. Smol, and J. Kristiansen. Cambridge University Press, Cambridge, UK.

Wetherell, D. F. 1958. Obligate phototrophy in *Chlamydomonas eugametos*. *Physiologia Plantarum* 11:260–274.

Wetzel, R. G. 1975. *Limnology*. W. B. Saunders, Philadelphia, PA.

Wetzel, R. G. 1996. Benthic algae and nutrient cycling in lentic freshwater ecosystems. Pages 641–667 in *Algal Ecology. Freshwater Benthic Ecosystems*, edited by R. J. Stevenson, M. L. Bothwell, and R. L. Lowe. Academic Press, New York, NY.

Wetzel, R. G., and G. E. Likens. 1991. *Limnological Analyses*, 2nd ed., Springer-Verlag, New York, NY.

Wetzel, R. G., and M. Søndergaard. 1998. Role of submerged macrophytes for the microbial community and dynamics of dissolved organic carbon in aquatic ecosystems. Pages 133–148 in *The Structuring Role of Submerged Macrophytes in Lakes*, edited by E. Jeppesen, M. Sondergaard, K. Christoffersen. Springer-Verlag, Berlin, Germany.

Wharton, R. A., D. T. Smernoff, and M. A. Averner. 1988. Algae in space. Pages 485–510 in *Algae and Human Affairs*, edited by C. A. Lembi and J. R. Waaland. Cambridge University Press, Cambridge, UK.

Whatley, J. 1993. Chloroplast ultrastructure. Pages 135–204 in *Ultrastructure of Microalgae*, edited by T. Berner. CRC Press, Boca Raton, FL.

White, A. W. 1976. Growth inhibition caused by turbulence in the toxic marine dinoflagellate *Gonyaulax excavata*. *Journal of the Fisheries Research Board Canada* 33:2598–2602.

Whittle, S. J., and P. J. Casselton. 1975. The chloroplast pigments of the algal classes Eustigmatophyceae and Xanthophyceae. I. Eustigmatophyceae. *British*

Phycological Journal 10:179–181.

Wiencke, C., and M. N. Clayton. 1990. Sexual reproduction, life history, and early development in culture of the Antarctic brown alga *Himantothallus grandifolius* (Desmarestiales, Phaeophyceae). *Phycologia* 29:9–18.

Wilce, R. T. 1990. Role of the Arctic Ocean as a bridge between the Atlantic and the Pacific Ocean: Fact and hypothesis. Pages 323–347 in *Evolutionary Biogeography of the Marine Algae of the North Atlantic*, edited by D. J. Garbary and G. R. Sough. NATO ASI Series G, Ecological Science Vol. 22, Springer-Verlag, Berlin, Germany.

Wilce, R. T., and J. R. Searles. 1991. *Schmitzia sanctae-crucis* sp. nov. (Calosiphoniaceae, Rhodophyta) and a novel nutritive development to aid in zygote nucleus amplification. *Phycologia* 30:151–169.

Wilcox, L. W. 1989. Multilayered structures (MLSs) in two dinoflagellates, *Katodinium campylops* and *Woloszynskia pascheri*. *Journal of Phycology* 4:785–789.

Wilcox, L. W., and G. L. Floyd. 1988. Ultrastructure of the gamete of *Pediastrum duplex* (Chlorophyceae). *Journal of Phycology* 24:140–146.

Wilcox, L. W., P. A. Fuerst, and G. L. Floyd. 1993. Phylogenetic relationships of four charophycean green algae inferred from complete nuclear-encoded small subunit rRNA gene sequences. *American Journal of Botany* 80:1028–1033.

Wilcox, L. W., L. A. Lewis, P. A. Fuerst, and G. L. Floyd. 1992. Assessing the relationships of autosporic and zoosporic chlorococcalean green algae with 18s rRNA sequence data. *Journal of Phycology* 28:381–386.

Wilcox, L. W., and G. E. Wedemayer. 1984. *Gymnodinium acidotum* Nygaard (Pyrrophyta), a dinoflagellate with an endosymbiotic cryptomonad. *Journal of Phycology* 20:236–242.

Wilcox, L. W., and G. J. Wedemayer. 1985. Dinoflagellate with blue-green chloroplasts derived from an endosymbiotic eukaryote. *Science* 227:192–194.

Wilcox, L. W., and G. J. Wedemayer. 1991. Phagotrophy in the freshwater, photosynthetic dinoflagellate *Amphidinium cryophilum*. *Journal of Phycology* 27:600–609.

Wilcox, L. W., G. J. Wedemayer, and L. E. Graham. 1982. *Amphidinium cryophilum* sp. nov. (Dinophyceae), a new freshwater dinoflagellate. II. Ultrastructure. *Journal of Phycology* 18:18–30.

Wilde, A., Y. Churin, H. Schubert, and T. Börner. 1997. Disruption of a *Synechocystis* sp. PCC 6803 gene with partial similarity to phytochrome genes alters growth under changing light qualities. *Federation of European Biological Societies Letters* 406:89–92.

Wildman, R. B., J. H. Loescher, and C. L. Winger. 1975. Development and germination of akinetes of *Aphanizomenon flos-aquae*. *Journal of Phycology* 11:96–104.

Wilkinson, A. N., B. A. Zeeb, J. P. Smol, and M. S. V. Douglas. 1997. Chrysophyte stomatocyst assemblages associated with periphytic, high arctic pond environments. *Nordic Journal of Botany* 17:95–112.

Wilkinson, C. R. 1992. Symbiotic interactions between marine sponges and algae. Pages 111–152 in *Algae and Symbioses*, edited by W. Reisser. BioPress, Bristol, UK.

Will, A., and W. Tanner. 1996. Importance of the first external loop for substrate recognition as revealed by chimeric *Chlorella* monosaccharide/H$^+$ symporters. *Federation of European Biological Societies Letters* 381:127–130.

Willen, E. 1991. Planktonic diatoms—an ecological review. *Archiv für Hydrobiologie Supplement* 89:69–106.

Willey, R. L. 1984. Fine structure of the mucocysts of *Colacium calvum* (Euglenophyceae). *Journal of Phycology* 20:426–430.

Willey, R. L., P. L. Walne, and P. A. Kivic. 1988. Origin of euglenoid flagellates from phagotrophic ancestors. *CRC Critical Reviews in Plant Sciences* 7:313–340.

Willey, R. L., and R. G. Wibel. 1985. A cytostome (cytopharynx) in green euglenoid flagellates (Euglenales) and its phylogenetic implications. *BioSystems* 18:369–376.

Williams, D. F., J. Peck, E. B. Karabanov, A. A. Prokopenko, V. Kravchinsky, J. King, and M. I.

Kuzmin. 1997. Lake Baikal record of continental climate response to orbital insolation during the past 5 million years. *Science* 278:1114–1117.

Williamson, R. E. 1979. Filaments associated with the endoplasmic reticulum in the streaming cytoplasm of *Chara corallina*. *European Journal of Cell Biology* 20:177–183.

Williamson, R. E. 1992. Cytoplasmic streaming in characean algae: mechanisms, regulation by Ca^{2+}, and organization. Pages 73–98 in *Algal Cell Motility*, edited by M. Melkonian. Chapman & Hall, London, UK.

Wilmotte, A. 1994. Molecular evolution and taxonomy of the cyanobacteria. Pages 1–25 in *The Molecular Biology of Cyanobacteria*, edited by D. A. Bryant. Kluwer Academic Publishers, Amsterdam, Netherlands.

Wilson, E. O., and W. H. Bossert. 1971. *A Primer of Population Biology*. Sinauer Associates, Stamford, CT.

Windust, A. J., J. L. C. Wright, and H. L. McLachlan. 1996. The effects of the diarrhetic shellfish poisoning toxins, okadaic acid and dinophysistoxin-1, on the growth of microalgae. *Marine Biology* 126:19–25

Winter, A., R. W. Jordan, and P. H. Roth. 1994. Biogeography of living coccolithophores in ocean waters. Pages 161–178 in *Coccolithophores*, edited by A. Winter and W. G. Siesser. Cambridge University Press, Cambridge, UK.

Winter, A., and W. G. Siesser. 1994. Atlas of living coccolithophores. Pages 107–160 in *Coccolithophores*, edited by A. Winter and W. G. Siesser. Cambridge University Press, Cambridge, UK.

Woelkerling, W. J. 1976. Wisconsin desmids. I. Aufwuchs and plankton communities of selected acid bogs, alkaline bogs, and closed bogs. *Hydrobiologia* 48:209–232.

Woelkerling, W. J. 1988. *The Coralline Red Algae: An Analysis of the Genera and Subfamilies of Non-geniculate Corallinaceae*. British Museum (Natural History), Oxford University Press, London, UK.

Wolf, F. R., and E. R. Cox. 1981. Ultrastructure of active and resting colonies of *Botryococcus braunii* (Chlorophyceae). *Journal of Phycology* 17:395–405.

Wolfe, G. V., and M. Steinke. 1996. Grazing-activated production of dimethylsulfide (DMS) by two clones of *Emiliania huxleyi*. *Limnology and Oceanography* 41:1151–1160.

Wolfe, G. V., M. Steinke, and G. O. Kirst. 1997. Grazing-activated chemical defense in a unicellular marine alga. *Nature* 387:894–897.

Womersley, H. B. S. 1984, 1987. *The Marine Benthic Flora of Southern Australia* (2 parts). Government Printer, Adelaide, Australia.

Wong, A., and T. Beebee. 1994. Identification of a unicellular, non-pigmented alga that mediates growth inhibition in anuran tadpoles: A new species of the genus *Prototheca* (Chlorophyceae, Chloroccales). *Hydrobiologia* 277:85–96.

Wood, A. M., and T. Leatham. 1992. The species concept in phytoplankton ecology. *Journal of Phycology* 28:723–729.

Wood, R. 1998. The ecological evolution of reefs. *Annual Review of Ecology and Systematics* 29:179–206.

Wood, R. D., and K. Imahori. 1965. *A Revision of the Characeae*, Vol. 1. Cramer, Weinheim, Germany.

Woods, J. K., and R. E. Triemer. 1981. Mitosis in the octaflagellate prasinophyte, *Pyramimonas amylifera* (Chlorophyta). *Journal of Phycology* 17:81–90.

Wray, J. L. 1977. *Calcareous Algae*. Elsevier, New York, NY.

Wright, P. J., J. A. Callow, and J. R. Green. 1995b. The *Fucus* (Phaeophyceae) sperm receptor for eggs. II.

Isolation of a binding protein which partially activates eggs. *Journal of Phycology* 31:592–600.

Wright, P. J., J. R. Green, and J. A. Calow. 1995a. The *Fucus* (Phaeophyceae) sperm receptor for eggs. I. Development and characteristics of a binding assay. *Journal of Phycology* 31:584–591.

Wujek, D. E. 1996. Chrysophyte ultrastructure and taxonomy: A mini review. Pages 21–36 in *Cytology, Genetics and Molecular Biology of Algae*, edited by B. R. Chaudhary and S. B. Agrawal. SPB Academic Publishing, Amsterdam, Netherlands.

Wujek, D. J., and J. E. Chambers. 1966. Microstructure of pseudocilia of *Tetraspora gelatinosa* (Vauch.) Desv. *Transactions of the Kansas Academy of Science* 68:563–565.

Wujek, D. J., and L. C. Saha. 1995 The genus *Paraphysomonas* from Indian rivers, lakes, ponds and tanks. Pages 373–384 in *Chrysophyte Algae: Ecology, Phylogeny and Development*, edited by C. D. Sandgren, J. P. Smol, and J. Kristiansen. Cambridge University Press, Cambridge, UK.

Wustman, B. A., M. R. Gretz, and K. D. Hoagland. 1997. Extracellular matrix assembly in diatoms (Bacillariophyceae). I. A model of adhesives based on chemical characterization and localization of polysaccharides from the marine diatom *Achnanthes longipes* and other diatoms. *Plant Physiology* 113:1059–1069.

Wustman, B. A., J. Lind, R. Wetherbee, and M. R. Gretz. 1998. Extracellular matrix assembly in diatoms (Bacillariophyceae). III. Organization of fucoglucuronoglactans within the adhesive stalks of *Achnanthes longipes*. 116:1431–1441.

Wynne, M. J. 1981. Phaeophyta: Morphology and Classification, in *The Biology of Seaweeds*, edited by C. S. Lobban and M. J. Wynne. Blackwell Scientific, Oxford, UK.

Wynne, M. J. 1983. The current status of genera in the Delesseriaceae. *Botanica Marina* 26:437–450.

Wynne, M. J. 1988. A reassessment of the *Hypoglossum* group (Delesseriaceae, Rhodophyta), with a critique of its genera. *Helgoländer Meeresuntersuchungen* 42:511–534.

Wynne, M. J. 1996. A revised key to genera of the red algal family Delesseriaceae. *Nova Hedwigia* 112:171–190.

Xia, B., and I. A. Abbott. 1987. Edible seaweeds of China and their place in the Chinese diet. *Economic Botany* 41:341–353.

Xiao, S., Y. Zhang, and A. H. Knoll. 1998. Three-dimensional preservation of algae and animal embryos in a Neoproterozoic phosphorite. *Nature* 391:553–558.

Xiaoping, G., J. D. Dodge, and J. Lewis. 1989. Gamete mating and fusion in the marine dinoflagellate *Scrippsiella* sp. *Phycologia* 28:342–351.

Xu, H. H., and F. R. Tabita. 1996. Ribulose-1,5-Bisphosphate carboxylase/oxygenase gene expression and diversity of Lake Erie plankton microorganisms. *Applied and Environmental Microbiology* 62:1913–1921.

Xu, M.-Q., S. D. Kathe, H. Coodrich-Blair, S. A. Nierzwicki-Bauer, and D. A. Shub. 1990. Bacterial origin of a chloroplast intron: Conserved self-splicing Group I introns in cyanobacteria. *Science* 250:1566–1572.

Yamagishi, Y., and K. Kogame. 1998. Female dominant population of *Colpomenia peregrina* (Scytosiphonales, Phaeophyceae). *Botanica Marina* 41:217–222.

Yeh, C.-C., S.-H. Wu, J. T. Murphy, and J. C. Lagarias 1997. A cyanobacterial phytochrome two-component

light sensory system. *Science* 277:1505–1508.

Yildiz-Fitnat, H., P. Davies-John, and A. R. Grossman. 1994. Characterization of sulfate transport in *Chlamydomonas reinhardtii* during sulfur-limited and sulfur-sufficient growth. *Plant Physiology* 104:981–987.

Yoon, H.-S., and J. W. Golden. 1998. Heterocyst pattern formation controlled by a diffusible peptide. *Science* 282:935–938.

Yopp, J. H., D. R.Tindall, D. M. Mille, and W. E. Schmid. 1978. Isolation, purification and evidence for a halophilic nature of the blue-green alga *Aphanothece halophytica* Fremy (Chroococcales). *Phycologia* 17:172–178.

Yoshida, T., and H. Mikami. 1996. *Sorellocolax stellaris* gen et sp. nov., a hemiparasitic alga (Delesseriaceae, Rhodophyta) from the East Coast of Honshu Japan. *Phycological Research* 44:125–128.

Yoshimura, Y., S. Kohshima, and S. Ohtani. 1997. A community of snow algae on a Himalayan glacier: Change of algal biomass and community structure with altitude. *Arctic and Alpine Research* 29:126–137.

Young, D. N., B. M. Howard, and W. Fenical. 1980. Subcellular localization of brominated secondary metabolites in the red alga *Laurentia snyderae*. *Journal of Phycology* 16:182–185.

Young, D. N., and J. A. West. 1979. Fine structure and histochemistry of vesicle cells of the red alga *Antithamnion defectum* (Ceramiaceae). *Journal of Phycology* 15:49–57.

Young, J. R., P. R. Brown, and J. A. Burnett. 1994. Paleontological perspectives. Pages 379–392 in *The Haptophyte Algae*, edited by J. C. Green and B. S. C. Leadbeater. Clarendon Press, Oxford, UK.

Young, J. R., J. M. Didymus, P. R. Bown, B. Prins, and S. Mann. 1992. Crystal assembly and phylogenetic evolution in heterococcoliths. *Nature* 356:516–518.

Zapata, M., and J. L. Garrido. 1997. Occurrence of phytylated chlorophyll *c* in *Isochrysis galbana* and *Isochrysis* sp. (Clone T-ISO) (Prymnesiophyceae). *Journal of Phycology* 33:209–214.

Zechman, F. W., and A. C. Mathieson. 1985. The distribution of seaweed propagules in estuarine, coastal, and offshore waters of New Hampshire, USA. *Botanica Marina* 18:283–294.

Zechman, F. W., E. C. Theriot, E. A. Zimmer, and R. L. Chapman. 1990. Phylogeny of the Ulvophyceae (Chlorophyta): Cladistic analysis of nuclear-encoded rRNA sequence data. *Journal of Phycology* 26:700–710.

Zeeb, B. A., K. A. Duff, and J. P. Smol. 1996. Recent advances in the use of chrysophyte stomatocysts in paleoecological studies. *Nova Hedwigia* 114:247–252.

Zehr, J. P., and L. A. McReynolds. 1989. Use of degenerate oligonucleotides for amplification of the *nifH* gene from the marine cyanobacterium *Trichodesmium thiebautii*. *Applied and Environmental Microbiology* 55:2522–2526.

Zehr, J. P., and H. Paerl. 1998. Nitrogen fixation in the marine environment: Genetic potential and nitrogenase expression. Pages 285–302 in *Molecular Approaches to the Study of the Ocean*, edited by K. E. Cooksey. Chapman & Hall, London, UK.

Zhang, X., M. Watanabe, and I. Inouye. 1996. Light and electron microscopy of grazing by *Poterioochromonas malhamensis* (Chrysophyceae) on a range of phytoplankton taxa. *Journal of Phycology* 32:37–46.

Zingmark, R. G. 1970. Sexual reproduction in the dinoflagellate *Noctiluca miliaris*. *Journal of Phycology* 6:122–126.

Taxonomic Index

Page numbers in boldface indicate illustrations. Numbers with t indicate tables.

Subject Index

Page numbers in boldface indicate illustrations. Numbers with t indicate tables.